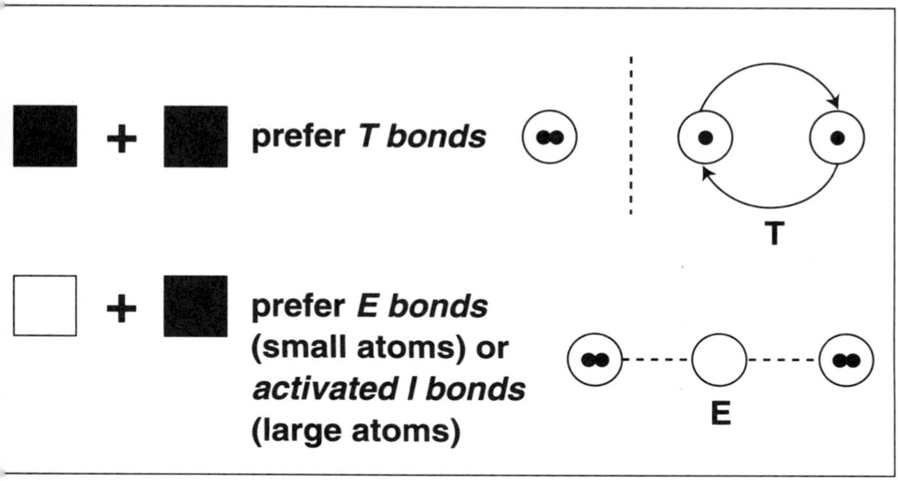
prefer *T bonds*
T
prefer *E bonds*
(small atoms) or
activated I bonds
(large atoms)
E

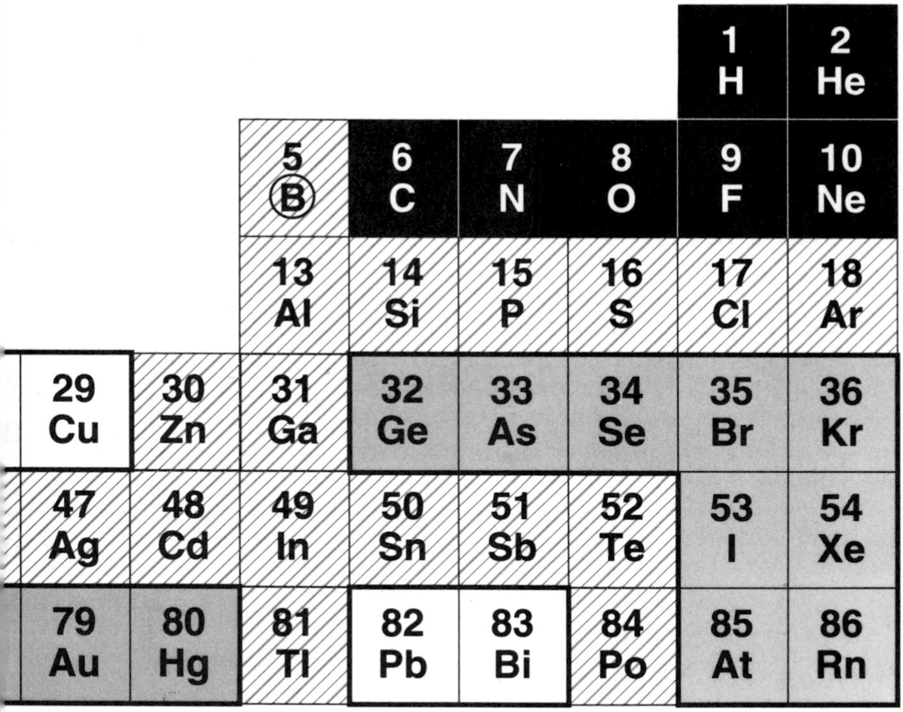
1 H
2 He
5 B
6 C
7 N
8 O
9 F
10 Ne
13 Al
14 Si
15 P
16 S
17 Cl
18 Ar
29 Cu
30 Zn
31 Ga
32 Ge
33 As
34 Se
35 Br
36 Kr
47 Ag
48 Cd
49 In
50 Sn
51 Sb
52 Te
53 I
54 Xe
79 Au
80 Hg
81 Tl
82 Pb
83 Bi
84 Po
85 At
86 Rn

Deciphering the Chemical Code

Deciphering the Chemical Code

Bonding Across the Periodic Table

Nicolaos D. Epiotis

VCH

Nicolaos D. Epiotis
Department of Chemistry
University of Washington
Seattle, WA 98195-1700

This book is printed on acid-free paper. ♾

Library of Congress Cataloging-in-Publication Data

Epiotis, N. D., 1944–
Deciphering the chemical code : bonding across the periodic table / N.D. Epiotis.
p. cm.
Includes bibliographical references and index.
ISBN 1-56081-946-4 (alk. paper)
1. Valence (Theoretical chemistry) 2. Chemical bonds. I. Title.
QD469.E65 1996
541.2'24—dc20 95-48282
CIP

Printed in the United States of America

ISBN 1-56081-946-4 VCH Publishers, Inc.

Printing History:
10 9 8 7 6 5 4 3 2 1

Published jointly by

VCH Publishers, Inc.
333 7th Avenue
New York, New York 10001

VCH Verlagsgesellschaft mbH
P.O. Box 10 11 61
69451 Weinheim, Germany

VCH Publishers (UK) Ltd.
8 Wellington Court
Cambridge CB1 1HZ
United Kingdom

This work is dedicated to the legendary Italian tenor

Franco Corelli

Preface

Dreams have elements of reality. In this context, this monograph argues that the constellation of chemical concepts that define the theory of chemical bonding (i.e., the *interpretation* of the experimental and computational data) is a dream, a mirage. A new approach, based on valence bond (VB) methodology, is devised, explained, and applied. New theory means new formulas which act as the most direct predictors of new experimental chemistry. This work is about new chemical formulas that lead to the self-consistent rationalization of old facts and the design of new chemistry. The crucial experimental and computational facts (selected from an ocean of data) are the pieces of a puzzle that defines the chemical code. This is an attempt to decipher the code.[1]

Why did this "new" theory, have to wait so long to surface? One reason is that nature has hidden well its secrets in a crypt called *electron–electron repulsion.* Computational chemists have, in the past, compromised with this enemy in the form of the zero differential overlap approximation in early semiempirical theory and in the subsequent wide adoption of the restricted Hartree–Fock (RHF) model. Nowadays, it is possible to carry out ab initio calculations that adequately deal with interelectronic repulsion. However, this ability has failed to forge new conceptual models. Chemists still think on the basis of an effective Hückel-type Hamiltonian, which assumes that classical Coulomb interactions cancel out. The few recognized cases in which Hückel molecular orbital (MO) theory fails because of the neglect of electron–electron repulsion have created the impression that the damage is of restricted nature.

We will argue that the truth is entirely different: The fundamental trends of chemistry cannot be understood on a self-consistent basis without *explicit* consideration of electron–electron repulsion. In other words, the popular models do not even contain the right underpinning factor. This is effectively why the second half of the twentieth century represents a long standstill in the development of valence theory. Indeed, modern one-electron MO models are most frequently restatements or refinements of old seminal ideas: for example, the Woodward–Hoffmann rules are application of Hückel's rule to transition states; hyperconjugation is a restatement of the "double bond, no bond resonance" concept of Pauling. With all the current efforts to sell chemistry to the public and the consequent search for wise public-relations spokesmen, it is worth stressing that conceptual chemical theory as we know it today has hardly made a step forward with respect to "Lewis–Hückel–Pauling theory" of the first half of the twentieth century. This work is the first post-Pauling theory of chemical bonding, in the essentials rather than in baroque details.

Knight[2] examines the evolution of chemistry and argues that the discipline has "a glorious future behind it!" Davenport[3] makes a point that is food for thought: *It is chemists who develop chemical theory operating far ahead of the physicists.* To quote,

> It was the 19th-century chemists, rather than the 20th-century physicists, who hammered out the essential macroscopic–microscopic duality of the physical world. It was the organic—and later the inorganic—chemists who elucidated the architecture of the molecular world using little more than elemental analysis, polarimetry, methods of purification . . ., and the assumption that the four valencies of carbon are directed to the corners of a regular tetrahedron. . . . With no assistance from electron theory or quantum mechanics and without the help of nuclear magnetic resonance, Fourier-transform infrared spectroscopy, high-performance liquid chromatography, or other such methods, Fischer could adumbrate the three-dimensional structure of the sugars. . . . All this at a time when many physicists, and even incipient physical chemists, were doubting the very existence of atoms. . . . The end of the 19th century witnessed the triumph of strictly chemical thought, an intellectual achievement on a par with Newtonian mechanics, Darwinian evolution, or, Einsteinian relativity. Alas, its story has never been coherently, let alone popularly, told. . . .

Does this make sense?

The Davisson–Germer experiment best defines the essence of quantum mechanics. The diffraction pattern of electrons scattered by a crystal suggests that the superposition principle applies to electrons (de Broglie waves) just as to electromagnetic waves. In Chapter 12, we will see that the concept of aromaticity (molecular stability) in VB formalism is a descriptor of the superposition of electron delocalization mechanisms as a function of geometry. Thus, the Davisson–Germer experiment and molecular stability are directly linked. As a result, the pre-quantum chemists implicitly discovered quantum mechanics before their physical colleagues.

The very theme repeats itself nowadays. While chemical physicists can compute with accuracy (some) molecules and transition states, the chemist has moved to studies in molecular replication, self-assembled monolayer films, enzyme-assisted synthesis, and other areas far beyond computability. At the same time, the description of the new chemistry remains phenomenological and one thing is crystal clear: Conventional concepts, which failed to predict or anticipate the recent spectacular developments in fullerene chemistry, cuprate superconductors, and so on, are no longer adequate to drive new research. How can we break the impasse? Theories that merely "explain" the known but cannot lead us past the intuition of the experimentalist are useless. How can we develop a predictive model? Models that are not refutable are useless. How can we ensure refutability?

To answer these questions, we go back to the theoretical apparatus that history has demonstrated to be the best-suited for chemistry: *nonorthogonal VB theory.* Chemists conceive of methane as "carbon bound to four hydrogens by four C—H bonds" rather than "four lower doubly occupied and four higher vacant MOs." Comparison of organic, inorganic, and physical chemistry textbooks of 30 years ago with the latest editions shows that Lewis formulas and "arrow pushing" still dominate.[4] Our first goal is to develop a physical interpretation of the full VB wavefunction. On this basis, we then proceed to replace the conventional Lewis formulas of "covalent-only" Pauling–Eyring VB theory by new formulas that expose what is "good" and what is "bad" about the molecule. *The hallmark of this work is the presentation of chemical formulas that have an associated error count.* Because formulas speak for themselves, we can rationalize old controversial trends and we can predict new chemistry in a fraction of a page. Finally, formulas are specific and, thus, refutable. Given the availability of canned computer programs, chemists have sought to apply the classical ideas of chemical bonding to molecules of ever increasing size. The problem lies elsewhere: What is the operationally significant formula of methane, cyclopropane, benzene, Diels–Alder transition state, ferrocene, and the lithium tetramer? An answer that captures the physics of the problem leads to unlimited applications to molecules of whatever size. Thus, the work is about methane, cyclopropane, benzene, and the rest of the seminal molecules and transition states of chemistry, but the story turns out to be very different from what one would normally expect.

The substance as well as the style of a scientific work are determined by the esthetic viewpoint of the author. I cannot envision a more apt description of my philosophy than this remark by the late Professor Coulson: "It is futile to obtain accurate numbers, whether by computation or experiment, unless these numbers can provide us with simple and useful chemical concepts; otherwise, one might as well be interested in a telephone directory."

Anyone who has even tangential contact with chemistry knows one fundamental truth: When it comes to molecules, anything is possible. The challenge is to put the right atoms together to accomplish a defined goal. This presupposes breaking the chemical code. The crucial question in chemistry is not *how* to do it. Nowadays, technology provides many options. An ab initio quantum chemical computation package can be bought and used in the same way one buys and uses (often as a "black box") an NMR spectrometer. Rather, the key question is *what* to select as a research target from a vast ocean of possibilities. Many gold prospectors had the right tools, but only few dug in the right place. The answer to the "what" question is concepts and ideas. This is what the monograph is all about. Much of the chemistry done nowadays is in the mode of data collection. This work says that there are huge opportunities if only one is willing to think patiently and without preconceptions.

The concepts developed here are founded on a large amount of data. Along the way, I had to be selective and inevitably unfair to many worthwhile contributions, which were left out either because of space limitations or because I was simply unaware of them. In this context, and unless otherwise stated, the cited thermochemical, spectroscopic, and structural data have been extracted from the following sources:

(a) Thermochemical data have been taken from the compilations by Lias et al.,[5a] by Cox and Pilcher,[5b,c] and by Benson.[5c] The heats of formation of perfluorinated rings have been obtained from the compilation of Smart.[6] Silicon thermochemical data come from Walsh's review.[7] Bond dissociation data of diatomics have been obtained from the excellent review by Morse[8] and the classic monograph of Gaydon.[9]

(b) Structural data have been taken from the collection by Callomon et al.[10]

(c) Atomic spectroscopic data have been extracted from Moore[11] and, in some cases, transition metal promotional energies have been obtained from Morse.[8]

(d) Atomic orbital radii have been obtained from Desclaux[12] and Morse.[8]

(e) Singlet–triplet energy gaps of a variety of molecules have been taken from literature compilations.[13] Valence orbital ionization energies have been taken from Hinze and Jaffé.[14]

It is worth stressing that this work would not have been possible without the data, each one a product of the physical and intellectual labor of many workers, that are collected in the invaluable sources just cited. For, if there is any work befitting the designation "theory stimulated by experiment," this is precisely it.

A word about the people who contributed to the making of the monograph. Kathleen Bennett shepherded the manuscript all the way to camera-ready form. Dr. Barbara Goldman, as thoughtful a science editor as one could ever find, and her expert team (especially C. Pecoul and B. Griffing), assisted in the preparation of the final copy and, ultimately, produced a book which is far superior to what I gave them in the first place. The initial thinking was stimulated by many encounters with diverse European scientists during a visit to Germany through an Alexander von Humboldt Senior U.S. Scientist award (1984–1986). It is not an accident that some of the reprints and preprints they gave me are featured herein. In Seattle, the organometallic chemists told me of many experimental facts I did not know. Finally, I am indebted to I.G. Csizmadia, P. Mezey, R.G. Pearson, N.J. Turro, and, especially, E.R. Davidson and R. West for their comments, admonitions, and suggestions.

References

1. Some of the key ideas have been published in the form of two monographs in the *Lecture Notes in Chemistry* series of Springer-Verlag, as well as in the *Journal of Molecular Structure (THEOCHEM)* and the *New Journal of Chemistry.*

2. D. Knight, *Ideas in Chemistry: A History of the Science,* Rutgers University Press, New Brunswick, NJ, 1992.

3. D.A. Davenport, *Chem. Eng. News* 32 (May 24, 1993).

4. (a) T.W.G. Solomons, *Organic Chemistry,* 5th ed., Wiley, New York, 1992. (b) P.H. Scudder, *Electron Flow in Organic Chemistry,* Wiley, New York, 1992. (c) D.P. Weeks, *Pushing Electrons,* Saunders College Publishing, New York, 1992.

5. (a) S.G. Lias, J.E. Bartmess, J.F. Liebman, J.L. Holmes, R.D. Levin, and W.G. Mallard, "Gas-Phase Ion and Neutral Thermochemistry," in *J. Phys. Chem Ref. Data,* Vol. 17, Suppl. 1, National Bureau of Standards, Washington, DC, 1988. (b) J.D. Cox and G. Pilcher, *Thermochemistry of Organic and Organometallic Compounds,* Academic Press, New York, 1970. (c) G. Pilcher and H.A. Skinner, in *The Chemistry of the Metal–Carbon Bond,* F.R. Hartley and S. Patai, Eds., Wiley, New York, 1983. (d) S.W Benson, *J. Chem. Educ.* 42, 502 (1965).

6. B.E. Smart, in *Molecular Structure and Energetics,* Vol. 3, J.F. Liebman and A. Greenberg, Eds., VCH, New York, 1986.

7. R. Walsh, *Acc. Chem. Res.* 14, 246 (1981).

8. M.D. Morse, *Chem. Rev.* 86, 1049 (1986).

9. A.G. Gaydon, *Dissociation Energies,* Chapman and Hall, London, 1968.

10. J.H. Callomon, E. Hirota, K. Kuchitsu, W.J. Lafferty, A.G. Maki, and C.S. Pote, in Landolt-Börnstein *Numerical Data and Function Relationships in Science and Technology,* Vol. 7, New Series, *Structure Data on Free Polyatomic Molecules,* K.H. Hellwege, Ed., Springer-Verlag, West Berlin, 1976.

11. C.E. Moore, *Atomic Energy Levels,* Vols. I–III, U.S. Department of Commerce, National Bureau of Standards, Washington, DC, 1958.

12. J.P. Desclaux, *At. Data Nuclear Data Tables* 12, 311 (1973).

13. N.D. Epiotis, *New J. Chem.* 13, 639 (1989).

14. J. Hinze and H.H. Jaffé, *J. Chem. Phys.* 84, 540 (1962).

Contents

Part IV The Cluster I Bond 691

PICTORIAL GLOSSARY OF NEW TERMS

Anti overlap dispersion: The anticooperative action of overlap and dispersion (induction).

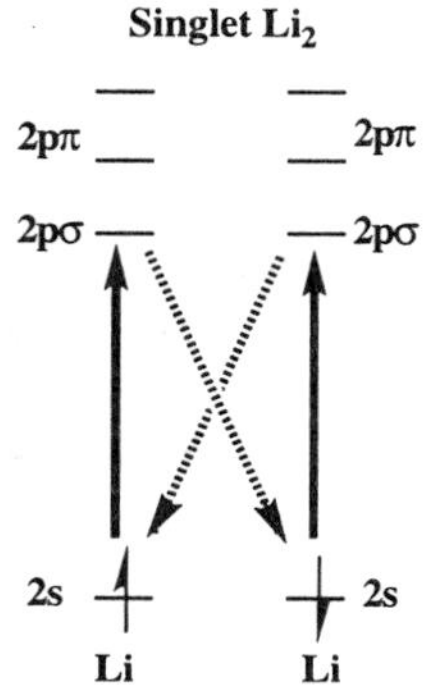

ANTI OVERLAP DISPERSION

Arrow: A pictorial representation of the transfer of a single electron.

Arrow train: The sequence of arrows that constitute the I formula. (EN = electroneutral, ELP = electropositive, ELNG = electronegative.)

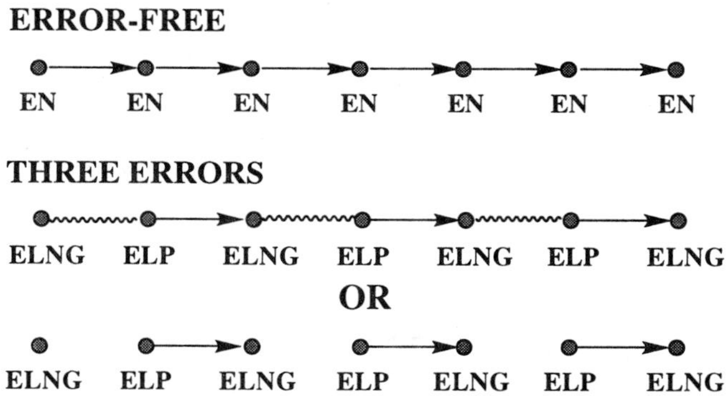

Arrow train, amphidromic: An error-free arrow train with right-left resonance; a nonpolar arrow train which spans a homonuclear set of atoms.

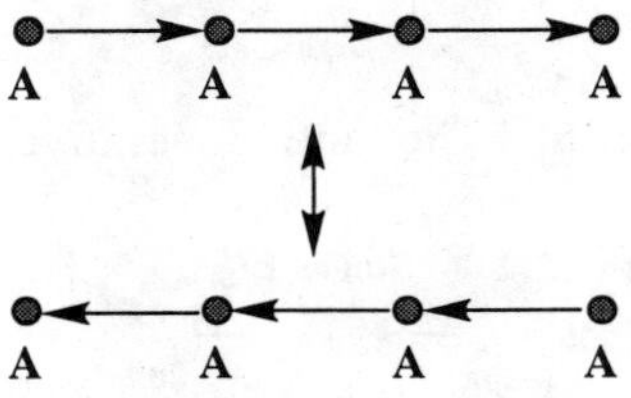

Arrow train, unidromic: An nonresonant, unidirectional error-free arrow train; a polar arrow train which spans a heteronuclear set of atoms.

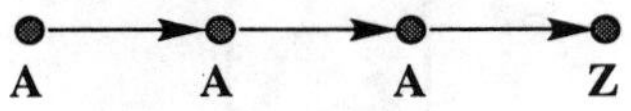

Arrow-train error: An arrow directed from a more electronegative to a more electropositive AO (vide supra).

Association diagrams: The diagrams which show how *one* hole, dot, or pair of a central atom ties up *two* ligand elements (holes, dots, or pairs) to generate one I bond.

PENTAGONAL PYRAMIDAL IF_7

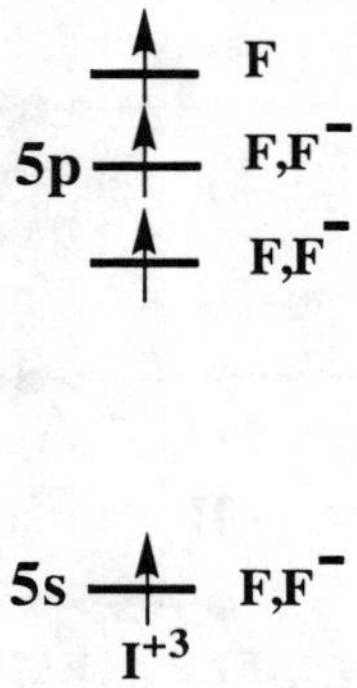

Association domain: The I-bound region of a molecule.

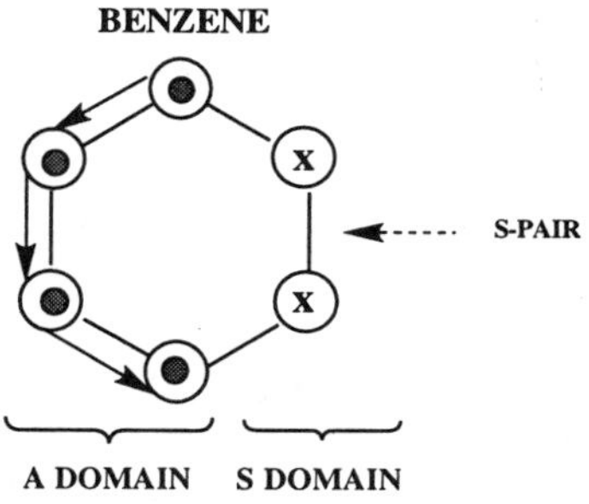

Association rule: The recipe for affixing arrows on holes, dots, and pairs.

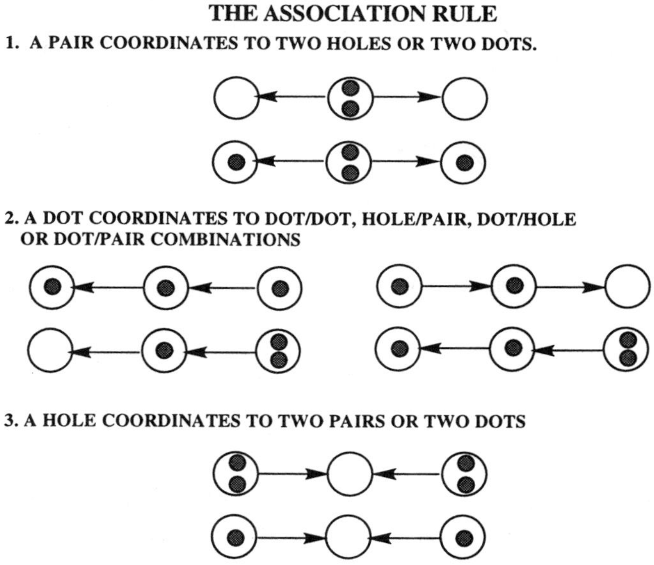
THE ASSOCIATION RULE
1. A PAIR COORDINATES TO TWO HOLES OR TWO DOTS.
2. A DOT COORDINATES TO DOT/DOT, HOLE/PAIR, DOT/HOLE OR DOT/PAIR COMBINATIONS
3. A HOLE COORDINATES TO TWO PAIRS OR TWO DOTS

Associative saturation: The association of every single pair, dot, or hole of a central atom with two ligands via one I bond.

Balanced allocation condition: The requirement that electrons are evenly distributed along a chain of metal atoms so as to diminish their Coulomb repulsion.

Banana formula (B formula): A Lewis formula in which each atom is sp^3 hybridized and in which multiple bonds are made by the overlap of sp^3 hybrids.

Bikrat: A molecule with two distinct domains: An association and a segregation domain. The two domains may or may not be in resonance (*See* Association domain).

Boundary pairs (B pairs): Pairs delocalized in both the radial and tangential AO arrays.

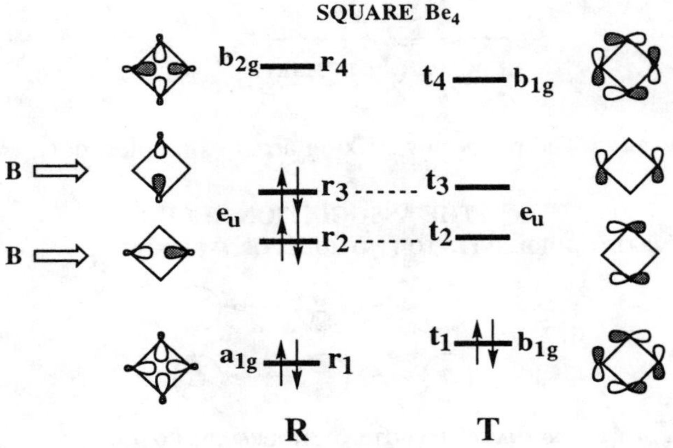

Codirectional arrow train: An arrow train in which all arrows point in the same direction.

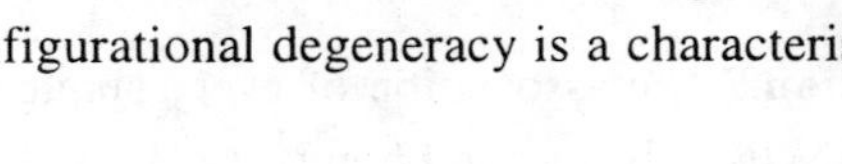

Configurational degeneracy: The situation in which a 2-e CT hop creates two degenerate or quasidegenerate valence bond configurations. One consequence is strong stabilization due to the interaction of the degenerate or quasidegenerate configurations. Configurational degeneracy is a characteristic feature of arrow codirectionality.

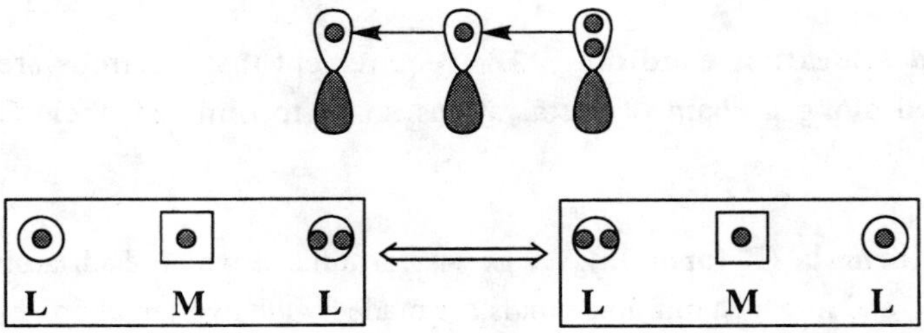

Contact repulsion: The Coulombic repulsion of two electrons in the same or different overlap regions.

Contradirectional arrow train: An arrow train in which all or some arrows point in opposite directions.

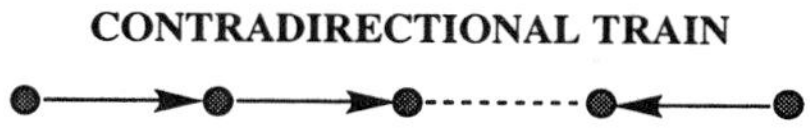

Coulomb exchange: The two-electron (Coulombic) interaction of two configurations connected by a 2-e hop across orthogonal orbitals.

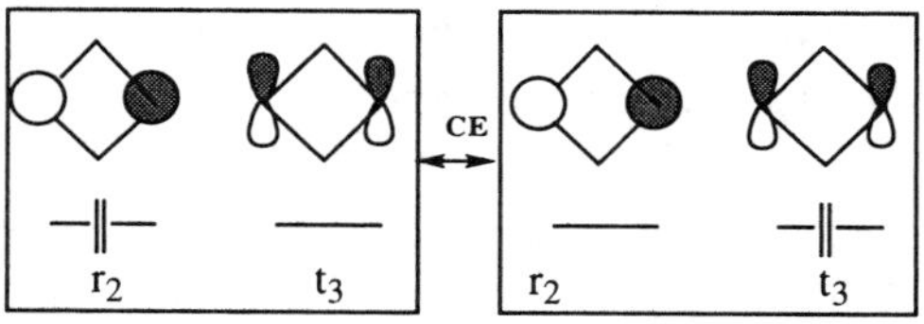

Crossed antiaromaticity: The crossing of an upper under a ground antiaromatic state under a suitable perturbation. This effectively eliminates antiaromaticity.

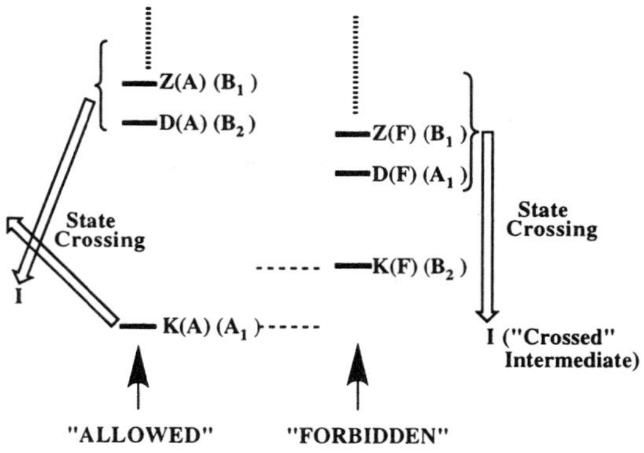

CT delocalization: Charge-creating transfer of electrons. CT delocalization is always acyclic (open loop).

2-e CT hop: The noncyclic, charge-creating transfer of two electrons.

Diastolic group: A group connected to another group by an outbound arrow.

EXAMPLES OF DIASTOLIC GROUPS

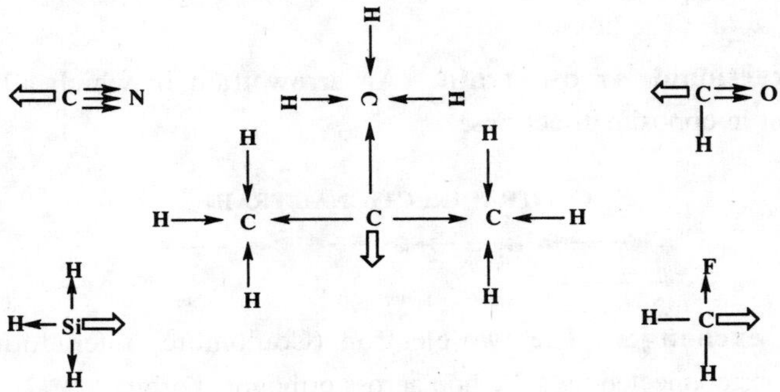

Docking rule: The formula for attaching ("docking") the correct ligands on the correct vacant AOs (starred versus unstarred).

E formula: The representation of the Coulombic attraction of pairs and holes in the absence of CT delocalization. The coordination of one pair to two holes is taken to be one E bond.

Exchange antibond (T antibond): An antibond represented by two head-to-tail arrows connecting two electrons of parallel spins and forming a closed loop.

Exchange bond (T bond): A bond represented by two head-to-tail arrows connecting two electrons of opposite spins and forming a closed loop. This can be taken to be equivalent to the Lewis electron-pair bond or the covalent bond.

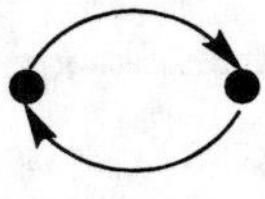

T-bond

Exchange delocalization: Charge-conserving transfer of electrons. Exchange delocalization is always cyclic (closed loop).

2-e exchange hop: The cyclic, charge-conserving transfer of two electrons.

Frustration: The destabilization accompanying the combination of atoms best suited for different mechanisms of bonding.

Geminal bond: The bond defined by two electrons occupying two orthogonal hybrids of one and the same atom. Hybrid AOs interact in spite of being orthogonal.

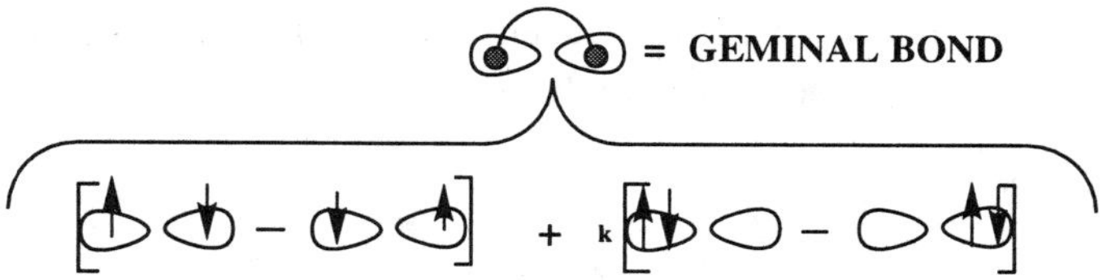

Hole-excessive reaction: A reaction in which the holes (vacant AOs) exceed the pairs (filled AOs).

Hole–Pair (HP) structure: The ionic VB configuration that contains, either in part or in toto, holes (vacant AOs) and pairs (filled AOs).

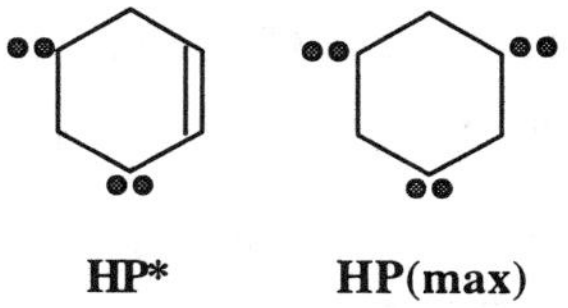

Hollow pairs: Pairs occupying radial AOs.

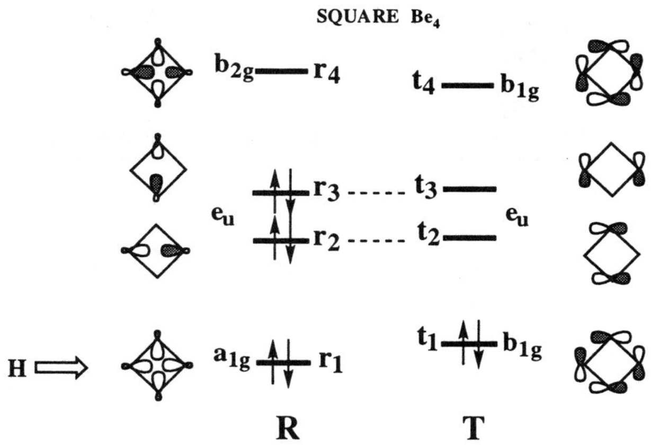

I bond: *See* Interstitial bond.

I descriptor: The representation of an ensemble of VB configurations by appending arrows on electrons and executing the arrows in all possible ways (one at a time, two at a time, and so on).

I formula: The formula of semimetal- or metal-containing molecules which shows how bonds are formed as a consequence of CT: Two arrows make up one I bond. The I formula is the principal I descriptor.

I FORMULA

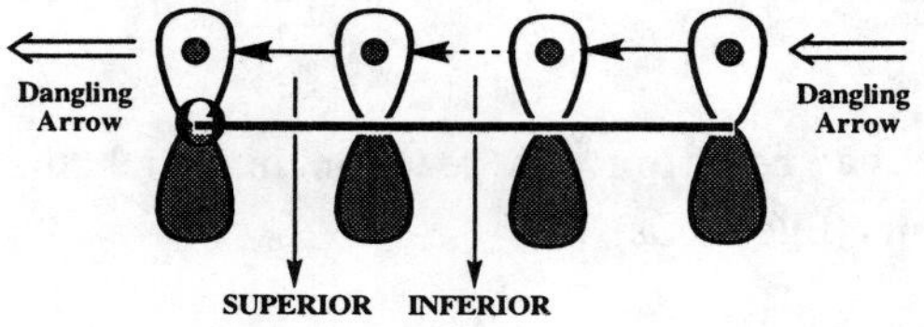

I hybrid: The linear combination of I descriptors.

Interbond CT delocalization: The CT delocalization of electrons from the space of one to the space of a second covalent bond.

EXAMPLES OF INTERBOND DELOCALIZATION

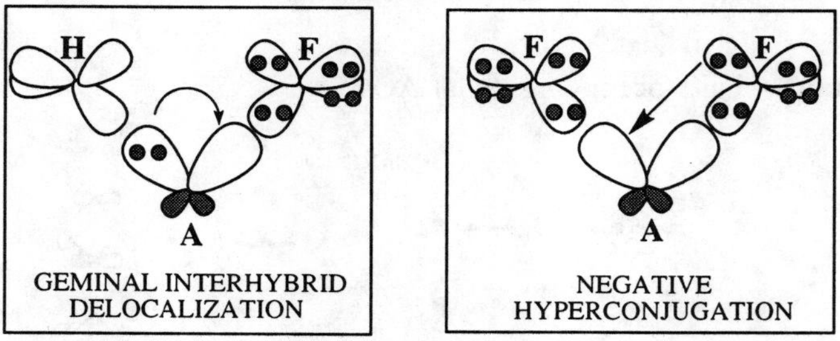

Interstitial bond (I bond): A bond represented either by two head-to-tail (superior) or two head-to-head (tail-to-tail) arrows forming an open loop. One I bond is formed by one 2-e CT hop and two ordered 1-e CT hops which are in resonance.

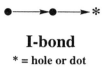

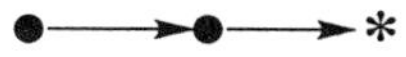

I-bond

*** = hole or dot**

Intrabond CT delocalization: The CT delocalization of two electrons making a covalent bond within the space of the two AOs defining the covalent bond.

Isoelectronic fragments: Those having the same number of valence electrons.

Isolobal fragments: Those having the same symmetry-wise set of valence AOs.

Isosynaptic fragments: Isoelectronic and isolobal fragments which bind by the same mechanism, T, I, or E.

Matched polyhedron: A polyhedron in which the number of valence AOs of the vertexial fragment matches exactly the order of the vertex.

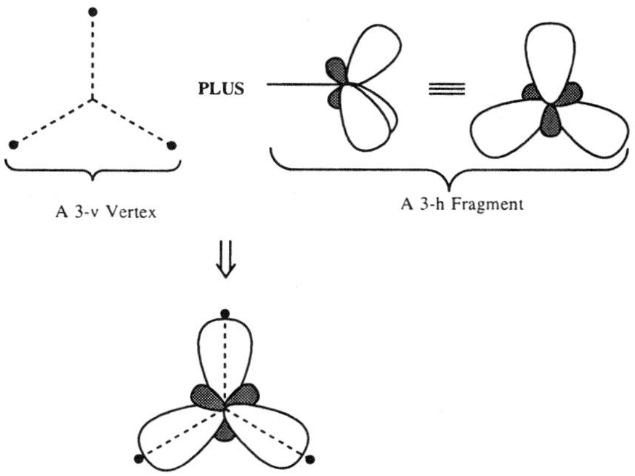

MOVB theory: VB theory which departs from fragment MOs.

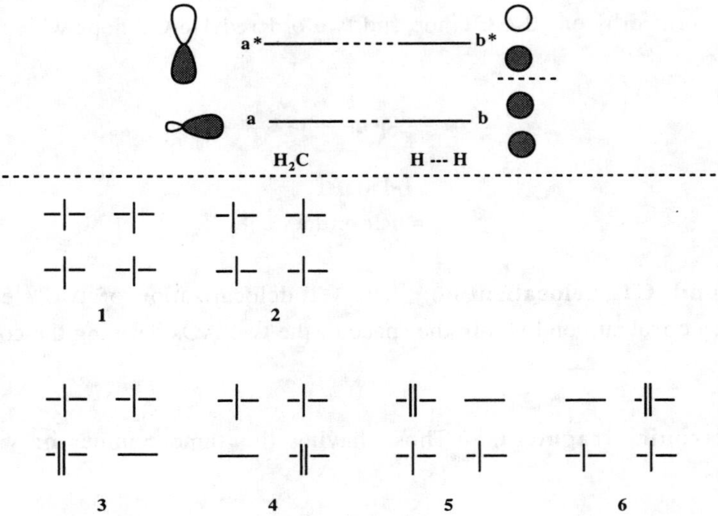

etc.

Multicatenation model: The view of a cluster as a polyhedron in which each face is spanned by a cyclic array of AOs (cyclic catena), each containing 4N + 2 electrons (aromatic count).

n-e CT: The simultaneous transfer of n electrons from one to another set of AOs (n = integer).

COLLECTIVE 3-e CT DELOCALIZATION

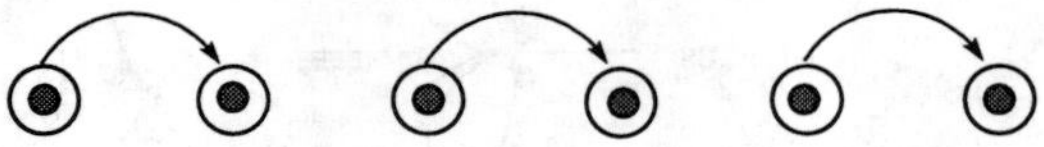

COLLECTIVE 6-e EXCHANGE DELOCALIZATION

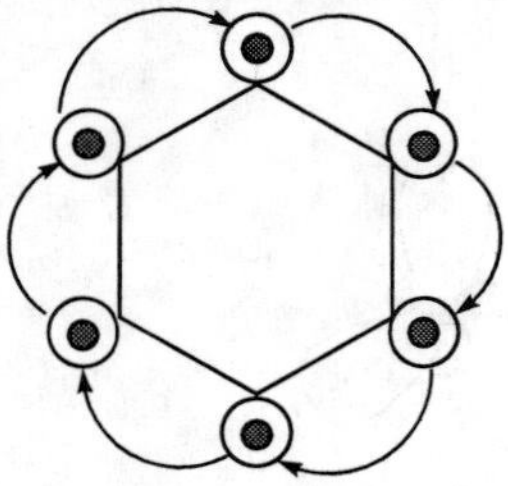

Nonlocal VB theory: VB theory which departs from perturbed (mixed) atom-centered AOs with the goal of facilitating the computation.

Nonorthogonal VB theory: VB theory which departs from the unperturbed atom-centered AOs familiar to every chemist.

Overlap dispersion (induction): The cooperative action of overlap and dispersion (induction).

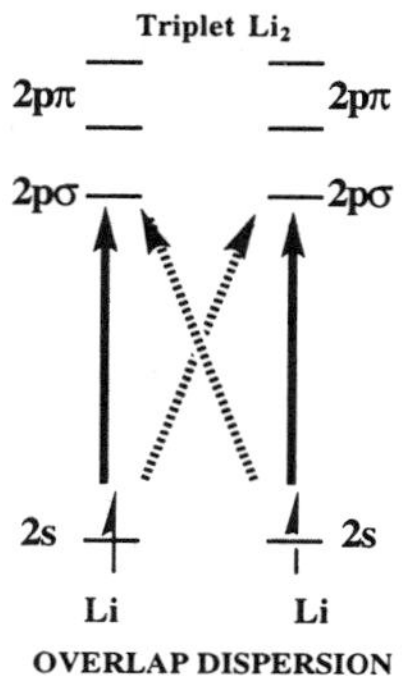

OX (Oxidative) delocalization: Synergic donation (oxidation) of *two* electrons.

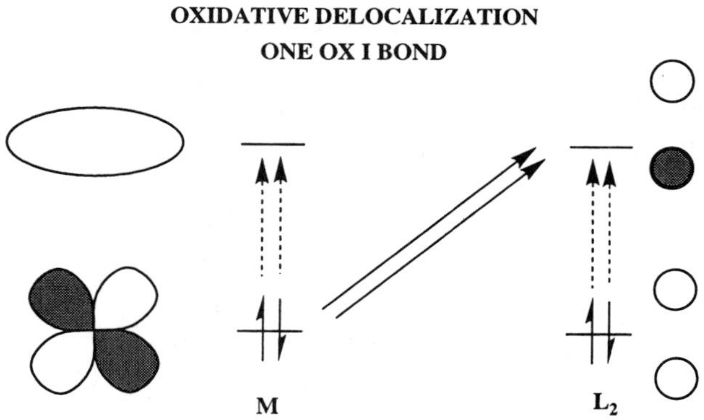

p error: A line replacing an arrow in the T formula indicating that optimal CT delocalization between two atoms has been arrested because it engenders strong exchange repulsion.

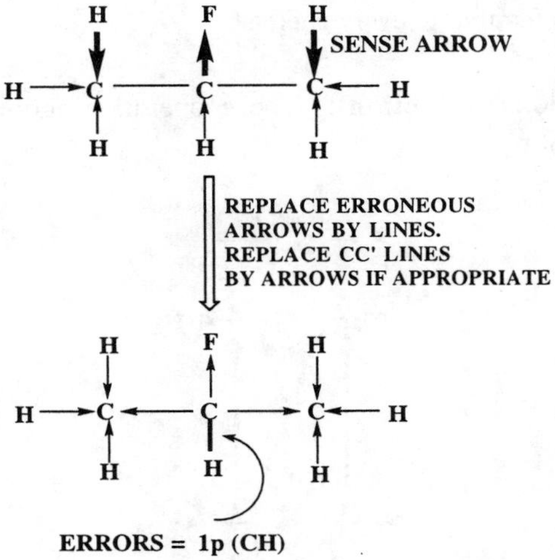

Pair-excessive reaction: A reaction in which the pairs (filled AOs) exceed the holes (vacant AOs).

Perfect pairing structure: The covalent VB configuration in which there is maximum coupling of electron pairs into exchange bonds.

Phantom bonding: An attractive interaction between two apparently nonbonded sites. It can be a consequence of I-bonding.

THE RESONANT I BOND

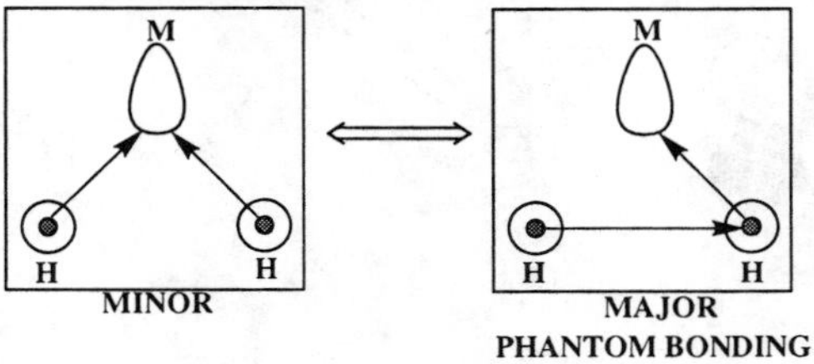

Prometals: A nonmetallic substance which can become metallic by appropriate perturbation.

Pseudoaromaticity: Aromaticity which fails to be expressed because high atom and AO electronegativity thwarts CT.

Quadrupolar diradical (zwitterion): A diradical (zwitterion) stabilized by quadrupolar interaction.

THE QUADRUPOLAR DIRADICAL

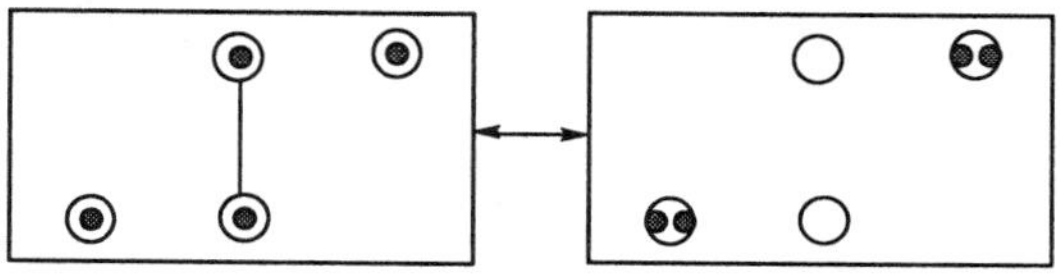

RED (Reductive) delocalization: Synergic acceptance (reduction) of *two* electrons.

Relay I bond: An I bond defined by two codirectional arrows. Its hallmark is configurational degeneracy, that is, strong configuration interaction.

ALLYL-TYPE
REL I BOND

METAL-TYPE
REL I BOND

Rotated molecule: A molecule in which a multiple bond is constructed by two overlapping sets of AOs forming a closed loop.

A [4] ANNULENOID MULTIPLE BOND

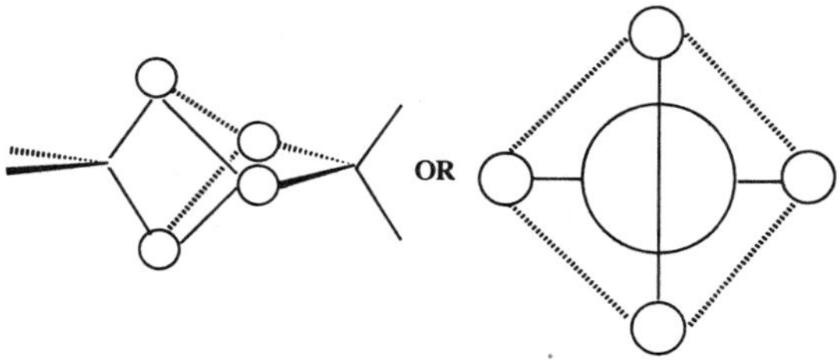

ROX (RedOx) delocalization: Synergic donation (oxidation) and acceptance (reduction) by a metal; that is, a forward/backward delocalization of *two* electrons.

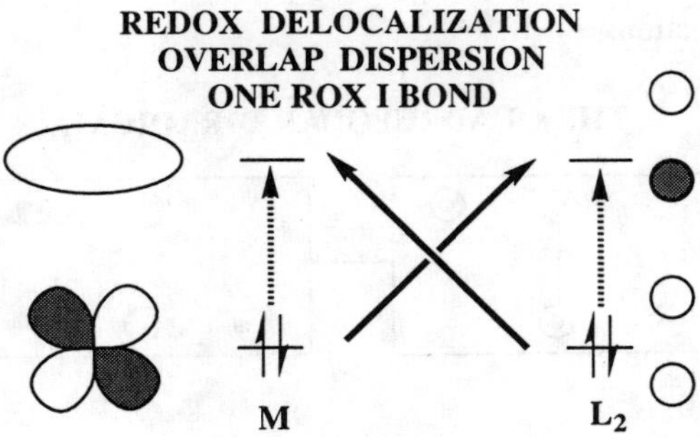

Segregation domain: The T-bound region of a molecule (*See* Association domain).

Shell model: The visualization of rings and clusters as composites of (interacting) radial and tangential arrays of AOs or MOs.

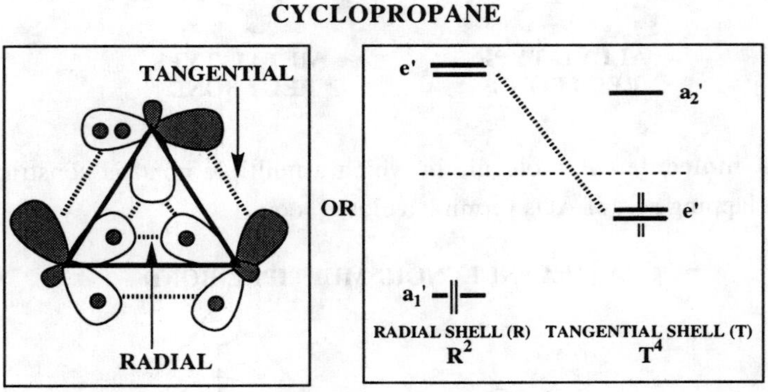

Starring procedure: The identification (by asterisks) of high-energy hybrids.

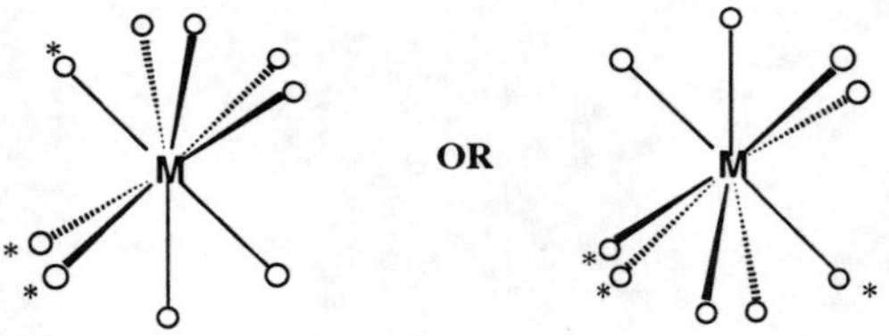

Systolic group: A group connected to another group by an inbound arrow.

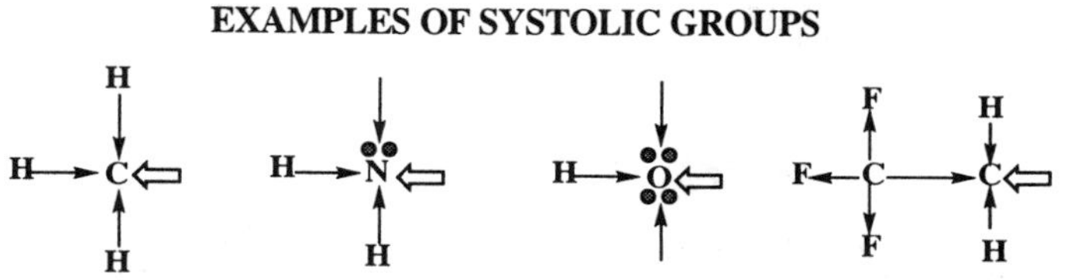

T bond: *See* Exchange bond.

T complement: The energy component due to exchange delocalization. This can be either bonding or antibonding. When more than a few pairs of electrons are present, the T complement is antibonding.

T formula: The formula of organic molecules in which an arrow represents intrabond CT delocalization and a line symbolizes a bond whose exchange repulsion thwarts CT delocalization. A T formula has an associated count of p- and v-errors.

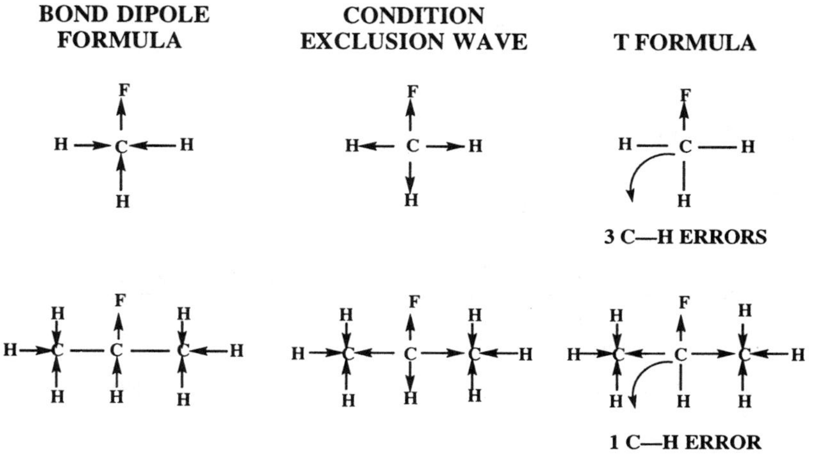

Tire chain model: The view of a cluster as a "chain," made up of occupied cyclic AO arrays, enshrouding a "tire," namely, the cluster core.

Toy model: The view of a cluster as a composite of rings. Each ring is constituted of atoms having a "90°-rotated" set of sp^2 AOs so that there is a stack of three different cyclic AO arrays: An upper set spanning one set of sp^2

hybrids, a lower set spanning a second set of sp^2 hybrids, and a middle "tangential" set spanning the p AOs. Lone pairs and exo ligands are attached to the exo set of sp^2 hybrids.

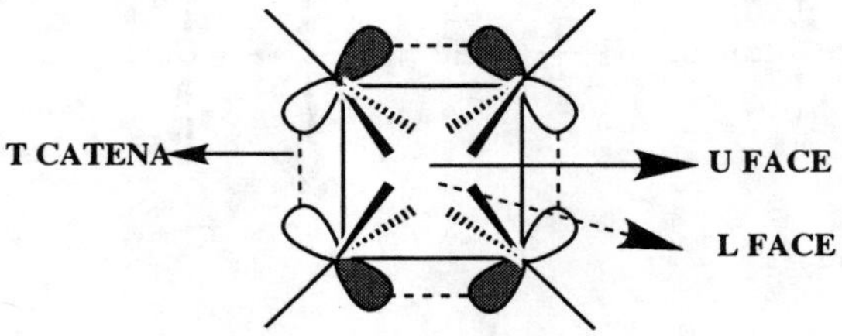

$U^xT^yL^z$

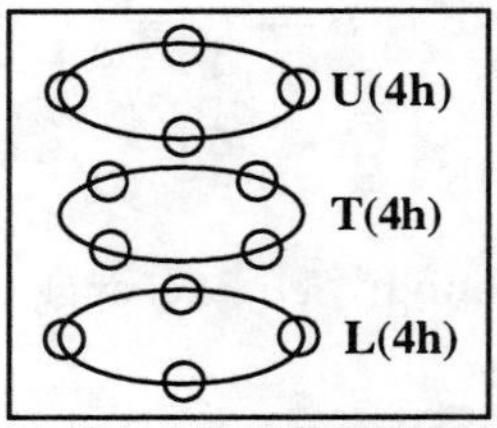

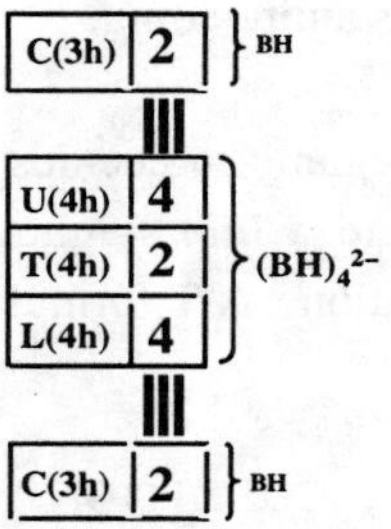

$(BH)_6^{2-}$ (Sq. Bipyr.)

v error: A pair of vicinal inbound (systolic) arrows.

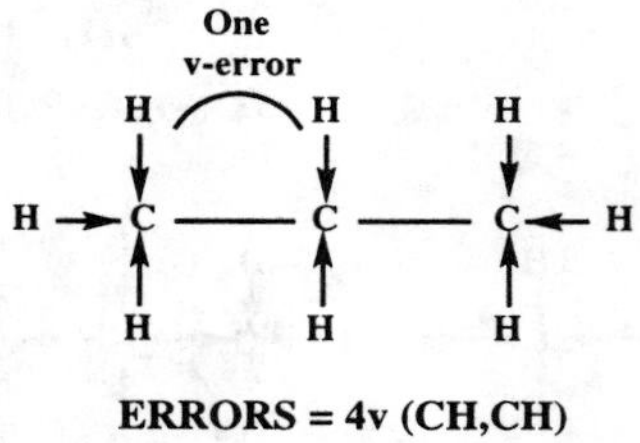

ERRORS = 4v (CH,CH)

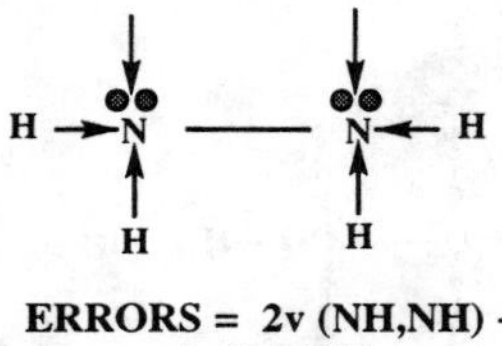

ERRORS = 2v (NH,NH) +
1v (LP,LP)

Vacuum pairs: Pairs occupying tangential AOs.

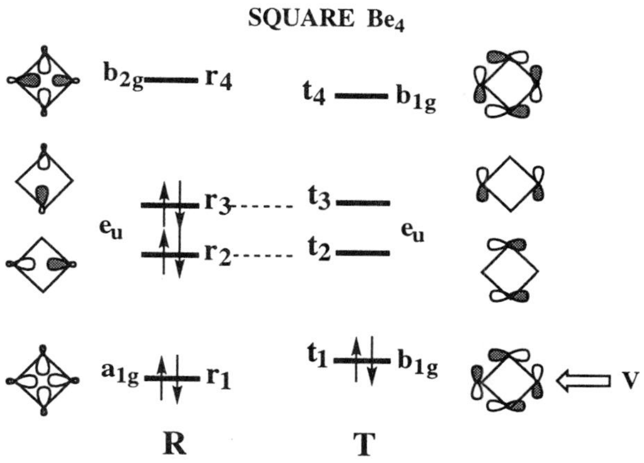

Valence shuttling: The ability of a metal to act with different combinations of valence AOs and valence electrons as a result of its capacity to rehybridize.

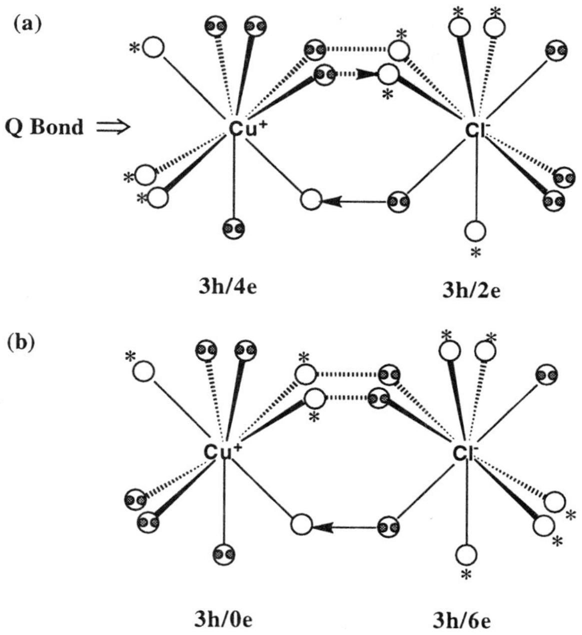

Part I

The VB Theory of Chemical Bonding

Chapter 1

Why Today's Theoretical Models Do Not Make Sense

Most chemists believe that chemical bonding is well understood and that all that remains to be done is a quantification of familiar concepts: If it ain't broke, don't fix it! After one look at the *total* picture, however, despair replaces optimism. To understand why, consider the three cornerstones of the conceptual framework of chemistry:

Magic numbers. The shape and stability of a molecule (or transition state) depend on the electron count: that is, there exist magic numbers in chemistry. The two most important examples are the inert gas rule and Hückel's rule.

Isoelectronic analogy. Atoms or fragments with the same number of valence electrons in the same type of symmetry orbitals behave similarly.

Covalent bonding. Molecular stability is determined by the number and quality of covalent bonds.

Here is the first important point of this work: Old and new experimental data refute each one of these central arguments. The first two tenets are destroyed by a very simple example: H_6 and Li_6 are isoelectronic, each having six valence electrons. It follows that they must have the same shape (isoelectronic principle), and this must be a regular hexagon (Hückel's rule). The available data are in disagreement: the stable form of H_6 is three H_2 molecules infinitely apart (neglecting van der Waals minima), and Li_6 is computed to be a piece of close-packed Li solid with a D_{3h} geometry.[1] The covalent/ionic model is refuted by the following long-known facts:

(a) H_2 has a stronger bond that H_2^+ but exactly the *opposite* is true of Li_2 and Li_2^+. This means that the relative strengths of two- and one-electron covalent bonds are not qualitatively invariant but depend on atom electronegativity.

(b) Cr_2, with a formal sextuple bond, has a *shorter bond but a smaller bond dissociation energy* (41 kcal/mol) than any of V_2 (57 kcal/mol), Ni_2 (48 kcal/mol), and Cu_2 (46 kcal/mol), which have fewer bonding and/or more antibonding electron pairs.

(c) The reaction $A_2 + B_2 \rightarrow 2\ AB$ is indeed *exothermic* when A = H and B = F only to become *endothermic* when A = H but B = Li. Again, the conclusion is that atom electronegativity must be the controlling factor.

To get a better idea of how helpless we are when it comes to the fundamentals, let us ask the preeminent pair of questions: What makes some molecules repel each other, and what makes some others stick? Ab initio calculations show that an H_4 rhombus lies 151 kcal/mol above two isolated H_2 while an Li_4 rhombus lies 15 kcal/mol below two isolated Li_2 molecules.[2] Note the spectacular difference (of the order of 166 kcal/mol) between the two systems! This means that H and Li bind by fundamentally different mechanisms.

One might think that problems of interpretation are endemic to metallic systems. Nothing could be further from the truth. The stereochemistry of "allowed" pericyclic reactions is consistent with a one-step mechanism, but the regiochemistry of the very same reactions seems consistent with a two-step process involving a diradical intermediate. Alcohols are much more stable than isomeric ethers, but no one has ever asked why. Incredible though it may sound, there has never been a serious attempt to understand what makes an organic molecule stable. Organic chemists are brought up with "steric effects"; but there are as many examples of molecules that prefer crowded geometries. A recent review of the "anomeric effect" dramatizes the urgency of the situation: the summary of the data and the alleged explanations of this *single* trend of organic stereochemistry takes up about 190 pages with no resolution in sight![3] If this is "theory," we may as well abandon all hope of understanding molecules and resign ourselves to data collection.

In summary, the failure to develop a unified model of chemical bonding is signaled by the very existence of interdisciplinary boundaries separating "organic chemistry," "inorganic chemistry," and so on. The organic chemist is well familiar with "aromaticity," "angle strain," and "hyperconjugation" but finds little use for Wade's rule. By contrast, clusters are central in the discipline of the inorganic chemist who is well familiar with the "borane analogy." On the other hand, "angle strain" and Hückel's rule provide very few insights when metal atoms come into play. Finally, many a physical chemist and many a solid state physicist concerned with metals are hardly aware of the concepts of their colleagues in the organic and inorganic fields. Rather, they use alien constructs like "cohesive energy," "electron–phonon coupling," and "Jellium model." This work explains why this is so by revealing the crucial variable that differentiates one chemical system from another: electron–electron repulsion. However, even if we were to strip this monograph of all theory, rationalizations, and predictions, the reader would still have the benefit of the formulation of a long litany of fundamental, unsolved problems from which the true theoretician cannot hide.

References

1. M.H. McAdon and W.A. Goddard III, *J. Non-Crystall. Solids* 75, 149 (1985).
2. (a) M. Rubinstein and I. Shavitt, *J. Chem. Phys.* 51, 2014 (1969). (b) H. Eckman, J. Koutecky, P. Botschwina and W. Meyer, *Chem. Phys. Lett.* 67, 119 (1979).
3. P.R. Graczyk and M. Mikolajczyk, *Top. Stereochem.* 21, 159 (1994).

Chapter 2

The Concepts of Nonorthogonal Valence Bond Theory

2.1 Lewis Formulas and Nonorthogonal VB Theory Are the "Experimentally Proven" Language of Chemistry

The early VB computations made use of a restricted set of covalent structures. Later on, some polar structures were included. By the time full-scale ab initio VB calculations of model systems (e.g., π-benzene) were made possible, MO theory, as a calculational approach, was firmly entrenched. The concurrent elevation of MO theory to the status of a conceptual theory created a huge vacuum: there has been little pedagogy with regard to the formal and conceptual aspects of VB theory. One of the few exceptions is a monograph by Sandorfy that gives a balanced description of the VB and MO methods and their applications to spectroscopy[1a]; the more recent monograph by Salem presents "parallel" MO and VB applications to organic chemistry.[1b] In this chapter, we lay the foundation for ultimately answering the question: How does chemical bonding arise when all the covalent and all the ionic VB structures are diagonalized with respect to the full Hamiltonian? This challenge has never been taken up before, for a very simple reason: Intuitively, one expects that the complexity of the solution will render impossible a simple chemical interpretation. Interestingly, this attitude, though seemingly reasonable, is the antithesis of the mathematician's philosophy: Simplify by complexifying! Henceforth, the term "VB theory" will be taken to mean "full ab initio nonorthogonal VB theory," to be clearly differentiated from "resonance theory" (a highly simplified and, as we will see, misleading version of VB theory) and other brands of VB theory such as orthogonal,[1c] generalized,[1d] or spin-coupled[1e] VB theory, in which the concept of a local, atom-centered AO is lost.

Why choose VB theory when practically every theoretical article in the literature (exceptions prove the rule) has chosen MO theory (and, more recently, local density functional theory)? Is there some desire to be different? In the beginning of this century and long before the advent of quantum mechanics, G.N. Lewis, following seminal contributions to the theory of carbon valence by Kekulé and Couper and by van't Hoff and LeBel, developed the concept of the electron pair bond. This paved the way to the concept of the Lewis formula as a descriptor of the electronic structure of a molecule.[2] In this, each atom uses a set of valence AOs that are nonorthogonal in an interatomic sense and identical or analogous to those of the atom in vacuo. The valence AOs contain the valence electrons. Bonding is effected by pairing the electrons. Thus, the formal underpinning of the Lewis formula is nonorthogonal VB theory.

An analogous development in inorganic chemistry was the proposal by Sidgwick of the electron pair coordinate bond,[3] an idea inspired by the Lewis concept. Every day, synthetic chemists plot strategies for making new molecules, mechanistic chemists formulate alternative pathways for a reaction, biochemists explore how molecules recognize each other, and computational chemists decide what should go next into the computer by scribbling Lewis structures of molecules and formulating options on this basis. Thus, it is hard to deny that the single most influential concept of chemical theory is the concept of the Lewis formula. [4] This implies that the theory best suited for chemistry is nonorthogonal VB theory.

Pauling, with his extensive knowledge of the facts of chemistry, proposed interpretations of diverse experimental findings using empirical VB theory and the notion of hybridization.[5] The emergent "resonance theory" provided chemists with a "kit" of useful concepts that could be used to rationalize interesting structure and reactivity trends. Ingold, Winstein, Bartlett, and Doering laid the foundation of physical organic chemistry[6] by carrying out systematic experimental studies of the reactivity of organic molecules and integrating the results into a conceptual framework that is still the backbone of every organic course at any level. Once again, mechanistic and stereochemical concepts were developed through manipulations of Lewis formulas rather than molecular orbitals. An analogous development occured in the area of inorganic chemistry where the "hard/soft acid/base" concept of Chatt and Pearson stands out.[7]

An example can be used to better make our point. In particular, the interpretation of the conformational preferences of organic molecules has been a oft-revisited topic since the very inception of conformational analysis. Acetaldehyde ($CH_3CH{=}O$) is an important molecule because replacing the three methyl hydrogens by three different groups, X, Y, and Z, generates a stereocenter vicinal to carbonyl. As a result, nucleophilic attack on $XYZCCH{=}O$ becomes a prototype of diastereofacial selectivity. This is how the biological world operates. It follows that the conformational preference of the parent acetaldehyde has broad implications. Now, much like propene, it is found that the preferred geometry has one methyl C—H bond eclipsing the vicinal C—O bond. Why?

Traditionally, the answer has been sought in simplified MO models by focusing on the stereochemical activity of the easy-to-handle π-type MOs or on selected frontier MOs.[8] VB theory suggests an entirely different "whole-molecule" view: by formulating the C═O double bond as a "double-banana" bond made up by overlapping sp^3-hybridized (rather than sp^2-hybridized) C and O atoms, and by using dummy atoms in the spirit of the Kahn–Ingold–Prelog *R/S* nomenclature system, we can formulate acetaldehyde as an ethane derivative. The resulting formula is called the "banana formula" (B formula):

B-Formula

We can now use what we have learned from ethane conformational isomerism to treat the conformational isomerism of every organic molecule on equal footing.

The explanation of the conformational preference of acetaldehyde is now straightforward: the stable form (C—H syn to C—O) is the "B-staggered" form and the higher energy conformer (C—H anti to C—O) is the "B-eclipsed" form in terms of the B formulas.

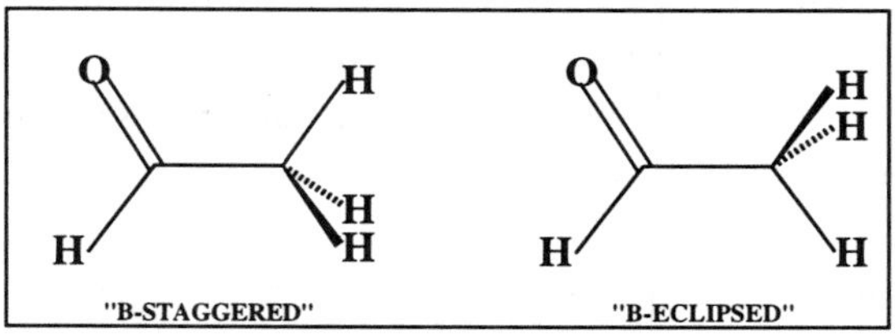

Let us pursue the issue. The planar molecules FHC═CHF and X═HC—CH═X (X = CH_2, NH, O) have isoelectronic σ frames. However, the former prefers the cis but the latter the transoid geometry. Writing the B formulas for the cisoid and transoid forms of the X═HC—CH═X reveals that these are "B-eclipsed" and "B-staggered," respectively.

The most important question remains: How does XYZC chirality determines the diastereofacial selectivity of XYZCCH═O? The answer is simple: The chiral methyl group selectively polarizes one of the two C—O banana bonds and renders it more susceptible to nucleophilic attack. We will return to this issue in Chapter 11 after we have seen how vicinal bonds in organic molecules interact with each other.

Bottom line: MO practitioners often rediscover the wheel by finding novelty in results that can be easily rationalized by conventional VB theory (why fluorination increases the inversion barrier of ammonia, what is the bonding of the [1.1.1]propellane, etc.).

With confidence that VB theory is the right *conceptual* tool, we now come to key questions: Has this tool been used properly in the past? Has the VB wavefunction been properly decoded? What is the difference between Heitler–London ("covalent-only") VB theory and a full VB treatment, not in numerics but in concepts? What is the difference between a full VB treatment and the popular Hückel MO theory? The answers to these questions define a path that ultimately leads to a new overview of chemical bonding. The essential point is that if VB theory is the "natural" theory of chemistry, failure to understand chemical bonding implies that VB theory has not been used properly — for example, that Lewis structures and Pauling covalent resonance are just the beginning of the story.

2.2 The Two Brands of Delocalization and the Three Mechanisms of Chemical Bonding

The gospel of conceptual VB theory is: Atoms (or fragments) are "promoted" ("excited") in return for "bond making," which produces a *net* "energetic profit" called the binding energy. We can envision the process as occurring by a sequence of three steps.

Step 1. Promotion of the ground atoms to their valence state at infinite interatomic distance. The energy requirement is called the atom promotional energy. The spin-adapted molecular configuration built from the promoted atoms is called the *reference configuration.* This configuration is a linear combination of *spin configurations.* A spin configuration is also called a *primitive configuration.*

Step 2. The promoted atoms are brought to the equilibrium interatomic distances while AO overlap is turned off. The energy lowering that accompanies this process is the *W, static interaction,* component of the total binding energy. This is the sum of the classical electrostatic interaction (EL) plus the dipole–induced dipole (induction, IND) and induced dipole–induced dipole (dispersion, DISP) interactions. The latter two constitute *deshielding* mechanisms: the exposed core of one atom or fragment attracts the electrons of a second atom or fragment.

$$W = \text{EL} + \text{IND} + \text{DISP}$$

The three different components of W are differentiated in a fundamental way: EL requires zero promotion, IND requires single promotion, and DISP requires double promotion. The static interaction is an *overlap-independent* interaction.

Step 3. Overlap is turned on, and this results in *delocalization* of the electrons. In nonorthogonal VB theory, there exist *two* mechanisms of delocalization:

Exchange delocalization, which involves exchanging electrons as follows (note how the two-electron exchange hop creates zero interatomic electron transfer):

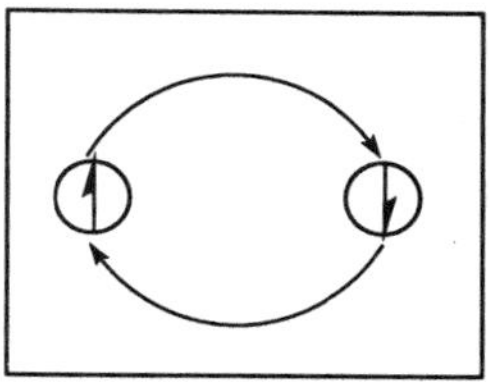

2-e EXCHANGE HOP

In organic molecules, the lowest energy exchange–stabilized, spin–adapted configuration is called the Lewis ("covalent") structure.

Charge Transfer (CT) delocalization, which involves interatomic electron transfer as follows:

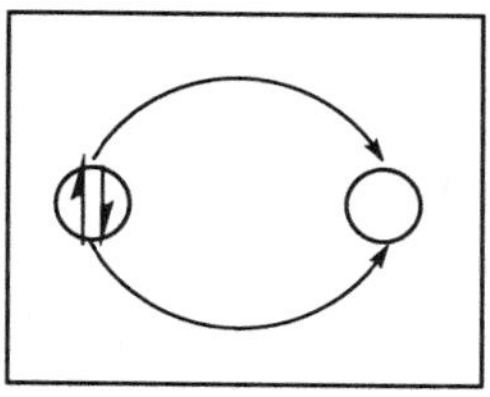

2-e CT HOP

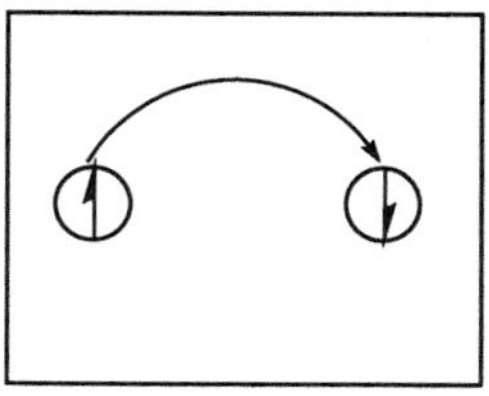

1-e CT HOP

Note how the two-electron CT hop is related to the two-electron exchange hop: The former preserves atomic charge, while the latter alters it by two units.

Focusing now on the physics of the problem, two important points need to be stressed. The first is that exchange and CT delocalizations are two delocalization modes that complement each other in a way that defines a continuum ranging from exclusive exchange to exclusive CT delocalization.

The second point is that two of the three mechanisms of static interaction, namely, induction and dispersion, can be viewed as intra-atomic, or, in general, intrafragmental CT delocalization in response to a permanent or induced electric field. The distinction between inter- and intrafragmental CT is arbitrary (because the definition of the fragments making up a system is arbitrary); thus it follows that there exists only one type of CT delocalization in which inter- is either assisted or antagonized by intrafragmental CT delocalization. This means that a meaningful theory of chemical bonding must involve a merger of the theories of interatomic and intermolecular bonding. We already anticipate a pedagogic problem: Organic and the inorganic chemists are subscribers to the covalent/ionic gospel (in either VB or MO language) who are uncomfortable with induction and dispersion. And yet, our description of what is commonly perceived as "covalent" will turn out to be not only "not covalent" but also heavily dependent on induction and dispersion as inseparable components of bonding due to CT delocalization.

The thesis of this work is that there exist three types of bonding:

(a) Bonding due to exchange delocalization, called *T-bonding.*

(b) Bonding due to CT delocalization complemented by dispersion and induction, called *I-bonding.* In homonuclear systems, the complementary action of dispersion is preeminent.

(c) Bonding due to electrostatic interaction complemented by induction and dispersion, called *E-bonding*.

In this light, *the novelty of this work lies in the recognition that the distinction between exchange, CT, and electrostatic bonding is operationally significant in a chemical sense.* For this reason, we must appreciate early on the key differences between the three bonding mechanisms:

(a) The feasibility of exchange delocalization is atom independent, that is, it does not depend on atom electronegativity because atomic charge is conserved. By contrast, the feasibility of CT delocalization is atom- dependent, that is, it *does* depend on atom electronegativity.

(b) Exchange delocalization can be either attractive or repulsive. Attractive exchange is equivalent to T-bond formation and repulsive exchange is equivalent to T-antibond formation. By contrast, CT delocalization is always attractive. CT delocalization is equivalent to I-bond formation. Whether the exchange delocalization is attractive or repulsive depends on the number of electrons and the total spin (see below). This is our first encounter with a dependence of stability on electron count and spin, a theme that reappears in the discussion of aromaticity. It follows that since exchange delocalization is a consequence of the Pauli principle, aromaticity should also have the same underpinning. This is what we will argue later on.

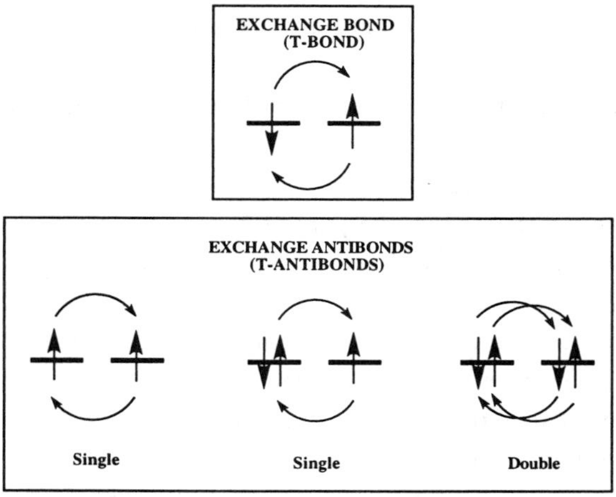

(c) Either exchange or CT delocalization can occur by a sequence of odd and/or even electron hops, the only exception being that a one-electron exchange hop is impossible. Energetically, the one-electron (1-e) hop of the odd family and the two-electron (2-e) hop of the even family are the most important. The overview of exchange and CT delocalization is shown schematically as follows.

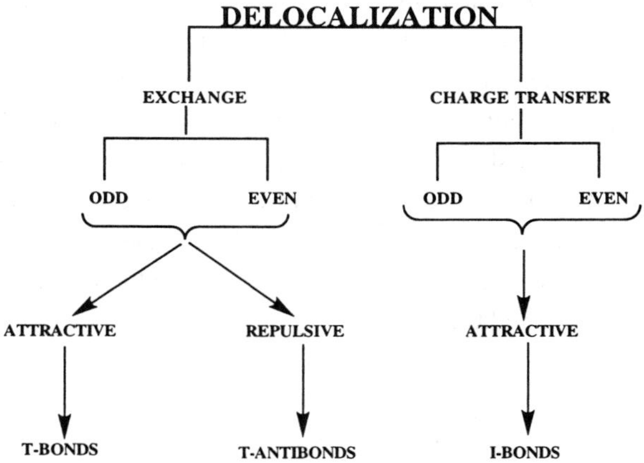

(d) Overlap-dependent delocalization (either exchange or CT) and overlap-independent E-bonding are two sides of the same coin. The feasibility of CT delocalization is atom dependent, as discussed above, and so is the feasibility of E-bonding. The condition for strong electrostatic bonding (EL component) is that the electron density is contracted rather than diffuse; that is, the best supporters of strong electrostatic attraction are small and compact anions and cations. The condition for strong inductive and/or dispersive (IND and DISP component) is that the atoms (or fragments) are polarizable.

Let us differentiate the three mechanisms from conventional models:

(a) Conventional covalent bonding focuses on T bonds. By contrast, the term "T-bonding" describes the interplay of T bonds and T antibonds. As we will see in Chapters 6–8, this will revolutionize our understanding of structural organic chemistry.

(b) I-bonding has no conventional counterpart, and yet this is the type of bonding of transition states and organometallics.

(c) Conventional ionic bonding assumes a classical interaction of ions formed by the combination of atoms of widely different electronegativities. By contrast, E-bonding can be dominated either by the electrostatic mechanism, which corresponds to ionic bonding, or by the inductive and dispersive mechanisms. Electrostatic E-bonding is expected in heteronuclear systems, while inductive/dispersive E-bonding is the norm in the homonuclear variety. As we will see in Chapter 23, this leads to a different interpretation of "hard" and "soft" behavior in ion combination reactions.

All in all, the T, I, and E mechanisms of chemical bonding are either partly or totally distinct from the age-old models that drive chemical thinking.

We will first focus attention on the two delocalization mechanisms of chemical bonding, seeking to answer a fundamental question: What is the physical meaning and what are the chemical implications of exchange and CT delocalization? Later on, and after some applications have convinced the reader that our target is real chemistry and not formalistic calisthenics, we will confront the next issue: How does intrafragmental CT (i.e., induction and dispersion) complement interfragmental CT?

2.3 VB Matrix Elements and the Consequences of Delocalization

The way in which the overlap interaction stabilizes a system has to do with the manner of interaction of two VB configurations defining exchange or CT delocalization. According to perturbation theory, the magnitude of the stabilization of a lower configuration as a result of its mixing with an upper one depends on two factors:

(a) The matrix element, U_{ab}, connecting two configurations, O_a (lower) and O_b (higher). This has the form:

$$U_{ab} = H_{ab} - H_{aa}S_{ab}$$

where H_{ab} are Hamiltonian and S_{ab} overlap integrals.

(b) The energy gap separating the two configurations, ΔE_{ab}.

The larger U_{ab} and the smaller ΔE_{ab} (in absolute magnitude), the stronger the interaction. In fact, we can think of the corresponding energy lowering as due to the characteristic matrix element U_{ab}, modulated by the delocalization promotional energy, ΔE_{ab}. There are two types of U_{ab}: one corresponding to exchange, U_{ab} (EX), and one corresponding to CT, U_{ab} (CT), delocalization. This defines the central problem of chemistry: What are the circumstances under which one or the other delocalization mechanism dominates, and what are the consequences? Before we consider this, the major issue, let us first take a look at how delocalization, regardless of type, effects chemical bonding.

Delocalization is the act of transferring electron density to the overlap region, and it is represented by the interaction of two overlapping VB configurations a and b, and this depends on $U_{ab} = H_{ab} - H_{aa}S_{ab}$, where H_{ab} is a sum of *atomic* one-electron h_{mn} terms, denoted by B_{ab}, plus a sum of two-electron terms, denoted by G_{ab}:

$$H_{ab} = B_{ab} + G_{ab}$$

S_{ab} is a sum of AO overlap s_{mn} terms. We can express each of h_{mn} in terms of energy contributions from the kinetic, T, and *one-electron* (nucleus–electron attraction) potential, F:

$$h_{mn} = T_{mn} + F_{mn}$$

We use two approximations to further simplify things. The first (F_{mn}) is the Mulliken[9] and the second (T_{mn}) the Ruedenberg/Cusachs[10] approximation. Note the pivotal difference between F_{mn} and T_{mn} with respect to the dependence on overlap.

$$F_{mn} = \frac{\left(F_{mm} + F_{nn}\right)s_{mn}}{2} = F_{mm}s_{mn} \qquad (F_{mm} = F_{nn})$$

$$T_{mn} = \frac{\left(T_{mm} + T_{nn}\right)s_{mn}^2}{2} = T_{mm}s_{mn} \qquad (T_{mm} = T_{nn})$$

From the virial theorem, we also have for a one-electron atom:

$$T_{mm} = -0.5\,F_{mm} = -\,h_{mm}$$

In addition, we have:

$$-\,h_{mm} = \mathrm{IE}_m \;@\; \mathrm{ELNG}$$

where IE_m is the ionization energy of orbital m and ELNG is the atom electronegativity as defined by Allen.[11a] The important result is that T_{mm} is proportional to IE_m and atom electronegativity.

$$T_{mm} \;@\; \mathrm{IE}_m \;@\; \mathrm{ELNG}$$

These expressions allow us to write the one-electron parts of the H_{ab} as sums of one-center terms F_{mm} and T_{mm} multiplied appropriately by s_{mn} overlap integrals. However, H_{ab} contains also bielectronic terms (which upon summation yield G_{ab}), and a useful approximation, again due to Mulliken, allows us to express repulsion integrals in terms of two-orbital repulsion integrals premultiplied by s_{mn} overlap integrals:

$$(ij|kl) = 0.25 s_{ij} s_{kl}\left[(ii|kk) + (ii|ll) + (jj|kk) + (jj|ll)\right]$$

We can now obtain simple expressions for the characteristic matrix element U_{ab} in terms of T_{mm}, F_{mm}, and G_{ab} terms. The kinetic contribution to bonding is the sum of the T_{mm} terms. The potential contribution is the sum of the F_{mm} and G_{ab} terms.

Having differentiated between two different modes of delocalization, exchange and CT, the key question becomes: What does delocalization, whether exchange or CT, accomplish? Traditionally, this question is answered by focusing on the CT delocalization responsible for the bonding of the prototypical H_2^+ ion. For example, Kutzelnigg tackled the problem of the origin of the one-electron bond of the H_2^+ radical cation within the framework of MO theory.[12] He noted that the quantity that *decreases* upon turning on AO overlap at the equilibrium interatomic distance is the electron kinetic energy, while the potential energy remains relatively constant. This appears to contradict the inference drawn from the virial theorem which predicts that bond formation is due to a twofold *decrease* of the potential energy outweighing a onefold *increase* of the kinetic energy. The apparent conflict was resolved by recognizing that the AOs may contract upon bond formation in order to optimize the total energy. This intra-atomic promotion, combined with the subsequent CT delocalization, satisfies the virial theorem while, at the same time, crediting bond formation to kinetic energy reduction.

At the VB level, the bonding of H_2^+ is due to the CT delocalization represented by the interaction of the two resonance structures H H^+ (O_a) and H^+H (O_b), where *a* and *b* denote configurations. Denoting the AOs of the first and second hydrogen atoms by 1 and 2, respectively, we can express the matrix elements that make up U_{ab} in atomic terms as follows:

$$H_{ab} = h_{12} \qquad H_{aa} = h_{11} \qquad S_{ab} = s$$

By using the approximate forms for the F_{mn} and T_{mn} discussed before, we obtain:

$$U_{ab} = s(s-1)T_{11}$$

Because T_{11} is positive and $(s-1)$ negative, the one-electron bond of H_2^+ must be due to reduction of kinetic energy. Note the key features of the expression:

Bonding disappears when the overlap is zero.

Bonding is maximal for $s = 0.5$.

When $s = 1$, we have one and the same AO, zero AO interaction, and no definable bond. Thus, it makes physical sense that U_{ab} goes to zero.

The "particle in a box" problem is the prototype of what happens as delocalization is turned on in our model two-electron diatomic. Specifically, one electron originally confined to one atom can now delocalize over both atoms. This is analogous to increasing the length of the box containing the particle. In turn, this lowers the kinetic energy of the particle in the box. From the chemical standpoint, the critical question is: How does the magnitude of the overlap interaction depend on the nature of the atoms? The answer is straightforward: The kinetic energy reduction is proportional to the orbital ionization energy and, hence, to orbital (and atom) electronegativity. Bare two-electron exchange bonds (i.e., bonds undisturbed by surrounding lone pairs or bond pairs) get stronger as atom electronegativity increases.[13] The same is true of three- and four-electron exchange antibonds.[14]

References

1. (a) C. Sandorfy, *Electronic Spectra and Quantum Chemistry*, Prentice Hall, Englewood Cliffs, NJ, 1964. (b) L. Salem, *Electrons in Chemical Reactions*, Wiley, New York, 1982. (c) R. McWeeny, *Proc. R. Soc. A*, 223, 306 (1953). (d) F.M. Bobrowicz and W.A. Goddard III, in *Modern Theoretical Chemistry*, Vol. 3, *Methods of Electronic Structure Theory*, H.F. Schaefer III, Ed., Plenum Press, New York, 1977. (e) D.L. Cooper, J. Gerrat, and M. Raimondi, *Chem. Rev.* 91, 929 (1991).
2. (a) G.N. Lewis, *J. Am. Chem. Soc.* 35, 1448 (1913); *ibid.* 38, 764, 768, 769, 772, 777, 779 (1916). (b) G.N. Lewis, *Valence and the Structure of Atoms and Molecules*, Chemical Catalog Company, New York, 1923.
3. (a) N.V. Sidgwick, *Electronic Theory of Valency*, Oxford University Press, London, 1927. (b) N.V. Sidgwick and H.M. Powell, *Proc. R. Soc.* 176A, 164 (1940). (c) R.J. Gillespie and R.S. Nyholm, *Inorg.Stereochem., Q. Rev.* 11, 341 (1951).
4. *Chem. Eng. News* 71(44), 25 (1993).
5. (a) L. Pauling, *J. Am. Chem. Soc.* 53, 1367 (1931). (b) L. Pauling, *The Nature of the Chemical Bond*, 3rd ed., Cornell University Press, Ithaca, NY, 1960.
6. No single monograph does justice to the works of these pioneers. Some classical studies and concepts are discussed in T.H. Lowry and K. Schueller-Richardson, *Mechanism and Theory in Organic Chemistry*, Harper & Row, New York, 1976.

7. (a) S. Ahrland, J. Chatt and N.R. Davies, *Q. Rev.* 11, 265 (1958). (b) R.G. Pearson, *J. Am. Chem. Soc.* 85, 3533 (1963).
8. Y. Jean, F. Volatron, and J. Burdett, *An Introduction to Molecular Orbitals*, Oxford University Press, New York, 1993.
9. R.S. Mulliken, *J. Chim. Phys.* 46, 497 (1949); 46, 675 (1949).
10. (a) K. Ruedenberg, *J. Chem. Phys.* 34, 1892 (1961). (b) L.S. Cusachs, *J. Chem. Phys.* 43, S157 (1965). (c) A discussion of extended Hückel MO matrix elements can be found in S.P. McGlynn, L.G. Vanquickenborne, M. Kinoshita, and D.G. Carroll, *Introduction to Applied Quantum Chemistry*, Holt, Rinehart and Winston, New York, 1972, Chapter 3.
11. (a) L.C. Allen, *J. Am. Chem. Soc.* 111, 9003 (1989). (b) G.-H. Liu and R.G. Parr, *J. Am. Chem. Soc.* 117, 3179 (1995).
12. W. Kutzelnigg, *Angew. Chem. Int. Ed. Engl.* 12, 546 (1973).
13. For a defense of VB theory, see D.J. Klein, *Pure Appl. Chem.* 55, 299 (1983).
14. A simple illustration of the chemical import of a three-electron antibond is the β-scission of radicals in which the cleaving C—C bond is spatially constrained to overlap in quasicyclic fashion with the radical center. For data, see M.-S. Lee, D.A. Hrovat and W.T. Borden, *J. Am. Chem. Soc.* 117, 10353 (1995).

Chapter 3

The New Valence Theory

3.1 The Chemical Classification of VB Configurations

A VB configuration has elements of three different types: holes (h), odd electrons, or, more succinctly, dots (r), and pairs (p). The latter are subdivided into three categories: lone pairs (p), bond pairs (bp), and geminal pairs (gp). Except for the geminal pair, all these entities are familiar to the chemist. While a full discussion is deferred until later, the essential point is that the two gp electrons reside in two hybrid AOs which have been constructed by mixing two (or more) canonical AOs, x and y. Interhybrid CT delocalization engenders a geminal bond (G bond) the strength of which is proportional to the energy gap separating x and y. Localization of the gp in the two hybrids in preparation for bonding requires expenditure of promotional energy.

The set of VB configurations necessary for describing even a "small" system is, by nature and size, intimidating. The conceptual breakthrough of this work starts with a classification that paves the way to a new valence theory. Specifically, VB configurations can be covalent or ionic, and these categories are subdivided as follows.

Each *covalent configuration* represents a different blend of exchange bonds and antibonds which is optimized at different molecular geometries. The structure that has the maximum number of bonds and minimum number of antibonds in a given geometry is called the perfect-pairing (PP) structure. A higher energy covalent structure, which represents an alternative coupling scheme that may attain lowest energy in a different geometry, is called a no-bond (NB) structure.

The *ionic configurations* are produced by starting with the PP configuration and turning q bond or geminal pairs into holes (vacant AOs) and pairs (filled AOs). These are called HP(q) configurations, where q is the ionicity index. In homonuclear systems, the facility of effecting such a transformation depends on atom electronegativity. Thus, q depends critically on atom electronegativity and, more generally, on AO electronegativity. In what will turn out to be an important matter, there are two types of HP(q) configuration:

(a) HP(q)* is defined as the ionic configuration that satisfies two conditions: it has the highest q value of all ionic configurations that make an appreciable contribution to the total VB wavefunction, but it also has the lowest energy among all ionic configurations of the same *q*. The ionicity index *q* of HP* is interpreted as the indicator of the number of pairs that become associated.

(b) HP(max) is the configuration that contains only holes and lone pairs. By assuming that HP* = HP(max), we have the oxidation state formalism of the inorganic chemist.

Examples are shown in Figure 3.1.

3.2 The Competition Between Exchange and Charge Transfer Delocalization; Covalency Represents a Restriction!

Hydrogen atoms combine to form H_2 molecules, which repel each other. By contrast, lithium atoms stick together to form a solid. What is the difference between the VB wavefunction of H_n and that of Li_n? The seemingly reasonable answer is: H has an active H1s but Li has active Li2s and Li2p AOs. Hence, the difference must lie in the number of active valence AOs. The availability of low-lying vacant 2p AOs in Li can promote "electron-deficienT-bonding" of the "two-electron/three-orbital" type first discussed by Longuet-Higgins in connection with the structure of diborane.[1] It is no wonder that this very seductive interpretation is the predominant viewpoint.

There is a different type of "chemical common sense" which says that in the absence a global overview of the problem, acceptance of theories, no matter how apparently convincing, is an invitation to disaster. So, here is an entirely different scenario which, in fact, is the thesis of this work: The difference between the H_n and Li_n wavefunctions lies in the balance of exchange and CT delocalization, the latter coupled to overlap-independent mechanisms of bonding.

We use H_2 as the illustrator molecule. There are four primitive electronic configurations, two diradical ($|1\bar{2}|$ and $|\bar{1}2|$) and two zwitterionic ($|1\bar{1}|$ and $|2\bar{2}|$), which combine to form the covalent, *R*, and the ionic, *Z*, symmetry-adapted configurations. For simplicity, we write only the form of the wavefunction,

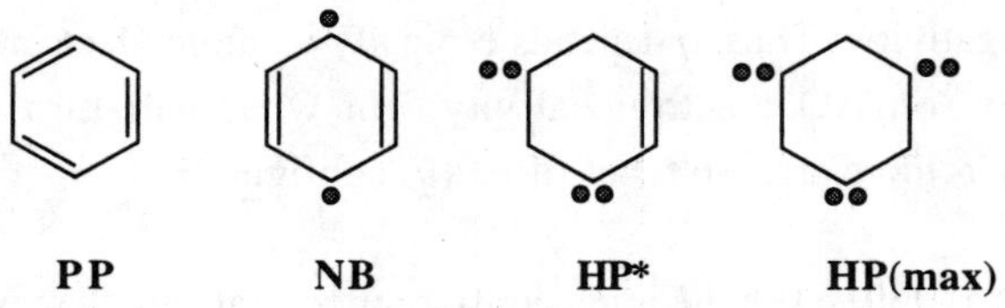

Figure 3.1 Prototypical VB configurations of benzene.

dropping the normalization constant. The notation $|123\ldots|$ symbolizes a determinant built from spin orbitals $1, 2, 3\ldots$, according to standard convention.[2]

$$R = |1\bar{2}| + |2\bar{1}|$$
$$Z = |1\bar{1}| + |2\bar{2}|$$

The interaction of the diradical configurations $|1\bar{2}|$ and $|2\bar{1}|$ defines a 2-e hop corresponding to exchange delocalization. The attendant energy lowering is denoted by A, where A is the *VB exchange integral.* We refer to the electron pair bond of H_2 as an *exchange bond.* When the sign of R is reversed (triplet rather than singlet H_2), and whenever there are more than two electrons in overlapping AOs, exchange delocalization produces *exchange antibonds.* Either two electrons with parallel spins (e.g., triplet H_2) or three electrons in two overlapping AOs (e.g., HeH) create an A-exchange antibond. Four electrons in two overlapping AOs create a 2A-exchange antibond (e.g., He_2). An exchange bond represents lowering and an exchange antibond represents raising of the kinetic energy as a result of overlap.

The interaction of the diradical $|1\bar{2}|$ with the higher lying ionic $|2\bar{2}|$ defines a 1-e hop that improves the bonding. This is now a case of 1-e CT delocalization. The exchange 2-e hop is much superior to the CT 1-e hop because of a smaller (zero) energy gap of the interacting primitive configurations, but let us for the moment disregard this property and compute the characteristic matrix elements U(CT*) and U(EX), where CT* indicates a 1-e CT hop and EX a 2-e exchange hop. The results are as follows:

1-e CT electron hop:

$$S_{ab} = s$$
$$H_{ab} = h_{12} + h_{11}s + (12|22)$$
$$H_{aa} = 2h_{11} + (11|22)$$
$$U\,(\text{CT*}) = s(s-1)\,T_{11} + 0.5\,s\,(J_{22} - J_{12}), \text{ where } J_{12} = (11|22), \text{ etc.}$$

2-e exchange electron hop:

$$S_{ab} = s^2$$
$$H_{ab} = 2h_{12}s + (12|12)$$
$$H_{aa} = 2h_{11} + (11|22)$$
$$U(\text{EX}) = Q = 2s^2(s-1)\,T_{11} + 0.5\,s^2\,(J_{22} - J_{12})$$

The major conclusions are:

(a) Exactly as in the monoelectronic case of H_2^+, bonding in H_2 is due to reduction of kinetic energy: that is, the negative $(s-1)T_{11}$ term premultiplied by either s (1-e CT hop) or $2s^2$ (2-e exchange hop).

(b) Because T_{11} is proportional to orbital electronegativity, it follows that the more electronegative an atom (or an orbital, in comparing two identical atoms), the stronger the kinetic energy reduction and the stronger the bonding.

(c) Setting $s = 0.5$, the value that maximizes the $s(s-1)$ term, we see that kinetic energy reduction is the same for the 1-e and 2-e hops because $2s^2 = s$. For smaller values, the 1-e hop is superior in reducing the kinetic energy. Operating in a direction exactly opposite to the kinetic energy term is electron repulsion which is s times smaller for the 2-e hop.

It follows that when overlap is appreciable (e.g., of the order of 0.5), the 2-e exchange will be *intrinsically* more favorable than the 1-e CT hop on account of smaller interelectronic repulsion. Hence, unless the energy gap separating two configurations dictates otherwise, 2-e delocalization, whether of the exchange or the CT type, can be, at the very least, competitive with 1-e CT delocalization.

The basis of the new valence theory is the recognition that *2-e delocalization of the CT type can be even more favorable than 2-e delocalization of the exchange type.* This is illustrated by the linear symmetrical H_3^+ prototype. The following 2-e CT hop can be thought of as a "relay 2-e hop" or "billiards 2-e hop."

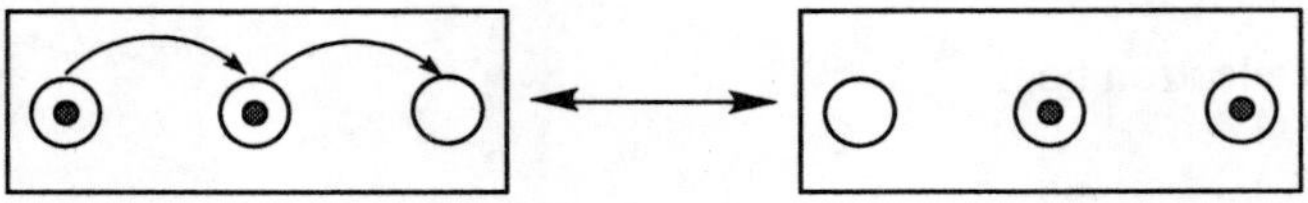

The U(CT) characteristic matrix element of H_3^+ produces the same kinetic energy reduction as the exchange U(EX) for H_2. On the other hand, the electron repulsion term (in brackets) now approaches zero (i.e., electrons delocalize without "feeling" each other):

$$U(\mathrm{CT}) = 2s^2(s-1)T_{11} + 0.25s^2[(J_{22} - J_{12}) - 2(J_{12} - J_{13})]$$

We have arrived at an important conclusion: Two arrows symbolizing an exchange 2-e hop and located within the same overlap region can be advantageously replaced by two arrows symbolizing a CT 2-e hop but now located in different overlap regions because of diminished interelectronic repulsion. This is illustrated in Figure 3.2, which is the pictorial representation of the central thesis of this work, namely: *Covalency represents a restriction imposed by the nature of the linked atoms, and this is a direct consequence of electron–electron repulsion.* The recognition that two arrows within the same overlap space define one T bond while two arrows within different overlap spaces define one I bond paves the way to the development of a new theory of chemical bonding.

The physical picture is simple: An electron pair switches from an exchange to a CT mode of delocalization to minimize intrapair interelectronic repulsion. For this to happen, the two electrons that make up the pair must avoid the "observer" electrons; that is, inter- must phase in with intrapair correlation. This becomes possible if the energy gap ΔE that separates a lower covalent from an upper ionic configuration that define the 2-e CT hop is small. The value of ΔE depends on orbital ionization energy (and atom electronegativity) because the latter is a measure of the self-repulsion integral J_{ii} which, in turn, is a measure of the energy of the ionic configurations relative to the lower lying covalent configurations. The magnitude of J_{ii} is approximately equal to the ionization

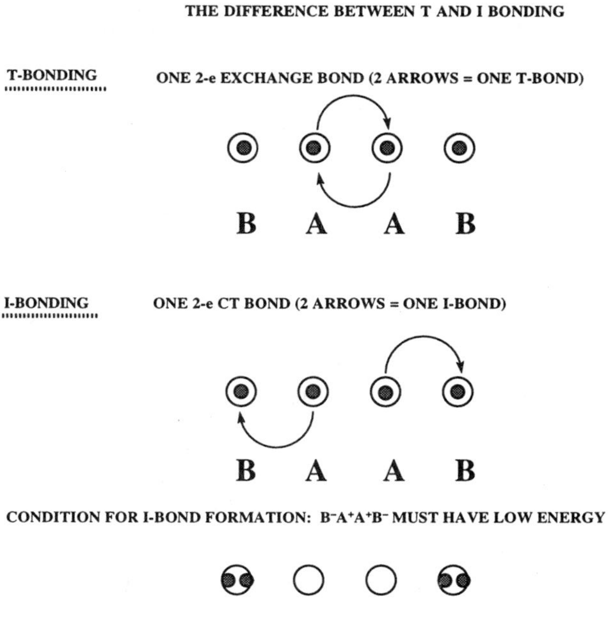

Figure 3.2 The crucial recognition: Two arrows make up either an exchange (T) or a CT (I) bond.

energy minus the electron affinity of the orbital *i*:

$$J_{ii} = \mathrm{IE}_i - \mathrm{EA}_i$$

Hence, low atom electronegativity means that electron pairs can switch from the exchange to the CT 2-e hopping mode because they are capable of avoiding the surrounding electrons (interpair correlation). The energy gap is the crucial index that tells us to what extent a T bond has acquired I-bond character.

The conceptual models of chemistry neglect electron–electron repulsion. As a result, they cannot perceive that covalency (i.e., T-bonding) is *not* the ideal type of bonding, and no one has been driven to seek an alternative type of bonding in which perfect pairing and covalency are either insignificant or secondary. We discovered this new type of bonding in the experimental facts (especially those summarized by the Chatt–Pearson hard/soft, acid/base construct). High level ab initio computations of small metal-containing systems, especially those of E.R. Davidson,[3] W.A. Goddard III,[4] and J.P. Malrieux,[5] provided reassurance that the conventional bonding models do not capture the physics of the problem and hinted that electron–electron repulsion plays a role beyond what is commonly assumed.

We have concluded that electron–electron repulsion makes a CT superior to an exchange 2-e hop. Exactly the same principle manifests itself in the comparison of the one-electron (1e) and three-electron (3e) bonds to the CT complement of the exchange bond. As illustrated below, the 1-e hop generating the 1e bond is free of electron–electron repulsion.

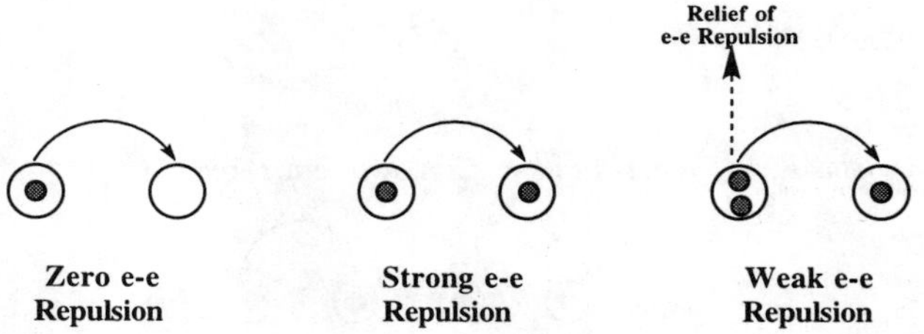

Similarly, the 1-e hop generating the 3e bond relieves interelectronic repulsion. By contrast, the 1-e hop complementing an exchange bond only introduces interelectronic repulsion. We caution that these conclusions are based on consideration of the $H_{ab} - H_{aa}S_{ab}$ matrix element, as it should always be the case in nonorthogonal theory. This leads to a prediction: As the electronegativity difference between metal and nonmetal increases, a 3e bond should improve relative to a 2e bond. The data given below are revealing. Assuming that Mn binds a univalent atom from the stable $4s^2 3d^5$ configuration by a 3e bond, while Cr binds a univalent atom from the stable $4s^1 3d^5$ configuration by a 2e bond, the counterintuitive variation of the MX (M = Cr, Mn) bond dissociation energies (BDEs, in kcal/mol) now makes sense. As X becomes more electronegative (thus promoting stronger CT), the 3e bond of Mn becomes superior.

CrF	92	CrCl	86	CrH	66
MnF	120	MnCl	85	MnH	55

CuCl	83	CuH	66
ZnCl	49	ZnH	20

3.3 The Configurational Profiles of the T, I, and E Mechanisms of Chemical Bonding; the Association Rule

In our discussion of matrix elements, we neglected the normalization constant, for convenience. This was done selectively for maximum simplification. On the other hand, the normalization constant in every nonorthogonal theory embodies a key concept: the in-phase (bonding) combination of two functions (whether AOs or primitive VB structures) generates a new wavefunction, which represents kinetic energy reduction. This new bonding function *Y* has a small normalization constant. Hence, it is incapable of interacting strongly with some other function, *X*, to further augmenT-bonding because the interaction matrix element $<YH|X>$ is premultiplied by the normalization constants of *Y* and *X*. By contrast, an antibonding function *Y** has a large normalization constant and is highly capable of interacting with some other function to further augmenT-bonding. For example, the normalization constants of the singlet and triplet Heitler–London functions of H_2 are $(2+2s^2)^{-1/2}$ and $(2-2s^2)^{-1/2}$. The greater the binding effected (i.e., the larger the overlap term entering the normalization constant), the weaker the "activity" of the function. This is why the mixing of the covalent and zwitterionic functions is actually extremely weak in H_2. Each effects the same bonding, and (in a frozen orbital approximation) each has the same small normalization constant.

In summary, covalent bonding functions have partly completed their mission and cannot be used further. By contrast, covalent antibonding functions have a large normalization constant and are ideally suited for effecting bonding by further interaction. This implies that a strong mixing of covalent and ionic configurations can occur only when two conditions are met:

The covalent configurations become quasi-degenerate.
The energy gap separating the covalent an ionic structures is diminished.

With this preamble, let us now consider an AB_2 three-membered ring which can be visualized as "A plus B=B" as our bonding prototype. When A = B = CH_2, we have the organic cyclopropane. When A = L_2Pt and B = CH_2, we have an organometallic metal–olefin complex. Chemists find no difference between the two other than the different blends of resonance structure involved. By contrast, we will argue that the two are bound by different mechanisms and that this difference is encoded in the stacking pattern of the valence VB configurations.

A VB configuration has a characteristic number of formal T bonds and T antibonds and a characteristic ionicity index. How does the spacing of VB configurations depend on exchange and interelectronic repulsion?

First, nonequivalent configurations of the same ionicity index differ in the exchange energy (i.e., in the number of exchange bonds and antibonds). Since the strength of an exchange bond or antibond depends on the orbital ionization energy IE_i, the spacing of configurations of fixed ionicity depends on orbital electronegativity.

In addition, configurations having the same exchange energy but differing in ionicity index are distinguished by the number of hole–pair couples. The energy required for turning two odd electrons into an hole–pair couple is the self-repulsion Coulomb integral J_{ii} which, as we have already seen, is approximately equal to the ionization energy minus the electron affinity of orbital *i*. Hence, the spacing of configurations of different ionicity index but constant exchange *also* depends on orbital electronegativity.

We can now see three possible scenarios:

(a) In homonuclear systems (e.g., A = B in our example) or in heteronuclear systems in which atomic electronegativities are not widely different, two things happen as the *average* orbital (and, by extension, atom) electronegativity decreases:

1. The spacing of configurations of fixed ionicity order decreases (i.e., the covalent configurations tend to become quasi-degenerate).

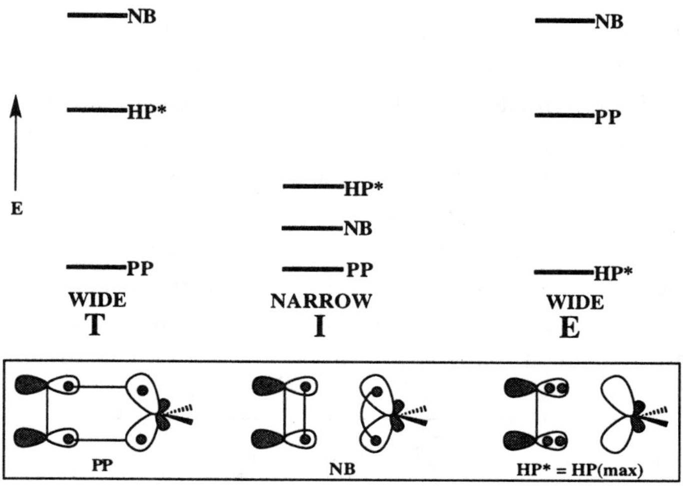

Figure 3.3 Three different stacking patterns of VB configurations characterized by wide or narrow bands of configurations. Diagram is schematic.

2. The spacing of configurations of different ionicity also decreases (i.e., the HP* configurations, and every HP configuration of lower ionicity index, tend to become quasi-degenerate with the covalent structures).

In other words, we have a transition from a *wide band* of VB configurations, in which the covalent configurations lie well below the ionic configurations, to a *narrow band* of VB configurations, where all are quasi-degenerate.

(b) In heteronuclear systems in which the component atoms differ greatly in electronegativity (e.g., A much more electropositive than B), we have a *wide band* of VB configurations in which the relative positions of PP and HP* have been inverted.

The three possible situations, represented schematically in Figure 3.3, define the three different mechanisms of bonding: T, I, and E. Each mechanism has a characteristic configurational profile, that is, a characteristic width of the band that encloses all valence configurations as well as characteristic widths of the covalent and HP(q) subbands. The band is wide in the case of T- and E-bonding and narrow in the case of I-bonding.

How are we to represent the three different types of AB_2 ring? It depends on whether the average electronegativity is high or low, and on the strength of the electronegativity variation.

High average electronegativity regime with zero or modest electronegativity variation. The PP lies far below all the other configurations with which it can still interact appreciably. The molecule is represented by the PP configuration. Each line represents a Heitler--London covalent bond. This is the first-order representation of a T-bound system, which will be modifiied in Chapter 6 to produce the revealing T-formula.

Low average electronegativity regime with zero or modest electronegativity variation. The configurations are quasi-degenerate, and bonding is now due principally to the interaction of all low energy covalent and ionic configurations, as shown in Figure 3.4.

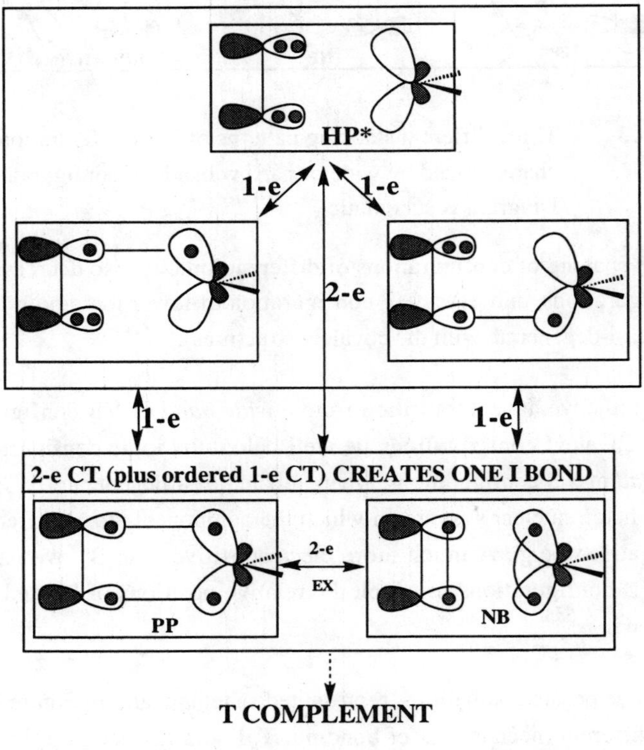

Figure 3.4 The interaction of the covalent and CT structures define one I bond. The exchange energy of each covalent configuration plus the exchange interaction of the covalent structures defines the T complement of the I bond created by the CT interaction. All CT structures can be generated by appending two arrows on the PP configuration and "executing" each arrow alone or in combination with the second.

The interaction of the covalent structures defines a set of resonating exchange bonds (T bonds) and exchange antibonds (T antibonds). The covalent resonance hybrid (T hybrid) is visualized as a "polyradical" in which the dots are "unlocked": that is, the different coupling schemes are quasi-degenerate. This sets the stage for omnidirectional CT delocalization which (unlike the previous case) is no longer restricted in directionality by a single coupling scheme. The interaction of the lowest energy ionic structures with the T hybrid defines the principal I descriptor, which is represented by appending arrows to the T hybrid, with the result that "execution" of one, two, . . ., arrows generates a unique ionic configuration. The allocation of arrows is made according to the *association rule:*

1. One arrow symbolizes the transfer of one electron.
2. A dot can have a maximum of one entering and one exiting arrow.
3. A pair can have a maximum of two exiting arrows.
4. A hole can have a maximum of two entering arrows.
5. Two arrows are "forbidden" from appearing in the same overlap region for this would produce an inferior exchange bond.

The principal I descriptor represents bonds created by CT delocalization but still having exchange character. The I-bound "A plus B═B" molecule is represented as follows:

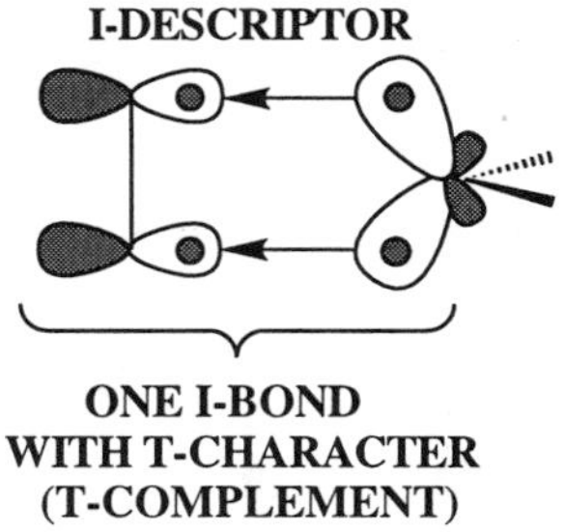

Strictly speaking, the reference basis is the T hybrid. However, for qualitative purposes, we can replace the latter by the lowest energy covalent structure(s). The arrows define a specific VB-CI that spans configurations produced by "executing" the arrows in singles and in pairs. The single execution is mandatory because a 2-e CT hop can be effected either directly or by two ordered 1-e CT hops. This configuration interaction (CI) is shown in Figure 3.4. Two CT arrows represent one I bond exactly analogous to the two exchange arrows that represent one T bond. A three-electron (one-electron) bond can be represented by drawing an arrow pointing from the pair (odd electron) to the odd electron (hole), and the single arrow is counted as one-half. The same goes for one two-electron bond resulting from CT delocalization. Thus, the number of I bonds can be either integral or half-integral. Each I bond has a T complement that, depending on the system, may be dominant. The T complement of an I bond is a crucial aspect of the bonding that is not explicitly shown but should always borne in mind. A low T complement implies strong association of electron pairs via CT delocalization (global minimum), while a high T complement implies instability with respect to segregated electron pairs (transition state). A molecule represented as an I-bound complex but having high T complement is called an *obligatory association complex*, to indicate instability with respect to fragmentation. A typical example is the transition state of a bimolecular organic reaction that is unstable with respect to the reactants infinitely apart. Finally, a more complete description can easily be accomplished by writing a hybrid of I descriptors, the I-hybrid, as shown in Figure 3.5.

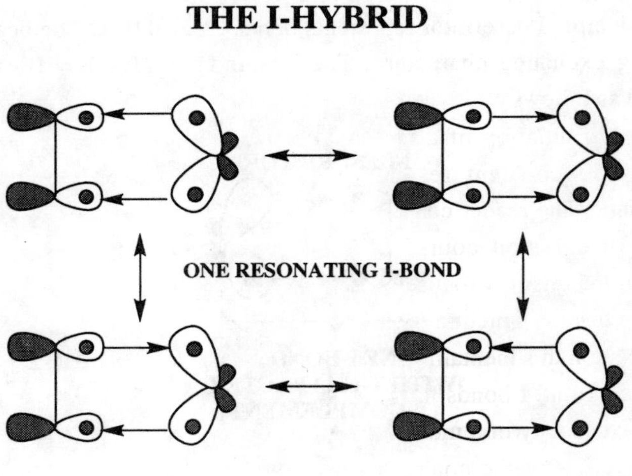

Figure 3.5 The resonating I bond. Henceforth, only principal I descriptors are shown; unless the pedagogy demands otherwise, equivalent I descriptors are omitted for brevity. In the general case of a polyelectronic system, *n* arrows represent the formation of *n*/2 I bonds defined by a VB-CI spanning all structures produced by the *n* arrows combined in singles, pairs, triplets, . . ., up to *n*-tuples.

Low average electronegativity regime but with strong electronegativity variation. The HP* lies far below the covalent configurations and fails to interact strongly with them. The molecule is represented, to a first approximation, by the HP* configuration. This is the representation of an E bound system.

The "arrow formulas" representing I-bonding are actually the expression of second quantization VB theory, since one arrow is a creation/annihilation operator; that is, one electron is created at the head and one electron is annihilated at the tail of the arrow. The conceptual revolution is manyfold, and this will become apparent as applications unfold. For the time being, we can get some first glimpses by focusing on the overlap-dependent T and I mechanisms:

(a) Since "bond due to overlap" means "exchange delocalization plus CT delocalization," one can equally well consider an overlap bond either as "exchange plus CT complement" or as "CT plus exchange complement." The former view is best for molecules made up of CH, C, N, O, and F. The second view is best for the rest of the elements. The "grand error" has been viewing the *whole* periodic table through the former lens (covalent bonding).

(b) As a result of (a), the vague notion of "delocalization" is replaced by concepts involving countable elements, namely, T bonds and I bonds. Of course, these are the limits of a T/I continuum, since every bond has both T and I character.

(c) The formulation of a molecule as either T- or I-bound hinges on the realization that covalent resonance is merely a preparatory step for CT delocalization. The reader can easily appreciate that this element of the theory puts us on a collision course with the traditional concepts of chemistry. Specifically, the chemist focuses on the lowest energy covalent structures. We call this the "frontier function model." For the sake of correct physics, however, we must abandon this mentality and consider all "active" configurations.

(d) Since T and I bonds have different chemical and physical implications, and since the capacity of a molecule to form bonds of either type depends on the electronegativities of the constituent atoms, we have a direct linkage of atom electronegativity and molecular properties. Thus, cyclopropane is "methylene connected by two T bonds to ethylene" but the Pt–ethylene complex is "Pt connected by one I bond to ethylene."

Here is the essence of our model: The movement of electrons into the overlap region reduces kinetic energy at the price of interelectronic repulsion. In other words, interelectronic repulsion in the overlap region is a form of promotion. When atom electronegativity is high, this type of promotion is justified; that is, the strong kinetic energy reduction outweighs the promotion expenditure. When atom electronegativity becomes low, the return (bonding due to kinetic energy reduction) hardly outweighs the investment (interelectronic repulsion in the overlap region). Now, a covalent bond is weak because of an unfavorable balance of kinetic energy reduction and electron–electron repulsion in the overlap region. However, the system has now a better way of effecting bonding, namely, by 2-e CT delocalization initiated from a "polyradical" state produced by exchange resonance.

3.4 I-Bonding and the Writing on the Wall

I-bonding represents the strategy by which atoms combine to reduce the kinetic energy with concurrent avoidance of interelectronic repulsion. This is consistent with the following facts.

The Hartree–Fock model is inadequate for treating metal-containing systems. This has been emphasized by Davidson.[6] Carter and Goddard computed the valence states of $RuCH_2^+$ and found that the $^2A_2$ ground state has conventional double bonding ("metal methylidene"), while a cluster of three degenerate excited states ($^4A_2$, $^4B_1$, $^4B_2$), lying about 13.0 kcal/mol higher, has σ-donor/π-acceptor bonding ("metal--carbene").[7] In a relative sense, the $^2A_2$ species has Coulomb correlation, and this can be described properly only at the CI level. By contrast, the $^4A_2$ species has exchange correlation, and this is adequately described at the HF level. Carter and Goddard obtained the following results by means of Hartree–Fock and generalized valence bond (GVB) computations. Clearly, correlation effects are important even in a molecule with just one metal atom and a metal–carbon "double bond."

Computation	ΔE ($^4A_2 - {}^2A_2$) (kcal/mol)
HF	– 20.1
GVB-PP	4.9
GVB-RCI	15.0

Covalency is not the best mechanism of bonding; that is, *covalency represents a restriction.* The $Ru{=}(CH_2)^+$ bond dissociation energy (in kcal/mol) computed by Carter and Goddard is -12 at the HF level, 27.6 at the Perfect Pairing GVB level and 68 at the GVB-CI level. Implictly, then, there exists another way of making bonds that is superior to perfect pairing. This is the I bond.

Because I bonds are formed by the collective CT interaction of *all* accessible covalent and ionic structures, one expects to find strong contribution from configurations that normally are unimportant in nonmetal systems. For example, covalent no-bond (NB) configurations must figure prominently in a wavefunction of a metal-containing molecule that, at first sight, appears to be bound by polar covalent bonds. In a recent paper, Cundari and Gordon computed a variety of transition metal alkylidenes in a multiconfiguration–self-consistent field (MC–SCF) frame and translated the results into a nonorthogonal VB wavefunction via MO.[8] Focusing on the metal–carbon formal double bond, we find that the PP "triplet metal plus triplet methylidene" and singly ionic (metal-to-carbon) configurations have the largest eigenvectors, as intuitively expected. However, it turns out that the "singlet metal plus singlet methylidene" NB configuration also has a very appreciable eigenvector (0.248, compared to 0.550 for the PP in the case of the Ti complex). Now, it is known that the eigenvectors of nonorthogonal functions do not accurately represent their energetic impact. Specifically, bonding functions (i.e., those having more exchange bonds than exchange antibonds), like the PP, are assigned disproportionately high eigenvectors relative to antibonding functions. This point was eloquently made by Gallup and Norbeck, who recomputed the eigenvectors of the VB wavefunction of BeH_2 in a way that properly exposes the true significance of each VB structure.[9] They found that the coefficients of the PP and NB structures shown below changed from to 0.474 (PP) and – 0.154 (NB) to 0.417 (PP) and 0.352 (NB)! We expect that when this problem is properly recognized and all nonorthogonal VB wavefunctions of metal-containing molecules are examined from this perspective, covalent structures other than the PP will turn out to play a key role.

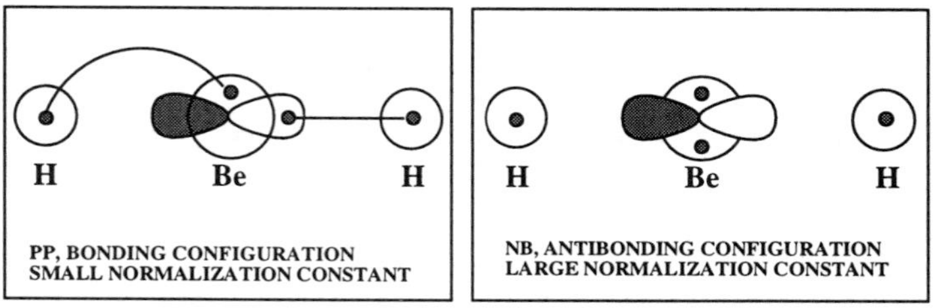

This work began when I became convinced that modern computational chemistry not only lacks the tools for analyzing chemical bonding but, far worse, paves the way to confusion because it offers the possibility of endless "achemical" reformulations that end up "confirming" the ideas of the 1930–1950 decades rather than breaking new ground. This should be clearly differentiated from the fantastic ability of modern computational chemistry to provide us with accurate numbers. In this light, I sought in the experimental literature a confirmation of the key postulate: *Covalent bonding (T-bonding) is a mere complement (T complement) of I-bonding in metal-containing species.* This point can be proven by experimental cases in which the PP has been eliminated and only the NB configuration is active. If our analysis is correct, we should see strong bonding, provided one or more component atoms are semimetals or metals. The conceptual experiment is shown in Figure 3.6.

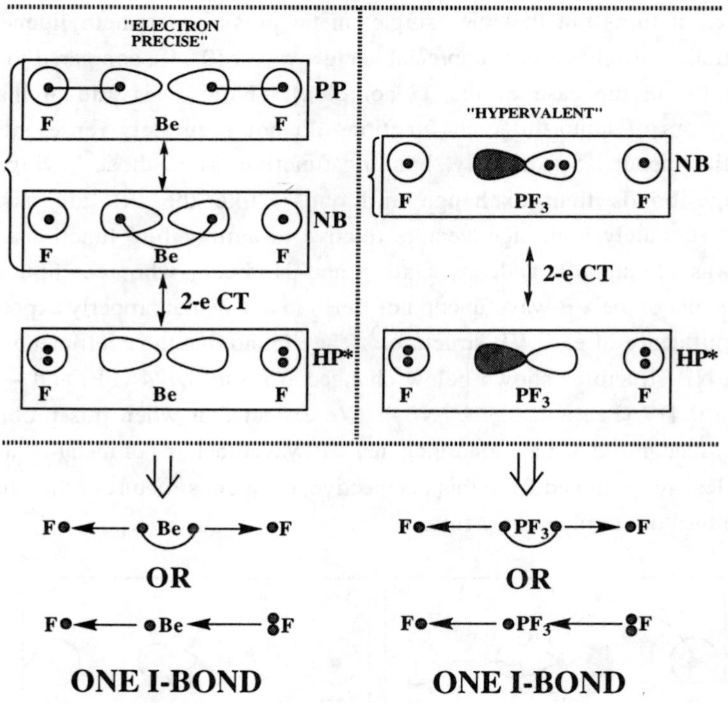

Figure 3.6 The CT resonance responsible for the binding of FBeF and $F(PF_3)F$. The 1-e CT hops are omitted. In each case, the 2-e CT hop is taken to imply the formation of a single I bond connecting a central atom or fragment with two ligands. The two molecules differ only in the T complement (i.e., exchange delocalization), as indicated in the lower portion of the figure.

Replacing a geminal pair of Be in F—Be—F by a lone pair of P in F—PF_3—F should wipe out bonding in the latter if perfect pairing is the key. However, bonding should be comparable if I-bonding is the common denominator. The data given below speak for themselves. Since we know that lower electronegativity of Be lends it greater capacity for CT delocalization than P and that PF_5 suffers enhanced nonbonded repulsion compared to PF_3, we can interpret the thermochemical data to mean that perfect pairing is just a mere complement of I-bonding. For example, the *difference* of the two sequential F_2 dissociations from PF_5 is only 64 kcal/mol, versus 190 kcal/mol for the first F_2 dissociation. Thus, the covalency complement is 64 kcal/mol when I-bonding is worth 190 kcal/mol. In other words, the covalency complement is about one-third of the actual I bond.

Reaction	BDE (kcal/mol)
F—Be—F $\rightarrow$ Be + 2F	+306
F—PF_3—F $\rightarrow$ F_3P + 2F	+190
F—PF—F $\rightarrow$ PF + 2F	+254

This turns the house upside down: All metal-containing molecules currently classified as "electron precise" (e.g., FBeF), "hypervalent" (e.g., FPF_3F), and "electron deficient" according to the covalent/ionic gospel must be actually I bound in a way that a *single* I bond associates three (rather than two) atoms or fragments!

At this early stage, it is helpful to state where all this is leading. The key idea is that two arrows make up one bond, and this can be of the exchange or the CT type. Once this has been visualized, all metal-containing molecules traditionally classified as "hypervalent," "electron precise," and "electron deficient" become analogous: the common denominator is now the I bond, which can link two ligands to one central atom as illustrated for our two apparently unrelated prototypes in Figure 3.6. The new challenge is to understand what makes an I bond strong and how an I bond differs from a conventional T bond.

The C—C is stronger than the Si—Si bond. This would lead one to infer that the C—H is also stronger than the Si—H bond, in accord with thermochemical data.[10] Thus, it is a big surprise to see that the situation reverses when H is replaced by a more electronegative halide. What is the explanation? Our argument has been that covalency represents a restriction. In other words, *all atoms would employ the I-bonding mechanism if only they could do so.* Metal atoms are qualified to do so, especially when combined with electronegative atoms, in which case CT delocalization becomes very favorable. Nonmetal atoms are too electronegative and can only rely on covalency. If electron repulsion were not differentiating T- from I-bonding, all nonmetal would be stronger than metal bonds to some reference atom simply because nonmetals are more electronegative and kinetic energy reduction, by itself, depends on atom electronegativity. The fact that Si—H is weaker than C—H but Si—Cl and Si—F are stronger than C—Cl and C—F bonds, respectively, is evidence of the superiority of I-bonding over T-bonding when CT becomes favorable. Replacing H by F allowed Si to fully exploit its capacity (low electronegativity) for I-bonding. By contrast, the (relatively) high electronegativity of C left it within the T domain of the bonding continuum. The superiority of the Si—halogen bonds is evidence of the intrinsic superiority of the I-bonding mechanism and has nothing to do with "d-orbital participation" as commonly envisioned.

Chemists have pondered "delocalization" and "resonance energy" for decades. Because of the neglect of electron–electron interaction, however, they never differentiated clearly between exchange and CT delocalization. In an implicit sense, they equate "delocalization" with "exchange delocalization" because they focus on the covalent resonance structures. In this way, they miss the crucial point that the type of delocalization that is dependent on atomic nature is the CT, not the exchange component. To see this point, consider what happens when we bring two H_2 molecules to a collinear geometry: The interbond exchange repulsion is counteracted, but only in part, by the exchange attraction that is represented by the interaction of the two familiar covalent structures. Because of the high electronegativity of hydrogen, CT delocalization is essentially turned off. The failure of two H_2 molecules to associate (neglecting long-range van der Waals minima) under standard temperatures and pressure conditions means that interbond exchange repulsion has won over exchange attraction. Furthermore, the failure of three H_2 molecules to form a globally stable H_6 means that covalent resonance, represented by the T descriptor, cannot cause association even when the resulting ensemble is "aromatic." The implication is clear: *Resonance involving the familiar covalent Kekulé structures is not the reason behind the stability of aromatic organic molecules.* If the stability of aromatic organic molecules were in face due to covalent resonance, aggregation would be independent of atom electronegativity. But, the very data cited in Chapter 1 clearly show that this is not the case. We now have the explanation: A T bond (exchange bond) links (by two arrows) two dots, and this causes bond pair segregation. By contrast, an I bond can associate (also by two arrows) three dots because of the operation of the association rule.

Most people compute and think in MO terms. In this context, the physical meaning of "correlation energy" is said to be the reduction of the ionicity of an MO wavefunction. This interpretation is wrong for, in a VB calculation, perfect pairing (i.e., zero ionicity) does not guarantee bonding and VB–CI is often necessary. The interpretation should be, instead: *Correlation energy represents the failure of the monodeterminantal MO calculation to reproduce the correct mechanism of bonding.* In other words, the calculation fails to place the molecule on the right spot along the T/I-bonding continuum. The magnitude of the failure is system dependent: Homonuclear systems require an MO-CI treatment, while heteronuclear systems often can be adequately treated by SCF-MO theory. The essential point is that what differentiates T- from I-bonding is the magnitude of interelectronic repulsion accompanying a correlated 2-e hop. In shifting from T to I-bonding, we replace "correlated exchange 2-e hops" by "correlated CT 2-e hops," which suffer from less interelectronic repulsion. These "correlated 2-e hops" have nothing to do with the "correlation energy" associated with MO theory.

3.5 The Map of Chemical Bonding

We now set out to construct a two-dimensional mapping of molecules that reveals two things: the mechanism of the bonding and the strength of the bonding. Our quest is governed by four guidelines.

(a) Kinetic energy reduction and, thus, bonding strength, is related to the sum of the ionization energies of the overlapping AOs and, as a result, to the sum of atom ionization energies and atom electronegativities. The larger the sum, the stronger the kinetic energy reduction and the stronger the bonding, everything else being equal. There are different, parallel scales of electronegativity, some based on BDEs (Pauling scale) and others based on ionization energies (Mulliken and Allen scales), because BDE depends on ionization energy.

(b) The mechanism of delocalization, T or I, also depends on the sum of atom electronegativities, everything else being equal. In homonuclear molecules, strong bonding implies T-bonding and weak bonding goes hand in hand with I-bonding. The origin of the crucial effect of electronegativity on the binding strength as well as on the bonding mechanism is emphasized in the diagram.

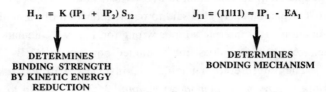

IP HAS *TWO* DIFFERENT CONSEQUENCES:

$H_{12} = K\,(IP_1 + IP_2)\,S_{12}$ $\qquad$ $J_{11} = (11|11) \approx IP_1 - EA_1$

DETERMINES BINDING STRENGTH BY KINETIC ENERGY REDUCTION

DETERMINES BONDING MECHANISM

(c) The condition for I-bonding is that the atoms support strong CT delocalization. For this to happen, the sum of atom electronegativities, x^+, should be small and the difference of the atom electronegativities, x^-, should be large. This dual condition cannot be met, however, because the limits are as follows: x^+ goes to zero as atom electronegativity goes to zero at which limit x^- is also zero. Thus, the "king of I-bonding" must involve some sort of compromise.

(d) E-bonding is classical Coulomb attraction complemented by induction and dispersion. The conditions for optimization of E-bonding are:

1. The interacting species are small and compact counterions, formed from precursor neutral atoms of widely different electronegativities by transfer of one or more electrons from the more electropositive to the less electronegative atom. This maximizes classical Coulomb attraction.
2. The interacting species are large polarizable neutral atoms or molecules. This maximizes the dispersive interaction.
3. One species is a polar molecule or ion and the other is a polarizable neutral molecule. This maximizes the inductive interaction.

If a lone pair is viewed as a bond involving an infinitely electropositive ghost ligand and a hole as a bond involving an infinitely electronegative ghost ligand, the preceding recipe predicts that E-bonding will tend to be prominent when a cation (or a deshielded nucleus) acts on an anion or a neutral atom carrying lone pairs. This means, for example, that H^- (one lone pair) will be stabilized far less than F^- (four lone pairs) by a nearby Li^+ via the E mechanism.

The proposed map of chemical bonding is shown in Figure 3.7. The two critical variables are the sum, x^+, and the difference, x^-, of atom electronegativities in a prototypical binary system. The two dimensional plot defines six domains, with the letters "w" (weak) and "s" (strong) referring to the bonding strength. The envelope of the pantheon conforms to the proper limits:

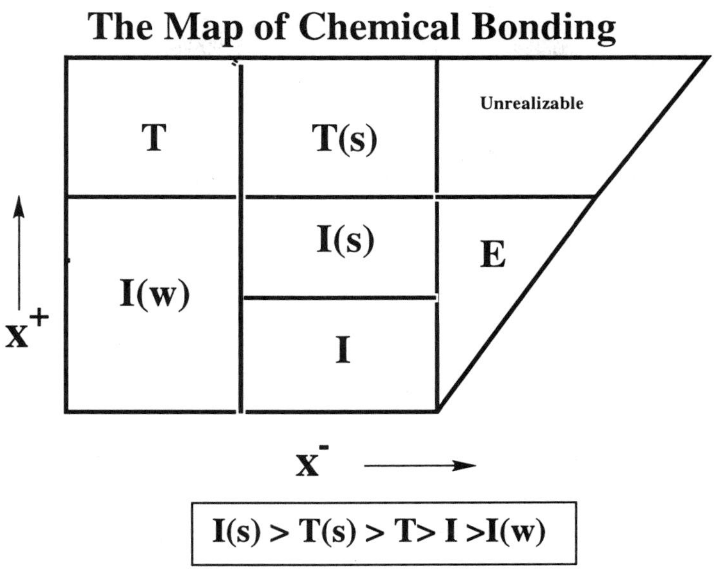

Figure 3.7 The map of chemical bonding. In a binary system, the three mechanisms of chemical bonding, T, I and E, are a function of the sum (x^+) and difference (x^-) of atom electronegativities. As x^- decreases, the strength of either T or I-bonding increases as CT delocalization becomes increasingly favorable relative to exchange delocalization (w = weak, s = strong). One key point is that the intrinsically superior I-bonding can become the supreme mechanism of bonding [I(s)] once CT delocalization becomes favorable while the sum of atom electronegativities is maintained high.

as x^+ goes to zero so does x^-. However, as x^+ approaches infinity, x^- can also do so. The nature of the elements precludes realization of this extreme.

As an example, consider the consequences as we move along the following series of binary molecules: $TiLi_4$, CH_4, CF_4, SiH_4, SiF_4, TiF_4. In the case of $TiLi_4$, the sum as well as the difference of electronegativities is low. This means two things: There is modestly efficient CT (small x^+, small x^-), and there is weak kinetic energy reduction (small x^+). The system falls in the I(w) domain, and the bonding is expected to be weak. Moving on CH_4, x^+ shoots up while x^- remains small. We are now in the T domain, and there is strong bonding because kinetic energy reduction is strong. Things improve as we go to CF_4 in the T(s) domain. This is a case of superior T-bonding explained by

Pauling's "polar covalence" concept. Going on to SiH_4, x^+ drops precipitously while x^- is kept modest. As a result, this molecule can benefit from neither strong kinetic energy reduction nor CT delocalization. This is called a "frustrated" molecule. It has two options, and both are bad: If it stays T-bound it gets modest kinetic energy reduction at the expense of interelectronic repulsion, while if it goes I-bound it needs a lot of promotional energy to support the requisite CT. SiF_4 is a very critical molecule for our theory. Specifically, in going from SiH_4 to SiF_4, both x^+ and x^- increase. The former index pushes the system toward the T domain, but the latter pushes it toward the I domain.

Which is the winner? Since I-bonding is intrinsically superior, the dominant factor must be the move in the I-bound direction. Hence, we expect that in SiF_4, strong kinetic energy reduction occurs via CT delocalization, which is now greatly favored because of the large difference of atom electronegativities. This molecule is now I-bound, and it resembles a metallic system in the mechanism of bonding even though it is much superior in the strength of bonding. In other words, SiF_4 is one molecule in which the compromise between the sum and difference of electronegativities becomes near-ideal. This molecule is an example of an inhabitant of the I(s) domain. Finally, TiF_4 is a classical E-bound molecule best formulated as $Ti^{4+}(F^-)_4$.

As pointed out in Chapter 1, the very facts of chemistry demand a new theory of chemical bonding. Silicones are $[(CH_3)_2SiO]_x$ polymers of great practical importance largely because they are water repellent. The conventional explanation of the bonding of SiF_4 and, by analogy, silicones is that the fluoride molecule is "polar covalent." However, this explanation fails to account for the hallmark of the silicon polymers: hydrophobicity. This very fact implies that the electrons of molecules in which Si binds the electronegative F and O are so well bound that they are able to resist an extramolecular perturbation, even an electrostatic one (as in the case of hydrogen bonding). This suggests some form of "superbonding," which we now have identified as optimal I-bonding.

3.6 The Affinity of Atoms and the Concept of Bonding Frustration

The first recognition that there exist laws of combination of atoms — that is, the first attempt to decipher the chemical code — is due to Chatt, Ahrland, and Davies back in the middle of this century, with subsequent key contributions made by Pearson. Our map of chemical bonding has similarities and differences with respect to that of Chatt et al. For example, the I domains have no counterpart in conventional models. Furthermore, the I(s) domain represents some type of superbonding. Is there any evidence that such a domain exists?

The members of a series of fluorides starting with LiF and ending with FF have the following bond dissociation energies (kcal/mol).

LiF	BF	CF	NF	OF	FF
137	181	127	92	55	37

It is evident that BF stands out of the correlation. Now, the increase in the BDE in going from LiF to BF can be attributed to the increasing x^+. Also, the decrease in going from BF to CF can be attributed to the introduction of a three-electron π antibond between one odd carbon electron and one fluorine lone pair. If this were the whole story, the subsequent addition of antibonding electrons in going from CF to FF would have comparable consequences, but this is clearly not the case. Our interpretation is that BF is actually a molecule with a very favorable I bond, as illustrated.

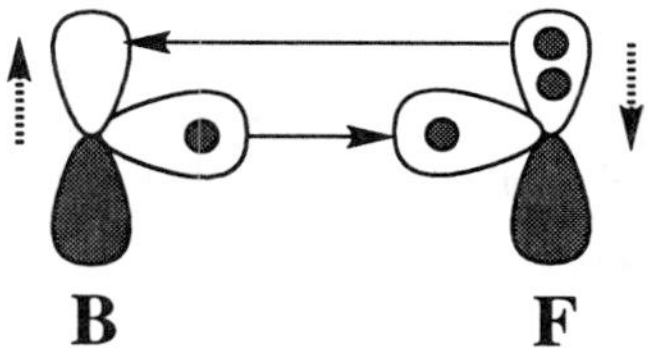

Specifically, two arrows are taken to signify a single I bond, and two head-to-tail arrows imply the formation of one I bond with assistance by dispersion (hatched arrows). The bottom line is that the σ bond of BF does not belong to the catalog of conventional theory. Rather, it is a strong bond due to CT which is assisted by its environment, the latter being the π space. Since BF_3 is known to be an inferior Lewis acid (compared to BCl_3, etc.), the π space must be engaged in some fashion. Thus, we meet, for the first time, a situation in which focusing on the apparently functional elements of a problem (e.g., the σ bond of BF) is simply not enough. The environment has a controlling influence on whether the functional elements are activated.

The preceding list of BDEs reintroduces the concept of bonding frustration. To appreciate this point, consider a question: Why does HF have a stronger bond than H_2? The VB answer is: Because it has a larger x^+ as well as a large x^-. Pauling attributed the stronger bond of HF to the second factor ("polar covalence").[11] As we will discuss in more detail in Chapter 6, this argument misses a crucial point: the first factor is actually more important! In other words, had it not been for lone pair–lone pair exchange repulsion, F_2 would have a much stronger bond than HF! The problem with F_2 is that the two-electron σ bond is antagonized by three four-electron exchange antibonds. As a result, F_2 is a frustrated molecule because one exchange bond is antagonized strongly by two exchange antibonds. Therefore, F_2 opts for a binding mode that is typically adopted by metal dimers. The C—H bond in CH_4 is stronger than the more polar Si—H bond in SiH_4 because C is more electronegative than Si. Again, SiH_4 is a frustrated molecule. Thus, either competing exchange antibonds or merely an incapacity to sustain strong CT creates a category of molcules that cannot become good citizens of any one of the five domains.

We now classify each atom with respect to its intrinsic ability to engender T-, I-, or E-bonding as follows.

(a) Because of the absence of "observer" holes or pairs, H acting alone is incapable of satisfactorily participating in anything but T-bonding. A proton acting alone has either T or E affinity. A case can be made that a proton attaches to neutral closed-shell molecules by the E mechanism, with the strength of attachment (base strength) being a function of the number of "observer" bond and lone pairs that are stabilized electrostatically by the proton.[12] A hydride acting alone has T affinity. The same holds true for alkyl groups. The situation changes when a hydrogen atom acts in concert with a hydrogen ion to define an I bond.

(b) Nonmetals with lone pairs are ideally suited for the E mechanism because the presence of a cationic ligand engenders a field that stabilizes the lone pairs. In other words, Li^+H^- has poor but Li^+F^- has excellent E-bonding because, in the latter, Li^+ sees four pairs (rather than one). This is clearly revealed by the enthalpies of the following two reactions:

$$LiF + CuCl \rightarrow LiCl + CuF \qquad \Delta E = +26 \text{ kcal/mol}$$

$$LiH + CuCl \rightarrow LiCl + CuH \qquad \Delta E = -40 \text{ kcal/mol}$$

Thus, F, OR, NR_2, and CH_3 (systolic groups) have E affinity when combined with alkalies. However, they also have T affinity when combined with either H or alkyl and I affinity when combined with semimetals and metals. Their T affinity stems from their ability to support strong *exchange delocalization* because of their high electronegativity. Their I affinity stems from their ability to support *CT delocalization* when combined with a more electropositive metalloid or metal. Furthermore, their lone pairs and/or bond pairs can match metal holes to activate I-bonding in a way that will be described later on.

(c) All semimetals and metals of the p block have I affinity but are T tolerant. The most electronegative semimetals (e.g., Cl) have I as well as E affinity.

(d) First transition series metals have E and third transition series metals have I affinity, always in a relative sense. The failure of the first transition metals to have strong I affinity is due to the contracted nature of the d AOs. However, in exceptional circumstances (e.g., ferrocene-type molecules), I-bonding can become preeminent.

(e) All alkalies have E affinity.

(f) Electropositive early transition metals are T intolerant but late transition metals are T tolerant.

When atoms or fragments having the same affinity combine, the resulting molecule is symbolized by the affinity symbol. When there is a mismatch, the resulting molecule is frustrated and is denoted by FR. An extreme form of frustration, which occurs when atoms having lone pairs combine with each other, is denoted by FR#. When there is a mismatch but there is toleration, the resulting molecule is designated TOL. The first-order laws of fragment combination are:

$$T + T = T \qquad E + E = E \qquad I + I = I$$

$$T + E = FR \qquad T + I = FR \qquad E + I = FR$$

By using this convention, we can visualize every chemical equation as the transformation of ideal (T, I, or S) and/or frustrated (FR) reactants to ideal and/or frustrated products. The following reaction offers an example:

$$LiH + CuCl \rightarrow LiCl + CuH \qquad \Delta E = -40 \text{ kcal/mol}$$

E + T	I + I	E + E	I + T
FR	I	E	TOL

The concept of bonding frustration is nicely illustrated by CrF_6, a molecule of uncertain existence.[13] Cr is neither electropositive enough nor small enough to produce an E-bound $Cr^{6+}(F^-)_6$. On the other hand, it is electropositive enough to shun T-bonding. This leaves I-bonding as the only option. However, as we will see later on, the "environment" fails to activate the I bonds. As a result, the best Cr can do is bind six Fs by falling back on a blend of T and E mechanisms; that is, it can form $F_3Cr^{3+}(F^-)_3$ which has three Cr—F T bonds and in which three fluoride anions are bound electrostatically to Cr^{3+}. Substituting W for Cr changes the "environment" and makes it capable of activating the I bonds. WF_6 is a well-known species.

In closing this preliminary round with chemical selectivity, we stress that a cut-and-dried separation of T-, I-, and E-bonding is impossible. What changes is simply the blend of these three mechanisms. In other words, what is chemically meaningful is the relative importance of kinetic energy reduction (I, T) and Coulombic bonding (E) as well as the mode of kinetic energy reduction (T or I). The crucial issue is that diminution of atom electronegativity has two consequences: diminution of the BDE, as well as change of the bonding mechanism from T to I or E. Finding evidence for the first trend is no problem, since the available thermochemical data speak for themselves. On the other hand, demonstrating a change of mechanism requires a reexamination of chemical selectivity across the periodic table. This monograph is primarily devoted to the latter task. Here, we briefly comment on the electronegativity–BDE correlation, since this is not widely appreciated.

In homonuclear molecules, there exists a vertical correlation between atom ionization energy (IE), a measure of electronegativity, and BDE, provided a second key variable is kept constant. This variable is the relative size of the valence AOs, which determines the relative efficiency of hybridization, hence the magnitude of the diatomic BDE. Specifically, in going from the first to the second row in the p-block of the periodic table, the ns AO contracts relative to the np AO. This is a fundamental consequence of atomic electronic structure and it has been discussed by Pyykkö and Kutzelnigg.[14] As a result, we expect a diminution of the BDE on two counts: contraction of the ns AO plus decrease of atom IE. This trend is uniformly observed, and the effect is large, assuming that C, N, O, and F bind from their sp^3-hybridized state and B from the sp^2-hybridized state while all the rest of the atoms bind from their unhybridized state. In going from the first to the second transition series, the (n – 1)d expands relative to the ns AO. As a result, AO overlap considerations alone dictate an increase in the diatomic BDE as we descend from first to second transition series within a column, as observed. The s contraction in the p block and the d expansion in the d block continue, but now abated, in descending below the second row in the p block and in going from the second to the third transition series (now because of relativistic effects). Thus, if we look at the periodic table and restrict our attention to the vertical trends starting with the second row in the p block and the second transition series in the d block, we see that the expected trend materializes with only one exception (Nb_2 vs. Ta_2). Even this may be only apparent owing to the uncertainty in the BDE of Ta_2.

3.7 What H_2 Has Been Trying in Vain to Tell Us

Pearson[15] and Parr[16] made a heroic effort to understand the electronic basis of the hard/soft distinction by using density functional theory. They concluded, in effective agreement with our analysis, that the sum (chemical potential) and the difference (hardness) of the highest occupied MO of one and the lowest unoccupied MO of a second fragment are the critical indices of bonding. Parr recognized that "both chemical potential and chemical hardness have to be used as coordinates of structure–stability diagrams to delineate and predict successfully crystal structure." However, the chemical meaning of these indices could not be deduced because the role of electron–electron repulsion in dictating the mechanism of chemical bonding is not transparent at the level of MO or density functional theory. The revolution promised by the hard/soft concept is a fundamental one: *Selectivity of atom combination implies the existence of more than one mechanisms of chemical bonding.* This is why the recognition of the hard/soft duality by Chatt[17] and Pearson[18] is one of the most important (but least appreciated) contributions to the theory of chemical bonding, ranking with the Lewis formula concept and Hückel's rule.

To illustrate how the explicit consideration of electron–electron repulsion revolutionizes our understanding of chemical bonding, we need nothing "bigger" than the prototype of chemical bonding itself: the hydrogen molecule.

Figure 3.8 shows three different ways of representing H_2 and identifies on one page the difference between conventional theory and the theory presented here:

In the Lewis representation of H_2, a line connects two "dots" to indicate an "electron pair bond." Exactly how bonding occurs is not stated. Hence, the Lewis formula is a stoichiometric formula.

In the Heitler–London representation, the mechanism of bonding is exchange. Bonding occurs as a result of kinetic energy reduction due to the 2-e exchange hop.

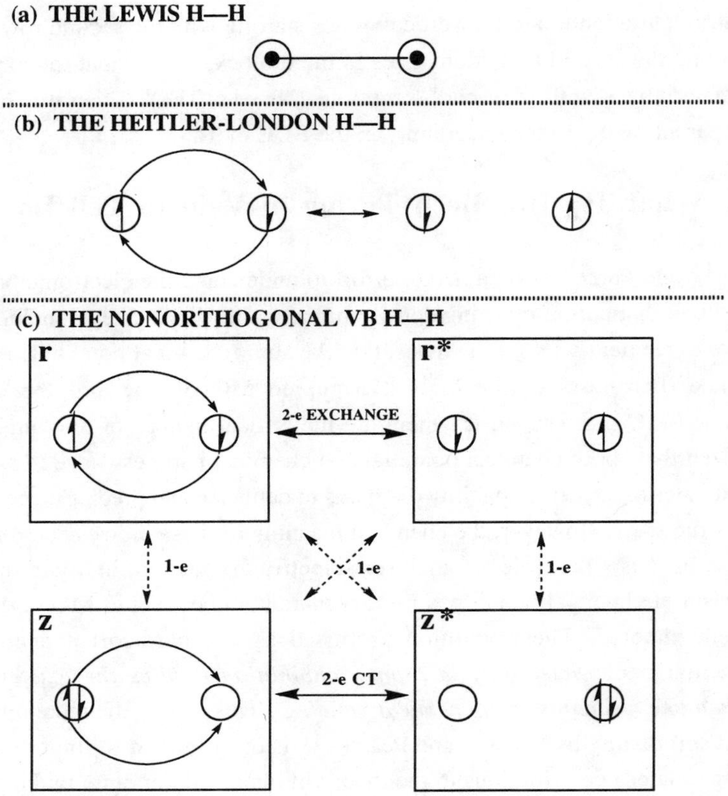

Figure 3.8 The three different interpretations of the H_2 molecule. The complete VB formulation suggests the existence of two different types of bond: exchange and CT.

The VB story is different. There are four primitive VB configurations: Two diradical (*r* and *r**) and two zwitterionic (*z* and *z**) configurations. The interaction matrix elements connecting *r* with *r** and *z* with *z** are identical:

$$\langle r|H|r^*\rangle = \langle z|H|z^*\rangle = 2\beta s + K$$

where β is the resonance, *s* the overlap, and *K* the Coulomb exchange repulsion integral. This means that there exist two different mechanisms of chemical bonding: 2-e exchange (T-bonding) and 2-e CT (I-bonding). The diradical and zwitterionic configurations are connected by 1-e hops, and these can be taken to represent the process by which an optimal blending of 2-e exchange and 2-e CT delocalization is obtained. The two distinct mechanisms of chemical bonding remained "undetected" because the H_2 molecule, with its large electronegativity, makes almost exclusive use of the exchange mechanism. The CT mechanism becomes activated only when two conditions are met:

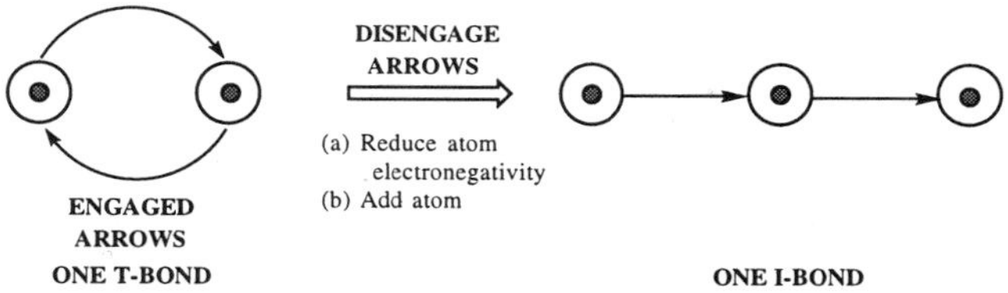

(1) Atom electronegativity becomes low, in which case the zwitterionic structures attain low energy (relative to the diradical structures); and

(2) more than two atoms participate in the bonding, in which case the 2-e CT hop becomes superior to the 2-e exchange hop because the bielectronic term (the Coulomb exchange repulsion integral K) is reduced, while the monoelectronic overlap term ($2\beta s$) is preserved.

The two conditions taken together predict that electropositive atoms will aggregate. As illustrated above, a pair of arrows defining a single T bond becomes "disengaged" to form an I bond connecting now three atoms. Clearly, the H_2 molecule provides us with all the concepts of homonuclear chemical bonding, but it projects one mechanism (T-bonding) and disguises the other (I-bonding).

There is an alternative way of formulating the bond of H_2 which makes transparent the relationship of T- and I bonds. Specifically, adopting the definition that one arrow signifies the transfer of one electron and that a 2-e hop can also be attained by a sequence of two ordered 1-e hops, we can see that the bond of H_2 can be represented by two arrows which form a closed loop. The arrows can be "executed" in pairs (2-e hop) or in singles (1-e hops). Over a basis set of primitive (rather than spin-adapted) VB configurations, these two arrows define the 4x4 CI depicted in Figure 3.8. Since two arrows are taken to represent one bond, we can see that the T bond of H_2 is actually an exchange (covalent) bond with ionic resonance. When the ionic structures lie high in energy, the T bond becomes approximately equivalent to an exchange bond, i.e., the T bond is only at the limit a Heitler-London bond. The advantage of this formulation is that it leads to the inescapable conclusion: If two arrows make a T bond by forming a *closed loop*, two arrows must make a qualitatively different bond by forming an *open loop*. The latter is the I bond.

Recent monographs have used the convention of the Ketelaar triangle to classify bonding into covalent, metallic, and ionic as a function of the sum and difference of electronegativities.[19] This represents an empirical statement of some of the basic facts of chemistry. The key point is that covalent, metallic, and ionic bonding are only distant relatives of T-, I- and E-bonding. As we will see, the two systems produce different concepts and predictions! For example, if we were to identify "metallic" with "I" bonding, then "metallic" bonding appears as early as in π aromatic systems (with high T complement) and, more importantly, as soon as one crosses below the first row to "heavy p-block nonmetals" or "metalloids." It also appears as early as in metal dimer, far in advance of metal solid. All this contravenes conventional wisdom. Thus, "metallic bonding," as commonly understood, has operationally nothing to do with I-bonding. It is best to think of the covalent/metallic/ionic gospel as the yield of intuition rather than theory, and the T/I/E pantheon as the expression of polydeterminantal theory with much chemical intuition thrown in.

References

1. H.C. Longuet-Higgins, *Q. Rev.* 11, 121 (1957).
2. (a) J.C. Slater, *Quantum Theory of Molecules and Solids*, McGraw-Hill, New York, 1963, Chapter 3. (b) C. Sandorfy, *Electronic Spectra and Quantum Chemistry*, Prentice-Hall, Englewood Cliffs, NJ, 1964.
3. E.R. Davidson, K.L. Kunze, F.B.C. Machado, and S.J. Chakravorty, *Acc. Chem. Res.* 26, 628 (1993).
4. M.H. McAdon and W.A. Goddard III, *J. Non-Crystal.Solids* 75, 149 (1985).
5. D. Maynau and J.P. Malrieu, *J. Chem. Phys.* 88, 3163 (1988).
6. E.R. Davidson, in *The Challenge of d and f Electrons*, D.R. Salahub and M.C. Zerner, Eds., American Chemical Society, Washington, DC, 1989, p. 153.
7. E.A. Carter and W.A. Goddard III, *J. Am. Chem. Soc.* 108, 2180 (1987).
8. T.R. Cundari and M.S. Gordon, *J. Am. Chem. Soc.* 113, 5231 (1991).
9. G.A. Gallup and J.M. Norbeck, *Chem. Phys. Lett.* 21, 495 (1973).
10. R. Walsh, *Acc. Chem. Res.* 14, 246 (1981).
11. (a) L. Pauling, *The Nature of the Chemical Bond*, 3rd ed., Cornell University Press, Ithaca, NY, 1960. (b) G.G. Balint-Kurti and M.

Karplus, in *Orbital Theories of Molecules and Solids*, N.H. March, Ed., Clarendon Press, Oxford, 1974, Chapter 6.

12. P.A. Frey, *Science* 269, 104 (1995).
13. J. Jacobs, H.S.P. Mueller, H. Willner, E. Jacob, and H. Buerger, *Inorg. Chem.* 31, 5357 (1992).
14. (a) P. Pyykkö, *J. Chem. Res.* 380 (1979); P. Pyykkö, *Phys. Scripta* 20, 647 (1979); P. Pyykkö and J.P. Desclaux, *Acc. Chem. Res.* 12, 276 (1979). (b) W. Kutzelnigg, *Angew. Chem. Int. Ed. Engl.* 23, 272 (1984).
15. R.G. Pearson, *Acc. Chem. Res.* 26, 250 (1993).
16. R.G. Parr and Z. Zhou, *Acc. Chem. Res.* 26, 256 (1993).
17. S. Ahrland, J. Chatt, and N.R. Davies, *Q. Rev.* 11, 265 (1958).
18. (a) R.G. Pearson, *J. Chem. Soc. Chem. Commun.* 65 (1968). (b) R.G. Pearson, Ed., *Hard and Soft Acids and Bases*, Dowden, Hutchinson and Ross, Stroudsburg, PA, 1973. (c) R.G. Pearson, *Inorg. Chem.* 27, 734 (1988). (d) R.G. Pearson, *Chem. Bri.* 27(5), 444 (1991).
19. (a) N.W. Alcock, *Bonding and Structure: Structural Principles in Inorganic and Organic Chemistry*, Ellis Horwood, New York, 1990. (b) N.C. Norman, *Periodicity and the p-Block Elements*, Oxford University Press, New York, 1994.

Part II

The T Bond

Chapter 4

The Shell Model of Rings

4.1 Angle Strain in Organic Rings

The protocol for the analysis of the bonding of an organic molecule is: Each species is viewed as a composite of atoms or fragments; the magnitude of the promotional energy needed for perfect pairing is estimated; and stability is determined by how effectively the electron pair bonds that are formed by perfect pairing are able to avoid interbond overlap repulsion. The problem of "angle strain" in organic rings is a pivotal one because it illustrates how the excitation factor and the overlap factor jointly determine the stability of organic molecules.

There is a more important reason for choosing angle strain as the first application of the VB concepts of this work. Specifically, our intention is to walk through the periodic table and show that different atoms support different mechanisms (T vs. I) of bonding. One key difference between organic molecules and metal- or semimetal-containing molecules is that the former "hate" but the latter "love" deltahedral arrangements: The organic tetrahedrane is unknown,[1] but tetrahedral $(BCl)_4$, P_4, and $[Ir(CO)_3]_4$ are well-characterized stable molecules. Be_4 is computed to prefer the tetrahedral geometry over the square alternative.[2] Many ascribe this preference to the availability of d orbitals. However, it is well established by ab initio computations that d AOs do not function as valence AOs in Be, B, or P! This implies that angle strain ("Baeyer strain")[3] must depend on some inherent atomic property, other than spatial overlap, which changes in a spectacular way in going from nonmetals to semimetals and metals. We will argue that this property is *atom electronegativity*, the variation of which causes a shift from T- to I-bonding, or vice versa.

The universally accepted interpretation of angle strain is that in cyclopropane, the internuclear CCC bond angle (60°) deviates from the ideal interorbital angle (109°) by 49°. The result is inefficient "banana bond" formation. Hence, the instability of cyclopropane is attributed to impaired overlap of the sp^3 hybrids.[4] However, these ideas fall short of explaining critical facts:

(a) The C—C bond (and C—H bond) of cyclopropane is *shorter* than that of cyclohexane (Table 4.1). The computed C—C bond of a tetrahedrane derivative

Table 4.1 Structural Data of Cycloalkanes

Molecule	Strain (kcal/mol)	rCC (Å)	rCH (Å)	m – H	R:T	Si Strain[a] (kcal/mol)
$(CH_2)_3$	27.6	1.510	1.089	+1	1:2	37.5
$(CH_2)_4$	26.4	1.555		0 to –2	2:2 to 3:1	12.9
$(CH_2)_5$	6.5	1.546	1.114	– 1	3:2	– 2.2
$(CH_2)_6$	0	1.536	1.1121	0	3:3	0

[a] Ab initio calculation of strain of cyclosilanes: P.v.R. Schleyer, in *Substituent Effects in Radical Chemistry*, H.G. Viehe et al., Eds., Reidel, New York, 1986, p. 69.

is one of the shortest known.[5] This has been pointed out by Greenberg and Liebman.[6]

(b) Although planar cyclobutane has an internuclear angle 30° *larger* than that of cyclopropane, the strain energies of the two molecules are nearly the same.

(c) Computations reveal that replacement of carbon by silicon *increases* the strain of cyclopropane, even though the Si—Si bond is much weaker than the C—C single bond and thus more indifferent to impaired overlap ("banana bonding").[7]

(d) Structural effects resulting from replacement of H by electronegative σ acceptors/π donors, such as F, are not accounted for by perturbation MO (PMO) models.[8]

The correct explanation of angle strain in cycloalkanes is, at the same time, easy and hard. It is easy because the root of angle strain is exactly what every organic chemist intuitively thinks: Deviation from 109° internuclear angles because of skeletal constraints. On the other hand, the problem is hard because it is not at all obvious how a strained molecule copes with angle strain.

The "easy part" can be exposed by reference to a system as simple as methane, provided one recognizes the following:

(a) In nonmetal systems, the key variables are exchange bonds and antibonds which, collectively, represent exchange delocalization. Because of the exchange phenomenon, it is unrealistic to focus on bonds and disregard the antibonds. In other words, it is misleading to use the term "strong bonding" to refer to electron pair bonds. The right term is "strong T-bonding," meaning that the *balance* of bonds and antibonds due to exchange delocalization is optimal.

(b) Optimal overall bonding represents the combination of optimal T-bonding at the expense of minimal atom promotional energy.

(c) Impaired overall bonding is manifested under two entirely different disguises:

1. Weak bonds and strong antibonds (i.e., weak T-bonding) at the cost of small atom promotion.
2. Strong bonds and weak antibonds at the cost of disproportionately high atom promotion. The key word here is "disproportionately." We describe this situation by saying that the molecule is forced to "squander" promotional energy.

Situations 1 and 2 represent two very different alternatives, neither of which is "good." We will argue that angle strain in cyclopropane represents choice 2 while angle strain in cyclobutane represents choice 1.

Turning our attention to methane, we note that the ideal geometry is tetrahedral because this involves the optimum blend of exchange-bond maximization and exchange-antibond minimization. Since these two antithetical overlap-dependent factors are interrelated, we can focus on interbond exchange repulsion in our pursuit of a concise explanation. In this context, consider transforming methane from T_d symmetry to C_{2v} by either "opening" or "closing" one HCH bond angle. Either of these changes brings geminal C—H bonds in closer proximity and raises geminal C—H/C—H interbond exchange repulsion without improving the C—H bonds themselves. This is the explanation of angle strain in cycloalkanes constrained to have an "open" or a "closed" CCC angle while the corresponding HCH angle is held fixed. In the former case, it is C—C/C—H interbond repulsion that is the cause of strain, while in the latter case the problem is the C—C/C—C interbond overlap repulsion.

This very simple argument actually defines the problem: The angle strain of cyclopropane and cyclobutane must be a manifestation of excessive geminal C—C/C—C interbond exchange repulsion. How do cyclopropane and cyclobutane cope with this difficulty? We will argue that they do so in diametrically opposite ways and that the bond length variation (i.e., short C—C bonds in cyclopropane and long C—C bonds in cyclobutane) is a mere reflection of two antithetical choices.

A successful interpretation of the physical significance of bond lengths is a conceptual minefield. For example, consider one plausible interpretation of the C—C bond length variation in cycloalkanes: Once it is recognized that orthogonal hybrids interact, cycloalkanes become effectively σ conjugated systems. Comparing the C—C bond of cyclopropane and cyclohexane is now analogous to comparing the C1-C2 π-bond orders of, say, 1,3-butadiene and 1,3,5,7-octatetraene. The shorter the polyene, the larger the individual bond order, since the occupied MOs have increasingly fewer nodes. For this reason alone, one expects a regular monotonic increase of the C—C bond length as the size of the cycloalkane increases. However, this trend does not materialize; rather, the short cyclopropane C—C bonds give way to very long cyclobutane C—C bonds, which become shorter again in going to cyclopentane and still shorter in going to cyclohexane (Table 4.1). Furthermore, the very shortness of the C—C bonds of tetrahedrane make an unmistakable suggestion: the bonds of cyclopropane (*not* the C—C bond dissociation energy of cyclopropane) are actually remarkably strong, and those of cyclobutane are remarkably weak, for reasons that cannot be exposed by reference to our methane model. To make further inroads, we must first appreciate the importance of atom promotional energy for molecular stability.

4.2 The Fallacy of Bond Length as Indicator of Stability

Isomer A is more stable than isomer B, but the interatomic distances in B are shorter than those in A. Why? In VB theory, the answer is automatic: B has intrinsically stronger bonds, but the price paid for arriving at the bonding configuration (i.e., the promotional energy) is disproportionately high. When the *less stable* isomer has *shorter* bonds, the reason behind its relative instability is higher promotional energy.

MO theory is conceptually murky because the interplay of promotion and bond making is not transparent. This is due to the delocalized nature of the MOs. As a result, there is a widespread impression that bond length correlates universally with molecular stability. In other words, some fail to differentiate between intrinsic bond strength, measured by the magnitude of the bond length, on one hand, and bond dissociation energy (BDE), on the other hand. The latter depends on the promotional energy (EXC) for preparing the fragments for bonding*and* the intrinsic bond dissociation energy (IBDE). For example, the BDE of A*— B* is IBDE – EXC.

$$A + B \xrightarrow[\text{EXC}]{\text{Fragment Promotion}} A^* + B^* \xrightarrow[-\text{IBDE}]{\text{Bond Making}} A^*\text{—}B^*$$

As an example, consider the following processes:

$$2CH_3 \text{ (quasi-planar)} \longrightarrow 2CH_3^* \text{ (pyramidal)} \longrightarrow H_3C^*\text{—}CH_3^*$$

$$2CF_3 \text{ (pyramidal)} \longrightarrow 2CF_3 \text{ (pyramidal)} \longrightarrow F_3C\text{—}CF_3$$

The promotional energy required for pyramidalizing the nearly planar methyl radical and bringing it to the geometry within H_3C—CH_3 is large. By contrast, very little promotional energy is required to prepare F_3C—CF_3. CF_3 itself is pyramidal.[9] As a result, the C—C BDE of C_2H_6 is *smaller* than that of C_2F_6 although the C—C bond length is *shorter* in the former.[10]

The double bond dissociation energies (DBDEs) of tetrafluoroethylene (to two ground singlet F_2C) and ethylene (to two ground triplet H_2C) are 69 and 172 kcal/mol, respectively.[11] Yet, the C=C bond of the former is shorter. The (relative) shortness of the C=C bond of tetrafluoroethylene is due to two factors:

(a) The fluorines draw p character, thus enhancing the s character of the carbon AO directed to the second carbon (Walsh's rule).[12]

(b) The smaller vicinal interbond repulsion of C—F bonds

On the other hand, F_2C is ground state singlet with a 51 kcal/mol singlet–triplet gap. By contrast, H_2C is ground triplet. Coupling two F_2C groups to form tetrafluoroethylene requires 102 kcal/mol excitation energy. Coupling two H_2C fragments to form ethylene requires zero orbital promotional energy. The DBDE of tetrafluoroethylene is so much smaller than that of ethylene because the former requires much larger fragment excitation than the latter. The small DBDE of $H_2Si{=}SiH_2$ (51 kcal/mol,[11] SiH_2 ground singlet with an 18 kcal/mol singlet–triplet gap) is also due to the appreciable promotional energy of the SiH_2 fragment plus the inability of the silicon σ hybrid AOs to engender strong overlap because of the contracted nature of the 3s AOs, which contributes to each hybrid.

We end by reiterating the key message: The BDE does not have to correlate with the bond length. Thus, the C—C BDE of cyclopropane can be lower than the C—C BDE of cyclohexane even though the C—C bond length of the former is shorter than that of the latter.

4.3 MOVB Is VB Theory with Symmetry Control Made Transparent

Up until now, we have presented all arguments in the language of nonorthogonal VB theory using prototypical systems containing few electrons. The concepts derived by reference to such "small" model systems can be extended to molecules of any size. However, one needs to be extra careful in making this transition because writing symmetry-adapted VB configurations (resonance structures) is more time-consuming and less amenable to pictorial representation than writing symmetry-adapted MOs. Hence, we can use a VB formalism tailored to large systems and then restate the emerging concepts in VB language.

MOVB theory is a fusion of the MO and VB recipes of constructing the wavefunction of a system.[13,14] We extracted the concept of the canonical, symmetry-adapted orbital from MO theory (to keep track of symmetry control) and the concept of "bond" from VB theory (to have an efficient bookkeeping procedure). The interrelationship of MO, MOVB, and VB theory has been discussed in the literature, and a translation from one to the other language can be easily made.[15] For the purpose of this work, all one needs to know is the basic recipe: A molecule is cut into two fragments consistent with utilization of maximal local symmetry (e.g., it is better to view water as "O plus H_2" than as "OH plus H"), and the canonical fragment orbitals are written. Next, *nonorthogonal* configurations are generated by allocating the valence electrons to the various orbitals in all possible ways.

The resulting configurations are grouped into "packets," and each packet is separately diagonalized to obtain packet substates. The ground state of each packet substate manifold is selected and denoted by W_{a0} (where a is the packet index and zero symbolizes the ground state of packet a), or, more conveniently, by W_i, where i is a numerical index. The W_i are diagonalized again to produce the final eigenstates. The final ground state is a linear combination of substates, each of them called a *bond diagram*. This represents one possible, symmetry-allowed way of connecting the two fragments by bonds and/or antibonds. For example, the bond diagrammatic description of the ground state of methane may be viewed as a composite of H_2C and H_2:

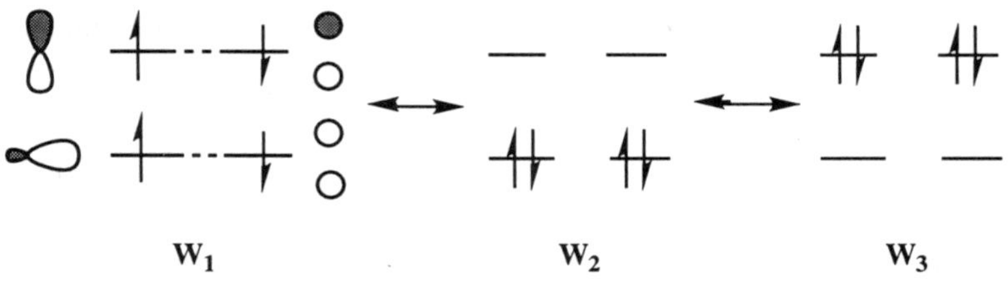

W_1 is a linear combination of 10 configurations generated by shifting electrons along the dashed lines starting with the parent configuration, 0_1, and W_2 and W_3 are each made up of one configuration, as shown in Figure 4.1. When the fragment orbitals are capable of strong overlap interaction, as in the case of methane, W_1 is the principal bond diagram and the perfect-pairing configuration

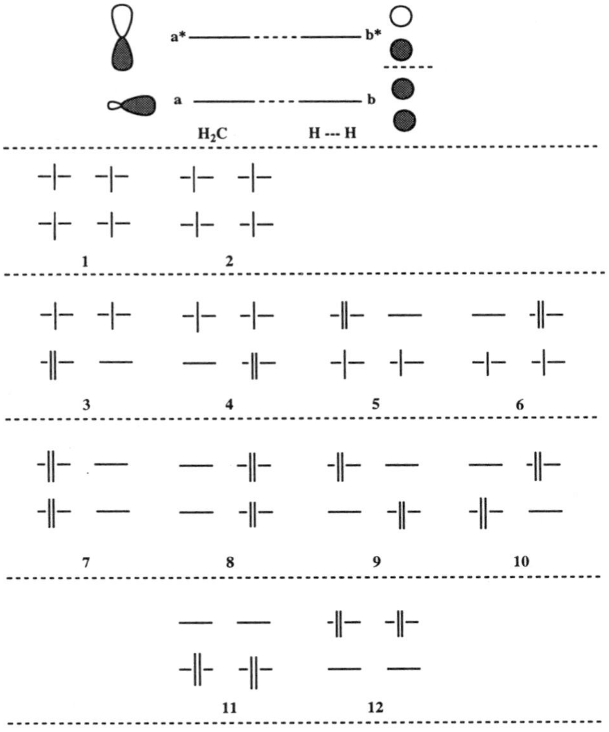

Figure 4.1 Bond diagrammatic representation of CH_4 viewed as "CH_2 plus H_2."

the dominant contributor. When C and H are replaced by metal atoms, there is no longer one principal bond diagram, and the interaction of several contributor bond diagrams becomes critical.

One can conceptualize the process of atoms or fragments combining to form a molecule as a "bank transaction": the fragments are excited (promoted) to a configuration suitable for bonding. This is the "investment." The excited fragments make bonds. This is the "return." The stability of the molecule is determined by the extent to which the return exceeds the investment ("profit"). The bond diagrammatic representation of $(H_2C)H_2$ allows us to monitor the following:

(a) The magnitude of the excitation energy. In our example, the magnitude of the excitation is, to a first approximation, the sum of the *a* to *a** plus *b* to *b** excitations.

(b) The way in which symmetry controls bond making. In our example, the bonds are formed by overlapping *a* with *b* and *a** with *b**.

(c) The intrinsic strength of each bond. This depends on the *a*—*b* and *a**—*b** overlap integrals and the ionization energies of the orbitals. Each bond is a *multicenter bond* formed by the overlap of AOs with MOs (MOs with MOs, in the general case) rather than by the overlap of AOs with AOs, as in VB theory.

4.4 Rings as Composites of Aromatic Shells

We focus on planar cycloalkane carbon rings, $(CH_2)_m$, symbolized by C_m, where *m* is the number of carbons, and seek to develop their VB wavefunction. We assume planarity for maximal simplification and because the key electronic features of the planar carbon rings are essentially maintained in the puckered derivatives. The carbon AOs available for σ C—C bond formation are designated n_r (radial) and $2p_t$ (tangential) and, since there are two valence electrons per methylene, there are three configurations, designated G, S, and D.

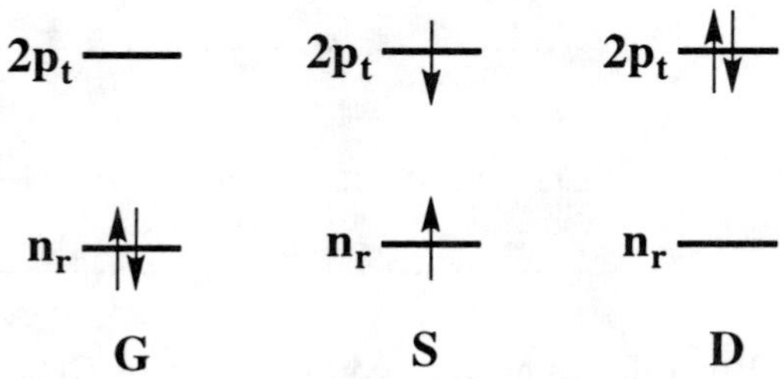

To construct the principal bond diagram for each carbocycle, we first write the symmetry MOs spanning the n_r AOs. This MO manifold defines *the radial, R, shell.* We repeat the procedure for the tangential AOs, and we obtain the manifold of the symmetry MOs spanning the tangential $2p_t$ AOs. This defines *the tangential, T, shell.* There is an important distinction between odd and even carbocycles: In *even* systems (*m* = even), the radial AOs of the R shell define a *Hückel* array and the same is true for the tangential AOs of the T shell. This is a case of *noncomplementary shells.* By contrast, in *odd* systems (*m* = odd), the radial AOs form a *Hückel* but the tangential AOs a *Möbius* array. We now say that the shells are *complementary.*

The next step is to deposit the valence electrons in the R and T MOs of lowest energy. Finally, we add dashed lines connecting MOs of the same

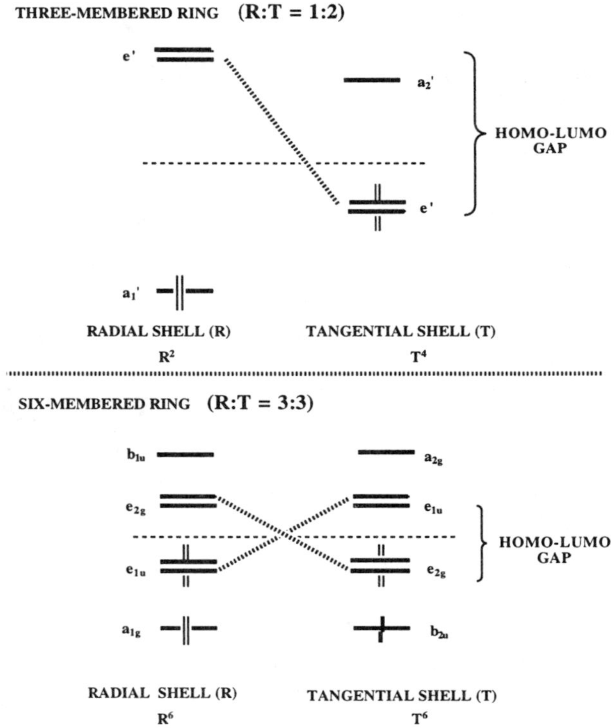

Figure 4.2 Shell diagrams for C3 and C6. Note how the diminution of radial AO overlap shrinks the HOMO–LUMO gap of the R shell in going from C3 to C6.

symmetry to indicate the delocalization from the R to the T and from the T to the R shells. This model is called the *shell model* of ring structure and the MOVB bond diagrams which are generated by this philosophy are called *shell diagrams.*

Here is now the key question: What is the signature of a ring with impaired bonding on its shell diagram? The answer is that angle strain is the consequence of one or both of the following attributes of the shell diagram:

Impaired intershell (R–T) delocalization. The criterion is a large HOMO(R)–LUMO(T) and HOMO(T)–LUMO(R) energy gap. These are referred to collectively as the intershell frontier MO (FMO) gap.

Occupancy of nonbonding R and T MOs.

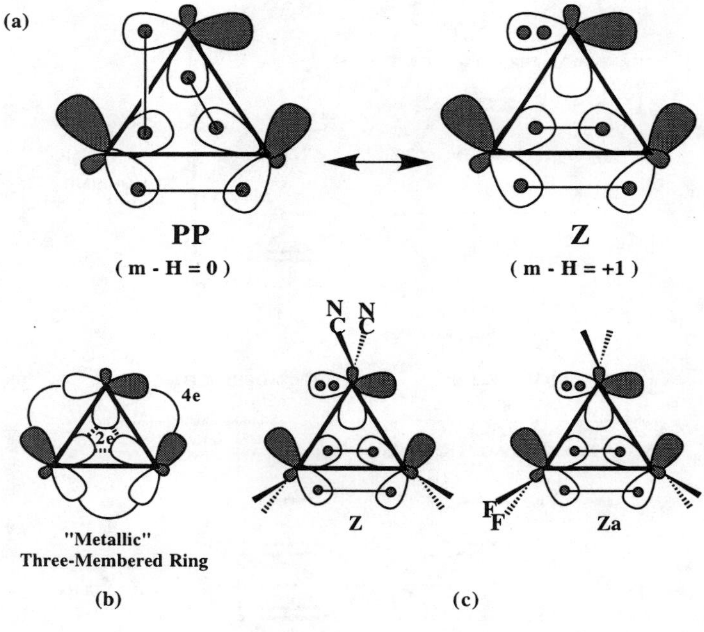

Figure 4.3 (a) The approximate VB wavefunction of C_3: *m*, number of carbon centers; *H*, number of radial (hollow) electrons. (b) A three-membered metallic ring is a doubly-aromatic R^2T^4 complex. (c) Optimal interaction of substituents with the carbenic resonance structure. In the case of the difluoro derivative, only one of two equivalent structures is shown.

The final step is the translation from the MOVB language of the shell diagram to the VB language of interacting nonorthogonal VB configuration. The simplest procedure has two steps:

(a) In the absence of intershell delocalization, the VB resonance structures must be consistent with the occupancy of the radial and tangential MOs as displayed in the shell diagram. For example, if there are two radial and four tangential electrons, VB structures that place three electrons in radial and three electrons in tangential AOs are excluded.

(b) When intershell delocalization is strong, pairs and holes that are symmetry-matched, in an intershell sense, engender intershell covalent bonds. In such a case, perfect-pairing VB structures that preserve atom electroneutrality and feature both intra- and intershell covalent bonds dominate the VB wavefunction.

4.5 The VB Formulas of the Three-Membered Rings

Figure 4.2 shows two shell diagrams. We begin by considering the diagram of C_3. Because of the R/T complementarity, each of R and T has only bonding pairs. On the other hand, the extremely strong spatial overlap of the radial AOs produces a high energy LUMO(R), which is separated by a large energy gap from the HOMO(T). As a result, intershell delocalization is thwarted. This means that C_3 approaches the limit of two *pseudoaromatic* shells, R and T, and it can be represented by the formula R^2T^4 . The term "pseudoaromatic" (rather than "aromatic") is crucial: it means that the radial and tangential AOs have high electronegativity, and aromaticity cannot express itself under these circumstances. We will return to the issue of the electronegativity dependence of aromaticity.

We can translate from MOVB to VB language by writing down the VB wavefunctions that correspond to the shell diagrams. This is done in Figure 4.3a, b.

(a) C_3 defends against strong C—C/C—C interbond overlap repulsion present in perfect-pairing structure PP by replacing one covalent bond by a radial hole and a tangential pair (i.e., by ejecting one electron from the triangular hollow to the tangential space) at the expense of Csp^x-to-C2p promotion. This generates the carbenic structure Z, which is a major contributor to the VB wavefunction (Figure 4.3a). The total VB wavefunction is interpreted as follows: C_3 has two covalent bonds plus a radial hole and a tangential pair which are further stabilized by pseudoaromatic delocalization . Furthermore, C_3 tends to the limit of "Hückel cyclopropenyl cation" plus "Möbius cyclopropenyl anion" (Figure 4.3b) in which CT delocalization has been largely arrested because of the relatively high electronegativity of the carbon AOs. This representation is suitable for a cyclopropane derivative in which carbon is replaced by a more electropositive semimetal or metal atom.

(b) Because perfect pairing is still important, C_3 cannot be appreciably stabilized by either π-donor or π-acceptor substituents. Only very powerful π acceptors (but not π donors) can effect stabilization.

(c) The effect of π-acceptor (effectively acting via a 2p hole) and π-donor (effectively acting via a 2p pair) substituents is determined by recognizing that only a tangential 2p AO of a ring can overlap strongly with a 2p AO of the substituent. The radial AO acts as an sp^x hybrid that is poorly aligned for overlap. This means that overlap considerations alone suggest that the response of a ring to a substituent will depend primarily on whether the carbenic Z contributor has a tangential pair or a tangential hole. In the case of C_3, the presence of a tangential pair decides it: π acceptors should be superior to π donors.

(d) C_3 controls the conformation of an attached π acceptor by the tangential pair and the conformation of an attached π donor by the radial hole. The two conformations of interest are the "bisected" and the "perpendicular" ones:

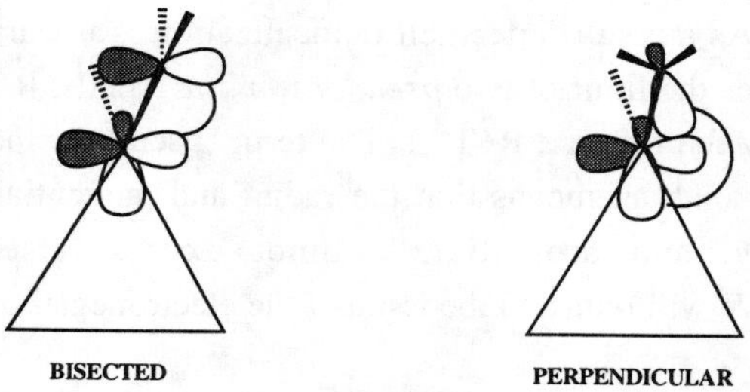

A ring tangential pair (hole) directs a π acceptor (donor) to adopt the bisected geometry. By contrast, a ring radial hole (pair) directs a π donor (acceptor) to adopt the perpendicular geometry with or without rehybridizing to an sp^3 geometry.

Thus in a single page (Figure 4.3), VB theory can be used to explain twotrends in cyclopropane chemistry:

(a) The tangential pair of Z dictates a bisected geometry to π acceptors. This is a rather trivial result because the PMO model does as well. By contrast, the radial hole of Z dictates a perpendicular geometry to π donors. In the case of cyclopropylamine, the NH_2 group adopts a perpendicular geometry in which the nitrogen sp^3 lone pair overlaps with the radial hole of Z.[16a]

(b) The order of stabilization of the carbenic G, S and D states by π donors (e.g., F, OR) and by π acceptors (e.g., CN) is:

$$\pi \text{ donors: } G > S > D$$
$$\pi \text{ acceptors: } D > S > G$$

As a result, the structural consequences of the two classes of substituents are differentiated by the carbenic contributor as illustrated in Figure 4.3c. π Acceptors should cause the "side" C1—C2 to be longer than the "distal" C2—C3 bonds and they should also promote preferential C1—C2 bond cleavage. The converse is true of π donors. Thus, in a fraction of a page, VB theory rationalizes well-known data[17] that are not explained by conventional MO models.[8a,b] And yet, the story is far from over: In Chapter 8, we will see that the same predictions are reached (for cyclopropane *and* propane) for the case of π donors even if one totally disregards the carbenic structure and assumes perfect pairing and intrabond CT delocalization.

As a final illustration, consider the electronic structure of the oxirane molecule, the mono-oxygen derivative of cyclopropane. In dealing with homoatomic rings (e.g., AAA), the singlet–triplet gap of A is the indicator of whether A can accommodate better the G or D closed-shell configuration. However, in the case of a heteroatomic ring, e.g., ABC, the singlet–triplet gap is the indicator of the preference of a fragment for either the open-shell S or the closed-shell, G or D, configuration. In going from H_2C to HN to O, the triplet is increasingly stabilized relative to the singlet, the G and D configurations tend to become degenerate, and the fragment is increasingly predisposed toward adopting the S configuration. It is the fragment that best accommodates the open-shell configuration. Because of the degeneracy of the two singly occupied AOs (and the greater preference for the triplet state), oxygen has a higher tendency than methylene for binding from the S electronic configuration. This means that the oxygen of oxirane will have high S character, while the two methylenes will each have S/D character. In turn, this means that π-donor substituents (which stabilize the G state of carbene) will destabilize the oxirane, while π-acceptor substituents (which stabilize the D state of carbene) will have the opposite effect. This is clearly seen in the epoxide–quinone–methide valence isomerization of benzofurans studied by Adam and his co-workers.[16b]

4.6 The Difference Between Cyclopropane and Cyclohexane

We can now contrast C_3 with C_6. From the shell diagram of the latter (Figure 4.2), it is apparent that C_6 fulfills neither of the conditions for instability stated earlier. Each of R and T have only bonding pairs, and the intershell FMO gap is modest simply because changing the CCC angle from 60° in cyclopropane to 120° in the planar form and to 109° in the chair form of cyclohexane diminishes greatly the radial without improving greatly the tangential AO overlap. In turn, this brings down the R LUMO and raises the R HOMO without greatly depressing HOMO(T) and raising LUMO(T). The result is shrinkage of the intershell FMO gap and strong intershell delocalization. The approximate VB wavefunction of C_6 (shown in Figure 4.4.) says that C_6 is a "classical" carbocycle bound, by six covalent bonds, which can be formed in inter- as well as in intrashell fashion.

We can now explain why cyclopropane has shorter C—C bonds despite being more strained than cyclohexane. Specifically, the approximate VB wavefunction of cyclopropane shown in Figure 4.3a says that the molecule has two "choices," both of them bad, corresponding to resonance structures PP and Z:

Form three C—C covalent bonds through the singlet-coupling of three S methylenes at the price of severe interbond repulsion. This is represented by PP.

Form only two C—C covalent bonds, while driving the third bond pair to the T shell so as to generate a doubly pseudoaromatic system built from two S and one D methylene. This is represented (in part) by Z.

Cyclopropane opts for a compromise, and both PP and Z are important contributors. The Z contributor (see Figure 4.3a) has only one radial bond pair in the triangular hollow. As a result, there is no exchange repulsion engendered by the electrons occupying the radial AOs. By contrast, the PP and PP* contributors of cyclohexane (see Figure 4.4) have intershell radial/tangential (PP) as well as intrashell (PP*) radial bond pairs that are either partly (the former) or totally (the latter) confined within the hexagonal ring hollow. As a result, there is radial interbond overlap repulsion which, however, is kept modest because of the large size of the ring hollow.

We conclude that the C—C bond lengths are a function of two variables:

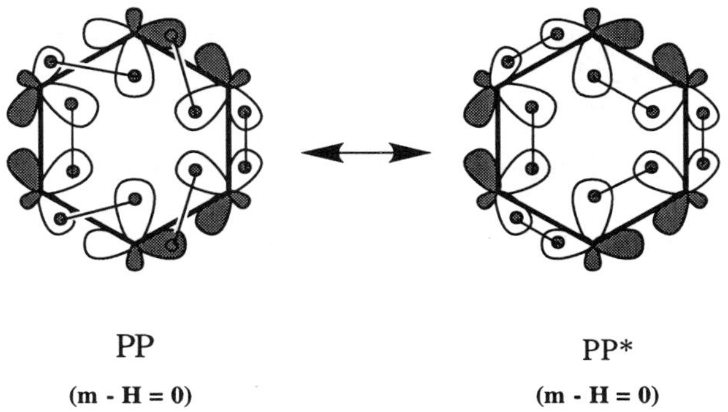

Figure 4.4 The approximate VB wavefunction of planal hexagonal C_6.

(a) The number of electrons (not pairs) in the R shell called *hollow electrons*, symbolized by *H*.

(b) The size of the C_m ring hollow. This is taken to be proportional to *m*. This analysis leads to the construction of the index $m - H$. Every reference MOVB configuration of the shell diagram and every VB configuration that has been generated by expansion of the former has an associated $m - H$ index. A positive value implies a ring hollow depleted of electron pairs ("depleted hollow") and predicts C—C bond shortening. A negative value implies an "enriched hollow" and predicts C—C bond lengthening. Cyclopropane has short C—C bonds because one of the principal VB contributors is Z, and this has $m - H = 3 - 2 = +1$. Cyclohexane has longer C—C bonds because it is effectively a classical covalent molecule in which each of the principal contributors, PP and PP*, has $m - H = 6 - 6 = 0$.

Cyclopropane is forced to depend for its stability at least partly on pseudoaromatic delocalization. By contrast, cyclohexane depends primarily on perfect pairing. Pseudoaromatic cyclopropane is made up by the combination of two S and one D methylenes; that is, each methylene in cyclopropane has 66% S and 33% D character. By contrast, covalent cyclohexane is made up of six S methylenes; that is, each methylene in cyclohexane has 100% S character. This means that the excitation energy per methylene is much higher in cyclopropane than in cyclohexane. Cyclopropane is strained and, at the same time, has short bonds because pseudoaromatic delocalization is too ineffective relative to the high investment of promotional energy. This leads to the strategy for rendering cyclopropane superior to cyclohexane: Replace the carbons by more electropositive atoms, in an attempt to turn intrashell pseudoaromaticity into aromaticity by activating CT delocalization without requiring a disproportionate amount of orbital promotion. As we will see, this is exactly what happens, when CH_2 is replaced by PH.

4.7 Orbital Promotion and Ring Strain

The hallmark of the isoelectronic, heavier congeners of carbon (A = Si, Ge, Sn, Pb) is twofold: they are more electropositive, and they also have a contracted ns AO. This means that AH_2 is less likely than CH_2 to invest in orbital promotion for the purpose of bond making (see Chapter 3). In other words, orbital promotion is unjustified when bond making is weak. This is why the tendency for the divalent state increases as we go from Si to Pb as atom electronegativity decreases and ns contraction increases. It follows that the best accommodation of AH_2 fragments is provided by rings that allow them to remain unpromoted while still being able to engender strong bonding. These are the rings with a large ratio of R to T pairs in the shell diagram. This leads to a selection rule: *Replacement of CH_2 by an AH_2 fragment having a large singlet–triplet gap preferentially stabilizes rings with high R:T ratios.* This means automatically that SiH_2 will be fundamentally different from the isoelectronic PH and S fragments because the latter have triplet ground states and require no orbital promotion. Needless to say, the same rule covers also X_2C species with large singlet–triplet gaps.

What are the rings with high and with low promotional energy requirement? To answer this question, all we have to to do is draw the Shell diagrams and determine the R:T ratio. Cyclopropane has R:T = 1:2, but cyclohexane has R:T = 3:3 (see Figure 4.2). This leads to the conclusion that replacing CH_2 by either SiH_2 or CF_2 (two fragments with a highly stabilized ground singlet state) should *destabilize* the three-membered ring. The (rounded-up) strain energies of cyclopropane, cyclobutane, and their perfluoro derivatives suggested by Smart are given below. Not only are they consistent with our expectations, but they also define a crucial problem: Why does perfluorination increase the strain of cyclopropane but diminish the strain of cyclobutane? Clearly, there must be an antipodal relationship of the two parent molecules. We will now see that the VB wavefunction of cyclobutane projects this in a very clean fashion.

Strain Energies (kcal/mol)

c-C_3H_6 : 28	c-C_4H_8 : 27
c-C_3F_6 : 49	c-C_4F_8 : 9

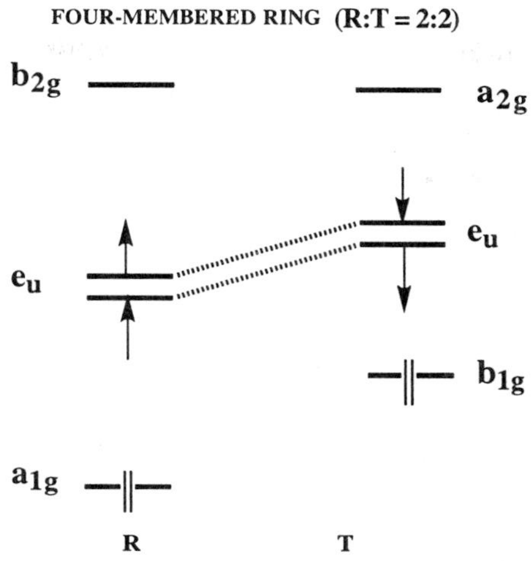

Figure 4.5 Shell diagram for square planar C_4.

4.8 The VB Formulas of Four-Membered Rings and the Difference Between Cyclobutane and Cyclopropane

The shell diagram of square C_4 is shown in Figure 4.5. Because of the lack of R/T complementarity, each of R and T has two singly occupied, nonbonding MOs. On the other hand, these nonbonding MOs are quasi-degenerate and, as a result, intershell delocalization is very strong. This means that C_4 is actually a composite of two triplet pseudoaromatic shells, R and T, which are coupled into an overall singlet that can be represented by the formula R^4T^4.

We can translate from MOVB to VB language by writing down the approximate VB wavefunction that corresponds to the shell diagram as shown in Figure 4.6. This diagram defines the hallmark of cyclobutane: It is the most "polarizable" of all rings because, depending on the relative importance of the three contributors, the $m - H$ index is free to attain any value between –2 and +2 and the R:T ratio can vary from 3:1 to 1:3!

Compared to C_3, C_4 suffers from less C—C/C—C interbond overlap repulsion in its PP structure. However, intrashell delocalization can no longer relieve exchange repulsion as in C_3. This occurs because intrashell delocalization involves the biscarbenic structures ZZ and ZZ*, which are interpreted as follows:

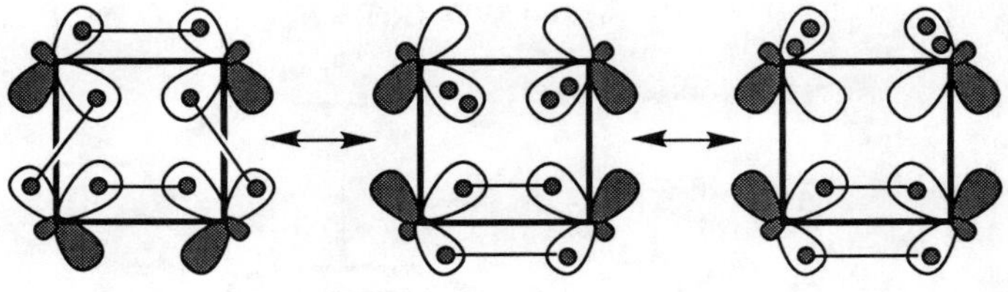

Figure 4.6 The approximate VB wavefunction of square planar C_4. Because of lower promotional energy (occupation of the lower energy sp^x radial AOs), ZZ makes a greater contribution than ZZ*.

In ZZ, two cis *radial pairs* are delocalized while experiencing mutual exchange repulsion. At the same time, there is complementary delocalization of two cis *tangential holes*.

In ZZ*, two cis *tangential pairs* are delocalized while experiencing mutual exchange repulsion. At the same time, there is complementary delocalization of two cis *radial holes*.

The intrinsic preference of C_4 for ZZ or ZZ* is determined by the interplay of interpair overlap repulsion (weaker in ZZ*) and promotional energy (more favorable in ZZ because the radial sp^x has lower energy than the tangential 2p). We formulate a hypothesis: *The relative weights of ZZ and ZZ* are primarily controlled by the promotional energy factor and, as a result, ZZ dominates.* This leads to the following predictions:

(a) The index $m - H$ is smaller than zero, and it can reach the limit of –2. Hence, the C—C bonds will be long (compared to the C—C bonds of cyclohexane).

(b) The R:T ratio is larger than 2:2 and it can reach the limit of 3:1. As a result, the four-membered ring will undergo the most precipitous stabilization when CH_2, with a triplet ground state, is replaced by AH_2, with a singlet ground state and a large singlet–triplet gap. However, as we will see, the four-membered ring will not turn out to be the most stable $(AH_2)_4$ ring simply because the five-membered ring also has a large R:T ratio. Although the stabilization of $(CH_2)_4$ exceeds that of $(CH_2)_5$ when C is replaced by Si and heavier congeners, the latter ends up being the most stable $(AH_2)_n$ ring simply because the reference cyclopentane started at lower energy than the reference cyclobutane.

(c) C_4 acts primarily with a *stabilized tangential hole* and a *destabilized tangential radial pair* toward a substituent. This means that C_4 will be stabilized preferentially by π donors because, on the basis of mere spatial overlap considerations, it acts preferentially by employing its overlap–superior tangential hole rather than the overlap–inferior radial pair.

(d) While ZZ may be relatively more important than ZZ*, cyclobutane is "polarizable" to the extent that it can easily readjust the weights of the close-lying ZZ and ZZ* as a response to a perturbation by either substituent or medium.

We can use these ideas to explain the following trends:

(a) The exceedingly long C—C bonds of cyclobutane can be explained in two equivalent ways:

1. They are the consequence of the occupancy of nonbonding R and T shell MOs which is not offset by intershell delocalization.
2. They are the result of cis radial pairs (dominant ZZ contributor) which engender exchange repulsion that is not offset by pseudoaromatic delocalization.

(b) Replacing CH_2 by either SiH_2 or CF_2 (i.e., singlet carbenic fragments with a large singlet–triplet gap) stabilizes the four-membered ring relative to the larger (six members) ring; that is, it reduces the strain of cyclobutane.[18] The effect is exactly the opposite of that found in the case of cyclopropane. The prediction that replacing C by Si stabilizes the four-membered ring, suggests that propellane-type as well as prismatic structures made up of heavier congeners of carbon are accessible synthetic targets. This is confirmed by the syntheses of the bicyclo[1.1.0]pentasilane[19] and the preparation of hexasilaprismanes[20] and octasilacubanes.[21] Prisms of heavier congeners of Si,[22] as well as stable tin derivatives[23] of the organic bicyclo[1.1.0]pentane,[24] are known.

(c) The key experimental result is that both π acceptors and π donors stabilize cyclobutane more than cyclopropane. Thermochemical data taken from the work of Fuchs et al.[25] and given in Table 4.2 show that substituents have virtually no effect on the stability of cyclopropane. Indeed, the π donor NH_2 weakly destabilizes cyclopropane, but the π acceptor CN has almost no effect. On the other hand, in a comparative sense, the effect of the same substituents on

Table 4.2 Effects of Substituents on the Stability of Cyclopropane and Cyclobutane

Substituents, Y	ΔH_r (kJ/mol)
Cyclopropyl-Y	
H	−114.8
CN	−115.2
NH_2	−117.8
Cyclobutyl-Y	
H	−110.4
CN	−100.3
NH_2	−102.5

cyclobutane is dramatic. *Equally important, substituents of both types stabilize cyclobutane.* This is a consequence of the "polarizability" of cyclobutane.

$$\text{c-}(CH_2)_n CHY + H_3C\text{—}CH_3 \rightarrow$$
$$(n-1)\, H_3C\text{—}CH_2\text{—}CH_3 + H_3C\text{—}CHY\text{—}CH_3$$

Here is now a crucial point: Assuming ZZ and ZZ* degeneracy (or quasi-degeneracy), radial/tangential character can be ascribed to each pair and each hole. This contrasts with the situation in cyclopropane, where the Z contributor features a radial hole and a tangential pair. Hence, unlike C_3, C_4 cannot dictate conformation to either π-donor or π-acceptor substituent: a radial pair dictates a "perpendicular" but a tangential pair dictates a "bisected" conformation of a planar π acceptor, and so on. A strong preference can be established only if ZZ far exceeds ZZ* in importance, an unlikely scenario. A survey of the data confirms that the preference for the "bisected" conformation by a π acceptor is much weaker in cyclobutane than in cyclopropane.[26] In turn, this means that the conformational preference of a substituent is not related to the stabilizing effect of the same substituent! The former depends on whether a pair or hole is constrained to occupy only one type of AO, radial or tangential. By contrast, the latter is independent of the type of orbital accommodating the pair or hole. Thus, the bisected preference of π acceptors is much greater in cyclopropane, but the stabilizing effect of the very same groups is much larger in the "polarizable" cyclobutane. Clearly, the VB theory of prototypical organic rings paints a picture totally at odds with that produced by conventional MO models over decades and explains all pieces of the puzzle, rather than some conveniently selected cases.

We end the discussion of the prototypical cyclic hydrocarbons by underscoring an implicit assumption: In comparing different rings, each methylene was assumed to be "frozen"; that is, the carbon hybridization is kept constant. In reality, the detailed hybridization is dictated by the perfect-pairing configuration. Since the overlap of the radial and tangential 2p AOs decreases as ring size increases, the 2p character of the endocyclic (skeletal) radial AOs progressively decreases as we sweep from cyclopropane to cyclohexane. This means that the exocyclic radial AO tying up the hydrogens gains increasingly 2p character in the same direction consistent with the NMR and IR spectroscopic characteristics of these molecules. Bottom line: To understand structural variations in carbocycles having fixed component atoms, focus on perfect pairing. On the other hand, to predict the effect of substitution on ring stability, focus on the dominant "carbenic" VB structures representing CT delocalization.

4.9 From Pseudoaromaticity to Aromaticity; the Electronic Structures of "Heavy" Tetrameric Rings

When first-row elementsare replaced by more electropositive, heavier congeners of the second or lower rows, pseudoaromaticity gives way to aromaticity in the sense that every stable ring becomes a coordination complex of two (interacting) aromatic R and T shells. The following formulas are proposed for four-membered rings of second-row atoms.

B_4^{+2}	$[R^2T^0P^0]R*^8$	P_4^{+2}	$[R^6T^2P^2]R*^8$
B_4	$[R^2T^2P^0]R*^8$	P_4	$[R^6T^2P^4]R*^8$
Si_4^{+2}	$[R^2T^2P^2]R*^8$	S_4^{+2}	$[R^6T^2P^6]R*^8$
Si_4	$[R^4T^2P^2]R*^8$	S_4	$[R^8T^2P^6]R*^8$

Because of the ns contraction, Al, Si, S, and Cl are all expected to bind from a ground $3s^23p^x$ valence configuration. Since the nonbonding 3s pairs are accommodated by the exo-radial AOs, each tetramer is expected to have a completely filled R* shell which, henceforth, is not shown explicitly. The formulas reveal directly whether a square ring will retain its integrity or distort to a rhombus or to a rectangle according to the following rules:

1. A high symmetry square is expected when every shell has an aromatic electron count.

2. Distortion of the square to either rhombus or rectangle is expected when one of the R and T shells has a formal antiaromatic electron count. Low atom electronegativity promotes rhombic distortion and high atom electronegativity supports the rectangular form.

3. An empty T shell motivates in-plane and an empty P shell an out-of-plane distortion. On the other hand, a partly or completely filled P shell enforces planarity because, in this way, exchange repulsion between σ and π pairs is averted.

The most notable features of the VB formulas are:

(a) B_4^{2+} is an analogue of Li_4^{2+}, which is expected to be a tetrahedral cage enclosing two singlet-coupled electrons. This stereochemical preference is revealed by the shell formula in the following way: the vacant T and P shells act on the R^2 shell to effect three-dimensional close-packing.

(b) Al_4 is expected to be a singlet butterfly because the vacant P shell will act on the R^2 and T^2 aromatic shells to drive the system toward a nonplanar geometry.

(c) Si_4^{2+} is the first semimetal ring in which each shell is filled and aromatic. Hence, this is expected to be a stable square ring.

(d) Square Si_4 is important because it illustrates the stereochemical consequences of an antiaromatic R or T shell when the atoms have low electronegativity. In such a case, rhombic distortion turns on the interaction of the upper D and Z states of the odd antiaromatic shell, R^4, so that state crossing occurs (i.e., the low lying Z crosses under the K state), and the system is stabilized. Hence, we expect Si_4 to be a rhombus. Another way of looking at it is to recognize that the σ system is isoelectronic to that of rhombic Be_2Li_2. Ab initio computations predict the isoelectronic Ge_4 to be a singlet rhombus.[27]

(e) Square P_4^{2+} represents the ideal situation in which every shell is aromatic.

(f) Square P_4 has an aromatic R/T σ frame and an antiaromatic four-electron P shell, which drives the system toward either a rhombic or a rectangular geometry.

(g) S_4^{2+} is the second example of a stable square ring in which each shell is aromatic; S_4^{2+}, Se_4^{2+}, and Te_4^{2+} have been prepared, and the experimental data are discussed in a review by Gillespie.[28]

(h) S_4 is the first tetramer having one completely full shell, namely, the in-plane radial R shell. This creates strong exchange repulsion in the ring plane which is pitted against the stabilizing influence of the aromatic T and P shells. This leads us to expect an in-plane deformation. Ab initio calculations indicate that the preferred geometry of S_4 is a cisoid chain.[29] We interpret this to mean that S_4 simply lengthens one square edge in order to relieve R-shell exchange repulsion while conserving, to the extent possible, the aromaticity of the T and P shells. Out-of-plane deformation is disadvantageous because it would engender exchange repulsion between the R/T and P pairs.

4.10 The Problem with "Sigma Aromaticity" and Non-local VB Treatments of Rings

Dewar posed the question: Why do cyclobutane and cyclopropane have nearly the same strain energies? He answered it by saying that the σ system of cyclopropane is isoconjugate to the π system of benzene, while the σ system of cyclobutane is isoconjugate to the π system of cyclooctatetraene (COT).[30] In other words, cyclopropane approaches σ aromaticity while cyclobutane tends toward σ antiaromaticity. By contrast, our conclusion was that cyclopropane and cyclobutane are both pseudoaromatic. Cyclopropane trades high promotional energy for strong intrinsic bonding, while cyclobutane does the exact reverse. This is why the two molecules end up with comparable strain energies. They are simply the opposite faces of the same coin.

Many chemists think that "orbital interaction" is synonymous with "orbital overlap" and that orthogonal hybrids do not interact. This is not the case at all. Consider the two sp hybrids, n_1 and n_2. The overlap inegral s_{12} is zero, but the resonance integral h_{12} is not zero at all. It is equal to the one-electron energy difference between a 2s and a 2p AO, the two AOs that make up the hybrid.

$$n_1 = \mathrm{s} + \mathrm{p}, \quad n_2 = \mathrm{p} - \mathrm{s}, \quad s_{12} = 0, \quad h_{12} = \mathrm{e(s)} - \mathrm{e(p)}$$

Consider now the hybrid orbital representation of cyclopropane in which we have two hybrids per methylene used to make a total of three σ C—C bonds. This is a cyclic system that is isoconjugate to the π system of benzene with one important difference: Instead of having six equivalent resonance integrals connecting overlapping π AOs, we have three resonance integrals, t_{ij}, connecting *overlapping hybrids* and three more integrals, h_{ij}, connecting *interacting hybrids of the same atomic center.* The six hybrid AOs interact in a cyclic fashion, and their interaction is "aromatic" if the t_{ij}s and h_{ij}s are chosen in such a way so that the product of their preceding signs is positive. The overlapping hybrids are as follows:

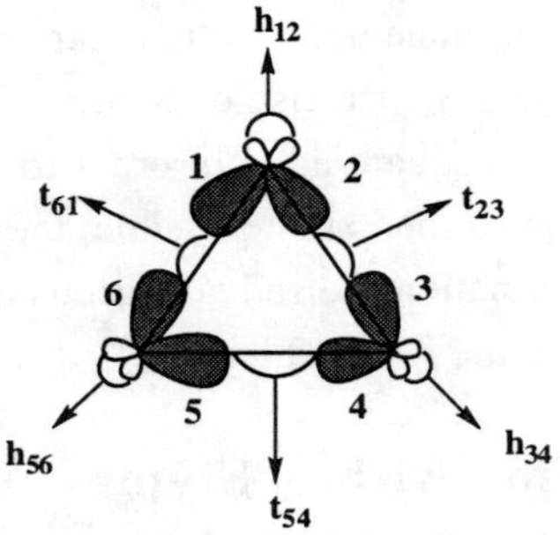

Every cycloalkane is *pseudoaromatic* because one can always choose the t_{ij}s so that the ring is either of the $4N + 2$ electron Hückel or the $4N$ electron Möbius variety. Odd rings belong to the former and even rings to the latter category. Thus, cyclopropane is Hückel pseudoaromatic and cyclobutane is Möbius pseudoaromatic. To have an aromatic Hückel array, in the case of cyclopropane, the product of the signs preceding the h_{ij}s and the t_{ij}s (which themselves are negative) must be positive. If we take all three t_{ij}s to have a positive sign, the condition becomes: The product of the signs of the hij's must be positive. Either all three have positive or two of the three have negative signs: that is, we need an even number of h_{ij} sign reversals to accommodate $4N + 2$ electrons. In the case of cyclobutane, the optimum number of negative h_{ij}s is either one or three; that is, we need an odd number of sign reversals to accommodate $4N$ electrons. In summary, cyclopropane is pseudoaromatic, not aromatic, and cyclobutane is pseudoaromatic, not antiaromatic!

All four prototypical rings, C_3, C_4, C_5 and C_6, have perfect pairing. Differentiating one from the other are the partly noncovalent resonance structures that involve either promotion or demotion of atoms relative to the perfect-pairing configuration. Later on, we will see that what differentiates aromatic and antiaromatic species are the ionic structures. VB approaches that disguise the action of noncovalent configurations are called nonlocal VB (NLVB) approaches because the basis orbitals are not the atom-centered nonorthogonal AOs that we use and to which every chemist can relate but, rather, the Coulson–Fischer type, which are delocalized over many centers. In other words, NLVB has "MO flavor."

Consider two local AOs, *a* and *b*, and two electrons that make a covalent bond. The nonorthogonal VB (NOVB) wavefunction is $(ab + ba) + g\ (a^{+}b^{-} +$

a^-b^+), where g is a mixing coefficient. We shift to NLVB by writing $a^* = a + kb$ and $b^* = b + ka$, where k is a mixing coefficient. The NLVB perfect-pairing structure $(a^*b^* + b^*a^*)$ *alone* is now an approximation of the *total* NOVB wavefunction because the NLVB-covalent structures contain the NOVB-covalent plus the NOVB-ionic structures. However, since the NLVB perfect pairing conceals the action of noncovalent resonance structures built from local AOs, crucial chemical information is lost. NLVB is good for computing numbers, and NOVB is good for interpreting and predicting trends. In fact, NLVB theory has much greater "confusion power" than the MO approach. The two are similar in their linear (LCAO) aspect, but the MO method is superior in making orbital symmetry control transparent.

Many methodologies for computing by NLVB fail to appreciate that once the atomic orbitals are prepared for computation (e.g., orthogonalized AOs, Coulson–Fischer AOs), the conceptual advantage of VB theory is lost. The spin-coupled (SC) methodology is particularly "dangerous" because the SC valence AOs are used in the same spirit as the nonorthogonal, atom-centered AOs to construct resonance structures. The results of the LC computations are, of course, "right" but "achemical". For example, everyone who knows that cyclobutane has notoriously long and cyclopropane notoriously short C—C bonds will be surprised to read the conclusion of Karadakov et al: "Orbital overlaps, orbital shapes, and correlation energies . . . indicate that the bonding along the carbon–carbon σ frameworks in cyclopropane and cyclobutane is rather similar, which represents a convincing theoretical explanation for their surprisingly close conventional ring strain energies."[31] The same objections follow the application of the SC method to benzene, "hypervalent" molecules, and other topics.

The hopelessness of trying to interpret chemistry with aged models is demonstrated in a recent review of organopolysilanes by West,[32] which highlights the uncertainty regarding the electronic structures of silicon rings. Since we have already seen that the carbocycles themselves cannot be understood by conventional models, discussing the purported "explanations" of the literature is an exercise in futility. Suffice it to say that none of these models proposes that fragment promotion is what differentiates one ring from another with respect to substituent effects. Why? Because the alleged "explanations" are "MO explanations," and the interplay of promotion and bond making is hidden under the delocalized MOs. To further underline this point, we focus on yet another piece of interesting data: chemical shift anisotropy and its implication for ring electronic structure.

The ring system that approaches most closely the limit of a bisaromatic complex is cyclotrisilane and its heavier congeners, because of the destabilization of the PP configuration and because the low atom electronegativity makes CT delocalization favorable. If $(H_2Si)_3$ is viewed as an R^2T^4 complex and if the Si–H bonds are tentatively assumed to be covalent, the allocation of the four silicon electrons is as follows:

2/3 endo-radial (R shell) plus 3/3 exo-radial (σ component of SiH_2) electrons
4/3 tangential (T shell)
3/3 π (π component of SiH_2) electrons

This means that the number of electrons along the three Cartesian axes fixed on a given Si molecule is 5/3 (radial), 4/3 (tangential), and 3/3 (π). This explains why the ^{29}Si chemical shift anisotropies are largest in Si_3 rings.[33] When perfect pairing becomes dominant, as in $(H_2Si)_6$, each of the endo radial (R shell) and tangential (T shell) AOs has one electron and the asymmetry is eliminated.

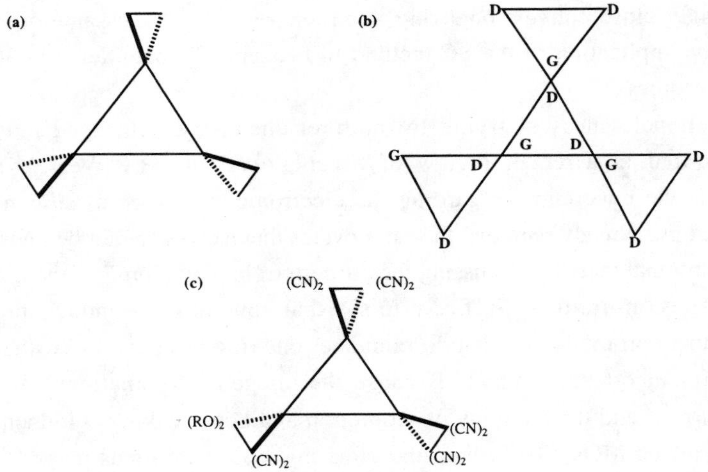

Figure 4.7 Design of a stabilized [3] rotane: G, ground methylene; D, doubly excited methylene.

4.11 Experimenting with Strained Systems

We have seen that cyclopropane can be envisioned as an aggregate of two S methylenes plus one D, or, by momentarily neglecting interelectronic repulsion, as an aggregate of one G and two D methylenes with electrons delocalizing to ensure that the radial orbitals always have two and the tangential orbitals always have four electrons. Since we know what kind of groups stabilize the G and D methylene configurations, we can predict the substitution pattern that is tailor-made for cyclopropane. An instructive application is the prediction of the substitution pattern of the highly strained [3]-rotane[34] shown in Figure 4.7a. The operative constraint is that each three-membered ring is made up of one G methylene and two Ds. This predicts the G and D distribution shown in Figure 4.7b. Given that G methylene is stabilized by π donors and D methylene by π acceptors, the final prediction of the most effective substitution pattern is given in Figure 4.7c.

Making highly strained rings is interesting, but so is the transformation of prototypical rings to molecules with counterintuitive properties. Are cycloalkanes able to transmit information? We can attempt to answer the question by recognizing that appropriate substitution of a small cycloalkane (cyclopropane or cyclobutane) can make it resemble a pseudoaromatic complex in which the PP configuration is not the dominant contributor and the carbons are connected by a formal double bond in the principal carbenic VB contributor (see Figures 6.3 and 6.6). The molecule shown below can undergo a suprafacial 1,5-homodienyl shift according to the Woodward–Hoffmann rules, provided the ring effectively acts as a pseudo–double bond. Getty and Berson have shown that stereospecificity is present when the ring is cyclopropane and persists when the ring is cyclobutane (though the reaction rate is reduced).[35]

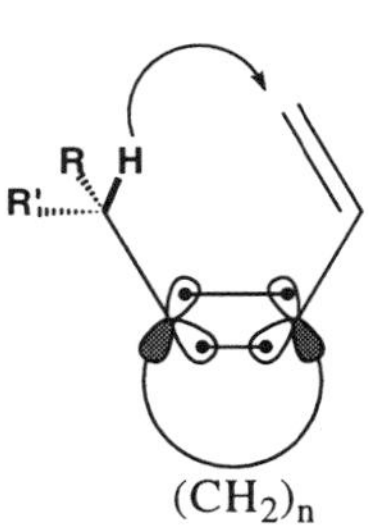

We can enhance the double-bond charater of cyclopropane and cyclobutane by enhancing the importance of the Z (Figure 6.3) and ZZ (Figure 6.6) contributors, respectively. The first requires that we place substituents that stabilize a D carbene — that is, π acceptors such as CN and CHO — and the second requires that we place substituents that stabilize a G carbene (singlet carbene) — that is, π donors such as F, OR, and NR_2. The following systems are proposed.

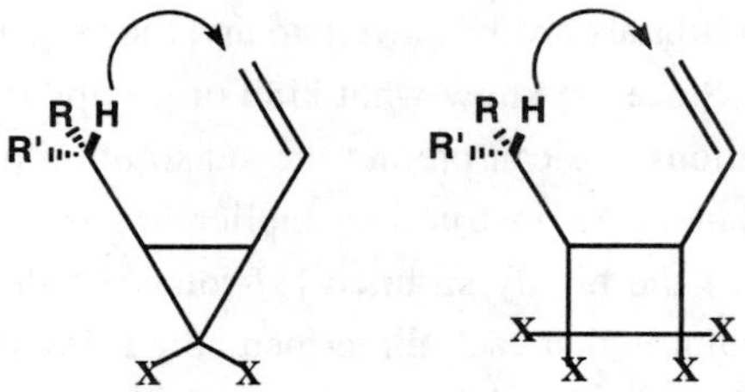

X = CN X = F

Chemists can visualize multiple mechanistic alternatives, but often the electron bookkeeping becomes obscure and useful analogies are lost. The role of theory is to expose these analogies so that new predictions can be formulated and tested. As an example, consider the transformation of a metal–cyclopropene complex to a metal-vinylcarbene M═CH—HC═CH_2. This seems like a very improbable scenario, but it constitutes a secondary pathway in the two-metal isomerization of cyclopropene studied by Grubbs and co-workers.[36] How can we look at cyclopropene to make the electronic events leading to the rearrangement transparent? The answer is given by Figure 4.3. The carbenic resonance structure Z says that the σ frame of cyclopropene has tangential nucleophilic (pair in the T shell) and radial electrophilic (hole in the R shell) character. The metal is linked to the cyclopropene by two formal carbon–metal bonds involving two metal and two pi carbon electrons. The rearrangement leading to M═CH—HC═CH_2 can now be visualized as follows. The methylene bearing the tangential pair rotates to permit the pair to attack one C—M bond, which attacks the radial pair, which, in turn, attacks the radial hole. This description leads to a prediction: To promote intramolecular rearrangement, enhance the importance of structure Z. This is done by replacing the methylene hydrogens by π acceptors (e.g., CN).

Ab initio calculations show that cyclobutyne lies in a shallow minimum with a very low barrier, leading to the much more stable cyclopropylidenecarbene isomer, and a huge barrier leading to butatriene.[37] These reactions are shown with the data in our usual units (kcal/mol).

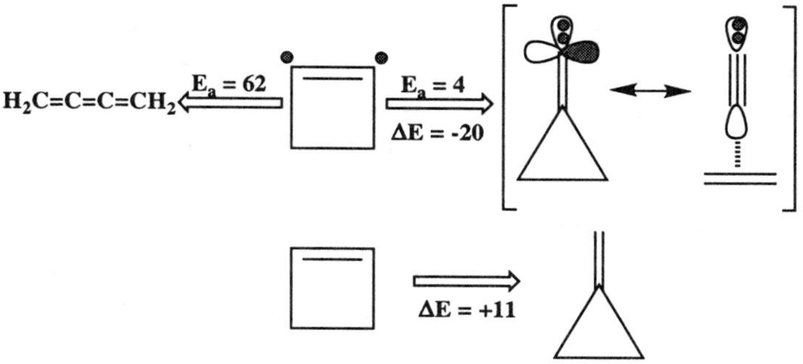

Why is the cyclobutyne ring contraction so exothermic when the corresponding cyclobutene reaction is endothermic? Why is the ring contraction barrier so low, and why is the (conrotatory) ring opening barrier so high? What can we do to raise the barrier of ring contraction so that a cyclobutyne derivative can become "makable"?

(a) The ring contraction is exothermic because of enhanced resonance stabilization of the cyclopropylidenecarbene.

(b) All indications are that the differential response of rings to perturbations is determined by the carbenic resonance structures. As a result, the rearrangement can be represented as follows:

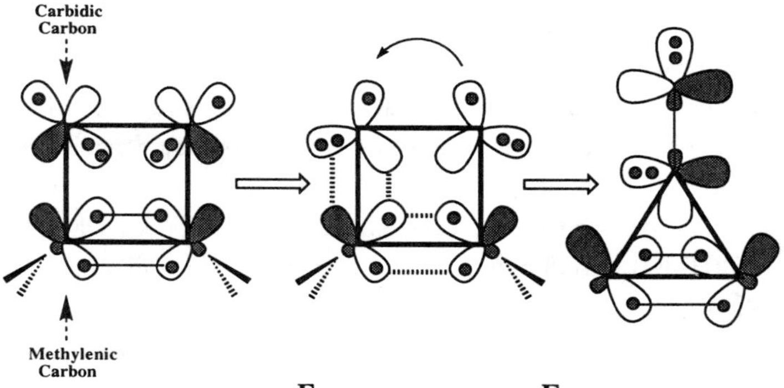

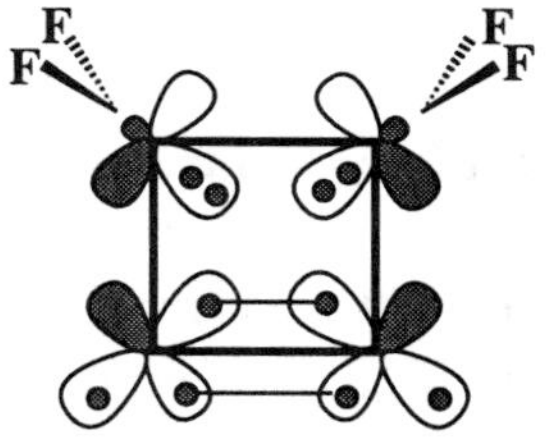

Assuming that the carbidic carbons are sp- while the methylenic carbons are sp^2-hybridized, the carbidic carbons are "nonbonded" and free to rehybridize and rearrange while the methylenic carbons are "double bonded" and unwilling to dissociate as a result of conrotatory ring opening. This explains the great disparity of the two barriers.

By what atoms should we replace the cyclobutyne hydrogens so that the barrier to rearrangement is increased and the cyclobutyne derivative becomes accessible? There are two different ways of proceeding:

(a) Use ab initio computations and try different substituent combinations until there is success.

(b) Use conceptual theory and make qualitative predictions that can be generalized to other ring contractions. These *specific* predictions can then be tested by experiment or ab initio calculation. The analysis presented above suggests the simple strategy of replacing H by F or the methylenic carbons by silica, to turn the carbidic carbons from "nonbonded" to "bonded." The reason is that difluorocarbene and silylene have a highly stabilized singlet ground state. Hence, the methylenic centers will acquire nonbonded (carbenic) character, dictating that the carbidic carbons acquire bonded (open-shell) character. As a result, the barrier of ring contraction will go up and the barrier of ring opening will go down (assuming that changing H to F does not destabilize the ring-opening transition state by F···F nonbonded repulsion). Unpublished results suggest that perfluorocyclobutyne has a higher ring contraction barrier than the hydrocarbon parent.[37]

The key lesson of this chapter is that the electronic structures of rings (and molecules, in general) cannot be understood without explicit consideration of the atom promotion/bond-making interplay. Specifically, the vast majority of experimentalists find their way to applied quantum mechanics via a π-electron Hückel MO (HMO) introduction. Because of the clarity and simplicity of the HMO method, the thinking of the eventual ab initio practictioner is dominated by the concepts of and the indices associated with this theory. There are two major problems:

(a) One trades simplicity for loss of physical reality (see Chapter 3).

(b) π-Electron HMO theory is unrepresentative and, thus, misleading. The total energy of the system is obtained from the Hamiltonian matrix, which is made up of diagonal and off-diagonal elements. The former are indicators of the atomic promotional energy needed for bonding, and the latter are indicators of intrinsic bond making. In π-electron theory all diagonal elements are constant

$\langle 2p|H|2p\rangle$. As a result, relative stability is due to relative bond making, leading many to think that bond distances are indicators of stability. Now, most molecules contain σ frames! This means that the diagonal elements of the energy matrix are no longer constant; that is, we now have $\langle s|H|s\rangle$, $\langle p|H|p\rangle$, $\langle d|H|d\rangle$, and so on, varying in relative importance from molecule to molecule. The reality now becomes what is preached in the gospel of VB theory: Stability is the result of an *interplay* of atom promotion (diagonal matrix elements) and bond making (off-diagonal matrix elements). This work would have not been possible if the author had remained hostage of "π-electron theory"-related intuition about chemical bonding.

References

1. A tetrasubstituted tetrahedrane has been prepared by H. Irngartinger, A. Goldmann, R. Jahn, M. Nixdorf, H. Rodenwald, G. Maier, K.D. Malsch, and R. Emrich, *Angew. Chem. Int. Ed. Engl.* 23, 993 (1984).
2. R.A. Whiteside, R. Krishnan, J.A. Pople, M.-B. Krogh-Jespersen, P.v.R. Schleyer, and G. Wenke, *J. Comput. Chem.* 1, 307 (1980).
3. (a) A. Baeyer, *Chem. Ber.* 18, 2269 (1885). (b) R. Huisgen, *Angew. Chem. Int. Ed. Engl.* 25, 297 (1986).
4. T.W.G. Solomons, *Organic Chemistry*, 5th ed., Wiley, New York, 1992.
5. The C—C bond lengths of the parent tetrahedrane have been taken from ab initio calculations: W.J. Hehre and J.A. Pople, *J. Am. Chem. Soc.* 97, 6941 (1975).
6. A. Greenberg and J.F. Liebman, *Strained Organic Molecules*, Academic Press, New York, 1978.
7. M.S. Gordon, *J. Am. Chem. Soc.* 1980, 102, 7419.
8. (a) R. Hoffmann, *Tetrahedron Lett.* 1970, 2907. (b) H. Guenther, *Tetrahedron Lett.* 1970, 5173. (c) Microwave study of 1,1-difluorocyclopropane: A.T. Peretta and V.W. Laurie, *J. Chem. Phys.* 62, 2469 (1975).
9. L. Kaplan, "The Structure and Stereochemistry of Free Radicals," in *Free Radicals*, Vol. II, J.K. Kochi, Ed., Wiley, New York, 1973, Chapter 18.
10. A critical review that exposes the conceptual problems posed by fluorine chemistry has been written by B.E. Smart in *Molecular Structure and Energetics*, Vol. 3, J.F. Liebman and A. Greenberg, Eds., VCH, New York, 1986.

11. (a) Thermochemical data for ethylene and perfluoroethylene taken from J.J. Low and W.A. Goddard III, *J. Am. Chem. Soc.* 108, 6115 (1986). (b) Thermochemical data for disilaethylene taken from R. Walsh, *Acc. Chem. Res.* 14, 246 (1981). G. Olbrich, P. Potzinger, B. Reimann, and R. Walsh, *Organometallics* 3, 1267 (1984).
12. A.D. Walsh, *Discuss. Faraday Soc.* 2, 18 (1947). See also: H.A. Bent, *Chem. Rev.* 61, 275 (1961)
13. N.D. Epiotis, "Unified Valence Bond Theory of Electronic Structure," in *Lecture Notes in Chemistry*, Vol. 29, Springer-Verlag, New York, 1982.
14. N.D. Epiotis, "Unified Valence Bond Theory of Electronic Structure" in *Lecture Notes in Chemistry*, Vol. 34, Springer-Verlag, New York, 1983.
15. P. Karafiloglou and G. Ohanessian, *J. Chem. Educ.* 68, 583 (1991).
16. (a) S.N. Mathur and M.D. Harmony, *J. Chem. Phys.* 69, 4316 (1978), and previous papers. (b) M. Sauter and W. Adam, *Acc. Chem. Res.* 28, 289 (1995).
17. For data review, see: T. Clark, G.W. Spitznagel, R.Klose, and P.v.R. Schleyer, *J. Am. Chem. Soc.* 106, 4412 (1984).
18. (a) Silicon effect: S. Nagase, M. Nakano, and T. Kudo, *J. Chem. Soc. Chem. Commun.* 60 (1987). (b) Fluorine effect: R.D. Chambers, *Fluorine in Organic Chemistry*, Wiley, New York, 1973.
19. Y. Kabe, T. Kawase, J. Okada, O. Yamashita, M. Goto, and S. Masamune, *Angew. Chem. Int. Ed. Engl.* 29, 794 (1990).
20. A. Sekiguchi, T. Yatabe, C. Kabuto and H. Sakurai, *J. Am. Chem. Soc.* 115, 5853 (1993).
21. H. Matsumoto, K. Higuchi, Y. Hoshino, H. Koike, Y. Naoi, and Y. Nagai, *J. Chem. Soc. Chem. Commun.* 1083 (1988).
22. L. Sita and I. Kinoshita, *J. Am. Chem. Soc.* 113, 1856 (1991).
23. L.R. Sita and R.D. Bickerstaff, *J. Am. Chem. Soc.* 111, 6454 (1989).
24. K.B. Wiberg and F.H. Walker, *J. Am. Chem. Soc.* 104, 5239 (1982).
25. (a) R. Fuchs, J.H. Hallman and M.O. Perlman, *Can. J. Chem.* 60, 1832 (1982). (b) R. Fuchs and J.H. Hallman, *Can. J. Chem.* 61, 503 (1983).
26. (a) F.H. Allen, *Acta Crystallogr.* B40, 64–72 (1984). (b) A. Greenberg and T.A. Stevenson, in *Molecular Structure and Energetics*, Vol. 3, J.F. Liebman and A. Greenberg, Eds., VCH, New York, 1986, Chapter 5.

27. G. Pacchioni and J. Koutecky, *Ber. Bunsenges. Phys. Chem.* 88, 242 (1984).
28. R. J. Gillespie, *Chem. Soc. Rev.* 8, 315 (1979).
29. G.E. Quelch, H.F. Schaefer III, and C.J. Marsden, *J. Am. Chem. Soc.* 112, 8719 (1990).
30. M.J.S. Dewar, *J. Am. Chem. Soc.*, 106, 669 (1984).
31. P.B. Karadakov, J. Gerratt, D.L. Cooper, and M. Raimondi, *J. Am. Chem. Soc.* 116, 7714 (1994).
32. R. West, in press.
33. J.D. Cavalieri, R. West, J.C. Duchamp and K.W. Zilm *J. Am. Chem. Soc.* 115, 3770 (1993).
34. R. Boese, T. Miebach, and A. de Meijere, *J. Am. Chem. Soc.* 113, 1743 (1991).
35. S.J. Getty and J.A. Berson, *J. Am. Chem. Soc.* 112, 1652 (1990).
36. R.T. Li, S.T. Nguyen, R.H. Grubbs, and J.W. Ziller, *J. Am. Chem. Soc.* 116, 10032 (1994).
37. R.P. Johnson and K.J. Daoust, *J. Am. Chem. Soc.* 117, 362 (1995).

Chapter 5

The Five-Membered Ring as a Key Piece of the Bonding Puzzle

5.1 The Electronic Structure of Cyclopentane

Tons have been written about cyclopropane, cyclobutane, and cyclohexane. By contrast, little attention has been focused on cyclopentane, a ring typically regarded as the poor relative of cyclohexane. We will now present the seemingly strange argument that the five-membered ring is the most interesting of them all simply because the periodic table has many more heavy than light (C, N, O, F, Ne) p-block elements!

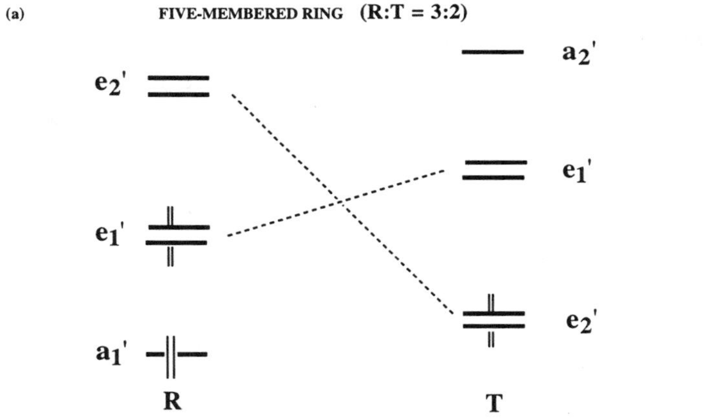

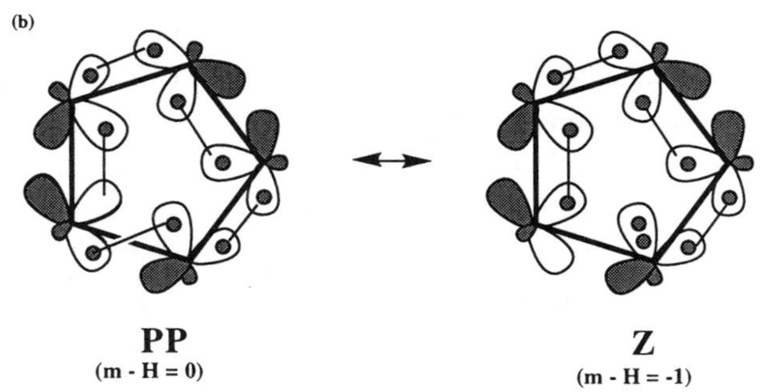

Figure 5.1 (a) shell diagram of cyclopentane. Note the high R:T ratio of bonding pairs. (b) Approximate VB wavefunction of cyclopentane.

The shell diagram of C_5 (Figure 5.1a) suggests that cyclopentane can be viewed as a close relative of cyclohexane (only bonding R and T pairs and modest intershell FMO gap). In VB terms, the large ring hollow renders the PP contributor dominant, much as in cyclohexane. On the other hand, comparison of the approximate VB wavefunctions of cyclopropane (Figure 4.3a) and cyclopentane (Figure 5.1b) suggests the following similarities and antitheses:

(a) Both are odd rings in which the R and T shells can attain a formally aromatic electronic configuration featuring one odd hole or pair in addition to bond pairs. This implies *configurational degeneracy* (e.g., one can write five equivalent VB configurations for the cyclopentadienyl-anion-type system of the R shell) and strong aromatic CT delocalization once the orbital electronegativity has been reduced. This stands in sharp contrast to the even rings of cyclobutane and cyclohexane.

(b) Cyclopentane is the "inverse relative" of cyclopropane insofar as the nature of the Z contributor is concerned: it inserts (rather than ejects) an electron in the ring hollow to generate a radial pair plus a tangential hole (rather than a tangential pair and a radial hole). Relocation of an electron from 2p to sp^x has the benefit of reducing promotional energy at the cost of increasing radial interpair exchange repulsion.

We can now appreciate the uniqueness of the C_5 ring:

(a) With the exception of C_4, C_5 has the highest R:T ratio of the prototypical cycloalkanes. Furthermore, if we count only *bonding* (and not nonbonding) R and T pairs, cyclopentane has the highest R:T* ratio of all prototypical cycloalkanes (asterisk denotes that only bond pairs are counted). This predicts that the most favorable accommodation of singlet heavy p-block groups with a large singlet–triplet gap is afforded by the five-membered ring. This is what ab initio calculations suggest (Table 4.1).

(b) The $m - H$ value of the Z contributor is –1. This predicts that cyclopentane is predisposed to have long C—C bonds, which is why the C—C bonds of cyclopentane are longer than those of cyclohexane. On the other hand, the C—C bonds are shorter than those of cyclobutane because of the greater weight of the PP contributor.

(c) Since the dominant contributor is PP, there will be a weak response to substituents. However, the presence of a tangential hole in Z will render π donors superior to π acceptors. Furthermore, since the hole and pair in the Z

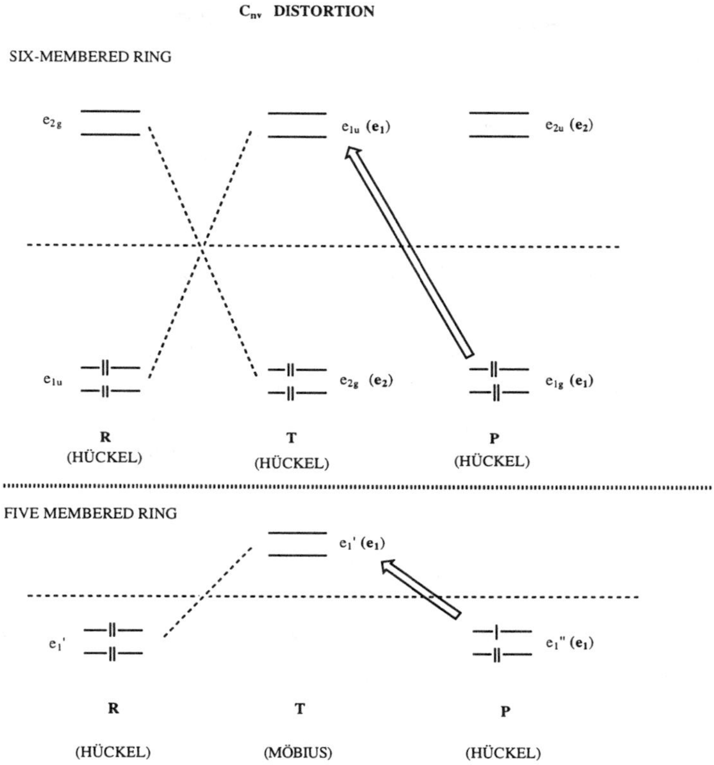

Figure 5.2 π–σ mixing in six- and five-membered rings as a result of C_{nv} deformation.

contributor are "unimodal" — since there is only a *tangential hole* and a *radial pair* in the singular Z contributor — they should exert a much stronger influence on the conformational preference of π acceptors and π donors than the "bimodal" holes and pairs of cyclobutane, everything else being equal. Unfortunately, perfect pairing is superior in cyclopentane, and this will tend to blunt the stereochemical impact of the Z contributor. Nonetheless, it would be interesting to study the conformational preferences of typical substituents in cyclopentane. The C_5 ring should direct π donors and π acceptors to adopt a "bisected" and a "perpendicular" geometry, respectively.

5.2 How Does Buckminsterfullerene Cope with Deplanarization Strain?

In discussions of the structure of buckminsterfullerene, C_{60}, it is implicitly assumed that this molecule facilely and nonspecifically absorbs the strain consequent upon deplanarization of benzene units. Is this a fair assumption? Comparison of the shell diagrams indicates a key difference between the σ skeletal frames of *planar* regular hexagonal C_6 and regular pentagonal C_5 rings. Specifically, the LUMOs of the σ-tangential MO manifolds differ in a fundamental way: those of C_6 are high but those of C_5 are low lying. This is a direct consequence of the even–odd distinction between the two rings. Also, and again for the same reason, the π HOMO of C_5 lies higher than that of C_6. The HOMOs and LUMOs of the radial (R), σ-tangential (T) and π-tangential (P) shells of C_6 and C_5 are shown in Figure 5.2. How do these differences express themselves when it comes to deforming D_{nh} planar rings toward formation of a spherical cluster? Such a deformation, called *"sphere deformation,"* causes reduction from local D_{nh} to C_{nv} symmetry in each ring. As a result, the HOMO(P) is now able to mix with the corresponding LUMO(T). The change in the symmetry labels accompanying D_{nh} to C_{nv} reduction is indicated in parenthesis in Figure 5.2. According to low order perturbation theory, the stabilization resulting from the HOMO–LUMO interaction is inversely proportional to their energy gap. This gap is much smaller in C_5 than in C_6. Hence, the distinguishing feature of a five-membered ring is that it promotes deformation required for spherical cluster formation. Thus, combining the strain-free five- and six-membered rings in a C_{60} cluster is energetically favorable because the C_5 rings "absorb" the damage done by abandonment of the planar graphitic motif. The tendency for sphere deformation is still greater for C_4 because the HOMO(P) and the LUMO (T) are, to a first approximation, quasi-degenerate. However, the strain energy of C_4 rings makes them unfavorable components of a cluster. An important consequence of the strong HOMO(P)–LUMO(T) mixing in the C_5 ring as a result of a sphere deformation is that it drives π-electron density on the concave surface.

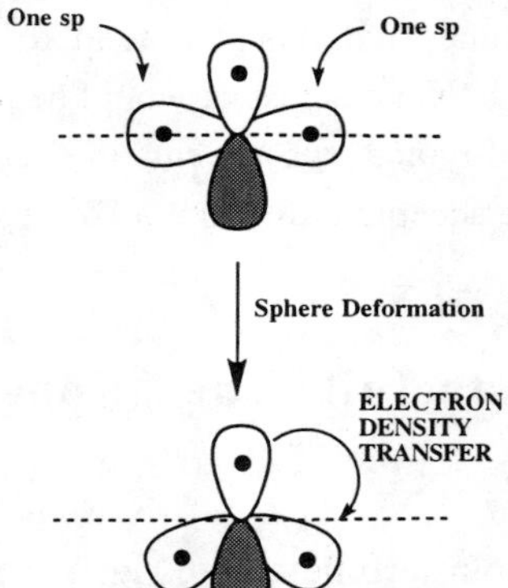

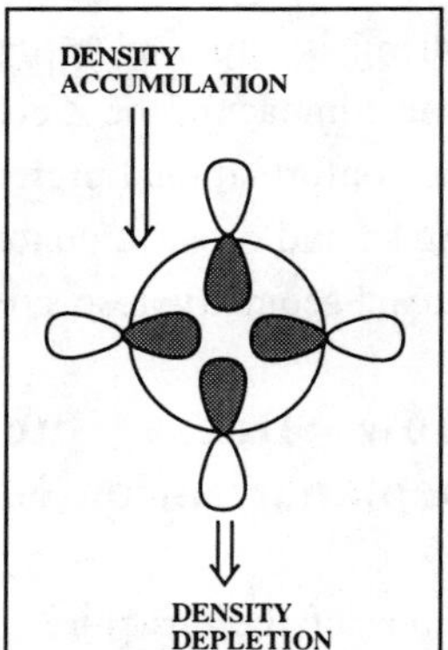

In turn, this sets the stage for the formation of onion fullerides.

ONION FORMATION IN FULLERIDES

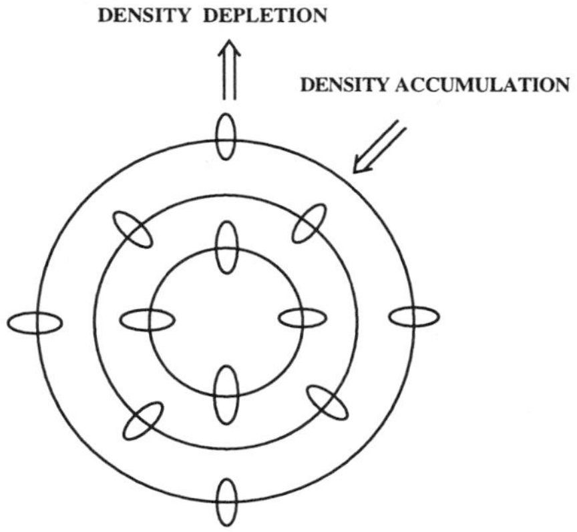

Electron-rich regions of one sphere attract electron-poor regions of an encompassing sphere.

A sphere deformation is related to a *puckering deformation* to the extent that both effect σ–π mixing. We can lump all deformations having such an effect under the title *S/P deformation.* We predict that a given ring will exhibit an increasingly greater tendency for S/P deformation as either the HOMO or LUMO of the P shell and the LUMO or HOMO of the R or T shell, respectively, tend to become degenerate. On this basis, the preference of rings for S/P deformation increases in the following order:

$$4 > 5 > 6$$

Note the position of the four-membered ring: acting with an R^6T^2 configuration, it has LUMO(T), which is degenerate with HOMO(P)!

The analyses presented in this work are based on basic principles and, as a result, the conclusions must be supported by basic molecules. An S/P deformation cannot be observed in ordinary π annulenes, such as the prototypical cyclobutadiene, cyclopentadienyl radical, and benzene, because the π bonds constrain the system to a planar geometry. The concepts remain immutable when the π bonds are replaced by carbon–hydrogen bonds in going from a cyclic polyene to a cycloalkane. The only difference is that the locking action of the π bonds has been eliminated. Now, we must be able to see the physical consequences of the driving force for S/P deformation.

Consider the case of planar hexagonal cyclohexane. The molecule is expected to pucker with the result that the unfavorable 120° CCC angles are turned into favorable 109° angles, which will maximize T-bonding and minimize T-antibonding at each carbon center. In addition, puckering relieves nonbonded H···H repulsion by replacing eclipsed by staggered H···H interaction. However, here are two observations that imply that something else must be superimposed:

(a) Planar cyclopentane can very well accommodate a tetrahedral carbon because the angle of a regular pentagon is 108°. Nonetheless, the molecule is known to undergo pseudorotation through a puckered conformation.[1] Puckering affects adversely the CCC angles.

(b) Cyclobutane is known to be puckered with a D_{2d} geometry even though puckering of the planar ring makes what is already bad (the 90° CCC angle) worse.[2]

(c) Ab initio computations indicate that the aromatic cyclobutadiene dication is puckered, with D_{2d} symmetry.[3]

We suggest that all these observations are the physical manifestations of the increasing driving force for σ–π mixing as we go from a six- to a five- and, ultimately, to a four-membered ring. This is a quantum mechanical effect that comes on top of the ordinary angle strain and torsional strain considerations.

5.3 The Relationship of Enthalpic Stability, Symmetry, and Entropy

While enthalpic considerations of stability take center stage, the role of entropy is obscure. One reason has been the inability to understand the factors that dictate enthalpic stability. This problem is now resolved in Chapters 4, 6–8 and 11. On the other hand, assuming that vibrational entropy is the differentiating element in isomeric comparisons, we isolate two factors:

Symmetry
Enthalpic stability

The higher the symmetry of the isomer, the lower the ability of the atoms to explore three-dimensional space by asymmetric long-amplitude vibrations, and the lower the $S°$ value. The higher the enthalpic stability of the isomer, the lower the ability of the atoms to explore three-dimensional space by *any* long-

Table 5.1 Thermochemistry of Structural Isomers

		ΔH_f° (kcal/mol)	S° [cal/(mol/deg)]	Methyl Rotors
1a	$CH_3(CH_2)_3CH_3$	–35	89	2
2b	$(CH_3)_4C$	–40	72	4
2a	$CH_3(CH_2)_6CH_3$	–50	112	2
2b	$(CH_3)_3CC(CH_3)_3$	–54	93	6
3a	Methylcyclopentane	–25	81	1
3b	Cyclohexane	–30	71	0
4a	$CH_3(CH_2)_2CH_2OH$	–66	87	1
4b	$(CH_3)_3COH$	–75	78	3
5a	CH_3CH_2CHO	–46	73	1
5b	CH_3COCH_3	–52	71	2
6a	*cis*–1,4–Dimethylcyclohexane	–42	89	2
6b	*trans*-1,4–Dimethylcyclohexane	–44	87	2
7a	*cis*-2–Butene	–2	72	2
7b	*trans*-2–Butene	–3	71	2
8a	$HCOOCH_3$	–81	85	1
8b	CH_3COOH	–104	68	1
9a	$CH_3(CH_2)_3CH_3$	–35	89	2
9b	$CH_3CH(CH_3)CH_2CH_3$	–37	82	3
10a	CH_3OCH_3	–44	64	2
10b	CH_3CH_2OH	–56	68	1
11a	$(CH_2)_3$	13	57	0
11b	$CH_3CH{=}CH_2$	5	64	1

amplitude vibrations, and the lower the $S°$ value. The three sets of thermochemical data presented in Table 5.1 teach us valuable lessons.

The first point is that only comparisons of isomers are clearly meaningful. In every case, we compare isomer a with isomer b. Here are the key trends.

(a) In items 1–5, the enthalpically favored molecule has higher symmetry and lower entropy. In other words, enthalpy and symmetry impose the same entropic outcome.

(b) In items 6–8, there is no significant symmetry distinction of the two isomers. The enthalpically favored molecule still has lower entropy. In other words, enthalpy dictates entropy.

(c) Items 9 and 10 are instances of the enthalpically favored molecule having lower symmetry. Now, there is a disjunction: In the alkane example, the enthalpically more stable molecule has lower entropy; that is, enthalpy dictates entropy. Replacement of carbon by an electronegative heteroatom reverses the trend; that is, symmetry dictates entropy.

(d) The well-known principle that restriction of conformational mobility lowers entropy is seen in item 11. This concept is part of the more general principle that higher symmetry implies lower entropy.

(e) There is no general correlation between entropy and the number of free methyl rotors. In fact, in the case of acyclic hydrocarbons, the isomer with the greater number of methyls has the lower enthalpy (see Chapters 6–8) and lower entropy.

We can now formulate a hypothesis: *Enthalpy and entropy operate antagonistically in hydrocarbons and synergistically in polar molecules when restrictions of internal rotation are kept constant.* In other words, hydrocarbons that are enthalpically favored are entropically disfavored. Note that this holds only for comparison of isomers that have the same conformational constraints. That is, we can compare only acyclic isomers or only cyclic isomers or only polycyclic isomers. For example, we can compare isomeric acyclic alkanes (as shown in Table 5.1) as well as cyclic alkanes (as shown below).

	$\Delta H_f^\circ/n$ (kcal/mol)	$S°/n$ [cal/(mole deg)]
$(CH_2)_3$	+4.2	19
$(CH_2)_4$	+1.7	16
$(CH_2)_3$	–3.7	14
$(CH_2)_3$	–4.9	12

Finally, in keeping with the spirit of our discussion, replacing first-row (electronegative) heteroatoms by (more electropositive) heavier congeners weakens the bonding (see Chapter 2) and raises the entropy.

	$S°$
CH_3OCH_3	63.7
CH_3SCH_3	68.3
$(CH_2)_8$	87.7
S_8	102.8

A rule emerges: *Preferential formation of an enthalpically disfavored hydrocarbon at high temperatures implicates entropy as the basis of the selectivity.*

5.4 The Mechanism of Formation of Fullerenes

Since nonplanar C_{60} is enthalpically unstable relative to all-planar graphite (loss of π aromaticity), its formation must be a kinetic, rather than a thermodynamic, phenomenon. Furthermore, since entropy and enthalpy are antithetically related (see Section 5.3) and since fullerene formation occurs under entropic conditions, VB theory makes a simple prediction: C_{60} must arise from a polycyclic carbon cluster that contains enthalpically unstable and thus entropically favorable three-, four-, and five-membered rings and that ultimately rearranges to C_{60}.

The mechanism of fullerene formation is still a mystery, though great progress has been made from the initial suggestion of the "Pentagon Road"[4] to the more recent demonstration that fullerenes can be formed by coalescence of polyyne carbon rings.[5] Whatever the detailed mechanism, our point is that forming more (enthalpically) unstable five-membered rings on the way to C_{60} is entropically more favorable than forming more (enthalpically) stable six-membered rings on the way to graphite. In other words, fullerenes are formed at high temperatures essentially because methylcyclopentane is entropically favored over the more stable isomeric cyclohexane. The same argument explains the "rule of five,"[6] namely, the preferential formation of products that seem to have been derived from biradical intermediates in which a five-membered ring has been formed (recently verified by Maradyn and Weedon[7]) as well as the preference of the 5-hexenyl radical to close to the cyclopentylmethyl radical.[8]

5.5 The Radical Ions of Heavy p-Block Rings

The concept of a heavy p-block ring as a composite of interacting aromatic R and T shells can be further expanded by writing down a formula that shows explicitly the π shell (P, spanning the π-type AOs), the exo shell (R*, spanning the exo-radial AOs) and any attached ligands. In this context, the complete formula of cyclopentasilane, $(SiH_2)_5$, where as always, the superscript denotes the electron count of the shell, is as follows:

$$R^6T^4[P^5R*^5(H_{10})^{10}]$$

The term in the brackets represents the bonding the five silica pairs with five pairs of hydrogens. Since we have I-bonding, the reference configuration for the complete CI implied by the term in the brackets (10 Si—H bonds) is the polyradical state ("covalent hybrid" or "T hybrid") which is made up of all independent covalent structures. In a fraction thereof, P^5 and $R*^5$ are isoconjugate to the nonaromatic π-cyclopentadienyl radical, and H_{10} corresponds to five H_2 molecules with a long bond. We can now ask the question: Which of the four shells is going to accept an odd electron to form an anion radical? The answer is clear: The P^5 will be transformed to an aromatic P^6 shell on addition of an electron. Furthermore, since a $(SiH_2)_n$ cyclopolysilane contains one P^n shell, only rings that fulfill the condition $n + 1 = 4N + 2$ can form stable anion radicals. In other words, the anion radicals of four-, six-, and seven-membered polysilane rings are expected to be unstable relative to the anion radical of the five-membered ring. This picture is consistent with the findings of West and collaborators.[9] The salient point is that the very switch of the bonding mechanism from T to I dictated that all metalloid rings be regarded in a way fundamentally different from the way that we look at nonmetal rings; that is, as composites of interacting shells that tend to behave as independent "molecules."

References

1. B. Fuchs, *Top. Stereochem.* 10, 1 (1978).
2. O. Bastiansen, K. Kveseth, and H. Moellendal, *Top. Curr. Chem.* 81, 99 (1979).
3. K. Krogh-Jespersen, D. Cremer, J.D. Dill, J.A. Pople, and P.v.R. Schleyer, *J. Am. Chem. Soc.* 103, 2589 (1981).

4. R.E. Smalley, *Acc. Chem. Res.* 25, 98 (1992).
5. (a) G. von Helden, N.G. Gotts, and M.T. Bowers, *Nature* 363, 60 (1993). (b) J. Hunter, J. Fye and M.F. Jarrold *Science* 260, 784 (1993). (c) N.S. Goroff *Acc. Chem. Res.* 29, 77 (1996).
6. (a) G.S. Hammond and R.S. Liu, *J. Am. Chem. Soc.* 89, 4930 (1967). (b) K.H. Carlough and R. Srinivasan, *J. Am. Chem. Soc.* 89, 4932 (1967).
7. D.J. Maradyn and A.C. Weedon, *J. Am. Chem. Soc.* 117, 5359 (1995).
8. A.L. Beckwith, *Tetrahedron* 37, 3065 (1981).
9. R.West and E. Carberry, *Science* 189, 179 (1975).

Chapter 6

The Valence Bond Formulas of Organic Molecules

6.1 A New Look at Hydrogen Fluoride

The H_2 molecule is the illustrator of homopolar covalent bonding. What happens when we go to a heteronuclear diatomic like HF? Pauling[1] proposed that replacing H by F in H_2 splits the degeneracy of the ionic structures and this creates a stronger covalent/ionic resonance because the energy gap separating the covalent, H—F, from the depressed ionic H^+F^- structure is diminished (relative to the energy gap separating H—H and H^+H^-). As a result, the H—F bond is said to be stronger than the H—H bond because of "polar covalence." This energy gap argument neglects the fact that the splitting of the degeneracy of the ionic configurations enhances one (e.g., H—H/H^+F^- interaction) but diminishes another covalent/ionic interaction (e.g., H—H/H—F^+ interaction). This leads one to suspect that the gigantic increase of the bond dissociation energy (BDE) in going from H—H to H—F is due to other reasons. This suspicion is supported by the experimental data. The "polar covalence" concept predicts a correlation between the difference of electronegativities (a measure of bond polarity) and the BDE of a diatomic molecule. Such a correlation does not exist! Indeed, instead of a correlation between difference of electronegativities and BDE, we have a correlation between electronegativity or sum of electronegativities and BDE.

HN is out of line simply because bond formation requires the destruction of the strong exchange correlation (a Coulombic phenomenon not to be confused with exchange bonds or antibonds) of three parallel spins in the ground N atom. The BDE data suggest that the primary determinant of BDE is not the energy gap separating covalent and ionic structures but, rather, the interaction matrix element that connects them. As we have already seen (Chapter 2), this is proportional to an AO resonance integral which, in turn, is proportional to orbital electronegativity and, by extension, to atom electronegativity. In short,

Table 6.1 The Bond Dissociation Energies (kcal/mol) of First-Row Hydrides

HLi	HB	HC	HN	HO	HF
56	78	80	74	102	135

the strength of a bond connecting A and B depends on *two* indices, namely, IE(A) + IE(B) and IE(A) – IE(B), and the former is more important than the latter. Thus, instead of "polar covalence," the strength of the H—F bond is due to "electronegativity-promoted CT delocalization."

Up to this point, we have treated every first-row hydride as if it were a simple heteroatomic analogue of H—H devoid of any extra valence pairs. Clearly, this is not the case. For example, H—H and H—F are distinct in two ways:

The sum and difference of electronegativities differ.
H—F has lone pairs, whereas H—H has none.

Focusing on the second difference, we now see yet another problem. If one were to completely disregard CT delocalization and assume that the two diatomics are simply bound by exchange, it would be necessary to conclude that the H—F could be weaker than the H—H bond because the former has one two-electron T bond antagonized by a three-electron T antibond while the latter has only one two-electron T bond. To see this point, consider what happens in the σ frame of HF as the two atoms approach each other. The H atom sees a singly occupied 2p and a doubly occupied 2s fluorine AO. If the F AOs were to remain frozen, there would be one H1s—F2p T bond and an offsetting H1s—F2s T antibond. To defend against bond pair–lone pair exchange repulsion, the F2s and F2p mix through H1s so that F approaches the limit of sp hybridization. This exacts promotion of one electron from F2s to F2p. The net benefit is bond formation with avoidance of exchange repulsion, but at the price of atom promotion. This means that the H—F bond must be weaker than the H—H bond because the former had to pay the penalty of promotion. This leads to the conclusion that the extremely strong H—F bond is not just the result of electronegativity-promoted CT delocalization but also a consequence of a mechanism by which HF avoids exchange repulsion and, thus, the need for strong atom promotion.

Is there a feature in HF that promotes avoidance of overlap repulsion, and, thus, reduces the requisite promotional energy? We get the answer by writing the VB resonance description of HF by keeping the F2s pair frozen and showing explicitly the σ pairs:

$$\mathrm{H{\cdots}F\!:} \longleftrightarrow \mathrm{H^{+}\ :\!F\!:^{-}} \longleftrightarrow \mathrm{H\!:^{-}\ F\!:^{+}}$$

The key point is that the H^+ :F:$^-$ structure has zero exchange repulsion because the σ pairs are in orthogonal AOs of F.

By contrast, the $H:^{-}\ F:^{+}$ structure has a full-blown four-electron T antibond. The H—F covalent Lewis structure is in between. Thus, the presence of lone pairs on F both narrows the energy gap separating the covalent H—F and the ionic $H^{+}F^{-}$ configuration by destabilizing the covalent structures and enhances covalent/ionic resonance beyond what is dictated by the sum and difference of electronegativity.

We have arrived at a key conclusion: Aside from electronegativity considerations, the strong bond of H—F is a consequence of CT delocalization, which is consistent with minimization of exchange repulsion; that is, a large fraction of the binding of HF is due to the contribution of a resonance structure that does not suffer from exchange repulsion. This state of affairs can be represented by writing down the Lewis structure and transforming each line to an arrow consistent with bond polarity to obtain the *bond/dipole formula.* In addition, fluorine has three lone pairs and each of them can be taken to imply the existence of an infinitely electropositive ligand. Finally, the hydrogen atom can be assumed to have three vacant AOs, and each one of them can be taken to imply the presence of an infinitely electronegative ligand. With these assumptions, we arrive at the following bond dipole formula for HF:

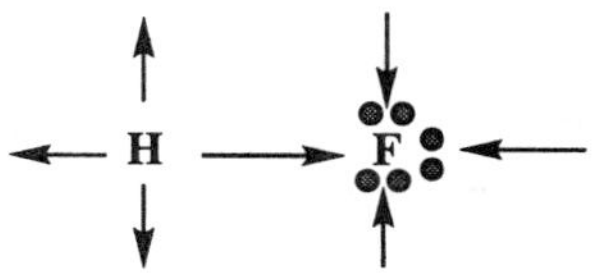

The lesson of this formula is pivotal for the discussions to come: *Optimal intrabond CT delocalization obtains if all arrows emanating from one atom either all point away from the central atom or all point towards it.*

The term "intrabond CT delocalization" means that two electrons within two AOs forming an exchange bond are allowed the additional benefit of CT delocalization but only within the two AOs that define the exchange bond. The preceding formula, which is the *VB descriptor* of HF, no longer represents a single VB configuration. Rather, it is a descriptor of a wavefunction produced by the superposition of two VB configurations, H—F: and $H^{+}:F:^{-}$. By performing a series of 1-e hops, *in an intrabond sense,* one goes from the perfect-pairing Lewis structure to the fully ionic configuration, which contains only unshared pairs and holes. In the case of HF, this ionic configuration is free of overlap repulsion. This feature is largely responsible for the strong bond of HF.

The metathesis (or bond switch, or atom redistribution) reaction shown below has served as the basis for the derivation of the Pauling electronegativity scale. The endothermicity was taken to be both a consequence of stabilization of the reactant side by "polar covalence" and a measure of the electronegativity difference of A and B. The "polar covalence" argument is now right because the sum of reactant atom electronegativities equals the sum of product atom electronegativities. The differentiation between reactants and products now falls exclusively to the province of the difference of the electronegativities of the bound atoms. In other words, the Pauling argument fails when the sum of electronegativities varies (H—H/H—F comparison) but works when it is kept constant (metathesis reaction).

$$2\ \text{A—B} \longrightarrow \text{A—A} + \text{B—B}$$

We put A = H and B = F and write the bond dipole formulas for reactants and products. The reaction is endothermic by 130 kcal/mol! Once again, we see a different scenario unfolding: Polar covalence is only one part of the story. The other part is exchange repulsion. The (twice) "in–out" pattern of the arrows on the reactant side is inferior to the "out–out" plus "in–in" pattern of arrows on the product side insofar as interpair exchange repulsion is concerned. Hence, we develop a simple hypothesis: The stability of isomeric organic molecules is determined by the exchange repulsion of intradelocalized bond pairs and lone pairs.

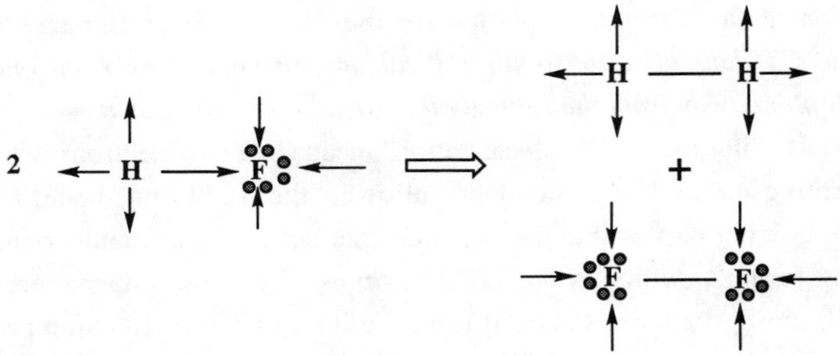

6.2 The Exclusion Rule

Consider the prototypical ligand redistribution reaction (6.1), where M is a divalent fragment (e.g., H_2C). Is this reaction exothermic, and if so why?

$$2\ \mathrm{F—A—H} \longrightarrow \mathrm{F—A—F} + \mathrm{H—A—H} \tag{6.1}$$

Assuming that electronegativity increases in the order F > A > H and treating the hydrogen atom as a univalent atom with three high lying vacant AOs and fluorine as a univalent atom with three doubly occupied AOs, we write the bond dipole formulas of reactants and products:

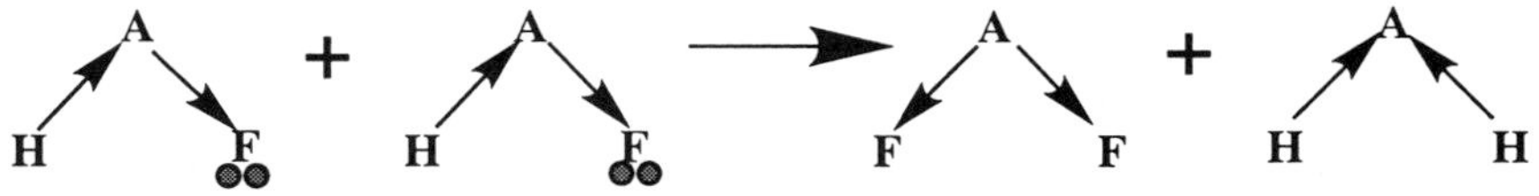

The reaction is predicted to be *exothermic* because optimal CT delocalization (i.e., delocalization that minimizes interpair exchange repulsion) can occur only in the products. To see this point, observe the ultimate localization of the electron pairs as a result of intrabond CT delocalization. On the product side, all pairs occupy either orthogonal orbitals of A (in AH_2) or nonbonded orbitals of F

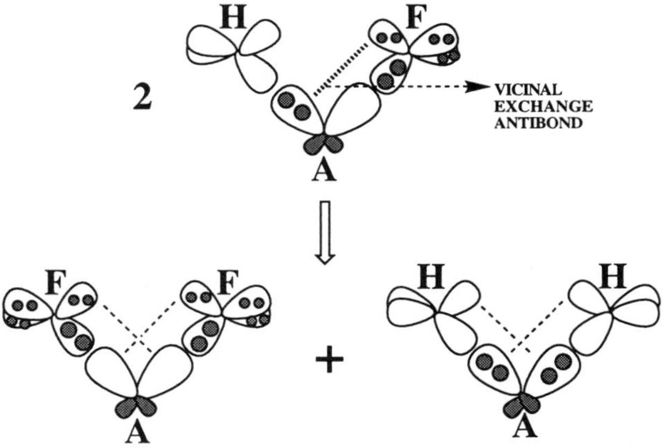

Figure 6.1 The interaction of A—H and A—F pairs in the reactants and products of the redistribution reaction 2 H—A—F ⟶ H—A—H + F—A—F. Pairs occupy noninteracting AOs in AF_2 and orthogonal AOs in AH_2 and escape exchange repulsion; that is, there exist no vicinal antibonds. Two pairs occupy vicinal AOs in AHF and create one vicinal antibond. The reactants and products are represented by the fully ionic configuration HP(max), which is reached by intrabond CT delocalization starting from the perfect-pairing configuration.

atoms which we can assume to have zero overlap (in AF_2). As a result, formation of a strong four-electron antibond is aborted. In other words, the two bonds in each of AF_2 and AH_2 are largely free of exchange repulsion. By contrast, on the reactant side, the electron pairs are allocated in such a way that two of them end up in vicinally overlapping valence AOs. This results in the formation of one strong four-electron antibond. Hence, the two bonds of H—A—F suffer from strong exchange repulsion. The argument is conveyed schematically in Figure 6.1.

We have generated a selection rule: If a molecule is represented by its Lewis formula with each lone pair, bond pair, and vacant orbital replaced by an arrow symbolizing the corresponding bond dipole, head-to-tail combinations of arrows at the same center are "forbidden," while head-to-head or tail-to-tail combinations are "allowed." In other words, minimization of interbond exchange repulsion occurs when all (bond-dipole) arrows at one atomic center have the same directionality. We refer to this as the *exclusion rule*, since it is derived from the Pauli exclusion principle.

6.3 The Vicinal Antibond and the T Formula

Figure 6.2 shows four tetratomic molecular fragments involving different combinations of atoms or groups (C = carbon, Z = atom more electronegative than C, ELP = electropositive group, ELNG = electronegative group). Which of the four combinations comes closest to being ideal? In the left column, we show the ordinary bond dipole formula *(BD formula)* and, in the middle column, we present the HP(max) configuration, which is generated by starting with the perfect-pairing configuration and effecting complete *intrabond* CT. We then count the geminal and vicinal four-electron and three-electron antibonds and enter the result under the HP(max) configuration. The symbol v^4 indicates a *vicinal* four-electron antibond, the symbol g^3 indicates a *geminal* three-electron antibond, and so on. Vicinal and geminal overlap are defined as follows:

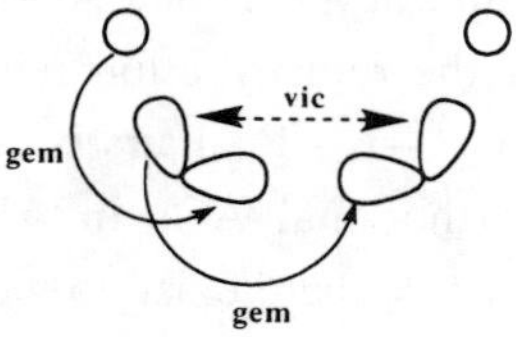

Optimal CT delocalization that is spared from exchange repulsion is represented by system I in Figure 6.2. The characteristic feature of the

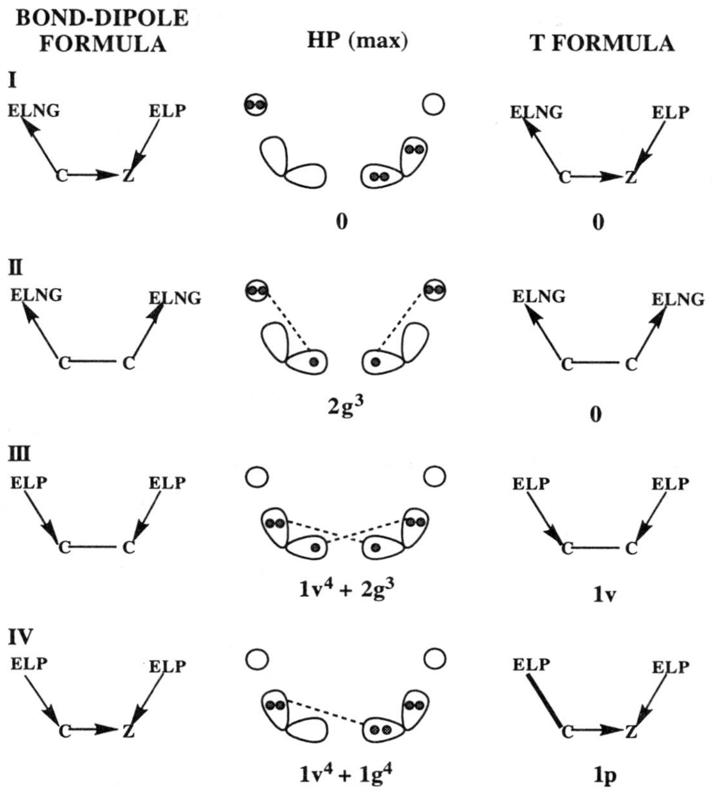

Figure 6.2 Four different atomic combinations represented by four different bond-dipole (BD) and T formulas. The symbol underneath each T-formula is the associated error-count.

corresponding bond dipole formula is that each atomic center has all arrows pointing either all "in" or all "out," with the direction alternating (in/out/in/out . . .) as we move along a chain of atoms. Arrows of the "out" type are called *diastolic* and arrows of the "in" type are called *systolic.* Furthermore, assuming that vicinal is much stronger than geminal AO overlap, we can neglect the geminal antibonds. In this case, systems I and II represent optimal while systems III and IV represent unfavorable CT delocalization.

Chemists are experts in counting bonds. We now suggest that a (vicinal) four-electron antibond must be elevated to the same pedestal as a (vicinal) two-electron antibond. We can count vicinal antibonds in the BD formula as follows: *A pair of vicinal systolic arrows, counted only once, represents one vicinal antibond.* Ethane has six C—H bond pairs, each represented by a systolic arrow. As a result, there are 3 × 3 = 9 vicinal CH—CH exchange-repulsive interactions but only three vicinal antibonds. The reason is that one CH bond interacts maximally with one (rather than all three) vicinal CH bond (anti-periplanar in staggered and syn-periplanar in eclipsed ethane). Note the break from convention: Ethane has six C—H plus one C—C bonds and three incipient CH—CH vicinal antibonds. The Pauli principle dictates that a closed-shell molecule has coexisting bonds and antibonds. The only exception is a two-electron system (e.g., H_2). Focusing exclusively on bonds is a costly error as we will soon find out.

In summary, molecules have two-electron bonds and four- (or three-) electron vicinal antibonds. In the prototypical B—A—A—B system, the strength of the (AB,AB) antibond is minimal when B is more electronegative than A (diastolic AB bonds) and maximal when B is more electropositive than A (systolic AB bonds). We discretize this continuum of antibond strength (for simplicity's sake) by neglecting diastolic and focusing on systolic antibonds, which are counted as vicinal errors. *The fingerprint of molecular instability is the presence of vicinal systolic arrows.*

We can now transform the BD formula of a molecule to another type of formula, called the *T formula*, which directly projects the bonding defect. This new formula is obtained as follows:

(a) We begin with the Lewis formula, which shows all valence pairs and valence holes. A lone pair counts as an infinitely polar systolic bond and a valence hole as an infinitely polar diastolic bond. Omission of lone pairs and holes is catastrophic. Any conclusions based on abridged T formulas that do not show explicitly the lone pairs and the holes should always be consistent with the action of these elements.

(b) The full Lewis formula is transformed to the BD formula by affixing bond dipoles to all bonds, including lone pairs and holes.

(c) Whenever a bond dipole is inconsistent with the exclusion rule, it is replaced by a line. For example, in the case of the H—A—F molecule, the presence of the three lone pairs on F and the A—F bond polarity requires an arrow pointing from A to H in order to satisfy the Exclusion rule. However, this requirement is incompatible with A—H bond dipole, which is directed from H to A. Hence, the A and H atoms are connected by a line and this represents a *polarity error,* symbolized by the letter p *(p-error).* Using T formulas, we can rewrite (6.1) as follows. Each T formula is associated with a number of errors shown underneath. Reaction (6.1) is exothermic because the reactants have two p-errors and the products have zero p-errors.

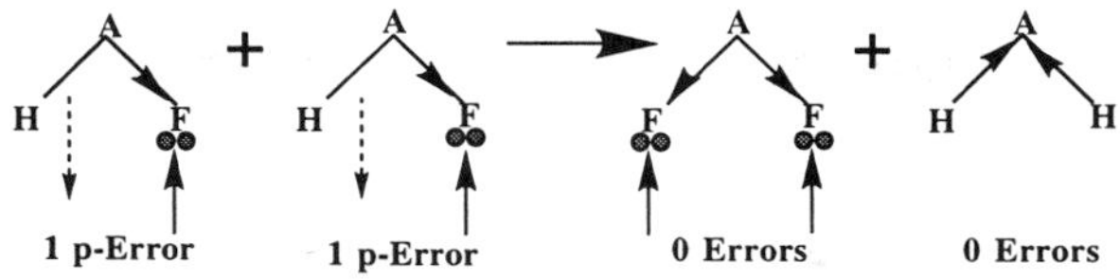

From models to the real world of molecules: Fluorine symbiosis on carbon cores is typified by the following redistribution reaction.

$$H_2CF_2 + H_2CF_2 \longrightarrow H_2CH_2 + F_2CF_2$$

Contrary to intuition, the process is exothermic by approximately 25 kcal/mol. Each bond of CF_4 and CH_4 gets an arrow because, in each case, optimal delocalization is compatible with bond polarity. In CH_2F_2, the electronegativity difference between C and F far exceeds that between C and H. Hence, each C—F bond carries an arrow but each C—H bond is represented by a line, which counts as one p-error. The reaction is predicted to be exothermic because there are now four p-errors on the reactant side and zero p-errors on the product side. By the way, the C—H bond is taken to be polarized in the C^-H^+ sense, reflecting the higher orbital ionization energy of a carbon sp^3 (14.6 eV, vs. 13.6 eV for H1s).

In summary, the condition for minimization of interbond exchange repulsion is that each atom within the BD formula must have all arrows pointing "out" or all pointing "in," and this means that arrow directionality must reverse in passing from one atom to its nearest neighbor. This optimal arrangement of bond dipoles is called an *exclusion wave*. To produce the T formula, bond dipole arrows that do not conform to the exclusion wave are

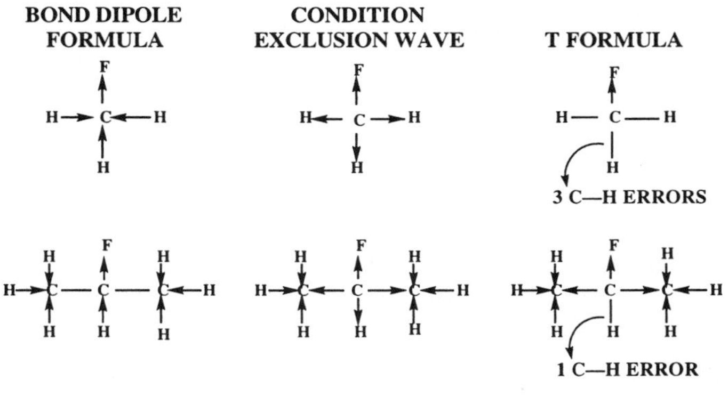

Figure 6.3 Two illustrations of the generation of the T formula.

turned into lines, and each line is counted as an error. Examples are shown in Figure 6.3.

6.4 The Error Count of the T Formula

Let us consider the series of the hydrofluoromethanes CH_3F, CH_2F_2, and CHF_3, and let us count the errors within each molecule using two different approaches:

Counting the p-errors of the T-formula.

Counting the vicinal antibonds of the BD formula, with each vicinal antibond representing one *vicinal error* symbolized by v *(v-error).*

The results are as follows.

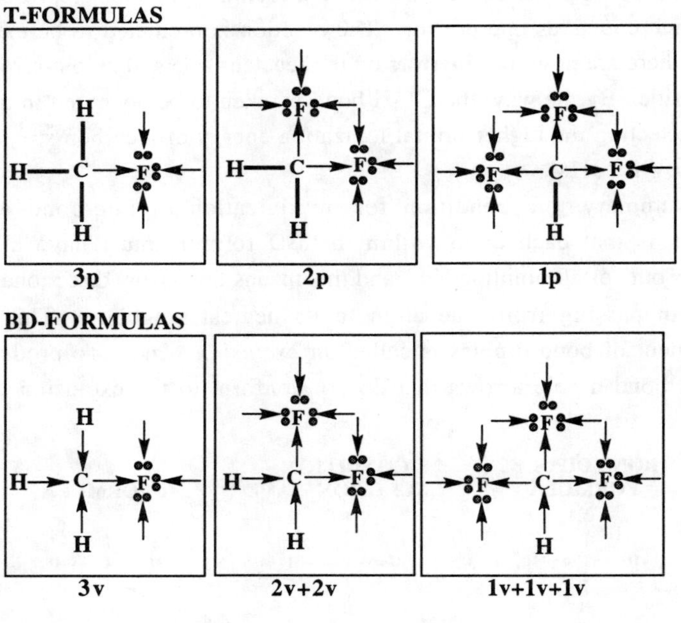

The stability of the hydrofluoromethanes is predicted to increase with increased fluorination according to the p-error count in the T formulas, since the number of p-errors gradually decreases. On the other hand, the stability is predicted to be roughly independent of the degree of fluorination according to the v-error count because this is bell-shaped but can be assumed to be roughly constant. Clearly, one has to decide which of the two error counts is more physically meaningful.

The key point is that the "truth" lies in the direction of the T formula simply because the count of v-errors in the BD formula assumes that bonds and antibonds are additive ("frozen bonding" approximation), while the count of p-errors in the T formula effectively assumes that bonds and antibonds interact.

To see this point, let us reconsider the derivation of the number of v-errors in the BD formula of CHF_3. Adding the first fluorine to the reference CH subfragment creates one vicinal CH—F(lone pair) antibond. Arresting CT delocalization in the C—H bond diminishes the strength of this vicinal antibond. The second fluorine now sees an adjusted C—H bond and the second vicinal antibond is largely avoided, while the C—H bond makes a further incremental adjustment. Entry of the third fluorine sees a C—H bond that is best prepared for avoiding vicinal exchange repulsion.

While energetic predictions require the use of the T formulas, structural predictions can be handled by either model. For example, both T and BD formulas predict that each fluorine suffers increasingly less exchange repulsion by the systolic C—H bonds as these become eliminated with increasing fluorination. For example, the *single* F sees three erroneous C—H bonds in the T formula of CH_3F, while *each* F sees only one erroneous C—H bond in the T formula of CHF_3. In each molecule, each C—H bond is erroneous because of the action of one vicinal lone pair on F. By the same token, the *single* F makes *three* vicinal antibonds with the *three* systolic C—H bonds in the BD formula of CH_3F while *each* F makes *two* vicinal antibonds with the *two* systolic C—H bonds in the BD formula of CHF_3.

We have come to the conclusion that T formulas (rather than Lewis formulas or BD formulas) are the most appropriate representations of organic molecules. In the general case, these will contain both p-type and v-type errors. By comparing the error counts in the T and BD formulas of the hydrofluoromethanes, we conclude that a p-error is equal to or greater than a v-error. This can be seen by assuming the "frozen bonding" approximation and considering the the two representations to be equivalent. This produces the result that 1p = 1v in fluoromethane, 1p = 2v in difluoromethane and 1p = 3v in trifluoromethane. This implies that, a p-error is equal or greater in size than a v-error.

To maintain simplicity and conciseness, we will assume that p- and v-errors have equal weights. What about the magnitude of such errors? The T formula of fluoromethane given above has three p-errors, and all three vanish if the fluorine lone pairs are artificially deleted in a computation. This sets up an interesting problem: Each of the isoelectronic H_3C—Z molecules, with Z = F, OH, NH_2 and CH_3, has three (p- or v-) errors. Since the three p-errors of fluoromethane are caused by the three fluorine lone pairs, they must be larger in magnitude than the three v-errors of ethane. Hence, the magnitude of the errors should vary in the order Z = F > OH > NH_2 > CH_3. However, this ordering assumes that vicinal AO overlap remains constant. In actuality, this varies in the reverse order because the AOs of Z contract as we go from CH_3 to F. By interpreting ab initio computations in the way illustrated in later chapters, we have concluded that error maximization occurs for Z = OH.

The concept of the T formula leads to a natural classification of chemical groups:

Groups having systolic arrows are called systolic (SY) groups because they preferentially connect to some other group by a systolic arrow.

Groups having diastolic arrows are called diastolic (DI) groups because they preferentially connect to some other group by a diastolic arrow.

A SY or a DI group can be error-free, or it may contain one or more errors. Examples of DI and SY groups are shown in Figure 6.4. The exclusion rule can now be reformulated as follows: *The most stable combination of SY and DI groups is the one in which these alternate* (i.e., SY-DI-SY-DI-SY-DI is more stable than SY-SY-SY-DI-DI-DI). An illustrative application is the following reaction, where CH_3 is SY and CF_3 is DI. The reaction is exothermic because the reactants are SY-SY + DI-DI and the products are SY-DI + SY-DI:

$$H_3C\text{—}CH_3 + F_3C\text{—}CF_3 \longrightarrow 2\ H_3C\text{—}CF_3 \quad \Delta E = -17\ \text{kcal/mol}$$

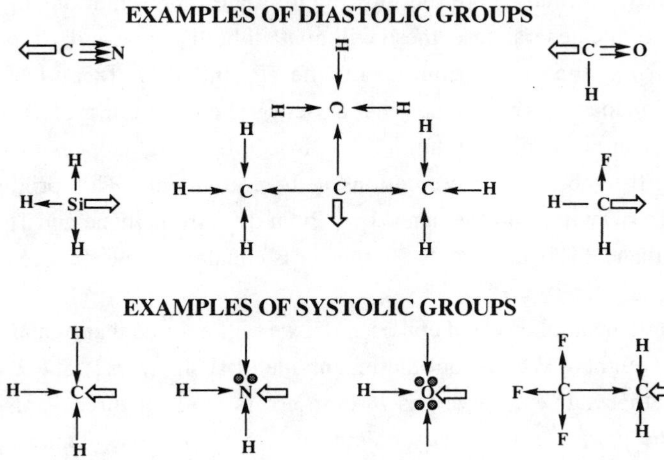

Figure 6.4 Examples of diastolic and systolic groups.

6.5 The Effect of Interbond Delocalization; Carbenic Resonance

Consider the fully ionic configuration that is produced by *intrabond* CT delocalization in each of reactants and products of the reaction shown in Figure 6.1. Starting from this configuration, we use an arrow to indicate the additional *interbond* CT delocalization that may occur in each species. This so-called secondary delocalization is illustrated in Figure 6.5.

Let us now consider how interbond delocalization influences the enthalpy of reaction (6.1). On the reactant side, this is tantamount to delocalization over two orthogonal hybrids causing the formation of a *geminal interhybrid bond (G bond)* the strength of which is proportional to the one-electron energy gap separating the two pure AOs from which the hybrids were constructed. In our case, the "precursor" AOs of A can be taken to be the 2s and 2p AOs. Hence, interhybrid CT delocalization on the reactant side effects A de-excitation, which is directly proportional the singlet-triplet gap of A denoted by DE(ST). In other words, the interhybrid CT delocalization effectively produces the resonance structure H^+ A: F:$^-$ which is now an important contributor in the resonance hybrid of H—A—F because it minimizes atom promotion (ground state A:). This type of resonance stabilization of a molecule is termed *carbenic resonance*.

$$H—(CF_2)\text{-}F \longleftrightarrow H^+\ (CF_2)\ F{:}^-$$

Carbenic resonance

EXAMPLES OF INTERBOND DELOCALIZATION

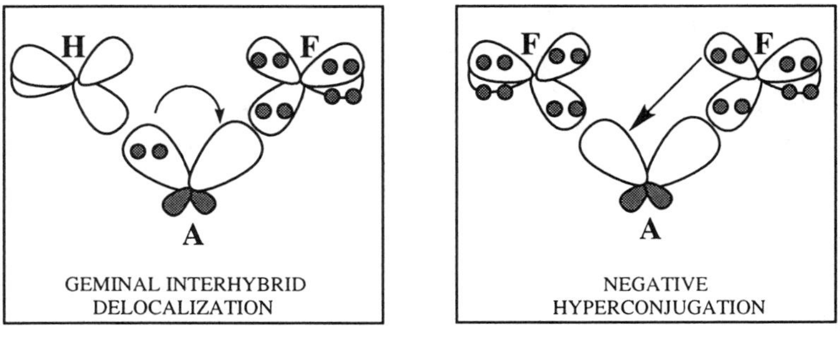

Figure 6.5 The two types of interbond delocalization.

Electron flow from H to C and concomitant electron flow from C to one F generates a ground CF_2 carbenic subfragment. Hence, carbenic resonance corresponds to stabilization via atom de-excitation.

The situation on the product side is familiar. Interbond delocalization in F—A—F is equivalent to "negative hyperconjugation," and interbond delocalization in H—A—H (no matter how weak it may be) is equivalent to "positive hyperconjugation." The stronger the vicinal AO overlap, the stronger the hyperconjugation. The final conclusion is that there are two types of interbond delocalization, one widely recognized (hyperconjugation) and the other totally neglected (carbenic resonance), which promote symbiosis and antisymbiosis, respectively. Hence, we can formulate the reasonable hypothesis that the two effects cancel out and that the enthalpy of reaction (6.1) is simply a reflection of the number of violations of the exclusion rule on the reactant and product sides.

Attachment of a diastolic (DI) and a systolic (SY) group on a central atom or fragment, A, violates the exclusion rule, hence, is characterized by instability which is partly alleviated by carbenic resonance. As a result, the molecule DI—A—SY tends to to undergo 1,1-elimination, which becomes increasingly facile as the DI—A and/or A—SY bonds become weaker and the singlet is increasingly stabilized relative to the triplet state of A:. Often, the two conditions cannot be satisfied simultaneously. For example, the first condition is met when A: bears one or two lone pairs, in which case the triplet dips below singlet A:. A typical DI group is acyl. A typical SY group is alkoxy. Consistent with our expectations, Barton, Hall and co-workers suggested that N,N´-diacyl-N,N´-dialkoxyhydrazines decompose by stepwise 1,1-elimination, yielding nitrene plus ester.[2]

A crucial point is that carbenic resonance materializes when two conditions are met:

(a) A divalent fragment with a large singlet–triplet gap, A, is attached to an electropositive diastolic group (e.g., H) and an electronegative systolic group (e.g., F). In the resonance structure H^+A: F:$^-$ *(carbenic zwitterion)*, the carbenic lone pair hybridizes so that it points toward the vacant H1s AO, while the carbenic hole hybridizes so that it points toward the fluoride lone pair. In this way, exchange repulsion of two formal bond pairs is averted. On the other hand, carbenic resonance is unimportant when A is attached to two identical atoms and groups. Now, the energy of the HA:H or FA:F *(carbenic diradical)* is high because of exchange repulsion of the A pair and the odd electrons of the two ligands.

(b) Because the carbenic–zwitterionic resonance sacrifices two exchange bonds for the benefit of atom de-excitation, it is particularly important when A is an electropositive metalloid or metal atom, which can support only weak kinetic energy reduction (see Chapter 2).

These two conditions are admirably met by a group like SiY_2—X (Y = F, OR, or NR_2) when this is attached to an electropositive atom. Satisfaction of the first condition alone qualifies CY_2—X under the same circumstances.

What are the structural consequences of carbenic resonance? It has been recently shown that *n*-perfluorobutane and *n*-tetrasilabutane have a distinct 90° conformational isomer ("ortho" isomer).[3] Our explanation is that carbenic resonance, expressed by the structure $F^-(CF_2)\ ^+(CF_2—CF_2)^+\ (CF_2)\ F^-$, renders the central F_2CCF_2 unit of *n*-perfluorobutane analogous to X_2BBX_2 (X = F, Cl). Depending on X, either the planar (X = F) or the 90°-twisted (X = Cl) ortho conformation is the global minimum.[4] Obviously, any substituent larger than F will tip the balance in favor of the ortho form. The counterintuitive ortho minimum is expected to be a hallmark of F—A—A—A—A—F molecules, where A is a carbenic fragment with a highly stabilized (relative to triplet) singlet ground state. For example, A can be cyclopropenylidene. The physical interpretation is that the *low energy* staggered (gauche and anti) forms best accommodate perfect pairing and the *dominant intrabond* CT delocalization while the *high energy* ortho form best accommodates the *secondary interbond* CT delocalization. In other words, the staggered geometries maximize T bonding (by minimizing vicinal T antibonding), while the ortho geometry minimizes atom promotion. Our argument can be proven invalid if it is shown that ortho forms exist as distinct minima in molecules made up of carbenic fragments that have a triplet ground state.

The preceding argument is consistent with another conclusion that, in turn, resolves a puzzle. It has been long known that perfluorination lengthens the carbon–carbon bond distance of ethane but shortens that of ethylene! We can now see that, because CF_3 is diastolic and F is systolic, perfluoroethane enjoys the benefit of carbenic resonance, $F_3C^+(F_2C:)\ F:^-$. As a result, the carbon–carbon interaction has partial no-bond character.

6.6 The T Formulas of Organic Molecules

We are now prepared to give a general recipe for writing formulas for organic molecules:

(a) The full Lewis structure is written, and arrows indicating bond dipoles are affixed on each bond pair, lone pair, and hole. Atoms of the same type (e.g., two carbons) are connected by lines, whether equivalent or nonequivalent.

(b) Every atomic center has four bonds. Each center is classified as either systolic or diastolic by reference to the *largest bond dipole* associated with it. This is indicated by an arrow that is called the *sense arrow* because it defines the character of the atomic center (systolic or diastolic). In first-row nonmetal atoms, sense arrows are associated with lone pairs (implying infinitely electropositive ligands), holes (implying infinitely electronegative ligands), and pairs of atoms differing greatly in electronegativity. In an alkane, the sense arrow at every carbon is the one corresponding to a CH bond dipole.

(c) Only systolic arrows are allowed on systolic centers and only diastolic arrows are allowed on diastolic centers. *Misdirected arrows are replaced by lines.* In other words, every center either has arrows pointing in the same direction as the sense arrow or it has lines. Equivalent atoms are connected by lines. Nonequivalent atoms of the same type (e.g., two nonequivalent carbons) are connected either by a line or by an arrow, depending on which choice minimizes the number or errors.

(d) The resulting T formula has an associated error count, which breaks down as follows. (1) Every line connecting two atoms of *different* type counts as a polarity error (p-error). *A line connecting two equivalent or nonequivalent atoms of the same type is not an error.* (2) Two vicinal *systolic* arrows count as a single vicinal error (v-error), the magnitude of which is proportional to the mgnitudes of the two systolic bond–dipoles.

(e) One p-error is assumed to carry the same weight as one v-error, and errors within each T formula have the following consequences: (1) The more the errors, the lower the stability of the structure; and (2) the presence of p-errors implies that intrabond CT delocalization in the erroneous bonds has been arrested. Erroneous bonds are expected to be longer and weaker. Furthermore, erroneous bonds weaken and lengthen error-free bonds on the same atomic center because of exchange repulsion. It should always be kept in mind that p-(indirectly) and v-errors (directly) are representatives of vicinal antibonds.

The revolution brought about by the T formula is easy to understand. Traditionally, the chemist focused on bonds. We are now focusing on (two-electron) bonds as well as (four-electron) antibonds. Lone pair–lone pair are stronger than bond pair–bond pair antibonds. Highly polar bonds are effectively "hole–lone pair" combinations; that is, they are nascent lone pairs. The sense arrow on a given atomic center identifies the bond dipole that can potentially engender the largest exchange repulsion, and this is why it determines the

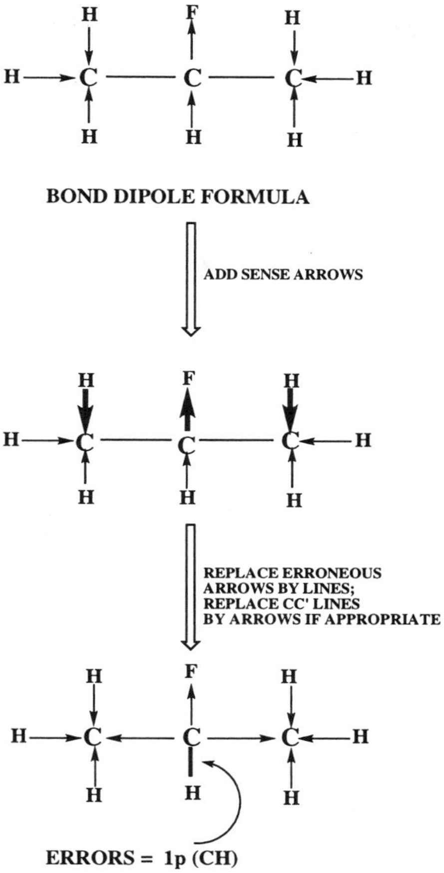

Figure 6.6 Step-by-step development of the T formula of 2-fluoropropane. Note how the sense arrows control the situation.

optimal arrow directionality. The prospects are now exciting: Since all explanations of molecular interactions have been based on consideration of bonds and interbond delocalization, these must be different and even antithetical to interpretations based on the T formulas, which effectively enumerate bonds as well as antibonds while assuming(to a first approximation) zero interbond CT delocalization (hyperconjugation).

The procedure for constructing a T formula is illustrated in a step-by-step fashion in Figure 6.6 using 2-fluoropropane as the example. Additional arrow formulas of typical organic molecules (Figure 6.7) are dscussed in turn.

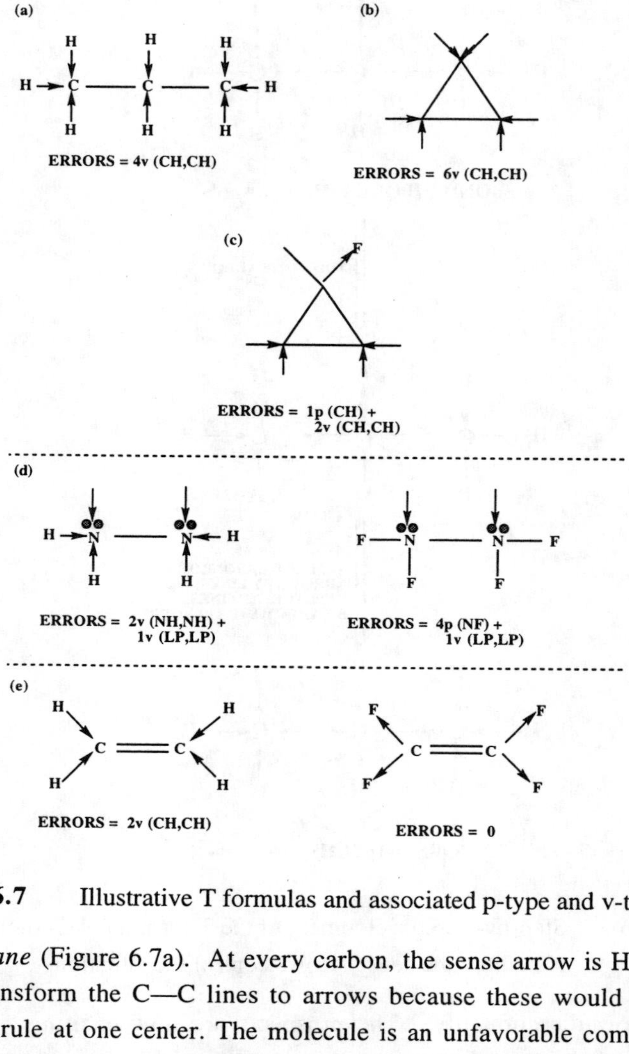

Figure 6.7 Illustrative T formulas and associated p-type and v-type errors.

Propane (Figure 6.7a). At every carbon, the sense arrow is H → C. We cannot transform the C—C lines to arrows because these would violate the exclusion rule at one center. The molecule is an unfavorable combination of three systolic groups, namely, two methyls and one methylene. There are four pairs of vicinal systolic arrows corresponding to four v-errors. Replacement of a hydrogen attached on the middle carbon of propane yields 2-fluoropropane and changes the formula drastically (Figure 6.6). Specifically, the methylene changes from systolic to diastolic because of the dominance of the C—F bond. This creates an alternation of systolic and diastolic groups. There is now only one line corresponding to one p-error involving the C—H bond of the CHF diastolic group. In addition, the vicinal errors of propane have been eliminated. The C—C bonds of this molecule are thus expected to be shorter than those of the parent hydrocarbon.

Cyclopropane (Figure 6.7b). This is made up of three equivalent systolic methylenes. Because of the equivalence of the carbon atoms, the lines are not errors. What remains are the six v-errors.

Fluorocyclopropane (Figure 6.7c). This represents a situation analogous to 2-fluoropropane. One p-error is created but four v-errors are eliminated compared with cyclopropane. The two v-errors occur across the line connecting the two systolic methylenes. This means that the "distal" C—C bond should be longer than the "side" C—C bond, as found by a microwave study of 1,1-difluorocyclopropane.[5]

Hydrazine (Figure 6.7d). This involves the combination of two systolic groups and suffers from three v-errors. The errors are serious because of the high polarity of the N—H bonds and the lone pair (infinite polarity). The problem remains in perfluorohydrazine but in a different disguise. The sense arrow of the nitrogen is determined by the lone pair. This requires that N and F be connected by a line. This creates four p-errors plus one serious v-error (lone pair–lone pair). This explains the great instability (highly positive heat of formation) of these molecules.

Ethylene (Figure 6.7e). Each CH_2 group is a systolic group. As a result, ethylene ends up with two v-errors. By contrast, perfluoroethylene involves a combination of two diastolic groups, and it is error-free. As a result, the central C—C bond is much shorter in tetrafluoroethylene. In an analogous fashion, ethane has three v-errors but perfluoroethane has no errors whatever. The rotational barrier of $F_3C—CF_3$ is virtually the same as that of ethane despite the presence of the additional fluorine lone pairs. The reason is that vicinal interbond exchange repulsion is actually smaller in the staggered and eclipsed conformers of perfluoroethane.

We end this chapter by stressing that the one thing not shown by the arrow formula is the atom or group promotion factor. This information should be independently supplemented and considered whenever appropriate.

References

1. L. Pauling, *The Nature of the Chemical Bond*, 3rd ed., Cornell University Press, Ithaca, NY, 1960.
2. M.V. de Almeida, D.H.R. Barton, I. Bytheway, J.A. Ferreira, M.B.Hall, W. Liu, D.K. Taylor, and L. Thomson, *J. Am. Chem. Soc.* 117, 4870 (1995).

3. B. Albinsson and J. Michl, *J. Am. Chem. Soc.* 117, 6378 (1995).
4. D.D. Danielson, J.V. Patton, and K. Hedberg, *J. Am. Chem. Soc.* 99, 6484 (1977).
5. A.T. Peretta and V.W. Laurie, *J. Chem. Phys.* 62, 2469 (1975).

Chapter 7

The Central Problem of Fluorine Symbiosis

7.1 Forty Years of Preoccupation with a Nonexplanation

It is difficult to envision a problem of greater importance for structural chemistry than that of constitutional (structural) isomerism: What is the optimum combination of a given set of atoms that leads to a highly stabilized molecule or transition state? Much is taught and discussed about conformational and geometric isomerism, but virtually nothing is said about constitutional isomerism — even though structural isomeric can be far stronger than rotational isomeric preferences. We will now see that this omission is not accidental: *Constitutional isomerism denies the very arguments that have been used to rationalize rotational isomerism.*

More than a half-century ago, Brockway discovered that the C—F bond length in fluoromethanes progressively shrinks as the number of fluorines increases.[1] The C—F bond dissociation energy increases also in the same direction.[2]

	r_{CF} (Å)	C—F BDE (kcal/mol)
CH_3—F	1.39	107
CH_2F—F	1.36	110
CHF_2—F	1.33	115
CF_3—F	1.32	116

Moreover, the following reaction is highly exothermic and demonstrates that fluorines prefer to be attached on the same atom rather than on different atoms (fluorine symbiosis).

$$2CH_2F_2 \xrightarrow{\Delta H_r = -25.0\ \text{kcal/mol}} CF_4 + CH_4 \qquad (7.1)$$

This symbiotic tendency is further exemplified by the energetic superiority (by 11 kcal/mol) of 1,1- to 1,2-difluoroethylene. So, here is a set of critical facts that cannot be explained by standard concepts, such as "steric effects."

Many years ago, Pauling proposed the concept of "double bond/no bond resonance,"[3] which can be illustrated by means of our CF_4 molecule:

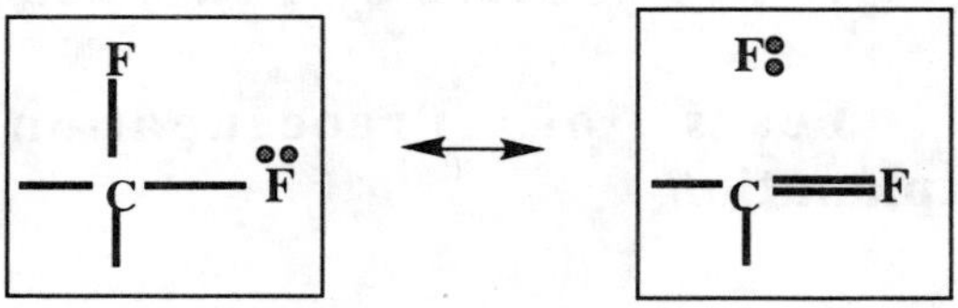

This model explains the energetic preference of placing two fluorines on the same center, but it fails to explain why the C—F bond in CH_2F_2 is shorter than that in CH_3F. The actual prediction of the Pauling concept is that the C—F bond length must *not* change as a result of "no bond/double bond resonance" since the average of zero and two is one! A surprising number of chemists rallied to support the Pauling concept despite its obvious flaw.[4]

The Pauling argument has been time and again reintroduced in the literature under different guises: "hyperconjugation," "doubly occupied nonbonding AO/vacant CF antibonding MO interaction," and so on. Other intuitive effects have been proposed to explain the phenomenon. Electrostatics, bond ionicity effects, rehybridization arguments, and our own "nonbonded attraction" from the PMO days of the author, were thrown at the problem. What has been the outcome? Chambers put it best: "The very range of explanations that have been offered is some indication in itself of the uncertainty which exists in explaining these data."[2]

Why are steric effects so popular with chemists? Because the concept is simple to utilize and its validity seems self-evident. But this is only a mirage. For example, one expects because of steric effects that 1,1-difluoroethylene will be less stable than the 1,2 form, and the opposite turns out to be true. This means that a key assumption underlying the "steric effect" construct must be invalid. In our isomeric difluoroethylenes example, the implicit assumption of the construct is that the σ and π bonds of the two isomers are invariant, and the only important difference lies in the way nonbonded groups interact. Is there some fundamental reason for the bonds of the 1,1 isomer to be more favorable than those of the 1,2? The T formulas give the answer.

7.2 T Formulas Explain Fluorine Symbiosis

The reaction shown below is exothermic by 13 kcal/mol. The T formulas provide a direct, explicit, and pictorial explanation: There are six C—H errors on the reactant side but only two C—H errors on the product size. These errors are p-errors. Hence, the reaction is expected to be exothermic, as indeed is found.

By contrast, ab initio calculations suggest that the following reaction is endothermic by only 1.1 kcal/mol.[5] We count two errors on the reactant and two errors on the product side. Hence, the near-thermoneutrality of the reaction is explained.

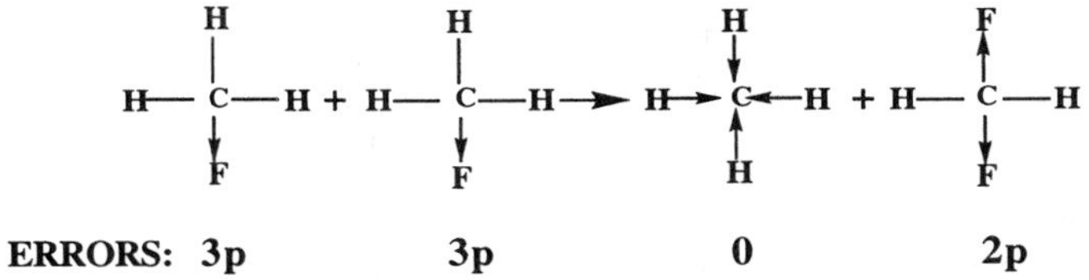

The trend of the enthalpy for the reaction 2H—A—F ⟶ H—A—H + F—A—F, as determined by ab initio computations,[5a] is summarized in Table 7.1.

Some time ago, J. Larson, using E. R. Davidson's programs, computed the enthalpy of the reaction with A = CH_2 at the extended Hückel (EH) MO, SCFMO, and SCFMO-CI levels. He found that all methodologies predict the reaction to be strongly exothermic. Since EHMO does not contain Coulombic effects, an explanation of the energetics based on electrostatics is rejected. Next,

Table 7.1 Error Count and Enthalpy of Redistribution Reaction

	C—H ERRORS		ΔH_r
A	REACTANT	PRODUCT	(kcal/mol)
CH_2	6	2	−14.0
CHF	4	3	− 8.5
CF_2	2	2	+1.1

Larson replaced F by a pseudoatom having the same electronegativity but devoid of lone pairs. As a result, the p-errors on the reactant side disappear. What remains now is nonbonded repulsion, which dictates that the reaction should be endothermic, as is found at all computational levels.[5b] The following reaction is exothermic by an astounding 25 kcal/mol. The reason is that there are four p-errors on the reactant side and zero errors on the product side.

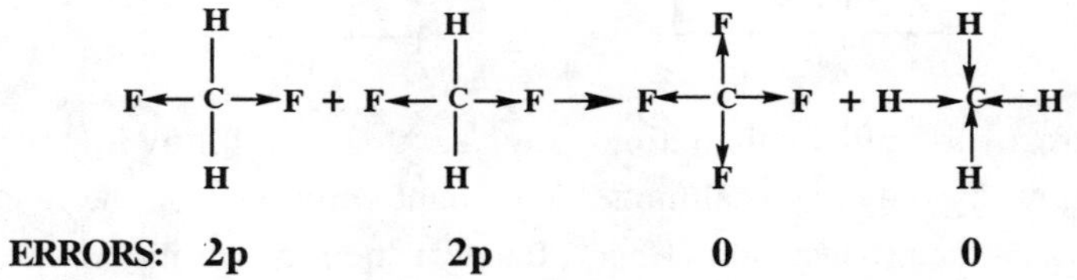

When we replace C by Si, there are zero errors on both sides simply because Si is much more electropositive than either F or H. Once again, the only force left to control the reaction enthalpy is nonbonded repulsion. If two nonbonded groups are classified as "large" (L) and "small" (S), the most unfavorable combination is LL, and this is what decides whether reactants or products are more stable. Hence, our prediction now is that the following reaction will be endothermic. Indeed, this turns out to be the case (ΔH_r = +9.3 kcal/mol).

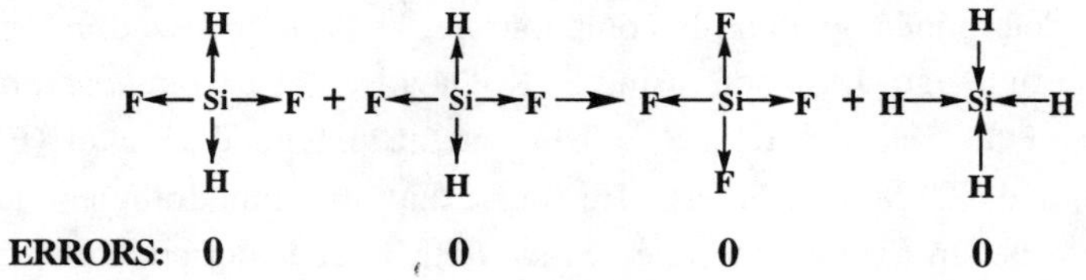

A related case is also shown:

$$2Me_2SnCl_2 \xrightarrow{\Delta H_r = +7\ \text{kcal/mol}} Cl_2SnCl_2 + Me_2SnMe_2$$

7.3 The Electronic Basis of Bond Contraction

Brockway found that the CF bond lengths decrease in the order CH_3F, CH_2F_2, CHF_3, and CF_4.[1] What property of the fluoromethanes decreases in the same fashion? The answer is simple: The number of errors in the corresponding T formulas:

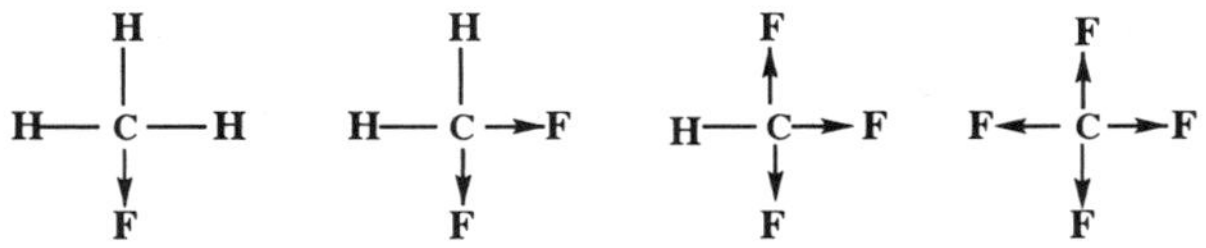

ERRORS: 3p 2p 1p 0p

A decreasing number of p-errors implies decreased exchange repulsion of each fluorine by the systolic C—H bonds. Increasing fluorination, starting with fluoromethane, progressively eliminates the p-errors and, as a result, the CF bonds become increasingly shorter and stronger.

Bond contraction is not the exclusive property of the fluoromethanes. Even though nonbonded overlap repulsion works in the opposite direction, the C—C bond of propane appears to be *shorter* than that of ethane. C—C bond elongation is observed as methylation increases and strong nonbonded repulsion becomes inevitable.

Replacement of the central carbon in the fluoromethane series by C═C produces the fluoroethylene series. Our argument remains the same: Sequential replacement of H by F deletes errors of both the p-type (compare fluoroethylene and 1,1-difluoroethylene) and the v-type (compare fluoroethylene and 1,2-difluoroethylene) and is expected to cause shortening of both the C═C and the C—F bonds! This is what spectroscopic data show (Table 7.2). Furthermore, Dixon et al. reported ab initio calculations showing that this trend is essentially reproducible at the SCF level.[5c] This is exactly what we expect, because the

Table 7.2 Bond Length Data Acquired by Electron Diffraction (ED) and Microwave (MW) Spectroscopy

	Bond Length (Å)		
Molecule	$r_{C=C}$	r_{C-F}	Method
$H_2C{=}CH_2$	1.377		ED
$HFC{=}CH_2$	1.329	1.347	MW
	1.333	1.348	ED
cis-HFC═CHF	1.324	1.335	MW
	1.331	1.335	ED
$F_2C{=}CH_2$	1.315	1.320	MW
	1.309	1.336	ED
$F_2C{=}CF_2$	1.311	1.319	ED

electronic basis of the bond contraction is T bonding, which relies primarily on overlap. Correlation effects become pivotally important when we cross the line between T and I bonding.

1,1-Difluoroethylene is more stable than *cis*-1,2-difluoroethylene by about 11 kcal/mol. The reason is that the former is error-free while the latter has one C—H p-error per carbon. The presence of these errors means that all bonds, namely, C—C, C—F, and C—H, emanating from each carbon will be longer in the cis isomer. This goes head-on against popular PMO models. Typically, a PMO explanation has some doubly occupied bonding orbital of fragment A interacting with some vacant antibonding orbital of fragment B. This stabilizing interaction is compared in two different geometries AB and A´B´, and an inference is drawn. The stronger the interaction, the more the *intra*fragmental bonds become weaker (because of depletion of electron density from a bonding MO and population of an antibonding MO) and the more the *inter*fragmental bonds are strengthened. To put it simply, in the conventional PMO models, orbital interactions shorten some bonds and lengthen others. By contrast, a comparison of two isomers, one error-free and the other error-full, means that *all* atoms "pack" better in the error-free isomer. Thus, we expect to find that in going from *cis*-1,2- to *cis*-1,1-difluoroethylene, all bonds become shorter! The following microwave data confirm these expectations.

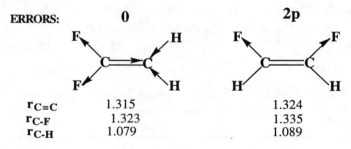

ERRORS: 0 2p

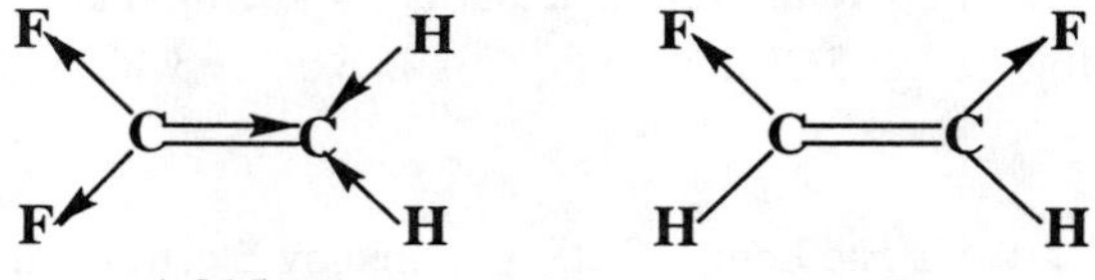

$r_{C=C}$	1.315	1.324
$r_{C\text{-}F}$	1.323	1.335
$r_{C\text{-}H}$	1.079	1.089

The underpinning of T formulas is the notion that molecular electronics is a *whole-molecule problem*. Further evidence is provided by recent ab initio computations,[6] which reveal the following trends:

(a) Both the RO—H and the R—OH bond dissociation energies increase as the degree of fluorination of CF_xH_{3-x}—OH increases and the number of C—H p-errors decreases.

(b) CH_4 has zero, CH_3F has three, CH_2F_2 has two, and CHF_3 has only one C—H p-error. Exactly paralleling this trend, the R—H BDE is highest in methane, decreases in going to fluoromethane, and then progressively increases.

T formulas can predict detailed structural differences. For example, consider the three nonequivalent vinylic hydrogens of propene. If we remove the approximation that a line connecting nonequivalent carbons is a nonerror and assume that this is actually a quasi-error, then the vinylic carbon bearing the methyl group becomes the one sustaining the greater number of errors. As a result, we expect that *all* bonds emanating from C2 will be weaker and longer. This predicts that the longest C—H bond must be the C—H_a bond. However, of the remaining two, C—H_c is cis to the methyl group and it experiences the nonbonded repulsion of the methyl C—H bonds. Thus, the predicted C—H bond lengths are C—H_a > C—H_c > C—H_b. The spectroscopic data, with all their uncertainties, support this scenario; the only difference is that C—H_c is essentially equal to C—H_a.

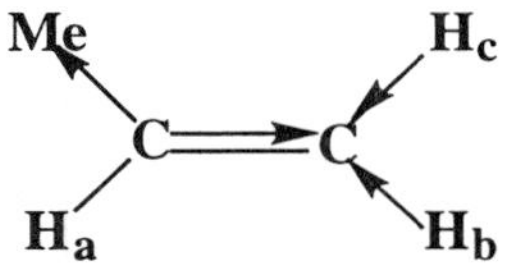

Bottom line: The Brockway bond contraction effect is clear-cut and fundamental. This means that the explanation must be basic and, thus, simple. This viewpoint is confirmed by the T formulas. One has only to count the number of errors in the fluoromethanes to understand the electronic basis of bond contraction: minimization of exchange repulsion.

7.4 Methyl Symbiosis and the Rejection of the Classical "Steric Effect"

We are now prepared to see why methyl acts qualitatively like fluorine.

The following reaction, which is exothermic by 2.6 kcal/mol, illustrates methyl symbiosis.

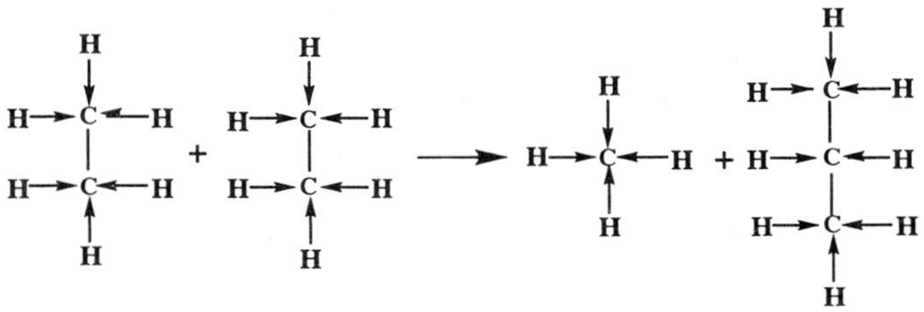

3v **3v** **0** **4v**

In alkanes, vicinal errors can be only of the v(CH,CH) type. Hence, each such error is symbolized by v not only in alkanes but also in alkane derivatives. Similarly, p(CH) is abbreviated as p. The reaction exothermicity is simply rationalized by saying that the products have fewer errors than the reactants. This is exactly the opposite of what "steric effects" would lead one to conclude.

A rule becomes apparent: *The most stable structural isomeric alkane is the one with a maximum number of methylated carbons.* A good example is the comparison of *n*-pentane and neopentane. Steric considerations suggest that the first isomer is energetically superior. Our analysis assigns eight errors to *n*-pentane and zero errors to neopentane; that is, neopentane is an error-free isomer. As a result, neopentane, the more "crowded" isomer, is expected to be the champion of stability in the C_5H_{10} structural isomer series.

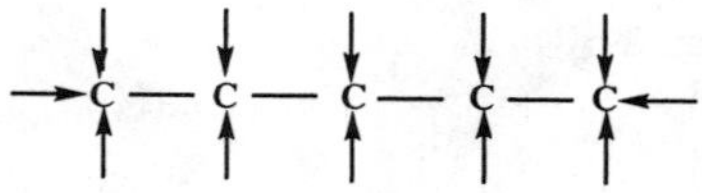

8v ERRORS

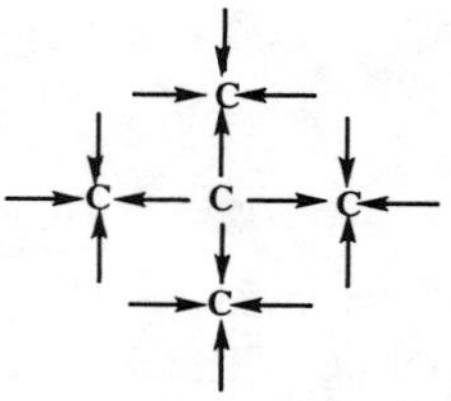

0 ERRORS

Neopentane is more stable than *n*-pentane by 5 kcal/mol. Neopentane and CF_4 are analogous molecules. Carbon dioxide is analogous to tetrafluoromethane.

In Table 7.3, we add some more examples from alkane structural isomerism that conform to the generalization: Methyls prefer to stick on the same carbon because, in this way, the number of errors is minimized (i.e., interbond overlap

Table 7.3 Relative Energies of Alkane Isomers

Isomer	E_{rel} (kcal/mol)
$CH_3—(CH_2)_4—CH_3$	4.6
$(CH_3)_2CH—CH(CH_3)_2$	1.9
$(CH_3)_3C—CH_2—CH_3$	0.0
$CH_3—(CH_2)_5—CH_3$	4.0
$(CH_3)_3C—(CH_3)_2CH$	0.0
$CH_3—(CH_2)_6—CH_3$	4.0
$(CH_3)_3C—C(CH_3)_3$	0.0

repulsion is minimized). Note how the "uncrowded" linear chain is actually disfavored!

Some are aware that "fluorines like to be attached to the same carbon" in fluorocarbons. Few have paid attention to the parallel statement that "methyls also like to be attached to the same carbon." The intuitive explanation given is that this preference is due to van der Waals attraction. By contrast, we say that both trends have a common origin: minimization of interbond overlap repulsion. Hence, we predict that, for example, the significantly larger stability of neopentane relative to *n*-pentane will be reproduced, for the most part, by a one-determinantal calculation of the restricted Hartree-Fock (RHF) type that does not contain correlation effects. A useful paper by Wiberg and Murcko[7] reports calculations of the relative energies of alkane rotamers and structural isomers. The superiority of iso- to *n*-butane and the superiority of neo- to *n*-pentane are qualitatively (not quantitatively) reproduced at the RHF level. However, these workers find that electron correlation is needed to reproduce semiquantitatively the relative energies of structural isomers.

The energetic superiority of *trans*- over *cis*-2-butene early on introduces chemistry students to the concept of "steric effects." They learn that the key difference between the two isomers is repulsive nonbonded interaction of the methyl groups. What often is *not* said is that isobutene is even more stable than *trans*-2-butene (by 1.2 kcal/mol). Why? Because isobutene is the error-free combination of $(CH_3)_2C$ (a DI group) and H_2C (an SY group). By contrast, 2-butene is the erroneous combination of two $(CH_3)HC$ SY groups. The SY-DI is the best and the SY-SY is the worst combination of two fragments. An interesting investigation of the chemical bonding morphology of the acyclic structural isomers of C_4H_8 from the momentum space perspective has appeared.[8] It was found that the measured and calculated momentum distributions are qualitatively different among the three butene isomers (i.e., C—C and C—H bonds in different environments are fundamentally different).

Another interesting example is the comparison of 1-hexene (least stable), 3-hexene, and tetramethylethylene (most stable). Again, the stability order is in contrast to what one would intuitively expect on the basis of "steric effects." On the other hand, organic chemists are quick to cite a golden rule that says, "The more highly substituted an alkene, the greater its stability." One way to explain this trend is by falling back on Walsh's rule and arguing that an sp^2-hybridized carbon prefers to direct its electronegative sp^2-hybrid AOs to electron-releasing groups. In contrast to our approach, methyl is traditionally regarded as a σ donor. Others subscribe to hyperconjugation involving the π-bond orbitals of the alkene and the π-type orbitals of each methyl group. Taken together, these interpretations lead to the conclusion that dimethyl ether should be more stable than ethanol. This is in direct contradiction of the data, presented shortly. The T formulas provide a simple rationalization: It is the number of errors that determines that 1-hexene (8v errors) is less stable than 3-hexene (7v errors) by about 2.3 kcal/mol. Tetramethylethylene (zero errors) is the most stable C_6H_{12} alkene, being 6.6 kcal/mol more stable than 1-hexene.

What is the most stable isomer of C_8H_{18}? The *t*-butyl group constitutes an error-free DI group, and combination of two of them leads to hexamethylethane as the proposed most stable isomer. This falls short by one isomer in the "stability derby." That is to say, it is more stable than all constitutional isomers except 2,2,4-trimethylpentane. Amazingly, hexamethylethane is roughly 4 kcal/mol more stable than *n*-octane.

7.5 The Antisymbiotic Action of Diastolic Groups

Since both methylene and methyl are SY groups, they can combine in an error-free fashion with diastolic CN groups. As a result, the redistribution reaction of the type shown below is predicted to have zero errors on both reactant and product sides. Now all that differentiates the two sides of the equation is nonbonded repulsion. This predicts that the reaction will be endothermic, as found ($\Delta H_r = 9.7$ kcal/mol). Antisymbiosis is the result of the disappearance of the selectivity due to intrabond CT delocalization and the emergence of "steric effect" as the only controlling factor.

$$2H\text{—}CH_2\text{—}CN \longrightarrow H\text{—}CH_2\text{—}H + NC\text{—}CH_2\text{—}CN$$

Errors: 0 0 0

A carbon-centered diastolic group, like CN, can be attached to another carbon either by an outbound arrow or by a line, depending on which choice minimizes the number of errors. In this light, a CN group placed on secondary carbon can "insulate" more polar CH bonds and prevent their vicinal repulsive interaction than a group placed on primary carbon. Tertiary placement is still better. The argument is exemplified below by means of the arrow formulas and supporting thermochemical data.

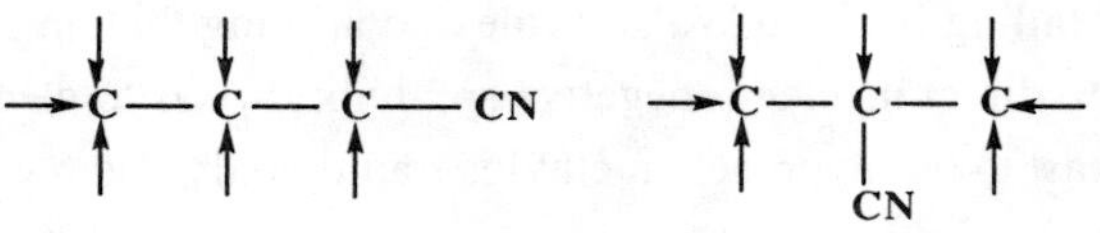

4v ERRORS 2v ERRORS

Z	(2-Propyl-Z -1-Propyl-Z), E_{rel} (kcal/mol)
CN	–1.2
COOH	–2.0

It is emphasized here how far apart we stand from traditional bonding theory. A chemist would rationalize the preceding energy data by saying that CN and COOH are σ-withdrawing groups that are best accommodated at C2 of propyl because 2-propyl has lower ionization potential (i.e., is a better σ donor) than 1-propyl. This is the Pauling "polar covalence" argument. By contrast, we say that each of these groups exerts its effect by "insulating" CH bonds from one another and, thus, preventing vicinal overlap repulsion! The best "insulation" is achieved when the group is placed in the midst of the largest number of CH bonds (i.e., when it occupies a tertiary site).

References

1. L.O. Brockway, *J. Phys. Chem.* 41, 185, 747 (1937).
2. R. D. Chambers, *Fluorine in Organic Chemistry*, Wiley, New York, 1973.
3. L. Pauling, *The Nature of the Chemical Bond,* 3rd ed., Cornell University Press, Ithaca, NY, 1960.
4. For expressed skepticism regarding the validity of negative hyperconjugation, see: (a) Ref. 2. As indicated in Section 7.1, with regard to the progressive CF bond contraction in the fluoromethanes, Chambers was uncertain more than 20 years ago of the usefulness of this concept (i.e., negative hyperconjugation). (b) B.E. Smart, in *Molecular Structure and Energetics*, Vol. 3, J.F. Liebman and A. Greenberg, Eds., VCH: New York, 1986, pp. 141–148. (c) C.L. Perrin and O. Nunez, *J. Am. Chem. Soc.* 108, 5997 (1986). (d) K.B. Wiberg and P.R. Rablen, *J. Am. Chem. Soc.* 115, 614 (1993). (e) The generality of the phenomenon of fluorine symbiosis (i.e., that all systolic groups are symbiotic) and the failure of the proposed models (such as hyperconjugation and our own through-space/through-bond coupling) to consistently explain the data has been recognized by A. Greenberg, S.D. Sprouse, and J.F. Liebman, unpublished work (1978).

5. (a) A.E. Reed and P.v.R. Schleyer, *J. Am. Chem. Soc.* 109, 7362 (1987). (b) N.D. Epiotis, "Unified Valence Bond Theory of Electronic Structure," in *Lecture Notes in Chemistry*, Vol. 34, Springer-Verlag, New York, 1983, Chapter 4. (c) D.A. Dixon, T. Fukunaga, and B.E. Smart, *J. Am. Chem. Soc.* 108, 1585 (1986).
6. W.F. Schneider, B.L. Nance, and T.J. Wallington, *J. Am. Chem. Soc.* 117, 478 (1995).
7. K.B. Wiberg and M.A. Murcko, *J. Am. Chem. Soc.* 110, 8029 (1988).
8. C.P. Mathers, B.N. Gover, J.F. Ying, H. Zhu, and K.T. Leung, *J. Am. Chem. Soc.* 116, 7250 (1994).

Chapter 8

Reinterpretation of Structural Organic Chemistry Through the T Formulas

8.1 Alcohols Versus Ethers, Ketones Versus Aldehydes, Primary Versus Tertiary Amines, and Carboxylic Acids Versus Esters

Why are ethers less stable than alcohols, and why are aldehydes less stable than ketones? We cannot find in the literature any systematic attempts to answer these important questions. On the other hand, the T formulas lead to a one-sentence explanation: Ethanol is roughly 12 kcal/mol more stable than dimethyl ether because the former has two and the latter six p-errors. Acetone is roughly 7 kcal/mol more stable than propionaldehyde because the former is error-free but the latter has a total of three errors.

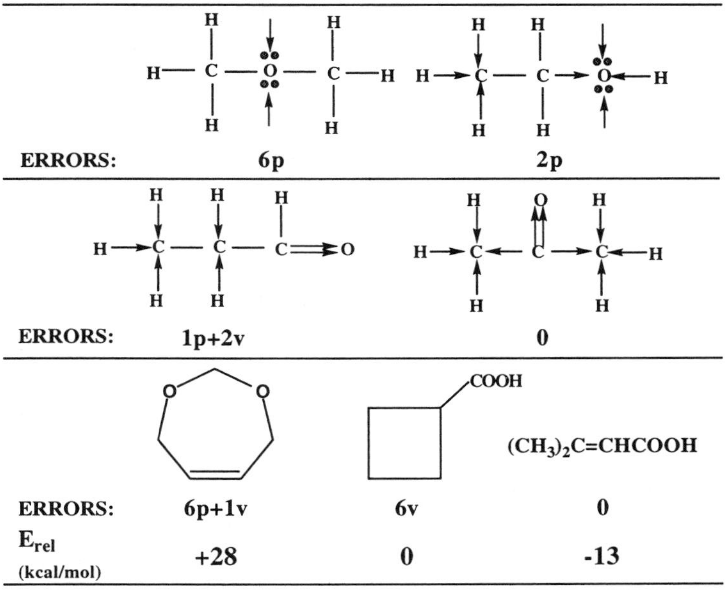

We now begin to see patterns that escaped us before: Moving an oxygen atom from within to the terminus of a $C_nH_{2n+2}O$ chain reduces the number of errors. We will encounter the same pattern in transition states described by I, rather than T, formulas. As shown above, a cyclic diether can be unfavorable enough to be greatly inferior to a ring-strained isomer. An unsaturated ether has a thermodynamic driving force for being converted to an aldehyde, as illustrated below. This is partly why the Claisen rearrangement of vinyl–allyl ethers is such a facile reaction.

	$O{=}CH{-}CH_2{=}CH_3$	$H_2C{=}CH{-}O{-}CH_3$
E_{rel} (kcal/mol)	0.0	21.0
Errors	1p + 2v	4p

The magnitude of the superiority of propionaldehyde over the methyl–vinyl ether confirms that a p-error must be weighed more than a v-error. For maximum simplification, however, we will continue to assume that polarity and vicinal errors are equally weighted, while always keeping in mind the quantitative nuances.

What makes alcohol superior to ether also makes carboxylic acid (e.g., CH_3COOH) superior to the isomeric ester (e.g., $HCOOCH_3$). The reader should go through the exercise of becoming convinced that this stability order reflects the number of errors in each isomer. In the following data (kcal/mol), note the stability trend and how increased aggregation of methyls on carbon improves stability.

$CH_3{-}(CH_2)_4{-}COOH$	$CH_3{-}(CH_2)_3{-}COOCH_3$	$(CH_3)_3CCOOCH_3$
0.0	10.1	5.8

Finally, tertiary are predicted to be less stable than isomeric primary amines, a situation analogous to the ether/alcohol comparison. Indeed, *n*-propylamine (2p + 2v errors) is roughly 11 kcal/mol more stable than trimethylamine (9p errors).

What happens when there is the same number of errors are found on the reactant and on the product side? This is the ideal situation for uncovering subsidiary effects. For example, the number of errors is conserved in the two reactions that follow. The first is exothermic because spatial overlap is stronger on the product side [the most strained $(CH_2)_n$ molecule is not cyclopropane but, rather, ethylene][1] and the second is also exothermic partly because of nonbonded attraction in dimethyl ether (see Chapter 11).

$$H_2C{=}O + H_2O \longrightarrow H_2C(OH)_2 \qquad \Delta H_r = -8.0$$

$$2H_3COH \longrightarrow H_3COCH_3 + H_2O \qquad \Delta H_r = -6.0$$

The most interesting case of isomers having a constant number of errors is the comparison of the keto and enol tautomers of acetaldehyde. The T formulas that follow reveal two things:

Both isomers have one erroneous C—H bond.

The keto form has only one but the enol form has two arrows connecting identical carbon atoms.

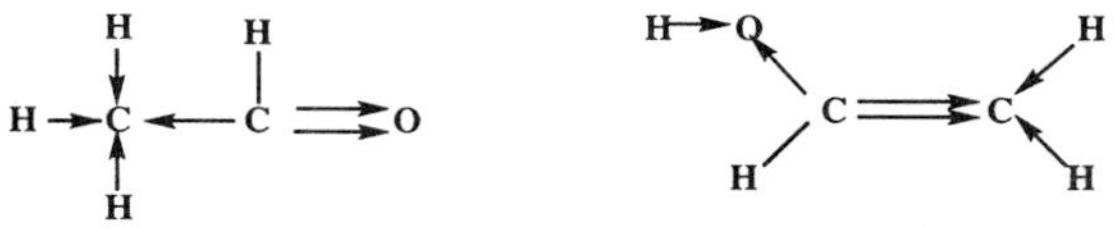

E_{rel} (kcal/mol)	KETO	ENOL
	0	+10

It follows that satisfaction of the exclusion rule requires a higher promotional energy in the case of the enol form (i.e., turning a homopolar C—C into a heteropolar C—C bond). In other words, the keto form is more stable because it has one extra polar bond appropriately oriented to satisfy the exclusion rule.

A biscoordinate oxygen atom prefers to occupy the terminal site of a carbon chain to minimize the number of errors, and the ethanol/dimethyl ether comparison is the simplest example. How far from conventional theory this takes us can be appreciated by considering a question I often put to visitors: Is vinyl methyl ether ($H_2C{=}CH{-}O{-}CH_3$) or vinyl carbinol ($H_2C{=}CH{-}CH_2OH$) the more stable? The vast majority of respondents look for lone pair–π-bond conjugation and proclaim the ether to be more stable. The alcohol is the more stable, however, by 6 kcal/mol. Simplifying, one can say that this happens because C—O bonds are weaker than O—H bonds. The traditional approach has been to focus on π conjugation. This work argues that it is the σ bonds that determine molecular stability and structure. This theme encompasses the sacred cow of π conjugation, namely, benzene (see Chapter 16).

8.2 T Formulas for Carbocations and Diazonium Ions

A carbenium ion is a neutral molecule with an infinitely electronegative ligand. Hence, the T formula of a carbenium ion has three arrows pointing away from the cationic center. This means that the tertiary butyl, which is error-free, will be much more stable than the isomeric primary *n*-butyl carbenium ion, which has 2p + 4v errors.

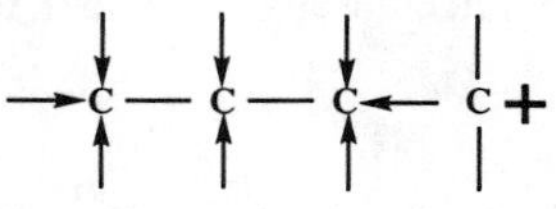

ERRORS: 2p + 4v

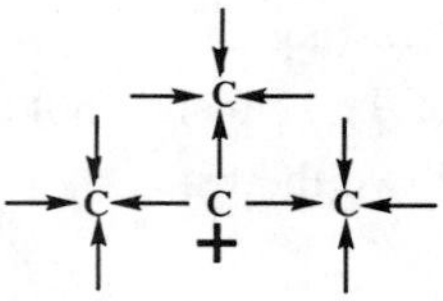

ERRORS: 0

The absence of a symbol on the tail of an arrow implies hydrogen. Note that if we think of methyl as analogous to fluorine (both systolic), and if we take the infinitely electronegative ligand to be fluorine, we come to the conclusion that the tertiary carbenium ion shown above is analogous to CF_4.

We now explore a domain of data at odds with popular notions. Specifically, it has been assumed that methyl groups stabilize carbocationic centers by hyperconjugation. This means that if we were to consider the graded series from CH_3^+ to CMe_3^+, we should predict a decrease in the charge on the cationic center. However, the preceding formula predicts exactly the opposite: that is, *increased methylation of a carbenium ion center should make it more positive!* After all, we argued that Me is analogous to F — that they are both systolic! This conclusion is supported by ^{13}C NMR data on $HCMe_2^+$ and CMe_3^+ obtained by Olah and co-workers.[2] The crucial point is that we no longer need hyperconjugation to explain the stability of carbocations.

A more interesting application concerns the effect of silicon on the stability of a carbocation. The formulas below lead to an immediate prediction: Placing a silyl group *alpha* to the cationic center will destabilize it because there is one serious polarity error involving two atoms of widely different electronegativity (Si,C). By contrast, placing the C_2Si *beta* will stabilize it because this polarity error disappears. The latter influence is the underlying cause for the preference of electrophiles to replace SiR_3 rather than H in the electrophilic aromatic substitution of Ph—SiR_3. The arenium ion formed by adding the electrophile on the carbon bearing the SiR_3 group is stabilized by the latter in beta fashion.[3a] Using this type of argumentation, we predict that α-silylation of a C—H bond will enhance its acidity (carbanion formation) while leaving the C—H bond dissociation energy (radical formation) relatively unaffected, as found by Bordwell and his co-workers.[3b]

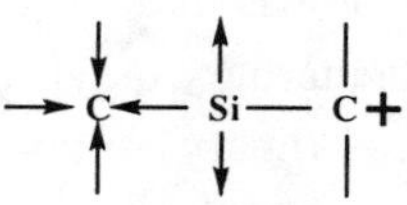

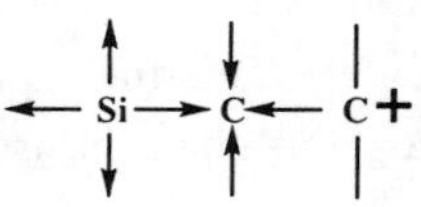

ERRORS: 2p(CH)+1p(SiC) 2p(CH)

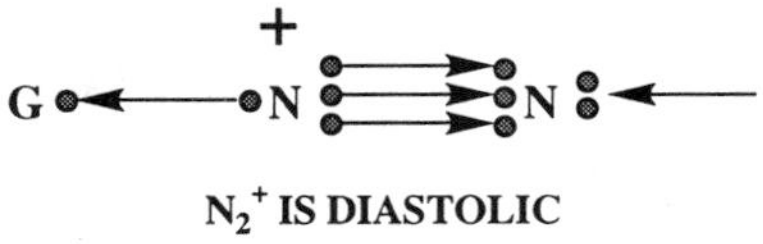

Figure 8.1 The diastolic N_2^+.

Addition of HCl to a vinyl silane results in stereospecific replacement of the silyl group by Cl via an intermediate carbocation, which adopts a conformation in which the C—Si bond eclipses the vacant C2p AO on the adjacent cationic carbon. This conformational preference can be rationalized by the very same principles on which the concept of the T formula is based: Minimization of exchange repulsion between the polar C—Si bond and the adjacent cationic center. Hyperconjugation is not needed!

Diazonium ions provide yet another surprise for conventional theory. Ab initio caculations show that the endothermicity of the nitrogen loss reaction shown below is roughly 42 kcal/mol for R = methyl and only 12 kcal/mol when R = ethyl![4]

$$R{-}N_2^+ \longrightarrow R^+ + N_2 \qquad (8.1)$$

What makes the diazonium ion unique is the presence of the diastolic N_2^+ functional group. Figure 8.1 explains why this functional group is bistolic by depicting the bonding in G—N_2^+ in which the N directly attached to G is unhybridized and the remote N is sp-hybridized. The consequences of the diastolic nature of N_2^+ are revealed by the T formulas of the reactants and products of reaction (8.1). The greater endothermicity of the methyldiazonium decomposition is due to two combined effects: reactant destabilization and product stabilization of the ethyl (relative to the methyl) derivative.

ERRORS: 1p 3p

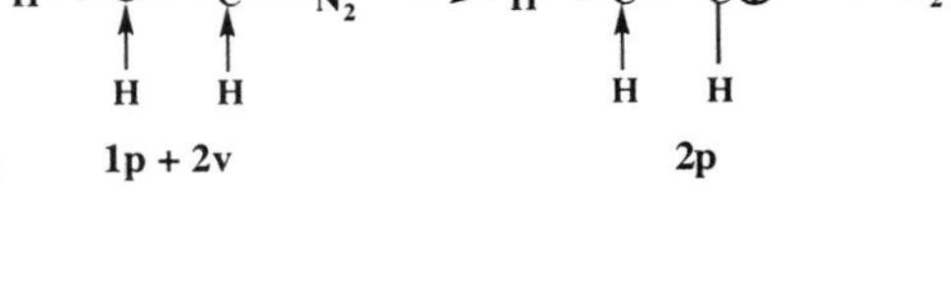

ERRORS: 1p + 2v 2p

Glaser et al.[4] resisted the temptation to pull an explanation from the conventional menu and append it to their calculations. Rather, they came to the conclusion that "the structural and topological features of the CN linkages in MeN_2^+ and EtN_2^+ are similar and their consideration alone cannot account for the large difference in their dissociation energies. . . . The characterization of the electron density in the bonding region alone is not sufficient to fully characterize all of the bond properties." This is precisely the spirit of the T formula: Stability is the result of what happens interactively in the *entire* molecule (how *all* bonds arrange themselves so as to minimize mutual exchange repulsion) rather than what happens at an isolated bond or subfragment.

Bottom line: Inserting an oxygen atom into the C—C bond of ethane to generate dimethyl ether is unfavorable because the two lone pairs of oxygen suffer the exchange repulsion of six adjacent systolic CH bonds. Inserting an oxygen atom into the C—H bond of ethane to generate ethanol is much better because the two lone pairs of oxygen suffer the exchange repulsion of only two adjacent systolic CH bonds. Finally, inserting an oxygen atom into the C—H bond of the cationic carbon of ethyl cation to generate CH_3CHOH^+ is the best because the two lone pairs of oxygen feel the exchange repulsion of one adjacent systolic CH bond. In addition, the electric field of the positive carbon stabilizes electrostatically the oxygen lone pairs. Stated in equivalent terms, the number of errors progressively dwindles as we go from the ether to the alcohol and to the carbenium ion. Thus, carbocation stability has to do with exchange repulsion. Is π conjugation important? The burden of proof lies on those who for decades argued on the basis of conjugation and hyperconjugation. We say that CO_2, CO_3^{2-} (the stable molecules of nature), and $C(OH)_3^+$ are error-free atom combinations and that any interbond delocalization (conjugation) is simply a bonus.

8.3 T Formulas and Angle Strain

In Chapter 6, we saw that we could explain the effect of π acceptors (DI groups) and π donors (SY groups) on the geometry and reactivity of cyclopropane. What happens if one assumes only perfect pairing and intrabond CT delocalization? The T formulas of the SY derivatives of propane, cyclopropane, and cyclobutane are shown in Figure 8.2. A SY group reduces the number of errors in *every* hydrocarbon. This improves the stability and changes the geometry of the substituted derivative relative to the parent. Furthermore, the effect is uniform. Specifically, an SY group is predicted to have the same effect on the lengths of the "side" C—C bonds of propane (C-2 substitution), cyclopropane, and cyclobutane.

Schleyer and co-workers calculated that replacing H by a SY group in cyclopropane and propane (substituent at C2), shortens the "side" C—C bonds in *both* cases.[5] Furthermore, the T formula of SY-substituted cyclopropane predicts that the "distal" should be longer than the "side" bond and, by implication, distal-bond should take precedence over side-bond cleavage, as is observed.[6]

Bottom line: The structure and reactivity trends of cyclopropane derivatives bearing systolic groups are due to the combined action of intra- and interbond CT delocalization.

8.4 Superexothermic Reactions

Theoretical treatments have traditionally focused on the explanation of what we may call "few kcal effects"! The energy difference between eclipsed and staggered ethane is of the order of 3 kcal/mol; the strengths of hydrogen bonds and intermolecular bonds are, on the average, of the order of a few kcal/mol; an "allowed" and a "diradical" transition state are often quasi-degenerate, and so on. We now have the opportunity to escape this uncomfortable zone by deliberately designing molecules that differ greatly in stability. That is, we can design reactions that are exothermic (endothermic) by dozens of kcal/mol. The strategy is simple: Connect the atoms in an error-free mode in one isomer and in an error-plagued mode in a second isomer.

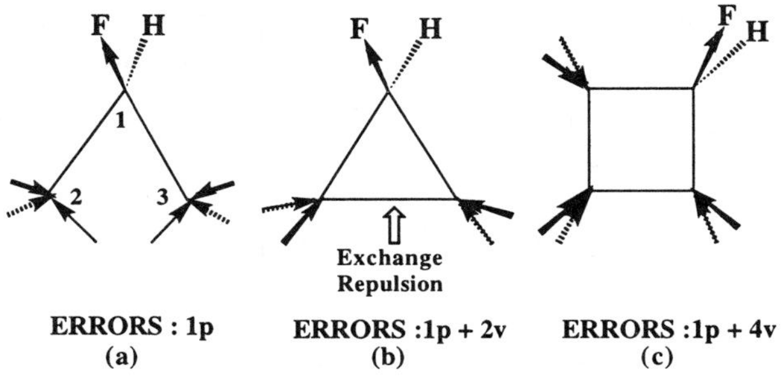

Figure 8.2 T formulas of (a) fluoropropane, (b) fluorocyclopropane, and (c) fluorocyclobutane. Observe the similarity between the acyclic and cyclic propane systems with regard to the two C—C bonds emanating from the carbon bearing the systolic group.

The best approach is to consider combinations of atoms that differ significantly in electronegativity and have lone pairs. In such systems, each error becomes energetically very costly and the error-free isomer turns out to be highly superior to an error-plagued one. In all the examples that follow, the reactants carry errors and the products are error-free. The reader should go through the exercise of identifying the number and types of errors on the reactant side. The reaction exothermicity (kcal/mol) is shown atop of the reaction arrow.

$$CH_3CH_2CH_2CH_3 + CH_3OH \xrightarrow{-14} (CH_3)_3COH + CH_4$$

$$CH_3NH_2 + H_2CO \xrightarrow{-30} HCONH_2 + CH_4$$

$$2CH_3NH_2 + H_2CO \xrightarrow{-45} H_2NCONH_2 + 2CH_4$$

$$H_3CNO \xrightarrow{-61} HCONH_2$$

$$HCNO \xrightarrow{-77} HNCO$$

$$CNH \xrightarrow{-80} HCN$$

8.5 Is There π Conjugation of Systolic Groups?

Every chemist anticipates that a π donor (e.g., F, OR, NR_2) prefers to be attached to a benzene ring to be able to reap the benefit of resonance stabilization. Thus the following reaction and its associated enthalpy data seem to be rather mundane. In our theory, however, these very π donor groups are systolic groups, and this suggests a totally different scenario: This simple reaction is endothermic, not because of π resonance but, rather, because there are more errors in the products than in the reactants.

$$\text{para } H_3C\text{—}C_6H_4\text{—}Y \longrightarrow H\text{—}C_6H_4\text{—}CH_2\text{—}Y$$

Y	ΔH_r (kcal/mol)
Me	2.7
NH_2	7.0
OH	6.0
$H_2C{=}CH$	10

In the following T formulas for reactants and products, we count every line connecting nonequivalent carbons as an error.

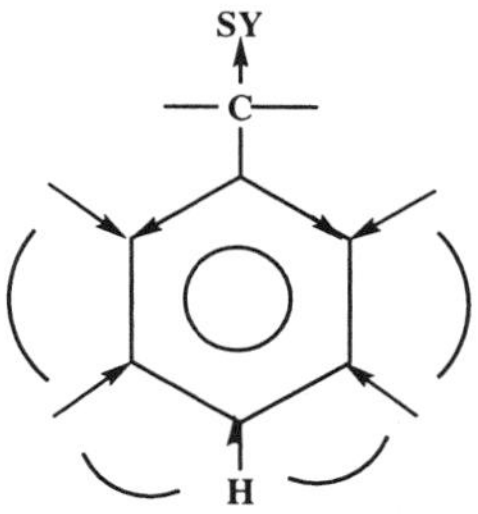

ERRORS = 7p + 4v

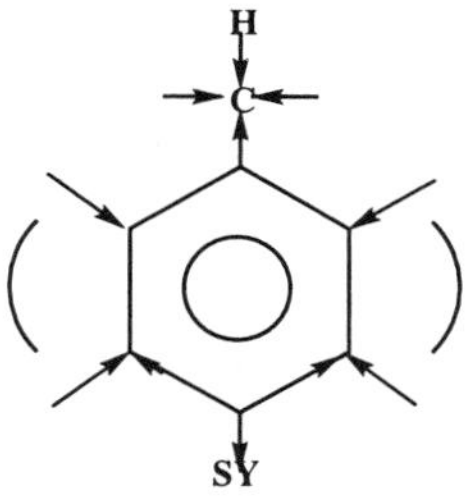

ERRORS = 4p + 2v

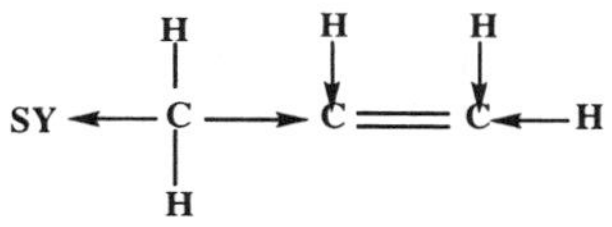

ERRORS = 4p + 2v

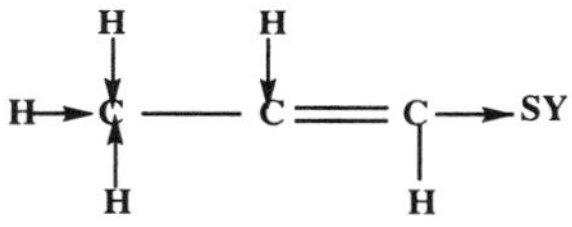

ERRORS = 3p + 1v

It is clear that σ bonding alone can explain the preference of systolic groups for direct attachment to the benzene nucleus or any other π-conjugated system (ethylene, 1,3-butadiene, cyclopentadienyl radical, naphthalene, etc.). In all cases, replacing H by a systolic group simply reduces the number of errors and stabilizes the system!

We will now argue that arrow formulas can explain the often unexpected relative stability of isomeric benzene derivatives bearing systolic groups. We have tabulated the most stable isomers of a variety of polysubstituted benzenes in Figure 8.3. In all cases, the most stable isomer is the one that minimizes the errors in the σ frame.

A proper perspective of the importance of σ binding can be achieved by observing the difference in stability between two pyridine isomers shown in Figure 8.4a. A mere displacement of OH away from the carbon ortho to the carbon meta to the nitrogen raises the energy by 12 kcal/mol. Recognizing that both H_3C and HO are systolic groups, we can now anticipate an extremely stable molecular skeleton, namely, 1,3,5-trihydroxytriazine and its triketo form (Figure 8.4b).

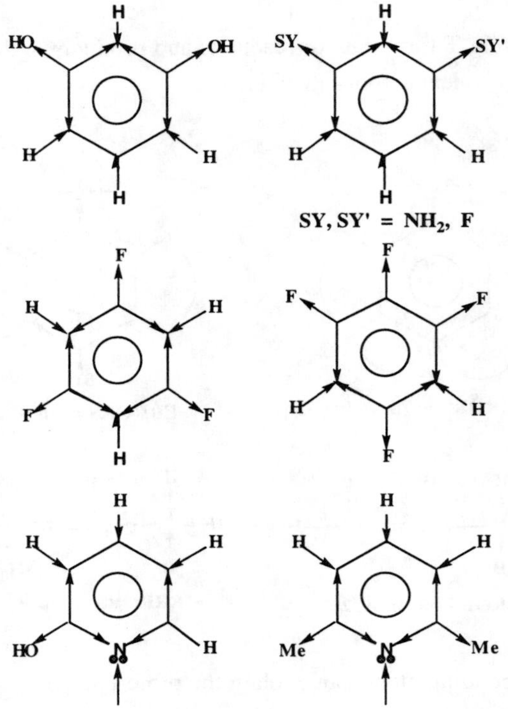

Figure 8.3 Stable isomers of multisubstituted benzenes. Note the partial or complete alternation of systolic and diastolic arrows.

8.6 New Concepts of Substituent Effects

T formulas enable us to visualize directly the impact of a substituent on the inherent stability of a molecule by simply counting the number of errors deleted or added as an atom in a given parent molecule is replaced by a substituent. Figure 8.5 gives some examples, which we discuss in turn.

(a) Methane is differentiated from ethane by systolic substituents. Replacing H by SY damages methane more than ethane. By contrast, replacing H by DI leaves the number of errors unaltered in both methane and ethane. SY substituents change the stability of alkanes. DI substituents have no effect.

(b) Two-carbon molecules like ethane and ethylene show a comparable response to substitution by SY groups: The number of errors is reduced by one. By contrast, alkynes are not affected by substitution. The reason is that two vicinal systolic arrows that are colinear generate negligible exchange repulsion. Hence, their presence does not constitute an error. This means that replacing hydrogens by SY groups is favorable in alkanes and alkenes but not in alkynes, a relationship that explains the following thermochemical data:

$$H_3C{-}CH_3 + HC{\equiv}CF \longrightarrow HC{\equiv}CH + H_3C{-}CH_2F \qquad \Delta H_r = -\,13.8 \text{ kcal/mol}$$

Errors: 3v 0 0 2p

$$H_2C{=}CH_2 + HC{\equiv}CF \longrightarrow HC{\equiv}CH + H_2C{=}CHF \qquad \Delta H_r = -\,16.7 \text{ kcal/mol}$$

Errors: 2v 0 0 1p

(c) Every hydrocarbon is stabilized when H is replaced by a systolic group. The effect of such a replacement is to reduce the number of vicinal errors (i.e., to reduce interbond exchange repulsion) in highly symmetrical polysubstituted systems as illustrated in Figure 8.6. For example, hexa-SY benzene is expected to be an extremely stable molecule because vicinal errors involving the C_b—H bonds have been eliminated by replacing diastolic H by systolic SY (C_b = benzene carbon; SY = F, OH, NH_2, CH_3). We now appreciate why methylation (as well as perfluorination) of carbon chains and rings leads to very stable systems: $C_2(CH_3)_4$ is the most stable C_6H_{12} alkene, $C_6(CH_3)_6$ is the most stable $C_{12}H_{18}$ polyene, and so on. Stabilization also occurs when three systolic groups alternate with three hydrogens on the benzene nucleus (Figure 8.7).

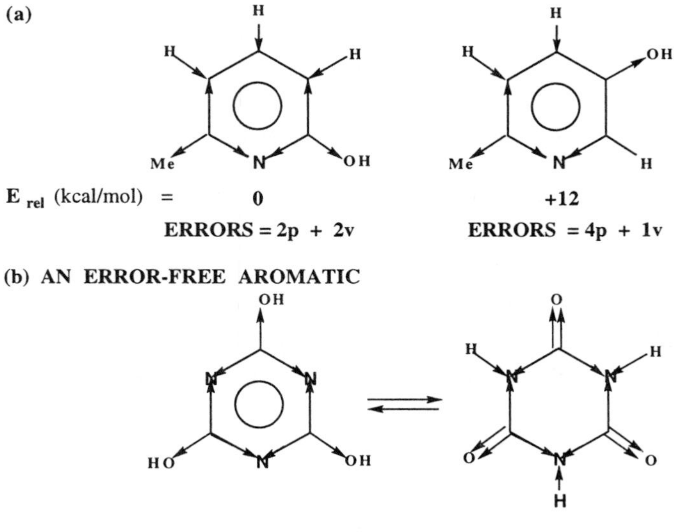

Figure 8.4 (a) Comparison of structurally isomeric pyridines. (b) A "perfect" aromatic molecule.

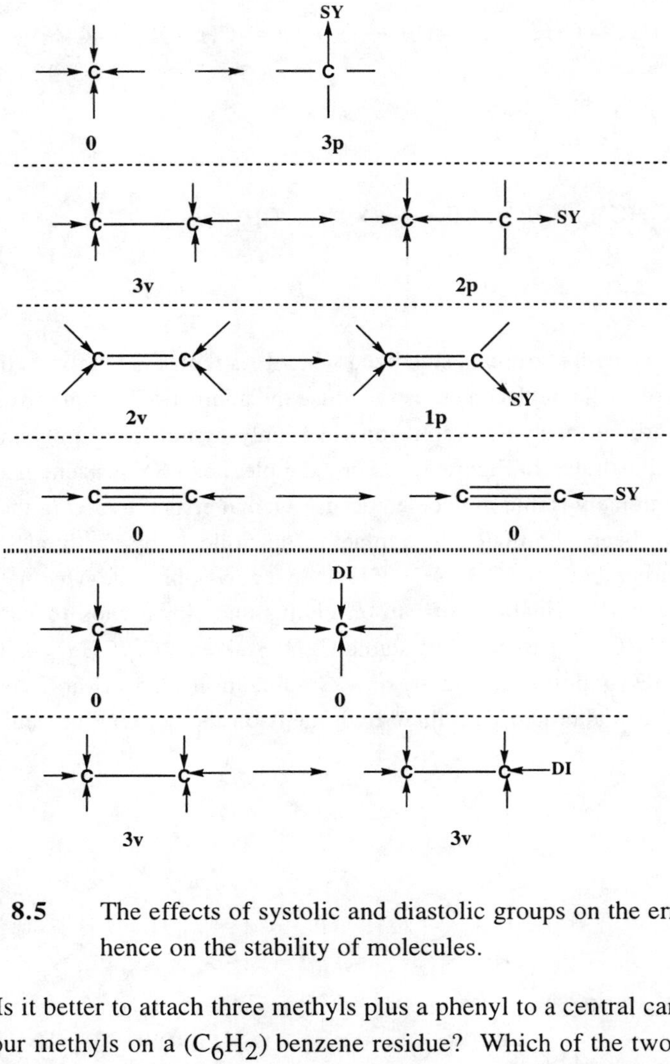

Figure 8.5 The effects of systolic and diastolic groups on the error count, hence on the stability of molecules.

(d) Is it better to attach three methyls plus a phenyl to a central carbon or to attach four methyls on a (C_6H_2) benzene residue? Which of the two $C_{10}H_{14}$ isomers is the more stable? It is easy to show that the former has six vicinal errors while the latter is error-free. This explains why 1,2,4,5-tetramethylbenzene is more stable by roughly 6 kcal/mol than *t*-butylbenzene and serves to reemphasize the need to look at the entire molecule rather than individual functional groups to deduce stability!

(e) Consider what happens when the chemist who has accepted the conventional ideas of π conjugation is asked to predict the most stable C_4H_6O isomer. If methyl is treated as isoconjugate to vinyl and one opts for maximal π conjugation, the predicted structure is H_3C—CH═CH—CH═O. By contrast, using the reasoning of (c) above, one immediately envisions a C_2 skeleton

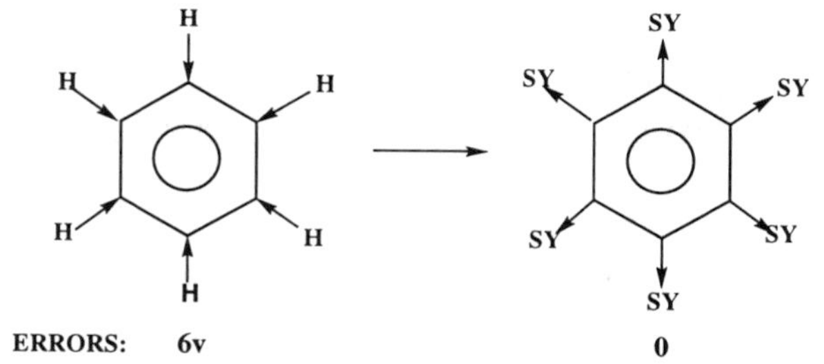

Figure 8.6 Stable aromatics produced by uniform substitution of H by SY groups.

connected to four systolic groups: two CH_3 groups and one double-bonded oxygen, the latter acting effectively as two systolic OHs. The predicted dimethylketene structure shown below is 7 kcal/mol more stable than the π conjugated aldehyde predicted by falling back on conventional models. Note that since ketene is an ethylene analogue, with hydrogens replaced by methyl groups and oxygen atoms, vicinal errors present in ethylene are eliminated.

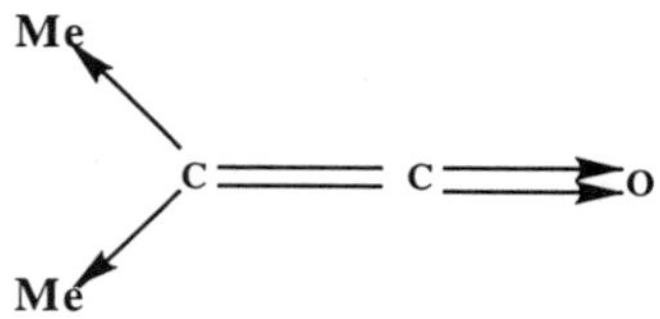

The "thinking man's" review of fluorine structural chemistry by Smart[7] points out some spectacular characteristics of perfluoro groups like trifluoromethyl (CF_3) and perfluorophenyl (C_6F_5). One example is the chairlike geometry of the pertrifluoromethylbenzene.[8] Another example is the ability of C_6F_5 to form an exceptionally strong bond to some group X: The

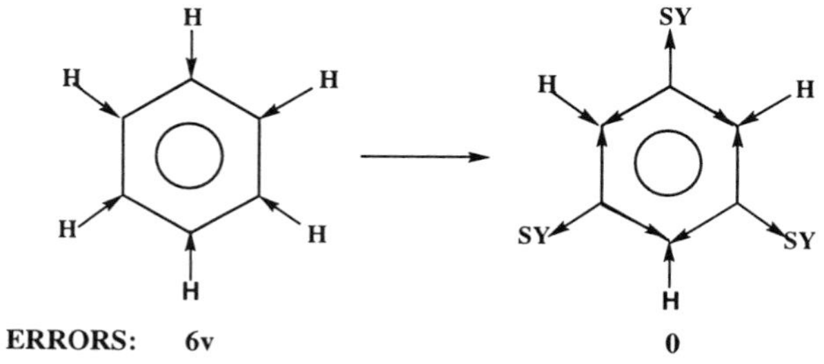

Figure 8.7 Stable aromatics produced by alternation of SY groups and (DI) hydrogens.

bond dissociation energy of C_6F_5—X exceeds that of C_6H_5—X by 41, 29, 34, 23 kcal/mol when X = H, F, Cl, and Br, respectively. We suggest that these and related trends are consequences of the diastolic nature of the fluorine-bearing carbon.

8.7 T Formulas and van der Waals Complexes

Propyne is a simple illustrator of the alkyne family and carbon acidity. Asked to predict the structure of the gas phase complex with argon, most chemists would envision coordination (of the hydrogen bond type) of the acidic hydrogen of propyne with argon. By contrast, the T formulas of organic molecules teach us that these units should be viewed as multipolar structures as projected by the corresponding T formulas. The T formula of propyne shown in Figure 8.8 speaks for itself in the sense that it identifies the regions of depleted and enhanced electron density. It is predicted that an inert gas atom will be attracted to the *diastolic* center (i.e., C2) of propyne. In an interesting microwave/infrared spectroscopic study, Watts and co-workers[9] found that the propyne–argon complex has the T-like geometry shown in Figure 8.8.

8.8 T Formulas and the Counterintuitive Structural Isomerism of Heavy p-Block Molecules

H—O—O—H and F—O—O—F are much more stable than H_2O—O and F_2O—O, respectively. However, changing O to S produces a preference that is hard to understand: The 1,2 and 1,1 isomers of the perfluoroderivatives are now very close in energy, with the best estimate going in favor of the 1,1 isomer. The following T formulas provide an immediate rationalization, once it is assumed that sulfur has one active and one inactive ("inert 3s") pair. The 1,1 isomer has zero errors while the 1,2 isomer has two serious errors (because the sense arrow goes to the active sulfur lone pair). This is more than enough for

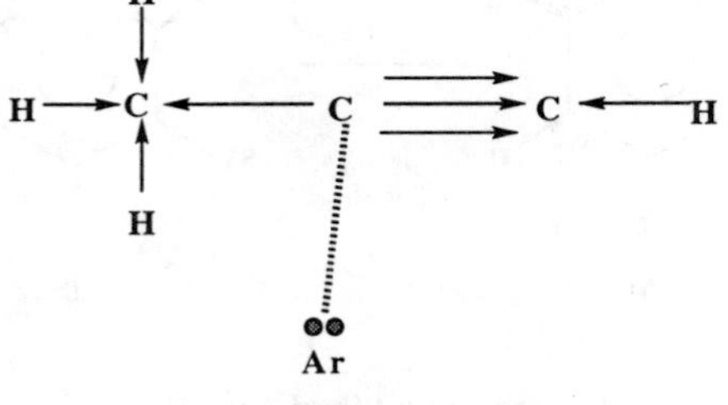

Figure 8.8 T formula of Ar—CH_3CCH.

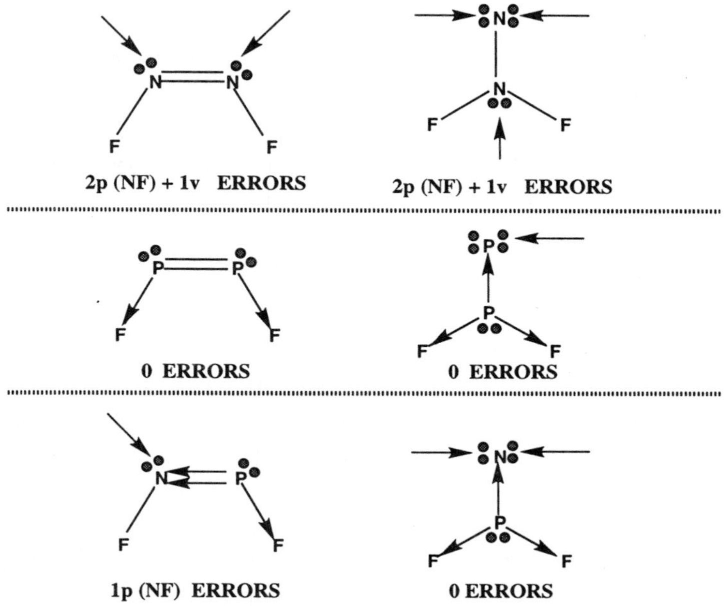

Figure 8.9 Arrow formulas of 1,2 and 1,1 structural isomers of derivatives of diimide. Note the unique situation of FNPF versus F_2PN.

counteracting the unfavorable charge separation and rendering the 1,1 isomer superior. Replacing F by H will return the preference to the 1,2 form because CT delocalization ceases to be important.

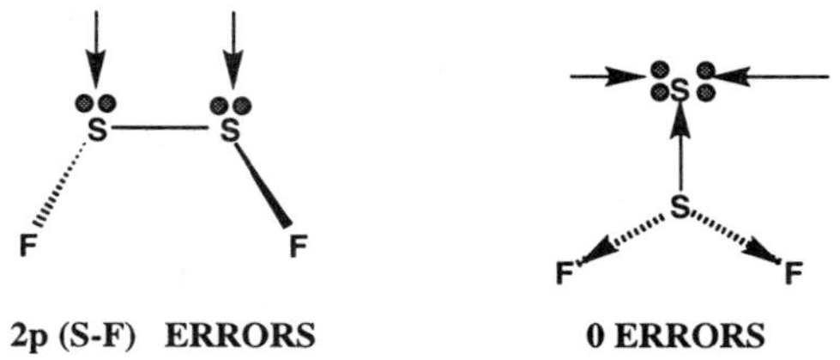

Nguyen et al.[10] obtained the surprising calculational result that F_2PN is more stable than *cis*-F-P=N-F by 70 kcal/mol. By contrast, *cis*-F—P=P-F is more stable than F_2P=P by 68 kcal/mol and *cis*-F—N=N—F is more stable than F_2N—N by 78 kcal/mol! These observations are cleanly rationalized by the T formulas shown in Figure 8.9 under the assumption that one P lone pair is inactive. Specifically, it can be seen that the 1,1 and the 1,2 isomers incur the same number of errors in both P_2F_2 and N_2F_2, with the situation changing dramatically in the case of the F_2PN system, in which the 1,1 has no errors but the 1,2 isomer does have errors.

8.9 On the Relationship of Protein Primary Structure and Folding

Ranganathan et al.[11] pointed out that "the rapid developments that have taken place pertaining to protein synthesis have not been matched with knowledge relating to the relationship between the primary sequence and its folding profile."[12] We now have the T formulas that inform us how many "mistakes" exist within a molecule with respect to bond polarity if exchange repulsion between bonds is minimized. We can use this notation to have a new look at proteins: Instead of classifying amino acids according to classical bonding concepts, we can subdivide them according to the number of errors in their side chains. The plan is to look at the primary structure of proteins with an eye to discerning whether there is a connection between the function of the protein and the way in which the errors are distributed along the protein. Why do we expect a correlation?

The arrow formula of a peptide linkage HN—CO— reveals that this is an error-free arrangement. An amino acid side chain that is also error-free combines with the two flanking peptide linkages to support an exclusion wave. This effectively charges the atoms by promoting the multipolar electronic configuration suggested by the T-formula. As a result, error-free side chains promote stronger intra- and intermolecular hydrogen bonding by the corresponding backbone stretches as well as by themselves, provided they carry a functional group capable of hydrogen bonding. The converse is true of a side chain with many errors. In short, the proposed linkage is between error distribution and capacity for hydrogen bonding.

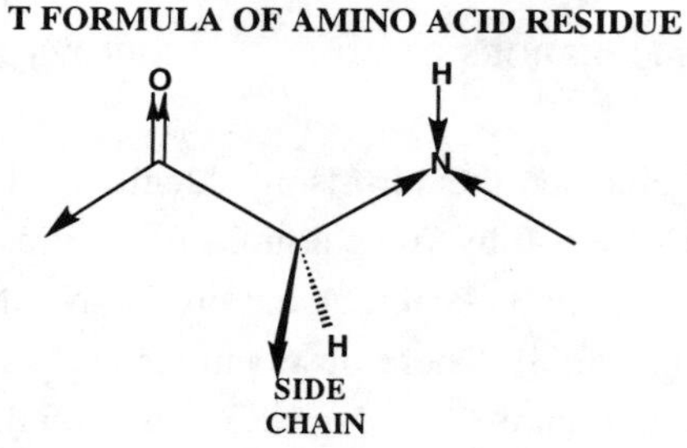

Figure 8.10 The alpha carbon of an amino acid is diastolic.

To write an arrow formula for a side chain, we must first understand whether the peptide linkage acts as a diastolic or systolic fragment. The answer is given in Figure 8.10: It acts as a systolic group. This means that the alpha amino acid carbon is diastolic. Hence, the side chain must be an error-free systolic group. One such example is the methyl side chain of alanine. By using our standard procedure, we can now enumerate the errors of each side chain and seek a correlation between the error distribution, the differential ability of the backbone to form hydrogen bonds, and the structure–function aspects of the protein. The methyl group of alanine is no longer viewed as a mere nonpolar side chain but rather as a promoter of hydrogen bonding of the flanking peptide linkages.

8.10 What Is Wrong with Glucose?

Photosynthesis is the uphill conversion of CO_2 plus H_2O to glucose. In other words, nature makes an unstable product that is effectively used as an energy storage device. Why is glucose unstable relative to CO_2 plus H_2? The T formulas of reactants and products reveal that glucose has errors while CO_2 plus H_2O has none! The same situation occurs in glycolysis. The cell now harvests energy released in the breakdown of glucose to two lactic acids.

$$\underset{\text{(D-glucose)}}{HCO{-}(CHOH)_4{-}CH_2OH} \longrightarrow 2\ \underset{\text{(L-lactic acid)}}{CH_3{-}CHOH{-}CO{-}OH}$$

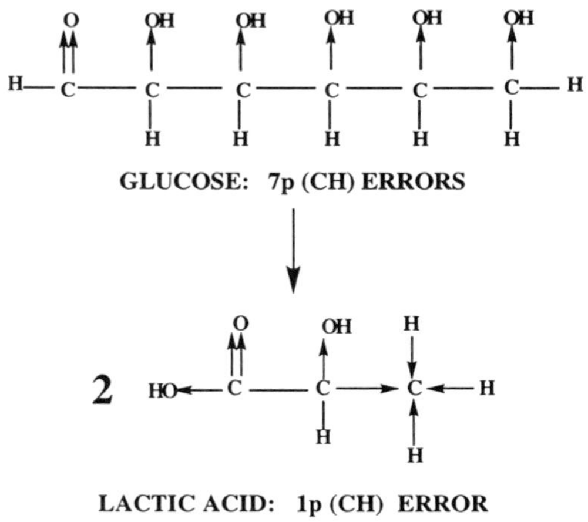

Figure 8.11 T formulas of glucose and lactic acid reveal the origin of the instability of the former relative to the latter.

Again, the reaction is exothermic by roughly 27 kcal/mol because the reactant has seven polarity errors while the product have only one (Figure 8.11). Note that although two lactic acids have one σ replaced by one π bond relative to the isomeric glucose, the former two are much more stable because of a nearly error-free combination of methyl, hydroxyl, and carbonyl groups.

In the past, research proposals seeking funding to explore possible uses of molecules to store energy have traditionally envisioned making strained organic molecules to act as energy storage devices. Nature teaches us a lesson: Glucose is a key "strained" (i.e., "error-full") molecule, and we now understand why. By using the T formulas, we can imitate the strategy of nature by constructing highly unstable yet makable molecules.

8.11 Heterocyclic Chemistry

The T formulas change our outlook of heterocyclic chemistry. Where we could see merely carbon frames "sprinkled" with diverse heteroatoms, we now see patterns which are readily interpreted to imply either instability or stability. As a simple first example, consider the even-center and odd-center cumulenes shown below. The arrow formulas make a straightforward prediction: Even systems are stabilized by SY (electronegative) and odd systems by DI (electropositive) groups, with the magnitude of the effect dropping off as size increases. This is supported by ab initio calculations of the odd ketenic system YCH=C=O. Gong and co-workers found that the enthalpy of the isodesmic reaction of YCH=C=O + $CH_3CH{=}CH_2$ to yield $CH_3CH{=}C{=}O$ + YCH=CH_2 plots linearly against Pauling electronegativity values of Y in the sense that a more electropositive Y produces a more positive reaction enthalpy.[13]

EVEN CUMULENES

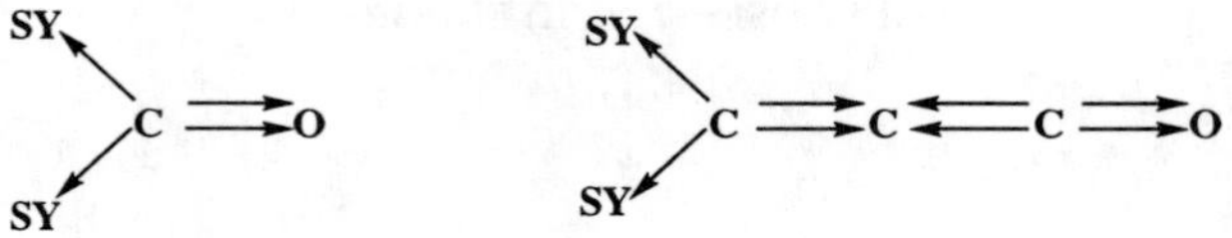

ODD CUMULENES

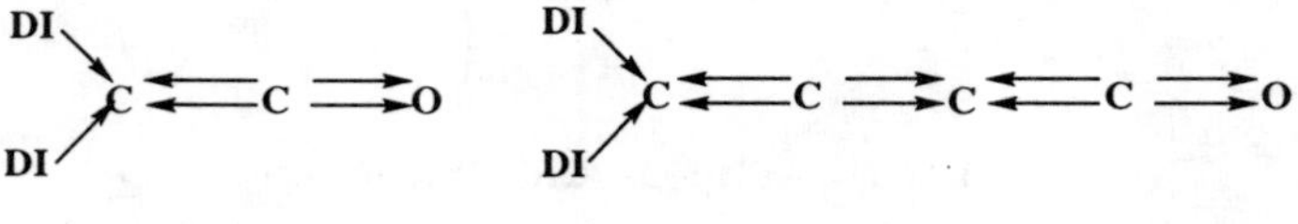

It does not take much searching in the realm of "aza-chemistry" to discover that diverse molecules ranging from simple derivatives of polynuclear hydrocarbons to pesticides and to the key biomolecules have the signature of stability: alternating carbons and nitrogens, or, more generally, alternating systolic and diastolic centers, a pattern that guarantees error-free bond making. Molecules I–IV are examples.

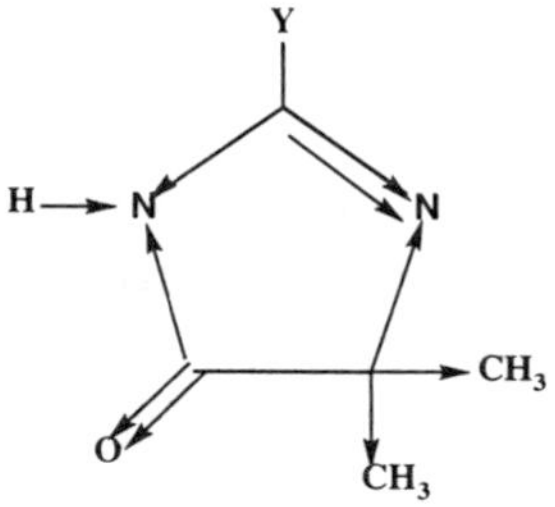

I

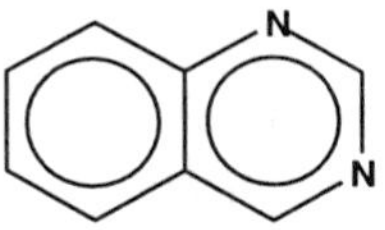

II

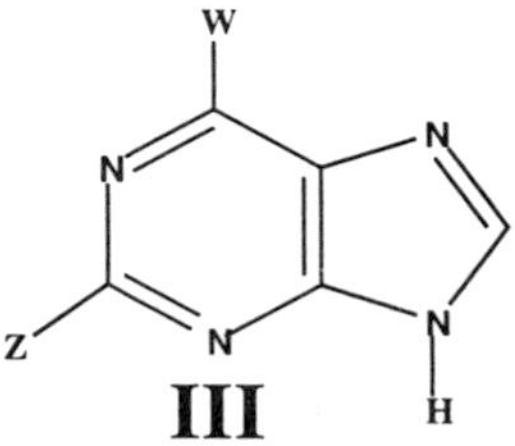

III

	W	Z
Adenine:	NH_2	H
Guanine:	OH	NH_2

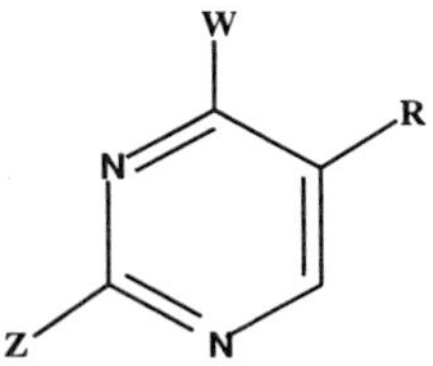

IV

	W	Z	R
Cytosine:	NH_2	OH	H
Uracil:	OH	OH	H
Thymine	OH	OH	Me

In our collection of aza compounds, molecule I is the active segment of imidazolinone herbicides developed by American Cyanamid.[14] Note how the termination of N/C/N alternation by two adjacent diastolic groups, CO and $C(CH_3)_2$, introduces no errors. (b) Molecule II is the most stable constitutional isomer among *ten* contestants listed in the survey of heats of formation reported by Lias et al. (c) The purine and pyrimidine bases of the nucleic acids are shown in their enol forms, III and IV. Note the strategic placement of the electronegative amino and hydroxyl groups on the diastolic centers in the five bases of the nucleic acids.

Here is one main challenge of heterocyclic chemistry: Given any set of heteroatoms, predict the global minimum. For example, we can think of the sequence of first-row atoms plus hydrogen and ask: What is the structure of HBCNOF? We now have the concepts that lead to an immediate prediction: Surround electropositive by electronegative atoms and vice versa. This guarantees an error-free molecule. The choice is obvious: The best structure must be HB═N—CO—F. Given a canned ab initio computer program, an undergraduate can now explore new areas of first-row chemistry by first predicting and then testing by calculation.

8.12 New Stable Covalent Solids

The recipe for producing stable organic molecules and, by extension, stable covalent solids, has been given in Chapters 6 and 7: Arrange atoms so that a less electronegative nonmetal is always surrounded by more electronegative nonmetals. This means starting with diamond and replacing carbons by the more electronegative N or O so that each of the remaining carbons is linked to four N or four O atoms. Recently, Lieber and co-workers synthesized a carbon nitride film (b-C_3N_4) that may be harder than diamond.[15] A related development is the synthesis and characterization of the cyanocarbon HTT shown here with X = CN.[16a]

Hexacyano Tris(imidazo) Triazine
(HTT, C_5N_{12})

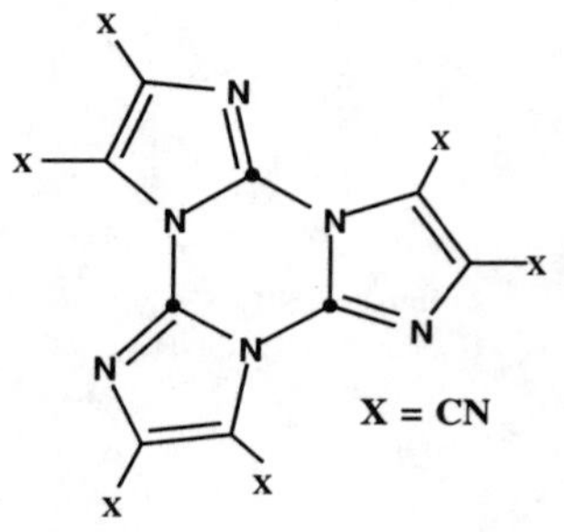

Again, note the characteristic pattern that guarantees minimization of interbond exchange repulsion and optimal T bonding: each of three carbons (dots) is linked to to three electronegative nitrogens. HTT, with a thermal decomposition point exceeding 400°C, has very high thermal stability. In addition, it has remarkable kinetic resistance to combustion. Further improvement can be made by using F instead of CF.

8.13 The Anomeric Confusion

The so-called anomeric effect has been taken to cover two entirely different problems because the hyperconjugation explanation seems reasonable for both:

A conformational isomerism problem. Heteroatom substituents prefer the axial, rather than the equatorial, position in saturated heterocyclic systems such as the pyranose form of a sugar.[17]

A structural isomerism problem. Electronegative atoms prefer to be attached on the same carbon. For example, many assume that the two problems are manifestations of the same electronic principle and focus attention on the energy difference between 1,4- and 1,3-dioxane, the latter being favored by 5 kcal/mol. Unfortunately, these formulations are simply the result of the bias exerted by conventional bonding theory developed in the middle of this century. Specifically, the problem is not the determination of why 1,3- is more stable than 1,4-dioxane. Rather, the real problem is: What is the most stable $C_4H_8O_2$ isomer and why? The T formulas give a direct answer: The acyclic carboxylic acid represents the most error-free combination of four carbons, six hydrogens, and two oxygens! This isomer is 34 kcal/mol more stable than 1,3-dioxane simply because the latter has more errors (Figure 8.12).

The superiority of the acyclic carboxylic acid over the 1,3-dioxane is attributed to a decrease of the kinetic energy in going from the latter to the former. Every time we say that isomer A is more stable than isomer B on account of T bonding, we automatically imply that the greater stability of A is

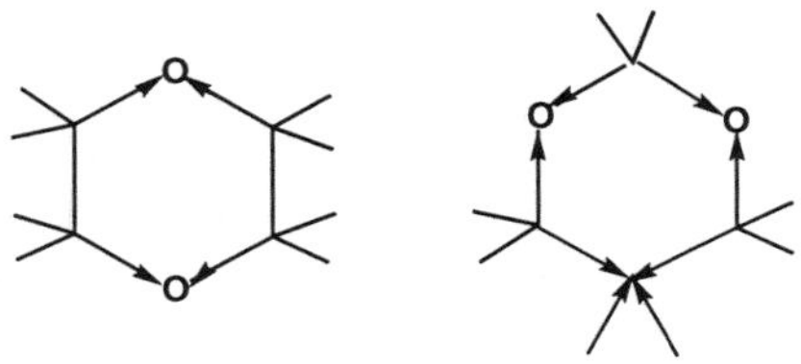

ERRORS:	**8p + 2v**	**6p + 0v**
E (rel.) :	**+39**	**+34**

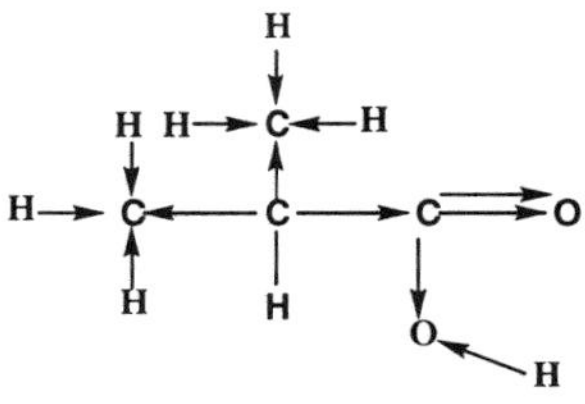

ERRORS:	**1p**
E (rel.) :	**0**

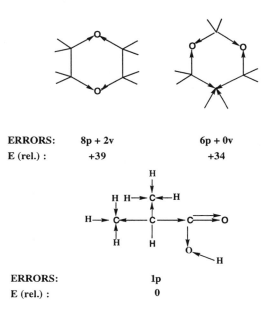

Figure 8.12 The "structural anomeric effect." Note how the carboxylic acid far exceeds the dioxane isomers in stability.

the result of reduced kinetic energy. This means that analyses that focus on potential energy aspects of the problem by employing electron density maps and other electrostatic models are misguided. In this connection, a recent theory of atoms in molecules by Bader[18] has been used by some to reexamine chemical bonding by looking at electron density maps, where atoms now appear as well-defined entities. According to Bader's theory, there exists a bond path (the path of maximum charge density connecting the atoms), and along it there is a bond critical point. In a homopolar bond, the critical point is at the bond center. As the electronegativity of one of the two atoms increases, the bond critical point moves toward the atom of lower electronegativity. In methane, the critical point of the C—H bond lies closer to H. This is consistent with expectations: H is less electronegative than C. However, the charge of H in methane is computed to be negative. This means that "modified" H acts as if it were more electronegative than C, in contradiction to the information provided by the critical point!

It is now coming to the point that every computational chemist has a favorite definition of "atom population." Unfortunately, these indices are dependent on the procedure and the basis set.[19] Some of the problems have been discussed in the recent literature.[20] However, beyond and above mathematical details, the problem with the "electron density map approach" is the erroneous impression that bonding is an electrostatic phenomenon. As discussed before, the crucial quantity insofar as bonding is concerned is the sum of kinetic and potential energies, and T- and I-bonding are principally due to a diminution of the first of these components. In other words, the total energy is equal to "pieces" of electron density multiplied by some factor X that translates density into kinetic plus potential energy, with the key role going to the first, rather than the second, component. Without the knowledge of factor X and without the ability to partition the total energy into its two components, an electron density map is chemically meaningless. Given the electron density map of cyclopropane, can one predict whether stannacyclopropane will be more or less strained? Given the electron density map of *n*-pentane or the populations of its hydrogens, can one explain why this molecule is less stable than neopentane? Given the electron density map of *anti* 1,2-difluoroethane, can one explain why the gauche isomer is more stable? Given the electron density map of tetrahedrane, can one predict the shape of $(SiH)_4$?

What is the best use of computations? This discussion begs the question. Given the tremendous advances of computational chemists, the answer is simple: Use calculations in lieu of experiment whenever justified and fall back on basic principles and the numerical data, whether experimental or computational, to develop concepts. The irony is that one cannot compute the right indices of chemical bonding before understanding chemical bonding! For example, the simple molecule CH_3F has three polarity errors. This means that CH_3F has two options for confronting the problem and neither is good:

Maintain the C—H bond polarity at the expense of exchange repulsion.

Reverse the C—H bond polarity (to avoid exchange repulsion) at the expense of orbital promotion (i.e., unfavorable) CT.

Adoption of the first strategy means that the H atoms will be positive, while adoption of the second strategy means that they will be negative. Now, examination of the computational literature reveals that some calculations get the first and others the second answer. Here is the point: Whether the H atoms are positive or negative is irrelevant to the stability issue. Whatever the case, CH_3F is an erroneous molecule.

The recurrent theme of this work is that thinking and, in particular, *global thinking* (i.e., consideration of the data in their totality) comes first, and computation or experiment comes second. Of course, in an age of specialization, command of the facts across interdisciplinary boundaries is hard. However, the penalty for slacking is also severe. For example, it is all too easy to propose that CF_4 and CO_2 are stable because of electrostatic attraction of the negative ligands and the positive central atom and, in fact, one can easily compute indices (e.g., Mulliken populations, charges) that make this assertion seem plausible. However, a global view of the accumulated data easily persuades one that such certainty is only a mirage. The electrostatic model fails to explain why perfluorination of ethylene is accompanied by shortening of both the C—F and the C═C bond, why the proximal is shorter than the distal C—C bond of 1,1-difluorocyclopropane, why methyl shows the same symbiotic tendency as fluorine, and so on.

"Theoretical organic chemistry" equals "Lewis plus Hückel plus Pauling." We have replaced Lewis/Pauling formulas by T formulas, the latter based on the Pauli exclusion principle. Some time ago, Lennard-Jones, in an address titled "New Ideas in Chemistry," commented that "it is the exclusion principle which plays the dominant role in chemistry. Its all-pervading influence does not seem hitherto to have been fully realized by chemists, but it is safe to say that ultimately it will be regarded as the most important property to be learned by those concerned with molecular structure"! We will next replace Hückel's rule by the VB theory of aromaticity. Finally, the valence rules for metal-containing molecules will take us completely out of the sphere of ideas of these true pioneers. By the way, neither Lewis nor Hückel ever received a Nobel Prize, and the same is true of the pioneers of physical organic chemistry Winstein, Ingold, Bartlett, and Doering.[21,22]

8.14 Molecular Stability Is a "Whole-Molecule" Problem

Chemists divide substituents into π donors (e.g., OH) and π acceptors (e.g., CH═O), reasoning that the former stabilize (or promote the formation of) carbenium ions and the latter stabilize carbanions by conjugation. This line of thought fails to explain an important trend of structural chemistry: The relative stability of RZ, where R is alkyl and Z a general substituent, is independent of the nature of Z. Specifically, the tertiary is much superior to the primary isomer irrespective of whether Z is π donor or π acceptor, as the following data illustrate.

	E_{rel} (kcal/mol)	
Z	$CH_3CH_2CH_2CH_2Z$	$(CH_3)_3CZ$
OH	9	0
CH═O	3	0

Since *t*-butyl is diastolic and *n*-butyl can be regarded as a derivative of the systolic methyl, it follows that the systolic OH prefers attachment to a tertiary carbon, while the diastolic CH═O prefers attachment to a primary carbon. Thus, *if one were to pay exclusive attention to the C—Z bond*, it would be necessary to predict that *n*-Bu-CH═O and *t*-Bu-OH are the most stable isomers. This local model fails because, independent of the nature of the substituent Z, *t*-Bu, which has zero v-errors, is far superior to *n*-Bu, which has six. Two conclusions follow:

(a) The most stable isomer of RZ, where Z is systolic, is always the tertiary form.

(b) The most stable isomer of RZ, where Z is diastolic, is either the primary or the tertiary form. The former occurs when the C—Z bond controls the situation, and the latter occurs when the alkyl fragment dominates the problem. For weak diastolic groups, the second is true, and this is why tertiary RZ is most stable irrespective of Z.

We can expose the frustration (i.e., the two antithetical effects) in RZ where Z is diastolic by choosing Z to be a strongly diastolic group. BH_2 is more electropositive than carbon and it has a 2p hole. This means that B is strongly predisposed to be connected to carbon with an outbound arrow. The limited thermodynamic data, with all their substantial experimental uncertainties, suggest that our expectation is met: In some cases the more highly branched isomer is most stable and in other cases the reverse is true. The relative energies (kJ/mol) of two different sets of isomeric trialkyl boranes are one example.

$(n\text{-Pr})_3B$	$(i\text{-Pr})_3B$	$(n\text{-Bu})_3B$	$(i\text{-Bu})_3B$
0	–15	–13	0

In summary, *two* arrows originating from two odd electrons and forming a closed loop symbolize the interaction of *four* VB configurations (two covalent and two ionic) and are counted as *one* T bond. This formulation respects the fact that one higher order electron-hop (e.g., a 2-e hop) can be also achieved by a series of lower order electron-hops (e.g., two 1-e hops) and the two "pathways" interact (see the crucial concept of VB configuration aromaticity in chapters 12 and 13). When the two atoms connected by one T-bond have different electronegativity, the T bond can be represented by a single arrow which "creates" the lowest energy ionic VB configuration. Minimization of exchange repulsion of T bonds requires that arrows emanating from a central atom are directed either all in or all out. Violation of this condition represents one type of error. As a result, the concept of the T bond leads to the concept of the T formula which, in turn, revolutionizes our understanding of organic structural isomerism. One key point is that a T bond is different from the Lewis electron-pair bond or the Heitler–London (covalent) bond which represent the interaction of the covalent VB configurations. This is why T formulas and Lewis formulas paint entirely different pictures of organic molecular electronics.[23]

References

1. This is why starands are more stable than ketonands: S.J. Cho, H.S. Hwang, J.M. Park, K.S. Oh and K.S. Kim, *J. Am. Chem. Soc.* 118, 485 (1996).
2. G. Olah, *J. Am. Chem. Soc.* 92, 4627 (1970).
3. (a) A.R. Bassindale and P.G. Taylor, in *The Chemistry of Organic Silicon Compounds*, S. Patai and Z. Rappoport, Eds., Wiley, New

York, 1969. (b) S. Zhang, X.-M. Zhang, and F.G. Bordwell, *J. Am. Chem. Soc.* 117, 602 (1995).

4. R. Glaser, G. Sik-Cheung Choy, and M.K. Hall, *J. Am. Chem. Soc.* 113, 1109 (1991).
5. T. Clark, G.W. Spitznagel, R. Klose, and P.v.R. Schleyer, *J. Am. Chem. Soc.* 106, 4412 (1984).
6. A. Greenberg and T.A. Stevenson, in *Molecular Structure and Energetics*, Vol. 3, J.F. Liebman and A. Greenberg, Eds., VCH, New York, 1986, Chapter 5.
7. B.E. Smart, in *Molecular Structure and Energetics*, Vol. 3, J.F. Liebman and A. Greenberg, Eds., VCH, New York, 1986.
8. M.H. Couldwell and B.R. Penfold, *J. Cryst. Mol. Struct.* 6, 59 (1976).
9. T.A. Blake, D.F. Eggers, S.-H. Tseng, M. Lewerenz, R.P. Swift, R.D. Beck, and R.O. Watts, *J. Chem. Phys.* 98, 6031 (1993).
10. M.T. Nguyen, H. Vansweevelt, T.-K. Ha, and L.G. Vanquickenborne, *J. Chem. Soc. Chem. Commun.* 1425 (1990), and private communication.
11. D. Ranganathan, N.K. Vaish, and K. Shah, *J. Am. Chem. Soc.* 116, 6545 (1994).
12. L.M. Gierasch and J. King, Eds., *Protein Folding, Deciphering the Second Half of the Genetic Code*, American Association for the Advancement of Science, Washington, DC, 1990.
13. L. Gong, M.A. McAllister, and T. Tidwell, *J. Am. Chem. Soc.* 113, 6021 (1991).
14. See *Chem. Eng. News*, March 7, 1994, p. 29.
15. C.M. Lieber, *Science* 261, 334 (1993).
16. (a) E.C. Coad, P.G. Apen, and P.G. Rasmussen, *J. Am. Chem. Soc.* 116, 391 (1994). See also (b) L.L. Bircumshaw, F.M. Tayler, and D.H. Whitten, *J. Chem. Soc.* 931 (1954). (c) L. Yu, M. Chen, and L.R. Dalton, *Chem. Mater.* 2, 649 (1990). (d) A.Y. Liu and M.L. Cohen, *Science* 245, 841 (1989). (e) M.R. Wixom, *J. Am. Ceram. Soc.* 73, 1973 (1990). (f) L. Maya and L.A. Harris, *J. Am. Ceram. Soc.* 73, 1912 (1990).
17. R.U. Lemieux and S. Koto, *Tetrahedron* 30, 1933 (1974).
18. R.F.W. Bader, *Acc. Chem. Res.* 18, 9 (1985).
19. (a) J. Cioslowski, *J. Am. Chem. Soc.* 111, 8334 (1989). (b) K.B. Wiberg and C.M. Breneman, *J. Am. Chem. Soc.* 112, 8765 (1990).
20. C.L. Perrin *J. Am. Chem. Soc.* 113, 2865 (1991).

21. K.J. Laidler, *Acc. Chem. Res.* 28, 187 (1995). This paper is obligatory reading for every chemist who maintains any connection with the world of scientific ideas.

22. The suspicion that intellectual satisfaction often takes a back seat to career advancement in modern academia was first confirmed during a discussion of a chemical bonding problem that I had with a famous visitor to our Department when I was an Assistant Professor. Specifically, the guest remarked that using "hyperconjugation" to name an effect that I then proposed would take me back to Pauling, while calling the same phenomenon "orbital interaction" would put a halo of novelty around my head. The message was: Rediscover the wheel but make it seem novel. The vast majority of "theoretical contributions" fall in this category.

23. Many a scientist have been castigated for advancing the right explanation when the popular view was the wrong explanation. Long ago, W.M. Schubert of the University of Washington suggested that certain solution trends popularly attributed to hyperconjugation were actually consequences of differential solvation: W.M. Schubert and W.A. Sweeney, *J. Org. Chem.* 21, 119 (1956); W.M. Schubert and J. Robins, *J. Am. Chem. Soc.* 80, 559 (1958), and subsequent papers.

Chapter 9

The Signature Concept

9.1 What Makes Bond Heterolysis Favorable?

The T formulas permit us to compare the inherent stability of two nonisomeric molecules (e.g., fluoromethane and tetrafluoromethane). We simply count the number of errors: fluoromethane has three and perfluoromethane zero. Consider now the relative stability of the carbocations generated from them by heterolysis of one C—F bond. Clearly, whatever happens at the reactant stage is simply accentuated at the level of carbocationic products. That is, there is an incipient hole on carbon at the reactant stage (due to the $C^{+}F{:}^{-}$ contributor resonance structure) and a full hole on positive carbon (C2p vacant AO) at the product level. The surrounding bonds are subjected to a stronger perturbation as we move from reactants to products, and this simply means that errors at the reactant stage are accentuated at the product level. This leads to the *signature concept:* The ease of heterolysis of a molecule is inversely proportional to the number of errors in its arrow formula. In other words, the stability of an ion is signed on the arrow formula of its precursor.

The signature concept is counterintuitive in the sense that chemists are taught that the more stable a molecule, the less resistant it is to a perturbation (e.g., the less reactive). The signature concept says precicely the opposite:

(a) An error-free R—H molecule is a strong acid because the R—H bond is protected from exchange repulsion with the rest of the pairs (bond pairs and lone pairs) in the entire molecule. Dissociation of a proton creates a lone pair in $R{:}^{-}$. Because $R{:}^{-}$ is also error-free, the lone pair is well accommodated because the observer bond pairs and lone pairs can best respond to the perturbation (departure of the proton). Note that this is a "whole-molecule" argument. One can simplify things and merely say that an error-free R—H is a strong acid because the lone pair of the product anion is spared of exchange repulsion.

(b) The reaction of R—H to form $R{:}^{-}$ plus H^{+} is called acid dissociation. Actually, it is merely one type of bond heterolysis. The same signature concept must apply to all heterolyses. In this light, an error-free R—Br molecule is prone to heterolytic cleavage to form a carbocation, R^{+}, and bromide, $Br{:}^{-}$, for exactly the same reasons that an error-free R—H is a strong acid. Departure of the bromide from R—Br enhances intrabond CT delocalization (already present in the reactant), which further minimizes of interpair exchange repulsion.

The first chapter outlined some reasons for the need for a new theory of chemistry. By now, it should be clear that many of the new VB concepts are greatly at odds with what is featured in the chemical literature, from undergraduate texts to research journals. One spectacular example can be drawn from the area of unimolecular solvolysis of R—L molecules, where L is a good leaving group: The more stable a carbocation, the less reactive it should be. The signature concept says precisely the opposite!

Consider the series of the isomeric primary, secondary and tertiary butanols. The predicted and observed order of thermodynamic stability is 3° > 2° > 1° and the same is predicted to be true of the corresponding protonated alcohols. The signature concept predicts that the more stable protonated alcohol will undergo the most facile heterolysis to yield the more stable carbocation. Thus, we have a relationship between reactant and intermediate stability which is opposite to that normally expected.

Next consider the stability of $H_2C—X^+$, where X is the systolic CH_3, OH, and F. The VB prediction is clear and "different": Assuming zero interbond delocalization ("conjugation"), the stabilizing ability of X should be proportional to its symbiotic tendency measured by the enthalpy of the following redistribution reaction, where we assume that F models the infinitely electronegative ligand of the carbenium ion.

$$2H_2CFX \longrightarrow H_3CF + H_3CX$$

In the absence of data, we have the following order extracted from early ab initio calculational results of Pople and his school[1] pertinent to the preceding reaction but with F replaced by X:

$$OH > F > CH_3$$

In this light, the statement "OH is the best stabilizer of a carbenium ion" (among OH, F and CH_3)[2] means that the infinitely electronegative ligand experiences the least exchange repulsion by the rest of the carbon ligands. This produces a counterintuitive rule: *The more stable the carbenium ion, the faster its capture by nucleophilic solvent,* the latter acting as the missing infinitely electronegative ligand. This explains the results of Richard et al., who found that the more stable $RHC—OEt^+$ reacts faster with H_2O than the less stable $RHC—pAn^+$ (pAn = p-anisyl).[3] We suggest that the stabilization of carbocations by what chemists call "π donors" may have less to do with conjugation and more to do with exchange repulsion of confined bond pairs and lone pairs. This is why carbenium ion stability and reactivity go hand in hand.

9.2 Acidity

The generation of an anion by the departure of proton creates a free electron pair that suffers on two counts:

(a) Exchange repulsion by the surrounding pairs.

(b) High interelectronic repulsion not compensated by nucleus–electron attraction.

The strategy for coping is Rydbergization of the anionic pair. As a result of (a) and (b), we come to view the conjugate base of a protonic acid as a hybrid of a valence (VL) and a nonvalence (RY) resonance structure, as shown schematically:

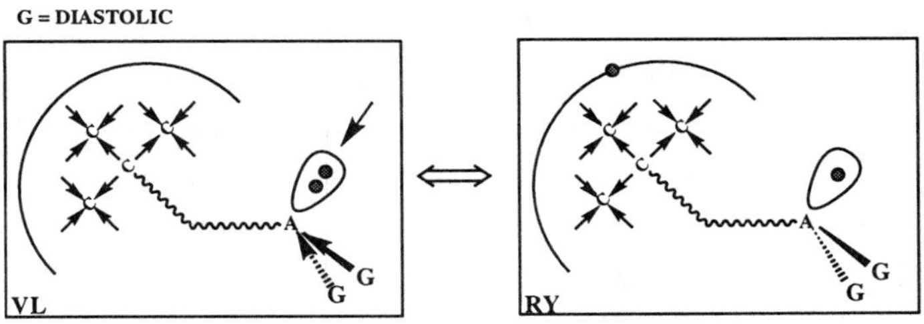

We can now formulate a recipe for anion stabilization:

To stabilize the VL contributor: The strategy is based on the recognition that the presence of a lone pair renders an anion a systolic group. Hence, this will be stabilized by either diastolic group. In other words, we must attach a substituent that has an arrow pointing toward the anionic center ("σ donor") in the T formula in direct contradiction of the notion of inductive effects.

To stabilize the RY contributor: The strategy is to attach either directly on the anionic center or on a neighboring atom a group with an "electron-deficient surface" or with low-lying vacant nonvalence AOs, which can act as a receptor of the Rydbergized electron. An excellent choice of the former type is the *t*-butyl group because its T formula shows it to be an error-free group with a "protonic surface." The latter is a consequence of the fact that three systolic methyls are ideally combined with a diastolic central carbon.

A large number of known experimental trends of gas phase acidity[4] can be explained by assuming preeminence of the VL contributor. We state three conclusions.

(a) The *systolic* H_3C, H_2N, HO and F groups are *not* good anion stabilizers despite being electronegative.

(b) Atom diastolicity increases going down a column for three reasons. First, electropositivity increases. Second, the ns lone pair, which ordinarily acts to render a group systolic, goes inactive (ns contraction). Third, the active nonvalence unoccupied space neutralizes the effect of any remaining valence electron pairs. Hence, a group should become increasingly capable of stabilizing an anion as we proceed from the first to lower rows within the same column. This is what the available gas phase and computational data show.

(c) Diastolic *t*-butyl is far superior to systolic methyl in acidifying water. In the gas phase, *t*-butanol is a stronger acid than methanol, simply because *t*-butoxide is error-free (zero errors) while methoxide is error-plagued (three errors).

All three conclusions can be reformulated using the signature concept. We use this version to explain other important trends. For example, why is a carboxylic acid a stronger acid than an alcohol? A better question is: Why does placement of a carbonyl next to OH enhances the acidity of the latter? The answer lies in the comparison of the pertinent structural isomers (e.g., H—CO—CH_2—OH vs. H—CH_2—CO—OH). The former has two polarity errors (the two CH bonds next to OH) and it is a weaker acid, while the latter has no errors and is a stronger acid. Why is methane a stronger acid than ethane? Because the first has zero but the latter three v-errors. Why is dihydrogen a stronger acid than many carbon acids discussed in textbooks? Because hydride fears no exchange repulsion.

The uncertainty regarding acidity can be best illustrated by the following menu of explanations:

Koppel et al. attribute the enhanced acidity of $(CF_3)_3CH$ relative to bicyclic analogues in which Bredt's rule applies to hyperconjugation.[5]

Bordwell and Satish say that resonance is important in determining acidity.[6]

Speers et al. say that reactant destabilization explains the acidity order $DMSO_2$ > DMSO > DMS.[7] This suggestion follows the invocation of inductive and field effects to explain enhanced acidity.[8]

Obviously, these many explanations cannot all be right. The common denominator of all these diverse experimental data is that, in every case, the superior acid is the one which is error-free! The point is illustrated by this explicit drawing of the arrow formulas of the champion acids.

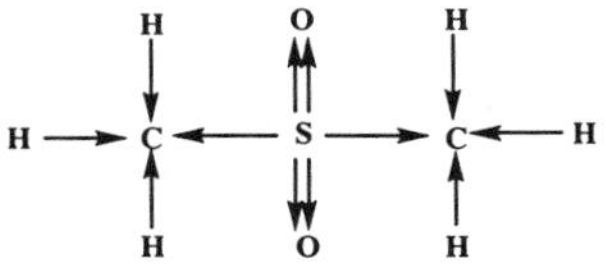

ERROR-FREE

COMPARE WITH

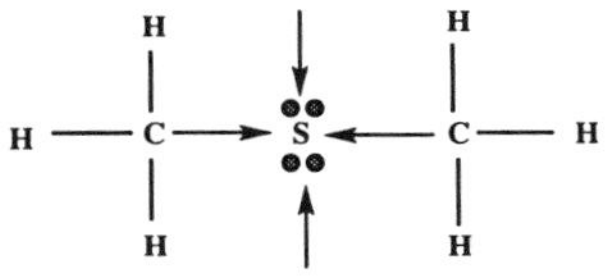

6 p (CH) ERRORS

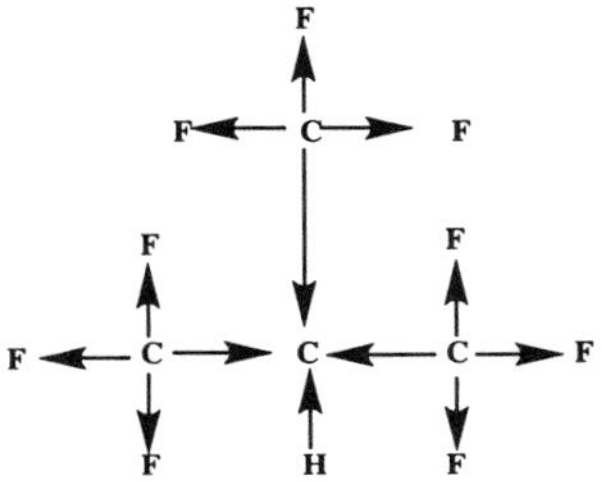

ERROR-FREE

The bicyclic analogue of $(CF_3)_3CH$ has C-C p-type errors (which we normally disregard for simplicity), and DMS has four v-type errors.

The signature concept assumes zero *interbond* (as distinct from intrabond) delocalization. The calculations (but not the interpretations) of Speers et al. show exactly this. To quote the authors: "We find no evidence...for an appreciable delocalization of charge from the anionic carbon in the anion."[7]

The importance of the RY contributor is inferred from a key observation of Brauman and Blair[4c]: *t*-Bu—CH_2—OH is a stronger acid than *t*-BuOH even though the former contains two p-errors in its arrow formula. The protonic surface of the t-Bu group stabilizes electrostatically the Rydbergized electron of the anionic center. These data suggest that the trends we rationalized using the signature concept materialize because the contribution of RY remains constant. Other pieces of the puzzle that suggest that RY is important are the enhanced acidity of the chloroacetic acids and the slightly higher acidity of methanol (with three p-errors) over water (with no errors).

The hypothesis that the RY contributor is an important determinant of carbanion stability can be further tested by focusing on the conformational isomerism of carbanion analogues such as amines. Although the effect will be weaker, we expect that electrostatic attraction of the nitrogen lone pair by the protonic surface of t-Bu may very well dictate an eclipsed geometry in *t*-Bu—

H_2C-—NX_2, where X are chosen to enforce sp^3 hybridization (rather than dehybridization) on N. Recent reports of eclipsed ethane derivatives of this type[9] support the hypothesis that *t*-Bu acts effectively as a charged sphere capable of stabilizing vicinal electron pairs.

Finally, we note that the signature concept explains why the error-free DI-CH_3, where DI is a diastolic NO_2, is a stronger acid (by 5 pK_a units) than the error-plagued 1,3-cyclopentadiene, which nevertheless yields an aromatic anion upon deprotonation. Thus, we have one more challenge to the conventional view that π acceptors stabilize carbanions by resonance. While CT delocalization (i.e., incipient I-bonding) may well be important, the acidification of methane by π acceptors can also be explained by assuming zero CT by simply recognizing that substituents of the latter type are diastolic groups and that replacing a hydrogen by a diastolic group in methane produces an error-free molecule.

9.3 A Mechanism for Site-Specific Chemical Modification of Nucleic Acids and Proteins

The signature concept finds a significant application in the field of site-selective reactivity of macromolecules with effector molecules such as antibiotics. Specifically, if a molecule is represented by its T formula and if this is taken to represent an exclusion wave, every perturbation that enhances the *intrabond* CT denoted by one or more arrows will enhance the entire exclusion wave. We propose that this is the electronic basis of intra- and intermolecular communication.

The characteristic feature of nucleic acids and proteins as well as of effector molecules is that they are made up of H, C, N, and O. Of these, amino-N and and carbonyl-O are strong systolic centers on account of both electronegativity and the presence of lone pairs. When the hydrogens attached to N or the lone pairs present in O are tied up by a hydrogen-bonding partner, the exclusion wave defined by the T formula is accentuated, with direct consequences for the reactivity of the molecule. We call this mechanism exclusion wave (EW) activation via H-bonding and we use the acronym EWAH to denote it.

When do we expect strong EWAH? Whenever oxygens and nitrogens of the effector molecule are linked by an exclusion wave. For this condition to exist, there must be an intervening *odd* number of carbon atoms. Our scenario is simple:

The nucleic acid or protein binds specifically the effector molecule through H bonds.

The H bonds enhance the exclusion wave already present in the effector.

A reactor site of the effector becomes activated and reacts.

A recent paper by Warpehoski and Harper describes the acid-dependent electrophilicity of cyclopropylpyrroloindoles.[10] Their studies were motivated by a desire to understandthe potent antitumor agent adozelesin, a highly selective alkylator of the N3 atom of adenine in certain sequences of the B form of DNA. How does this happen in vivo? Our scenario is indicated below. Note that the nitrogens and oxygen of the effector are separated by an odd number of intervening carbons.

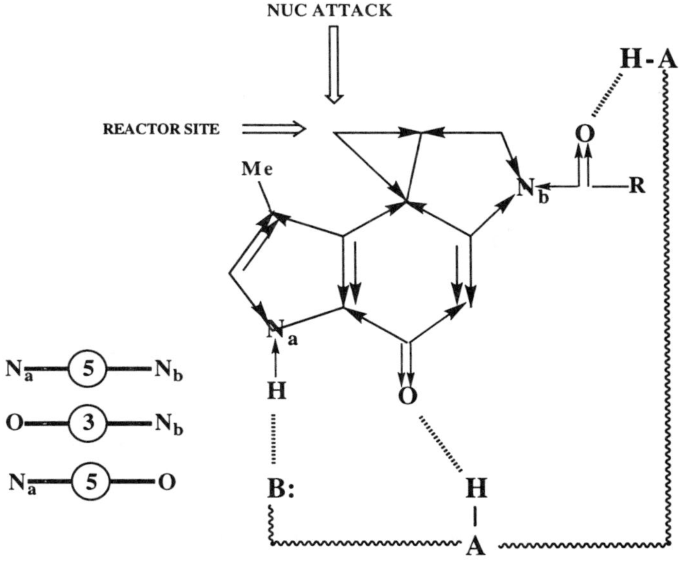

The first step is selective hydrogen-bonding of adozelesin with a DNA sequence: Tying up any of its nitrogens and oxygens by hydrogen bonds triggers an exclusion wave represented by the arrows. This activates the reactor site (i.e., the cyclopropane ring) in a way that directs the nucleophile to the *diastolic* cyclopropane carbon. Thus, we have a linkage between specific macromolecule–effector interaction and specific effector reactivity. In the work of Warpehoski and Harper, activation is effected by carbonyl protonation rather than by cooperative H-bonding and its consequences are precisely the same as those just described.

If this scenario is what nature has designed, the door is opened to the deliberate construction of effectors capable of strong, cooperative EWAH with selected sequences of macromolecules in which not only the spacing of O and N is right but the entire molecule is error-free (adozelesin is not) and the reactor site may be any one of multiple reactive fragments capable of responding to an exclusion wave.

9.4 Nucleic Acids and Electron Transfer

The signature concept says that if a molecule is represented by its T formula and if this is taken to represent an exclusion wave, every perturbation that enhances *intrabond* CT (denoted by one or more arrows) will enhance the entire exclusion wave. An electron moving down a polymer can be regarded as a perturbation that follows the path that constantly enhances the exclusion wave, with some drop-off with distance from the perturbing electron.

Consider the nitrogen bases of DNA as an array of stacked π aromatic systems that can promote electron transfer (ET). The key feature of the nitrogen bases is their relatively error-free T formulas. Two examples are given.

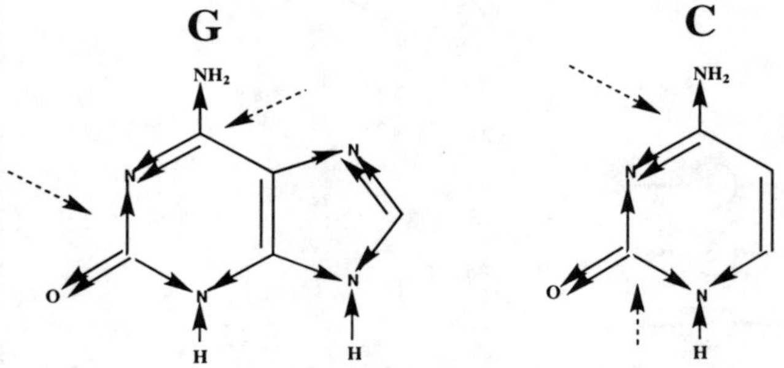

DASHED ARROWS INDICATE SITES OF PREFERRED ATTACHMENT OF ONE ELECTRON

The existence of a multipolar σ frame sets the stage for a cooperative mechanism of ET via the DNA nitrogen bases (N bases). This can be viewed as a perturbation that enhances the multipolar character of the N base to which the electron was added. This enhances H-bonding of the affected base pair. In turn, this causes a local constriction of the duplex, which enhances the H-bonding and the multipolar character of the next base pair. As a result, the electron can hop to the next base, which is ready to receive it because of its augmented multipolar character. The argument is illustrated schematically as follows.

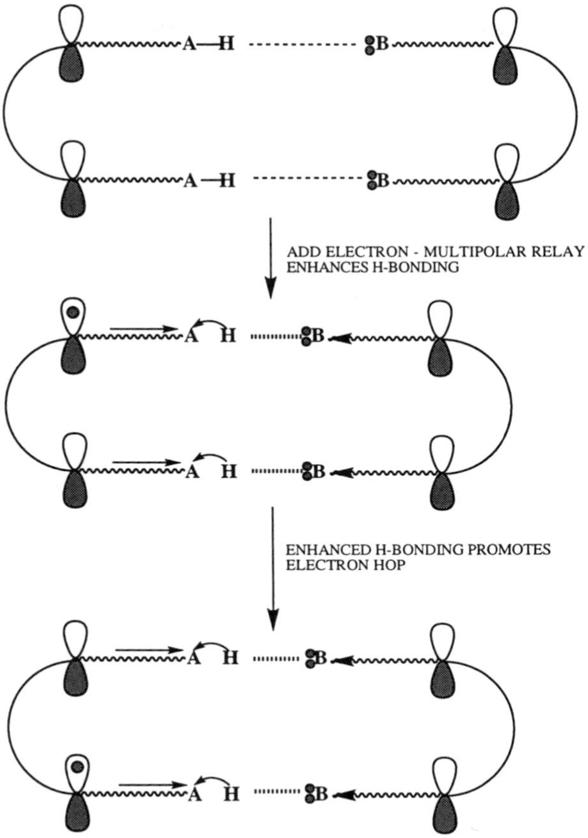

Recent work suggests that DNA can act as a molecular wire.[11]

References

1. L. Radom, W.J. Hehre, and J.A. Pople, *J. Am. Chem. Soc.* 93, 289 (1977). The amino group is left out because it seems to have a π-conjugation component that puts it ahead of hydroxyl as a carbenium ion stabilizer.
2. S. Hoz and J.L. Wolk, *Tetrahedron Lett.* 31, 4085 (1990).
3. J.P. Richard, T.L. Amyes, V. Jagannadham, Y.-G. Lee, and D.J. Rice, *J. Am. Chem. Soc.* 117, 5198 (1995).
4. (a) E.A. Brinkman, S. Berger, and J.I. Brauman, *J. Am. Chem. Soc.* 116, 8304 (1994). (b) M. Born, S. Ingemann, and N.M.M. Nibbering,

J. Am. Chem. Soc. 116, 7210 (1994). (c) J.I. Brauman and L.K. Blair, *J. Am. Chem. Soc.* 92, 5986 (1970).

5. I.A. Koppel, V. Pihl, J. Koppel, F. Anvia, and R.W. Taft, *J. Am. Chem. Soc.* 116, 8654 (1994).
6. F.G. Bordwell and A.V. Satish, *J. Am. Chem. Soc.* 116, 8885 (1994).
7. P. Speers, K.E. Laidig, and A. Streitwieser, *J. Am. Chem. Soc.* 116, 9257 (1994).
8. D. Holtz, *Prog. Phys. Org. Chem.* 8, 1 (1971).
9. J.E. Anderson, D. Casarini, A.I. Ijeh, L. Lunazzi, and D.A. Tocher, *J. Am. Chem. Soc.* 117, 3054 (1995).
10. M.A. Warpehoski and D.E. Harper, *J. Am. Chem. Soc.* 116, 7573 (1994).
11. C.J. Murphy, M.R. Arkin, Y. Jenkins, N.D. Ghatlia, S. Bossmann, N.J. Turro, and J.K. Barton, *Science* 262, 1025 (1993).

Chapter 10

T Formulas and Molecular Association

10.1 Promotional Energy and the Connection of Interatomic and Intermolecular Bonding

Here are two critical reactions that reintroduce important VB concepts:

$$C(CH_3)_4 \longrightarrow C + 2H_3C{-}CH_3 \qquad \Delta E = +191 \text{ (kcal/mol)}$$
$$Pb(CH_3)_4 \longrightarrow Pb + 2H_3C{-}CH_3 \qquad \Delta E = -6 \text{ (kcal/mol)}$$

These data illustrate a central dogma: An electronegative nonmetal atom opts for becoming excited to reap the benefit of forming the maximum number of T bonds. The willingness to become excited is due to the ability of the electronegative nonmetal atom to engender strong overlap interaction with some other atom. The preceding equation involving the nonmetallic carbon illustrates this state of affairs. Note the extremely high reaction endothermicity. By contrast, a metal atom lies at the antipodes. Its low electronegativity does not allow it to engender strong overlap interaction. As a result, it opts to remain de-excited, even though, in this way, it misses the opportunity to form the maximum number of bonds. This is the case with the metallic lead in the other equation.

Intermolecular bonding of hydrocarbons is one of the weakest forms of bonding brought about by the inductive and dispersive components of static bonding. The third component — namely, classical Coulomb interaction — is assumed to be negligible. By direct analogy with the interatomic case, we can now formulate a hypothesis: *The enthalpy of hydrocarbon association depends primarily on the promotional energy required by induction and dispersion.* The lower the promotional energy, the stronger the enthalpy of association. There is now a twofold question:

What is the optimal promoted state of a hydrocarbon?
How do we determine the associated promotional energy?

The T formulas give answers to both questions.

10.2 T Formulas, Molecular Promotion, and Enthalpy and Entropy of Association

The T formula of methane is shown in Figure 10.1a. We now envision the molecule as a tetrahedron inscribed in a sphere so that each hydrogen atom lies on the surface of the sphere. There are four units of *formal* negative charge at the center and four units of *formal* positive charge at the surface of the sphere, as depicted in Figure 10.1b. Assuming that the intermolecular bonding of methane depends primarily on the field exerted by the sphere surface atoms, we symbolize

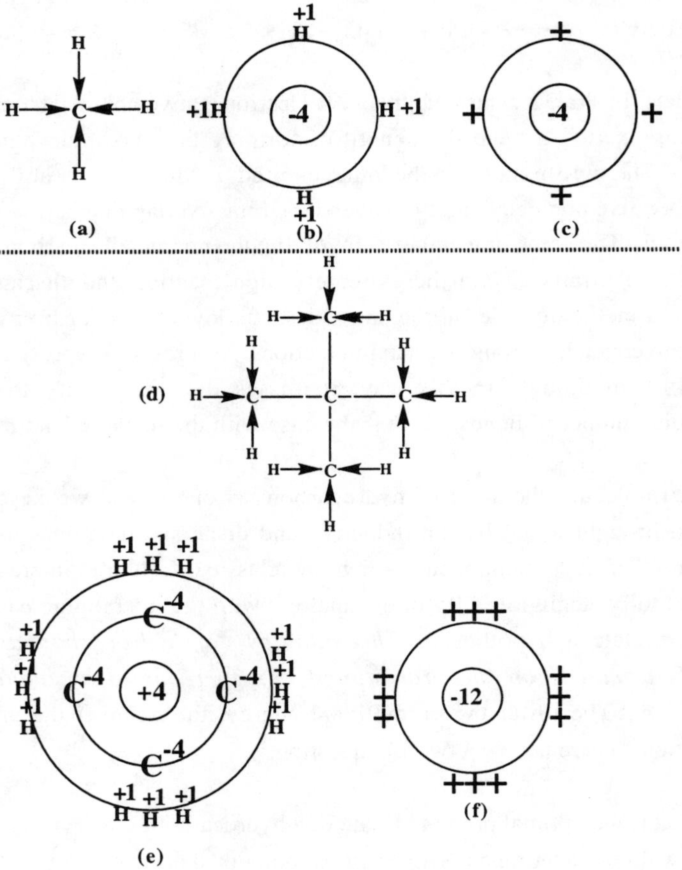

Figure 10.1 Representations of methane and neopentane according to the concepts of Chapters 6–8.

ground methane as a sphere with a positively charged surface. This is done by drawing a circle and putting the total number of unit charges deduced from the arrow formula at the center, as in Figure 10.1c. The same procedure can be applied to neopentane (Figure 10.1d–f).

How can one methane attract a second methane? Although a variety of scenarios could be offered, we formulate a mechanism that contains the features we believe to be essential. Specifically, we assume that one methane presents itself as an *unpromoted*, uniformly positive sphere (protonic hydrogens), while the second enters as a *promoted*, uniformly negative sphere (hydridic hydrogens). The result is strong omnidirectional Coulomb attraction. The promotion is effected by starting with the T formula of methane and reversing all arrows. This is equivalent to introducing *four polarity (4p) errors*, which represent the energetic price of promotion. The ground and promoted methanes are shown in Figure 10.2a, whereas Figure 10.2b represents their Coulombic interaction. Exactly the same strategy can be used by two neopentane molecules to attract

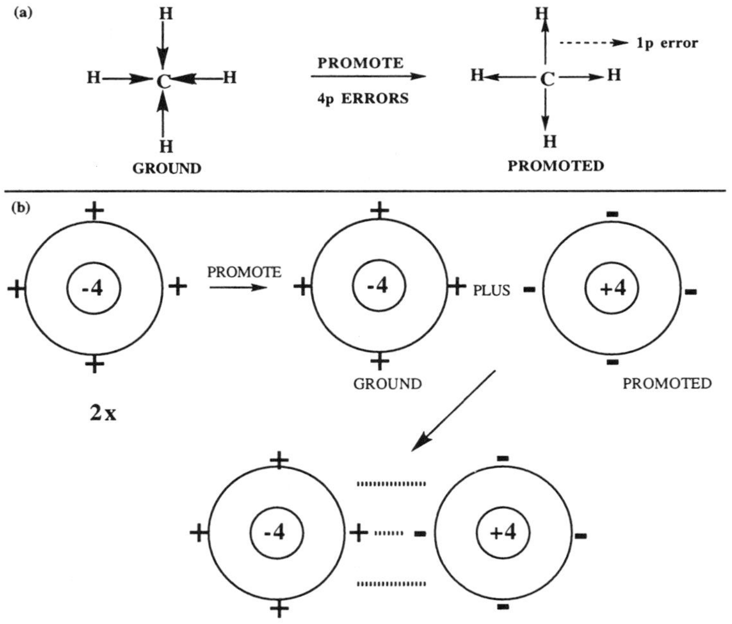

Figure 10.2 Promotion and intermolecular bonding of two methane molecules. The promotional energy is proportional to 4p (four polarity) errors corresponding to four outbound arrows from the more electronegative C to the more electropositive H.

each other. The requisite promotion is now 12p, as illustrated in Figure 10.3. As in the case of methane, the goal of promotion is to turn a positive (protonic) to a negative (hydridic) sphere surface so that one ground and one promoted molecule can attract each other.

Let us now shift the focus to a comparison of two hydrocarbon isomers and inquire which of the two associates with the least promotion. The best example is the neopentane/*n*-pentane comparison. Ground neopentane has an error-free T formula, while ground *n*-pentane has a T formula with eight vicinal (8v) errors, as shown in Figure 10.4. Intuitively, one expects that the more stable isomer will be more "polarizable." This is exactly what we find by replacing "polarizability" by "promotion." Specifically, promotion of *n*-pentane introduces 12p errors but removes 8v errors (Figure 10.4). This means two things:

Promotion costs less in *n*-pentane.

Because promotion is the yardstick for association, n-pentane must have a larger heat of fusion and a larger heat of vaporization. This is in accord with experiment.

	Fusion		Vaporization	
	ΔH (kcal/mol)	ΔS [cal/(deg mol)]	ΔH (kcal/mol)	ΔS [cal/(deg mol)]
n-Pentane	2.0	14.0	6.3	21.0
Neopentane	0.8	3.0	5.4	19.2

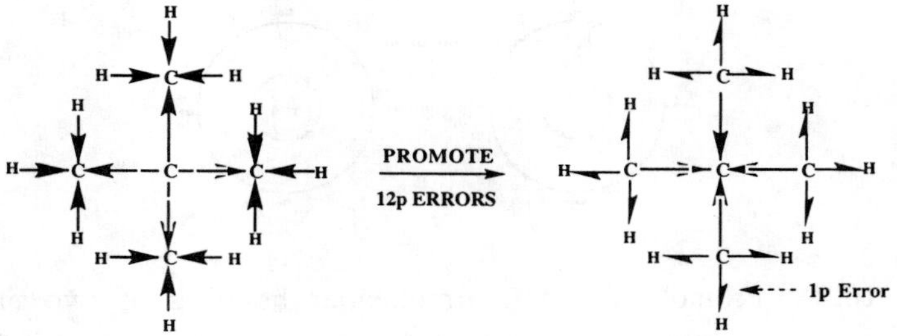

Figure 10.3 Promotion of neopentane is equivalent to creation of 12p errors.

We have a rule: *The greater the error count of an isomer, the lower its relative stability, the lower the promotional energy required for intermolecular bonding, and the more negative the enthalpy of association.*

In going from gas to liquid to solid, there is an associated "freezing" of internal rotational degrees of freedom and an accompanying diminution of entropy. This vibrational entropy loss (internal modes of rotation are counted as vibrational modes) becomes more prevalent as the shape of the molecule changes from spherical to linear.[1] Rotational entropy (i.e., the entropy ascribed to the rotation of the molecule as a whole) works in the same direction: A spherical molecule can spin about any one of three Cartesian axes while a linear cylindrical molecule can spin only about the cylinder Cartesian axis in its lattice site. As a result, rotational–vibrational entropy, or, simply, entropy, favors aggregation of spherical over linear isomers. Recognizing that the melting points (MPs) and the boiling points (BPs) of isomers are determined by the free energy of fusion and vaporization — that is, by the interplay of enthalpy and entropy — we arrive at the following scenario:

(a) In comparing isomers that have radically different shapes, entropy dominates, so that the *higher-symmetry* isomer has the higher MP because the solid is entropically stabilized. In such comparisons, the variation of MP and BP in a pairwise comparison of isomers is *discordant.*

(b) In comparing isomers with modestly different shapes, enthalpy dominates so that the *more stable* stable isomer has lower MP as well as BP. In contrast to (a), the variation of MP and BP in a pairwise comparison of isomers is *concordant.*

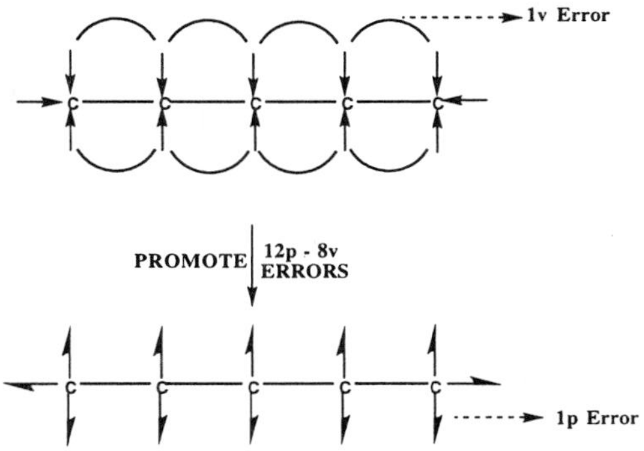

Figure 10.4 Promotion of n-pentane is equivalent to creation of 12p - 8v errors. Vicinal errors in the ground state (indicated by arcs) are undone in the promoted state.

10.3 The Relationship of Isomer Stability, Isomer Shape, Melting Points, and Boiling Points

"The stronger the intermolecular bonding, the higher the melting point of a solid or the boiling point of a liquid." This nearly universal credo is seemingly supported by correlations between MP, BP, enthalpy of sublimation, and enthalpy of vaporization.[2] Here are two contradictory interpretations of thermochemical data:

(a) Linear alkanes (e.g., *n*-pentane) have higher boiling points than spherical alkanes (e.g., neopentane) because the surface-to-volume ratio is minimum in a sphere and van der Waals bonding becomes increasingly favorable as molecular surface area increases.

(b) Spherical alkanes have higher melting points than linear alkanes because they "pack" better, like hard spheres.

Table 10.1 Physical Data for Isomeric Alkanes Illustrating VB Concepts

		MP (°C)	BP (°C)	E_{rel} (kcal/mol)
1a	Neopentane	–17	10	0.0
1b	n-Pentane	–130	36	5.0
2a	Hexamethylethane	101	106	0.0
2b	n-Octane	–56	126	4.0
3a	Propene	–185	-48	0.0
3a	Cyclopropane	–127	-33	8.0
4a	*trans*-2-Butene	–106	1	0.0
4a	Cyclobutane	–50	13	8.0
5a	Tetramethylethene	–74	73	0.0
5b	*trans*-2-Hexene	–133	68	5.0
6a	Hexamethylbenzene	165	265	0.0
6b	1-Phenylhexane	–62	226	5.0

The VB concepts can be illustrated using the data of Table 10.1 For example, items 1 and 2 show a discordant variation of MPs and BPs because we are comparing linear with spherical isomers. Enthalpic considerations predict that the less stable isomer will make stronger intermolecular bonds on account of lower promotional energy. The prediction is reversed in the case of the solid–liquid transition because entropy dominates enthalpy according to our hypothesis (vide supra) and stabilizes preferentially the higher symmetry spherical isomer in the solid state. This entropic stabilization is a consequence of shape, rather than stability. If entropy were to remain constant, *n*-pentane (*n*-octane) would have higher MP *and* higher BP than neopentane (hexamethylethane) because the enthalpy of association is more negative for the less stable isomer. This trend is called *entropic discordance.*

Items 3 and 4 show a *concordant* variation of MPs and BPs because we are comparing linear and cyclic isomers. Enthalpic considerations predict that the less stable isomer will make stronger intermolecular bonds on account of lower promotional energy. Since enthalpy now dominates entropy according to our hypothesis, increased stability correlates with lower MP as well as with lower BP. Since, however, the lowering of the MP is significantly greater than the lowering of the BP, it is possible that entropic effects (promoting a higher MP for the cyclic isomer) still play a role. This trend is called *enthalpic concordance.*

Items 5 and 6 are exceptional cases in which the lower energy isomer has both a higher MP and a higher BP. In this condition, called entropic concordance, entropy becomes all-important, dictating both MP and BP, that is, Trouton's rule breaks down.

In summary, MPs and BPs have to do with the interplay of enthalpy and entropy. In turn, enthalpy is related to the interplay of molecular promotion and intermolecular bond making and, thus, to isomer stability. On the other hand, vibrational entropy is related to isomer shape, and rotational entropy differentiates sharply between highly symmetrical and unsymmetrical isomers.

Next, we seek to develop a relationship between isomer stability and melting and boiling points in hydrocarbon derivatives in which systolic methyl groups are replaced by other systolic groups (e.g., NH_2, OH, F). Indeed, if we were to tentatively assume that classical dipolar interaction and hydrogen bonding introduced by the electronegative NH_2, OH, and F are of secondary importance, we would end with the same predictions regarding the variation of the MPs and BPs of isomeric amines, alcohols, and fluoroalkanes that we made for isomeric alkanes:

Table 10.2 Physical Data for Isomeric Alcohols

Isomer	ΔH_f (kcal/mol)	MP (°C)	BP (°C)
$H_3CCH_2CH_2OH$	–61	–127	97
$H_3CCH(OH)CH_3$	–65	–90	82
$H_3CCH_2CH_2CH_2OH$	–66	–90	118
$(CH_3)_3COH$	–75	26	82

The least stable isomer has the higher BP. This has enthalpic origin, which is traced to reduced promotional energy of the least stable isomer.

The most symmetrical (and more stable) isomer has the highest MP. This has entropic origin, which is traced to the quasi-spherical shape of the most stable isomer.

The data of Table 10.2 confirm these expectations. The trend for the isomeric alcohols parallels the trend for the corresponding alkanes obtained by replacing OH by CH_3.

The heats of formation and the MPs and BPs of the isomeric hexanes given in Table 10.3 lead us to appreciate the sharp dependence of entropic stabilization of the solid phase on shape. As expected, the BPs increase as stability decreases. The MP data are incomplete but still instructive: an effectively spherical shape cannot be attained unless there is a highly symmetrical distribution of alkyl groups that populates the surface of an inscribing sphere with a high density of hydrogen atoms. The branched hexanes c and d of Table 10.3 fail to meet this condition. As a result, entropy no longer dominates enthalpy, and the less stable straight chain isomer ends up having a higher MP. This confirms our hypothesis: Entropy dominates enthalpy only in comparisons of isomers that have radically different shapes.

Bottom line: To understand intermolecular bonding, we must first determine whether a pairwise comparison of isomers is entropy-discordant, enthalpy-concordant, or entropy-concordant with respect to the MP/BP variation.

Table 10.3 Physical Data for Isomeric Hexanes

Isomer	ΔH_f (kcal/mol)	MP (°C)	BP (°C)
a. $H_3C(CH_2)_4CH_3$	–40	–95	68
b. $H_3CCH_2CH(CH_3)CH_2CH_3$	–41		63
c. $H_3CCH(CH_3)CH_2CH_2CH_3$	–42	–154	60
d. $H_3CCH(CH_3)CH(CH_3)CH_3$	–42	–129	58
e. $H_3CC(CH_3)_2CH_2CH_3$	–44		50

Only after this has been done can we evaluate whether a trend is enthalpy- or entropy-dominant and how this status relates to molecular electronic structure.

We have focused on the association of hydrocarbons because these pose the most challenging problem to electronic structure theory. Intermolecular bonding of polar molecules is due to overlap-independent static bonding, namely, classical Coulomb attraction plus a combination of induction and dispersion. Because only the latter two components require strong molecular promotion and because the first dominates, the interplay of promotion — "bond making" — is no longer crucial, and the problem is much simpler, in a qualitative sense. On the other hand, the issue of overlap-dependent CT bonding involving interaction of lone pairs with σ bonds has been controversial. An argument in favor of preeminence of the CT mechanism could be made under the following conditions:

The molecular deformation energy is high.

Magic numbers arise in cyclic structures.

The hypersurface has one deep minimum (i.e., association is strongly stereoselective).

An overview of experimental and computational data on bulk liquids and small molecular clusters[3] suggests that these three conditions are *not* met. This leads us to concur with the viewpoint championed by Buckingham:[4] Hydrogen bonding is mainly a classical electrostatic phenomenon. While most of the binding comes from classical attraction, weak stereoselectivity can still be the result of minimization of exchange repulsion plus maximization of CT. If this is the case, the global minimum must feature an aromatic or, as a compromise to static bonding, a nonaromatic array of atoms. What limits this type of investigation is lack of data on the structures of small molecular clusters.

References

1. K.B. Wiberg, *Laboratory Technique in Organic Chemistry*, McGraw-Hill, New York, 1960, pp. 77–79.
2. M.S. Westwell, M.S. Searle, D.J. Wales, and D.H. Williams, *J. Am. Chem. Soc.* 117, 5013 (1995).
3. (a) S.S. Xantheas and T.H. Dunning Jr., *J. Chem. Phys.* 99, 8774 (1993); S.S. Xantheas, *J. Chem. Phys.* 100, 7523 (1994). S. Scheiner, *Acc. Chem. Res.* 27, 402 (1994).
4. (a) A.D. Buckingham in *Physical Chemistry — An Advanced Treatise*, D. Eyring, D. Henderson, and W. Jost, Eds., Academic Press, New York, 1970. G.J.B. Hurst, P.W. Fowler, A.J. Stone, and A.D. Buckingham, *Int. J. Quantum Chem.* 29, 1223 (1986).

Chapter 11

Why "Crowded" Rotational Isomers End Up Being Global Minima

11.1 The Anti Effect and the Syn Effect

The current popularity of one-electron orbital models as interpretative tools owes much to the Woodward–Hoffmann rules but also to numerous applications of the perturbation MO (PMO) model[1] to structural organic chemistry in the late 1960s and early 1970s.[2] Why is the gauche form more stable than the anti form of 1,2-difluoroethane? Why is *cis*- more stable than *trans*-1,2-difluoroethylene? Why is the barrier to pyramidal inversion of phosphine greater than that of ammonia? Why is isobutane more stable than *n*-butane? Seduced by the simplicity of the PMO method, we sought to answer these questions without first developing the necessary conceptual tools and by falling back on old ideas: steric effects, hyperconjugation, aromaticity and second-order Jahn–Teller (SOJT) effect. We will now argue that the available data are inconsistent with these concepts. Then, we will explain the counterintuitive structural patterns of organic chemistry by using the new VB concepts.

A mainstay in the arsenal of conceptual tools is the "steric effect": Nonbonded groups prefer to be as far as possible away from each other. The textbook example is 2-butene: trans is more stable than cis. However, isobutene, where the two methyls are closest together, is the most stable isomer, and the cis form of XHC═CHX, where X is a systolic group (like F and OR), is more stable than the trans form!

Next, consider "negative hyperconjugation." Assuming that a trans interaction of a σ C—H bond with a vicinal C—F antibond is most favorable when the two are in a trans arrangement,[2] one may explain why the cis form of 1,2-difluoroethylene is more stable than the trans. However, the same argument predicts that cis must have longer C—F and shorter C═C bonds than the trans. Spectroscopic and ab initio computational results show the opposite.[3] Another prediction is that lone pairs should be oriented trans to vicinal vacant antibonding orbitals. For example, hydrogen peroxide (HOOH) is expected to have a dihedral angle of 60° (sp^3 oxygens, staggered geometry). Instead, this angle is roughly 120° (i.e., HOOH is eclipsed). Hydroxylamine (H_2N—OH) is expected to prefer a "W," but is computed to have a "Y" conformation.[4] Whereas cis is expected to be more stable than trans HN═NH, thermochemical data show the latter to be more stable by 4 kcal/mol.

Yet another model is the SOJT approach to molecular stereochemistry.[5] Here, the recipe is: Start with a reference (e.g., linear) geometry and compute the MOs. A geometrical distortion that causes strong mixing of the HOMO and the LUMO is favored. For example, linear acetylene and planar ethylene are predicted to prefer a trans over a cis deformation. Addition of two electrons creates two molecules isoelectronic to diimide and hydrazine, respectively, and the prediction is reversed — both should prefer cis over trans bending. This is the opposite of what is found, namely: trans is strongly preferred over cis HN═NH. Pearson has presented SOJT applications and pointed out failures of the model.[5]

Finally, consider the notion of aromaticity. We once argued that *cis*- and *trans*-1,2-difluoroethylene are quasi-aromatic and nonaromatic, respectively, and this is why cis is more stable than trans.[6] This argument requires a respectable F2p—F2p nonbonded overlap which, however, does not materialize.

Are structural trends due to the combination of many different "effects," and is trying to decipher them hopeless? Or, could it be that everything is a consequence of a general principle which we failed to recognize because we have been seeing things through the wrong glasses?

The published literature contains many articles dealing with the "gauche effect,"[7] the "anomeric effect,"[8] and the "σ conjugation effect."[2] All of them, whether implicitly or explicitly, target one and the same phenomenon: σ donor bonds (e.g., H—C, Si—C) or lone pairs prefer to be anti to σ acceptor bonds (e.g., C—F, N—F). We call this the "anti effect." For example, the preference of 1,2-difluoroethane for the gauche form[9] has been identified as the "gauche effect." This term focuses on the two acceptor bonds and identifies their preference for adopting a gauche orientation. Alternatively, one could formulate the gauche isomer as the one having a donor C—H bond anti to an acceptor C—F bond. Figure 11.1 shows representative examples of the "anti effect." The common denominator is the presence of donor bonds or lone pairs anti to acceptor bonds. Is the trans periplanar arrangement of donor and acceptor bonds the origin of all these effects, or have we been led astray by this seemingly convincing pattern?

While the data defining the "anti effect" have been celebrated in the journals, virtually no attention has been paid to a different and even more unexpected stereochemical trend: the preference of certain molecules for adopting a geometry in which lone pairs eclipse vicinal donor bonds. We refer to this as the "syn effect." Hydrogen peroxide (HOOH), diimide (HNNH), and hydroxylamine (NH_2OH) are some illustrative examples (Figure 11.1). The gauche hydrazine can be regarded as a distorted form of one eclipsed conformation where the syn effect obtains. Clearly, the very existence of these preferences suggests that the models used to account for the "anti effect" might have missed the broader picture that accommodates both effects.

11.2 1,2-Difluoroethane and the Gauche Rule

Certain 1,2-disubstituted ethanes have conformational preferences that defy intuition: for example, those in which the more "crowded" staggered isomer turns out to be the most stable form. A good example is 1,2-difluoroethane, where the gauche is favored over the trans form by about 1 kcal/mol.[9] We can view 1,2-difluoroethane as "C_2 plus H_4F_2" and we can construct the principal MOVB bond diagrams of *gauche*- and *anti*-difluoroethane (DFE) as shown in Figure 11.2. We now see that only in the gauche isomer can the C_2 fragment use the two highest energy π-type MOs to make two MOVB electron pair bonds with the two MOs spanned by the two fluorines. *This is not possible in either the anti or the syn periplanar form.* Hence, the gauche form is more stable because the upper orbitals of the C_2 fragment are permitted to "drain" their electrons into the low-lying AOs of the two fluorines. The depopulation of the

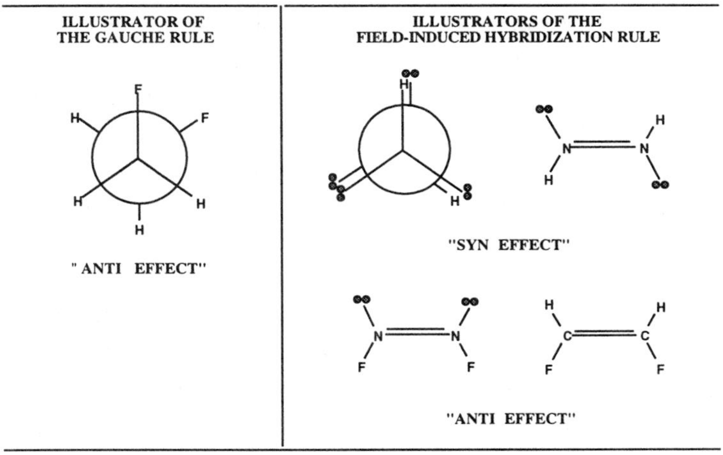

Figure 11.1 Illustration of the anti and syn effects. The anti effect identifies vicinal anti placement of a C—H bond or lone pair with respect to a polar A—F bond. The syn effect identifies vicinal syn placement of a lone pair with respect to a polar A—H bond (A = atom of variable electronegativity).

antibonding MOs of an open shell C_2 fragment by intrabond CT delocalization that drives electron density to electronegative ligands has two effects:

It enhances the C—F bond polarity (Pauling's "polar covalence").

It contracts the C—C bond length. Translated into VB language, this means that in going from the anti to the gauche isomer, the C—F bonds shorten as a result of increased "polar covalence" and the C—C bond also shortens because of decreased vicinal C—F/C—F exchange repulsion.

Let us contrast these results with the predictions of the hyperconjugation model. The latter argues that gauche is favored over anti because the placement of two donor C—H bonds anti to two acceptor C—F bonds causes favorable delocalization from the occupied orbitals of the former to the unoccupied orbitals of the latter. Hence the prediction that the C—F bonds should lengthen and the C—C bond should shorten as well. Thus, the energetic predictions of the two models are the same but the structural predictions are different. Ab initio computations, reported in a very useful paper by Kirby and co-workers,[10] show that the C—C, C—F, and C—O bonds of F—H_2C—CH_2—OY (Y = Me, CHO, NO_2) all shorten in going from trans to gauche isomer.

1,1,2,2-Tetrafluoroethane is predicted by the hyperconjugation model to adopt the gauche geometry. Our analysis leads to different conclusions simply because the maximum number of fluorines that can tie up high-lying singly occupied MOs of the C_2 fragment is two. To see this point, consider the C_2 septet, that is, the C_2 fragment which is excited so that it can make six bonds with six ligands using the odd electrons in orbitals w_2 to w_5. The C_2 MOs are shown in Figure 11.2 in the following order of increasing energy: $w_1 < w_2 < w_{3x}$, $w_{3y} < w_4 < w_{5x}$, $w_{5y} < w_6$. If we think of the ligands as being σ acceptors or σ donors, we may ask the question: How do the ligand orbitals best match with the C_2 orbitals? To answer this question, we use the same reasoning as before. We first observe that the π antibonding w_{5x} and w_{5y} lie far above the rest of the singly occupied MOs. Hence, we conclude that the upper w_{5x} and w_{5y} will engage the most electronegative ligands while the remaining lower four C_2 MOs will accommodate the remaining four ligands without exhibiting any strong preference, since they are quasi-degenerate. It follows that the preferential placement of six different groups on a C_2 fragment to form an ethane derivative is given by the following rule: Place the two strongest σ acceptor groups gauche and accommodate the remaining four ligands according to conventional nonbonded repulsion considerations.

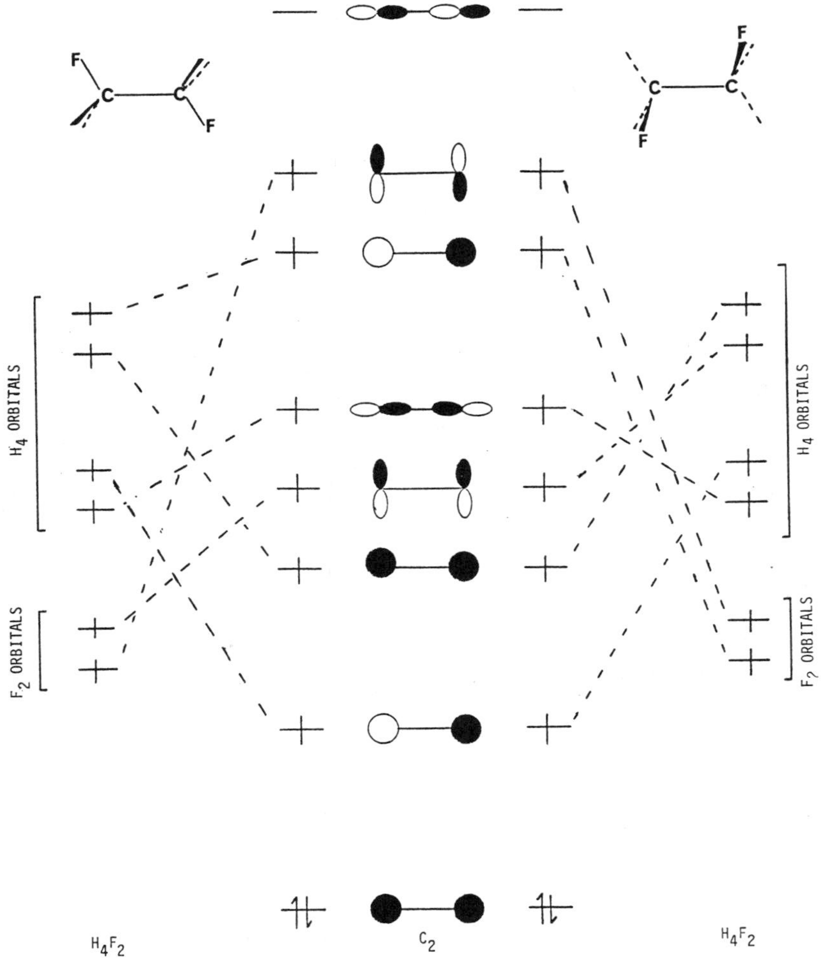

Figure 11.2 The MOVB bond diagrams of *trans*- and *gauche*-1,2-difluoroethane.

We now return to the case of 1,1,2,2-tetrafluoroethane. If the hyperconjugation model were valid, the most stable isomer would be 1. If the VB analysis is correct, the most stable isomer should be 2. This places two fluorines gauche and treats the remaining two as bulky analogues of hydrogen. Experiment shows that isomer 2 is indeed the global minimum on the rotational surface.[11]

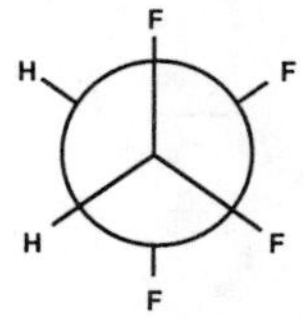

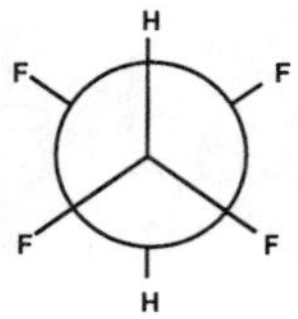

1 **2**

Obviously, if the four ligands beyond the two most electronegative gauche ligands have comparable bulk, minimization of interbond overlap repulsion is best achieved by placing the two least electronegative units anti to the two gauche ligands. This will now permit some small interbond delocalization (hyperconjugation).

We can now formulate the *gauche rule*: The lowest energy conformer of a hexaligated diatom (ethane-type molecule) places the two most electronegative ligands, A/A, gauche and the best two donor ligands, D/D, anti to A/A provided the remaining two ligands, L/L, have bulk comparable to D/D. This rule is restricted to ethane-type molecules, for which one has to choose between periplanar (anti) and nonperiplanar (gauche) staggered geometries.

11.3 The Anti Hyperconjugation of Donor and Acceptor Bonds Is Not Supported by the Data

Consider the seemingly reasonable and appealingly simple explanation of conformational isomerism by the hyperconjugation model: anti placement of donor and acceptor bonds apparently explains *both* the gauche preference of $O_2NO—H_2C—CH_2—F$ and the trans preference of $O_2NO—H_2C—CH_2—SiH_3$ determined by ab initio computations.[10] It makes an additional specific prediction: Since the magnitude of the effect is expected to increase as the donor and acceptor ability of the bonds increases, the trans stereoelectronic preference should be larger and the C—C bond shorter in the silyl species because this has a much better donor bond (CSi) than the former (CH). However, the results of the ab initio computations are at odds with these expectations.

With only one exception, the C—C bond is shorter in $F—H_2C—CH_2—OX$ than in $SiH_3—H_2C—CH_2—OX$ for a fixed (trans or gauche) conformation (X = Me, CHO, NO_2). Observe the very important trend presented in Table 11.1:

Table 11.1 Conformational Dependence of Bond Lengths

		$r_{C—C}$ (Å)	
X	Conformation	$FH_2C—CH_2OX$	$H_3SiH_2C—CH_2OX$
Me	g	1.509	1.524
Me	t	1.515	1.523
CHO	g	1.507	1.521
CHO	t	1.514	1.519
NO_2	g	1.509	1.521
NO_2	t	1.517	1.517

(b) The gauche preference of $F—H_2C—CH_2—OX$ is greater than the trans preference of $SiH_3—H_2C—CH_2—OX$, although by not much.

The C—C bond variation is critical because it is pertinent to a prediction made before. The comparison of two different constitutional isomers or two different molecules is always made by reference to their T formulas and associated concepts. We argued that the combination of two diastolic (DI) fragments can be as good a combination as one diastolic and one systolic (SY) fragment because the DI-DI combination is particularly effective in minimizing vicinal repulsion. This was illustrated by the geometries of $F_2C{=}CF_2$, a combination of two DI CF_2 groups, and $F_2C{=}CH_2$, a combination of SY = CH_2 with DI = CF_2. The C=C bond is shorter in $F_2C{=}CF_2$ (1.311 Å) than in $F_2C{=}CH_2$ (1.316 Å). Increasing fluorination of a central carbon fragment leads to shortening of the C—F bonds because electron pairs are displaced toward the outer fluorines and exchange repulsion within the carbon core is minimized. The hyperconjugation model leads to diametrically opposite conclusions.

The following reaction constitutes a test of the interaction of two groups. According to the hyperconjugation model, switching from Y being σ acceptor to Y being σ donor should render the reaction enthalpy less negative or more positive. Indeed, ab initio calculations find this to be the case. But, the very same calculations find that the C—C bond of $Y—CH_2CH_2—OX$ is *shorter* when Y is a σ acceptor! So, hyperconjugation gets the energetics right but the structures wrong! Hence, this model cannot be right.

	YH_2CCH_2OX +	H_3CCH_3	⟶	H_3CCH_2OX +	YH_2CCH_3
Y = F	4p	3v		2p	2p
Y = SiH_3	2p	3v		2p	3v

The T formulas disperse the confusion. The group OX is an SY group. When Y is chosen to be SY (e.g., F), the reaction is predicted to be *exothermic* because the products have three fewer errors than the reactants. By contrast, when Y is chosen to be DI (e.g., SiH_3), the reaction is predicted to be *thermoneutral* because the errors is the same on both sides of the equation. This is consistent with the trend for greater endothermicity as F is replaced by SiH_3, revealed by the ab initio computations. When X = NO_2 (i.e., when OX is a powerful SY group), the reaction with Y = F is strongly exothermic (–4.8 for the trans and –4.1 kcal/mol for the gauche reactant), while the reaction with Y = SiH_3 is essentially thermoneutral (+0.4 for the trans and –0.1 kcal/mol for the gauche reactant).

11.4 Cooperative and Anticooperative Action of Overlap and Induction

Until this point, we have regarded the static (classical Coulomb interaction, induction, and dispersion) and delocalization mechanisms of bonding as acting independently. Conventional wisdom is that static effects come in the form of classical Coulomb repulsion (dipole–dipole repulsion), and this condition has been used to explain trends in rotational isomerism ranging from the preference of alkenes for the trans geometry all the way to the anomeric effect. We now present a very different scenario which, in turn, illustrates how far away from chemical sense the orbital models of electronic structure have taken us. First, we realize the obvious: In a typical heteronuclear organic molecule neither all bonds are covalent nor are all bonds ionic! In other words, we have incomplete intrabond charge transfer. As a result, the field due to one polar bond can act to direct hybridization of a vicinal nonpolar bond. In this way, the static and CT delocalization mechanisms of bonding interact. We will argue that this coupling, called *field-induced hybridization (FIH)*, is important in the conformational analysis of molecules having polar bonds and/or lone pairs and that it constitutes the explanation of numerous counterintuitive trends of structural organic chemistry.

Consider two open-shell fragments approaching each other on the way to bond formation. The first (fragment A) has two (orthogonal) orbitals, a and a*, and a total of three electrons; the second (fragment B) has a single orbital b with one electron. Each orbital of B can interact by symmetry with the sole orbital of A. The goal is to maximize exchange bonding and minimize exchange antibonding while, at the same time, the classical Coulomb interaction of the fragments remains favorable. The exchange optimization is principally due to

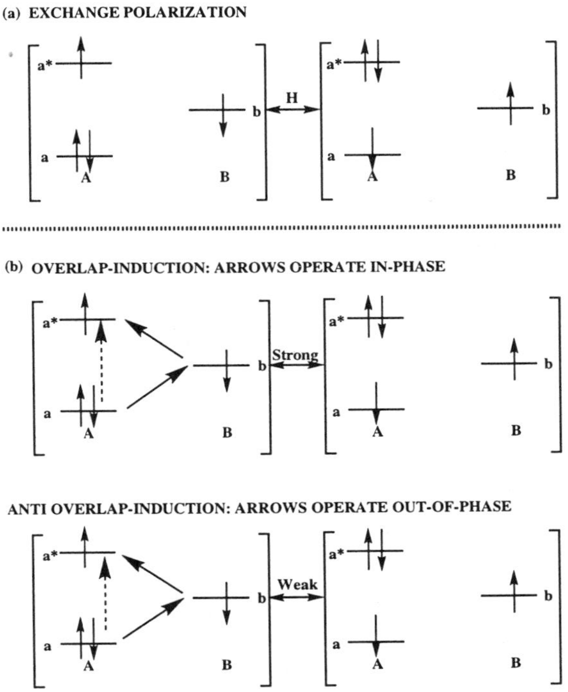

Figure 11.3 (a) The concept of exchange polarization. Mixing of the two configurations is equivalent to hybridizing a and a* with the three electrons allocated to best respond to the singly occupied b in an overlap sense. (b) The concepts of overlap induction and anti overlap induction. Solid (overlap) and dashed (induction) arrows can operate either in phase or out of phase to create a/a* hybridization or dehybridization, respectively.

the interaction of the two configurations shown in Figure 11.3a. This mixing, which is called exchange polarization, causes hybridization of orbitals a and a* in fragment A, with the result that one pair is directed away from B while the odd electron is directed towards B so as to generate an exchange bond (T bond or covalent bond). There are now two possibilities:

The electric field of a vicinal polar bond *assists* the exchange polarization. In this case, we have strong hybridization of a and a*.

The electric field of a vicinal polar bond *antagonizes* the exchange polarization. In such a case, hybridization is arrested.

Figure 11.4 provides further illustrations of these concepts by using the model systems A—Li and A—I, where A is some arbitrary fragment, and focusing on the σ system. Depending on the charge on A (e.g., depending on the active electric field of a vicinal polar bond), we have either in-phase (a, d) or out-of-phase (b, c) action of exchange polarization and classical Coulomb interaction.

We can now appreciate what will become a recurrent theme: *The transformation of one VB configuration to another can be effected by distinct electron hops, all of which produce the same end result albeit by different mechanisms.* In our case, there exist *two* different mechanisms for going from the first to the second configuration of Figure 11.3a:

By a 2-e interfragmental CT hop (assisted by 1-e CT hops). This is the overlap mechanism.

By a 1-e intrafragmental CT hop. This is the induction mechanism, and it depends on the field exerted by the environment.

Both mechanisms are illustrated in Figure 11.3b, with solid arrows signifying the overlap and dashed arrows the induction component. Depending on the field, we can have either overlap induction (hybridization) or anti overlap induction (dehybridization).

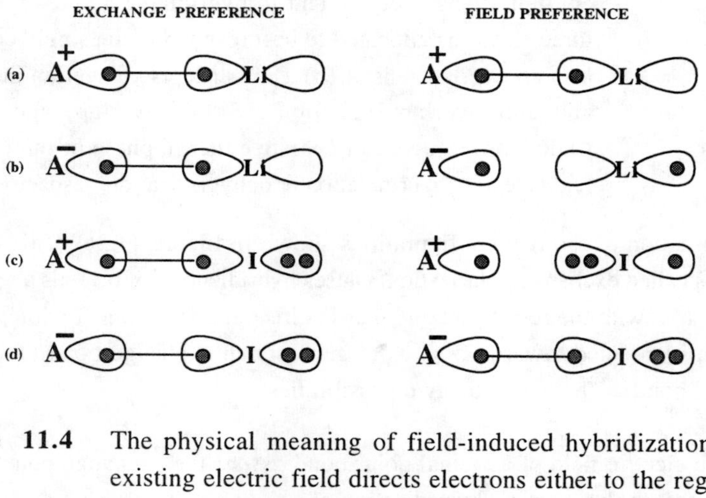

Figure 11.4 The physical meaning of field-induced hybridization. The existing electric field directs electrons either to the regions of space in which exchange bonding is possible (i.e., hybridization) or to regions of space in which there can be only exchange antibonding or only nonbonding (i.e., dehybridization).

11.5 Geometric Isomerism: Graveyard of Conventional Explanations

The surprising preference of 1,2-difluoroethylene for the cis form has provoked much thought and generated many would-be explanations. Since these are frequently revisited in the literature, an explicit presentation of the pros and cons is appropriate.

(a) One explanation, once favored by this author, was that the cis form is quasi-aromatic. We call this the "nonbonded attraction" explanation. In VB terms, it amounts to delocalization of the π- and σ-fluorine lone pairs into the π and σ components of the C═C double bond. This model predicts shortening of the C—F bonds and lengthening of the C═C bond as we go from trans to cis, as observed experimentally. The problem with this interpretation is that nonbonded overlap of fluorine AOs is not at all great. Furthermore, the FCC angle opens up in going from the trans to the cis isomer, suggesting nonbonded repulsion rather than attraction of the fluorines.

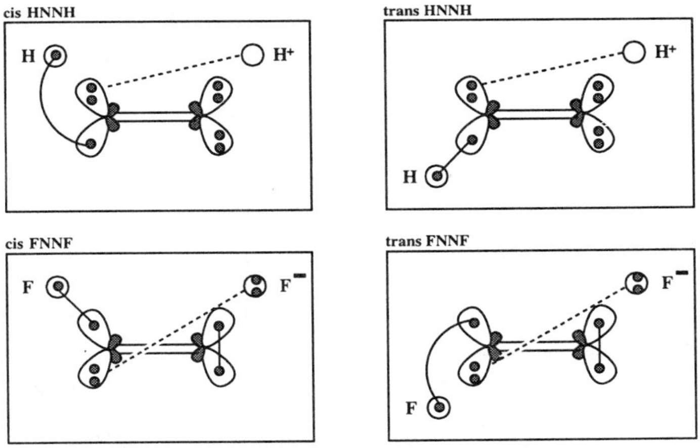

Figure 11.5 The field-induced hybridization rationalization of the trans preference of HNNH and the cis preference of FNNF. Note how the electric field of the formal proton or fluoride controls the stereochemistry of the pair and the odd electron at the remote nitrogen. In HNNH, one proton attracts the distal lone pair and dictates trans entry of the other hydrogen. In FNNF, one fluoride repels the distal lone pair and dictates cis entry of the other fluorine.

(b) The anti interaction of a C—H bond pair with a low-lying C—F antibonding hole (i.e., hyperconjugation: "anti effect") seems like a plausible explanation. However, this model predicts shortening of C=C and lengthening of C—F bonds in passing from trans to cis. This is exactly opposite to what is observed experimentally: The C=C bond lengthens and the C—F bond shortens.

In 1,2-difluoroethane, there is a choice between gauche and periplanar arrangement of the two fluorines. The electronegative ligands create a preference for gauche because only in this geometry are two high energy π-antibonding MOs of the C_2 fragment available for binding the two fluorines. In the case of 1,2-difluoroethylene, the molecule is forced to have periplanar fluorines and the high energy π-antibonding MO of the C_2 fragment is equally available to only one fluorine in both cis and trans. Hence, the greater stability of *cis*-1,2-difluoroethylene must have a different origin from the greater stability of gauche 1,2-difluoroethane.

The problem becomes more interesting when the focus shifts to diimide, HNNH, and its fluoro derivative, FNNF. Now, the preference depends on ligand electronegativity: HNNH prefers trans but FNNF prefers cis![12] Furthermore, in contradistinction to 1,2-difluoroethylene, the NF bond is longer and the N=N

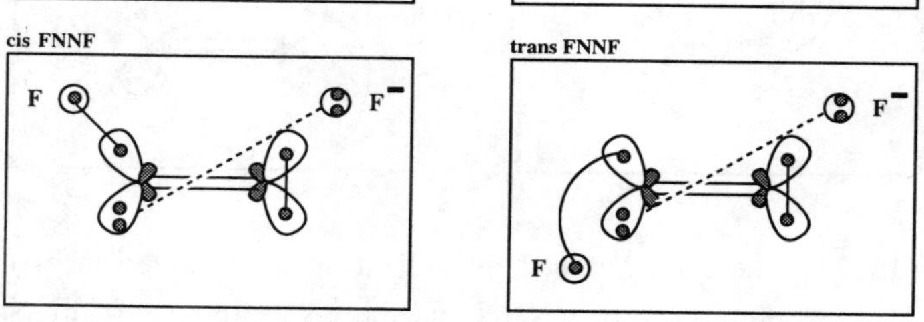

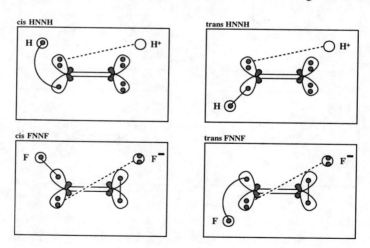

Figure 11.6 The action of exchange polarization and the field of a vicinal polar bond cause hybridization in *trans*-HNNH (a) and *cis*-FNNF (d) and dehybridization in *cis*-HNNH (b) and *trans*-FNNF (c). Symmetry adaptation is obtained by writing equivalent resonance structures.

bond shorter in the cis isomer![9,12] In summary, geometric isomerism is an ancient problem that has practically exhausted all possible explanations! And yet, none is satisfactory because none comes close to explaining both the energetics *and* the structural aspects of the problem.

The HNNH and FNNF geometric isomers are the simplest illustrators of our VB approach. The explanation of why the first prefers trans while the second

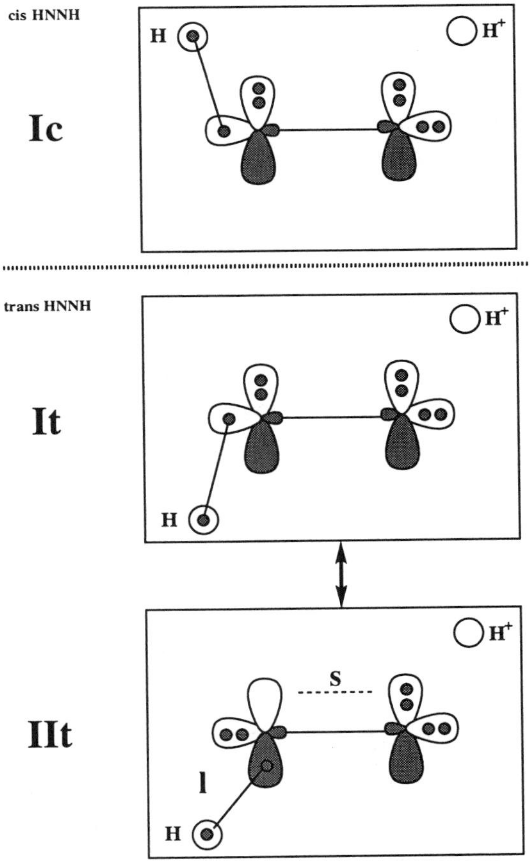

Figure 11.7 Resonance formulation of the cis and trans isomers of HNNH. The difference in NH and NN bond lengths is due to the differentiating structure IIt. The letters l and s, respectively, denote the bond-lengthening and bond-shortening effect of the differential contributor. These diagrams, and those of Figures 11.8 and 11.9, were constructed under the assumption that C2p has lower energy than Csp^x.

opts for cis is based on the field induced hybridization concept, and it is shown in Figure 11.5.

Electron transfer within one polar bond as part of intrabond CT delocalization creates either a hole or a pair on the ligand, which creates an electric field that directs the stereochemistry of the lone pair on the vicinal nitrogen. Exchange and field are phased in only in *trans*-HNNH and *cis*-FNNF. By contrast, they are in conflict in *cis*-HNNH and *trans*-FNNF. This means that that hybridization can occur only in *trans*-HNNH and *cis*-FNNF. By contrast, *cis*-HNNH and *trans*-FNNF are condemned to be dehybridized, as illustrated in Figure 11.6.

The structural consequences of hybridization and the absence thereof in the geometric isomers of HNNH are elucidated in Figure 11.7. The difference between the two isomers is contributor IIt. This shortens the N=N bond because it removes an N2p—N2p four-electron antibond and it lengthens the NH bond, which is now formed by utilization of the more electropositive N2p AO. Similarly, the difference between the cis and trans geometric isomers of the FNNF isomers is contributor IIc, which shortens the N=N bond [because the formal fluoride and the remote nitrogen have reduced exchange repulsion (3- vs. 4- exchange antibonds in IIc and It, respectively)] and lengthens the NF bond (because this is now formed by utilization of the more electropositive nitrogen

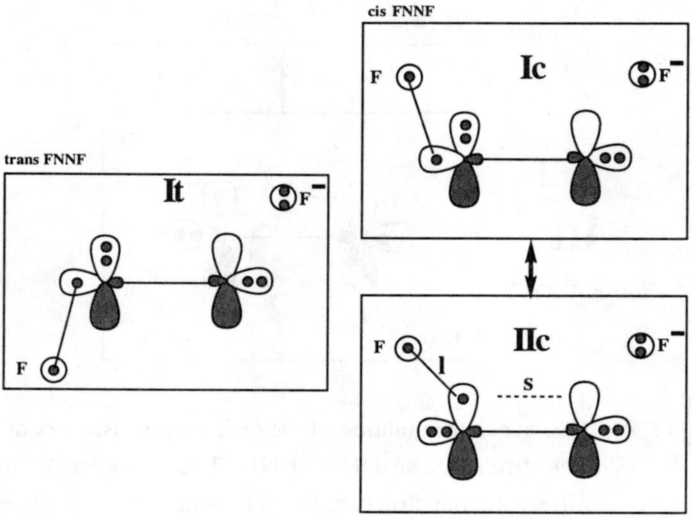

Figure 11.8 Resonance formulation of the cis and trans isomers of FNNF. The difference in NH and NN bond lengths is due to the differentiating structure IIc.

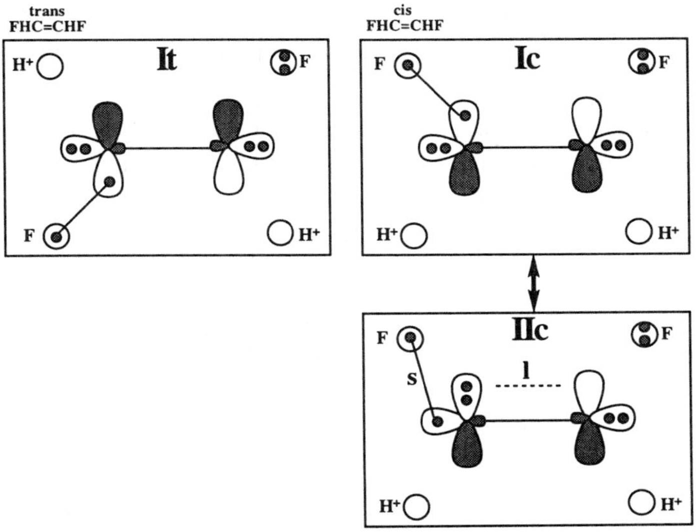

Figure 11.9 Resonance formulation of the cis and trans isomers of FHC=CHF. The difference in CF and CC bond lengths is due to the differentiating structure IIc.

2p AO). These considerations are illustrated in Figure 11.8.

To a first approximation, 1,2-difluoroethylene is 1,2-difluorodiimide with the nitrogen lone pairs being replaced by CH bond pairs. However, there is one important difference: In diimide and in its difluoro derivative, the interaction of the two sp^x AOs is strong because the N—N distance is short. As a result, the one-electron energy of sp^x is higher than that of 2p. In going to the ethylene derivatives, the situation is reversed because the longer C—C distance diminishes the sp^x—sp^x interaction. As a result, the more electropositive 2p has higher energy than the more electronegative sp^x. In turn, this reverses the trends in geometry: The differentiating VB structure IIc causes a lengthening of the C=C bond and a shortening of the CF bond in the cis isomer (Figure 11.9). In all cases, we have followed Walsh's rule: The more electronegative ligands tie up the higher energy orbitals of the central fragment. In 1,2-difluoroethylene, this means that the fluorines are preferentially bound to the C2p AOs.

Is the nonbonded attraction explanation of cis preferences dead? No, because this does assume a principal role in molecules in which the key electronic effects described in this and the preceding chapters remain constant. The best example is the cisoid geometry of methyl vinyl ether and its derivatives discussed in Ref.

2. Another example is the planar (D_{2h}) geometry of N_2O_4. The isoelectronic B_2F_4 is also planar but only slightly more stable than the "perpendicular" form.[13]

We now condense the analysis to bare essentials. What determines the relative stability of rotamers having lone pairs and polar bonds is not the intuitively appealing 1,4 nonbonded interaction but, rather, the overlooked 1,3 interaction as illustrated below for the two FNNF geometric isomers:

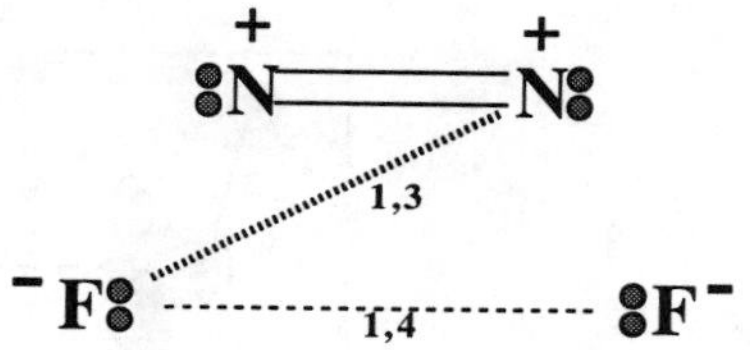

11.6 The Field-Induced Hybridization Rule

We can transform the essence of the VB analysis of geometric isomerism into a predictive rule: A weaker donor ligand directs a stronger vicinal donor ligand cis, a weaker acceptor ligand directs a stronger acceptor ligand cis, and an acceptor ligand directs a donor ligand trans.

According to the field-induced hybridization rule, active lone pairs are taken to imply the presence of infinitely strong donor ligands and active holes the presence of infinitely strong acceptor ligands; heavy p-block atoms have inert ns lone pairs, which are considered to be stereochemically inactive and are treated as though they did not exist all. The FIH rule can be restated for LNNL (L = H, F) as follows: σ donors direct an active vicinal lone pair cis and σ acceptors direct an active vicinal lone pair trans.

Ab initio computations reported by Nguyen et al.[14] give us another chance to test our model. Specifically, replacing N by P is equivalent to replacing an active by an inactive lone pair. This predicts that FIH control of stereochemistry will persist in LNNL as well as in LPNL, but it will vanish in LPPL. Thus, we expect that LPPL will be a very unlikely prototype of nonbonded repulsion: There is simply no other operative effect save for the nonbonded repulsion of the two PL bonds simply because the P lone pairs can be treated as "nonexistent"! This is exactly what calculations of the cis and trans isomers of FNPF and FPPF reveal: The first prefers cis but the latter trans.

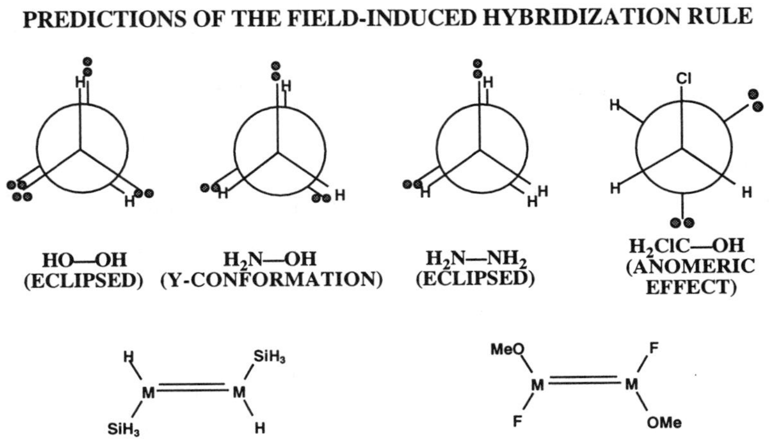

Figure 11.10 Illustrations of the field-induced hybridization rule. Si and H are more electropositive and O and F more electronegative than M.

Figure 11.10 provides illustrations of the FIH rule. In the case of HOOH, the rule predicts an eclipsed geometry with an HOOH dihedral angle of 120°. In this conformation, one lone pair at one oxygen center is eclipsed by a bond pair of the second oxygen center and vice versa (two BL interactions, where B = bond pair and L = lone pair). There is an additional pair of eclipsed lone pairs (LL interaction). The observed HOOH dihedral angle is roughly 120°. The hyperconjugation model predicts a gauche structure with a 60° dihedral angle. Turning to hydroxylamine, the rule predicts an eclipsed "Y" geometry where every lone pair or bond pair of one center is eclipsed by a bond pair or lone pair, respectively, of the other center. By contrast, the hyperconjugation model predicts a staggered "W" geometry with three anti lone pair–σ* interactions.

Hydroxylamine is an important molecule because it is the simplest representative of the conformational preference of the ligands for the N—O bond. The antitumor antibiotic calicheamicin γ-1 has a hydroxylamine-type glycosidic linkage. Kahne et al.[15] recognized the fundamental difference of the rotational profiles of the methyl derivatives of H_3C—OH and H_2N—OH and proposed that the N—O bond organizes the two halves of the molecule into a shape that complements the shape of the minor groove of the DNA. Apparently, this is a prerequisite for the DNA double-strand cleavage effected by this antibiotic.

The anomeric effect refers to the tendency of an electronegative substituent next to the oxygen of a pyranose ring to occupy the axial position.[16,17] This counterintuitive trend, which has mesmerized stereochemists since its discovery, has been repeatedly reviewed,[18] and still attracts great interest.[19,20] The FIH explanation is straightforward: The action of a formal chloride in CH_2ClOH, a simple prototype of the "anomeric effect," dictates the anti localization of one oxygen lone pair, and this is possible only in the gauche conformation. In a pyranose ring, this can be realized only if the C—Cl bond next to oxygen is axial. The electron distribution in the O—C—Cl array anticipated by the FIH model is consistent with the electron density shifts in going from 1,4-dioxane to *trans*-2,5-dichloro-1,4-dioxane found by Koritsanszky et al.[21]

Hydrazine is an interesting case. Our model predicts an eclipsed conformation with an LNNL dihedral angle of 120°, whereas the hyperconjugation model predicts a gauche geometry with an LNNL dihedral angle of 60°. The experimental result is right in the middle: the dihedral angle is roughly 90°. We suggest that the eclipsed geometry is rendered unfavorable by the steric repulsion of two eclipsed NH bonds, a feature absent in HOOH and NH_2OH.

Further support for FIH model comes from the observation that replacement of a first-row heteroatom (e.g., N) by a heavier congener (e.g., P) destroys a "counterintuitive" and establishes an "intuitive" stereochemical preference. For example, replacing N by P in hydrazine and its derivatives either diminishes or abolishes gauche preference, with the effect increasing as we move down the nitrogen column. Thus, tetramethylhydrazine is "counterintuitive" gauche but the arsenic derivative is "intuitive" trans. The data have been summarized elsewhere.[22] One feature that differentiates N from P and heavier congeners is the contraction of the ns relative to the np valence AO as we go down a p-block column. This causes s-p dehybridization. As a result, FIH ceases to be important and the molecule adopts the conformation that minimizes nonbonded repulsion, much like ethane.

We have argued that structural isomerism is controlled by minimization of interbond exchange repulsion and rotational isomerism in molecules containing polar bonds and lone pairs by FIH. Vicinal interbond CT (i.e., hyperconjugation) has been deemed unimportant and inconsistent with stereochemical trends across the board. In other words, we have a hierarchy of stereoselection, and hyperconjugation has yet to weigh in. The last chance is the differentiation of nonequivalent bonds within a fixed conformer. For example, Bohlmann[23a] noted that the C—H bond anti to an sp^3 N lone pair in conformationally defined amines is longer and weaker than the two gauche C—H bonds.

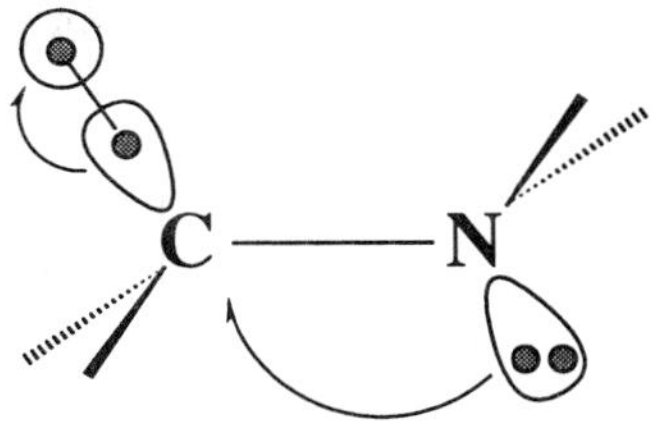

The conventional explanation has been hyperconjugation. While this weak selectivity can indeed be signaling the effect of vicinal interbond CT,[23b] there is an alternative explanation: The sp^3 lone pair generates a field that *selectively* polarizes the anti bond. This corresponds to induced intrabond CT delocalization and once again removes the necessity of invoking any interbond CT delocalization. Replacing N by P is expected to turn the directional Nsp^3 into a spherical P3s lone pair, diminishing or eliminating induction, the Bohlmann effect, and the anomeric effect. This is what happens when O is replaced by S in 1,3-dioxanes.[24] In essence then, stereoselectivity may arise from the Coulombic interaction of oriented holes and pairs without interbond CT much as it has in the case of the (weakly) directional hydrogen bond.

The presence of more than one σ-acceptor ligand in an ethane-type system takes us away from the range of applicability of the FIH rule to the range of applicability of the gauche rule. The hallmark is the obligatory placement of *one pair* of electronegative ligands gauche. It is illustrative to compare HOOH and FOOF. In contrast to HOOH, FOOF is expected to have the two fluorines gauche, much like 1,2-difluoroethane, with a 60° FOOF dihedral angle, assuming sp^3 hybridization of the oxygens. A change of hybridization to sp^2 allows the fluorines to get further away from each other while the FOOF dihedral angle becomes 90°, the experimentally observed angle.

The two rules merge into a single succinct recipe: A diastolic and a systolic bond prefer trans while two *nonequivalent* diastolic or two *nonequivalent* systolic bonds prefer cis alignment. A lone pair (hole) is taken to be the remnant of an infinitely systolic (diastolic) bond. Exceptions can be anticipated on the basis of the detailed discussions presented in this chapter.

11.7 Diastereofacial Selectivity

We have seen that structural organic chemistry can be interpreted by assuming, to a first approximation, zero vicinal interbond delocalization (hyperconjugation). Furthermore, we documented the proliferation of MO models of fundamental structural trends, most frequently based on hyperconjugation, and we interpreted this as proof of what we asserted in the very beginning: MO models explain everything and nothing. MO theory is good for computing but not for thinking. A recent paper by Wipf and Kim[25] points out the inadequacy of the general understanding of how facial selectivity of nucleophilic addition to carbonyl comes about and suggests an electrostatic basis for diastereofacial selectivity consistent with the foundational notion that the electrons of organic molecules stay locked in the bonds. A chiral substituent L effects a selective polarization of the two faces of LRC=O by a dipolar mechanism as illustrated.

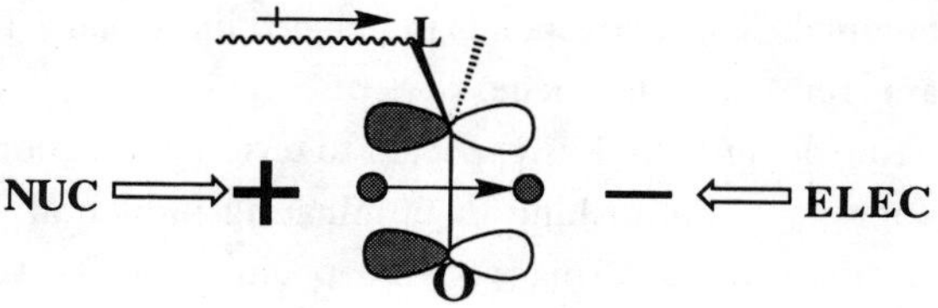

The diplacement of the π electrons by the L dipole creates a depleted (electrophilic, denoted by +) and an enriched (nucleophilic, denoted by –) face. Electrophiles attack preferentially the (–) and nucleophiles the (+) face. The essential points are that the axial polarization of the π system comes about by the mixing of valence and/or nonvalence AOs with the π-type p AOs and that this is a consequence of the electric field of either the substituent L or the electric field of a polar C—L bond. In the latter case, L must be a first-row electronegative systolic group (CH_3, NH_2, OH and, best of all, F). We designate this the *axial π-polarization* model because the polarization occurs along the axis of a p AO.

We can now make some predictions about the stereochemistry of nucleophilic addition to carbonyl:

(a) Using the B-formula representation of a carbonyl molecule in Newman projection form (see Section 2.1) as illustrated following (b), we predict that the entering nucleophile follows a trajectory in which the NUC—C—O angle is greater than 90°.

(b) The facial selectivity of nucleophilic attack is predicted by assuming that the reactive conformation is the B-staggered conformation shown below, where S (= small) and ELNG and ELP are atoms or groups more and less electronegative than C, respectively. The diastereoselection rule is that nucleophilic attack will occur trans and electrophilic attack cis to ELNG.

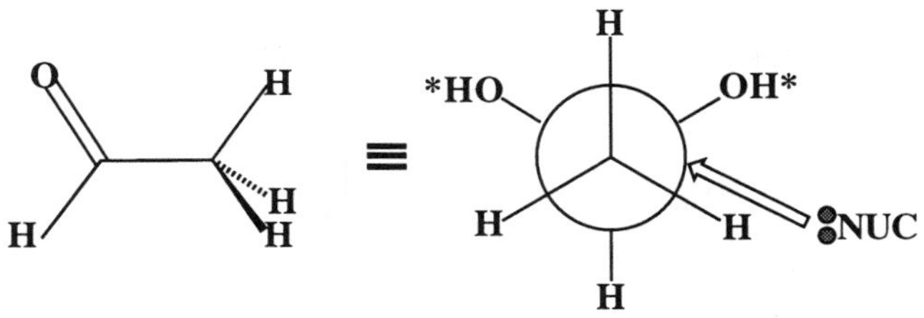

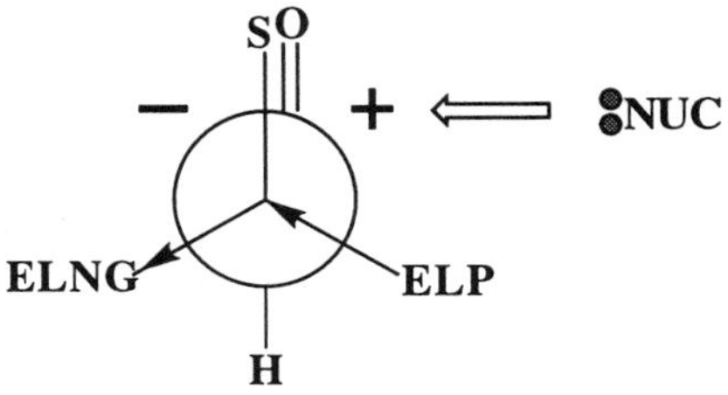

Placing S = Me, ELNG = Ph and ELP = H, we can rationalize the classic work of Cram, as summarized by various authors[26] and the later work of Reetz[27] on methylmetal addition to chiral carbonyl. Putting S = H, ELNG = Cl and ELP = H explains the computational results of Anh.[28]

That transmission of information to the carbonyl is effected by the electrostatic interaction of bond dipoles, as marvelously illustrated by the 4,4-disubstituted cyclohexadienone system studied by Wipf and Kim.[25] When X = alkyl, the dominant C—OR bond dipole on the upper face of the planar ring polarizes the π-electron pair toward the upper face. In turn, the π pair polarizes the upper component of the double-banana C═O bond. This predicts prefential nucleophilic attack on the lower face (backside with respect to the selectively polarized banana C—O bond). In other words, the diastereoselection rule is that nucleophilic attack should occur trans with respect to the dominant outbound bond dipole at C4. That chirality is transmitted by the interaction of bond dipoles and not by hyperconjugation is revealed by placing X = CF_2—CF_3. In this case, the dominant bond dipole at C4 is the remote C—F bond dipole(s) and facial selectivity is reversed.[25]

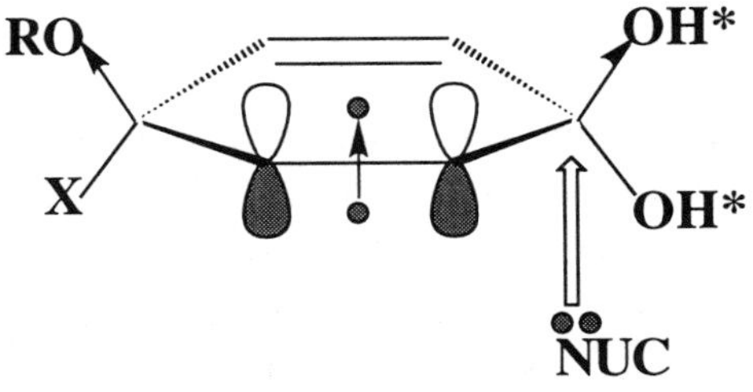

The diastereoselectivity model presented above is pertinent to chiral molecules of the type XYZC—CR═A, where A is an atom more electronegative than C. When A is less electronegative than C (or even equally), the reactive conformation may well be the B-eclipsed (rather than the B-staggered) form if two of X, Y, and Z carry lone pairs that can associate with the electron-depleted A terminal atom. In such an event, the diastereoselection rule is reversed: Nucleophilic attack will occur cis and electrophilic attack trans to X = ELNG. Recent work by Evans and co-workers suggests that chiral propenes (A = CH_2) reacting with the electrophile BH_3 obey the latter rule.[29]

The axial π-polarization model explains also the diastereofacial selectivity of the Diels–Alder reactions of model dienes such as I, IIc and III shown here. The placement of the π electrons in the white p lobes is taken literally to signify axial polarization by the field of the C—F dipole.

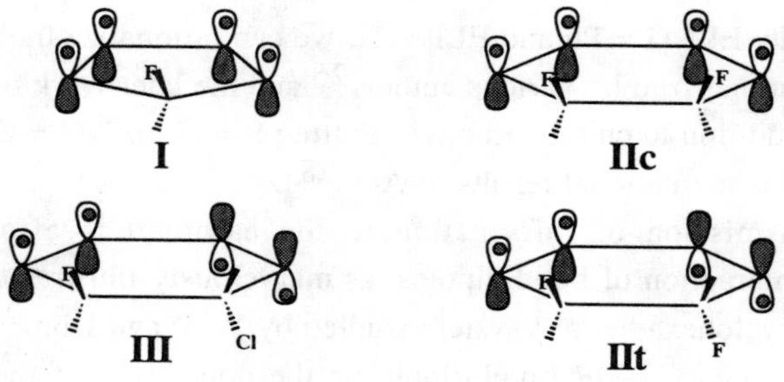

The predictions are:

(a) An electrophilic dienophile will attack I, IIc, and III syn with respect to F. π-Acceptor dienophiles are the most common reagents. The data summarized and contributed by Paquette and co-workers[30] are in agreement, especially the behavior of the *N*-methyltriazolinedione dienophile which, because of lone pair exchange repulsion, should have always entered anti rather than sometimes syn, as observed.

(b) A nucleophilic dienophile will attack anti with respect to F. There are no available pertinent data.[31]

Let us now go beyond simple intuition inspired by some key experimental findings by Paquette and co-workers, who determined the relative reactivity of dispiro analogues of IIc and IIt in which oxygen, rather than fluorine, is incorporated into a five-membered ring that is attached in spiro fashion to the 1,3-cyclohexadiene nucleus. Knowing that trans is more stable than cis in planar 1,2-disubstituted rings (such as cyclopropane), one would be tempted to predict a higher reactivity of IIc. By contrast, it is found that the oxospiro derivative of IIt is much more reactive than that of IIc! In other words, two trans-oriented C—F dipoles apparently polarize a diene in a way that is superior to polarization effected by two cis-oriented C—F dipoles. Why is this so? Instead of deferring the answer until after the discussion of VB aromaticity, we state the result here, to forewarn the reader that what follows in Chapter 12 and after is not a mere reprisal of the standard aromaticity concepts in VB language.

As we will see shortly, an aromatic is differentiated from an antiaromatic [even] annulene by the action of association-promoting VB configurations which, in turn, promote I-bonding. These are VB structures that contribute either exclusively or predominantly to the aggregated system of n bonds [(2n) annulene] in comparison to the hypothetical segregated system of n bonds infinitely apart. One crucial VB configuration of this type is the one in which there is hole–pair alternation. Stabilization of this results in stabilization of either an aromatic molecule or an aromatic transition state.

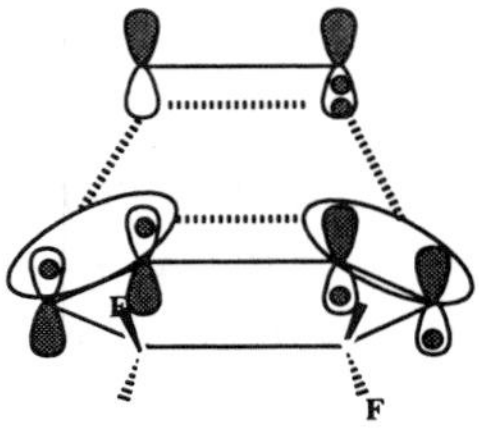

The diagram above explains simply how the transaction of two adjacent C—F dipoles in *trans*-5,6-difluoro-1,4-cyclohexadiene polarizes the π system in a way that effects hole–pair alternation.

References

1. M.J.S. Dewar, *The Molecular Orbital Theory of Organic Chemistry*, McGraw-Hill, New York, 1969.
2. N.D. Epiotis, W.R. Cherry, S. Shaik, R.L. Yates, and F. Bernardi, *Top. Curr. Chem.* 70, 1–242 (1977).
3. D.A. Dixon, T. Fukunaga, and B.E. Smart, *J. Am. Chem. Soc.* 108, 1585 (1986).
4. W.H. Fink, D.C. Pan, and L.C. Allen, *J. Chem. Phys.* 47, 895 (1967).
5. R.G. Pearson, *Symmetry Rules for Chemical Reactions*, Wiley-Interscience, New York, 1976.

6. N.D. Epiotis, *J. Am. Chem. Soc.* 95, 3087 (1973).
7. S. Wolfe, *Acc. Chem. Res.* 5, 102 (1972).
8. C. Romers, C. Altona, H.R. Buys, and E. Havinga, *Top. Stereochem.* 4, 39 (1969).
9. For a tabulation of the relative stability of X—CH_2CH_2—X rotamers and derivatives, see A. Yokozeki and S.H. Bauer, *Fortschr. Chem. Forsch.* 53, 71 (1975).
10. R.D. Amos, N.C. Handy, P.G. Jones, A.J. Kirby, J.K. Parker, J.M. Percy, and M.-D. Su, *J. Chem. Soc. Perkins Trans.* 2 549 (1992).
11. P. Klaboe and J.R. Nielsen, *J. Chem. Phys.* 32, 899 (1960).
12. G.T. Armstrong and S. Marantz, *J. Chem. Phys.* 38, 169 (1963). $ONOO^-$ and the conjugate acid are isoelectronic and isostructural to FNNF: J.-H.M. Tsai, J.G. Harrison, J.C. Martin, T.P. Hamilton, M. van der Woerd, M.J. Jablonsky, and J.S. Beckman, *J. Am. Chem. Soc.* 116, 4115 (1994). On the other hand, Bohle et al. calculate the trans $ONOO^-$ isomer to be more stable, with cis preference recovered in the sulfur derivatives: S. Bohle, B. Hansert, S.C. Paulson, and B.D. Smith, *J. Am. Chem. Soc.* 116, 7423 (1994).
13. D.D. Danielson, J.V. Patton, and K. Hedberg, *J. Am. Chem. Soc.* 99, 6484 (1977). For ab initio computations and references, see T. Clark and P.v.R. Schleyer, *J. Comput. Chem.* 2, 20 (1981).
14. M.T. Nguyen, H. Vansweevelt, T.-K. Ha, and L.G. Vanquickenborne, *J. Chem. Soc. Chem. Commun.* 1425 (1990), and private communication.
15. S. Walker, D. Gange, V. Gupta, and D. Kahne, *J. Am. Chem. Soc.* 116, 3197 (1994).
16. J.T. Edward, *Chem. Ind. (London)* 1102 (1955).
17. R.U. Lemieux, in *Profiles, Pathways, and Dreams*, J.I. Seaman, Ed., American Chemical Society, Washington, DC, 1990.
18. E. Juaristi and C. Cuevas, *Tetrahedron* 48, 5019 (1992).
19. (a) G.R.J. Thatcher, Ed., *The Anomeric Effect and Associated Stereoelectronic Effects*, ACS Symposium Series 539, American Chemical Society, Washington, DC, 1993.
(b) K.B. Wiberg and M. Marquez, *J. Am. Chem. Soc.* 116, 2197 (1994).
20. U. Ellervik and G. Magnusson, *J. Am. Chem. Soc.* 116, 2340 (1994).
21. T. Koritsanszky, M.K. Strumpel, J. Buschmann, P. Luger, N.K. Hansen, and V. Pichon-Pesme, *J. Am. Chem. Soc.* 113, 9148 (1991).

22. N.D. Epiotis, "Unified Valence Bond Theory of Electronic Structure," in *Lecture Notes in Chemistry*, Vol. 34, Springer-Verlag, Berlin and New York, 1983, p. 223.
23. (a) F. Bohlmann, *Angew. Chem.* 69, 641 (1957); *Chem. Ber.* 91, 2157 (1958). (b) K.B. Wiberg and P.R. Rablen, *J. Am. Chem. Soc.* 115, 614 (1993).
24. E. Juaristi, G. Cuevas, and A. Vela, *J. Am. Chem. Soc.* 1116, 5796 (1994).
25. P. Wipf and Y. Kim, *J. Am. Chem. Soc.* 116, 11678 (1994).
26. (a) J. Martens, *Top. Curr. Chem.* 125, 165 (1984). (b) J.D. Morrison and H.S. Mosher, *Asymmetric Organic Reactions*, Prentice-Hall, Englewood Cliffs, NJ, 1971.
27. M.T. Reetz, *Pure Appl. Chem.* 57, 1781 (1985).
28. N.T. Anh, *Top. Curr. Chem.* 88, 145 (1980).
29. D.A. Evans, A.M. Ratz, B.E. Huff, and G.S. Sheppard, *J. Am. Chem. Soc.* 117, 3449 (1995).
30. L.A. Paquette, B.M. Branan, R.D. Rogers, A.H. Bond, H. Lange, and R. Gleiter, *J. Am. Chem. Soc.* 117, 5992 (1995).
31. The VB model should not be confused with an "electrostatic" model of facial selectivity proposed by S.D. Kahn and W.J. Hehre [*J. Am. Chem. Soc.* 109, 663 (1987)]. According to these authors, syn attack of an electrophilic dienophile on, say, I is due to the presence of the heteroatom lone pairs rather than the action of the C—F dipole on the π system of the diene.

Part III

The Molecular I Bond

Chapter 12

What Is Aromaticity and How Has It Been Misinterpreted?

12.1 Cooperativity and Anticooperativity of One- and Two-Electron Delocalization

It is hard to imagine a theoretical principle that has fascinated organic chemists as much as Hückel's rule,[1] pivotally extended by Heilbronner,[2] and applied by Dewar, Zimmerman, Woodward and Hoffmann, and others.[3] Since aromaticity implies stabilization of a composite system relative to its component parts in isolation, and since the electronic basis of association is I-bonding, VB theory leads to the conclusion that *Hückel's rule is a necessary but not sufficient condition for association.* This, in turn, requires a basic revision of our thinking about molecules, in general. We begin with the case of a three-orbital cyclic system in which the AOs overlap in phase and they contain a total of two electrons. Examples are H_3^+ and the π system of cyclopropenyl system. These systems are called "Hückel aromatic" because they contain $4N + 2$ electrons within a Hückel cyclic AO array. The designation of the AOs and the three equivalent covalent VB configurations which are the dominant contributors to the total wavefunction are shown in Figure 12.1. With these definitions, we have:

$$<1|2>=<2|3>=<3|1>=s$$

$$<a|H|b>=<b|H|c>=<c|H|a>=H$$

$$H @ \left[\left(h_{13} + h_{11}s_{13} \right) + \left(h_{12}s_{23} + h_{23}s_{12} \right) \right] + (13|22) + (12|23)$$

$$H @ F(1-\mathrm{e}) + F(2-\mathrm{e}) + G(1-\mathrm{e}) + G(2-\mathrm{e})$$

H represents the action of two different CT delocalization mechanisms, namely, a 1-e and a 2-e hop, which operate in-phase. As a result, *H* is very large in absolute magnitude.

We assumed that all AOs overlapped in phase, that is, all overlap integrals are positive. Now, let us change the sign of one overlap integral, say the sign of s_{12}. This is equivalent to converting the Hückel to a Möbius AO array. We get:

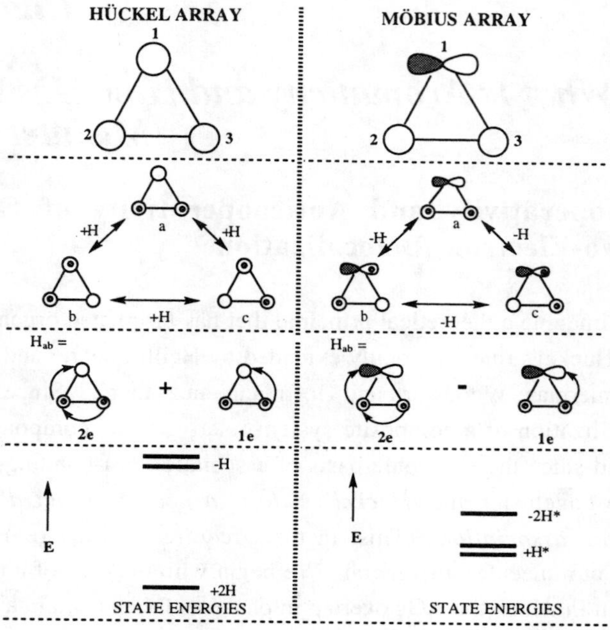

Figure 12.1 The VB interpretation of aromaticity and antiaromaticity in odd annulenes. There are two reasons for the superiority of the aromatic species: aromatic interaction of the low energy VB configurations and the fact that H(aromatic) is larger than H(antiaromatic) because the former represents the synergic (additive) while the latter the anticooperative (subtractive) effect of 2-e and 1-e CT delocalization.

$$H = \left[\left(h_{13} + h_{11}s_{13}\right) - \left(h_{12}s_{23} + h_{23}s_{12}\right)\right] + (13|22) - (12|23)$$

H represents now the anticooperative action of a 1-e and a 2-e hop, that is, the two delocalization mechanisms operate out-of-phase. As a result, *H* is very small in absolute magnitude.

By working in similar fashion, we arrive at the conclusion that the general form of the VB matrix element, H_{ab}, is:

$$H_{ab} = k\left(P_0 s^0 + P_1 s^1 + P_2 s^2 + \cdots P_n s^n\right)$$
$$+ \text{(corresponding repulsion integrals)}$$

Here, k is a constant, P_i a one-electron negative term, and s the AO overlap integral. There exist two situations:

(a) In Hückel cyclic AO arrays containing $4N + 2$ electrons, most, but not all, signs in the first set of parentheses in the preceding equation are positive. As a result, H_{ab} is large and interaction is strong. Such a system is called aromatic. The presence of any negative signs at all is due to interpair exchange repulsion. In Möbius cyclic AO arrays containing the same number of electrons, the fraction of negative signs is larger. Hence, H_{ab} is smaller and interaction much weaker. Such a system is called antiaromatic.

(b) In systems containing $4N$ electrons, the situation is reversed. Most signs are now positive in the case of Möbius cyclic AO array. A larger fraction of signs is negative in the case of a Hückel cyclic AO array.

Fischer and Murrell[4] and Mulder and Oosterhoff[5] were the first to recognize that at the level of VB theory, aromaticity and antiaromaticity are consequences of permutational symmetry. These important but unrecognized theoretical contributions set the stage for a reinterpretation of the VB wavefunction which, in turn, leads to concepts that have either little to do with what has been produced by MO theory or are even antithetical to it.

This is half of the VB aromaticity story. The other half concerns the way in which the sign of each H_{ab}, in addition to its magnitude, controls the strength of the cyclic interaction of the VB configurations. Assuming that each H that is preceded by a positive sign is a negative overall quantity, there exist two possible cyclic VB CIs.

(a) A cyclic CI in which the product of the signs of the H_{ab}s is positive. In this case, the VB configurations form a cyclic Hückel array, and this mode of interaction produces a highly stabilized ground state. This occurs in the case of a $4N + 2$ electron Hückel AO array or a 4N electron Möbius AO array (aromatic systems). An example is the interaction of the covalent configurations of the cyclopropenyl cation (Figure 12.1). We say that this system is *configurationally aromatic.*

(b) A cyclic CI in which the product of the signs of the H_{ab}s is negative. In this case, the VB configurations form a cyclic Möbius array and this mode of interaction produces a high energy doubly degenerate ground state. This occurs in the case of a 4N electron Hückel AO array or a $4N + 2$ electron Möbius AO array (antiaromatic systems). An example is the interaction of the covalent configurations of cyclopropenyl cation where one 2p has been replaced by a delta-type d AO (Figure 12.1). We say that this system is *configurationally antiaromatic.*

In an *n*-electron system, and in the frame of nonorthogonal VB theory, we can have all possible delocalization modes starting with one-electron (proportional to s^1) and ending with *n*-electron delocalization (proportional to s^n). We can subdivide these delocalization mechanisms into odd and even electron hops and examine whether these cooperate or anticooperate. For qualitative discussion, we can simply focus on the dominant representatives of the odd and even classes: the 1-e and 2-e delocalization mechanisms.

There can be no more general principle of chemical bonding than the notion of cooperativity and anticooperativity. Aromaticity is an expression of cooperativity. The important point is that orbital aromaticity (i.e., the conventional aromaticity concept) differs fundamentally from configuration aromaticity. What is, for example, orbitally nonaromatic at the EHMO level may be configurationally aromatic at the VB level. For example, the strong bond of the orbitally nonaromatic ground state triplet O_2 is the result of configuration aromaticity, as illustrated in Figure 12.2.

Our VB treatment of aromaticity is actually nothing other than an illustrator of high order, many-electron perturbation theory over VB structures. The device

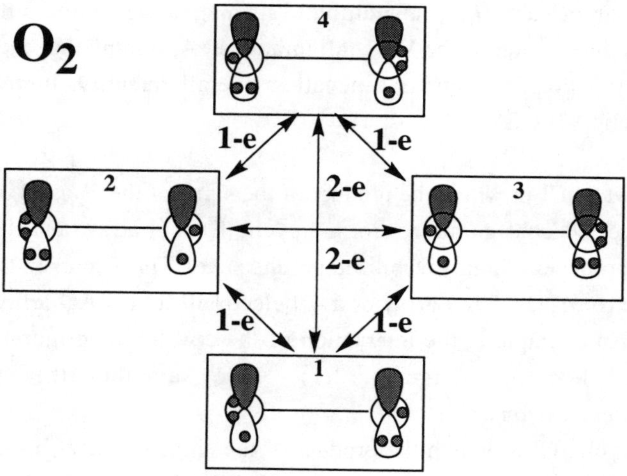

Figure 12.2 The 2-e CT hop defined by the interaction of the 1 and 4 VB configurations plays a key role in determining the stability of triplet O_2, in addition to the Hückel-type aromatic CI brought about via 1-e CT hops.

of the Hückel and Möbius configuration rings simply saved us the trouble of writing out long expressions for nth-order perturbation corrections over a complete Hamiltonian. Configuration aromaticity tells us how one configuration interacts with a second via a third configuration, and so on.

Bottom line: VB theory is not only the best conceptual road to understanding chemistry but also the best illustrator of the counterintuitive world of quantum mechanics. The interaction of two polyelectronic VB configurations represents an electronic reorganization that can be executed via more than one "path." What counts is whether the various paths act cooperatively or anticooperatively. Aromaticity and antiaromaticity represent the two opposite limits. Exactly the same ideas are now finding application in laser chemistry. Two different excitation paths leading to the same reaction product can interfere constructively or destructively, and this duality can be exploited to modulate the amount of product.[6]

12.2 Collective Delocalization and the VB Derivation of Hückel's Rule

Recall that there exist two types of collective delocalization, exchange and CT, as illustrated. The former conserves atom neutrality while the latter creates charge separation. It will be appreciated that exchange delocalization can always be brought about by cyclic CT delocalization and that the latter can be viewed as a form of disrupted exchange delocalization. We will now illustrate how permutational symmetry determines whether collective delocalization of the exchange type is favorable in an even annulene. Permutational symmetry controls collective CT delocalization in an analogous manner.

COLLECTIVE 3-e CT DELOCALIZATION

COLLECTIVE 6-e EXCHANGE DELOCALIZATION

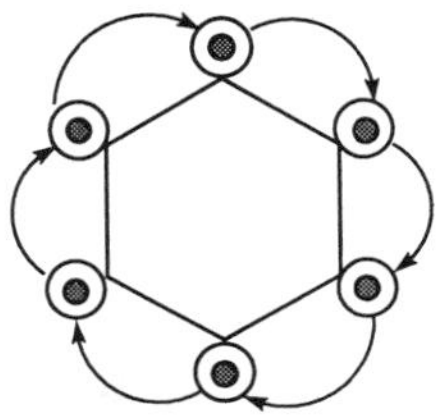

Consider an n-electron, even alternant annulene such as the π system of square cyclobutadiene. There are four primitive VB configurations which when combined according to standard procedures yield the following equivalent Kekulé structures, K_1 and K_2 (shown omitting normalization):

$$K_1 = \left|\bar{1}2\bar{3}4\right| - \left|\bar{1}\bar{2}34\right| - \left|12\bar{3}\bar{4}\right| + \left|1\bar{2}3\bar{4}\right|$$
$$K_2 = \left|\bar{1}2\bar{3}4\right| - \left|\bar{1}23\bar{4}\right| - \left|1\bar{2}\bar{3}4\right| + \left|1\bar{2}3\bar{4}\right|$$

Let us focus attention on the energy of K_1 and, in particular, to the energy due to the spin alternant primitives a and a^*

$$a = \left|\bar{1}2\bar{3}4\right|$$
$$a^* = \left|1\bar{2}3\bar{4}\right|$$

Now, $<a|H|a> = <a^*|H|a^*>$ and this is equal to the classical electrostatic energy of the system plus the "diagonal" exclusion repulsion of electrons of the same spin plus a higher order attractive overlap term. The spotlight falls on the interaction element $<a|H|a^*>$. Our first task is to line up the spin orbitals in the second determinant so that they match the spin orbitals of the first in terms of spin. We have

$$<\left|\bar{1}2\bar{3}4\right|H\left|1\bar{2}3\bar{4}\right|> \rightarrow <\left|\bar{1}2\bar{3}4\right|H\left|\bar{2}3\bar{4}1\right|>$$

The transposition of $\left|\bar{1}2\bar{3}4\right|$ to $\left|\bar{2}3\bar{4}1\right|$ is an odd cyclic permutation with length k-1 = 3. The integral $<\left|\bar{1}2\bar{3}4\right|\mathrm{H}\left|\bar{2}3\bar{4}1\right|>$ is made up of four terms corresponding to one zero, two single, and one double exchange of electrons having the same spin on the right side of the operator. The zeroth-order exchange term represents collective exchange delocalization of the highest order (i.e., fourth order), and it is proportional to the product of the overlap integrals of neighboring AOs, s_{ij}.

$$<\left|\bar{1}2\bar{3}4\right|H\left|\bar{2}3\bar{4}1\right|> \quad @ \quad s_{12}\, s_{23}\, s_{34}\, s_{41}$$

Since the sign preceding this term is negative, collective delocalization acts to *destabilize* the system, provided the product of the s_{ij} overlap integrals is positive (i.e., provided there are an even number of phase changes of the overlapping AOs). This occurs because the spin alternant primitives in the K_1

Kekulé structure, *a* and *a**, have the same sign. The reason for this is that singlet coupling of *one electron pair* requires that a *single* transposition of the spin label (while the order of the spatial AOs remains fixed) results in sign reversal:

$$|\bar{1}2| \xrightarrow{\text{spin label transposition}} -|1\bar{2}|$$

It follows that the two spin alternant primitives of every even annulene have either the same or opposite signs depending on $(-1)^p$, where p is the number of electron pairs. This leads to a prediction: Even annulenes with p = even ($4N$ electrons) are destabilized while even annulenes with p = odd ($4N + 2$ electrons) are stabilized in a relative sense, provided the product of the AO overlap integrals of neighboring atoms is positive. This is equivalent to Hückel's rule. Our analysis demonstrates that Hückel's rule is an expression of the Pauli principle subject to orbital topology. Orbital topology is critical because orbitals have phases. In turn, the phase properties of an AO are manifestations of atomic structure. *Hence, Hückel's rule is an expression of the Pauli principle and atomic structure.*

Heilbronner's rule of Möbius aromaticity and antiaromaticity is the counterpart of Hückel's rule: When the product of the AO overlap integrals of neighboring AOs is negative (odd phase changes of overlapping AOs), stabilization occurs when the spin alternant primitives have the same sign. As a result, Hückel's rule is reversed.

The VB picture of aromaticity is collective delocalization in which each electron hop is assisted by the simultaneous hopping action of all remaining electrons. This is a "nonclassical" consequence of electron indistinguishability and atomic structure. Hence, there can be no more fundamental principle of electronic structure than the concept of aromaticity. When formulated in VB theoretical terms, this principle unlocks many secrets. When formulated in Hückel MO theoretical terms, it can lead to confusion. We will discuss this issue shortly. Henceforth, VB aromaticity will be distinguished from HMO-aromaticity either explicitly or implicitly in all discussions.

We extracted the Hückel selection rule by focusing on n-e exchange delocalization. Since this is a function of s^n, the stabilization conferred on the system is small. This leads us to confront the central issue: How does aromatic stabilization via CT delocalization come about? As discussed before, CT delocalization requires a dense set of configurations. *Aromaticity cannot be turned on unless this condition is met.*

12.3 Hückel's Rule as a Necessary But Not Sufficient Condition for Molecular Stability

What is the connection between aromaticity and pair association? The answer is most clearly obtained from inspection of the VB wavefunctions of the antiaromatic Hückel (symbolized by *K*) and the aromatic Möbius (symbolized by *M*) cyclobutadienes shown in Table 12.1. This is an important table to which we will be referring time and again. The essential features are:

(a) A ground aromatic system is one in which the critical, *association-promoting* ALT and HRP (hole, radical, pair) configurations can participate. An association-promoting ionic configuration is defined as the configuration that makes either an exclusive or dominant contribution to the aggregated (relative to the segregated) ensemble of electron pairs.

(b) A ground antiaromatic system is one in which either some or all of the association-promoting ALT and HRP configurations are not "allowed" to participate by symmetry. In the case of the Hückel cyclobutadiene, neither the HRP nor the ALT configurations can participate by symmetry!

The emerging conclusion is that a homonuclear aromatic complex has access to interactions that could render it stable relative to the noninteracting component bonds. However, *the seminal point is that this potentiality will come to fruition only if the orbital electronegativity is either modest or low!* By contrast, the game is already lost for the ground antiaromatic complex: Symmetry precludes the action of the very configurations that are necessary for promoting aggregation, namely, the HRP and ALT configurations. Thus, regardless of orbital electronegativity, a ground antiaromatic complex is condemned to be destabilized relative to the noninteracting components. The message is clear: *The terms "aromatic," "nonaromatic" and "antiaromatic" based on electron count alone are meaningless.* In the absence of consideration of the key variable of orbital eletronegativity, it is fruitless to make any prediction regarding the tendency of a system of bond pairs to aggregate or segregate by falling back on the aromaticity yardstick.

The overview developed in this section is simple: Aromatic and antiaromatic systems represent, respectively, the best- and worst-case scenarios for I-bonding. The latter represents the extreme at which 2-e CT delocalization is effectively turned off. To see clearly this point, note that the interaction of the covalent with the ALT and the HRP configurations defines two different types of 2-e CT hop which are turned off in the antiaromatic complex. Thus,

Table 12.1 The Three Singlet States (K, D, Z) of Hückel and the Ground State of Möbius (M) 4e-4c Systems

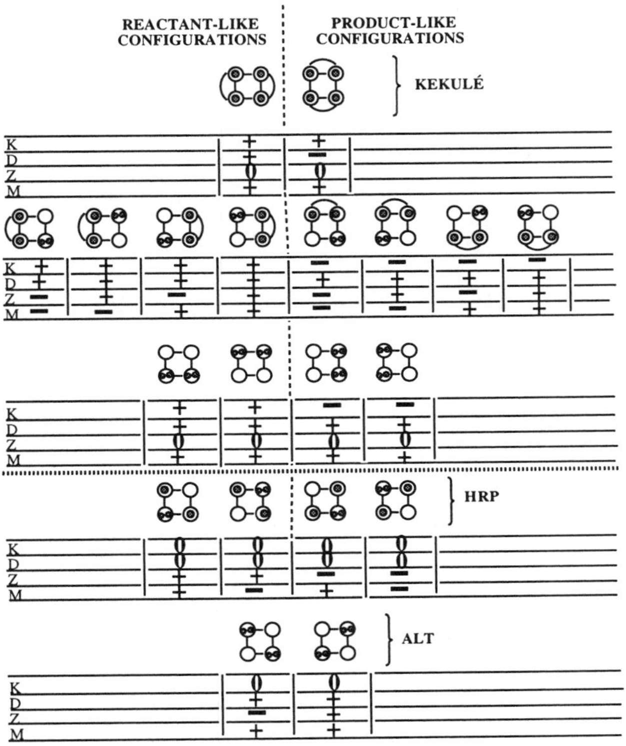

the very concept of aromaticity is implicitly related to the second dimension of chemical bonding: I-bonding. Since this represents kinetic energy reduction with concomitant minimization of interelectronic repulsion, only CI-type calculations can cope with aromatic systems. In this connection, the all-carbon [18]annulene prefers an acetylenic structure at the SCF level only to adopt the symmetrical D_{18h} aromatic cumulene form when electron correlation is included.[7]

References

1. For early review and applications of HMO theory, see A. Streitwieser Jr., *Molecular Orbital Theory for Organic Chemists*, Wiley, New York, 1961.

2. E. Heilbronner, *Tetrahedron Lett.* 1923 (1964).
3. For review, see N. Epiotis, S. Shaik, and W. Zander, in *Rearrangements in Ground and Excited States*, Vol. 2, P. deMayo, Ed., Academic Press, New York, 1980.
4. H. Fischer and J.N. Murrell, *Theor. Chim. Acta* 1, 464 (1963).
5. (a) J.J.C. Mulder and L.J. Oosterhoff, *J. Chem. Soc. Chem. Commun.* 52 (1970). (b) W.J. van der Hart, J.J.C. Mulder, and L.J. Oosterhoff, *J. Am. Chem. Soc.* 94, 5724 (1972).
6. (a) M. Shapiro, J.W. Hepburn, and P. Brumer, *Chem. Phys. Lett.* 149, 451 (1988); M.Shapiro and P. Brumer, *Int. Rev. Phys. Chem.* 13, 187 (1994). (b) V.D. Kleiman, L. Zhu, X. Li, and R.J. Gordon, *J. Chem. Phys.* 102, 5863 (1995). (c) Related review: B. Kohler, J.L. Krause, F. Raksi, K.R. Wilson, V.V. Yakovlev, R.M. Whitnell, and Y. Yan, *Acc. Chem. Res.* 28, 133 (1995).
7. V. Parasuk, J. Almlöf and M.W. Feyereisen, *J. Am. Chem. Soc.* 113, 1049 (1991).

Chapter 13

VB Aromaticity Is Different from Hückel Aromaticity

13.1 Aromaticity in Acyclic Systems; The Concept of Overlap Dispersion

We have already seen that there are two criteria for aromatic stability: the right electron count ($4N + 2$ for Hückel and $4N$ for Möbius strips) and low orbital ionization energy. HMO theory, the backdrop of Hückel's rule, errs in failing to respect the second criterion. We will now see that once electron repulsion is admitted in the form of bielectronic repulsion integrals, aromaticity and antiaromaticity are manifest even in acyclic systems!

Consider an ML system in which each of the components, M and L, has two nondegenerate orbitals and two electrons. The interaction of the ground (no bond-type) and the two independent open-shell, doubly excited configurations defines Overlap dispersion (Figure 13.1a). The interaction of configurations I and II is effected by a 2-e hop that can be executed in two distinct ways as indicated in Figure 13.1b.

The hatched pair of arrows signifies a CT 2-e hop proportional to ab^* overlap times ba^* overlap, while the solid pair signifies a 2-e hop representing the interaction of two transition dipoles. This means that 2-e CT delocalization operates in phase with dispersion. The interaction matrix element H_{12} takes the following form, where the leading term of OV is a *negative* term proportional to $s_{ab^*} \times s_{ba^*}$ and the leading term of DISP is the *positive* $(aa^*|bb^*)$ repulsion integral.

$$H_{12} = \text{OV} - \text{DISP}$$

Overlap dispersion comes about because the sign of the preceding equation is *negative*, in which case the 2-e hop indicated by the hatched arrows (overlap) is assisted (algebraic addition) by the 2-e hop indicated by the solid arrows (dispersion). When the number of electrons changes from four to either two or six, the sign changes to positive and we have *anti overlap dispersion* .

Next, we observe that the two independent open-shell configurations II and III differ in the sense that II involves inter- while III involves intrafragmental

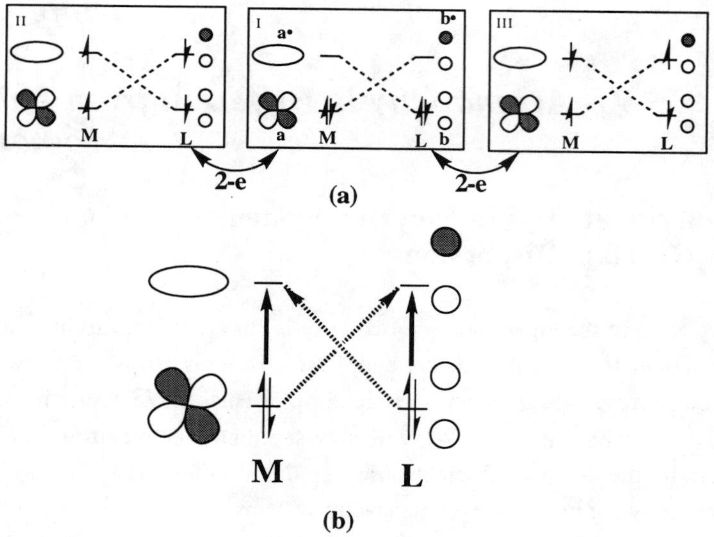

Figure 13.1 The concept of overlap dispersion. (a) CI due to a 2-e hop which can occur by two distinct routes, one inter- and the other intrafragmental. (b) The interaction matrix element, which represents overlap dispersion (i.e., the cooperative action of 2-e CT delocalization and dispersion); 2-e CT delocalization is denoted by solid arrows and dispersion by hatched arrows.

pairing. The interaction of the two fragments in II is attractive because of the presence of two bonds. By contrast, the interaction in III is repulsive because of the presence of a four-electron exchange antibond. As a result, the energy minimum of the two configurations occurs at different interfragmental distances. We can think of II and IIIas the "tight" (minimum at short distance) and "loose" (minimum at long distance) configurations, respectively. Both II and III interact with the no bond configuration I in an overlap dispersion sense. The only difference is that the I–II interaction gives greater weight to the overlap term while the I–III interaction does exactly the opposite. This means that overlap dispersion can occur either at two different geometries, a tight and a loose, or at some compromise geometry of intermediate nature. This raises the possibility of valence isomerism as a consequence of overlap dispersion: The "tight" isomer will have primarily perfect pairing assisted by dispersion and the species will resemble, but only superficially, a covalent molecule. By contrast, in the "loose" isomer, the overlap component of overlap dispersion will eliminate all or part of the inherent four-electron antibond present in configurations I and III, leaving the dispersion term to act as the binding force. This species will resemble a van der Waals complex which, however, is strongly bound because exchange repulsion between A and B has been largely eliminated by the overlap component of overlap dispersion. Hence, one anticipates the existence of a pseudocovalent and a pseudo–van der Waals complex as distinct isomers having comparable stability!

13.2 The Physical Basis of Overlap Dispersion and Anti Overlap Dispersion

By focusing on the way in which 1-e and 2-e hops combine in the off-diagonal matrix elements connecting nonorthogonal electronic configurations of different types, one generates rules for interfragmental interaction. In general, when each of two fragments has a pair of orthogonal orbitals, the electron counts and spin multiplicity consistent with overlap dispersion are:

Two or six electrons overall triplet-coupled.
Four electrons overall singlet-coupled.

A typical example of the first category is the π bonding of the O_2 molecule, as illustrated below. This explains why the restricted Hartree–Fock bond dissociation energy of O_2 is only one-fourth the experimental value.[1] An example of the second category is the binding of organometallic complexes, which has already been exemplified (see Figure 13.1) and will be discussed extensively later.

O_2

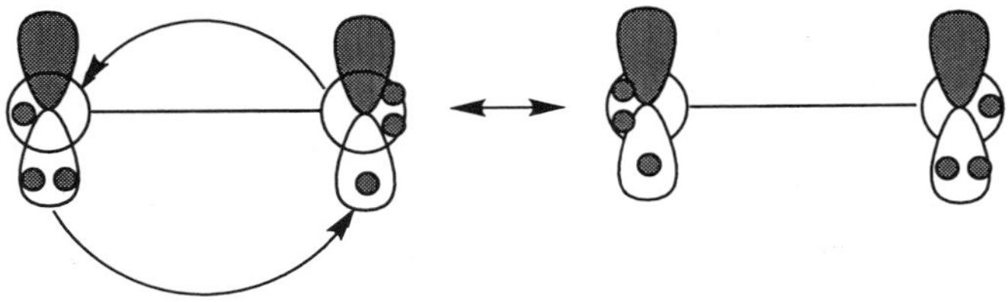

What is the physical meaning of overlap dispersion and anti overlap dispersion in singlet molecules? The answer is that the electron count decides whether two atoms can deshield so that one attracts the other by dispersion while still maintaining perfect pairing. In other words, the electron count decides

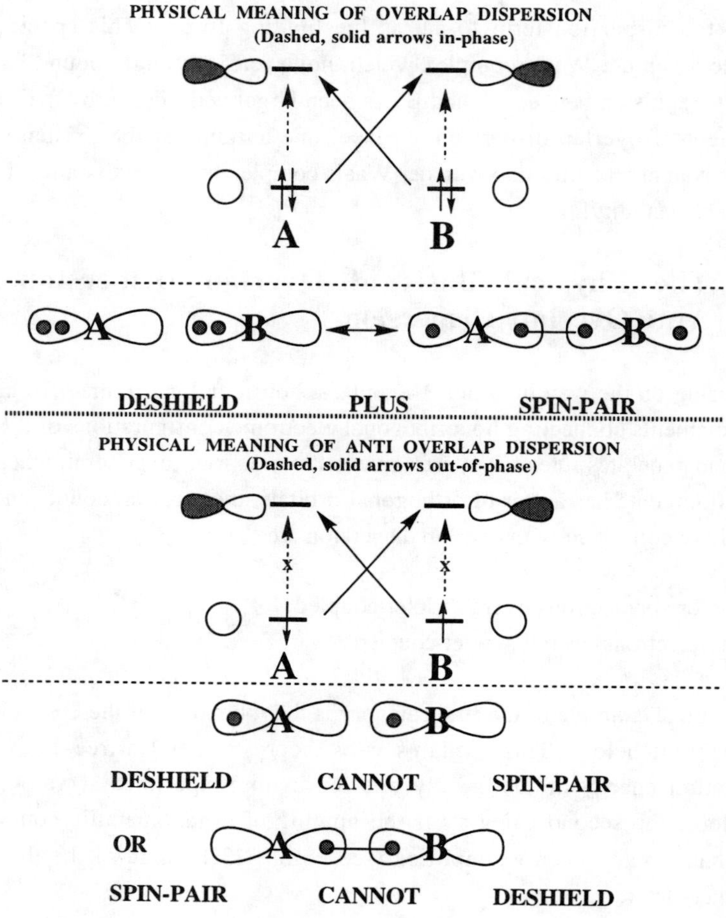

Figure 13.2 Cooperative and anticooperative action of inter- and intrafragmental CT.

whether intrafragmental CT delocalization (producing induction and dispersion) can be phased in with interfragmental delocalization of the exchange or CT type. In the case of 4 ($4N$ number) electrons, we have cooperativity, but, in the case of the 2 or 6 ($4N + 2$ number) electrons, we have anticooperativity. In the case of odd 3 or 5 electrons, we have an intermediate situation. The argument is illustrated schematically in Figure 13.2.

We can get an early appreciation of the diverse implications of the concept of overlap dispersion by recalling that the two commonly invoked mechanisms of energy transfer, namely, the Förster dipole–dipole mechanism[2] and the electron exchange mechanism,[3] are not two independent processes but, rather, two distinct pathways connecting the same two terminal configurations, as illustrated in Figure 13.3. This means that when the overlap of the two chromophores is non zero, one mechanism promotes the other (overlap dispersion). Exactly how strong the effect will be depends on the the validity of the "four-orbital/four-electron" approximation. Some recent studies aimed toward mimicking photosynthetic energy transfer are already providing crucial information.[4]

The concept of overlap dispersion had its birth in the original treatise of VB aromaticity when we recognized that a single 2-e hop in a Hückel antiaromatic system (e.g., π cyclobutadiene) can be executed in two distinct ways corresponding to two attractive overlap terms which, however, subtract algebraically. Going to a diatomic (polyatomic) molecule with two orthogonal AOs (MOs) per atom (fragment) generates a situation in which the positive overlap term becomes a negative dispersion term. We now have *algebraic addition* and overlap dispersion. Two distinct 2-e hops connect the same association-promoting ALT configuration with one of the covalent Kekulé structures in cyclobutadiene:

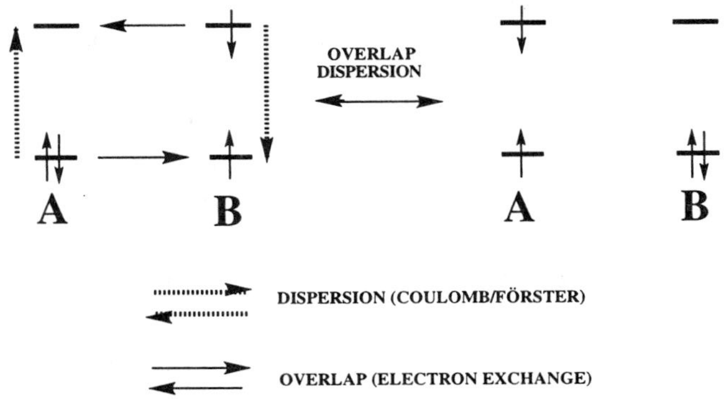

Figure 13.3 Energy transfer by the mixing of two locally excited configurations. When overlap is zero, the mixing is of the excitonic type. When overlap is nonzero, the mixing is due to the in-phase action of two terms: the Coulombic (excitonic) interaction and electron exchange. Energy transfer cannot be discussed by assuming independence of the two mechanisms. The two can combine either in or out of phase, as discussed in the text. The chromophore electron count may determine the mechanism as well as the efficiency of energy transfer!

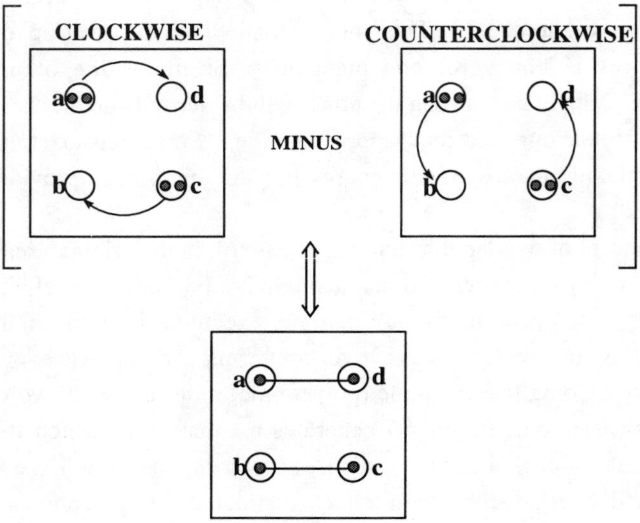

In this work, development of the theoretical themes takes a back seat to chemical applications of the fundamentals. However, we cannot forego the opportunity to mention the way in which "chirality" enters into the VB energy matrix. Specifically, the preceding example illustrates that CT delocalization can occur via clockwise and counterclockwise electron hops either in phase (aromatic case) or out of phase (antiaromatic case). Thus, the VB theory of aromaticity relies implicitly on the notion of handedness.

13.3 Overlap Induction and Anti Overlap Induction

Induction and dispersion are two polarization mechanisms that can either assist or antagonize CT delocalization. This is a consequence of a fundamental phenomenon that has remained unrecognized because nonorthogonal VB theory has been used, though very infrequently, as a computational rather than as a conceptual tool. The fundamental point is that there exist more than one "pathway" for going from one VB configuration to another and, depending on the orbital basis, these different "paths" may be overlap dependent as well as overlap independent (induction and dispersion). As a result, the concept of overlap induction is exactly analogous to the concept of overlap dispersion, and the same is true of their "anticooperative" forms, namely, anti overlap induction and anti overlap dispersion. Again, whether we have cooperativity or anticooperativity

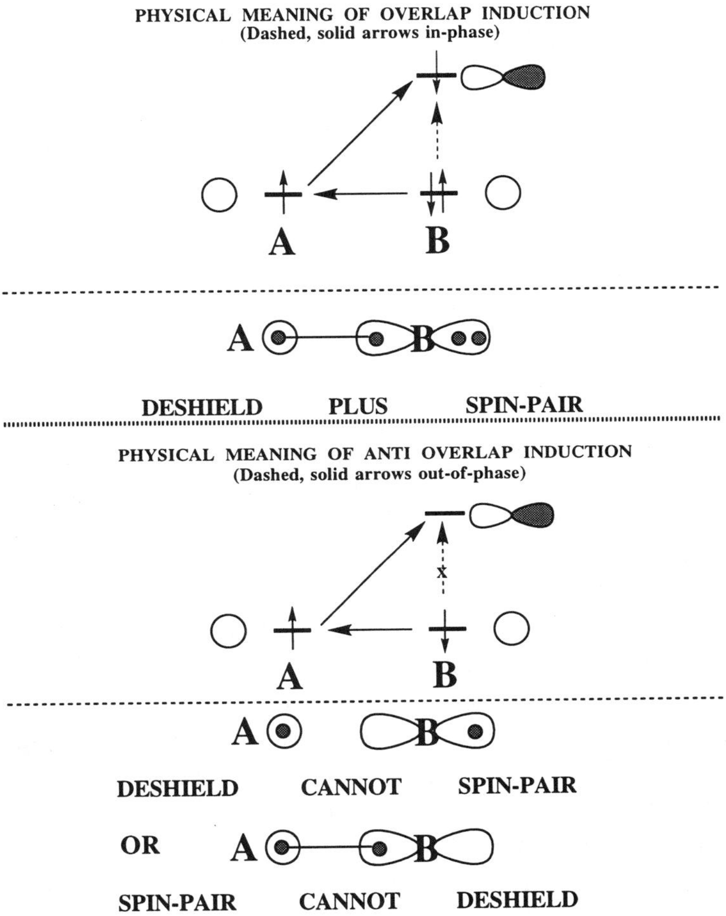

Figure 13.4 Overlap and anti overlap induction.

depends on the number of electrons. Figure 13.4 explains schematically the physical meaning of overlap induction and anti overlap induction.

13.4 VB Aromaticity as the Mechanism of Strong Electron Pairing

The mechanism of stabilization of the two-electron/three-center (2e-3c) π-aromatic cyclopropenyl cation involves the cooperative action of 1-e (solid arrow) and 2-e (two dashed arrows) delocalization as illustrated in Figure 13.5, in

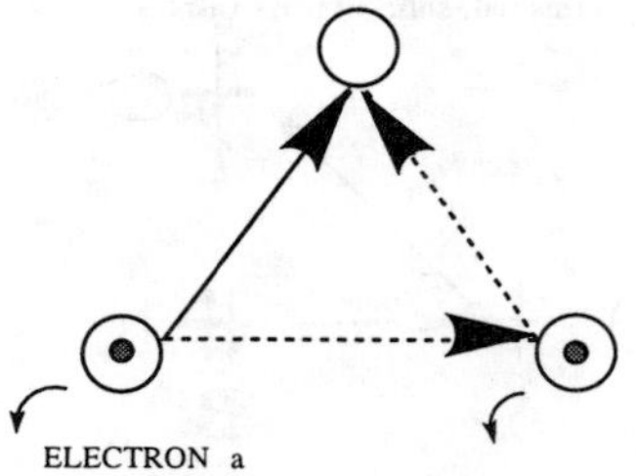

Figure 13.5 The singlet coupling of two electrons by a hole.

which electron a moves by effectively "dragging" electron b. This movement arises because electron a can either hop directly to the hole or it can collide, in a billiards sense, with electron b in a way that directs it to the hole. Because electrons are indistinguishable, we can say that actually electron a can arrive at the hole via two different routes, provided it is singlet-coupled to electron b. This means that the presence of the hole (needed for delocalization) has effected the singlet-coupling of two odd electrons in the 2e-3c aromatic π system.

We can reformulate ground state VB aromaticity as the mechanism by which two odd electrons are singlet-coupled by an effector system. In the case of the 2e-3c (Hückel) aromatic system, the effector was a hole. In a 4e-3c (Möbius) aromatic system, the effector is a pair. Alternative designations are possible. For example, our 2e-3c aromatic system can be viewed as the coupling of a radical and a hole by another radical subject to the condition that the two radicals be singlet-coupled.

We depart the subject carrying the following idea: Coupling of different elements such as holes, radicals, and pairs can be brought about by effector holes, radicals, or pairs. This coupling is most effective when our system is an aromatic system. We will use these ideas to tackle the high T_C superconductivity problem.

References

1. H.F. Schaefer, *J. Chem. Phys.* 54, 2207 (1971).
2. (a) T. Förster, *Ann. Phys.* 2, 55 (1948).
   (b) T. Förster, *Disc. Faraday Soc.* 27, 7 (1959).
3. (a) D.L. Dexter, *J. Chem. Phys.* 21, 836 (1953). (b) For an excellent review of energy transfer and applications to organic photochemistry,

see N.J. Turro, *Modern Molecular Photochemistry*, Benjamin Cummings, Menlo Park, CA, 1978, Chapter 9.

4. D. Gust, T. Moore, and A. Moore, *Acc. Chem. Res.* 26, 198 (1993).

Chapter 14

Resonance Theory and Hückel MO Theory Are Unphysical Versions of VB Theory

14.1 Covalent Resonance Cannot Avert Bond Segregation

Asked to describe the electronic structure of aromatic benzene, the chemist writes the two equivalent Kekulé resonance structures and claims stabilization. Since, however, two equivalent Kekulé structures also can be written for antiaromatic cyclobutadiene, it must be that what makes a molecule aromatic or antiaromatic has little to do with covalent resonance! We will now argue that π-conjugated systems provide us with the first glimpse of the I bond, but in incipient rather than full-blown form. In other words, aromatic molecules provide the first piece of evidence in support of the proposition that covalency represents a restriction.

Applied mathematicians worry about implementing a calculation of matrix elements in the computer. We worry over what matrix elements mean for chemistry. The π system of 1,3-butadiene (BU) is described by the two well-known covalent structures shown in Figure 14.1. Their interaction defines exchange bond (solid lines) resonance as well as exchange antibond (hatched lines) resonance.

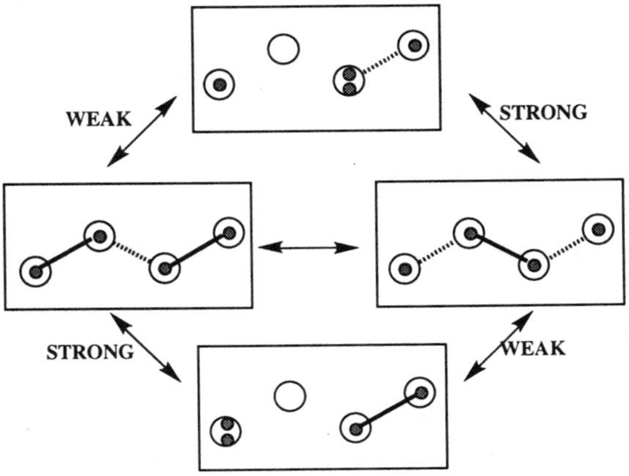

Figure 14.1 Selective interaction of covalent an ionic configurations of π-1,3-butadiene.

Resonance theory is based on a fundamental misinterpretation of the physical meaning of the covalent resonance hybrid, denoted CRH. Specifically, it is thought that covalent resonance is the agent of stability (i.e., association). By contrast, our argument is that this type of resonance loses out to the exchange repulsion (existing in each resonance structure) and that the latter drives the system to segregation. In other words, the two π bonds defined by the PP structure move away from each other so as to delete the hatched antibond in preference to staying associated for the benefit of covalent resonance. If the σ frame were absent, the two π bonds of BU would fly apart.

The interaction of two or more bond pairs was treated nearly half a century ago by "covalent-only" VB theory of the Eyring–Pauling type.[1] In Sandorfy's superb monograph on the electronic spectra of organic molecules,"[2] one finds the following results:

The two π bonds of BU are destabilized by $0.5A$, where A is the VB resonance (or exchange) integral, relative to two π bonds infinitely apart.

The three π bonds of aromatic benzene are also destabilized by $0.5A$ relative to three π bonds infinitely apart!

The message is straightforward: Even when given the best chance, as in the case of the unconstrained H_6 system, covalent resonance cannot effect association. Three H_2 molecules infinitely apart are more stable than hexagonal "aromatic" H_6. In other words, covalent resonance *alone* fails to overwhelm interpair exchange repulsion. Nature itself suggests that this argument is correct: If covalency were the exclusive factor, chemical and physical properties of molecules would be independent of the nature of the constituent atoms; for example, three A_2 molecules would always be more stable than an A_6 hexagon.

14.2 The Physical Meaning of Covalent Resonance

Ab initio calculations show that the interaction of the covalent structures of benzene is weak.[3] So, if covalent resonance cannot produce stabilization, does it serve any useful purpose at all? What is the correct physical interpretation of the CRH? The answer is that this is merely the launching pad of a superior type of bonding in which I bonds replace T bonds. The I bonds are produced by (assisted) 2-e delocalization, with the CRH acting as the reference state. Covalent resonance is merely a necessary step for guaranteeing CT delocalization conforming to the association rule. This CT delocalization is represented in the following way, with the arrows symbolizing the "allowed" 1-e, 2-e, and 3-e CT hops:

I-FORMULA

The three arrows that combine to form 1.5 I bonds can act only because the reference state is a resonance hybrid of covalent structures, each of which is now interpreted as defining a "CT delocalization corridor." Let us see what this term actually means.

The PP configuration of BU can interact with two singly ionic structures that differ in the placement of the zwitterion. One ionic structure localizes the zwitterion where the PP has a covalent bond. As a result, there is strong interaction of the two VB configurations. The other ionic structure places the zwitterion where the PP has a covalent antibond and, as a result, it is incapable of strong interaction with the PP. The interaction matrix element is a fraction of the preceding one. We say that the PP can promote strong CT delocalization only between atoms that are linked by exchange bonds within the PP. This is called intrabond CT delocalization. Thus, we conclude that CT along *each* of the three pairs of atoms can originate only from a linear combination of covalent structures. Thus, we come to view the CRH as the departure point for the formation of the superior I bonds which, by their very nature, cause association.

Will the two π bonds of BU associate? The three arrows in our diagram define 1.5 I bonds with T character (covalent character). The latter can be either dominant or secondary, depending on atom electronegativity. Carbon is an electronegative nonmetal. Hence, T character (covalent resonance) is dominant, I bonding (CT delocalization) is weak, and association is averted. To associate, π bonds have to go beyond covalency. If they are unsuccessful, they fail to associate.

What is the best chance for producing I bonds? It is afforded by an aromatic system in which the interaction of the covalent and ionic structures is enhanced by the mechanisms discussed in Chapter 12. However, even benzene cannot remove the major stumbling block: Carbon is too electronegative; therefore, the ionic configurations lie high (with respect to the covalent structures), and association is averted again. However, the substantial I character of the bonds endows benzene with the special stability we all know. In other words, I-character works to neutralize a part of the destabilization of the π system caused by the exchange antibonds. Benzene has special stability because of significant intrusion of a new type of bond: the I bond. This point is amplified in the next chapter.

14.3 Why Are There Reaction Barriers?

If we denote the exchange repulsion of the PP configuration by P_{EX}, the exchange attraction due to resonance by R_{EX} (exchange resonance), and the attraction due to the interaction of the covalent structures with higher lying ionic configurations by R_{CT}, the barrier height of a bond switch reaction such as the degenerate valence isomerization of three H_2 molecules via an "aromatic" hexagonal transition state is $P_{EX} - R_{EX} - R_{CT}$. Unless R_{CT} is very large, the barrier will be preserved, although diminished relative to that calculated in the hypothetical absence of CT delocalization. The salient point is that the connection between barrier height and the nature of the reactants rests primarily on CT delocalization (i.e., on R_{CT}): The higher the orbital ionization energy (IE) of the constituent atoms of a homonuclear ensemble, the less favorable is CT delocalization, the lower the magnitude of R_{CT}, and the higher the reaction barrier. A barrier arises when the intrinsically superior I bonds fail to replace the inferior T bonds. In other words, *barriers exist because the reactants cannot remove the covalency restriction.*

14.4 Hückel MO Theory and the Association Catastrophe

Whether said explicitly or implied, the concepts of modern applied quantum chemistry (*not the actual computations*) are based on extended Hückel MO theory.[4] Open any chemical journal, find a theoretical paper dealing with molecular electronic structure, and be certain that in the vast majority of cases, an *orbital interaction* will be invoked to explain some experimental observation or computational result.

Consider the description of the single bond of H_2 by the HMO method but now translating from MO to VB language. To do so, one may use Hückel VB (HVB) theory, which is VB theory with the integral approximations of HMO theory. HVB and HMO are equivalent theories (HVB = HMO). In HVB theory with neglect of overlap, exchange delocalization is wiped out and only CT delocalization remains; that is, the interaction of the diradical (HH) with the zwitterionic (H^+ $H:^-$) configuration. Because of neglect of electron–electron repulsion, these two configurations are *degenerate*, and their interaction defines CT delocalization. This is proportional to the AO resonance integral *b*, which, in turn, is proportional to the first power of overlap. This is why the stability of a system in HMO theory is measured in units of *b*, in contrast to Heitler–London VB theory, where stability is measured by the VB exchange integral *Q* proportional to *bs*.

Extended Hückel VB theory is VB theory with the integral approximations of EHMO theory (EHVB = EHMO). Now, the AO overlap integral survives but the bielectronic repulsion integrals are still set equal to zero. As a result, binding still depends principally on CT delocalization much as in HMO theory. The net result is that both HMO and EHMO methods reproduce binding as a result of kinetic energy reduction without any regard for the consequences of electron–electron repulsion (which is neglected). This is referred to as *unrestrained CT delocalization* or *hyperdelocalization*. It follows that EHMO theory is a construct in which artificially strong CT delocalization (brought about as a result of neglect of interelectronic repulsion) hopefully compensates for the missing static mechanisms of bonding (S-bonding). It reproduces a band within the T/I continuum that is displaced toward the I end of the bonding continuum. As a result, EHMO theory has a greater chance of success with metallic than with nonmetallic systems. In other words, *EHMO theory is a "metallic" theory*. This contradicts the intuition of many, who think that the more CI is needed for the description of a system at the MO level, the greater the illegitimacy of EHMO method. By contrast, our expectation is that EHMO theory should actually have greater difficulty with nonmetal systems! This is revealed by the observation that an aromatic or nonaromatic array of bond pairs always has lower energy than the isolated components irrespective of the orbital ionization energy. At the HMO level with neglect of overlap, even an antiaromatic array can be as stable as the isolated components! As a result, EHMO theory makes two predictions that contravene the experimental reality:

It predicts universal aggregation of nonmetals as well as metals. EHMO predicts that regular hexagonal H_6, π benzene, and Li_6 are all global minima. In general, homonuclear systems with more bonding than antibonding pairs will tend to aggregate because of the artificial degeneracy of covalent and ionic configurations.

It predicts that all arrays (aromatic, nonaromatic, and even antiaromatic) are more stable than the corresponding diradical species.

These two facts represent the *association catastrophe*.

One way out of this difficulty is to force HMO theory to artificially inhibit CT delocalization. For example, Longuet-Higgins and Salem[5] compared a uniformly bound polyene, U, in which all bond lengths are equal, to the isomer having bond alternation, R. They chose the HMO resonance integral for the U species to be b while those corresponding to the "short" and "long" bonds of the bond-alternant R species to be b(s) and b(l) so that the following condition is met, assuming absolute values of bs:

$$b = [b(\mathrm{s})b(\mathrm{l})]^{1/2}$$

However, this means that

$$\frac{b(\mathrm{s}) + b(\mathrm{l})}{2} > b$$

In other words, the arithmetic is greater than the geometric average.

As a result, bond alternation is favored simply because the resonance integrals have been made effectively larger. Since HMO energies depend exclusively on the resonance integrals, this prediction follows: Every π system should exhibit bond alternation if the resonance integrals are chosen as above and the σ bonds are neglected. When the preference of the σ bonds for the symmetrical U form is parametrized in the calculation, a compromise is reached and, depending on the parameters, bond alternation sets in at a given number of atoms.

The choice of variable b's *simulates* bond alternation but cannot alter the incorrect physics of HMO theory. This is seen in the case of the allyl cation. Now, VB theory predicts bond uniformity because there is no longer interbond exchange repulsion to enforce segregation! By contrast, choosing the resonance integrals as above will result in the prediction of bond segregation in the frame of HMO theory. A "short/long" combination of bonds will always be superior to a "uniform intermediate" combination of bonds because of the inequality shown above. However, there is a still larger problem. Instead of two, there are actually four choices of polyene bond lengths: "short/long," "uniform long," "uniform intermediate," and "uniform short." In a "perfect" ab initio calculation, the system chooses among these four possibilities. The problem of HMO theory is neglect of interelectronic repulsion: Because of hyperdelocalization, HMO theory *always* predicts that nonaromatic and aromatic systems will prefer the "uniform short" mode.

We conclude that chemical bonding is dependent on electron repulsion and that failure to reproduce this properly leads to a fictitious world in which the smooth transition from the T- to the I-bonding mechanism disappears.

Whenever the predictions of EHMO theory are qualitatively right, there has been a happy error cancellation. This methodology, the very foundation of conventional argumentation, is *neither quantitatively nor qualitatively* a proper descriptor of chemical bonding. Of course, this is not an indictment of the EHMO method alone. The VB equivalent of EHMO theory is EHVB theory, and the same criticism applies there. Every theory (MO-type, VB-type, etc.) that makes the same integral approximations as EHMO theory will also fail. The "VB explanations" of the literature are actually "EHVB = EHMO explanations" masquerading as "VB explanations." Many will argue: Why not use the wrong model if it produces the right result? The answer is that such a philosophy leads to the development of local models that work here but not there. This is the defining characteristic of a chemist: the ability to pick the right model for the right problem! By contrast, in this work, we are interested in the right model for all problems. As we shall see, by getting things right for the right reason, we will reap the benefit of new recipes for real-life chemistry.

"Thou shall not speak ill of a theory unless you fully understand it." In the mid-1960s, I was impressed by some adherents of ab initio methods who delighted in criticizing the Woodward–Hoffmann rules and the EHMO method, in general, as being "theoretically deficient." When asked to pinpoint what exactly was physically wrong with this methodology beyond mere numerical accuracy, they could offer only vague, formalistic litanies. One favorite pronouncement was: "CI is important"! But, of course, what is CI? At the end, they would retreat to the position that EHMO was qualitatively right but quantitatively deficient. This credo shows up in the extensive use of EHMO-type models by the very people who carry out ab initio calculations and, at the same time implicitly or explicitly criticize the EHMO method. One has the "right" to criticize the EHMO method only after electron–electron repulsion has been *explicitly* incorporated in the deliberations. Once this is done at the VB level, the distinction between T- and I-bonding becomes obvious, and everything is open to reinterpretation.

References

1. (a) H. Eyring, J. Walter, and G.E. Kimball, *Quantum Chemistry*, Wiley, New York, 1944. (b) L. Pauling and E.B. Wilson, *Introduction to Quantum Mechanics*, McGraw-Hill, New York, 1935.
2. C. Sandorfy, *Electronic Spectra and Quantum Chemistry*, Prentice-Hall, Englewood Cliffs, NJ, 1964.
3. J.M. Norbeck and G.A. Gallup, *J. Am. Chem. Soc.* 96, 3386 (1974).

4. (a) R. Hoffmann, *J. Chem. Phys.* 39, 1937 (1963). (b) R.B. Woodward and R. Hoffmann, *The Conservation of Orbital Symmetry*, Verlag Chemie, Weinheim, 1970.

5. H.C. Longuet-Higgins and L. Salem, *Proc. R. Soc.* A251, 172 (1959).

Chapter 15

The Association Rule and the I Formula

15.1 The Association Rule and the Count of I Bonds

The conceptual breakthrough is that CT, much like exchange, creates "bonds" that can be counted by assuming that two arrows, representing the transfer of two electrons, define one CT bond, which is called an I bond. A molecule is looked upon as a resonance hybrid in which the ensemble of all independent covalent structures is interpreted as a "polyradical" substate, which sets the stage for CT delocalization. The interaction of the ionic structures with the covalent structures (as well as with each other) is interpreted to be equivalent to formation of I bonds, which are identified by affixing arrows on the polyradical substate (or, on the lowest energy covalent structure) subject to the association rule: *A hole gets two in-bound arrows, a dot gets one in- and one out-bound arrow, and a pair gets two out-bound arrows.* The association rule sets the stage for counting the number of I bonds: *Two arrows make one I bond.*

The wavefunction of a heteronuclear metal–metal "double bond" shown in Figure 15.1 is a simple illustration of our approach. The molecule is a resonance hybrid of I descriptors, and each I descriptor is a resonance hybrid of the covalent PP and NB configurations and three ionic configurations

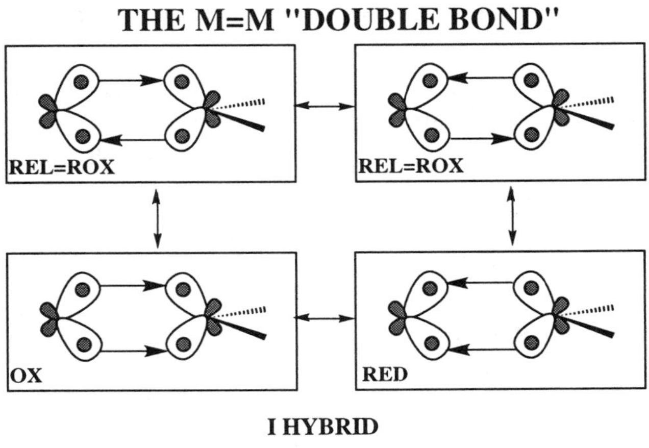

Figure 15.1 The metal–metal "double bond" is actually a single resonating metal–metal I bond

representing one 2-e CT hop plus two ordered 1-e CT hops. Each I descriptor represents one I bond (two arrows). The superposition of all I descriptors, called the I hybrid, represents a resonant I bond. By construction, the I-bond has T character. This is called the *T complement* of the I bond, and it is ever-present because all individual VB configurations have an exchange contribution (either net attractive or net repulsive). At the T end of the bonding continuum, we have T bonds with some I complement. At the I end, we have I bonds with some T complement. In between, we have I bonds with variable T complement, and the latter increases with atom electronegativity.

A descriptor with *n* arrows represents the complex CI of all configurations obtained by 1-e, 2-e, . . ., n-e CT hops, and the number of these configurations is proportional to the sum of binomial expansion coefficients. A seemingly simple-looking I descriptor with four arrows represents a highly complex CI. Our hypothesis is that 1-e and 2-e CT hops dominate the wavefunction and that 2-e CT hops are charge-creating variants of the charge-conserving 2-e exchange hops. Thus, we have succeeded in reducing a complicated problem to simple pictures because we captured the physics of the problem: *CT resonance represents the "replacement" of T bonds by I bonds, which have lower interelectronic repulsion.*

15.2 The I Formula; Superior and Inferior Arrows

The lowest energy configuration of a molecule is made up of three elements: holes, dots (odd electrons), and pairs. The *I formula* shows how these basic elements are connected by arrows subject to the association rule. The arrows represent the CT delocalization, which engenders I bonds with T complement. We use *superior arrows* to indicate the presence of covalent bonds and *inferior arrows* to indicate their absence. As a result, the I formula is an *arrow train* made up of solid and dashed arrows appended according to the association rule. Each pair of arrows symbolizes one I bond with a T complement, which can be large (two superior arrows) or small (two inferior arrows). As an example, we give the I formula of the π system of acrolein.

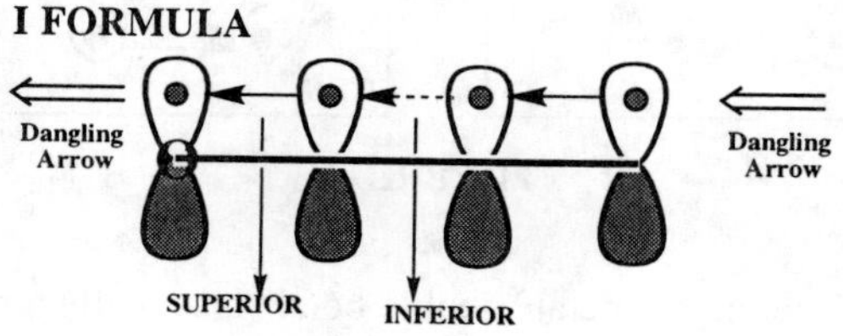

The I formula is a key development. It replaces the representations used by chemists to convey the electronic structures of "delocalized" species in which dashed lines indicated unspecified delocalization with unidentified physical consequences. Instead, we now have countable I bonds and, as we will see, very specific ways of appraising them.

In what follows, we will often be forced to differentiate the effects of primary (major) and secondary (minor) CT delocalization. To minimize complexity, we will be consistently using the superior/inferior notation along with a specification of the meaning. In the absence of any such specification, the conventions just outlined apply.

15.3 The Strength of the I Bond; Contact Repulsion

The strength of a single I bond depends on three factors:

(a) Orbital overlap.

(b) The efficacy of CT (which depends on the ionization potential of one atom and the electron affinity of the other).

(c) The "mission" of an I bond is to produce superior bonding (relative to a T bond) by minimizing interelectronic repulsion. It follows that a hallmark of the I bond is the amount of electron–electron repulsion associated with it. This is a function of the orientation of the arrows representing the I bond. In general, the greater the angle of two arrows affixed on the same atom, the more favorable the I bond they represent.

Factor c differentiates our theory from the conventional models. Figure 15.2 shows one exchange and two CT 2-e hops. Each 2-e hop, whether of the exchange or CT type, effects the *same* reduction of kinetic energy, which is proportional to $Cs_{ij}s_{kl}$, where C is a factor proportional to orbital electronegativity and s_{ij} is the AO overlap integral of AOs i and j. On the other hand, the reduction of kinetic energy cannot avoid the penalty of *Coulomb repulsion* measured by the bielectronic repulsion integral $(ij|kl)$. This specific

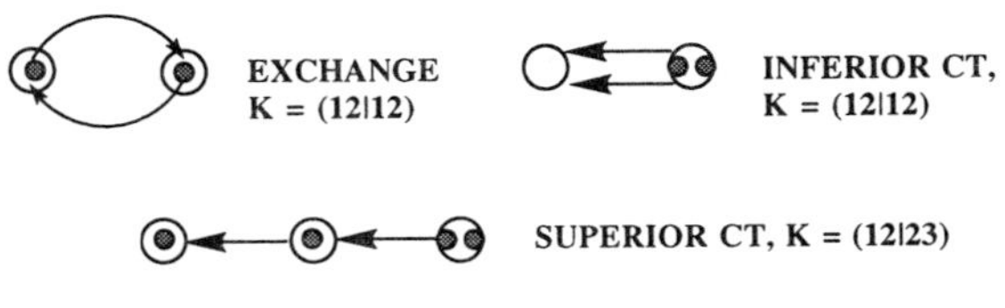

Figure 15.2 Different 2-e hops and the associated contact repulsion.

type of repulsion that attends a CT hop is called *contact repulsion.* The key point is that the CT is intrinsically superior to the exchange 2-e hop because it involves much smaller contact repulsion, provided the CT 2-e hop spans two different overlap regions, or, better still, two different interatomic regions. Specifically, an exchange 2-e hop, represented by two "head-to-tail" arrows within one and the same overlap region, gets the same contact repulsion penalty as a CT 2-e hop, which is represented by two "head-to-head" arrows within one and the same overlap region.

The intrinsic superiority of the CT 2-e hop is expressed only when the two arrows span, at the very least, two different overlap regions. Coulomb repulsion is minimized when the *ij* and *kl* centroids of charge are furthest away from each other. This means that two arrows converging on an atom engender minimum contact repulsion if they are collinear (i.e., if the angle they subtend is 180°).

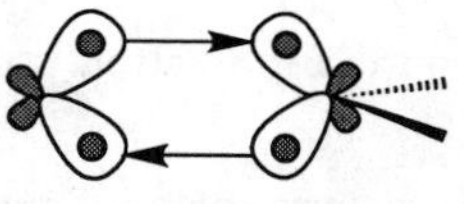

0° ARROWS
HIGH CONTACT REPULSION

109° ARROWS
LOW CONTACT REPULSION

Since two arrows symbolizing one I bond must subtend the largest angle compatible with maintainance of AO overlap, I-bound molecules will tend to have geometries that maximize the distance of bonding electrons.

Why are the formal double bonds (and multiple bonds, in general) connecting semimetals or metals notoriously unstable?[1] The explanation of this important trend is simple: The "double bond" is only a single I bond with the two defining arrows representing two close-lying overlap densities. The angle subtended by the two arrows is diagnostic of the magnitude of contact repulsion. A "double bond" represents the limit of zero angle and maximum contact repulsion.

Bottom line: I-bonding would be the universal principle, were it not for the difference in promotional energies required by exchange and CT 2-e hops. The very nature of the exchange process (i.e., zero net electron transfer) guarantees that this is more favorable according to the promotional energy yardstick. By contrast, a CT 2-e hop causes charge transfer, and this is justified only when the electronegativity of the donor atom is low. On the other hand, the key point is that without the restraining action of the promotional energy, *all* molecules would be bound by the intrinsically superior I bonds.

15.4 Aromaticity, Arrow Directionality, and the Map of Chemical Bonding

CT delocalization is more important in an associated than a segregated set of n bonds because ionic configurations that fail to contribute to the wavefunctions of the isolated components are now important participants. Ionic configurations that contribute either mainly or exclusively to the wavefunction of the aggregated system are called *association-promoting configurations.* Given this, we have an early test of the concept of the I bond: Aromatic should be differentiated from antiaromatic complexes by the magnitude of CT delocalization "allowed" in the former and "forbidden" in the latter. In turn, this implies that crucial association-promoting configurations are rendered inactive (by symmetry) in antiaromatic species. This is exactly the case.

Consider the prototypical degenerate exchange of two bonds via a symmetrical square Möbius–aromatic complex shown at the top of Figure 15.3. This can be considered to be a valence isomerization of the "segregated" S via the

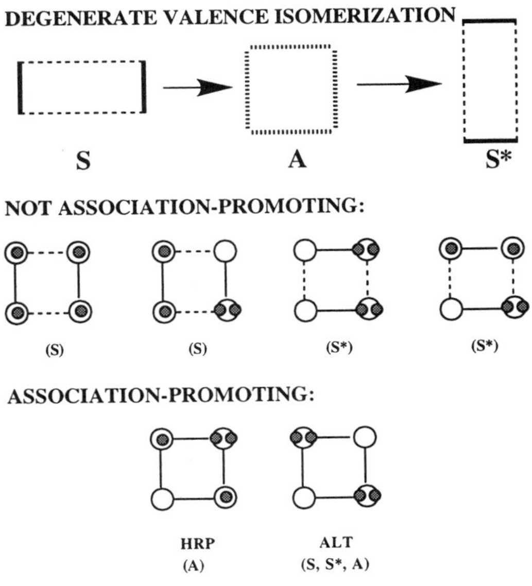

Figure 15.3 Valence isomerization via a Möbius–aromatic complex.

"associated" A, with the latter being either a saddle point or a global minimum. The stability of A relative to S depends on the association-promoting configurations that contribute mainly or exclusively to A. These are two equivalent ALT and four equivalent HRP configurations of the types shown in Figure 15.3. Note the multipolar character of ALT and the way in which one odd electron is sandwiched by a hole and a pair in the HRP configuration. The importance of the multipolar ALT configurations for association is indicated by the elimination, in going from even- to odd-electron systems, of the ALT configurations, which is partly the reason for the very weak or nil stereoselection (i.e., no aromatic/antiaromatic distinction) found in radicals.[2] The key point is that these crucial ionic configurations are eliminated from the wavefunction of the correspondinding Hückel–antiaromatic species, as shown explicitly in Table 12.1. This is effectively tantamount to the partial elimination of I bonds as projected by the I formulas of the prototypical aromatic, nonaromatic, and antiaromatic molecules shown in Figure 15.4.

In particular, note the *codirectional* arrows in the aromatic case and the two missing arrows in the antiaromatic case. This absence can be taken to mean that the antiaromatic molecule has *contradirectional* arrows, as also illustrated in Figure 15.4. Application of the association rule eliminates two arrows, and we obtain the final formula. It follows that *arrow directionality* (co- versus contradirectionality) is an important determinant of I-bond strength and the stability of a molecule. Codirectionality means cooperativity, and contradirectionality means anticooperativity. Note that the I formulas represent

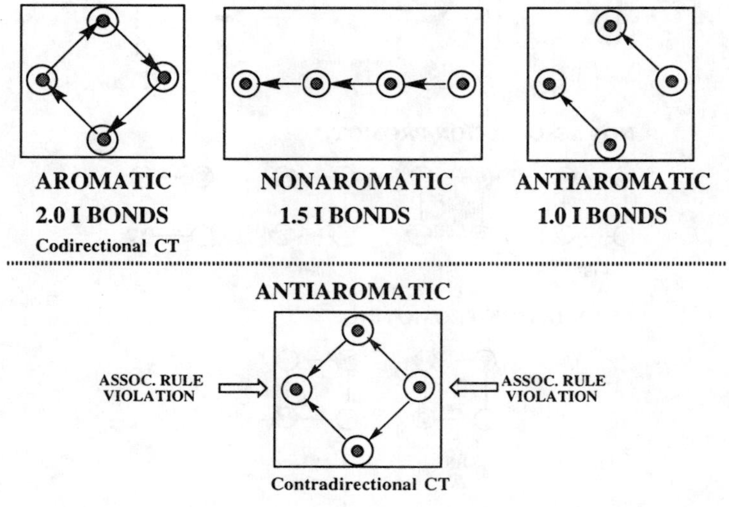

Figure 15.4 I formulas of prototypical complexes.

CT delocalization by arrows that are "executed" in all possible combinations. Aromatic systems are distinguished by a codirectional arrow train in which execution of all arrows effects a cyclic permutation that represents exchange delocalization of highest order.

We now examine how the concept of aromaticity relates to the map of chemical bonding presented earlier. To do so, we consider the case of a Möbius aromatic four-electron system and we inquire into the nature of bonding as the electronegativities of the atomic centers are varied subject to the *arrow restriction condition:* An arrow can enter only an electronegative radical and it can exit only an electropositive radical.

Figure 15.5 shows the two limiting mechanisms of association corresponding to I- and E-bonding as well as two crucial compromise situations. First, in a homonuclear system made up electroneutral (EN) atoms (i.e., atoms with modest electronegativity), the degeneracy of the ALT configurations is preserved and we have *amphidromic I-bonding* represented by a right (R) and a left (L) codirectional arrow train. The presence of degenerate R and L trains is crucial for I-bonding.

In the second compromise situation, alternation of ELNG and ELP atoms splits strongly the degeneracy of the two ALT configurations. We now have multipolar *heteronuclear E-bonding* represented by the lowest energy ALT configuration. Any measurable CT delocalization is no longer codirectional but, rather, contradirectional. This means that a molecule cannot have its pie and eat

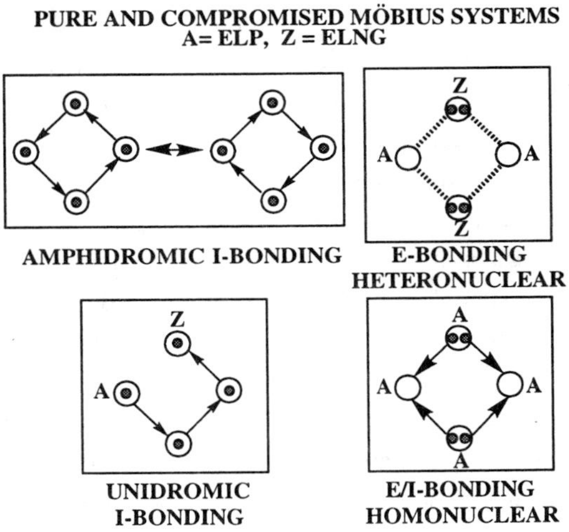

Figure 15.5 The limiting cases of I- to E-bonding and intermediate situations.

it too: Choosing the E mechanism means abandonment of the superior I mechanism involving codirectional CT delocalization.

Here is the message: To benefit from I-bonding, maintain the ALT degeneracy. On the other hand, to switch to E-bonding, split the ALT degeneracy. In-between cases represent compromised bonding, as exemplified by the following systems

(a) A system derived from the I-bound homonuclear parent, $(EN)_4$, by replacing two adjacent atoms by one electronegative (ELNG) and one electropositive (ELP) atom. As a result, the R/L train degeneracy is lifted but we still maintain one codirectional arrow train which is, however, defective because the arrow exiting the ELNG and pointing to the ELP atom (to maintain arrow codirectionality) represents unfavorable CT delocalization, hence, is eliminated. This is *unidromic I-bonding*, which stands between the limits of perfect I- and E-bonding.

(b) A system derived from the I-bound $(EN)_4$ by replacing each EN by an ELP atom. Once again, the system stands between the perfect I and E limits. E-bonding is inferior because electron transfer in homonuclear systems is intrinsically inferior to that in heteronuclear systems. I-bonding is also inferior because the charge alternant configuration can promote only contradirectional CT delocalization. The simplest example, Li_4, is discussed in detail in Chapter 35.

We have differentiated between I- and E-bound cyclic complexes and in-between derivatives. What is the representative of T-bonding? The answer is simple: This is the $(EN)_4$ transoid diradical in which neither I- nor E-bonding is a factor. Thus, we now have a confirmation of the map of chemical bonding by one of the simplest molecular prototypes.

We can now make a disjunction between the VB model and conventional bonding theory. Asked to identify a favorable transition state (in actuality, an obligatory association complex), the organic chemist would point to an "aromatic" transition state "allowed" by the Woodward–Hoffmann rules. This ceases to be the case here. There are two ways of achieving stability, namely, via I-bonding or E-bonding. Furthermore, any in-between choice is defective and the complex is "frustrated." As we will see, this viewpoint has important practical consequences. There are two different ways (I and E) of designing a facile reaction, but, until now, only one has been intensely explored.

15.5 The Physical Meaning of Arrow Codirectionality; Configurational Degeneracy

Aromaticity has spoken: Arrow codirectionality represents superior chemical bonding and arrow contradirectionality represents inferiority in this respect. This is a general bonding principle that finds its apotheosis in cyclic systems but loses none of its power when it comes to acyclic systems. The advantage of arrow codirectionality is that the configurations defining the CT delocalization preserve atom electroneutrality to the maximum extent possible because an exiting arrow at one odd-electron site is matched by an entering arrow at the same site. This produces CT configurations that lie not far above the perfect-pairing structures. As a result, the low energy configurations describing the system tend to be either degenerate or quasi-degenerate, and the I bond(s) tend to be strong because of *configurational degeneracy*. By contrast, arrow contradirectionality means that the configurations defining the I bond(s) are separated by a wide energy gap. As a result, the I bonds themselves are weak. In short, arrow directionality is one measure of the strength of the constituent I bonds having to do with the energy spacing of the interacting configurations, as depicted here.

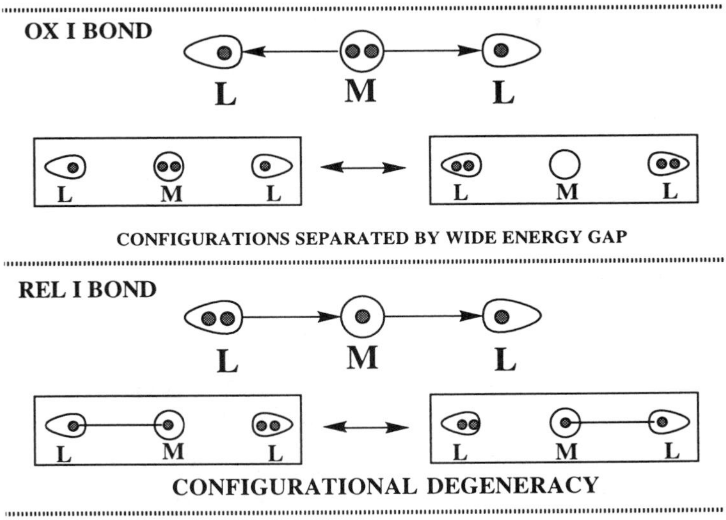

The four-electron LML system can become bound by a strong I bond if M is sufficiently electropositive and L sufficiently electronegative to promote the LM^+L^- configuration. In other words, the M(0) metal can make only a weak (contradirectional) OX (oxidative) I bond, but the M(I) metal can make a strong (codirectional) REL (relay) I bond.

The term "configurational degeneracy" does not strictly imply degeneracy of VB configurations but, rather, their location within a narrow energy range. The best realization of this state of affairs is effected by a homonuclear system in which the component atoms have modest electronegativity. In other words, optimization of configurational degeneracy requires that highly electronegative or electropositive atoms be excluded from an array of electroneutral atoms. Optimal or limited configurational degeneracy is a requirement for strong I-bonding.

The best example of the consequences of configurational degeneracy is provided by the n-center/($n\pm1$)-electron odd linear polyenes or aromatic annulenes. The presence of an odd hole or pair allows us to write a set of low-energy degenerate VB configurations connected by 2-e CT hops that define I-bonding and endow the associated system with great stability. For this reason, cycloheptatrienyl is computed to be more stable than benzyl cation.[3a]

15.6 I Formulas and Arrow Directionality Errors

Let us recall the philosophy of the T formulas: The bond dipole formula is written by affixing arrows consistent with bond polarity on the perfect-pairing configuration. We then apply the condition that each center must satisfy the exclusion rule: All arrows are either in-bound (systolic center) or out-bound (diastolic center). The bonds that fail to meet this criterion are *"erroneous bonds."* Turning the arrows of these aberrant bonds into lines generates the T formula. The stability of the molecule is a function of the number of recorded errors.

We can now proceed in an analogous fashion to generate the most revealing formula of an I-bound (rather than a T-bound) molecule. Specifically, we write the I formula of the association complex assuming arrow codirectionality and subject to the *arrow restriction condition:* Arrows exiting electronegative atoms or entering electropositive atoms are "forbidden." Such misdirected arrows either are replaced by a wiggly line or are simply deleted. We assume that arrows connecting atoms of the same or similar electronegativities are "allowed." Arrows terminating at atom holes or originating from atom pairs are always "allowed" irrespective of electronegativity considerations. Finally, in assessing stability, not only the number is important; the magnitude of the errors must be considered as well. The presence of multiple errors implies that either T- or E-bonding is a superior alternative.

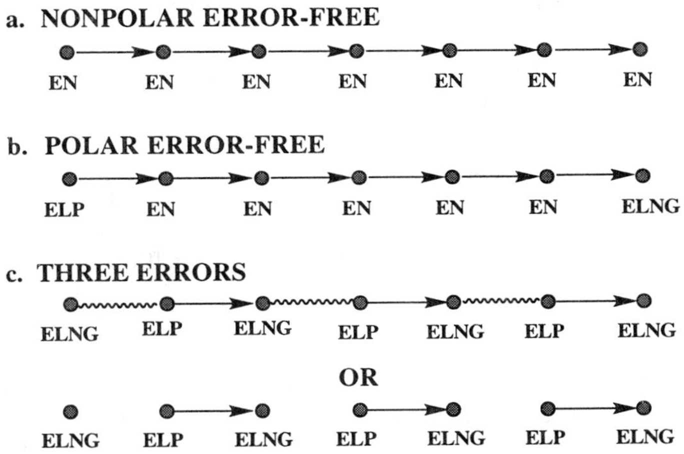

Figure 15.6 Error-free and error-plagued arrow trains.

Error-free arrow codirectionality, a counterintuitive view of molecular electronic structure, is projected in Figure 15.6. The condition for realizing an error-free codirectional arrow train is that electronegative (ELNG) and electropositive (ELP) atoms be placed at the ends of an array of electroneutral (EN) atoms. This generates a miniature conductor of electricity. By contrast, alternation of ELNG and ELP centers produces an error-full arrow train, which is incapable of supporting strong I-bonding. As a result, the error-free species owes its stability principally to I-bonding while the error-full species can only rely on E-bonding for its existence.

The simplest illustration of the concept of the erroneous arrow train is the stability of the allyl radical. Rendering each terminal atom either electronegative or electropositive deletes one codirectional arrow; that is, it introduces one error as illustrated (by convention, we show only one of two equivalent I descriptors: right and left arrow train).

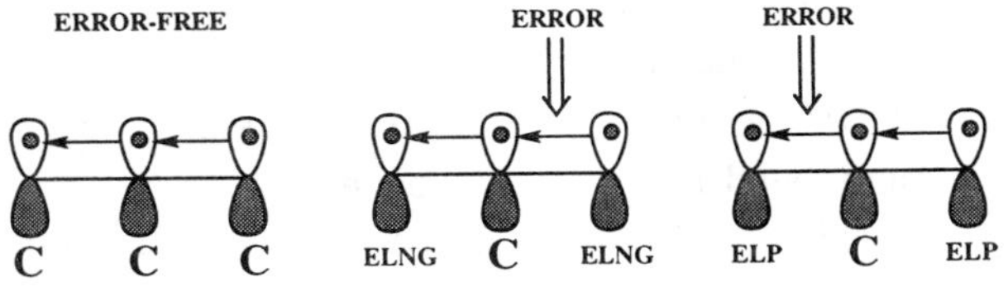

This is confirmed by ab initio calculations[3b,c] and experiment.[3d] The same is expected to be true of allyl anions and cations. Does the error in the I formula of the π system of the formyloxy radical, OCHO (ELNG—CH—ELNG case) mean

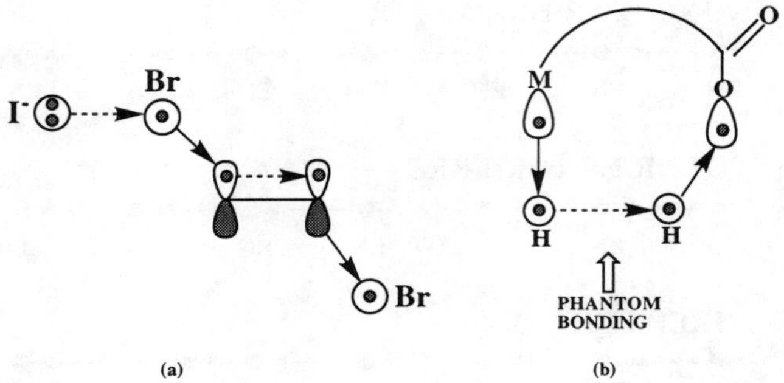

Figure 15.7 (a) The I formula of the transition state of the dehalogenation of a vicinal dibromide. (b) H—H phantom bonding.

that this is an intrinsically unstable species? Quite the contrary! All that it means is that it is incapable of supporting CT delocalization (CT resonance) and it will opt for an alternative bonding mechanism for which the alternation of electronegative (O) and electropositive (C) atoms is ideal: E-bonding. At the limit of pure E-bonding, OCHO is predicted to be unsymmetrical (lower than C_{2v}) as a consequence of the unsymmetrical $O{:}^{-}CH^{+}O$ resonance structure that defines E-bonding. Furthermore, such an E-bound OCHO is expected to be intrinsically more stable than the I-bound H_2CCHCH_2 because carbon is too electronegative to support CT and strong I-bonding. In other words, the π allyl radical has a strong T complement that would drive the π bond and π radical components to segregation (because T antibonding dominates T bonding) had it not been for the action of the σ frame which is ultimately responsible for the C_{2v} geometry. Later on, we will see that radical substitution reactivity is consistent with this conclusion.

The idea that low energy transition states and stable molecules have terminal ELNG and ELP groups and an interior chain of EN atoms explains why only a small change will serve to incapacitate two reactants. For example, the dehalogenation of vicinal dibromides by iodide occurs because one Br has come close to being EN, thus allowing the generation of an error-free codirectional arrow train as shown in Figure 15.7a. Replacing Br by Cl improves the electrophilic terminus but destroys the interior chain.

The way in which the notion of the error-free codirectional train changes our view of molecules is exemplified by the simple organometallic species in which the metal is directly attached to an axial hydride and indirectly connected to a carboxylic acid residue (Figure 15.7b). Whenever we see ELNG or ELP atoms within an organometallic complex, we immediately think of them as terminals of a potentially error-free codirectional arrow train. As a result, the innocent-looking organometal of Figure 15.7b can be formulated as a complex of H—H with one metal and one oxygen atom. The key feature of this complex is the appearance of bonding electron density where one would normally expect to find none (i.e., between the two H atoms). This phenomenon is called *phantom bonding*, and it is a direct consequence of I-bonding. We emphasize that the incipient bonding of the apparently nonbonded H atoms is *not* the result of covalent resonance. The recent works of Crabtree and co-workers and the studies of Ramachandran and Morris strongly point to a direction that will undoubtedly come onto the screen of organometallic chemists before long.[4]

15.7 The Heteroatom Exclusion Rule

The heteroatom exclusion rule says:

> Only electroneutral radicals are capable of full association.
>
> Electronegative or electropositive radicals are either completely excluded from the association domain of a molecule or they are relegated to terminal positions.
>
> Electronegative (electropositive) radicals can become associated with only one donor (acceptor) radical and a second radical that is either equally or more electronegative (electropositive).

The rule is confirmed by the following data:

(a) Nitrogen is excluded from the aromatic benzene nucleus as shown below. The unwillingness of heteroatoms to be parts of carbon aromatic systems (thus damaging configurational degeneracy) is illustrated by the observation that a nonaromatic ketone can be more stable than an isomeric furan derivative, an aromatic molecule.

(b) Benzyl (BZ) is computed to be less stable than cycloheptatrienyl (CHP) cation. However, the stability order turns around, and BZ becomes more stable than CHP cation, when C is replaced by Si in CHP and in the benzylic position of BZ.[3a] In other words, the winner is the isomer that has the heteroatom excluded from the aromatic array. The same journal issue that reported the change in stability order just cited also contains a different computational study of the isomers of $C_6H_5O_2$. The most stable isomer is the one in which the two oxygens act as substituents of a benzene nucleus, one as OH and the other as oxygen radical. In other words, both oxygens are excluded from the benzene nucleus.[5]

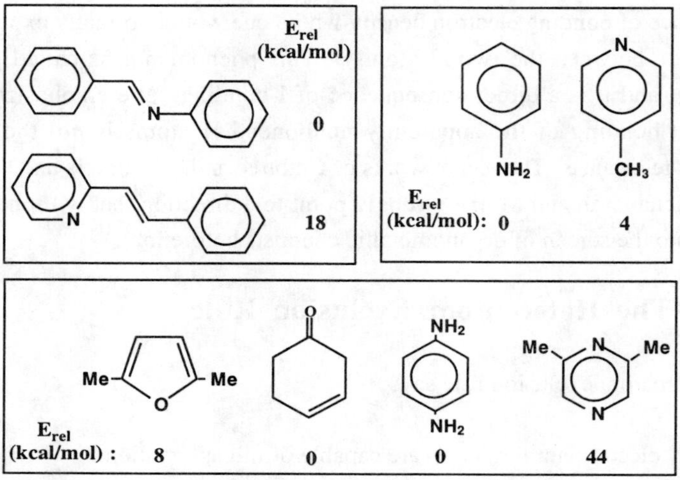

(c) Replacement of C by O in a linear polyene diminishes the association tendency.

$$CH_3\text{—}CH\text{=}CH\text{—}CH\text{=}CH_2 \xrightarrow{\Delta H_r = +7} H_2C\text{=}CH\text{—}CH_2\text{—}CH\text{=}CH_2$$

but

$$CH_3\text{—}CH\text{=}CH\text{—}C(OH)\text{=}O \xrightarrow{\Delta H_r = 0} H_2C\text{=}CH\text{—}CH_2\text{—}C(OH)\text{=}O$$

$$CH_3\text{—}CH\text{=}CH\text{—}C\equiv N \xrightarrow{\Delta H_r = +1} H_2C\text{=}CH\text{—}CH_2\text{—}C\equiv N$$

Many more examples can be found.

15.8 The Difference Between the T and I Formulas

The simplest illustration of the difference between the T and I formulas is provided by a triene system. The T formula departs from the perfect-pairing (PP) configuration and allows for intrabond CT delocalization. By contrast, the I formula departs from a linear combination of all covalent structures and allows for CT in all possible directions but always limited by the association rule.

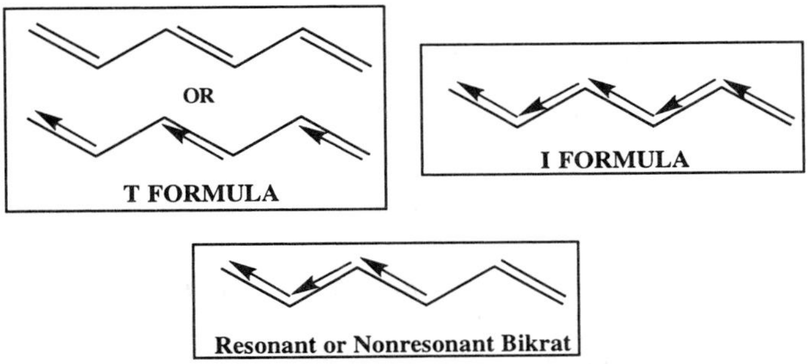

In particular, note how the I formula indicates instability (or, reactivity) by virtue of the capacity of the two terminal atoms to receive one inbound and one outbound arrow, called *dangling arrows.* Note also how execution of the arrows in the I formula leads to VB resonance structures that are never considered in conventional chemical deliberations (e.g., CT involving three adjacent odd electrons).

While T and I formulas are different, they often converge to the representation of the same dominant factors in some cases. For example, CF_4 is represented by a T formula but $SiBr_4$ is represented by an I formula. The two molecules have one thing in common: diastolic CT delocalization, which minimizes interpair exchange repulsion. This commonality is physically more important than the disjunction: CF_4 is dominated by the PP configuration while the bonding of $SiBr_4$ involves the action of all independent covalent configurations. Hence, we can use the T formulas to determine the relative stabilities of isomeric metal-containing molecules much as we did before in the case of organic molecules. This makes physical sense: Exchange repulsion of bond pairs never disappears, even when the mechanism of the formation of the bond pairs changes.

15.9 The HRP I Bond and the Captodative Stabilization of Radicals

Figure 15.8 presents a reformulation of the association rule that makes explicit how the concept of the I bond changes the conceptual landscape of chemistry. Note that one hole, radical, or pair can engage two other different fragments by means of a *single I bond!* Let us examine some consequences. Two illustrative applications show how the three-electron I bond leads to reinterpretation of old facts.

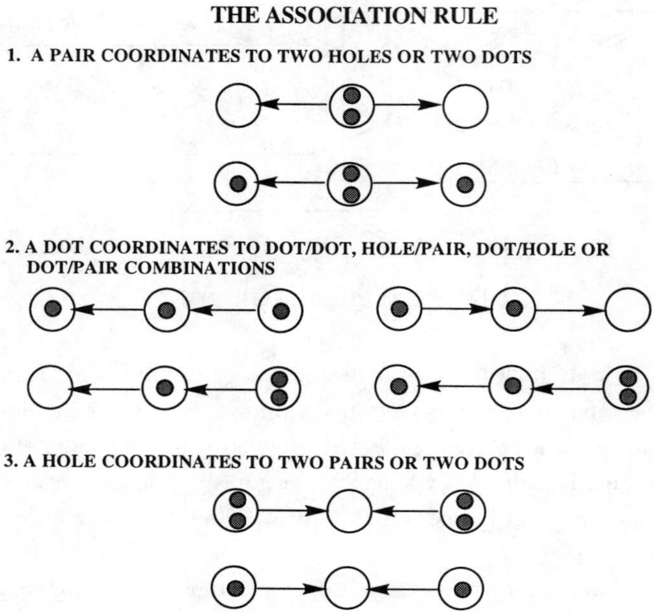

THE ASSOCIATION RULE
1. A PAIR COORDINATES TO TWO HOLES OR TWO DOTS
2. A DOT COORDINATES TO DOT/DOT, HOLE/PAIR, DOT/HOLE OR DOT/PAIR COMBINATIONS
3. A HOLE COORDINATES TO TWO PAIRS OR TWO DOTS

Figure 15.8 Restatement of the association rule.

(a) The regioselectivity of radical additions to olefins and radical eliminations.[6] One example is shown.

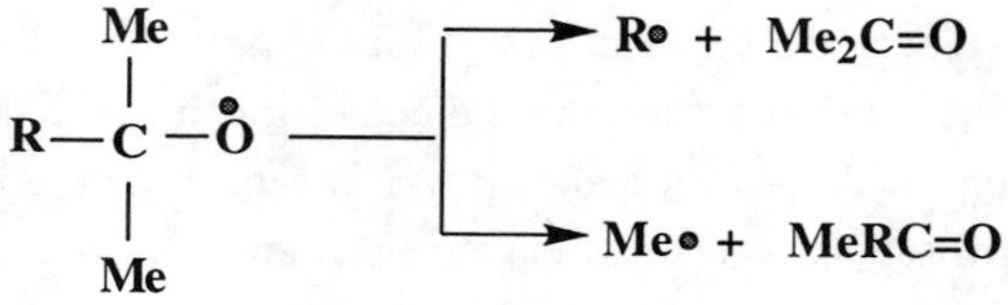

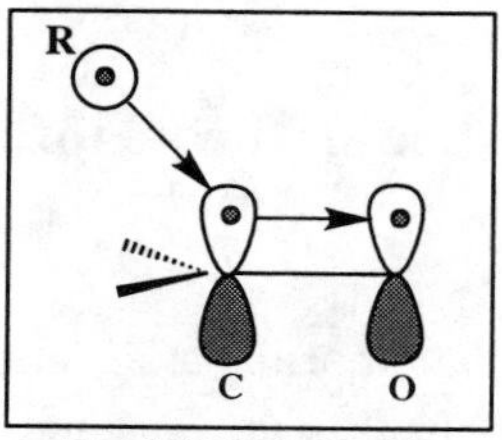

TRANSITION STATE

The arrow directionality imposed by the relative atom electronegativities predicts that the radical has cationic character and explains why *t*-butyl has the highest elimination aptitude and why this disappears when the -butyl is replaced by 1-norbornyl. I would like to reassert here my credo that the best theoretician is the experimentalist. Radical chemists thought of "polar effects" in radical transition states a long time ago and developed arguments that are parallel to what we say here.[7] The only difference is that now these "polar effects" are manifestations of a unifying principle that extends from organic transition states to metals: CT causes I-bonding, and this stabilizes either obligatory (e.g., transition states) or genuine association complexes.

(b) The preference for formation of exocyclic rather than the thermochemically more stable cyclic radicals upon intramolecular radical addition to a double bond.[8,9] The 1.5 I bond is so directed in the closure to the five-membered ring that an electron pair is driven out of the ring and the system is spared of exchange repulsion, according to the following scheme:

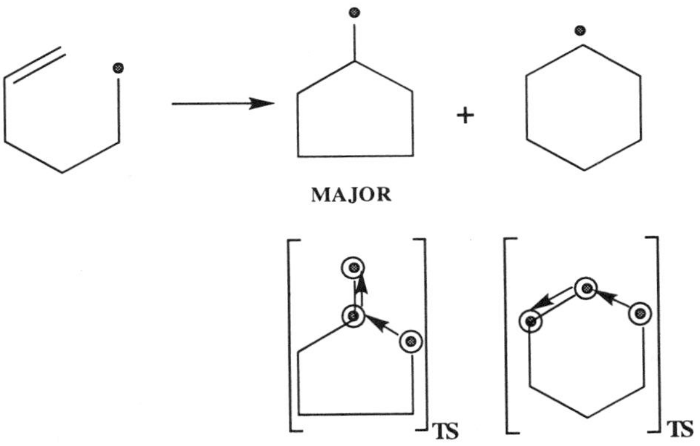

As discussed in Chapter 5, entropy acts in the same direction.

Every chemist can write the two equivalent structures of allyl anion. We now know that these define one I bond that is relatively strong because of configurational degeneracy. However, one I bond is also defined by one odd-electron sandwiched by a hole and a pair. This is now weaker because configurational degeneracy no longer obtains. These considerations are illustrated in Figure 15.9a. Note how a 2-e CT hop engenders the allylic-type VB structure in both instances. In general, one I bond can have two, three, or four electrons! The crucial difference between the even-electron (2,4) and odd-electron (3) systems is that only the former have the benefit of configurational degeneracy. Configurations that are incapable of engendering an I bond are shown in Figure 15.9b.

The "hole-radical-pair" (HRP) I bond leads to many important insights because it suggests the recipe for stabilizing a radical: Flank it with one π-donor (DO) group and one π-acceptor (AC) group. This is the captodative stabilization

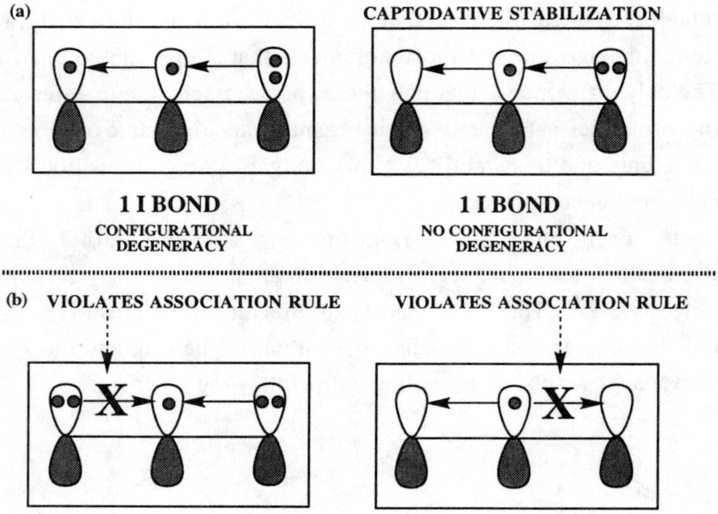

Figure 15.9 (a) Two I bonds differing in the electron count and the electronic basis of the captodative effect. (b) Reference configurations that fail to produce an I bond.

mechanism discovered by Viehe and co-workers.[10] These investigators found that a π radical is best stabilized by a combination of a π donor (pair) and a π acceptor (hole) rather than by two π donors or two π acceptors. Moving on to the frontier of biochemical research, we find the proposal that ascorbic acid mediates N-dealkylation and N-deoxygenation of amine oxides by a mechanism that involves electron transfer from the former to the latter.[11a] Thus, we have claims that vitamin C can act against aging and chronic disease because of its capacity to act as a strategic electron donor. The formula of the singly ionized ascorbic acid reveals that this radical cation is simply a captodative radical!

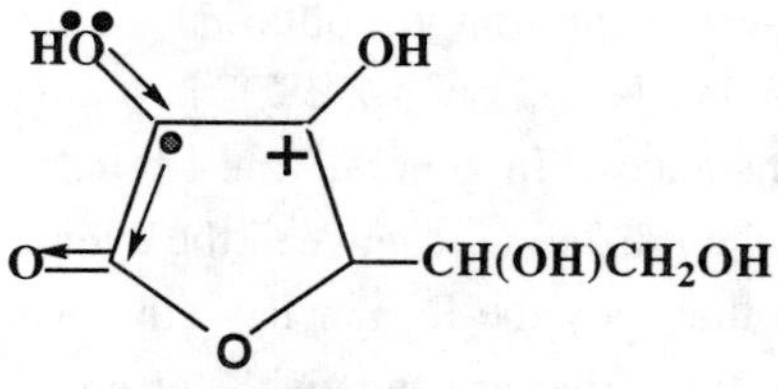

Ionized Ascorbic Acid
A Captodative Radical

The captodative effect is one of the many manifestations of I-bonding and, as a fundamental principle, it leads to immediate predictions. For example, each of the following radicals (the neutral 4-carboxypyridinyl radical and its radical-cationic and radical-anionic derivatives) is a captodative radical. In the radical cation, the odd electron is localized so that the pyridine moiety has six π electrons. The donor and acceptor fragments are circled. The predictions match the findings of Tripathi et al.[11b]

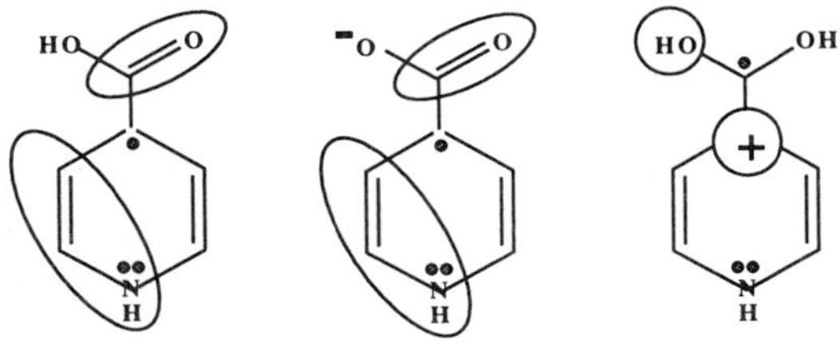

ALL THREE ARE CAPTODATIVE RADICALS

We can use the concept of the I bond and the consequent notion of captodative stabilization to design radical traps of the following types.

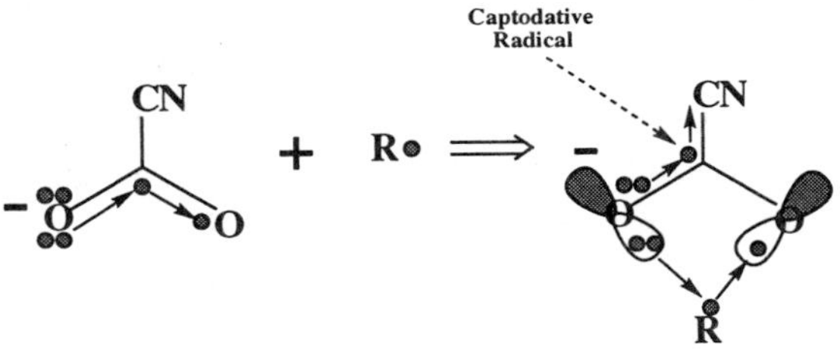

15.10 The I Bond Fells the Concept of the Magic Number

Because one I bond is compatible with anything from two to four electrons, there must exist different stable organometallic complexes that have the same number of I bonds and yet very different magic numbers (MNs) ranging from a maximum to a minimum MN as illustrated in Figure 15.10. This explains at once why there are so many "unprecedented" papers on metal-containing systems in the chemical literature.

Since the work of Lewis in the early twentieth century, the concept of "atomic valence" has been the unchanging centerpiece of chemistry. VB theory brings now a fundamental modification: Carbon still has a valence of 4 because four odd electrons in promoted carbon can make four covalent bonds. On the other hand, the *maximum* valence of the elements that are isoelectronic with

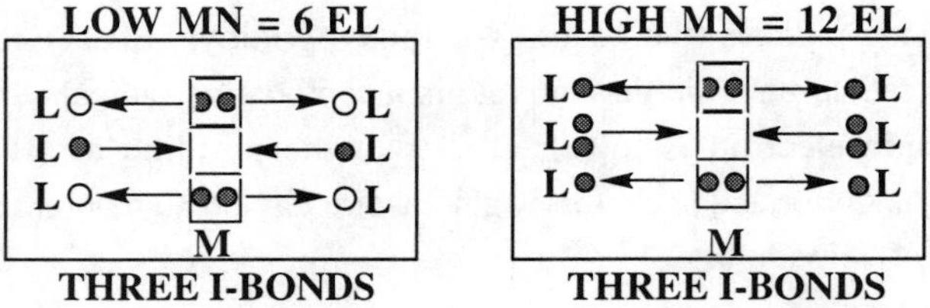

Figure 15.10 One I bond — many magic numbers!

carbon (i.e., Si, Ge, Sn, and Pb) is 8!!! Each of these atoms is now viewed as having one pair, one hole, and two dots, and the *maximum* number of ligands that can be engaged is eight (two per hole-dot-pair element) rather than four!

In the preface, I called the accepted conceptual theory of chemistry a mirage. It is now possible to become more specific. The failure of the chemist to recognize the existence of two distinct types of overlap-dependent bonds lies in the failure to treat electron–electron repulsion in an explicit manner. *Interatomic* electron–electron repulsion is a form of promotion. Chemists failed to recognize that interelectronic repulsion is as important as promoting carbon from $2s^2 2p^2$ to $2s^1 2p^3$ to make methane. Once this is seen, it becomes evident that the covalent bond has a problem; namely, strong electron–electron repulsion in the overlap region, resulting because two electrons delocalize in proximity to each other. Strong electron–electron repulsion is justified only when strong kinetic energy reduction is possible. This is what happens when atom electronegativity is high. When atom electronegativity is low and kinetic energy reduction small, covalency must be abandoned in favor of a better alternative. One can now anticipate that some other kind of bond must take the place of the covalent bond in metals. This is the I bond.

At the EHMO level of theory, and because of neglect of electron–electron repulsion (and Coulomb interaction, in general), there is no longer a distinction between T- and I-bonding. Instead, there is only one brand of bonding, which may be called "covalent bonding." As recently as 1994, this was the message passed to chemists by EHMO practitioners in the form of "urgent communications."[12] Organometallic chemists have been persuaded to view their molecules as "metal derivatives of organic molecules."[13] Hoffmann and his school have argued that "chemistry and physics meet at the solid state,"[14] meaning that orbitals overlap in organic molecules and in solids and, thus, the two domains obey the same rules of bonding. We suggest that the experimental facts argue otherwise: First-row chemistry is the exception. If there was only one type of bonding, the hard/soft dichotomy would not exist, Li_2^+ would not have a stronger bond than Li_2, and so on, ad infinitum.

The central thesis of this work is schematically presented in Figure 15.11. The original nonquantum formulation of the electron pair bond by Lewis actually misled chemists into thinking that a single bond is "a single line connecting two dots." On the other hand, quantum mechanics says that it is "two arrows that effect the exchange of two dots." Once this is realized, a conceptual breakthrough follows: Two arrows can operate in either an exchange or a CT mode. The former corresponds to a T bond and the latter to the I bond. Thus, the "covalent/ionic" spectrum is a product of chemical intuition, while the T/I spectrum is the result of the quantum formulation of bonding.

15.11 The Electronics of Radical Ions

The best way to illustrate new ideas is by application to areas in which no organized thinking has been possible before. One such domain is the chemistry of radical ions. What is the preferred structure(s) of AH_4^+, where A is C or a heavier congener? The key recognition is that a radical cation does not have

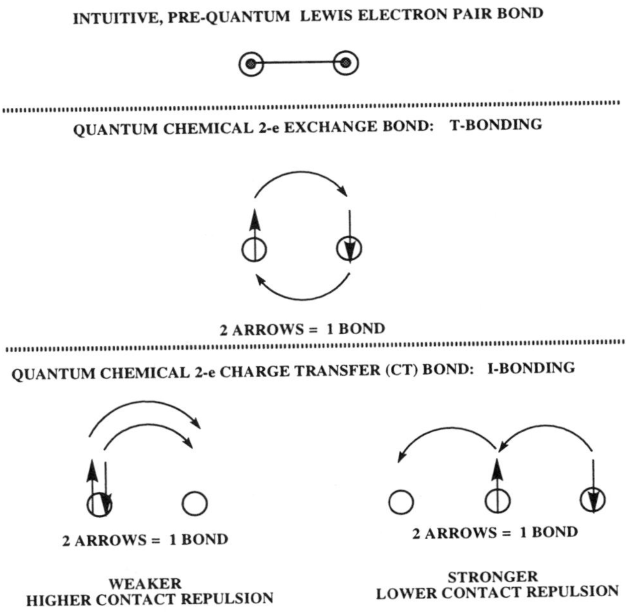

Figure 15.11 The history of the I bond.

enough electrons to satisfy perfect pairing. Hence, it must rely on I-bonding. There exist two alternatives:

The radical cation employs a superior two-electron REL bond, which has the advantage of configurational degeneracy.

The radical cation employs a three-electron I bond, which no longer has the advantage of configurational degeneracy.

The two possible forms are shown in Figure 15.12a. Note that the I(2) form has an ionized hydrogen and a REL bond involving the carbon electron with the lowest ionization energy, since this is best for delocalization in the 1s hole of the formal H^+. The I(3) form has an ionized carbon, with the hole residing in the higher energy C2p AO. Furthermore, this form can be formalized as a captodative $H^+(CH_2)^+H{:}^-$ radical. In the absence of any difference in the promotional energy required for making the I bonds, the I(2) form would be clearly superior. However, as in many instances before, promotional energy and bond making vary in opposite directions in going from one isomer to another. The C_{2v} [I(2)] form has a stronger two-electron I bond at the expense of (formal) ionization of the more electronegative H1s AO. At the antipodes, the T_d [I(3)] form has a weaker three-electron I bond with the benefit of (formal) ionization of the more electropositive C2p AO. Because double occupancy of the σ-type AO of the carbenic H_2A fragment can occur only in the I(3) form (via CT delocalization), the tetrahedral form is expected to become increasingly stabilized as we descend the carbon column and Ans becomes increasingly contracted. For the same reason, the C_{2v} [I(3)] form is expected to become the global minimum of the AH_4^- radical anion in which A is a heavy congener of carbon because the contracted nature of the Ans AO dictates double occupancy (Figure 15.12a). All these expectations are in accord with the data summarized by Almond.[15]

Shifting attention to the reactivity of ion radicals, we note that a substituted alkane bearing multiple π-donor substituents that can stabilize the cationic center as well as the radical site can be formulated as a captodative radical in which a stabilized cationic hole and one substituent π-type pair flank the radical site. As a result, nucleophilic attack on a delocalized radical cation is predicted to occur selectively on the atom bearing a π-donor group (e.g., methyl). One possible example is the preferential nucleophilic attack of the methyl-substituted carbon of 1,1-diphenyl-2-methylcyclopropane cation studied by Dinnocenzo and co-workers (Figure 15.12b).[16]

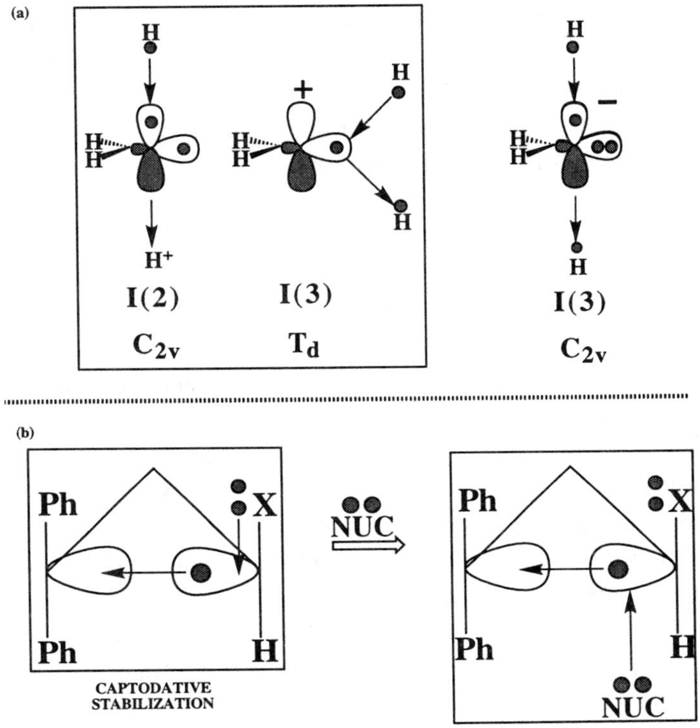

Figure 15.12 (a) The two forms of the radical cation and the predominant form of the radical anion of AH_4 (A = C, Si, Ge, Sn, Pb). (b) Regioselectivity of the nucleophilic attack on a substituted cyclopropane radical cation.

A clear disjunction between aromatic and antiaromatic complexes exists only in even-electron systems. By contrast, this distinction is lost in odd-electron systems. What is the physical reason? Recall that an aromatic molecule has exalted while an antiaromatic molecule has impaired CT delocalization and that the latter depends crucially on the association-promoting configurations. One of them is the ionic configuration in which holes and pairs alternate. Ionic configurations having full alternating holes and pairs exist only in even-electron systems. In addition, even-electron ionic systems have one more advantage over neutral odd-electron systems, and this is configurational degeneracy that promotes CT delocalization. Thus, allyl ion resonance is superior to allyl radical resonance because these correspond to 2-e CT and 2-e exchange delocalization, respectively, and CT is inherently superior to CT delocalization when configurational degeneracy is kept constant. Furthermore,

CT delocalization is unfavorable in the allyl radical because configurational degeneracy is no longer possible. Stated equivalently, both species have one I bond but the ion's bond is superior to that of the radical. Or, again, the ion lies toward the I end and the radical toward the T end of the bonding continuum. Note that in contradistinction to the problem of radical ion isomerism discussed at the beginning of this section, promotional energy considerations are no longer important, since the orbital energies are kept constant (C2p AOs).

We can now see that the geometry of radical ions presents yet another case of bonding selectivity much like the underpinning of the hard/soft distinction discussed in Chapter 19. Specifically, we predict that a radical ion will prefer to segregate into "ion plus radical" so that each subsystem optimizes a different modality of bonding. The ionic domain features I- and the radical domain T-bonding. For example, the radical cation of cyclobutadiene is expected to have the C_{2v} (rather than D_{2h}) rhombic structure shown below, if only the π system is considered. However, the constraining action of the σ frame may well impose a D_{2h}geometry. The same considerations within the context of the shell model (see Chapter 4) lead to prediction of a C_{2v} rhombic structure (R^4T^3 electronic configuration) for the radical cation of cyclobutane itself driven by an effectively antiaromatic R shell, which is now matched with a nonaromatic T shell.[17] A segregation of ion and spin domains in radical cations has been observed by Roth and co-workers.[18]

Bottom line: We are making predictions based on physical principles rather than by invoking an arbitrarily prescribed set of orbital interactions. Once one realizes that there exist three types of bonding (T, I, E), the recipe is simple: Fragments bound by the same mechanism associate if the mechanism is I or E. Fragments bound by different mechanisms segregate. This principle unifies the electronic structure of metals (see Chapter 19) and the electronic structure of the apparently unrelated organic π-radical cations.

References

1. A.H. Cowley, *Acc. Chem. Res.* 17, 386 (1984).

2. (a) E.N. Marvell, *Thermal Electrocyclic Reactions*, Academic Press, New York, 1980. (b) S. Olivella, A. Solé, and J.M. Bofill, *J. Am. Chem. Soc.* 112, 2160 (1990).

3. (a) A. Nicolaides and L. Radom, *J. Am. Chem. Soc.* 116, 9769 (1994). The refusal of Si to be a part of an aromatic ring also signifies the general distaste of heavy p-block atoms for double bonds, yet another consequence of I-bonding. (b) D. Feller, E.R. Davidson, and W.T. Borden, *J. Am. Chem. Soc.* 106, 2513 (1984). (c) K.B. Wiberg, J.R. Cheeseman, J.W. Ochterski, and M.J. Frisch, *J. Am. Chem. Soc.* 117, 6535 (1995). (d) F.G. Bordwell, G.-Z. Ji, and X. Zhang, *J. Org. Chem.* 56, 5254 (1991).

4. (a) J.C. Lee Jr., E. Peris, A.L. Rheingold, and R.H. Crabtree, *J. Am. Chem. Soc.* 116, 11014 (1994). (b) R. Ramachandran and R.H. Morris, *J. Am. Chem. Soc.* (1994).

5. A.M. Mebel and M.C. Lin, *J. Am. Chem. Soc.* 116, 9577 (1994). The global minimum is easily predicted by following two simple guidelines: Preserve a benzene nucleus and stabilize a formal radical center by the captodative effect (see Chapter 25).

6. (a) C. Walling and A. Padwa, *J. Am. Chem. Soc.* 85, 1593 (1963). (b) F.D. Greene, M.L. Savitz, F.D. Osterholtz, H.H. Lau, W.N. Smith, and P.M. Zanet, *J. Org. Chem.* 28, 55 (1963).

7. (a) J.K. Kochi, in *Free Radicals*, Vol. II, J.K. Kochi, Ed., Wiley, New York, 1973, p. 685. (b) T. Zytowski and H. Fischer, *J. Am. Chem. Soc.* 118, 437 (1996).

8. For summary and original references, see B. Giese, *Angew. Chem. Int. Ed. Engl.* 24, 553 (1985).

9. N.J. Turro, *Modern Molecular Photochemistry*, Benjamin-Cummings, Menlo Park, CA, 1978.

10. H.G. Viehe, Z. Janousek, and R. Merenyi, *Acc. Chem. Res.* 18, 148 (1985).

11. (a) K. Wimalasena and S.W. May, *J. Am. Chem. Soc.* 117, 2381 (1995). (b) G.N.R. Tripathi, Y. Su, and J. Bentley, *J. Am. Chem. Soc.* 117, 5540 (1995).

12. (a) W.P. Anderson, J.K. Burdett, and P.T. Czech, *J. Am. Chem. Soc.* 116, 8808 (1994). (b) L.C. Allen and J.F. Capitani, *J. Am. Chem. Soc.* 116, 8810 (1994). These authors conclude that "the term 'metallic bond' should be dropped from the vernacular because it is fully encompassed by molecular orbital and band theory and the broader concept of covalent bonding." This is what most chemists think: All

brands of MO theory have the same physical meaning (i.e., HMO is qualitatively the same as MO-CI across the board). The thesis of this work is precisely the opposite.

13. R. Hoffmann, *Angew. Chem. Int. Ed. Engl.* 21, 711 (1982).
14. R. Hoffmann, *Angew. Chem. Int. Ed. Engl.* 26, 846 (1987).
15. M.J. Almond, *Short-Lived Molecules*, Ellis Horwood, New York, 1990, Chapter. 8.
16. J.P. Dinnocenzo, D.R. Lieberman, and T.R. Simpson *J. Am. Chem. Soc.* 115, 366 (1993).
17. At present, the available evidence points toward a D_{2h} rhombus. As usual, we have the thorny problem of a fast interconversion of C_{2v} rhombuses simulating a D_{2h} shape. (a) K. Ushida, T. Shida, M. Iwasaki, K. Toriyama, and K. Nunome, *J. Am. Chem. Soc.* 105, 5496 (1983). (b) P. Jungwirth, P. Carsky, and T. Bally, *J. Am. Chem. Soc.* 115, 5776 (1993).
18. H.D. Roth, *Acc. Chem. Res.* 20, 343 (1987).

Chapter 16

The Benzene Problem

16.1 Orbital Ionization Energy Is the Indicator of Association and Electronic Structure; The Cohesion Rule

Consider the simple case of the aggregation of three bond pairs (e.g., three A_2 molecules infinitely apart) in a hexagonal geometry (e.g., D_{6h} A_6 complex) defined by the following equation. The binding energy per pair is called the cohesive energy (CE) and this can be negative (segregation), zero or positive (aggregation).

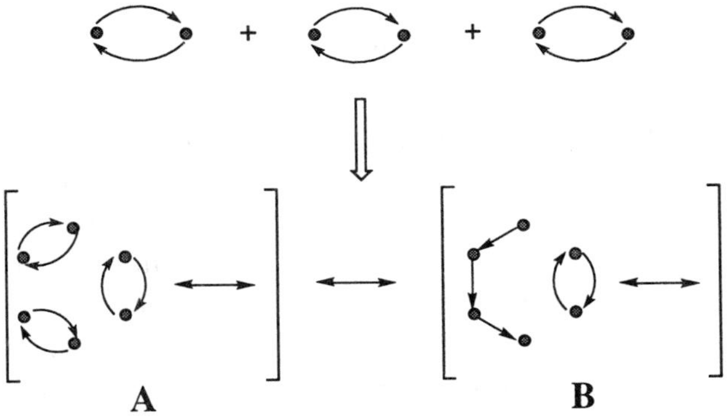

What is the crucial index of association, and what does this imply for the electronic structure of the aggregated state?

The transformation of the segregated to the preceding aggregated state introduces two conditions of delocalization:

Exchange delocalization absent in the segregated state. This is a composite of a repulsive (due to exchange repulsion of the bond pairs) and an attractive part (due to exchange resonance).

CT delocalization absent in the segregated state. This is only attractive, and is the equivalent of the generation of interfragmental CT bonds.

Assuming now that exchange repulsion dominates exchange attraction, we see that the only way of stabilizing the aggregated state is by CT delocalization, which depends on AO ionization energy (AOIE). This produces a simple picture of association: The transformation of the segregated to the aggregated state is seen as a process in which T bonds, symbolized by engaged arrow pairs, are transformed to I bonds by disengagement of some or all engaged arrow pairs.

Let us now consider the foregoing transformation in some detail. The aggregated state is an aromatic [*n*]-annulene that is a resonance hybrid of an exchange (A) and a CT (B) VB descriptor. (Remembering that a VB descriptor is the ensemble of VB configurations implied by the arrows.) The interaction described by A causes overall destabilization on the tested assumption (stated shortly) that interpair repulsion dominates exchange resonance. The interaction described by B causes attraction, which is, however, limited by the AOIE. In our example, the value of the AOIE is such so that only two of the three arrow pairs are disengaged to form 1.5 I bond. By generalizing to an aromatic [*n*]-annulene, we realize that the equation linking AOIE with CT stabilization, hence the fraction of disengaged arrow pairs, must have the following form:

$$\Omega = f(\text{AOIE}) = f(\text{ELNG})$$

Where Ω is the fraction of disengaged arrow pairs and f(AOIE) is some unspecified function of AOIE and, as a result, atom electronegativity (ELNG). The message is: the lower the AOIE and atom electronegativity, the larger the Ω and the greater the stability of the aggregated state. In our example $\Omega = 2/3$. Lowering the assumed AOIE value will drive Ω to the limit of unity. When Ω is significantly smaller than unity, exchange dominates CT delocalization and segregation is favored. When Ω approaches unity, the converse is true. The next crucial issue is: What does the value of Ω tell us about the size dependence of association? In other words, if n bond pairs, with $n = 3$, prefer segregation because Ω is significantly less than unity, what will happen to n bond pairs where $n = 4N + 2$ (aromatic case)?

The answer is founded on one reasonable assumption: The value of Ω is independent of n. In other words, the AOIE determines the *fraction* of disengaged arrow pairs for any [*n*]-annulene. For example, in the example above, $\Omega = 2/3$ means that $6 \times 2/3$ exchange arrow pairs disengage to produce the association of the same number of centers in one resonating I descriptor. In an aromatic [*n*]-annulene, the value of $\Omega = 2/3$ means that only $2n/3$ of the atomic centers become associated in one resonating I descriptor. After the resonance interaction of the equivalent I descriptors, each of which features $2n/3$ associated centers, has been counted, the conclusion is that the aggregated state is still destabilized relative to the aggregated state simply because Ω is 2/3 (i.e., much less than unity).

What is the dependence of the magnitude of the destabilization of the aggregated state for a given Ω on n in an aromatic [n]-annulene? We can easily answer this question provided we have a clear understanding of the essence of T- and I-bonding. Specifically, a single exchange bond (T bond) is defined by two arrows, and so is a single CT bond (I bond). The stabilization is proportional to the AO overlap integral in one interatomic space times the resonance AO integral in either the same (exchange) or different (CT) overlap region. *This means that both T and I bonds have cooperative character.* Enhancing the overlap interaction represented by one arrow enhances the entire bond. The consequences of this series of enhancements are depicted schematically below. It can be seen that Ω determines whether segregation or aggregation is favorable. Once this has been decided, the cooperative nature of exchange and CT bonding ("one arrow helps the other") implies that increasing n will make things either increasingly worse ($\Omega << 1$) or better ($\Omega = 1$) for the aggregated state!

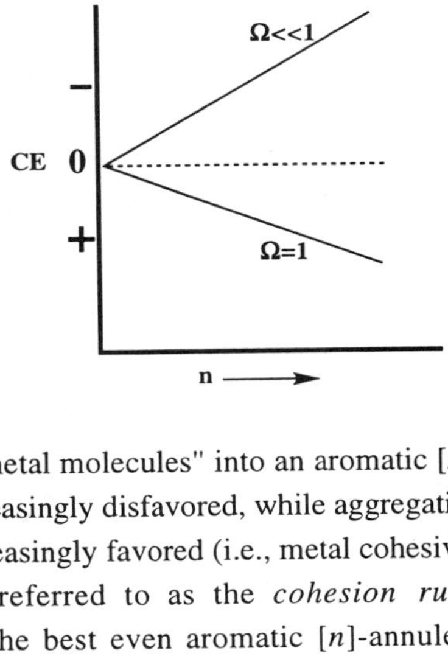

Aggregation of "nonmetal molecules" into an aromatic [n]-annulenoid geometry is predicted to be increasingly disfavored, while aggregation of "metal molecules" is predicted to be increasingly favored (i.e., metal cohesive energy increases) as n increases. This is referred to as the *cohesion rule*. One of its many implications is that the best even aromatic [n]-annulene is the one with the minimum value of n because the electronegativity of C is high. As a result, the interaction of the π-bond pairs is net repulsive. We will now reinvestigate the very conceptual basis of chemistry: benzene, resonance, and aromaticity.

16.2 Why Is Benzene Hexagonal?

The iconoclastic answer is that had it not been for the constraining action of the σ frame, benzene would have had D_{3h} geometry, rather than D_{6h}. In other words, the "aromatic" ensemble of the three π bonds drives the system away from, rather than toward, a regular hexagonal geometry. Indeed, the electronic structure of benzene, a molecule that has been computed to death, is the clearest illustrator of the difference between "chemical thinking" and "chemical engineering thinking." I am aware of only one paper in the literature, a computational study of the bond exchange of three H_2 molecules by Dixon, Stevens, and Herschbach,[1] that identifies the problem: Symmetry considerations alone do not suffice to explain molecular energetics. The realization that H_6 is not a global minimum despite being aromatic calls for reexamination of all the chemist holds dear: aromaticity, resonance, etc. The absence of any serious attempt to solve these problems for nearly a half-century carries the unflattering implication that chemists either cannot recognize fundamental contradictions in their conceptual apparatus or are capable of living with them happily forever.

Why is it that H_6 opts for a geometry in which three H_2 molecules are infinitely apart while Li_6 is "close packed"? The difference lies in nothing other than orbital electronegativity: H1s is quite electronegative while Li2s is very electropositive. The orbital ionization energies are 13.6 and 5.4 eV, respectively, a huge difference. The electronegativity of C2p is not very different from that of H1s (11.2 and 13.6 eV, respectively). Hence, the three π-bonds of benzene must repel each other much as the three σ bonds of the three H_2 molecules do. Orbital electronegativity determines the nature of the VB configuration stacking, hence the type of bonding. The stabilization of the aromatic complex relative to the noninteracting components vanishes when atom electronegativity is high, in which case the ionic structures lie high in energy relative to the covalent structures (sparse set of VB configurations). In other words, both H_6 and π-benzene lie toward the T limit, while Li_6 lies toward the I limit of the bonding continuum.

That the π system of benzene, acting alone, would have driven the molecule away from a regular hexagonal geometry is a conclusion that can be reached without any theoretical deliberations, by mere examination of chemical data:

Three ethylenes do *not* aggregate to form a stable trimer isoconjugate to π-benzene. Such a complex (or a derivative thereof) serves as a transition state to ethylene trimerization to form cyclohexane. Ditto for acetylene.

1,3-Butadiene does *not* coordinate to ethylene. Rather, the complex is a transition state of the Diels–Alder reaction.

In general, unconstrained organic π systems do not form aromatic coordination compounds that are global minima. In the absence of the constraining action of the σ frame, the π bonds of benzene would do exactly what the π bonds of three ethylenes do, namely: They would opt to stay apart. Organic chemistry is said to be the chemistry of the covalent bond without any qualitative distinction between σ and π carbon–carbon bonds. If σ bonds do not aggregate (the example of the three H_2 molecules), neither should π bonds. The VB interpretation is that this occurs because the orbital electronegativities of the Csp^3 and C2p AOs are similar.

Here is the age-old conceptual error the benzene precipitated on us: In writing two equivalent Kekulé structures and saying that π resonance lies behind the stability of benzene, one represents the attractive part while totally disregarding the repulsive part of exchange delocalization. In other words, people counted the exchange bonds and discounted the exchange antibonds. Unfortunately, the latter overwhelm the former. The only way that the complex can become a global minimum is via CT resonance. However, the complexity of matrix elements prevented theoreticians of the 1950s from going beyond "covalent-only" VB theory by incorporating the ionic structures. As a result, the chance for recognizing the physical significance of ionic resonance was missed. With the increasing popularity of MO theory, the evolution of conceptual VB theory stopped, but the power of the method still kept it a mainstay in the education of every chemist. Unluckily, what we inherited is not VB theory but, rather, resonance theory (i.e., "covalent-only VB theory"), and this is why the I bond was not discovered earlier than 1995! It can be reproduced only by full-scale ab initio VB, and it is due to the CT (rather than exchange) interaction of all VB configurations.

What does the "resonance energy" of benzene determined from heats of hydrogenation mean? The π bonds that form an aromatic array have much greater I character than bonds forming an antiaromatic array, hence engender much less interbond exchange repulsion. In other words, CT delocalization is much stronger in aromatic than nonaromatic species and least of all, antiaromatic species. As a result, CT delocalization counteracts a greater portion (but only a portion) of the exchange repulsion in an aromatic system (relative to an antiaromatic or a nonaromatic system). The π system of benzene falls in the T extremum of the bonding pantheon, but it has incipient I-bonding. All thermochemical data can be explained on the hypothesis that π-aromatic systems are less destabilized (rather than more stabilized) than alternative nonaromatic or antiaromatic species. The conventional discussions of benzene fail to recognize that "stability" means either "greater stabilization" or "smaller destabilization." HMO theory pushed chemists to adopt the first interpretation. The right explanation of the remarkable benzene stability is the second interpretation: The π bond of cyclohexene suffers much greater exchange repulsion than one π bond of benzene from the surrounding bonds because CT delocalization (I-bond formation) is at a maximum in the latter.

The confusion about "resonance stabilization" continues unabated in the literature.[2] To banish it, I adopt a dialectic approach.

Question: What does "π-resonance stabilization" mean?

Answer: It means "reduction of interpair exchange repulsion" because of CT delocalization. However, *net exchange repulsion persists.*

Question: Toward what geometry is the allyl anion driven by the π system?

Answer: Toward π pair plus π bond (minimization of interpair exchange repulsion).

Question: Why does the allyl anion have C_{2v} symmetry?

Answer: Because the σ frame says so.

Question: Why is the allyl anion planar?

Answer: Because rotating one methylene by 90° effectively increases interpair exchange repulsion by turning off the counteracting CT delocalization.

In summary, exchange delocalization is either net repulsive or net attractive. With the exception of a two-electron bond, it is repulsive! Hence, exchange drives every polyelectronic system to segregation of bond pairs. On the other hand, CT delocalization is always attractive. Because certain VB configurations (the association-promoting configurations) contribute either mainly or exclusively to the aggregated form, CT delocalization always favors association of bond pairs. This is what ab initio calculations by physical organic chemists show.[2b] The key point is that exchange dominates CT delocalization in aromatic systems, so that what seems "stabilized" is actually "antistabilized," but less so than nonaromatic or, worst, antiaromatic systems. Physical organic chemists have made a living with the notions of conjugation and hyperconjugation by embracing MO theory. The problem is that MO theory does not make an explicit and transparent differentiation of exchange and CT delocalization. Hence, what is simple in VB theory becomes confusing in MO theory. This work shatters the illusion of conjugation and hyperconjugation: If three H_2 molecules do not form a stable "aromatic" H_6 hexagon, why should one expect "aromatic" conjugation in benzene and inferior "nonaromatic" hyperconjugation in organic molecules ever to be important? Computing with the conventional MO theory of chemistry is done all the time; using it to understand molecular electronic structure tends to be unproductive.

Bottom line: Chemistry needs more thinking and less computing. For example, the "double-bond no-bond resonance" ("anionic hyperconjugation") concept goes back to Pauling. It was popularized among organic chemists by Roberts in the 1950s and it has been the focus of multitudinous computational studies from the 1960s to present. For nearly half a century no serious attacks on the justification of the concept were published. The same is true of the concepts of "resonance stabilization," "aromaticity," and so on. Shaik and collaborators provide the only exception to the rule; that is, they were the first to rethink the meaning of the "aromatic sextet."[3] The "theoretical" pages of *JACS* are devoted to papers speaking in favor or disfavor of the ideas of the 1930s: resonance, conjugation, hyperconjugation, van't Hoff's rule, Hund's rule, and so forth. Once the disjunction between exchange and CT delocalization is made, and once it is realized that the former is repulsive and that only electropositive atoms can support the latter, the conventional explanations of organic chemistry lose their force.

16.3 The VB Wavefunction of Benzene; Association and Segregation Domains in Molecules

Benzene defines the quintessential puzzle of chemistry. All appearances indicate that the problem has been resolved. And yet the two preceding sections say the opposite. If it is true that the three π bonds would opt for segregation if the σ frame were not there, the VB wavefunction of π-benzene must bear the sign of instability. We already know where to look for the answer. Association of bonds is effected by I bonds generated by CT delocalization with the HP* configuration defining the association domain. If HP* = HP(max), the wavefunction would be the following one, and the three π bonds would associate.

THE ELECTRONIC STRUCTURE OF BENZENE FOR HP* = HP(max)

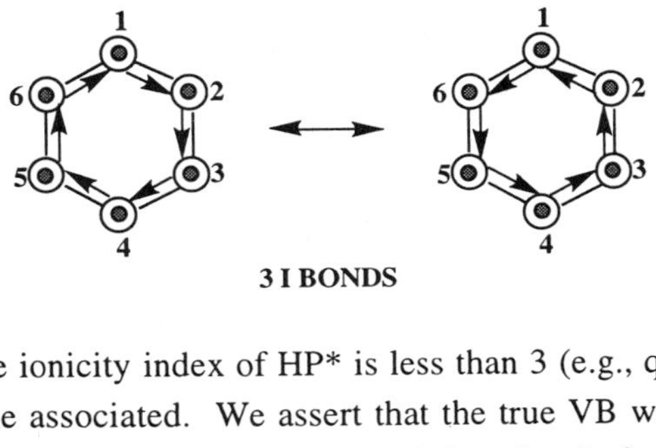

3 I BONDS

By contrast, if the ionicity index of HP* is less than 3 (e.g., q = 2), only two π bonds can become associated. We assert that the true VB wavefunction of π-benzene is actually the one represented by the I formula using the superior/inferior arrow convention and shown in Figure 16.1. We can rephrase the argument using the ideas of Section 16.1: The value of Ω is much less than unity because carbon is significantly electronegative. We assert that the value of Ω is 2/3 corresponding to an ionicity index of 2. At once, Ω tells us two things:

The π system of benzene is actually destabilized.

Since only two of the three π-bond pairs are disengaged, two π electrons remain coupled by exchange-shunning association with the other four π electrons. The key features of the π VB wavefunction of benzene form the basis for all following applications to ground aromatic molecules as well as aromatic transition states.

(a) The π system is a hybrid of I_o (o = ortho) and I_p (p = para) I descriptors.

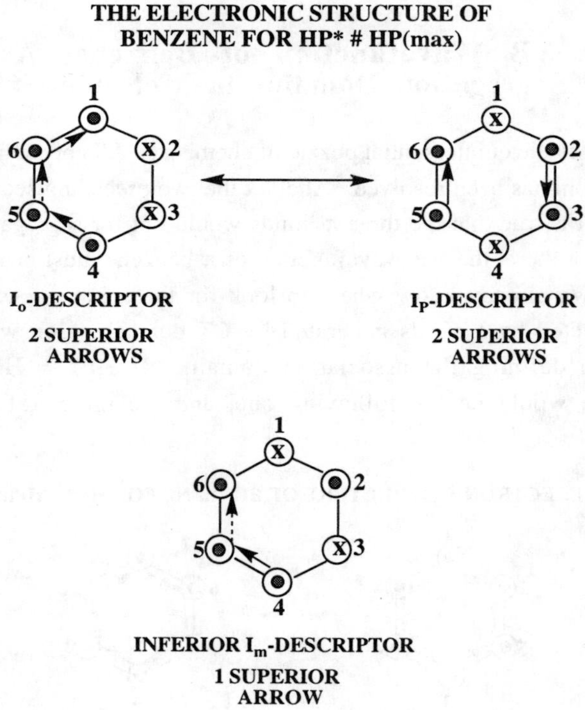

Figure 16.1 The VB formula of benzene. The molecule is a resonance hybrid of "2 times 6" ortho and "2 times 3" para I descriptors, each of which has resonating association and segregation domains. The meaning of "2 times 6" is that there are six ortho descriptors for a right-handed and six more for a left-handed codirectional arrow train.

(b) Each I descriptor has an association domain (A domain) and a segregation domain (S domain). The latter is localized either ortho or para. Metalocalization is inferior because it has only one superior arrow. In other words, ortho/para localization of the S domain allows for strong T complement ("covalent character") of I-bonding in the A domain. Each I descriptor, called a *bikrat* (Gr: two domains), consists of two distinct domains that are associated with secondary 2-e CT delocalization (not shown).

(c) The A domain is bound by a codirectional arrow train. Since there can be right- and left-handed codirectional arrow trains, there are 2×6 ortho and 2×3 para I descriptors. The A domain involves the association of two electron pairs.

(d) The S domain is a pair of antiferromagnetically coupled electrons that cannot support CT delocalization. Each S electron acts effectively as a radical, which can be further stabilized in three ways: by a radical-stabilizing substituent, by the A domain, or by both combined. In a heterobenzene, the S pair localizes on electronegative or electropositive (relative to carbon) heteroatoms to allow a homonuclear set of electroneutral carbons to accommodate the A domain. This guarantees configurational degeneracy required for strong I-bond formation.

(e) Following the usual chemical convention, the molecular formula shows only one of the equivalent descriptors.

(f) π-Benzene is a *resonant bikrat*. The nonresonant analogue is the soliton.[4] A resonant bikrat is the consequence of strong CT delocalization, and the latter is maximized in aromatic systems.

16.4 The Counterintuitive Predictions of VB Benzene

Chemical journals bombard their readers with unending computations of six-electron transition states (Diels–Alder, Cope, Claisen, etc.). Precious little has been done with benzene because people assume that it is well understood. However, this attitude makes little sense: The π system of benzene is isoconjugate to the "active" component of the Diels–Alder transition state. Because of the absence of the constraining action of the σ bonds, the Diels–Alder six-electron complex is a transition state rather than a global minimum consistent with the arguments presented earlier. Thus, whatever "problems" exist in the Diels–Alder, Cope, Claisen, etc., transition state, they must also be present in benzene! In fact, we will argue that our VB wavefunction of benzene is proven right by the multitude of experimental results pertinent to the Diels–Alder transition state.

The wavefunction of π-benzene shown in Figure 16.1 leads to immediate predictions:

(a) The S pair is ortho/para localized. Because the I_o descriptor has 1.5 I bonds while the I_p descriptor has 1.0, ortho- is superior to para-localization of the S pair. Since the S pair is essentially a stretched π bond (i.e., two singlet-coupled radicals), it will be stabilized by radical-stabilizing groups. Since both π donors and π acceptors stabilize radicals, the prediction is that ortho- and para-disubstituted benzenes should be much more stable than the meta variety unless some other factor overrides this preference. This "other factor" is the effect of substituents on the stability of the σ frame: Both systolic and carbon-centered diastolic groups prefer meta attachment because this minimizes the number of errors (see Chapters 7–9). Typically systolic groups are conventional π donors, and typical diastolic groups are conventional π acceptors. Hence, we predict that a cancellation of the two opposing effects will lead to a small differentiation of the *m* and *p* isomers with nonbonded repulsion disfavoring the *o* isomer.

(b) The heteroatom exclusion rule leads to two predictions:

1. Electropositive (e.g., Si) or electronegative (e.g., N and O) heteroatoms are excluded from the six-electron domain.
2. Electropositive (e.g., Si) or electronegative (e.g., N and O) heteroatoms constrained to be part of the six-electron domain are excluded from the A domain and accommodate the S pair.

(c) Heteroatoms accommodating the S electrons in the principal I contributor, accommodate also the A domain in the minor I contributors. Since CT delocalization responsible for the stability of the A domain and the molecule as a whole is impaired as atom electronegativity increases, it is expected that replacement of the benzene hydrogens by more electronegative heteroatoms (e.g., F) will turn off the I character of the bonds and will diminish the resonance energy of benzene. The same thing will happen if a perturbation either directly or indirectly causes carbon pyramidalization and, thus, introduction of C2s character in the original C2p.

Although traditionally benzene has not been examined from this angle, the random experimental data found in the literature are in line with these counterintuitive expectations. Meta placement of two electronegative heteroatoms is strongly favored in saturated six-membered rings (apply the T formulas of Chapters 7–9). However, this preference disappears in the benzene analogues, with the ortho becoming stabilized much more than the para.

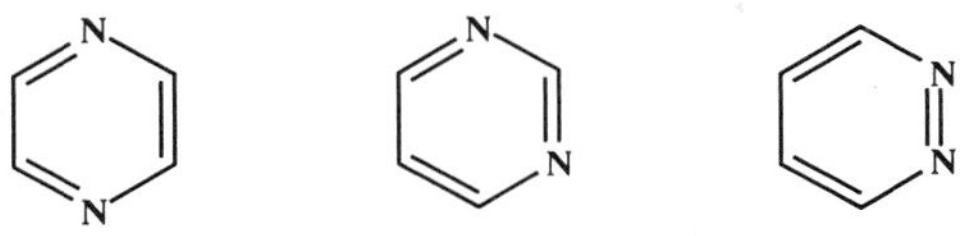

E_{rel}: **0** **0** **22**

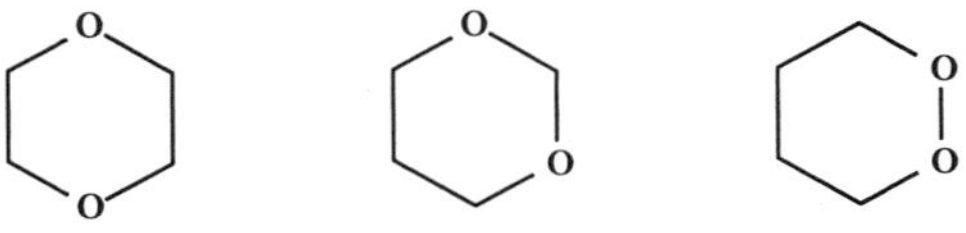

E_{rel}: **5** **0** **45**

The thermochemical data of Table 16.1 show that a π donor and a π acceptor prefer meta/para and not ortho/para placement, as one would have predicted by falling back on the conventional concepts of π conjugation in the case of the CH_3O/NO_2 pair.

Some confirmations of the heteroatom exclusion rule were discussed in Chapter 15. We can further test the prediction that the S electrons localize preferentially on the heteroatoms of a heterobenzene by using C_5H_5E (E = P, As, Sb, or Bi) as the illustrator molecule. Realizing that the A domain is more stabilized than the S domain and assuming that one of the S electrons localizes on E leads to the prediction that C_5H_5E can act as either an ortho- or para E-centered bikrat, which in combination with a dienophile will yield a 4 + 2 cycloadduct in which E will be at one union site, as is found to be the case.[4d]

The dependence of aromaticity on orbital electronegativity is a key element of the VB theory of chemical bonding. The counterintuitive prediction that replacing H by F will diminish the resonance energy of benzene is confirmed by the following thermochemical data (kcal/mol).

$$\text{cyclohexene} + H_2 \xrightarrow{\Delta H_r = -28} \text{cyclohexane}$$

Table 16.1 Heats of Formation (kcal/mol) of the Isomeric C_6H_4XY

X,Y	Ortho	Meta	Para
CH_3O, NO_2	–17	–22	–21
CH_3O, CH_3O	–53	–58	–56

$$\text{benzene} + 3\text{H}_2 \xrightarrow{\Delta H_r = -49} \text{cyclohexane}$$

Resonance Energy = 35 kcal/mol

$$\text{perfluorocyclohexene} + \text{F}_2 \xrightarrow{\Delta H_r = -110} \text{perfluorocyclohexane}$$

$$\text{perfluorobenzene} + 3\text{F}_2 \xrightarrow{\Delta H_r = -338} \text{perfluorocyclohexane}$$

Resonance Energy = –8 kcal/mol

More evidence in support of the proposal that electronegativity determines whether aromatic delocalization is activated comes from results of semiempirical and ab initio computations[5]:

(a) Because of the conversion of π bonds to σ bonds, both the "antiaromatic" dimerization and the "aromatic" trimerization of acetylene to cyclobutadiene and benzene, respectively, are exothermic. Replacing HC by N increases the orbital ionization energy of the 2p AO from 11.3 eV (C2p) to 13.9 eV (N2p). As a result, both reactions are rendered endothermic. The computed energy change is colossal: 156 kcal/mol for the dimerization and 295 kcal/mol for the trimerization! Of course, one reason is vicinal lone pair repulsion. The other is impairment of π CT delocalization due to the enhanced orbital electronegativity.

(b) The computed exothermicity difference between the "aromatic" trimerization and the "antiaromatic" dimerization of acetylene is 122 kcal/mol. This reflects the strong preference for formation of an aromatic benzenic π system. By contrast, the endothermicity difference between the corresponding processes for the nitrogen derivatives is computed to be –17 kcal/mol! This means that there is no longer any significant aromatic driving force operating in favor of "nitrogen-benzene."

When a benzene C2p is transformed to a more electronegative Csp^3 AO by pyramidalization, aromatic stabilization is expected to decrease. A fullerene is a curved sheet of graphite produced by pyramidalization at every carbon. Recent investigations have identified small ring currents and bond alternation in fullerenes (i.e., loss of aromatic character).[6a] More recent magnetic resonance studies using ^{3}He atoms as probes in endohedral fullerenes have suggested the existence of some aromatic character.[6b] All these results simply mean that graphite is displaced toward the I end and fullerene toward the T end of the T/I bonding continuum, always in a relative sense. C_{70} has a stronger ring current than C_{60} because the former approaches more closely the planar graphitic limit. Saunders put his finger on the problem in 1994: "I believe that fullerenes require us to extend our concept of aromaticity from the black and white — aromatic or antiaromatic — to the notion that aromaticity is a continuous property and that C_{60} is roughly halfway between aromatic and antiaromatic, leaning slightly toward aromaticity."[6c] Haddon, too, had invoked a gray area.[6d] The distinguishing feature of an aromatic molecule is its potentiality for strong CT delocalization (i.e., I-bonding). This potentiality becomes reality only when the AO electronegativity is low. The essential point is that diminished ring currents in formal aromatic systems confirm the validity of our analysis: Aromaticity is a function of AO electronegativity!

16.5 Aromatics as Polybenzenes

The story of coronene has been told in a recent instructive and entertaining monograph.[7] This $C_{24}H_{18}$ hydrocarbon represents most of the challenges one faces when it comes to describing extended two-dimensional π systems. The key question is: What is the operationally significant formulation of this molecule? Two choices are as follows:

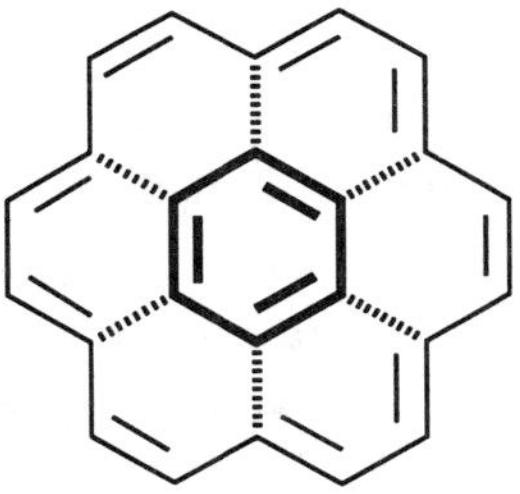

[6] + [18]

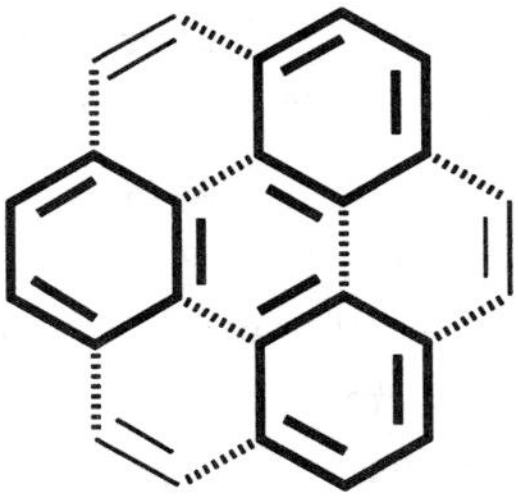

3x[6] + 3[2]

One molecule is a [6] π-annulene (π-benzene) "hub" plus an [18] π-annulene "wheel" connected by spokes via the hatched lines, which represent the interaction of the two subfragments. The other molecule is three [6] π-annulenes plus three π-bonds.

For each dissection, the total π energy, $E(T)$, is:

$$E(T) = E(F) + IE$$

where $E(F)$ is the total π energy of the chosen fragments and IE the interaction thereof. Clearly, $E(T)$ is dissection invariant while $E(F)$ and IE depend on the fragmentation mode. This provides now a clear solution to our dilemma: The operationally significant formulation of a two-dimensional π system is the one for which $E(F)$ is maximum. In other words, as IE approaches the limit of zero, the chosen fragments become an increasingly more faithful representation of the composite molecule. The key point now is that the relative magnitudes of $E(F)$ and IE depend on the level of theory one uses to compute them.

The HMO wavefunctions of π-conjugated systems have been tabulated by Heilbronner.[8] According to this brand of theory, the fragment energies (in beta units) of the two dissection modes of coronene are:

$$E(6 + 18) = 8.0 + 23.0 = 31.0$$

$$E(3 \times 6 + 3 \times 2) = 24.0 + 6.0 = 30.0$$

The results predict that the "chemistry-revealing" formulation of coronene is the "hub/wheel." However, we recall the fundamental failure of HMO theory: unrestrained CT delocalization. in the absence of such restraint, coronene will tend to resemble the 6 + 8 combination of aromatic annulenes because in the situation described by HMO theory, all π-bond pairs can be turned into zwitterions to support CT delocalization.

The VB story is different: Only a fraction of the π-bond pairs can be turned over into zwitterions. That is, CT delocalization is limited by the ionization potential of the C2p AO. According to the cohesion rule of Section 16.1, the optimum even [n]-annulene has $n = 6$ (i.e., n is a minimum because net destabilization increases with n). As a result, all two-dimensional planar π systems are actually "polybenzenes:" composites of π-benzene fragments plus "π leftovers." The final conclusion is that the "chemistry-revealing" formulation of coronene is the "polybenzene" one. This predicts that the shortest (most doublelike) bonds of the molecule will be the HC═CH bonds on the outer rim, as is found. These are also predicted to be the sites of highest reactivity.

The comparison of anthracene and phenanthrene is also instructive.

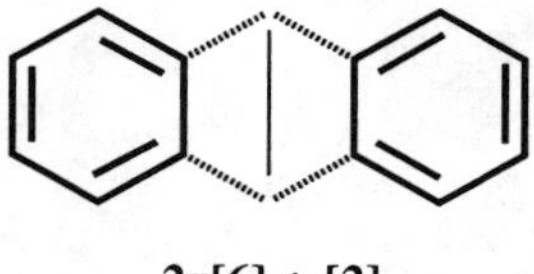

2x[6] + [2]

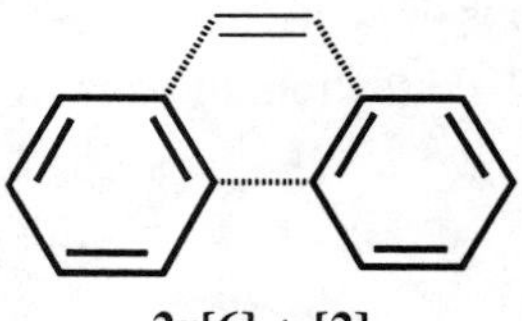

2x[6] + [2]

Now, the chemical formulation of each is "2 × [6] + [2]." The difference comes to the extent that the two-center annulene (i.e., the lone π bond) involves the nonadjacent C9 and C10 atoms in anthracene but the adjacent C9 and C10 in phenanthrene. Hence, the latter must be the most stable isomer, as in fact is found. In addition, both should react preferentially at C9 and C10, as is found. This approach predicts that curved will always be superior energetically to linearly fused aromatic rings.

As a final example, the bowl-shaped [5] circulene (or corannulene) can be formulated in either of the following ways:

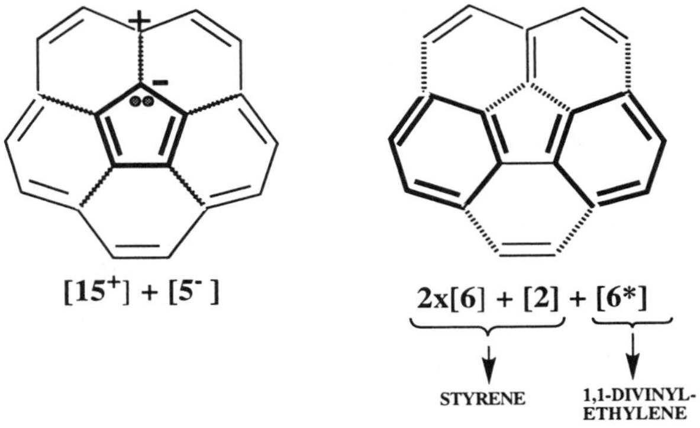

The X-ray crystal structure makes it clear that the polybenzene (right), rather than the annulenoid (left), depiction comes closer to the truth.[9,10] Furthermore, the very colorlessness of the molecule also argues against the latter formulation. The ^{1}H and ^{13}C NMR spectrometric data are reminiscent of those for benzene. The "π styrene plus π-1,1-divinylethylene" formulation also explains why this molecule is willing to accept up to four more electrons.[11] The tetraanion is a urealike six-electron fragment plus two resonance-stabilized benzyl anions:

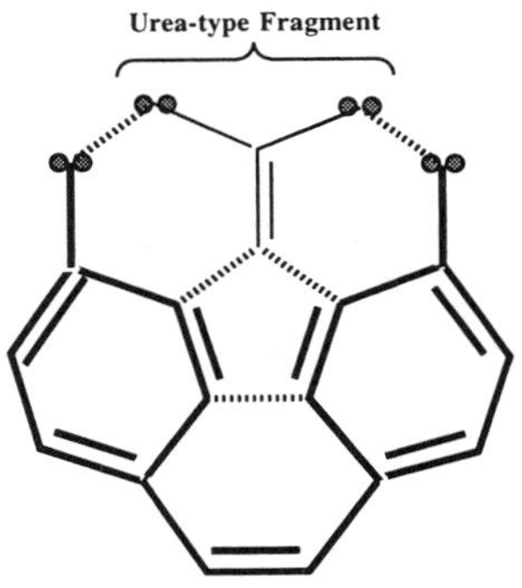

The essential point is that the "benzene rule," first envisioned by Clar a long time ago,[12] is the manifestation of the balance between atom promotion (i.e., turning bond pairs into zwitterions) and CT delocalization. Because HMO theory fails to reproduce this balance properly, planar benzenoid hydrocarbons resemble either annulenes or "annulenes within annulenes." VB theory changes all this and restores benzene to its proper place as the fundamental repeating unit of all planar benzenoid hydrocarbons.

16.6 Buckminsterfullerene as Polydicyclopentadienyl

The electronic structure of buckminsterfullerene, C_{60},[13] seems simple enough: C_{60} appears to be a piece of curved graphite. However, according to VB theory, this is not the case: The *planar* graphitic system has aromatic character because of the modest ionization potential of the C2p AO. By contrast, C_{60} has pyramidalized carbons in which the exo radial carbon AOs have enhanced electronegativity. Thus, the aromaticity of "curved graphite" is impaired. As a result, ordinary graphite can be regarded as "aromatic," while C_{60} is really a less stable quasi-nonaromatic analogue of graphite. This model is supported by the reduced "ring currents" found in C_{60}.[6d] In turn, this raises the question: What is the chemically meaningful formulation of this molecule? Restricting our attention exclusively to the π systems, we focus on two possible choices:

C_{60} is "polybenzene;" that is, it is formally $(BZ)_{10}$ (BZ = benzene residue).

C_{60} is "polycyclopentadienyl;" that is, it is formally $(CP)_{12}$ (CP = cyclopentadienyl residue).

There are two possibilities:

(a) If the effect of curvature on the energy of the formal π system is neglected (i.e., if C_{60} is effectively a piece of graphite), the molecule should behave like a "polybenzene."

(b) If the effect of curvature is selective and strong (i.e., if it stabilizes five- over six-membered rings while diminishing the fragment interaction), the molecule should behave like "polycyclopentadienyl." However, since two CPs can combine to form the more stable dicyclopentadienyl, the molecule should behave like "poly-DCP," where DCP stands for the closed-shell dicyclopentadienyl fragment :

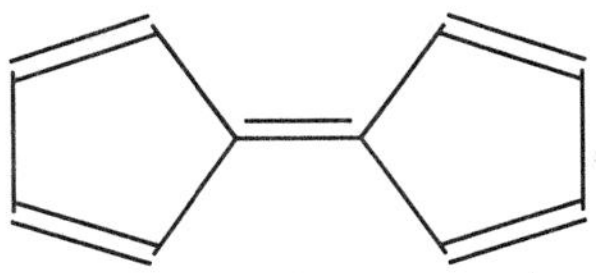

DCP

The answer was given effectively in Chapter 5: Buckminsterfullerene avoids strain due to graphitic curvature because of the capacity of the five-membered ring to eschew strong bonding in preference for atom de-excitation. Thus, the characteristic feature of the molecule is the localization of electron density in the five-membered rings in preparation for favorable C2p to C2s demotion. As a result, we visualize the C_{60} molecule as $(DCP)_6$. There are 12 CP fragments at the 12 vertices of an icosahedron. Six icosahedral edges (sides) connect connect six different pairs of CPs to produce the six DCPs. These are called *structural edges*. The remaining edges represent the interaction of one CP belonging to one DCP with a neighboring CP belonging to a different DCP. These are called *interaction edges*. Different allocations of the six structural edges constitute either resonance structures in the form of maximum symmetry or valence isomers of reduced symmetry.

The DCP model of buckminsterfullerene leads to several predictions:

(a) C_{60} can accept six electrons to form a highly stable septet because, by doing so, one of the two cyclopentadienyl rings of each DCP is converted to a stable aromatic cyclopentadienide anion.

(b) C_{60} can react like DCP (e.g., as dienophile and dipolarophile). Because of the presence of the reducible cyclopentadienyl rings, DCP is also expected to act as an electrophile.

(c) The formal single and double bonds of DCP can be classified according to the type of ring junction as (6,6) or (5,6). The central exocyclic double bond is (6,6), while the endocyclic single and double bonds are (5,6). Of these, only the (5,6) π bonds are involved in the direct inter-DCP interaction. Hence, addition of a carbene is expected to occur preferentially in the "uninvolved" (6,6) π bond. By contrast, insertion of a carbene to a σ bond to form a fulleroid is expected to occur preferentially in a (5,6) σ bond.[14]

(d) Addition of four moles of F_2 to DCP generates a $DCPF_8$ residue that now has only one double bond. One possible $DCPF_8$ structural isomer is designated A:

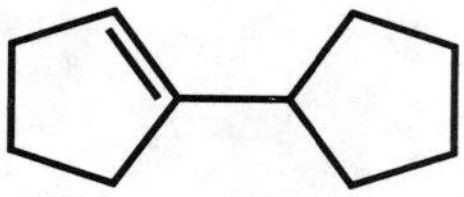

$DCPF_8$ - A

We can assemble three $DCPF_8$ subunits in the following manner to form a truncated trigonal pyramidal structure in which the three edges connecting the truncated apex to the triangular base are the three $DCPF_8$ subunits.

"UPPER HALF" OF $C_{60}F_{48}$ IS THREE $C_{10}F_8$ IN LOCAL C_3 SYMMETRY

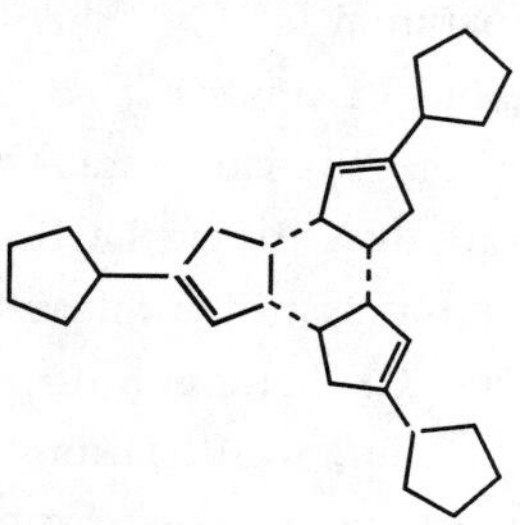

Repeating the procedure generates a second inverted truncated trigonal pyramid. Meshing the two pyramids generates a spherical cluster that corresponds to the chiral $C_{60}F_{48}$ molecule elucidated by Gakh et al.[15] The best way to visualize the shape is to take a soccer ball and color it appropriately. We predict the existence of other stable $C_{60}F_{48}$ isomers which are hexamers of $DCPF_8$ structural isomers. The important point here is that $C_{60}F_{48}$ can be formulated as a hexamer of $DCPF_8$-A much as C_{60} was formulated as a hexamer of DCP.

Space does not permit an extensive discussion of these predictions. Suffice it to say that the existing experimental data are consistent with this picture. Wudl and Fagan realized the operational significance of the "pyracyclene" formulation of C_{60},[16] and a survey of fullerene reactivity was published in 1994.[17]

16.7 Physical Interpretation of Ring Current

The upfield shift of the proton signal of the hydrogens directly attached to an aromatic ring has been attributed to "aromatic ring current" and has been used in the diagnosis of aromaticity. VB theory gives a direct physical interpretation of ring current. Specifically, π aromaticity is the consequence of incipient I-bonding due to kinetic energy reduction occurring with correlated CT delocalization. This is equivalent to saying that the π electrons are capable of delocalizing inside the ring perimeter to best enjoy nucleus–electron attraction while avoiding each other. The reason for the aromatic proton deshielding is twofold:

(a) Because of CT delocalization, the π electrons become interstitialized in the overlap regions. As a result, both the hydrogen nucleus and the hydrogen electron see less of a π electron. The latter is beautifully shown in the VB calculations of Norbeck and Gallup,[18] where going from a single VB configuration to the symmetry-adapted set (e.g., from one adjacent singly ionic to the 12 equivalent adjacent singly ionic configurations) produces a stabilization that is largely due to the reduction of σ–π repulsion.

(b) For the same reason given in (a), there is an additional "inside shift" of π-electron density.

In antiaromatic rings, the impairment of CT delocalization has exactly the opposite effect.

The problem becomes more interesting when we consider cyclopropane. According to the shell model, this is a pseudoaromatic R^2T^4 system (see Chapter 4) in which the four tangential electrons of the T shell are confined outside the ring perimeter and stay in proximity to the C—H bonds. As a result, the cyclopropane hydrogens are notoriously shielded. The situation becomes normal in cyclohexane because intershell interaction is now established, and the T-shell electrons can delocalize into the R shell ("inside shift").

The paradigm of the notion of "nominal fallacy" is the textbook explanation of the ^{1}H NMR chemical shifts of acetylene and ethylene (or formaldehyde and aldehyde). Both the relative shielding of the protons of the former and the deshielding of the protons of the latter are attributed to ring current. However, there exist two stereochemically distinct ring currents: one in the plane containing the multiple bond and one in the plane normal to the multiple bond. The former causes deshielding and the latter shielding. Hence, the key issue is to understand why the ring current travels normal to the triple bond but parallel to a double bond. The answer is given by the concept of overlap dispersion. In the case of the double bond, one σ and one π bond are coupled by overlap dispersion. Each component amounts to a 2-e CT hop, which can be represented by a pair of arrows (e.g., the overlap is symbolized by two dashed and the dispersion by two solid arrows). The four arrows define codirectional CT as illustrated below; that is, overlap dispersion is really "acyclic aromaticity" involving four electrons circulating in a closed loop in the plane containing the double bond. By contrast, in the case of the triple bond, overlap dispersion couples the weaker two of the three bonds. These are the two π bonds. The four-arrow loop extends perpendicular to the triple bond axis; that is, four electrons circulate in a plane normal to the triple bond.

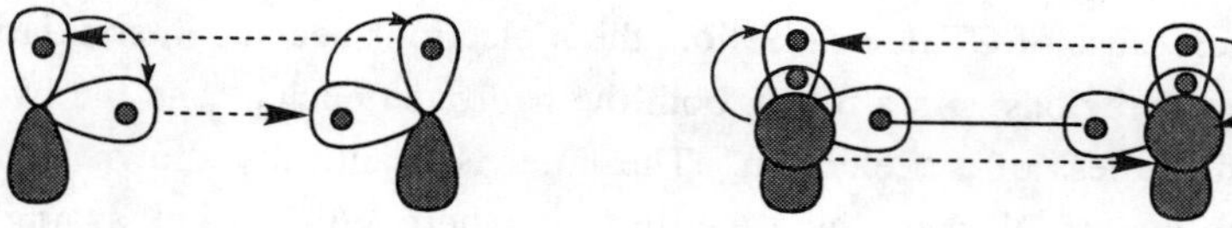

The same type of analysis predicts that the phosphorus chemical shift of HCP along the C_{inf} axis will be low (shielding) while that normal to the CP "triple bond" will be high (deshielding), as found by ab initio calculation and experiment.[19]

16.8 Can a π-Attractive Benzene Be Made?

"New theory" means "new beginning." More than a century after Kekulé suggested a formula for benzene, we find ourselves wondering whether a "stable π-benzene" can be made! To be more precise, can we make a benzene molecule in which the three π bonds have net attraction? The problem is defined in Figure

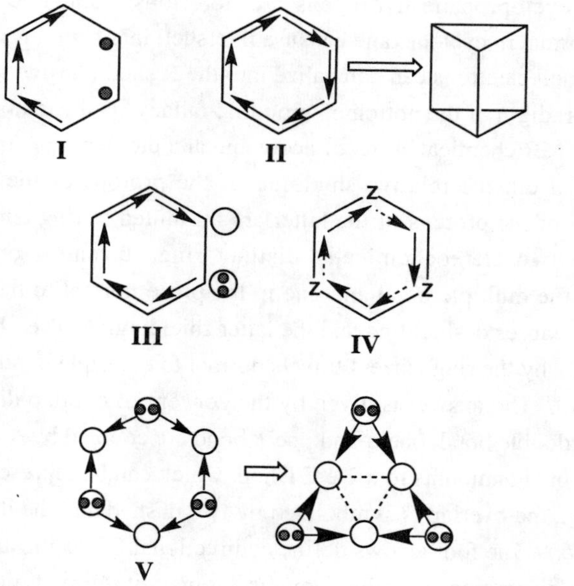

Figure 16.2 Six-electron cyclic systems of five different types; open arrows indicate the distortion modes.

16.2, which shows five different systems that are isoconjugate either to benzene or to the π system of benzene.

System I is the real bikratic benzene which has an association and a segregation π domain. Replacing C by Si leads to a decrease of the electronegativity of the np AO and to a fully associated system II, which, however, obeys different rules! Contact repulsion dictates that the prismane valence isomer is now superior. We will return to this issue in greater detail in Chapter 24. By the time we arrive at V, typified by the Li hexamer, the orbital electronegativity has dropped so low that the bonding is directed by the multipolar ALT configurations. These cause a close-packing distortion that is discussed in detail in Chapter 35.

We can now appreciate the problem: The "window of opportunity" for making π-attractive hexagonal benzene is far too small. Starting with carbon–benzene and replacing carbon by more electropositive atoms brings us too fast inside the realm of I-bonding in which the planar hexagonal structure becomes disfavored.

References

1. D.A. Dixon, R.M. Stevens, and D.R. Herschbach, *Faraday Discuss. R. Soc.* 62, 110 (1977).
2. (a) A. Gobbi and G. Frenking, *J. Am. Chem. Soc.* 116, 9275 (1994). (b) E.D. Glendening, R. Faust, A. Streitwieser, K.P.C. Vollhardt, and F. Weinhold, *J. Am. Chem. Soc.* 115, 10952 (1993).
3. Shaik has argued independently and on the basis of computations that the π bonds of benzene repel each other: S. Shaik, P.C. Hiberty, J.-M. Lefour, and G. Ohanessian, *J. Am. Chem. Soc.* 109, 363 (1987). For recent developments, see Y. Haas and S. Zilberg *J. Am. Chem. Soc.* 117, 5387 (1995).
4. (a) W.P. Su, J.R. Schrieffer, and A.J. Heeger, *Phys. Rev. Lett.* 42, 1698 (1979). (b) W.P. Su, J.R. Schrieffer, and A.J. Heeger, *Phys. Rev. B*, 22, 2099 (1980). (c) Review of solitons in conducting polymers: J.L. Bredas and G.B. Street, *Acc. Chem. Res.* 18, 309 (1985). See also L.M. Tolbert, *Acc. Chem. Res.* 25, 561 (1992). (d) A.J. Ashe III, *Acc. Chem. Res.* 11, 153 (1978).
5. (a) H. Kollmar, F. Carrion, M.J.S. Dewar, and R.C. Bingham, *J. Am. Chem. Soc.* 103, 5292 (1981). (b) J.S. Wright, *J. Am. Chem. Soc.* 96, 4753 (1974).

6. (a) For a compact review of this ever-expanding frontier and pertinent references, see P.J. Fagan, J.C. Calabrese, and B. Malone, *Acc. Chem. Res.* 25, 134 (1992). (b) M. Saunders, H.A. Jimenez-Vazquez, R.J. Cross, S. Mroczkowski, D.I. Freedberg, and F.A.L. Anet, *Nature* 367, 256 (1994). (c) M. Saunders, *Chem. Eng. News* 72(9), 40 (1994). (d) Haddon had previously concluded that "C_{60} is of ambiguous aromatic character": R.C. Haddon, *Science* 261, 1545 (1993)

7. F. Voegtle, *Fascinating Molecules in Organic Chemistry*, Wiley, New York, 1992, Chapter 3.

8. E. Heilbronner, *Tetrahedron Lett.* 1923 (1964).

9. W.E. Barth and R.G. Lawton, *J. Am. Chem. Soc.* 93, 1730 (1971).

10. R.C. Haddon, *J. Am. Chem. Soc.* 109, 1676 (1987).

11. (a) A. Ayalon, M. Rabinovitz, P.-C. Cheng, and L.T. Scott, *Angew. Chem. Int. Ed. Engl.* 31, 1636 (1992). (b) M. Baumgarten, L. Gherghel, M. Wagner, A. Weitz, M. Rabinovitz, P.-C. Cheng, and L.T. Scott, *J. Am. Chem. Soc.* 117, 6254 (1995).

12. D. Clar, *Aromatische Kohlenwasserstoffe*, Springer-Verlag, Berlin, 1952.

13. R.E. Smalley, *Acc. Chem. Res.* 25, 98 (1992).

14. P.M. Warner, *J. Am. Chem. Soc.* 116, 11059 (1994).

15. A.A. Gakh, A.A. Tuinman, J.L. Adcock, R.A. Sachleben, and R.N. Compton, *J. Am. Chem. Soc.* 116, 819 (1994).

16. (a) See ref. 6a. (b) F. Wudl, *Acc. Chem. Res.* 25, 157 (1992).

17. For recent survey of fullerene reactivity, see A. Hirsch, *The Chemistry of Fullerenes*, G. Thieme Verlag, New York, 1994.

18. J.M. Norbeck and G.A. Gallup, *J. Am. Chem. Soc.* 96, 3386 (1974). The VB wavefunction of benzene produced by these workers may shock organic chemists because the predominant contributors are the orthozwitterionic, rather than the Kekulé, structures. The crucial point is that this does *not* mean that benzene is a zwitterion. The wavefunction can be interpreted exactly as we did in Section 16.2, except that the A domain has a zwitterionic, rather than a covalent, reference configuration and the S domain has two odd electrons. In this picture, ionic configurational degeneracy in the A domain creates strong association of two of the three π bonds.

19. K. Eichele, R.E. Wasylishen, J.F. Corrigan, N.J. Taylor, and A.J. Carty, *J. Am. Chem. Soc.* 117, 6961 (1995).

Chapter 17

The Pericyclic Transition State

17.1 The Four-Electron Rule and the Bikratic Transition State

In a reversal of chemical common sense, the six-electron transition states of pericyclic reactions discovered by Diels and Alder, Huisgen, Cope, and Claisen and studied by the pioneer physical organic chemists Doering and Bartlett confirm the electronic structure of benzene just described (Section 16.7):

Substituents end up either ortho or para irrespective of whether they are π donors or π acceptors.

Electronegative atoms either are excluded from the reactive sites or accommodate the two electrons of the S pair.

Fluorination of the dienophile destroys transition state aromaticity and converts a one-step reactioninto a stepwise Diels–Alder reaction.

In this chapter, our main intent is to summarize the critical data not only to provide a definitive interpretation of the "six-electron problem" but also to project the work that must be done before we are certain that new surprises do not await us around the corner. The VB analysis does vindicate the concept of transition state aromaticity advanced long ago by Dewar and Zimmerman, but it must be clearly understood that VB aromaticity and bonding, in general, are very different from the corresponding notions of resonance theory, HMO theory, and any models that do not have as a centerpiece the explicit treatment of electron repulsion.

Recalling that a codirectional arrow train can be either right- or left-handed, the transition state of the 4+2 cycloaddition of a diene and an olefin is a resonance hybrid of the following I descriptors.

2×3 reactant-like ortho descriptors.
2×3 product-like ortho descriptors.
2×3 reactant/product-like para descriptors.

The operationally significant wavefunction of the 4+2 transition state is one of the three shown in Figure 17.1, where we continue our usual practice of writing only one of the equivalent contributors. In the absence of radical-stabilizing substituents, a reactant-like (product-like) transition state is expected in the case of a highly exothermic (endothermic) reaction (Hammond's postulate). When the reaction enthalpy is modest, the transition state must be the in-between state III (i.e., the one with 1,4-diylic character), as found by calculations not only of the Diels–Alder transition state but also of the related Cope and Claisen transition states.[1] The key point is that the transition state is neither a diradical nor some undefined resonance hybrid. It is a resonant bikrat.

The crucial point in Figure 17.1 is the notion that the number of associated electron pairs is limited by the orbital ionization energy. Our hypothesis is that in a cyclic aromatic array of C2p AOs, the orbital electronegativity of a C2p AO can promote the association of only *two* bond pairs. For example, in a [6]-annulene, *each* equivalent I descriptor has four associated centers (2 × 6/3). In other words, the highest order ionic configuration contributing to the VB wavefunction [i.e., the HP* configuration (see Chapter 3)], is the quadrupolar rrhphp structure (r = radical, h = hole, p = pair), which is interpreted to mean that

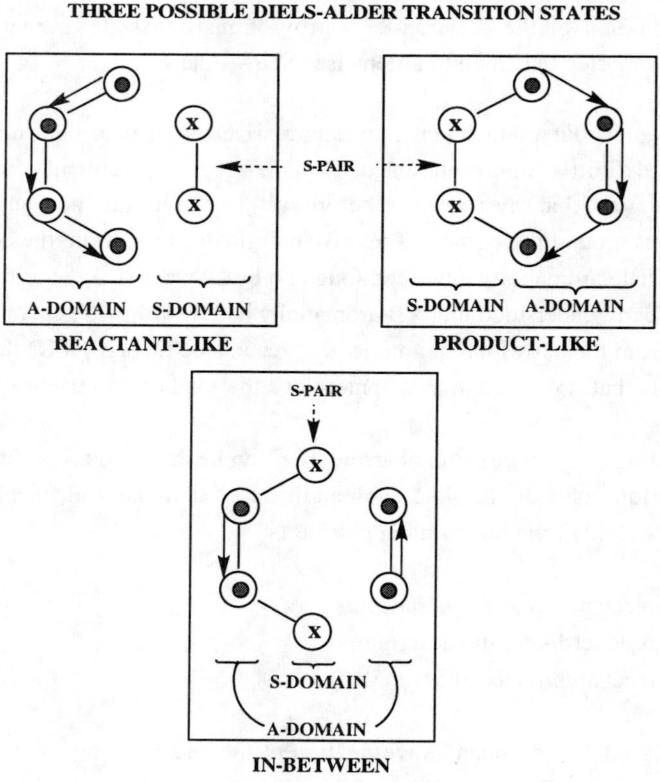

Figure 17.1 The three possible bikratic Diels–Alder transition states.

only two pairs can associate. In turn, this suggests a general rule for organic transition states: The A domain of every neutral [*n*]-annulenoid organic transition state contains four electrons while the S domain contains n – 4 electrons with the two domains being either in resonance (aromatic case) or out of resonance (antiaromatic case). For the most common case of an aromatic, six-electron organic transition state (e.g., the Diels–Alder reaction transition state), the A domain spans four centers with four electrons, while the S domain spans two centers with two electrons. By reference to the A domain, this is called the *four-electron rule.* It should be noted that the number of electrons assigned to the A domain (four) is appropriate for *neutral* carbon systems. In *ionic* carbon systems in which the inductive action of an anionic pair or a cationic hole gives added impetus to bond-pair association, the number of electron pairs assigned to the A domain may exceed two. For example, the A domain of the aromatic cyclopentadienyl anion may involve three electron pairs, meaning that there is no S domain.

17.2 The Dilemma of the Regioselectivity of the Diels–Alder Reaction and the 1,3-Dipolar Cycloaddition

In 1968 Firestone[2] and Huisgen[3] discussed the mechanism of 1,3-dipolar cycloadditions from apparently antithetical points of view. In actuality, both investigators recognized crucial pieces of evidence but they proceeded to reconcile them in different ways. Firestone proposed that the reaction involves a "spin-paired diradical intermediate," which may lie in a shallow energy well. Huisgen supported a one-step mechanism with no secondary minimum. Indeed, controversies surrounding the depth of a given minimum abound in chemistry, the "nonclassical ion" controversy being a good example.[4]

The fuel for the diradical proposal of Firestone was the observation that in the great majority of dipolar cycloadditions, the preferred regioisomeric product can be rationalized by postulating a "best diradical" intermediate and recognizing that *all* groups stabilize radicals, albeit to different extents. This concept can be used to explain an impressive general trend: The regioselectivity of the reaction of a dipolarophile (e.g., benzonitrile *N*-oxide) with a monosubstituted ethylene or acetylene is independent of the π-donating or π-accepting nature of the substituent; that is, the reactants combine preferentially in the same way, regardless of whether the substituent is a π acceptor or π donor. Firestone argued as follows: "For one or the other group of substituents, this orientation must be wrong for concerted cycloadditions. Both groups, however, stabilize a diradical intermediate. . . ."

The Firestone formulation is founded on a crucial and apparently inexplicable fact: The postulated "best diradical" in the 1,3-dipolar cycloaddition of ArC≡N—O and G—HC═CH_2 has one odd electron on the oxygen (rather than the carbon) atom of the 1,3-dipole. This is illustrated in Figure 17.2a. A problem with this scenario can be seen by comparing the allegedly "better" and "worse" diradicals. As seen in Figure 17.2a, the first arises from C—C and the second from C—O linkage of the 1,3-dipole and the olefin. The C—C and C—O bonds are comparable relative strengths and uncertain depending on the model system, as indicated by the following BDEs (kcal/mol):

CH_3CH_2—CH_2CH_3	82	CH_3—CH_3	88
CH_3O—CH_2CH_3	80	CH_3—OH	92

Thus, the "best diradical" model predicts nothing in actuality. With today's ab initio computational facilities, a better appraisal can be made.

In any event, the key challenge presented by the 1,3-dipolar cycloaddition

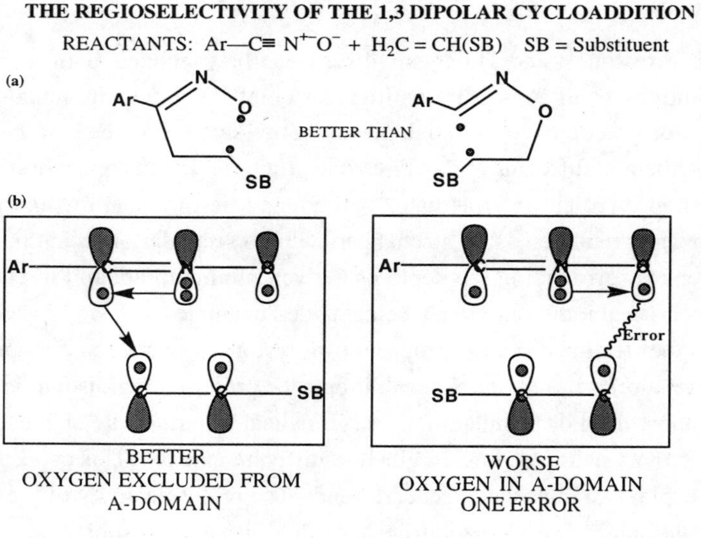

Figure 17.2 (a) Alleged "best diradical" rationalization of the regioselectivity of the 3 + 2 cycloaddition and comparison of alleged "best" and "worst" diradicals. (b) VB explanation of the regioselectivity of the 3 + 2 cycloaddition. Oxygen inclusion in the A domain disrupts arrow train and violates the heteroatom exclusion rule.

regioselectivity is: Why is the odd electron localized on the *more electronegative* oxygen despite a thermochemical preference for localization on the electroneutral carbon of the 1,3 dipole? What does this fact imply about the electronic nature of the transition state? Clearly, the only way to explain these crucial data is by assuming that the mechanism of bonding of the pericyclic transition state differs from that of reactants, products, and diradical intermediates, which is what we have argued all along. Since we know that the transition complex is I-bound while reactants, products, and diradical intermediates are T-bound, the regioselectivity of pericyclic reactions must be a consequence of I-bonding.

The idea that the regioselectivity of a pericyclic reaction can be rationalized and predicted by the "best diradical" concept was not new in 1968. It had already been used by Bartlett to rationalize the regioselectivity of the Diels–Alder reaction.[5] It was argued that regardless of whether two groups are π donors or π acceptors, they end up either ortho or para in the major regioisomer as if this arose from the "best diradical" intermediate. This is illustrated in Figure 17.3. In contrast to the 3+2 cycloaddition case, the alleged "best diradical" is now unequivocally the best. Huisgen, Grashey, and Sauer, in their classic review of cycloaddition reactions, also recognized that the regioselectivity of the Diels–Alder reaction is independent of the nature of the substituent.[6]

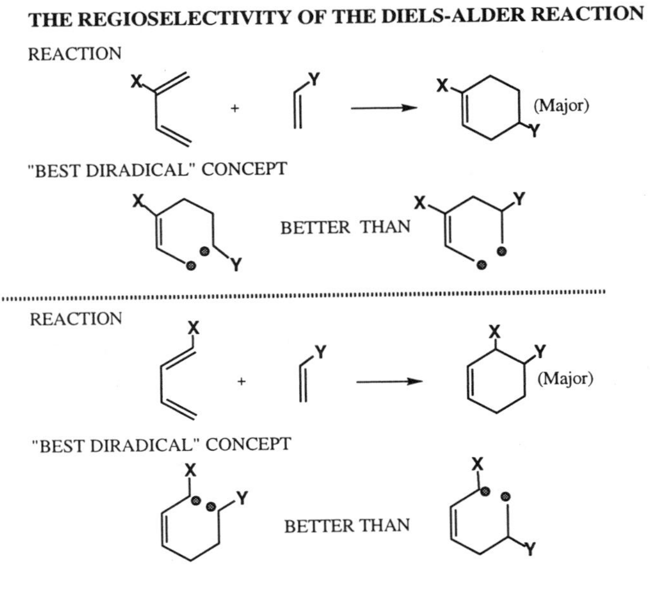

Figure 17.3 "Best diradical" rationalization of the regioselectivity of the 4 + 2 cycloaddition.

In summary, we have the following situation: The regiochemistry of "allowed" pericyclic reaction defies a self-consistent rationalization by recourse to diradical models. Even when such a construct succeeds unambiguously (e.g., the Diels–Alder reaction), most other mechanistic criteria point to a one-step mechanism. Alas, what many believe to have been a successful response to this challenge, namely, the frontier orbital–perturbation MO interpretation of regioselectivity proposed some years ago by Houk[7a] and the author[7b] is an illusion. The solution to the problem of pericyclic regioselectivity is most clearly seen in the following 1,3-dipolar cycloaddition:

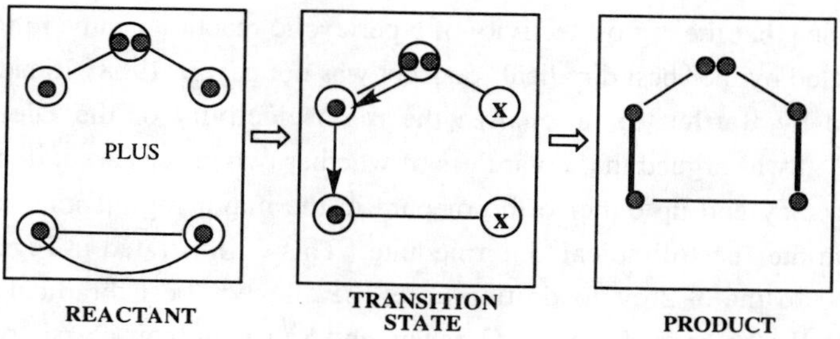

The four-electron π system of the 1,3 dipole is formalized as "dot-pair-dot" according to calculations by Hiberty and Leforestier.[8] The location of the electron pair fixes the A domain defined by the codirectional arrow train that satisfies the four-electron rule: Only two electron pairs can be associated in nonresonant fashion at the transition state. Figure 17.2b gives a direct explanation of why oxygen ends up carrying one odd electron in the putative diradical formulated for explaining the regioselectivity: Oxygen is excluded from the A domain because its presence there causes the interruption of the codirectional arrow train. This is the first application of material importance of the heteroatom exclusion rule formulated in Chapter 15.

17.3 Electronegative Atoms Control the Regioselectivity of Pericyclic Reactions

One of the most remarkable trends in "allowed" cycloadditions is that if a choice exists, the reactive atoms are almost always carbons, rather than electronegative nitrogen or oxygen. For example, the strong dienophiles of the Diels–Alder reaction have C—C as well as C—X (X = N,O) π bonds (e.g., $H_2C{=}CH{-}CH{=}O$). Why does the diene attack the C=C, rather than the C=O, π bond with extremely high site selectivity?

We first inquire whether the FO-PMO model successfully accounts for this trend. All we have to do is look up the AO coefficients of the LUMO of a representative dienophile (e.g., acrolein). For a reasonable choice of heteroatom parameters, the HMO method yields the LUMO coefficients given in Scheme 17.1a. The sum of the LUMO densities at C1 and C2 is smaller than that at C3 and O, though by not very much. Thus, the FO-PMO model predicts that the diene should attack acrolein with virtually no selectivity.

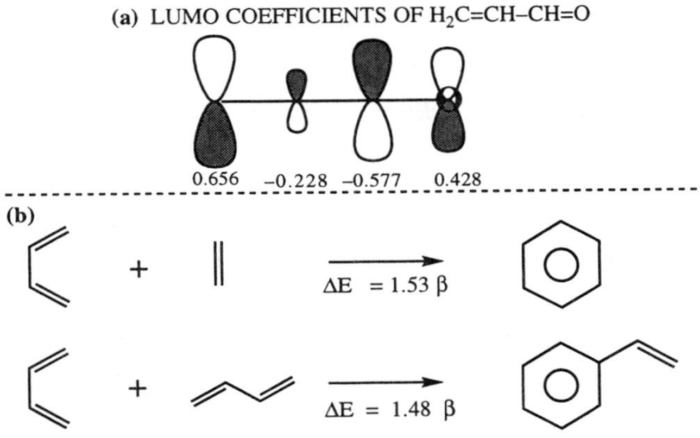

Scheme 17.1

Once again, the VB explanation is contained in the heteroatom exclusion rule discussed in Chapter 15: Electronegative and electropositive atoms are excluded from the A domain. Furthermore, since the A and S domains resonate, the rule implies that these atoms are also excluded from the reactive sites. An equivalent statement is that heteroatoms are more suitable to be "observers" rather than "participants" in either benzene or the isoelectronic Diels–Alder transition state. One illustration of the rule is the observation that the acrolein oxygen ("square" orbital) is excluded from the six-center transition state complex; that is, acrolein acts as a dienophile and not as a diene.

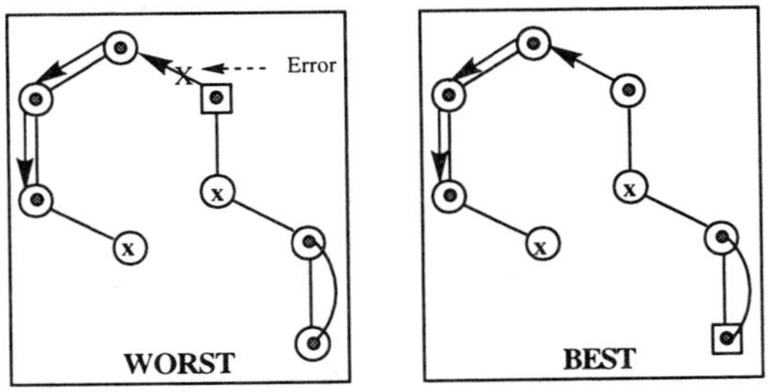

Carbonyl acts better as a substituent of a dienophile rather than as a component of a diene, and all this happens to prevent the disturbance of maximal association in the resonating A domain. Still more interesting is the observation that when electronegative oxygen is constrained to participate in a Diels–Alder or a 1,3-dipolar cycloaddition, it invariably ends up in the S domain. Consider, for example, the reaction of acrolein with a monosubstituted ethylene. In the major regioisomeric product, the A domain spans the electroneutral carbons, leaving the more electronegative atoms to accommodate the S domain. This means that the bikrat has one odd electron on oxygen, leading, in turn, to the prediction that when acrolein reacts with an alkene bearing a substituent, the alkenic carbon bearing that substituent will combine preferentially with the oxygen of the acrolein (Figure 17.4).

One remarkable trend of the Diels–Alder regioselectivity, in full accordance with the heteroatom exclusion rule, is the nearly inviolable preference for a carbon bearing a substituent, whether π donor or π acceptor, and an electronegative heteroatom to be located ortho or para in the major regioisomeric 4+2 cycloadduct. A good source of many illustrative regioselectivity trends of this type is the monograph of Boger and Weinreb.[9a] An interesting case can be found in the work of Sorensen and his co-workers, who found that the dimerization of oxyallyl [$CH_2{=}C(CH_2)O$] through a six-electron suprafacial transition state yields selectively the regioisomer having two ring oxygens para (1,4-dioxane derivative).[9b] In other words, oxygen is excluded from the A

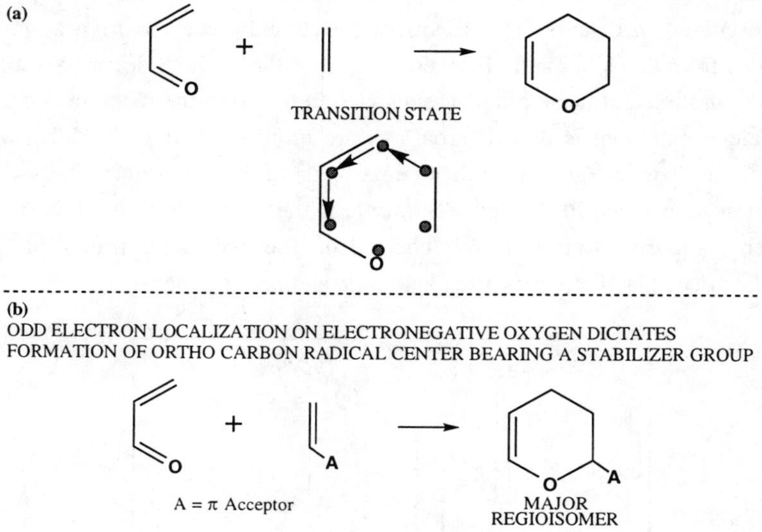

Figure 17.4 Regioselectivity of acrolein plus alkene cycloaddition.

domain but not from the reactive sites. Full adherence to the heteroatom exclusion rule predicts that the major product should be the 1,4-cyclohexadione isomer. It follows that either steric repulsion hinders formation of the latter or the reaction is a combination of two oxyallyl zwitterions via an E-selective transition state (i.e., a complex in which the binding of the two reactants is electrostatic).

The regioselectivity of the 1,3-dipolar cycloadditions can now be explained in a satisfactory way. Calculations by Hiberty and Leforestier have shown that, in most cases, the electron pair of the 1,3 dipole resides on the central atom; that is, the 1,3 dipole has a "dot-pair-dot" configuration.[8] The exceptions may very well be artifacts due to the modeling of the large aryl substitutents normally borne by the 1,3 dipoles. Focusing attention on the common 1,3 dipoles with the atom sequence CNZ, where Z is more electronegative than C, we have a choice between a right-handed and a left-handed codirectional arrow train. As shown before (see Figure 17.2), one incurs more errors than the other. Taking into consideration the restriction placed on the complex by orbital electronegativity, we arrive at the final description of the bikratic transition state, which features an S domain that spans Z and the carbon of the dipolarophile bearing a substituent, whether π donor or π acceptor.

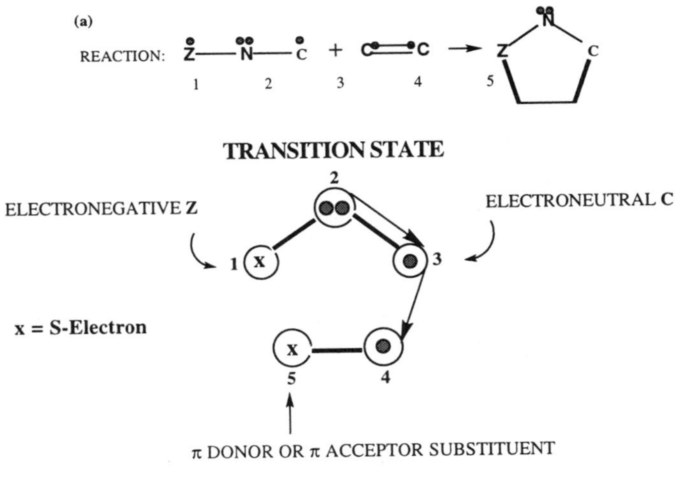

In the case of nitrile ylides, Z is also carbon but it bears radical-stabilizing alkyl groups. A meaningful deviation from the foregoing regioselectivity pattern is the 3+2 cycloadditions of azide 1,3 dipoles (Ar—NNN), which tend to adopt the "pair-dot-dot" rather than the common "dot-pair-dot" configuration according to the Hiberty–Leforestier calculations.

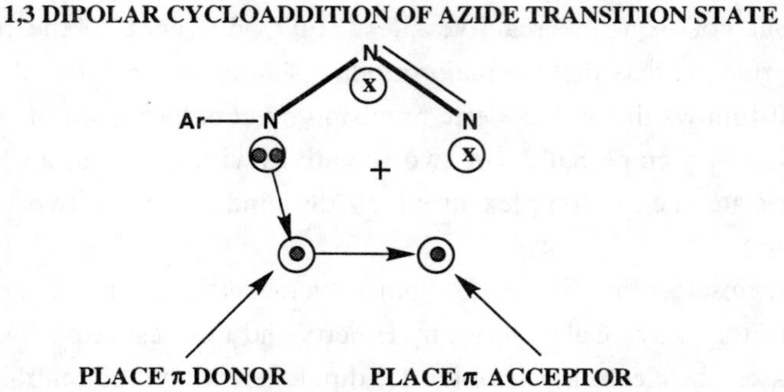

The S pair localizes on two adjacent (ortho) electronegative nitrogens. As a result, the alkene and one of the azide nitrogens end up defining the four-electron A domain. This configuration leads us to predict that substituents of the olefin will act to stabilize the A domain. The optimal placement is straightforward: π donors should be placed at the tail and π acceptors at the head of the arrow spanning the olefinic π bond. Thus we have a rationalization of the striking observation that regioselectivity does depend on the nature of the alkene substituent in the case of azide 1,3 dipoles, and we can suggest a rule: The diaza derivatives of the all-carbon pericyclic transition states will exhibit unsymmetrical regioselectivity in contrast to the regioselectivity of the latter, which is predicted to be invariant. In both cases, the optimal placement of two substituents is predicted to be ortho/para. However, in the unsymmetrical mode, one must be a π donor (systolic group) and the other a π acceptor (diastolic group) while, in the invariant mode, each substituent can be either a π donor or a π acceptor (i.e., any radical-stabilizing group). We will return to a more specific discussion of the implications of the azide 3+2 cycloaddition shortly. For the time being, we underscore the key point: The counterintuitive dependence of the regioselectivity of the 3+2 cycloaddition on the nature of the 1,3 dipole is entirely consistent with the concept of the resonant bikratic transition state.

We are developing chemical theory that relies heavily on the power of analogy. We say that pericyclic organic transition states are obligatory (saddle points) I-bound complexes, while ground transition metal complexes (global minima) are natural I-bound species. Since the nature of bonding remains roughly constant, we expect uniform selectivity. Organic chemists know that the dienophilic part of α,β-unsaturated carbonyl molecules is the C=C unit. The same subunit must show preferential association with a transition metal center, and this is observed.[10a,b] Organometallic chemists know that olefins are better π acids than aldehydes.[10c]

Bottom line: The strength of the I bond depends crucially on configurational degeneracy and, more specifically, on ionic configuratioinal degeneracy. For this reason, electroneutral atoms seek to be parts of a homonuclear A domain because this guarantees degeneracy of left- and right-handed codirectional arrow trains (i.e., maximal degeneracy of ionic configurations). The importance of ionic configurational degeneracy is clearly suggested by Norbeck and Gallup's ab initio VB calculation of benzene cited in Section 16.7, according to which the π system resembles a hybrid of the degenerate orthopolar structures. In other words, the prerequisite for strong I-bonding is narrow ionic subbands, a condition that is met ideally in homonuclear systems. Replacing electroneutral by highly electropositive or electronegative atoms turns a homo- into a heteronuclear system, broadens the ionic subbands, and frustrates I-bonding.

17.4 The Effect of Substituents on the Rates of Pericyclic Reactions Is an Unfinished Story

It has been long known that Diels–Alder reactions are accelerated when the diene is made a good donor and the dienophile a good acceptor. However, whether this

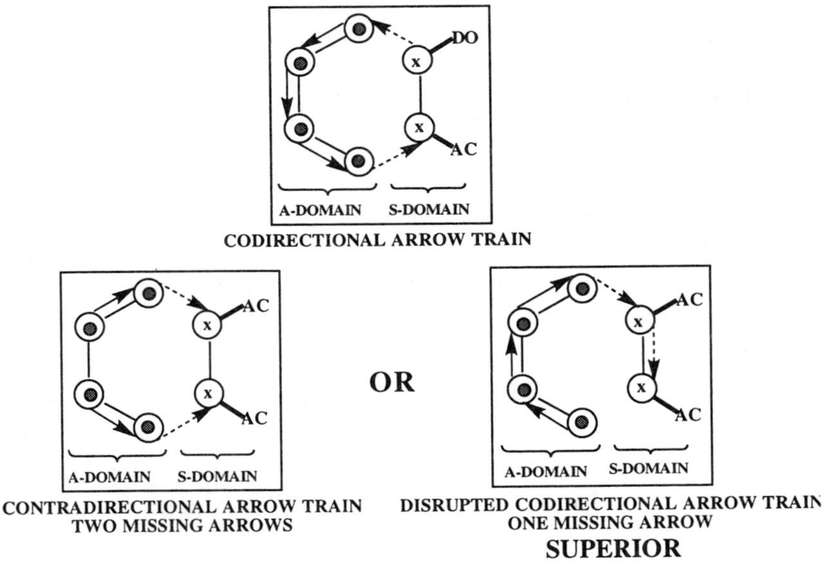

Figure 17.5 Explanation of the superiority of donor–acceptor combination of reactants in the 4+2 cycloaddition. Solid arrows, primary CT delocalization; dashed arrows, secondary CT delocalization.

is a transition state effect or a ground state effect has not been studied systematically, partly for lack of an appropriate theoretical motivation. We now have such a motivation because VB theory presents a vivid picture of the transition state. The strategy for stabilizing the resonant bikrat is straightforward: The substituents of the carbons accommodating the (ortho or para) S electrons must be chosen to activate a codirectional arrow train. As shown in Figure 17.5, there are two choices:

A donor–acceptor combination triggers an uninterrupted codirectional arrow train.

An acceptor–acceptor (or donor–donor) combination triggers either a contradirectional arrow train or a (superior) codirectional arrow train with one error (one missing arrow).

This rationalizes the well-known acceleration of the 4+2 cycloaddition when one component bears donor and the other acceptor groups.

However, it is also clear that the two substituents will stabilize the transition state complex in either of the following situations:

The diene bears the donor and the dienophile the acceptor (or vice versa), and the regioselectivity is ortho or para ("inverse electron demand" Diels–Alder reaction).

Both donor and acceptor substituents are borne by either the diene or the dienophile, and their orientation is ortho or para.

In other words, an ortho or para donor–acceptor pair of substituents can act to stabilize either a product-like or a reactant-like transition state. Hence, the two isomeric reactions shown in Figure 17.6 are expected to have transition states of comparable stability.

And yet, the accumulated data show that reaction I is faster! This suggests that the famous "donor–acceptor" rate effect does not reside in the transition state but reflects ground stabilization of reactants in which the donor–acceptor combination is locked within the same molecule. In other words, ortho/para allocation of one donor and one acceptor has a comparable effect on the transition states of reactions I and II but stabilizes more the reactants of reaction II. The same argument explains why diene plus tetracyanoethane (TCNE) will react faster than dicyanodiene plus dicyanoolefin. The implication is that there is a basic flaw in the popular frontier orbital argument that the 4+2 cycloaddition

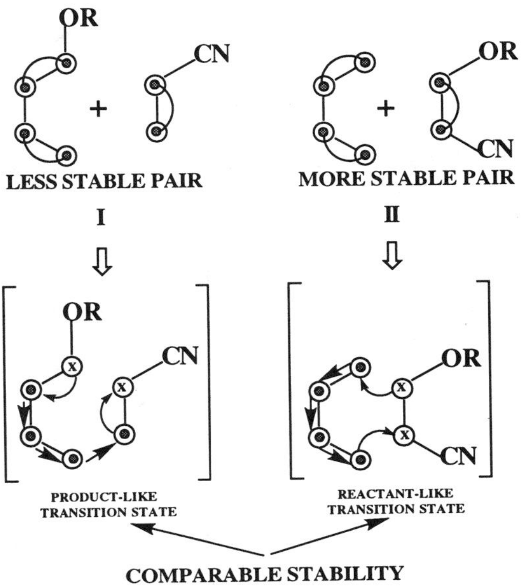

Figure 17.6 Substituent effects on the rates of 4+2 cycloadditions may be due to differential ground reactant stabilization.

becomes accelerated because of a diminished HOMO-LUMO gap when a donor is matched with an acceptor reactant. Indeed, this makes sense: The theory on which our arguments are based has nothing to do with the covalent/ionic canon to which the simple models hark.

We can suggest systematic studies to probe the transition state structure. The strategy is to pin the S pair (ortho or para) by electronegative heteroatoms and probe the A domain by appropriate placement of substituents. By using the reaction system of 2,3-diaza-1,3-butadiene plus olefin shown in Figure 17.7, we lock the S pair on the nitrogens and use substituents on the olefin to probe the electronic structure of the A domain.

A codirectional arrow train predicts rate acceleration by unsymmetrical 1,1-disubstitution (or even monosubstitution) of the olefin by π donors relative to the case of symmetrical 1,2-disubstitution after ground reactant stabilization has been factored out. Sauer and co-workers found that a monoalkoxy reacts faster than a 1,2-dialkoxy olefin with 1,2,4,5-tetraazabenzene derivatives.[11] This result confirms the ortho/para placement of one π donor and one π acceptor (unsymmetrical regioselection) as the best approach to stabilizing the A domain.

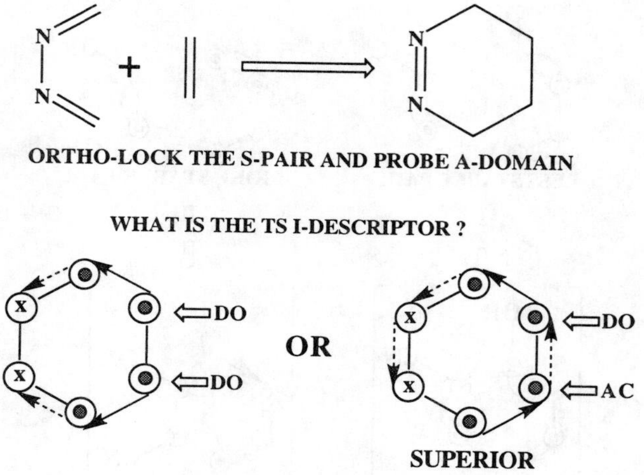

Figure 17.7 Pinning the S pair and probing the A domain: solid and dashed arrows, primary and secondary CT delocalization, respectively, open arrows, optimal placement of π donors and π acceptors.

17.5 Hydrogen Transfer and Bridging as Diagnostic of the I Bond; The Hydrogen Rules

The time has now come to ask a basic question: Assuming constant electronegativity, what atom types are intrinsically suitable as members of the A domain and which befit the S domain? For example, what is the affinity of hydrogen and alkyl groups for the A domain? For a radical to be included in the A domain, it must easily fulfill the requirement of one in- and one outbound arrow dictated by the association rule. A spherical H1s AO is ideally suitable for the A domain because it places no stereochemical constraints on the two atoms to which it must be connected by the two arrows. By contrast, an sp^x alkyl AO falls short of this ideal situation. This leads to two conjugate rules:

H participates in preference to R (R = alkyl) in the A domain of an organic pericyclic transition state.

Superiority of H over R either as a migrating group or as a bridging group implies I-bonding, with the characteristic feature that H has one inbound and one outbound arrow:

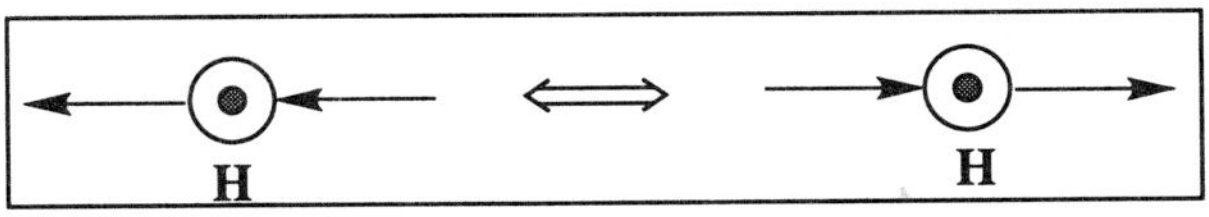

OR

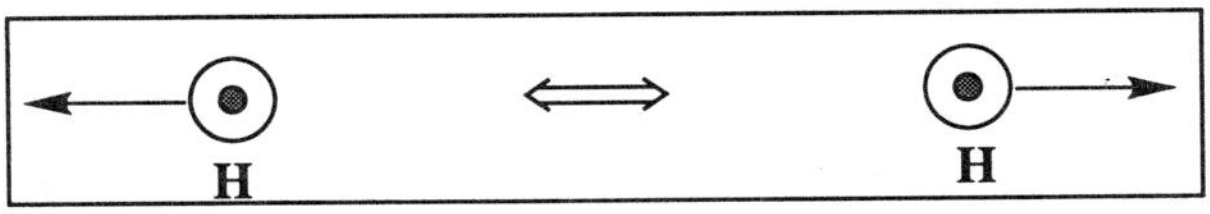

These two rules are baptized the hydrogen rules.

It goes without saying that the first yardstick of migratory aptitude in thermally "allowed" sigmatropic shifts is the strength of the bond of the migrating group. For example, silyl migrates in preference to hydrogen in fluctional 5-silyl-1,3-cyclopentadiene. However, when bond strength does not produce a strong discrimination, other factors must be considered. A theme that recurs across the periodic table is the superiority of hydrogen over alkyl as a transferring atom as well as a bridging atom. This is seen in three prototypical cases:

Thermally "allowed" sigmatropic shifts (i.e., H migrates in preference to R).
Homolytic substitution (i.e., H- is preferred over R-abstraction).
In methyl derivatives of diborane, H has higher bridging ability than R.

We suggest that these are consequences of the superior ability of H to support I-bonding.

Here is now the opposite face of the coin: The superiority of alkyl as a transferring or bridging group implies that arrow codirectionality, the hallmark of I-bonding, cannot be achieved. In turn, this implies that the mechanism of bonding has changed from I to E type. We expect to find this state of affairs in two categories of complexes:

"Forbidden" transition states.
Ground complexes involving atoms of widely different electronegativities.

The two limiting cases are depicted in Figure 17.8. And indeed, the data speak convincingly in favor of this scenario. By way of showing that we are on the right track, we point out, for starters, that only H-bridged diboranes are stable, while alkyl-bridging of the aluminum analogues is commonplace.

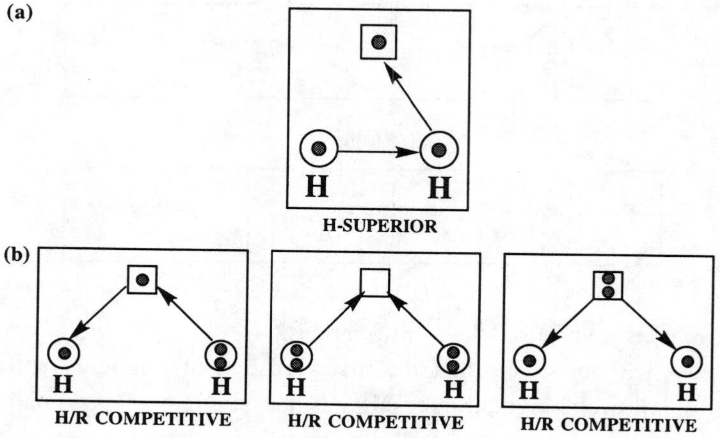

Figure 17.8 (a) The nature of bonding implied by the bridging superiority of hydrogen over alkyl. (b) The nature of bonding implied when hydrogen and alkyl have comparable bridging abilities.

17.6 Pinning the S Pair in Molecular Rearrangements

The Cope rearrangement transition state has one ortho or one para S pair coordinated to a four-electron A domain. Much as in the case of the regioselectivity of the Diels–Alder reaction, rate enhancement by substituents oriented ortho or para (relative to meta) is compatible with a concerted process in which the transition state is a resonant bikrat and is not evidence for a diradical intermediate. Doering et al.[12] recently compiled the activation enthalpies and entropies for the thermal Cope rearrangements of phenyl-substituted hexa-1,5-dienes. VB theory says that in a concerted reaction, two phenyls ortho (i.e., 3,4) or para (i.e., 1,4 or 2,5) will be superior to two phenyls meta (i.e., 1,3 or 2,4). Following the admonitions of Doering et al. regarding the data reliability (1,3 case) and the introduction of steric effects (3,4 case), the activation enthalpies are consistent with this expectation. For example, 2,5 reduces the activation enthalpy by roughly 3 kcal/mol compared to 2,4 placement of two phenyls.

Replacing one CH_2 by O in 1,5-hexadiene takes us to the Claisen rearrangement. The heteroatom exclusion rule predicts that the electronegative oxygen will accommodate one of the S electrons. The other S electron should be either ortho- or para-localized. Indeed, a spectacular drop of the activation energy is observed when a trimethylsiloxy group is attached on the carbon ortho to oxygen.[13,14]

A greater challenge comes in the case of a 1,5-hydrogen sigmatropic shift. The hydrogen rules predict that H will have the highest migratory aptitude because the spherical nature of the H1s AO is ideal for I-bonding, and that as a result, H will be part of the A domain. Thus the transition state of a 1,5-sigmatropic shift should have one of the following structures:

TWO POSSIBLE BIKRATIC TRANSITION STATES OF A 1,5-SIGMATROPIC SHIFT

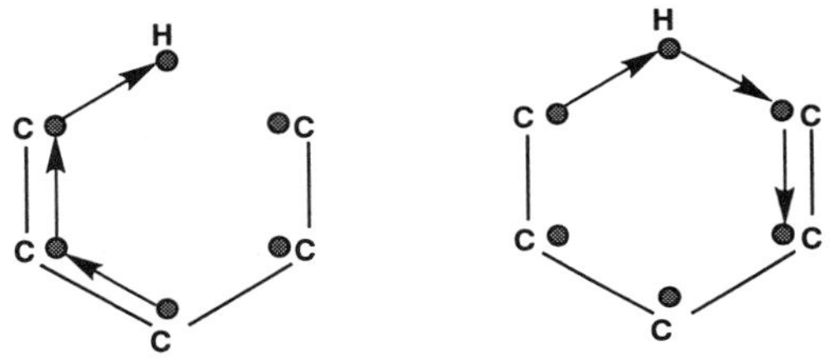

Once again, two substituents will effect maximal acceleration when placed ortho or para.

There is a second reason why hydrogen migrates in preference to alkyl and other electronegative groups like NR_2 and OR even though a C—H bond is stronger than a C—C bond. It has to do with the superiority over H of all other substituents when it comes to stabilizing the S-electrons of a bikrat. In other words, the migratory aptitude in "allowed" thermal pericyclic reactions is partly determined by the left-behind *bystander groups.* This view is supported by the thoughtful studies of singlet carbene rearrangements by Nickon.[15] In these systems, hydrogen migration permits radical-stabilizer bystander groups to exert the maximum possible effect on the singlet carbenic center, which is essentially a diradical (two odd electrons in two hybrid carbon AOs), at the transition state.

A dramatic change occurs when we look at the "suprafacially forbidden" 1,3-sigmatropic shift. Now, the hydrogen rules predict no migratory preference for H. It is found that the regioselectivity of the rearrangement is a function of the IP (donor) – EA (acceptor) difference between the (formal radical) migration framework and the (formal radical) migrating group. For example, when the migration framework is chosen according to the criterion of highest EA, the migrating group is the one with the lowest IP.[16] This donor–acceptor ("push–pull") relationship is necessary for engendering a stabilized antiaromatic transition state via state-crossing much as in the case of "push–pull" cyclobutadiene.

What is the electronic structure of an "allowed" pericyclic transition state? The real problem was first put on the table by the Huisgen–Firestone exchange discussed earlier in connection with the 1,3-dipolar cycloaddition. The Diels–Alder reaction is a related issue. Nowadays, exactly the same debate is met in connection with the Cope rearrangement.[17] More experiments and computations will resolve nothing. The problem lies in the theory of chemical bonding that is used to interpret the results. Within the context of VB theory, the language of the chemist, the resolution is clear:

(a) The "allowed" transition state is *not* a diradical. In other words, it is not represented by the perfect-pairing covalent structure with one missing covalent bond.

(b) The "allowed" transition state is *not* a fully delocalized "aromatic" array. The VB structures that can contribute are limited by atom electronegativity (i.e., a subset of ionic structures is effectively excluded).

(c) The "allowed" transition state is a bikrat. In the case of a six-electron all-carbon transition state, triply ionic configurations are effectively excluded. The substantial electronegativity of carbon effectively limits the VB configuration set to the covalent, the singly ionic, and the doubly ionic. This means that only two of the three electron pairs can associate. If the triply ionic configuration were important, all three pairs would associate and the transition state would be an I-bound global minimum rather than a transition state. In other words, the S pair is the *defect* that makes the transition state a transition state!

17.7 Orbital Electronegativity Control of the Mechanism of the Diels–Alder Reaction

We have been preoccupied with "allowed" pericyclic transition states in an effort to show that their apparent diradicaloid nature is a consequence of their resonant bikratic character. How can we turn a one-step "allowed" process into two steps involving a diradical intermediate? To define the strategy, we must first understand what makes a diradical stable. To this extent, consider the combination of two ethylenes to form a cyclobutane cycloadduct (2+2 reaction) or a 1,4 diradical (1+1 reaction). According to VB theory, the reaction enthalpy is determined by the interplay of fragment promotion and fragment bond making:

(a) In the case of the 2+2 reaction, the promotion step is the deformation that produces an ethylenic geometry identical to that found in the product. This promotional energy requirement ΔP which can be further broken down to Δs, the energy for C—C stretch while maintaining planarity, plus $\Delta p(2)$, the bispyramidalization energy. Substituents that stabilize radicals by π

conjugation (e.g., strong π donors like amino and strong π acceptors like cyano or carbonyl) reduce Δs but also increase Δp. Substituents which are not effective π donors or π acceptors but which favor bispyramidalization because of their electronegativity (e.g., F) are the best choices for minimizing ΔP because they exert little effect on Δs while causing a strong reduction of Δp. The second component of the promotion/bond-making interplay is the energy return due to twice C—C bond formation, ΔB. Hence, the energetics of the transformation of reactants to products can be discussed by reference to the index $\Delta B + \Delta s + \Delta p(2)$.

(b) In the case of the 1+1 reaction, exactly the same considerations hold true, only now the bis is replaced by a monopyramidalization step $\Delta p(1)$. The reason is that the product 1,4 diradical is a composite of an odd σ bond that requires pyramidalization plus two odd electrons that are free to select the right geometry. Note the key departure from conventional thinking: A 1,4-diradical is not "two dots" but, rather "one σ bond plus two dots," and its stability is determined by what transpires at each component rather than by what occurs exclusively at the radical sites.

We now have a rule: A perturbation that promotes pyramidalization by diminishing Δp while keeping Δs and ΔB relatively constant will stabilize the cycloadduct as well as the diradical intermediate. One well-known perturbation that promotes pyramidalization is replacement of H by F.[18] The greater exothermicity (by kcal/mol) of the 2+2 cyclodimerization of tetrafluoroethylene over that of ethylene confirms the rule. By contrast, cyclodimerization of tetracyanoethylene is not expected to be more favorable than that of the parent ethylene because CN will reduce Δs (good radical stabilizer) but it will also increase $\Delta p(2)$.

We have learned how to appraise the effect of substituents on the stability of a diradical intermediate relative to the reactants. We next consider the effect of substituents on an "allowed" organic pericyclic transition state such as the Diels–Alder reaction of 1,3-butadiene and ethylene. The stability of the latter depends critically on CT delocalization, which, in turn, depends on carbon orbital electronegativity. Two perturbations that enhance orbital electronegativity are pyramidalization and σ-electron withdrawal, and both are brought about by replacing H by F. Hence the same type of substitution that stabilizes a diradical destabilizes a pericyclic transition state. We will come to diradical stabilization shortly

Here is the crucial point: There is an antithetical relationship between molecular distortion and CT delocalization in the sense that pyramidalization is beneficial for T-bond formation either in the diradical intermediate or in the product, but it is deleterious for I-bond formation in the transition state of an "allowed" multicenter reaction. This antithetical relationship can be nicely exploited: Substituents that promote pyramidalization will tend to stabilize the diradical while destabilizing the pericyclic transition state. The ideal situation is reached when, in addition, the substituents render the carbon atoms more electronegative by direct σ-electron density withdrawal, an effect that further discourages CT delocalization. At this limit, we expect a complete breakdown of the Woodward–Hoffmann rules. The best strategy is now self-evident: Perfluorination will turn a one-step "allowed" process into two steps involving a diradical intermediate.

It was known[19] (even before the classic work of Bartlett[20]) that fluoroalkenes react with 1,3-butadiene and its derivatives to yield mixtures of cyclohexene and vinylcyclobutane products, whereas ethylene yields nearly 100% cyclohexene product. The observation of mixtures of 2+2 and 4+2 adducts is indicative of a stepwise mechanism via a diradical intermediate. The uniqueness of the fluorine in promoting pyramidalization and raising the electronegativity of the atom to which it is attached is seen by comparison to other substituents which, though superior radical stabilizers, fail to abort the Diels–Alder reaction. For example, one might have expected tetracyanoethylene would have a much greater chance of promoting a stepwise 4+2 cycloaddition than tetrafluoroethylene simply because a cyano group is a more efficient radical stabilizer than fluorine.[21] This is not the case, since TCNE is known to be a superb dienophile. Similarly, one might also have expected captodative[22] substitution of dienes and dienophiles to stabilize a diradical intermediate that could close to cyclobutane and cyclohexene adducts. Once again, this is not the case: Captodative substitution leads to favorable Diels–Alder cycloaddition with no evidence for cyclobutane product formation.[23]

We just argued that tetrafluoroethylene cycloadditions are "anomalous" because of impairment of CT delocalization in the "allowed" transition state. Consider now the potential closure of the 1,6-dienyl diradical intermediate to either a four- or a six-membered ring. The corresponding transition states are inevitably "quasi-pericyclic"; that is, the forming bond cannot avoid overlapping with the remaining "active" electron pairs because of the mere skeletal constraints. CT delocalization no longer acts in favor of closure to the six-membered ring, however, because the indirect action of the fluorines has turned it off! Thus, the two transition states involved in the ring closure of the diradical are analogous to "stretched cyclobutane" and "stretched cyclohexane." Which of the two is more favorable? In Chapter 4, we saw that cyclobutane can form weak bonds but with the benefit of low atom promotion. By contrast, cyclohexane owes its stability to stronger bonds at the price of higher atom promotion. It follows that "stretched cyclobutane" is energetically superior to "stretched cyclohexane" because it owes proportionately more of its stability to low fragment excitation. This explains why closure to the less thermodynamically four-membered ring is favored.

The preferential formation of 2+2, rather than 4+2, adducts in the cycloaddition of 1,3-butadiene and fluorinated olefins has been known since 1949. Nearly half a century later, Dolbier and co-workers confirmed that this is a result of the fundamental electronic consequences of fluorination discussed here, namely, the intramolecular 4+2 cycloaddition of a fluorinated 1,3,5-hexatriene (i.e., the "allowed" conrotatory closure to a 1,3-cyclohexadiene) is bypassed and the major products can be rationalized by postulating a transoid-1,4-diradical intermediate.[24] We reiterate: The success of the Woodward–Hoffmann rules is accidental; that physical truth is that carbon orbital electronegativity in many organic systems is not high enough to turn off CT delocalization and render every multicenter reaction stepwise!

Further evidence that orbital electronegativity determines the stability of an "allowed" pericyclic transition state comes from the work of Piers et al.,[25a] who discovered that the rearrangement of the vinylcyclopropane derivative shown in Figure 17.9 is directed by X in the following way. When X = OMe, the generation of a reactive AO over the σ C—O bond is avoided. By contrast, when X = $SiMe_3$, the rearrangement occurs over an underlying C—Si bond. We suggest that this preference is the consequence of the greater electropositivity of the AO of the carbon bearing the Si group.

A related example of how atom electronegativity controls reaction mechanism is the cycloaddition of 1,1-difluoroallene to butadiene and 1,3 dipoles. It is expected that the unfluorinated double bond with the more electropositive carbons will be the one to combine with the diene in a 4+2

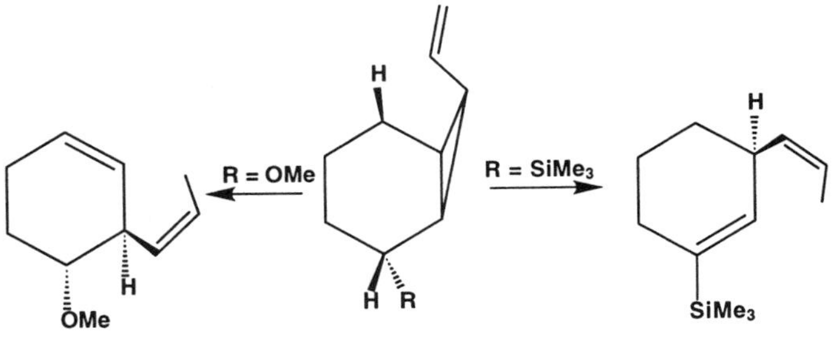

Figure 17.9 The regioselectivity of the vinylcyclopropane rearrangement is determined by the electronegativity of the substituents.

cycloaddition, and this is exactly what Dolbier found.[25b] No HOMOs or LUMOs!

These ideas define a general strategy: To promote single-bond cleavage at the expense of fragment loss, replace hydrogens by fluorines. Perfluorination will always favor fragmentation via a diradical rather than via a pericyclic transition state. The ring opening of perfluorocyclopropane leading to 1,3-dihalo derivatives is a case in point.[25c]

Now we come to a tantalizing problem. According to conventional wisdom, both F and CF_3 are termed "σ withdrawing." By contrast, VB theory says that the two are fundamentally different: F is systolic (i.e., a σ acceptor) but CF_3 is diastolic (i.e., a σ donor). Nonetheless, one can envision that CF_3 can also act as a Coulombic inductor to enhance the effective electronegativity of the atom to which it is attached. This scenario implies that F and CF_3 can act in a similar fashion, albeit for different reasons, to arrest CT delocalization on the carbon skeletons to which they are attached. This could explain why pertrifluoromethylation (rather than perfluorination) produces the following results:

Stabilization of Dewar benzene and prismane relative to aromatic benzene by 30 kcal/mol compared to methylation.[25d]

Stabilization of quasi-antiaromatic cyclooctatetraene.[25e]

Since the relative importance of CT delocalization increases in the order aromatic > nonaromatic > antiaromatic, pertrifluoromethylation destabilizes an aromatic (relative to a nonaromatic) while it stabilizes an antiaromatic (relative to a nonaromatic) molecule or transition state by virtue of destabilizing the nonaromatic (relative to the antiaromatic) isomer. This suggests that to obtain the sought-after square cyclobutadiene as a global minimum, one should replace the four Hs by four Fs or four CF_3 molecules.

Perfluoroethylene (TFE) is the ideal dienophile for promoting a two-step Diels–Alder reaction. Each olefinic center is highly electronegative (due to the systolic action of the fluorines) and has a large driving force toward pyramidalization. What happens when we replace TFE by $R_2Si{=}SiR_2$? The situation now becomes ambiguous: The pyramidalization tendency remains, but the atomic centers are electropositive and, thus, more amenable to CT delocalization. It is not surprising that the experimental evidence is "mixed":

Silaethylene derivatives react with 1,3-butadiene to yield principally cyclobutane products.[26a]

Disilaethylene derivatives seem to follow the normal Diels–Alder route.[26b]

Woodward and Hoffmann declared that there should be no exception to a principle so fundamental as the conservation of orbital symmetry, other than cases for which credible alibis exist (e.g., skeletal constraints, steric repulsion). We can now see the flaw: Transition state aromaticity depends on orbital electronegativity. Hence, whatever the successes of the Woodward–Hoffmann rules and related one-electron models, they are an accident. If the electronegativity of the carbon were much higher than it is, aromaticity would be so diminished that there would be no gas phase concerted reactions. Stepwise reactions involving diradical intermediates would be the rule!

17.8 The Case Against the Hückel–PMO Model of Reactivity

While I was a graduate student in love with the frontier orbital Hückel PMO (FO-PMO) method, I wrote a paper attempting to explain the regioselectivity observed in concerted 4s+2s cycloadditions (s = suprafacial).[7b] I suggested that to predict the major regioisomer, one should take the following steps:

Identify the dominant HOMO-LUMO interaction.

Look up the AO coefficients of the HOMO and the LUMO at the union sites.

Connect the two sites, one in the first and one in the second reactant, having the highest FO electron density.

That paper was not accepted for a long time because one referee made the point that the FO electron densities at the two reactive sites of the substituted 1,3-butadiene and the substituted ethylene were not very different. In other words, the magnitude of the proposed effect was very small. Additional consideration of orbitals other than the FOs ("subjacent" or "superjacent" orbitals) could then upset the prediction. About the same time, Houk published a similar analysis, and many subsequent papers argued along essentially the same lines.[7a] We now recognize that these papers identified a correlation between some computed quantity and experimental data and nothing more!

The Hamiltonian one uses in the FO-PMO model is the effective one-electron Hamiltonian of the EHMO method. So, the FO-PMO model can be thought of as being either:

(a) A mere *heuristic model*, which identifies correlations between reactivity trends and MO theoretical indices such as frontier orbital densities, and HOMO-LUMO gaps.

(b) A *shortcut* to obtaining the same result that full-scale EHMO theory would have yielded after one had gone through the motions of calculating explicitly molecules and transition states, including all valence orbitals and electrons. In other words, in the minds of the experimentalists and the theoreticians who have embraced the Hoffmann EHMO-type approach (explicitly or implicitly), the following "equations" hold:

FO-PMO model = EHMO theory

EHMO = "Perfect CI-Type theory"

Denying the foregoing "equalities" is tantamount to rejecting today's conceptual framework of chemistry because it is largely based on EHMO-type ideas.

Does the FO-PMO model equal EHMO theory? If the former is not a good approximation of the latter, and since EHMO theory is only an empirical framework, the FO-PMO model is an exercise in futility. In a Diels–Alder reaction, the best combination of reactants insofar as reaction rate is concerned, involves the pairing of a diene bearing π-donor groups and an olefin bearing π-acceptor groups. This is explained by saying that the HOMO-LUMO gap of the two reactants decreases relative to that in the case of unsubstituted reactants. According to the principles of second-order perturbation theory with an effective Hückel Hamiltonian, a smaller HOMO-LUMO gap is taken to mean stronger stabilization of the reaction complex. At this level, and neglecting overlap, the stabilization due HOMO-LUMO interaction is given by the following expression:

$$SE = \frac{H^2}{\Delta E}$$

where SE is the stabilization energy, H the HOMO-LUMO resonance integral, and ΔE the HOMO-LUMO gap.

The "energy gap explanation" is based on the assumption that upon varying the diene and/or the olefin, H remains relatively constant and what changes is ΔE. Let us compare ethylene (ET) and tetracyanoethylene (TCNE) reacting with 1,3-butadiene (BU). If we plug numbers in the preceding equation, it turns out that what really changes much faster in going from the BU/ET to the BU/TCNE reaction complex is H, not ΔE. Although TCNE has indeed the lower LUMO, the latter is delocalized on the cyano groups, causing the C2p AO coefficients to be small. As a result, H is much smaller in the TCNE than in the ET reaction. Thus, H and ΔE work in opposite directions, and the winner is the H variation. Hence, if we were to take the FO-PMO model literally, we would have to conclude that ET is a better dienophile than TCNE, which is opposite to what is found experimentally.

The intuitive approach to substituent effects on pericyclic transition states goes as follows: An unsaturated substituent (e.g., vinyl or cyano) attached on a diene or a dienophile should slow down the Diels–Alder reaction because conjugation is sacrificed in the benzenelike transition state. At first hearing, this makes sense: The more the "active" π electrons are stabilized by substituents, the less available for reaction they are. And this is precisely what we just found by application of second-order perturbation theory: Delocalization of the active electrons in the substituents diminishes the matrix element H, lowers the stabilization of the transition state, and slows down the reaction, contrary to the HOMO-LUMO argument! This is confirmed by explicit comparison of the cycloaddition of ethylene plus 1,3-butadiene with the 4+2 cycloaddition of two 1,3-butadienes. We can calculate the π energies of the reactants and compare them to the π energies of the model transition states, benzene and vinylbenzene. The results, given in Scheme 17.1b and taken from the compilation of π HMO calculations by Streitwieser and Brauman,[27b] show that a vinyl group *reduces* the stabilization of the transition state model relative to the reactants. Hence, HMO predicts that unsaturated substituents will decelerate the Diels–Alder reaction even though the HOMO-LUMO gap is diminished!

We now have a twofold problem:

(a) The intuitive HOMO-LUMO gap explanation of the experimental data is opposite to the numerical predictions of low order perturbation theory and HMO theory. Hence, *the HOMO-LUMO gap argument does not parallel the theory on which it is based.* It is a "correlation," not a theory.

(b) HMO theory predicts that cyclodimerization of two BU molecules will be slower than the cycloaddition of BU plus ET. The exact opposite is found experimentally.[28]

What is the VB explanation of the rate enhancement due to the vinyl substituent? Very simply, the effect is due to the vinyl group, an excellent radical stabilizer, which stabilizes the S domain of the bikratic transition state complex. The mechanism of bonding at the transition state (defective I-bonding)

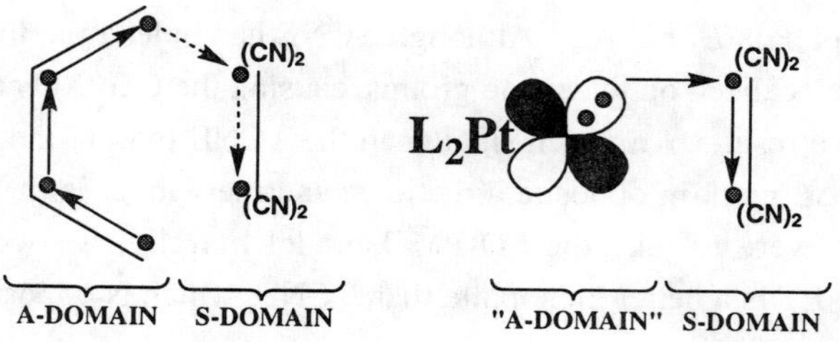

Figure 17.10 The analogy of diene–TCNE transition state complex and L_2Pt–TCNE ground organometallic complex: solid arrows, primary CT localization; dashed arrows, secondary CT delocalization.

is different from the mechanism of bonding at the reactant stage (T-bonding). As a result, substituents are more important at the transition state because they act on the defects (i.e., the electrons of the S domain). Because of unrestrained CT delocalization, numerical HMO theory cannot properly describe the difference in bonding between a transition state and reactants and fails to make the right prediction.

How do we replace the heuristic HOMO-LUMO argument that apparently rationalized the superiority of TCNE as a dienophile? The VB explanation is straightforward: The diene–TCNE transition state complex is a destabilized version of, for example, the L_2Pt–TCNE organometallic complex, and it owes its superiority to two factors:

Each S-electron is stabilized by two CN groups, which are superb radical stabilizers.

Further stabilization (albeit weak) is effected by CT delocalization from the A to the S domain, called oxidative (OX) CT delocalization.

The comparison is illustrated in Figure 17.10.

Table 17.1 shows that the effect of cyano groups on activation free energy is far greater than on reaction free energy. The effect of solvent on reaction rate is consistent with an essentially nonpolar Diels–Alder transition state.

Bottom line: The FO-PMO model of chemical reactivity should have been a brief interlude in the development of chemical theory. The problem, however, has to do with the entire framework of one-electron theory. As a result, the EHMO-type models persist. Indeed, I know of only one article that presents a critical view of FO-PMO model in terms of chemical specifics.[29]

Table 17.1 Activation ($\Delta G\#$) and Reaction (ΔG) Free Energies for the Cycloaddition of Cyclopentadiene to Electron Acceptor Dienophiles

	Free Energies (kcal/mol)	
Dienophile	$\Delta G\#^a$	ΔG
Ethylene	28.7[b]	–12.0
Acrylonitrile	24.1	–13.0
Fumaronitrile	21.3	–14.0
Maleonitrile	21.2	–14.0
1,1-Dicyanoethylene	17.8	–14.0
Tricyanoethylene	16.4	–14.0
Tetracyanoethylene	13.7	–15.0

[a]Except as noted, from J. Sauer, H. Weist, and A. Mielert, *Chem. Ber.* 97, 3183 (1964)
[b]J.J. Gajewski, *J. Am. Chem. Soc.* 101, 4393 (1979)

17.9 Excited State Analogues of Ground Pericyclic Reactions

VB configurations can be subdivided into covalent and ionic, and one can build covalent and ionic substates by diagonalizing separately the two sets. As we move upward in each substate manifold, the nodes increase and so does the interaction of covalent and ionic substates of the same symmetry. Accordingly, one can think of low-lying covalent excited states as ground molecules in which the bonds have a different orientation and CT delocalization is stronger. This implies that the T formula of an excited covalent molecule can suggest the mode of reaction. In the case of benzene, the lowest excited state is the out-of-phase combination of the two Kekulé structures. This translates into a species with two long 1,3 bonds, one of which must have high zwitterionic character. This photointermediate can be viewed as a combination of one "long" bond and one "long" zwitterion (LBLZ intermediate) which rationalizes:

(a) The conversion of benzene to benzvalene.[30]

(b) The conversion of *transoid* butadiene to bicyclobutane. (Figure 17.11a)[31]

(c) The di-π methane rearrangement[32] of an unconstrained 1,4-diene, a highly stereospecific reaction despite the peculiar, strained carbon skeleton of the product (Figure 17.11b). The active system is not a diene, as one might have intuitively assumed, but a π bond plus a vicinal conjugated single C—C bond. A key feature of the reaction is its remarkable regioselectivity.[33] According to our model, electron-releasing, R, groups stabilize the hole, and either electron acceptor, W, or π-conjugated groups, U, stabilize the pair of the "long" 1,3 zwitterion. The predicted regiochemistry (i.e., the preference of U or W groups to become incorporated in the cyclopropane residue) is in accord with the experimental facts.

The reaction of photoexcited benzene with an alkene yields highly strained photocycloadducts with pronounced regioselectivity. We suggest that this is effectively a thermal 4+2 cycloaddition of B_{2g} benzene plus olefin, which is assisted by the 1,3 zwitterion hole and one pair. The reaction complex shown in

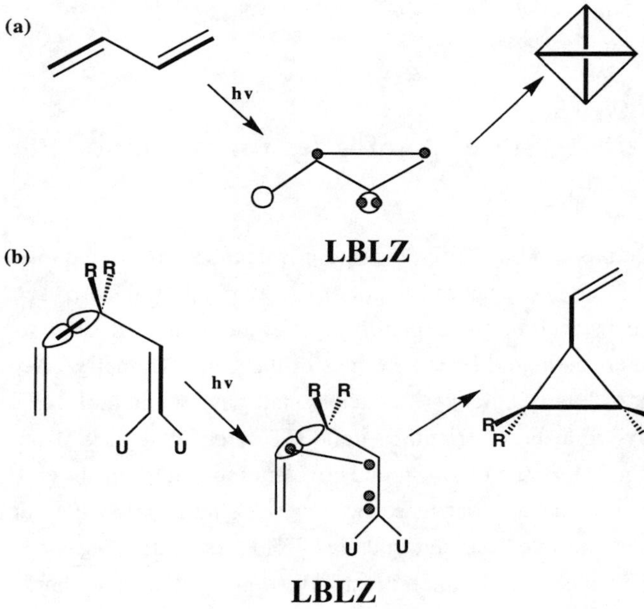

Figure 17.11 Two intramolecular photorearrangements involving intermediacy of the LBLZ species. (a) The intermediate resembles the highly stabilized cyclopropylcarbinyl cation. (b) The di-π methane rearrangement; the polarization determines the reaction regioselectivity [R = π donor or alkyl, U = π-conjugated group (polyenyl, aryl, etc.)]

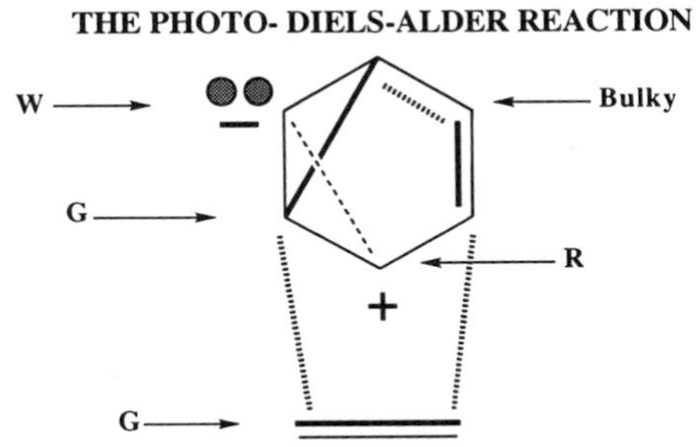

Figure 17.12 The meta-(4 + 2) photocycloaddition of arene and alkene. The role of photoexcitation is to produce the LBLZ intermediate that adds to the olefin. Substituents stabilize the reaction complex and control regioselectivity as indicated: R, π donor (electron-releasing group), W, π acceptor (electron-withdrawing group), G, radical stabilizer.

Figure 17.12 illustrates how the "long" zwitterion controls the regioselectivity of the reaction. Wender et al.[34] summarized the theoretical and experimental history and the present status of the problem. Their paper contains a table of stereo- and regiochemical data that can be rationalized by the LBLZ intermediate suggested by the work of Cornelisse and his group.[35]

The hole or the pair of the 1,3 zwitterion can itself partake in the reaction on the excited surface. In this light, we can identify a case that constitutes the ("long") 1,3-dipolar cycloaddition analogue of the Diels–Alder reaction. This is

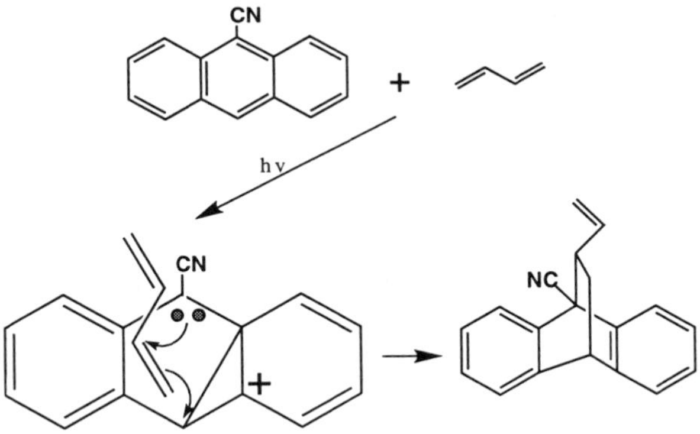

Figure 17.13 The formulation of a "forbidden" suprafacial 4+2 photocycloaddition as an effectively thermal 1,3-dipolar cycloaddition of LBLZ and an olefin.

the case of the "forbidden" 4+2 photocycloaddition of 9-cyanoanthracene and a 1,3-butadiene derivative.[36] The reaction can now be formulated as an effective thermal 1,3-dipolar cycloaddition (Figure 17.13).

A photochemical reaction is initiated on an excited hypersurfaces, and decay to ground state via conical intersections is a big part of the problem. Recently, Bernardi and Robb and their co-workers presented systematic computational studies of the problem, to which the interested reader is referred.[37]

References

1. (a) M. Bearpark, F. Bernardi, M. Olivucci, and M.A. Robb, *J. Am. Chem. Soc.* 112, 1732 (1990). (b) H.Y. Yoo and K.N. Houk *J. Am. Chem. Soc.* 116, 12047 (1994).
2. R.A. Firestone, *J. Org. Chem.* 33, 2285 (1968). See also R.D. Harcourt, *Tetrahedron* 34, 3125 (1978).
3. R. Huisgen, *J. Org. Chem.* 33, 2291 (1968).
4. H.C. Brown, *The Nonclassical Ion Problem*, Plenum Press, New York, 1977; *Pure Appl. Chem.* 54, 1783 (1982).
5. P.D. Bartlett *Q. Rev. Chem. Soc.* 1970, 24, 473.
6. R. Huisgen, R. Grashey, and J. Sauer, in *The Chemistry of Alkenes*, S. Patai, Ed., Wiley-Interscience, New York, 1964. Recent Diels–Alder review: J. Sauer and R. Sustmann, *Angew. Chem. Int. Ed. Engl.* 19, 779 (1980).
7. (a) K.N. Houk, *J. Am. Chem. Soc.* 94, 8953 (1972). (b) N.D. Epiotis, *J. Am. Chem. Soc.* 95, 5624 (1973).
8. P.C. Hiberty and C. Leforestier, *J. Am. Chem. Soc.* 100, 2012 (1978).
9. (a) D.L. Boger and S.M. Weinreb, *Hetero Diels–Alder Methodology in Organic Synthesis*, Academic Press, New York, 1987. (b) A.P. Masters, M. Parvez, T.S. Sorensen, and F. Sun, *J. Am. Chem. Soc.* 116, 2804 (1994).
10. (a) W.D. Harman, W.P. Schaefer, and H. Taube, *J. Am. Chem. Soc.* 112, 2682 (1990). (b) Y. Wang, F. Agbossou, D.M. Dalton, Y. Liu, A.M. Arif, and J.A. Gladysz, *Organometallics* 12, 2699 (1993). (c) O. Blum and D. Milstein, *J. Am. Chem. Soc.* 117, 4582 (1995).
11. J. Sauer, private communication.
12. W. von E. Doering, L. Birladeanu, K. Sarma, J.H. Teles, F.-G. Klaerner and J.-S. Gehrke, *J. Am. Chem. Soc.* 116, 4289 (1994).
13. R.E. Ireland and R.H. Mueller, *J. Am. Chem. Soc.* 94, 5897 (1972).

14. R.E. Ireland, R.H. Mueller, and A.K. Willard, *J. Am. Chem. Soc.* 98, 2868 (1976).
15. A. Nickon, *Acc. Chem. Res.* 26, 84 (1993).
16. N.D. Epiotis, *Theory of Organic Reactions*, Springer-Verlag, Berlin, 1978, especially pp. 205–214.
17. K.J. Shea, G.J. Stoddard, W.P. England, and C.D. Haffner, *J. Am. Chem. Soc.* 114, 2635 (1992).
18. (a) L. Kaplan, in *Free Radicals*, Vol. II, J.K. Kochi, Ed., Wiley, New York, 1973. p. 361. (b) H. Sakurai, in *Free Radicals*, Vol. II, J.K. Kochi, Ed., Wiley, New York, 1973. p. 741.
19. D.D. Coffman, P.L. Barrick, R.D. Cramer, and M.S. Raasch, *J. Am. Chem. Soc.* 71, 490 (1949).
20. (a) L.K. Montgomery, K. Schueller, and P.D. Bartlett, *J. Am. Chem. Soc.* 86, 622 (1964). (b) P.D. Bartlett and L.K. Montgomery, *J. Am. Chem. Soc.* 86, 628 (1964). (c) P.D. Bartlett, *Q. Rev. Chem. Soc.* 24, 473 (1970). (d) R.D. Chambers, *Fluorine in Organic Chemistry*, Wiley, New York, 1973, pp. 179–189.
21. (a) A. Beckwith and D.H. Roberts, *J. Am. Chem. Soc.*108, 5893 (1986). (b) F.G. Bordwell and J.-P. Cheng, *J. Am. Chem. Soc.* 113, 1736 (1991).
22. H.G. Viehe, Z. Janousek, and R. Merenyi, *Acc. Chem. Res.* 18, 148 (1985).
23. (a) L. Stella, in *Substituent Effects in Radical Chemistry*, H.G. Viehe et al., Eds., Reidel, New York, 1986, p. 361. (b) G. Coppe-Motte, A. Borghese, Z. Janousek, R. Merenyi, and H.G. Viehe, in *Substituent Effects in Radical Chemistry*, H.G. Viehe et al., Eds., Reidel, New York, 1986, p. 371. (c) D. Doepp, and J. Walter, in *Substituent Effects in Radical Chemistry*, H.G. Viehe et al., Eds., Reidel, New York, 1986, p. 375.
24. W.R. Dolbier Jr., K. Palmer, H. Koroniak, H.-Q. Zhang, and V.L. Goedkin, *J. Am. Chem. Soc.* 113, 1059 (1991).
25. (a) E. Piers, A.R. Maxwell, and N. Moss, *Can. J. Chem.* 63, 555 (1985). (b) W.R. Dolbier Jr., *Acc. Chem. Res.* 24, 63 (1991). (c) Z.-Y. Yang, P.J. Krusic, and B.E. Smart *J. Am. Chem. Soc.* 117, 5397 (1995). (d) D.M. Lemal and L.H. Dunlap Jr., *J. Am. Chem. Soc.* 94, 6562 (1972). (e) L.F. Pelosi and W.T. Miller, *J. Am. Chem. Soc.* 98, 4311 (1976).
26. (a) A.G. Brook, K. Vorspohl, R.R. Ford, M. Hesse, and W.J. Chatterton, *Organometallics* 6, 2128 (1987). (b) Unlike

perfluoroethylene, disilaethylene gives only 4+2 adducts upon reaction with a diene: H. Sakurai, Y. Nakadaira, and T. Kobayashi, *J. Am. Chem. Soc.* 101, 487 (1979); G. Raabe and J. Michl, *Chem. Rev.* 85, 419 (1985).

27. (a) C.A. Coulson and A.S. Streitwieser Jr., *Dictionary of Π Electron Calculations*, Freeman, San Francisco, 1965. (b) A.S. Streitwieser Jr. and J.I. Brauman, *Supplemental Tables of Molecular Orbital Calculations*, Pergamon Press, Oxford, 1965.

28. (a) M.R. Wilcott, R.L. Cargill, and A.B. Sears, *Prog. Phys. Org. Chem.* 9, 25 (1972). (c) That the S pair dictates a fine distinction between one- and multistep cycloaddition is clearly brought out by F.-G. Klaerner, B. Krawczyk, V. Ruster, and U.K. Deiters, *J. Am. Chem. Soc.* 116, 7646 (1994).

29. (a) B.K. Carpenter in *Advances in Molecular Modeling*, Vol. 1, D. Liotta, Ed., JAI Press, Greenwich, CT., 1988, p. 41. (b) We have universally assumed that stereoselectivity is an expression of the electronic features of the reaction hypersurface. Whether dynamic effects can engender stereoselectivity is a pivotal question which we have circumvented; it is addressed by B.K. Carpenter, *J. Am. Chem. Soc.* 117, 6336 (1995).

30. K.E. Wilzbach, R.J. Ritscher, and L. Kaplan, *J. Am. Chem. Soc.* 89, 1031 (1967).

31. (a) R. Srinivasan, *J. Am. Chem. Soc.* 84, 4141 (1962); 85, 4045 (1963). (b) R. Gleiter and W. Sander, *Angew. Chem. Int. Ed. Engl.* 24, 566 (1985).

32. H.E. Zimmerman, in *Rearrangements in Ground and Excited States*, Vol. 3, P. de Mayo, Ed., Academic Press, New York, 1980.

33. H.E. Zimmerman and B.R. Cotter, *J. Am. Chem. Soc.* 96, 7445 (1974).

34. P.A. Wender, L. Siggel, and J.M. Nuss, *Org. Photochem.* 10, 357 (1989).

35. See especially A.W.H. Jans, B. van Arkel, J.J. van Dijk-Knepper, H. Mioch, and J. Cornelisse, *Tetrahedron* 40, 5071 (1984).

36. N.C. Yang, K. Srinivasachar, B. Kim, and J. Libman, *J. Am. Chem. Soc.* 97, 5006 (1975).

37. M.J. Bearpark, M. Olivucci, S. Wilsey, F. Bernardi, and M.A. Robb, *J. Am. Chem. Soc.* 117, 6944 (1995).

Chapter 18

Torquoselectivity

The groups of Frey,[1] Dolbier,[2] and Kirmse[3] made an important contribution by uncovering a stereochemical trend that is counterintuitive and poses a challenge to theory. It has been found that 1 undergoes thermal conrotation in which the systolic F atoms rotate "outside" while the bulkier diastolic CF_3 groups rotate "inside."[2]

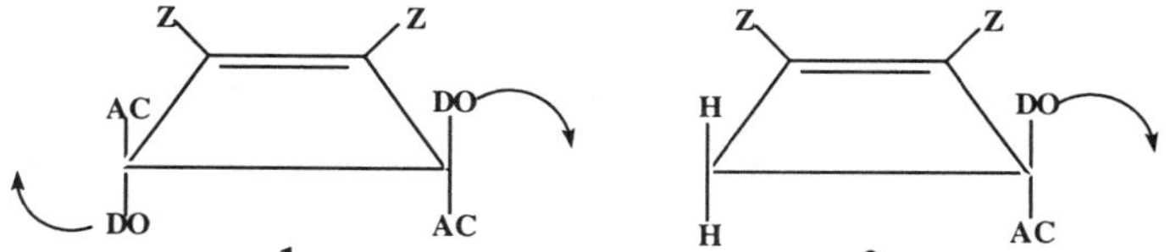

1 2

AC = CF_3
DO = F

A related important fact that has lain unappreciated in the literature is preference of methyl to rotate outside while the larger ethyl rotates inside in 2.[1] Finally, systematic investigation of the thermal conrotatory ring opening of cyclobutene derivatives such as 1 and 2 and ab initio calculations[4] have produced the following generalization: DO groups rotate preferentially "out" while AC groups rotate preferentially "in." Why?

According to HMO theory, a nonaromatic acyclic three-center/four-electron (3c-4e) system is more stable than the corresponding cyclic antiaromatic one. Rondan and Houk[4] proposed a model comprising a heteroatom lone pair of a DO substituent (doubly occupied NBAO) and a breaking bond between C_3 and C_4 (occupied HOMO, unoccupied LUMO), which would define such a 3c-4e system. Inside rotation of DO generates an antiaromatic array with strong destabilizing NBAO-HOMO and weak stabilizing NBAO-LUMO overlap, while outside rotation creates a *nonaromatic* complex with weaker NBAO-HOMO and stronger NBAO-LUMO interaction. Replacing DO by AC changes the electron count by two. Inside rotation of AC is now superior because of the generation of a 3c-2e *aromatic* array.

This model leads to two predictions:

(a) Because of the stronger NBAO-LUMO interaction during outside rotation of DO, the distance between the 3rd and 4th C atoms must be longer at the transition state.

(b) The preference for outside rotation of DO is independent of the number of active electron pairs, hence it is independent of the mode of electrocyclization (con- vs. disrotation) and the size of the unsaturated ring. To be specific, there should be outside rotation of DOs in the electrocyclic ring opening of cyclobutene, 1,3-cyclohexadiene, 1,3.5-cyclooctatriene, and so on.

The first prediction is at odds with the calculational results of the same authors: The C3–C4 distance is shorter and the transition state is tighter in the case of outside conrotation of the DOs (see Table IV of ref. 4). This begins to remind us of the "hyperconjugation adventure," specifically recalling how the inconsistent energetic forecasts with the geometric forecasts (see Chapter 11). The second prediction is contradicted by experimental evidence. *Torquoselectivity* (i.e., a preference for either "outside" or "inside" rotation of substituents) *is not found beyond four-membered rings.*[5,6]

In an electrocyclic reaction, a clear-cut σ/π separation of "active" MOs at the reactant and product stages gives way to σ–π mixing at the transition state, with the result that the detailed evolution of bonding is obscured. In other words, one can draw a symmetry MO correlation diagram, but exactly what happens in between reactants and products cannot be discerned. VB theory removes these ambiguities: We can formulate the electrocyclization reaction so that we always keep track of what reactant bond pairs become product bond pairs. However, once this has been done, a completely different picture of the electrocyclization reaction emerges: Rather than being an intramolecular cycloaddition, it turns out to be an *intramolecular atom transfer reaction* in which two atoms are transferred from the π to the σ system, with the simultaneous transformation of one σ to one π bond. Coupled with our understanding of ring electronic structure (see Chapter 4), this leads to a very clear picture of what happens along the reaction coordinate and defines new dimensions of reaction stereoselectivity.

According to our VB model, 1,3-butadiene and cyclobutene are each viewed as a planar C_4H_2 fragment plus two hydrogen pairs connected to C1 and C4 (with variable C1–C4 distance). The transformation of the latter to the former is envisioned as the rotation of the two hydrogen pairs, which starts with each HCH plane making a 90° dihedral angle with the carbon skeleton plane (cyclobutene stage) and ends with the planar diene in which the same dihedral angle is 0°. There are eight critical AOs assigned to the carbon fragment and four AOs to the four rotating ligands, two of which are assumed to be DO and two AC. The active AOs of the C_4H_2 fragment and the four ligands are the four

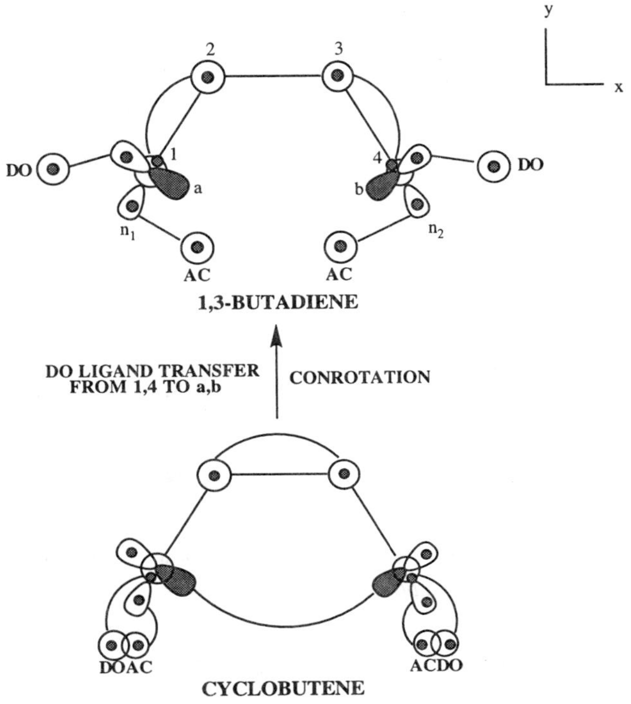

Figure 18.1 The critical AOs involved in the electrocyclic transformation of cyclobutene to 1,3-butadiene.

2p π AOs, each denoted by a number, and one sp^x hybrid radial AO (n_1, n_2) plus one *tangential* 2p AO (a,b) on each of the terminal carbons (Figure 18.1).

The key thing to note is that two of the four ligands remain always bound to two radial (n_1, n_2) AOs while the remaining two are transferred from the 2p π (1,4) to the tangential 2p (a,b) AOs in going from cyclobutene to 1,3-butadiene. To decide which two ligands (DO or AC) bind preferentially to the radial AOs, we recall Walsh's rule: Electronegative DO groups prefer to bind to the more electropositive 2p AOs, while the AC groups are left to be accommodated by the more electronegative radial AOs. In turn, this means that the ring opening is a transfer of two DO groups from the π to the σ frame of a four-carbon chain.

As we saw in Chapters 6 through 8, the conventional DO groups are actually the systolic F, OH, NH_2 and CH_3. The conventional AC groups are the diastolic CN, SiH_3, and so on. Will the systolic DO or the diastolic AC groups rotate out? The transition state of the conrotatory transformation of cyclobutene to 1,3-butadiene is actually an eight-electron/eight-orbital Möbius aromatic system (Figure 18.2). It can be envisioned as the addition of two

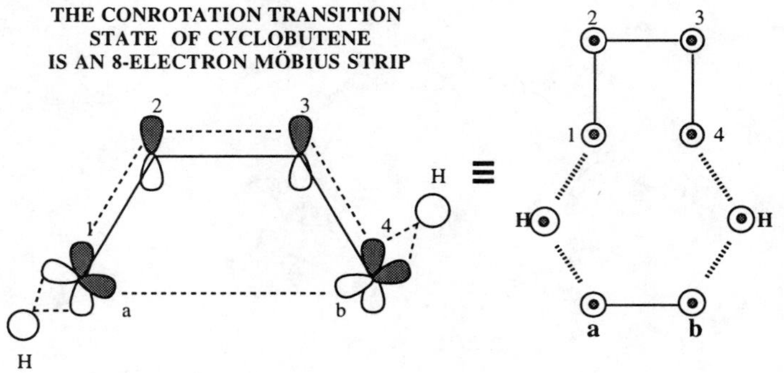

Figure 18.2 Two views of the eight-electron/eight-orbital Möbius aromatic transition state of cyclobutene ring opening.

hydrogens, which were originally bound to the 2p π AOs at the cyclobutene stage, to a cyclobutene tangential bond made by the overlap of the tangential 2p AOs identified as *a* and *b* in Figure 18.2. The key point is that torquoselectivity is the consequence of the presence or absence of an "observer C—C pair" within the ring hollow. If such an observer pair is present, systolic DO groups like F will preferentially rotate out to avoid exchange repulsion of the formal fluoride pair with the observer C—C pair. By contrast, diastolic groups like CF_3 will preferentially rotate in, so that the formal trifluoromethyl cation hole interacts attractively with the observer C—C pair. In other words, σ donation from carbon toward the rotating (systolic) group drives it out, while σ withdrawal by carbon away from the rotating (diastolic) group drives it in. Note the sharp distinction between the systolic/diastolic of VB theory and the σ-donor/σ-acceptor dichotomy of conventional theory. No one calls CF_3 (a methyl cation) a σ donor!!! According to VB theory, it is just this (i.e., diastolic).

As discussed in Chapter 4, four- as well as three-membered rings have σ-pseudoaromatic character. In cyclobutene, the major biscarbenic contributor has an observer endo radial bond next to the active tangential bond (Figure 18.3). This immediately explains the following experimental observations:

(a) Smaller systolic groups, such as F, will rotate out, while larger diastolic groups, such as CF_3, will rotate in.

(b) Methyl rotates out in preference to ethyl because the former is an error-free systolic group while the latter is intermediate between systolic (methyl) and diastolic (*t*-butyl).

(c) Torquoselectivity is a function of the double bond character of the reactive centers of the ring. That is, torquoselectivity should vanish in rings

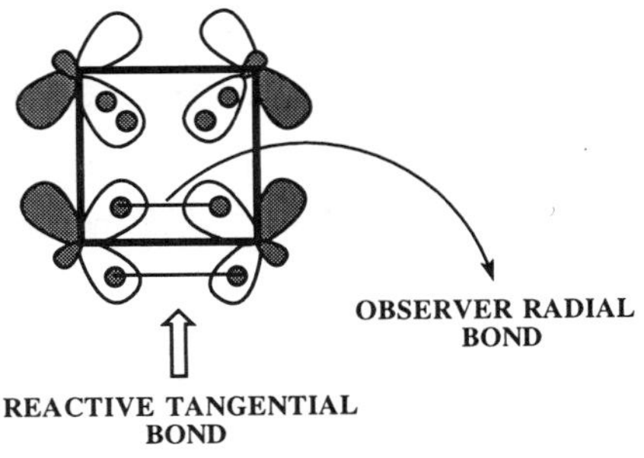

Figure 18.3 Reactive tangential and observer radial bonds in the biscarbenic VB resonance structure of the σ frame of cyclobutene.

with no significant pseudoaromatic delocalization (i.e., rings larger than four-membered rings in which perfect pairing dominates). Experimentally, it is found that formal six-electron pericyclic reactions fail to show the torquoselectivity exhibited by the four-electron electrocyclization.[5,6]

(d) Replacing the methylene hydrogens of cyclobutene by systolic groups (e.g., F) will raise the activation energy for conrotatory ring opening by increasing the exchange repulsion of the inside-rotating C—F bond and the reactive C—C bond. Indeed, the most spectacular experimental result in this area is that perfluorination raises the barrier by roughly 15 kcal/mol, a part of which may be due to conventional F···F nonbonded repulsion.[7]

(e) Replacing the methylene hydrogens by diastolic groups (e.g., CN) will lower the activation energy for conrotatory ring opening by decreasing the exchange repulsion of the inside-rotating C—CN bond and the reactive C—C bond. There are no available data to test this prediction.

(f) Replacing the methylene hydrogens by groups that are de facto incapable of rotating inside will lower the activation energy for conrotatory ring opening. A substitution of this type is to replace the two hydrogens by one oxygen to generate a cyclobutenedione, which opens to bisketene. Tidwell and co-workers found by ab initio computations that this reduces the activation enthalpy by 5–6 kcal/mol.[8]

(g) Since the transition state must be bikratic, the S electrons must localize ortho (1,2) or para (1,4). As a result, placement of radical stabilizers away from the sites of electrocyclization (i.e., at C1 and C2 of cyclobutene) should accelerate the reaction unless steric effects (exchange repulsion with the rotating methylenes) dictate otherwise. 1,2-Dimethylation of cyclobutene raises the barrier by 3–4 kcal/mol, and this can be attributed to a steric effect. On the other hand, 1,2-diphenylation leaves the barrier essentially unchanged, implying that some other stabilizing factor is at work. The best case is provided by the ring opening of cyclobutenedione, where steric effects are absent because there is no definable rotation of the C3 and C4 oxo groups. Ab initio computations now show that replacing the vinylic hydrogens by two silyl groups reduces the activation enthalpy by nearly 3 kcal/mol.[8]

We depart the organic transition states with a comment on the mechanism of enzyme action,[9] an area of research in desperate need of new ideas beyond "entropy" and "propinquity."[10] The issue is not whether these notions are credible but, rather, how an enzyme implements them. Specifically, the concept of "propinquity" has been verified repeatedly in the past but under disguises one would not normally recognize. The exothermic 2+2+2 cycloaddition of three ethylenes to yield cyclohexane can be represented by a double-minimum (reactants/products) reaction profile. Constraining the three olefins into proximity of one another is equivalent to entering the (ground) hypersurface well above the unperturbed reactants and exiting it well above the unperturbed products. Photochemical studies probing the entry and exit of (excited) hypersurfaces have been carried out, notably by Turro and co-workers.[11] Thus, there is nothing novel or surprising in the notion that skeletal constraints that align reacting molecules serve to lower reaction barriers. In the conventional organic experiments, the constraint is effected by the σ frame of a rigid ring or cage. The real challenge is to understand the mode of constraining employed by the enzyme. We now know that there are three choices: T-, I-, or E-bonding, with each imposing different selectivity on the enzyme–substrate binding.

References

1. H.M. Frey and R.K. Solly, *Trans. Faraday Soc.* 65, 448 (1969).
2. W.R. Dolbier Jr., H. Koroniak, D.J. Burton, A.R. Bailey, G.C. Shaw, and S.W. Hansen, *J. Am. Chem. Soc.* 106, 1871 (1984).
3. W. Kirmse, N.G. Rondan, and K.N. Houk, *J. Am. Chem. Soc.* 106, 7989 (1984).
4. N.G. Rondan and K.N. Houk, *J. Am. Chem. Soc.* 107, 2099 (1985).
5. W.R. Dolbier Jr., A.C. Alty, and O. Phanstiel IV, *J. Am. Chem. Soc.* 109, 3046 (1987).
6. W.R. Dolbier Jr. and O. Phanstiel IV, *Tetrahedron Lett.* 29, 53 (1988).

7. The activation parameters for a variety of thermal rearrangements and decompositions are drawn from the review by M.R. Wilcott, R.L. Cargill, and A.B. Sears, *Prog. Phys. Org. Chem.* 9, 25 (1972).
8. A.D. Allen, J. Ma, M.A. McAllister, T.T. Tidwell, and D. Zhao, *Acc. Chem. Res.* 28, 265 (1995).
9. G.A. O'Doherty, R.D. Rogers, and L.A. Paquette, *J. Am. Chem. Soc.* 116, 10883 (1994).
10. F.C. Lightstone and T.C. Bruice, *J. Am. Chem. Soc.* 116, 10789 (1994).
11. N.J. Turro, *Modern Molecular Photochemistry*, Benjamin/Cummings, Menlo Park, CA, 1978, p. 428.

Chapter 19

Alkali Dimers as Coordination Compounds

19.1 Chemical Formulas for Li_2 and Be_2

Invented before the advent of quantum chemistry, the concepts of the Lewis bond pair and the Lewis molecular electronic formula are probably the most important concepts of chemistry. We communicate, and we design experiments and computations, by falling back on usage of Lewis structures. Is the Lewis concept applicable to metallic systems? One popular view is that the same type of Lewis structure represents H_2 and Li_2:

H—H Li—Li

This impression has been reinforced by the very numerous papers of EHMO practitioners as well as ab initio computational chemists who have embraced the one-electron model of chemical bonding. Everything from organic molecules (i.e., molecules comprised of nonmetal atoms) to metal solids is thought to be bound by the familiar bonding mechanisms: covalent, ionic, and in between (i.e., polar covalent). We will now challenge these perceptions by writing formulas that directly project the difference between nonmetallic and metallic bonding.

According to VB theory, metal-containing systems are bound subject to one requirement: minimization of promotional energy. There are two types of promotion: orbital promotion and Coulomb repulsion (in the form of exchange loss in going from atoms to molecules plus contact repulsion). This means that metal atoms form I bonds by making a minimum investment in hybridization. In this light, the bonding of Li_2 can be described as follows.

(a) We mix the valence 2s AO with the 2p AOs to form three conical, sp^3-type hybrids in which the dominant contributor is the 2s AO.

(b) The valence electrons are allocated so that deshielding occurs. In other words, the placement of the electrons reproduces the combined effect of induction and dispersion. Thus we end up with two sets of sp^3-like AOs, which are staggered with respect to each other and have the two electrons occupying anti-periplanar or gauche hybrids. This construction is shown in Figure 19.1.

(c) Starting with the aforementioned "electron geometry," 2-e CT delocalization, symbolized by the joint action of two arrows, effects further bonding. This is the equivalent of one resonating I bond. The two resonant I descriptors have comparable weights because each one preserves electroneutrality.

We can investigate which of the two contributors is relatively more important by exposing their respective physical consequences. To do so, we replace Li by H_2B. It can now be seen that two different geometries of H_2BBH_2 correspond to the two different ways of making an I bond (Figure 19.1). If the two I bonds are comparable, the two geometries must be nearly degenerate (i.e., the molecule H_2BBH_2 must have essentially free rotation about the BB bond). A high level ab initio computation can test this conclusion.

In summary, we have arrived at an important conclusion: Li_2 is bound by a single resonating I bond rather than by a covalent bond. One pair is connected to two holes by two arrows spanning two different overlap regions. Hence, Li_2 is the simplest illustrator of the concept of the I bond. Li_2 is not appreciably bound at the SCF level.[1a] This is evidence that the one-electron formula of Li_2 is different from the many-electron formula of the same molecule. The physical and chemical consequences of all this will become increasingly clear as our story unfolds.

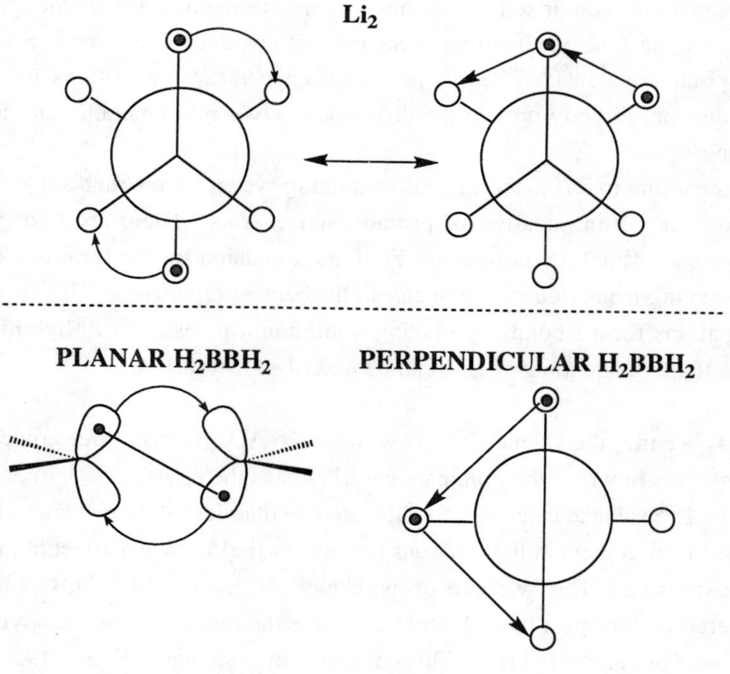

Figure 19.1 "Electron stereochemistry" and bonding of Li_2 and the related H_2BBH_2.

Be_2 can be envisioned as the combination of two Be atoms, each of which is sp^3-hybridized and contains two singlet coupled electrons in two of the four hybrids. The dimer is bound by one I bond (two arrows equal one I bond) as illustrated in section 19.2 (see Figure 19.2). We draw attention to the hallmarks of our formulas: Initial deshielding, obtained by allocating electrons so that one electron of one Be faces a hole of the neighbor Be ("anti electron geometry"), sets the stage for 2-e CT delocalization.

19.2 Li_2 and Be_2 as Illustrators of Overlap Dispersion

What is the best way of delocalizing the two valence electrons of Li_2 starting with the ground perfect-pairing configuration? We focus attention on intra- and interatomic 2-e delocalization as shown in Scheme 19.1. The matrix elements H_{ij}, are depicted in Scheme 19.2a. Note how in H_{13} the double solid arrow representing dispersion undoes (has opposite sign) what the double hatched arrow

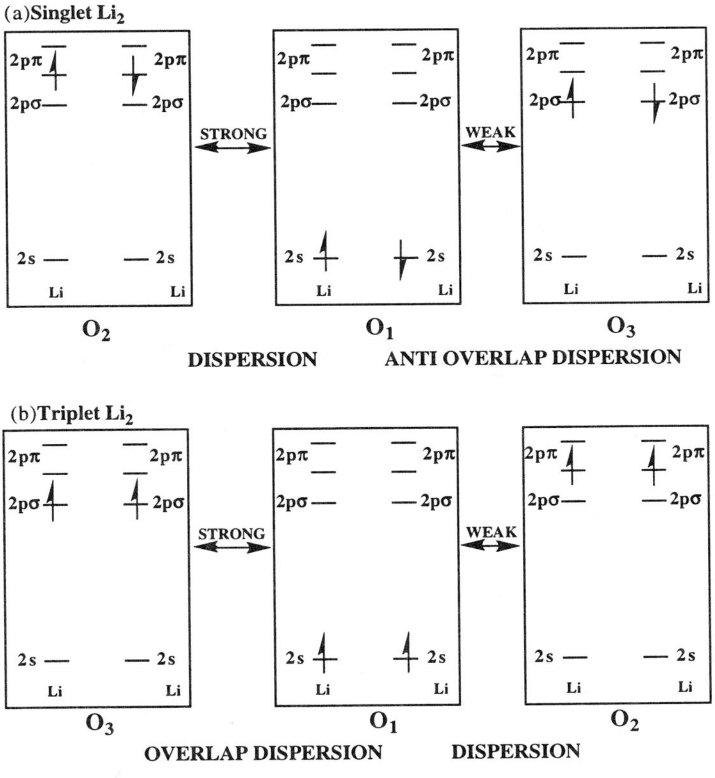

Scheme 19.1

representing overlap does. In other words, we have anti overlap dispersion. Thus, an interatomic 2-e CT hop in the σ space leads to anti overlap dispersion, while an intra-atomic 2-e hop in the π space results in pure dispersion. We call the latter "strong dispersion" because it is actually a sum of two bielectronic exchange integrals, and we note that it is improved by 2s-2p hybridization in the σ space. The net result of the entire complex CI is expressed by the formulas shown in Figure 19.1.

The situation reverses in the triplet state, where now delocalization in the σ space corresponds to overlap dispersion, while delocalization in the π space amounts to "weak dispersion" because it is a difference of two bielectronic

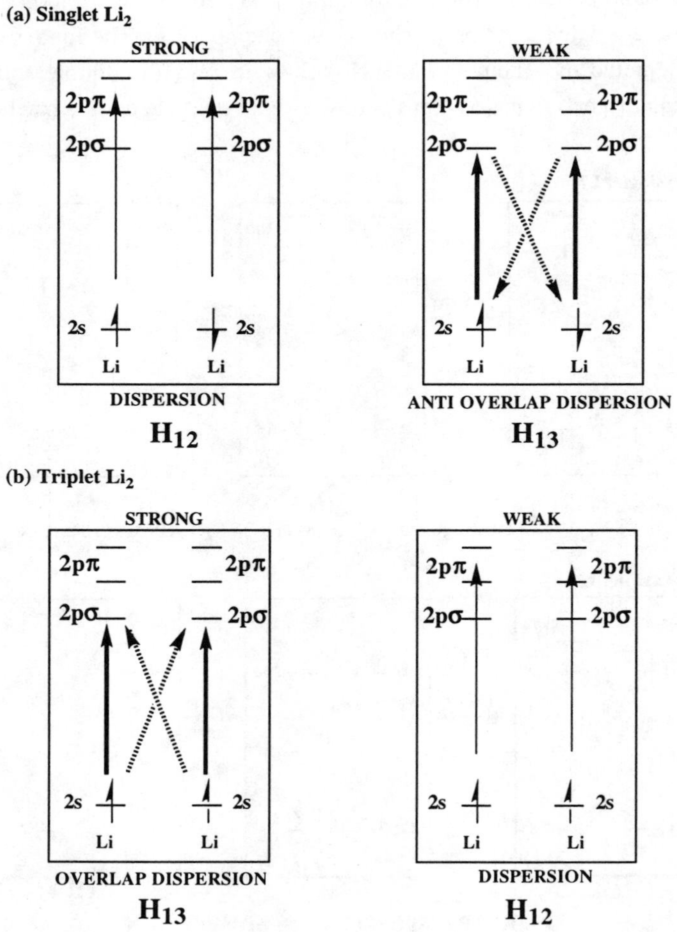

Scheme 19.2

exchange integrals. The matrix elements H_{ij} are shown pictorially in Scheme 19.2b. Note how, in H_{13}, the double solid arrow representing dispersion now reinforces what the double hatched arrow representing overlap does. As a result, the triplet Li dimer can be thought of as a species in which the exchange repulsion of two odd electrons having the same spin is partly canceled by 2-e CT delocalization leaving the dispersion term of overlap dispersion to effect bonding. This van der Waals molecule can also be visualized either as "triplet Li anion plus Li cation" or as "electron plus Li_2 cation radical."

Overlap dispersion is also behind the proposed formulas for Be_2. In the absence of hybridization, the exchange repulsive interaction of the two Be2s pairs renders the approach repulsive. Once we allow mixing of the 2s with the 2p orbitals on the same center, we have overlap dispersion due to the interaction of configurations **1** and **2** shown in Figure 19.2a. Overlap dispersion is described by moving two electrons at a time, starting with the $(s^2)(s^2)$ configuration. As a result, the magnitude of the stabilization depends on the energy proximity of the $(s^2)(s^2)$, the $(s^1p^1)(s^1p^1)$, and the $(p^2)(p^2)$

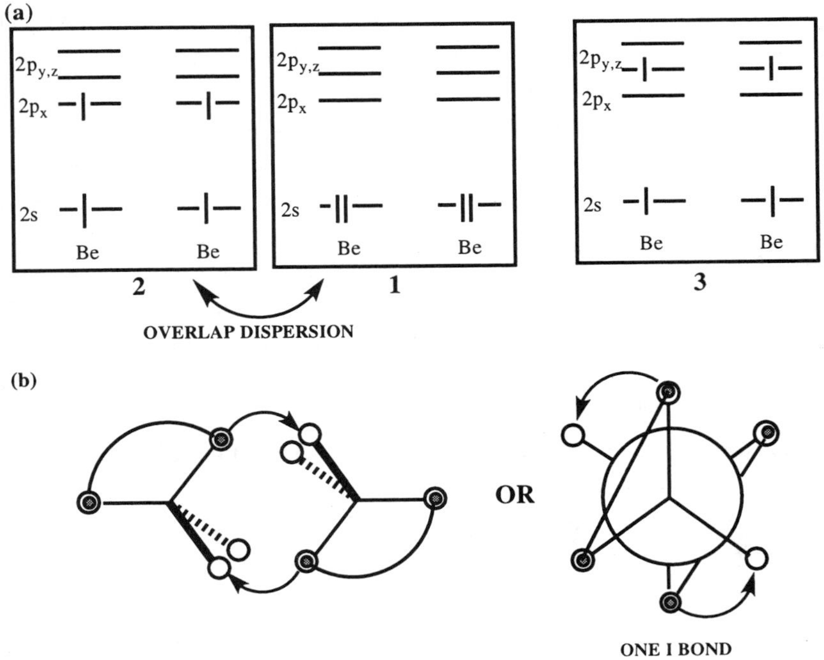

Figure 19.2 (a) The critical CI in Be_2. (b) The VB formulas of Be_2. Each Be has a geminal (intra-atomic) bond of two electrons. The two atoms are connected by a single I bond (two CT arrows).

configurations. The narrower the energy range that is spanned, the stronger the interaction, the more favorable the effective hybridization, and the stronger the bonding. The interaction of configurations 1 and 3 is a pure dispersive interaction that requires the mixing of $(s^2)(s^2)$ with $(p^2)(p^2)$ and is expected to be weaker than the overlap dispersive interaction.

19.3 Computational Evidence Relevant to the VB Formula of Li_2

The difference between Li_2 and H_2 is now apparent. In the former, there are two electrons occupying two Coulomb holes; that is, we have a gas pair, or a pair of *interstitial electrons*, coordinated to two Li cations. By contrast, the interelectronic distance of the electron pair of the H_2 molecule can take every conceivable value because interelectronic repulsion is more than compensated by the strong reduction of kinetic energy due to the high electronegativity (always in a relative sense) of the hydrogen atom. The difference between isolated Li_2 and metallic solid Li can also be understood. Each Li_2 is pictured as a flexible quadropole made up of two cationic cores and two electrons. The quadropoles associate in a way that maximizes their electrostatic interaction to form a solid.

Williams wrote a nice paper explaining solid state architecture by applying the concept of the molecular electric quadropole moment to the structure of solid benzene.[2] Equivalently, Li_2 units aggregate so that each gas pair becomes coordinated to several Li cations and each Li cation becomes coordinated to several gas pairs. All in all, we come to view a metal solid as the result of association of M_2 subunits due to electrostatic attraction between metal (M) cations and interstitial electrons. The conclusion is inescapable: Metallic bonding is already present in the metal dimer, and solid formation is a mere manifestation thereof. Since the notion that solid Li has metallic bonding while Li_2 has covalent bonding marks a substantial deviation from conventional thinking, it is good to ask: Is there evidence supportive of these ideas?

Ab initio computational chemists are well aware that a proper description of metal-containing molecules requires extensive configuration interaction. The challenge is to understand what CI does and translate this understanding into a chemical formula that can guide the experimentalist into new areas of research. The explicit VB calculation of Li_2 and related model systems by Pelissier and Davidson[3] shed considerable light on the problem. The most important calculational results relevant to our arguments are as follows.

(a) The ground covalent configuration of Li_2 accounts for only part of the bonding. The CI which, in our language, generates the gas pair, is responsible for roughly one-third of the binding in the alkali dimers. As Pelissier and Davidson wrote: " Double excitations of the form $s_A p_A / s_B p_B$ are essential to explain the bond in the metal diatomics." This contrasts with the case of H_2, where the covalent configuration is responsible for more than 90% of the binding.

(b) In the absence of the double excitations responsible for gas pair formation, 2s–2p mixing in the σ space is more pronounced in H_2 than in Li_2 despite the much lower energy of a p AO in Li than in H! This is exactly what anti overlap dispersion in the σ space is expected to cause: negligible 2s–2p hybridization within each Li atom.

Another important paper is that by Konowalow and Fish.[4] These authors report ab initio computed potential energy curves for the 26 lowest lying states of Li_2, giving their asymptotic limit. Going back to Scheme 19.1, we recognize that configuration O_1 is the parent of the ground $1\Sigma_g^+$ state derived from the Li2s + Li2s asymptote, while O_3 and O_2 are the parents of the $4\Sigma_g^+$ and $5\Sigma_g^+$ excited states, both derived from the Li2p + Li2p asymptote. These two states have very different dissociation energies (to the 2p + 2p asymptotic limit) and equilibrium bond distances. Specifically, the $4\Sigma_g^+$ state features a single undistorted deep minimum with D_e = 7773 cm^{-1} and R_e = 3.621 Å while $5\Sigma_g^+$ is a highly distorted curve with D_e = 1804 cm^{-1} and R_e = 5.121 Å. Our explanation goes as follows: O_3, the principal contributor of $4\Sigma_g^+$, cannot interact with O_1 because of anti overlap dispersion. Hence, O_3 yields an energy curve that shows no effect from any interaction with a lower energy state, such as $1\Sigma_g^+$, that would cause it to be "pushed up" and become deformed near the R_e of the ground $1\Sigma_g^+$ state. This is why $4\Sigma_g^+$ has a large D_e and a tight R_e. Experimental spectroscopic results of Bernheim et al.[5] are compatible with this picture. By contrast, O_2, the principal contributor of $5\Sigma_g^+$, interacts strongly with O_1 by dispersion. As a result, this state is "pushed up" by $1\Sigma_g^+$ about the R_e of the latter. This is why $5\Sigma_g^+$ ends up having a small D_e and a very long R_e.

Finally, with regard to triplet Li_2, Konowalow et al.[6] propose a "most likely" potential energy curve of the lowest triplet state of Li_2 which has a minimum at about 7.8 a_o and a dissociation energy of about 340 cm^{-1}. We attribute the binding of Li_2 to overlap dispersion.

19.4 Physical Meaning of the Failure of Pauling's "Polar Covalence" Concept

Many years ago, Pauling pointed out that "polar covalence" is responsible for the exothermicity of the following recombination reaction:

$$AA + ZZ \rightarrow 2AZ$$

This trend obtains when A and Z are nonmetals having different electronegativity. Examples of the "Pauling trend" are given in Table 19.1.

However, recall the lessons of Chapters 6 to 8: What is ultimately the most important determinant of stability is the error count in the T formula. We are reminded of this by a spectacular example of an error-plagued aromatic molecule (1-nitropyrrole) losing by a large margin to a near-perfect (only one error) nonaromatic isomer:

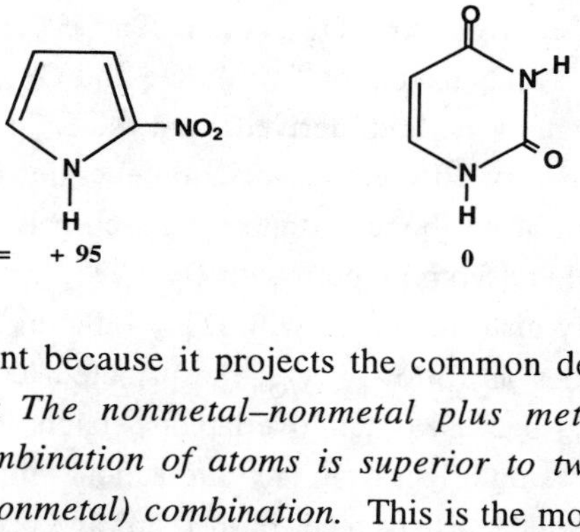

E_{rel} (kcal/mol) = + 95 0

Table 19.1 is important because it projects the common denominator in subsections B, C, and D: *The nonmetal–nonmetal plus metal–metal (or semimetal–semimetal) combination of atoms is superior to twice a metal–nonmetal (or semimetal–nonmetal) combination.* This is the most compelling evidence in favor of the proposal of the T/I/E map of chemical bonding.

In 1958 Ahrland, Chatt, and Davies[8] made a contribution of great significance: They pointed out that metal ions (Lewis acids) can be divided roughly into two classes depending on their ability to coordinate with specific donor atoms or groups (Lewis bases). Specifically, "class a" metal ions bind most strongly with first-row donors (N, O, F) while "class b" ions form complexes of high stability in combination with donors from lower rows of the periodic table. In the early 1960s, Pearson made a related important contribution: He recognized that if neutral metals and metal ions were classified as "hard" and "soft," the available thermochemical data imply that "soft likes to bind soft and hard likes to bind soft." Equally significant, Pearson recognized that the hard/soft acid/base (HSAB) concept led to thermochemical predictions diametrically opposite to those of the Pauling "polar covalence" concept.[9] Typical examples of the "HSAB trend" are given in Table 19.1. Pearson commented: "The HSAB principle is phenomenological in nature. This means that there must be underlying theoretical reasons which explain the chemical facts which the principle summarizes. It seems certain that there will be no one simple theory."

So, what theory has emerged to rationalize a phenomenon of such importance? Practically all attempts are founded on the same central assumption: *There is only ionic and covalent bonding.* The only deviation

Table 19.1 Thermochemistry of Bond Interchange

Reaction[a]			ΔE (kcal/mol)
A. Pauling Trend			
F—F + H—H	$\longrightarrow$	H—F + H—F	−129
HO—OH + H—H	$\longrightarrow$	H—OH + H—OH	−83
H_2N—NH_2 + H—H	$\longrightarrow$	H—NH_2 + H—NH_2	−45
H_3C—CH_3 + H—H	$\longrightarrow$	H—CH_3 + H—CH_3	−11
B. Hard Acid/Soft Base Trend			
COF_2 + $HgBr_2$	$\longrightarrow$	$\underline{COBr_2}$ + HgF_2	+85 (−66)
CO + PbS	$\longrightarrow$	$\underline{CS}$ + PbO	+71 (−64)
H_2O + CaS	$\longrightarrow$	$\underline{H_2S}$ + CaO	+37 (−25)
HF + NaI	$\longrightarrow$	$\underline{HI}$ + NaF	+32 (−76)
C. Metal Trend			
Li_2 + H_2	$\longrightarrow$	$\underline{LiH}$ + $\underline{LiH}$	+13
Na_2 + H_2	$\longrightarrow$	$\underline{NaH}$ + $\underline{NaH}$	+26
Cs_2 + H_2	$\longrightarrow$	$\underline{CsH}$ + $\underline{CsH}$	+30
D. Suggestive Trend			
O_2N—NO_2 + Cl_2	$\longrightarrow$	2$\underline{Cl—NO_2}$	+4
NC—CN + I_2	$\longrightarrow$	2$\underline{I—CN}$	+16
HC≡C—C≡CH + Cl_2	$\longrightarrow$	2$\underline{Cl—C≡CH}$	+17
H_3C—CH_3 + I_2	$\longrightarrow$	2$\underline{I—CH_3}$	+12

[a]Underscored molecules are "frustrated" combinations of nonmetal with semi-metal or metal atoms.

from this theme is found in a contribution by Pitzer, who argued that London forces, depending on the product of the polarizabilities of the combining groups, may account for the preference of a polarizable (soft) acid to combine with a polarizable (soft) base.[10]

We can now provide a theoretical basis for the HSAB concept by considering the combination of univalent and isoelectronic atoms in the gas phase. In this way, there are no longer HOMOs and LUMOs, the combining atoms are isoelectronic, with the result that orbital symmetry considerations are irrelevant (i.e., the valence AOs have the same symmetries) and there are no solvent effects to muddle the picture.

The critical thermochemical data that involve the simplest nonmetal and metal atoms and demonstrate counter-Pauling behavior define the "metal trend" (subsection C, Table 19.1). In the reaction shown below, lithium can be called "soft" and hydrogen "hard." Hence, this reaction is the simplest embodiment of the HSAB concept.[11] The terms "soft" and "hard" are now seen as inventions that serve the same purpose of differentiating between more than one mechanism of chemical bonding. Our interpretation of the endothermicity of the reaction is straightforward: The two H atoms taken together optimally engage in T-bonding ("homonuclear T-bonding"), while the two Li atoms taken together engage in I-bonding ("homonuclear I-bonding"). The LiH combination involves "frustrated" bonding.

$$\underset{T}{\text{H—H}} + \underset{I}{\text{Li—Li}} \longrightarrow \underset{FR}{2\ \text{Li—H}} \qquad \Delta E = +13\ \text{kcal/mol}$$

Here is now the key realization: The thermochemical data we have classified as demonstrative of the HSAB trend as well as other data exemplifying Pauling behavior are actually demonstrators of the three domains of the map of chemical bonding as illustrated in the reactions that follow. The reaction endo- or exothermicity depends on whether reactants or products are frustrated according to the rules of atom affinity given in Chapter 3.

HSAB Reaction

$$\underset{T}{COF_2} + \underset{I}{HgBr_2} \longrightarrow \underset{FR}{COBr_2} + \underset{FR}{HgF_2} \quad \Delta E = +85\ \text{kcal/mol}$$

Anti-HSAB Reaction

$$F\text{—}F + Li\text{—}Li \longrightarrow 2Li\text{—}F \qquad \Delta E = -211 \text{ kcal/mol}$$

FR# I E

The preceding data validate the map of chemical bonding presented in Chapter 3. All the counter-Pauling redistributions produce metal–nonmetal combinations (molecules underlined) that are frustrated. The data that follow remind us of the lessons of Chapters 7 to 9: As the systolic character of a heavy p-block fragment increases (e.g., as we sweep from SiH_3 to Cl), its combination with systolic methyl becomes increasingly frustrated:

	ΔE (kcal/mol)
$CH_3\text{—}CH_3 + I_2 \longrightarrow 2CH_3I$	12
$CH_3\text{—}CH_3 + HS\text{—}SH \longrightarrow 2CH_3SH$	5
$CH_3\text{—}CH_3 + H_2P\text{—}PH_2 \longrightarrow 2CH_3PH_2$	–4
$CH_3\text{—}CH_3 + H_3Si\text{—}SiH_3 \longrightarrow 2CH_3SiH_3$	–13

The dogma "metals like to bind metals and nonmetals like to bind nonmetals" takes its place next to the well-known rule "like dissolves like." Indeed, inorganic chemists have long been aware that molecules made up of metallic and nonmetallic fragments tend to decompose to "metal molecules" plus "nonmetal molecules." For example, metal chalcogenolates are known to decompose to chalcogenides according to:

$$[M(AR)_2]_n \longrightarrow (MA)_n + nAR_2$$

where M = Zn, Cd, and Hg and A = S, Se, and Te.

Generalizing, we have the equation below where M is a red atom, A a green atom, and R an organic alkyl:

$$M(AR)_4 \longrightarrow A\text{=}M\text{=}A + 2RAR \longrightarrow A\text{=}M\text{=}A + A\text{—}A + 2R\text{—}R$$

While thermochemical data for the first step are not available, the existing evidence demonstrates clearly the proclivity of the intermediate molecules to decompose to red–green, green–green and blue–blue combinations. Note the switchover of the reaction enthalpy as A tends to become a metal.

$$2H_3C\text{—}CH_3 + A\text{—}A \longrightarrow 2H_3C\text{—}A\text{—}CH_3$$

A	ΔE (kcal/mol)
O	-47.8
S	-8.5
Se	13.2

Finally, the enthalpies of the following binary redistribution reactions provide further confirmation of the I-bond concept.

A = alkali, alkaline earth, transition metal, boron, F and congeners

$$A\text{—}A + H\text{—}H \longrightarrow 2AH$$

A = C and congeners

$$A\text{—}A + 3H\text{—}H \longrightarrow 2AH_4$$

A = N and congeners

$$A\text{—}A + 3H\text{—}H \longrightarrow 2AH_3$$

A = O and congeners

$$A\text{—}A + 2H\text{—}H \longrightarrow 2AH_2$$

The results shown in Table 19.2 define a remarkable trend: The enthalpy shifts from endo- to exothermic as A tends to become a nonmetal. A related example is the metal–organic chemical vapor deposition (MOCVD) reaction, crucial for industry and technology:

$$Ga(CH_3)_3 + PH_3 \longrightarrow GaP + 3CH_4$$

Table 19.2 The Enthalpies (in kcal/mol) of the Metathesis Reaction $A_2 + nH_2 \longrightarrow 2AH_n$

Li: +17	Be: +13	B: +14	C: −128	N: −22	O: −58	F: −130
Na: +26	Mg: +18	Al: +6	Si: −45	P: −32	S: −41	Cl: −44
K: +29				As: −14		Br: −24
Rb: +36		In: +9		Sb: +14	Te: +16	I: −3
Cs: +31				Bi: +57		

Sc: +44 Ti: +37 V: +79 Cr: +62 Mn: +61 Fe: +32 Co: +34 Ni: +33 Cu: +27 Zn: +64

The following derivative reaction (for which reliable thermochemical data exist) is exothermic by 6 kcal/mol even though the reactants have one more formal covalent bond than the products!

$$GeH_4 + SeH_2 \longrightarrow GeSe + 3H_2$$

We conclude that the thermochemical data are convincing evidence that there are two types of chemical bonding in molecules. What is conventionally called "metallic bonding" in metal solids is simply the apotheosis of I-bonding already present in a metal dimer.

19.5 Why Does Li_2^+ Have a Stronger Bond than Li_2?

Both H_2 and Li_2 have two electrons in a totally symmetric σ_g MO. Yet, they are completely different: The two-electron bond of H_2 is much stronger than the one electron bond of H_2^+. By contrast, homonuclear alkali dimer cations are more tightly bound than their parent neutral dimers.[12] Heteronuclear dimers, with the exception of NaLi, reverse the trend. Even so, the difference in the bond dissociation energies of the neutral and the cation radical dimer is very small. Equally important are the geometrical trends: Specifically, the weaker Li_2 bond is shorter than the stronger Li_2^+ bond. These data have been

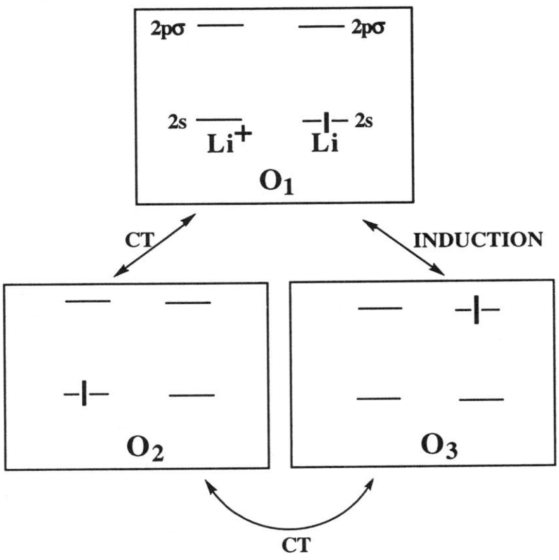

Figure 19.3 Overlap induction in Li_2^+.

summarized in a paper by Buckingham and Rowlands.[13] Are these trends exceptional, or do they reflect fundamental differences between nonmetal and metal bonding?

The fundamental break of this work from conventional theory is exemplified by Li_2: A covalent bond connecting the two Li atoms is now deemed unfavorable because Li is a metal atom and covalency represents a restriction. The ideal situation for metals is CT-bonding because reduction of kinetic energy can effect bonding only if it is not counteracted in the overlap region by strong electron–electron repulsion (contact repulsion).

What about the Li_2 cation radical? This is the simplest illustrator of configuration aromaticity in which one charge-transfer (CT) hop (O_1—O_2 interaction) is assisted by another electron hop of the induction type (O_1—O_3 interaction) to create a third CT hop (O_3—O_2 interaction) in the σ frame (Figure 19.3). This is a version of overlap induction: The positive charge on Li^+ causes the mixing of the 2s and 2p AOs of the Li atom, drawing the single electron into the overlap region between the two Li cores. This promotes an electron hop from the polarized Li toward the Li cation. The overlap term is expected to be small because of the high electropositivity of Li. Hence, $Li_2^{\dot{+}}$ owes most of its binding to induction. As a result, it can be described as a "gas electron sandwiched by two Li cations."

We can now compare Li_2 and $Li_2^{\dot{+}}$. Considered from the vantage point of VB theory, the BDE's and the internuclear distances of Li_2 and $Li_2^{\dot{+}}$ give an unmistakable signal: Li_2 is less strongly bound because it is more promoted. Consider two neutral Li atoms approaching each other. Each Li ejects one electron into the interstitial space. This corresponds to double quasi-ionization, a form of promotion. The resulting gas pair, with the electrons staying far apart to minimize their mutual repulsion, fastens the two Li cations together. By contrast, when one neutral Li atom and one Li cation approach each other, only one Li needs to eject one electron into the interstitial space. Thus, only *single* quasi-ionization is required. However, we now have two Li cations fastened together by only a single gas electron. Hence, the shorter bond length of Li_2 goes hand in hand with a greater instability that is due to larger promotion.

The Li_2 bond is weaker than the $Li_2^{\dot{+}}$ bond for another reason as well, and it has to do with a feature of atomic structure not yet considered explicitly. Specifically, inspection of the Desclaux tables of atomic radii of maximum electron density R_{max} shows that in transition metals, the valence nd AOs have R_{max} values not greatly exceeding those of the nonvalence np AOs. As a result, using the metal, valence AOs to form bonds with some ligand L makes formation of (3-e or 4-e) M···L antibonds involving the nonvalence metal electrons inevitable. If we were to assume that filled nonvalence Li1s pairs are active, we could conclude that two interstitial electrons in Li_2 form twice as strong an antibond (4-e antibond) as one interstitial electron does in Li_2^+ (3-e antibond) with one nonvalence pair. Hence, nonvalence AOs can potentially become the source for a smaller destabilization of the metal cation radical compared to the neutral dimer. This point has been made by Pelissier and Davidson[3] as well as by Buckingham and Rowland.[13] However, this argument fails to explain why Li_2 has a shorter bond than Li_2^+. By the way, the greater binding energy of Li_2^+ relative to Li_2 was first computed by James,[14] decades before the experimental determination.

The VB conclusion that the higher atom promotional energy causes Li_2 to have a bond that is weaker but shorter than that of Li_2^+ is echoed in a different comparison: According to Bondybey, Be_2 has a shorter but far weaker bond than Li_2![15] The reason is that the former requires effective promotion of four electrons, while the latter needs only two.

	r_e (Å)	D_e (cm^{-1})
Be_2	2.45	790
Li_2	2.67	8440

Many people realize that electron–electron repulsion can be a significant factor when a bond is weak. What has not been recognized is that covalency represents a restriction, that is, the molecule can find some other superior mechanism for bonding. In the case of Li_2, this is formation of an I bond. Since, however, the 2s-to-2p excitation is disproportionately high, Li_2 has two options and both are bad: to make a T bond at the expense of high interelectronic repulsion or to make an I bond at the expense of high promotional energy. Indeed, the M_2/M_2^+ comparison is the earliest warning sign that there is something wrong with conventional bonding theory, not in the details but in the fundamentals.

Bottom line: As illustrated in this chapter, as well as Chapters 4 and 27, only VB theory leads to a clear interpretation of BDEs. As more thermochemical data accumulate to complement structural determinations, the chemist will find that the beloved "bond strength/bond length" correlation is simply an accident. There will be nearly as many cases of strong bonding going hand in hand with long bond distances as cases that conform to the conventional wisdom. The inexorable process has already started.[16]

19.6 The Analogy Between Interatomic and Intermolecular Bonding Selectivity and the Mechanism of Enzyme Action

The uniqueness of an enzyme lies in its ability to specifically solvate its substrate by a composite of covalent and noncovalent interactions. This ability must be a net consequence of the very structure of the enzyme, and the latter is determined by the tendency of hydrophobic residues to aggregate in the protein interior, while the hydrophilic residues preferentially occupy "surface" sites exposed to aqueous solvent. This can be formalized by the first equation below, nonpolar and polar groups or molecules are represented by N and P, respectively. The crucial point now is that the three different modes of pairwise interaction are differentiated by the component of static bonding, which is characteristic of each combination:

The characteristic feature of the P—P interaction is classical Coulomb attraction (EL). This requires no promotion.

The characteristic feature of the NP—P interaction is induction (IND). This requires promotion on the NP partner.

The characteristic feature of the NP—NP interaction is dispersion (DISP). This requires promotion on the both NP partners.

P—NP + P—NP	⟶	P—P	+	NP—NP
H—Li + H—Li	⟶	H—H	+	Li—Li
Monopromoted:		Nonpromoted:		Dipromoted:
Medium Bonding		Strong Bonding		Weak Bonding

This formulation exposes the common denominator between the electronic structure of molecules and the electronic structure of solid–liquid and liquid–liquid solutions. The solute–solvent redistribution above is predicted to be exothermic for the very same reasons that the reaction of two HLi groups to yield H_2 plus Li_2 is exothermic. This is why "like dissolves like." The textbook example is the water–dioxane system, which shows a positive deviation from Raoult's law.

The connection between atomic bonding and phase bonding is simple: "Strong" plus "weak" binding is superior to twice "medium-frustrated" binding. In other words, fragments predisposed toward one mechanism of bonding prefer to bind with fragments of the same type; that is, the "hard/soft" concept manifests itself not only in gases but also in solutions. Much as in the interatomic paradigm, the case of intermolecular interaction offers two extreme choices: classical Coulomb attraction with zero promotion or dispersion with strong promotion. Bonding via induction represents an in-between "frustrated" situation. Polar fragments opt for one extreme (classical Coulomb attraction) and nonpolar fragments for the other (dispersion). The combination of polar and nonpolar molecules represents an in-between situation, which can be formulated as "frustrated" static bonding because the polar can act inductively on the nonpolar unit but the reverse is not possible.

The iceberg model[17] projects entropy as the primary determinant of the free energy of mixing a polar and a nonpolar phase. VB theory says precisely the opposite: Liquid–liquid and solid–liquid interaction are dictated primarily by enthalpy, and "like attracts like." The situation ought not to be confused with the case of gas–liquid interaction. The enthalpy of solution of a nonpolar gas in a polar liquid is almost always negative simply because the NP—NP bonding in the preceding equation is zero! An important paper by Ishikawa et al.[18] makes the key point that bilayer formation is enthalpically driven.

The VB treatment of solute–solvent redistribution suggests that the uniqueness of enzyme catalysis is due to polar groups at the active site which are unresponsive to the surrounding nonpolar environment (i.e., they are "desolvated") because the very shape of the enzyme was dictated by optimization of the remaining polar–polar and nonpolar–nonpolar interaction according to the selectivity principle espoused above. The amino acid side chains, which provide the crucial substrate "solvation," are active because they are themselves "desolvated" with respect to the remainder of the enzyme. The orientation of the polar side chains at the active site is the strategic error in the minimization of the total nonbonded interaction of an enzyme! In short, the mechanism of enzyme catalysis is equivalent to strategic phase transfer catalysis, so well known to organic chemists.

References

1. (a) P.J. Bertoncini, G. Das, and A.C. Wahl, *J. Chem. Phys.* 52, 5112 (1970). (b) G. Das, *J. Chem. Phys.* 46, 1568 (1967). (c) An account of early ab initio calculations of "small" molecules is given by H.F. Schaefer III, *The Electronic Structure of Atoms and Molecules*, Addison-Wesley, Reading, MA, 1972.
2. J.H. Williams, *Acc. Chem. Res.* 26, 593 (1993).
3. M. Pelissier and E.R. Davidson, *Int. J. Quantum Chem.* 25, 723 (1984).

4. D.D. Konowalow and J.L. Fish, *Chem. Phys.* 84, 463 (1984).
5. R.A. Bernheim, L.P. Gold, and C.A. Tomczyk, in *Comparison of Ab Initio Quantum Chemistry with Experiment for Small Molecules*, R.J. Bartlett, Ed., Reidel, New York, 1985, p. 325.
6. D.D. Konowalow, R.M. Regan, and M.E. Rosenkrantz, *J. Chem. Phys.* 81, 4534 (1984).
7. This is the basis for the construction of Pauling's electronegativity scale: L. Pauling, *General Chemistry*, 3rd Ed., W.H. Freeman, San Francisco, CA 1970.
8. S. Ahrland, J. Chatt, and N.R. Davies, *Q. Rev.* 11, 265 (1958).
9. R.G. Pearson, *Science* 151, 172 (1966).
10. K.S. Pitzer ,*J. Chem. Phys.* 23, 1735 (1955).
11. R.G. Pearson, *Chem. in Br.* 444 (1991).
12. Y.T. Lee and B.H. Mahan, *J. Chem. Phys.* 42, 2893 (1967).
13. A.D. Buckingham and T.W. Rowlands, *J. Chem. Educ.* 68, 282 (1991).
14. H.M. James, *J. Chem. Phys.* 3, 9 (1935).
15. V.E. Bondybey, *Chem. Phys. Lett.* 109, 436 (1984).
16. R.D. Ernst, J.W. Freeman, L. Stahl, D.R. Wilson, A.M. Arif, B. Nuber, and M.L. Ziegler, *J. Am. Chem. Soc.* 117, 5075 (1995).
17. For review, see N. Matubayasi, *J. Am. Chem. Soc.* 116, 1450 (1994).
18. Y. Ishikawa, H. Kuwahara, and T. Kunitake, *J. Am. Chem. Soc.* 116, 5579 (1994).

Chapter 20

Arrow Trains and I-Bond Activation

20.1 The Types of Arrow Train

The I formula and the associated arrow trains form the basis for all subsequent discussions. We have already seen that co- are superior to contradirectional arrow trains. We now recognize that more than one codirectional train can coexist in a molecule. Representative situations in which one train optimally complements the other are shown in Figure 20.1.

We will encounter all of these situations in actual molecular examples. For example, the Shell model of rings is the illustrator of two different arrow trains within the concentric R and T shells. When the chirality of the motion is the same, the two trains are called cochiral and each is symbolized by the Cahn-Ingold-Prelog R/S convention. When it is opposite, the two trains are called antichiral.

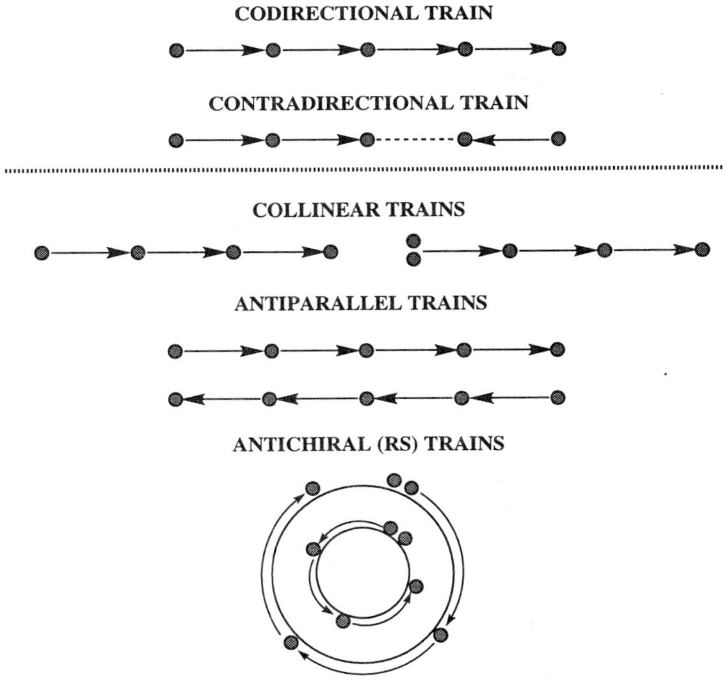

Figure 20.1 Types of arrow trains.

20.2 Activation of Codirectional Arrow Trains

A codirectional arrow train represents the minimum satisfactory way of making I bonds. The different types of perturbation that improve, or activate, the train are shown in Figure 20.2. We discuss them in turn.

Oxidation or reduction by external agents (Figure 20.2A) causes configurational degeneracy and enhancement of I-bonding. Internal redox (Figure 20.2B), which has the same effect as external oxidation (reduction), becomes possible either when the train is homonuclear and the atoms have low electronegativity or when the train is terminated at highly electronegative atoms. One "functional" arrow train can become activated by an appropriately directed "observer" arrow train, as shown in Figure 20.2C. According to the concept of overlap dispersion, under certain conditions, two antiparallel arrow trains running on the same atomic chain can effect I-bonding by a combination of inter- and intra-fragmental CT. This represents an optimal situation and becomes the criterion by which we will be judging the quality of I-bonding.

The reader may know the experience of the computational chemist: Organic molecules are easier to compute than metal-containing molecules and ground

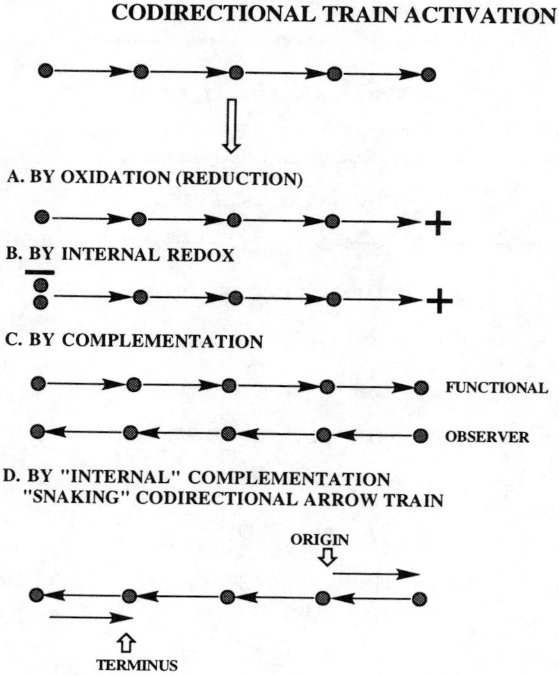

Figure 20.2 Mechanisms of activation of a codirectional arrow train.

states are much easier to compute than excited states. The difficulties arise from the requirement of extended AO bases and the necessity for CI. Translated in VB language, "Hard MO computation" implies that the system in question involves the hard-to-describe I-bonding in which electrons avoid each other by correlating their motions (this is the meaning of the n-e hop with $n > 1$). We can now see the linkage between metallic ground states and excited states of π-conjugated organic molecules. That metal-containing systems are I-bound has been already discussed, and it is a topic to which we will return shortly. On the other hand, a π-excited organic molecule is an *antiresonant zwitterion* in which the cationic and anionic ends activate a codirectional arrow train and institute I-bonding in the way illustrated in Figure 20.2B. This I-bonding is "defective" because the hole and the pair interact in an antibonding sense and are not allowed to approach each other. Hence, the number of arrows is curtailed relative to a ground *resonant zwitterion* analogue. Nonetheless, we can still think of the zwitterionic first excited state of a long polyene as an I-bound π system.

20.3 The Three Choices of BeF_2 and the Superior Relay Bond

There exist two types of I bond connecting two ligands to a metal: OX and ROX.

An OX ("oxidative") bond involves 2-e CT from metal to ligand. The two electrons move in the same direction. Using canonical orbitals, the OX I bond is described by the CI shown in Figure 20.3a. The intrafragmental 2-e hops prepare the polyradical state from which 2-e CT delocalization can occur with facility. In RHF theory, the "preparation step" is left out (dashed arrows are inoperative).

A ROX ("redox") bond, on the other hand, involves 2-e CT occurring in opposite directions: One electron hops from metal to ligand while a second electron moves in the opposite direction. The critical CI is shown in Figure 20.3b. The ROX bond is produced by overlap dispersion in which the solid (overlap) and hatched (dispersion) arrows act synergistically. In RHF theory, the dashed and hatched arrows are inoperative. The term ROX is reserved for an I bond within a molecule formulated as a composite of two fragments (e.g., M plus L_2). The equivalent term REL (for "relay CT") designates the same I bond within the same molecule formulated as a composite of three fragments (e.g., M plus K plus L). The hallmark is two head-to-tail arrows representing codirectional CT delocalization.

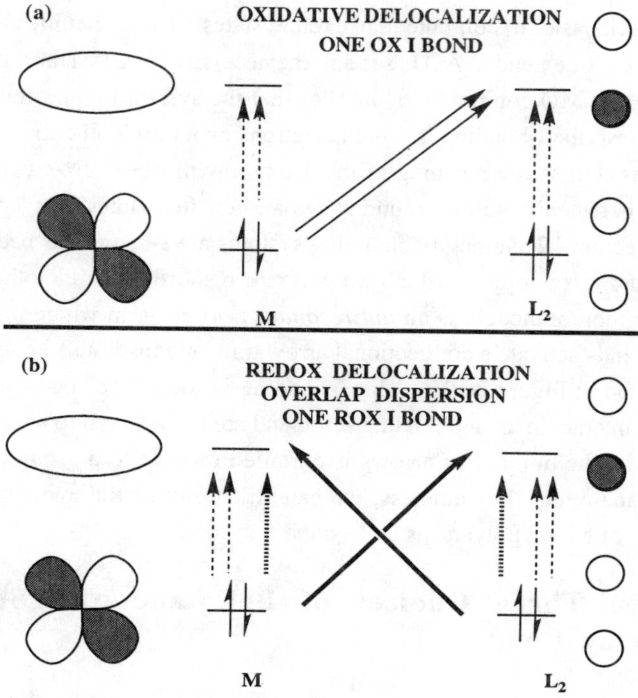

Figure 20.3 The OX and ROX (REL) I bonds.

ROX delocalization represents "assisted" while OX delocalization is "unassisted" CT delocalization, where the quoted terms refer to the action of dispersion. However, assistance by dispersion (overlap dispersion) vanishes unless the total number of electrons is four. Otherwise, we get either anti overlap dispersion (two or six electrons) or unassisted CT (because the dispersion term approaches zero: three or five electrons). The key point is that when the condition spelled out above is met, the ROX mode is the one intrinsically favored because it has the larger interaction matrix element H_{ab}, where a and b are the interacting configurations. However, the strength of a bond depends not only on H_{ab} but also on the inverse of the energy gap, ΔE_{ab}. As we go from a homo- to a heteronuclear system (and with increasing difference of electronegativity), the energy gap factor increasingly favors the OX mode. Hence, the relative strength of ROX and OX bonds depends critically on the nature of the component fragments.

The consequences of the preceding reasoning are better appreciated when we inquire whether a given single 1-e CT hop that occurs within some "functional" space is assisted by "observer" holes and pairs. The problem is formulated and

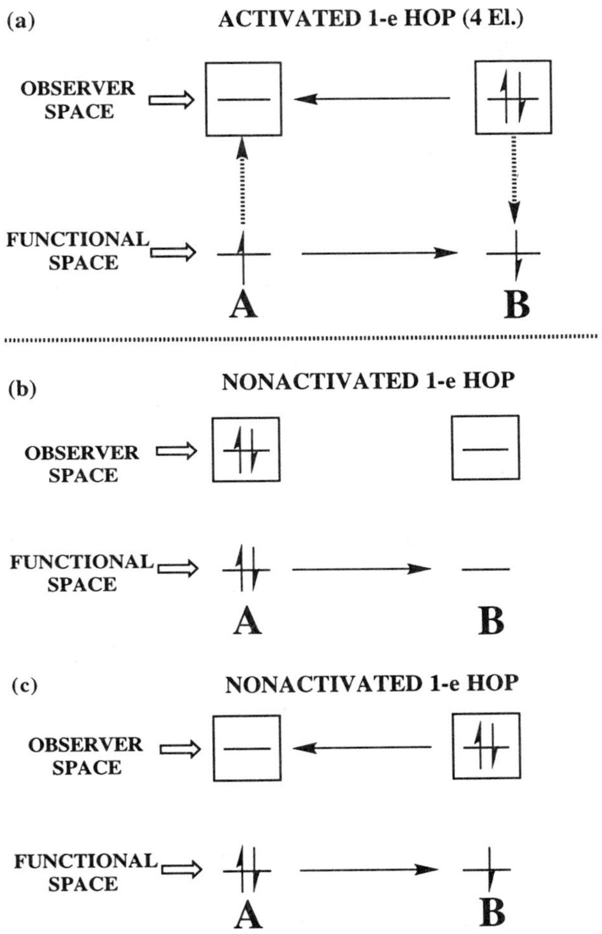

Figure 20.4 Assistance of CT delocalization in the functional space by observer pairs and bonds.

solved in Figure 20.4. We envision a 1-e hop occurring under the "observation"of surrounding holes and pairs that differ in orientation with respect to the 1-e hop. It can be seen that a forward 1-e hop from atom A to atom B can be assisted only if A has an observer hole and B and observer pair, to ensure that a reverse electron hop can complement the forward electron hop. If the observer hole and pair are reversed in orientation, there is no assistance by dispersion (Figure 20.4b). The same thing is true if one or both observer elements (hole or pair) is absent or if the number of electrons is two, three, five, or six (Figure 20.4c).

Two ordered 1-e CT hops act in tandem with a 2-e CT hop to make up one I bond. The I bond itself becomes favorable when each 1-e CT hop becomes activated. The physical meaning of this is simple: One cannot turn the charge-conserving exchange 2-e hop into a charge-creating 2-e CT hop without first ensuring that the surrounding electron density will not render the CT ineffectual. In other words, I-bond formation demands an electronic environment that is tolerant of CT delocalization. If it is not, I-bonding is turned off and the only remaining choice of a metal system is E-bonding.

In this light, let us consider the bonding of FBeF. Because Be is electropositive, small kinetic energy reduction does not justify promotion to the divalent $2s^12p^1$ configuration. Thus, we should inquire as to the best possible way of making bonds starting with ground atomic configurations. Normally, the inorganic chemist uses oxidation state formalism and formulates the molecule as $F^-Be^{2+}F^-$. It is said that Be is actually Be(II), where the numeral in parentheses denotes the oxidation state. To make sense of the experimental data, various researchers have postulated different degrees of covalency.

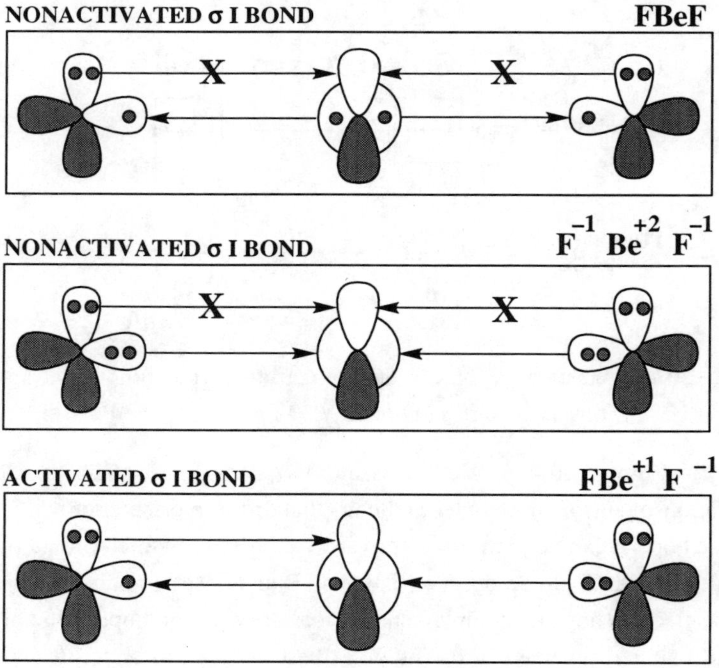

Figure 20.5 The three bonding choices of FBeF. An activated codirectional arrow train can be established only if the system can attain the reference $F{:}^-Be^+F$ configuration.

The VB scenario shown in Figure 20.5 is different. Specifically, Be has three choices, and one is superior. It can bind as either Be(0) or Be(II) with one nonactivated, OX and RED, respectively, σ I bond. Note that RED is simply the reverse of OX delocalization, and the term that is applicable depends on the (implicitly) defined frame of reference, which, by convention, is taken to be the metal center (or central atom). Be(II) binds effectively by the E mechanism, at which limit only an inferior contradirectional CT delocalization is possible.

The third choice, to bind as Be(I) with one activated REL bond, offers the best of the three worlds: It is a codirectional I bond with T complement (covalency) and E complement (Coulomb attraction). We conclude that BeF_2 must be actually FBe^+F^-.

The example discussed in connection with Figure 20.5 suggests a fruitful way of thinking about ML_n, where M is metal and L are ligands. We can envision each ligand as a composite of two different types of elements:

A functional hole (Lewis acid), pair (Lewis base), or dot (radical).
Observer holes, pairs, and dots.

The bonding is now seen as principally due to the action of the functional element(s), with its "quality" being dependent on the assistance provided by the observer elements. It is now crucial that this assistance be measured not by the conventional yardstick of overlap but, rather, according to the criterion of overlap dispersion. If the I bond is thought of as a correlated form of the T bond, the assisted I bond must be viewed as a supercorrelated form of bonding. We have produced simple pictures that capture the essence of the "correlation of correlation," and we are now ready to apply these concepts to real chemistry.

20.4 The Difference Between F_3B—NH_3 and H_3B—CO

The first of the two title molecules is a "high school chemistry" molecule. The second is a prototype of metal–carbonyl bonding. Both groups are strongly bound, yet very different. However, the difference does not lie in the conventional explanations. It is a mere difference in the directionality of two arrows, as illustrated. As a result, F_3B—NH_3 has an unassisted I bond and opts for E-bonding while H_3B—CO has an assisted I bond and opts for I-bonding.

NONACTIVATED DATIVE BOND. E-BONDING

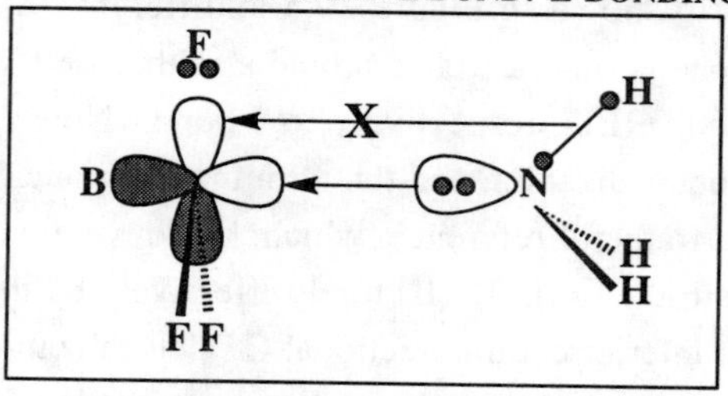

ACTIVATED DATIVE BOND. I-BONDING

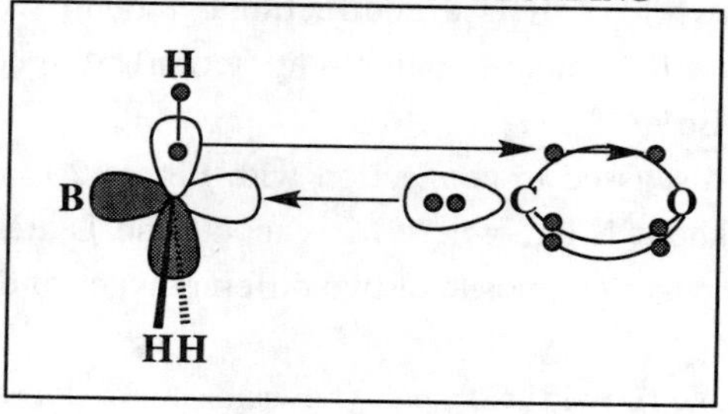

One of the first indications that the so-called coordinate bond is not what most writers have in mind comes from a Morokuma energy decomposition in which one assumes additivity of classical Coulombic attraction, exchange repulsion, and charge transfer. While the Morokuma scheme predicts F_3B—NH_3 to be bound by Coulomb forces, it shows term nonadditivity in the case of H_3B—CO.[1] This suggests a "nonclassical" mechanism of bonding which we have identified as overlap dispersion (i.e., activated I-bonding). This suggests a variety of strategies for creating strong I complexes. For example, the best match of CO is not H_3B but rather $(R_3Si)_3B$.

We can now appreciate that the terms "Brønsted acid" and "Lewis acid" are misleading. The proper classification of acids and bases is according to the mechanism of bonding, T, I, or E. An examination of the literature reveals that "Lewis complexes" or "charge-transfer complexes" involving semimetals or metals are actually either I- or E-bound coordination compounds. The synthetic chemist finds it useful to devise strategies by matching pairs with holes, but this approach alone cannot tell us which type of complex will materialize. The work of Veith[2] and co-workers is especially important because it teaches us that the rules of coordination chemistry extend all the way in the p block and the carbon family which, unfortunately, most chemists believe to be the domain of the covalent bond.

20.5 The Classification of Nucleophiles, Electrophiles, and Radicals According to I Activation

An atom cannot support strong I-bonding unless it meets a minimum criterion: It must have observer holes and pairs that indirectly promote I-bonding of the functional valence electrons. This means that nucleophiles (as well as nucleophilic radicals) and electrophiles (as well as electrophilic radicals) fall in two categories:

(a) An *I-permissive* nucleophile has observer holes (e.g., Cl anion, with low-lying 3d holes) and an *I-permissive* electrophile has observer pairs (e.g., CO, with σ and π pairs). These are minimum qualifications that do not guarantee activation. To induce activation, an I-activated nucleophile must have its observer holes matched with observer pairs of the partner electrophile and an I-activated electrophile must have its observer pairs matched by observer holes of the partner nucleophile.

(b) An I-nonactivated nucleophile (e.g., fluoride) has no observer hole, and an I-nonactivated electrophile (e.g., Li cation) has no observer pairs.

In other words, if one functional electron moves forward, some other observer electron must move in an opposite direction to promote the forward motion of the functional electron. This pairing provides assistance by dispersion; it ensures that CT delocalization can occur without engendering undue repulsion with the surrounding electrons. Thus, the electronic configuration of an atom uniquely determines its capability to bind by either the I or the E mechanism.

In the following illustrations of I-activated nucleophiles (N), radicals (R), and electrophiles (E), observer holes and electron pairs are indicated by squares.

The reader will appreciate that the differentiation of "functional" and "observer" elements is somewhat arbitrary. We can easily bypass this difficulty by treating explicitly functional as well as observer elements and by stating the observer elements whenever we attempt to simplify things by focusing on the functional elements. For example, in $Cr(CO)_6$, two CO fragments can make one OX bond with one Cr 3d pair because the action of the CO pairs in conjunction with the Cr 4p holes clears the way for I-bonding.

20.6 The d-Block Test of I-Bond Activation

One way of testing our ideas of selectivity is to exploit the well-known trend of d expansion down a p-block column. Observation of enhancement of bonding in replacing a first- by a third-row congener implies I-bond activation. To see this point, consider the case of CrF_6. There are three pairs in three axial σ AOs of an octahedral sd^2 Cr atom making three I bonds with the six fluorine atoms. Whether the σ I bonds are activated depends on the interaction of the fluorine lone pairs with the three Cr3d π-type holes. In the case of Cr, the d orbitals are contracted and I-bond activation fails. CrF_6 is an "uncertain" molecule.[3] Replacing Cr by W effects I-bond activation because the d AOs expand and can now overlap with the fluorine lone pairs. WF_6 is a stable molecule. This test gives three types of information:

It suggests that transition metal fluorides are I-bound molecules.
It reveals that CrF_6 is frustrated.
It rules out the ionic model, because W is more electronegative than Cr and yet WF_6 is more stable than CrF_6.

A review of descriptive inorganic chemistry in the light of our concepts is a fascinating pastime. We now understand, inter alia, why $HgCl_2$ is a discrete, stable species and $Au(CN)_2^-$ is an exceedingly stable molecule while their first-transition row counterparts are inferior and the fluorides entirely different. The common denominator is that Cl^- and CN^- (but not F^-) are I-permissive nucleophiles, which become I-activated when matched with electrophiles having active observer d pairs. This explains the observation that late transition metals prefer to bind to heavy p-block atoms. We can write specific formulas that differentiate among isoelectronic species as follows:

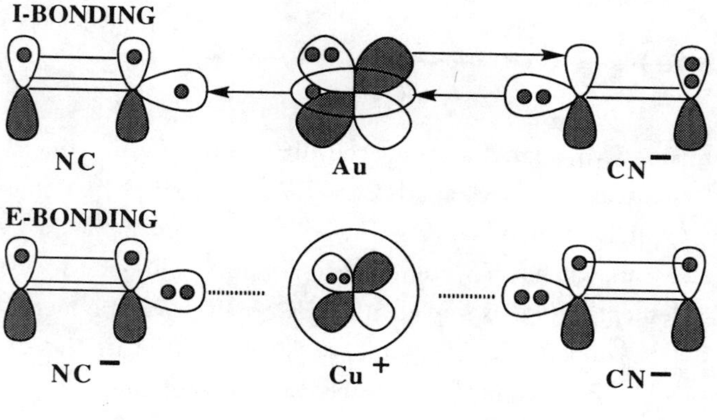

What is the valence of a transition metal? This question is posed to students of general descriptive chemistry who have no recourse other than memorize the facts. Inorganic textbooks itemize the observed valence states of transition metals and identify the most common one. The key observation now is that in moving from the first to the third transition series, the most common valence state (oxidation state) floats to the top of the list. For example, Cr is most often +3 but W is most frequently +6. The VB interpretation is that the d expansion enables the third-series metal to engage in I-bonding in a way that involves all or most of its valence electrons. At the limit, each metal pair can tie up two radicals or two Lewis acids.

20.7 The Electronic Structure of Borazine

In the typical approach to the borazine molecule, calculations are made, indices are reported, and the discussion centers on how "aromatic" the easily identifiable π system is. All this has little to do with the problem! Borazine owes its stability to the action of all electrons of all atoms including the "innocent" and disregarded R ligands attached on the boron atom.

Before facing borazine, one must graduate from RBZZ, a simple derivative of BH_3, where R is alkyl and Z a univalent electronegative ligand. The triangular diagram in Figure 20.6a reveals how the three atoms are attached on the central B atom and, in particular, which two atoms become part of a REL bond and which one atom makes the covalent bond with the odd-out electron of B. Since the more electronegative ligands must tie up the more electropositive AOs, we obtain the formula shown in Figure 20.6a. Since nitrogen atom acts as "Z,Z," the σ frame of borazine is constructed as shown in Figure 20.6b. Note two key features:

The presence of three (resonating) covalent B—N bonds. In the absence of the π system, these are unactivated σ covalent bonds.

The action of the "innocent" R groups. These determine the stability of the σ frame because they are integral parts of the REL bond responsible for a fraction of the bonding of the σ frame.

All arrows are codirectional.

Turning on the π system causes activation of the covalent σ bonds and generates a codirectional arrow train the spans the entire σ frame. This is complemented by a π codirectional arrow train (Figure 20.6c). We now have the explanation of the stability of borazine in one quarter of a page.

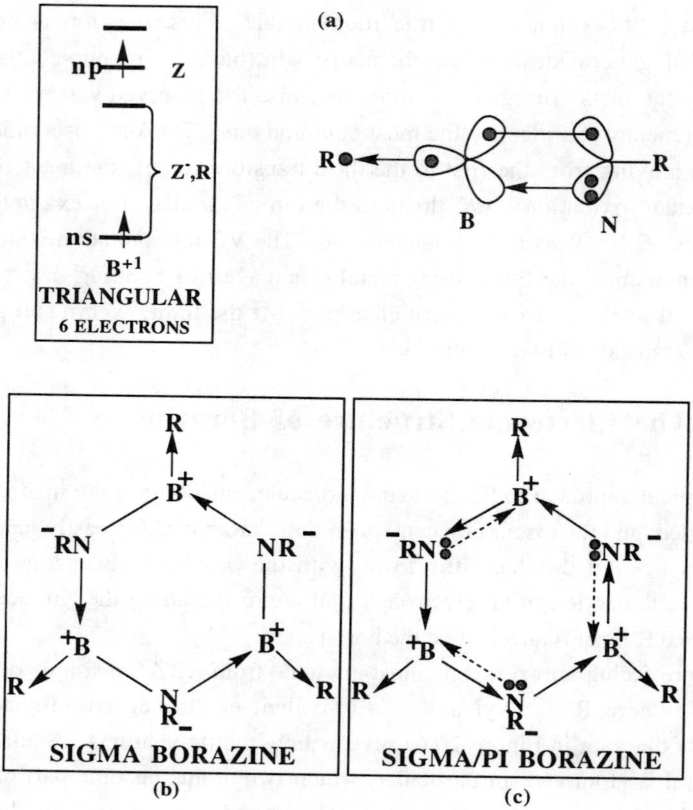

Figure 20.6 The electronic structure of borazine; the formula is the principal I descriptor.

References

1. K. Morokuma, *Acc. Chem. Res.* 10, 294 (1977).
2. M. Veith, *Angew. Chem. Int. Ed. Engl.* 26, 1 (1987).
3. J. Jacobs, H.S.P. Mueller, H. Willner, E. Jacob, and H. Buerger, *Inorg. Chem.* 31, 5357 (1992).

Chapter 21

The Relay I Bond as the Foundation of Organometallic Bonding

21.1 Allyl Resonance and the Primitive Relay Bond

Organic chemists learn resonance theory from the prototypical allyl ions. Two equivalent resonance structures imply strong resonance stabilization. But, what is the physical meaning of "resonance stabilization"? This question cannot be answered without considering explicitly the role of interelectronic repulsion. With interelectronic repulsion explicitly included, the π-allyl ion becomes the simplest example of a *primitive* I bond: Two codirectional arrows define the ensemble of four interacting VB structures, which, in turn, define one resonating I bond. The defining CI is shown in Figure 21.1a, where L represents the outer

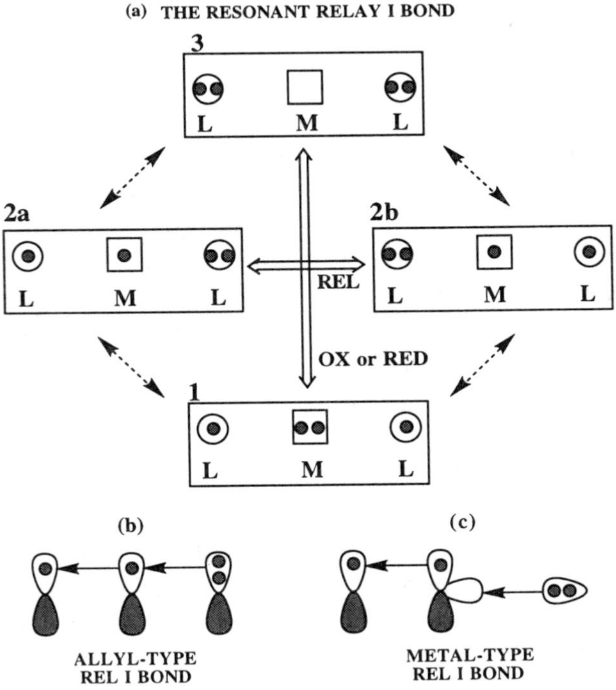

Figure 21.1 (a) 4 × 4 CI defining a REL bond. (b) Arrow representation of the REL bond. (c) REL bonds span either three AOs or four AOs, with the two pairs differing in symmetry.

and M the interior carbon of the allyl fragment. Alternatively, M can be a metal and L a ligand. When the allyl-type structures are dominant, we have a relay I bond (REL bond), which is symbolically represented as illustrated in Figure 21.1b. Because the 2-e CT hop defined by the two equivalent allyl structures can be executed either directly or by a sequence of two 1-e CT hops, the strength of the REL bond is dependent on all four structures, hence on atom electronegativity. This is why the strength of the REL bond of A_3^- depends on the electronegativity of A.

A REL bond has three hallmarks:

(a) Arrow codirectionality.

(b) Configurational degeneracy (i.e., the degeneracy of the two allyl-type configurations), which promotes strong CI.

(c) High T complement (i.e., the presence of a T bond in each allyl-type structure), which makes it ideal for semimetals and metals of relatively high electronegativity. In other words, the REL bond is suitable for systems falling in the midrange of the T-to-I continuum. One good example of a REL bond is that featured in the simple SNS^+ molecule.[1]

The REL bond of the allyl anion has nothing to do with the conventional concepts of monodeterminantal MO theory. Two arrows define one I bond, and each can be appended to orbitals of the same or different spatial symmetry. Thus, the REL bond shown in Figure 21.1c is simply just one more variant of the bonding of the π-allyl anion which, however, cannot be described by one-electron theory. There can be three types of REL bond depending on the overlap spaces spanned by the two arrows: σ/σ, σ/π, and π/π. Indeed, the allyl anion has the worst REL bond, namely, a π/π REL bond, which owes its existence only to the constraining action of the σ system. The REL bond ought to be differentiated from the "allyl resonance" of the organic chemist: Instead of two covalent VB structures, there must be four making appreciable contribution. In other words, the REL bond is actually a resonating I bond with predominant REL bond character. Hence, a semimetal or metal (by virtue of its tolerance of ionization and reduction) is much more suitable than carbon for anchoring a REL bond. The operational consequences are both simple and counterintuitive: Enhancing the binding ability of any one of the three centers must enhance the apparent bonding of all centers because one REL bond connects three atoms! We will find proof of this concept in three apparently unrelated research areas: the mechanism of vitamin B_{12} action (Chapter 25), the trans influence in transition metal complexes (Chapter 31), and the reactivity of transition metal complex ligands (Chapter 33).

A REL bond can be formed when one metal electron pair combines with two radical ligands and electron transfer from metal to ligand is favorable. The original metal pair can occupy either a single AO (lp for "lone pair") or a pair of AOs (rr for "radical/radical). Whatever the case, we end up with two REL bonds differing only in the T complement. The two species are analogous when viewed from our perspective, only to become "hypervalent" and "electron precise" when viewed from the lens of conventional theory:

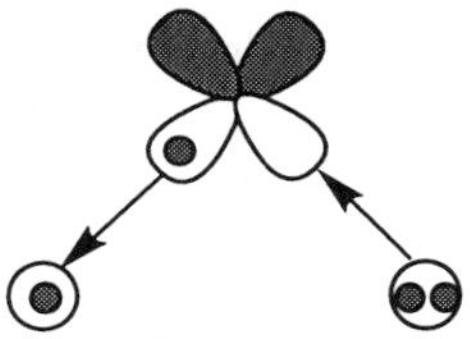

ONE REL I BOND

ONE REL I BOND

Ab initio calculations[2a] have confirmed the REL character of the Schrock metal carbene.[2b] The presence of bulky substituents, which can coordinate with holes or pairs, suggest that even silicon–silicon double bonds may actually be REL bonds.[3]

Bottom line: Zwitterionic resonance (such as that manifested in the ab initio VB computation of benzene by Norbeck and Gallup, mentioned in Chapter 16) and ionic resonance (such as that manifested in semiempirical VB computations of oxocarbons by Herndon)[4] are now interpreted to signify the presence of I bonds. To discover the I bonds, one must include ionic structures (indeed, everything) in the VB wavefunction. Nonorthogonal VB theory was initially used as an appendage of spectroscopy, and the complexity of the matrix elements discouraged the practitioners from carrying out VB computations over a complete set of resonance structures. As a result, insofar as concepts are concerned, VB theory actually means "covalent VB theory" because there has never been an opportunity to even ponder the physical significance of the ionic structures. In this work, we discovered the chemical meaning of these structures not by carrying out ab initio VB calculations but by following first principles and by "listening" to the experimental data.

21.2 REL-Bond Activation

Even the σ/σ variant of the REL bond of the π system of the allyl ion is, by itself, far from perfect; that is, it is primitive. The reason is that the REL bond is unactivated. In the case of the π/π REL bond of the allyl anion, the surrounding σ pairs cannot get out of the way of the delocalizing π electrons. Thus, this type of bonding will be unsuitable for electropositive metals, which are intrinsically incapable of supporting strong kinetic energy reduction via overlap. Molecules made up of electropositive metals will have to rely on either activated I bonds or E-bonding. I-bond activation is effected by overlap dispersion, which turns out to be the breakthrough concept that differentiates not only T- and I-bonding but also nonactivated and activated I-bonding. This set of distinctions becomes the key for decoding chemical bonding across the periodic table.

As an example, consider the case of the linear ClHgCl molecule. We transfer one electron from Hg to Cl and we generate the Cl^-Hg^+—Cl configuration. CT delocalization generates a primitive REL bond. Furthermore, since Cl^- is an activated nucleophile, the REL bond is half-activated. The alternative is to transfer two electrons and generate an inferior (contradirectional) RED I bond, which, in addition, is nonactivated. We opt for the first choice and we say that the electronic structure of ClHgCl is best represented by following the I formula, featuring a REL bond:

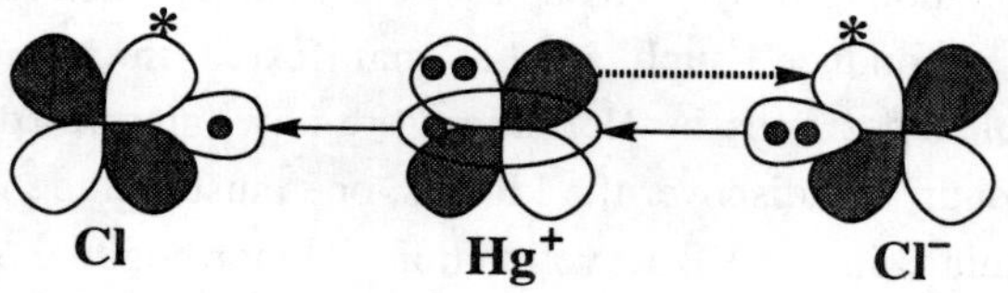

As mentioned before, the REL bond does not have to span three overlapping AOs; that is, the two arrows do not have to connect orbitals of the same spatial symmetry. The same result can be accomplished if one arrow spans two AOs of one symmetry and the second arrow two AOs of another. The simplest example is the Fischer carbene, shown here.[5]

THE REL I BOND

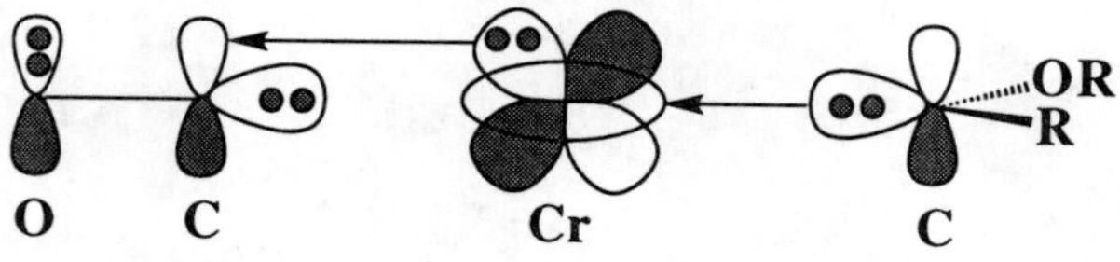

The I formula explains why the Fischer carbene actually has carbocationic character.

21.3 The Difference Between EHMO and VB Theory

The hypothetical beryllocene sandwich $BeCp_2$, with assumed D_{5h} geometry, is a convenient illustrator of the difference between EHMO-type arguments and polydeterminantal concepts. Specifically, the standard EHMO analysis of the problem goes as follows:

We view the molecule as "Cp plus Be plus Cp."

We feed electrons in the lowest energy fragment orbitals. This generates the $Cp^{-}Be^{2+}Cp^{-}$ configuration.

Orbitals of the same symmetry are allowed to interact to produce the final MOs of the system.

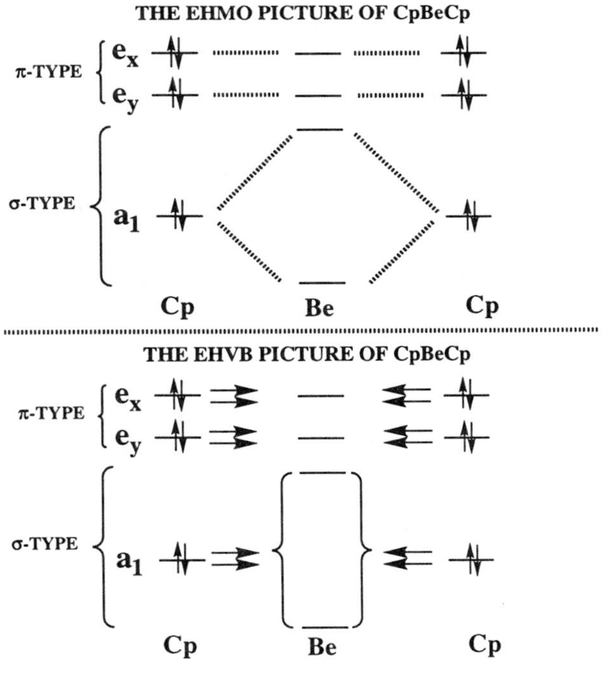

Figure 21.2 EHMO orbital interaction diagram and EHVB (= EHMO) configuration interaction diagram of symmetrical beryllocene.

The essence is conveyed by the EHMO interaction diagram in Figure 21.2, where the hatched lines indicate the symmetry-allowed interactions. Note the plethora of hatched lines and the impression that all interactions are simply treatable by overlap and energy gap considerations.

We can say exactly the same thing in a different language if we replace EHMO by EHVB theory. Starting from the same reference configuration, electrons are allowed to delocalize as indicated by the arrows affixed on each pair (Figure 21.2). The indicated CT delocalization respects Pauli's principle, but a 2-e hop can involve two electrons occupying either the same or two different orbitals. In other words, electrons do not feel each other. There is a total 12 arrows. The situation improves when we go to SCFMO theory, which incorporates electron–electron repulsion but not Coulomb correlation. The

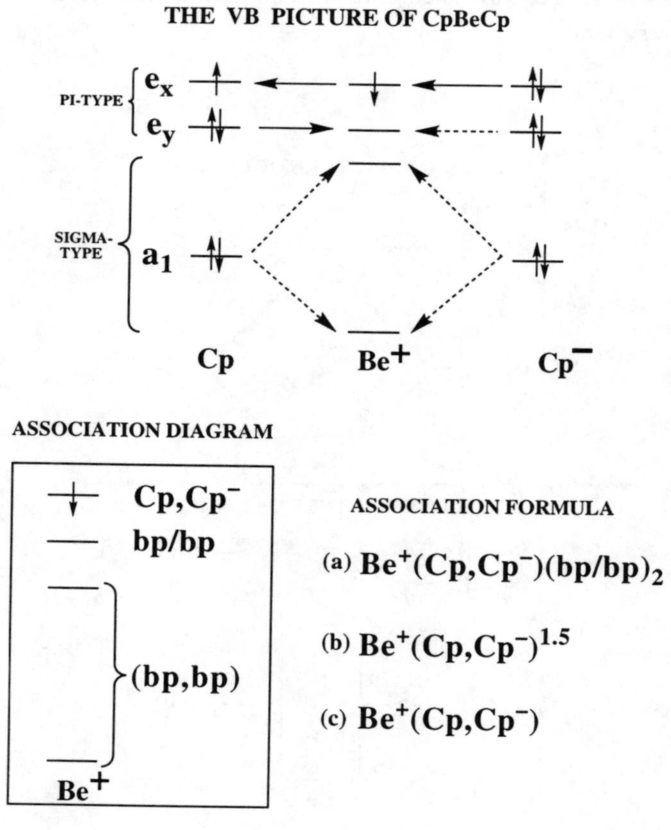

Figure 21.3 VB description, A diagram, and A formulas of beryllocene: bp, bond pair

major difference is that the reference configuration may change to the less ionic $Cp^{-}Be^{+}Cp$. Beyond this, the management of the electrons remains the same.

A radically different picture emerges when we implement "perfect" VB theory, as shown in Figure 21.3. As is common practice, only one of two equivalent VB descriptors is displayed. We now see only three codirectional arrows plus five more dashed arrows that, for all practical purposes, can be neglected. Behind the fundamental difference between the EHVB and VB versions of symmetrical beryllocene lie the VB concepts themselves:

Contact repulsion forbids the appearance of two arrows in the same interfragmental overlap space. This is the basis of the concept of the I bond. As a result, only one arrow exits a pair.

Overlap dispersion dictates codirectional arrow trains.

The fundamental difference between the EHVB (= EHMO) and VB scenarios can be expressed very simply: We have gone from a twelve- to a three-arrow picture! We simplified by complexifying. Is there a physical basis for this? Coulomb correlation means effective restriction of the space available for electron delocalization. In turn, this means that there are fewer interactions that matter after all counterproductive interactions have been eliminated.

We have arrived at a simple conclusion: We say that CpBeCp has 1.5 I bonds connecting the three fragments. According to our formulation, one Cp anion starts a three-arrow codirectional train that ends at Be after it has bound all three fragments. As always, there are two equivalent arrow trains, one right- and the other left-handed, but we show only one. The essence of the bonding is that a central atom dot couples with a ligand pair and a ligand dot to produce an activated REL bond. We can immediately predict that slippage of the Cp anion will replace a π by a σ solid arrow, and this will create a stronger REL bond. This is why CpBeCp is not a symmetrical complex, but, rather, a σ/π sandwich. Why is it not a σ/σ complex? Because this would preserve two codirectional arrows but largely destroy the third π-type arrow of the train.

We can now dispense with a variety of problems of metal chemistry in limited space. For example, linear FBeF, C_{5v} CpBeR (R = Me) and D_{5h} CpBeCp are three molecules in which the three fragments are connected by one activated REL bond as shown in Figure 21.4. Cp is effectively equivalent to a halogen constrained to bind as a π radical. The three molecules differ in the nature of the REL bond: σ/σ, σ/π, and π/π, respectively. Symmetrical beryllocene mutates to a geometry that transforms the π/π into a superior σ/π

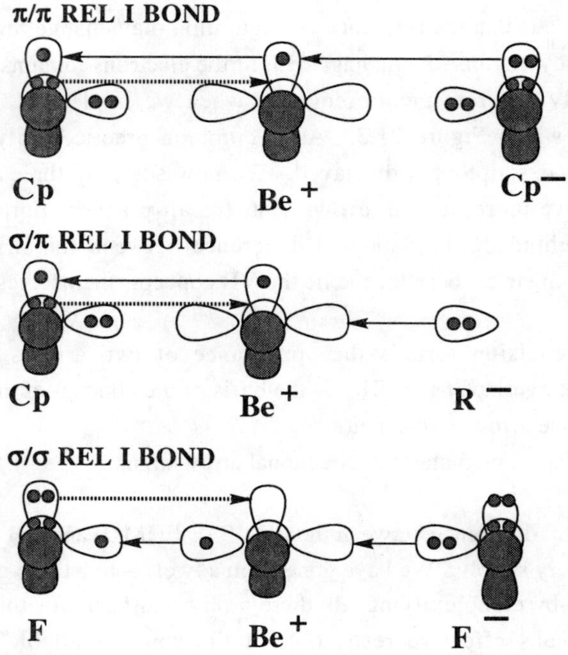

Figure 21.4 Three different molecules, each bound by a different type of REL bond.

REL bond. This is the "slipped-Cp" geometry in which one Cp is π-pentahapto and the other σ-monohapto.[5]

21.4 The Association Diagram and the Association Formula

Let us now return to the symmetrical CpBeCp and consider ways of representing the bonding in a compact but illuminating way. There are two choices: the association diagram and the association formula.

The *association diagram* (A-diagram) method illustrated in Figure 21.3 displays the elements of the central atom [holes, dots (odd electrons), and pairs] and lists next to them the ligand elements that combine with them to form REL bonds. Each Cp enters with one dot plus two bond pairs. The presence of one ligand radical (RD) plus one ligand Lewis base (LB) next to a metal dot implies a single REL bond. In our case, the ligand RD is the Cp radical and the ligand LB is the Cp anion. The REL bond is activated by an I-bond generated by the coupling of one metal hole with two ligand pairs (or one metal pair with two ligand holes). This counts as half because only one of its two arrows can be codirectional, with the arrows symbolizing the REL bond, as suggested by the slash separating the two ligand elements that form the activating I bond.

The essence of the A-diagram can be expressed by the *A formula*, which lists in order the pairs of ligands that become coupled by the central atom through a REL bond. Three different varieties are shown in Figure 21.3: In (a), all I bonds are shown explicitly. In (b), we show only the REL bond, using a superscript to specify the total I-bond character by following the usual convention that two arrows equal one I bond. Since a primitive REL bond is defined by two arrows and a fully activated REL bond by four arrows (always codirectional), the total I-bond character of a REL bond can be 1.0, 1.5, or 2.0. In (c), the third A formula, we simply indicate the REL bond that glues the fragments together without specifying the activator I bond or the I-bond character.

We have developed a weapon: We can visualize organometallic complexes as the pairwise association of ligands by the central metal. Let us see what we can get out of this by using the simplest possible A formulas.

21.5 The Conservation of REL-Bonding

The simplest A formula of CpBeMe is Be^{+}(Cp,Me^{-}). Note that the Me plays the role of the LB because the Cp is an I-permissive electrophilic radical that combines with an I-permissive nucleophilic radical, namely, Be^{+}. One arrow must go in the singly occupied MO of Cp if another arrow is to exit a doubly occupied MO and enter the Be2p hole. Now, examination of Zr complexes reveals that these are "twice CpBeMe" complexes. In other words, the prototypical $ZrCp_2R_2$ has the A formula Zr^{2+}(Cp,Me^{-})(Cp,Me^{-}). We can now attempt to understand how a complex of this type will react.

The reaction sequence that follows is the predicted mechanism of the transformation of the initial $ZrCp_2H_2CO$ complex to the final $ZrCp_2R(OCH_3)$.[7] The mechanism respects only one condition: The LB component of one REL I-bond must always be maintained. In this way, the integrity of both REL I bonds is also maintained. Thus, the reaction is a sequence of transformations that create new LB components for one REL I bond. These components change from hydride to dihydrogen to hydroxycarbene to hydroxymethyl anion to methoxyanion!

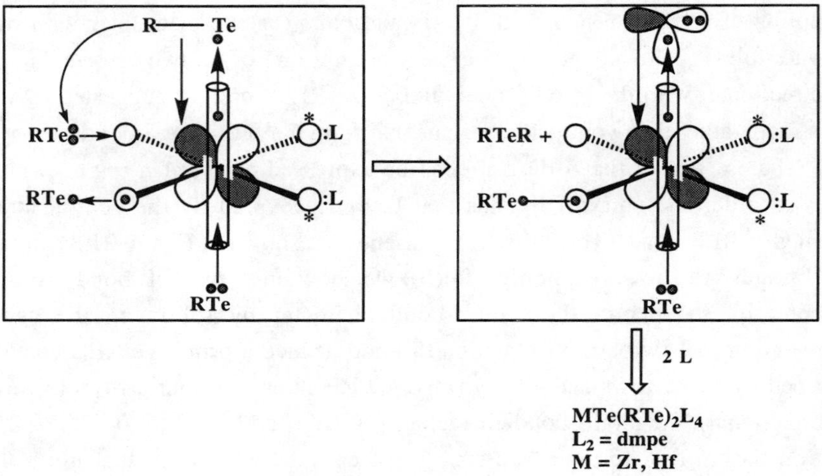

Figure 21.5 The reaction of two REL bonds produces one stronger REL bond. The central metal is Zr or Hf.

$$Zr^{2+}(Cp^*,H^-)(Cp^*,H^-)CO \xrightarrow{H_2} Zr^{2+}(Cp^*,H^-)(Cp^*,H_2)CHO^- \longrightarrow$$

$$Zr^{2+}(Cp^*,H^-)(Cp^*,H^-)CHOH \longrightarrow Zr^{2+}(Cp^*,H^-)(Cp^*,CHOH)H^- \longrightarrow$$

$$Zr^{2+}(Cp^*,H^-)(Cp^*,CH_2OH^-) \longrightarrow Zr^{2+}(Cp^*,H^-)(Cp^*,CH_3O^-)$$

A corollary of the principle just enunciated is that every perturbation will tend to occur, including bond heterolysis, which enhances the strength of a preexisting REL bond. The simplest example is the proton dissociation of a "heavy" halomethane, as discussed later on. However, a still more interesting case is that depicted in Figure 21.5. It can be seen that departure of a formal alkyl cation from one RTe group liberates an electron pair, which enhances the strength of the axial REL bond. The alkyl cation combines with the Lewis base component of the diequatorial REL bond to produce RTeR.[8]

References

1. S. Parsons and J. Passmore, *Acc. Chem. Res.* 27, 101 (1994).

2. (a) T.R. Cundari and M.S. Gordon, *J. Am. Chem. Soc.* 113, 5231 (1991). (b) R.R. Schrock,*Acc. Chem. Res.* 12, 98 (1979).
3. (a) R. West, *Angew. Chem. Int. Ed. Engl.* 26, 1201 (1987). (b) G. Raabe and J. Michl, *Chem. Rev.* 85, 419 (1985).
4. W.C. Herndon, *Theochem* 12, 219 (1983).
5. K.H. Doetz, Ed.,*Transition Metal Carbene Complexes*, Verlag Chemie, Weinheim, 1983.
6. (a) A. Almenningen, A. Haaland, and J. Lusztyk, *J. Organomet. Chem.* 170, 271 (1979). (b) J.S. Pratten, M.K. Cooper, M.J. Aroney, and S.W. Filipczyk, *J. Chem. Soc. Dalton Trans.* 1761 (1985), and references therein.
7. (a) P.T. Wolczanski, R.S. Trelkel, and J.E. Bercaw, *J. Am. Chem. Soc.* 101, 218 (1979). (b) P.T. Wolczanski and J.E. Bercaw, *Acc. Chem. Res.* 13, 121 (1980).
8. V. Christou and J. Arnold, *J. Am. Chem. Soc.* 114, 6240 (1992).

Chapter 22

The Gordian Knot of Chemistry: What Is a Base and What Is a Nucleophile?

22.1 I- Versus E-Selective Transition States

The hypothesis of this work is that there are three mechanisms of chemical bonding: T, I, and E. The transition state of a chemical reaction is an obligatory association complex that is bound by either the E or the I mechanism. In effect, E-bonding can be viewed as a limiting form of I-bonding in which superior arrow codirectionality has given way to inferior arrow contradirectionality by electron transfer from the more electropositive to the more electronegative atom, with minimization of promotional energy. For example, CaF_2 minimizes the one-electron energy by adopting the $F^-Ca^{2+}F^-$ ionic configuration from which only contradirectional CT delocalization is possible. Impairment of E- or I-bonding (as in "forbidden" pericyclic reactions) forces the transition state to adopt the third mechanism of chemical bonding, namely, T-bonding. Since, however, T-bonding is incompatible with association, it leads to the formation of a diradical or dipolar intermediate.

The hallmark of an I-selective transition state is a codirectional arrow train of maximum length. By contrast, the hallmark of an E-selective transition state is the interaction of two formal ions, each of which is associated with the maximum number of polarizable bonds via induction (indicated by hatched arrows). The best opportunity for studying chemical selectivity is afforded by reactions having T-bound reactants and products because these permit one to compare the consequences of T-bonding in the valleys and E- or I-bonding in the mountain passes. T-bound reactants *and* products are present in organic reactions of neutral molecules (e.g., in pericyclic reactions).

As a first illustration of the disjunction between I- and E-bonding, consider the transition state of the addition in the case of Z, which bears an electron pair, to a C=A double bond, where A is some atom of variable electronegativity. There exist two choices, as shown in Figure 22.1, which depend on the nature of A and Z.

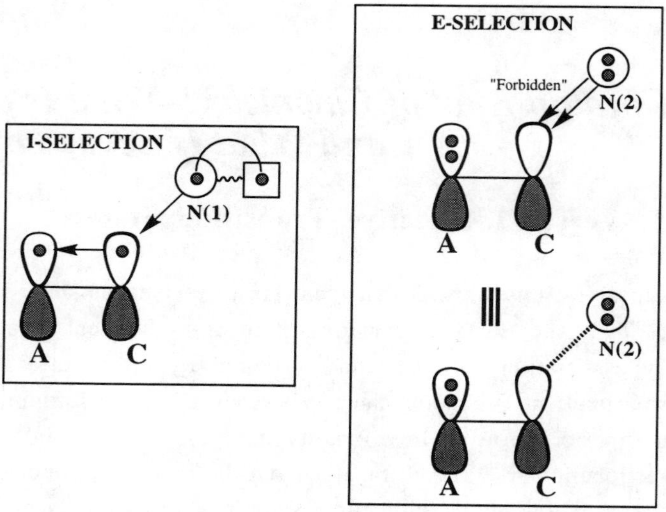

Figure 22.1 The I and E mechanisms of nucleophilic addition to a C—A π bond. A is more electronegative than C. The I mode involves CT delocalization involving codirectional arrows. The E mode involves formation of a multipolar array of holes and pairs.

(a) If Z (and A) is I-activated, the transition state is an I-bound species with a codirectional arrow train starting at Z and terminating at the remote end of the π-conjugated system. Z is represented by N(1) to indicate that it acts effectively as an *odd electron*, and the formula of the transition state is an I formula. The reaction is said to be under I selection. The hallmark of an I complex is arrow codirectionality.

(b) If Z (and A) is I-nonactivated, the transition state is an E-bound species in which the Z pair experiences the electric field of a deshielded carbon nucleus. Z is represented by N(2) to indicate that it acts effectively as an *electron pair*, and the formula of the transition state is an E formula. The reaction is said to be under E selection. The hatched line exiting the pair implies one E-bond.

We are delineating two concepts for which the reader is most unprepared. The first is the notion that either actual or obligatory (e.g., transition states) association complexes have two choices, namely, I- and E-bonding. The second is that unless the components are properly assembled, the complex becomes frustrated because it fails to meet the requirements of either I- or E-bonding. While the simplest illustration of these ideas is the nucleophilic addition to carbonyl discussed earlier, a better connection with the experimental reality is made when we focus on bond switch reactions such as metal oxidative addition and cycloadditions. The I/E spectrum and the corresponding I and E formulas tell us that there exist *two* different ways of creating a facile Diels–Alder reaction: by combining either three homonuclear or three heteronuclear bonds. The first choice produces an I- while the second choice produces an E-type transition state, as indicated in the corresponding formulas:

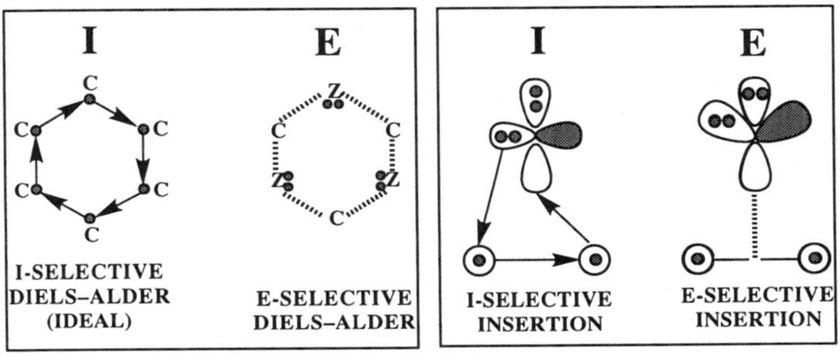

The experimental literature suggests that E-bound six-electron transition states can attain very low energy:

(a) Head-to-tail trimerization of formaldehyde has a much lower barrier[1a] than the trimerization of three ethylenes[1b] despite its lower exothermicity.

(b) Danishefski and co-workers found that a "nucleophilic" 1,3-dioxydiene can combine with an aldehyde in the presence of zinc chloride or boron trifluoride catalyst and under mild conditions to form a substituted 2,3-dihydro-*g*-pyrone.[2a] The corresponding aza-4+2 cycloaddition is also facile.[2b] In this and the preceding case, the carbon substituents are ideally placed to stabilize one of the two hexapolar, hole–pair alternant VB structures. In other words, the substitution pattern is ideal for alternant E-bonding.

(c) Evans and Golob found that charged π donors (e.g., O^-), cause pronounced acceleration of the Cope rearrangement.[3] Overman and his co-workers have made related observations.[4] These are diagnostic of a switch in the mechanism of bonding from I to E.

We return to the case of the formaldehyde trimerization to point out how conventional concepts can lead us astray. Specifically, one could argue that the low energy transition state of the reaction reflects strong HOMO-LUMO interaction of the three fragments. This would appear to be a reasonable explanation to anyone unaware of the hallmark of the "ordinary" (I-selective)

Diels–Alder reaction and benzene itself, namely, the exclusion of oxygen from the aromatic domain. This fact says that one can either shoot for error-free arrow codirectionality (I selection) or go for either alternant or nonalternant multipolar bonding (E selection). The first strategy has something to do with HOMOs and LUMOs and the best option is the establishment of an all-carbon array aided by substituents (e.g., 1,3-butadiene plus TCNE). The second strategy has to do with static bonding, which lies outside EHMO theory.

The nucleophilic addition to a π bond discussed earlier is an analogue of nucleophilic addition to a σ bond commonly known as nucleophilic substitution. The gas phase S_N2 reaction[5] affords the best opportunity for studying chemical selectivity. The first step is the classification of nucleophiles (and electrophiles) according to their intrinsic ability to act as I- or E-selective species.

22.2 Unimodal, Bimodal, and Ambident Nucleophiles and Electrophiles

At first it seems unnecessary to use both "base" and "nucleophile" because each entity carries an electron pair. The terms are not redundant, however, because a nucleophile is an I-activated while a base is an I-nonactivated reagent. The former is I selective and the latter E selective. Clearly, to do justice to the concept, we need a term other than "base" or "nucleophile," but we will leave this to the linguists. In the remainder of the work, we will simply differentiate between I-selective and E-selective nucleophiles, keeping in mind that the latter correspond to "bases" in conventional nomenclature. The same policy will be followed with respect to the acid/electrophile duality.

Once we have crossed the threshold of polydeterminantal VB theory, we can no longer live blissfully in ignorance. Reagents once thought to be analogous are now known to be very different:

A *unimodal E nucleophile* (e.g., F^-) has small size, lone pairs, and no observer holes. Since it cannot support I-bonding, it must fall back on the E mechanism for which it is mostly suitable because of its small size.

A *unimodal I nucleophile* (e.g., Cl^-) has large size, lone pairs, and also observer holes. However, this is not enough to qualify it. In addition, the fragment with which it is matched must have observer pairs. In other words, there are only I-permissive nucleophiles. For them to become I nucleophiles, the partner must be right.

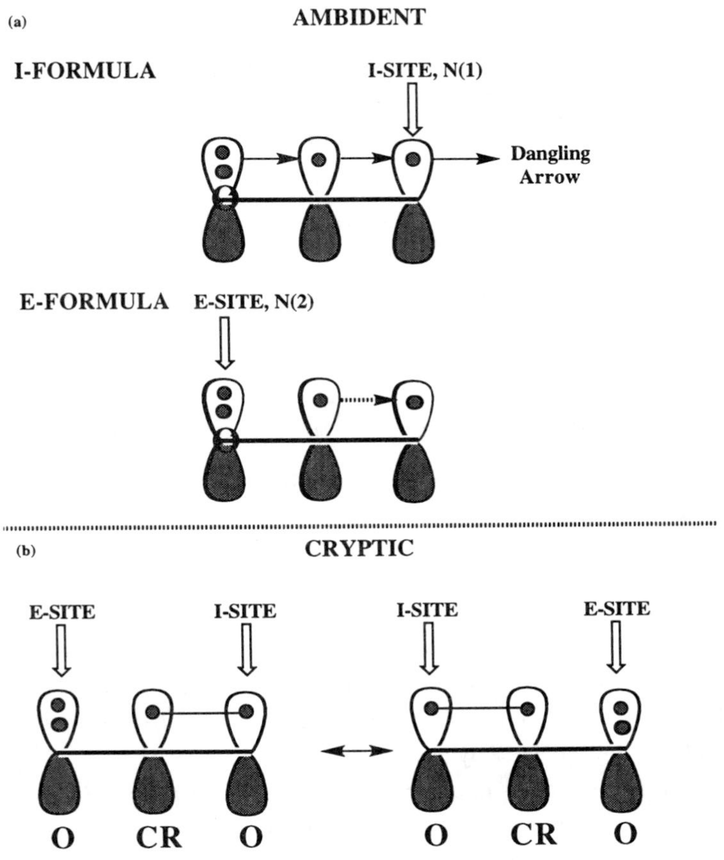

Figure 22.2 I and E sites of (a) ambident enolate anion and (b) cryptic carboxylate anion.

A *bimodal nucleophile* (e.g., CN^-) has small size, lone pairs, and observer holes, hence can act as either E or I nucleophile.

One and the same nucleophile having two different reactive sites, one E and one I, is called an *ambident nucleophile*. In the case of π-conjugated ambident nucleophiles, the I site is the terminus of a codirectional arrow train initiated by the nucleophilic pair. On the other hand, the E site is the one in which the nucleophilic pair is surrounded by the maximum number of polarized π bonds. Thus, we have a principle of maximum bond accumulation exactly as in Alder's rule. We will see that this is no accident. These considerations are illustrated in Figure 22.2a, using the enolate anion as an example. To determine the I site, we write down the lowest energy VB configuration (pair on oxygen) and append

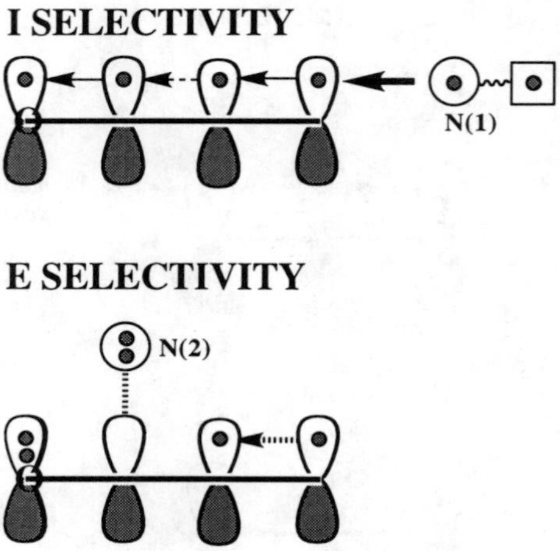

Figure 22.3 I and E selectivity of acrolein.

arrows according to the association rule and respecting arrow codirectionality. The I site is the terminus of the longest codirectional arrow train, and it is identified by a dangling arrow. To determine the E site, we simply localize the nucleophilic pair on the more electronegative atom (i.e., oxygen). This pair is now a part of a multipolar array having hole–pair alternation.

A third variety of a nucleophile, called a *cryptic nucleophile,* has indistinguishable I and E sites. This means that molecules of this type (e.g., the carboxylate anion shown in Figure 22.2b) can readily adjust to the requirements placed by a partner electrophile. However, its very nature denies us the opportunity of differentiating between I and E selectivity.

We can go through the same motions with an electrophile. As an example, we show the I and E sites of the ambident electrophile acrolein in Figure 22.3.

One implicit assumption of Figures 22.2 and 22.3 becomes standard practice. Specifically, we have seen that organic transition states cannot support the association of more than two electron pairs; moreover, they have resonating A and S domains, with the latter having $n - 4$ electrons, where n is the total number of active electrons. As we proceed to examine transition states containing atoms of all different sorts, the four-electron rule may be relaxed. Even when this is not the case, however, I selectivity is always associated with the longest codirectional arrow train that one can construct. To see why this is true, recall that such an arrow train derives the greatest benefit from A- and S-domain resonance.

22.3 The Rule of Complementation

Nucleophiles react with electrophiles according to the following selection rule: *E combine with E sites and I combine with I sites.* In other words, I + I plus E + E is superior to twice E + I. This is exactly analogous to "T + T plus I + I is better than twice T + I." In other words, atoms predisposed toward T-bonding (nonmetals) prefer to combine with each other in the ground state. Atoms predisposed toward I-bonding (metals) also prefer to combine with each other in the ground state. Similarly, atoms predisposed toward CT delocalization combine with each other while atoms favoring static interaction also combine with each other in the transition state. But this is nothing else but one manifestation of the hard/soft acid/base concept! As an illustration, consider the two superior modes of combination of the enolate anion and acrolein:

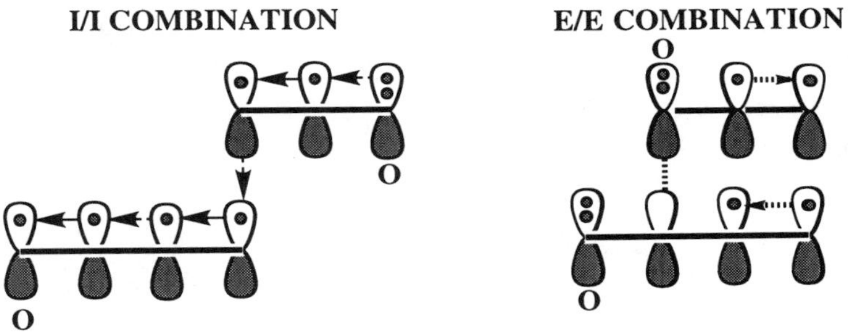

The key result of this work is that there is nothing like a singular "hard/soft" comparison. Rather, there exist three different types of "hard/soft" comparison, namely, T/I, T/E, and I/E, because there are three mechanisms of chemical bonding, T, I and E. Most of the examples of "hard/soft" selectivity originally provided by Pearson and augmented by Klopman and others are E/I comparisons. In Chapter 19, we gave T/I comparisons (e.g., the $Li_2 + H_2$ to LiH + LiH metathesis). The problem as we now define it is immense and the possibilities unlimited, given the data that have accumulated.

22.4 The Disguised Selectivity Of The Gas-Phase S_N2 Reaction

We now begin the application of the theory to diverse reactions, starting with a clear-cut example of the degenerate second-order nucleophilic substitution (S_N2) reaction, which encapsulates the whole problem of reaction selectivity. What is the reaction efficiency when X = F versus X = Cl?

$$X^- + CA_3{-}X \longrightarrow X{-}CA_3 + X^-$$

The answer is: It all depends on X and A!

Fluoride is an ideal E nucleophile because of its compact size and the presence of observer pairs. By contrast, chloride is very unsuitable for E-bonding. As a result, both the transition state and the enthalpy of the first step of the gas phase S_N2 shown in Figure 22.4 will be more favorable for X = F. In other words, the ΔU energy well will be larger (more negative) when X = F. On the other hand, chloride is an I-permissive nucleophile. Because it has observer holes, it can engage in I-bonding, provided it is matched with a partner having observer pairs. This depends on the nature of A. Specifically, when Y = H, the H—C bond is polarized toward C and the H—C pair activates X = Cl. As a result, the transition state of the second step will be more favorable when X = Cl. In other words, the ΔQ barrier will be smaller when X = Cl. The final result (when A = H) is shown in Figure 22.5a. We see that the reaction efficiencies of the thermoneutral fluoride and chloride reactions can end up being the same because of the antithetical effect of the electronic structure of the nucleophile on the enthalpy of the first step and the activation enthalpy of the

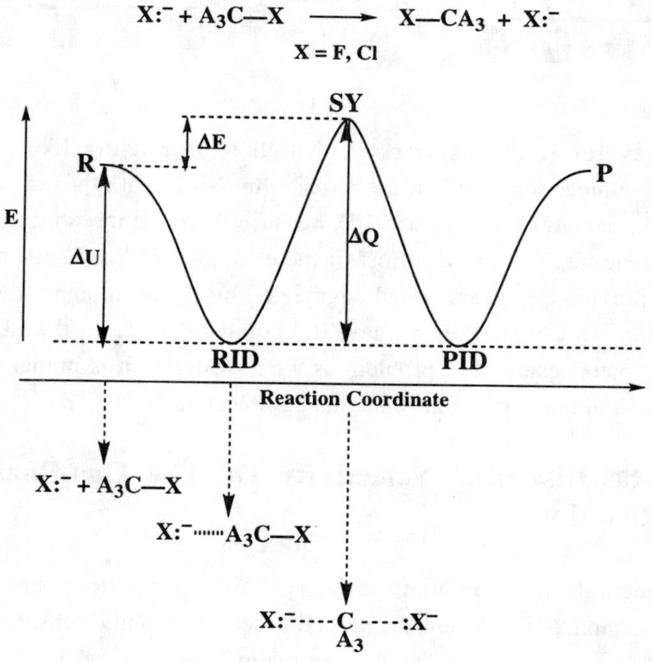

Figure 22.4 The energy profile of the gas phase S_N2 reaction.

second step. In a *relative* sense, the first step measures the ability of the nucleophile to engender E-bonding, and fluoride wins. On the other hand, the second step measures the ability of the nucleophile to generate I-bonding, and now chloride wins! As a result, the efficiencies of the two gas phase S_N2 reactions can be comparable for diametrically opposite reasons, as discovered by de Puy and co-workers.[6a]

Let us now change A from σ donor to σ acceptor. Now, the chloride will become I-deactivated, and it will be forced to the intrinsically unfavorable E-bonding. At the same time, the more positively charged carbon will respond more favorably to the electric field of the compact fluoride. Hence, we now expect a very different result: The fluoride should be far more efficient than the "frustrated" chloride reaction as depicted in Figure 22.5b.

The demonstration that the degenerate fluoride and chloride gas phase S_N2 reactions have comparable efficiencies can be simply explained by saying that two different reaction complexes have attained the same energy by fulfilling entirely different sets of conditions: The symmetrical $F(CH_3)F^-$ complex is E selective while the $Cl(CH_3)Cl^-$ is I selective, and the two have comparable stabilities for entirely different reasons. This underlines a general principle:

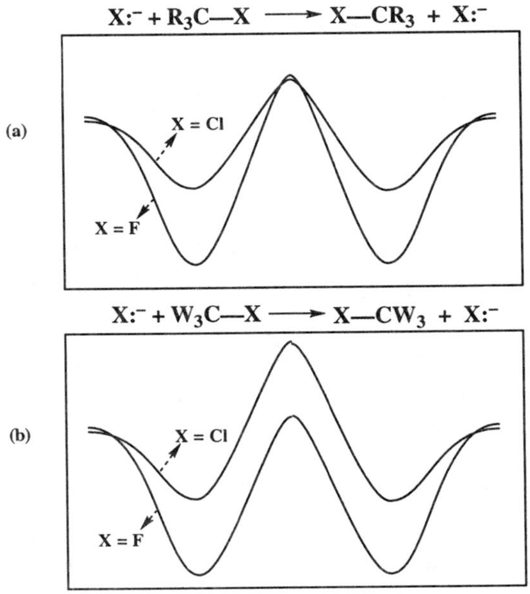

Figure 22.5 (a) Explanation of the equal efficiencies of the F^- and Cl^- reactions (R = electron releasing). (b) Predicted hypersurfaces when "observer" ligand W is σ withdrawing.

There exist two different ways of realizing a low energy transition state, namely, by employing I-bonding or by employing E-bonding.

Are there other cases in which I- and E-selective transition states end up having comparable energies? One possible example is provided by the ab initio calculations of Wong et al., who found that the transition states of cyanide addition at the E sites (C═O) and the I sites (C═C) of acrolein are nearly isoenergetic even though the product of the I attack is far more stable than the product of the E attack! Thus, in one shot, these calculations illustrate the difference between T-bonding (products), I-bonding (conjugate addition), and E-bonding (carbonyl addition).[6b]

We can extract an important lesson from this discussion. Assuming that all nucleophilic substitutions of the type $X:^-$ plus A—B to yield X—A plus $B:^-$ in which the attacking reagent is an electronegative, first row atom (e.g., F or O) go through an E-selective transition state of the type $X:^-A^+B:^-$ in which pair exchange repulsion is minimized, we have a selectivity principle: Given a choice, an electronegative nucleophile will attack preferentially the A—B bond with the more electronegative B. For example, fluoride will attack the C—F and not the C—H bond of methyl fluoride. That this is not a trivial matter is revealed when we refocus on radical rather than nucleophilic substitution. The transition state now has the form $XA^+B:^-$, and the same selectivity principles apply. These explain a mystery: Tert-butoxy radicals abstract hydrogen from $(RCH_2)_2NH$ exclusively from nitrogen even though abstraction from the alpha carbon would be more exothermic.[6c,d] This contrathermodynamic kinetic preference can now be understood: The transition state is E-bound while reactants and products are T-bound.

We can now answer the question posed in the title of this chapter: A base is an E-selective and a nucleophile is an I-selective reagent each bearing an electron pair ("pair reagent"). Whether a "pair reagent" will exhibit E or I selectivity depends on the "hole reagent" with which it reacts. Hence, the basicity/nucleophilicity dichotomy can be discussed only by reference to the acidity/electrophilicity dichotomy: Bases react preferentially with acids (E-bonding, hard/hard bonding) and nucleophiles react preferentially with electrophiles (activated I-bonding, soft/soft bonding).

22.5 The Electronic Basis Of Ligand Apicophilicity

The gas phase S_N2 reaction taught us an important lesson: The elementary step that converts an ion–dipole complex to its symmetrical form (the RID-to-PID conversion in Figure 22.4) is the simplest illustrator of I selectivity in a

Table 22.1 Relative Energies (kcal/mol) of Trigonal Bipyramids

Molecule	Apical Atoms	E_{rel} (kcal/mol)
PCl_2F_3	F,F	0.0
	Cl,Cl	11.5
BUT:		
PH_2ClF_2	F,Cl	0.0
	F,F	7.0

thermoneutral reaction. We now focus attention on the factors that stabilize the I-bound symmetrical transition state (SY in Figure 22.4). VB theory predicts a linkage of the action of the equatorial ligands and the binding of the apical ligands in any trigonal bipyramidal molecule that is isoelectronic to the organic S_N2 transition state. What are the facts?

Holmes and his group used ab initio calculations to address the question of the apicophilicity dependence on the equatorial group in pentacoordinate phosphorus.[7] The data of Table 22.1 are drawn from their work. To quote the authors:

> In both the phosphorus and silicon series, molecules constructed of highly electronegative atoms have fluorine more apicophilic than chlorine. As the collective electronegativity of the equatorial atoms is reduced by replacement with hydrogen atoms, a crossover occurs with chlorine becoming more apicophilic than fluorine.

Figure 22.6 shows how the apicophilicity of Cl is linked to the electronegativity of a variable equatorial atom.

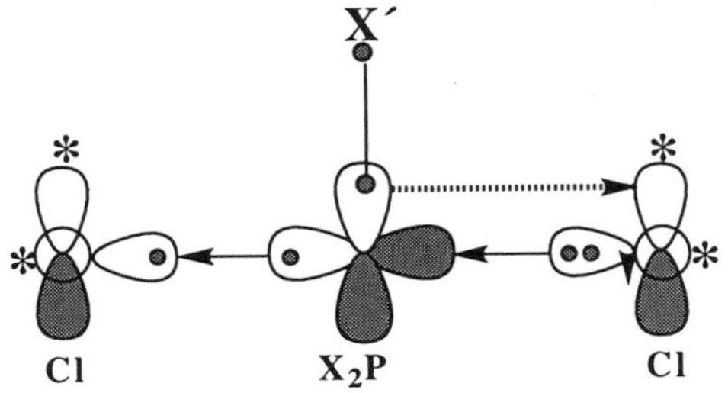

Figure 22.6 The apicophilicity of Cl depends on the σ-donor power of X´ in pentacoordinate phosphorus. When X´ is a σ-donor, Cl apicophilicity is high.

22.6 Nucleophilic Addition to Carbonyl as an S_N2 Reaction

The return to the world of chemical formulas leads us to reexamine old problems from a new angle. Since a carbon–carbon multiple bond can be formulated by using sp^3-hybridized carbons, nucleophilic addition to carbonyl is nothing other than a regioselective S_N2 reaction in which the leaving group is oxygen. When the carbonyl group bears a good leaving group, such as chloride, there exist two distinct S_N2-type transition states which are mere stereoisomers:

> One leads to direct substitution. This is the I-selective transition state.
> The other, the E-selective transition state, leads to initial addition.

The two plausible transition states are shown in Figure 22.7.

How can we force the reaction to follow the direct substitution route? Choose one equatorial group that is a σ-donor and a nucleophile and a leaving group having I affinity. Wilbur and Brauman found that the reaction of CN^- with 3,5-difluorobenzoyl chloride[8] in the gas phase involves direct substitution instead of the conventional addition/elimination mechanism. According to the T formula, the aryl group acts as a σ-donor. This is a consequence of inbound and outbound arrow alternation around the benzene nucleus. Furthermore, cyanide has strong I affinity, and the same is true of chloride. These properties suggest the following scenarios:

(a) Carbonyl chlorides and imino chlorides, as well as vinyl chlorides, may go by direct substitution into the gas phase. A recent ab initio study has found this to be the case for vinyl chlorides.[9]

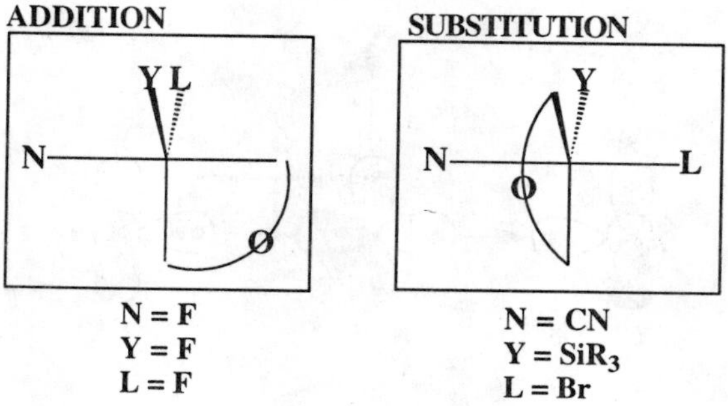

Figure 22.7 I- and E-selective attack of YLC═O. The first produces direct substitution and the latter direct addition.

(b) Solvation always favors the E-selective transition state. It will stabilize the addition transition state, and this will reproduce the conventional mechanism of solution chemistry.

(c) We can force the addition mechanism in the gas phase by starting with the Wilbur–Brauman system and making the following replacements: (1) cyanide by fluoride and (2) 3,5-difluorobenzoyl by fluorine.

Bottom line: There is no such thing as "understanding" *one* reaction, in gas or solution phase. There is only chemical selectivity. This is the choice among the T, I, and E mechanisms of bonding. The degenerate gas phase S_N2 reaction (nucleophile = X^-, leaving group = X) is one of the simplest systems that allows for a clear disjunction of E and I selectivity.[10] Depending on the precise nature of nucleophile/leaving group and carbon substituents, we have a continuum of hypersurfaces that range between the two limiting shapes shown in Figure 22.8.

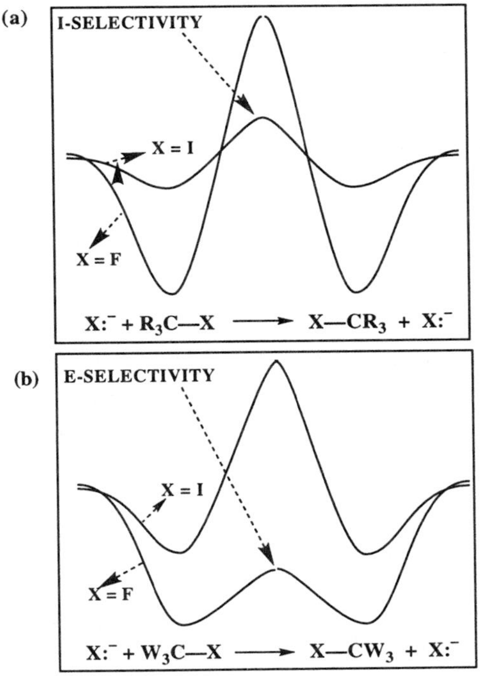

Figure 22.8 The two limiting hypersurfaces of the gas phase S_N2 reaction. Depending on the choice of the carbon substituents (R = σ donor, W = σ acceptor), either the F^-/F reaction (E selectivity) or the I^-/I reaction (I selectivity) can become superior.

The role of conceptual theory is to drag experimentalists and computationalists away from the beaten path. The classical S_N2 reaction involves a lone pair/bond pair switch. Equally fascinating but far less popular is the bond pair/bond pair, S_N2-type switch encountered in boron chemistry. A borane of the type RR´R´´B is "activated" by a nucleophile, N, to yield RR´R´´BN and one of the B—R bonds acts now as a nucleophilic bond pair to attack another bond pair in a reaction with multiple synthetic applications[11] and, still more important, many implications for transition metal chemistry, since an unsaturated metal can be viewed as a borane analogue. In short, the chemical applications of the concepts discussed here go far beyond the narrow (!) limits of the S_N2 reaction, but space limitations force us to end at this point.

22.7 Cationic and Anionic Bond Metatheses

The lessons of the gas phase S_N2 reaction apply equally well to cationic and anionic metatheses defined by the following equations.

Anionic metathesis (e.g., nucleophilic substitution, nucleophilic addition)

$$N:^{-}\cdots A\text{—}B \longrightarrow [I] \longrightarrow N\text{—}A\cdots :B^{-}$$

Cationic metathesis (e.g., electrophilic substitution, electrophilic addition)

$$E^{+}\cdots A\text{—}B \longrightarrow [I] \longrightarrow E\text{—}A\cdots B^{+}$$

In either case, the reactants and products can be regarded as valence isomers, and the intermediate complex denoted by [I] can be either a transition state or a global minimum. If we assume that the active AOs can define only a cyclic Hückel system, [I] is likely to be a (linear) transition state in the anionic case but a (cyclic) global minimum in the cationic case, assuming that spatial overlap considerations do not dictate otherwise. All this is true for the gas phase. In a polar solvent, preferential stabilization of reactants/products creates a double-minimum reaction profile, with solvated [I] now being a transition state.

Consider now the textbook case of acid-catalyzed nucleophilic addition of an alcohol to carbonyl to form a hemiacetal or hemiketal. The reaction of proton (the catalyst) plus carbonyl can be viewed as a cationic metathesis reaction in which one valence isomer is "proton electrostatically attached to carbonyl oxygen," $C{=}O\cdots H^{+}$, and the second valence isomer is "carbocation electrostatically attached to an O—H bond," C^{+}—O—H. The first species has a

COH angle of 180° and the second a COH angle of 120°. An intermediate species [I] with some intermediate angle can be either a global minimum or a transition state, neglecting, for the moment, the possibility that [I] is actually an inflection point.

Let us now forget whether we have a single- or a double-minimum situation in the gas phase and ask the question: Which of the two valence isomeric complexes is stabilized more by water solvent? The answer depends on what exactly we mean by "solvent." How many water molecules, and in what geometry? Let us take the minimalist approach of using only one water molecule to simulate the solvent. Furthermore, let us assume arbitrarily that the single water molecule is coordinated to the H atom in each of the two complexes. Computations indicate that the two forms lie close in energy with the carbenium ion form being somewhat more stable, as expected.[12] This result, which implies that the nature of the active intermediate in electrophilic or nucleophilic additions or substitutions depends on the detailed structure of the solvent, has profound implications. Assuming that protonation of carbonyl leads to C^+—O—H places the spotlight on "resonance." Assuming that protonation leads to C=O···H^+ shifts the spotlight on "overlap-independent intermolecular bonding." Depending on the answer, a different type of thinking and a different pedagogy are required.

Bottom line: The staple reactions of organic chemistry are actually cationic or anionic metatheses. This means that there exist choices of valence isomeric intermediates that have not been considered before. Which intermediate is active in solution depends on the nature of the solvent. Depending on the ultimate answer, a complete rethinking of mechanistic preconceptions may be needed. For example, is proton acting as an E-selective (C=O···H^+ active) or as a T-selective (C^+—O—H active) catalyst in solution? Our design of selective catalysis depends on the answer. High level ab initio computations of solvated cationic and anionic metatheses (necessitated by the likelihood that the valence isomers may be separated by a small energy difference) can help us define the catalysis strategies which, after all, is what matters when it comes down to "real chemistry."

22.8 Solution Reactivity as a Problem of Linkage Isomerism

Consider the case of two solvated ions, namely, an alkyl cation and a halide anion. How many distinct energy minima are there? If we treat the solvent (symbolized S) as a homogeneous cluster of solvent molecules, we recognize that a solvated ion is nothing but an ambident reactant capable of interacting with some partner via either the solvent or the ionic end:

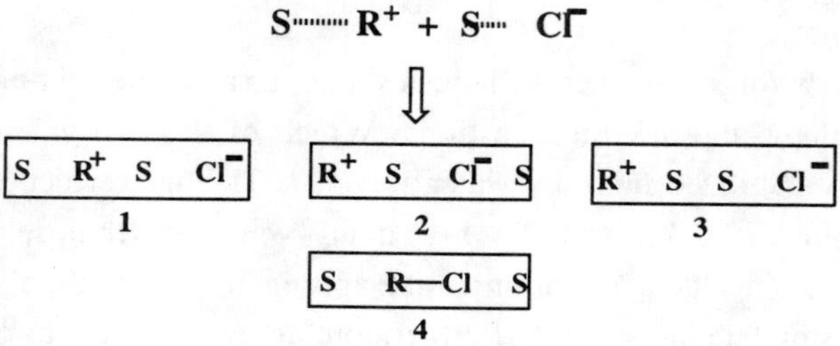

Species 1 and 2 are unsymmetrical solvent-separated "tight" ion pairs, while species 3 is a symmetrical solvent-separated "loose" ion pair. By contrast, species 4 is a solvated molecule resulting from the collapse of two counterions to either a T- or an I-bound molecule.

We can now look at the classical addition of HCl to an olefin (symbolized by E) as a process in which an ambident $S\cdots Cl^-$ combines with a proton or a carbocation in two distinct regiochemical ways: either by the S end via the E mechanism or by the Cl^- end via the T or I mechanism. The situation can be schematized as follows:

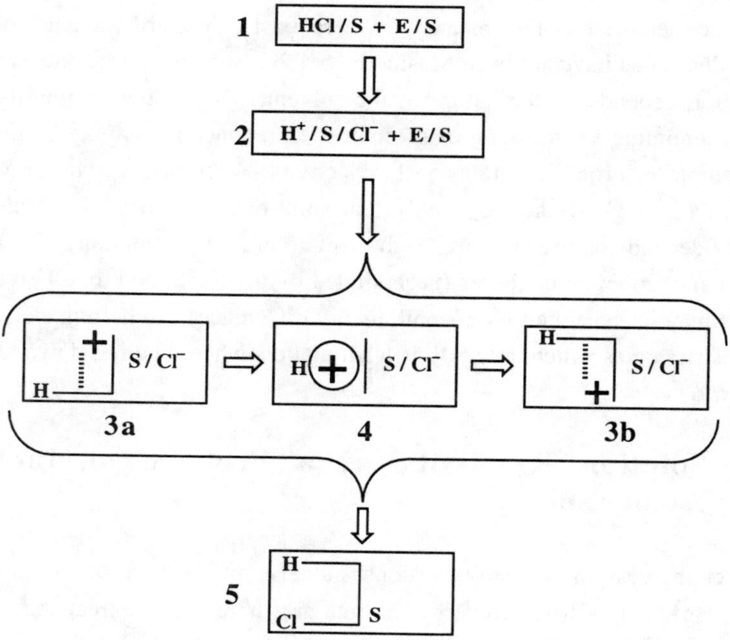

The key point is that 1 and 2 are *linkage isomers*, which are bound by different mechanisms: HCl/S involves T-bonding of H^+ and Cl^- but $H^+/S/Cl^-$ involves E-bonding of H^+ and S. The same is true of 3 or 4 and 5. Species 3a,b are "classical" cations and 4 a "nonclassical" cation. Whether 4 is a local minimum or a transition state for the interconversion of 3a and 3b depends on the detailed electronic structure of the olefin and the nature of S. If we strip away the solvent and the chloride anion, we expect species 3a, b or 4 to be the intermediates in the gas phase addition of proton to olefin. In the latter case, either 3 or 4 is the global minimum while, in the solution reaction, the low energy species are the reactant 2 (if S is a hydroxylic solvent) and the product 5, with 4 now being either a transition state or a high-lying intermediate.

An increase in stereochemical complexity occurs when the solvent, S, hosts a solute, Q, which can act as nucleophile, electrophile, or both in the interaction of RCl with S(Q). Given that now the solvent S(Q) itself is ambident since it can act on RCl with either the S or the Q "end," the number of possible distinct intermediates in the so-called nucleophilic substitution of RCl increases as follows:

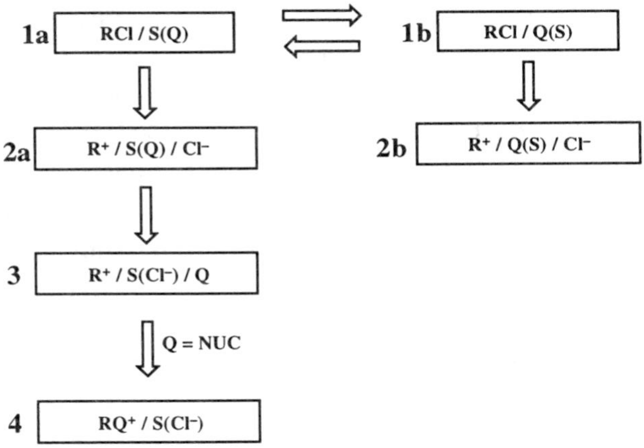

In summary, "solvent reorganization" can be discussed in the context of ambidency and the trichotomy of the mechanism of chemical bonding.[13]

References

1. (a) R.L. Burnet and R.P. Bell, *Trans. Faraday Soc.* 34, 420 (1938).
(b) A. Ioffe and J. Perkin, *J. Chem. Soc.* 2, 2101 (1992).

2. (a) S. Danishefski, J.F. Kerwin Jr., and S. Kobayashi, *J. Am. Chem. Soc.* 104, 358 (1982). (b) K. Ishihara, M. Miyata, K. Hattori, T. Tada, and H. Yamamoto, *J. Am. Chem. Soc.* 116, 10520 (1994).
3. D.A. Evans and A.M. Golob, *J. Am. Chem. Soc.* 97, 4765 (1975).
4. L. Overman, *Acc. Chem. Res.* 25, 352 (1992).
5. (a) J.I. Brauman, J.A. Dodd, and C.-C. Han, in *Nucleophilicity*, J.M. Harris and S.P. McManus, Eds., American Chemical Society, Washington, DC, 1987. (b) S.T. Graul, M.T. Bowers, D.M. Cyr, L.A. Posey, G.A. Bishea, C.-C. Han, M.A. Johnson, J.L. Wilbur, and J.I. Brauman, *J. Am. Chem. Soc.* 113, 9696, 9697, 9699 (1991).
6. (a) C.H. DePuy, S. Gronert, A. Mullin, and V. Bierbaum, *J. Am. Chem. Soc.* 112, 8650 (1990). (b) S.S. Wong, M.N. Paddon-Row, Y. Li, and K.N. Houk, *J. Am. Chem. Soc.* 112, 8679 (1990). (c) Y.Maeda and K.U. Ingold, *J. Am. Chem. Soc.* 102, 328 (1980). (d) For an interesting correlation of energies of activation of hydrogen abstractions with experimental observables in which the authors point out various mysteries of the field, see A.A. Zavitsas and C. Chatgilialoglou, *J. Am. Chem. Soc.* 117, 10645 (1995). One spectacular case is the much higher activation of the identity hydrogen abstraction, A plus H—A, when A = CH_3 compared to A = RO. In a relative sense, the transition state of the first reaction (A = CH_3) is I-bound while the transition state of the second reaction (A = RO) is E-bound, with the latter coming much closer to ideality because of the very nature of the constituent atoms (i.e., carbon is hardly a good promoter of I bonding when in combination with atoms of similar electronegativity, while oxygen is an excellent promoter of E bonding when in combination with more electropositive atoms).
7. J.A. Deiters, R.R. Holmes, and J.M. Holmes, *J. Am. Chem. Soc.* 110, 7672 (1988).
8. J.L. Wilbur and J.I. Brauman, *J. Am. Chem. Soc.* 116, 9216 (1994).
9. M.N. Glukhovtsev, A. Pross, and L. Radom, *J. Am. Chem. Soc.* 116, 5961 (1994).
10. Recent computations of the degenerate S_N2 reaction of H_3CX produced method-dependent absolute numbers. However, the qualitative trends are exactly as predicted: The well depth, ΔU, as well as the barrier height, ΔQ, increase in absolute magnitude in the predicted order F > Cl > Br > I with only two reversals (F/Cl) in the cases of Hartree-Fock and MP2 computations: L. Deng, V. Branchadell, and T. Ziegler, *J. Am. Chem. Soc.* 116, 10645 (1994). For related ab initio studies of

proton transfer, see S. Gronert, *J. Am. Chem. Soc.* 115, 10258 (1993). For an excellent treatment of ion-molecule interaction, see R. Zahradnik, *Acc. Chem. Res* 28, 306 (1995).

11. S.E. Thomas, *Organic Synthesis. The Roles of Boron and Silicon*, Oxford University Press, New York, 1991.
12. S. Scheiner, *Acc. Chem. Res.* 27, 402 (1994).
13. A recent interesting paper deals with the problem of solvent reorganization and enthalpy–entropy compensation by treating explicitly different solvent–solute geometries: E. Grunwald and C. Steel, *J. Am. Chem. Soc.* 117, 5687 (1995).

Chapter 23

The Difference Between Kinetics and Thermodynamics

23.1 Ambident Reactivity and Kinetic Control in Ion Combination Reactions

When two ions are combined in solution and the nucleophile (electrophile) is ambident while the electrophile (nucleophile) is unimodal, the reaction is represented as follows:

nucleophile (E/I) + electrophile (E or I) + solvent (E)

where the symbols in parentheses indicate the reactant selectivities.

The solvent is an ensemble of molecules held together by static bonding. Hence, the solvent is counted as an E-selective reagent. If the electrophile is also E-selective, the VB prediction is clear: Attack will occur at the E site of the ambident nucleophile. One example is the C3 (interior) protonation of the cyclohexadienyl anion formed in the Birch reduction of benzene.[1] All three of the conventional routes for explaining data fail:

Product stability. The major product isomer is the one that is thermodynamically the less stable. Hence, the reaction is an example of *kinetic control.*

Frontier orbital electron density. Atoms C1 and C3 of the pentadienyl anion have equal NBMO coefficients. Hence, on statistical grounds alone, protonation should have occurred at C1.

Charge density. There is no significant differentiation of C1 and C3.

The I and E sites of the cyclohexadienyl anion are shown in Figure 23.1. The I and E formulas speak for themselves. The nucleophilic I site is the one at the terminus of the longest codirectional arrow train. On the other hand, the nucleophilic E site is the one in which the nucleophilic pair is best stabilized by the "environment" by a Coulombic mechanism. Note that unlike the case of the enolate anion, the nucleophilic pair could in principle localize at any center because all the centers have the same electronegativity, since all are carbons.

Replacing a π bond by a π-isoconjugate methylene fragment of methyl and replacing the anionic pair by a cationic hole generates a cationic analogue of the

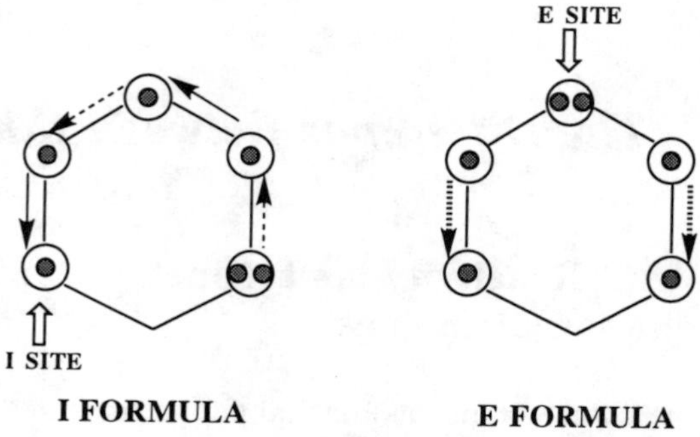

Figure 23.1 The E and I sites of the cyclohexadienyl anion.

cyclohexadienyl anion, namely, the $(CH_3CHCHCH_2)^+$ intermediate of the electrophilic addition of HBr to 1,3-butadiene. This captures a nucleophile by the E site, which is the hole flanked by one double bond and the methyl group, to produce the thermodynamically less stable 1,2-addition product. Once again, we have kinetic control.

Exactly the same arguments hold in the electrophilic addition to benzene to produce the cyclohexadienyl cation itself. The para form has an E-selective hole, while the ortho form has an I-selective terminus. Every systolic group carries one or more pairs and is E selective. For example, a methoxy group is E selective. Hence, para attack of anisole by an electrophile generates a superior E + E combination. This explains a puzzle: Why is para attack favored over the ortho form in electrophilic aromatic substitution (after the statistical factor has been eliminated)?[2] In HSAB parlance, the VB answer is that para is favored over ortho because the paracyclohexadienyl cation is "hard" and so-called π donors are also "hard."

The final example is the competition between second-order elimination (E2) and S_N2 reactions of an alkyl halide. As shown in Figure 23.2, the selectivity is controlled by the nature of the attacking reagent and the leaving group: I nonactivated atoms or groups (e.g., F) select the E2 mode, but I-permissive atoms or groups (e.g., I) select the S_N2 mode, provided the alkyl halide can activate the reagent. As the last step, replacing the alkyl halide by a vicinal dibromide and using the I-permissive I^- as the attacking reagent leads to vicinal dehalogenation by an I-selective transition state. Hence, the two different alkene-producing reactions have two qualitatively different transition states. We stress here the "case-by-case" nature of the theory: In ionic reactions, hydrogen acts as E-selective proton (e.g., E2 reaction), but in reactions of neutral molecules (e.g., sigmatropic shifts), it acts as an I-selective spherical atom.

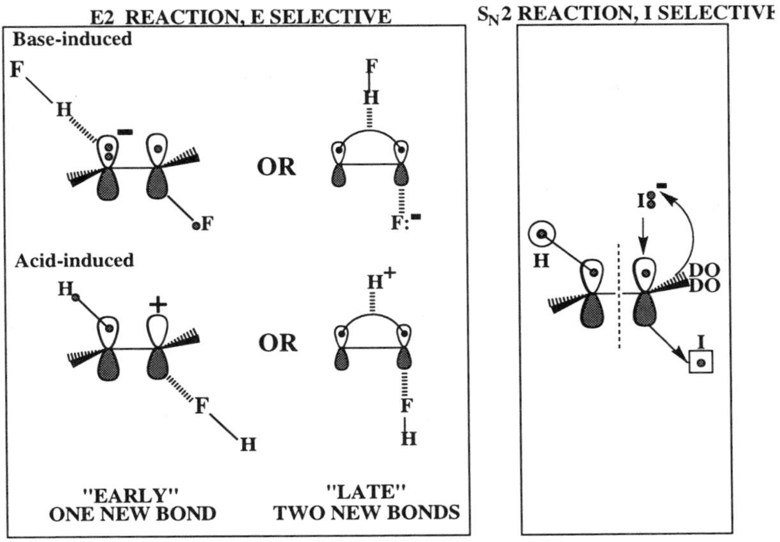

Figure 23.2 I- and E-selectivity in the attack of an alkyl halide by a group bearing an electron pair. The E2 transition state is primarily held together by the electrostatic interaction of ions and/or polar molecules. The S_N2 transition state is held together by codirectional CT delocalization.

Klopman was the first to suggest that the selectivity of organic reactions requires the action of two different physical variables. He proposed that "soft" means frontier orbital control and "hard" means charge control.[3] One might be tempted to equate "charge control" with E-bonding and "FO control" with I-bonding. However, E-bonding means the cooperative action of the electrostatic *and* inductive plus dispersive mechanisms (see Chapter 2) rather than the action of the first alone. When it comes to homonuclear organic transition states, the inductive component becomes dominant, and this provides for a clean disjunction between E- and I-bonding. For example, preferential para attack in electrophilic aromatic substitution is said to be due to FO control; that is, a C3 hole in cyclohexadienyl cation is regarded as "soft." This is diametrically opposite to our conclusions: The orientation of the Birch reduction of benzene and the ortho/para distinction in electrophilic aromatic substitution are due to the concept of maximum accumulation of π bonds about an "inductor" charge predicted by the E mechanism of chemical bonding. Of course, FO control and I-bonding are in no way related, since the former is an EHMO-type index and, at this level of theory, there is no distinction between T- and I-bonding.

23.2 Kinetic Control of the Diels–Alder Reaction

An illustrator of the map of chemical bonding is the Diels–Alder reaction:

The diradical intermediate represents T-bonding.

The symmetrical transition state represents (partly impaired) I-bonding (bikratic transition state, see Chapter 17).

In a relative sense, the exo mode is an I-selective while the endo mode an E-selective transition state. In other words, the former is displaced towards the I end and the latter towards the E end of the bonding continuum. In the exo geometry, exchange repulsion of the I bonds is minimized while, in the endo geometry, the (overlap-independent) Coulombic attraction of the I bonds is maximized.

The endo selectivity of the Diels–Alder reaction[4] is a prime illustrator of kinetic versus thermodynamic control of reactions: What is more stable at the transition state (endo orientation of reactants) becomes less stable at the product stage, and the reaction is said to be under kinetic control. The endo addition rule of Alder and Stein requires that the reaction partners arrange themselves in parallel planes and addition proceeds from the geometry that has the maximum accumulation of double bonds. The phenomenon is quite diverse: The groups showing endo preference are so varied that no single specific explanation such as "inductive effects," "charge transfer," "secondary orbital effects," or "primary overlap considerations," will be successful everywhere. Huisgen et al.[4] pointed this out by saying that even "innocent-looking" dienophiles, such as cyclopropene, cyclopentene, and propene, favor endo cycloaddition!

Our explanation is that endo preference reflects the mutual polarization of an "observer" homopolar π bond by the A domain or the polarization of the A domain by a hetoeronuclear (polar) π bond (or some π-isoconjugate group: e.g., a methylene). A comparison of the exo and endo transition states of the 4+2 dimerization of 1,3-butadiene is shown in Figure 23.3a. The hallmark of the endo complex is a dienophile whose nominally inactive π bond (the "observer" bond) interacts with the C_2–C_3 fragment of the A domain in an acyclic fashion. This is why maleic anhydride is the quintessential illustrator of the endo rule. Note the head-to-tail arrangement of one superior arrow in the A domain and the hatched arrow in Figure 23.3a.

Things get more interesting when we compare the reaction of the geometric isomers of a 1,2-disubstituted ethylene in which the two substituents are π acceptors, AC (e.g., CH=O). Treating AC as a pseudoatom with a vacant π

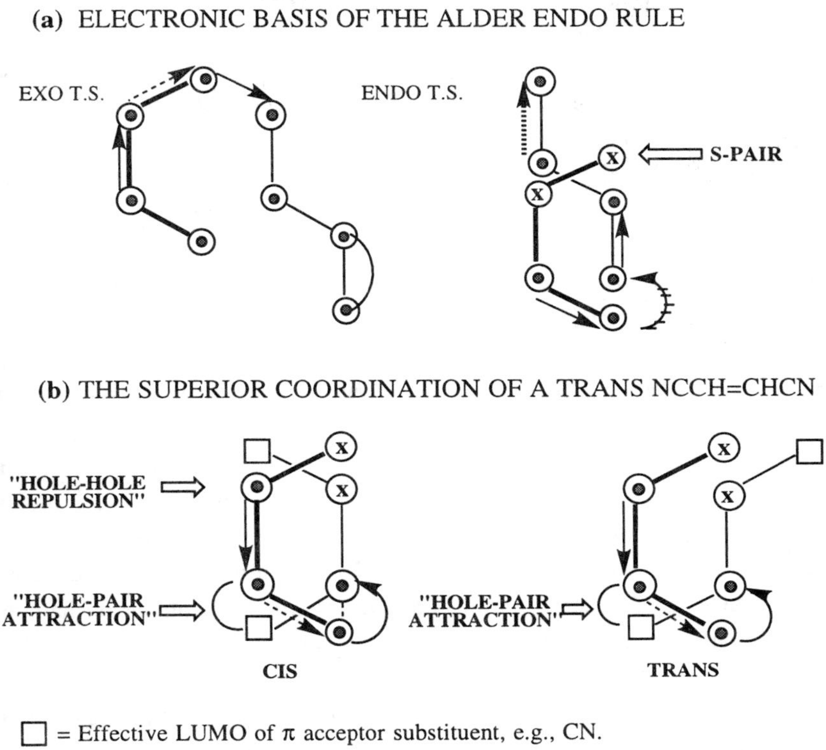

Figure 23.3 (a) Exo and endo transition states of the Diels–Alder reaction; hatched arrow indicates polarization. (b) The differentiation of the geometric isomers of 1,2-dicyanoethylene at the Diels–Alder transition state.

AO and recognizing that the A domain has hole character at the tails and pair character at the heads of superior arrows, we see that endo attack by the cis isomer occurs at the expense of one hole–hole destabilizing interaction (Figure 23.3b). Such a source of instability is absent in the trans form. Indeed, the two geometric isomers of 1,2-dicyanoethylene react with dienes with comparable rates even though the trans is more stable than the cis isomer. Hence, the transition state involving the trans isomer must have lower energy. The magnitude of the superiority of the trans isomer depends on how strong a π acceptor the AC group is. The superiority of the trans over the cis dienophile (at the transition state) is accentuated in the case of 1,3 dipolar cycloadditions,[4] and the interpretation remains the same.

The conventional view is that the S_N2/E2 selectivity observed in the attack of a first-row base (e.g., OH^-) on an alkyl halide is unrelated to the exo/endo selectivity of the Diels–Alder reaction. We paint a very different picture: The two processes are linked in the following way: The *kinetically preferred* (E2 and endo product) is the *thermodynamically unfavorable* product which arises from an E-selective transition state. The distinction between kinetic and thermodynamic control is a consequence of the distinction between T, I, and E bonding. This simple example casts a pointer on how different the VB theoretical concepts compared to the accepted view.

23.3 Homolytic Substitution; The "Forgotten" Reaction

Every student of organic chemistry is exposed to the bimolecular nucleophilic substitution at carbon (S_N2 reaction). By contrast, the homolytic S_H2 analogue is hardly to be found in textbooks. One reason is that it disconfirms the concepts taught by the former reaction. Specifically, it is found that the second most prevalent S_H2 reaction, after attack on hydrogen, is attack on halogen bonded to carbon. In other words, the most favorable array of atoms in the S_N2 reaction is (NUC)(R_3C)Br^-, where NUC^- is the attacking nucleophile and Br the leaving group, but the most favorable array of atoms in the S_H2 reaction is (R_3C)Br(RAD) instead of (RAD)(R_3C)Br, where RAD is the attacking radical and R_3C the leaving radical. Furthermore, there are pronounced substituent effects: Methyl radical attacks F_3C—I roughly 500 times faster than H_3C—I.[5a] One can hardly find a better example of kinetic control: Making a C—C bond by attacking C by means of a carbon radical is thermochemically superior by roughly 33 kcal/mol to making a C—I bond via an attack on I. And yet, methyl radical attacks iodine in preference to carbon!

The solution of the puzzle is that I-permissive nucleophiles as well as I-permissive radicals intrinsically prefer attack on heavy halogen than on carbon because by attacking heavy halogens, they become activated by the observer pairs and holes as illustrated in Figure 23.4. The frequent superiority of the S_N2 substitution reaction (nucleophilic attack on carbon) to alkyl reduction (nucleophilic attack at halogen) is simply due to the low electronegativity of carbon, which effectively thwarts configurational degeneracy. For example, in the nucleophilic attack of methyl bromide by bromide at bromine, the structure $H_3C:^-$ Br—Br lies far above H_3C—Br $Br:^-$. Note how the "observer" hatched arrows activate the "functional" REL bond. One hatched arrow taken in combination with one codirected solid arrow (head-to-tail sequence) produces

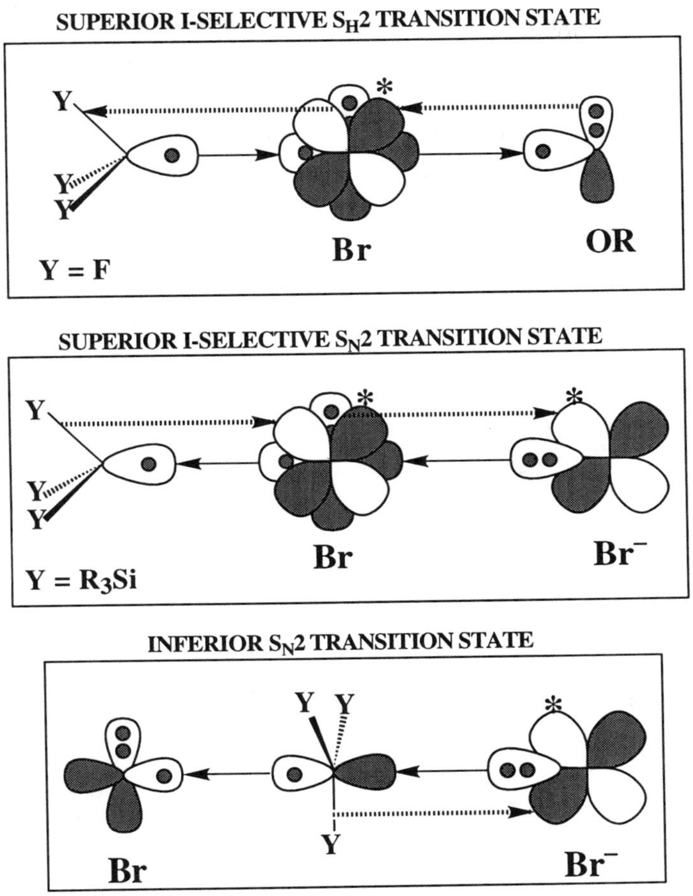

Figure 23.4 The length of a (looping) codirectional arrow train determines transition state stability.

induced dipole–induced dipole bonding in addition to bonding due to CT delocalization.

A superior complex is always distinguished by the presence of an extended codirectional arrow train. The emerging viewpoint is that when Y is appropriately chosen, both halogen atom abstraction by a radical (S_H2 on halogen) and halonium ion abstraction (or, equivalently, alkyl reduction) by a nucleophile (S_N2 on halogen) can become favorable reactions. When the carbon atom of an alkyl bromide becomes capable of accommodating an anionic electron pair, nucleophilic attack should occur at Br rather than at C. This is what happens in the dehalogenation of vicinal dibromides by iodide anion, a textbook case. Furthermore, reduction of alkyl halides by I^- and RS^- nucleophiles via nucleophilic substitution at halogen has been proposed.[5b]

We can use the new VB concepts to design new reactions. For example, we saw before that a REL bond is preserved in the course of chemical transformations. The A formula of PCl_3 is ClP(Cl,Cl^-). It follows that we can replace Cl^- by an olefin. Subsequent attack of the departed chloride on the attached Cl corresponds to nucleophilic abstraction of halide, leading to formation of a three-membered ring made up of two carbons and one phosphorus:

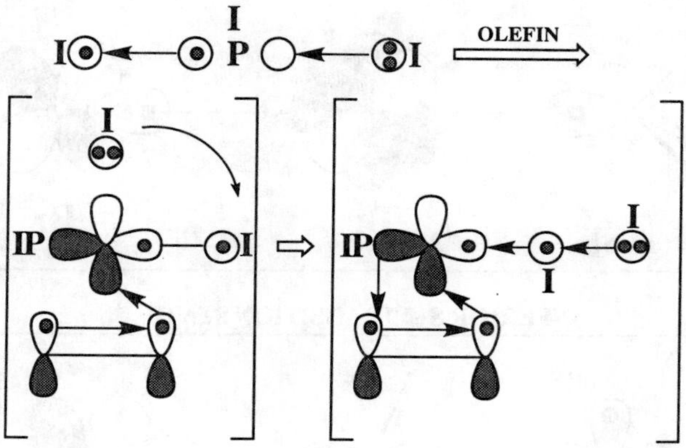

This type of "cyclopropanation" by PCl_3 has been discovered by Lappert and co-workers.[6a]

Bottom line: The bond linking two Cl atoms to PCl_3 to form the "hypervalent" PCl_5 is not a "three-center/four-electron bond." Rather, it is an I bond of the REL type, which has the advantage of configurational degeneracy. An I bond spanning three atoms can have four, three, or two electrons, as illustrated below. Furthermore, a single I bond is nothing but a special case of a codirectional arrow train that is equivalent to *n* I bonds. This means that systems like C can exist either as distinct intermediates or as low energy transition states. Ab initio calculations uncovered the possibility of nucleophilic attack by chloride anion at the radical center of CH_2—CH_2Cl.[6b] Radical substitution at a heavy halogen is normally a double-minimum reaction in which an I-bound, three-electron complex is the transition state. However, appropriate substitution may transform the latter to a global minimum. One recent example is the detection of a "hypervalent" iodine radical intermediate.[6c] The concept of the (even- or odd-electron) I bond casts a pointer to the new areas of fruitful experimentation or computation.

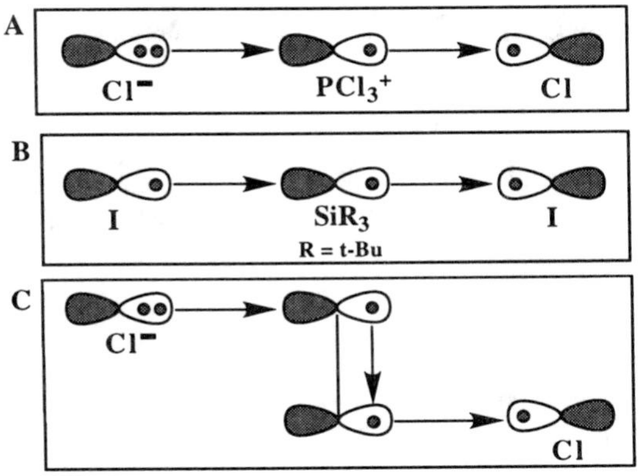

A note of caution: The three-electron REL bond associating the three fragments at the S_H2 transition state has high (or dominant) T complement simply because configurational degeneracy cannot be attained. As a result, the reaction approaches the limit of uncoupling and recoupling spins (allyl valence isomerization). Depending on the relative magnitude of the T complement, we have two limiting cases:

Valence isomerization, as in homolytic substitution or homolytic addition to π bonds, which occurs when the T complement is strong.

Electron transfer, which occurs when the T complement is weak, always relative to CT.

The I-bond formalism is operationally significant across the board because the stability of an odd-electron complex always depends on CT delocalization, which, in turn, is dependent on the nature of the component atoms. Always keep in mind that exchange delocalization is, to a first approximation, independent of *relative* atom electronegativity because atom neutrality is conserved; that is, exchange does not create electron flow.

23.4 The Enolate Paradigm

Definitions of "nucleophile" and "leaving group" are possible in a symmetrical S_N2 reaction where these entities are identical. On the other hand, either the in- or the outgoing group can function as the nucleophile or the leaving group. In other words, there exist three possible transition states:

$$N: + A—L \rightleftharpoons [TS] \rightleftharpoons N—A + L:$$

TRANSITION STATES

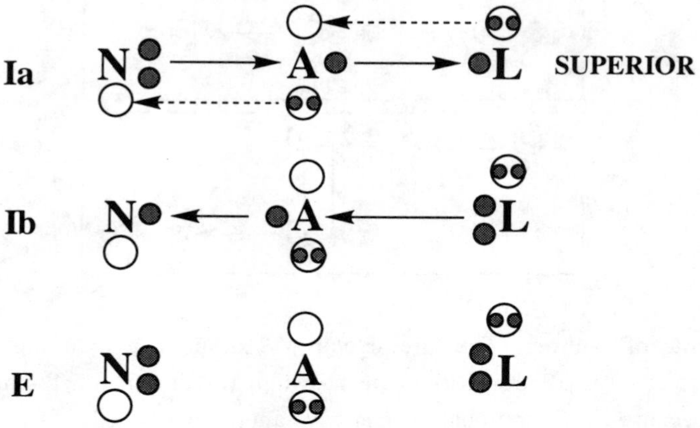

Ia and Ib are actually resonant I descriptors, but one or the other must be dominant, depending on the situation.

The key question is: Which group, N or L, functions as a nucleophile and which acts as a radical in an I-type transition state? The answer is straightforward: The mode of action of N and L is the one consistent with the formation of the longest codirectional arrow train.

This brings us to a disjunction of conventional models and VB theory. The nearly universal assumption is that the S_N2 transition state can be viewed as a three-center/four-electron problem. By contrast, the very nature of VB theory dictates that the "observer" pairs and holes must be included. The reason is that I-bond formation requires that the surrounding electrons get out of the way. As a result, the explicit treatment of the entire occupied and unoccupied space is mandatory. In this light, let us return to the example given in Section 23.3 and assume that the central fragment is SiR_3, which has three valence bond pairs and a low-lying unoccupied space simulated by one vacant d-type orbital. A heavy p-block central atom of this type can now activate an I-permissive nucleophile (i.e., a nucleophile with a vacant AO such as Cl^-) and an I-permissive electrophilic radical (i.e., an electronegative radical with a lone pair such as RO). Therefore, **Ia** is far superior to **Ib** because it can support a codirectional arrow train made up of four arrows. In other words, the primary REL bond is fully activated. This is illustrated as follows.

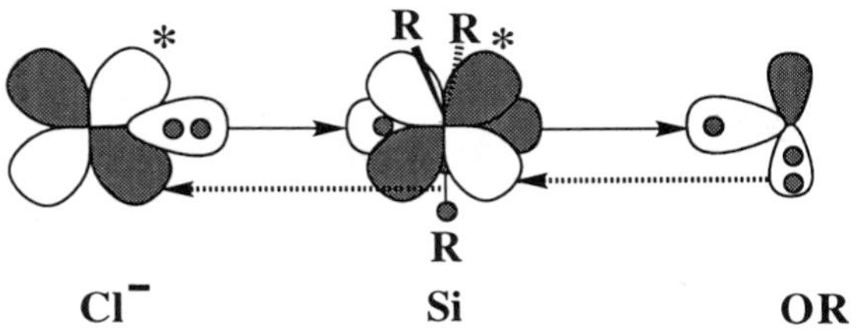

Under these conditions, we can unambiguously predict that N will act as a nucleophile and L as the radical in an I-type transition state (Ia in the diagram at the beginning of this section).

In this light, consider the reactions of enolate anion, ENL^- ($H_2C{=}CH{-}O^-$), with $Me_3Si{-}F$. The prediction is clear: ENL will act as an oxygen (rather than carbon) nucleophile in an E-type transition state because the E site of ENL^- is oxygen and the leaving group is a small and compact electronegative F, which has E affinity. The situation changes when F is replaced by Cl. We now discover that we can build a full codirectional arrow train, exactly as above, if Cl acts as a nucleophile and ENL as an electrophilic O-centered radical! In other words, Cl dictates that the crucial feature of the enolate fragment is not the carbon–carbon double bond but, rather, the oxygen lone pairs that render oxygen an I-permissive electrophilic radical. Later on (see Section 28.2), we will see that systolic groups (i.e., atoms bearing lone pairs and/or systolic bond pairs such as F, OH, NH_2 and CH_3) have a characteristic high affinity for heavy p-block atoms because their pairs can combine with metalloid nonvalence unoccupied space to activate valence I bonds connecting the systolic group and the metalloid atom. Thus, the reason for the kinetic preference for O-silylation is the same as the reason for the high thermodynamic stability of the vinyl silyl ether product.

The final result is that enolate anion will act as an effective O nucleophile in its reaction with halotrimethylsilane, as is well known. The key message is that we have gone beyond the point where allusion to "accepted" terms and concepts is productive because it connects the reader to what is already known. Specifically, we just argued that the identity of the I and E sites of a nucleophile change as a function of the substrate. Thus, a monolithic classification of molecules or regions of molecules as "soft" or "hard" is unphysical. The problem becomes simpler when enolate anion reacts with an organic molecule containing only nonmetal atoms, in which case the I and E sites of the enolate are substrate-invariant. In this case, the rule is that this ambident nucleophile will attack electroneutral carbon via its C terminus (amphidromic arrow train, I-bonding) but it will attack positively-charged carbon via its O terminus (charge alternation, E-bonding). Ab initio calculational studies by Houk and co-workers, as well as experimental gas phase data coming from the laboratories of Squires, Nibbering, and Brauman and co-workers, summarized by Zhong and Brauman,[7] are consistent with this scenario. The contrathermodynamic preference for O-attack is unequivocal evidence for an overlap-independent E-bound transition state which is fundamentally different from the overlap-dependent T-bound reactants and products.

"New theory" means "new beginning." There is hardly a simpler, more basic reaction than the S_N2 substitution, and hardly a synthetically more important nucleophile than the enolate anion. Indeed, these are the cornerstones of organic chemical pedagogy. And yet, the picture painted here has little to do with what one finds in the literature. Clearly, we must start from scratch in our effort to get things right for the right reason. The intricacies of ambident reactivity of enolate anion with silicon, phosphorus, and sulfur halides and their heavier congeners (silyl halides undergo O- but selenyl halides C-attack) in the presence or absence of coordinating metal atoms (e.g., lithium enolates) can easily be the subject of an entire monograph when the problem is viewed from the VB standpoint. Such attempts are for the future.

23.5 Electron Transfer

The best evidence for two types of transition states, one I selective and the other E selective, comes from the studies of electron transfer by Taube.[8] Here are the analogies.

Inner sphere ET corresponds to either I-selective halogen abstraction by radical (one-electron transfer) or I-selective halonium ion abstraction by nucleophile (two-electron transfer).

Outer sphere ET corresponds to E-selective ET.

Is there a basic quantum mechanical model system that illustrates two distinct modes of ET? One possible choice is the E′ degenerate ground state of π-cyclopropenyl anion in D_{3h} geometry. Upon Jahn–Teller distortion, this can be transformed to either an obtuse triangle (the A_1 species) with a π system that resembles the allyl anion or an acute triangle (the B_2 species) having the π characteristics of the π systems of ethylene anion plus carbon. According to ab initio calculations, the A_1 species is a minimum on the pseudorotational surface.[9] The B_1 species is a saddle point lying appreciably above A_1. A VB interpretation of these key features of the hypersurface has been published.[10]

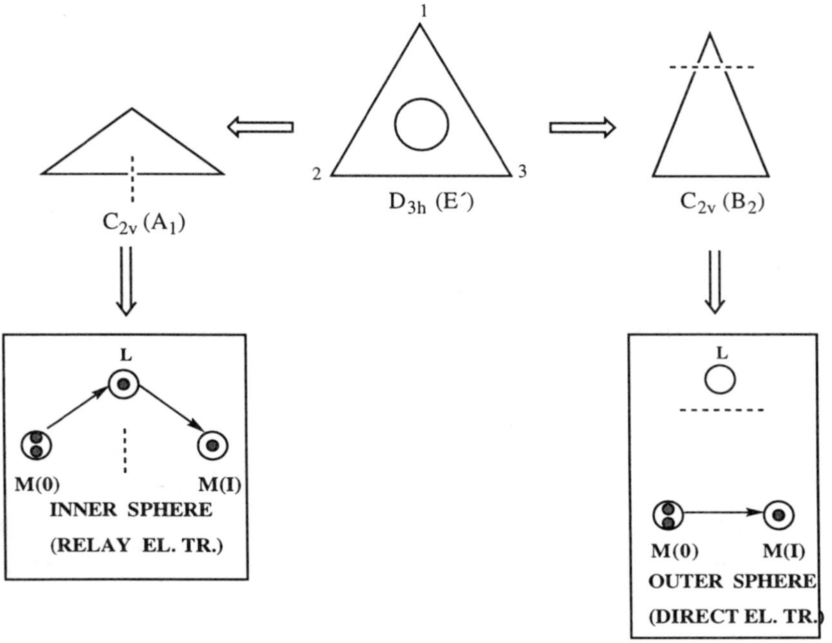

Figure 23.5 The two mechanisms of ET are illustrated by the Jahn–Teller distortion of cyclopropenyl anion.

Figure 23.5 illustrates the above considerations. It also shows the dominant ET mechanism within the two species. If we replace the first C by L (= ligand), the second C by M(0) — that is, a neutral metal — and the third C by M(I) — that is, a monopositive metal — we see that the ET in the A_1 species corresponds to inner sphere electron transfer via the bridging L ligand, while the delocalization in the B_2 species corresponds to outer sphere electron transfer, where L has effectively "withdrawn" and plays the role of mere observer. Thus, ET can be viewed as a giant Jahn–Teller distortion of some reference symmetrical species. According to this analogy, the inner is distinguished from the outer sphere mechanism to the extent that the former involves indirect ET by means of a "billiards mechanism" corresponding to the REL bond: One electron of the reducer metal strikes an electron of the bridging ligand and the latter is transferred to the oxidizing metal. By contrast, the outer sphere mechanism involves direct ET from metal to metal. The transition states of the following prototypical inner and outer sphere reactions are depicted in Figure 23.6.

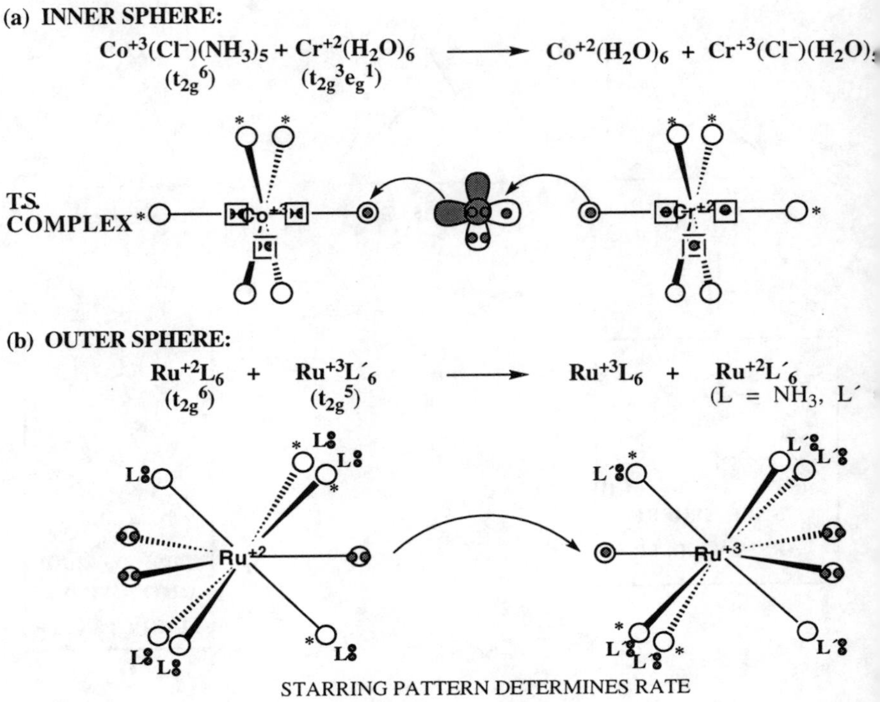

Figure 23.6 Prototypical inner and outer sphere ET. The starring pattern is crucial in both cases.

Inner sphere:

$$(NH_3)_5(Cl^-)Co^{3+} + (H_2O)_6Cr^{2+} \xrightarrow{H^+} (H_2O)_6Co^{2+} + (H_2O)_5(Cl^-)Cr^{3+} + 5\ NH_4^+$$

Outer sphere:

$$RuL_6^{2+} + RuL'^{3+}_6 \longrightarrow RuL_6^{3+} + RuL'^{2+}_6 \quad \text{(fast)}$$

The rates of inner sphere reactions and their dependence on the type of donor and acceptor orbitals (σ vs. π) have been discussed in the literature.[11] For example, the preceding inner sphere ET reaction is formally a homolytic abstraction of Cl from $Co{-}Cl^{2+}$ by Cr^{2+}, and the same considerations as those

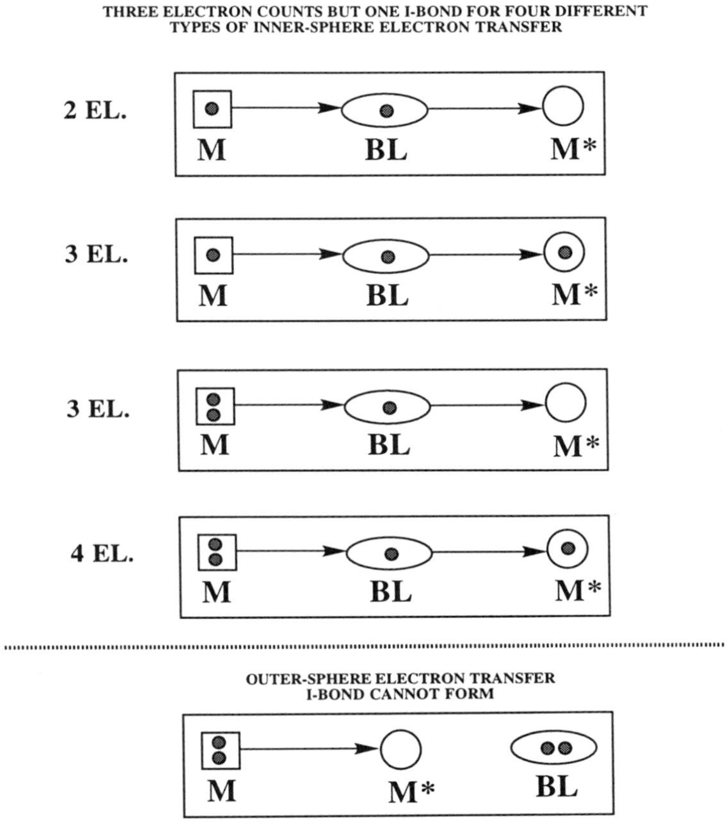

Figure 23.7 Three different electron counts (two, three, and four electrons), each corresponding to just a single REL bond in the transition state of inner sphere ET. The strength of the REL bond depends on configurational degeneracy, which is the element that differentiates the four situations. Outer sphere ET occurs when an I bond cannot be established.

discussed in Section 23.4 apply. Inner sphere ET proceeds via a transition state in which the electron count varies but the number of REL bonds (one) connecting the three fragments is conserved. This phenomenon, illustrated in Figure 23.7, underlies the difference between EHMO-type and VB explanations.

The final drawing of this chapter summarizes the strategy the material exemplifies: Chemical reactions of different types expose the disjunction of different bonding mechanism types. Opposing thermochemical and kinetic preferences in ionic and radical reactions tell us something about the difference between T-bonding (thermochemical preference) and either E- or I-bonding (transition state preference). The disjunction between E- and I-bonding is provided by the regioselectivity of chemical reactions.

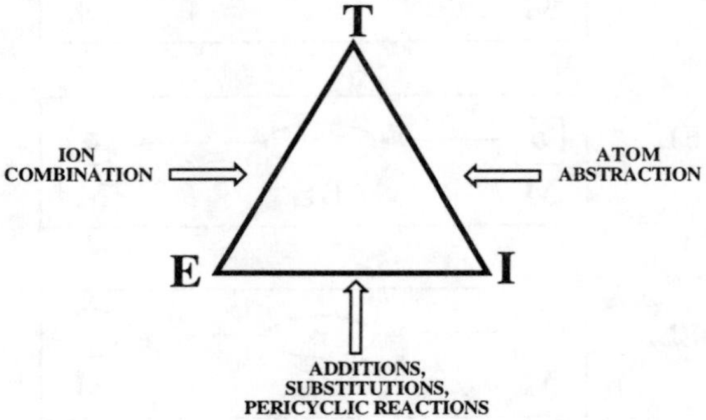

References

1. A.J. Birch and H. Smith, *Q. Rev.* 12, 17 (1958).
2. R. Taylor, *Electrophilic Aromatic Substitution,* Wiley, New York, 1990.
3. G. Klopman, *J. Am. Chem. Soc.* 90, 223 (1968).
4. R. Huisgen, R. Grashey, and J. Sauer, in *The Chemistry of Alkenes*, S. Patai, Ed., Wiley-Interscience, New York, 1964.
5. (a) M. Poutsma, in *Free Radicals*, Vol. II, J.K. Kochi, Ed., Wiley, New York, 1973, Chapter 14.
   (b) N.S. Zefirov and D.I. Makhonkov, *Chem. Rev.* 82, 615 (1982).
6. (a) A.G. Avent, M.F. Lappert, B. Skelton, C.L. Raston, L.M. Engelhardt, S. Harvey, and A.H. White, in *Heteroatom Chemistry*, E. Block, Ed., VCH, New York, 1990, Chapter 15. (b) H. Zipse, *J. Am. Chem. Soc.* 116, 10773 (1994). (c) J.T. Banks, H. Garcia, M.A. Miranda, J. Perez-Prieto, and J.C. Scaiano, *J. Am. Chem. Soc.* 117, 5049 (1995).
7. M. Zhong and J. I. Brauman, *J. Am. Chem. Soc.* 118, 636 (1996).
8. H. Taube, *Electron Transfer Reactions of Complex Ions in Solution*, Academic Press, New York, 1970.
9. E.R. Davidson and W.T. Borden, *J. Chem. Phys.* 67, 2191 (1977).
10. N.D. Epiotis, *Nouv. J. Chim.* 8, 421 (1984).
11. K.F. Purcell and J.C. Kotz, *An Introduction to Inorganic Chemistry*, Saunders College Publishing, Philadelphia, 1980.

Chapter 24

The Chemical Code Cracks in the p-Block

24.1 Chemical Bonding Across a Colored Periodic Table

We are interested to find out what happens when nonmetal atoms like H, C, N, O, and F combine with heavier semimetal and metal atoms. The determinant of the mechanism of bonding is the atomic ionization potential (IP). This dictates a breakdown of the periodic table by colors as illustrated in Figure 24.1 for up to radon. Our table meets the following conditions:

(a) Along a column, the IP decreases in going from a black to a grey and, finally, to a white atom.

(b) Two columns, corresponding to groups 4 and 13 (Ti and B groups, respectively) are made up of atoms of comparable IP (except B) and they form hatched transitory zones.

(c) Along a row, any one black atom has a higher IP than any one grey atom, which has higher IP than any one white atom. The only exception is C, which has lower IP than grey Kr and Xe.

(d) Colored zones are separated by hatched transitory (borderline) zones with the exception of Ti, which separates two white zones. An atom in a borderline zone resembles in color the atom below (column) or the atom to its left (row).

(e) The problem with the anomalous C, Ti, and B can easily be remedied if we define different areas of white and black. However, we stick with only three basic colors in order to maximize simplicity. The reader should note that semimetals and metals are differentiated into grey, white, or hatched atoms in a

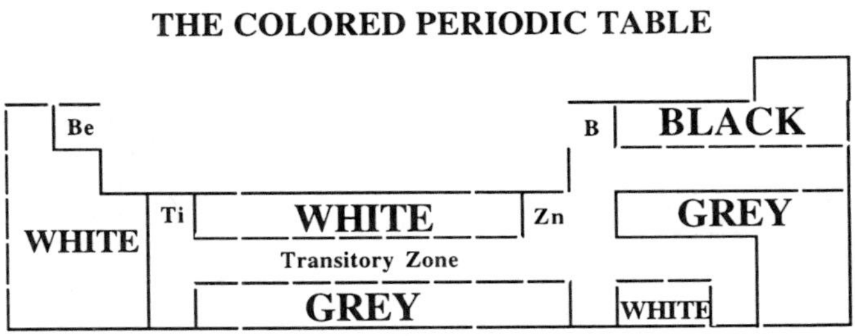

Figure 24.1 The colored periodic table.

way that is totally disconnected from the conventional classification of the elements.

The bonding mechanisms which are appropriate for every domain of the colored periodic table are shown in Figure 24.2. In every case, we combine atoms of the given color with black nonmetal atoms. For example, T-bonding is appropriate when black are combined with black atoms. I-bonding is appropriate when grey are combined with black atoms. I (activated)- or E-bonding is appropriate when white atoms combine with black atoms. When the combining atoms fail to satisfy the requirement of the optimal mechanism, they become frustrated. The organometallic pantheon shown in Figure 24.2 defines the following rules of chemical bonding.

(a) Black atoms choose T-bonding.

(b) Grey atoms combine with black atoms by either unactivated I-bonding of the REL type, or, still better, activated I-bonding. The tolerance for the unactivated form of REL-bonding is attributable to the high electronegativity of the grey atoms, which allows them to benefit from the high T complement of REL bonds.

(c) White atoms are too electropositive to tolerate unactivated I-bonding. Thus, the only two options are E- or activated I-bonding. Activation is effected by complementation of the primary arrow train by a secondary arrow train. This is equivalent to bonding by overlap dispersion. Small size favors the E and large size favors the I mechanism.

The stage is set: We expect a number of counterintuitive trends to arise from the very subdivisions of the colored periodic table[1] and the ascribed optimal mechanisms. Because the data available for testing our model are usually

THE MECHANISM OF BONDING OF ORGANOMETALLICS

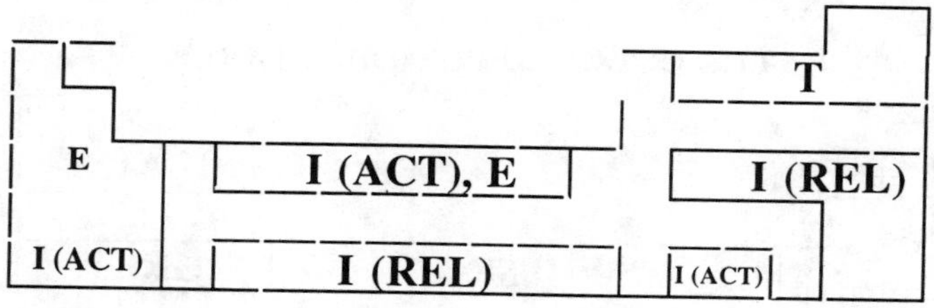

Figure 24.2 The mechanism of bonding of the atoms of the periodic table with first-row nonmetals from C to F (REL = relay I-bonding).

limited, we often illustrate the key concepts by reference to borderline atoms even though these may not the best test cases.

24.2 The VB Model of Organometallic Complexes

The VB model of a prototypical organometallic complex of the form ML_n, where M is metal (or semimetal) and L is a nonmetal ligand, is as follows:

(a) Each heavy p-block atom binds from its ground electronic configuration.

(b) According to the association rule, two arrows are allowed to exit a doubly occupied AO. However, two arrows are also allowed to exit two different singly or doubly occupied AOs. As a result, the elements of the theory must be pairs, dots, and holes that are combined either in singles or doubles. The former are called *lone elements* (symbolized by the prefix l, for "lone") and the latter *double elements*. There exist six distinct pairwise combinations of the three lone elements:

THE NINE ELEMENTS OF THE GAME

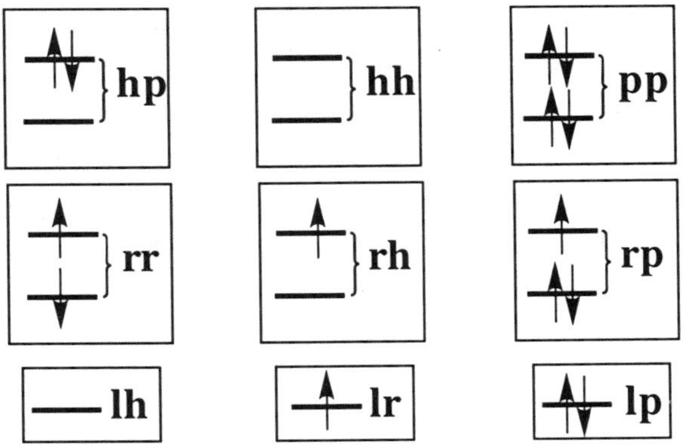

The key point is that each M enters as a composite of any combination of nine elements.

(c) Each ligand binds as a Lewis acid (LA), a Lewis base (LB), or a radical (RD).

(d) One metal hole, pair, or radical (whether lone or double) ties up two ligand elements to form one I bond according to the association rule. An LA acts with one hole, an LB with one pair, and an RD with one dot. Because of configurational degeneracy (arrow codirectionality), REL bonds are superior and are always formed in preference to either OX or RED bonds by effecting electron transfer from metal to ligand (or vice versa), if necessary.

(e) Observer ligand pairs and holes combine with available metal holes and pairs to activate the REL bonds. Whenever the latter are anchored by metal dots, the number of activator I bonds is halved.

(f) The ideal inorganic complex must conform to the bonding mechanism specified in the organometallic pantheon (Figure 24.2).

The breakthrough concept here — the realization that the binding of two ligands by a single I bond can be accomplished by using either a single or a double atomic element — reveals what has been wrong with the notion of universal covalency. Specifically, all semimetal- and/or metal-containing molecules thought to be "electron precise" are actually I-bound, with each ligand pair being associated with a double element. On the other hand, all molecules thought to be "hypervalent" or "electron deficient" are actually also I-bound, only now either some or all ligand pairs are associated with lone elements. Thus, from the standpoint of VB theory, all molecules (except those comprising H,C, N, and O) are "normal valent," with the common denominator being the I bond. The proof of the statement lies in the myriads of molecules that illustrate the possible combinations of lone and/or double elements with ligand pairs.

As discussed before, a prediction of complex stability, geometry, reactivity, and photochemistry can be made on the basis of the association diagram (A diagram), which shows how M binds a radical (RD) and a Lewis base by one dot to make a REL bond, the optimal type of an I bond. The sequence of REL bond formation is subject to the priority rules:

Higher energy AOs bind ahead of lower energy AOs.

Double elements bind ahead of lone elements (because of the higher T complement).

The geometry of the complex is predicted by adherence to the stereochemical rules. Three of these are relevant to the p-block elements:

An ns AO binds two ligands in cis geometry in all cases except a triatomic. Severe nonbonded steric repulsion with the remaining ligands results from trans binding because of the contracted nature of the ns AO.

An np AO has the flexibility of binding two ligands either cis or trans, depending on which mode produces the lowest nonbonded steric repulsion.

Two np AOs bind two ligands in an angular fashion.

An A diagram displays the elements of the central atom and lists next to them the ligands that combine with them to form I bonds. For example, SF_2 is a molecule in which one sulfur radical/hole (rh) double element makes one REL bond to one fluorine radical and one fluoride anion. Its A diagram is shown in Figure 24.3a. Note that we do not have two "polar covalent bonds" but, rather, one REL I bond with T complement. The prediction of geometry is straightforward: Because the rh is ($3p_x$,$3p_y$), the molecule is bent at an angle approaching 90°. SF_4 further clarifies the A diagram convention, as illustrated in Figure 24.3b. We now have two REL bonds formed by one rh and one lr. Figure 24.3 also shows the A diagrams of other well-known "hypervalent" molecules. Exactly how favorable each REL bond is (or, better, exactly how significant the REL character of the I bond is) depends on the metal–ligand electronegativity difference.

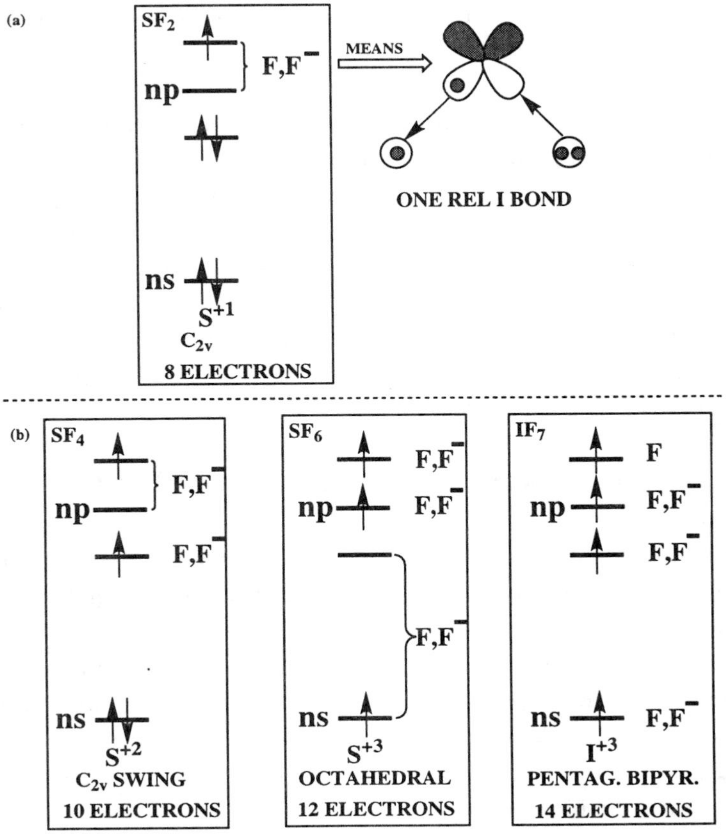

Figure 24.3 A diagrams of prototypical semimetal fluorides.

The A formula is the compact representation of the A diagram that shows the pairs of ligands that become associated by an element of the central atom. Each associated pair of ligands is enclosed in parentheses, and an asterisk is used to differentiate a double element from a lone element. Although emphasis will be placed on stable ground, singlet molecules, we add that a singlet rr binds either two electronegative or two electropositive dots by a REL bond but a triplet rr binds either two pairs or two holes by two correlated 3-e or 1-e bonds. Here are some examples:

The ground, angular SF_2 is written $S^+(FF^-)^*$. By contrast, the linear form is $S^+(FF^-)$. This means that the two F atoms are associated by an S3p dot.

Ground SF_4 is written $S^{2+}(FF^-)^*(FF^-)$.

Ground SiF_4 is written $Si^{2+}(FF^-)^*(FF^-)^*$.

The ground, angular SLi_2 is written $S^-(LiLi^+)^*$. By contrast, the linear form is $S^-(LiLi^+)$. This means that the two Li atoms are associated by an S3p dot.

This section is pivotal because it says that the "electron-precise" SF_2 and the "hypervalent" SF_4 and SF_6 molecules all belong to the same family of I-bound molecules. The only difference between the bonding of the two F atoms by S (in SF_2) and by SF_2 (in SF_4) is that the former benefit from a higher T complement; that is, we can bind the two F atoms, in part, by two exchange bonds in $S(F_2)$ but by only one in $SF_2(F_2)$. That SF_2 is not a covalent molecule (such as, e.g., H_2CF_2) is evidenced by its decomposition to S_8 plus SF_4. In other words, SF_4 is "more normal" than SF_2 despite what the labels "electron precise" and "hypervalent" imply!

24.3 Linking p- and d-Block Atoms by REL Bonds

The terms "coordination compound," "inorganic complex," and so on, are associated with a conceptual theory of bonding that is fundamentally flawed. Hence, we will use the term "I-complex" in order to project the notion that these molecules have bonds that are different from the covalent bonds of organic molecules. The first application of the theory instructs us how to use A diagrams to predict stable molecules and how to show explicitly their bonding. Two categories of molecules involving the combination of p-block and d-block metals are shown in Figure 24.4.

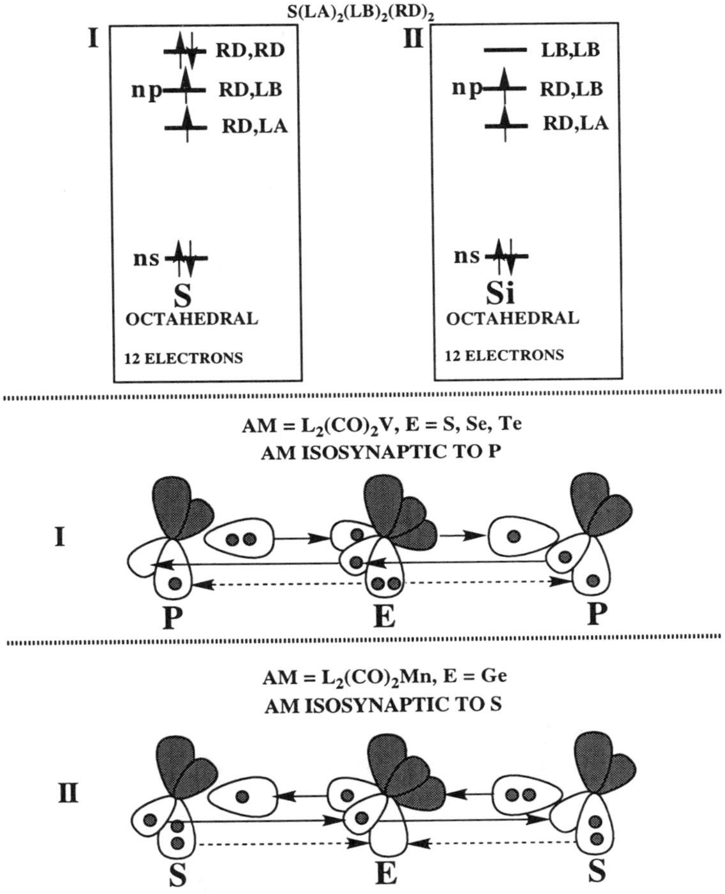

Figure 24.4 The binding of two transition metals and a heavy p-block element by two complementary REL bonds; dashed arrows indicate a secondary contradirectional I bond.

The VB rationalization is as follows:

(a) Each I complex is formulated as a central Si or S atom combining with three *pairs* of formal LA, LB, or RD ligands by three REL bonds.

(b) The action of the (six) LA, LB, or RD ligands is duplicated by two fragments that combine these six elements. For example, we can generate the analogue of the octahedral S(RD,LA)(RD,LB)(RD,RD) by combining S with two fragments, one of which acts as RD/RD/RD and the other as LA/LB/RD. A fragment of this type must have three conical valence AOs and three electrons. In other words, the sought-after fragment is P or a heavier congener or an isolobal organometallic fragment, such as $Cp(CO)_2Mn$:

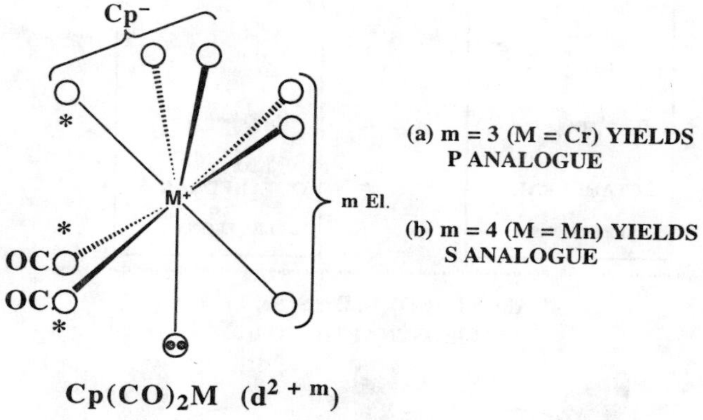

(c) The appropriate fragments are combined and the electrons are reshuffled to generate the maximum number of complementary REL bonds.

If we denote the $Cp(CO)_2$ ligand ensemble by A, we find in the literature many combinations of AM groups with heavy p-block atoms. In the following examples, taken from the works of Selegue, Herrmann, Weiss, and Legzdins reviewed by Herrmann,[2] A* means that permethyl-Cp has replaced Cp.

AM Fragment	*m*	Complex
ACr	3	ACr═E═CrA (E = S, Se)
A*Mn	4	A*Mn═E═MnA* (E = Ge, Sn, Pb)

The bonding of these molecules is illustrated in Figure 24.4, where the $Cp(CO)_2$ and the $L_2(CO)_3$ (L = phosphine) ligand ensembles are regarded as effectively equivalent. In each case, there are two complementary (antiparallel) REL bonds plus an inferior contradirectional I bond, all formed with minimum atomic promotion.

24.4 The Story of the Inorganic Rings

One of the most critical problems of molecular electronics is the stability of the prototypical X_n rings, where X = SiH_2, PH, and S. The puzzle is as follows:

(a) Among the cyclosilanes, the three-membered ring is the most unstable and far more so than cyclopropane among the cycloalkanes. On the other hand, ab initio calculations reveal that the most stable ring is the five-membered ring.

(b) Among the cyclophosphanes, the three membered ring shows unusual stability. Once again, the five-membered ring is the most stable.

(c) The sulfur rings show a progressive increase in stability as size increases up to S_8.

This is a bewildering, "nonlinear" puzzle that we can now solve by recognizing two key variables:

The singlet–triplet energy gap of X, E(S) – E(T). This is negative for X = SiH_2, mildly positive for X = PH, and very positive for X = S.

The electronegativity of the heteroatoms. This is an indicator of the ability of the ring to support strong CT delocalization.

The resolution comes in the form of the shell model of ring electronic structure (Chapter 4) and the concept of the arrow train.

We begin with the cyclosilanes. Because SiH_2 is ground singlet with a large singlet–triplet gap, replacement of CH_2 by SiH_2 stabilizes the rings with the highest ratio of radial over tangential pairs (R:T ratio). These are, in order, the four- and the five-membered rings. Because of the greater stability of the reference cyclopentane over cyclobutane, the most stable cyclosilane turns out to be the cyclopentasilane. On the other hand, the least stable cyclosilane is the one derived from the cycloalkane with the smallest R:T ratio. This is cyclopropane. In short, ring selection in cyclosilanes is governed by orbital promotion.

The game changes when we go to phosphorus rings and clusters. Since the valence electrons occupy spatially degenerate P3p AOs (PH is ground triplet and P is ground quadruplet), orbital promotion ceases to be the controlling factor. The best ring must now be the one with superior CT delocalization in the R and T shells. The odd-membered rings are superior because the R and T shells are odd annulenes with one activator hole or pair. In other words, the three- and five-membered rings can form activated arrow trains in both R and T shells, as illustrated in Figure 24.5.

(a) In each case, we have two codirectional arrow trains, one in the R and one in the T shell, that can be cochiral or antichiral. In actuality, the two forms are resonance I descriptors.

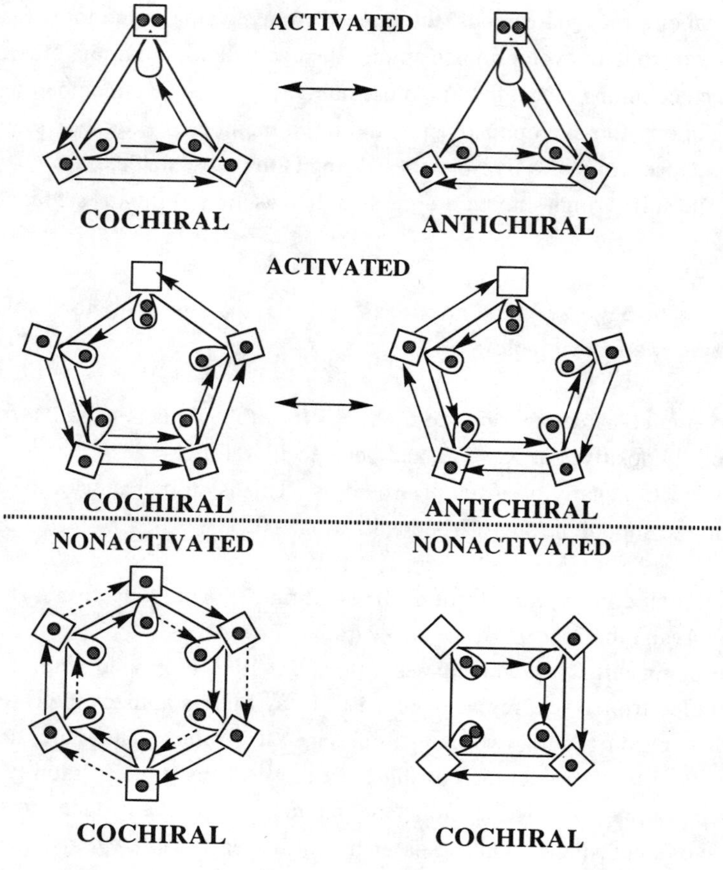

Figure 24.5 Odd rings and even rings have activated and nonactivated codirectional arrow trains, respectively, in the R and T shells.

(b) Maximum association is always effected by the cochiral contributor in which one codirectional arrow train "drags" the other.

(c) In the odd-membered rings, activation of each shell is effected by one pair or one hole; that is, each shell benefits from configurational degeneracy. Neither of the even rings is activated: the four-membered ring because of the presence of two pairs or holes, and the six-membered ring because it still takes substantial energy to turn a bond pair into a zwitterion.

Among the most important and least appreciated contributions to the theory of chemical bonding are the works of von Schnering[3a] and Baudler[3b] and their co-workers. The major conclusion of these studies is codified in Baudler's rules, one of which essentially identifies the hallmark of phosphorus chemistry: *Phosphorus prefers making five-membered rings.* The series of counterintuitive structures that conform to this statement provides direct support for the concept of the I bond simply because there exists no conventional alternative for rationalizing the observations. Ab initio calculations (and, of course, chemical experience) show that the pentagonal dodecahedral P_{20} is unstable,[4a,b] but this is only an apparent violation of Baudler's rules. To understand the problem, consider the key difference between NH_3 and PH_3: The latter has a much higher inversion barrier than the former because of "heterovalent hybridization" (i.e., the failure of the P3s and P3p orbitals to hybridize because of the contracted nature of the P3s), which causes the phosphorus to use predominantly the three P3p AOs for binding the three ligands while confining one pair to the "impotent" P3s AO ("inert P3s pair effect"). As a result, planar tricoordination of P is inherently unfavorable. The pentagonal dodecahedral form of P_{20} is unstable because each P approaches planar tricoordination.

The situation changes again when we go to the sulfur rings. Now, the increased atom electronegativity partly arrests CT delocalization, and these ringsbegin to resemble the cycloalkanes with one exception: The contracted nature of the 3s pair eliminates part of the nonbonded ligand repulsion present in the cycloalkanes. As a result, stability increases with increasing ring flexibility (i.e., with ring size), up to some limit that seems to be the S_8 ring.

We can now appreciate the difference between nonmetal (e.g., cycloalkanes) and semimetal rings as revealed by the shell model: The former are pseudoaromatic species in which perfect pairing (which hinges on intershell overlap) dominates. By contrast, the latter are "double-aromatic" R^mT^n complexes, where *m* and *n* represent aromatic electron counts. The crucial ring systems that project the disjunction between the T bond and the I bond are the cyclophosphanes. They demonstrate that the best CT delocalization does not occur in the six-membered, benzene-type $(PH)_6$, but, rather, in the forgotten $(PH)_5$! CT delocalization is most favorable when there is arrow train activation by a hole, a pair, or both. Furthermore, though differing in activation, all semimetal rings are essentially "double aromatics." Recalling that aromaticity represents the best-case scenario for I-bond formation, the rings and cages of p-block semimetals and metals are superb illustrators of I-bonding. This is evidenced by counterintuitive electron counts that do not conform to the ideas of covalency. The aromatic six-electron radial array of the four-membered ring, which recurs in inorganic stereochemistry, is one example:

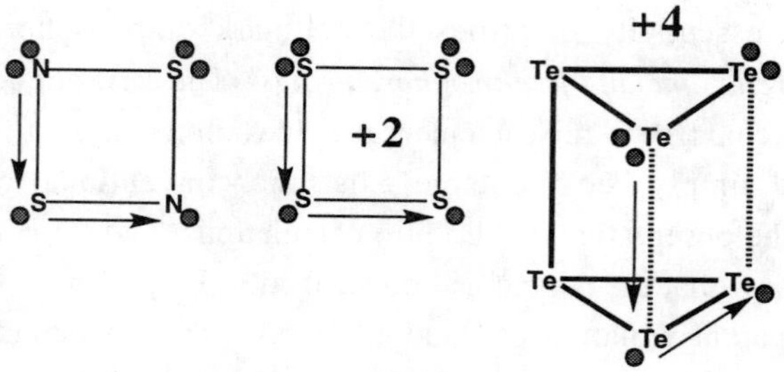

24.5 PF_3 and Its Derivatives

The A diagrams of Figure 24.6 constitute a new interpretation of the electronic structures of phosphorus fluorides and their derivatives. We note several features of these compounds.

(a) PF_3 is not a covalent molecule but a pyramid featuring one REL bond plus one covalent bond.

(b) As mentioned in Section 24.4, the P3s AO is contracted and the P3s pair tends to be "inert." As a result, inversion of PF_3 must occur via a T-shaped transition state that leaves the inert P3s pair uninvolved. Dixon and Arduengo discovered that heavy congeners of N prefer a T-shaped planar transition state in the pyramidal inversion of MF_3.[5]

(c) PF_5 can be viewed as a derivative of PF_3 and yet another application of the concept of I-bonding. Instead of one REL plus one T bond, we now have two REL and one T bond, and one of the REL bonds is formed by a lone (rather than a double) element. One can envision two possible ways of making the bonds:

1. In the equatorial isomer, P(II) uses a ($3p_x$,$3p_y$) rh to bind one F/F^- pair cis (diequatorial binding). This leaves one 3s lr to cis-bind a second F/F^- pair (axial–equatorial binding). The fifth fluorine is attached by a T bond.

2. In the axial isomer, P(II) uses a (3s,$3p_y$) rh to bind one F/F^- pair cis (diequatorial binding). This leaves one $3p_z$ lr to trans-bind a second F/F^- pair (diaxial binding). The fifth fluorine is attached by a T bond.

The key difference between the equatorial and axial forms is that, in the EQ isomer one 3s lr effects cis- while in the AX isomer one 3p lr effects trans-

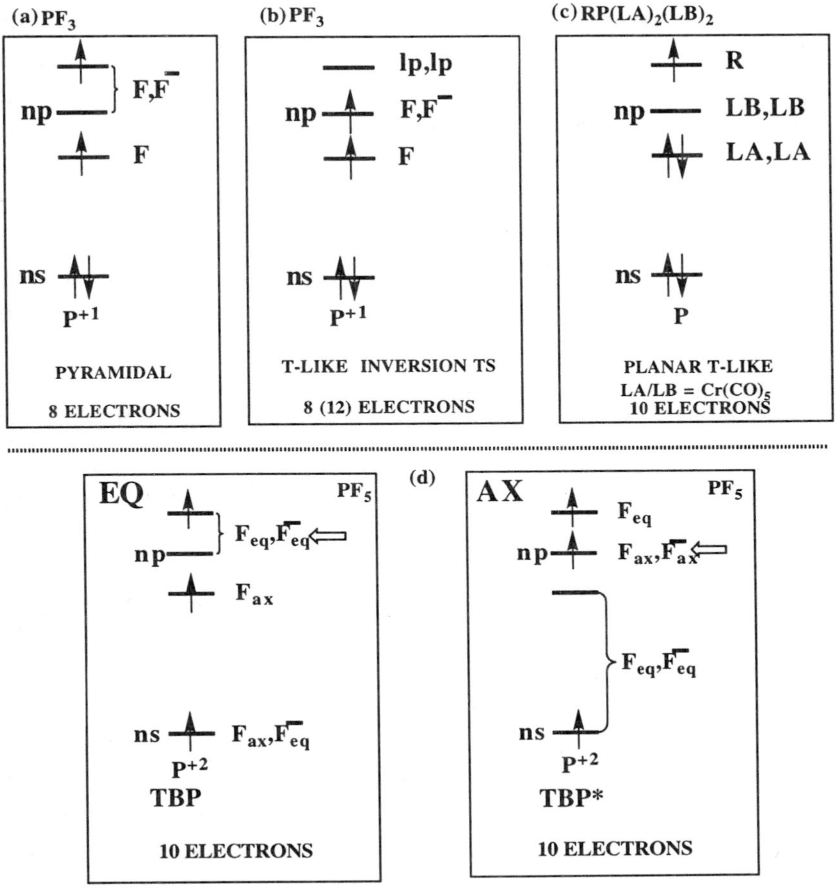

Figure 24.6 (a,b) A diagrams of PF_3. (c) A diagram of $RP[Cr(CO)_5]_2$. (d) Two different constructions of PF_5.

binding of an F/F⁻ pair. Hence, the axial isomer should be more favorable because it evades strong interligand interelectronic repulsion. The most electron-rich fluorines of the AX form are the two diaxial ones. This rationalizes the well-known ability of electronegative ligands to go axial in pentacoordinate phosphorus.

We can easily design counterintuitive phosphorus structures in which P rejects the pyramidal geometry of PF_3 or the trigonal bipyramidal geometry of PF_5. One option is to enforce a T-shaped geometry by using two LA/LB ligands as illustrated in Figure 24.6c. This should produce a planar molecule in which the three ligands relax into a C_{2v} geometry normally thought to be consistent with sp^2 hybridization. Examination of the literature reveals the existence of a transition metal derivative of PF_3 in which an RP fragment uses an lp and an lh to bind two $Cr(CO)_5$ fragments, each of which acts with a σ-lh and a π-lp. P is trigonal planar in $PR[Cr(CO)_5]_2$ and the electron count is short by two compared to the the pyramidal PF_3.[6] $PR[Cr(CO)_5]_2$ has two I bonds while PF_3 has only one. The P3s pair is not at all involved in the bonding of the heterometallic complex, and the trigonal planar arrangement of the heavy atoms is not at all a manifestation of sp^2 hybridization, as one might have assumed. Without the R group (steric repulsion), the CrPCr array would be linear.

24.6 Ligand Attraction In Square Pyramidal $BiPh_5$

The following stereochemical trends are discerned in the series AF_5 and APh_5, where A comprises P and heavier congeners.

The pentafluorides prefer a trigonal bipyramidal (TBP) geometry irrespective of A.[7]

The pentaphenyls switch from TBP preference when A = P to square pyramidal (SP) preference when A = Bi.[8]

The message is clear: Phenyls have an intrinsic preference for the SP geometry, but this can be expressed only when the bonds connecting A to Ph have become so long that nonbonded phenyl exchange repulsion is no longer a problem. The explanation requires a little imagination: Four equatorial phenyl groups are ideal for engendering *two* REL bonds with the $6p_x$ and $6p_y$ odd electrons of zerovalent P:

The $6p_x$ odd electron associates with a phenyl radical and a phenyl anion denoted by G^{-}·

The $6p_y$ odd electron associates with a phenyl radical and a phenyl cation denoted by D^{+}·

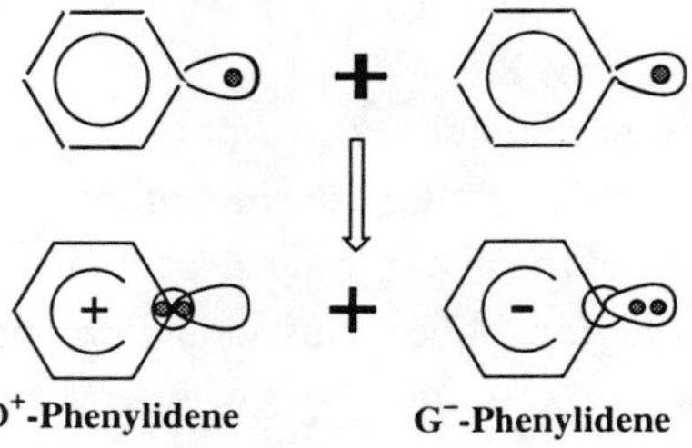

Thus, one pair of phenyl radicals is turned into one cation and one anion in which an overlying π system stabilizes the σ hole of the cation or the σ pair of the anion. This association yields a pair of ionic phenylidenes. Coulomb attraction of the nonbonded phenyl rings is engendered by cis placement.

The resulting structure is shown in Figure 24.7. There are two *cis*-Ph groups and *cis*-G^- and -D^+ phenylidenes, with the former acting as σ-LB and the latter as σ-LA. The four groups are linked to the central metal atom by two REL bonds. The Bi6s pair is totally uninvolved.

There are three important implications of this analysis.

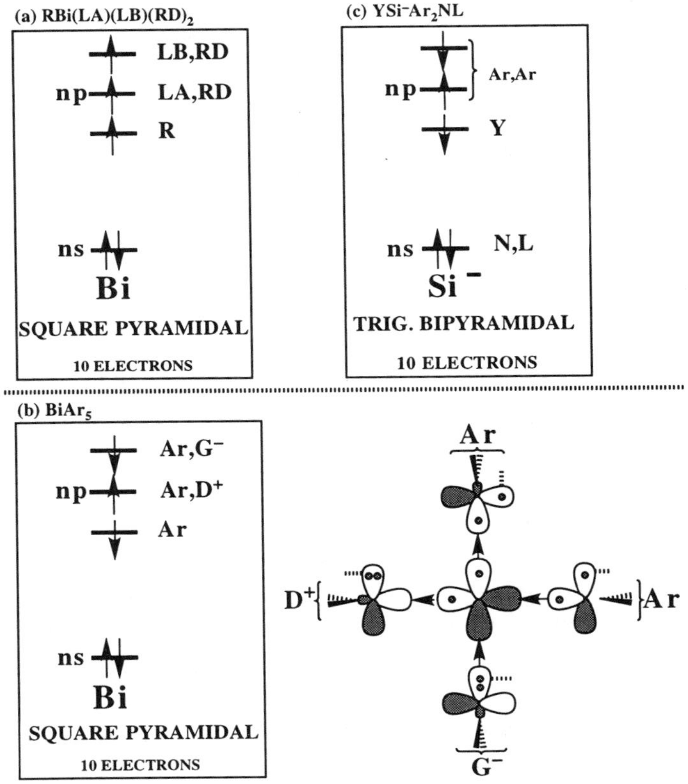

Figure 24.7 (a) A diagram of $RBi(LA)(LB)(RD)_2$ (LA = Lewis acid, LB = Lewis base, RD = radical). (b) The electronic structure of $BiAr_5$ (Ar = Aryl); two Ar ions have been converted to G^- and D^+. (c) The A diagram of N(Si^- Ar_2Y)L trigonal bipyramid (N^- = nucleophile, L = leaving group).

(a) Polyenyl and aryl groups attached on metals can function as "carbenes in disguise." Appropriate placement of substituents can further accentuate the carbenic function of apparent monovalent groups and lead to the discovery of new fields of fruitful research activity. This is one more implication of the concept of *carbenic resonance* discussed in Chapter 6.

(b) Two (rather than four) aryl groups will also stabilize zerovalent P by forming two complementary REL bonds through the intervention of their aromatic π systems as follows:

THE EQUATORIAL ENSEMBLE OF THE N-($Si^{-}Ar_2R$)-L TRIGONAL BIPYRAMID

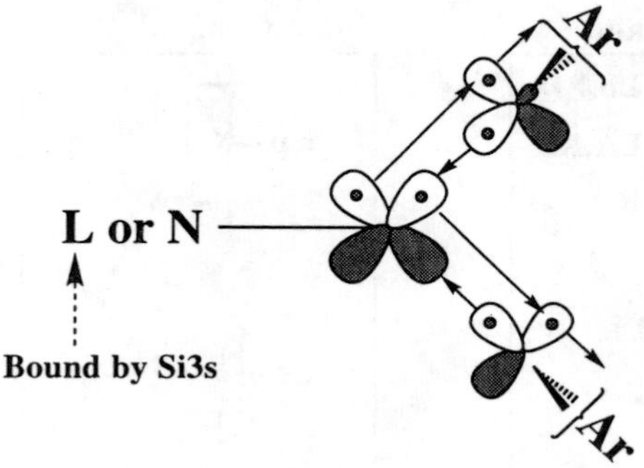

As a result, the aryl groups will relegate two electronegative atoms to the P3s pair for REL bonding. This pair of electronegative atoms will now bind in a cis geometry, leading us to predict a reversal of the stereochemistry of the S_N2 reaction at silicon,[9a] as discussed in Section 24.7.

(c) A carbanion with an overlying π system, such as ethoxyvinyllithium (EVL), is best represented as "singlet carbene stabilized by an anionic substituent." This explains why EVL, in contradistinction to alkyllithium, metallates a parasubstituted anisole ortho to its methoxy group.[9b] The vacant C2p AO of the "carbenic" EVL coordinates with the methoxy lone pairs.

24.7 Why and When Can S_N2 Reactions at Silicon Occur by Retention?

Corriu and his collaborators found that the system NpPhRSiL (Np = naphthyl, Ph = phenyl, R = alkyl, L = leaving group) undergoes replacement of L by N (N = nucleophile) by retention when L = H, F, and OR and N is also an electronegative first-row atom or group. In addition, NpPhRSiF reacts by retention, but NpPh(OMe)SiF reacts by inversion.[9a]

The presence of two aryl groups immediately suggests the explanation: The two aryl groups stabilize zerovalent P in the way described in Section 24.6. Because of the action of the two aryls, the formal Si^- 3s pair binds in *cis* geometry two electronegative radicals: N, the formal nucleophile radical cation, and L, the formal leaving atom. This binding mode predicts axial entry and equatorial departure and retention of configuration in the case of bimolecular nucleophilic substitution at silicon. The transition state is depicted schematically below. All this happened because the $P3p_x$ and $P3p_y$ AOs became tied up by the two aryls.

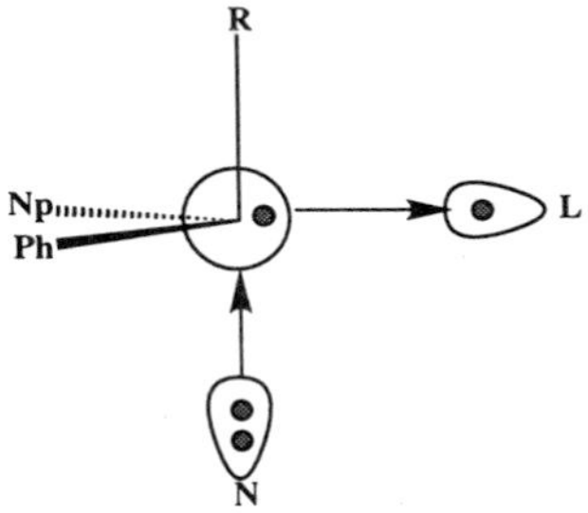

The superiority of the trigonal bipyramidal structure in which the two electronegative atoms are diaxial can be reestablished by a perturbation that turns the $P3p_x$ and $P3p_y$ odd electrons into an "internal zwitterion," thus destroying the two REL bonds with the two aryls. Now, the pair component of the internal zwitterion assumes the role of binding the two most electronegative atoms, while the two aryls are relegated to the 3s pair by means of replacing the axial R with OMe. The latter group has a π-type oxygen lone pair that has the effect just described. As a result, retention gives way to inversion because now N and L are associated in a diaxial sense with a 3p pair, rather than in an axial/equatorial sense with a 3s pair.

24.8 Why Heavy p-Block Atoms Prefer to Make Single-Normal Rather Than Multiple-Banana Bonds

"Suppression of details may yield results more interesting than a full treatment. More importantly, it may suggest new concepts. Pure quantum mechanics alone, in all its details, cannot supply a definition of, e.g., an acid or a base or a double bond." This quotation is attributed to E. Schrödinger. So, what is a "double bond"? In VB terms, a double bond can be formulated in either of two ways:

As one σ plus one π bond (σ/π formulation).
As two banana bonds (banana formulation).

At the limit of a complete treatment, the two formulations are equivalent. However, the banana formulation has the distinct conceptual advantage of rendering multiple-banana and single-normal bonds directly comparable. Recalling that "impaired T-bonding" means reduced exchange bonding and enhanced exchange antibonding, we can formulate two rules:

(a) Irrespective of atomic composition, multiple bonds suffer from strong exchange and contact repulsion.

(b) In going from nonmetal to metal multiple bonds, we shift from T- to I-bonding. This means that the importance of contact repulsion in determining stability is greatly enhanced and, thus, multiple bonding becomes increasingly unfavorable.

In short, multiple bonding is unfavorable in every system. This is why the polymerization of ethylene to polyethylene is exothermic, as well as every analogous process involving conversion of banana to normal bonds. However, multiple bonding is much more unfavorable in metal-containing molecules.

Schaefer et al.[10] showed that double bonds connecting semimetals in the alkene derivatives can be weaker than single bonds connecting the same atoms in the alkane derivatives, as indicated by the following list of molecules and their bond dissociation energies (kcal/mol).

$H_3Ge{-}GeH_3$	64.3	$H_2Ge{=}GeH_2$	36.9
$H_3Ge{-}SiH_3$	67.0	$H_2Ge{=}SiH_2$	44.8
$H_3Si{-}SiH_3$	69.4	$H_2Si{=}SiH_2$	52.0

The correct explanation exists only in the context of many-electron theory: One I bond connecting one pair of electropositive atoms is intrinsically unfavorable because the strong Coulomb repulsion of the overlap densities (contact repulsion) is tantamount to disproportionate interatomic promotional energy that offsets kinetic energy reduction. Ring formation, cage formation, or bridging improves the bonding by moving the overlap densities apart and minimizing contact repulsion.

In attempting to explain the relative instability of the Si=Si double bond, the argument embraced in the literature[11] goes as follows: $R_2S{=}SiR_2$ is weakly bound (relative to two ground R_2Si groups) because the promotional energy investment for preparing R_2Si for bonding ($3s^23p^0$ to $3s^13p^1$ orbital promotion) is disproportionately high as a result of the weak bonding effected by the contracted Si3s AO. This argument is fine if all that needs to be explained is the tendency of heavy congeners of carbon to adopt the divalent state, a well-recognized trend of inorganic chemistry. On the other hand, it constitutes no explanation for the preference for single over multiple bonding. R_2Si will tend to be divalent in $R_2Si{=}CH_2$ as well as in H—(R_2Si)—H; that is, both molecules suffer from the inert pair effect.

There is a second way of seeing that the orbital promotion argument is faulty: One needs to promote R_2Si to make $R_2Si{=}SiR_2$. However, it is not necessary to promote P to make P_2. And yet, the PP triple bond has all the earmarks of instability displayed by the SiSi double or triple bond! Thus, the "hatred" of semimetal and metal atoms (across the periodic table) for multiple bonds cannot be due to atom promotion. VB theory, in which covalency is now seen as a restriction, gives a direct and simple answer: Contact repulsion declines as the geminal internuclear angle of atom A (i.e., the angle formed by the two lines emanating from A in the Lewis formula) changes from zero to 109° in going from one formal double to two formal single bonds.

We can make a disjunction between this argument and the conventional models by simply writing the Lewis formula and proceeding as follows:

(a) If the connected orbitals are electronegative, the lines are taken to imply T-bond pairs.

(b) If the connected orbitals are electropositive (always in a relative sense), electron pairs are added to the lines to represent an equal number of I bond pairs.

The key point is that the criteria for evaluating T and I pairs are entirely different because the strength of the T bond depends primarily on overlap while the strength of the I bond is critically dependent on contact repulsion. After all, it is electron–electron repulsion that differentiates the two types of bond.

Nagase computed the relative energies of the valence isomers of $(AH)_6$ with A = C, Si, Ge, Sn, and Pb.[12] Restricting our attention to the benzene and prismane isomers, the results in Table 24.1 leave no doubt that there is a "night and day" difference between the bonding of nonmetals (carbon) and metals (germanium, tin, lead). Note how the extraordinarily unstable carbon prismane becomes the most stable isomer when C is replaced by Si. Note how the legendary benzene is superb for C but highly unfavorable for the semimetallic Si and the metallic Ge, Sn, and Pb. Carbon–benzene wins because of absence of

Table 24.1 Relative Energies of $(AH)_6$ Isomers

	E_{rel} (kcal/mol)	
A	"Benzene"	"Prismane"
C	0.0	+ 128
Si	0.0	–8
Ge	0.0	–14
Sn	0.0	–31
Pb	0.0	–67

strain of the σ T pairs and the offsetting of exchange repulsion of the π pairs by CT delocalization. Carbon–prismane loses because of extreme strain (interbond exchange repulsion) of the σ T pairs. By contrast, stannaprismane wins because of minimization of the contact repulsion of the I pairs. In actuality, the situation is even better in the case of the prism: The pairs move from the edges to the faces as described by the multicatenation cluster model, discussed later on. Stannabenzene loses because of maximization of contact repulsion of the I pairs (Figure 24.8).

24.9 CF_4, SiF_4, SF_4, and $SiLi_4$; Counterintuitive Differences and Similarities

The molecule SiF_4 seems to be a simple analogue of CF_4 by conventional yardsticks. From our vantage point, however, they are entirely different species.

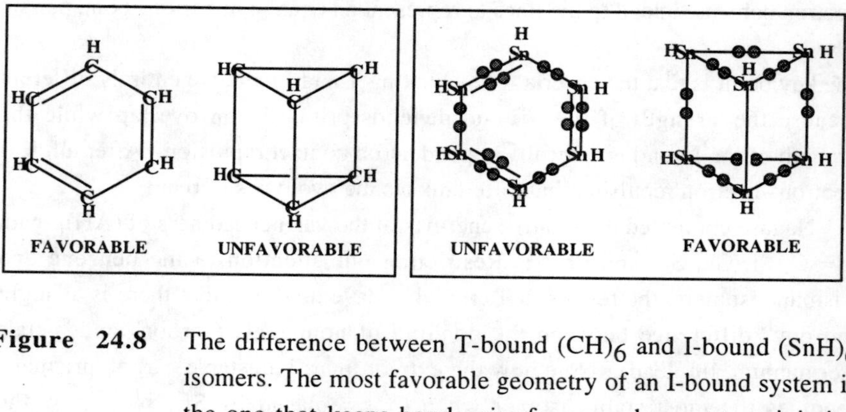

Figure 24.8 The difference between T-bound $(CH)_6$ and I-bound $(SnH)_6$ isomers. The most favorable geometry of an I-bound system is the one that keeps bond pairs far enough apart to minimize contact repulsion. As a result, benzene is the most stable $(CH)_6$ isomer but prismane is the most stable $(SnH)_6$ isomer!

The electronegative carbon engages in T-bonding but the electropositive silicon engages in I-bonding. CF_4 has four T bonds, but SiF_4 has two REL bonds, as illustrated by the A diagram of Figure 24.9a. That binding two ligand pairs by two double elements is only one of a large number of possible combinations of Si and ligand pairs is illustrated by the following experimental structures.[13]

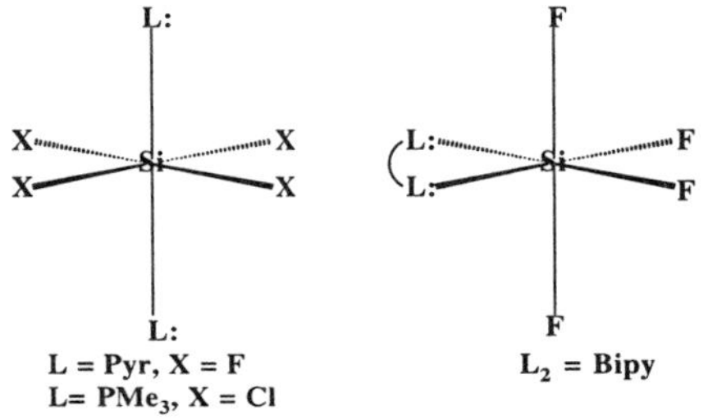

The corresponding A diagrams of Figure 24.9c show that the two LBs are bound either by single (axial lh) or double (equatorial hh) elements of two different isomeric Si configurations. It should be pointed out that whereas organic chemists tend to think of Si, Ge, Sn, and Pb as tetravalent analogues of carbon with a strong predisposition for the divalent state, each heavy congener of carbon can also be penta- and hexacoordinate, two modes of bonding that clearly set it apart from carbon itself. Examples can be found in the excellent monograph by Elschenbroich and Salzer[14a] and in the recent literature.[14b,c]

The silylenic nature of silanes is exposed by computations showing that the R_3Si—SiR is competitive with the $R_2Si{=}SiR_2$ isomer, with the carbenic form becoming further stabilized as we move down the column. This leads us to the formulation of a simple illustrator of antiferromagnetism due to an I bond, which may be the species involved in the disiladioxetane rearrangement observed by West et al.[15]

ANTIFERROMAGNETIC COUPLING IN $(R_2Si)_2O_2$

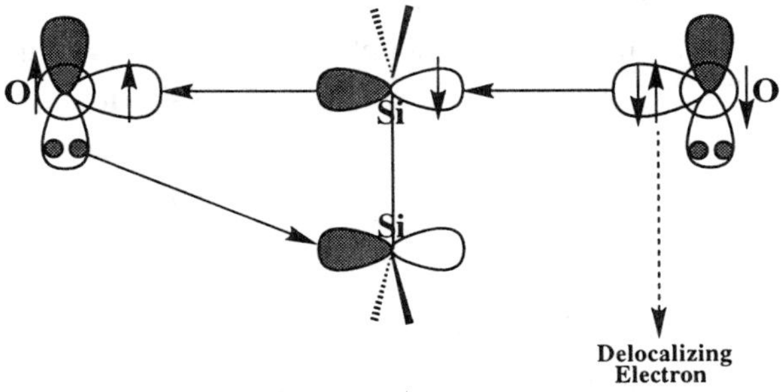

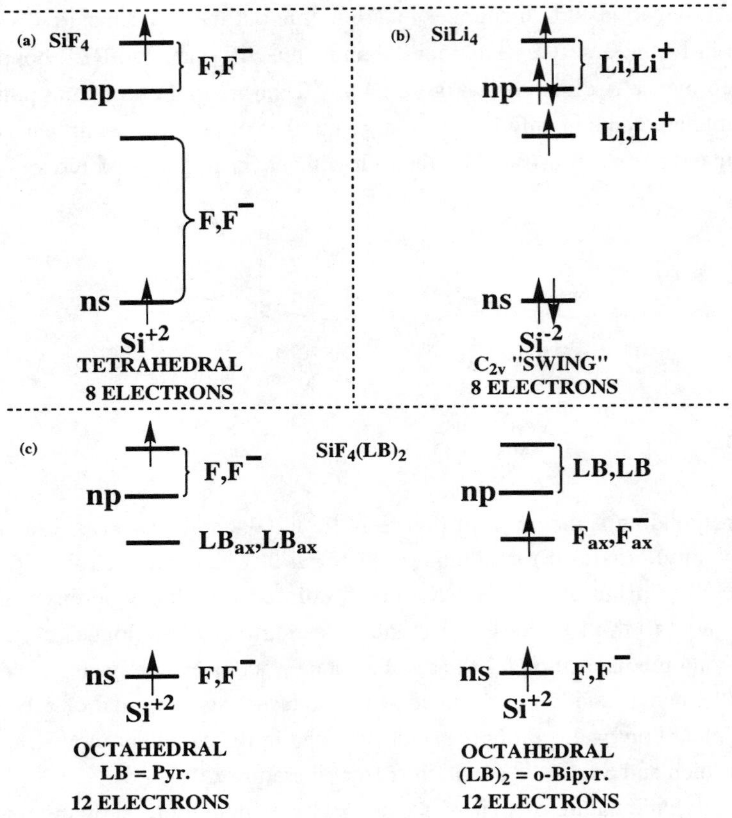

Figure 24.9 A diagrams of SiF_4 and derivatives. Note how one central atom element (hole, pair or dot) associates with two ligands (or two ligand elements).

In SiF_4, Si uses two rh's (two double elements) to tie up two F/F$^-$ pairs. In $SiLi_4$, the situation changes: Si uses one rp (one double element) and one lr (one single element) to tie up two Li/Li$^+$ pairs. The reason that the perlithio derivative abandons bonding by two double elements (two rp's) is simple: Not only does the Si3s AO lie far below the valence Li AOs, it is also contracted. The change in the bonding alters the geometry from C_{2v} or T_d in the case of SiF_4 to the C_{2v} structure found in SF_4 (compare Figures 24.3 and 24.9). The reason is that the number and stereochemistry of I bonds defined by Si (in $SiLi_4$) and by S (in SF_4) is conserved: One four-electron REL bond is replaced by a two-electron REL bond involving the same 3p AO of the central atom. This explains the ab initio computational results of Schleyer and co-workers.[16a]

24.10 The Sequential Bond Dissociation Energies of Halosilanes

The hallmark of I-bonding is that ligands bind in pairs. What happens when we have an odd number of ligands? In this case, the odd-out ligand is bound either by a covalent bond or by a three-electron or one-electron bond. An odd bond of this type should be weaker than one I bond, leading us to predict a sawtooth variation of sequential bond dissociation energies. For example, the first ($n = 3$) and third ($n = 1$) M—L BDE in MX_n—X, with $n = 0$–3, should be larger than the second ($n = 2$) and fourth ($n = 0$), respectively. The problem with these predictions is that the promotional energy of M is also dependent on the number of ligands. This muddles the picture, but enough is left to suggest that we are on the right track.

The difference in the bonding between alkanes and silanes is clearly projected by bond dissociation energy data reviewed by Walsh (Table 24.2).

Table 24.2 BDE Data (kcal/mol) for Alkanes and Silanes

BOND	BDE	BOND	BDE
SiH_3—H	90	CH_3—H	105
SiH_2—H	64	CH_2—H	111
SiH—H	84	CH—H	100
Si—H	70	C—H	81
$SiCl_3$—Cl	111	CCl_3—Cl	71
$SiCl_2$—Cl	66	CCl_2—Cl	67
SiCl—Cl	114	CCl—Cl	92
Si—Cl	91	C—Cl	80
SiF_3—F	160	CF_3—F	130
SiF_2—F	123	CF_2—F	88
SiF—F	155	CF—F	123
Si—F	132	C—F	129

Source: R. Walsh, *Acc. Chem. Res.* 14, 246 (1981).

The map of chemical bonding back in Chapter 3 makes these predictions:

(a) When CT is largely arrested (electroneutral X), the CX_n—X must be larger than the SiX_n—X BDE. SiX_n is a frustrated molecule. This is what happens when X = H.

(b) When CT is favorable (electronegative X), the SiX_n—X must now be larger than the CX_n—X BDE, provided I-bond formation is possible. This condition is met when $n = 3$ or 1 because single-atom cleavage no longer reflects the breakage of a covalent bond but rather the removal of one component of one I bond. As a result, the first and third BDEs of SiX_n—X reflect I-bond destruction, while the second and fourth reflect covalent bond cleavage. Thus, in the absence of any other competing effects, we expect the following correlation: The first and third BDEs of SiX_n must be larger than the corresponding CX_n BDEs because I-bonding is intrinsically superior to covalent bonding. Furthermore, the effect must vanish when we compare instead the second and fourth BDEs. The foregoing data have tantalizing features: The advantage of the Si—X over the C—X BDE in MX_n—X (M = C, Si; X = halogen) disappears in all cases except one (underlined in Table 24.3) when n attains the critical values of 2 or 0.

	BDE (kcal/mol) for:		
n	(SiF) – (CF)	(SiCl) – (CCl)	(SiH) – (CH)
3	30	40	–15
2	<u>35</u>	–1	–47
1	32	22	–16
0	3	11	–11

Since Cl is much more electropositive than F, $SiCl_4$ is a better model of I-bonding than SiF_4. In this light, the preceding thermochemical data make still better sense: The first and third BDEs of $SiCl_4$ are clearly superior to those of CCl_4, with the effect tending to disappear in the second and fourth BDEs. This goes hand in hand with a second observation: The sum of the first and the second BDEs far exceeds the sum of the third and fourth BDEs of $SiCl_4$. Completing the picture are thermochemical data pertaining to the following metathesis reaction.

$$MH_4 + MX_2 \longrightarrow MX_4 + MH_2$$

When M is a metallic congener of C and X is a heavy congener of F, the reaction is predicted to be endothermic because two halogens can become bound to M by an activated REL bond in MX_2 but not in MX_4. The activation is provided by two halogen pairs in conjunction with one np hole of M.

Here are some facts:

When M = Si and X = F, the reaction is close to thermoneutral, confirming the position of Si in the borderline region of the colored periodic table. However, when F is replaced by Cl, the reaction becomes endothermic by 30 kcal/mol.

Calculations by Kaupp and Schleyer suggest that the reaction endothermicity can reach up to 100 kcal/mol in the case of M = Pb and X = F![16b]

Bottom line: The nonmonotonic (sawtooth) variation of sequential BDEs within one MX_n—X system can always be explained by the argument that the first ligand promotes ("hybridizes") the central atom for a more effective binding of the second ligand, and so on, irrespective of the mechanism of bonding.[17] On the other hand, comparison of BDE variations in two different systems provides clues to the nature of bonding.

24.11 I-Conjugate Molecules

What is the shape and stability of the cluster BLi_5? The A diagram shown in Figure 24.10 speaks for itself. The shape is predicted to be the one found in the apparently unrelated ClF_5 (square pyramid with Cl at the center of the square). ClF_5 is derived from BLi_5 by simply turning boron holes into pairs. Since the same AOs are involved in the I-bonding, the stereochemistry is retained. Two molecules of this type are called *I-conjugate:* They have the same stereochemistry and the same number of I bonds; they differ only in the number of electrons defining the I bonds. This exercise shows that the C_{2v} $SiLi_4/SF_4$ comparison establishes a trend that links "hypovalent" (or "electron deficient") and "hypervalent" molecules. Thus, we have another piece of evidence that these molecules are actually "normal-valent" species obeying the same rule which, however, we could not discern before.

A more important test case is the structure of the Li_9^+ cluster. The A diagram of Figure 24.10 is interpreted as follows:

The three 2p dots of Li^{3-} form three REL bonds with three Li/Li^+ pairs along the three diagonals of an octahedron.

The last Li/Li^+ pair is bound by the 2s dot of Li^{3-} which, according to the stereochemical rules, binds cis.

This interpretation predicts a bicapped octahedron that upon slight readjustment becomes a dodecahedron with a guest Li^{3-}. But, the dodecahedron is an exemplary deltahedron, and it is the structure adopted by $(BH)_8^{2-}$ with a magic number of $(2 \times 8) + 2 = 18$. Our lithium cluster has only eight electrons!

Things get even more interesting when we consider the I-conjugate of Li_9^+,

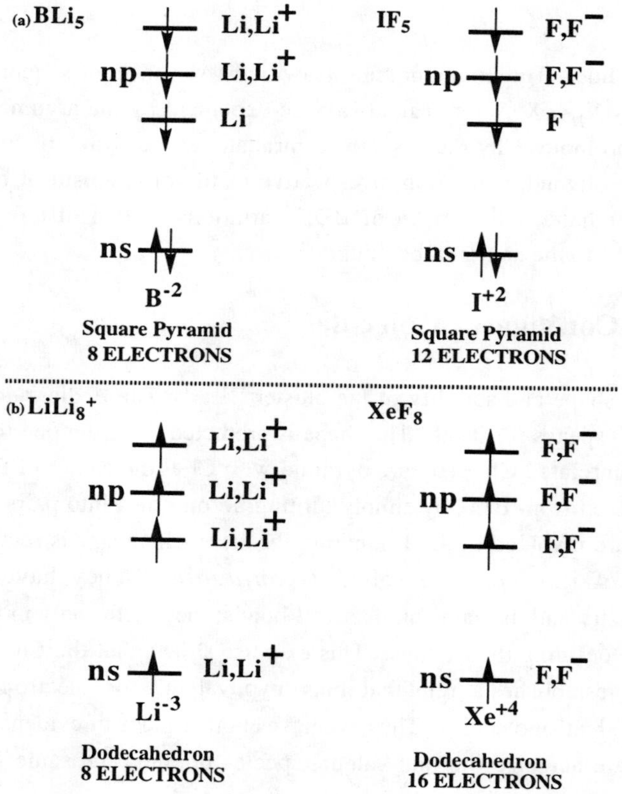

Figure 24.10 I-conjugate molecules have different electron counts but the same number of I bonds and the same shape.

namely, XeF_8. In XeF_6, each of three F/F^- pairs makes one REL bond with one Xe5p dot. Theoretically, one additional F/F^- pair could be bound by the Xe5s dot to form a dodecahedral F cluster with a xenon guest! Of course, this does not happen because cage "dilation" would be necessary to minimize nonbonded steric repulsion. In turn, this would impair the I bonds, and the molecule would disintegrate. For a suitable choice of cage atoms isoelectronic to F and a guest atom isoelectronic to Xe, we fully expect that a stable dodecahedron can be made! If $(BH)_8^{2-}$ can be made, every analogue conforming to the same bonding rules (but not the same electron count) should be realizable. Is there any good reason to think that inert gas halides and closoboranes are related? There is hardly a suspicion in the literature that this can be so. In defiance of ordinary intuition, we will now show that they are the closest of relatives.

24.12 The Octahedral $(BH)_6^{2-}$, XeF_6 and $(AuL)_6C^{2+}$ Are the Same Stories with Slightly Different Endings

There is a very simple, yet unusual, way of visualizing the $(BH)_6^{2-}$ octahedron: It is made up of a central *ghost atom*, created by a subspace of six BH, plus six "residual" BH fragments. The procedure is as follows:

(a) Each BH binds with one radial and two tangential AOs plus two valence electrons. This is the standard assumption from which all treatises depart.

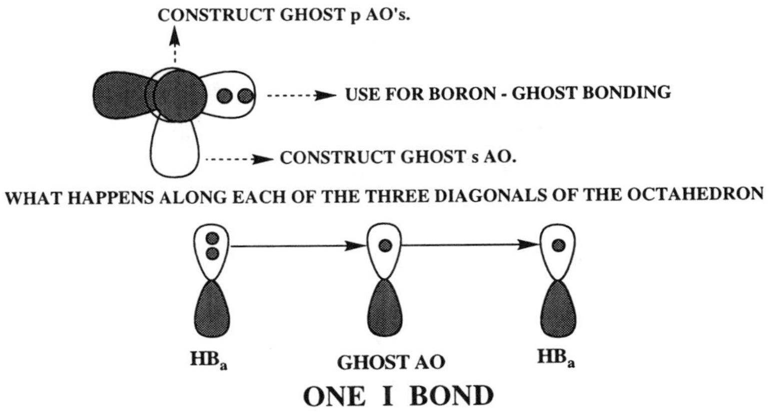

Figure 24.11 Ghost orbitals and the bonding of $(BH)_6^{2-}$.

(b) We collect one of the two tangential 2p AOs from each of the six BH units and build three ghost p AOs (p_x, p_y, p_z). Each ghost p AO is the in-phase combination of two diagonal BH tangential 2p AOs. The remaining three (out-of-phase) combinations are reserved for electron correlation.

(c) We collect the six radial AOs and we build one totally symmetric ghost s AO. The remaining combinations are again reserved for electron correlation.

(d) We end up with one tangential pair per BH plus one s-type and three p-type ghost holes. Reallocation of the electrons along one diagonal produces one REL bond, as illustrated in Figure 24.11. There are three REL bonds along the three octahedron diagonals. Two additional electrons are placed in the ghost s AO. The story is over!

The final ghost-electronic structure of $(BH)_6^{2-}$ is shown in Figure 24.12. Note the pairwise association of BH units with the "ghost" via a REL bond. The drawing of the ghost AOs is schematic. The prescription is that p_a (ghost) $= p_a(B) + p_a(B')$, where B and B′ are trans-located borons.

Comparison with the electronic structure of XeF_6 (Figure 24.13) makes the analogy evident. $(BH)_6^{2-}$ and XeF_6 have the same number of I bonds and the same skeletal electron count! The only physical difference between the two species is one that cannot be shown explicitly: The 2V-electron spherical cage is "tight" in XeF_6 but "loose" in $(BH)_6^{2-}$. Hence, the former deforms to better accommodate the "guest" s pair.

Schmidbaur and collaborators prepared and characterized a dipositive $(AuL)_6$ octahedron centered by carbon and isoelectronic species differing in the guest

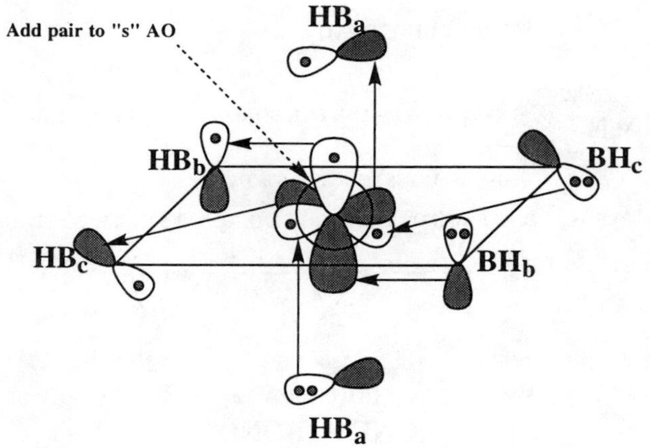

Figure 24.12 The electronic structure of $(BH)_6^{2-}$ and its derivatives.

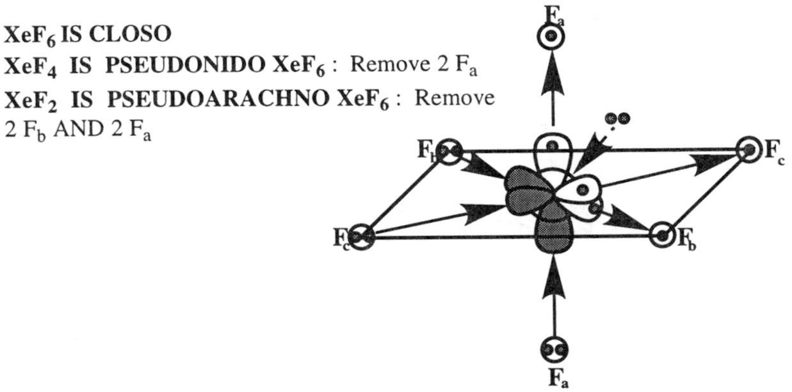

Figure 24.13 The electronic structure of XeF_6.

atom.[18] In this eight-electron octahedron, a formal tetraanionic carbon has one C2s pair plus three C2p pairs, each of which makes one I bond with two *trans*-LAu holes. Thus, this cluster is entirely analogous to a XeF_6 octahedron; the only difference is that in a formal sense, each Xe5p pair ties two fluorine atoms (radicals) while each C2p pair ties two LAu Lewis acids. The XeF_6 octahedron becomes distorted because of the exchange repulsion of the ligand odd electrons with the 5s pair. This effect disappears in $(LAu)_6C^{2+}$ (holes replace odd electrons), which stays undistorted.

Our understanding of the underpinnings of the electronic structures of apparently unrelated octahedra combined with the ghost atom model of cluster electronic structure allows us to add one more member to the series, namely, Ar_6.[19] Calculations suggest that inert gases form van der Waals complexes, which often prefer deltahedral structures. We can now see that the Ar_6 octahedron can be rationalized exactly like the $(BH)_6^{2-}$ octahedron with one difference: The ghost AOs are built from diffuse unoccupied nonvalence (rather than valence) AOs of the Ar atoms. As a result, an added electron pair can no longer be accommodated by the ghost atom. Since the radial AOs overlap more strongly with the ghost AOs, the Ar_6 octahedron is viewed as six Ar atoms linked by three I bonds to the central ghost atom through the utilization of one σ pair each. Hence, octahedral Ar_6 is a $2V$-electron deltahedral cluster.

Chemists perceive as analogous two molecules that have similar shapes. In other words, geometry becomes a yardstick of similarity. However, one may also argue that two molecules are analogous if they have the same number of I bonds. Now, it is the electron count that becomes the measure of similarity. In this light, we can now add the 14-electron linear stannocene $SnCp_2$ (where Cp is cyclopentadienyl) to the list of the title 14-electron clusters. As illustrated, *linear* stannocene has three I bonds plus one Sn5s lone pair:

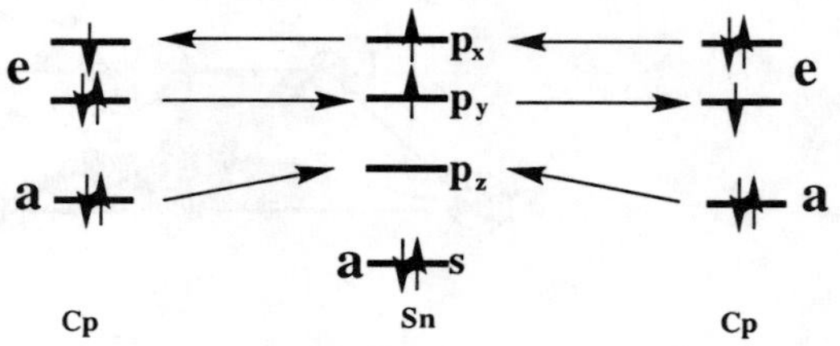

The relationship between XeF_6 and $SnCp_2$ becomes now self-evident: In both cases one lone pair of the central atom is encapsulated by a cage defined by the ligands. Because of the small cage size, each molecule undergoes a distortion, with the result that the lone pair spills out of the cage. XeF_6 abandons the octahedral shape and $SnCp_2$ bends.

Finally, let us focus on the physics that encompasses the bonding of all five (apparently unrelated) molecules: SF_6, XeF_6, $(BH)_6^{2-}$, Ar_6, and $SnCp_2$. In each case, three I bonds connect a central (real or ghost) atom to a set of ligands located on the surface of a sphere. Each I bond can be pictured as electron density along a sphere diameter. Optimization of I-bonding requires minimization of contact repulsion. This means maximal spacing of the sphere diameters representing the I bonds. By the same token, the optimal geometry is a polyhedral arrangement that approaches a sphere in which the ligands are spaced as far as possible from each other on the surface of the sphere. This minimizes interpair repulsion, in general. The polyhedron that approaches a sphere is the deltahedron. This explains why deltahedral preference is the common denominator of diverse molecules which, however, have one common characteristic: They are made up of semimetal or metal atoms.

24.13 The Skeleton in the Closet of Inorganic Chemistry

According to the theory outlined here, alkali halides are E-bound only when the metal atom is small and highly electropositive and activated I-bonding is not possible. More than any other chemists, solid state researchers make extensive use of the concepts of covalency and ionicity only to realize that these ideas are suspect, at best! For example, they use heats of formation to determine the electronegativity difference of two atoms and, on this basis, determine whether a binary solid is covalent or electrovalent (i.e., ionic). The usual prescription follows, in which x_A is the electronegativity of atom A, Q is the heat of formation of the compound (kJ/mol), and n is the number of bonds connecting a central atom A with some ligands B.

$$\Delta x = x_A - x_B = \left(\frac{Q}{n}\right)^{1/2}$$

For example, the electronegativity difference of P and I in PI_3 is computed by putting Q = 45.6 kJ/mol and n = 3. Using this procedure, one calculates that $\Delta x(SiO_2)$ is roughly equal to $\Delta x(Na_2O)$, even though these two species are commonly regarded as covalent and electrovalent, respectively! I am aware of two thoughtful monographs on solid state stereochemistry[20] in which the authors recognize that point-charge models are approximate versions of some other elusive bonding mode.

The belief that "chemical bonding is understood" cannot be maintained in the face of three facts:

(a) Despite Ba being much more electropositive than Be, BeF_2 is linear but BaF_2 is bent.[21] In addition, BaO (BDE = 133 kcal/mol) is much more strongly bound than BeO (BDE = 106 kcal/mol). Finally, Be solid is hexagonal close-packed (hcp), while Ba solid is body-centered cubic (bcc).

(b) Li_2O is linear but Cs_2O is bent.[22]

(c) Consideration of the bond dissociation energies of the alkaline earth oxides reveals a striking pattern: There is a decrease in going from Be to Mg followed by a relative flattening and a final pronounced rise in going from Sr to Ba. A similar trend, though less pronounced, is met in the case of the BDEs of alkali fluorides. Clearly, the BDEs do not show the same variation as the metal ionization potentials. Thus, ionic bonding is ruled out.

	BeO	MgO	CaO	SrO	BaO
IP(M), eV	9.3	7.6	6.1	5.7	5.2
BDE, kcal/mol	106	94	100	97	133

	LiF	NaF	KF	RbF	CsF
IP(M), eV	5.4	5.1	4.3	4.2	3.9
BDE, kcal/mol	137	114	117	120	123

The organometallic pantheon of Figure 24.2 provides the explanation, which relies on the recognition that as we go down the alkali and rare-earth column, the atom size increases and the $(n-1)$d orbitals are stabilized relative to the ns and np AOs, ultimately dropping below the np AOs. The key point is that all alkalies are white atoms that have only two options: E-binding or activated I-binding. With the possible exception of FBeF, the smaller alkalies adopt the E strategy. On the other hand, the larger alkalies must fall back on activated I-bonding. Because of the presence of quasi-degenerate s, p, and d valence AOs, these atoms are best suited for the latter type of bonding. Hence, the common denominator of all three counterintuitive experimental results is the efficacy of activated REL I-bonding, as illustrated:

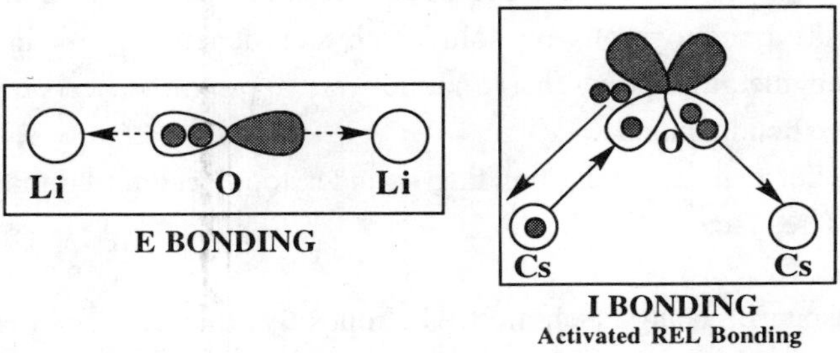

While it is easy to explain the angular shape of FBaF by invoking sd hybridization and covalency, it is impossible to rationalize the angular shape of CsOCs without invoking metal–metal bonding. We suggest that this is not the case. Rather, the angular shape of CsOCs is a consequence of activated REL-bonding: The activator arrow connects a nonmetal lone pair with a pd hybrid of the metal atom. An alternative interpretation will be presented later, when we discuss a counterintuitive scheme of multiple bonding.

24.14 Phantom Bonding

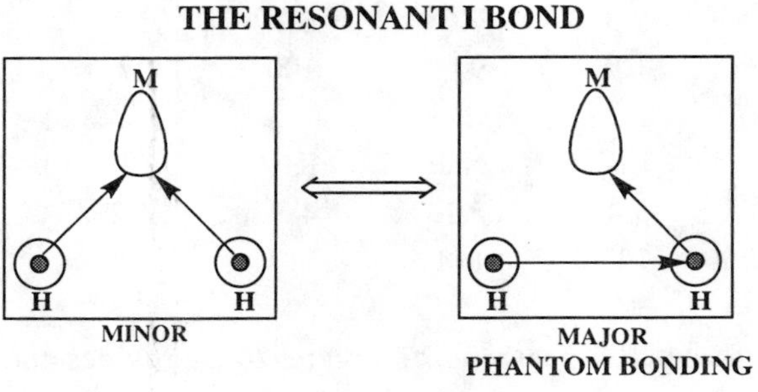

Figure 24.14 Origin of phantom bonding.

Phantom bonding is defined as the accumulation of bonding electron density in molecular regions where one would expect to find none on the basis of the conventional covalent/ionic model. This "extra bonding" is the result of I-bond resonance. The quintessential example of phantom bonding is the metal–dihydrogen complex. Because of the spherial nature of the H1s AO, H becomes a good participant in I-bonding despite its relatively high electronegativity. Figure 24.14 shows how metal–H_2 binding occurs in such a way so that one resonant I bond can simultaneously effect H—H and M—H bonding. Note that the major contributor is the one involving a codirectional arrow train. Furthermore, consider the implications: Every two-electron/three-center array of the ABB type, where B is more electronegative than A, will be tolerant to adoption of either an "open" or a "closed" geometry because of the superiority of arrow codirectionality. If arrow contradirectionality were superior, there would be a strong preference for the closed form. This observation about geometry preference rationalizes the geometry of monoalkynyl derivatives of Al and its heavier congeners.[23]

The most interesting chemical fact that supports the notion of the resonant I bond is the methylation effect on bond lengths and bond strengths. Specifically, it is known that replacing H by Me lengthens and weakens the C—C bond of ethane. By contrast, the same replacement shortens and strengthens the Si—Si bond of disilane![24]

	BDE (kcal/mol)
H_3C—CH_3	88
Me_3C—CMe_3	70
H_3Si—SiH_3	74
Me_3Si—$SiMe_3$	81

ONE RESONANT REL I BOND [I FORMULA]

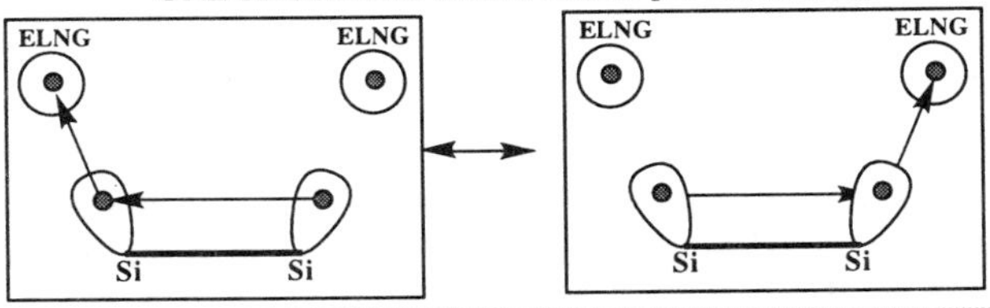

PERFECT-PAIRING PLUS INTRABOND CT DELOCALIZATION [T FORMULA]

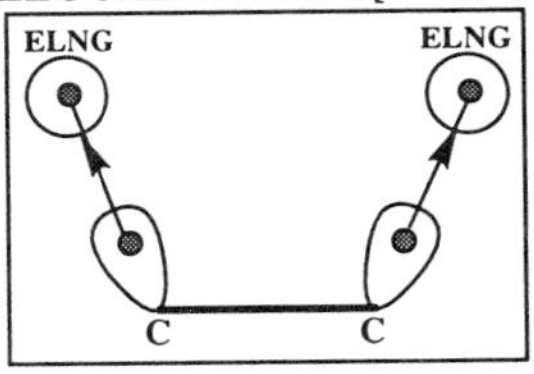

The explanation is that Me is a systolic group, which is superior to H as a σ acceptor. Replacing H by Me enhances CT and the relative importance of the REL-bonding. In turn, this enhances phantom bonding, as illustrated above.

A substantial amount of structural data defines the following theme: Nonbonded semimetal or metal atoms come closer together when linked to electronegative nonmetals. In this case, strong CT favors REL-bonding, which, in turn, creates phantom bonding. Some of the best evidence comes from the investigations of West and co-workers, who found that the silicon–silicon bond distance in small siloxane rings is abnormally short compared to cyclic or acyclic polysilanes.[25]

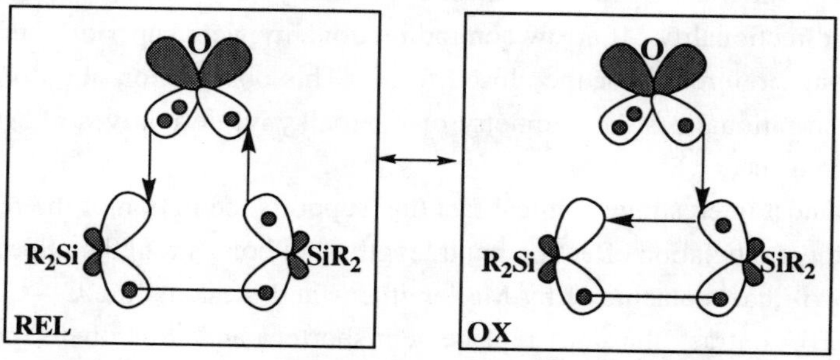

Phantom Bonding

VB theory gives a simple and straightforward answer: The molecule above is not an oxirane-type molecule with two covalent C—O bonds but, rather, an association molecule in which $R_2Si{=}SiR_2$ is connected to O by a single resonating I bond having REL character. This is why the two silicas stay "abnormally" close together.

24.15 The Different Types of "Ethylene"

Heavy congeners of carbon form compounds in which a nonmetal atom or group acts as a bridging ligand to generate an MLLM array. Using the heteroatom exclusion rule and the ideas about the hydrogen transfer, we can connect the nature of L to the nature of the bonding mechanism:

When L = F, we have electron transfer from M to F, arrow codirectionality is inhibited, and the MLLM array is E-bound.

When L = H, we have formation of two complementary REL I bonds, each connecting three centers.[26]

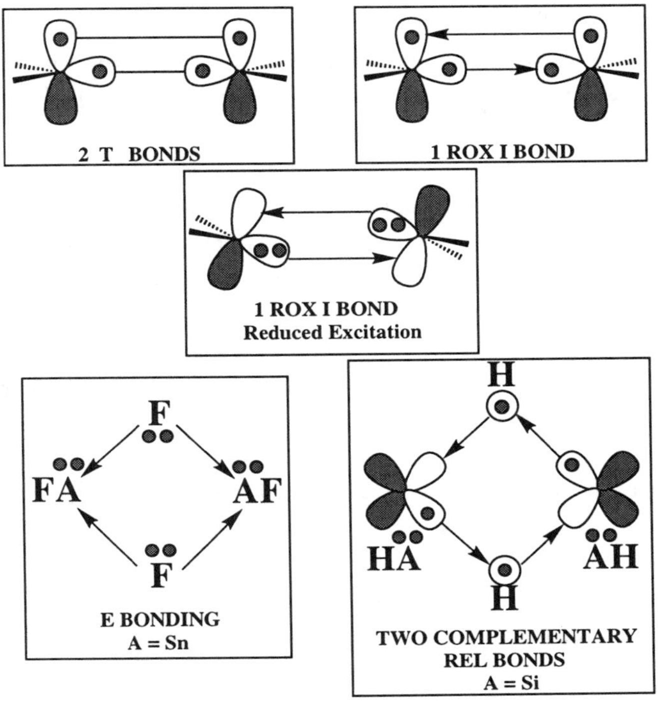

Figure 24.15 Different types of X_2AAX_2 structure illustrating T-, I-, and E-bonding.

With these guidelines, we can write a representative spectrum of chemical structures and identify the bonding mechanism as shown in Figure 24.15.

The explanation of trans bending in disilaethylene dates back to Pauling and has been recast in MO language. However, all explanations miss the point: Ethylene and disilaethylene involve a different mechanism of chemical bonding. In particular, computational chemists have attempted to understand the origin of trans deformation and double bridging of heavy p-block derivatives of ethylenes. One correlation, which follows from the Pauling double-carbene formulation, is between the degree of trans bending and the singlet–triplet gap of each methylene fragment. The more destabilized the triplet relative to the singlet, the greater the tendency for trans bending. Superficially, this works in going from the Si to the Sn derivatives. However, a key question is left unanswered: Why are tetrafluoro-, tetraalkoxy-, and tetraaminoethylene "normal," although CF_2, $C(OR)_2$, and $C(NR_2)_2$ have a more stabilized singlet than R_2Si and $C(NR_2)_2$ has a more stabilized singlet than even R_2Sn? The answer is that in ethylenic derivatives, we have T-bonding and conventional organic structures. Substituting the hydrogens (by F, OR, etc.) changes the C=C bond dissociation energy by changing the requisite methylene promotional energy, but it does not change the nature of the bonding. Thus, the gross shape of the unsubstituted parent is retained. By contrast, reducing the electronegativity of C (by replacing it by Si, Ge, etc.) takes us to another bonding realm. The trans bending of planar $Y_2Si{=}SiY_2$ is the mechanism by which fragment de-excitation is achieved at the expense of reducing the T complement (covalency) of the I bond connecting the two silylenes.

24.16 Symbiosis on Semimetals and the Strength of p-Block Acids

The distinguishing features of heavy p-block elements are contracted ns pairs and np lone pairs that can be polarized by the action of the nonvalence unoccupied space.

It follows that in transforming the I formula to a T formula by applying the exclusion wave condition (as we showed in Chapter 6), the only difference between first-row p-block nonmetals and the heavier congeners is that the lone pairs no longer determine the sense arrow in semimetals or metals. Rather, it is the bond linking the two atoms of highest electronegativity that determines the group peristolicity. In the following illustrative diagrams, note how easy it becomes to explain the inexplicable — for example, why F_2SS and DMSO (1,1 isomers) are more stable than the conventional 1,2 isomers. This means that the concepts of Chapters 6 to 8 can be easily extended to the entire p block. With this assumption, we can now understand acidity in the different light.

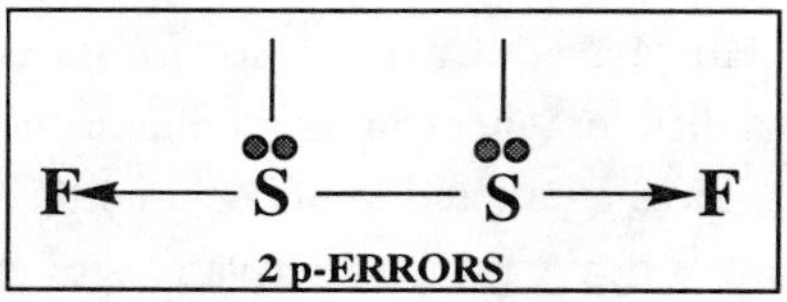

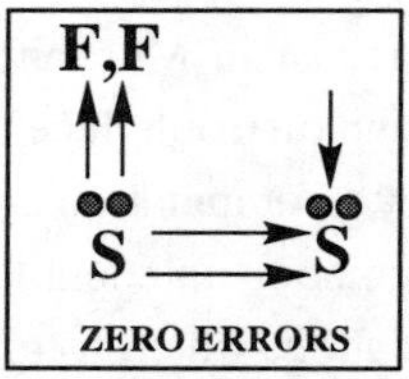

R=Me

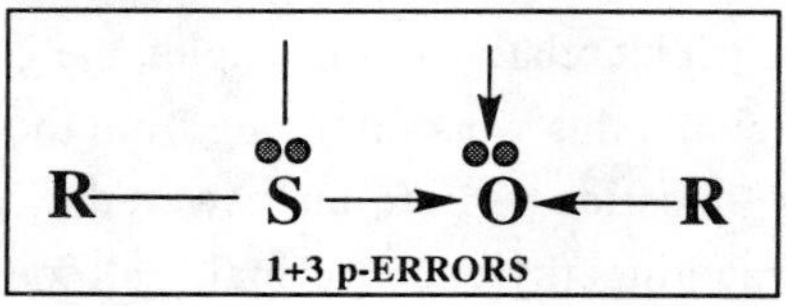

R=Me

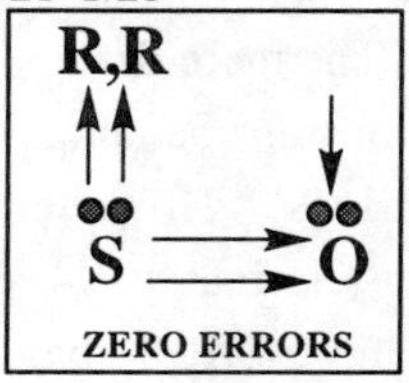

Cl—OH is a weak but O_3Cl—OH is an exceedingly strong acid. It is argued that increasing the number of oxygens makes the central atom more positive and the acid stronger. We argue that the added oxygens simply eliminate the errors and, according to the signature concept, enhance acidity. In other words, the key is that the oxygen pair released by proton departure suffers less exchange repulsion as the number of the oxygen ligands increases. Perchloric acid is strong for the same reason that CF_4 and CO_2 are stable!

The A diagrams and the corresponding schematic T formulas of Cl—OH and O_3Cl—OH shown in Figure 24.16 reveal that the former has three errors while the latter has none. The errors are the lines emanating from the lone pairs. These are ideally systolic arrows which the action of the Cl—O bond has turned into lines. The analogy between Cl—OH and CH_3—F and the analogy between O_3C—OH and CF_4 is also projected in Figure 24.16. In short, acidity has to do with exchange repulsion.

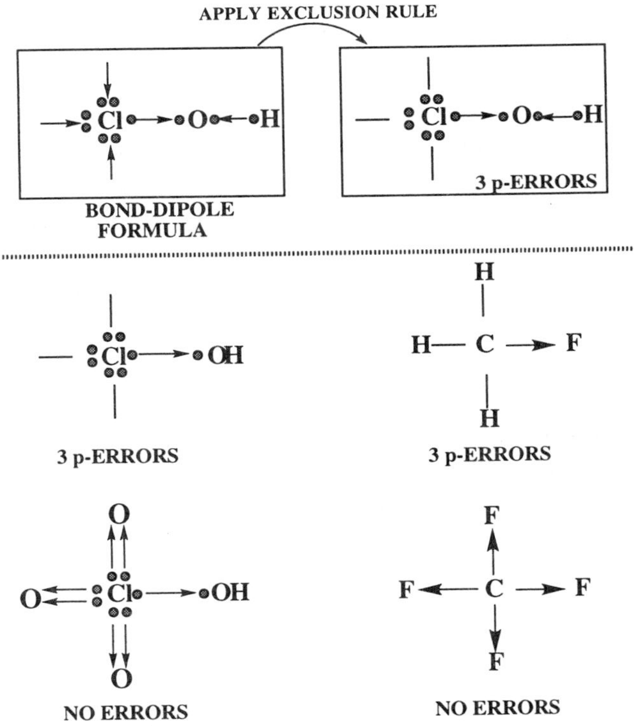

Figure 24.16 T formulas of oxyacids and analogy between chlorine oxyacids and fluoromethanes.

An interesting problem arises in the sulfur series of oxy acids: Triflic acid (HO—SO_2—CF_3) is much stronger than sulfuric acid (HO—SO_2—OH), even though CF_3 is less electronegative than OH! Once again, our experiences in Chapter 6 provide the clue: CF_3 can act as a systolic group because, in this way, there is atom relaxation signaled by a resonance structure featuring the highly stabilized singlet CF_2 carbene unit. Our explanation of the exceptional acidity of triflic acid is carbenic resonance (see Chapter 6).

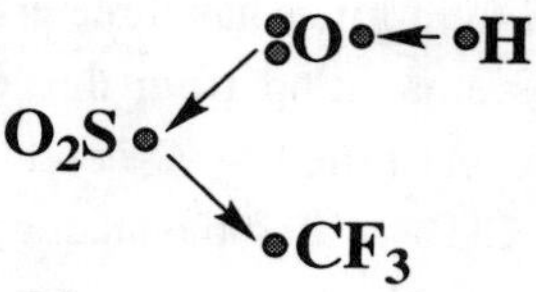

TRIFLIC ACID

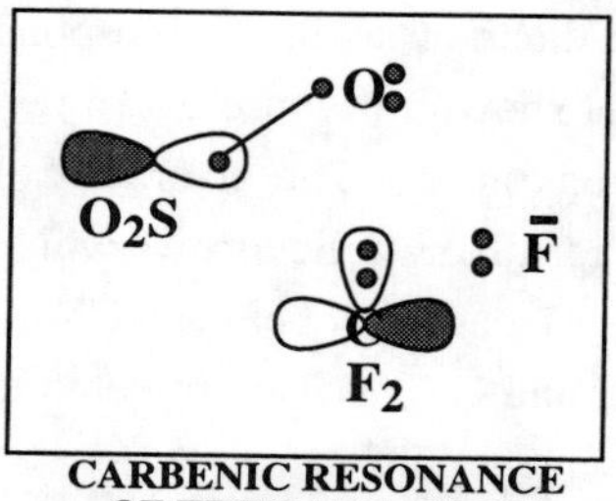

CARBENIC RESONANCE OF TRIFLATE ANION

24.17 Associative Saturation and Unsaturation

The value of theory lies in suggesting what has not been done that can be done. The association rules and the A diagrams have taught us that a given semimetal or metal atom has a minimum and a maximum valence. The applications presented above suggest that the surface of inorganic chemistry has barely been scratched! For example, we saw that Xe can take up to eight electronegative radicals. XeF_8 is called *associatively saturated*, or, more briefly, A-saturated. The even-electron xenon fluorides form a series: XeF_2, XeF_4, XeF_6, and XeF_8 (at the moment, only the first four members are known), and the first three are A-unsaturated.

Next, we can see that replacing the electronegative F radicals by powerful Lewis acids should yield the corresponding $Xe(LA)_n$ series, with $n = 2, 4, 6, 8$. At present, these molecules are unknown. They will be realized when the appropriate strong LA is found.

In contrast to XeF*n* (n = even), all members of the $(IF)F_n$ series have been isolated. The A-saturated member is the pentagonal bipyramidal IF_7. Exactly the same deltahedral shape is adopted by the apparently unrelated $(BH)_7^{2-}$. The A diagram is shown in Figure 24.3b.

SF_2 is conventionally referred to as "electron precise" and SF_4 as "hypervalent." And yet, the former apparently normal molecule decomposes to S_8 plus SF_4, an apparently abnormal molecule! We now know why: SF_2 is associatively unsaturated relative to SF_4 and both are bound by I bonds. Conventional theory has simply produced an illusion with which chemists have lived for decades.

24.18 The Folly of "Hypervalency," "Hypovalency," and "Electron Precision"

Sulfuric acid is a "high school molecule". Sulfur hexafluoride is a "college molecule." And yet the two are interrelated by the concept of the I bond: Sulfuric acid can be thought of as a derivative of the "classical" $(HO)_2S$ in which each of the 3p and 3s lone pairs is capped as shown by an oxygen atom through a REL bond.

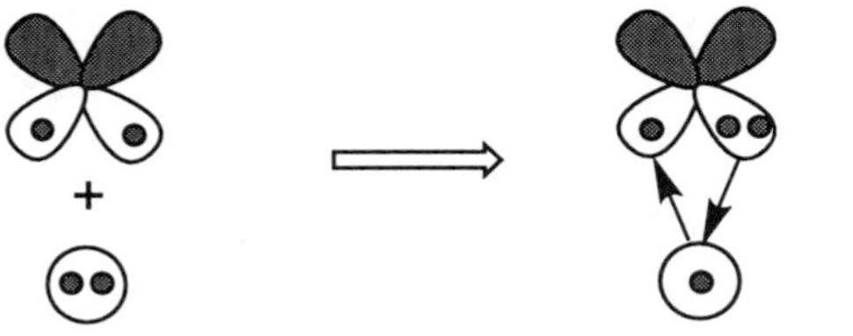

Similarly, sulfur hexafluoride can be thought of as a derivative of the "classical" F_2S in which each of the 3p and 3s lone pairs is capped by a pair of fluorine atoms through a REL bond. The key point is that each REL bond is described by a complex VB CI over the normally chosen set of AOs in which individual configurations interact, regardless of whether they overlap. Rather than being a mere detail, the Coulomb exchange (zero-overlap) interaction of the following type is the one responsible for I-bond (REL-bond) formation.

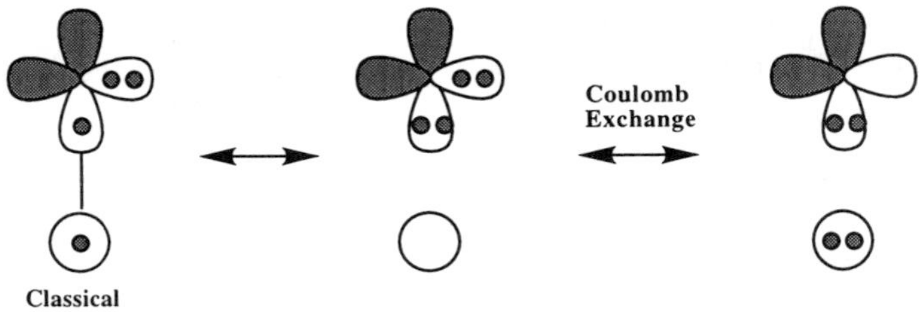

According to VB theory, hypervalent molecules are not "hypervalent" and electron-precise molecules are not "electron precise." Rather, they are all I-valent molecules. For example, according to conventional theory, SiO_2 is an isoelectronic analogue of CO_2 having four covalent bonds and represented by the Lewis structure O═Si═O. VB theory paints a different picture: Because of its much lower electronegativity, Si cannot justify a 3s-to-3p promotion (even though the 3s–3p gap in Si is smaller than the 2s–2p gap in C) for the purpose of forming four weak (in a relative sense) covalent bonds that suffer from strong interelectronic repulsion. Instead, Si stays unpromoted and choses a different strategy: Rather than four T bonds, it makes two I bonds, which involve lower interelectronic repulsion. One oxygen ties up two singly occupied Si3p AOs and the second oxygen caps the Si3s lone pair. Transferring one 3p and one 3s electron to the two oxygens generates two REL bonds, and the molecule is represented as $Si^{2+}(O^-)[O^-]$ in which the parentheses imply a REL bond spanning three AOs and the bracket a REL bond spanning four AOs. Superposition of the two equivalent resonance structures $Si^{2+}(O^-)[O^-]$ and $Si^{2+}[O^-](O^-)$ is the representation of the symmetrical linear molecule. The formula now explains the fundamental difference between CO_2 and SiO_2: The latter polymerizes to $(SiO_2)_x$ because it has silicon holes and oxygen pairs that set the stage for further I-bonding, and the molecule has a large quadrupole moment that sets the stage for strong Coulomb attraction. The actual oxidation state of Si is II rather than the formal IV. We end by saying the obvious: The Si3p AO that is not populated in our scheme does play some role. However, the structure of the molecule is closer to $Si^{2+}(O^-)[O^-] \longleftrightarrow Si^{2+}[O^-](O^-)$ than to $Si^{2+}[O^-][O^-]$.

We can use the "oxygen cap" procedure to gain an immediate understanding of the electronic structures of "hypervalent" and "electron-precise" metalloids in a way that differentiates them from isoelectronic first-row molecules:

(a) The strong electrophilic action of SO_3 can be understood by starting with singlet SO in which each atom is sp^2-hybridized and capping the two S lone pairs by two REL bonds. Each capping oxygen induces one-electron transfer from semimetal to oxygen to form the REL bond. If we assume that the actual oxidation state of a central atom approximately equals the number of oxygen-capped lone pairs (where a lone pair is a pair occupying a single-valence AO), the sulfur atom of SO_3 is S(II). The +2 charge on S in addition to the partly vacant $S3p_z$ AO qualifies SO_3 as a strong electrophile.

(b) Acidity increases in the order $HClO_4 > H_2SO_4 > H_3PO_4$, reflecting the number of oxygen-capped lone pairs: 3, 2, and 1, respectively. As a result, the charge of the central atom increases and so does the acidity to the hydroxylic group(s).

(c) The conventional Lewis structure of trialkyl phosphate, $(RO)_3P{=}O$, implies that the P=O bond is roughly twice as strong as the P—OR bond. This explains why $(HO)_3P$ is actually unstable with respect to $(HO)_2HPO$. However, ab initio computations suggest that d orbitals are unimportant and pentacovalency an illusion (Chapter 28). The association formula RO—P(RO,RO)(O) resolves the conflict. Each set of atoms in parentheses is linked to phosphorus by one REL bond and, in addition, there is one odd RO—P T bond. In essence, each of the two ROs in parentheses is linked by one-half I bond to P, while O is linked by one full I bond to P. Furthermore, the RO—P T bond is inferior to the P—O I bond. This explains why the P—O bond is effectively stronger than the RO—P bond in the absence of d-orbital involvement. As we will see in Chapter 28, the nonvalence unoccupied space of semimetals, simulated by the d AOs, has important chemical consequences, which become visible only in systems in which the valence space has played out its part. However, for most problems involving metal–containing molecules, focusing on the valence I bonds is sufficient.

It must now be apparent that we have entered a new era of "back-of-the-envelope" chemistry in which new stable molecules can be discovered by using the new formulas and procedures described in this work. For example, we can take transition metals in their ground electronic configurations, cap the electron pairs (either geminal pairs or lone pairs) by oxygens, allow for geometry relaxation, and come up with the likely closed-shell global minimum. By using this approach in conjunction with test calculations and test experiments, we hope to make metal chemistry as predictable on the back of an envelope as organic chemistry.

Bottom line: We have a simple procedure for rationalizing and predicting the structures of metal or semimetal oxides. One oxygen atom caps a pair of electrons, whether the two occupy two AOs (geminal pair) or one (lone pair). Either way, one forms one I bond, the only difference lying in the T complement. Thus, "hypervalent" fluorides, oxides, nitrides, and carbides are perfectly normal I-valent molecules when viewed from our present vantage point. On the other hand, "electron-precise" molecules are not what the name implies but, rather, simple analogues of the "hypervalent" molecules; that is, they are also perfectly normal I-valent molecules. The common denominator is the I bond, which is symbolized by two arrows. Since two arrows can connect four, three, or two AOs (although the latter possibility is eliminated because it does not have any advantage relative to a T bond), there are two types of I bond: Those spanning four AOs (e.g., oxygen atom capping a metal geminal pair) and those spanning three AOs (e.g., oxygen atom capping a metal lone pair). "Hypervalent" molecules have some I bonds of the first type and some of the second type (e.g., SF_4). "Electron-precise" molecules either have I bonds of the first type only (e.g., SF_2) or, if the frame of reference becomes the unpromoted configuration of the metal atom (e.g., SiF_4 and SiO_2), are actually "hypervalent." In summary, there is nothing "abnormal" about "hypervalency" or "hypovalency," and there is nothing "normal" about "electron precision"! The two statements are reconciled by the recognition that the real anomalies are the bonding theories of the past, as they have carried over into the present. Once the third dimension of bonding, namely, I-bonding, is seen, all molecules fall on the T/I/E map of chemical bonding, and the real game of chemical selectivity begins.

References

1. (a) R.J. Puddephatt and P.K. Monaghan, *The Periodic Table of the Elements*, 2nd ed., Clarendon Press, Oxford, 1986. (b) N.N. Greenwood and A. Earnshaw, *Chemistry of the Elements*, Pergamon Press, New York, 1986. (c) In recent years there has been an interest in periodic systems of molecules and interesting ideas have been put forth by R. Heferlin and G.W. Burdick, *Abstracts X Mendeleev Discussion*, St. Petersburg, Russia, 1993, and M. Randic, *J. Chem. Educ.* 64, 713 (1992).
2. W.A. Herrmann, *Angew. Chem. Int. Ed. Engl.* 25, 56 (1986).
3. (a) H.G. von Schnering, in *Homoatomic Rings, Chains and Macromolecules of Main-Group Elements*, A.L. Rheingold, Ed., Elsevier, Amsterdam, 1977. (b) M. Baudler and K. Glinka, *Chem. Rev.* 93, 1623 (1993).
4. (a) M. Haeser, U. Schneider, and R. Ahlrichs, *J. Am. Chem. Soc.* 114, 9551 (1992). (b) M. Haeser, *J. Am. Chem. Soc.* 116, 6925 (1994).
5. D.A. Dixon, A.J. Arduengo and T. Fukunaga, *J. Am. Chem. Soc.* 108, 2461 (1986). See also P. Schwerdtfeger, P.D.W. Boyd, T. Fischer, P. Hunt, and M. Liddell, *J. Am. Chem. Soc.* 116, 9620 (1994), and Y. Yamamoto, X. Chen, S. Kojima, K. Ohdoi, M. Kitano, Y. Doi, and K. Akiba, *J. Am. Chem. Soc.* 117, 3922 (1995).
6. G. Huttner, J. Borm, and L. Zsolnai, *J. Organomet. Chem.* 263, C33 (1984).

7. The presence of fluorines drives mixed-ligand molecules to a TBP geometry. $BiPh_3F_2$ is TBP with F atoms axial: A. Schmuck, D. Leopold, S. Wallenhauer, and K. Seppelt, *Chem. Ber.* 123, 761 (1990). This is further evidence that the SP preference of $BiPh_5$ is due to a distinctive property of the phenyl groups.
8. (a) PPh_5 and $AsPh_5$ are TBP but $SbPh_5$ is SP: P.J. Wheatley and G. Wittig, *Proc. Chem. Soc.* (London) 251 (1962); P.J. Wheatley,*J. Chem. Soc.* 2206 (1964); C.P. Brock and D.F. Webster, *Acta Crystallogr. B* 32, 2089 (1976); P.J. Wheatley, *J. Chem. Soc.* 3718 (1964); A.C. Beauchamp, M.J. Bennett, and F.A. Cotton, *J. Am. Chem. Soc.* 90, 6675 (1968). (b) $BiPh_5$ is SP: A. Schmuck, J. Buschmann, J. Fuchs, and K. Seppelt, *Angew. Chem. Int. Ed. Engl.* 26, 1180 (1987). (c) For review, see: K. Seppelt, in *Heteroatom Chemistry*, E. Block, Ed., VCH, New York, 1990, Chapter 19, p. 335.
9. (a) R.J.P. Corriu, C. Guerin, and J.J.E. Moreau, in "*The Chemistry of Organic Silicon Compounds*, S. Patai and Z. Rappoport, Eds., Wiley, New York, 1989. (b) M. Shimano and A.I. Meyers, *J. Am. Chem. Soc.* 116, 10815 (1994).
10. R.S. Grev, H.F. Schaefer III, and K.M. Baines, *J. Am. Chem. Soc.* 112, 9458 (1990).
11. M.W. Schmidt, P.N. Truong, and M.S. Gordon, *J. Am. Chem. Soc.* 109, 5217 (1987).
12. S. Nagase, referred to by A. Sekiguchi and H. Sakurai, in *The Chemistry of Inorganic Ring Systems*, R. Steudel, Ed., Elsevier, Amsterdam, 1992.
13. (a) V.A. Bain, R.C.G. Killean, and M. Webster, *Acta Crystallogr.* 25B, 156 (1969). (b) H.E. Blayden and M. Webster, *Inorg. Nuclear Chem. Lett.* 6, 703 (1970). (c) A.D. Adley, P.H. Bird, A.R. Fraser, and M. Onyszchuk, *Inorg. Chem.* 11, 1402 (1972). (d) Hexacoordinate Si may be involved in a silatropic ene pathway of the aldol-type reaction of an aldehyde with ketene silyl acetals: K. Mikami and S. Matsukawa, *J. Am. Chem. Soc.* 116, 4077 (1994).
14. (a) C. Elschenbroich and A. Salzer, *Organometallics*, VCH, New York, 1989. (b) D. Dakternicks, K. Jurkschat, H. Zhu, and E.R.T. Tiekink, *Organometallics* 14, 2512 (1995). (c) The same is true of heavy congeners of N, O, and F. For example, see: Y. Yamamoto, R. Nadano, M. Itagaki, and K. Akiba, *J. Am. Chem. Soc.* 117, 8287 (1995).

15. K.L. McKillop, G.R. Gillette, D.R. Powell, and R. West, *J. Am. Chem. Soc.* 114, 5203 (1992).

16. (a) A.E. Reed, P.v.R. Schleyer, and R. Janoschek, *J. Am. Chem. Soc.* 113, 1885 (1991). (b) M. Kaupp and P.v.R. Schleyer, *J. Am. Chem. Soc.* 115, 1061 (1993).

17. A sawtooth variation can be observed even in the case of metal–ligand bonding by the E mechanism: C.L. Haynes, P.B. Armentrout, J.K. Perry, and W.A. Goddard, III, *J. Phys. Chem.* 99, 6340 (1995).

18. (a) F. Scherbaum, A. Grohmann, B. Huber, C. Krueger, and H. Schmidbaur, *Angew. Chem. Int. Ed. Engl.* 27, 1544 (1988). (b) H. Schmidbaur, *Pure Appl. Chem.* 65, 691 (1993).

19. D.J. Wales, *J. Am. Chem. Soc.* 112, 7908 (1990).

20. (a) J.A. Duffy, *Bonding, Energy Levels and Bands in Inorganic Solids*, Wiley, New York, 1990. (b) D.M. Adams, *Inorganic Solids*, Wiley, New York, 1974.

21. (a) BeF_2: L. Warton, R. Berg, and W. Klemperer, *J. Chem. Phys.* 40, 3471 (1974). (b) BaF_2: V. Calder, D.E. Mann, K.S. Seshardi, M. Allavena, and D. White, *J. Chem. Phys.* 51, 2093 (1969).

22. (a) A. Buechler, J.L. Stauffer, and W.J. Klemperer, *J. Chem. Phys.* 1963, 39, 2299. (b) R.C. Spiker Jr. and L. Andrews, *J. Chem. Phys.* 58, 713 (1973).

23. (a) G.D. Stucky, M.M. McPherson, W.E. Rhine, J.J. Eisch, and J.L. Considine, *J. Am. Chem. Soc.* 96, 1941 (1974). (b) P. Jutzi, *Adv. Organomet. Chem.* 26, 217 (1986).

24. R. Walsh, *Acc. Chem. Res.* 14, 246 (1981).

25. R. West, in *The Chemistry of Inorganic Ring Systems*, R. Steudel, Ed., Elsevier, Amsterdam, 1992, Chapter 4.

26. For *ab initio* computations, see G. Trinquier and J.-C. Barthelat, *J. Am. Chem. Soc.* 112, 9121 (1990), and references therein.

Chapter 25

VB Selection Rules for Metal Hybridization

25.1 Perfect Pairing Versus Coulomb Hybridization

Unlike organic chemists, inorganic and organometallic chemists never embraced the VB method and the concept of hybridization. One major reason is the apparent lack of a single hybridization scheme that suffices for rationalizing the multitudes of organometallic structures. This is in stark contrast to the experience of the organic chemist: Using sp^3-hybridized carbons, one can rationalize not only the structures of alkanes and their derivatives but also those of alkenes and alkynes. Because of the resurfacing controversy surrounding the role of d orbitals in the binding of main group atoms, it has been problematic to decide exactly what type of hybridization is "permitted" in metallic systems.

Hybridization is the process by which the valence AOs of atoms are prepared for optimal bonding. The concept was first put to work by Pauling in connection with *covalent bonding*. We now recognize that there exist three different types of bonding. As a result, hybridization is motivated by entirely different requirements in each case.

In T-bound systems, hybridization leads to AOs that are optimal for the formation of exchange bonds (T bonds) by the perfect pairing of odd electrons. This is called *perfect-pairing hybridization*. In the case of electronegative nonmetal atoms, minimization of promotional energy takes a back seat to strong perfect pairing simply because the large kinetic energy reduction justifies a large promotional energy investment.

In I- and E-bound systems, hybridization occurs in a way that engenders optimal *atom deshielding* by the overlap-independent mechanisms of induction and dispersion. This creates a match of holes and pairs of two interacting atoms, with the result that a fraction of the attractive energy is due to the static Coulomb interaction of the deshielded atoms. Starting with the deshielded electronic configuration, 2-e CT delocalization engenders further attraction that is due to kinetic energy reduction. When the two components override the inevitable exchange repulsion of the components, we have net bonding. The hybridization that aims to optimize both overlap-independent and overlap-dependent mechanisms of attraction is called *Coulomb hybridization*.

25.2 The Concept of the Geminal Bond and the Starring Procedure

To illustrate the central idea, we consider the Be atom. Our first goal is to describe the ground state configuration of this atom, $2s^2$, by using a set of two sp hybrids. These are produced as indicated in the following hybridization correlation diagram.

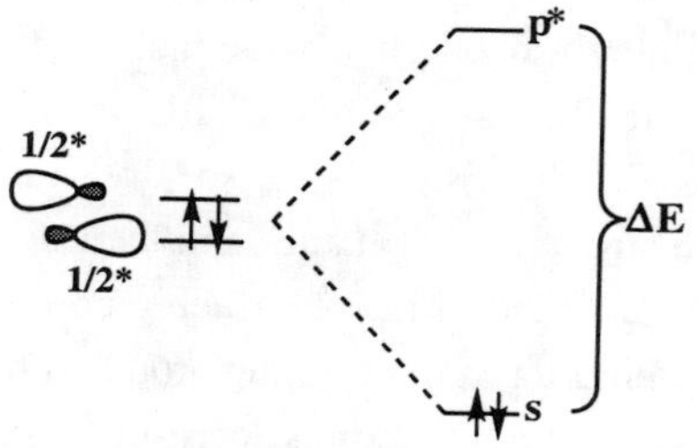

The lower s mixes with the upper p_x, which is *starred* to indicate its higher energy, to form two hybrid sp orbitals. In the absence of an external perturbation, the two hybrids are equivalent, and this means that each hybrid carries half of the asterisk. In the presence of an asymmetrical perturbation, the hybrids become nonequivalent and the asterisk is confined to one of them. This *starring procedure* immediately informs us which hybrid will act preferentially as a receptacle of electrons: *Electrons can occupy only unstarred hybrids in the reference configuration.*

The ground state of Be can be represented as a resonance hybrid of VB configurations over hybrid AOs. Because most people are used to canonical AOs (i.e., the Be2s and Be2p AOs rather than two Be sp hybrids), we show the equivalence of configurations built from canonical and hybrid AOs in Figure 25.1. If we use hybrid sp AOs, ground Be is represented by S + kZ^- with k assuming some intermediate value between zero and one. Whatever this value, the important thing is that interhybrid delocalization (i.e., mixing Z^- into S) is equivalent to the formation of an interhybrid bond which has the following meaning: The greater the mixing of Z^- into S (i.e., the greater the value of k), the greater the atom de-excitation. We call this a *geminal bond (G bond)*, and we represent it thus:

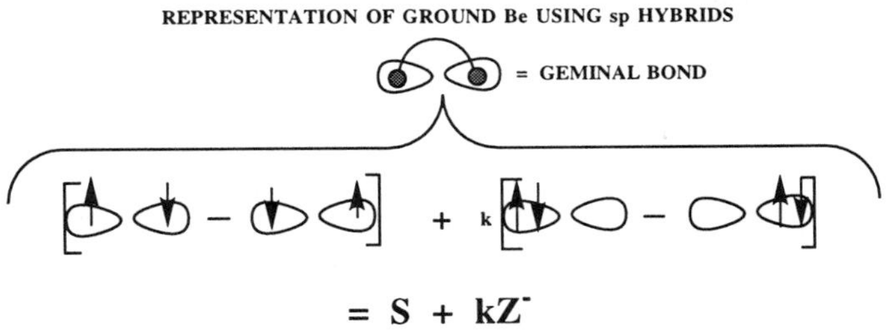

The pair that makes up the G bond is called the G pair (gp). As we discussed before (see Chapter 4), the strength of the G bond is gauged as follows: If two canonical AOs, x and y, combine to form two hybrids, x' and y', the strength of the G bond defined by the hybrids x' and y' is proportional to the energy gap separating x and y. While the mixing of Z^- into S represents atom de-excitation and defines the strength of the G bond, the mixing of Z^+ into S represents atom excitation for the purpose of binding, via overlap dispersion, with some other fragment.

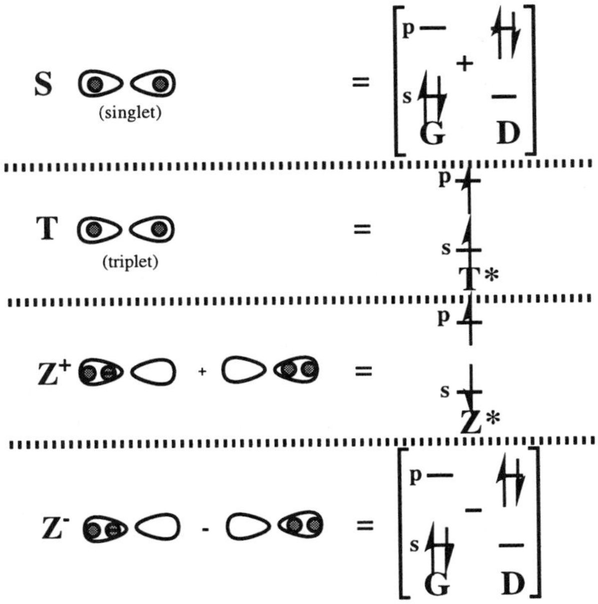

Figure 25.1 The equivalence between configurations built from hybrid and configurations constructed from canonical AOs.

Two different hybrids belonging to the same atom can combine to form a G bond much as two hybrids belonging to different atoms can overlap to form an interatomic bond. The question becomes: How many hybrids combine to form a single G bond? For example, how many of the four sp^3 hybrids should we link in a pairwise fashion to indicate the existence of one delocalized G bond? Clearly, each sp^3 hybrid has arisen from the mixing of one 2s and three 2p orbitals, and atom de-excitation (the physical consequence of the G bond) can occur only from one 2p to one 2s. Hence, only two sp^3 hybrids combines to form a G bond. The result is generalized as follows: If a set of n equivalent lower energy AOs mixes with a set of m higher energy AOs to produce $n + m$ hybrids, the maximum number of G bonds is n if $n < m$. The formulas that show explicitly the presence of a G bond for the prototypical C, N, and O atoms are as follows:

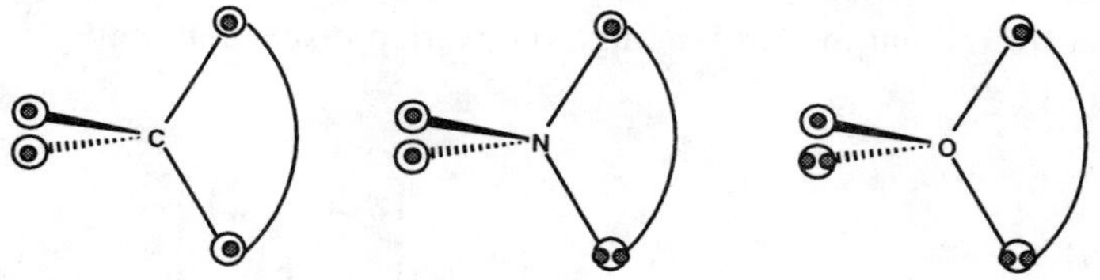

The lesson here is that the G bond is not restricted to two electrons. As with interatomic bonds, we can have one-, two-, and three-electron G bonds.

Can G bonds produce interatomic bonding? The answer is that the gp's act as if they were ordinary bond pairs, and they obey the same rules of association: Two gp's associate in trans, rather than cis, fashion to produce 1.5 I bonds, as illustrated by the Ti dimer in Figure 25.2. The trans association of the two intra-atomic G bonds allows the two atoms to approach each other so that additional bonds (i.e., two π bonds) can also be formed. The efficacy of s/d hybridization depends on atom promotional energies, a matter well-familiar to VB practitioners.

We are now prepared to outline a recipe for metal hybridization which we will be using routinely in the remainder of this work:

(a) Hybrids are generated by mixing nominally valence and nonvalence AOs. The nonvalence AOs are actually "nonvalence functions," but this issue is tentatively bypassed (see Chapter 28) to keep things as simple as possible.

(b) Hybrids that are "descendants" of nonvalence AOs are *starred*. The valence and nonvalence AOs of different classes of atoms are defined in Table 25.1.

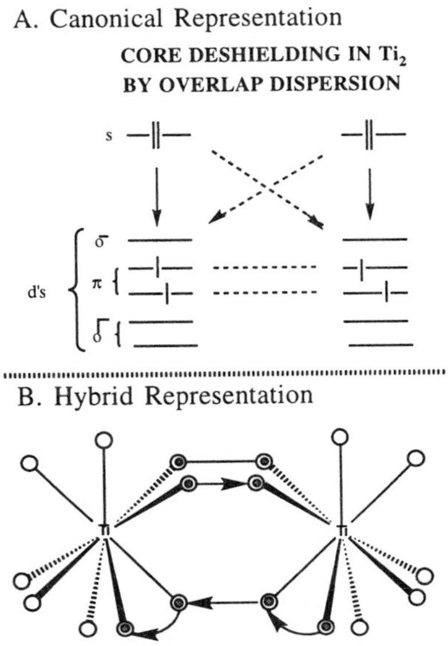

Figure 25.2 Overlap dispersion in the σ space of the Ti dimer causes association of the geminal bonds, and this is responsible for a major fraction of the binding energy.

(c) There are as many stars as nonvalence AOs. Symmetry-equivalent hybrids have identical "star character." When two hybrids are nonequivalent, a star indicates the hybrid that has higher nonvalence character.

(d) Occupation of starred hybrids is "forbidden."

(e) The smaller the number of starred hybrids, the lower the metal promotional energy even if only unstarred hybrids are occupied.

(f) The ns AOs of transition metals and the np AOs of alkalies and alkaline earths represent borderline cases and, when necessary, they are identified by a dagger (#) to indicate their higher energy relative to the (n – 1)d or ns AOs, respectively.

Table 25.1 Valence and Nonvalence AOs in Four Classes of Atoms

Type of Atom	Valence AOs	Nonvalence AOs
Zerovalent transition metals	(n – 1)d, ns	np
High-valent transition metals	(n – 1)d	ns, np
Alkali and alkaline earths	ns, np	nd
Heavy p-block	ns, np	nd

As an example, consider an sp^3-hybridized Zn atom in the C_{3v} field generated by one K and three L ligands (Figure 25.3). Disregarding the (n – 1)d AOs and assuming that Zn acts with one valence 4s and three nonvalence 4p AOs means that three of the four sp^3 hybrids are "descendants" of the three nonvalence 4p orbitals and they must be starred. An unsymmetric environment has more than one physically distinct star allocation. In one of them, two stars go to hybrids facing the L ligands and one to the hybrid pointing toward the K ligand. This produces a formula that requires that each hybrid have 2/3 "star character" to be consistent with C_{3v} symmetry.

Symmetry adaptation of starred hybrids is not a trivial matter, and it requires some attention. Specifically, the linear combination of AOs into equivalent hybrids (isovalent hybridization) represents a restriction that is undone during an actual computation. For example, the three conical AOs have been derived from one σ and two π AOs, and the former one is differentiated from the latter two. As a result, symmetry adaptation is actually effected by "star resonance" as

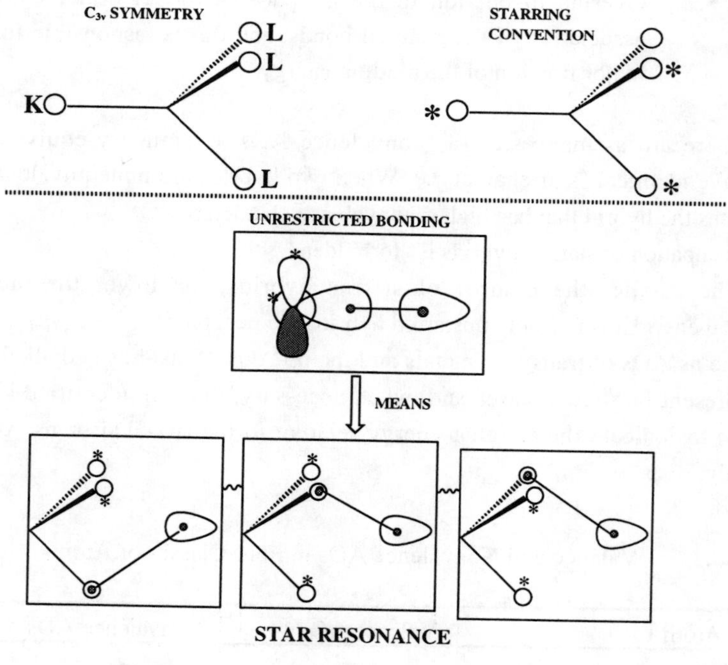

Figure 25.3 Application of the starring procedure to molecules. Equivalent hybrids have the same "star character" because of "star resonance."

illustrated in Figure 25.3. Of course, when the σ and π AOs become quasi-degenerate, isovalent hybridization is no longer a serious restriction and each equivalent hybrid has the same fractional star character. Henceforth, symmetry adaptation is implied in our drawings much as symmetry adaptation is implied when we symbolize a symmetrical molecule by one of many equivalent resonance structures. This convention simplifies the drawings and the discussion.

Since the separation of AOs into valence and nonvalence is not based on some absolute criterion, the implications of the starred hybrids for atom excitation energy differ from atom to atom. Recognizing that the np AOs that are "nonvalence" in the transition block become "valence" in the p block, it is clear that starred hybrids of the late transition metals are much more effective in promoting CT delocalization than starred hybrids of the early transition metals. This should be always kept in mind whenever we consider the energetic implications of the star character of a transition metal atom.

25.3 The Hybridization of Transition Metals

The very concept of the I bond implies that the electronic structures of molecules can be accounted, to first approximation, by assuming that only ns and np AOs in p-block atoms and only (n – 1)d and ns AOs in transition metals are active. When it comes to transition metals, there are three important s/d hybridization schemes: ds (Figure 25.4), d^2s (Figure 25.5), and d^3s (Figure 25.6). These will play in organometallic chemistry the same roles taken by sp, sp^2 and sp^3 carbon hybridization in organic chemistry.

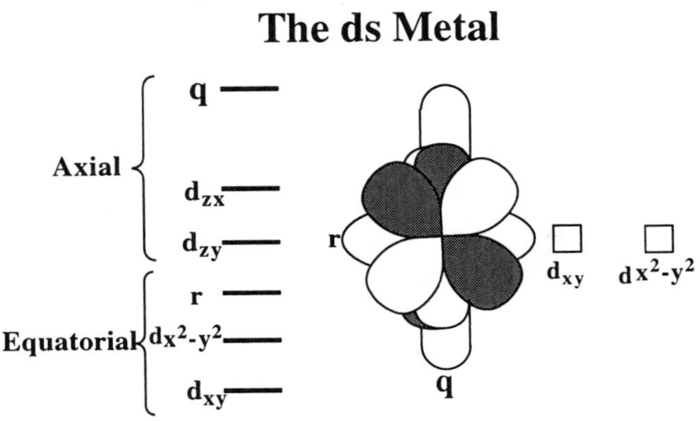

Figure 25.4 The ds hybridization mode.

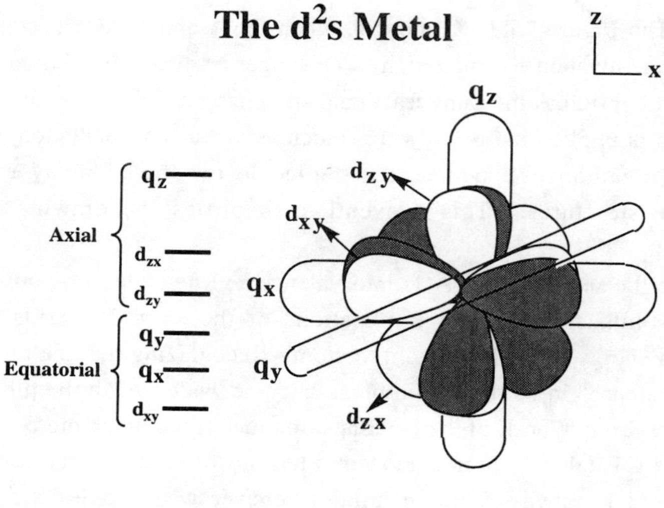

Figure 25.5 The d^2s hybridization mode.

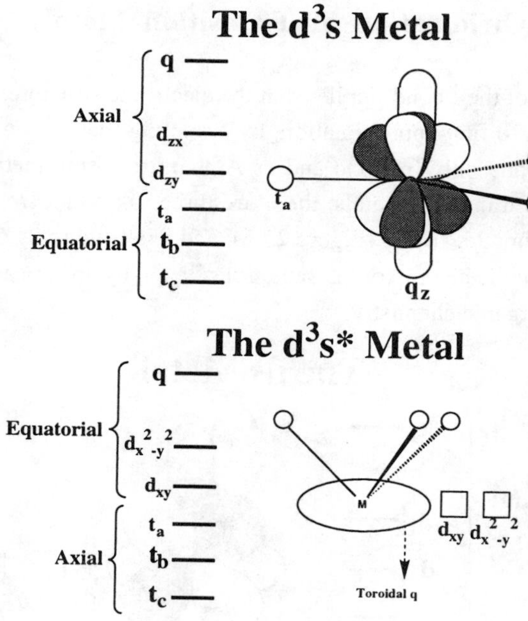

Figure 25.6 Two equivalent d^3s hybridization modes. The d^3s^* mode exposes the ability of one equatorial AO (the toroidal q AO) to overlap with neighboring AOs as a result of its large radial extension (a consequence of the s character of the q AO).

We illustrate our approach by reference to the octahedral d^2s hybridization. We place a transition metal atom at the origin of a Cartesian coordinate systems and we first mix the 4s, $3d^2{}_z$, and $3d_{xy}$ metal AOs. We obtain three σ-type d^2s hybrids, designated by q_i, which are aligned with the three Cartesian coordinate axes. Each of the three has roughly cylindrical symmetry about each axis. The remaining three d orbitals, d_{zx}, d_{zy}, and $d_{x^2-y^2}$, are π-type. This means that two groups placed along a diagonal of the octahedron share one σ-type q_i and one π-type d AO. Two of the three d^2 hybrids define one G bond. The single dagger goes to any one of the three equivalent d^2s hybrids.

One important variant, the pseudo-octahedral d^4s hybridization, is extremely useful in dealing with metal atoms having local C_{3v} symmetry. There is one set of three *equatorial* (EQ) and one set of three *endo conical* (ENDO) hybrids:

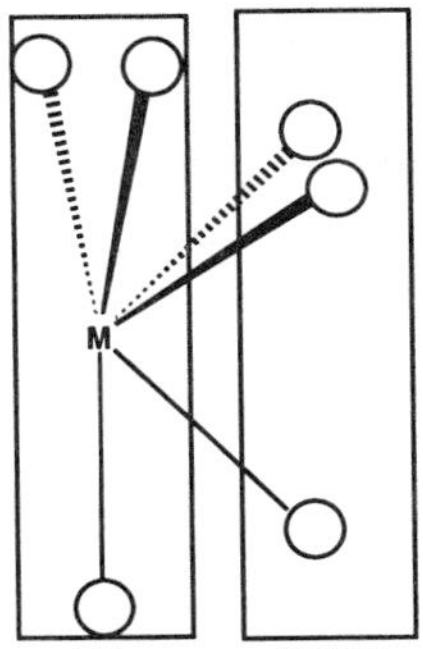

One EQ and one ENDO hybrid define a G bond, and the single dagger goes to either EQ or ENDO hybrids.

Nonvalence AOs play a pivotal role in directing metal orbitals so that atom deshielding sets the stage for CT delocalization. In the case of transition metals, we can complement the (n – 1)d and ns valence with np nonvalence AOs to generate nine hybrids. Five (those parented by the five d orbitals) will be unstarred, one will be daggered (the one parented by the single s), and three will be starred (those parented by the three p orbitals). The principal G bond is the one formed by one unstarred hybrid and the daggered hybrid, but this will be shown explicitly only when necessary. The most common types of d^nsp^3 metal hybridization schemes for a zerovalent metal are discussed in Sections 25.3.1 to 25.3.4.

25.3.1 Octahedral d^2sp^3 Hybridization

We derive three σ-type d^2s hybrids, designated by q_i, and three π-type d_{zx}, d_{zy}, and $d_{x^2-y^2}$ as in the case of octahedral d^2s hybridization. Each of the three q_i AOs can mix with a 4p AO that is aligned along the same Cartesian axis generating two new hybrids, designated by h and h*:

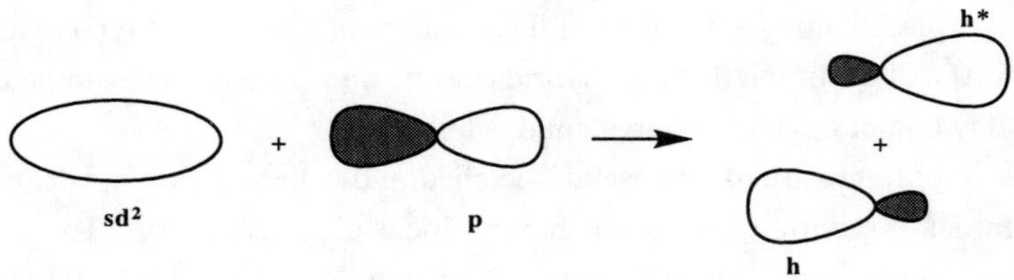

25.3.2 Pseudo-octahedral d^5sp^3 Hybridization

The hybrid orbital ensemble is shown in Figure 25.7. Ordinarily, this is called "trigonal prismatic" or "trigonal antiprismatic" hybridization. However, we prefer the term "pseudo-octahedral" for two reasons: first, to relate the d^5sp^3 to the d^2sp^3 octahedral hybridization scheme and, second, as a reminder of spatial relationships present in the octahedron. The nine hybrids are subdivided into two sets:

(a) The *conical* set comprised of three AOs along the negative direction of one Cartesian axis (EXO subset) and three AOs directed along the positive direction of the same axis (ENDO subset). The two sets define a double cone.

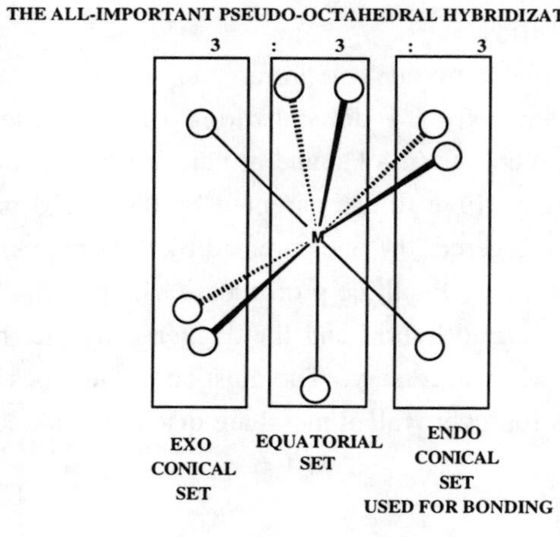

Figure 25.7 The pivotal d^5sp^3 pseudo-octahedral hybridization.

These six hybrids are formed by combining one σ-type $d_{z^2}s$ hybrid lined up with the *z* axis with one σ-type $2p_z$ and by combining, as illustrated, the two π-type d_{xz} and d_{yz} with the two π-type p_x and p_y AOs.

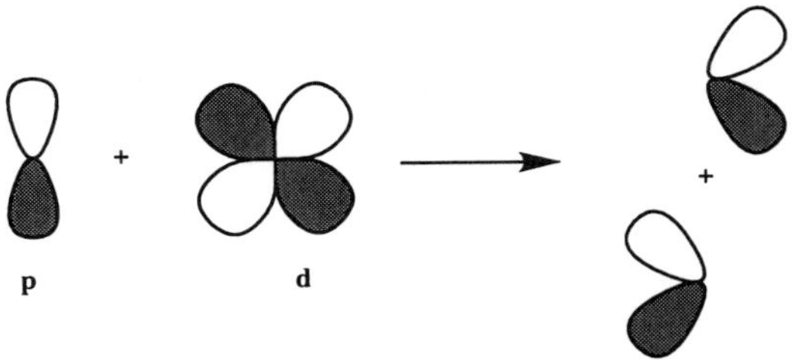

(b) The *equatorial* set of three coplanar hybrids is generated by mixing one $d_{z^2}s$ hybrid and two δ-type $d_{x^2-y^2}$ and d_{xy} AOs. These three hybrids are directed to the corners of an equilateral triangle. Alternatively, these three hybrids are equivalent to a disk-like σ $d_{z^2}s$ plus $d_{x^2-y^2}$ and d_{xy} AOs of δ symmetry.

We must now assign three stars and one dagger and leave five hybrids unstarred. The rule is that *the stars go to any three conical hybrids (but never to an equatorial hybrid), while the dagger goes to either a conical or an equatorial hybrid.* The remaining five hybrids remain unstarred.

Here are some important properties of the conical and equatorial hybrids.

(a) Because the p contribution goes exclusively to the conical set, there exist two choices: Either one of the two conical subsets (EXO and ENDO) takes up all three stars (e.g., ENDO/EXO***) or one gets one star and the other two stars (e.g., ENDO*/EXO**). This sets up the stage for *star isomerism*, as depicted in Figure 25.8a.

(b) The EXO and ENDO sets can be either "staggered" or "eclipsed" relative to each other (Figure 25.8b). The eclipsed form describes prismatic and the staggered antiprismatic (pseudo-octahedral) coordination of the metal.

(c) The three hybrids of each conical set are isolobal to one radial σ- and two tangential 2p- π AOs. Thus, depending on the situation, one may interchangeably use the EXO or ENDO sp^3-like or sp^2-like or sp-like set (Figure 25.8c).

(d) In a typical situation, the ENDO conical subset is used to effect metal–metal bonding. The EXO conical subset, also comprising a triplet of hybrids, is typically used to house nonbonding electron pairs or to accommodate ligands. When using one of the conical subsets to effect bonding, a metal fragment is designated a 3h (trisynaptic) fragment to indicate that three hybrid AOs are available for bonding. The symbol "h" is inspired by the term "hapticity," which is related to our term "synapticity."

(e) Because the ns AO is a valence AO [though of higher one-electron energy than the (n – 1)d AOs], the daggered equatorial hybrid can be occupied. As a result, the equatorial set is typically used to accommodate up to three nonbonding electron pairs.

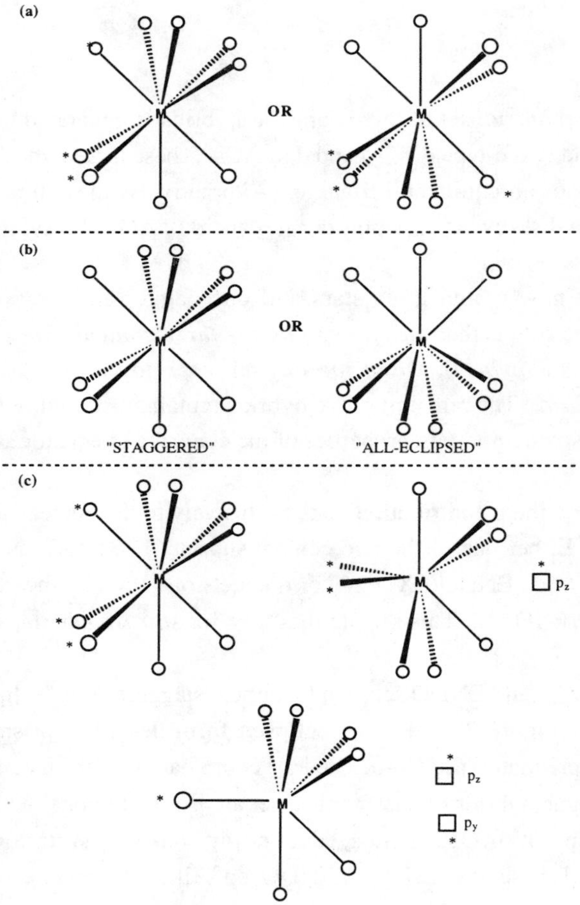

Figure 25.8 (a) Pseudo-octahedrally hybridized centers differing only in the location of the stars ("star isomers"). (b) The staggered and eclipsed arrangements of the exo and endo conical sets. (c) Equivalent representations of the three exo starred hybrids.

The pseudo-octahedral hybridization scheme, otherwise designated 3:3:3 (where the numbers indicate the ENDO:EQUATORIAL:EXO orbital ratio), is convenient for a vast number of complexes that have axial symmetry including pure metal dimers, sandwiches like ferrocene, and complexes with mixed ligands [e.g., $(PR_3)_3M(CO)_3$]. The important point is that a d^5sp^3 pseudo-octahedral transition metal is a triple-deck piece with two trisynaptic (3h) faces (corresponding to the two conical sets) plus an ensemble of three trigonal equatorial AOs.

While we will focus on metal clusters in a later chapter, it is prudent to give an early example of how we plan to use the concept of the pseudo-octahedral transition metal. Specifically, consider the recently discussed issue of the tetranuclear Ru(IV) aqua ion.[1] What is the likely global minimum? We start with pseudo-octahedral Ru atoms in which the unstarred axial endo set effects I-bonding, the equatorial AOs accommodate the metal lone pairs, and the starred axial exo set binds E-selective ligands (e.g., water) by the E mechanism. When the metal is in a high formal oxidation state, we must worry about the dagger. This goes to the equatorial set, which now has two (low energy) undaggered and one (high energy) daggered AO.

We now seek to produce the most stable tetraruthenium oxo complex. This is done by one-electron oxidation of each Ru metal to yield an ion with seven electrons, with four filling two undaggered equatorial AOs and three going to the unstarred endo set. Next, four Ru^+ ions combine to form a tetrahedron analogous to P_4. Next, six oxygens are formally inserted into the six tetrahedral edges. Finally, each Ru^+ accommodates conically the water molecules via the starred exo set. The resulting closed-shell molecule has an adamantane-type structure with the formula $[Ru_4O_6(H_2O)_{12}]^{4+}$. This comes close to the suggested structure.[1]

25.3.3 Quasi-prismatic and Square Antiprismatic Hybridization

(c) Figure 25.9 illustrates 4:2:3 (quasi-prismatic) and 4:1:4 (square antiprismatic) d^5sp^3 hybridization. The first scheme is derived from the pseudo-octahedral model by combining one d-δ AO of the equatorial set with one conical set. The second scheme is derived by combining one delta-type AO of the equatorial set with the ENDO conical set and the second delta-type AO of the equatorial set with the EXO conical set. In this way, we obtain new EXO and ENDO sets, each made up of four AOs. The single equatorial AO of the 4:1:4 model is a toruslike σ orbital normal to the z axis. The starring pattern shown in Figure 25.9 is only one of two possible as discussed earlier. Again, a d^5sp^3 square antiprismatic transition metal is a triple-deck piece with two tetrasynaptic

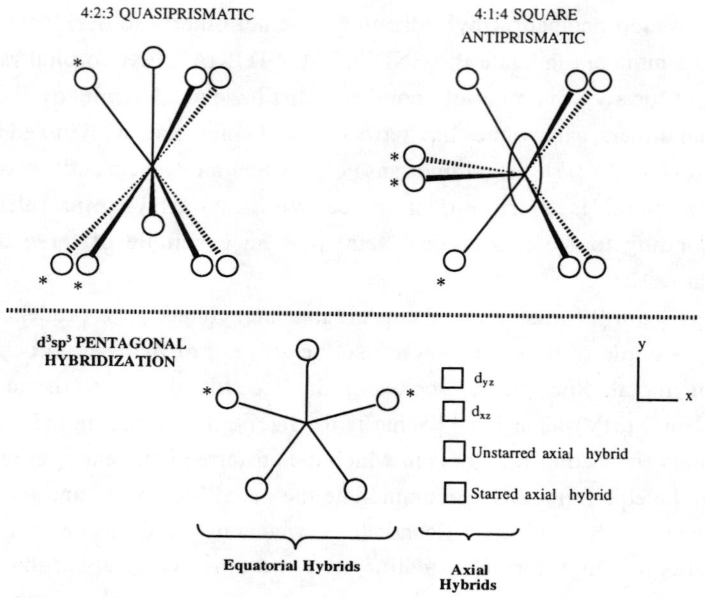

Figure 25.9 Different hybridization modes of transition metals.

(4h) faces and one toroidal AO. Similarly, a 4:2:3 d^5sp^3 transition metal is a triple-deck piece with one tetrasynaptic (4h) and one trisynaptic (3h) face and an ensemble of two orthogonal equatorial AOs.

The d/s/p hybridization modes discussed in Sections 25.3.1 to 25.3.3 are optimal for binding a metal center to axially disposed ligands. For example, the pseudo-octahedral mode is optimal for iron in ferrocene. The characteristic feature is that the in-plane p_x and p_y AOs mix with the out-of-plane d_{xz} and d_{yz} AOs, and the resulting hybrids are directed above the *xy* plane and along the *z* axis. The other extreme is the d/s/p hybridization modes that are optimal for bonding a metal center to equatorially disposed ligands. The characteristic feature now is that the in-plane p_x and p_y AOs mix with the in-plane d_{xy} and d_{x2-y2} AOs, whereupon the resulting hybrids lie in the *xy* plane. The most common hybridization mode of this type is the pentagonal d^3sp^3 hybridization (Figure 25.9). There are five hybrids (generated by combining one sd_{z2} hybrid with p_x, p_y, d_{x2-y2}, and d_{xy}) directed to the five vertices of a regular pentagon situated on the *xy* plane, and two of them have major p character. Normal to the ring extend the d_{xz} and d_{yz} plus the second sd_{z2} hybrid AOs aligned with the *z* axis. Finally, the additional p_z AO can mix with the *z*-aligned sd_{z2} hybrid to generate two collinear hybrids pointing in opposite directions along the *z* axis. A derivative of the pentagonal

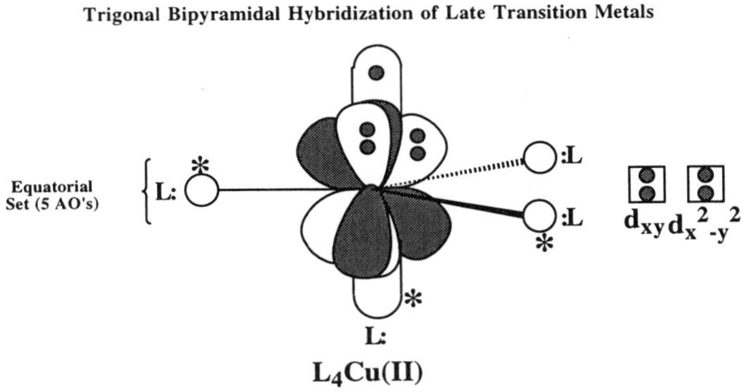

Figure 25.10 The dsp^3 trigonal bipyramidal hybridization scheme of Cu(II).

hybridization mode that essentially involves dehybridization in the equatorial plane is the dsp^3 scheme, appropriate for the treatment of trigonal bipyramidal late transition metals such as Cu(II). This is shown in Figure 25.10.

How do the metal electron count and the number of ligands determine the hybridization mode? While guidelines can be given, the very nature of I-bonding suggests that when dealing with a molecule, it is best to consider several alternatives, conscious that these may correspond to quasi-degenerate valence isomers. Nonetheless, since this work marks a transition to a different type of thinking regarding transition metal complexes, an illustration of the modus operandi is appropriate. Thus, d^6 metals with six ligands are expected to adopt the d^5sp^3 octahedral hybridization, but d^4 metals with the same number of ligands are expected to adopt the 4:2:3 quasi-prismatic mode, with one of the three exo conical starred AOs deleted. One example is the molecule $OsH_2X_2(PR_3)_2$ (X = Cl, Br, I).[2] The two H and the two PR_3 groups are accommodated by the four endo unstarred hybrids, the two metal pairs go to the two equatorial hybrids, and the Cl's dock on the two exo starred hybrids.

We close by noting that the s/p/d hybridization of heavy main group atoms is analogous to that of transition metals. The only difference is the number of starred hybrids: We now have four unstarred (descendants of the valence ns and np AOs) and five starred hybrids (descendants of the five d AOs).

25.4 The Mode of Action of the B_{12} Cofactor

We can illustrate the concept of transition metal hybridization by considering the way in which the axial base affects the rate of cleavage of a *trans*-Co(II)—R bond in alkylcobalamins (R—Cbl). This also takes us into the heart of the concept of the I bond as illustrated by I with X = R:

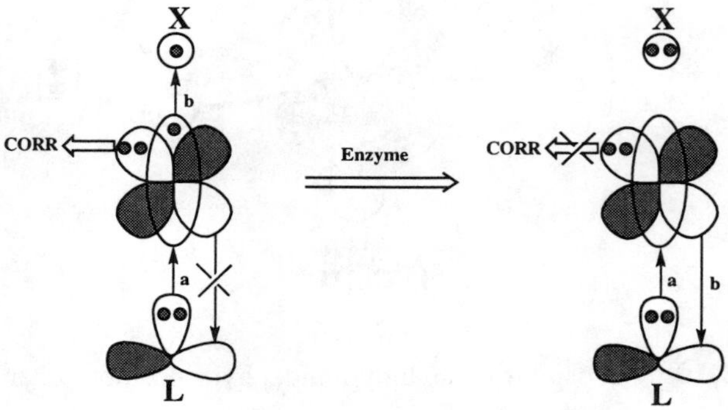

I II

A d^2s hybridized Co(II) binds four σ nitrogen pairs of the corrin ring by the two equatorial hybrid holes (see Figure 25.5) via two I bonds. One lone pair stays in the equatorial d_{xy} AO and two pairs go to the two π-type axial AOs; from there they can delocalize into the π system of the corrin. This leaves a singly occupied σ-type axial AO to bind the R group and the trans base L. It is seen that the axial ensemble of the base L, the Co(II) metal center, and the R group is linked together by a single REL bond. It follows that the Co(II)—R bond should become stronger as the basicity of L increases. Brown and Brooks[3] sumarized the key pieces of data that support this hypothesis:

In a series of X-ray R—Cbl structures, the Co—C bond lengths decrease monotonically with the axial Co—L bond lengths.

The enthalpy of activation of Co—C bond homolysis increases approximately linearly with increasing L basicity in a series of isosteric cobaloximes.[4]

In a rigid octaethylporphyrin system, the enthalpy of activation of Co—C bond homolysis increases linearly with phosphine base basicity.[5]

In summary, the available crystallographic data and the pivotal kinetic studies of Halpern and his group provide direct evidence in support of the concept of the REL bond: Improving the CT delocalization in one interatomic space [L to Co(II)] improves CT delocalization in a consecutive interatomic space [Co(II) to R].[5] In other words, "improving" either of the two arrows which, taken together, represent one I bond, strengthens the entire I bond, hence enhances the overall bonding. The implications of all this for the mechanism of activation of R—Cbl by the required enzyme have been discussed by Brown and Brooks, who concluded that the "base-on" effect is primarily steric.[3]

What happens when the alkyl group is replaced by a highly electronegative radical like H_2O^+, as in aquacobalamin perchlorate (H_2O—Cbl^+)? The axial σ REL bond is now destroyed, only to be replaced by an L—Co(III) ROX bond, as illustrated by II at the beginning of this section, with X = H_2O^+.
The result is an exceedingly short axial Co(III)—L distance, as one d pair of Co(III) becomes disengaged from the corrin ring, which now becomes free to to undergo an upward (toward H_2O) folding deformation to avoid the exchange repulsion with the σ lone pair of the now tightly bound L. This picture is supported by a recent crystallographic study by Kratky, Kreutler, and co-workers[6] and leads to the following scenario for the mechanism of action of B_{12}: The associated enzyme turns off the π acceptor action of the corrin, which most likely is due to the action of the dipoles of the acetamide side chains. This effects a transition from a REL bond spanning the DMB—Co(II)—R axial array (DMB = dimethylbenzoimidazole base) to a ROX bond connecting DMB and Co(III) accompanied by an upward-folding deformation that expels alkide anion, rather than alkyl radical, as illustrated above with X = R. In short, the crucial part of the B_{12} machinery that is capable of responding to the environment is the corrin π system!

References

1. A. Patel and D.T. Richens, *Inorg. Chem.* 30, 3792 (1991).
2. D.G. Gusev, R. Kuhlman, J.R. Rambo, H. Berke, O. Eisenstein, and K.G. Caulton, *J. Am. Chem. Soc.* 117, 281 (1995).
3. K.L. Brown and H.H. Brooks, *Inorg. Chem.* 30, 3420 (1991).
4. (a) F.T.T. Ng, G.L. Remple, and J. Halpern, *J. Am. Chem. Soc.* 104, 621 (1982). (b) F.T.T. Ng, G.L. Remple, C. Mancuso, and J. Halpern, *Organometallics* 9, 2762 (1990).
5. M.K. Geno and J. Halpern, *J. Am. Chem. Soc.* 109, 1238 (1987).
6. C. Kratky, G. Faerber, K. Gruber, K. Wilson, Z. Dauter, H.-F. Nolting, R. Konrat, and B. Kreutler, *J. Am. Chem. Soc.* 117, 4654 (1995).

Chapter 26

. . . and It Shatters in the d Block

26.1 The Multifaceted Transition Metal Ligands

The traditional view is that ligands can be subdivided in categories such as π acceptors or σ donors. While frequently justified as a matter of convenience, this simple approach is actually doomed by the very concept of the I bond: One central atom can bind two ligands by one I bond, and exactly how one ligand enters I-bonding depends on the identity of its partner. For example, what type of ligand is ethylene? There are three possible answers, and the correct one depends on the partner ligand L and the metal M:

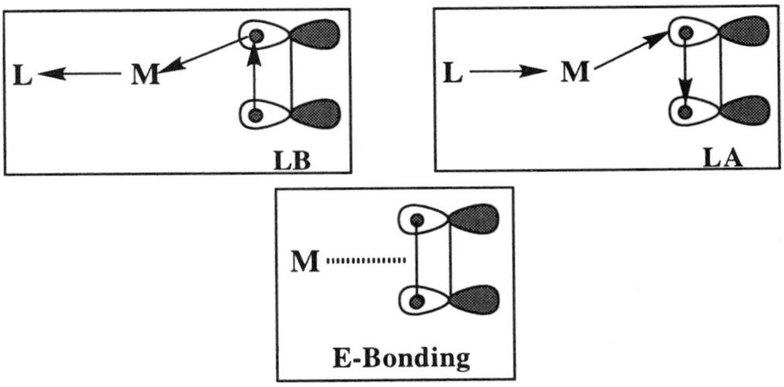

It follows that every single organometallic complex represents a real challenge in identifying the mode of action of the various ligands. Since we already know when to expect E-bonding, and since we have identified (activated) REL-bonding as the supreme form of I-bonding, our task is not hopeless. In addition, certain ligands are, by construction, predisposed to act as acids, radicals, or bases that can enter into activated I-bonding through the agency of observer holes, dots and pairs. For example, CO can be legitimately classified as π acid that is activated by a σ pair toward I-bonding with a π base, which, in turn, is activated by a σ hole toward I-bonding.

One major decision we will have to make regularly is whether a metal-containing molecule subscribes to E- or I-bonding. The criterion has already been established: If polarization by induction or dispersion sets the stage for codirectional CT delocalization, we have activated I-bonding. On the other hand, if polarization sets the stage for contradirectional I-bonding, we have frustrated bonding, which may be dominated either by CT delocalization or polarization.

Figure 26.1 shows how dispersion and induction can set the stage for I-bonding via CT delocalization.
In one case, dispersion promotes codirectional CT delocalization and REL I-bonding while, in the other case, induction promotes contradirectional CT delocalization and either I-bonding or E-bonding. Every model that neglects the first step (e.g., EHMO theory) will end up predicting that both complexes owe their stability to "delocalization." This misses the essence: The two complexes have widely different stabilities because of the way in which polarization complements CT delocalization.

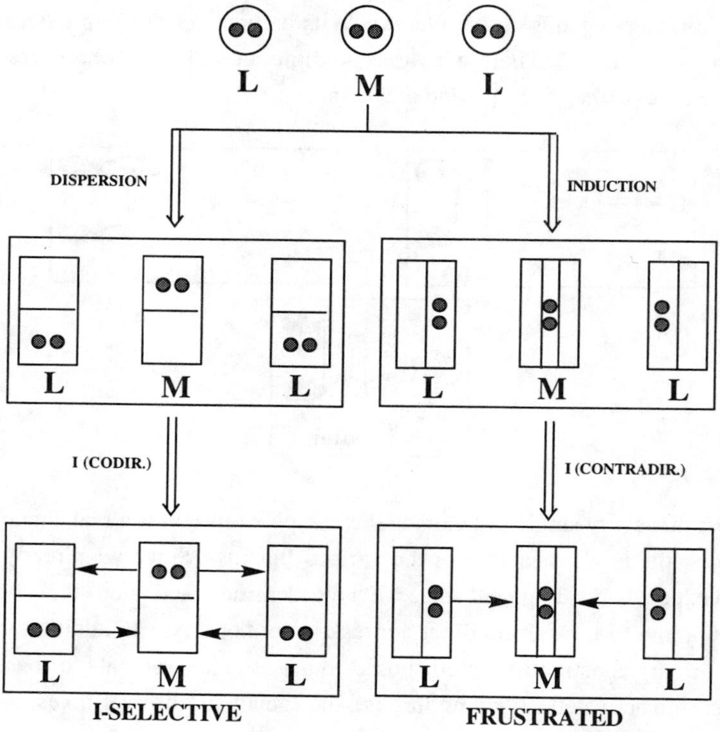

Figure 26.1 The complementation of static bonding by CT delocalization. In one case, dispersion sets the stage for codirectional CT delocalization. This represents activated I-bonding. In the other case, induction can promote only inferior contradirectional CT delocalization. This represents frustrated bonding.

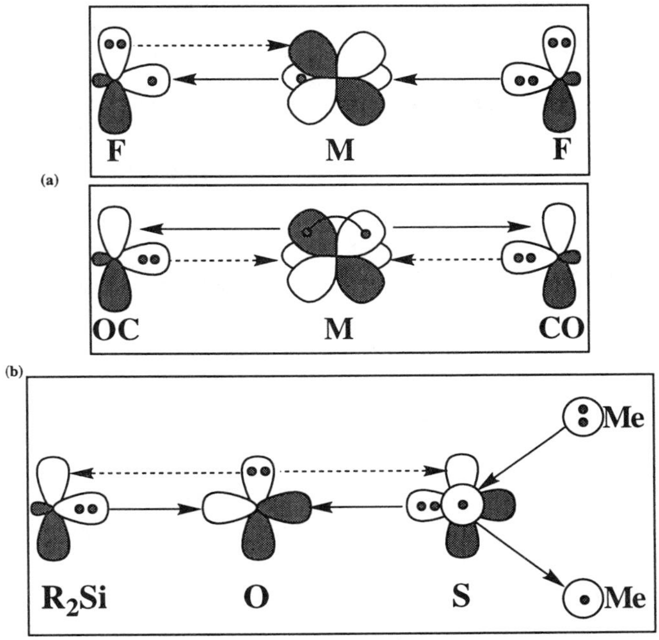

Figure 26.2 (a) The difference in bonding between FMF (in MF_6) and OCMCO in $[M(CO)_6]$ (M = Cr, Mo, W). (b) The analogous R_2Si—O—SMe_2 complex involved in the oxygen transfer from DMSO to silylene.

We will work out a variety of problems that illuminate the central concepts. All one has to remember are three guidelines:

Steric promotion is fully justified.

Antiparallel codirectional arrow trains imply activated I-bonding and pronounced stability.

E-bonding can dominate when codirectional arrow trains cannot be established.

26.2 The Bonding of Transition Metal Fluorides

One fundamental problem of transition metal chemistry is: Why is octahedral $Cr(CO)_6$ stable and octahedral CrF_6 a "doubtful" species?[1] The answer is revealed by looking at the bonding of LCrL (L = CO, F) along one one of the diagonals of the octahedron. Using superior arrows to indicate the primary action of the ligands and inferior arrows to indicate the activation of the primary I bond by the observer elements, we can see that COCrCO has superior bonding because the primary π-type OX I bond is activated by an observer RED I bond (two dashed arrows). By contrast, only half of the REL I bond of FCrF is activated (one dashed arrow). The argument is shown diagrammatically in Figure 26.2a.

The next key question is: Why is WF_6 stable while CrF_6 is doubtful? Once again, the answer has to do with the activating dashed arrows: The d AOs expand as we go from Cr to W, and the action of the dashed arrows is enhanced. This implies that CrF_6 is actually a frustrated molecule because the contracted d orbitals render REL-bond activation impossible. How does a first transition series metal deal with the frustration? Our hypothesis is that the metal adopts mixed bonding: Part of the bonding is T- and part is E-like. For example, Cr donates three electrons to three F atoms and makes three T-bonds with the remaining three. This is called T/E bonding and is represented by the conventional formula $F_3Cr^{3+}(F^-)_3$. By contrast, WF_6 has three activated REL bonds and is represented by the A formula $W^{3+}(F,F^-)_3$.

WF_6 is not covalent. Rather, it is I-bound, and the choice of geometry is dictated by two factors:

Optimization of I-bonding (i.e., activation of REL bonds through the action of ligand lone pairs).

Minimization of interligand nonbonded repulsion.

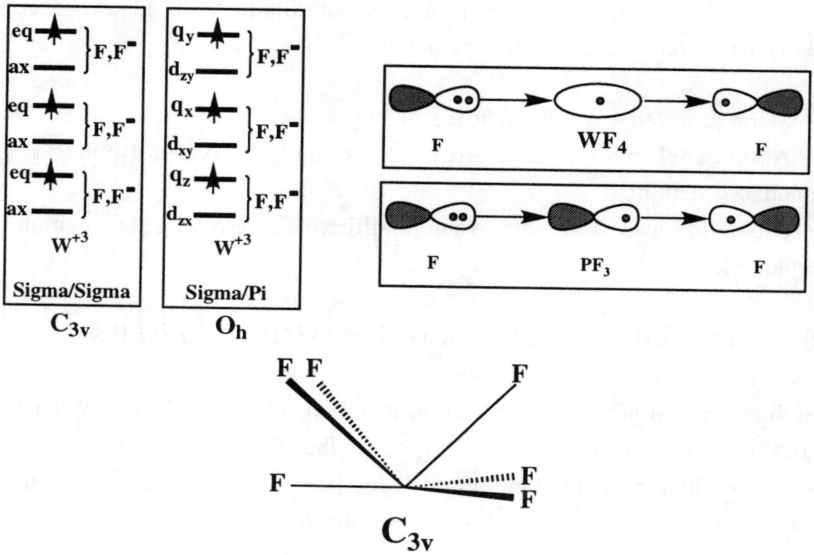

It turns out that the two conditions are fulfilled by the octahedral structure but only at the expense of loss of overlap, hence, loss of covalency (T complement). In other words, a candidate σ/σ C_{3v} geometry is abandoned in favor of the σ/π O_h structure, as illustrated. The term "σ/σ" denotes the type of one-to-one overlap match of ligand and metal AOs. The C_{3v} form involves an sd^5 set of metal atom hybrids, and it is easily derived starting from the pseudo-octahedral set and eliminating three starred conical hybrids (see Chapter 25). As just illustrated, W binds three F/F^- pairs by a σ/π mode exactly as PF_5 does. In the VB theory of bonding, the unsuspected common denominator is the I bond. The true oxidation number of W in WF_6 is III and not VI!

What would we predict, had we adopted the covalent model? Assuming that the six valence electrons of W are allocated to the six valence AOs (five 3d and one 4s) and assuming that F acts as a conventional σ radical, the predicted structure would be the C_{3v} geometry shown above, not the octahedral geometry determined by ab initio calculations.[2] EHMO theory "understands" only covalency, with no regard for interelectronic repulsion. Thus, it ought to predict the C_{3v} geometry as the global minimum of CrH_6. Indeed, EHMO calculations of the isoelectronic TiH_6^{2-} predict symmetry reduction that transforms the octahedral, O_h, to a C_{3v} form.[3a] By contrast, SCFMO-type calculations predict retention of the intuitively-expected octahedral shape because they do contain, albeit crudely, interelectronic repulsion.[3b]

How can we best enforce the C_{3v} geometry in a WF_6 derivative? The conventional view is that MX_6 (M = Cr, Mo, W) has an octahedral geometry because this is the geometrical arrangement that minimizes nonbonded ligand repulsion. In this way, one sees no motivation for abandoning this geometry in favor of a lower-symmetry structure irrespective of the nature of M and X. We now have a very different scenario. *The octahedral geometry is I-optimal because it brings each AO of an sd^2 metal into maximum overlap with two trans ligand AOs.* The same is true of any geometry that features linear XMX arrays. An I bond of the REL type associates three AOs. On the other hand, an I-optimal is not necessarily a T-optimal geometry; that is, the best I-bonding may necessitate diminished overlap and T complement. Remember: I-bonding is a function of both inter- and intrafragmental CT, hence is not solely dependent on overlap. T-bonding (perfect pairing) is.

We can now answer the question posed earlier: To enforce the C_{3v} geometry in a WF_6 derivative, replace F by H, since the latter is both incapable of REL-bond activation (no lone pairs) and smaller than F. This expectation is consistent with recent ab initio calculational findings by Albright et al. and Schaefer et al. These results have been summarized and interpreted (by Pauling

VB theory) by Landis and co-workers.[3c] We further expect that CrH_6, unlike WH_6, will be octahedral (and we can only predict trends) again for two reasons: The lower electronegativity of Cr pushes it further toward the I end of the bonding continuum, and the smaller size of Cr accentuates ligand nonbonded repulsion, which is best relieved in the octahedral geometry. Thus, the apparent covalency of third transition series hydrides is only an exception. Furthermore, the bonds of WH_6 are not covalent but, rather, I bonds with high T complement that decides the shape. An ab initio VB computation is expected to show that perfect pairing makes only a small fractional contribution to the bonding.

The VB view of the "simple" WF_6 has turned out to be very different from the conventional view: It is a molecule that opts for low T complement in exchange for optimum I-bonding and minimization of nonbonded repulsion. Had it been covalent, the structure would have been entirely different. Are there more examples of "aborted covalence"? $CoCl_2$ is linear.[3d] Upon reflection, this fact refutes the concept of covalency: Assuming that Co uses the 3d and 4s AOs to bind, one would predict an angular shape because two s-d hybrids make a 90° angle. The observed geometry implies that the system is willing to sacrifice covalency to attain an I-optimal linear geometry and to avoid steric repulsion. One way of accomplishing these goals is to transfer one electron to one Cl and use one 4s odd electron to bind the Cl/Cl^- couple. In other words, instead of using a (3d,4s) double element, a system can use a (4s) lone element to effect binding. MoO_3 has D_{3h} geometry,[3e,f] rather than C_{3v}, for the same reasons. Finally, we can now appreciate that the "skeleton in the closet of inorganic chemistry" (Section 24.13) is an unfinished story. The angular BaF_2 and OCs_2 must be I-bound with a high T complement. Since optimal I-bonding occurs in the linear geometry (maximum overlap of each one central metal AO with two trans ligand AOs), we expect that changing F to I and O to Te may restore the linear shape!

Since the common denominator of heavy p-block and transition elements is I-bonding, we next seek the p-block analogue of the bonding of $W(CO)_6$ along one octahedral diagonal. It turns out that one complex that results from replacing the two nonmetallic COs by the semimetallic silylene and sulfur and the metallic W by the nonmetallic O is the species that serves either as a low-energy transition state or as a bona fide intermediate in the reaction of R_2Si and $Me_2S{=}O$ (DMSO) to yield (oxygen transfer) $R_2Si{=}O$ and Me_2S.[4a] As can be seen in Figure 26.2b, the bonding of the latter complex is entirely analogous to that of COWCO.

By virtue of analogical reasoning, one formula suggests innumerable experimental and computational projects. Writing a formula for the linear fragment COWCO is tantamount to a suggestion of innumerable novel synthetic targets. For example, combining ethylene and two oxygen atoms is expected to yield the T-bound oxirane plus triplet oxygen. Replacing O by S (and, still better, by RP) and C by Si is expected to throw the system to the I domain of bonding. The consequence is that the three fragments can now aggregate to form the following complex.

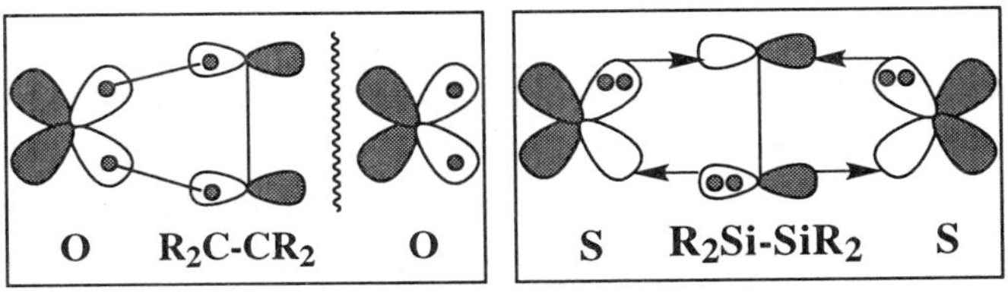

Note how two T bonds have been replaced by two I bonds. Note also that the I-bonding is the "reverse" of what obtains in COWCO: The central disilaethylene fragment acts as a σ donor and π acceptor and the two sulfur ligands act as σ acceptors and π donors. To obtain a "nonclassical" complex of this type, the two ligands must have a triplet ground state, which can be easily promoted to the D-type carbenic configuration (see Chapter 4), and the central fragment must have a polarizable double bond. The first condition is met by RP and S (and their heavier congeners) but not by R_2Si. The distinction among R_2Si, RP, and S is identical to that made in Section 24.4. The second condition is met by any olefin in which the carbons have been replaced by heavier congeners, or, conceivably, by ethylene itself, with the hydrogens replaced by polarizable substituents (e.g., R_3Si, etc.). Ab initio computations must be directed to calculating new molecules rather than recalculating old molecules and regurgitating old concepts.

We have argued that WF_6 and, still better, WR_6 and $W(OR)_6$, are I-bound molecules. It follows that a layer structure of the type $[(OR)_2W_2O_2]_n$, with the OR—W—OR array (bound by one I bond) perpendicular to the plane, is the polymeric, I-bound analogue of WF_6. Since the π system of benzene is incipiently I-bound, it follows that $[(OR)_2W_2O_2]_n$ must have graphitic character. Recently, Herrmann and co-workers prepared a derivative of the related $[(CH_3)(O)ReO_2]_n$ (which simply has one excessive bond pair).[4b,c] The polymer is actually $[H_{0.5}(CH_3)_{0.92}(O)ReO_2]_n$. It looks like graphite and undergoes pressure-induced depolymerization to $(CH_3)(O)ReO_2$.

Finally, let us connect with the most recent experimental literature: How can we describe the geometry and electronic structure of L_2PtAr_4, where L is a two-electron donor? Using the concept of the REL bond and the association formula (see Figure 21.3), we predict a cis geometry featuring three REL bonds, in which the metal bears three π-type electron pairs. The association formula is $Pt^{+}(Ar^{-},Ar)(L,Ar)_2$. The intramolecular analogue of this species, $Pt(C_6Cl_5)_4$, has been characterized Fornies et al.[4d] Two Cl atoms act as two Ls.

26.3 A Closer Look at Metal Carbonyls

The carbonyl ligand acts as a functional π acceptor with one observer σ pair. The latter activates the I bonds made by the π holes. Two CO fragments combine with one metal pair to define one OX I bond, which is activated by the σ pair, forms one antiparallel RED I bond with two metal holes. The primary acceptor function of CO is indicated by the available thermochemical data. Incorporation of a fragment X: into a bond A—A to produce A—X—A requires one-electron excitation from the HOMO to the LUMO of X:. The larger the HOMO–LUMO gap, the larger the promotional energy, and the weaker the bonding. From this standpoint, inserting a CO is highly unfavorable compared to inserting ethylene into a hydrocarbon like ethane because the singlet–triplet gaps (measures of the HOMO–LUMO gaps) of ethylene and carbon monoxide are roughly 100 and 160 kcal/mol, respectively. This means that CO disfavors perfect pairing. That CO acts primarily as a π acceptor is made evident by the carbonyl/ethylene comparison: Incorporation of ethylene into ethane is far superior to incorporation of carbonyl in the same molecule. However, the situation changes dramatically when ethane is replaced by the electron-rich hydrazine. The turnover is 33 kcal/mol!

	ΔE (kcal/mol)
$CH_3CH_3 + H_2C{=}CH_2 \longrightarrow CH_3CH_2CH_2CH_3$	–23
$CH_3CH_3 + CO \longrightarrow CH_3COCH_3$	–5

But

$NH_2NH_2 + H_2C{=}CH_2 \longrightarrow NH_2CH_2CH_2NH_2$	–40
$NH_2NH_2 + CO \longrightarrow NH_2CONH_2$	–55

Finally, the synthesis of highly reduced metal carbonyls such as $V(CO)_5^{3-}$ and $Cr(CO)_4^{4-}$ leaves no doubt about the inherent π-acid nature of CO.[5]

The most revealing ab initio work in the area of organometallics has been done by the groups of B.O. Roos and E.R. Davidson. These authors have consistently argued that chemical bonding is much more than what lies within the simple, EHMO-type models. A complete active space (CAS) SCF-CI calculation of the nickel–ethylene complex by Roos and co-workers showed that the dominant electronic configurations are those that are not normally considered in the simple models. For example, the only important reference configuration that describes interfragmental CT is the one corresponding to delocalization of a Ni d pair to the π^* MO of ethylene (2-e CT). The total electron density donated in the σ frame (a_1 space) and accepted in the π frame (b_2 space) of ethylene was 0.42 and 0.61 electrons, respectively.[6a] Similar trends were found in the case of the NiCO complex.[6b] Davidson and co-workers[7] found that "the MO model (meaning monodeterminantal MO theory) cannot actually predict that $Cr(CO)_6$ exists, let alone give a reasonable estimate of the bond energy." *Extramolecular correlation* is responsible for the binding, and a major part of it corresponds to what the authors call "dynamic shielding." This is a seminal paper that puts the discussion of organometallic bonding on the right course and is fully consistent with one central theme of this work: overlap dispersion and I-bonding in metal-containing molecules.

Baerends and Rozendaal[8] computed the RHF binding energy of $Cr(CO)_6$ in two steps: First, they interacted the σ CO pairs with the metal holes while turning off backbonding to obtain the dative stabilization energy, DE. Second, they interacted the metal pairs with the CO LUMOs while turning off dative bonding to obtain the backbonding stabilization energy, BE. Finally, they let dative and backbonding delocalization to occur together. The resulting DBE is 50% larger in absolute magnitude than the sum of DE and BE. This synergic effect is the result of reduction of interelectronic repulsion by partial preservation of atom electroneutrality and it is a strong hint that metal–carbonyl binding has a significant ROX delocalization component. This picture breaks down completely at the EHMO level at which the energy lowering upon orbital interaction is additive.

The electronic structure of the prototypical $Cr(CO)_6$ is represented by the following A diagram and A formula. Each pair of *trans*-CO groups sees a σ hole and a π pair. This means that a total of four arrows connect the two *trans*-CO with Cr; that is, we have two I bonds (indicated by superscripts on the A formula). There exist three different equivalent ways of visualizing the two *interacting* I bonds as illustrated above and the choice depends on the problem at hand. The RED/OX formulation is convenient when the response of the carbonyls to different metals (e.g., Cr, Mo, W) is considered. The REL/REL formulation is more suitable when ligand basicity is varied. Finally, the ROX/ROX formulation is best when we compare the affinity of two trans ligands for the central metal (trans influence). Here, we adopt the last formulation.

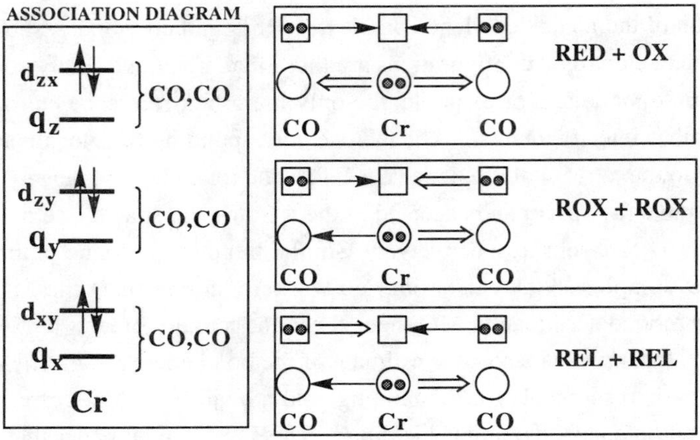

ASSOCIATION FORMULA:

$$Cr\ (CO,CO)^2(CO,CO)^2\ (CO,CO)^2$$

COMPARE WITH

$$\left[Cr\ (F,F)^{1.5}\,(F,F)^{1.5}(F,F)^{1.5} \right]$$

We can now use our concepts to rationalize some puzzling computational and experimental results. First, the mean carbonyl dissociation energy (CDE*) with respect to the promoted metal atoms stays nearly constant along the series $Cr(CO)_6$, $Fe(CO)_5$, $Ni(CO)_4$.[9a] To understand how this comes about, all we have to do is determine the ratio of the coordination number (CN) over the total number of I bonds (IB) at two limits:

(a) The high-excitation limit at which each CO acts as a ROX ligand to the maximum extent possible. This is equivalent to saying that activation by the σ pair is as important as binding by the π holes. ROX action requires that each of the metals and one carbonyl undergo single excitation. The promotion price is high, however, because of the huge HOMO–LUMO gap (singlet–triplet gap) of CO.

(b) The low-excitation limit at which each CO acts as an OX ligand. This is equivalent to admitting zero activation of the I bonds by the σ pairs for the benefit of reduced atom excitation. The energy gap separating the metal HOMO from the carbonyl LUMO is much smaller than the HOMO-LUMO gap of CO.

Note that the fragment excitations (internal CO promotion and metal to CO charge transfer) on which we focus come after the metal atom itself has been promoted to a closed-shell configuration suitable for binding the carbonyls.

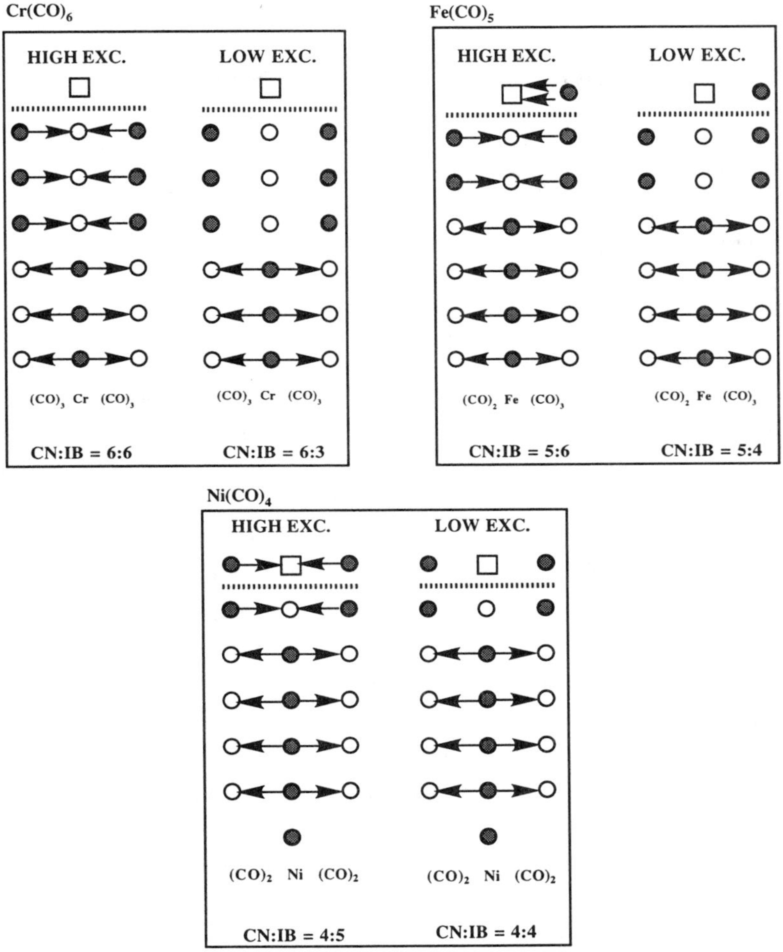

Figure 26.3 The matching of one metal pair (hole) with two ligand holes (pairs) to produce one I bond (CN = coordination number, IB = number of I bonds). Each metal has five (n – 1)d and one ns valence AO.

Assuming that each metal atom uses exclusively valence 3d and 4s holes and pairs and each carbonyl uses one (σ) pair and two (π) holes, the bonding of the prototypical metal carbonyls at the high- and low-excitation limits is shown schematically in Figure 26.3, and the crucial indices appear in Table 26.1.

The conclusion is straightforward: Both the "high" and the "low" CN:IB ratios predict the same order of binding: $Cr(CO)_6 < Fe(CO)_5 < Ni(CO)_4$. The lower the CN:IB ratio, the greater the stability. On the other hand, if we count the percentage of the intrinsically superior ROX I bonds at the high-excitation

Table 26.1 I Bonds, I-Bond Ratios, and Dissociation Energies for Three Metal Carbonyls

Carbonyl	I Bonds (high)	CN:IB ratios High	CN:IB ratios Low	CDE* (kcal/mol)
$Cr(CO)_6$	6ROX	6:6	6:3	50
$Fe(CO)_5$	4ROX,2OX	5:6	5:4	51
$Ni(CO)_4$	2ROX,3OX	4:5	4:4	46

limit (i.e., if we assume that overlap dispersion is the essence of metal–carbonyl bonding), we arrive at exactly the opposite conclusion: $Cr(CO)_6 > Fe(CO)_5 > Ni(CO)_4$. The truth is in between: Both the CN:IB ratio and the quality of I bonds matters. As a result, the two effects cancel out, and the values of CDE* are nearly the same. Note the recurrence of the VB theme: Bonding has to do with the *interplay* of fragment excitation and fragment bond making. The optimal situation always involves a compromise between these two antithetical variables.

We can again make use of our concepts to see that intervention of only one 4p AO (indicated by the empty squares above the hatched lines in Figure 26.3) is required for further improvement of the bonding, and this occurs partly in the case of $Fe(CO)_5$ and fully in the case of $Ni(CO)_4$. Intervention of one p in the case of $Fe(CO)_5$ is not very helpful because it produces a "forbidden" I bond (two arrows in the same overlap space). This means that $Fe(CO)_5$ has a defective carbonyl σ pair. It is this defect that drives $Fe(CO)_5$ to dimerize to form $Fe_2(CO)_9$. We now have two doubly bridged $Fe(CO)_4$ units plus an odd ninth carbonyl that associates with two Fe4p holes. In this way, the "forbiddenness" of the I bond is removed, as illustrated:

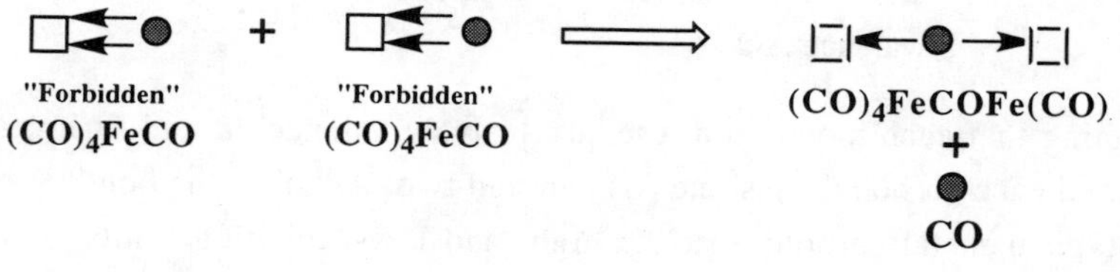

We can now see that the favorite 18-electron rule of the organometallic chemist works accidentally for the Cr, Fe, and Ni carbonyls [by assuming that a transition metal acts like a Lewis acid with (n – 1)d, ns, and np valence AOs] because it gives a proper account of the docking holes. That the success is accidental can be easily revealed by asking the question: Would $Ni(CO)_4$ be bound if the CO LUMOs (disregarded by the rule) were not present? VB theory gives a negative answer. Zerovalent metals are not σ Lewis acids but π Lewis bases. The problem shows up in the case of $Ti(CO)_n$ because the rule cannot perceive how bonding actually comes about once the carbonyls have docked on the metal holes, and it predicts the existence of the unknown $Ti(CO)_7$. The VB story is different. Because Ti is highly electropositive, we expect preeminence of OX delocalization. In general, we can make as many OX I bonds as there are metal pairs. Assuming exclusive OX I-bond formation and since one OX I bond can bind two ligands, Ti can bind only a maximum of 4, Cr can bind a maximum of 6, Fe can bind maximum 8, and Ni can bind maximum 10 carbonyls, provided there exist docking holes. Fe and Ni do not have the requisite number of docking holes and fail to attain maximum coordination.

Finally, we note that SF_2 and SF_4 can be viewed as structural derivatives of the octahedral SF_6, and the same holds true for $Mo(CO)_3$ and $Mo(CO)_5$ relative to octahedral $Mo(CO)_6$. The latter trend can be explained by assuming that ROX bonding is significant in transition metal carbonyls, in which case each of the six COs is effectively bound by one I bond to the central Mo in $Mo(CO)_6$. On this basis, the *gross* structures of the $Mo(CO)_n$ complexes can be derived by sequential removal of the CO vertices of the $Mo(CO)_6$ octahedron. The same trend encompasses all transition metal carbonyls.

Pauling proposed a VB theory of organometallic bonding implicitly based on Sidgwick's concept of the coordinate-covalent bond. The electronic structure of the octahedral $Mo(CO)_6$ was derived by assuming d^2sp^3 hybridization of the metal so that three d AOs accommodate the six valence electrons while six d^2sp^3 hybrids tie up the six σ-type CO pairs. In other words, Pauling assumed that one needs six metal holes to tie up six ligand pairs. The new VB theory says that one needs only half this number because the bonding is not of the coordinate-covalent type. Rather, it is of the I type governed by the association rule: One hole can tie up two pairs. As a result, the participation of the p AOs is no longer strictly needed, other than for polarization purposes.

With this background, we can now move into an area that only the most sophisticated experiments and calculations may enter, where more disjunctions between "old" and "new" theory can be made: the sequential bond dissociation energies of metal complexes. According to the covalency canon, a π acid ligand K (e.g., CO) placed trans to another π acid ligand L must reduce Pt—L bonding because one π-type d AO originally reserved for L must now be shared by K and L. By contrast, VB theory says that M—L is bound by one and (linear) K—M—L by two ROX bonds (see the COWCO in Section 26.1) when M has high electronegativity. If the two ROX bonds are assumed to be independent and if nonbonded repulsion is neglected, the first and second BDEs of $M(CO)_2$ must be equal, to a first approximation. The single ROX I bond of MCO is, of course, independent, and it is represented by two arrows confined in the same interatomic space. On the other hand, the two ROX I bonds of $M(CO)_2$ are not independent. Their interaction entitles us to consider them to be either two coupled ROX or two coupled REL bonds. Since a REL bond is represented by two arrows confined in two different interatomic spaces (lower contact repulsion), it follows that the (CO)M—CO must be stronger than the M—CO bond. In other words, one ROX bond enhances a second trans ROX bond. When M is Cu^+ or Ag^+, this expectation is met, as shown in an important work by Armentrout and co-workers.[9b] As the number of ligands increases past two, the requirement of ns-to-np promotion for polarization plus nonbonded repulsion causes a progressive diminution of the $(CO)_xM$—CO ($x > 1$) BDEs.

The situation changes dramatically when the metal binds π acid ligands by OX bonds, a situation that occurs when the metal has low electronegativity (e.g., M = Ni). Now, both NiCO and $Ni(CO)_2$ have a single OX bond. Although this form of bonding is superior in the biliganded metal (two arrows assigned to two different interatomic spaces), the singular CO in NiCO is still bound by one (inferior) OX bond while each CO in $Ni(CO)_2$ is bound effectively by one-half (superior) OX bond. As a result, the (CO)Ni—CO must be either as strong as (at best) or weaker than the Ni—CO bond, as is indicated by computations.[9c] The same is expected to be true in molecules of the type MO_x, where M is electropositive, as confirmed by the following tabulation; O can be taken to be effectively analogous to CO, with one difference being that one π hole in CO is replaced by a π pair in O.

	BDE (kcal/mol)	
Molecule	M—O	OM—O
TiO_2	159	144
ZrO_2	192	142
CrO_2	103	98
WO_2	161	144

In summary, our model predicts a *nonmonotonic* variation of the sequential $M(CO)_2$ BDEs only when ROX delocalization is predominant (e.g., in all cationic metal carbonyls). This appears to be the case with the exception of $Ni(CO)_2^+$.[9b]

The game is far from over! The increase in BDE in going from a mono- to a dicarbonyl complex of a transition metal also can be attributed to the necessity for promotion of the metal for binding the first ligand. The second ligand attaches to the already promoted metal with virtually no penalty. This is the conventional explanation of why the $X(SiX_2)$—X is stronger than the (SiX_2)—X bond. Atom X induces the mixing of the Si $3s^23p^0$, $3s^13p^1$, and $3s^03p^2$ configurations at the expense of Si promotional energy. This is called "hybridization," or, equivalently, "deshielding," necessary for the formation of the (SiX_2)—X bond. The fourth X is then bonded to the already deshielded atom without the penalty of promotional energy. In the case of Si, the atom promotional energy requirement is sharp because of the wide separation of the Si3s and Si3p AOs. Attachment of one CO to a transition metal such as Ni effects deshielding via the mixing of the $3d^24s^0$, $3d^14s^1$, and $3d^04s^2$ and sets the stage for a more exothermic attachment of a second CO. However, the promotional factor is not as important in determining the sequential BDEs because the $(n - 1)$d and ns AOs of transition metals are, by comparison to the ns and np AOs of semimetals, quasi-degenerate. Hence, promotional energy considerations cannot fully explain the sequential BDE variation in transition metal complexes. If this explanation were correct, the M—L bond would always turn out to be weaker than the LM—L bond, and this is not the case: The trend is dependent on the nature of M as discussed earlier. Clearly, the accurate determination of the sequential BDEs of ML_x is a crucial issue that will have definite repercussions on the conceptual theory of chemical bonding.

Bottom line: Organometallic chemists envision metal–carbonyl (and related metal–ligand) bonding as *synergistic bonding*. This term is actually a descriptor of ROX bonding, which, in turn, is only one of the three possible modes of metal–ligand bonding: OX, ROX, and RED. Synergistic bonding (i.e., ROX bonding) is a manifestation of overlap dispersion and is the most intrinsically favorable mode of I-bonding. *The key point is that the great common denominator of organometallic bonding is not synergistic bonding but, rather, I-bonding.* Depending on the nature of the metal and the ligand, any one of the three modes of bonding can predominate. Thus, there is nothing like "covalent, synergistic and ionic bonding."[10a] There is only I-bonding and this comes under three disguises. What is termed "synergistic" bonding by most organometallic chemists is actually prevalent ROX I-bonding. What is called "covalent" or "ionic" bonding is simply prevalent OX or RED I-bonding. This is why octahedral ML_6 complexes exist irrespective of the nature of L; that is, L can be an OX (e.g., CO, PF_3) or a ROX (e.g., PR_3, S).[10b,c]

26.4 VB Theory Makes Complex Molecules Easy to Understand

Most molecules inhabiting the real world of experimental metal chemistry have complex structures. Understanding how such molecules are put together, let alone predicting their existence, can be a difficult task. Can we rationalize the D_{2d} shape of CrO_8^{3-}?[11a] Since an electropositive metal is combined with electronegative oxygens, the situation must be analogous to that encountered in "hypervalent" molecules (PF_5, SF_4, ClF_3, etc.). Hence, we envision this molecule as a composite of Cr^+ plus four superoxide anions connected by four mutually complementary REL bonds. This represents activated I-bonding, and it is the only way first transition series metals can bind, aside from the E mechanism. The resulting structure has a D_{2d} geometry, and half of it is shown in Figure 26.4.

26.5 The Electronic Structure of Metallocenes

The hallmark molecules of organic chemistry are methane, benzene, and the Diels–Alder transition state, and we have already proposed new chemical formulas for them. We have done the same for the hallmark molecules of organometallic chemistry: $Ni(CO)_4$, $Fe(CO)_5$, and $Cr(CO)_6$. We now add to the list two more important prototypes: ferrocene[11] and dibenzene chromium.[12a] These two molecules further expose the fictitious world of "organometallic bonding," which has dominated the literature for decades.

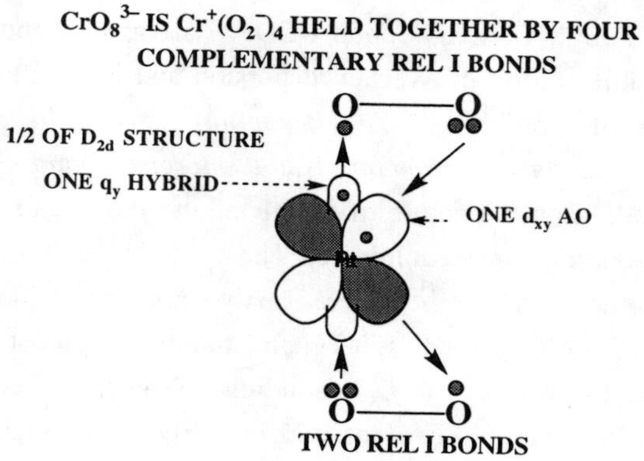

Figure 26.4 Electronic structure of CrO_8^{3-}. Only half of the D_{2d} structure is shown.

The A diagrams of the prototypical ferrocene and dibenzene chromium shown in Figure 26.5 illustrate the central principle that one metal element binds two ligand elements. Ferrocene is formalized as "Fe plus 2Cp." Each Cp is made up of three elements: an odd electron, identified by the fragment symbol (Cp) plus two bond pairs. Dibenzene chromium is formalized as "Cr plus 2 Bz." Bz has three elements, namely, three bond pairs. Each molecule has three I bonds. Note how the six ligand elements combine in pairs to match three metal elements in structuring three I bonds. In addition, each molecule has a six-electron equatorial mantle.

The A diagrams reveal also the key difference between the two molecules. In the case of ferrocene, transfer of one electron from iron to Cp generates one REL bond connecting the Fe(I) with the two rings. This is activated by the I bond produced by the combination of one 3d hole with two Cp bond pairs. The essence of ferrocene bonding is conveyed by the I formula shown in Figure 26.6.

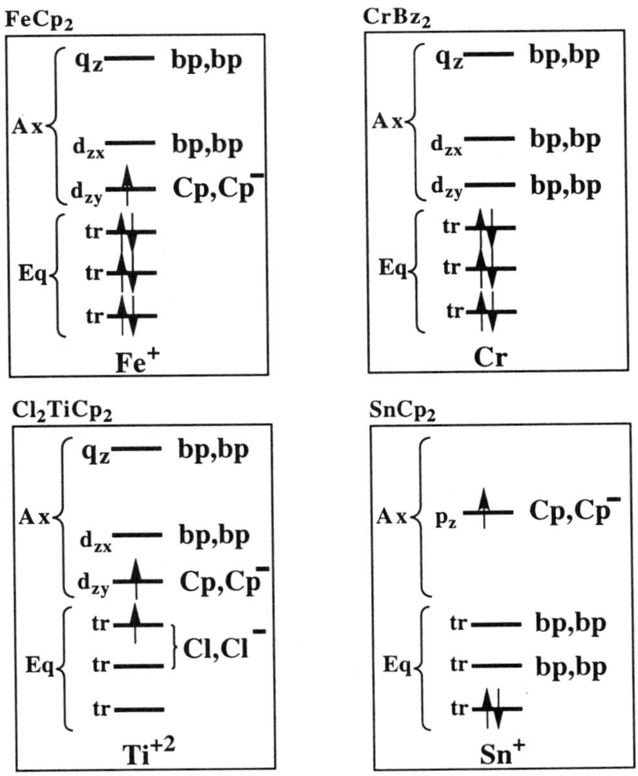

Figure 26.5 A diagrams of prototypical transition organometallics: bp, bond pair.

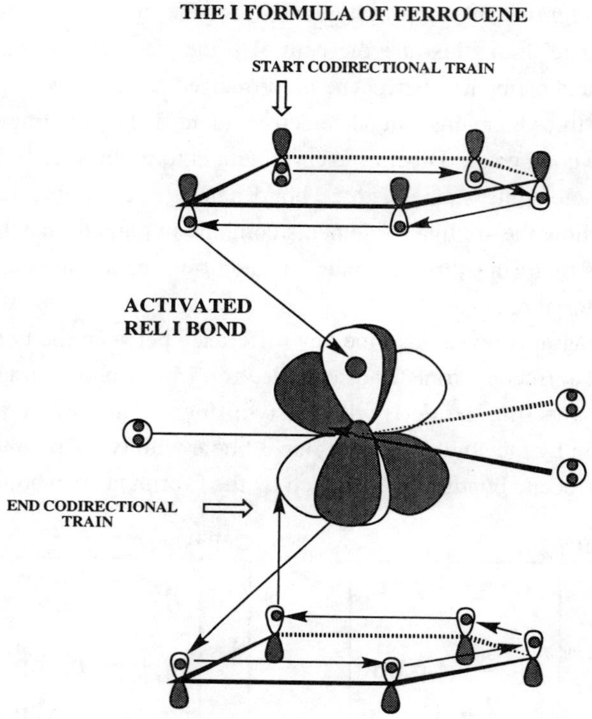

Figure 26.6 Electronic structure of ferrocene. Note arrow codirectionality.

By contrast, the benzene HOMOs are too low and the benzene LUMOs are too high in energy. As a result, electron transfer to generate REL I bonds is thwarted. The molecule has two weak contradirectional RED I bonds. The I formula is shown in Figure 26.7. As a result, ferrocene has superior activated I-bonding, while dibenzene chromium is a frustrated molecule that relies greatly on E-bonding [i.e., Coulomb attraction of the exposed (through induction) Cr core and the benzene π electrons]. This explains why dibenzene chromium is prone to oxidation rather than nucleophilic addition to the benzene rings.

We have seen that benzene does not have a fully associated π system because the C2p AO electronegativity is too high for such an arrangement. In other words, the triply ionic configuration with hole–pair alternation is inaccessible. As a result, benzene has only a partly associated π system (Figure 26.8a,b). This means that a metal can act to promote either full association or complete segregation. How can this occur? In a metallocene, the central metal enters with sd^3 hybridization in which three triangular equatorial hybrids accommodate the "nonbonding" electrons (a misnomer, as we shall soon see) plus two π-type d and one σ-type sd^3 hybrid, which are axially directed toward the π ligands. In the cases of ferrocene and dibenzene chromium, the three metal equatorial pairs can act in two distinct ways either to enhance or to abolish the association of the π bonds of the ligands, as illustrated in Figure 26.8c. This predicts three things:

(a) There should exist two different varieties of benzene metal complexes:

1. A nonalternant form, involving *associated benzene* with uniform C—C bond lengths and with mobile D_{3h} symmetry (twofold symmetry axes through vertices). Rapidly interconverting valence isomeric D_{3h} forms of this type can simulate D_{6h} symmetry. This seems to be the case in dibenzene chromium.

2. An alternant form, involving *segregated benzene* with C—C bond alternation with frozen D_{3h} symmetry (twofold symmetry axes through sides).

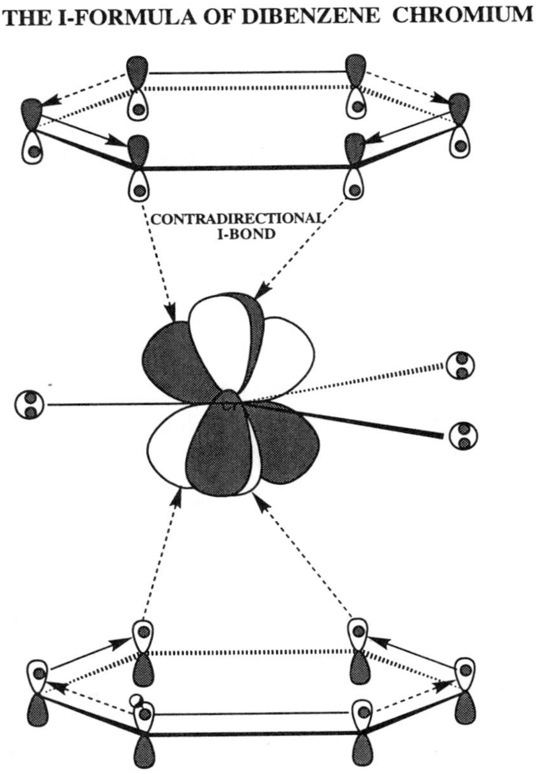

Figure 26.7 Electronic structure of dibenzene chromium. Note arrow contradirectionality.

This is exemplified by $(CO)_3CrBz$.[12b] Using the pseudo-octahedral metal hybridization scheme, we see that the $(CO)_3Cr$ fragment acts with three endo unstarred hybrids that are vacant and three equatorial pairs that are staggered with the endo set and eclipsed with the exo starred hybrids, which are used for the docking of the three carbonyls. The eclipsing occurs because the equatorial pairs delocalize into the π unoccupied MOs of the carbonyl, which are perfectly lined up for acceptance of equatorial metal electron density. Stacking the $(CO)_3Cr$ fragment on top of benzene gives an automatic explanation of the bond alternation, as illustrated:

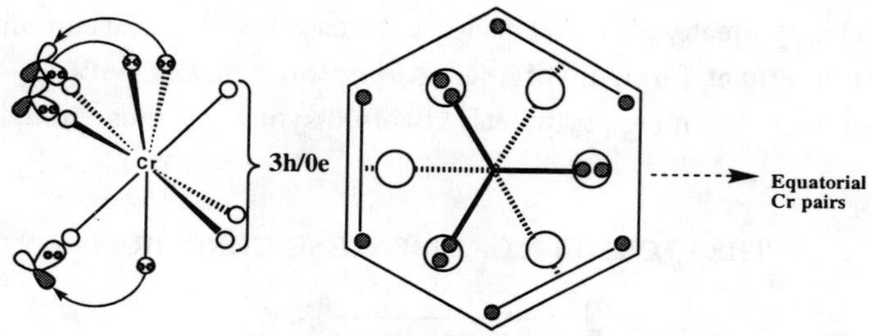

$(CO)_3CrC_6H_6$

Cr is pseudooctahedrally hybridized using the three endo conical hybrids (hatched lines) to bind benzene and the three exo conical starred hybrids (not shown) to tie up the CO σ pairs.

(b) The nonalternant form activates the benzene toward nucleophilic addition and predicts meta attack on anisole, as is observed.

(c) The metal equatorial pairs communicate with the ligands, and the most likely mode of communication is E-bonding, which is visualized as Coulomb attraction due to the matching of the equatorial pairs with induced ligand π holes as in the nonalternant form of dibenzene chromium. Hence, progressive removal of electrons from the equatorial mantle should diminish the dissociation energy and lengthen the metal–ligand distance. Indeed, as the number of equatorial electrons decreases in going from ferrocene to vanadocene, the M—C distance in MCp_2 lengthens.[11b]

M	Number of Equatorial Electrons	*r*(M—C) (Å)
Fe	6	2.064
Mn	5	2.144
Cr	4	2.169
V	3	2.280

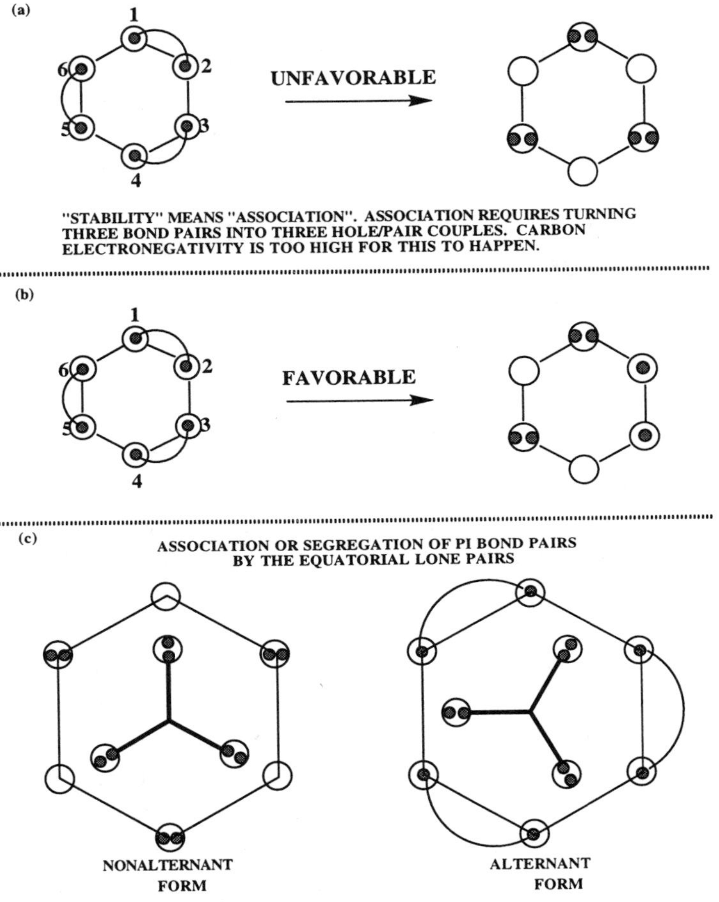

Figure 26.8 The electronic structure of benzene in isolation and upon coordination to $Cr(CO)_3$. There are two limiting structures of $C_6H_6Cr(CO)_3$ corresponding to two different orientations of the three triangular equatorial hybrids of $Cr(CO)_3$.

The MO formalism has dominated organometallic chemistry, and thus attention never focused on the action of the metal "lone pairs" simply because it was not possible to see where these pairs were located! The equatorial pairs of the $(CO)_3M$ group play an important role as indicated by two additional facts:

(a) The iron equatorial pairs of $Fe(CO)_3$ cause retention of stereochemistry in the solvolysis of 5-dienyl dinitrobenzoates.[13a]

(b) Replacing the Cp hydrogens by lithium atoms is expected to produce a stable ruthenocene or osmocene derivative by virtue of equatorial metal pairs delocalizing into Li2p holes, as is observed.[13b] We choose Ru and Os because they possess d AOs of greater radial extension than those of Fe. As a result, the equatorial lone pairs can overlap better with the Li2p holes.

The crucial role of the three equatorial hybrids of pseudo-octahedral or sd^3 metals (see Chapter 25) which point to the corners of an equilateral triangle can be further appreciated by considering the case of WR_6, where R is H or alkyl [i.e., ligands that are devoid of lone pairs, hence which can form only either unactivated (H) or weakly activated (alkyl) REL bonds]. Using an sd^3 W, we can form three REL bonds using the three axial metal AOs, each of which can be visualized as linking a formal "upper" R anion with a formal "lower" R radical via a singly occupied axial AO of formal W^{3+}. Whether the R-anion/R-radical pair is eclipsed or staggered is determined by the "observer" equatorial belt. Since a hole–radical or hole–pair interaction is attractive (whether primarily static or CT), each R-anion/R-radical pair must be eclipsed to afford each component maximum attractive interaction with one interposed equatorial hole. Note that each REL bond is "observed" by one equatorial hole, yielding the prediction of a *trigonal prismatic* structure for WR_6 consistent with recent experimental findings by Haaland's group[14a] and ab initio calculations.[3c] Exactly the same result is expected when the equatorial holes are replaced by pairs. Now, eclipsing of the R-anion/R-radical pair minimizes the repulsive interaction of each component with one equatorial pair. This rationalizes the well-known near-eclipsed conformation of crystalline ferrocene found by Seiler and Dunitz.[14b]

In summary, C_{3v} WH_6 represents a tradeoff of strong I-bonding (octahedral geometry) for high T complement (see Section 26.2), while prismatic $W(alkyl)_6$ is an example where strong I-bonding traded for static stabilization by observer holes. The "trading" occurred simply because the ligands were incapable of forming activated REL bonds. If lone pairs were present (replace alkyl by halogen), they would tie up the equatorial holes, giving rise to activated REL-bonding and transforming the prism to octahedral form.

Uranocene[15] provides another example of how one formula speaks a million words. The original proposal was that the stability of this molecule is due to f-orbital participation.[16] With the concept of the I bond at hand, the VB rationalization is different. Specifically, the f AOs are too contracted to bind efficiently and, as a result, both the bonding and the accommodation of any metal lone pairs are due to the action of the valence d and s AOs. This is

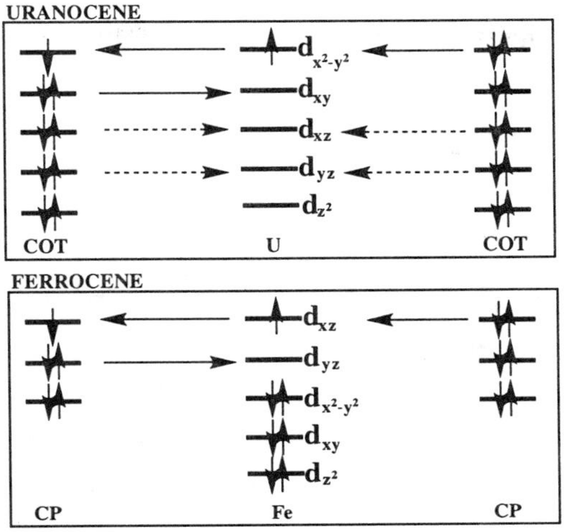

Figure 26.9 Uranocene and ferrocene have the same number of primary codirectional I bonds (solid arrows).

consistent with the known octahapto cycloctatetraene (COT) coordination by transition metals devoid of f-valence AOs [e.g., $(COT)(C_4H_8O)ZrCl_2$].[17]

The A diagrams shown in Figure 26.9 speak for themselves: When uranocene is formulated as $(COT^-)_2U^{2+}$, the bonding is analogous to that of ferrocene; the only difference is that "nonbonding" pairs of the former are replaced by "nonbonding" holes of the latter. Ferrocene and uranocene have very different electron counts, but the same number of crucial I bonds. These properties explain why uranocene has special stability among actinide sandwiches. If f-AO participation were the answer, all actinide metallocenes with four valence electrons or more should be stable because all can adopt an electronic configuration in which two f holes of the formal U^{4+} can match one NBMO of each formal COT^{2-}.

A "theoretician" once complained that experimental colleagues included his computations in joint papers without reading them! The reason is easy to understand. The MOs of ferrocene have appeared in books, journals, and conference proceedings. What has been learned? We suggest that, rather than improving understanding, these publications have perpetrated the illusion that drawing MOs means understanding molecular electronic structure. But, for every molecule in a given geometry, one can always and easily produce the symmetry-adapted MOs! MOs are only tools to be used for the development of physical models. They are not ends in themselves. There is no such thing as "chemical bonding explained by orbitals."

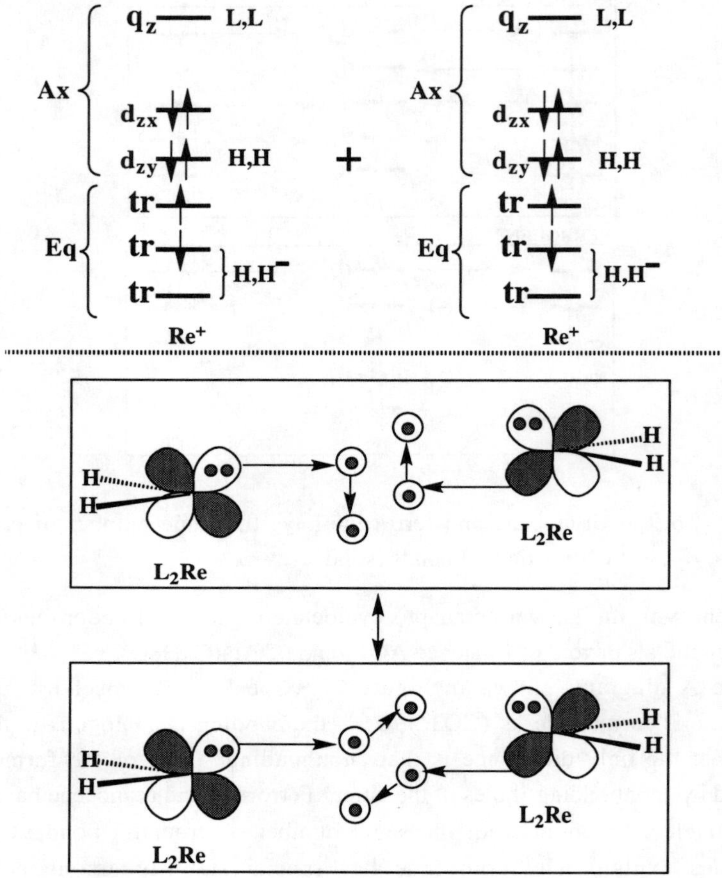

Figure 26.10 Electronic structure of $L_4Re_2H_8$ seen as a dimer of L_2ReH_4.

26.6 The Surprising Rhenium Polyhydrides

The tricapped trigonal prismatic ReH_9^{2-} (Re^{2-}-centered H_6 trigonal prism capped on the there square faces by three H atoms) is a classic structure of inorganic chemistry. It illustrates that the best a hydrogen atom can do is to form a REL bond by combining with a third-row transition metal and a second hydrogen atom. The metal adopts the sd^3 hybridization scheme and enters as Re^+ with one electron (one "dot") per AO. The three equatorial hybrids make one REL bond plus one T bond with three hydrogens, while the three axial dots associate with one hydride and one hydrogen atom each to form three axial REL bonds as shown:

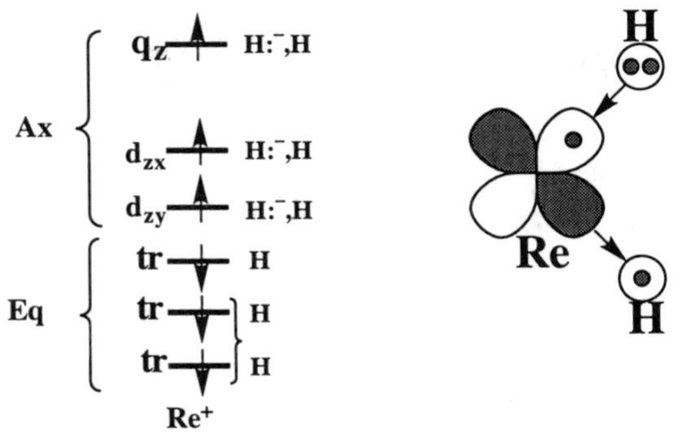

Bau et al.[18] determined the structure of $L_2H_2Re(H_4)ReH_2L_2$. We visualize this species as the product of the dimerization of two $L_2H_2ReH_2$ monomers described by the A diagram of Figure 26.10. The association of two $L_2H_2ReH_2$ fragments is the result of I-bond resonance. The formula given in Figure 26.10 speaks for itself.

26.7 The Formula of $(PR_3)_2Pt(H_2C{=}CH_2)$

The title molecule[19] has been beaten to death by investigators attempting to reach a "theoretical understanding" of the bonding of organometallic complexes. VB theory dispenses with the problem in half a page and the answer is very different. Specifically, the two sets of ligands are connected to a ds-hybridized Pt (see Figure 25.4) as follows:

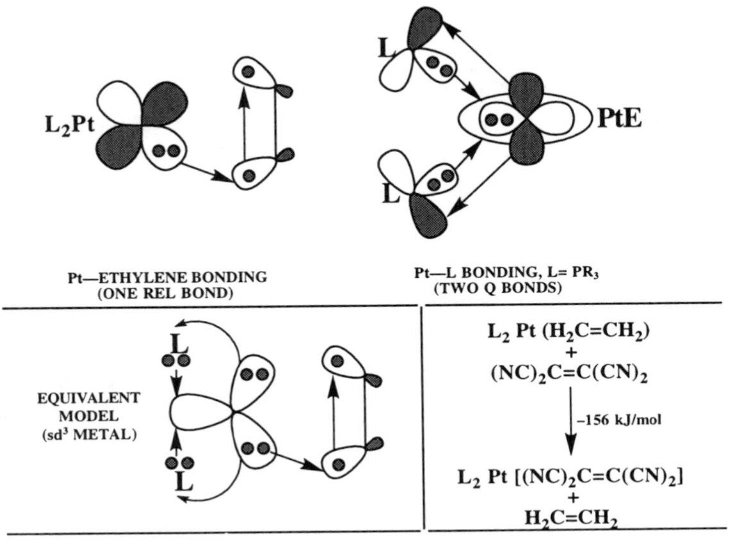

The hallmarks of this arrangement, which represents the principal I descriptor of the molecule, are fourfold.

(a) Each phosphine is linked to Pt by one ROX-type I bond involving the low-lying unoccupied space of P simulated by a vacant p AO. When the nonvalence space participates in the formation of such a bond, it is called a Q bond (see Chapter 28). The key point is that one valence hole and one valence pair are dedicated to the two phosphines.

(b) The Möbius arromatic array involving the antisymmetric metal pair and an effective ethylene diradical (stretched ethylene) constitutes one REL bond linking the metal to the ethylene. This predicts that the ethylene will have radical anion character. This explains the bonding of the analogous $(Ph_3P)_2COClIr(O_2)$ complex, where the O_2 ligand has an O—O distance such as that found in superoxide anion.[20] It also suggests that the stability of the complex should increase as the π acceptor ability of the alkene increases in the presence of phosphine ligands, as found. Of course, when the alkene becomes a very good acceptor and/or the metal becomes highly electropositive, the REL bond will be replaced by an OX bond (two arrows directed from metal to alkene). The key point is that only one valence pair is dedicated to the two phosphines.

(c) In the principal I descriptor, the metal binds the two phosphines by a hole–pair combination and the ethylene by a pair. The situation reverses in the secondary descriptor, which is equivalent to the famous Dewar–Chatt model of metal–olefin bonding embraced by chemists almost without exception. Hence, VB theory paints a very different picture for as "simple" a molecule as $(PR_3)_2Pt(H_2C{=}CH_2)$.

(d) The complex is planar because it has an equatorial σ hole and an axial σ pair (along with two π-type axial d_{xz} and d_{yz} pairs; see the ds-hybridized metal in Figure 25.4). As a result, going from the planar to the perpendicular (tetrahedral) geometry introduces exchange repulsion between the ethylene π-bond pair and the axial metal σ pair.

Rationalizations of the geometry of the complex and the effect of increasing the π acceptor ability of the alkene on complex stability are not very interesting because these trends conform to the intuition of the inorganic chemist. With respect to these topics, the superiority of the VB treatment over published treatments lies in the pictorial nature of our easily understood, half-page explanations, in which simple pictures represent the conclusions of high level theory. Much more interesting are the intricacies of the bonding, which can be exposed only by an electron-by-electron, orbital-by-orbital approach. In particular, the π bond of ethylene is formed by the combination of two carbon π radicals, each symbolized by RAD. There exist two possible association formulas:

(a) $Pt(PR_3)_2$(RAD, RAD). Each phosphine is bound by one ROX bond and the π system of ethylene by one REL bond. We say that each ligand is *self-associated.*

(b) $Pt(PR_3$, RAD$)_2$. Each phosphine is bound in tandem with one RAD site by one REL bond.

In the case of ethylene, the two carbon π radicals are constrained into proximity by the σ bond and situation a represents the best option. However, things change when the σ constraint is removed (e.g., when ethylene is replaced by two alkyl groups). Preservation of a REL bond now requires electron transfer from metal to one alkyl group as illustrated in II below. When the metal atom has a relatively high ionization potential, this is unfavorable. Hence, the molecule $(PR_3)_2PtR_2$ adopts the $Pt(PR_3, R)_2$ mode of bonding (III), in which the ds AO hole, originally dedicated to the binding of the phosphines, and the equatorial d pair, originally dedicated to binding the olefin, mix to produce two equatorial hybrids (see the d^2s hybridization scheme in Figure 25.5), with one electron in each one.

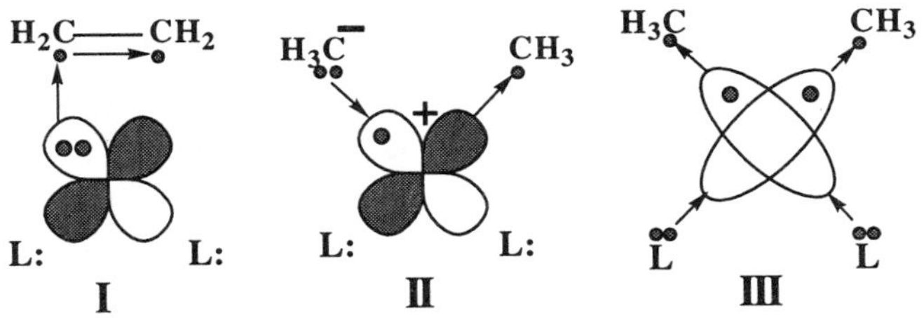

What structural variations of the ligands shift the preference from mechanism I to mechanism III? The first thing to note is that in I, each phosphine acts as a ROX (π acid/σ base) ligand while in III it acts as a RED (σ base) ligand. Furthermore, codirectional CT effects a one-electron transfer from metal to ethylene in I but a two-electron transfer from phosphines to ethylene in III. It follows that every perturbation of ethylene that renders it a better acceptor (π acid) and every perturbation of phosphine that renders it a better donor (σ base) will shift the mechanism of bonding from I to III. The enhancement of ethylene acceptor ability can be brought about in a variety of ways: pyramidalization, either by skeletal constrain or by perfluorination, as well as by replacing the hydrogens by π acceptor substituents (e.g., CN). As the effective LUMO of the olefin is depressed, the s character of Pt directed to the olefinic carbons will increase, while that directed to the phosphine ligands will decrease, causing the *J*(Pt—C) coupling constant to increase and the *J*(Pt—P) coupling constant to decrease.

Fortunately, a number of studies of derivatives of the $(PR_3)_2Pt(H_2C{=}CH_2)$ complex have been reported, and these provide a good test of our model.[21–23] For example, platinum complexes of the type L_2PtZ, where L is a phosphine ligand and Z a pyramidalized olefin, have been synthesized.[21] The following NMR coupling constants (in Hz) conform to our expectations.

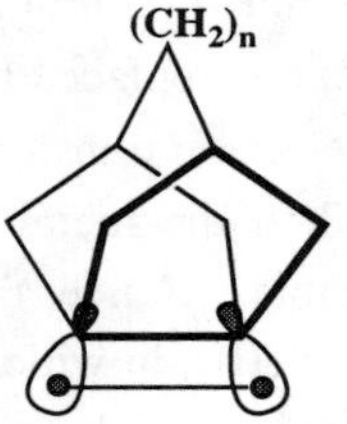

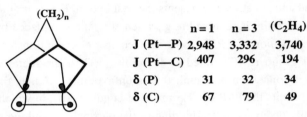

	n = 1	n = 3	(C_2H_4)
J (Pt—P)	2,948	3,332	3,740
J (Pt—C)	407	296	194
δ (P)	31	32	34
δ (C)	67	79	49

Here is a big puzzle: While the coupling constants testify to an increasing communication between the phosphorus and carbon Pt ligands as the olefin becomes an increasingly better acceptor, the nearly constant chemical shift of phosphorus seemingly suggests the opposite![22,23] What is the matter? To answer this question, we must have a clear understanding of the consequences of the RED, ROX, and OX modes of action of a ligand insofar as the electron density around the nucleus, hence its shielding, is concerned:

(a) RED action means that the ligand resembles a cation with a deshielded nucleus.

(b) OX action means that the ligand resembles an anion with a shielded nucleus.

(c) ROX action has two components. The overlap component represents the mutually canceling donor–acceptor action of the ligand. On the other hand, the dispersion component requires that the ligand act as an internal zwitterion. Now, we know that zwitterions acquire Rydberg character in order to cope with electron–electron repulsion. The best example is the Rydbergized zwitterionic singlet excited state of ethylene, and analogous considerations are being applicable to atoms. Recognizing that a Rydbergized zwitterion tends to the limit of "radical cation plus electron," we come to the conclusion that ROX action means that the ligand resembles a radical cation with a deshielded nucleus.

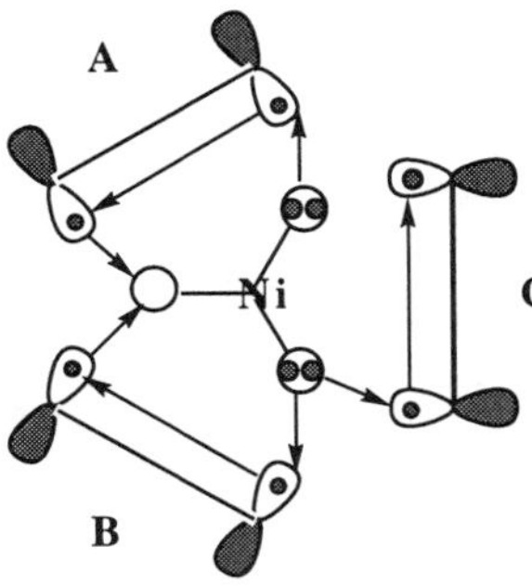

Figure 26.11 The electronic structure of $Ni(H_2C{=}CH_2)_3$; A, B, and C identify three ethylenic units.

Thus, as the mode of action of a ligand changes from OX to ROX to RED, we expect initial deshielding in going from OX to ROX, which levels off in going from ROX to RED. Since shifting from type-I to type-III bonding is tantamount to changing the action of phosphine from ROX to RED, it follows that the phosphorus chemical shift should remain relatively constant, as is found. If this analysis is correct, we have a way of using NMR spectroscopy to determine the mode of action of one ligand as a function of the metal and the observer ligands.

Bottom line: We reject the accepted view that $(PR_3)_2Pt(H_2C{=}CH_2)$ is an "inorganic cyclopropane" or a "π complex" or in between. The mechanism of bonding is variable but neither mechanism I nor mechanism III corresponds to the Dewar–Chatt model. We will find further evidence that the mechanism changes from I to III when ethylene is replaced by two alkyls in Section 32.2, where we discuss reductive elimination of metal alkyls, a reaction that clearly reveals the communication of trans ligands. With a good grasp of the electronics of the prototypical complex discussed in this section, we can easily understand the bonding of related complexes. For example, the much-discussed all-planar $Ni(H_2C{=}CH_2)_3$ complex[24] is a relative of $Pt(PR_3)_2(H_2C{=}CH_2)$. The VB formula shown in Figure 26.11 speaks for itself.

26.8 Organometallic Photochemistry and the Harpoon Mechanism

The lowest excited state (LES) corresponding to a ground T bond is an *antiresonant zwitterion* (i.e., the out-of-phase combination of the two zwitterionic structures).[25] When it comes to organometallics, T bonds are

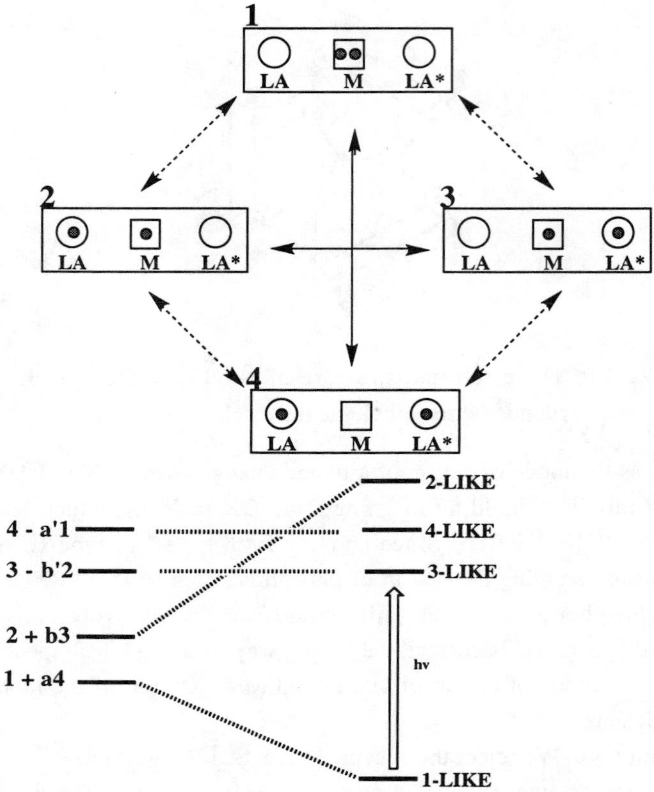

Figure 26.12 The low-lying states of a two-electron/three-orbital I bond.

replaced by I bonds. For example, consider the typical situation: A metal pair binds two holes of two different trans π-acceptors. What is the nature of the LES, and where does it localize? The 4 × 4 CI shown in Figure 26.12 leads to a manifold of states. The lowest energy excited state of the I bond is seen to be analogous to that of the T bond with the difference that a bond pair replaces a restricted pair in the antiresonant zwitterion. Specifically, the lowest excited state involves ML CT to the inferior acceptor, LA*. The resulting bond pair interacts in an antibonding fashion with the LA hole.

Figure 26.13 shows the reactive potentialities of the LES. The inferior LA* either induces the dissociation of the superior LA or it pivots and reacts with the superior LA. This is called the harpoon mechanism because the metal is envisaged as having harpooned the inferior ligand to be able to transfer it to the other component of the original I bond. The scenario here entails the exploitation of photoexcitation to identify, by observing the mode of ligand

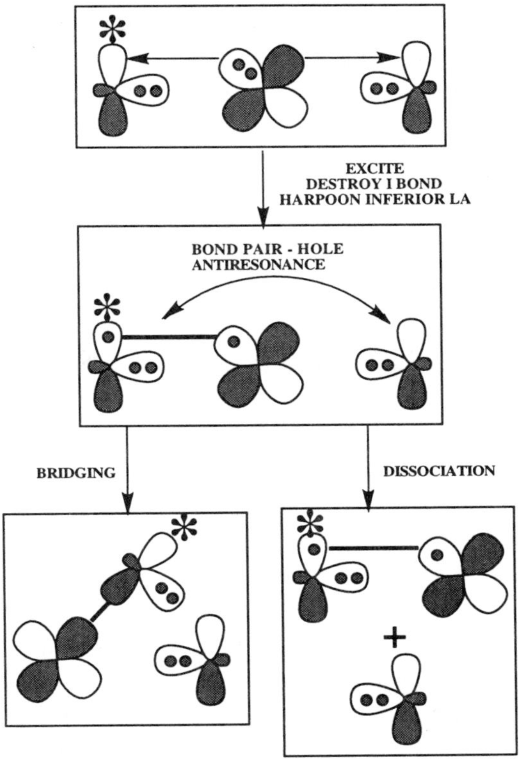

Figure 26.13 The reactions of one excited I bond.

rearrangement in the primary photochemical event, the ligands that are coupled by the metal into an I bond. Recent advances in spectroscopy now make it possible to identify the harpooned ligand and to follow its fate. Two cases are noteworthy:

Excitation localizes on the *inferior* π acceptor 4,4′-bipyridyl ligand (Bipy) in $ClRe(CO)_3Bipy_2$. CO and Bipy are coupled by a metal d pair.[26]

Upon long-wavelength excitation, one of the corner carbonyls rearranges to a bridging position in the $Ru_3(CO)_{12}$ cluster. This implies a CO···M···M I bond.[27]

An interesting situation arises when a homonuclear I bond coexists with a π system. In such a case, the resonance interaction of the corresponding excited states can produce a low-lying excited state that renders the compound colored, as exemplified by the trithiapentalene at the end of this section. The molecule has a 10-electron π system analogous to that of naphthalene, but its hallmark is the I bond spanning the three collinear sulfurs. An interesting computational paper by Spanget-Larsen et al. demonstrated that color is the consequence of a low-lying excited state which, in MO terms, is the result of interaction of the π-π* and σ-σ* electronic configurations.[28]

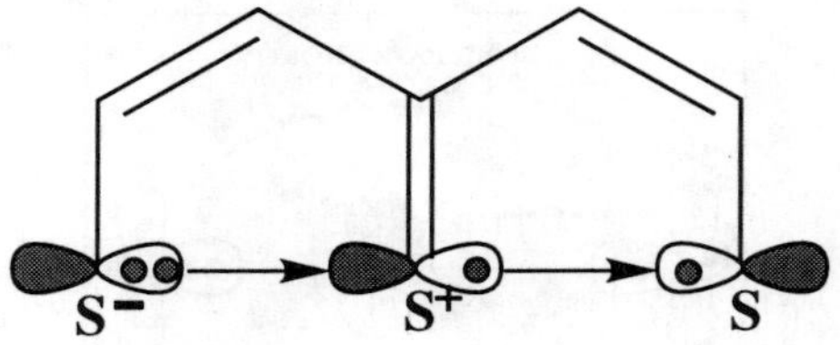

26.9 How Theoretical Inorganic Chemistry Went Astray

There have been four major attempts to gain an understanding of organometallic bonding:

Crystal field theory (CFT). This is based on the notion that bonding is effectively an electrostatic phenomenon.[29]

Pauling–Eyring VB theory (not to be confused with the ab initio VB theory of this work). The key ideas are "perfect pairing" and "covalency."

MO theory of the EHMO variety[30a] [frequently coming under the heading ligand field theory (LFT)].[30b] In this conceptual relative of the Pauling VB theory, "perfect pairing" is replaced by "unrestrained CT delocalization" and "covalency" is simulated by the mixing of covalent and ionic VB configurations.

SCFMO theory. This differs from EHMO theory to the extent that electron repulsion is treated explicitly, albeit in a crude way.

The common denominator of the four models, insofar as the essence of chemical bonding is concerned, is that none of them assigns the role of the protagonist to electron–electron repulsion. [31] This is contrary to the thesis of this work.

Is the "coordinate bond" a simple variant of the covalent electron pair bond? Sidgwick answered affirmatively ("covalent coordinate bond"), and the inorganic chemists followed. Later on, the issue was reexamined in the form of the CFT and LFT models, but the investigators failed to see what the very facts of metal chemistry actually hinted. Four prototypical molecules, namely, SF_6, WF_6, $W(CO)_6$, and IrF_6, bring the problem to focus. Chemists treated the octahedral SF_6 and WF_6 as if they were entirely different beasts. They called the former

Table 26.2 Inconsistency of Conventional Theory as Applied to Systems Containing Three I Bonds

	SF_6	WF_6	$W(CO)_6$	IrF_6
Conventional Explanation	"d-Orbital effect"	None	18-Electron rule	None
Spin	Low	Low	Low	High
Type of conventional bonding	"Hypervalent"	Ionic	Covalent	Ionic
Number of I bonds	3	3	3	3

"hypervalent" and, to explain the bonding of the latter, fell back on either the ionic model and CFT or the overlap model and LFT. In addition, $W(CO)_6$ was taken to suggest an analogy with organic chemistry: If there is an 8-electron rule for organic molecules, why not an 18-electron rule for transition metal complexes? This seemingly reasonable scenario is based on one fundamental assumption: universal covalency according to which one hole binds only one pair.

VB theory paints a different picture: According to the association rule, a hole can bind two pairs. Valence AOs of metal atoms can bind a maximum of twice as many ligands because of the very nature of the I bond: Two arrows emanating from a central atom can connect with a maximum of two ligands. The essence of our argument is projected in Table 26.2: Whereas there exists no apparent common denominator among the four illustrator systems in conventional theory, all four are related at the level of VB theory; that is, three I bonds connect twice as many ligands to the central atom.

WF_6 and IrF_6 disobey the 18-electron rule while $W(CO)_6$ obeys it. In fact, inorganic chemists have guidelines for anticipating success and failure of the 18-electron rule. Our point is that partial failure is not a case of an exception confirming the rule. Rather, it is indication that the house is upside down. The commonality of the CFT and one-electron LFT models lies in their failure to capture the physics of the problem: Bonding due to kinetic energy reduction is possible only if interelectronic repulsion is minimized. This occurs as exchange T bonds are replaced by CT I bonds. The difference between T- and I-bonding shows up only when interelectronic repulsion is explicitly considered. No electron–electron repulsion, no theory of chemical bonding!

Since the publication of the Lewis concept of the electron pair bond and the Heitler–London VB treatment of the hydrogen molecule, chemists have held fast to one central dogma: The best two atoms can do is make a "covalent bond," meaning a "Heitler–London covalent bond," or an "exchange bond." We have challenged this thesis: The best thing for two electrons is CT (rather than exchange) delocalization, either in two different diatomic overlap regions or — still better — in different interatomic overlap regions. This is the most efficient way of reducing kinetic energy by keeping down interelectronic repulsion. The worst-case scenario is that of the covalent bond: Two electrons delocalize in the same overlap region! To be specific the bonding of H_2 (T bond) is far more inefficient than that of Li_2 (I bond), even though the much higher electronegativity of H compared to Li causes much greater kinetic energy reduction in H_2 than in Li_2. This is why the bond dissociation energy of H_2 is much larger than that of Li_2. Here is the societal analogy: A rich man (H_2) does not mind extravagance (strong interelectronic repulsion) because he has a lot of money (strong kinetic energy reduction). By contrast, a poor but wise man (Li_2) is cautious because of his misfortune. Thus, he uses his resources more efficiently, minimizing his expenses (interelectronic repulsion) to compensate for the dearth of income (weak kinetic energy reduction).

One of the first things the MO theorist learns upon being introduced to the Hartree–Fock method is the expression of the total energy of a closed-shell molecule in terms of the LCAO spin orbitals *m* and *n*:

$$E = \sum_{m} E_m + \sum_{m<n}\sum J_{mn} - \sum_{m<n}\sum K_{mn}$$

The second double sum is over spin orbitals having the same spin. The physical interpretation of K_{mn} is that it represents exchange correlation: Two electrons having the same spin avoid each other because of the operation of the Pauli principle, and their mutual repulsion is reduced. This is the restricted MO interpretation of the physical meaning of the bielectronic exchange integral K_{mn}. Once we go to the unrestricted VB theory, the interpretation of K_{ab} (where *a* and *b* are nonorthogonal overlapping AOs) is entirely different: It represents the penalty of interelectronic repulsion that must be paid if two electrons delocalize in the same overlap region. This makes transparent the covalency restriction: A better bond is one in which bielectronic repulsion represented by $K_{ab} = (ab|ab)$ is replaced by repulsion represented by $(ab|cd)$!

All the foregoing considerations should not distract from the major issue: Even if EHMO and related MO models were physically acceptable, they would still be MO theories! To impact the experimental frontier, the theoretician must understand a molecule atom by atom and electron by electron. Only explicit formulas showing all valence orbitals and electrons can do the job. Trying to pin the "explanation" of the electronic structure of a molecule on one or few selected orbitals is an exercise in futility. MO theory can produce numbers but it cannot produce a chemically meaningful formula.

As an example of how VB formulas allow us to view molecules as analogues of macroscopic objects, consider the case of the "molecular brake" synthesized by Kelly et al.[32] In their model, a triptycene "wheel" is subjected to the braking action of a dimethoxybipyridyl group that upon coordination to Hg^{2+} acts as a "brake." VB theory suggests analogies in organometallic molecules. A pseudo-octahedral metal can act with a conical three-tooth (3T) gear to bind ligands under the influence of an equatorial 3T gear that is either "locked" or free to rotate. The molecule Cp*RhLXY ($L = PR_3$; X,Y = univalent ligand) is an illustrator of the key ideas.

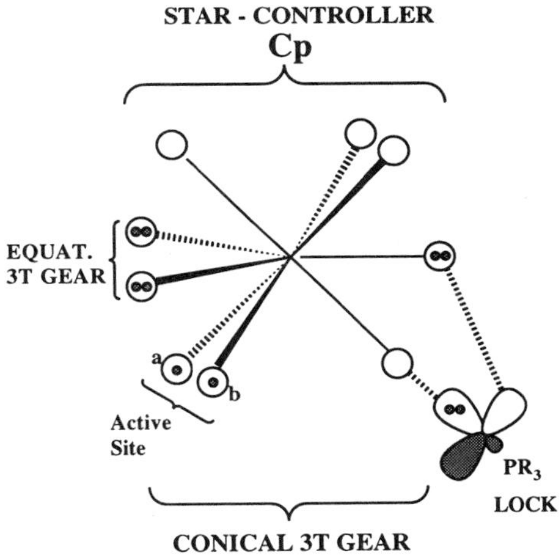

A ROX ligand, like PR_3, can act as a lock of the two 3T gears, with the strength of the locking action depending on the star character (not shown explicitly) of the hole on which L is docked. In turn, this depends on the number of starred holes tied up by the star controller ligand which, in this example, is Cp* (permethylcyclopentadienyl). Changing from a ROX (σ donor/π acceptor) to a RED (σ donor/π donor) ligand is equivalent to removing the lock.

In the same issue of *JACS* that published the molecular brake report, there is a paper dealing with the catalytic hydrogenolysis of the strong aryl–aryl bond by Cp*RhLXY (X = H, Y = Ph).[33] When this complex reacts with biphenylene in the presence of H_2 and benzene, the major product is the one in which H and Ph are replaced by 2,2′-biphenyl. How is this molecule formed? We can base a scenario on the electronic structure of the molecule as represented above rather than searching through stacks of occupied and unoccupied MOs.

Bottom line: Covalency represents a restriction because there is a superior alternative, namely, the I bond, which has two electrons delocalizing in different interatomic spaces rather than the same one. Organic chemists represent organic transition states with "dashed lines" to indicate delocalization, and coordination chemists follow exactly the same practice in representing ground organometallic complexes. These dashed lines signal that coordination complexes and transition states are related to each other but are different from organic molecules. The latter are T- but the former are I-bound. The dashed lines are an unwitting representation of the I bonds which we can now count and appraise.

26.10 Apparently Covalent Metal–Metal Bonds Are Not Covalent at All

Metals often do not do what chemists expect them to do, and for this reason, organometallic chemistry is both fascinating and frustrating, because the expectations of chemists are based on the covalent/ionic model and nature did not intend that metals obey the covalency rules.

For example, the prototypical dimer $(CO)_5Mn—Mn(CO)_5$ is thought to possess a covalent Mn—Mn bond because this formulation fulfills the 18-electron rule. VB theory paints an entirely different picture: There are no covalent bonds when it comes to organometallics, and the apparent Mn—Mn bond is just a component of an extremely favorable REL bond as illustrated below. Each Mn is assumed to be octahedrally hybridized. Only one of two equivalent I descriptors of the resonant REL bond is shown.

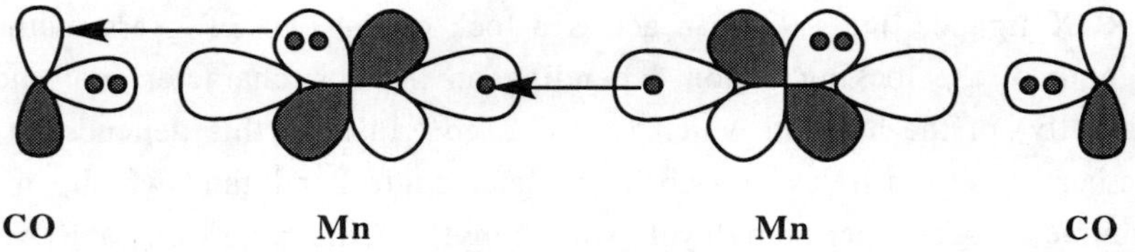

The experimental test of this proposal is simple: A metal–metal bond in a zerovalent bimetallic carbonyl cluster will not behave chemically as a covalent, two-electron bond; for example, it will fail to undergo the addition reactions expected of such bonds. Bullock and Casey studied $(CO)_3(Cp—)Mo—Mn(—$

$PR_2)(CO)_4$ and found that the assumed Mo—Mn covalent bond does not add H_2.[34] Bursten and co-workers found by Raman spectroscopy that the metal–metal bond is strongly coupled to a metal–carbonyl bond in $[CpFe(CO)_2]_2$.[35]

A well-known chemist once confessed that the most important thing he learned from his postdoctoral supervisor was how to get his papers published. This illustrates the mindset of a great many university professors. Instead of confronting a major problem with broad ramifications, they worry about how to appear knowledgeable in their writing of papers (grant proposals) so that the latter are accepted for publication (funded). Since one can be knowledgeable only when dealing with solved problems, the process guarantees the ad nauseum rediscovery of the wheel, supported by taxpayers' money. There is an astronomical number of combinations of atoms into molecules. Do they all justify a separate publication or grant proposal? The theory presented here and, in particular, the guarantee of multiple magic numbers by the very nature of the I bond (Section 15.10), suggests that the time might have come to abandon "per-molecule research" in favor of "viewpoint research," in which only comprehensive statements about a central problem are deemed satisfactory.

Bottom line: Incremental science is safe and arguably useful. At the same time, it is boring and inefficient. The cornerstone of incremental science is the peer-review system. Have the countless metal complexes reported in journals led to new ideas about the laws that govern how atoms are put together to form molecules? Have they taught us how to make things that count (catalysts, a conductor, a high T_c superconductor, etc.)? The answer we give is that the "coordinate bond" has actually little or nothing to do with the time-honored ideas of covalency and ionicity. We are as much at the beginning (but at a new beginning) as we were decades ago![36,37]

References

1. J. Jacobs, H.S.P. Mueller, H. Willner, E. Jacob, and H. Buerger, *Inorg. Chem.*31, 5357 (1992).
2. K. Pierloot and B.O. Roos, *Inorg. Chem.* 31, 5353 (1992).
3. (a) A. Demolliens, Y. Jean, and O. Eisenstein, *Organometallics* 5, 1457 (1986). (b) A.D. Cameron, G. Fitzgerald, and M. Zerner, *Inorg. Chem.* 27, 3437 (1988). (c) C.R. Landis, T. Cleveland, and T.K. Firman, *J. Am. Chem. Soc.* 117, 1859 (1995). WH_6 is not covalent and WF_6 is not ionic. This is revealed by the related $W_6O_{19}^{2-}$ cluster [V.W. Day and W.G. Klemperer, *Science* 228, 533 (1985)], which is

visualized as follows: A W_6 octahedron accommodates 12 edge and 6 vertex oxygens plus an encapsulated oxygen dianion. In essence, one W═O and the four surrounding oxygen atoms (which act as monovalent ligands with respect to the W atom) are analogous to $F_4W{=}O$, which is analogous to WF_6. Replacing two *cis*-W atoms by two Nb anions produces *cis*-$Nb_2W_6O_{19}^{4-}$, which reacts with different methylating agents with very low selectivity. In other words, a methyl cation attacks indiscriminantly the NbONb, NbOW, and WOW oxygens. The ionic model would predict that NbONb oxygen, which is flanked by two pentapositive niobiums, should react faster than WOW oxygen, which is flanked by two hexapositive tungstens. This lack of selectivity implies exactly what the I-bond concept denotes: interstitialization of the bonding electrons so that their mutual repulsion is minimized. In other words, the 6 metal atoms and the 12 bridging oxygens form a core about which the 48 $[(4 \times 6) + (2 \times 12)]$ metal–oxygen bonding electrons are symmetrically distributed. This tends to make all bridging oxygens equivalent insofar as local electron distribution is concerned. (d) M. Hargittai and I. Hargittai, *J. Mol. Spectrosc.* 108, 155 (1984) (e) W.D. Hewett Jr, J.H. Newton, and W. Weltner Jr., *J. Phys. Chem.* 79, 2640 (1975). (f) D.W. Green and K.M. Ervin, *J. Mol. Spectrosc.* 89, 145 (1981).

4. (a) Y. Nakadaira, T. Kobayashi, T. Otsuka, and H. Sakurai, *J. Am. Chem. Soc.* 101, 486 (1979). (b) W.A. Herrmann and R.W. Fischer, *J. Am. Chem. Soc.* 117, 3223 (1995), and following papers. (c) *Chem. Eng. News* 73(15), 32 (1995). (d) J. Fornies, B. Menjon, R. Sanz-Carrillo, M. Tomas, N.G. Connely, J.G. Crossley, and A.G. Orpen, *J. Am. Chem. Soc.* 117, 4295 (1995).
5. J.E. Ellis, *Adv. Organomet. Chem.* 31, 1 (1990).
6. (a) P.-O. Widmark, B.O. Roos, and P.E.M. Siegbahn, *J. Phys. Chem.* 89, 2180 (1985). (b) C.W. Bauschlicher Jr., P.S. Bagus, C.J. Nelin, and B.O. Roos, *J. Chem. Phys.* 85, 354 (1986).
7. E.R. Davidson, K.L. Kunze, F.B.C. Machado, and S.J. Chakravorty, *Acc. Chem. Res.* 26, 628 (1993).
8. (a) E.J. Baerends and A. Rozendaal, *NATO ASI* C176, 159 (1986). (b) T. Ziegler, J.G. Snijders, and E.J. Baerends, in *The Challenge of d and f Electrons*, American Chemical Society, Washington, DC, 1989, Chapter 23. (c) E.J. Baerends and A. Rozendaal, in *The Challenge of Transition Metals and Coordination Chemistry*, A. Veillard, Ed., Reidel, New York, 1986.

9. (a) E.R. Davidson, in *The Challenge of d and f Electrons*, American Chemical Society, Washington, DC, 1989, Chapter 11. (b) F. Meyer, Y.-M. Chen, and P.B. Armentrout, *J. Am. Chem. Soc.* 117, 4071 (1995). (c) L.A. Barnes, M. Rosi, and C.W. Bauschlicher Jr., *J. Chem. Phys.* 93, 609 (1990). Given experimental uncertainties, the problem of the sequential BDEs of $Ni(CO)_4$ must still be considered to be an open issue.

10. (a) J.S. Thayer, *Organometallic Chemistry*, VCH, New York, 1988. This marvelous descriptive book highlights the key molecules of organometallic chemistry and the puzzles they define. (b) V.J. Murphy and G. Parkin, *J. Am. Chem. Soc.* 117, 3522 (1995). (c) M.L. Buil, M.A. Esteruelas, F.J. Lahoz, E. Onate, and L.A. Oro, *J. Am. Chem. Soc.* 117, 3619 (1995).

11. (a) F.A. Cotton and G. Wilkinson, *Advanced Inorganic Chemistry*, Wiley, New York, 1988. (b) A. Haaland, *Acc. Chem. Res.* 12, 415 (1979).

12. (a) E. Keulen and F. Jellinek, *J. Organomet. Chem.* 5, 490 (1966). (b) B. Rees and P. Coppens, *Acta Crystallogr. B,* 29B, 2516 (1973). See also D.E. Koshland, S.E. Myers, and J.P. Chesick, *Acta Crytallogr. B*, 33B, 2013 (1977). (c) S.G. Kukolich, *J. Am. Chem. Soc.* 117, 5512 (1995).

13. (a) N.A. Clinton and C.P. Lillya, *J. Am. Chem. Soc.* 92, 3065 (1970). See also W.R. Roush and C.K. Wada, *J. Am. Chem. Soc.* 116, 2151 (1994). (b) A. Bretschneider-Hurley and C.H. Winter, *J. Am. Chem. Soc.* 116, 6468 (1994).

14. (a) For data summary, see A. Haaland, H. P. Verne, H. V. Volden, and J. A. Morrison, *J. Am. Chem. Soc.* 117, 7554 (1995). (b) P. Seiler and J. D. Dunitz, *Acta Crystallogr. B* 35B, 1068, 2020 (1979).

15. A. Avdeef, K.N. Raymond, K.O. Hodgson, and A. Zalkin, *Inorg. Chem.* 11, 1083 (1972).

16. A. Streitwieser Jr., in *Organometallics of the f-Elements*, T.J. Marks and R.D. Fischer, Eds., Reidel, Dordrecht, 1979, p. 149.

17. D.J. Brauer and C. Kruger, *Inorg. Chem.* 14, 3053 (1975).

18. R. Bau, W.E. Carroll, R.G. Teller, and T.F. Koetzle, *J. Am. Chem. Soc.* 99, 3872 (1977).

19. P.T. Cheng and S.C. Nyburg, *Can. J. Chem.* 50, 912 (1972).

20. L. Vaska, *Acc. Chem. Res.* 9, 175 (1976).

21. A. Nicolaides, Ph.D. Thesis, University of Washington, 1992.

22. G. Pellizer, M. Graziani, and B.T. Heaton, *Polyhedron* 2, 657 (1982).

23. M.A. Bennet, *Pure Appl. Chem.* 61, 1695 (1989).
24. K. Fischer, K. Jonas, and G. Wilke, *Angew. Chem. Int. Ed. Engl.* 12, 565 (1973).
25. L. Salem, *Electrons in Chemical Reactions,* Wiley, New York, 1982. For application of the VB two-electron/two-orbital model to bimetallic complexes, see D.G. Nocera, *Acc. Chem. Res.* 28, 209 (1995).
26. A.D.R. Gamelin, M.W. George, P. Glyn, F.-W. Grevels, F.P.A. Johnson, W. Klotzbuecher, S.L. Morrison, G. Russel, K. Schaffner, and J.J. Turner, *Inorg. Chem.* 33, 3246 (1994).
27. B.F.-W. Grevels, W.E. Klotzbuecher, J. Schrickel, and K. Schaffner, *J. Am. Chem. Soc.* 116, 6229 (1994).
28. J. Spanget-Larsen, C. Erting, and I. Shim, *J. Am. Chem. Soc.* 116, 11433 (1994).
29. S.F.A. Kettle, *Coordination Compounds*, Nelson, London, 1971.
30. (a) R. Hoffmann, *Science* 211, 995 (1981). (b) F.A. Cotton, *Chemical Applications of Group Theory*, 3rd ed., Wiley, New York, 1990.
31. C. Elschenbroich and A. Salzer, *Organometallics*, VCH, New York, 1989.
32. T.R. Kelly, M.C. Bowyer, K.V. Bhaskar, D. Bebbington, A. Garcia, F. Lang, M.H. Kim, and M.P. Jette, *J. Am. Chem. Soc.* 116, 3657 (1994).
33. C. Perthuisot and W.D. Jones, *J. Am. Chem. Soc.* 116, 3647 (1994).
34. R.M. Bullock and C.P. Casey, *Acc. Chem. Res.* 20, 167 (1987).
35. M. Vitale, K.K. Lee, C.F. Hemann, R. Hille, T.L. Gustafson, and B.E. Bursten, *J. Am. Chem. Soc.* 117, 2286 (1995).
36. At long last, someone decided to strike at the monster of peer review by launching a new journal with different rules of acceptance [*Chem. Eng. News* 73 (21), 26 (1995)]. Since incremental science is easily recognizable but new ideas are murky at first sighting, a "novelty journal" may be a utopian enterprise. However, the experiment is both necessary and interesting.
37. The concept of overlap dispersion (induction) and its relevance to metal chemistry was published nearly a decade ago but failed to capture any attention. One exception is the far-ranging review of the electronic structures of transition metal compounds by Professor Tsipis of the University of Thessaloniki: C.A. Tsipis *Coord. Chem. Rev.* 108, 163 (1991).

Chapter 27

The Transmutation of Endothermic into Exothermic Reaction

27.1 The Design of Frustrated Molecules

The major goal of chemistry is the preparation and exploitation of molecules that "live" in high energy wells. In other words, the molecule should be makable and yet unstable. Understanding chemical bonding means understanding how to play this game. There exist only two well-known strategies for making unstable molecules, and one is limited:

Antiaromatic molecules are unstable but also unmakable because they "escape" (by Jahn–Teller distortion) to more stable, unremarkable molecules.

Strained organic molecules are accessible but do not lend themselves to easy transformation and cannot act as catalysts.

In Chapter 8, we showed how a new understanding of the chemical bond leads to a recipe for the manufacture of high energy organic molecules that are represented by error-full T formulas. We can achieve high levels of instability by simply stringing atoms together in the wrong sequence. These unstable molecules are makable because they are not thwarted by Jahn–Teller distortion. Understanding the T bond can lead to the design and exploitation of organic instability. We are now ready to use our understanding of the I bond to design and exploit inorganic instability.

Chemistry is the science dedicated to the study of the transformation of matter. Typically, a chemist seeks cheap starting material, which can be converted to precious products. Nature stands in the way. The key question becomes: Can we modify reactions involving H_2, O_2, N_2, and other abundant molecules in a way that turns them from unfavorable to favorable? Our strategy entails the following points.

(a) Every reference reaction is pictured as a transformation of molecules labeled as T, I, E, and FR to indicate the nature of bonding and stability. Endothermic reactions have more FR products than reactants.

(b) To either decrease the number of FR products or increase the number of FR reactants, we seek to render a transformation favorable by metal intervention.

(c) Two guidelines for generating a frustrated molecule are:

1. Stable closed-shell molecules that have an intrinsic affinity for E-bonding must be combined with metals that are not suitable for E-bonding, such as third transition series metals.
2. Hydrogen (which is unsuitable for E-bonding in hydridic form and, at best, can form only nonactivated REL I bonds) must be combined with first transition series metal that have affinity for E-bonding.

(d) Substitution can turn an FR to either an I or an E product. Hence, there are two different ways of "curing" an unfavorable reaction, and the one that is superior depends on metal availability.

27.2 How to Turn H_2 into an I- or E-Selective Nucleophile

As a first example, consider the following synthetic target: Starting with H_2, produce a hydrogen nucleophile that can displace chlorine from an alkyl halide bearing either σ-donor (e.g., R_3Si) or σ-acceptor (e.g., F) groups. The latter has practical applications because it would constitute an effective way of destroying harmful chlorofluorocarbons such as CF_3Cl.[1] At first sight, metal intervention appears to be a doomed strategy because the H—H bond is much stronger than the M—H bond. The enthalpy of the following reference reaction is predicted, and found, to be very positive. The reason is that both products are frustrated (see Chapter 3).

$$\underset{T}{\text{H—H}} + \underset{I}{\text{Ni—Cl}} \longrightarrow \underset{FR}{\text{Ni—H}} + \underset{FR}{\text{H—Cl}} \qquad \Delta E = 29 \text{ kcal/mol}$$

Our strategy is twofold: We seek to decrease the number of FR products and/or increase the number of FR reactants. For the first case, we write

$$\underset{T}{\text{H—H}} + \underset{I}{L_nOs^+(\text{Cl,Cl}^-)} \longrightarrow \underset{I}{L_nOs^+(\text{H,Cl}^-)} + \underset{FR}{\text{HCl}}$$

where the atoms in parentheses combine with the metal fragment to form one REL bond. The overall transformation is expected to be more favorable than the reference reaction because two optimal molecules (I plus T) are transformed to two products only one of which is frustrated.

To increase the number of FR reactants, we write

$$\underset{T}{H{-}H} + \underset{FR}{Cl^{-}L_nFe^{+2}Cl^{-}} \longrightarrow \underset{FR}{Cl^{-}L_nFe^{2+}H^{-}} + \underset{FR}{HCl}$$

where each iron species is now FR because the contracted d AOs of Fe render it unsuitable for I-bonding. In addition, the size of Fe and Cl makes them unsuitable for E-bonding. Hydride is also unsuitable for E-bonding.

Once we have ClMH at hand, the question remains: Will it be able to act as a hydride transfer reactant? There are two possible successful transition states:

The I-selective transition state, obtained with M = Os and Y = R_3Si in Y_3CCl.

The E-selective transition state, obtained with M = Fe and Y = F in Y_3CCl. These two different transition states are depicted as follows.

I-SELECTIVE TRANSITION STATE (Y = SiR_3)

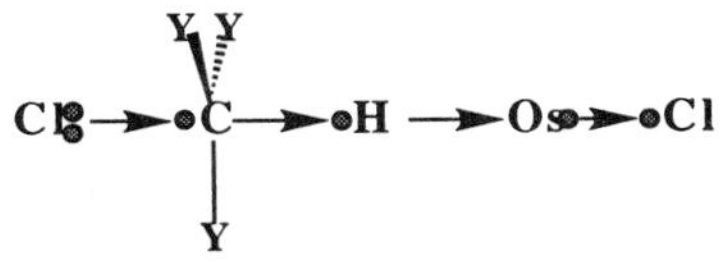

E-SELECTIVE TRANSITION STATE (Y = F)

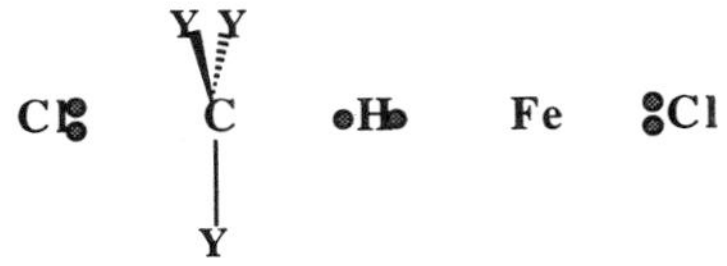

27.3 How Transition Metals Cleave Nitrogen

Why does NNO react with two H_2O molecules to yield NH_4NO_3? One major reason is that the reactants and products have zero errors in their T formula. The enthalpy of this redox process is –5 kcal/mol.

$$\underset{T}{NNO} + \underset{T}{2H_2O} \longrightarrow \underset{T}{NH_4NO_3}$$

The situation changes dramatically when we go to another reference reaction:

$$\underset{T}{NNO} + \underset{FR}{6HCl} \longrightarrow \underset{T}{2NH_3} + \underset{FR^*}{2Cl{-}Cl} + \underset{FR^*}{Cl{-}O{-}Cl}$$

Because three products are superfrustrated (lone pair repulsion), the reaction is endothermic by 110 kcal/mol! We can turn this unfavorable reference reaction to a favorable transformation by replacing O by metal M and choosing M so that the reactant NNM is frustrated while the product ClMCl is optimally E- or I-bound.

$$\underset{FR}{NNM} + \underset{FR}{6HCl} \longrightarrow \underset{T}{2NH_3} + \underset{FR*}{2Cl{-}Cl} + \underset{I}{ClMCl}$$

What should M be? Clearly, we must select M that has I but not E (or T) affinity so that the reactant NNM will be frustrated while the product ClMCl will be I-bound because Cl too has I affinity. Typical transition metals having low E affinity and high I affinity are those of the third transition series.

The reference reaction above can also be viewed as a sum of two elementary reactions: a metathesis plus a redox reaction.

Metathesis:

$$\underset{FR}{NNM} + \underset{FR}{4HCl} \longrightarrow \underset{T}{NH_3} + \underset{FR}{HNCl_2} + \underset{I}{ClMCl}$$

Redox:

$$\underset{FR}{NHCl_2} + \underset{FR}{2HCl} \longrightarrow \underset{T}{NH_3} + \underset{FR*}{2Cl{-}Cl} \quad \Delta E = -6 \text{ kcal/mol}$$

As implied by the oxidation numbers, the metathesis reaction is not a redox process. Only the conversion of $NHCl_2$ to NH_3 is a redox reaction. We can see that, as in the preceding example of the reaction of NNO with H_2O, there is no problem with the thermochemistry of the redox process. The problem may arise in the metathesis reaction, and it is insurmountable in the case of NNO. Once O is replaced by an M that turns NNM into FR and ClMCl into I, the enthalpy of the metathesis will be favorable and so will be dinitrogen reduction.

Every strongly bound, closed-shell molecule made up of first-row nonmetal atoms such as C, N, O, and F can interact with a metal only by the E mechanism. To produce weak E-bonding, we must select a "large" third-row transition metal, M, so that NNM will be frustrated. Furthermore, to ensure frustrated E-bonding, we must adorn the third-row metal with ligands that are high in the trans-influence and trans-effect series. In this way, all or some of the metal orbitals directed to N_2 will have high star character. As a result, (frustrated) E-bonding will be assured because CT delocalization involving high energy starred AOs is effectively "forbidden." The way in which the star character of metal AOs determines the mechanism and the strength of bonding is a central theme of our theory, and the concept is illustrated extensively in Chapter 31.

What are the facts? Our discussion has been motivated by the works of Henderson,[2] Schrock,[3] and Leigh,[4] and we use their systems as illustrators of the VB concepts. There exist many monometallic nitrogen complexes in which dinitrogen binds end-on, as well as many bimetallic complexes in which the dinitrogen acts as a bridging ligand. Reduction to hydrazine and ammonia requires four and, respectively six electrons, and these must be provided by the metal(s) and/or additional reagents. The Henderson complex $[Nb(S_2CNEt_2)_3]_2N_2$ yields hydrazine and $[Nb(S_2CNEt_2)_3]_2Cl_2$ upon treatment with HCl, and this implies sequential reduction to diazene and hydrazine by the two metals, each of which can offer two electrons. On the other hand, the Schrock complex $(Cp^*Me_3Mo)_2N_2$ produces ammonia. The heteronuclear complex in which one Mo is replaced by W gives 100% while the homonuclear (Mo or W) species gives only half of ammonia when protonated under reducing conditions. Clearly, the Schrock complex falls near our optimal design. It has second and/or third transition series metals and, most importantly, ligands (alkyls and Cp*) that have high affinity for unstarred AOs.

Figure 27.1 shows how (Cp^*Me_3Mo) connects to one nitrogen atom of dinitrogen either in the monomeric $(Cp^*Me_3Mo)N_2$ or in the bridged dimeric $(Cp^*Me_3Mo)_2N_2$. Since Cp and alkyl are far superior to N_2 in the trans-influence and trans-effect series (see Chapter 31), they tie up the maximum number of unstarred conical AOs dictating that N_2 must dock on a starred endo hybrid of a four-legged piano stool via one σ-nitrogen lone pair. This produces only weak E-bonding due to induction. The equatorial d has the right spatial orientation for overlapping with one π component of the triple N_2 bond. The equatorial "s" has the right spatial orientation for overlapping with the second π component of the triple N_2 bond. As a result, we have metal–ligand bonding due to overlap dispersion. Since the σ component of N_2 binds by the E mechanism, the CT component of the interaction of the equatorial metal AOs and the π system of N_2 is minimized compared to the dispersion component. As a result, the overall metal-N_2 bonding is effectively Coulombic, featuring a π-quadrupolar N_2. The key point is that the initial protonation at nitrogen does not involve nitridic but, rather, azenic nitrogen, which, in turn, owes its

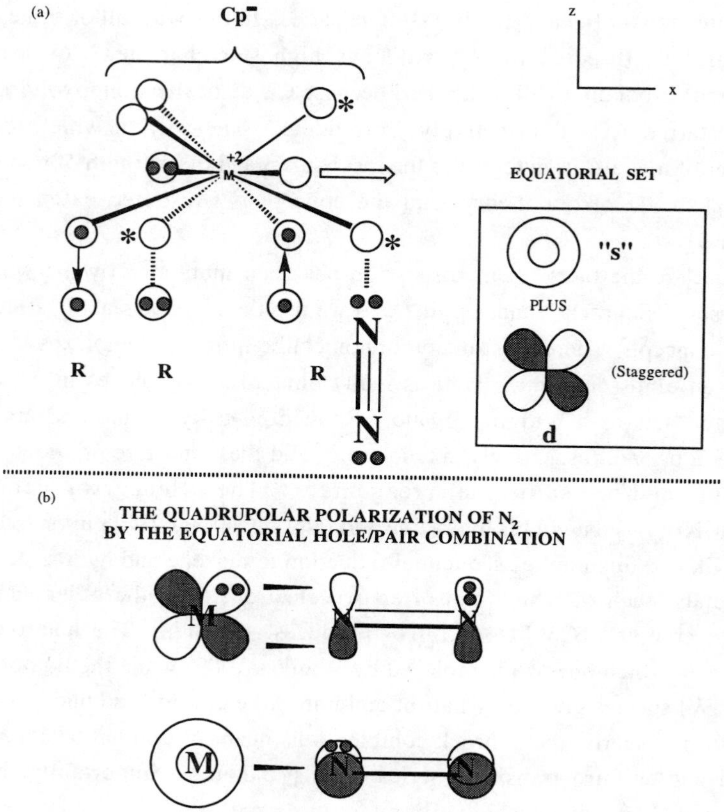

Figure 27.1 The polarization of N_2 by the $CpMR_3$ fragment (M = Mo, W).

existence to the "observer" ligands of the metal because they were the ones to determine the star allocation, hence the mechanism of metal–nitrogen bonding.

With T-, I-, and E-bonding as viable alternatives, it is obvious that we are just starting to learn how to piece together molecules to obey our commands. One lesson is that third transition series metals can bind to closed-shell organic molecules by turning them into multipolar species. The hallmark of this binding mode is that a large amount of promotional energy has to be expended to polarize strong T bonds, with the result that weak metal–ligand bonding goes hand in hand with a pronounced rearrangement of the ligand electron density. This answers a key question raised by Leigh: Why does *trans*-$[V(N_2)_2dmpe_2]^-$ have an exceedingly low value of ν(NN) and an activated dinitrogen which is, nonetheless, easily pumped away at room temperature? The significance of all this is that observation of a low ν(NN) and a "long" NN bond in dinitrogen complexes is consistent with either E- or I-bonding, and the former is more likely when the dissociation energy is low.

Our conclusion is that formation of diimide, hydrazine, and ammonia from MNN or MNNM complexes involves initial protonation of a π-quadrupolar dinitrogen. To make ammonia, one should choose a metal capable of weak E-type metal–nitrogen bonding. To make a metal nitride (which does not produce ammonia), however, a metal capable of strong I-type metal–nitrogen bonding is called for. The essential point is illustrated by two equations:

$$\mathrm{MNNM} \longrightarrow \mathrm{2MN}$$

$$\mathrm{MN + 3HCl} \longrightarrow \mathrm{MCl_3 + NH_3}$$

The first reaction is rendered exothermic when M can bind N strongly, while the second reaction is rendered exothermic when M can bind N weakly. What conditions must M fulfill so that the first reaction is realized? The answer is straightforward: It must be a second- or third-row transition metal (expanded d orbitals) fragment acting as a 3h/3e piece (three conical valence AOs/a total of three electrons), with none of the three active AOs starred. One possible choice is L_3Mo or L_3W (L = OR or NR_2). The electronic structure of $L_3MoNNMoL_3$ and the mode of cleavage to two L_3MoN complexes are shown in Figure 27.2. The the intermediate complex has two key features.

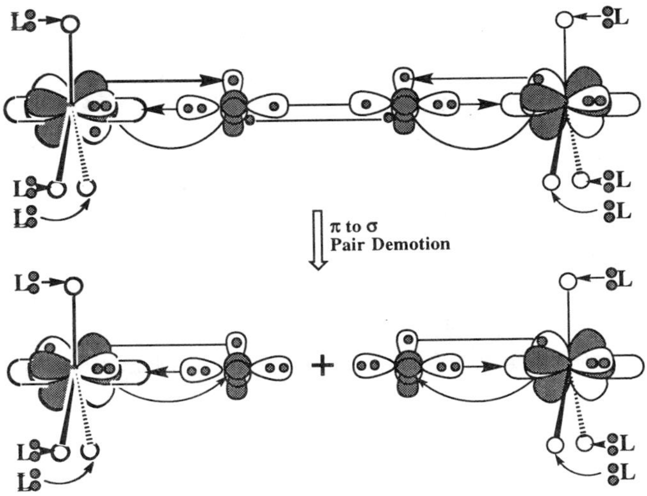

Figure 27.2 The electronic structure of $L_3MoNNMoL_3$ (L = NRAr).

The metal center binds one nitrogen by two 2-e bonds and one 3-e "bond." The seventh electron is the one that effects the cleavage of the central N—N bond.

A 3-e "bond" is actually a 3-e antibond that may become a bond by virtue of CT delocalization. Hence, the stability of the complex depends on the nature of the two three-electron interactions indicated by the curved lines. In essence, this is what qualifies a metal, rather than a nonmetal, as an "activator" of the N_2.

This discussion was motivated by the intriguing results of Cummins and Laplaza:

L_3Mo (L = NRAr) reacts with N_2O to produce $L_3Mo{\equiv}N$ and $L_3Mo{\equiv}NO$; that is, N_2O can indeed transfer N rather than O (as one might have expected (see Chapter 34). This suggested that L_3Mo can cleave N_2.[5]

And indeed, it was found that L_3Mo (L = NRAr) cleaves N_2 via an intermediate $L_3MoNNMoL_3$ complex to produce $L_3Mo{\equiv}N$.[6]

M_2N_2 complexes of the bridged type MNNM that fail to cleave dinitrogen do not contain a 3h/2e metal fragment and/or must use starred valence AOs. For example, $(Cp^*Me_3Mo)_2N_2$ has a 3h/2e M, but at least one of the three valence AOs directed to nitrogen is starred.

27.4 The Real Game of Chemical Bonding Is Just Beginning

The thesis of this work is that the existing models of chemical bonding are greatly inadequate, and this is why a good synthetic chemist has little sympathy for theory. Such a chemist simply understands chemistry much better because experiences with molecules have taught, for example, that fluoride is different from chloride, although both are called halogens. In other words, the experimentalist has an intuitive feeling about *chemical selectivity*. Understanding chemical bonding means understanding chemical selectivity. Some examples of the complexity of the problem follow.

The formal fragment SCN^- can combine with the formal fragment R^+ to form either RSCN or RNCS. Now, the first isomer involves superior I-bonding, while the second isomer best supports T- or E-bonding. Depending on the nature of R, the major product of nucleophilic attack of SCN^- on R^+ may involve either the S or the N end of the ambident nucleophile. N selectivity can now be rationalized as the result of E- or T-bonding! While T-bonding is a

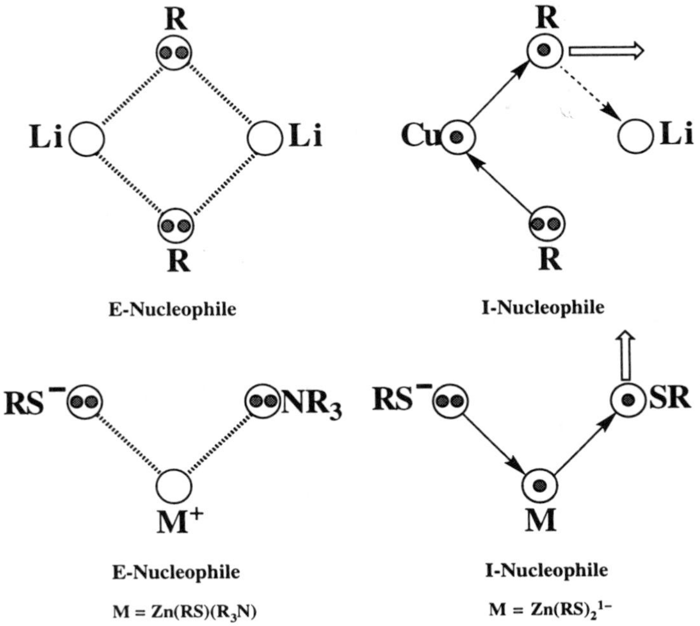

Figure 27.3 The differing bonding mechanisms used by lithium and by copper at the transition state of nucleophilic addition to an enone.

hallmark of ground organic reactants or products and I- and E-bonding are the two likely choices of a transition state (which is an obligatory association complex), the issue of ambident selectivity cannot always be resolved because two different modes of bonding often have the same stereochemical or regiochemical consequences.

A different problem is posed by the carbonyl group. Specifically, enolate anion adds to simple carbonyl compounds and gives conjugate addition to a,β-unsaturated carbonyl compounds (Michael addition) via the I site; that is, it acts as a carbanion. However, addition to carbonyl is equivalent to an S_N2 reaction in which the leaving group is effectively OY and one of the equatorial groups is also effectively OY (O is decomposed to two OYs, where Y is a dummy atom). That is, carbonyl is an E-selective substrate predisposed toward reacting with E-selective nucleophiles. This implies that the carbonyl group is "tolerant" to I-selective reactants such as enolate anions. We envision an ultimate classification of reagents into "sharp" and "blunt" depending on how selective they are with regard to partners.

We can now see that isoelectronic molecules traditionally thought to be analogous are actually very different. For example, mild reduction of aldehydes and ketones by $NaBH_4/H_2O$ implies that hydride derived from $NaBH_4$ acts as an I-selective reagent. By contrast, the requirement for the exclusion of water from $LiAlH_4$/ether implies that hydride derived from $LiAlH_4$ acts as an E-selective reagent. The way in which the metal determines E/I selectivity is illustrated by the behavior of conjugated enones, which undergo 1,2-additions with RLi but 1,4-additions with R_2CuLi. Although other mechanisms are possible (e.g., the oxidative addition path involving the R_2CuLi dimer), the argument here is that Cu differs from Li in the way that it determines the mechanism of bonding, E versus I, at the transition state of the addition, as indicated in Figure 27.3.

Partly because complexity thwarts selection and partly because of lack of guiding theoretical principles, experimental research is randomly targeted. Meaningful questions are rarely heard, and when they are asked, they stand out: "Why zinc and four cysteines?" asks Lippard and his group.[7] Why is the active site of the Ada protein $[Zn(S\text{-}Cys)_4]^{2-}$ and not $[Zn(S\text{-}Cys)_2(N\text{-}His)_2]$? Our answer is based on the assumption that through evolution, nature knows perfectly well the rules of chemical selectivity. In the first and complexes sulfur is an I- and an E-selective nucleophile, respectively, and the substrates on which Ada acts are I selective. The difference is portrayed in Figure 27.3.

Bottom line: For a reaction to be successful, the individual components must be carefully selected. Fluoride is different from chloride, carbon is different from silicon, and so on, and we are starting to learn how to put atoms together to achieve specific goals. Unfortunately, this game cannot be entered at the level of simple theory. The I bond effectively disappears at the monodeterminantal level. Because of the interaction of CT delocalization and static bonding, the theory of molecular bonding and the theory of intermolecular forces merge. Finally, covalent bonding expressed by the T formulas contradicts the accepted explanations of chemical trends.

References

1. D.M. Heinekey, private communication.
2. J.R. Dilworth, R.A. Henderson, A. Hills, D.L. Hughes, C. MacDonald, A.N. Stephens, and D.R.M. Walton, *J. Chem. Soc. Dalton Trans.* 1077 (1990), and following papers.
3. R.R. Schrock, R.M. Kolodziej, A.H. Liu, W.M. Davis, and M.G. Vale, *J. Am. Chem. Soc.* 112, 4338 (1990).

4. G.J. Leigh, *Acc. Chem. Res.* 25, 177 (1992).
5. C.E. Laplaza, A.L. Odom, W.M. Davis, C.C. Cummins, and J.D. Protasiewicz, *J. Am. Chem. Soc.* 117, 4999 (1995).
6. C.E. Laplaza and C.C. Cummins, *Science* 268, 861 (1995).
7. J.J. Wilker and S.J. Lippard, *J. Am. Chem. Soc.* 117, 8682 (1995).

Chapter 28

The Importance of Nonvalence Functions

28.1 The Great d-Orbital Participation Confusion

The problem of chemical bonding is twofold. The first issue is the recognition of the T/I/E trichotomy. The I bond removes the necessity for any ad hoc hypotheses, such as the constructs of "d-AO participation" in the p block and "p-AO participation" in the d block. The second issue is that I bonds are activated by observer pairs and holes, and the latter can be nonvalence polarization holes. The lack of understanding of how observer pairs and holes function has a dire consequence: One cannot see how I-bonding becomes activated. As a result, one cannot see how the "observer" (as distinct from the "functional") elements of the problem decide whether the system will choose I- or E-bonding.

Nothing speaks more eloquently about the nonexistence of a conceptual theory of chemical bonding than the controversy of the "d-AO participation" in p-block molecules. PF_5 is the prototypical "hypervalent" molecule.[1a] Most chemists think they understand the bonding of PF_3 and focus on how the fourth and fifth fluorine atoms become attached to phosphorus to produce PF_5. Here are the conventional explanations:

PF_5 has four covalent and a fifth "ionic" P—F bond; that is, it is really $PF_4^+F^-$, an analogue of $NH_4^+Cl^-$. The problem is that $PF_4^+F^-$ is actually observed as an ion pair species distinct from PF_5.

P is dsp^3-hybridized and has five covalent bonds directed to the five F atoms. The problem is that ab initio computations show that d AOs do not act as valence orbitals; this is demonstrated shortly.

PF_5 is planar PF_3 with a four-electron/three-center (4e-3c) bond spanning the axial FPF subunit. The problem is that the prototypical illustrator of such bonding, the H_3^- system, is actually unbound with respect to H_2 plus H^-.

Our interpretation is that PF_5 is neither "covalent" nor "ionic." Rather, it is an association complex that is described by the A diagram presented in Chapter 24 (see Figure 24.6d). No d AOs or d functions are needed to explain the bonding of PF_5. The stalemate has been broken by the concept of the I bond, whereby a single valence pair can bind two radicals.[1b]

Four different groups attempted to resolve computationally the d-orbital controversy, only to generate more of it. Magnusson[2] finds that in p-block molecules, d functions (rather than d orbitals) do little more than simply tighten the "bonds" that are formed by the valence AOs. This is an important contribution because it eliminates the "d-AO alibi" as the explanation of "hypervalency." We expect that an analogous study will find that in d-block molecules, p functions (rather than p orbitals) also do little more than simply tighten the "bonds" formed by the valence AOs. Such a result will remove the 18-electron alibi so dear to organometallic chemists. From this positive contribution by Magnusson, things go downhill. On the basis of ab initio MO calculations, Reed and Schleyer[3] conclude that the bonding in OCF_3^- (isoelectronic to ONF_3) is of the same type found in OPF_3. By contrast, on the basis of ab initio GVB computations in the perfect-pairing (PP) approximation, Messmer[4] concludes that nitrogen forms four (presumably covalent) primary bonds in ONF_3 while phosphorus forms six (presumably covalent) bonds in OPF_3 oriented roughly toward the corners of an octahedron. Finally, on the basis of spin-coupled ab initio VB calculations, Cooper et al.[5] end up agreeing more with Reed and Schleyer than with Messmer, despite their interpretation of the transcription of SCFMO-derived localized MOs to spin-coupled VB theory to imply six bonds in SF_6. All this controversy evaporates once it is realized that *one* central atom can be linked to two ligands by a single I bond. ONF_3 has four covalent bonds. By contrast, OPF_3 (analogous to PF_5) has two I bonds and one T bond according to the A diagram. People computing by VB methodologies still fail to appreciate that once the atomic orbitals are prepared for computation (orthogonalized AOs, Coulson–Fischer AOs, etc.), the conceptual advantage of VB theory is lost. PCl_5, ICl_3, and $SeCl_4$ were prepared by Davy, Gay-Lussac, and Berzelius in 1810, 1814 and 1818, respectively. If, in 1995, theorists are uncertain about the bonding of these molecules, then, surely, there must be something wrong with theory!

In summary, d orbitals are *not necessary* for explaining the bonding of "hypervalent" molecules. Chemists were pushed to the d-orbital participation model by a valence theory founded on the notion that one pair ties up one hole (covalent bond). On this basis, SF_4 is "hypervalent." We have replaced the old by a new valence theory, namely, that one pair can tie up either two holes or two radicals (I bond). As a result, the necessity for invoking the d-orbital participation model vanishes. "Hypervalent" SF_4 becomes "normalvalent" SF_4!

Much has been sought in "polarization" as a potential explanation of chemical trends. However, this term is so poorly defined that it explains everything and nothing! There is one-electron exchange polarization (an overlap phenomenon) as well as one- and two-electron Coulomb polarization (induction and dispersion) and, to make things worse, we have shown that these mechanisms interact!

The issue can be reformulated: What are the observable chemical consequences of Coulomb polarization acting either alone or in tandem with overlap? We will now give specific answers by specific formulas featuring specific bonds. We represent the activity of the nonvalence unoccupied space of p-block and d-block atoms by d and p AOs (functions), respectively. This is the context in which the hybridization schemes of the sp^3d^5-type (p-block atoms) and the d^5sp^3-type (d-block metals) discussed in Chapter 25 should be interpreted. We will argue that polarization functions play a critical role in determining chemical trends by engendering partly or fully "static bonds." The time has come for putting to work the starring procedure developed earlier in connection with the VB concept of hybridization.

28.2 Fluoromethane Is "Organic" but Methyl Iodide Is "Metallorganic;" The Q Bond and Pair Affinity

A fourth-row atom, such as iodine, has the same number of valence electrons as its first row congener, F, but, it is fundamentally different from it: I is less electronegative than F. This implies that I has a more active unoccupied space than F. *The two properties are implicitly related.* It is exactly because of the lower electronegativity and the implicitly related presence of a low-lying, nonvalence unoccupied space (simulated by d AOs or d functions) that iodine is displaced toward the I limit of the bonding continuum. This means that fluoromethane is a fluoro derivative of nonmetallic methane but iodomethane is a methyl derivative of semimetallic iodine. Bonding is fundamentally different for the two species. CH_3F is bound principally by perfect pairing, while CH_3I is bound by overlap dispersion and overlap induction (over canonical orbitals), represented by a formula that shows explicitly the deshielding effected by induction and dispersion. Specifically, the iodine lone pairs bend away from the CH_3 fragment, exposing the C—H bond pairs to a deshielded I core with further "transverse deshielding" (dispersion) improving the situation. This is shown in Figure 28.1, from which we can draw three conclusions.

Iodide is an I-activated *nucleophile*; that is, it has nonvalence observer holes that can be matched by valence observer pairs of an I-activated electrophile partner.

CH_3I is formalized as $CH_3^+I^-$. There is one hatched arrow connecting a polarized valence C—H pair with a nonvalence (or polarization) starred hole. In

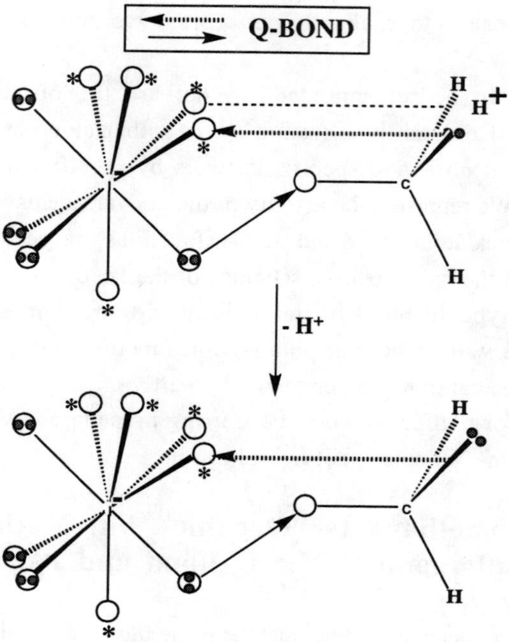

Figure 28.1 Incipient Q bond in CH_3I acidifies methyl hydrogens.

addition, there is a dashed line that indicates a resonating hatched arrow. The molecule has incipient carbenic character.

The C—I bond indicated by the solid arrow would have been a normal covalent bond (T bond) had it not been for the codirectional (head to tail) CT that can be effected by the hatched arrow.

There are two equivalent ways of interpreting Figure 28.1, and both are based on the fact that two CT arrows make up one I bond:

(a) The hatched arrow is "observer," the solid arrow is "functional," and the former activates the latter. As a result, the single solid arrow counts as half an I bond.

(b) Two codirected solid arrows define one I bond in which overlap is assisted by dispersion. One solid and one hatched codirected arrow count for one *Q bond*, in which dispersion is assisted by overlap. In other words, the I and Q bonds are differentiated by the relative importance of two coactive mechanisms of electron delocalization: interfragmental CT (overlap) and intrafragmental CT (dispersion). The Q bond has major quadrupolar character.

Our CH_3I formula suggests that heavy p-block atoms should have a strong affinity for vicinal pairs because these can combine with the unoccupied space of the heavy p-block atom to form the component of the Q bond represented by the hatched arrow. The first application of the principle of *heavy p-block pair affinity* is the following metathesis, where X is a p-block group.

$$2\text{H—CH}_2\text{—X} \longrightarrow \text{H—CH}_2\text{—H} + \text{X—CH}_2\text{—X}$$

When X is a light first-row systolic group (e.g., F, OH, NH_2, CH_3), the reaction is highly exothermic. When X becomes a heavy first-row congener (e.g., X = M = Cl, SH, PH_2, SiH_3), the activity of the nonvalence functions tends to render the reaction endothermic as the starred holes interact with C—H bond pairs toward forming a Q bond. This means increased stabilization of the reactant side, where each M is bound to one methyl.

The experimental data speak clearly: When X = F, the reaction is exothermic by 14 kcal/mol. On the other hand, when X = Cl, the reaction turns out to be thermoneutral. Schleyer and co-workers[6a] computed the exothermicities of the reaction for a series of first-row and corresponding second-row substituents, X = NH_2, OH, F and PH_2, SH, and Cl. It was found that replacing each first row by its second row analogue reduces the reaction exothermicity by approximately 10 (NH_2/PH_2), 17 (OH/SH), and 18 (F/Cl) kcal/mol!

A further interesting illustration is the contrast betweem ethanol and thioethanol. In particular, ethanol is more stable than dimethyl ether by about 12 kcal/mol. It is found that replacing O by S comes close to rendering the reaction thermoneutral.

$$\text{H}_3\text{C—S—CH}_3 \xrightarrow[\;]{\Delta H_r = \;-2.1\text{ kcal / mol}} \text{H}_3\text{C—CH}_2\text{—S—H}$$

A final set of data leaves no doubt that first-row are qualitatively different from heavier congeners insofar as bonding mechanism is concerned. Specifically, the combination of systolic first-row group (e.g., NH_2, OH, F) with some other systolic first-row group is unfavorable because it creates vicinal errors in the arrow formula. By contrast, the combination of the isoelectronic PH_2, SH, and Cl with systolic first-row groups is predicted to be favorable because of the generation of Q bonds. Hence, all reactions in which A is a first-

Table 28.1 The Affinity of Second-Row Atoms for Systolic Groups

$$\underset{T}{A(CH_3)_n} + \underset{FR}{ZH_n} \longrightarrow \underset{T}{AH_n} + \underset{I}{Z(CH_3)_n}$$

A	Z	n	ΔH_r (kcal/mol)
C	Si	4	–26
N	P	3	–31
O	S	2	–18
F	Cl	1	– 4
F	I	1	– 9

$$\underset{T}{AF_n} + \underset{FR}{ZH_n} \longrightarrow \underset{T}{AH_n} + \underset{I}{ZF_n}$$

C	Si	4	– 189
N	P	3	– 208
O	S	2	– 130
F	Cl	1	– 55

row and Z a heavier congener (see, e.g., Table 28.1), must be exothermic because one of the reactants is frustrated while both products approach optimal bonding. Indeed, in all cases, A prefers to bind H and Z prefers to bind to first-row systolic groups (e.g., CH_3 and F) by a long shot!

Bergman and co-workers noted that complexation energies of large alkanes (e.g., cyclohexane, neopentane) bearing unsaturated transition metal fragments are of the order of 10 kcal/mol, while methane complexes have not been observed in solution.[7] A simple rationalization is now possible: Since metal–alkane bonding comes about by overlap dispersion, the more polarizable the alkane, the larger the dispersion component and the stronger the overall binding. Since polarizability increases with size, it is reasonable that the metal–alkane bond should become stronger as alkane size increases. The generality of the phenomenon is illustrated by the following exothermic reaction, which illustrates that metalloids prefer to be linked to large alkanes in comparison to methane (Cy = cyclohexyl).

$$\text{Cy—H} + CH_3\text{—I} \longrightarrow \text{Cy—I} + CH_3\text{—H} \qquad \Delta H_r = -4 \text{ kcal/mol}$$

Bottom line: The metal-alkane bond is not a covalent bond. It is an I bond that has an overlap-independent component.

The message of this section is that if the active unoccupied space of heavy p-block elements is incorporated into the description, either some or all of the I bonds connecting a central heavy atom to a set of ligands bearing lone pairs become Q bonds and the molecule is effectively a metal residue. This is why the physical and chemical properties of all the fluorides of, say, the second row, are so different from those of the first row irrespective of whether the molecule is classified as "electron precise" or "hypervalent" according to standard convention. To be specific, we no longer see a fundamental difference between, say, SF_2, SF_4, and SF_6, other than unlike degrees of associative saturation and numbers of I and Q bonds.

28.3 Why Heavy p-Block Atoms Acidify Vicinal C—H Bonds

Heavy p-block groups have been known for some time to be potent stabilizers of carbanions. This has been reconfirmed by recent gas phase measurements and ab initio computations.[6b,c] For example, the acidity of silylmethane is greatly enhanced relative to methane. The explanation is (what else?) anionic hyperconjugation; that is, delocalization of the carbon lone pair to the vacant σ-antibonding MO of Si—H. This argument is defeated by the observation that *all* heavy second-row Y groups acidify the hydrogens of H_3C—Y including chlorine which cannot hyperconjugate. The pertinent data are as follows.

Y	ΔH° (acid) (kcal/mol)
H	417
Me_3Si	387
Me_2P	384–391
MeS	393
Cl	400

Next, one resorts to inductive stabilization of the carbanionic center by the electronegative chlorine. Again, the attempt is defeated by the observation that fluoromethane is a much weaker gas phase acid than chloromethane.[8] So, what is the common denominator of SiH_3 and Cl in acting as stabilizers of carbanions in contradistinction to the much more electronegative and yet inferior CH_3 and F? Why is I—CH_3 a stronger proton acid than any of the other halomethanes?

The answer is given directly by Figure 28.1: Deprotonation of I—CH_3 improves the Q bond in the sense that the molecule changes from "incipient iodide plus carbene" to "iodide plus carbene" as a carbon bond pair is replaced by a carbenic lone pair. Another way of saying the same thing is that both iodine (radical) and iodide (anion) are I-activated nucleophiles, while both methyl and carbene are I-activated electrophiles.

There have been many attempts to understand the origin of the acidification effected by heavy p-block atoms. A convincing answer cannot be given in the context of conventional theory because "polarization" means all and nothing. Rather, we have shown countable bonds, which, however, are not the covalent bonds the chemist envisions. The same story can be told for all molecules that can be classified as ylides. The simplest ylid is methyl chloride. Another ylid is the metal–phosphine complex! In reality, they are all Q-bound species.

28.4 The Hierarchy of Metal Bonds; I Bonds, Q Bonds, and E Bonds

One of the conceptual advantages of using VB formulas built from hybrid AOs is that the static mechanisms of bonding (i.e., the consequences of induction and dispersion) are explicitly projected. This reminds us that what is overlap dispersion when canonical AOs are used becomes overlap when hybrid AOs are used because the dispersion component has already been taken care by occupancy of the hybrids in a way that minimizes interelectronic repulsion. In other words, the solid arrows are preserved, but the dashed arrows in the canonical representation disappear in the VB formula because the allocation of the electrons to the hybrids has already taken care of the electron correlation. Furthermore, since the nonvalent character of starred holes cannot be quantified, the best way of viewing a Q bond is as one I bond with impaired CT. In short, the difference between I and Q bonds is quantitative. These are crucial invariants: two CT arrows make one CT bond, and arrow codirectionality (head-to-tail arrows) always implies superior bonding.

A hatched arrow connects a pair and a starred hole and, taken together with a codirectional solid arrow, defines a Q bond. When a hatched arrow is not complemented by a codirectional solid arrow, it is turned into a hatched line and counts as an E bond. Thus, the final menu of metal bonding, in order of strength, becomes: I, Q, E. A Q bond is better than two E bonds because it is formed by combining two pairs and one unstarred plus one starred hole, while the (two) E bonds are formed by assembing two pairs and two starred holes. This is the hierarchical order of metal bonds. It says that CT delocalization can occur only if the electrons correlate their motion (dispersion) and electron transfer preserves electroneutrality to the maximum extent possible (codirectional CT).

28.5 The VB Heme Model

Dioxygen is transported by hemoglobin (heme)[9] and is broken down by cytochrome P450 to mono-oxygen, which then inserts into an alkane bond.[10] What are the common denominators, and what are the differences in the binding of dioxygen and mono-oxygen by these biomolecules? What is the "minimum model" that contains the key features of the biologically active species that we will use in attempting to duplicate nature?

To answer these questions, we focus attention on the interaction of heme with prototypical ligands. We adopt the *axial polarization model* shown in Figure 28.2 where we start with pseudo-octahedral Fe(II), discussed in Chapter 25, and we resolve the three equatorial hybrids into two diagonal sd^2 hybrids, q_i, plus an equatorial d_{xy} AO. The four porphinato nitrogen lone pairs are tied up by the two diagonal equatorial q holes forming two I bonds. The d_{xy} acts as a receptacle of two or one iron electron in Fe(II) or Fe(III), respectively. There

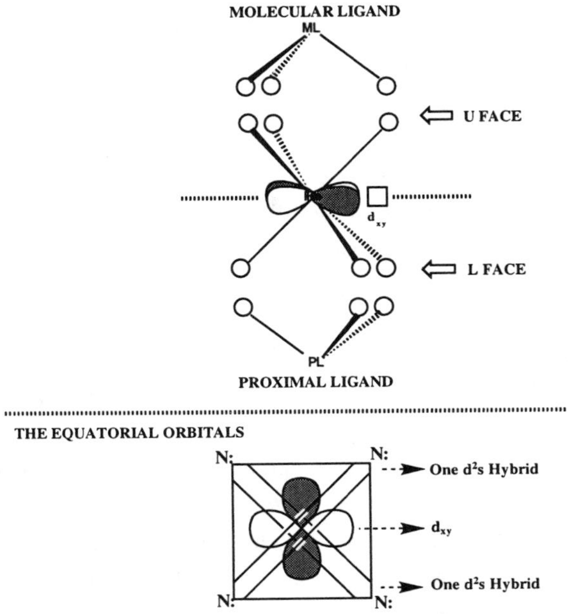

Figure 28.2 Orbital ensemble for the elucidation of the electronic structure of hemoglobin.

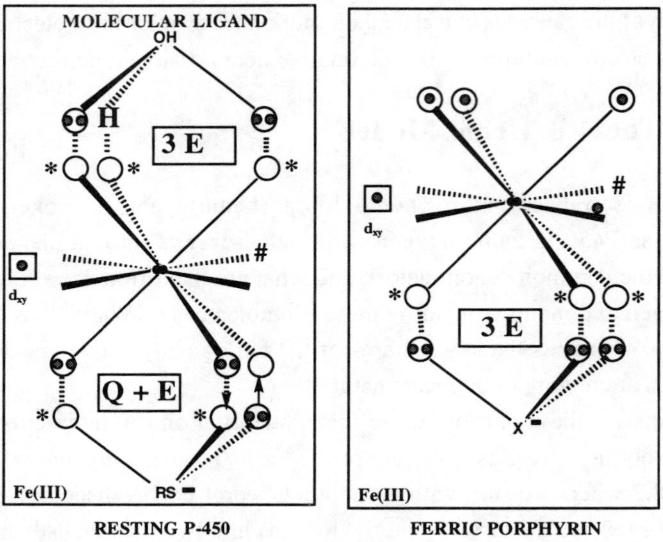

Figure 28.3 Iron coordination in resting cytochrome P450 and in ferric porphyrine (heme).

remain two conical sets of three hybrids, each of which can be used to tie up two apical ligands: the *proximal ligand* (PL: e.g., histidine, cysteinyl anion) and the *molecular ligand* (ML: e.g., dioxygen, oxygen, carbon monoxide). The conical set that ties up the ML ligand is designated U (upper) and the one tying up the PL ligand is called L (lower). The U and L sets can be thought of as two *triangular faces*. The binding of the ML and PL ligands by the U and L faces, respectively, depends on the synapticity (i.e., the number of binding AOs) and number of valence electrons of each element.

Assuming that each ML and PL can be formalized as a trisynaptic fragment (three conical hybrids) and given that both U and L are trisynaptic (three endo conical AOs each), we have a simple game: Match 3h/*me* U with 3h/*ne* ML and 3h/*je* L with 3h/*ke* PL, where *j*, *k*, *m*, *n* are integers and "3h/*ne*" is the standard representation of the synapticity (3h) and number of valence electrons (*ne*) of a fragment. This is called the *valence state* of a fragment.

As a first example of how we use these concepts, Figure 28.3 shows the representations of the low spin ligated heme in the resting state of cytochrome P450[11] and the high spin ferric porphyrin (heme), which has no sixth ligand.[12] Each of the equatorial lines represents one equatorial hole. The rules are as follows.

There are three starred axial AOs plus one daggered AO that is either axial or equatorial. We assume that the dagger always remains equatorial. Removal of this assumption unfolds a whole set of intricacies, which are bypassed in this initial presentation.

Starred or daggered AOs cannot be occupied.

The function of the signed AOs is to polarize the d orbitals, to prepare them for eventual I-bonding and, in addition, to make E bonds themselves. These are indicated by hatched lines, assuming that the water hydrogens are acidic enough to act as effective starred holes.

In this light, there are two things to note. First, in the high spin form, the proximal ligand acts to eliminate electron density from the L face, and is connected to the deshielded Fe(III) by three E bonds. Minimization of interelectronic repulsion is achieved by placing an electron in one of the two unstarred equatorial AOs directed toward the nitrogens of the porphyrin. This engenders exchange repulsion and causes the iron to move toward the proximal ligand. For this to happen, the upper tripod with the three odd electrons must fold, much like an umbrella. Folding, which is controlled by the macrocycle, constitutes another dimension of the mode of action of high spin hemes.

The second important point is that in the low spin form, the iron is envisioned as Fe(II) and acts with a 3h/4e upper face. The low spin is induced by the electric field of a water molecule acting as the sixth ligand. However, it may very well be that the pairing of electrons is effected by the overall electric field of the cavity rather than that of an individual water or water cluster.[13]

28.6 The Mechanism of Dioxygen Binding by Heme

The mode of heme–dioxygen binding has attracted attention since time immemorial. VB theory provides a half-page interpretation in Figure 28.4. The PL is an R_3N nitrogen base (or a first-row isoelectronic system) that can be formalized either as 1h/2e (focusing on the lone pair exclusively) or as 3h/6e (focusing on the lone pair plus the two π components of the three σ N—R bonds). Dioxygen enters in polarized singlet form in which the cationic oxygen is sp^2-hybridized. The resultant bond is primarily electrostatic, with incipient I-bonding indicated by the arrows. The crucial point is the less-than-perfect match of the multipole of the Fe(II) U face and the multipole of the sp^2-hybridized oxygen: Each has two pairs and one hole, and the former engender significant exchange repulsion that renders the overall bonding weak. A multipole is

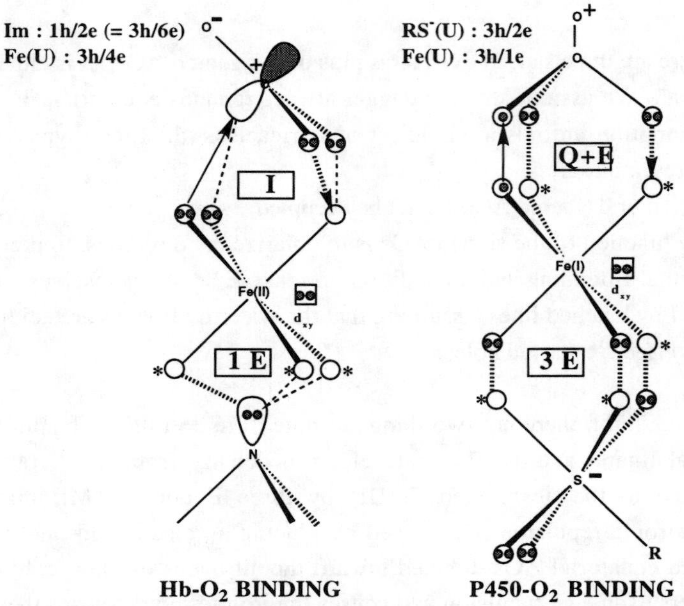

Figure 28.4 Oxygen binding by hemoglobin (Hb) and by cytochrome P450.

represented as an array of holes and pairs. The scenario makes good physical sense: Hemoglobin is an oxygen transporter because it can bind O_2 weakly. Stronger I-binding would render hemoglobin ineffective as transporter.

The way in which we envision metal–ligand bonding is further illustrated by showing the matchings of the U face of iron and the face of the cationic oxygen when the latter is either sp^2- or sp^3-hybridized.

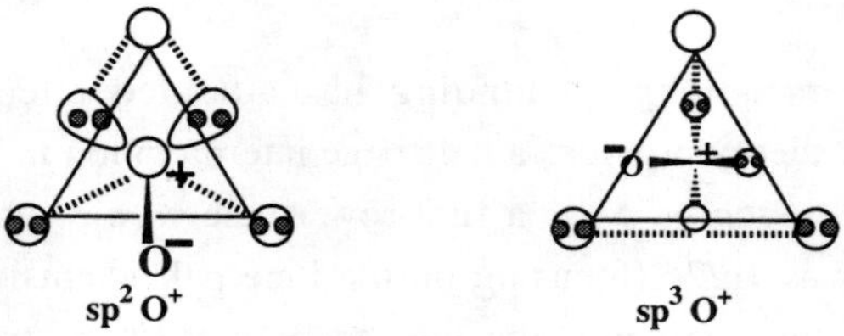

Actually, the sp^3representation is closer to the truth. Hence, the Fe(II)—O_2 binding in hemoglobin is best envisioned as the result of the match of a 3h/4e with a 2h/2e fragment. Changing the PL from a first- to a second-row base changes the valence state of the PL from 3h/6e to 3h/2e, and this translates to a change of the Fe(II)—O_2 bonding and geometry, as shown in Figure 28.4.

How can we "kill" the metal–oxygen bond and destroy the capacity of iron to transport oxygen? Since O_2 acts as a functional acceptor, the answer is straightforward: Diminish metal-to-ligand CT delocalization by increasing the metal electronegativity. This can be done by oxidizing Fe(II) to Fe(III). Methemoglobin is not an efficient transporter of O_2.

28.7 Valence Shuttling of Metals and Semimetals

Everything said up to this point rests on one fundamental assumption: The proximal ligand of heme (PL) ties up three starred holes of the L face of iron and remains passive to what is happening in the U face! While this appears to be the most likely scenario for the systems we have discussed, the role of theory is to expose unexplored potentialities, leading to the reexamination of old data and the design of new and more telling experiments. Of substantial importance is the possibility of valence shuttling, which gives the PL the flexibility of determining the electron count of the L face (to which it binds), thus determining indirectly the electron availability in the U face to which the ML binds. For example, hemoglobin and cytochrome P450 use an iron in the center of a porphyrin ring to bind oxygen, which is subsequently utilized in fundamentally different ways. In hemoglobin, the axial PL is the nitrogen atom of a histidine side chain, while, in P450, it is a thiyl residue of a cysteine side chain. Can the difference in function be due to the difference between N, a nonactivated nucleophile, and S, an I-permissive nucleophile?

If we assume zero participation of the nonvalence unoccupied space of chloride, we would conclude that the sp^3-hybridized anion acts like a 3h/6e species to bind Cu^+. However, if we assume sp^3d^5 hybridization, the chloride can bind Cu^+ by acting either as 3h/6e or 3h/2e as shown in Figure 28.5. The shift in valence from 3h/6e to 3h/2e is called *valence shuttling*. Heavy p-block atoms are not monolithic but, rather, flexible with respect to valence, and this flexibility may have interesting chemical consequences. In the case at hand, the 3h/2e mode of binding is superior on account of arrow codirectionality.

Is there any evidence that chloride binds preferentially by the 3h/2e mode? The formation of layers by cyanuric chloride implies that nitrogen lone pairs are interacting with chloride holes.[14,15] Recently, Brooks and collaborators obtained evidence for preferential electron transfer to the Br end of oriented CF_3Br.[16] Given that phosphine acts like a 3h/2e piece much as CO does, is there any evidence that chloride anions (or any isoelectronic species) are 3h/2e species interchangeable with phosphines? One telling fact is that chloride can act like a gold ligand in gold clusters much as phosphine does. Another

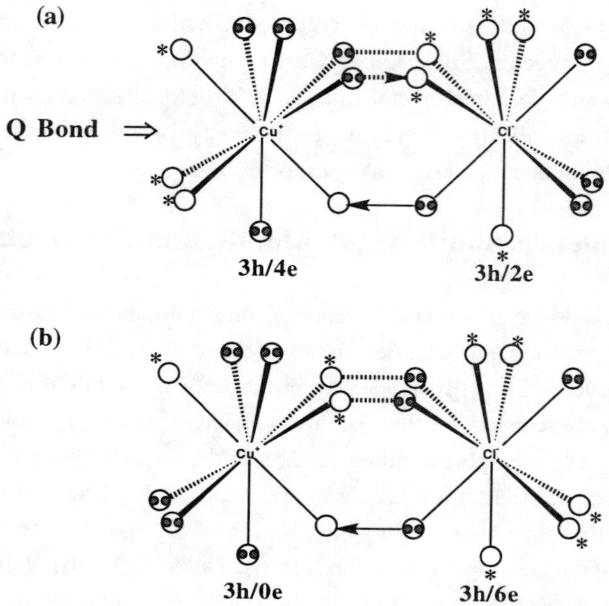

Figure 28.5 Two different modes of CuCl bonding. Mode a is superior.

interesting piece of data is the difference in how the axial ligands Z (Z = methyl or chlorine) act to select the most stable spin state of the octahedral TiZ_2L_4, where L is an equatorial phosphine ligand. It is found that the methyl derivative is diamagnetic (singlet)[17] while the chlorine derivative is paramagnetic (triplet)[18a] even though chloride is regarded as a stronger π-donor ligand that is expected to cause larger splitting between the equatorial $d_{x^2-y^2}$ and the axial d_{zx} and d_{zy}. Simpson et al. explained this by noting that the chlorine derivative has more contracted Ti3d AOs (because of the greater electron-withdrawing character of the chlorines). This property accentuates the effect of electron–electron repulsion in the d shell, rendering the chloro derivative a triplet.[18b]

There is an alternative explanation, however, as follows. The transformation from low spin to high spin is equivalent to transferring one electron from the in-plane $d_{x^2-y^2}$ AO, which has minimal overlap with the axial ligand AOs, to the out-of-plane d_{zx} or d_{zy} AO, which has maximal overlap with axial ligand AOs. This is unfavorable when each axial ligand is a 3h/6e piece but favorable when each is a 3h/2e piece. The axial chlorines act like 3h/2e pieces to induce the arrival of one electron in one axial Ti3d AO, which forms half an E-bond with a starred hole of Cl.

Let us now return to the problem of dioxygen binding in hemoglobin, introduced in the preceding section. How does the geometry of Fe(II)—O_2 binding reflect the nature of the bonding? How does the proximal ligand affect the bonding? How does the π system of the enclosing porphyrin (POR) control the bonding? How does the electric field of the environment (ligands plus protein taken as a whole) determine the bonding? After innumerable attempts, MO theory has provided slightly more than an illusion of understanding. The reasons are straightforward: MO theory can project T-bonding (exclusive occupancy of bonding MOs) and T-antibonding (occupancy of bonding and antibonding MOs) but it cannot project either I- or E-bonding. That is, it fails to give a simple and explicit account of precisely the mechanisms of bonding that operate in metal-containing molecules. The VB picture of Fe(II)—O_2 bonding in hemoglobin and related systems is both simple and operationally significant: A multipolar U face of Fe(II) presents itself to a multipolar "face" of a ligand, and binding occurs either because of electrostatic attraction alone (E-bonding) or because of the latter supplemented by CT delocalization (I-bonding). A multipolar face is made up of holes and pairs, and the action of the PL and the porphyrin is to effectively determine the number of holes and pairs on the U face of Fe(II) by shuttling among the 3h/2e, 3h/4e and 3h/6e valence states. In addition, the electric field of the environment acts to stabilize polar over nonpolar ligand forms.

We can now give a simple account of the dependence of geometry on electronic structure using a few simple pictures. In each of the cases presented in Table 28.2, O_2 acts as singlet O_2 in which each oxygen is sp^3-hybridized; open circles represent holes and solid circles pairs; single AO occupancy is indicated by a dotted circle. As can be seen, four different geometries of Fe(II)—O_2 correspond to four different face–multipole matches. The best matches, corresponding to formal triple bonds, are those in the two linear geometries. A key point is that the bent structure of Fe(II)—O_2 is the result of the action of the 3h/6e PL.

Furthermore, we have a set of predictions:

(a) The nature of the PL (i.e., whether it can act as 3h/2e, 3h/4e, or 3h/6e) differentiates three of the four geometries but fails to differentiate the linear from the end-on structure. The latter two are differentiated by the environment.

(b) The valence state of the PL determines the multipolar nature of the U face of Fe(II); that is, whether it can act as 3h/0e, 3h/2e, or 3h/4e. The same type of control is exerted by the π system of porphyrin (POR). In other words,

Table 28.2 The Dependence of Geometry on the Match of Metal and Oxygen Faces

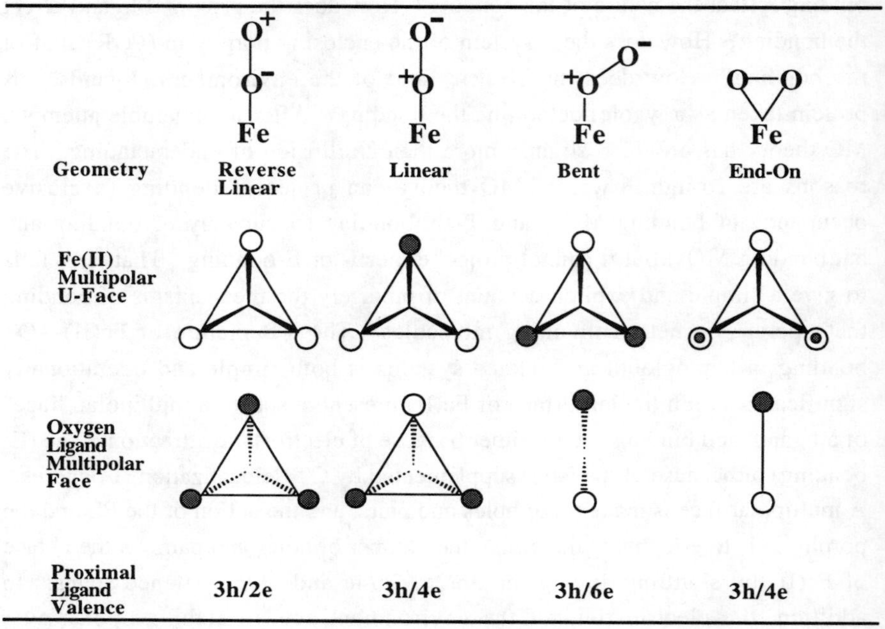

either one or both pairs of the U face of Fe(II) can be delocalized in the POR envelope. Thus for Fe(II)—O_2, both geometry and binding are functions of *both* PL and POR.

(c) The reverse linear geometry is expected when the PL is Cl^-, RS^-, PR_2^- or SiR_3^-. When PL is histidine, we have the hemoglobin story. When it is cysteinyl anion, we have the cytochrome P450 story. Thus we can predict that the Fe(II)—O_2 binding in cytochrome P450 approaches the reverse linear geometry that is ideal for a reduction of O_2 at the remote oxygen atom. It is known that one-electron reduction plus addition of two protons is necessary for generating the active Fe(IV)=O species.

(d) When O_2 binds to Fe(II) of hemoglobin, it does so as a 2h/2e ligand. This means that a singlet carbene can also become weakly bound to Fe(II).

(e) The same treatment can be extended to Fe(II)—CO complexes. The differentiation of the two geometries depends exclusively on the electric field of the environment, as indicated in Table 28.3.

Table 28.3 The Dependence of Geometry on the Match of Metal and Carbonyl Faces

Geometry	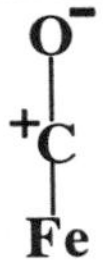 **Linear**	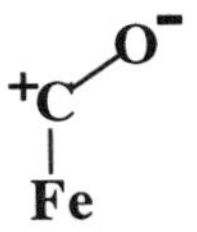 **Bent**
Fe(II) Multipolar U-Face	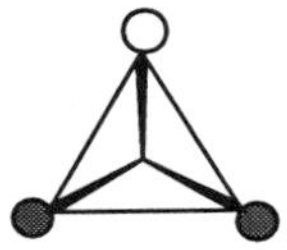	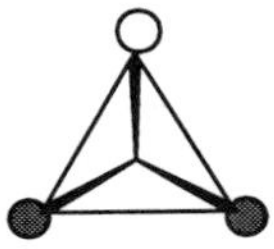
Carbonyl Ligand Multipolar Face	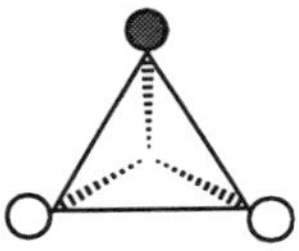	
Proximal Ligand Valence	**3h/6e**	**3h/6e**

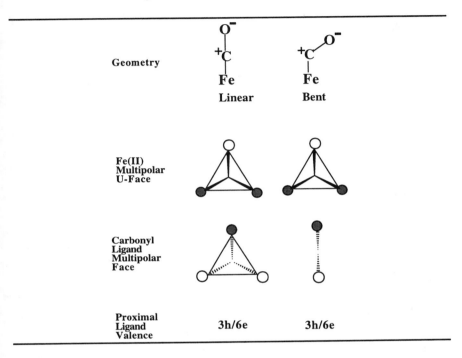

Furthermore, in the presence of a 3h/6e PL (e.g., histidine, H_2O and derivatives, NH_3 and derivatives), the multipolar interaction is "perfect" in linear Fe(II)—CO and "imperfect" in Fe(II)—O_2. We would expect changing the ligand from 3h/6e (e.g., R_3N) to 3h/4e (RS^-) is expected to destabilize Fe(II)—CO and stabilize Fe(II)—O_2, and recently this was found.[19]

We now have an overview (Table 28.4) of how the proximal base alone or in combination with protein determines the binding of a molecule by a metal embedded in a corrin ring.

Why has nature chosen porphyrins? The answer is that the four symmetrically disposed nitrogen atoms best fit the two orthogonal σ-type equatorial holes of the d^2s metal. Synthetic isomers (porphycene, corrphycene, N-confused porphyrin), first made by E. Vogel and co-workers, are handicapped by their asymmetry. Thus, they are incapable of supporting two I bonds each connecting two trans nitrogens to the central metal.

Table 28.4 Effect of Proximal Base on Binding

Protein	Molecule	Trans Base	Metal–Molecule[a]
Hb	O_2	Im	ImFe(II)—O_2 ("Impaired 6-e triple bond")
P450	O_2	RS^-	RS—Fe(II)—O_2 ("6-e triple bond")
B_{12}	R	DMB–Enzyme	DMBCo(III) + $R:^-$ (No Bond)

[a]Quotation marks imply formal multiple bonds connecting molecule and metal.

28.8 Strategies for Electrostatic Complex Stabilization

The synthesis of molecular aggregates is at center stage in the chemical literature,[20] hence the need to expand our understanding of the enthalpic driving force for molecular association beyond the bounds of conventional models (hydrogen-bonding, point-charge electrostatic attraction, van der Waals bonding). We begin the process by considering the simplest case of molecular association, namely, the so-called π complexes. The action of unoccupied starred AOs derived from the nonvalence space can be used advantageously in designing stable complexes. For example, alkali metals are known to form complexes with cyclopentadienyl (Cp) and benzene (Bz) rings.[21] The question is: How should we design a system to enforce strong E-bonding? The answer is simple: Match alkali holes with carbon pairs. For example, the strategy for stabilizing an E complex of Li^+ and Cp^- is to localize three Cp π pairs on the five-membered ring so that they best fit the unoccupied AOs of lithium cation. Furthermore, a pair prefers to be localized on a carbon that is σ-bonded to an atom or group that has starred holes. Such a group is called a pair promoter. Silyl, phosphinyl, thiyl, and chlorine groups are pair promoters.

The following example is taken from the literature.[22] Note how the SiR_3 groups act as tags of the π pairs and how the Li^+ ideally caps them. The complex is soluble even in nonpolar solvents like hexane, and it exists as a monomer in the gas phase. That the formal Li^+ has three sp^3 hybrids tied up by the ring pairs is indicated by existence of a 1:1 complex with the monodentate quinuclidine as a monomeric species in the solid state, in solution, and in the gas phase. An X-ray structure shows that the distance of the lithium atom from the centroid of the Cp is considerably shorter than the analogous Li–TMEDA complex, which has one $SiMe_3$ missing. This molecule teaches us how to effect strong E-bonding by using silyl substituents to orient pairs on a π-conjugated system. X-ray structures of silicon derivatives show that lithium–aromatic complexes place the Li cation in the vicinity of carbon atoms bearing trialkylsilyl groups.[23]

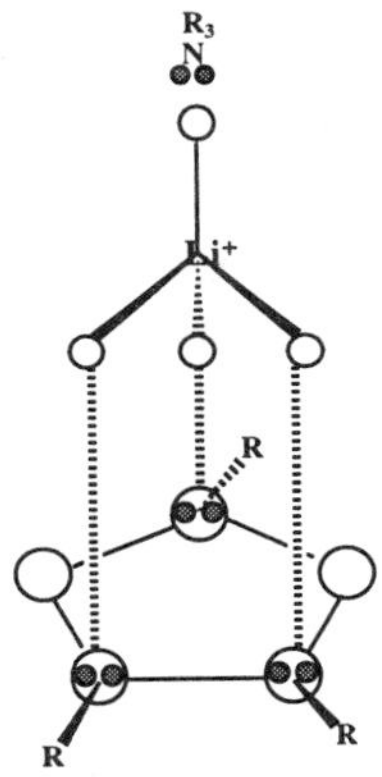

R = $SiMe_3$

28.9 The Structures of the Inert Gas Fluorides

In planar XeF_4, two F/F$^-$ pairs are bound by two in-plane Xe5p dots. However, in addition, there exists a Xe5s lone pair, which engenders exchange repulsion with the four fluorines. Intervention of a nonvalence Xe5d_{z^2} AO (function) polarizes the in-plane Xe5s lone pair out of the molecular plane and spares it from exchange repulsion with the four fluorines. Octahedral XeF_6 has the same problem: If the Xe5s lone pair is to excape exchange repulsion with the six fluorines, it must be polarized by one Xe nonvalence hole. By symmetry, this must be an $(n + 1)$s nonvalence hole.

It follows that an octahedral MX_6 molecule, where M is isoelectronic to Xe and X is a halogen, will show an increased tendency to adopt an undistorted octahedral structure as the np → $(n + 1)$s promotion, which relieves exchange repulsion, becomes increasingly favorable relative to the np → nd promotion, which fails, by symmetry, to direct the ns lone pair away from the fluorines while preserving the octahedral geometry. In other words, the geometry of MX_6 is a function of the $(n + 1)$s – nd energy gap, ΔE. As Table 28.5 shows,

Table 28.5 Diminution of Relative Promotional Energy (ΔE) According to Postion on a p-Block Column

	Energy (cm^{-1})		
ATOM	np → (*n* + 1)s	np → nd	Δ*E*
Ar	93,143	111,667	18,524
Kr	79,972	96,772	16,800
Xe	67,068	79,771	12,703

(*n* + 1)s lies below *n*d, with the gap decreasing as one descends down a column of the p block. This means that the chance of observing an undistorted MX_6 octahedron improves upon moving *upward* along a p-block column.

Christe and Wilson [24] found that BrF_6^- is octahedral but IF_6^- is a distorted octahedron on the time scale of vibrational spectroscopy, though both are fluxional. Furthermore, it had been known that gaseous XeF_6, solid IBr_6^- and solutions of SbX_6^{2-} and $BiBr_6^{2-}$ are distorted octahedra[25] while SeX_6^{2-} is octahedral.[26] Thus, fourth-row XeF_6 analogues tend to be distorted octahedra while third-row KrF_6 analogues tend to be undistorted. The only exception seems to be the octahedral TeX_6^{2-}.

How can we best visualize the distortion of the XeF_6 octahedron: as a lone pair squeezing through the walls of a cage or as a lone pair rupturing the walls? Since the 5s AO accommodating the Xe pair has the lowest energy, the latter view is more justified.

The preceding discussion can be connected to the apparently unrelated problem of metal ion solvation in the following way. A distorted MX_6 octahedron (M = inert gas atom or isoelectronic ion) may be viewed as an octahedron in which one M electron pair has penetrated the octahedral surface ("surface cluster") while the undistorted counterpart can be viewed as an octahedron in which one M electron pair is accommodated in the interior ("interior cluster"). It turns out that a differentiation between surface and interior clusters has been made in clusters of the type $Na(NH_3)_n$ where, instead of one electron pair, the metal has an active odd electron.[27]

28.10 Supramolecular Chemistry as "Organic Organometallic" Chemistry

Dibenzene chromium can be formulated as a promoted pseudo-octahedral Cr atom with three equatorial pairs and two sets (exo and endo) of three conical holes with each set matching three benzene π pairs. Can we identify an *organic molecule*, X, that is a close analogue of this promoted Cr atom that will give a stable supramolecular complex $(C_6H_6)_2X$?

Acidic hydrogens of organic molecules can be envisioned as analogues of metal starred holes that can dock on lone pairs or bond pairs to generate E bonds. The chair form of cyclohexane has six equatorial C—H bond pairs and two sets of three axial C—H bond pairs, one above and one below the mantle of carbon atoms. If we assume that the axial C—H bond polarity is C^-H^+, it is immediately obvious that the two sets of axial formal protons define an ensemble of holes that are stereochemically disposed as those of the promoted Cr save for a slip displacement of one set relative to the other. Hence, chair cyclohexane is the sought-after molecule X. The only problem is that the protonic character of the axial hydrogens is not strong enough. How can we improve the situation? The T-formulas suggest an answer: Replace each equatorial H by a diastolic group so that the axial C—H bonds can support unimpeded CT delocalization from H to C. One possible choice is the SiR_3 group. Hence, we expect that all-equatorial 1,2,3,4,5,6-hexatrialkyl-silylcyclohexane will bind two benzenes to form a slip-sandwich organic supermolecule.

References

1. (a) J.I. Musher, *Angew. Chem. Int. Ed. Engl.* 8, 54 (1969). (b) The most powerful weapon of any scientist is intuition. The late Angelo Mangini, professor of organic chemistry at the University of Bologna, had this in his arsenal. He thought that d orbitals do not act as valence AOs in heavy p-block elements and, in particular, not insulfur. He convinced distinguished ab initio computational chemists, like I.G. Csizmadia of Toronto and F. Bernardi of Bologna, to investigate the problem. We now have a theory that explains why sulfur forms compounds characteristically different from those of oxygen without ever requiring the action of d orbitals. To do so, we had to abandon the traditional notions of covalency. The maestro is probably smiling,

since he knew all along! F. Bernardi, I.G. Csizmadia, A. Mangini, H.B. Schlegel, M.-H Whangbo, and S. Wolfe, *J. Am. Chem. Soc.* 97, 2209 (1975).

2. E. Magnusson, *J. Am. Chem. Soc.* 112, 7940 (1990).
3. A.E. Reed and P.v.R.Schleyer, *J. Am. Chem. Soc.* 112, 1434 (1990).
4. R.P. Messmer, *J. Am. Chem. Soc.* 113, 433 (1991).
5. D.L. Cooper, T.P. Cunningham, J. Gerratt, P.B. Karadakov, and M. Raimondi, *J. Am. Chem. Soc.* 116, 4414 (1994).
6. (a) P.v.R. Schleyer, E.D. Jemmis, and G.W. Spitznagel, *J. Am. Chem. Soc.* 107, 6393 (1985). (b) E.A. Brinkman, S. Berger, and J.I. Brauman, *J. Am. Chem. Soc.* 116, 8304 (1994). (c) K.B. Wiberg and H. Castejon, *J. Am. Chem. Soc.* 116,10489 (1994). These authors, along with other physical organic chemists, have embraced the notion of hyperconjugation as the explanation of the carbon acidity enhancement by second-row heteroatoms.
7. B.A. Arndtsen, R.G. Bergman, T.A. Mobley, and T.H. Peterson, *Acc. Chem. Res.* 28, 154 (1995).
8. M. Born, S. Ingemann, and N.M.M. Nibbering, *J. Am. Chem. Soc.* 116, 7210 (1994).
9. M.F. Perutz, G. Fermi, B. Luisi, B. Shaanan, and R.C. Liddington, *Acc. Chem. Res.* 20, 309 (1987).`
10. *Cytochrome P450: Structure, Mechanism, and Biochemistry*, P.R. Ortiz de Montellano, Ed., Plenum, New York, 1986.
11. R. Sligar, *Biochemistry* 15, 5399 (1976).
12. W.R. Scheidt, *Acc. Chem. Res.* 10, 339 (1977).
13. D. Harris and G. Lowe, *J. Am. Chem. Soc.* 115, 8775 (1993).
14. K. Xu, D.M. Ho, and R.A. Pascal Jr., *J. Am. Chem. Soc.* 116, 105 (1994).
15. J.A. Zerkowski, J.C. McDonald, C.T. Setop, D.A. Wierda, and G.M. Whitesides, *J. Am. Chem. Soc.* 116, 2382 (1994).
16. G. Xing, T. Kasai, and P.R. Brooks, *J. Am. Chem. Soc.* 116, 7421 (1994).
17. J.A. Jensen, S.R. Wilson, A.J. Schultz, and G.S. Girolami, *J. Am. Chem. Soc.* 109, 8094 (1987).
18. (a) G.S. Girolami, G. Wilkinson, A.M.R. Galas, M. Thorton-Pett, and M.B. Hursthouse, *J. Chem. Soc. Dalton Trans.* 1339 (1985). (b) C.Q. Simpson II, M.B. Hall, and M.F. Guest, *J. Am. Chem. Soc.* 113, 2898 (1991).

19. D. El-Kasmi, C. Tetreau, D. Lavalette, and M. Momenteau, *J. Am. Chem. Soc.* 117, 6041 (1995).
20. (a) G.M. Whitesides, E.E. Simanek, J.P. Mathias, C.T. Seto, D.N. Chin, M. Mammen, and D.M. Gordon, *Acc. Chem. Res.* 28, 37 (1995). (b) N. Branda, R.M. Grotzfeld, C. Valdes, and J. Rebek ,*J. Am. Chem. Soc.* 117, 85 (1995).
21. P. Jutzi, *Adv. Organomet. Chem.* 26, 217 (1986).
22. P. Jutzi, E. Schlueter, S. Phl, and W. Saak, *Chem. Ber.* 118, 1959 (1985).
23. E. Lukevics, O. Pudova, and R. Sturkovich, *Molecular Structure of Organosilicon Compounds*, Ellis Horwood, Chichester, 1989.
24. K.O. Christe and W.W. Wilson, *Inorg. Chem.* 28, 3277 (1989).
25. R.J. Gillespie, *Molecular Geometry*, Van Nostrand Reinhold, London, 1972.
26. W. Abriel, *Acta Crystallogr. B* B42, 449 (1986).
27. K. Hashimoto and K. Morokuma, *J. Am. Chem. Soc.* 117, 4151 (1995) and references therein.

Chapter 29

The Electronic Structures of Metal Dimers

29.1 Orbital Rotation and Electron Stereochemistry in Light and Heavy Main Group Diatomics

"Simple" diatomics have been investigated ad nauseum. What could be less inspiring? Read on: A new theory aims precisely to turn the "boring" into "exciting" or "controversial." We will now write explicit formulas for diatomics that succeed in explaining important counterintuitive trends that have been locked in the skeleton closet. We begin by tentatively assuming that each main group atom, regardless of shade (see Chapter 24), can be treated at the minimal AO basis set level. Each atom is regarded as sp^3-hybridized, with three hybrids making up the endo set and the fourth being the sole member of the exo set. Interatomic bonds are made by using the endo set. We have a choice of making bonds by eclipsed or staggered orbital overlap, as indicated in Figure 29.1.

In the Newman projection formulas, the electrons placed in the "back" exo-sp^3 hybrid are not shown, but they are the same in number as those of the "front" exo-sp^3 hybrid in even-electron homonuclear diatomics. Finally, to assist the reader who is used to MO theory, we indicate the correspondence of the occupancy of the hybrids of the Newman projection to the occupancy of the

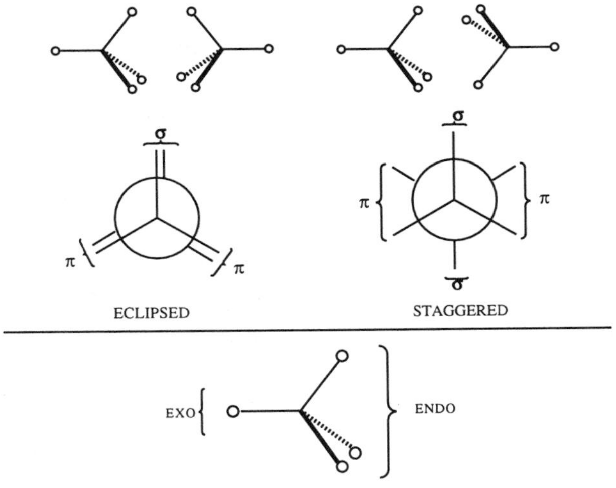

Figure 29.1 The disposition of hybrids in the bonding of p-block dimers.

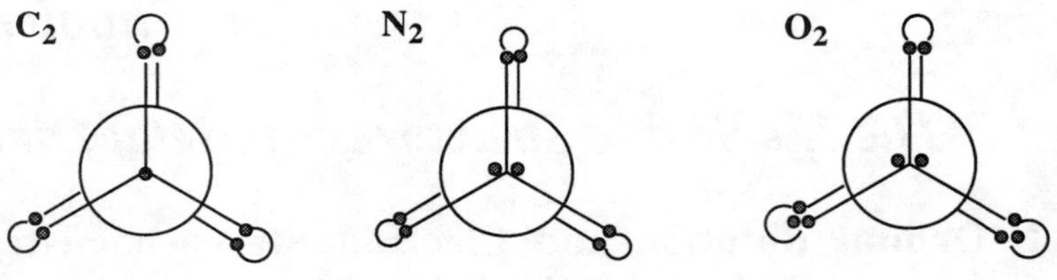

Figure 29.2 T-bonding of electronegative p-block dimers involves eclipsed hybrids.

MOs of the diatomic: Putting electrons in the hybrids marked o is analogous to putting electrons in the o MOs localized between the two centers (o_g and o_u, etc.).

We now embark on "electron conformational analysis" of prototypical diatomics by recognizing two extremes: When the atoms making up the diatomic are black atoms of the first row, we have T-binding and the usual maximum overlap considerations. As a result, the right formulas are the eclipsed ones because these allow strong overlap of the hybrids at the expense of interelectronic repulsion. When we shift to the grey or white atoms of the second or lower rows, T- gives way to I-bonding: Deshielding due to dispersion and subsequent 2-e CT delocalization is best achieved by placing the electrons in a staggered conformation. The hallmark of staggered bonding is 2-e delocalization involving vicinal hybrids with a dihedral angle of 60°, as we have already seen in the case of Li_2. The transition from eclipsed to staggered bonding is an example of "orbital rotation" that marks a transition from T-bonding (favored by electronegative atoms that seek to make strong T bonds by maximizing AO overlap) to I-bonding (favored by electropositive atoms that seek to make I bonds that involve the combined action of 2-e CT plus dispersion).

The eclipsed first-row dimers and the staggered second-row dimers are shown in Figures 29.2 and 29.3, respectively. One exceptional case is F_2, which is regarded as being analogous to Cl_2. The staggered forms are grouped in a way that exposes hole–pair symmetry. For example, Li_2 and Cl_2 are related in the sense that placing electron pairs in the holes of Li_2 generates the electronic configuration of Cl_2. Be_2 and Cl_2 are related in the same way, only now hole–pair symmetry is destroyed in the exo orbitals because there is one exo electron in each Be but two exo electrons in each Cl. Here are some noteworthy trends of Figures 29.2 and 29.3:

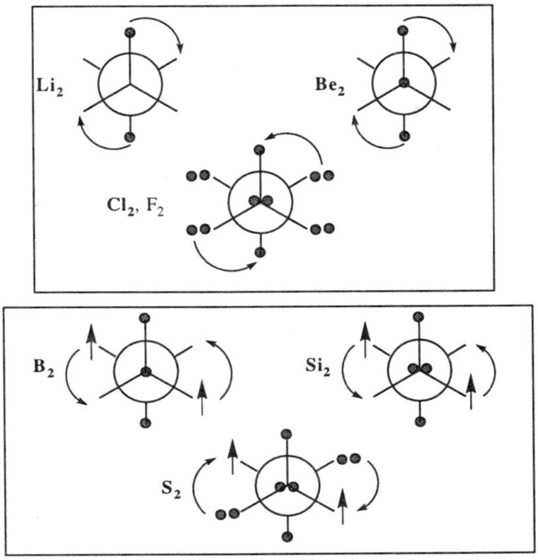

Figure 29.3 I-bonding of electropositive p-block dimers involves staggered hybrids. Dimers are grouped so that pair–hole symmetry is apparent.

(a) Because C is black, it opts for making three C—C bonds through the eclipsed conformation; that is, we have two promoted tetravalent carbons combining to form C_2. By contrast, grey silicon opts for minimization of atom excitation at the expense of making fewer bonds by remaining in its ground divalent configuration. The staggered arrangement of the electrons effects deshielding, and a fraction of the bonding is the result of just that. Note also how the staggered conformation places the two odd electrons with the same spin trans to each other, minimizing their mutual overlap repulsion. Further bonding is effected by codirectional 2-e CT delocalization of the trans triplet pair indicated by the two arrows. Codirectional delocalization of the trans singlet pair plays a secondary role and is not shown. In summary, Si_2 can be regarded as two ground triplet silicon atoms combining into an overall triplet Si_2 in a way that promotes static interaction and I delocalization.

(b) Si_2 and S_2 have hole–pair symmetry in the sense that replacing the holes by pairs in the formula of the first diatomic gives the formula of the second diatomic. In a one-electron sense, this amounts to replacing one-electron/two-center by three-electron/two-center bonds. Since a one-electron bond is stronger than a three-electron bond (because the latter suffers from three-electron exchange repulsion), we expect that Si_2 and C_2 will have stronger

Table 29.1 Bond Dissociation Energies (kcal/mol) of Second- and Third-Row Diatomics

C_2	144	Si_2	75
N_2	225	P_2	116
O_2	118	S_2	101
F_2	37	Cl_2	57

C_2	144	Si_2	75	Ge_2	65	Sn_2	46	Pb_2	19
O_2	118	S_2	101	Se_2	74	Te_2	54	Po_2	44

bonds than S_2 and O_2, respectively. In MO language, O_2 and S_2 are expected to have smaller bond dissociation energies because they carry two π-antibonding electrons. Now, the key fact is that while C_2 has a stronger bond than O_2, the reverse is true in the Si_2/S_2 comparison: S_2 has a larger BDE! This means that our formulas have failed to reproduce a key qualitative trend: A group IVB heavy dimer (Si and congeners) has a smaller BDE than a group VIB heavy dimer (S and congeners) of the same row (Table 29.1). Again, we have the recurring theme: *Organic bonding concepts are the exception and not the rule.*

(c) Li_2 and F_2 are related by hole–pair symmetry. In MO terms, the presence of four π-antibonding electrons in the latter might suggest that Li_2 has a larger BDE than F_2. However, in the hypothetical absence of the fluorine lone pairs, the BDE of F_2 would be much larger than that of Li_2 simply because F is much more electronegative than Li. The key question now is: How is it ever possible for F_2 to be bound, even though, in a formal Lewis sense there is only one σ two-electron bond but two four-electron antibonds? The answer is that F_2 adopts a mechanism of bonding in which deshielding is essential to the binding. In other words, F_2 is an "obligatory" I-bound species. This is why one-determinantal MO theory fails to predict binding. In an interesting paper, Shaik et al. found by ab initio VB computations that the stability of F_2 is due to "resonance" rather than perfect pairing.[1] The term "resonance" is effectively a composite descriptor of deshielding plus 2-e CT delocalization.

(d) While the minimum s/p basis staggered formula of F_2 does justice to this molecule, the analogous representation is ill-suited for Cl_2. For example, one and the same formula for two isoelectronic species of different shades fails to reveal why Cl_2 has a *stronger* bond than F_2 while every other first-row dimer has a larger BDE than its second row counterpart. Note the emerging counterintuitive trend: Grey atoms with np lone pairs form stronger bonds than black isoelectronic relatives (F/Cl comparison), or they lose by a small margin

(O/S comparison). By contrast, in the absence of np lone pairs, the BDEs of black are far greater than those of grey diatomics (N/P and C/Si comparisons).

What is the cause of the failure of our minimum s/p basis formulas to account for these fundamental bonding trends? Going down a column in the main block, electronegativity decreases. This implies the appearance of low-lying orbitals; that is, there is an accessible nonvalence unoccupied space. Clearly, we must include d orbitals in the basis set describing heavy main group dimers, but always in conjunction with the starring procedure.

29.2 Star Isomers, Electron Repulsion, and the VB Model of Metal Diatomics

Consider the prototypical A_2 diatomic, where A is a heavy p-block atom (e.g., any of Al, Si, P, S, or Cl) hybridized in an sp^3d^5 pseudo-octahedral fashion (see Chapter 25). There are five stars corresponding to the admixed five d orbitals, and three of them are assigned to the equatorial set. As a result, occupancy of the equatorial set is forbidden. This means that the valence electrons should be assigned to the unstarred endo and exo conical AOs. Furthermore, since bonding is effected using the endo set of conical hybrids, each A can enter by adopting any one of three star-isomeric ensembles of AOs, differing only in the number of endo-starred AOs. The three possibilities are shown in Figure 29.4.

A fourth possibility, namely, that all three endo hybrids are starred, is

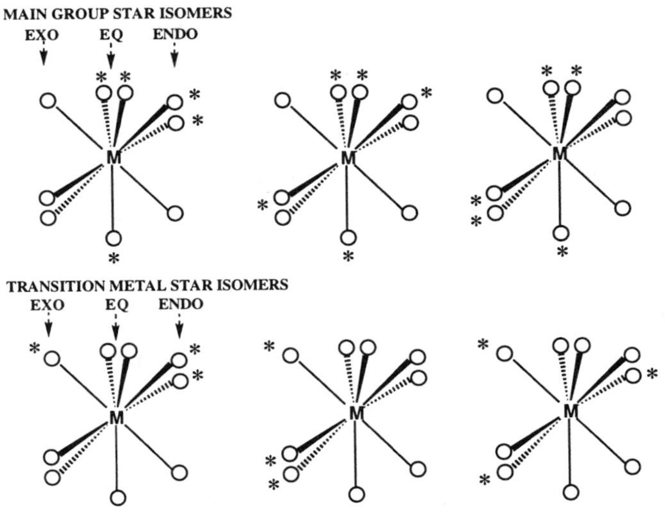

Figure 29.4 Main group and transition metal star isomers.

disregarded: It cannot lead to bonding because starred AOs can never be occupied. We identify each star isomer by the exo:endo ratio of stars. Thus, the 2:0 form is the one in which there are two starred exo and zero starred endo hybrids. Hence, the first critical decision is whether the choice of the hybridization is dictated by the nature of A or by some other factor. We will argue that a metal or semimetal atom has a distinct preference for the star-isomeric form that is consistent with a balanced allocation of electrons within the atom that effects minimization of intra-atomic electron–electron repulsion.

Each atom of a diatomic has local cylindrical symmetry with the three sets of hybrids (exo, equatorial, endo) defining three domains, as illustrated.

PSEUDOOCTAHEDRAL ATOM WITH CYLINDRICAL SYMMETRY

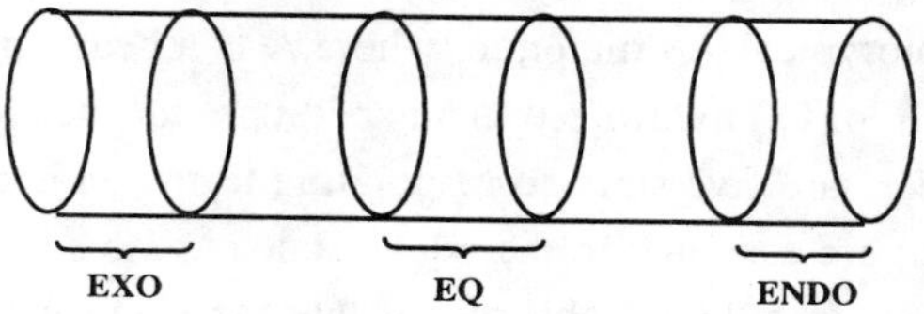

Minimization of atom promotional energy requires that two conditions be jointly met:

The valence electrons must be uniformly distributed to the exo and endo domains rather than being locked in one of them. This minimizes the potential energy by minimizing interelectronic repulsion.

The valence electrons must occupy unstarred orbitals. This minimizes the kinetic energy.

From these statements, it can be seen that the optimal starring pattern is the one consistent with the condition that in even-electron systems there are as many electrons in the exo as in the endo domain, while in odd-electron systems the odd electron may be assigned to either the exo or endo set. In this way, the ratio of exo to endo electrons becomes fixed within a narrow range.

We next consider how the electron distribution that minimizes promotion will be altered by the requirement for maximal interatomic bonding. Since the endo set can, at best, make a formal triple bond, the maximum number of endo electrons should be three. Hence, the best compromise of promotion and bond making is the *balanced allocation condition*, which states that the number of endo electrons cannot exceed three and the difference in the occupancy of exo and endo sets cannot exceed one electron. Exactly the same condition applies to transition metals. Again, the hybridization most suitable for bonding is the sp^3d^5 pseudo-octahedral mode. There are three stars corresponding to the admixed three p orbitals, and they are assigned to the conical exo and endo sets of hybrids. This means that the valence electrons can be assigned to unstarred endo and exo conical AOs as well as to unstarred equatorial AOs. Furthermore, since bonding is effected using the endo set of conical hybrids, each A can enter by adopting any one of three star-isomeric ensembles of AOs differing only in the number of endo starred AOs. We identify each star isomer by the exo:endo ratio of stars. Unlike the case of main group heavy atoms, there are three, rather than two, domains that can accommodate the valence electrons: exo, equatorial, and endo. Each of the exo and endo sets has an average of 1.5 unstarred AOs, while the equatorial set has three unstarred AOs. It follows that the equatorial set can accommodate twice as many electrons as the two conical sets, everything else being equal. However, this in no way alters the balanced allocation condition with respect to the distribution of the valence electrons in the conical sets.

29.3 The Formulas of Heavy Main Group Diatomics

The I bonds connecting two metal atoms respect two conditions:

The balanced allocation condition. This tells us how many electrons are actually in the bonding region, thus permitting the minimization of promotional energy, which includes intra-atomic interelectronic repulsion.

The principle of arrow codirectionality. This says that the best I bonds are those that are components of a codirectional arrow train.

Furthermore, the hierarchy of individual bond types is as follows:

The strongest bonds are I bonds (two codirectional solid arrows), followed by Q bonds (one solid and one hatched codirectional arrow). Triplet I bonds are much weaker because the subset of ionic configurations that place two electrons in the same AO is eliminated.

The weakest bonds are those not spanned by arrows. These are either T bonds (solid line) or E bonds (hatched line).

It must now be obvious that the "multiple bonds" of diatomics will bear no resemblance to the multiple bonds of organic chemistry. In fact, they will hardly be multiple! For example, the "triple bond" of P_2 turns out to be one I plus one T bond.

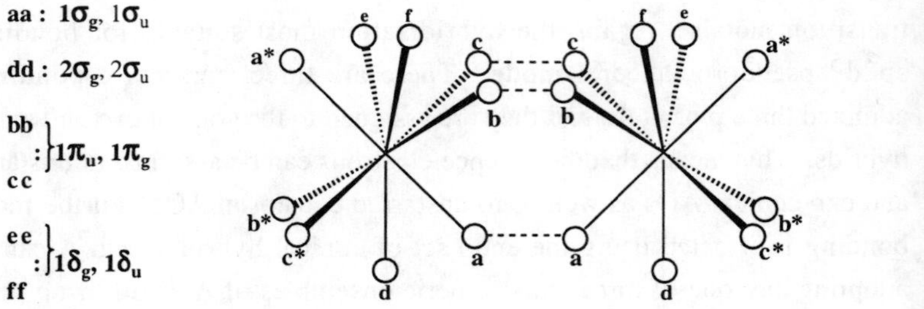

Figure 29.5 The correspondence between dimer electronic configurations built from pseudo-octahedral hybrids and from canonical MOs.

Most readers are familiar with MO theory. Hence, to be able to compare our conclusions with what experimentalists and computational scientists have found, it is important to translate the dimer configurations over hybrid AOs to dimer configurations over MOs. To this extent, the correspondence is spelled in Figure 29.5. With this background, the formulas of heavy main group diatomics shown in Figure 29.6 speak for themselves. For example, "aa : $1\sigma_g$, $1\sigma_u$" means that the electrons deposited in the two a AOs act like those occupying the $1\sigma_g$, $1\sigma_u$ MOs.

The crucial comparison is that of Si_2 and S_2. Whereas Si_2 has one triplet I bond plus one T bond, the balanced allocation condition causes S_2 to be an "exo triplet," relinquishing the endo space to electron pairs that associate with starred holes. The result is the formation of one Q bond plus one E bond. Hence, the related (hole–pair symmetry relatives) dimers Si_2 and S_2 are completely different in their binding. This major difference reminds us that we are just beginning to understand chemical bonding, since even a semiquantitative estimate of the relative strengths of E, T, triplet I, and Q bonds is beyond our reach. However, we can assume that a singlet Q is superior to a triplet I-bond and that both T and E bonds are weak in semimetals. On this basis, we have an explanation of why S_2 is more strongly bound than Si_2.

Bottom line: Heavy p-block atoms can hybridize in a way that imparts electrostatic character to the resulting interatomic bond. Because of the accessibility of a low-lying unoccupied space (simulated by the d AOs), a heavy p-block diatomic, such as Cl_2, can easily arrive at a deshielded electronic configuration that promotes electrostatic multipolar bonding with some organic molecule. This always involves the concomitant loss of Cl—Cl bonding with the benefit of reduction of atom promotion, as illustrated for the case of the ethylene–Cl_2 complex, where Cl—Cl bonding has been diminished (compare with Figure 29.6) while Cl hybridization has been "reduced" to sp^3d:

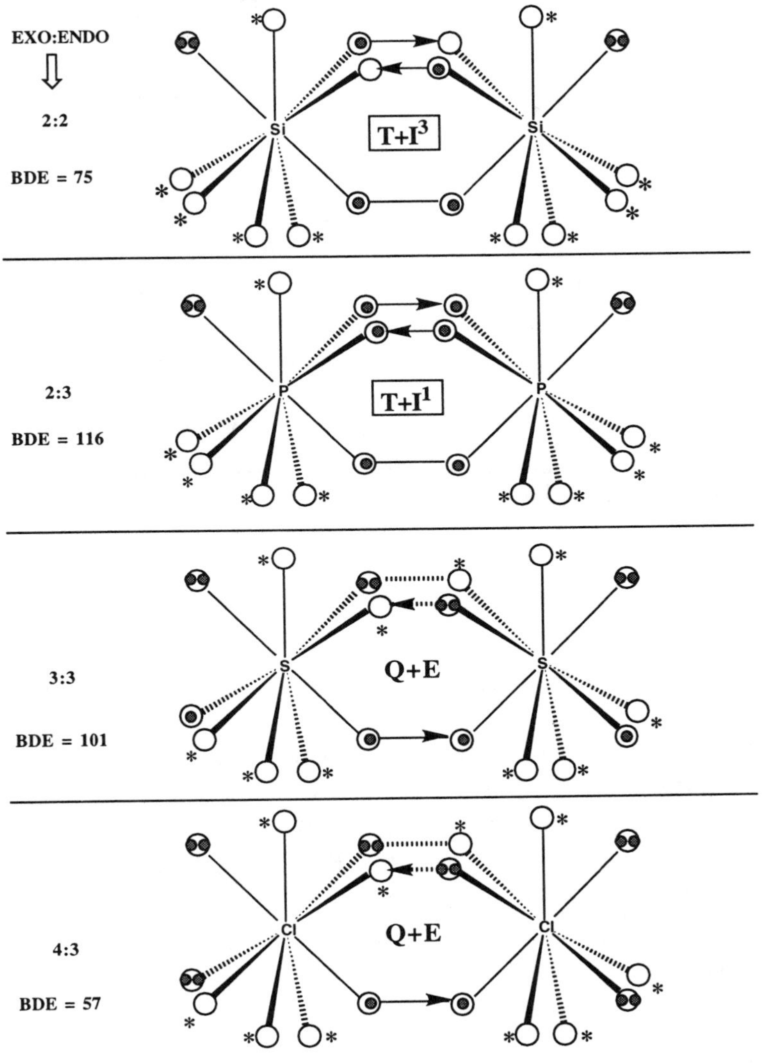

Figure 29.6 Formulas of heavy p-block dimers. Second-row diatomics are represented by VB "repulsion formulas" over an s/p/d basis. Each atom is pseudo-octahedrally hybridized. The bonding is now effected subject to the condition that there be a balance of electrons in the endo and exo polar sets. This ensures minimization of interelectronic repulsion and causes S_2 to have six and Si_2 to have only four electrons in the overlap region.

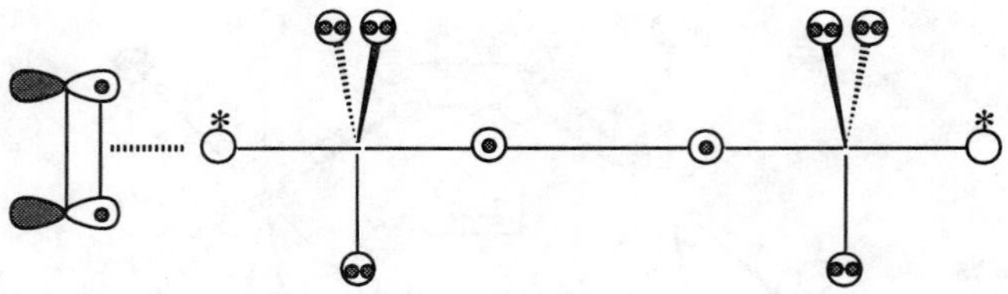

The promoted molecule has now strong quadrupolar character along the molecular axis, which can be abstracted as two positive charges at the atomic centers sandwiching two negative charges in the interatomic region. This suggests an analogy between H—Cl and Cl—Cl and vindicates the electrostatic multipolar model advanced by Legon to explain the observation that the stereochemistry of B···H—Cl is analogous to that of B···Cl—Cl (B = organic σ or π base).[3] The same model succeeds in rationalizing structures of $B\cdots M^{+q}$ complexes (M = metal).[4,5]

References

1. S. Shaik, P. Maitre, G. Sini, and P.C. Hiberty, *J. Am. Chem. Soc.* 114, 7861 (1992).
2. R. Huisgen, R. Grashey, and J. Sauer, in *The Chemistry of Alkenes*, S. Patai, Ed., Wiley-Interscience, New York, 1964.
3. A.C. Legon, *Chem. Phys. Lett.* 237, 291 (1995).
4. P.C. Kearney, L.S. Mizoue, R.A. Kumpf, J.E. Forman, A. McCurdy, and D.A. Dougherty, *J. Am. Chem. Soc.* 115, 9907 (1993).
5. S. Roszek, and K. Balasubramanian, *Chem. Phys. Lett.* 234, 101 (1995).

Chapter 30

VB Formulas For Transition Metal Dimers

30.1 Star Isomers of Cr_2

Transition metal dimers can be conveniently described by assuming that each atom is pseudo-octahedrally hybridized. Cr has four choices of arranging stars in the exo and endo hybrids to effect the best bonding with a second Cr atom: The endo (bond-forming) set has zero, one, two, or, three stars. Because only unstarred AOs are occupied in the reference configuration, the last choice is inconsistent with bonding and is no longer considered. The three possible star-isomeric Cr_2 diatomics are shown in Figure 30.1.

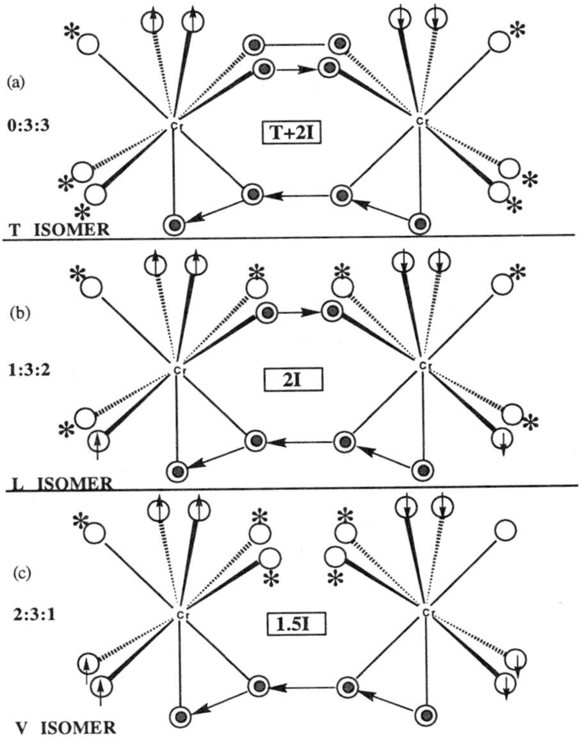

Figure 30.1 Star isomers of Cr_2 differing in the nature of the Cr—Cr bond. The EXO:EQ:ENDO allocation is indicated for each isomer.

Metal–metal bonding is always subject to the rules arrived at in Chapter 29:

The bonds are of the I type. One I bond is symbolized by one pair of arrows. Ideally, we must have only arrows in the bonding region. In practice, the odd number of overlapping pairs of endo hybrids make this impossible. Frequently, one odd bond is left out and this is either of the T or the E type.

The balanced allocation condition fixes the number of electrons available for metal–metal bonding.

Arrows must be codirectional.

The new element is that arrows can connect one endo and one equatorial hybrid because these interact strongly. These intra-atomic arrows effect bonding by dispersion and are counted in the enumeration of the intermetallic I bonds.

The linkage of two Cr atoms can be effected in three qualitatively different ways (see Figure 30.1): The first possibility is to connect the two atoms by using endo hybrids all of which are unstarred [isomer T (= tight)]. This produces a T+2I bond. The four arrows (three σ and one π) make up two I bonds. Note that we no longer have the classical triple bond of acetylene. This star isomer, however, gives us a large number of electrons (six) in the overlap region and a highly unbalanced distribution of electrons within each Cr: three in the endo and three in the equatorial set.

The second possibility is to connect the two Cr atoms by using endo hybrids of which only two are unstarred [isomer L (= loose)]. This produces a 2I bond. This type of bonding (i.e., double I-bonding) has two positive features: reduction of the number of electrons in the overlap region and a balanced allocation of electrons within each atom: One exo, three equatorial, and two endo electrons is an electron distribution that obeys the balanced allocation condition. The physical meaning of all this is that a weak T bond has been effectively annihilated by electron–electron repulsion.

The final possibility is to use endo hybrids of which only one is unstarred (isomer V). This produces a 1.5I bond. Much like the L star isomer, the V form fulfills the balanced allocation condition, but it has fewer metal–metal bonding electrons.

In summary, the three singlet Cr_2 species of Figure 30.1 are star isomers that differ only in Cr—Cr bond length: isomer T is a tightly bound analogue of As_2, isomer V is a weakly bound analogue of Be_2 (but without the promotional energy problem of the latter), and isomer L is in between. In going from the T to the L and V isomer, bonding due to overlap is sacrificed for the benefit of reduced interelectronic repulsion. However, we have yet to reach the limit, as represented by the following high spin form in which only a single I bond (of the type encountered in Li_2) links the two atoms:

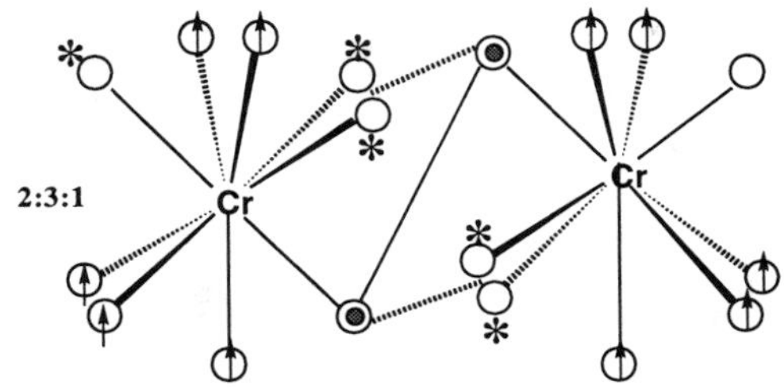

HIGH-SPIN ISOMER

Are the singlet T, L, and V forms distinct energy minima (i.e., star valence isomers)? Is there only one minimum? Irrespective of the answer, which form has the lowest energy? Which one, that is, represents the best compromise between electron–electron repulsion (a form of promotion) and bond making?

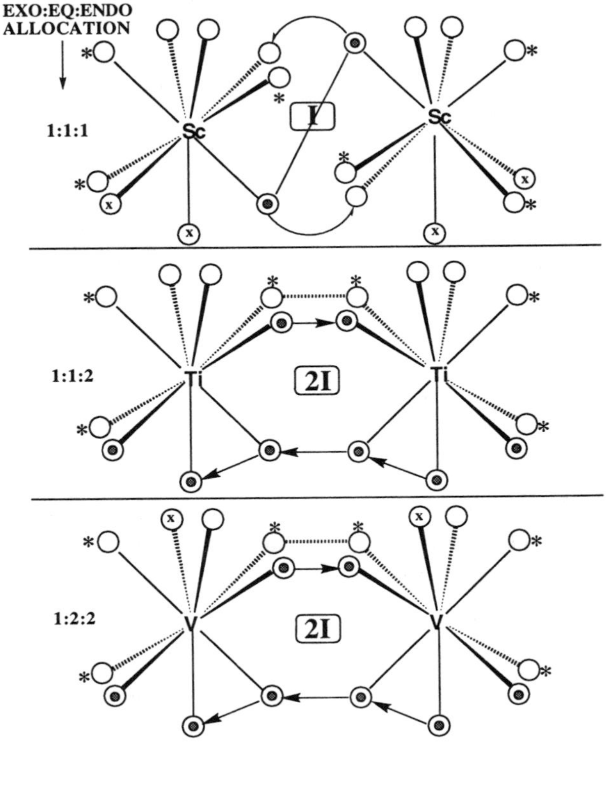

Figure 30.2 VB formulas of Sc_2, Ti_2, and V_2. X-electrons are high-spin coupled. Dot-electrons are low-spin coupled.

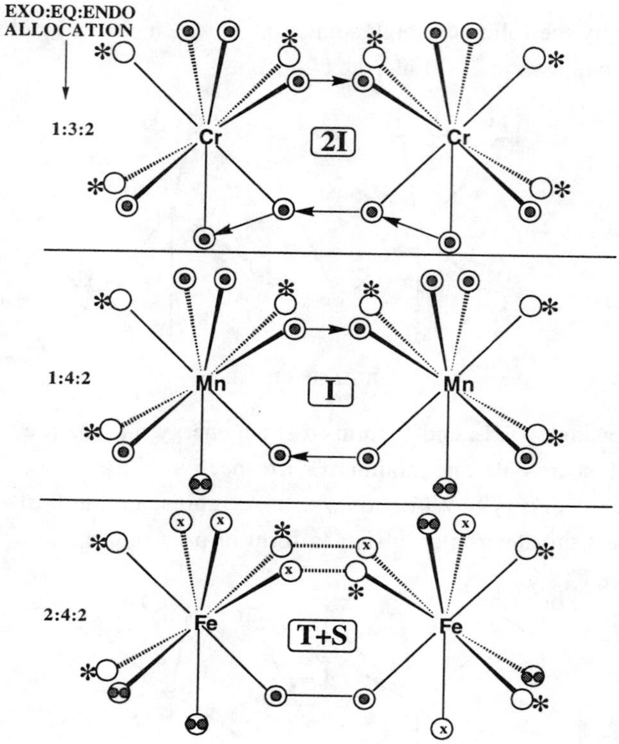

Figure 30.3 VB formulas of Cr_2, Mn_2, and Fe_2. X-electrons are high-spin coupled. Dot-electrons are low-spin coupled.

VB theory raises more questions than it can presently answer. We emphasize that the preceding discussion has nothing to do with the "stretch-isomerism" ideas of EHMO theory.[1] Observe the cascade of physical effects that produce the final picture: Interelectronic repulsion determines the star character of the endo hybrids, and this determines the number and type of bonds that can link the two atoms. The EHMO model neglects interelectronic repulsion and, as a result, star isomerism, which physically translates into "stretch isomerism" because the only observable structural difference among the T, L, and V singlets, the bond length, does not exist at the EHMO level.

Some very interesting computational results have been obtained by Goddard and Goodgame in their studies of the Cr_2 and Mo_2 dimers.[2] They found two different minima, which they describe as "covalent" and "antiferromagnetically coupled" dimers. These may correspond to two of the three star-isomeric structures of Figure 30.1.

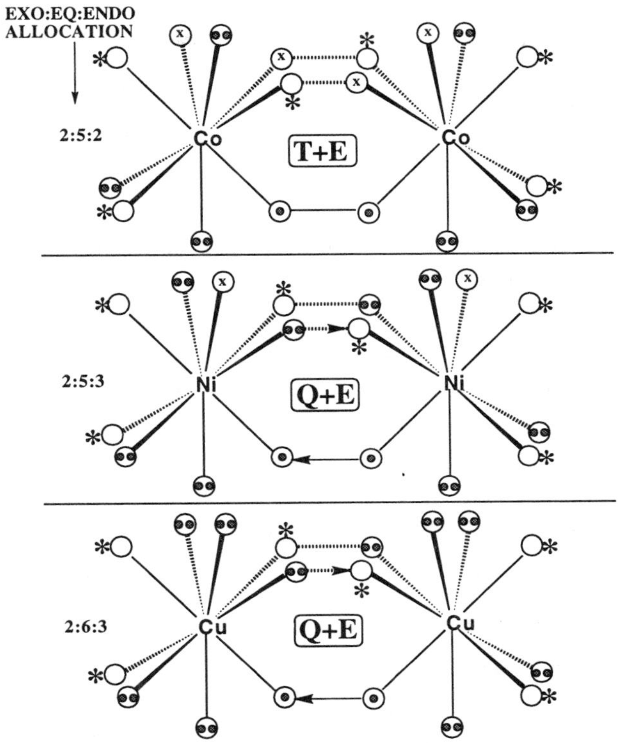

Figure 30.4 VB formulas of Co_2, Ni_2, and Cu_2. X-electrons are high-spin coupled.

30.2 VB Formulas for Transition Metal Dimers

The MO configurations of the ground states of the transition metal diatomics that are consistent with experiment and ab initio computations can be found in the critical review of Morse,[3] which we also cited in the preface. By translating from MO to VB language, we can write the VB formulas of the transition metal dimers, as shown in Figures 30.2 to 30.4. The rather amazing result is that the MO configurations deduced from experiment and the corresponding VB formulas meet the balanced allocation condition with no exception!

Now, armed with concrete formulas to help us organize our thinking, we attempt an interpretation of the seemingly unintelligible data obtained from studies of transition metal dimers. Here are some of the key features of the formulas.

In the case of Sc_2, the optimum allocation is expected to be 1:1:1, yielding the intuitively unexpected high spin structure shown in Figure 30.2. The bond dissociation energy of the Sc_2 dimer has been estimated at a surprisingly high 1.65 eV. Our formula shows that we have a formal single bond entirely analogous to that described for Li_2, save in one important aspect: The endo pair effectively sees two unscreened Sc cores because one electron in each atom is deposited in an exo hybrid. This is the main reason why the BDE of *quintuplet* Sc_2 is larger by about 0.5 eV than that of K_2. The remaining four electrons become high-spin coupled and act as nonbonding electrons. Of the four, two are nonbonding equatorial σ electrons. The other two are π electrons representing two superstretched one-electron π bonds.

Figure 30.5 emphasizes that the binding of Ti_2 is due largely to the association of two G bonds.

Returning to Figure 30.2, we see that the V_2 dimer has the Ti_2 structure plus two δ-type one-electron bonds, which are not counted because δ overlap is judged small.

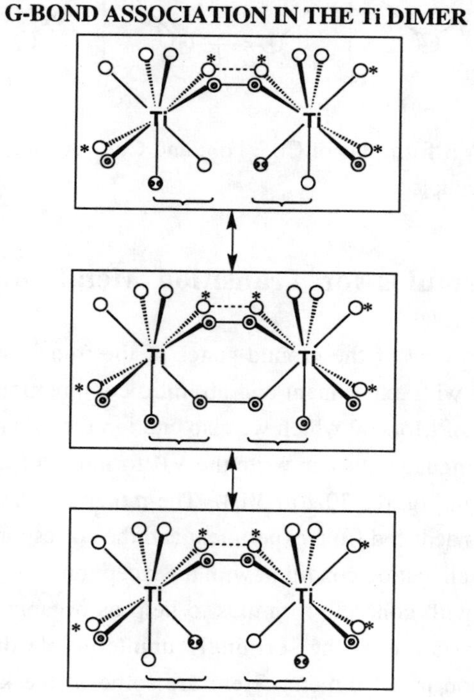

Figure 30.5 G bond association in Ti_2.

The Cr_2 dimer with maximum bonding is the T form (Figure 30.1a). According to conventional theory, this diatom has a sextuple bond. According to VB theory, it has two I bonds plus one T bond because one additional δ I bond is not counted. If this bond scheme in fact obtained, Cr_2 would be more strongly bound than V_2. As is known, however, the bond dissociation energy of Cr_2 is lower than that of V_2, Ni_2, and Cu_2 even though the latter molecules have a much lower apparent bond order at the level of EHMO theory. We can see the resolution of the problem: The singlet form of Cr_2 is actually the "compromise" L form (Figure 30.1b), which has two I bonds, exactly like V_2. Hence, the superiority of V_2 insofar as binding is concerned must be due to more favorable promotional energy. We will return to this issue shortly.

The Mn_2 molecule (Figure 30.3) represents a sudden downturn of the fortunes of the transition metal diatom. The balanced allocation condition can be satisfied only if there is just one I bond linking the two atoms. Note how the 1.5σ I bond has disappeared because of the double occupancy of one equatorial hybrid within each metal atom.

In all dimers from Sc_2 to Mn_2, an exo unstarred hybrid is singly occupied. This means that an electron can be delivered to the exo sets of these dimers, causing minor bonding changes. From Fe_2 on, the situation changes. Fe_2 (Figure 30.3) and Co_2 (Figure 30.4) have three important features:

(a) Since each of the odd electrons in the endo space is coordinated to a starred hole, they couple into a high spin pair, inducing high spin coupling of the remainder of equatorial electrons. Observe the pattern: Each coordinated endo odd electron is there because one pair opted to go exo rather than enjoy coordination to a starred hole. All this is a consequence of the balanced allocation condition.

(b) The exo unstarred hybrid is filled. Hence, delivery of an electron to form the corresponding anion radical is now possible only in the endo overlap region, where there are two half-filled unstarred AOs. As a result, such a delivery is expected to strengthen the metal–metal bond.

(c) Fe_2 and Co_2 are distinguished from all other dimers in having equal number of exo and endo electrons; that is, they obey perfectly the balanced allocation condition. Hence, their high spin character is, again, the consequence of minimization of interelectronic repulsion. That is, they refuse to form ordinary covalent triple bonds because this would violate the balanced allocation condition.

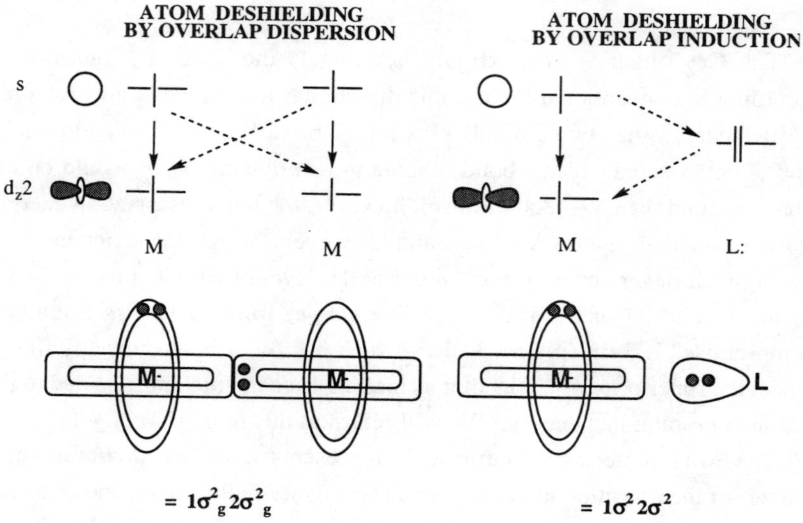

Figure 30.6 Atom deshielding by the cooperative overlap dispersion and overlap induction mechanisms.

In going from Co_2 to Ni_2 (Figure 30.4), addition of two electrons enhances interatomic bonding. We now have one Q and one E bond, but there is no longer any vacancy in the overlap region. Reduction of Ni_2 to form the radical anion is now unfavorable because there is no longer a singly occupied unstarred hybrid coordinated to a starred hole in the overlap region, as in Co_2 and Fe_2.

To complete our discussion of Figure 30.4, we note that ab initio computations indicate that a significant fraction of the bonding of Cu_2 is due to the action of the d electrons. This now creates a puzzle: Intuitively, one would expect that a copper atom (with a single s electron and 10 d electrons) could form only a weak bond with a second copper atom because of the anticipated exchange repulsion of the filled d subshells. At the very least, one would expect the Cu—Cu bond to be weaker than the K—K bond, where the presumed repulsive effect of the d shells is absent. Quite the opposite is true: The Cu—Cu bond is stronger than the K—K bond and also stronger than the nominally sextuple bond of Cr_2! The reason is revealed by our formula: Cu_2 has one Q plus one E bond, much like Ni_2. In addition, virtually no energy is needed to promote the Cu atom to its valence configuration because no exchange stabilization is lost, as is the case when a high spin atom is forced to adopt a low spin electronic configuration in the process of bond making.

30.3 The Bond Dissociation Energies of Transition Metal Dimers

Up until now, we have tentatively assumed that transition metal atoms can attain a pseudo-octahedral hybridization with equal facility. In other words, we have assumed that the promotional energy required for mixing the $(n-1)$d with ns and np AOs is constant along a transition row. Obviously, this is only an approximation. If we tentatively forget about the action of the unoccupied nonvalence space and consider the bonding of transition metal diatomics to be due exclusively to d and s electrons, the most significant electronic rearrangement is the one that results from the mixing of the d and s AOs and causes σ deshielding by dispersion and/or induction (Figure 30.6).

Since any remaining π- or δ-type electrons cannot engender bonding unless σ deshielding allows the atoms to approach each other, the bond strength will depend in a major way on the relative energies of three critical configurations: s^2d^{m-2}, s^1d^{m-1}, and d^m. The efficacy of d-s mixing in zerovalent transition metals depends on the energy gaps separating these configurations. We recognize three classes of transition metals: A, B, and C.

Class A transition metal atoms, which have either an s^2d^{m-2}(e.g., Mn) or an s^0d^m (e.g., Pd) ground electronic configuration that lies at least 1 eV below s^1d^{m-1}. The efficacy of d-s hybridization and attendant core deshielding is expected to be very low.

Class B transition metal atoms, which have an s^1d^{m-1} ground electronic configuration that lies at least 1 eV below s^2d^{m-2}(or s^0d^m if lower than s^2d^{m-2}). The other considerations from class A apply here.

Class C transition metal atoms, which have s^2d^{m-2}, s^1d^{m-1}, and s^0d^m configurations that lie within a range smaller than 2.5 eV. When two metal atoms combine to form a diatomic (or when a metal atom combines with some other fragment), these low energy configurations interact strongly to effect σ deshielding.

Atoms that do not belong to any of these three categories represent borderline cases. One can determine from the data of Table 30.1 the extreme to which each of them leans.

The bond dissociation energies (eV) of the transition metal dimers are Sc_2, 1.65; Ti_2, 1.2; V_2, 2.5; Cr_2, 1.8; Mn_2, 0.8; Fe_2, 0.9; Co_2, 1.0; Ni_2, 2.1; Cu_2, 2.0. We can now attempt to understand the bewildering variation of the

Table 30.1 Electronic Configurations (eV) for the Three Classes of Transition Metals

Class A[a]		Class B[b]		Class C[c]	
Sc	1.43	Cr	1.00	V	2.47
Mn	2.14	Cu	1.49	Ni	1.74
Y	1.36	Mo	1.47	Nb	1.32
Hf	1.69	Rh	1.63	Ru	1.09
Ta	1.04	Ag	3.97	Rh	1.63
Re	1.76	Au	1.74	Pt	0.64
Pd	0.95[d]				

[a] $E(s^1d^{m-1}) - E(s^2d^{m-2})$.
[b] $E(s^2d^{m-2}) - E(s^1d^{m-1})$.
[c] Bracketing range of $E(s^2d^{m-2})$, $E(s^1d^{m-1})$ and $E(s^0d^m)$.
[d] $E(s^1d^{m-1}) - E(s^0d^m)$.

BDEs by simply counting the number and types of metal–metal bonds in Figures 30.2 to 30.4. Assuming that promotional energies remain constant, we can predict the following trend:

The BDE should increase in going from Sc_2 (one I bond) to Ti_2 (two I bonds) and then remain constant in the series Ti_2, V_2, Cr_2 (which all have two I bonds).

A steep drop should occur at Mn_2 (only one I bond).

The BDE should rise hardly at all in going from Mn_2 to Fe_2 because one I bond is traded for one inferior T bond plus one E bond. Note that one electron coordinated to a starred hole counts for 0.5 E bond.

Fe_2 and Co_2 must have comparable BDEs, since both have one T bond plus one E bond.

A steep increase must occur in going from from Co_2 to Ni_2, since one inferior T bond is replaced by one Q bond, which must have the same BDE as Cu_2 because both dimers have one Q bond plus one E bond.

The VB predictions of BDEs fit the data, with two diatomics falling off the expected trend: Both Sc_2 and V_2 are bound far more strongly than is suggested by the mere bond count. What is the explanation? Why does the BDE more than double as we go from Ti_2 to V_2, and why does V_2 have a stronger bond than Cr_2? The answer is straightforward: V, being a class C atom, is superior to Ti and Cr in promoting s-d hybridization. In other words, V can arrive at the valence configuration necessary for bonding without investing as much promotional energy as either Ti or Cr. Cr_2 has a larger BDE than Ti_2 on account of an extra-weak δ I bond. The promotional energy factor is what renders V_2 and Nb_2 the champion diatomics, having the strongest bond within the first and second transition series, respectively. This argument also explains why VO, NbO, and TaO have BDEs that exceed by 40 kcal/mol or more (!) those of CrO, MoO, and WO, respectively, even though two adjacent metals (e.g., V and Cr) do not differ greatly in electronegativity. Ever since the preparation of inorganic complexes containing metal–metal bonds by Cotton and his associates,[4] the interest in the topic has promoted a great number of calculational investigations of model and actual systems aimed toward understanding the nature of these bonds. The investigations have been guided, at least insofar as interpretation was concerned, by covalent/ionic notions: Carbon–carbon multiple bonds (e.g., the double bond of ethylene and the triple bond of acetylene) and metal–metal multiple bonds are analogous, but the latter can assume any status from double to sextuple; the only difference between inorganic and organic homopolar bonds is the degree of covalency. Our point is simple: The experimental data do not support this scenario. "Bond" means "delocalization." The "covalent bond" represents "exchange delocalization." The "I bond" represents "CT delocalization." Organic and metal "multiple bonds" are fundamentally different. Covalency is ideal for nonmetals but represents a restriction in metals.

Organic chemists are uncomfortable with induction and dispersion simply because these static mechanisms of bonding are not prominent in the covalent world of organic monomers. In addition, the theoretical underpinning of this overlap-independent type of bonding is not as easily accessible as that of bonding due to overlap.[5] VB theory, however, can give a pictorial account of all types of bonding which can be immediately understood by the nonexpert. In metal systems, the covalent bonds (T bonds) give way to I, Q, and E bonds, which differ in the relative importance of CT and static bonding, with I-bonding having CT dominance and Q- and E-bonding having static dominance. We now enumerate the "deviant" trends of bond dissociation energies that can be explained by the VB concept of static bonding either as a distinct or an inseparable component of I- and Q-bonding. In all comparisons, the first mentioned diatomic or molecule has more or superior I, Q, and/or E bonds.

S_2 has a stronger bond than Si_2.
Cl_2 has a stronger bond than F_2.

Ni_2 and Cu_2 have stronger bonds than Fe_2 and Co_2 (Figures 30.3, 30.4).

Fe_2^- is bound roughly 0.75 eV more strongly than Fe_2. Similarly, Co_2^- is bound roughly 0.45 eV more strongly than Co_2.[6] By referring to our formulas for Fe_2 (Figure 30.3) and Co_2 (Figure 30.4), we expect that the added electron will end up in a half-vacant unstarred orbital of Fe_2 and Co_2 which is coordinated to a starred hole. In other words, the extra electron is delivered to the overlap region between the two atoms to produce an additional one-half E bond.

What about the bond length variations in the transition metal dimers? When the added electron increases the E-bond count, bond shortening is not necessarily expected, as it is when an added electron increases either the T- or I-bond count. Hence, the absence of significant bond shortening in the more strongly bound species is evidence in favor of the concept of the Q and E bond. It turns out that this is the case in the Fe_2/Ni_2, Co_2/Co_2^-, and Fe_2/Fe_2^- comparisons.[2]

Interesting trends of BDEs of small and large molecules that can now be illuminated by VB theory abound. To mention only one, the BDE of Fe_2^+ lies between 2.43 and 2.92 eV and the BDE of Fe_2 between 0.83 and 1.32 eV. Analogously, the BDEs of Ni_2^+ and Ni_2 are 3.30 and 2.07 eV, respectively.[6] What happens is that removal of one electron increases the effective electronegativity of each metal and moves the cationic dimer toward the T limit of the bonding continuum. As a result, the balanced allocation condition is no longer valid and the electron density becomes rearranged, whereupon each cationic metal dimer is driven toward the limit at which three electrons go to the three unstarred endo conical AOs to form a formal metal–metal triple bond, and the rest are placed in the equatorial hybrids. Thus, intermetallic bonding is strengthened because of the increase in atom electronegativity, which, in turn, causes enhanced perfect pairing.

In the foregoing discussions, we focused on first-row transition metal dimers. How does the picture change as we move to the third-series congeners? A good illustration is provided by the comparison of Cr_2 and W_2. Chromium has lower electronegativity than tungsten, as well as more contracted d AOs. These two properties imply that the orbital electronegativity of the valence Cr AOs is much lower than that of the W AOs. In turn, this means that the W_2 dimer will be predisposed to seek a greater T complement (i.e., to improve the covalency of the I bonds). Thus a switch in preference from the L to the T star isomer of Figure 30.1 is predicted. W_2 has now a metal–metal bond that begins to resemble the triple bond of acetylene. In addition, it has three weaker bonds formed by the overlap of the equatorial hybrids. This explains why W_2 as well as W solid have the strongest bonds of all third transition series metals.

The fundamental difference between first and third transition series metals insofar as binding is concerned is displayed schematically on the cover of the Puddephatt–Monaghan monograph[8]: To a first approximation, the first series has two maxima (V and Co) and a secondary minimum (Mn). This means that when it comes to metals, there is no simple bond strength/bond order correlation. The third series has a single maximum (W).

30.4 Metal Monocarbonyls; The Simplest Organometallic Complexes

The metal monocarbonyls provide more indications that the one-electron picture of metal bonding is qualitatively (leave alone quantitatively) inadequate. The interaction of a carbonyl with a metal center is almost invariably discussed in the literature by reference to the Dewar–Chatt–Duncanson model of σ dative bonding from ligand to metal and π back-donation from metal to ligand.[9] This implies that the best-case scenario for metal–carbonyl bonding is to have a pseudo-octahedral metal adopt a 3h/4e electronic configuration (three endo conical hybrids with two electrons), to match an effective 3h/2e carbonyl electronic configuration (three sp^3 hybrids with two electrons, or one σ pair plus two π holes) in the process of making a formal triple bond. This is not at all what happens.

Table 30.2 The Occupancies of the Exo, Equatorial, and Endo Sets of the Third-Series Metals[a]

Metal	Exo (x)	Equatorial (y)	Endo (z)
Sc	0	1	2
Ti	0	3	1
V	0	3	2
Cr	1	3	2
Mn	1	4	2
Fe	1	3	4
Co	1	4	4
Ni	0	6	4
Cu	1	6	4

[a] $|y-z| \leq 2$

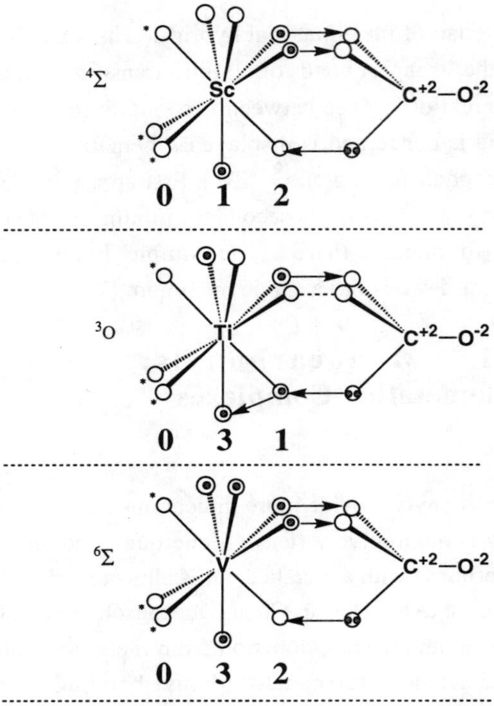

Figure 30.7 Monocarbonyl complexes of Sc, Ti and V.

A critical review of the structures of metal carbonyls by Weltner and Van Zee[10] is our major source of information. By translating from MO to VB language, we obtain the formulas shown in Figures 30.7 to 30.9. Because CO acts as an acceptor, arrow contradirectionality is allowed for the benefit of minimizing promotion. The occupancies of the exo, equatorial, and endo sets given in Table 30.2 leave no doubt that even in the prototypical organometallic MCO, minimization of interelectronic repulsion is of paramount importance:

(a) In five out of nine complexes, there is a single occupancy of the exo set. This tells us that the corresponding metals (Cr, Mn, Fe, Co, Cu) are striving toward the balanced allocation observed in the metal dimers.

(b) If it were merely a matter of delocalization of metal pairs from the endo hybrids to the CO holes, we would have an aufbau principle according to which we would fill the two available endo hybrids with the maximum of two pairs before placing electrons in the equatorial set. As can be seen in Table 30.2, the complexes from ScCO to MnCO do not bear this out. Rather, there is balanced allocation of electrons to the three pseudo-octahedral domains in the sense that

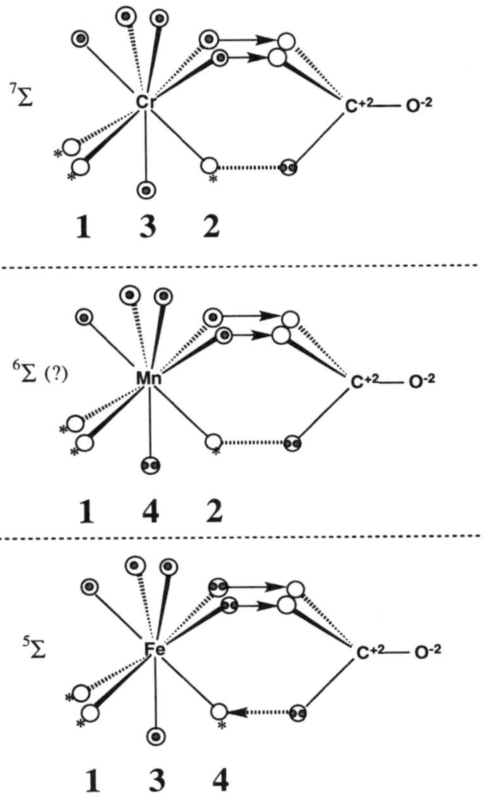

Figure 30.8 Monocarbonyl complexes of Cr, Mn and Fe. Unpaired electrons are high-spin coupled.

the number of electrons in the equatorial set never exceeds by more than 2 the number of electrons in the endo set. It is even more meaningful that the equatorial exceed the endo electrons in direct contradiction of the usual overlap notions (only endo electrons can delocalize to the CO holes in the back-bonding sense). We attain the maximum of two endo pairs only at FeCO and thereafter.

The simple transition metal dimers and monocarbonyls teach us concepts that can be extended to complex clusters. For example, Ni_2 and NiCO are open- and closed-shell fragments, respectively. In the most general sense, the ability of a ligand to reduce the magnetism of a metal atom is the result of a shift of the mechanism of bonding from I to T along the continuum as the average atom electronegativity increases. This trend is clearly seen in ligated high nuclearity clusters. Pacchioni and Roesch found that surface and bulk metal atoms are differentiated with respect to "metallic character".[11]

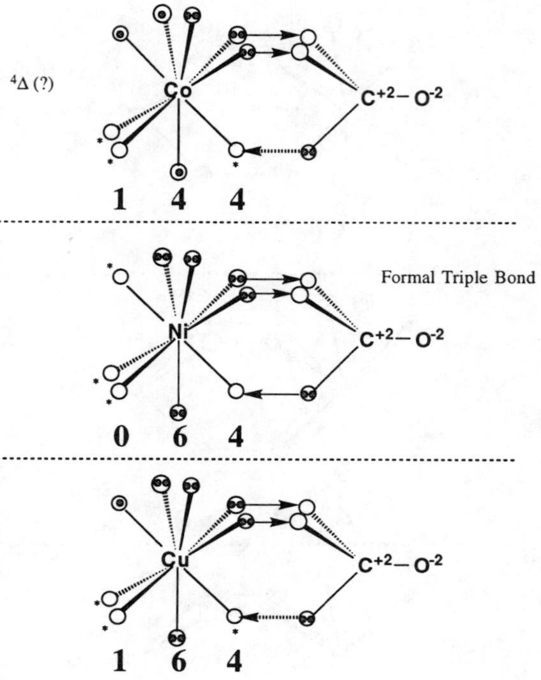

Figure 30.9 Monocarbonyl complexes of Co, Ni and Cu.

Recent experiments begin to shed light on how CO ligands modify the structures of bare metal clusters. For example, Co_2^+ is unreactive toward alkanes,[12] but $Co_2(CO)^+$ reacts efficiently.[13] The metal ion dimer is expected to have the electronic structure shown in Figure 30.10.

There exists no unstarred hole that can combine with a metal pair to effect addition to a C—C or C—H alkane σ bond. Coordination of one CO creates unstarred holes which, in cooperation with (unstarred) pairs, can lead to oxidative

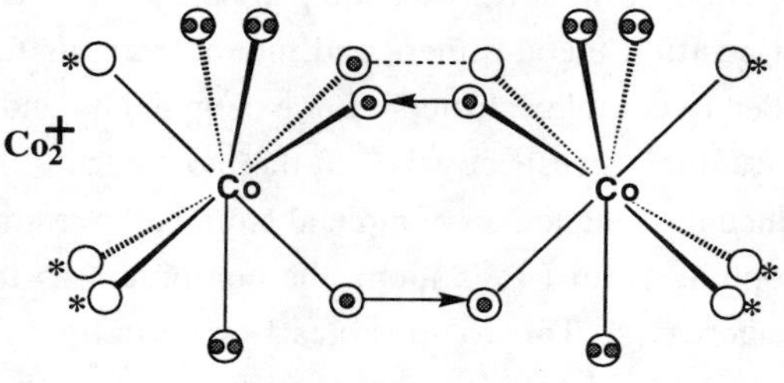

Figure 30.10 Proposed electronic structure of Co_2^+.

addition to an alkane bond. The following formula is the low spin isomer that can react efficiently with the alkane as a hole–pair reagent.

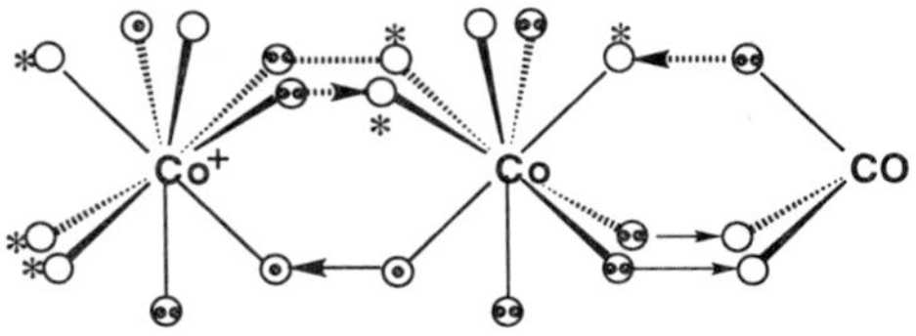

References

1. K. Yoon, G. Parkin, and A.L. Rheingold, *J. Am. Chem. Soc.* 113, 1437 (1991).
2. M.M. Goodgame and W.A. Goddard III, *Phys. Rev. Lett.* 48, 135 (1982).
3. M.D. Morse, *Chem.Rev.* 86, 1049 (1986).
4. F.A. Cotton and R.A. Walton, *Multiple Bonds Between Metal Atoms*, Wiley, New York, 1982.
5. By contrast, ab initio computational chemists make a living by handling electron–electron repulsion and, as a result, the way in which induction and dispersion effect deshielding in metallic systems, is well recognized. For clear descriptions, see (a) E.R. Davidson, in *The Challenge of d and f Electrons*, American Chemical Society, Washington, DC, 1989, Chapter 11. (b) P.-O. Widmark, B.O. Roos, and P.E.M. Siegbahn, *J. Phys. Chem.* 89, 2180 (1985).
6. D.G. Leopold and W.C. Lineberger, *J. Chem. Phys.* 85, 51 (1986).
7. P.J. Brucat, L.-S. Zheng, C.L. Pettiette, S. Yang, and R.E. Smalley, *J. Chem. Phys.* 84, 3078 (1986).
8. (a) R.J. Puddephatt and P.K. Monaghan, *The Periodic Table of the Elements*, 2nd ed., Clarendon Press, Oxford, 1986. (b) N.N. Greenwood and A. Earnshaw, *Chemistry of the Elements*, Pergamon Press, New York, 1986.
9. (a) M.J.S. Dewar, *Bull. Soc. Chim. Fr.* 79 (1951). (b) J. Chatt and L.A. Duncanson, *J. Chem. Soc.* 2939 (1953).
10. W. Weltner Jr. and R.J. Van Zee, in *The Challenge of d and f Electrons*, American Chemical Society, Washington, DC, 1989.
11. G. Pacchioni and N. Roesch, *Acc. Chem. Res.* 28, 390 (1995).
12. R.B. Freas and D.P. Ridge, *J. Am. Chem. Soc.* 102, 7129 (1980).
13. R.B. Freas and D.P. Ridge, *J. Am. Chem. Soc.* 106, 825 (1984).

Chapter 31

The Starring Concept and the Docking Rule

31.1 The Formulas of Organometallic Fragments

Coordination compounds are association complexes of elementary subfragments. We are now in a position to write explicit formulas for metal-containing fragments and assemble them into a composite system by countable I, Q, and E bonds. The guidelines for depicting an organometallic piece are as follows.

Starting with a d/s/p AO basis, we generate the active metal AOs, which are subdivided into starred and unstarred hybrids. The starred AOs are never occupied by metal electrons. They are the "descendants" of canonical nonvalence AOs that have already played out their role in effecting deshielding in addition to yielding hybrids that are appropriately directed to support CT delocalization.

The choice of the hybridization scheme depends on the number and spatial arrangement of the ligands. For example, the ubiquitous $Fe(CO)_3$ group can appear in 3:3:3 or a 4:2:3 form of endo:equatorial:exo hybrids. By using the 3:3:3 hybridization scheme (see Chapter 25), Fe acts as a trisynaptic, two-electron fragment denoted 3h/2e. This is how $Fe(CO)_3$ acts in the vast majority of cases.

Two fragments are called isolobal if their valence AOs are of the same number and symmetry type. For example, the metallic $Fe(CO)_3$ and the nonmetallic HC^+ are isolobal. On the other hand, two fragments are called isosynaptic if they are isolobal and, in addition, they bind by the same delocalization mechanism, T or I. Two isolobal semimetallic or metallic fragments are also isosynaptic [e.g., $Fe(CO)_3$ and Sn]. The same is true of two isolobal (or, isoelectronic) nonmetallic fragments. On the other hand, one metallic and one nonmetallic fragment that are isolobal are not isosynaptic; for example, $Fe(CO)_3$ and HC^+ are isolobal but not isosynaptic. This exposes the very different physics reproduced by EHMO and ab initio polydeterminantal theories. The EHMO model says that $Fe(CO)_3$ is isolobal and isoelectronic to HC^+ and asserts that the former will bind as the latter fragment. By contrast, VB theory says that although the two fragments have similar sets of valence symmetry orbitals, these effect bonding in different ways. HC^+ binds by the T and $Fe(CO)_3$ by the I mechanism, always in a relative sense.

The most common types of organometallic fragments fall into three classes:

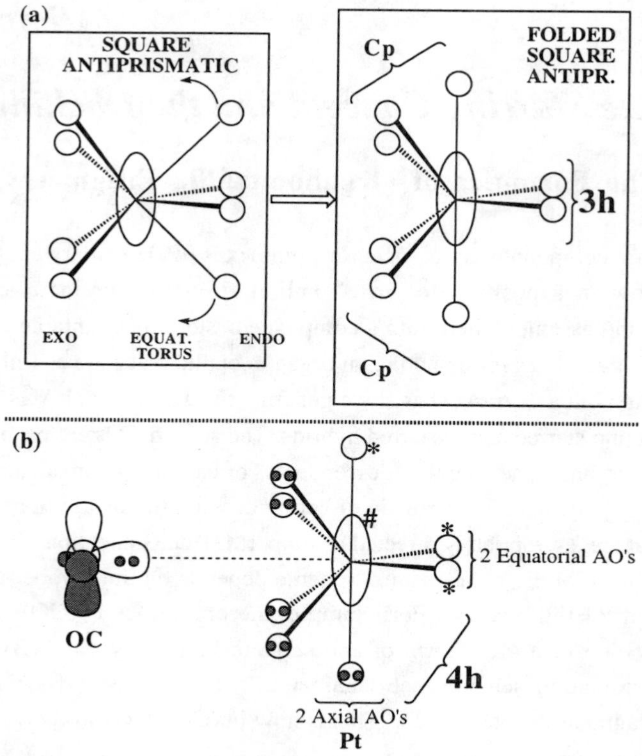

Figure 31.1 (a) The Cp_2M fragment. (b) The COPt fragment.

(a) The trisynaptic L_3M fragment involving a pseudo-octahedral metal and exemplified by $(CO)_3Fe$.

(b)The angular C_{2v} Cp_2M fragment shown in Figure 31.1a. This is produced by folding back two hybrids of the antiprismatic 4:1:4 form and it acts as a trisynaptic fragment with three in-plane hybrids. This fragment is part of ansa-metallocenes which are both theoretically and practically important, as the elegant works of Brintzinger and Kaminsky and their associates have demonstrated.[1a] The L_2CpM fragment of "piano stool" complexes is a derivative of the Cp_2M fragment.

(c) The simple LM fragment (L = PR_3 or CO and M = Ni, Pd, or Pt) shown in Figure 31.1b. This acts as a tetrasynaptic fragment in which the equatorial set of the two empty valence AOs sets the stage for in-plane coordination, while the axial set of one hole and one pair sets up for columnar association.

$NaReO(RC \equiv CR)_2$

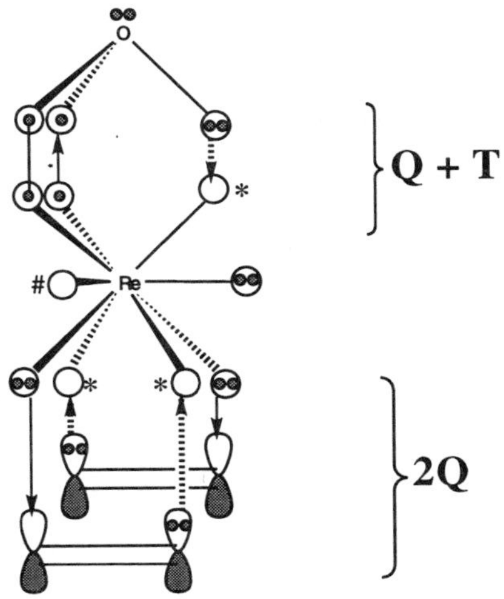

Figure 31.2 Coordination of anionic Re with O and two acetylenes.

The best approach to gaining familiarity with the VB method is by drawing explicit formulas for diverse organometallics. In this way, one comes to appreciate the versatility of the component organometallic fragments. Three examples are whimsically chosen from the literature.

We begin with the rhenium oxo complex $NaReO(RCCR)_2$, to illustrates the coexistence of "hard" nonmetallic oxygen and "soft" π ligands on the same metal atom. The molecule undergoes a variety of reactions,[1b] which can be rationalized by reference to the formula shown in Figure 31.2. Each acetylene is bound to the metal by a Q bond.

Our next case is the R_3P-Pd fragment, which acts like a 4h/2e fragment (see Figure 31.1b) that upon one-electron oxidation can combine with two ligands, each of which contributes two formal σ pairs, to produce the planar dibridged structure shown in Figure 31.3. Bridging is effected by using the starred equatorial holes of Pd. Solid arrows, indicating I bonds, terminate on starred AOs because the np AO activity tends to be valencelike toward the end of the transition series. The Pd—Pd bond is analogous to that in the prototypical Li—Li molecule. The linear R_3P—Pd—Pd—PR_3^{2+} fragment recurs in the literature with variable bridging ligands.[2a]

Finally, each CO in $Fe(CO)_3$ acts as 2h/2e. Replacing each CO by singlet O that also can act as 2h/2e generates the isoelectronic FeO_3, and replacing Fe by Cr yields the 3h/0e CrO_3 which is isosynaptic to Re^+O_3. The latter is present in $CpReO_3$.[2b,c] Isolobal organic pieces are interchangeable in

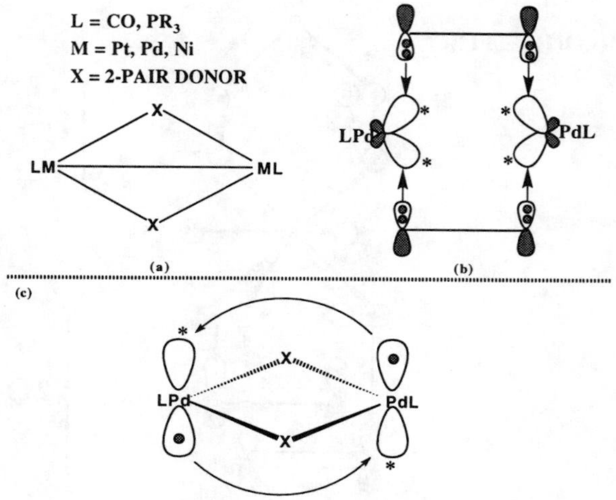

Figure 31.3 The electronic structure of $LPdX_2PdL$.

organometallic fragments, provided they belong to the same ligand category: CO and singlet O are both OX (π-acceptor) ligands.

A recent review[3a] of matrix isolation studies of $Fe(CO)_4$ defines interesting problems, which we can now confront with the aid of simple formulas. Using the octahedral hybridization scheme and the starring procedure and suppressing the six π-type d electrons, we can write three-dimensional formulas for singlet and triplet $Fe(CO)_4$ as follows:

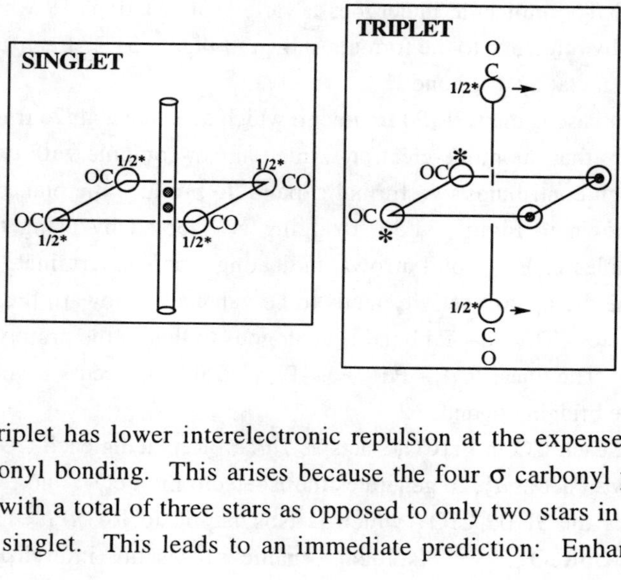

The C_{2v} triplet has lower interelectronic repulsion at the expense of weaker metal-carbonyl bonding. This arises because the four σ carbonyl pairs tie up four holes with a total of three stars as opposed to only two stars in the case of the planar singlet. This leads to an immediate prediction: Enhancement of

Table 31.1 Sequential Bond Dissociation Energies (kJ/mol) of Metal Carbonyls[3c]

M	$M(CO)_4$—CO	$M(CO)_3$—CO	$M(CO)_2$—CO	M(CO)—CO	M—CO
Fe	232 (6:5)	19 (4:4)	135 (4:3)	96 (3:2)	96 (3:1)
(EL:LG)[a]					
Co^+	75	75	82	152	174

[a]Ratio of unstarred holes and pairs (elements) to carbonyl groups (ligands)

bond-making by replacing Fe by Ru or Os (d orbital expansion) will stabilize the singlet relative to the triplet. The bending mode of the two axial carbonyls and the possibility of pseudorotation in the triplet form are obvious from mere inspection of the formulas. The difference between axial and equatorial Fe—C bond lengths (shorter axial because of lower star character of the axial holes), the difference between C_{2v} $Fe(CO)_4$ (axial π-acceptor CO bending toward two equatorial radicals) and C_{2v} SF_4 (axial π-donor F bending away from equatorial lone pair), and related stereochemical trends are now simple exercises of the VB theory of chemical bonding.

The planar singlet can combine with two electronegative atoms (e.g., two halogens) to form an octahedral complex in which the axial pair binds the two atoms by one REL bond analogous to that in F—PF_3—F. The C_{2v} triplet can mutate to a singlet and bind a closed-shell fragment by overlap dispersion. This is how the fifth ligand actually binds and this is why the first sequential bond dissociation energy (SBDE) of $Fe(CO)_5$ is the largest.[3b]

A key problem is defined by a recent important paper from Armentrout's group,[3c] which reports the SBDEs of $Fe(CO)_5$ and the isoelectronic $Co^+(CO)_5$. The data are reproduced in Table 31.1. There are two puzzling trends in going from $M(CO)_5$ to M(CO) for the metals represented in Table 31.1:

The trend is nonmonotonic in the neutral Fe series but monotonic in the ionic Co^+ case.

If $Fe(CO)_4$ is regarded as "anomalous" and excluded from consideration, the SBDEs decline in the Fe series but rise in the Co^+ series.

What is the physical meaning of these two trends?

We can attempt to understand the nonmonotonic behavior of the neutral $Fe(CO)_5$ complex by assuming octahedral hybridization and noting how many metal valence holes and pairs act to accommodate (overlap with) the equatorial and axial CO fragments. The unstarred holes and pairs are called *elements* (EL). The CO groups are ligands (LG). Assuming, for the moment, bond independence, the larger the EL:LG ratio of a given $Fe(CO)_n$ molecule, the stronger the SBDE. The computed ratios are shown in Table 31.1. We begin with $Fe(CO)_5$ in which two axial π-type pairs and one axial σ-type hole accommodate the two axial CO ligands, leaving one σ-type hole, one σ-type pair, and one π-type pair in the equatorial plane to accommodate the three equatorial CO ligands. The EL:LG ratio is 6:5. Since there are more axial than equatorial elements and more equatorial than axial ligands, the first SBDE reflects the departure of an equatorial CO, which produces triplet $Fe(CO)_4$ in which two high spin coupled odd electrons occupy two valence, unstarred equatorial AOs. This diminishes interelectronic repulsion at the expense of bond making: The EL:LG drops to 4:4 and, in addition, there is strong nonbonded repulsion of the four CO ligands. As a result, the second SBDE is very low. Note that in this triplet species, EL equals 6 (the total number of valence AOs) minus 2 (the two AOs accommodating the two odd electrons).

Dissociation of the second CO produces $Fe(CO)_3$. Assuming a triplet triangular species, the EL:LG ratio shoots up to 4:3 with concomitant reduction of nonbonded repulsion, and the third SBDE rises. From here on, the EL:LG ratio increases to 3:2 in triplet linear $Fe(CO)_2$ and to 3:1 as one electron pair goes into a δ-type AO, which cannot interact with the CO LUMO. According to the EHMO model, therefore, the SBDEs should rise beyond a minimum value at $Fe(CO)_4$. Instead, the SBDE first rises and then declines! In addition, the first SBDE with a 6:5 EL:LG ratio is far greater than the third, fourth, and even the fifth SBDEs, despite the greater EL:LG ratios of the latter three, reaching a value of 3:1 in Fe(CO)! What is going on?

The answer is that I-bonding is a cooperative phenomenon. One I bond is symbolized by two arrows. "Improving" one arrow, effectively "improves" the next arrow because the entire I bond becomes stronger. This means that attachment of one ligand promotes the binding of a second ligand, and so on, just as in the case of metal clusters, in which cohesive energy increases with cluster size. Thus, if promotional energy and nonbonded repulsion were to remain constant, the VB prediction would be "SBDEs of metal complexes will show progressive diminution"! The same trend is expected in either I- or T-bound molecules in which the SBDE trend is controlled by the promotional energy and, specifically, the loss of exchange correlation as high spin groups are (partly) converted into low spin ensembles of electrons for the purpose of bond making. For example, the SBDEs of ammonia are as follows:

NH_2—H	108 (kcal/mol)
NH—H	97 (kcal/mol)
N—H	75 (kcal/mol)

Interestingly, NF_3 displays the opposite trend as a result of the predominance of steric repulsion. The trend is restored in PH_3 *and* PF_3. The importance of loss of exchange correlation in determining BDEs has been stressed by Goddard and his school. This is one more example of VB theory revealing clearly what is obscure (though present) in MO theory.

The preceding analysis suggests a litmus test: Diminishing SBDEs, with whatever anomalies, are indicative of I-bonding in "metal molecules" and T-bonding in "nonmetal molecules." When neither T- nor I-bonding is operative, the trend is dictated by steric repulsion, and the SBDEs must increase. This type of situation is expected in E-bound molecules in which classical Coulombic attraction, which requires no promotion, makes a dominant contribution. As seen in Table 31.1, replacing Fe by Co^+ changes dramatically the trend from anomalous decline [$Fe(CO)_4$ is the exception] to steady ascent as we go from $M(CO)_5$ to M(CO).

Examination of the available SBDEs reported in the literature for a variety of metal complexes reveals that many fall in either of two categories:

Anomalous descending trend implying I-bonding, where "descending" means that a more ligated metal has a stronger metal–ligand bond than a less ligated metal.

Anomalous ascending trend implying E-bonding, where "ascending" means that a less ligated metal has a stronger metal–ligand bond than a more ligated metal.

For example, the trends (kcal/mol) of TiO_2 (144, 159) and ZrO_2 (142, 192) are ascending but those of PtO_2 (120, 94) and MoO_3 (145, 145, 134) are descending exactly as expected: Early electropositive transition metals are predisposed toward E-bonding, while middle and late transition metals are predisposed toward I-bonding, in a relative sense. Of course, there is a problem: Many of the thermochemical data have huge uncertainties. As with much of the rest of this work, we are at the beginning rather than at the end! However, the key point is that when promotional energy variations are small [e.g., when the $ns^2(n-1)d^m$ and the $ns^1(n-1)d^{m+1}$ configurations are quasi-degenerate], a counterintuitive descending trend, either unbroken or anomalous can only imply bonding cooperativity, that is, I-bonding. Furthermore, shifting from zerovalent metals to metals having a formal cationic charge to metal cations changes the mechanism of bonding from I to E. We will revisit this issue time and again.

31.2 The Docking Rule of Metal–Ligand Combination

The electronic structures of organometallic molecules can be derived by assembling metal and ligand pieces to achieve a perfect match. The metal pieces have (unstarred) pairs and/or dots as well as unstarred and/or starred holes. There can be no progress in understanding organometallic electronic structure unless we understand the *specificity* of metal–ligand interaction. In a typical situation, a transition metal makes available to a set of ligands six low energy hybrid orbitals; five are unstarred and one is daggered. In addition, there are three high energy starred hybrid orbitals. The key question is: What type of ligand docks on an unstarred and what kind of ligand docks on a starred metal AO if the starred holes have been exhausted? What determines the affinity of a ligand for a starred or an unstarred metal AO?

Two critical aspects of the electronic structure of a ligand decide its mode of combination with a metal. We begin by assuming temporarily the conventional oxidation state formalism, classifying ligands into closed-shell acid (OX), basic (RED) and acid/base (ROX) ligands. An OX ligand has a low-lying LUMO and a RED ligand has a high-lying HOMO, but both have large HOMO-LUMO gaps. By contrast, a ROX ligand has a high energy HOMO and a low energy LUMO and, thus, a small HOMO-LUMO gap. Here is now the first important point: Either OX or RED ligand acts with one AO while ROX acts with two AOs. Hence, the former two can form only one covalent bond (T bond) while the latter can form one I bond. Since metal–ligand bonding falls in the I domain of the mechanistic continuum, ROX ligands are intrinsically superior to either OX or RED ligands. This means that the metal ought to accommodate ROX ligands in advance of OX and RED ligands.

Our second important point is that every closed-shell ligand has a σ pair that must dock on a metal hole, and selectivity of docking depends on the mode of action of the ligand. This point has three ramifications.

(a) The binding action of the OX ligand is principally due to its LUMO hole matching metal pairs. The σ pair of an OX ligand does not care whether the metal hole on which it docks can sustain strong overlap interaction. It does not, that is, care whether the metal hole is starred or unstarred. Hence, OX ligands have high affinity for unstarred holes. CO is an example of an OX ligand (with ROX-character).

(b) The binding action of a ROX ligand is due to the *simultaneous* interaction of its σ HOMO with metal holes and its valence π LUMO with metal pairs (overlap dispersion). As a result, a ROX ligand has an overlap-dependent and an overlap-independent (dispersion) component. When the ROX ligand has a nonvalence LUMO, it forms a Q bond with the metal (see Chapter 28). When it has an effective valence LUMO, it forms an I bond, as discussed later. Hence, ROX ligands have high affinity for unstarred holes. H_2C and PH_3 are examples of ROX ligands.

(c) One might have guessed that the the σ pair of a RED ligand that is a strong proton base must dock on a metal σ hole to engender strong overlap interaction with it. However, since metals are linked to ligands by I bonds, proton basicity no longer qualifies a RED ligand for attachment on an unstarred hole because a covalent must be replaced by an I bond, and two CT arrows in the same overlap region are "forbidden." Hence, RED ligands, such as first-row acids (HF, H_2O, H_3N, H_4C, etc.) and their conjugate bases dock on starred holes and bind by the E mechanism. The only exception is the exceedingly strong base CH_3^- (and H^-). Unlike its isoelectronic first-row relatives, CH_3^- has a strong affinity for an unstarred hole; that is, it opts for T- rather than E-bonding because the resulting T bond between M and C is strong. In other words, the oxidation state formalism fails in the case of H_3C because this ligand acts as a radical rather than as a formal anion.

This classification leads to the *docking rule*, which is: OX and RED ligands dock on starred metal holes (if unstarred holes have been exhausted) and ROX ligands dock on unstarred metal holes. The exceptional cases of R^- and H^- are accommodated by the statement that radicals (like R and H) dock on singly occupied metal AOs, which are, by definition, unstarred. It should always be kept in mind that unstarred holes are tied up in advance of starred holes, which are used as docking sites only when unstarred holes have been exhausted. Thus, the docking rule can be reformulated as follows: ROX ligands have priority over OX and RED ligands for unstarred metal holes.

Here is now the key point: The docking rule is the product not only of a new formulation of hybridization (starring procedure) but, more importantly, of a new theory of chemical bonding, which replaces the covalent metal–ligand bond by the interstitial metal–ligand bond. According to our formulations, ROX ligands have higher priority relative to π acids (OX ligands) and σ bases (RED ligands) for an unstarred docking hole; that is, the concept of the I bond dictates

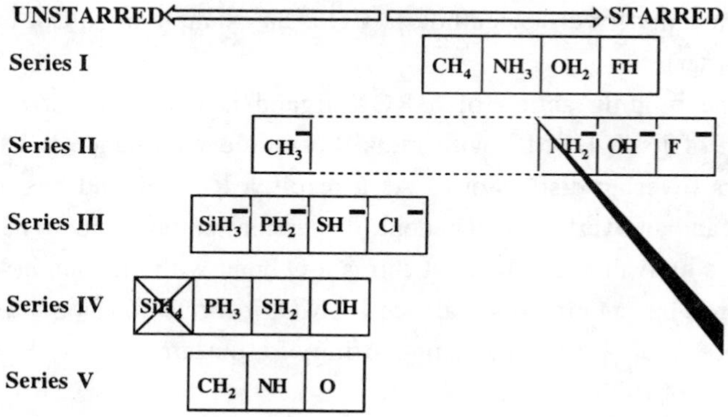

Scheme 31.1 The relative docking preference of typical ligands.

the concept of metal–ligand selectivity. When metal is replaced by nonmetal, the selectivity reverses because the mechanism of bonding changes from I to either E or T: RED NH_3 is a stronger proton base than ROX PH_3. In other words, RED has a greater affinity than ROX for the unstarred H1s AO of a nonmetal (T-bonding) and the positive charge of the proton (E-bonding). Note the remaining puzzle: Is H_3N—H a T bond or an E bond? If it is an E bond, a Mulliken population analysis must reveal an essentially intact N2s lone pair.

Scheme 31.1 presents a schematic differentiation of the action of prototypical ligand families according to the docking rule. Here are the essential points:

(a) In every series, the affinity for unstarred holes increases with proton basicity.

(b) Anionic series II (III) is displaced relative to neutral series I (IV) because anionic rely more than neutral ligands on the electrostatic mechanism of bonding. Hence, they have greater tolerance for an unstarred docking hole. Remember that because of the high energy of a starred AO, CT delocalization (hence I-bonding) from the ligand to the starred AO approaches zero. This is a tentative hypothesis because there are two opposing factors at work: F^- is a stronger base than HF but also more capable of E-bonding with a unipositive metal. The first attribute qualifies F^- for a stronger overlap interaction with the metal and predicts higher affinity for an unstarred hole. On the other hand, the second attribute suggests that F^- has a higher affinity for a starred hole. We hypothesize that the second factor dominates.

(c) Cl^- and PH_3 are fundamentally different from F^- and NH_3. According to conventional models, all four are classified as σ donors. According to VB theory, however, the first two are ROX and the latter two are RED ligands.

(d) The affinity of an unstarred holes in series III and IV increases in going from right to left of the row, not only because of increasing proton basicity but also because of a change from a nonvalence to an effectively valence LUMO. To see this point, compare the M—Cl to the M—PH_3 I bond. As illustrated, Cl has only a nonvalence LUMO simulated by a 3d AO, while PH_3 can bind the metal d pair as part of relay bonding that includes the "innocent" H atom. As a result, phosphine can use the valence space to make an I bond with the metal but Cl can make only an inferior Q bond with the metal.

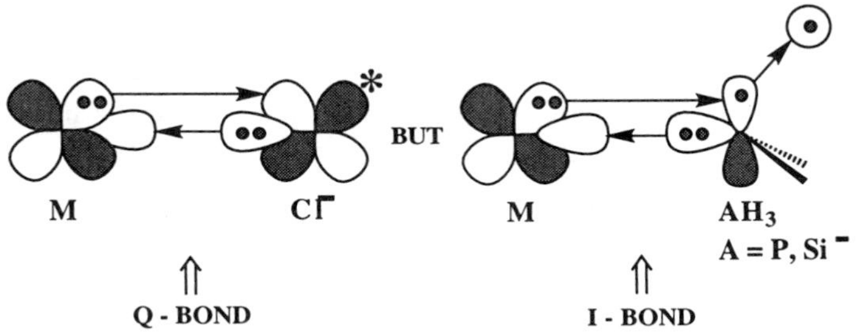

PH_3 is superior to HCl as a ligand because its one σ pair and two π-type P—H bonds (that effectively act as two π-type LUMOs) render PH_3 this an ideal 3h/2e fragment capable of making a formal triple bond (actually 1I + 1T bond) with a metal.

The docking rule is based on an implicit understanding that the mode of action of a ligand is actually dependent on the nature of the metal. For example, CO is an OX ligand with respect to a neutral metal, but it can become either a ROX ligand (with affinity for an unstarred hole) or even a RED ligand (with affinity for a starred hole) with respect to a metal in modest or high oxidation state. How do we decide which is the better option? Since RED action implies E-bonding while ROX action implies I-bonding (overlap dispersion), and since only the former depends (in part) on an interaction that does not require fragment promotion (the classical Coulomb interaction component of E-bonding), the prediction is clear: the more promotable the formal metal ion, the more favorable the ROX mode. Since promotability ("polarizability") is linked to the ion ionization potential, a rule follows: As the IP of a formal or actual metal ion increases, the mechanism of metal–ligand bonding changes from ROX (I-bonding) to RED (E-bonding). It turns out that the IP of transition metal monocations increases monotonically along a transition series, with exceptional depressions occurring when ionization leaves behind a half- or fully occupied d shell (e.g., Mn^+ to Mn^{2+} and Zn^+ to Zn^{2+}). The same considerations apply to oxygen atom and to every ligand that is capable of acting as π-acid/σ-base combination. Thus, the docking rule makes different predictions regarding the affinity of ligands for unstarred and starred holes, as well as the mechanism of bonding, depending on the oxidation state of the metal.

While the classification of most conventional ligands of organometallic chemistry may be intuitively apparent, the mode of action of π-conjugated ligands is not obvious: What metal holes are appropriate matches for the three π pairs of cyclopentadienyl anion? To find the answer, we determine the number of high energy and low energy π-occupied MOs of Cp^- and assume that a corresponding number of ligand pairs will dock on unstarred and starred holes, respectively. For example, Cp^- has two high-lying (degenerate) MOs and one low-lying occupied MO. This is taken to mean that two of the three pairs of Cp^- will dock on unstarred holes and one will dock on a starred hole.

Pseudo-octahedral Cr can bind six ligands via three exo starred and three endo unstarred holes, while placing three pairs in the three unstarred equatorial hybrids. The starred σ AOs of an octahedral or pseudo-octahedral set of hybrids are all cis-disposed. As a result, ideal ligation of pseudo-octahedral Cr requires three *cis* OX ligands docking at the three exo starred holes and three *cis* ROX ligands docking at the three endo unstarred holes. In this way, we reap an extra benefit: A ROX, acting as RED, and an OX ligand combine to form a REL bond involving the array ROX(RED)—M—OX. Increasing the basicity of the ROX ligand and the acidity of the OX ligand is expected to enhance the strength of the REL bond. The selectivity of ligand attachment is illustrated as follows:

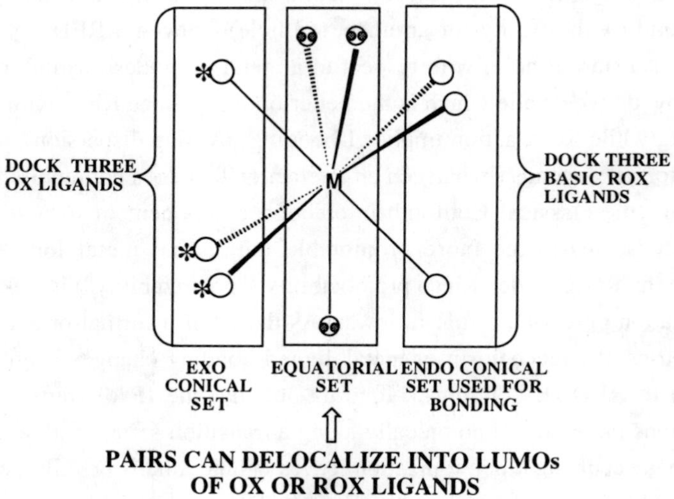

The concept of the REL bond in conjunction with the docking rule rationalizes cleanly the following pieces of data given (HMB = hexamethylbenzene):

$$Cr(CO)_6 + Cr(HMB)_2 \longrightarrow 2(CO)_3Cr(HMB) \qquad \Delta E = -12.0 \text{ kcal/mol}$$

$Cr(CO)_6$ readily loses three carbonyls to form the facial $L_3Cr(CO)_3$ cluster ($L = PR_3$); that is, the reaction is exothermic.

$$CrL_6 + Cr(CO)_6 \longrightarrow 2\textit{fac}\text{-}L_3Cr(CO)_3$$

The fac isomer is more stable than the mer isomer of $L_3Cr(CO)_3$ because only the former accommodates all three π acceptor (OX) carbonyls on cis-disposed starred hybrids and all three σ-donor/π-acceptor phosphines on cis-disposed unstarred hybrids of the octahedral AO ensemble. A trans orientation of carbonyls or phosphines occurs only when steric effects come into play (e.g., when the phosphine R groups are bulky).

Since PR_3 is isolobal to CO, why are there not stable phosphine analogues of the known metal–carbonyl clusters? Our theory, with the docking rule as the specific underpinning, dispenses of the problem very briefly:

$Fe(CO)_3$ is apt to form clusters because the carbonyls dock on the starred holes, leaving the low-energy unstarred holes available for cluster bonding.

By contrast, $Fe(PR_3)_3$ fails to form stable clusters because the phosphines, which are ROX ligands, dock on unstarred holes, leaving the high energy starred holes available for cluster bonding.

As a result, the inherent stability of transition metal–phosphine clusters is low.

A nice illustration of the docking rule is provided by the work of Armentrout and co-workers[4a] and Strauss and collaborators.[4b] The first group found that an H_2O slightly strengthens while CO strongly weakens an $Fe—H^+$ bond. We can explain this observation by assuming that CO acts as an OX ligand with significant ROX character, in which case the electronic structures of $COFe^+$ and H_2OFe^+ are those shown below. The key difference is that CO must dock on an unstarred hybrid because it acts (partly) as a ROX ligand when attached on a metal ion. By contrast, H_2O always docks on an unstarred orbital. It follows that attachment of a hydrogen atom to the metal ion will produce an angular $COFe—H^+$ but a linear $H_2OFe—H^+$. Avoidance of the strong nonbonded repulsion in the former can be achieved only by docking CO on a starred hybrid to attain a linear geometry. Remember that starred AOs can never accommodate metal electrons. Either choice (i.e., angular geometry obeying the docking rule or linear geometry disobeying it) is bad, and this is why the metal–hydrogen bond dissociation energy of COFe—H^+ is lower by roughly by 1 eV (!) than that of H_2OFe—H^+. By the way, the predicted electronic structures of the two complexes are exactly those determined calculationally by Bauschlicher and co-workers upon translating from MO to VB language.[4c]

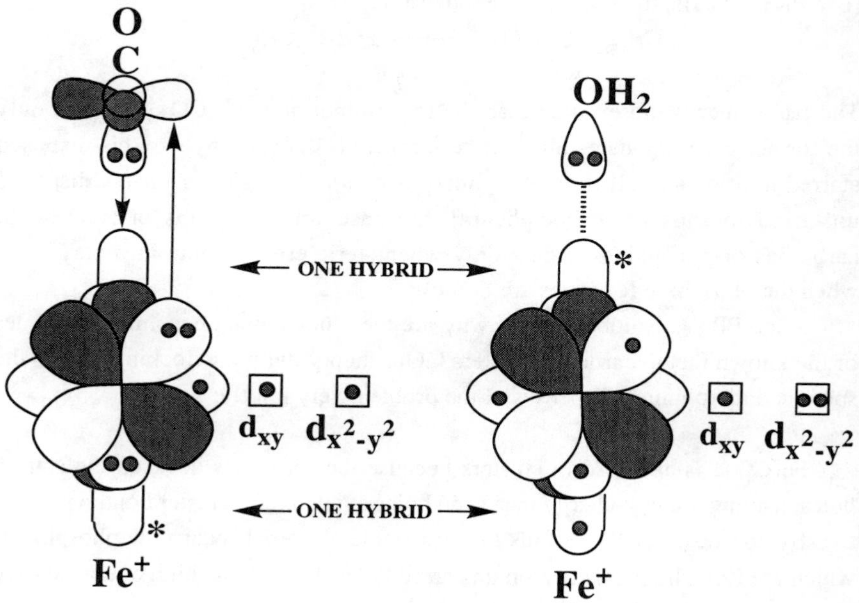

Changing from Fe^+ to Ag^+ (IP = 16.2 and 21.5, respectively), is expected to change the mode of action of CO from ROX to RED, and this result has been reported.[4b] Because of the absence of competing ligands, each CO in $Ag(CO)^+$ and $Ag(CO)_2^+$ can actually dock on one unstarred axial hole, but the binding mechanism remains E-bonding, as indicated by vibrational spectroscopic data. The key points are that the mode of action of a ligand depends on the electronic structure of the metal and that when attached to a metal, CO can embrace any one of the three modes of bonding.

Finally, the mere recognition that groups, G, like H_2C, HN, and O (and their heavier congeners) are ROX ligands with high affinity for unstarred AOs not only is consistent with our analysis of the bonding of polyphenyl derivatives of phosphorus and congeners (see Chapter 24) but also suggests new synthetic possibilities. Specifically, all we have to do to get a formal M═G complex is to mix metal salts of the type MX_n (X = halogen) with alkenes and arenes, which are substituted to further promote a "carbenic" valence state, and hope for a good yield (i.e., hope that alternate coordination paths have higher activation energies). A typical reaction involving a ROX ligand gives an illustration of our thinking:

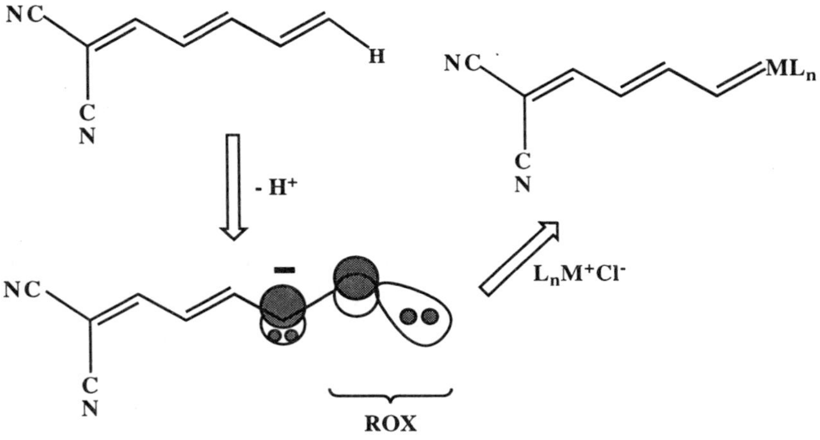

The I-bond concept says that ROX ligands are best suited for I-bonding. This means that M═G molecules will turn out to be ubiquitous. A recent commentary on porphyrin research[4d] highlights the discovery that N-confused porphyrins ionize by proton loss from the interior sp^2 carbon to form complexes with nickel. The reporter noted, correctly, the absence of a published rationale for this surprising propensity to form metal–carbon bonds. In this context, the analysis of the bonding of $BiPh_5$ in Chapter 24 belongs to dreamland. And yet, all these unexpected trends are not accidental but a consequence of the basic physics of metal bonding: I-bonding.

The game of chemical bonding has just begun. Only now, theory has direct and immediate operational consequences: for example, presenting specific recipes for the discovery of new molecules. Thus, the organic molecules that can form metal complexes by acting as "carbenic" ROX fragments after heterolytic cleavage of a C—H or a C—Cl bond are shown below. Obviously, instead of attaching AC (DO) groups on the appropriate carbon, we can also replace the latter by an electronegative (electropositive) heteroatom.

METAL-ALKYLIDENE/CARBENE PRECURSORS

(AC = π Acceptor, DO = π Donor)

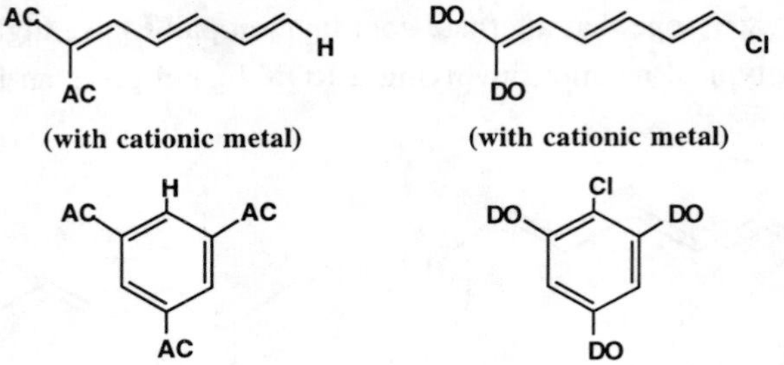

We have broken the chemical code but the decoding of chemical information will not be completed until many scientists have worked for many years. The numerous applications given here merely scratch the surface of a gigantic problem. To see this clearly, consider that the organometallic chemist thinks of, say, CO as a π-acid/σ-base (i.e., ROX) ligand that can enter into "synergic" bonding as defined by the Dewar–Chatt–Duncanson model. By contrast, we say that CO is an OX ligand with ROX character. A better example of a ROX ligand is singlet carbene, but this is only one of a number of different types of ROX ligand. In other words, ROX ligands are atoms or fragments that can be attached to a metal via two arrows: one ingoing, the other outgoing. Thus, ROX ligands can be closed-shell or open-shell species as follows:

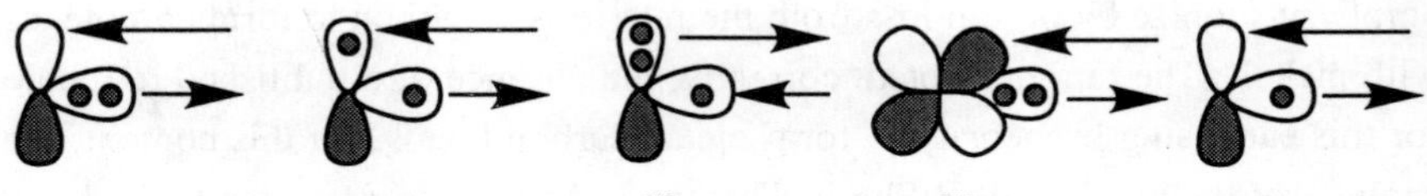

F_2C H_2C Cl Cl^- Na

Consider now a decision regarding the mode of binding of chlorine on metal. One alternative is to use Cl as a valence ROX ligand and bind by the I-mechanism. A second option is to oxidize the metal and bind the two ions by the E mechanism. Still a third option is to combine the two ions by a Q bond by using the chloride anion as a ROX ligand acting with one valence pair and one d hole simulating the low-lying nonvalence unoccupied space. What is best depends not only on the metal but also on the rest of the ligands with which Cl can cooperate to establish an I bond or a Q bond. Our discussions assume that Cl opts for the third mode; we realize, however, that situations in which it acts in either the first or third mode can be engineered. Clearly, the game of chemical bonding is just beginning.

31.3 The trans Influence

The "trans influence" and the "trans effect" occupy a central place in theoretical inorganic and organometallic chemistry. The puzzlement as to the origin of these phenomena is evident in monographs and papers. The phenomenon of "trans influence" has multiple mechanistic and synthetic implications, and yet there is no theory that explains satisfactorily the known facts.[5a] This state of affairs is symptomatic of the absence of a unified theory of chemical bonding, and this is nowhere more evident than in organometallic chemistry, where covalency is merely an illusion.

Consider the cis and trans geometric forms of the complex PtX_2L_2 (X = Cl, Br, I; L = PR_3):

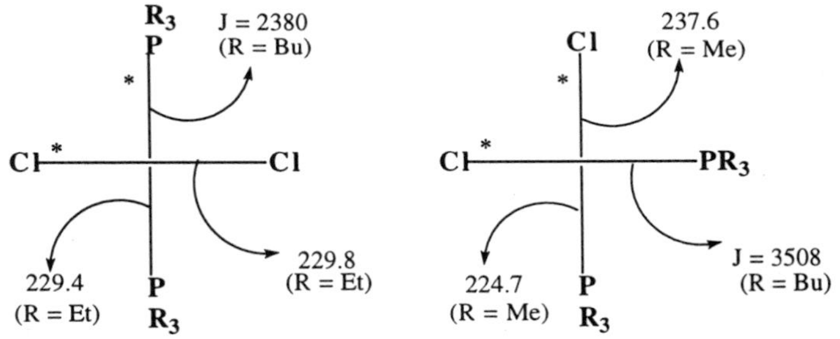

There are three interconnected observations:

The cis isomer is more stable.

The Pt—Cl bond distance is longer in the cis isomer.

The *J*(Pt—P) NMR coupling constant (in Hz), which measures the s component of the Pt hybrid AO extended to the phosphine, is greater in the cis case.

All these data[5a] taken together are interpreted to mean that PR_3 is a superior ROX ligand compared to Cl^- exactly as anticipated in Scheme 31.1. As a result, PR_3 ties up the s-rich unstarred σ hybrids forcing Cl^- to use the p-rich starred hybrids. Consequently, the cis isomer has higher s character in the Pt—P bond and greater p character in the Pt—Cl bond. Since the binding ability of a (more electronegative) s-rich AO is superior to that of a (more electropositive) p-rich AO, the Pt—P bonds are shorter and the Pt—Cl bonds are longer in the cis isomer. A rule follows: *A transition metal directs selectively s character toward superior ROX ligands and p character towards RED, OX, or inferior ROX ligands.*

Everything said above pertains to the *trans influence*, which is defined as the effect of L on the trans metal–ligand bond length and vibrational frequency. According to structural data, Purcell and Kotz[6] cite the following order within the framework of oxidation state formalism:

$$R^- \approx H^- > \text{carbene} \approx PR_3 > CO, RNC, C{=}C, Cl^-, NH_3$$

First, note that the formal R^- and H^- are at the top! This is because H and R combine with the metal as radicals that tie up singly occupied unstarred metal AOs. Note also that the ROX carbene and PR_3 are also high on the list. Each has a σ-donor/π-acceptor duality. On the other hand, the trans effect is the effect of a coordinated group L on the rate of substitution of a group Y trans to L.[5b] For the particular case of Pt(II) complexes, the labilizing effect of ligands is in the following order:

$$CN^-, CO, \text{olefins} > \text{phosphines, arsines}, H^- > CH_3^-$$
$$> NO_2^-, I^-, SCN^- > Br^-, Cl^- > \text{amines}, NH_3, OH^-, H_2O$$

It is immediately obvious that groups with low-lying LUMOs rise in the hierarchy as we move from the trans influence to the trans effect. We suggest that the classical interpretation is correct: In the nucleophilic substitution of square planar transition metal complexes, the incoming nucleophile faces exchange repulsion by metal d-π lone pairs on the way to forming a pentacoordinate transition state or intermediate. The presence of ligands with low-lying π LUMOs allows the metal to respond by enhancing metal to ligand CT delocalization. As a result, the incoming nucleophile is spared of exchange repulsion. Thus, the increased prominence of CN^-, CO, and olefin is a signal that these groups act primarily as OX ligands.

VB theory resolves the controversies surrounding the problem of the trans influence and trans-effect as follows:

(a) The existing "π-only" and "σ-only" theories are one-electron constructs based on the assumption that the sharing of one metal π-type pair and one σ hole by two trans ligands is unfavorable. This is not true because two trans ROX ligands can still be bound by two I bonds (four arrows). When the two ligands are different, the I bond connecting the dominant ligand to the metal is stronger than the I bond linking the inferior ligand to the metal because the dominant ligand docks on the σ hole of lower star character.

(b) The concept of the REL bond predicts that in a (CO)ML array where L is, at least, a σ-donor RED ligand, increasing the π-acceptor action of CO increases the σ donation (i.e., basicity of L), and increasing the basicity of L increases the π-acceptor action of CO. This means that, when CO becomes a better π acceptor or L a better σ donor, the CO vibrational frequency, ν_{CO}, *decreases* as the MC vibrational frequency, ν_{MC}, *increases.* Furthermore, this correlation must be universal (because it is a consequence of 2-e CT), and it should be present irrespective of the nature of the L ligand. In other words, we expect three linear plots on the ν_{CO}/ν_{MC} plane for the three situations shown in Figure 31.4. The important thing is that the lines should be parallel or nearly so; that is, "adding" or "subtracting" orbitals from L can never annihilate the 2-e CT hop, which exists even when L disappears. The best evidence for this type of behavior has been provided by Spiro and co-workers, who studied the axial "OC—Fe(II)—L" subsystem of diverse iron–porphyrins.[7] The situation depicted by Figure 31.4a obtains with L = neutral imidazole; Figure 31.4b shows the situation with five-coordinate iron–carbonyl complexes; and in Figure 31.4c, L = anionic thiolate. That *increasing* the basicity of L *lowers* the ν_{CO} frequency of a trans carbonyl was observed long ago by Birgogne in $Ni(CO)_3X$ and

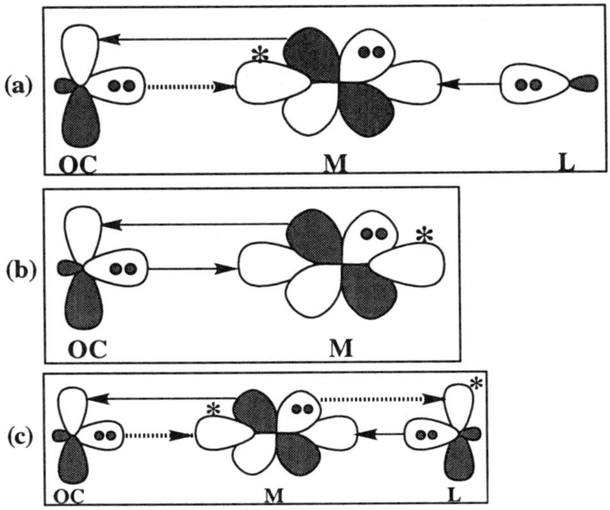

Figure 31.4 The common denominator of the three complexes is that π donation requires σ acceptance (solid arrows). This corresponds to the formation of one REL bond.

$Ni(CO)_2X_2$ complexes[8] and by Angelici in *cis*-$Mn(CO)_3L_2Br$.[9] This propelled the "σ-only" challenge to the "π-only" theory.

(c) Structural data reveal that $(PhO)_3P$ competes better than Ph_3P with a trans carbonyl in $Cr(CO)_5L$,[10] and this has been taken as support of the "π-only" model. However, according to our interpretation, this can be taken to mean that $(PhO)_3P$ is superior either as a ROX or as an OX ligand. The stability of $Ni(PF_3)_4$ and $Ni(CO)_4$ while only $Pt(PF_3)_4$, but *not* $Pt(CO)_4$, is known suggests that the key difference between PF_3 and CO is not the π-OX ability but, rather, the σ/π-interdependent ROX ability. By analogy, we conclude that the superiority of $(PhO)_3P$ over Ph_3P in the competition with CO is due to difference in ROX delocalization ability. The π-acceptor ability of X_3P is expected to decrease as X electronegativity increases, while the P3s lone pair remains relatively unperturbed. Thus, the ability of X_3P to support ROX delocalization is expected to increase in the same direction.

(d) In several octahedral complexes, the phosphine trans to Cl has a shorter bond than either of two trans phosphines. Furthermore, this difference remains essentially constant for a variety of metal oxidation states.[11] This is inconsistent with a "π-only" theory. Our explanation is that these structural trends are predominantly consequences of ROX, rather than OX, delocalization, which is tolerant to changes in oxidation state because it neither oxidizes nor reduces the metal. This suggests a general diagnostic procedure: Binding that remains unaffected by metal oxidation is ROX- or REL-binding.

An ambident ligand can bind to a metal by either its "soft" (e.g., sulfur) or "hard" (e.g., nitrogen) end. Typically, the former acts with a σ pair and a π hole (ROX ligand) while the latter acts with only a σ pair (RED ligand). In addition, the "hard" center is negatively charged. Presented with a central metal having a σ hole and a π pair as well as a trans ligand that is either "soft" or "hard," the choices are as follows:

When the central metal is in a high oxidation state, E-bonding dominates and the ambident ligand binds with its "hard" end.

In all other cases, and assuming that steric effects do not act prohibitively, the ambident ligand binds with its "soft" end irrespective of the nature of the trans ligand simply because this is the combination that maximizes the number of I bonds.

A typical ambidentate nucleophile is SCN^-, in which the N end acts as σ donor and the S end as σ donor/π acceptor. It has been shown that, in square planar complexes of the type V—ML_2—K (where V is a variable ligand, K is the ambident ligand trans to V, and L are constant ligands), the ambident nucleophile NCS^- shows increased preference for binding with the sulfur end as the metallic character of V increases (e.g., in going from Me_3P to Me_3As to Me_3Sb).[12] We interpret this to mean that there is a cooperative interaction of two trans donor–aceptor ligands that increases as the "metallicity" of V increases. This cooperativity is required by the REL bond concept. The more V resembles a metal (i.e., the greater its capacity to form I bonds by CT delocalization), the greater the disposition of the ambident ligand to act via the semimetallic (rather than the nonmetallic) end in response to V. In turn, this tendency increasingly outweighs the steric effect, which works against the larger semimetal atom. The important point is that if steric effects were nonexistent, S would always bind in preference to N to a low valent metal regardless of what the trans ligand is. This is borne out by the following (seemingly puzzling) experimental results: SCN^- binds with the sulfur end when V = Me_3Sb (two I bonds) as well as when V = Me_3N (1.5I bonds).[12] Binding with the nitrogen end would engender 0.5I bond less in each case because of the absence of an active amine LUMO.

"An uneasy truce now prevails in the area with those supporting and those opposing π-bonding each occupying certain undisputed territories, but with a large zone of no-man's land between them." This comment is taken from the excellent overview of the trans-effect contained in the monograph *The Chemistry of Phosphorus* by Emsley and Hall.[13] The uncertainty over what the trans effect really signifies has the same origin as the uncertainty with respect to the "hard–soft" concept, the uncertainty about "d-orbital participation," the uncertainty regarding the electronic structure of a metal dimer, and so on. All these uncertainties are manifestations of one and only one thing: the nonexistence of an electronic theory of bonding across the periodic table.

31.4 The Alkylidene/Carbene Distinction

In 1986, Carter and Goddard published a paper[14a] that is important in three ways:

(a) It illustrates the inadequacy of the Hartree–Fock model to deal with metal-containing systems.

(b) Still more important, it implies that there exists another form of bonding that is superior to perfect pairing because one needs to go beyond perfect pairing (PP) VB to get the bond energy right. In fact, PP-GVB is already ahead of nonorthogonal PP-VB because the former contains some CT delocalization.

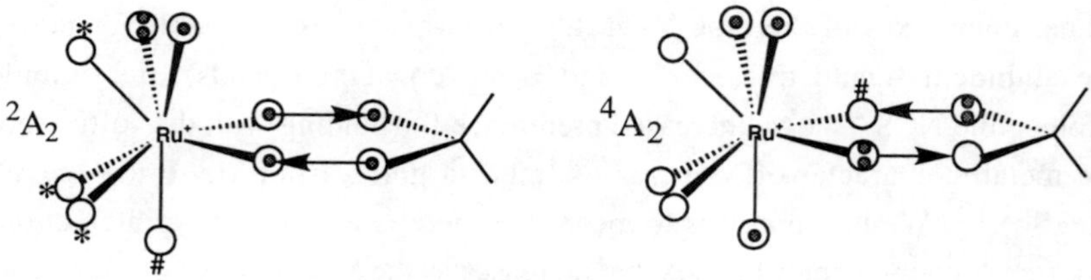

Figure 31.5 Two spin isomers differ in the reference configuration because they are also dagger isomers.

(c) It shows that the $^2A_2$ ground state of $RuCH_2^+$ has conventional double bonding ("metal methylidene"), while a cluster of three degenerate excited states ($^4A_2$, $^4B_1$, $^4B_2$), lying about 13.0 kcal/mol higher, has σ-donor/π-acceptor ("metal–carbene") bonding. This says that the mechanism of bonding is dependent on spin multiplicity.

Keeping in mind that a dagger can go to either the equatorial or the conical set of hybrids of a pseudo-octahedral metal, the explanation of the Carter–Goddard results is given schematically in Figure 31.5. Note that in both species, we have a single I bond; the difference is that the ground configuration has been changed. Converting the metal from doublet to quadruplet renders relocates a dagger from an equatorial to an endo AO because each equatorial AO must now be occupied by a single electron and a daggered AO cannot be occupied in the ground configuration. We call this the "chase the dagger" game.

A second illustration is the mode of bonding of Cp_2M, where M is W or Re^+, to a methylene group to form a "metal carbene." Assuming that M is pseudo-octahedrally hybridized, the linear form of Cp_2M has each Cp tying up one set of three conical AOs of the "staggered" metal (see Figure 25.8), with the remaining four electrons going to the triangular equatorial set. On the other hand, the bent form has the two Cp groups affixed on six AOs, four conical and two equatorial, of the "all-eclipsed" metal, leaving three AOs arranged in the shape of a trident to accommodate the four electrons. Thus, the first point is that a linear-to-bent transformation of a metallocene is unfavorable largely because it amounts to a rearrangement of nonbonding electrons that enhances their mutual interelectronic repulsion,[14b] as illustrated.

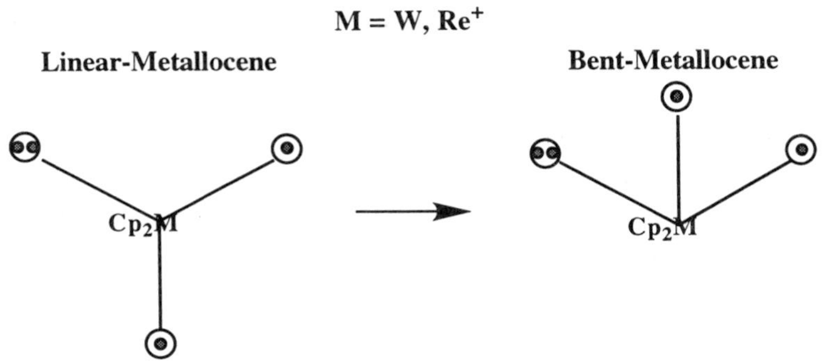

Consider now the problem of allocating three stars and, if the ns has significantly higher one-electron energy than the (n – 1)d, one dagger to the nine AOs tying up the two Cp fragments and the CH_2 in $Cp_2Re^+CH_2$, a molecule made by Heinekey and co-workers.[14c] In linear Cp_2Re^+, the three stars go to the axially oriented Cp groups and we can assume that the situation remains unaltered in going to bent $Cp_2Re^+CH_2$. On the other hand, the dagger is subject to the condition that it is not occupied in the reference configuration, and it can go to either the conical or the equatorial set of the pseudo-octahedral metal. This means that in the case of $Cp_2Re^+CH_2$, there exist two possibilities: the trident binding the CH_2 either has no dagger or has a dagger in either the central or one terminal AO. The first type of binding predicts the aklylidene structure I, and the second the carbene structure II:

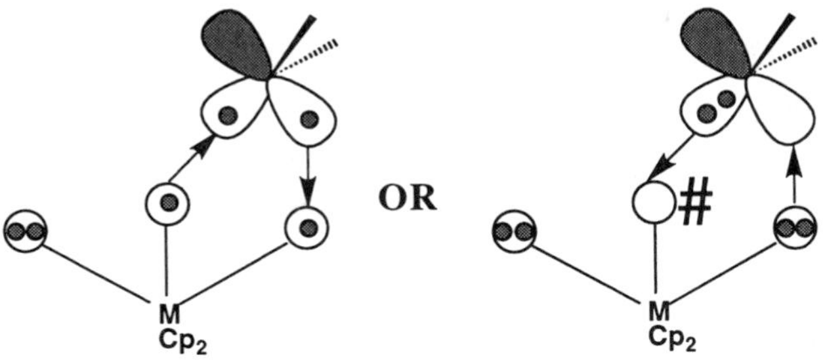

I II

The most likely scenario is that the cationic rhenium complex is analogous to the tantalum Schrock alkylidene (replace lone pair by alkyl group), with the two having structure I with the polarity of the formal double bond reversed. We expect structure I to yield nucleophilic or electrophilic addition to the formal M═C "double bond," depending on polarity. On the other hand, we expect structure II to lead to methylene transfer. The challenge is to transform the cationic rhenium complex from alkylidene to carbene. Every perturbation that enhances metal–Cp bonding will tend to shift the preference toward structure II, and vice versa. In other words, the higher the binding ability of the Cp, the greater the utilization of the low energy unstarred metal AOs for binding the Cp, and the greater the dagger character of the metal AO binding the methylene. Since "dagger character" means "ns character," we can use NMR coupling constants to probe the s character of the hybrids directed to the Cp fragments and the methylene. In short, the M—Cp binding strength must correlate with the electronic structure and, thus, the type of reactivity of the methylene fragment. Of course, the conventional steric effect continues to muddy the picture.

We have argued that the alkylidene/carbene nature of ligands like O, NH, and CH_2 is determined by the dagger character (or star character) of the σ-type AO of the metal extended to the ligand and that carbenic character implies reactivity of the atom transfer type. What is the relative dagger affinity of these three prototypical ligands? The answer is given in Scheme 31.1: All three groups, being ROX ligands, have a high affinity for an unstarred hole. However, on account of relative basicity, methylene has the lower and oxygen the higher dagger affinity. This means that the *relative preference* of carbenic-type bonding increases in the order O > NH > CH_2. In other words, when CH_2 and O are placed trans on a reducible (high valent or cationic) metal center competing for the starred or daggered σ AO, the oxygen "wins." Oxygen, that is, has the highest predisposition to bind as a "carbenic" (oxenic) fragment among isoelectronic first-row relatives. This explains the counterintuitive tendency of formal M=O bonds to transfer oxygen to phosphines (by formal nucleophilic attack) to form phosphine oxides. One example is intermolecular oxygen transfer to phosphine by $(Bipy)_2Ru(O)PEt_3^{2+}$.[14d] Each Bipy can be formalized as a "double NH." In this light, the complex has one O and one NH trans in an octahedral environment, and oxygen becomes transferred. Of course, our formulation of Bipy as "double NH" is not a very good one because of its aromaticity makes it unreactive. However, the very fact that oxygen is transferred implies oxenic character which, in turn, is explained by the presence of a trans ligand with an affinity for an unstarred hole substantial enough to rival that of oxygen atom.

Unfortunately, the competition of formal divalent groups for the star or dagger cannot be easily seen because all these groups have high affinity for an unstarred hole, thus forcing other ligands to starred holes. Any two of O, NH, CH_2 will end up cis, rather than trans, in an octahedral complex so that they do not have to share a star. For example, Bryan and Mayer prepared a *cis*-oxo alkylidene tungsten complex that permits each of the two functionalities to dock on an unstarred σ AO in preference to a trans halide.[14e] In such a case, nucleophilic attack by phosphine is expected to occur at carbon rather than oxygen because both bind in the same "alkylidene" fashion and carbon is less electron rich than oxygen.

Let us now illustrate the hoped-for incremental approach to understanding chemical bonding. Miller and Grubbs developed the ruthenium complex $Cl_2L_2Ru{=}CHR$ (L = PCy_3) which catalyzes ring-closing metathesis reactions.[14f] This fact argues in favor of an alkylidene formulation. By replacing Cl by R_3Si (i.e., by increasing the affinity of the observer ligands for an unstarred hole: see Scheme 31.1), we chase the dagger to the metal AOs binding the divalent carbon. This is expected to change CHR to the carbenic form and transform the complex to a cyclopropanation reagent. Whether the desired limit is reached must be determined by experiment.

31.5 The Metal–H_2 Valence Isomers Are Star Isomers

The combination of singlet carbene with ground H_2 to yield methane is the prototypical case of four-electron *valence isomerism*. The situation changes when we cross into the organometallic domain: Kubas has provided evidence for the existence of two bound valence isomeric forms of $W(CO)_3(PR_3)_2H_2$, the dihydrogen and the dihydride form, which exist in equilibrium.[15] This raises two questions:

(a) Looking at the diagonal of the $W(CO)_3(PR_3)_2H_2$ octahedral complex, which contains the H_2 ligand, we have the prototypical trans array of OC—W—(H_2) ("dihydrogen"), which can exist in equilibrium with OC—W—H_2 ("dihydride"). By contrast, there is only a single isomer of OC—C—H_2, namely, $O{=}C{=}CH_2$. What is the difference between "organic" and "organometallic" chemistry?

(b) The dihydrogen–dihydride equilibrium is established in both neutral and cationic complexes, effectively eliminating the argument that the dihydride requires strong π donation from the metal to the σ-antibonding MO of H_2. Thus, the inadequacy of a "π-only" model shows up again, as in the case of the trans influence. Why is this so?

A popular interpretation — that the two valence forms are simply resonance hybrids with different resonance structure eigenvectors — is flawed by the failure to differentiate between resonance structures and electronic states. Two different minima represent two different ways of binding by a codirectional arrow train that is directed either toward H_2 ("dihydride") or away from H_2 ("dihydrogen"). Because the direction of the arrow train is star dependent, the dihydrogen and dihydride isomers are actually *star isomers*:

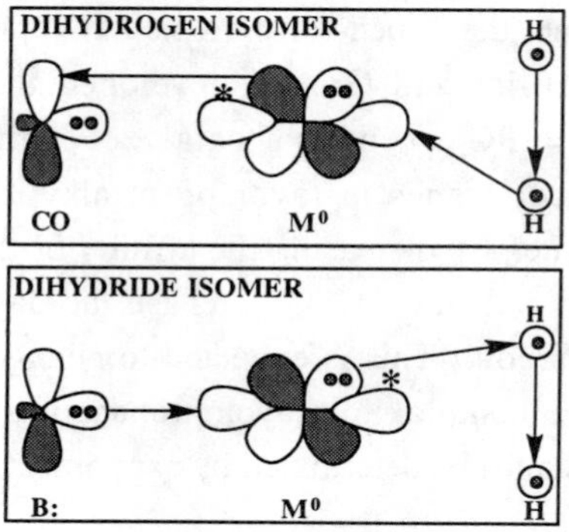

The wavefunctions come close to the intuition of the organometallic chemist in the sense that a trans π acid is predicted to stabilize the dihydrogen, while a trans σ base is expected to stabilize the dihydride form. On the other hand, the way in which H—H bonding becomes part of the metal–H_2 bonding is clearly revealed by the arrows. Furthermore, the REL bond (or, more precisely, the 1.5 REL bond: three codirectional arrows) connecting the two trans ligands via the metal conserves the charge of the latter. Thus, we have an arrow train that features one arrow entering and a second arrow exiting the metal center, leading to charge conservation on the metal. As a result, oxidation of the metal can have only a secondary effect because both "dihydride" and "dihydrogen" are bound by a REL bond that is independent (to a first approximation) of metal oxidation state. This is one of the (many) counterintuitive consequences of I-bonding.

The next key point is that star allocation limits the number of observable valence isomer equilibria. Unfortunately, it is not easy to differentiate between degenerate isomerization of two (or more) unsymmetrical valence isomers from a single symmetrical valence isomer. However, in a recent report, Hartwig and de Gala managed to distinguish between the two situations by using NMR isotopic perturbation methods.[16] Specifically, they were able to show that the there exists a symmetrical form and two interconverting unsymmetrical forms of $Cp_2NbH(BR_2)H$. Recalling that each Cp ties up one star, we explain the results as shown in Figure 31.6. The two forms are star isomers (i.e., they differ in the allocation of the single star).

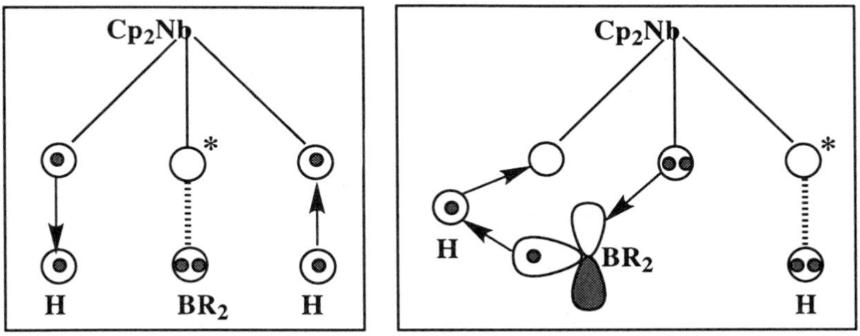

Figure 31.6 Star isomers of $Cp_2NbH(BR_2)H$.

Another case in which two valence isomers are star isomers has been reported recently. It involves a pentagonally hybridized metal, which can act as axial 3h/4e by using two occupied π-type d AOs (d_{zx} and d_{zy}) and one vacant sd^2 axial hybrid to tie up two 3h/2e ligands (e.g., PR_3, RS^-, or Cl^-). The five remaining in-plane AOs (the ninth AO, namely, p_z, is neglected) hybridize to form five hybrids directed to the corners of a regular pentagon with two of the five holes being starred. One can accommodate three hydrogens, one chloride, and one metal pair in two different ways, as shown (note the odd-out covalent bond and the codirectional character of the REL bonds):

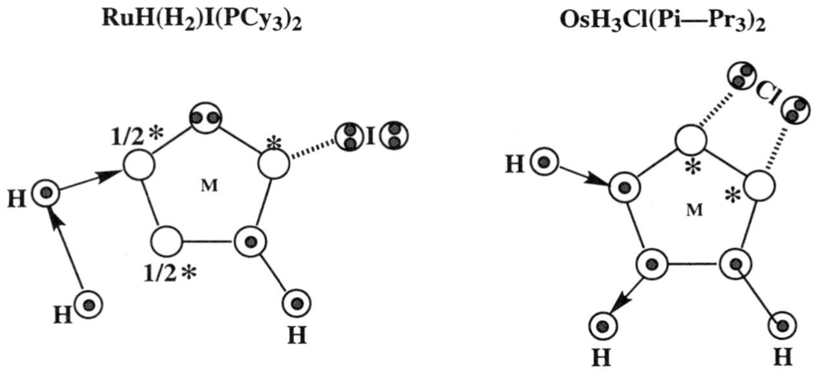

In one form, H_2 acts as a donor (RED) and in the other as a donor–acceptor (ROX) ligand. These two different situations are realized in two 16-electron complexes which were recently studied by Chaudret et al.[17] and by Caulton et al.[18]

31.6 Magnetic Exchange Coupling as a Probe of Star Allocation

The application of simple MO models to organometallic chemistry has created the impression in some quarters that understanding the electronics of metal-containing molecules is straightforward: One can always base an argument on some MO (or MOs) that becomes stabilized in the course of a transformation.

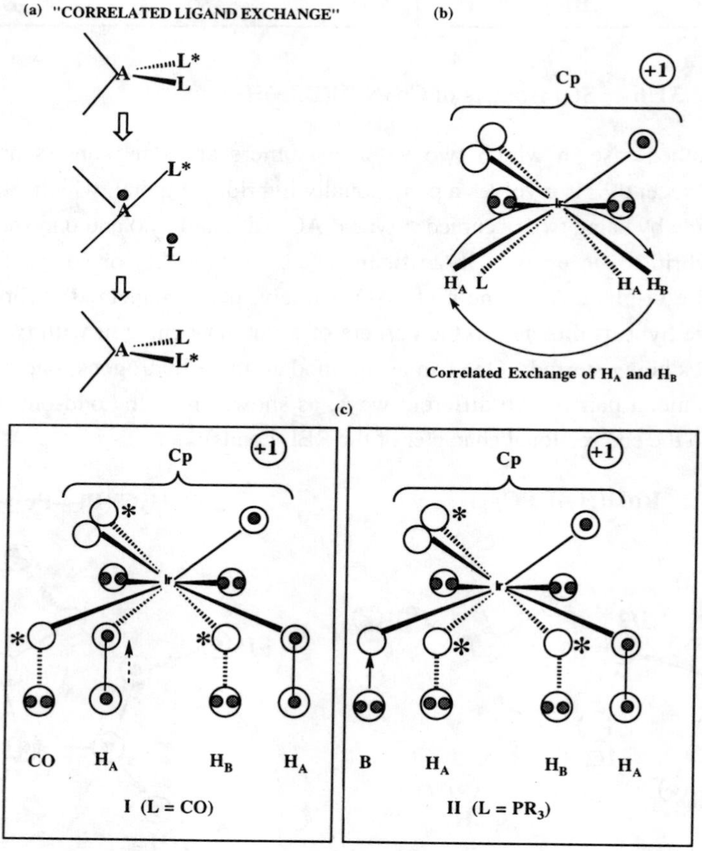

Figure 31.7 (a) The analogy of two reactions involving the 180° rotation of one triatomic molecular subunit. In each case, the transition state is a four-electron antiaromatic complex with three close-lying singlet states. The saddle point is a diradical. (b) Correlated ligand exchange in $LIrCpH_3^+$. (c) Control of star character and metal–hydrogen bonding by the observer ligand: star isomers of $LIrCpH_3^+$.

We say quite the opposite: Molecular electronics is a "game of angstroms." When VB theory cannot make a categoric prediction of a trend, it can set the alternatives for the experiment (or the "perfect" calculation) to decide. We illustrate this philosophy by reference to magnetic exchange coupling in transition metal hydrides.

The cis-trans isomerization of a 1,2-disubstituted ethylene and the enantiomerization of a chiral methane derivative can be visualized as the result of correlated rotation, as illustrated in Figure 31.7a. The second reaction can be formulated as a ligand exchange reaction. The lower the A—L bond dissociation energy, the lower the barrier. The higher the temperature, the faster the reaction. When the two ligands have magnetically active nuclei, the latter are effectively coupled as a result of the reaction itself. This is one possible interpretation of magnetic exchange coupling recently observed in transition metal hydrides.[19] This effect is expected to generalize to any L_nMYY molecules in which L are observer ligands, M is a semimetal or metal, and Y are magnetically active nuclei.

Let us now recall that the electronic basis of the trans effect (Section 31.3): An observer ligand deshields the metal by accepting the π-type metal "lone pairs" so that the incoming nucleophile is spared of exchange repulsion. Furthermore, let us recall the electronic basis of the preference for planar geometry by X_2PtL_2 (formal d^8) complexes (Section 26.7): In the planar geometry, the metal–ligand bonds stay orthogonal to three axially directed metal lone pairs (one σ- and two π-type), and exchange repulsion is minimized. Ligand exchange via a tetrahedral transition state has a significant barrier because the tetrahedral species suffers from bond pair/lone pair exchange repulsion. This analysis suggests the following general strategy for observing strong exchange coupling in L_nMYY:

The M—Y bonds must be weak.

The L must be high on the trans effect (not trans influence) series: for example, CO.

The VB formula of $LIrCpH_3^+$ reveals that correlated exchange of the two nonequivalent hydrogens in $LIrCpH_3^+$ (Figure 31.7b), where L is a variable ligand, runs into the equatorial lone pairs of Ir. Choosing the observer ligand L to be a strong π acid stabilizes the transition state of this transformation and maximizes the H_A—H_B coupling constant. It has been observed that shifting from L= PR_3 to L = $P(OR)_3$ increases the coupling by an order of 100 Hz.[21]

There exists a second alternative: Magnetic exchange coupling is *not* the effective result of a chemical reaction as described above but rather the result of the through-space interaction of the magnetic nuclei controlled by the nonbonded interaction of the Ir—H_A and Ir—H_B bonds. The limit of the atomic through-space interaction is the collapse of two nonbonded to two bonded hydrogens (dihydrogen form) and the form of the magnetic interaction may be rotatory exchange equivalent to rotational tunneling. We refer to this process as the vibrational mechanism of ligand magnetic interaction. Is this scenario compatible with the dependence of J_{AB} on the nature of the ligand?

Two cationic iridium trihydride complexes differing only in the allocation of the stars are shown in Figure 31.7c. Cp is associated with two unstarred holes and one starred hole for reasons discussed earlier. When L is a π-acidic ligand [e.g., CO, P)OR)$_3$, etc.], it docks on a starred hole, which means that only one of the three hybrids directed to the three hydrogens can be starred. This acts to tie up a hydride. Remember the rule: Only unstarred hybrids can be occupied in the reference (or ground) configuration. As a result, the M—H_A and $M^+H_B^-$ bonds attract each other by induction, leading to two predictions:

A high Ir—H stretching frequency (because the metal AOs directed to the hydrogens have low star character).

A low H_A—Ir—H_B scissoring frequency.

The result of the second prediction is H—H attraction, the limit of which is dihydrogen formation, and a large J_{AB} as a result of the coupling of the scissoring vibration (which allows the nuclei to come close to each other) with a rotation that effects nuclear exchange. By contrast, a σ-basic L ligand (e.g., PR_3) docks on an unstarred hybrid, dictating that two of the three hybrids directed toward the three hydrogens must be starred. Therefore, not only H_B but also H_A has hydridic character. This leads to opposite predictions: H—H repulsion, the limit of which is dihydride formation, and a small J_{AB}. These predicted trends are consistent with the experimental facts. Note that our scenario credits a scissoring vibration for magnetic exchange coupling. A stretching vibration is predicted to have the opposite effect exactly as one expects from the position of σ donors (higher) and π acceptors (lower) on the trans-influence scale (see Section 31.3).

The final conclusion is that there exist two conceivable mechanisms of magnetic exchange coupling: The vibrational mechanism operates along the entire temperature range. By contrast, the chemical exchange mechanism becomes operative at temperatures significantly above 0K. As a result, the plot of J_{AB} versus temperature must be flat in the vicinity of 0K and it should rise beyond a certain value at which the chemical exchange mechanism becomes activated.

31.7 The Dilemma of the Metal-Metal "Multiple Bonds"

The well-known contributions of Cotton and his co-workers[21a,b] to the chemistry of the metal–metal bond are summarized in a monograph[21b] that poses an important question: How are the Cotton complexes bound? The two choices are illustrated by reference to the prototypical dimer $Re_2Cl_8L_2^{2-}$ (L = H_2O) in Figure 31.8. There are two extreme possibilities:

The oxidation state is Re(I), the four chlorines at one metal center are held by two REL bonds, and there are two ROX metal–metal bonds. A ROX bond is the diatomic version of a REL bond. This is I-bonding.

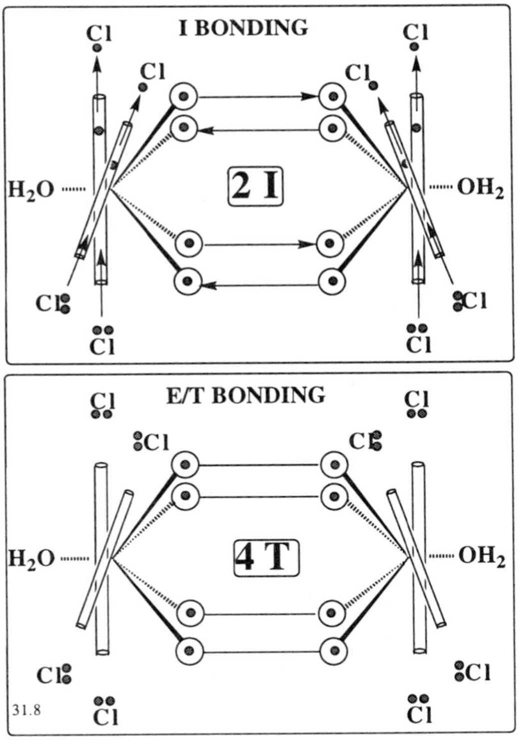

Figure 31.8 Two different mechanisms of bonding of $Re_2Cl_8(H_2O)^{2-}$.

The oxidation state is the intuitively expected Re(III) and the metal–chloride bonds are ionic. The metal–metal bonds are covalent.

What are the guidelines for interpreting molecules of this type:

According to the heteroatom exclusion rule, bridging by electronegative nonmetal implies E-bonding.

A complex made up of first transition series metals bearing nonbridging electronegative nonmetal ligands employs T-type metal–metal bonding and E-type metal–ligand bonding. In other words, we have mixed bonding.

A complex made up of third transition series metals bearing nonbridging semimetal ligands employs ROX-type metal–metal bonding as well as REL-type metal–ligand bonding. In other words, the entire complex is uniformly I-bound.

Because molecular electronic structure has been viewed from an entirely different perspective, meaningful comparisons are hard to locate in the literature. For example, what happens in borderline cases, such as the $(RO)_3WW(OR)_3$ complex of Chisholm and co-workers?[22] In essence, the meaningful investigation of molecular electronic structure (i.e., the issue of selectivity) has yet to begin!

31.8 The Docking Rule and the Dependence of Metal-Metal Bonding on the Nature of the "Observer" Ligands

Three groups have contributed three types of metal dimer that differ in structure and magnetism even though the only difference between the two is replacing $2Cl^- + 2H_2O$ by $4R_3P$! Our interpretation is three-pronged.

(a) In the Cotton dimer $Re_2Cl_8L_2^{2-}$ ($L = H_2O$),[1] each Re uses unstarred AOs to effect metal–metal bonding. The possibilities were discussed in the preceding section.

(b) Replacing two Cl^- and two H_2O by four PR_3 units leads to a Walton dimer,[23] which is paramagnetic. The paramagnetism is an indication that CT delocalization at the bridging sites is defective. The four phosphines, acting like ROX ligands, tie up the four unstarred holes, leaving four starred equatorial holes to accommodate two bridging chlorines. Since the delocalization of the chlorine electrons to the high-lying starred metal orbitals is effectively zero, the six chlorine atoms oxidize the two metals, and the resulting chlorides bind by

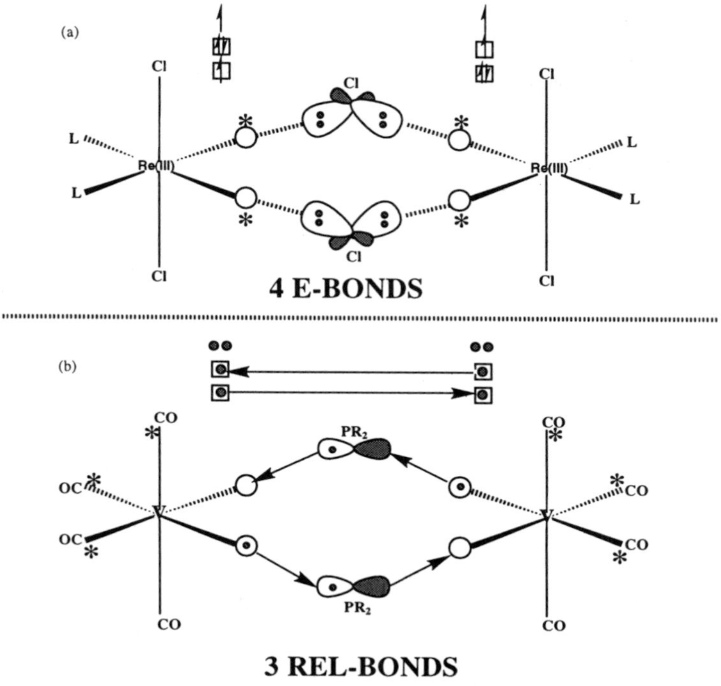

Figure 31.9 The star character of the bridging sites determines the mechanism of bridging, the nature of the metal–metal bonding, and the molecular spin.

predominant classical Coulomb attraction. The two bridging chlorides form four electrostatic E-bonds, the metal centers are kept wide apart, and the result is a high spin dimer (Figure 31.9a).

(c) The Vahrenkamp dimer[24] has three carbonyls docking at three starred holes of each metal center. This makes a total of four unstarred endo AOs and two electrons available to two R_2P bridging ligands. The result is the formation of two metal–ligand REL bonds, which are exactly analogous to those responsible for the bridged structure of diborane. In turn, the two REL bonds induce one additional metal–metal ROX (= REL) bond (Figure 31.9b).

We have discovered the mechanism of information relay in a molecule, and we have expressed it by simple formulas. External ligands determine indirectly the mode of action of internal bridging ligands by controlling the star allocation. Replacing E by I bonds effects a switch from ferromagnetism to antiferromagnetism. The latter has a "through-space" and a "through-bond" component. The diagram that follows illustrates how two complementary REL bonds (i.e., two I bonds described by a codirectional arrow train) cause "through-bond" antiferromagnetic coupling. One of the most recent examples is the ferric wheel.[25]

ANTIFERROMAGNETIC COUPLING BY TWO I-BONDS

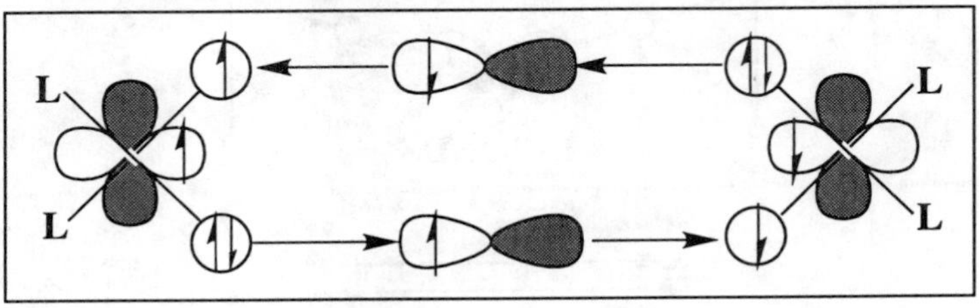

Comparison of this analysis with the EHMO analysis of the same problem by Hoffmann and collaborators[26] will lead to an appreciation of the chaos that separates this work from the conventional models. At present, chemists are "true believers" in "universal" covalency: All homonuclear bonds are "covalent," from the Cotton clusters to zerovalent metal carbonyl clusters to bare metal clusters! This notion, a cornerstone of chemical thinking, whether stated in "old VB" or "new MO" terms, is contrary to our conclusions, namely: There are three limiting mechanisms of bonding, T, I, and E, and organometallic chemistry walks a tightrope between the three extremes. Without the appropriate tools (e.g., the star formulas and the docking rule), no progress can be made in understanding how small hybridization changes produce spectacular structural, physical, and chemical differences even between isoelectronic species.

31.9 The Size Dependence of Cluster Reactivity

In an important experimental work, Smalley and co-workers found transition metal clusters that exhibit a size-dependent reactivity toward the hydrogen molecule.[27] Their results indicate the existence of three categories of transition metal clusters:

(a) Those that give oxidative addition to H_2 regardless of size. An example is Ni. Here, there is steady increase in the activity as cluster size increases.

(b) Those that fail to give *facile* oxidative addition regardless of size. An example is Cu.

(c) Those that show a pronounced sensitivity to cluster size. Small clusters are reactive, intermediate clusters are unreactive, and solid surfaces are again reactive. Specifically, neutral as well as ionic niobium clusters comprising 8, 10, 12, and 16 atoms are unreactive toward dissociative chemisorption. The same is true of neutral cobalt and iron clusters with six to nine atoms.

Trying to rationalize these data by saying that even-atom clusters are closed-shell singlets, hence unreactive, implies that open-shell metal atoms are capable of abstracting hydrogen atoms and are consequently reactive. This is not necessarily so because the enthalpy of hydrogen abstraction is very unfavorable as the following example suggests.

$$\text{Cu} + \text{H—H} \longrightarrow \text{Cu—H} + \text{H} \qquad \Delta H = +\,37 \text{ kcal/mol}$$

The H—H bond is worth 103 kcal/mol while M-H bond dissociation energies from M = K to M = Zn range from 20 to 61 kcal/mol only.[28] Thus, the hydrogen abstraction is highly endothermic and the barrier high.[29]

A small cluster has low cohesive energy and, as a result, it can rearrange and expose an individual metal atom to H_2. In going from this limit to a solid surface, the major changes are the increase in the cohesive energy and the drop of the ionization potential. The decreasing IP implies an increasing ability of the metal to donate electron density to the incoming H_2. The mystery is: What makes intermediate clusters unreactive? These represent cases in which the cohesive energy is not low enough to permit easy deformation of the metal, nor is the IP low enough to promote the metal-to-H_2 CT required for dissociative chemisorption. As a result, these species must satisfy more refined requirements for reactivity.

Our model of intermediate-size cluster reactivity is as follows:

(a) Dissociative chemisorption occurs by oxidative addition of H_2 to one metal atom and requires one active metal pair and one active metal hole.

(b) Medium-size clusters are deltahedra constructed from pseudo-octahedrally hybridized transition metals atoms. The endo set of hybrids binds the atom to the "bulk," while the exo set points to the vacuum and is available for binding a molecule or atom. The exo hybrids play the role of "dangling orbitals."

(c) The combined endo and exo sets have three unstarred and three starred AOs. Starred AOs can be used only for coordination, but they are (relatively) unreactive. Hence, the distribution of the stars to the two conical sets determines whether the surface atoms will be reactive.

(d) In principle, oxidative addition can be effected using two equatorial orbitals, or one equatorial and one exo, or two exo orbitals with a total of two

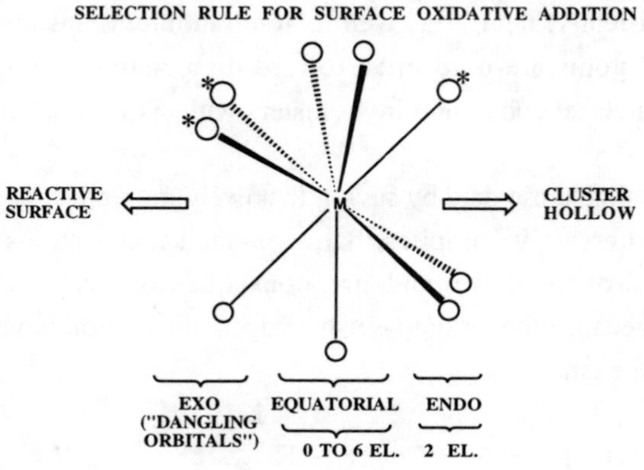

Figure 31.10 The conditions for reactivity of a surface metal atom.

electrons. The first two possibilities are rejected on the basis of exchange repulsion of the incoming H_2 with the endo-bonding pairs.

(e) Reaction can be effected by utilization of two exo hybrids under the condition that *at least* one is unstarred. In other words, the minimum requirement is that the endo hybrids are ENDO* and the exo hybrids are EXO**. When strong metal–metal bonding occurs, the situation changes to ENDO/EXO*** and the cluster becomes unreactive.

(f) The electron configuration of the surface atom is derived by first allocating a total of two electrons to the endo set (for the purpose of deltahedral metal–metal bonding, as discussed shortly), followed by filling the equatorial set of hybrids (three pairs) and then proceeding with the exo set. This configuration can accommodate only one pair because it has only one unstarred AO (starred AOs are never occupied in the ground configuration).

The reactivity conditions for a surface atom (i.e., the electron count and the starring pattern) are shown schematically in Figure 31.10. As a consequence, transition metals with up to eight electrons (two endo plus six equatorial) — that is, transition metals from Sc to Fe (and their congeners) — can be ideally accommodated in a deltahedral cluster in a way that renders them unreactive. Since the first transition metal with an exo pair is nickel, the prediction is that nickel will be reactive while all earlier transition metals will be unreactive. The difference between unreactive Nb and reactive Ni is illustrated in Figure 31.11. The former has no reactive exo pairs while the latter does.

Siegbahn and co-workers[30] recognized the significance of Smalley's results and attempted a resolution. They computed medium-size clusters of cobalt in the range of Co_6 to Co_{10} and found that these have closed shells. As a result, they concluded that the closed-shell character of the cluster is responsible for lack of reactivity. The VB interpretation is different: The surface atoms of these clusters cannot project exo pairs to the incoming molecule (without an undue loss of metal–metal bonding) and thus they are incapable of reaction.

31.10 Steps Versus Terraces

One of the most fascinating puzzles is the chemoselectivity of surfaces. Do steps, terraces, and the other various domains of a metal surface differ in their affinity for different molecules? If so, what determines the selectivity? Since we can now write formulas for metal atoms and metal ensembles, we can envision some new ways of thinking about the problem.

Everything that we said about intermediate-cluster reactivity in Section 31.9 hinged on an important implicit assumption: The occupancy of the equatorial metal AOs is not subject to any constraint. We will now argue that such a constraint does exist, and it is one important factor that differentiates clusters from single-crystal faces and surfaces. The idea is illustrated by considering two limiting coordination modes:

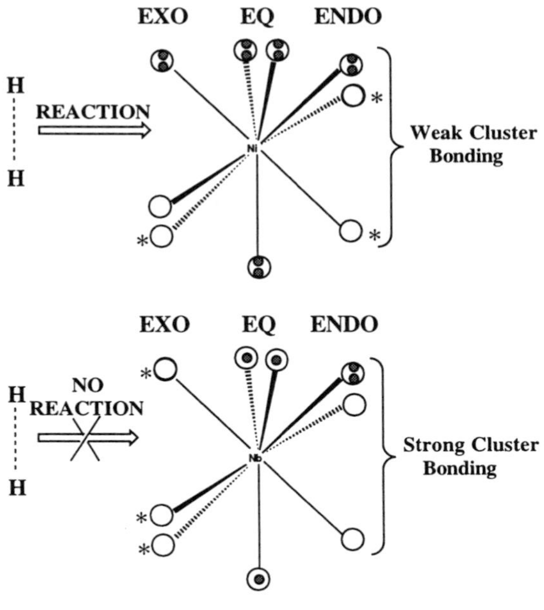

Figure 31.11 Reactive and unreactive surface metal atoms.

A linear array of metal atoms represents the situation encountered on very large clusters as well as on terraces of surfaces.

A curved array of metal atoms represents the situation encountered on small or medium-size clusters as well as on steps of surfaces.

How does the metal atom adjust in going from the first to the second geometry? Our hypothesis is that the primary difference between terrace and step atoms lies in the way the equatorial hybrids overlap. These orbitals are assumed to be tangent to the curve defined by a string of metal atoms, as illustrated in Figure 31.12. As a result, equatorial–equatorial overlap depends on the curvature in the sense that it is maximal when the curvature is zero. The situation is entirely analogous to what happens to the in-plane tangential 2p AOs as we go from a three- to an infinite-membered carbon polygon. Strong overlap of equatorial hybrids on the terraces requires that these orbitals be only partly filled, to prevent strong exchange repulsion. By contrast, overlap of the same hybrids is weakened

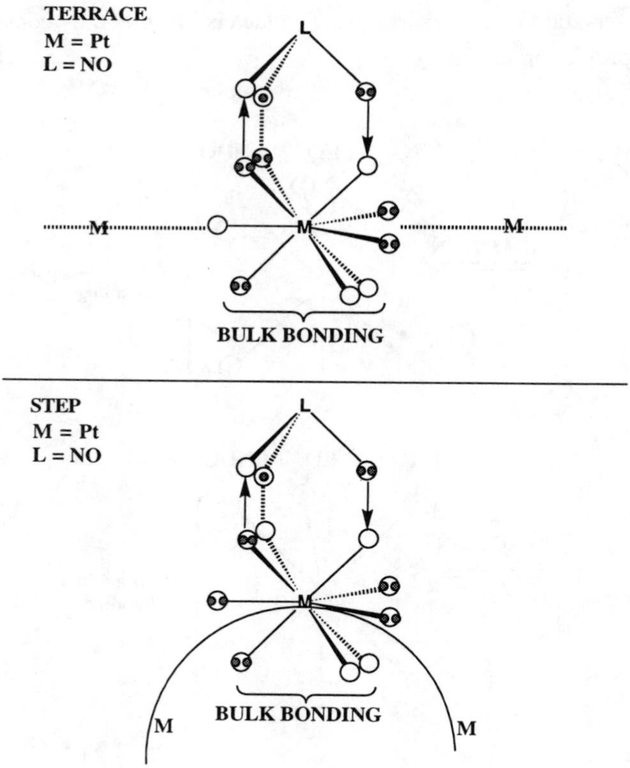

Figure 31.12 Proposed electronic structures of step and terrace metals.

on steps, which can now attain full occupancy. This is why our formulation of the reactivity of intermediate-size clusters assumed no constraint on the occupancy of the equatorial hybrids.

As we have already discussed, the endo set of three conical hybrids is dedicated to metal–metal bonding between the surface and bulk atoms. This set is occupied by two electrons, much as in the case of borane clusters. The exo set of three "dangling orbitals" is used to bind molecular ligands. Every molecule or atom is pictured as also binding with a set of three conical AOs and an appropriate number of valence electrons. Optimal surface molecule bonding occurs when the electron occupancy of the dangling orbitals is compatible with the electron occupancy of the molecular valence AOs. In general, if we assume that the surface metal has p and the molecule q electrons, the rule for strong binding is that $p + q$ must be 6, the number corresponding to a formal triple bond. The key point now is that the number of exo metal valence electrons, p, is determined by the number of electrons the metal can accommodate in the equatorial set. From what has been said, we expect that more electrons will be allocated to the exo set as we go from steps to terraces and full equatorial occupancy becomes less favorable.

For example, consider the difference between a terrace (T) and a step (S) Pt atom. To ensure that equatorial–equatorial overlap is not prohibitively repulsive, the T atom must leave at least one equatorial AO vacant. This means that the exo set, which must now accommodate two pairs plus one hole, will now be unable to interact in a bonding fashion with a molecule having more than two electrons. By contrast, the S atom can have all the equatorial hybrids full. This allows the exo set to accommodate only two electrons. As a result, a molecule having up to four electrons can now become bound.

The situation was illustrated in Figure 31.12 using Pt and NO as metal and molecule, respectively. Note how one odd electron of NO must be coupled to a pair of T-Pt generating a 3-e exchange antibond. This gives way to 0.5E bond in the case of S-Pt. Hence, we predict that the Pt surface will have higher reactivity toward NO (and CO) on steps than on terraces. In addition, we predict that this difference will vanish as the number of electrons in the molecule falls below three! Thus, we conclude that step versus terrace reactivity is a function of the electron counts of both metal and molecule.

Control of surface selectivity does not depend solely on electron count. Isoelectronic metals can be differentiated in the following way: In the case of transition metals, the promotional energies of $(n-1)$d to ns and the $(n-1)$d to np determine the efficacy of hybridization. When the latter is reduced, the differentiation between steps and terraces just described ceases to exist. Now,

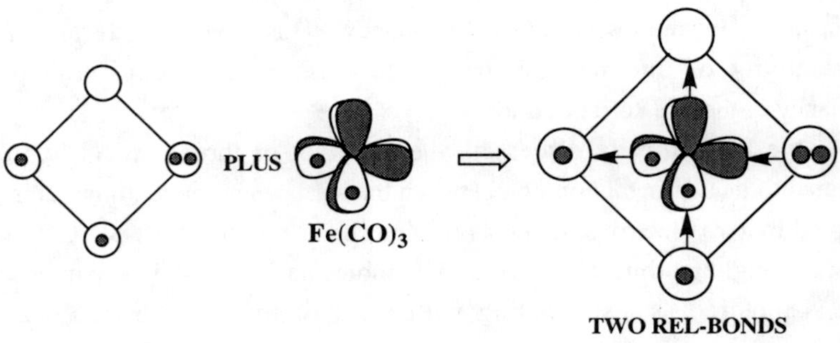

Figure 31.13 Electronic structure of the cyclobutadiene–$Fe(CO)_3$ complex.

terraces become the less sterically encumbered sites for molecular association. This may explain why NO adsorbs on Pt steps but on Pd terraces.[31] The distinguishing feature of Pd is the notorious stability of the d^{10} configuration, which discourages strong metal hybridization.

31.11 The sd^3 Metal

While polarization is important, the essence of organometallic chemistry is the REL bond. Organic chemistry is the chemistry of the covalent bond and the tetrahedral carbon. Trigonal and linear carbon are derivative forms. Organometallic chemistry is the chemistry of the REL bond and the sd^3 metal (Figure 25.6). Other hybridization modes can be thought as derivatives of this basic form.

The stable iron tricarbonyl–cyclobutadiene (CB) complex[32] is one simple illustrator of the VB view of organometallic structures. One operationally significant question is: Why does $Fe(CO)_3$ cap CB rather than adding to it? The $Fe(CO)_3$ fragment, with one daggered σ and two singly occupied π-type d electrons, combines with a zwitterionic form of CB to yield a complex having two REL bonds, as illustrated in Figure 31.13. Changing from $Fe(CO)_3$ to the isolobal HC^+ turns REL bonds to T bonds, and this produces an entirely different global minimum, which involves optimal perfect pairing rather than REL-bonding. For example, covalent $C_5H_5^+$, the result of inserting HC^+ into a σ bond of CB, is more stable than the associated CB—HC^+.

One more illustration of the action of $(CO)_3Fe$ as an sd^3 iron with two electrons is the following reaction sequence reproduced with permission from ref. 33a (copyright 1992 American Chemical Society).[33]

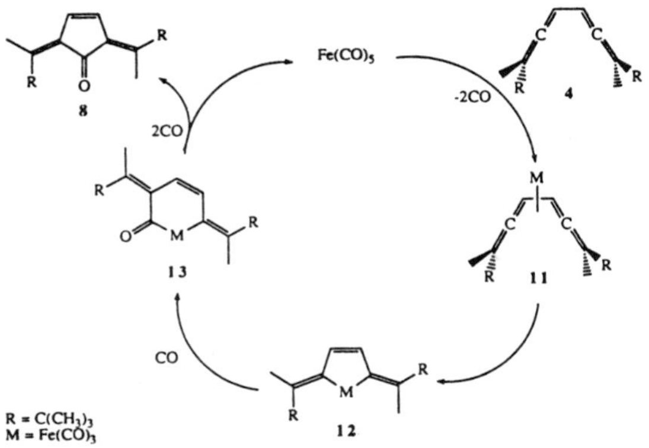

The VB interpretation is given in Figure 31.14. Note how one Fe hole disengages from the C_2—C_3 butadienic π bond pair in the $Fe(CO)_3$–1,3-butadiene complex to become associated with the X lone pairs in the metallacyclopentene isomer. When Y = CH_2, the metallacyclopentene is apparently not formed. When Y is replaced by X bearing a lone pair or a π-bond pair, the intermediate metallacyclopentene is formed and proceeds to the ultimate cyclopentenone product by addition of CO.

The sd^3 metal dispels the seductive illusion of the "18-electron rule." Of the nine hybrids, only six unstarred AOs can effect I-bonding. Furthermore, the sd^3 metal implies neither an 18-electron nor a 12-electron rule. Rather, it predicts that the metal will be able to bind twice as many ligands as active AOs (6) with the electron count ranging from a minimum of 12 (6 holes binding 12 radicals) to a maximum of 24 (6 pairs binding 12 radicals)!

The failure of the 18-electron rule to capture the physics of metal bonding becomes exposed when different electron counts for one and the same ligand are required for satisfying the rule. For example, is acetylene a two- or a four-

M = $Fe(CO)_3$

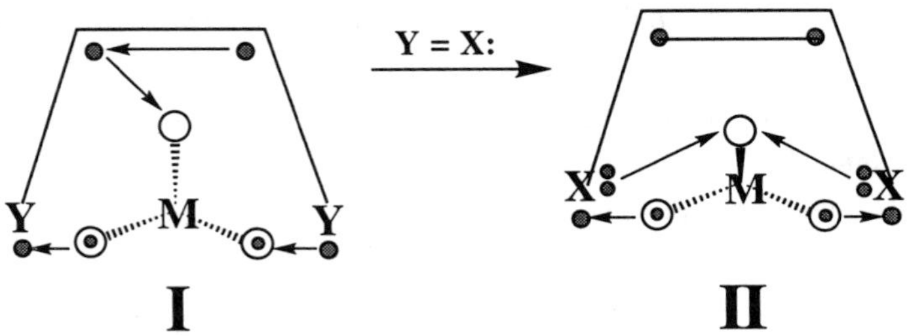

Figure 31.14 "Dragging" of $Fe(CO)_3$ by terminal groups of a 1,3-diene bearing lone pairs or π pairs.

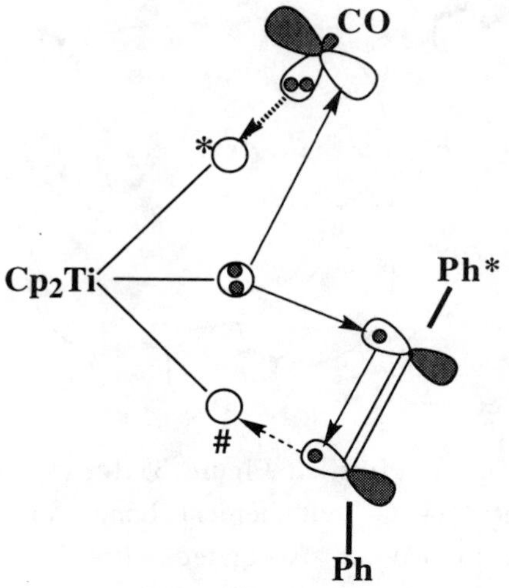

Figure 31.15 Electronic structure of $Cp_2TiCO(PhCCPh)$.

electron donor? A two-electron count is required by $Cp_2TiCO(PhCCPh)$[34a] but a four-electron count is required by $(PhCCPh)_3WCO$.[34b] VB theory paints a different picture: In either complex, the acetylenic ligand acts as a two-electron ligand with the "observer" π bond responding to (polarizing) the active π bond. The only thing that matters is whether we can link each ligand to the metal by a codirectional arrow train, that is, by one REL bond. As can be seen from Figures 31.15 and 31.16, this condition is met by both complexes.

As a final illustration, an sd^3 Os can tie up two alkylidenes and two alkyl groups to form pseudotetrahedral $Os(CH\text{-t-Bu})_2(R)_2$ as follows:

One equatorial hybrid and one π-type d_{xz} with a total of two electrons tie up one alkylidene by one I bond.

A second equatorial hybrid and a second π-type d_{yz} pair with a total of two electrons tie up a second alkylidene by one I bond.

The third equatorial hybrid is doubly occupied and ties up two alkyl radicals by one I bond.

The remaining σ-type axial hybrid accommodates the Os lone pair.

This molecule can be considered to have two formal metal–alkylidene double bonds, a formal analogue of an organic allene. Hence, it is called the "allenic" isomer. By working in the same way, we can derive the electronic structure of the "acetylenic" isomer $Os(C{-}t\text{-}Bu)(CH_2{-}t\text{-}Bu)(R)_2$. Which of the two

THE STRUCTURE OF WL_3CO WHERE L = Diphenyl Acetylene

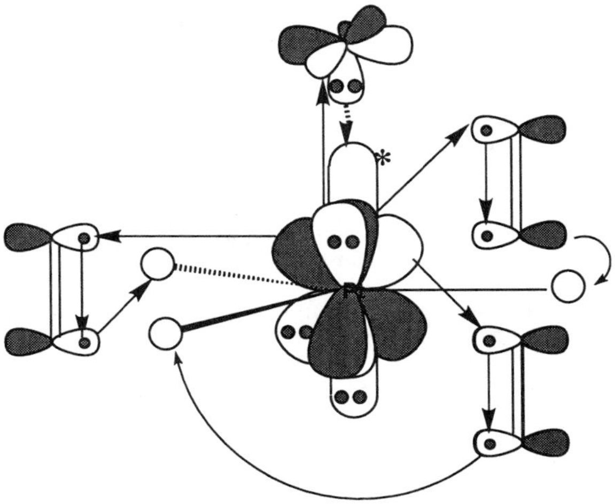

Each π–metal pair connects to two ligand π-LUMOs and each acetylene π-HOMO pair connects to two unstarred σ metal holes. Carbonyl docks on a single starred hole. One σ metal pair is a lone pair.

Figure 31.16 Electronic structure of $(PhCCPh)_3WCO$.

isomers is more stable? Assuming predominant ROX delocalization, attachment of two alkylidenes by two I bonds is more favorable than attachment of one alkylidyne by 1.5 and alkyl by 0.5 I bonds, because only in the former case can ROX delocalization occur with preservation of metal electroneutrality. In other words, making the "allenic" isomer requires less promotion. Replacing one R by OR effectively increases the metal electronegativity and shifts the bonding toward the T end of the spectrum. Since the T-bound CH_3CCH is thermodynamically more stable than the T-bound H_2CCCH_2 (by 1 kcal/mol), we expect the preference to shift in favor of the "acetylenic" isomer. All this discussion has been motivated by a recent contribution by Schrock and coworkers, who prepared and studied Os(CH—*t*-Bu)$_2$(R)$_2$.[35] The essential point is that VB formulas and underpinning concepts have moved the discussion of molecular electronic structure to another plane at which all orbitals and all electrons matter, and where molecular stability depends on the precise mechanism of bonding which in turn depends on the detailed electronic structure of the metal and the ligands.

31.12 The Pseudo-Octahedral Metal

While the sd^3 metal reveals the essence of organometallic bonding by differentiating I- from T-bonding, the pseudo-octahedral metal (see Figure 25.7) has two different advantages:

(a) It shows explicitly the role of polarization identified by the starred holes. Matching one starred metal hole with a ligand valence pair counts as one (electrostatic) E bond.

(b) It shows readily why organometallic complexes adopt the shape they do by reverting to a formalism in which we imagine that a T bond is a single covalent bond, an I or Q bond a double covalent bond, and an E bond a single covalent bond. This is the basis of the occasional success of the EHMO model, which treats the three different types of bond (T, I, E) as if they were all of the same covalent type.

The nitrogen atom of ammonia directs three sp^3 hybrids to the corners of a triangle to tie up three hydrogens. We say that the "hapticity" of N is 3 (i.e., there are three valence AOs reserved for binding some other fragment), and we use the designation 3h to symbolize this state. Furthermore, the three AOs are conically arranged; that is, N is a 3h conical fragment symbolized by $3h_c$. In this light, let us now consider how we can cap a pseudo-octahedral metal by a ligand that acts effectively as a 3h fragment. A typical example is the Cp^- ligand, which can be envisioned as having three pairs arranged in a triangular fashion and ready to tie up three triangularly disposed metal holes. The capping can proceed as follows:

(a) Either the staggered or the all-eclipsed form can be capped in a linear fashion, as illustrated. In either case, this leaves out three triangular equatorial hybrids that can serve as reservoirs of metal lone electrons. This is the situation in linear metallocenes.

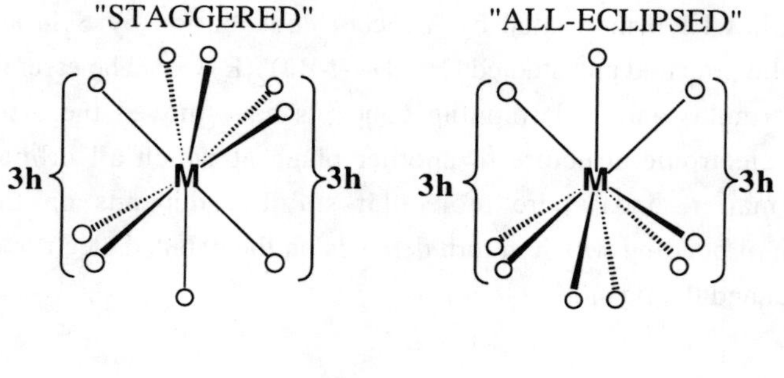

(b) Either the staggered or the all-eclipsed form can be capped in bent fashion, leaving out a set of three hybrids having different spatial orientations. One constitutes a conical and the other a trident 3h set, as illustrated. Hence, we can differentiate, in principle, between a *conical* and a *trident* bent Cp_2M or any isoelectronic bent XYM fragment in which M is a pseudo-octahedral metal. Note the tricky part: In the trident form, the two angular Cp ligands are attached to two 3h sets: One is made up of the three solid AOs and the other of the three hatched AOs.

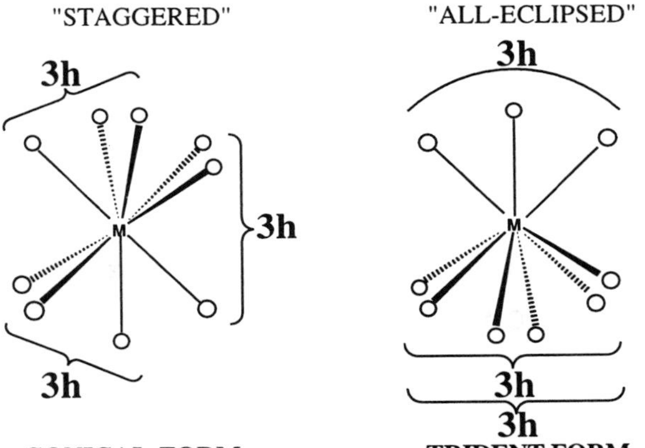

CONICAL FORM $\mathbf{3h_c3h_c3h_c}$

TRIDENT FORM $\mathbf{3h_c3h_c3h_t}$

We now have a simple way of predicting and visualizing stable complexes. Specifically, 3h ligands will combine with 3h sites of a pseudo-octahedral metal to form formal triple bonds having a total of six electrons. For example, a 3h/6e cyclopentadienyl anion will connect to a 3h/0e site of a pseudo-octahedral metal by means of a formal triple bond. A 3h/2e carbonyl will attach to a 3h/4e site by a formal triple bond, etc. Of course, this is the ideal situation. Triple bonds with five rather than six electrons are also possible. Sections 31.12.1 and 31.12.2 present some representative examples of the metal fragments or molecules we have been discussing.

31.12.1 Trident Molecules

The simplest illustrator of a trident metal is the $Cp_2Ta(CH_2)Me$ Schrock alkylidene.[36] The characteristic feature is that the HCH alkylidene plane is normal to the MeTaC (of CH_2) plane. The two planes would become coplanar if the pseudo-octahedral Ta were conical. Two related complexes are

$Cp_2Ti(CH_3CO)Cl$[37] and $Cp_2TiCO(PhCCPh)$.[34a] The complex $[C_3H_3N$(*t*-Bu)B(Ph)]COFe(μ-CO)$_2$FeCO[C_3H_3N(*t*-Bu)B(Ph)][38a] illustrates how the trident AOs accommodate the bridging ligands (with Cp replacing the actual heteroatomic isoelectronic rings). The number 3 denotes a formal six-electron triple bond connecting Fe with Cp and with CO. Each Fe dedicates one electron to the binding of Cp (Fe acts like 3h/1e and Cp like 3h/5e to produce a triple bond), as well as four electrons to the binding of CO (Fe acts like 3h/4e and CO like 3h/2e to produce a triple bond). This leaves three electrons dedicated to bridged intermetallic bonding. A related example is the apparently different Cp(N-*t*-Bu)Mo(μ-S)$_2$Mo(N-*t*-Bu)Cp.[38b]

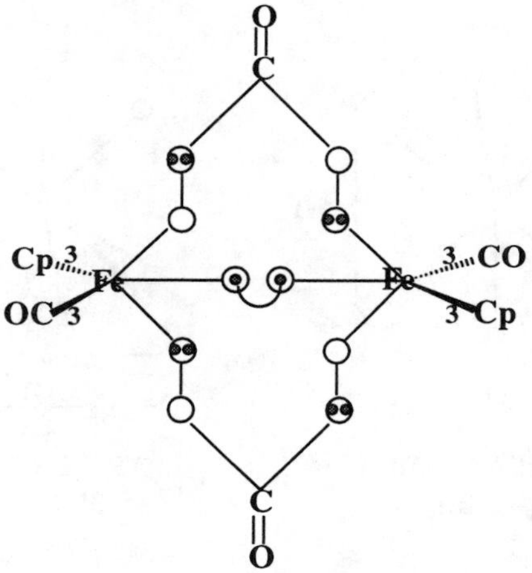

On the other hand, the complex (CpCORh)(μ-CH_2)(RhCOCp)[39] fits either the trident or the conical description, as illustrated. (For clarity, the deformation required for optimizing the two-electron bond and minimizing the four-electron antibond is not shown.)

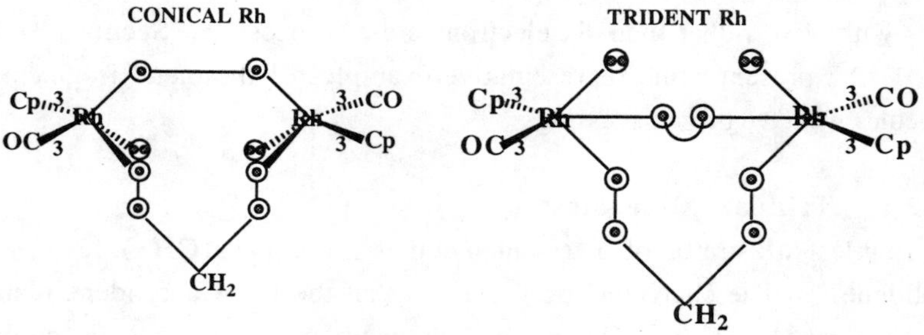

31.12.2 Conical molecules

Surprisingly, the best illustrator of the conical class is the textbook molecule Cp_4Zr.[40] The conical pseudo-octahedral metal predicts that triangular Cp_3Zr^+ should be stable. Indeed, this is so. This fragment is squashed to a pyramidal geometry by one electrostatically bound *monohapto* Cp^- in Cp_4Zr (= $Cp_3Zr^+Cp^-$). The way in which the conical AOs accommodate bridging ligands, which is not self-evident, is illustrated using (CpCORu)(μ-CMe)(μ-CO)(CpCORu)[41] as an example. Note how the two Ru conical sets dedicated to bridging are staggered. Moreover, there is no direct Ru—Ru bond, there is one formal Ru_2—carbyne five-electron triple bond, and the carbyne is in low spin form.

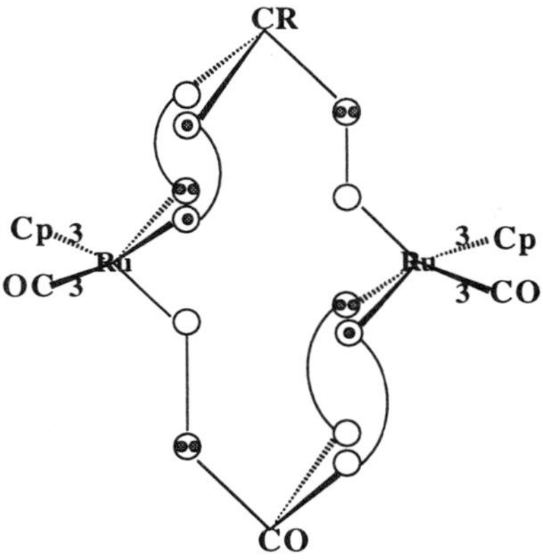

More examples can be found in tetraallyl dirhenium[42a] (five- rather than six-electron Re–allyl formal triple bonds) and $(H_2C{=}CH{-}HC{=}CH_2)_2MnCO$[42b] (one formal triple bond has five electrons).

The ability to write down a stereochemical representation of metal fragments that shows the orientation of the metal lone pairs in space makes it easy to anticipate their mode of action. Using the pseudo-octahedral model, it is easy to see that the similarities and differences between $Co(CO)_3$ and $MoCp(CO)_2$. Each can act as a 3h/3e fragment but there are two differences:

$Co(CO)_3$ has three while $MoCp(CO)_2$ has only one equatorial lone pair. If one were to focus on the available metal lone pairs, recognize that these bind primarily by interacting with the lone pairs, and compute the ratio of lone pairs to carbonyls, one would conclude that metal pairs are less available in the Mo fragment and any superiority thereof has some other electronic basis.

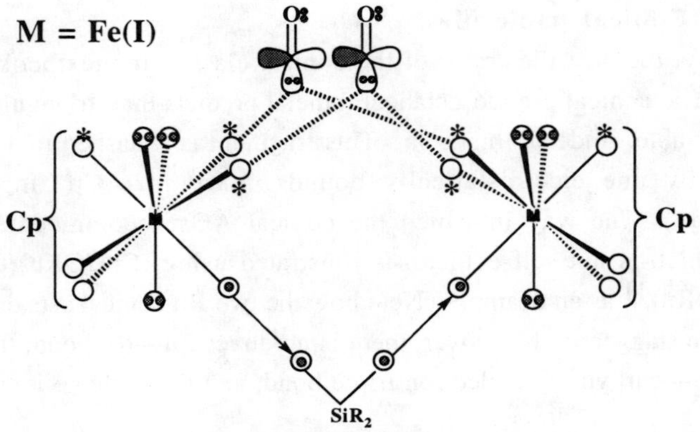

I(singlet)

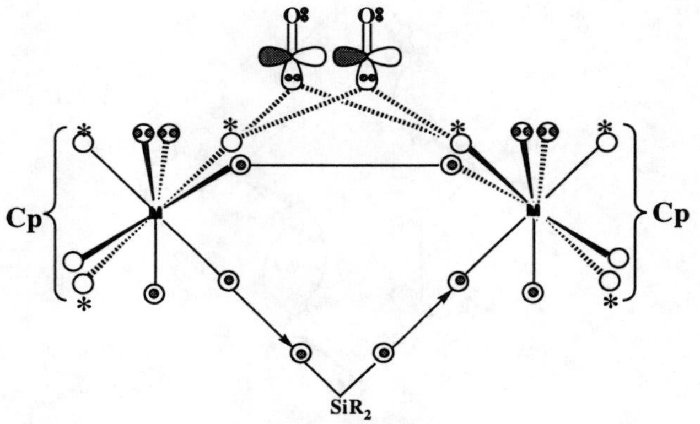

II(triplet)

Figure 31.17 Star-isomeric choices available to $CpFe(\mu\text{-}CO)_2(\mu\text{-}SiR_2)FeCp$.

(b) In combining with RC carbyne fragments to form organometallic tetrahedranes of the type $(RC)_2M_2$ (metallotetrahedranes), the Mo—C bond is much more polar than the Co—C bond on account of the Cp/CO differentiation.

It has been shown that $MoCp(CO)_2$ acting as a metallotetrahedrane vertex on a carbocationic center attached to a metallotetrahedrane carbon stabilizes the latter more than a $Co(CO)_3$ vertex.[43a] We suggest that this is a consequence of the metal–carbon bond polarization in the metallotetrahedrane cluster. Experiments of this type will help us understand the workings of metal pairs, until now confined to the dungeon of chemical consciousness.

Chemical theory is the exploration of the avenues available to atoms for bonding. Once such pathways have been identified, the synthetic chemist will follow them to the end. An illustration is the newly synthesized CpFe(μ-CO)$_2$(μ-SiR$_2$)FeCp, which has a triplet ground state and a very short Fe—Fe distance.[43b] Each Fe adopts a pseudo-octahedral hybridization and ties up the Cp by three exo conical AOs while reserving three conical AOs for bridging and/or Fe—Fe bonding. The remaining metal electrons go to the equatorial sets. There are now two different star-isomeric choices (Figure 31.17).

Each Cp is bound by one starred and two unstarred exo hybrids. This means that there are two unstarred endo hybrids and one unstarred. As a result, we have the singlet isomer I in which each Fe has a completely full equatorial belt. There is no Fe—Fe bond.

Each Cp is bound by two starred exo hybrids and one unstarred. This means that there are one unstarred and two unstarred endo hybrids. As a result, we have the triplet isomer II in which each Fe has an incompletely full equatorial belt with five electrons. The two odd equatorial electrons couple ferromagnetically. Furthermore, there is now a full Fe—Fe bond.

The issue now is: Why is II the preferred species? The answer is that Fe—Cp bonding has been sacrificed for the benefit of Fe—Fe bonding because Fe—Cp bonding is, any way, impaired by the proximal bulky silylene substituents. The two spin isomers are actually star isomers.

Bottom line: A transition metal binds with *six* valence AOs (sd^3 metal) or, in more detail, with six unstarred and three starred AOs (pseudo-octahedral metal). The unstarred AOs make I bonds and the starred AOs make E bonds. By contrast, the 18-electron rule of the organometallic chemist assumes that the metal binds with *nine* AOs, which can make a maximum of nine covalent bonds. The flaw is revealed by the mere recognition that there are as many metal complexes that disobey as there are metal complexes that obey the rule if one looks across the board rather than at, say, metal carbonyls. The best thing for the reader is to try to derive the electronic structure of a metal complex by first using the sd^3 and then the pseudo-octahedral sd^{5}p^3 scheme. There should be no difference in the conclusions other than the consequences of polarization.

Where would organic chemistry be now if, instead of VB formulas, we had to rely on MO diagrams? The answer can be given by pondering where inorganic chemistry would be now if, instead of MO diagrams, we had sd^3 metals and pseudo-octahedral metals. Has the inorganic chemist, against all chemical common sense, been prejudiced against VB theory? I know of only one publication that presents a drawing of a pseudo-octahedral metal.[44]

31.13 The Electronic Structure of the FeMo Nitrogenase Cofactor

The sd^3-hybridized Fe atom (Figure 25.6) can be reformulated as follows:

The three axial AOs are equivalent to a set of endo conical valence AOs.

The three equatorial AOs resolve into two δ-type d AOs plus a toroidal, s-type AO, which can be assigned the dagger by assuming that it has predominant s character.

This type of Fe can adopt the following electronic configuration, and it can bind (via the conical AOs) much like a trivalent P atom.

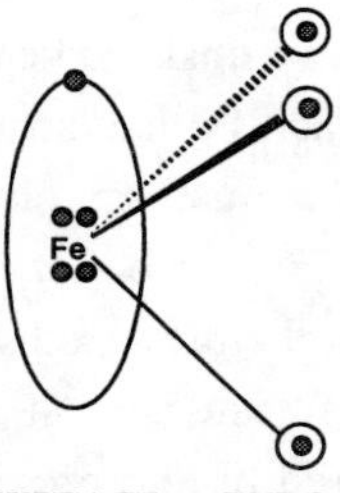

We can now construct the following clusters:

(a) An Fe_6 prismane that has, in addition, a "triple bond" made up of six toroidal electrons.

(b) An Fe_6S_9 prismane derivative, generated by insertion of nine S atoms to the nine prism edges.

(c) The cluster of Figure 31.18, where pseudo-octahedral Mo and Fe can cap the two triangular faces of Fe_6S_9 by matching their conical holes with S lone pairs. This is the VB depiction of the electronic structure of the FeMo cofactor of nitrogenase recently described in the literature.[45] The starring pattern of the conical AOs of the capping Mo and Fe is not shown explicitly, but it is critically dependent on the nature of the ligands L.

What is the mechanism of dinitrogen reduction? Recent work by Coucouvanis and co-workers[46] involving model systems implicates the Mo site. The nature of the L ligands determines the star character of the exo and endo conical AOs of Mo, and it was found that PR_3 and CO ligands suppress the

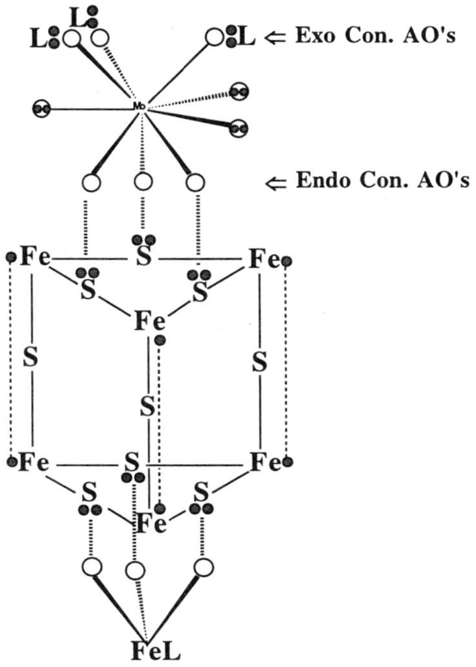

Figure 31.18 The FeMo nitrogenase cofactor in VB diagram form.

reaction of the $[MoFe_3S_4]^{3+}$ core with acetylene. We interpret this to mean that the rate-determining step of the reduction is the breaking of one formal S—Mo hole–pair bond (indicated by hatched line) to effect partial uncapping of the iron prism triangular face. The ligand L controls the facility of the reduction by indirectly determining the star character of the endo conical Mo holes: The more the L ligands of Mo prefer unstarred AOs, the lower the star character of the exo conical Mo AOs, the higher the star character of the exo conical Mo AOs, and the weaker the S—Mo binding.

31.14 The Activation of Stable Molecules by Transition Metals

Finally, we use the concept of the sd^3 metal in conjunction with the discussion of the cleavage of N_2 by $(NRAr)_3Mo$ to delineate a recipe for the activation of the prototypical, highly stable C_2, N_2, O_2, and F_2 by organometallic fragments of the type X_3M in which M is sd^3-hydridized and X is a monovalent group such as RO, R_2N, or R_3C. We illustrate our approach by reference to the N_2 target:

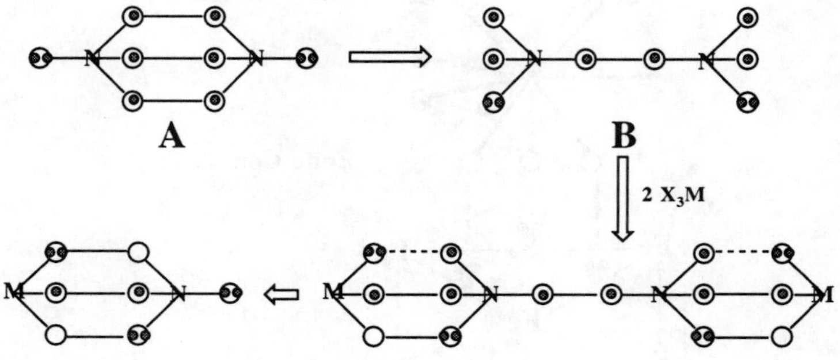

(a) The ground state N_2 molecule is viewed as a composite of two sp^3-hybridized nitrogen atoms connected by a triple banana bond, which is produced by the overlap of two sets of three endo conical nitrogen AOs, as in form A in the illustration above.

(b) Each nitrogen is inverted to produce form B, which has three exo conical AOs at each nitrogen.

(c) The three exo conical AOs of each nitrogen are matched by three conical AOs of a metal fragment.

(d) The magic number required for complete rupture of N_2 and formation of X_3MN is 7, corresponding to two 2-e and one 3-e formal metal–nitrogen bonds.

Going to the general case of an A_2 target (A = F, O, N, C) and denoting the number of exo A electrons of the inverted form of A_2 by Q and the number of metal electrons by R, we have the condition:

$$Q + R = 7$$

This leads to the following recipe, using transition metals of the second series as examples:

Target	Activator Fragment	Metal Valence
CC	X_3Tc	3h/4e
NN	X_3Mo	3h/3e
OO	X_3Nb	3h/2e
FF	X_3Zr	3h/1e

This defines the strategy for the full activation of "resistant," stable molecules made up of H, C, N, O, and F.[47] For example, ethane is isoelectronic to F_2. Because the stability of the X_3MAAMX_3 complex depends on the nature of the 3e interaction and because the latter depends on the efficacy of CT, the best activators may not be the neutral fragments given above but, rather, charged isoelectronic derivatives with the specified metal valence.

References

1. (a) M.K. Leclerc and H.H. Brintzinger, *J. Am. Chem. Soc.* 117, 1651 (1995). (b) E. Spaltenstein, R.R. Conry, S.C. Critchlow, and J.M. Mayer, *J. Am. Chem. Soc.* 111, 8741 (1989).
2. (a) For review and recent examples, see P. Braunstein, C. de M. de Bellefon, S.-E. Bouaoud, D. Grandjean, J.-F. Halet, and J.-Y. Saillard, *J. Am. Chem. Soc.* 113, 5282 (1991). (b) W.A. Herrmann, R. Serrano, and H. Bock, *Angew. Chem. Int. Ed. Engl.* 23, 383 (1984). (c) W.A. Herrmann, M. Floel, J. Kulpe, J.K. Felixberger, and E. Herdtweck, *J. Organomet. Chem.* 335, 297 (1988).
3. (a) M. Poliakoff and E. Weitz, *Acc. Chem. Res.* 20, 408 (1987). (b) P.C. Engelking and W.C. Lineberger, *J. Am. Chem. Soc.* 101, 5570 (1979). (c) S. Goebel, C.L. Haynes, F.A. Khan, and P.B. Armentrout, *J. Am. Chem. Soc.* 117, 6994 (1995).
4. (a) B.L. Tjelta and P.B. Armentrout, *J. Am. Chem. Soc.* 117, 5531 (1995). (b) P.K. Hurlburt, J.J. Rack, J.S. Luck, S.F. Dec, J.D. Webb, O.P. Anderson, and S.H. Strauss, *J. Am. Chem. Soc.* 116, 10003 (1994). (c) L.A. Barnes, M. Rosi, and C.W. Bauschlicher Jr., *J. Chem. Phys.* 93, 609 (1990) and earlier papers. (d) *Chem. Eng. News* 73(26), 30 (1995).
5. (a) L.M. Venanzi, *Chem. Bri.* 4, 162 (1968). (b) F. Basolo and R.G. Pearson, in *Prog. Inorg. Chem.* 4, 381 (1962).
6. K.F. Purcell and J.C. Kotz, *An Introduction to Inorganic Chemistry*, Saunders College Publishing, Philadelphia, 1980.
7. G.B. Ray, X.-Y. Li, J.A. Ibers, J.L. Sessler, and T.G. Spiro, *J. Am. Chem. Soc.* 116, 162 (1994).
8. M. Birgogne, *J. Inorg. Nuclear Chem.* 26, 107 (1964).
9. R.J. Angelici, *J. Inorg. Nuclear Chem.* 28, 2627 (1966).

10. (a) H.J. Plastas, J.M. Stewart, and S.O. Grim, *J. Am. Chem. Soc.* 91, 4326 (1969). (b) H.S. Preston, J.M. Stewart, H.J. Plastas, and S.O. Grim, *Inorg. Chem.* 11, 161 (1972)

11. (a) J. Chatt, L. Manojovlic-Muir, and K.W. Muir, *J. Chem. Soc. Chem. Commun.* 655 (1971). (b) L. Aslanov, R. Mason, A.G. Wheeler, and P.O. Whimp, *J. Chem. Soc. Chem. Commun.* 30 (1970).

12. S.J. Anderson and R.J. Goodfellow, *J. Chem. Soc. Dalton Trans.* 1683 (1977).

13. J. Emsley and D. Hall, *The Chemistry of Phosphorus*, Halsted Press, Wiley, New York, 1976.

14. (a) E.A. Carter and W.A. Goddard III, *J. Am. Chem. Soc.* 108, 2180 (1986). (b) This is why the angularly constrained (C_5H_4-CMe_2-C_5H_4)WHMe, in which two Cp residues are linked by CMe_2, fails to undergo reductive elimination to yield methane plus bent metallocene, while the unconstrained Cp_2WHMe does produce methane plus linear metallocene: L. Labella, A. Chernega, and M.L.H. Green, *J. Organomet. Chem.* 485, C18 (1995). (c) D.M. Heinekey, B.M. Schomber, and C.E. Radzewich, *J. Am. Chem. Soc.* 116, 4515 (1994); C.E. Radzewich, Ph.D. Thesis, University of Washington, 1996. (d) M.E. Marmion and K.J. Takeuchi, *J. Am. Chem. Soc.* 108, 510 (1986). (e) J.C. Bryan and J.M. Mayer, *J. Am. Chem. Soc.* 112, 2298 (1990). (f) S.J. Miller and R.H. Grubbs, *J. Am. Chem. Soc.* 117, 5855 (1995).

15. G.R.K. Khalsa, G.J. Kubas, C.J. Unkefer, L.S. Van Der Sluys, and K.A. Kubat-Martin, *J. Am. Chem. Soc.* 112, 3855 (1990).

16. J.F. Hartwig and S.R. de Gala, *J. Am. Chem. Soc.* 116, 3661 (1994).

17. B. Chaudret, G. Chung, O. Eisenstein, S.A. Jackson, F.J. Lahoz, and J.A. Lopez, *J. Am. Chem. Soc.* 113, 2314 (1991).

18. D.G. Gusev, R. Kuhlman, G. Sini, O. Eisenstein, and K.G. Caulton, *J. Am. Chem. Soc.* 116, 2685 (1994).

19. K.W. Zilm, D.M. Heinekey, J.M. Millar, N.G. Payne, S.P. Neshyba, J.C. Duchamp, and I. Szczyrba, *J. Am. Chem. Soc.* 112, 920 (1990).

20. (a) D.M. Heinekey, J.M. Millar, T.F. Koetzle, N.G. Payne, and K.W. Zilm, *J. Am. Chem. Soc.* 112, 909 (1990). (b) Review: D.M. Heinekey and W.J. Oldham Jr., *Chem. Rev.* 93, 913 (1993).

21. (a) F.A. Cotton, *Acc. Chem. Res.* 11, 225 (1978). (b) F.A. Cotton and R.A. Walton, *Multiple Bonds Between Metal Atoms*, Wiley, New York, 1982.

22. (a) M.H. Chisholm, D.L. Clark, K. Folting, J.C. Huffman, and M.J. Hampden-Smith, *J. Am. Chem. Soc.* 109, 7750 (1987). (b) M.H.

Chisholm, D.L. Clark, and M.J. Hampden-Smith, *J. Am. Chem. Soc.* 111, 574 (1989).

23. (a) J.A. Jaecker, W.R. Robinson, and R.A. Walton, *J. Chem. Soc. Dalton Trans.* 698 (1975). (b) T. Mimry and R.A. Walton, *Inorg. Chem.* 16, 2829 (1977).
24. (a) H. Vahrenkamp, *Chem. Ber.* 111, 3472 (1978). (b) H. Vahrenkamp, *Angew. Chem. Int. Ed. Engl.* 17, 379, 1978.
25. K.L. Taft, C.D. Delfs, G.C. Papaefthymiou, S. Foner, D. Gatteschi, and S.J. Lippard, *J. Am. Chem. Soc.* 116, 824 (1994). The concept of superexchange can be used to explain antiferromagnetic coupling through bridging ligands [for references, see L. Noodleman, *J. Chem. Phys.* 74, 5737 (1981)], but it is "achemical" to the extent that it does not reveal how the coupling depends on the observer ligands and the nature of the metal.
26. S. Shaik, R. Hoffmann, C.R. Fisel, and R.H. Summerville, *J. Am. Chem. Soc.* 102, 4555 (1980).
27. (a) M.D. Morse, M.E. Geusic, J.R. Heath, and R.E. Smalley, *J. Chem. Phys.* 83, 2293 (1985). (b) M.E. Geusic, M.D. Morse, and R.E. Smalley, *J. Chem. Phys.* 82, 590 (1985). (c) J.L. Elkind, F.D. Weiss, J.M. Alford, R.T. Laaksonen, and R.E. Smalley, *J. Chem. Phys.* 88, 5215 (1988).
28. P.B. Armentrout, in *Bonding Energetics in Organometallic Compounds,* American Chemical Society, Washington, DC, 1990.
29. P. Madhavan and J.L. Whitten, *J. Chem. Phys.* 77, 2673 (1982). See also G. Mills and H. Jónsson, *Phys. Rev. Lett.* 72, 1124 (1994), and references therein.
30. I. Panas, P. Siegbahn, and U. Wahlgren, in *The Challenge of d and f Electrons,* American Chemical Society, Washington, DC 1989.
31. Q. Gao, R.D. Ramsier, H.N. Waltenburg, and J.T. Yates Jr., *J. Am. Chem. Soc.* 116, 3901 (1994).
32. (a) For review of the transition metal chemistry of cyclobutadiene, see P.M. Maitlis, *J. Organomet. Chem.* 200, 161 (1980). (b) Theoretical MO considerations were first presented by H.C. Longuet-Higgins and L.E. Orgel, *J. Chem. Soc.* 1969 (1956). (c) A recent $Fe(CO)_3$–cyclobutadiene analogue to which the VB concepts are applicable is $[E_7M(CO)_3]^{3-}$ (E = P, As, Sb and M = Cr, Mo, W): S. Charles, B.W. Eichhorn, A.L. Rheingold, and S.G. Bott, *J. Am. Chem. Soc.* 116, 8077 (1994).

33. (a) B.E. Eaton, B. Rollman, and J.A. Kaduk, *J. Am. Chem. Soc.* 114, 6245 (1992). (b) M.S. Sigman, C.E. Kerr, and B.E. Eaton, *J. Am. Chem. Soc.* 115, 7545 (1993).

34. (a) G. Fachinetti, C. Floriani, F. Marchetti, and M. Mellini, *J. Chem. Soc. Dalton Trans.* 1398 (1978). (b) R.M. Laine, R.E. Moriarty, and R. Bau, *J. Am. Chem. Soc.* 94, 1402 (1972).

35. A.M. LaPointe, R.R. Schrock, and W.M. Davis, *J. Am. Chem. Soc.* 117, 4802 (1995).

36. R.R. Schrock, *Acc. Chem. Res.* 12, 98 (1979).

37. G. Fachinetti, C. Floriani, and H. Stoeckli-Evans, *J. Chem. Soc. Dalton Trans.* 2297 (1977).

38. (a) J. Schulze and G. Schmid *Angew. Chem. Int. Ed. Engl.* 19, 54 (1980). (b) L.F. Dahl, P.D. Frisch, and G.R. Gust, *J. Less-Common Met.* 36, 255 (1974).

39. W.A. Herrmann, C. Kruger, R. Goddard, and I. Bernal, *J. Organomet. Chem.* 140, 73 (1977).

40. J.N. Francis, A. McAdam, and J.A. Ibers, *J. Organomet. Chem.* 29, 131 (1971).

41. D.L. Davis, A.F. Dyke, A. Endesfelder, S.A.R. Knox, P.J. Naish, A.G. Orpen, D. Plaas, and G.E. Taylor, *J. Organomet. Chem.* 198, C43 (1980).

42. (a) F.A. Cotton and M.W. Extine, *J. Am. Chem. Soc.* 100, 3788 (1978). (b) G. Hunter, D. Neugebauer, and A. Razavi, *Angew. Chem. Int. Ed. Engl.* 14, 352 (1975).

43. (a) M. Kondratenko, H. El Hafa, M. Gruselle, J. Vaissermann, G. Jaouen, and M.J. McGlinchey, *J. Am. Chem. Soc.* 117, 6907 (1995). (b) H. Tobita, H. Izumi, S. Ohnuki, M.C. Ellerby, M. Kikuchi, S. Inomata, and H. Ogino, *J. Am. Chem. Soc.* 117, 7013 (1995).

44. R.H. Crabtree, *The Organometallic Chemistry of the Transition Metals*, Wiley, New York, 1988, p.313.

45. (a) M.K. Chan, J. Kim, and DC Rees, *Science* 260, 792 (1993). (b) J.T. Bolin, A.E. Ronco, T.V. Morgan, L.E. Mortenson, and N.H. Xuong, *Proc. Natl. Acad. Sci. USA* 90, 1078 (1993).

46. L.J. Laughlin and D. Coucouvanis, *J. Am. Chem. Soc.* 117, 3118 (1995), and following paper.

47. A combination of X_3Cr and X_3V cleaves NO: A.L. Odom, C.C. Cummins, and J.D. Protasiewicz, *J. Am. Chem. Soc.* 117, 6613 (1995).

Chapter 32

The Mechanism of Oxidative Addition

32.1 How Bare Metals Cleave Covalent Bonds

A transition metal has a diffuse ns and contracted $(n-1)$d AOs. When the metal inserts into H—H, it makes two bonds to the two H atoms in two phases, illustrated in Figure 32.1.

In the reactant (R) phase, H—H interacts with a vacant sd hybrid.

In the product (P) phase, the disrupted H—H acts as an anion radical which interacts with a singly occupied contracted $(n-1)$d AO.

The two stages of bond making are differentiated in a simple way: The R phase obeys the hydrogen rules (Chapter 17) because one H has one inbound and one outbound arrow. The P phase disobeys the hydrogen rules. Assuming that the transition state occurs at some intermediate stage, we have an important prediction: Metal insertion into H—H will be kinetically favored over insertion into R—R bonds (R = alkyl), with the effect disappearing at the product level. In other words, the activation energy must be substantially lower for H—H than for R—R, while the reaction enthalpies must be comparable.

Blomberg et al. have studied the oxidative addition of a variety of transition metals to single bonds using ab initio CI computations.[1] They found that metals (iron, cobalt, nickel, rhodium and palladium) add oxidatively to H_2 with either very low or no activation energy. By contrast, oxidative addition to the

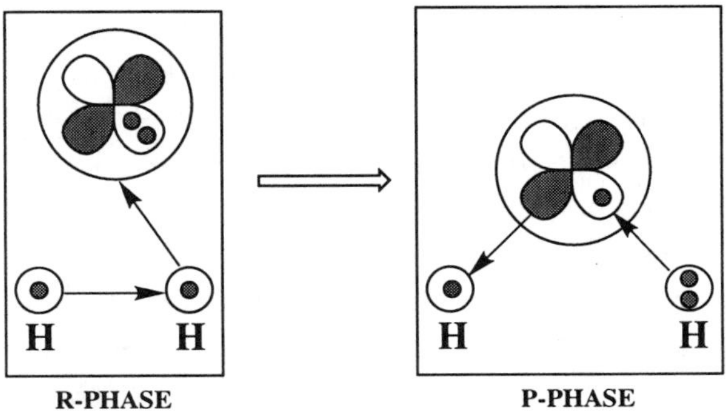

Figure 32.1 The R and P phases of oxidative addition of Ni to H—H.

Table 32.1 Calculated Enthalpies and Activation Energies (in kcal/mol) for Prototypical Insertions and Diels–Alder Type Reactions

Reaction	ΔE	$\Delta E^{\#}$	Ref.
Insertion			
$Ni + H_3C{-}CH_3 \longrightarrow H_3C{-}Ni{-}CH_3$	–1.5	+40	(a)
$Ni + H{-}H \longrightarrow H{-}Ni{-}H$	–9.6	+3	(b)
Diels–Alder-Type			
$3H_2 \longrightarrow 3H_2$	0	+69	(c)
$H_2C{=}CH{-}HC{=}CH_2 + H_2C{=}CH_2 \longrightarrow$ cyclohexene	–32.4	+43	(d)

[a]M.R.A. Blomberg, P.E.M. Siegbahn, U. Nagashima, and J. Wennerberg, *J. Am. Chem. Soc.* 113, 424 (1991).

[b]M.R.A. Blomberg and P.E.M. Siegbahn, *J. Chem. Phys.* 78, 986 (1983).

[c]D.A. Dixon, R.M. Stevens, and D.R. Herschbach, *Faraday Discussions Chem. Soc.* 62, 110 (1977).

[d]R.E. Townshend, G. Rammuni, G. Segal, W.J. Hehre, and L. Salem, *J. Am. Chem. Soc.* 98, 2190 (1976). 1976

C—C bond of ethane has a much higher barrier. As shown in Table 32.1, the activation energy for oxidative coupling of H_2 is a paltry 3 kcal/mol, in contrast to 40 kcal/mol for $H_3C{-}CH_3$. The reaction enthalpies, however, are not very different. The situation seems to persist even when neutral are replaced by cationic metal atoms.[2] Blomberg et al. rationalized the exceptional facility with which transition metals bond to H—H by using the intuitive argument of the inorganic chemist that "H1s can make bonds in all directions."

The computational data collected in Table 32.1 show that the cycloaddition of butadiene and ethylene is much more exothermic than the insertion of Ni to H_2. However, the barrier of the first reaction is larger by roughly 36 kcal/mol! That there is no relation between reaction exothermicity and barrier height is also made apparent by a comparison of the three-bond switch of three H_2 molecules with the Ni + H_2 reaction.

Why is Ni so reactive toward H_2? Our answer is that the transition state is crossed in the R phase of the reaction but induction is as important as I-bonding. In a typical bond switch reaction, an ascending reactant curve is crossed by a descending product curve. A reaction barrier exists because the transition state complex suffers from net nonbonded exchange repulsion that is not overcompensated by CT delocalization. This formulation of the origin of the

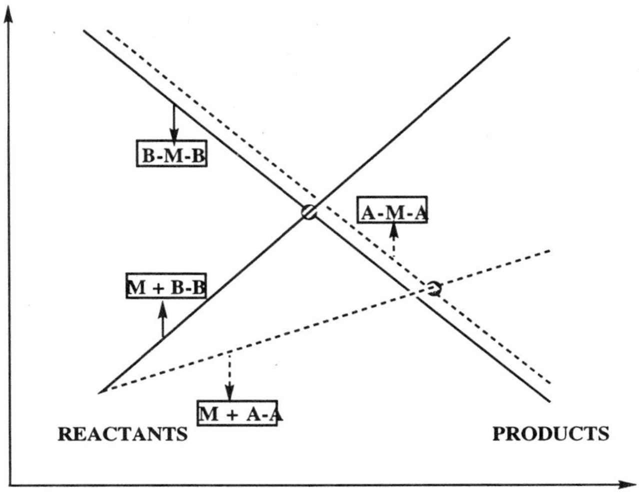

Figure 32.2 The VB representation of the insertion of Ni into H—H.

reaction barrier suggests the strategy for lowering it: Find a way of reducing exchange repulsion of the active reactant electron pairs that will serve to reduce the steepness of the ascending reactant curve and cause the crossing with the product curve to occur at lower energy.

Since only comparisons are physically meaningful, the problem is defined schematically in Figure 32.2. For constant reaction enthalpy, the barrier of the insertion of a metal atom into a σ bond depends on the deshielding of the metal along the reaction coordinate. In the R phase of the reaction, the σ pair of the attacking H—H sees a σ Ni hole (4s or $3d_{z^2}$) and a σ Ni pair ($3d_{z^2}$ or 4s).

Overlap induction (see Chapter 11) effects metal deshielding in the following way: The Ni4s mixes with the $Ni3d_{z^2}$ AO to generate two hybrids, one vacant AO pointing in the direction of the H_2 and one doubly occupied AO normal to the HNiH reaction plane. As a result, the ascending curve describing the Ni + H_2 ensemble crosses the descending H—Ni—H product curve at low energy. CT delocalization adds to the stability of the reaction complex at the crossing point.

As discussed earlier, the efficacy of metal deshielding depends critically on the energy gap separating the $ns^2(n-1)d^8$, $ns^1(n-1)d^9$, and $ns^0(n-1)d^{10}$ configurations. This rationalizes the finding that Fe and Co are unreactive while Ni is reactive with linear alkanes.[3] The difference among the three metals lies in the energy gaps separating the $4s^23d^8$ and $4s^13d^9$ configurations. They increase in the order Fe > Co > Ni. Hence, 4s-3d mixing and, consequently, deshielding is very inefficient with Fe and very efficient with Ni and this accounts for their very different reactivity.

The final conclusion is that the transition state of a thermoneutral or exothermic oxidative addition features a deshielded metal that acts as a weak Lewis acid. This has the following implications:

(a) Since the oxidative addition transition state is largely dependent on overlap induction, it does not have to conform to the Woodward-Hoffmann rules. A splendid example is the stereospecific retrocycloaddition of diosmacyclobutane discovered by Norton and his group.[4] The attacking alkene induces Os═Os deshielding in the fashion illustrated below and in a way exactly analogous to the deshielding of Ni induced by H—H.

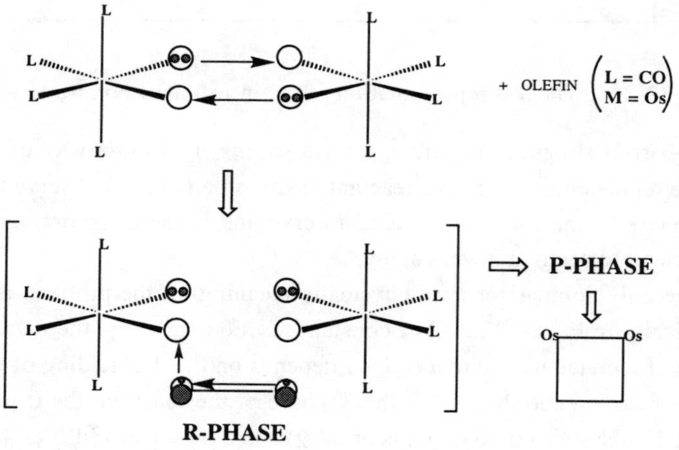

(b) The requirement for inductive deshielding means that "observer ligands" play a key role. Bowers and co-workers[5] have studied the insertion of Sc^+ into H_2 as a function of the number of H_2 ligands and they argued that the high-pressure kinetic data could be rationalized by postulating that sequential "electrostatic binding" of H_2 molecules diminishes the activation energy for insertion into one of them. What happens to the reaction enthalpy and the reaction barrier if one coordinates three H_2 molecules to reactant Sc^+ as well as to product H—Sc^+—H? In other words, what is the difference between the following reactions?

$$Sc^+ + H\text{—}H \longrightarrow H\text{—}Sc^+\text{—}H$$

$$(H_2)_3Sc^+ + H\text{—}H \longrightarrow (H_2)_3H\text{—}Sc^+\text{-}H$$

If all that is involved is electrostatic binding of the H_2 molecules to a cation, one would hardly expect any significant change in enthalpy. In fact, the better accessibility of the cationic center on the reactant side would tend to render the second reaction less exothermic. The alternative scenario is that the H_2 ligands cause metal promotion by induction. Specifically, the three "observer" H_2 ligands rearrange the electron density on the metal, causing the two active metal electrons to end up in AOs that are properly oriented for matching the AOs of the H—H reactant. In such a case, *both* the reaction barrier and the reaction enthalpy should be reduced in going from the first to the second reaction above because only the transition state and the products have incipient or complete Sc^+—H bonds. This is what is suggested by the experimental results of Bowers et al. (negative temperature dependence of reaction rate). Finally, on the premise that the three H_2 molecules promote the Sc^+ to a favorable hybridization state from which it can insert, and assuming this to be the pseudo-octahedral hybridization mode, we surmise that the structure of $(H_2)_3H$—Sc^+—H is not the one computed by Bowers et al., but rather, the following:

IN $HScH(H_2)_3^+$ EACH H—H LIGAND MATCHES ONE EQUATORIAL AND ONE EXO HOLE

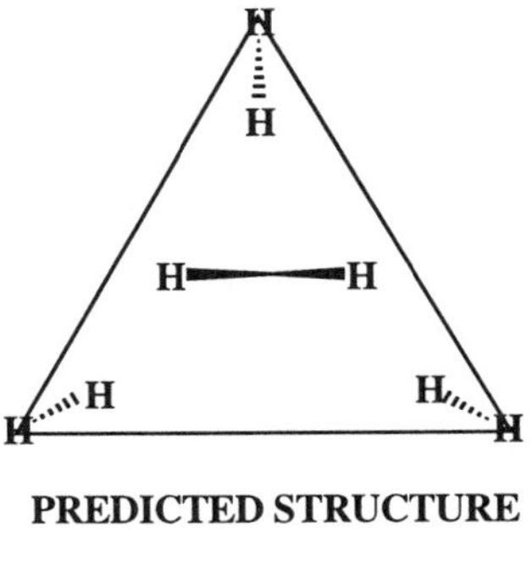

PREDICTED STRUCTURE

32.2 The Mysteries of the Oxidative Addition of Transition Metal Complexes

Collman and Hegedus exposed the puzzle of the oxidative addition of organometallic complexes in their marvelous monograph.[6] The key elements of the problem, drawn from the works of Collman, Stille, and their co-workers, are as follows:

(a) Addition of methyl halide to a square planar d^8 complex occurs kinetically in trans fashion even when the thermodynamically favored product is the one resulting from cis addition.

(b) Reductive elimination of two alkyl groups in PdR_2L_2 requires prior rearrangement of *trans*-alkyls to *cis*-alkyls. Thus, if cis elimination/addition is required and if the cis product is more stable, why is there ever trans elimination/addition, as suggested by the mode of addition of methyl halide to a square planar d^8 complex described in (a)?

(c) Reductive elimination in PdR_2L_2 requires prior departure of one L.

To this list, one additional key piece of data is that provided by the work of Puddephatt and co-workers: Metallacyclobutanes of the type $L_2X_2Pt(H_2CCHRCH_2)$ undergo reductive elimination to yield cyclopropane but they also undergo facile skeletal isomerization.[7a]

The VB explanation is simple:

(a) A square planar complex (geometry I) yields trans additions, but a stereoisomeric C_{2v} complex (geometry II) yields cis addition, which has as a prerequisite the departure of one L ligand, as illustrated. The crucial difference between I and II is the type of metal pairs reserved for binding the addend A—B; in I, the binding is effected by one metal lone pair (two electrons in one AO). In II, it is effected by one metal geminal pair (two electrons in two AOs).

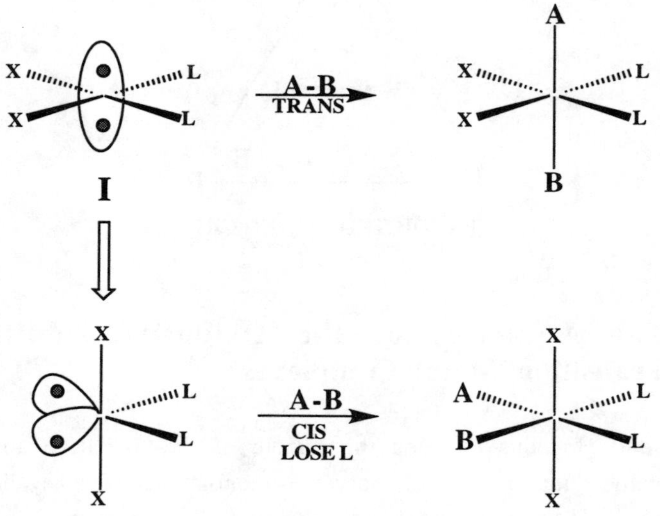

II

(b) Three reactive trajectories corresponding to three possible mechanisms of adding A—B to square planar complex I, two concerted and one stepwise (S_N2-like), are as follows:

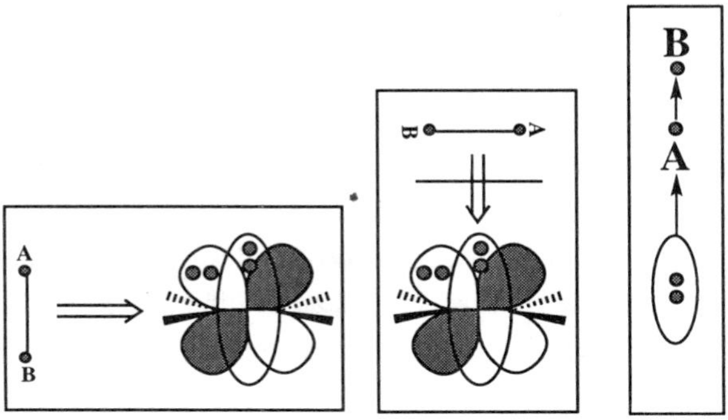

1 2 3

REBOUND MECHANISM

Focusing on the concerted mechanisms 1 and 2, we see that incipient cis addition, which maintains A—B bonding, is possible in either case because the d pair can engage A—B by one I bond in either 1 or 2 (see Section 26.7). However, A—B runs into the σ pair in 2, which becomes unfavorable compared to 1. As a result, kinetic trans addition is the result of a *rebound mechanism* according to which A—B becomes first engaged by an axial π-type d pair, and from there it rebounds to an axial σ-type hybrid pair to form the trans adduct. The key point is that the transition state of the process is a hexacoordinate species like the following, which is not connected with the cis adduct.

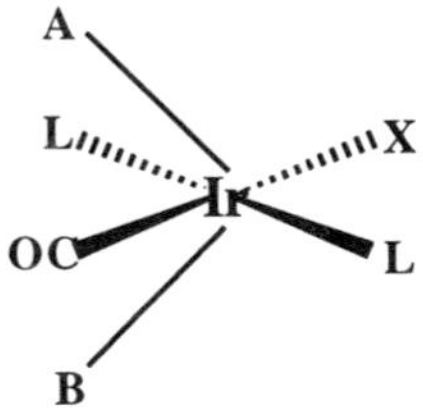

Transition State

In other words, once on the "rebounder" axial d pair, A and B can end up only trans on the axial σ pair. The mechanism predicts retention of configuration in A and B, which explains why trans is favored over cis addition without the requirement of L departure.

(c) Although cis addition maintains A—B bonding, it has two requirements: The square planar complex must adopt the C_{2v} geometry II, and one trans L ligand must depart before addition can occur. We explain by considering the reverse process of reductive elimination of a metallacyclobutane complex of the type $L_2M(H_2CCHRCH_2)$ as illustrated in Figure 32.3.

There are two crucial points. First, the spherical nature of the 1s AO makes it ideal for omnidirectional overlap, and this is why addition/elimination of H_2 is facile. The transition state has strong I-bonding. When H is replaced by alkyl, addition/elimination no longer occurs readily, and the complex must fall back on some overlap-independent mechanism of stabilization at the transition state. The inductive mechanism requires that the metal center act as a dipolar entity to induce a matching dipole on the two alkyl groups. This means that one of the two equatorial AOs directed to the alkyls must become doubly occupied and the other must be vacant. Then, to avert the formation of a four-electron antibond,

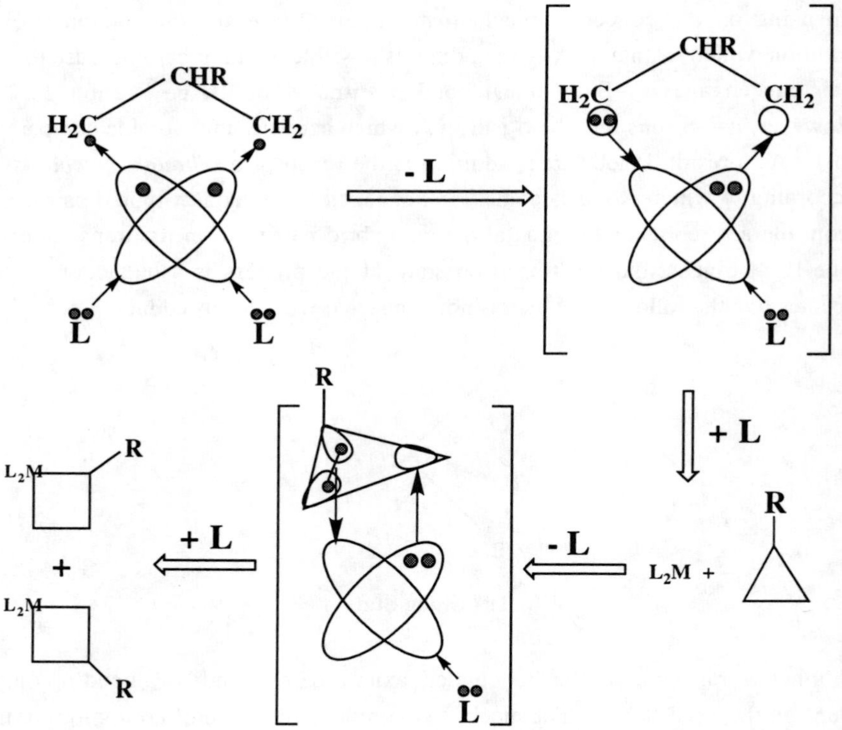

Figure 32.3 Reductive elimination of an $L_2M(H_2CCHRCH_2)$-type metallocyclobutane complex ($M = X_2Pt$).

the ligand attached on the doubly occupied equatorial hybrid must depart.

If the scenario of the "dipolar Pt" is correct, a second consequence is evident: A dipole of this type makes a perfect fit with the R shell of cyclopropane (see Chapter 4) by a formal "double bond"! In turn, this intermediate can decay to either of two isomeric metallacyclobutanes.

Thus, we see that trans-ligand departure and skeletal isomerization of the complex have a common origin: the ability of the metal to attain a dipolar state, which is simply described by VB formulas.

Space limitations preclude a discussion of the S_N2-type mechanism (which has more obvious consequences) and the radical chain mechanism (which we have not presented). A beautiful illustration of an S_N2-type reductive elimination of R—I and a concerted elimination of R—R from the same pentacoordinate cationic intermediate produced by ligand departure from a hexacoordinate octahedral precursor has been given by Goldberg and co-workers.[7b] With respect to the radical chain alternative, it ought to be mentioned that the crucial thing is attack by a metal radical (resulting from the addition of some initiator or propagator radical) on the halogen, X, of R—X, for reasons already discussed in Chapter 23.

32.3 Reductive Elimination and "Carbenes in Disguise"

We have already seen that groups of the type —AY, where A is an electropositive divalent fragment with a highly stabilized singlet ground state and Y is an electronegative nonmetal, act as carbenes in disguise when attached to carbon. The reason is carbenic resonance (R—A—Y in resonance with $R^+A{:}Y^-$), a concept introduced in Chapter 6. Since a divalent fragment A is a ROX ligand optimally designed for metal coordination, it follows that —AY groups, as well as appropriately substituted polyenyl and aryl systems (see Chapter 24), are carbenes in disguise when attached to metals. One consequence is that metal ligands that appear to be innocent alkyl derivatives, such as trifluorosilyl or trialkoxysilyl, are actually carbenes in disguise that will refuse to act as ordinary alkyl analogues. A clear demonstration of the concept of carbenic resonance can be found in the work of Lichtenberger and Rai-Chaudhuri.[8a] When $SiCl_3$ is attached to metal, as in $CpFe(CO)_2SiCl_3$, it acts as a "carbene in disguise" tying up one of the three metal π-type d pairs. This shows up by means of photoelectron spectroscopy as selective stabilization of the d_{xy} and d_{yz} AOs, that is, two (due to symmetry adaptation) of the three d AOs making up the "t_{2g} band" relative to $CpFe(CO)_2SiMe_3$ (where carbenic resonance is inoperative and the unoccupied space of Si has already played out its role). Stabilization of the third component ($d_{x^2-y^2}$ AO, see Figure 4 of Blomberg et al., ref. (a) of Table 32.1), suggests that the unoccupied space of Cl (the comparison being between SiR_3 and $SiCl_3$) also plays a (smaller) role, a trend anticipated in Chapter 28. We suggest that one consequence of carbenic resonance is the controversial short metal–silicon distances in metal–trichlorosilyl complexes in which the metal has d pairs. A summary of pertinent data can be found in the paper of Lichtenberger and Rai-Chaudhuri.[8a]

A second example is provided by the work of Milstein and co-workers, who studied the reductive elimination of $(Me_3P)_3IrH(CH_3)SiX_3$.[8b] They found that when X = OR, only methane is eliminated, whereas all three possible elimination reactions are observed when X = Et. We suggest that this is a consequence of the binding of $Si(OR)_3$ as a carbene in disguise, because $Si(OR)_2$ is expected to have a very large singlet–triplet gap by analogy to CF_2 and SiF_2.

We have seen that carbenes of two types have high metal affinity (high trans influence) because they act as ROX ligands: actual carbenes and carbenes in disguise. We can turn the problem inside out: Can a metal induce carbenic character in a "tenacious" organic molecule that has no carbenic character when in isolation? In other words, are there carbenes in disguise as a result of metal action? Carbon dioxide is a highly stable, error-free (T-formula) molecule, which is the paradigm of successful T-bonding. Is there a metal that can combine with CO_2 and effectively transform it to the carbenic diradical (a limiting situation, of course) as illustrated here?

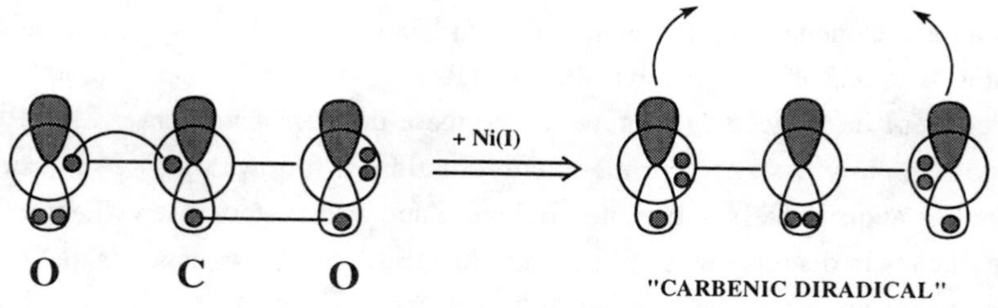

Ab initio calculations by Sakaki[9] suggest that this is possible. Specifically, a ds-hybridized Ni(I) (see Figure 25.4) can form an octahedral-type complex with four E-bound equatorial amines and one axial fluoride. The sixth axial position is then taken up by the CO_2. There are five electrons in the equatorial set (odd electron in the toroidal ds hybrid or in the d AO pointing toward four equatorial amines). The axial set is made up of two π-type d pairs and a σ-type hole, which are optimal for interaction with the carbenic diradical form of CO_2 in the following way:

The axial hole ties up the carbenic pair.

The d_{xz} pair interacts with the two formal oxygen radical sites in typical OX-delocalization fashion.

The d_{yz} pair interacts with the carbenic hole.

Clearly, the carbenic diradical acts as an acceptor of metal electron density. As a result, only the Ni(I) but neither the $Ni(I)^+$ nor the $Ni(II)^+$ complex is found to be bound. The message is clear: To successfully attach CO_2 to a metal, the axial π-type d pairs must be available. This requirement rules out utilization of "soft" ligands (e.g., phosphines). "Hard" ligands that bind by the E mechanism are needed. In turn, the last condition implies that the metal must have cationic character. However, cationicity must not be high enough to arrest donation to CO_2.

A closing comment on the virtual nonexistence of a conceptual theory of chemical bonding, despite the appearances that the opposite is true. Boron forms a tremendous variety of compounds but only when combined with the right elements. For example, BX_3 is unreactive as a Lewis acid when X = F, OR, and NR_2. By contrast, BCl_3, which has full metalloid character, yields interesting products upon reaction with normally inert organic molecules, as indicated in the second of the reactions that follow.

$$Cl_2B{-}BCl_2 + H_2C{=}CH_2 \longrightarrow Cl_2B{-}H_2C{-}CH_2{-}BCl_2$$

$$Cl_3B + Ph_4Sn \longrightarrow Cl_2B{-}Ph + Ph_3CSn{-}Cl$$

Now, the first (cis-stereoselective) reaction is the analogue of the $L_nOs{=}OsL_n$ plus ethylene (cis-stereoselective) retrocycloaddition discovered by Norton's group.[4] The only difference is that the polarization AOs are valence B2p in the boron but nonvalence Os6p in the osmium case. We suggest that many "anti-Woodward–Hoffmann" reactions of this type remain to be discovered. The second reaction is a bond switch process that either goes in one step through an E-selective transition state or proceeds in a stepwise fashion via an ylid-type intermediate, which can be formalized as $Ph_3CSn^+{-}Cl^-{-}BCl_2Ph$. The bonding of this species is analogous to the bonding along one diagonal of of $C(CO)_6$; that is, analogous to COCrCO (see Chapter 26). The only difference is that the vacant AO of chloride anion is effectively a starred d AO and, instead of two I bonds, we have two Q bonds linking the central Cl ion with the tin and boron ligands. The phenyl groups play an important role: They can act as formal phenyl cations or anions because these are actually "disguised excited carbene stabilized by a cationic π system" and "disguised ground carbene stabilized by an anionic π system," respectively.

Here is the point. The "theoretical understanding" of boron chemistry is based on the concept of the 3c-2e bond with diborane serving as the illustrator. Since, however, the designated illustrator is not at all a strong explainer, in the teaching of chemistry, the hydroboration reaction and the two prototypes mentioned above are presented as "material to be memorized."

References

1. (a) M.R.A. Blomberg, P.E.M. Siegbahn, U. Nagashima, and J. Wennerberg, *J. Am. Chem. Soc.* 113, 424 (1991). (b) M.R.A. Blomberg, U. Brandemark, and P.E.M. Siegbahn, *J. Am. Chem. Soc.* 105, 5557 (1983). (c) M.R.A. Blomberg and P.E.M. Siegbahn, *J. Chem. Phys.* 78, 986 (1983).
2. P.A.M. van Koppen, M.T. Bowers, E.R. Fisher, and P.B. Armentrout, *J. Am. Chem. Soc.* 116, 3780 (1994).
3. (a) D. Ritter and J.C. Weisshaar, *J. Am. Chem. Soc.* 112, 6425 (1990). (b) J.C. Weishaar, *Acc. Chem. Res.* 26, 213 (1993).
4. (a) R.T. Hembre, C.P. Scott, and J.R. Norton, *J. Am. Chem. Soc.* 109, 3468 (1987).
5. J.E. Bushnell, P.R. Kemper, P. Maitre, and M.T. Bowers, *J. Am. Chem. Soc.* 116, 9710 (1994).
6. J.P. Collman, L.S. Hegedus, J.R. Norton, and R.G. Finke, *Principles and Applications of Organotransition Metal Chemistry*, University Science Books, Mill Valley, CA, 1987.
7. (a) R.J. Puddephatt, *ACS Symp. Ser.* 211, 353 (1983). Review: V.K. Jain, G.S. Rao, and L. Jain, *Adv. Organomet. Chem.* 27, 113 (1987). (b) K.I. Goldberg, J. Yan, and E.M. Breitung, *J. Am. Chem. Soc.* 117, 6889 (1995).
8. (a) D.L. Lichtenberger and A. Rai-Chaudhuri, *J. Am. Chem. Soc.* 113, 2923 (1991). (b) M. Aizenberg and D. Milstein, *J. Am. Chem. Soc.* 117, 6456 (1995).
9. S. Sakaki, *J. Am. Chem. Soc.* 112, 7813 (1990).

Chapter 33

The Regioselectivity of Nucleophilic Addition to Coordinated Ligands

33.1 Five Experiments in Search of Interpretation

From all appearances, nucleophilic addition to coordinated ligands has failed to convince organometallic chemists that something is amiss! And yet, it is here that some of the clearest evidence for the concept of the I bond is found. Sections 33.1.1 to 33.1.5 present examples that have been discussed in the literature with varying amounts of self-deception.

33.1.1. Olefin Activation by Metals

Consider ethylene capped by a group Z that has two valence AOs and two valence electrons. When Z is CH_2, we have an "organic cycloalkane." When Z is ML_2, we have an "organometallic π complex." Now, regardless of what Z is, it is obvious that it will tie up (to a larger or smaller extent) the π system of ethylene. Thus, Z should deactivate the π bond of ethylene toward nucleophilic addition. However, the experimental results say otherwise: When Z is a metal fragment, regardless of whether charged or neutral, it often activates the olefin toward nucleophilic addition.[1] For example, it has been shown that olefinic complexes of $Fe(CO)_4$ undergo nucleophilic addition.[2] It has been suggested that activation is due to the willingness of the metal to "slip" in order to expose the olefin to the nucleophile.[3] But if all that molecular fragments had to do to be reactive was just . . . slip, then the existence of selectivity principles in chemistry would be very much in doubt. Everything can slip!

VB theory makes a clear prediction: To become activated, an olefin must be linked to a metal fragment by a REL bond, to permit the attacking nucleophile or electrophile to enter the transition state in a way that extends a codirectional arrow train already present in the olefin complex. The situation is illustrated in Figure 33.1. Note the requirement that the metal act with two AOs (2h metal) with one electron acceptance from (anionic metal) or donation to (cationic metal) the olefin. This explains several disparate observations:

The activation of olefins toward nucleophilic attack by $Fe(CO)_4$. This occurs because this fragment can act as a 2h anionic piece. $H_2Fe(CO)_4$ has one of the highest pK_a values among metal carbonyl hydrides.

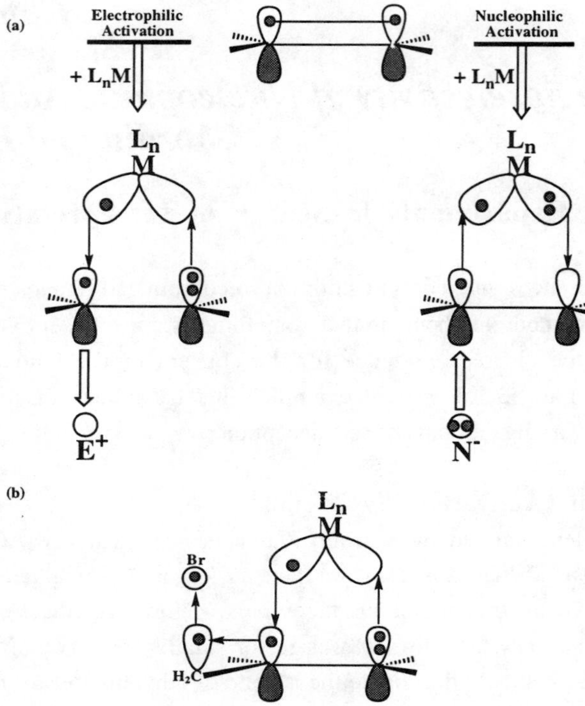

Figure 33.1 (a) Transition states of activated nucleophilic and electrophilic attack on coordinated olefin. Attack enhances the best codirectional arrow train already present in the substrate. (b) Solvolysis of an allyl halide–type molecule is effectively electrophilic attack on a coordinated olefin.

The great enhancement of solvolytic rates when a ferrocenyl group is attached to the carbon bearing a good leaving group.[4a] The situation is effectively that illustrated in Figure 31.1b.

The conversion of α,α-dihaloketones to iron-stabilized oxyallyl cations by $Fe(CO)_4$. Now, the latter acts as a 2h cationic piece.[4b]

33.1.2. Green's Rules

Green's rules[5] pertain to reactions in which the deviation from the "expected" is spectacular. Unfortunately, the considerable implications of these empirical rules for chemical bonding have not been properly appreciated. In synoptic form, the rules are:

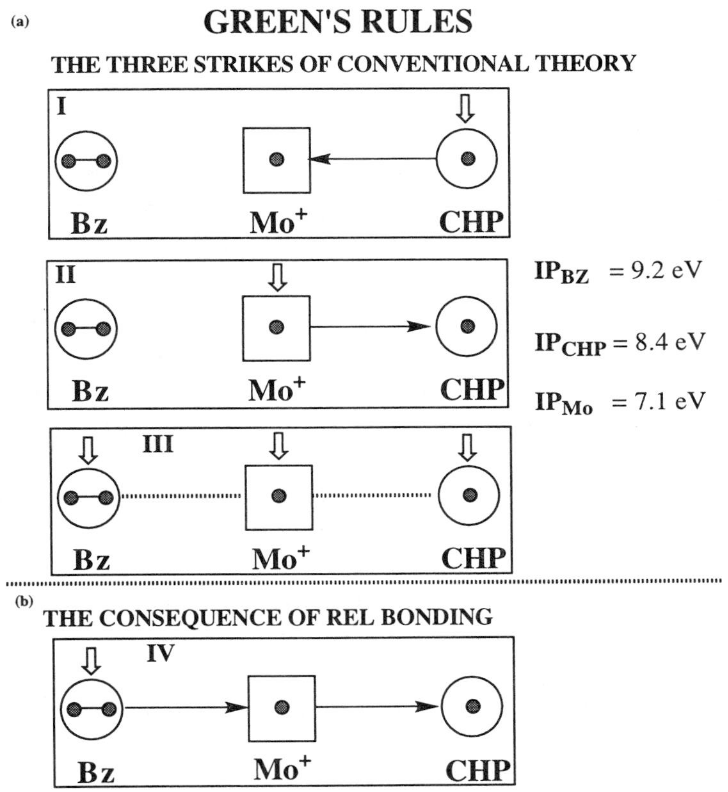

Figure 33.2 (a) Failure of conventional "arrow pushing" (resonance theory) and equivalent MO models to rationalize Green's first rule. (b) Regioselectivity of nucleophilic attack is determined by REL-bonding (codirectional arrow train). Solid arrows indicate predicted site of nucleophilic attack.

Even polyenes (e.g., butadiene) react before odd (e.g., allyl).

Open polyenes (e.g., butadiene) react before closed (e.g., benzene).

Terminal carbons are preferred in even polyenes, but interior carbons are selected in odd polyenyls, unless the metal becomes electron poor.

These rules are applicable to cationic complexes in which the metal carries a unit of positive charge.

Green's first rule says that nucleophilic attack occurs preferentially on an even rather than on an odd polyene. What would be predicted by a chemist predict who did not know the result? The cationic complex can be formulated as EM^+O, where E stands for the even and O for the odd polyene and M^+ for the cationic metal. The pattern of interaction of the frontier MOs is crucial: The HOMO of E, the semioccupied MO of M^+, and the SOMO of O have the same symmetry. According to conventional thinking, there is a covalent bond linking M^+ and O. Here is the key observation: The preference for E over O nucleophilic attack persists even when the difference in the IPs of the neutral E and O is very large. For example, attack on ethylene (IP = 10.5 eV) takes precedence over attack on cyclopentadienyl (IP = 8.9 eV) when the two are coordinated to $(CO)_2Fe^+$. Attack on benzene (IP = 9.2 eV) takes precedence over attack on cycloheptatrienyl (IP = 6.4 eV) when the two are coordinated to Mo^+. Thus, even a formal cycloheptatrienyl cation seems to be unfavorable!

Since there are three conventional ways of managing the electrons, as shown in Figure 33.2, and none can rationalize the observations, there must be something wrong with the intuitive resonance formulation of the EM^+O metal complex.

Scenario I of Figure 33.2a is seemingly the most logical: Electron delocalization should occur from the most electropositive ligand to the metal cation. In other words, one would normally represent the $[BzMoCHP]^+$ complex [E = benzene (Bz), M = Mo, and O = cycloheptatrienyl (CHP)] as follows:

$$BzMo^+CHP \longleftrightarrow BzMoCHP^+$$

This predicts a preferential attack by the nucleophile on the odd polyenyl, contrary to Green's rule.

Scenario II is close to the heart of EHMO practitioners and chemists comfortable with oxidation state formalism. The CHP radical has a one-electron SOMO that lies lower in energy than the *neutral* metal d AOs. One may now propose that the major contributing resonance structure is $BzMo^{2+}CHP^-$. Once again, this scenario fails because it predicts attack on the metal.

In scenario III, an act of desperation that leads nowhere, hatched lines to represent some undefined delocalization that predicts everything and nothing.

33.1.3. Nucleophilic Attack on Butadiene Complexes.

In $(CO)_3FeBU$ (BU = 1,3 butadiene), the pseudo-octahedral Fe of $Fe(CO)_3$ extends to the diene three conical AOs containing two electrons. Careful experiments that monitored kinetically controlled product formation by Semmelhack's group have revealed that nucleophilic attack occurs at C2 of 1,3-

butadiene.[6] The FO-PMO model fails to describe the selectivity of nucleophilic attack on coordinated polyenes. Specifically, nucleophilic attack is predicted to occur on C1 of butadiene in $(CO)_3FeBU$ because the LUMO of the complex is essentially the BU LUMO.[7] An unfortunate response to failure of the FO-PMO model is to look for some other orbitals to get out of trouble. Typically, filled–filled interactions are postulated, despite incompatibility with the following general principle: Maximization of delocalization and minimization of overlap repulsion go hand in hand in EHMO theory, and this is why qualitative trends are unaffected if overlap is neglected. An aromatic species has maximal CT delocalization and minimal exchange repulsion, an antiaromatic complex lies at the antipodes, and a nonaromatic complex is in between.

33.1.4. Nucleophilic Attack on Cationic Cyclohexadienyl Complexes

In cationic $(CO)_3Fe$–cycloxadienyl complexes, nucleophilic addition is regiospecific (terminal C-atom) and stereospecific (exo face). Pearson, who has given an account of the synthetic utility of this complex, points out that the FO-PMO model fails to produce a clear picture.[8]

33.1.5. Ligand Coupling

Recently, there has been considerable progress in the synthesis of organic rings by utilization of organometallic complexes. One example is (6 + 4) cycloaddition effected by photolysis of a $Cr(CO)_3$–triene metal complex.[9–12] The way in which an organometallic fragment "directs" a cycloaddition of two organic molecules, A and B, can be visualized as follows:

One partner, A, is coordinated to the organometallic fragment L_xM, which acts as an *n*h/*m*e piece.

The L_xM fragment is activated, either thermally or photochemically, by loss of one L ligand (or more). The resulting $L_{x-1}M$ fragment has different synapticity and a different number of electrons available for binding A.

The activated $L_{x-1}M$ binds a second molecule, B.

A and B combine in a ligand coupling reaction.

A simple example of how loss of a ligand causes a change in synapticity and electron count of the metal fragment is given in Figure 33.3. Loss of a carbonyl from $Cr(CO)_3$ unties one equatorial pair of pseudo-octahedral Cr. As a result, the metal is now free to abandon 3h/0e in favor of a 4h/2e configuration. Two

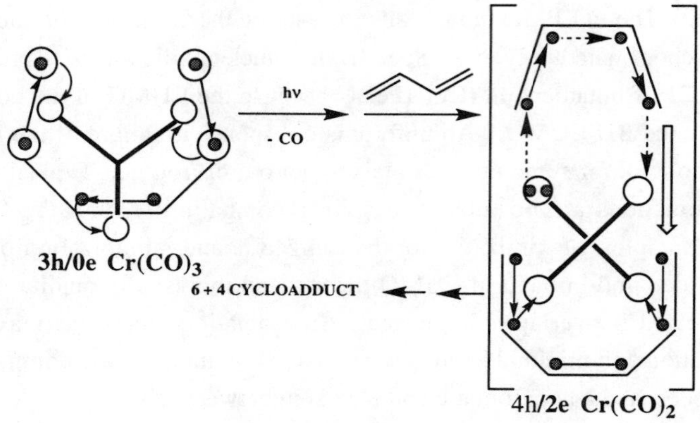

Figure 33.3 Ligand coupling brought about by a codirectional arrow train initiated by a metal pair.

unstarred holes act to tie up one polyene, and one hole plus one pair tie the other.

33.2 Error-Free Arrow Codirectionality as the Common Denominator of Facile Nucleophilic Additions

The common denominator of all five problems described in Section 33.1 is readily projected by the I formulas of the reactive complexes. In each case, an error-free codirectional arrow train links an actual or formal nucleophile with the olefin and the metal.

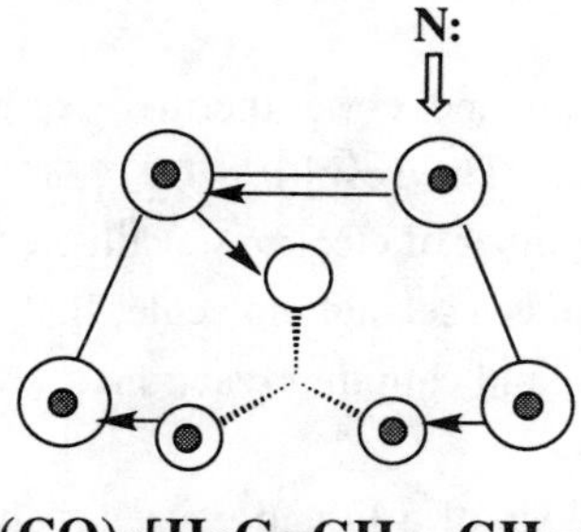

$Fe(CO)_3[H_2C{=}CH{-}CH{=}CH_2]$

Figure 33.4 Nucleophilic attack on $(CO)_3Fe$(1,3-butadiene) complex. Attack enhances the best codirectional arrow train already present in the substrate.

Having illustrated the $Fe(CO)_4$–olefin complex undergoing nucleophilic attack in Figure 33.1 [$L_nM = (CO)_4Fe$], we show in Figure 33.4 how the $(CO)_3$FeBU complex undergoes the same experience. In both cases, the nucleophile initiates a codirectional arrow train that terminates in a metal hole.

The I formula of $CpFeBz^+$ shown in Figure 33.5 is identical to the I formula of ferrocene with Cp^- replaced by Bz. The I formula of the cationic $(CO)_3$Fe–cyclohexadienyl complex is shown in Figure 33.6. The 6+4 cycloaddition of the $(CO)_3$Cr–(diene)(triene) complex (Figure 33.2) is effectively an intramolecular nucleophilic addition analogous to nucleophilic addition to the cationic $(CO)_3$Fe–cyclohexadienyl complex. The crucial point is that nucleophilic attack of each of the two cationic metal complexes [and, by analogy, the $(CO)_3$Cr–(diene)(triene) complex reaction] goes through a transition state featuring an arrow train initiated by a nucleophilic pair and terminated in a metal (or ligand) hole.

THE I-FORMULA OF $CpFe^+Bz$

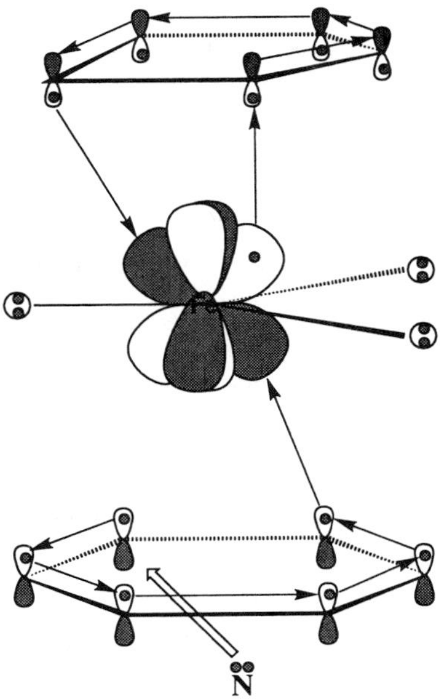

Figure 33.5 Selectivity of nucleophilic attack on $CpFeBz^+$. Attack enhances the best codirectional arrow train already present in the substrate. Double arrow indicates the preferred site of nucleophilic addition.

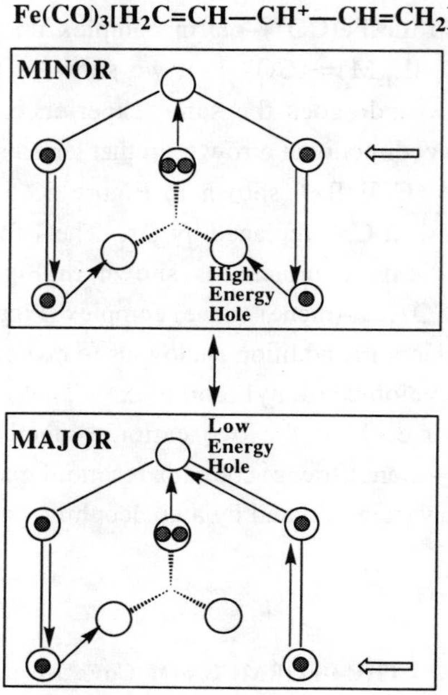

Figure 33.6 Selectivity of nucleophilic attack on $(CO)_3Fe(H_2C{=}CH{-}CH^+{-}HC{=}CH_2)$. Major contributor of the complex has two codirectional arrows linking a π bond with a carbon hole rather than with a (higher energy) metal hole. Attack enhances the best codirectional arrow train already present in the substrate. Preferred site of nucleophilic addition in each case is indicated by double arrow.

As a final illustration of the consequences of activated arrow codirectionality, we show below the transition state for the coupling of two olefins or acetylenes by CoCp in Figure 33.7.[13] Once again, note how a terminal hole and a terminal pair activate a codirectional arrow train which effects ligand coupling.

33.3 Green's Rules Prove the Concept of the REL Bond

The I formula of a prototypical transition state in which one can determine the positional selectivity in both E and O ligands of a cationic transition metal complex is shown in Figure 33.8. The preferred site of nucleophilic attack is

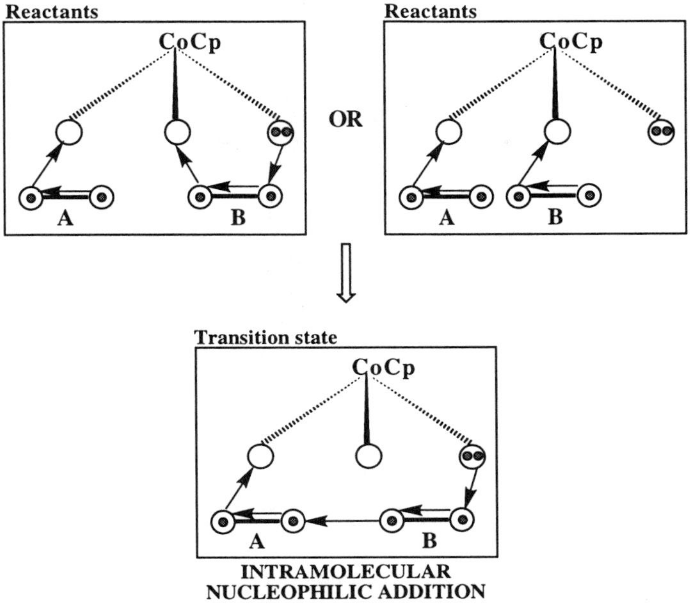

Figure 33.7 Coupling of two olefins (acetylenes) by CpCO.

the one consistent with formation of the longest, error-free codirectional arrow train. This rationalizes Green's third rule.

Finally, in the model EM^+O system, E is predisposed to act as donor and O as acceptor. A closed O has a higher electron affinity than an open O (e.g., Cp vs. 1,3-pentadienyl) and an open E has a lower ionization potential than a closed E (e.g., 1,3,5-hexatriene vs. benzene). This rationalizes Green's second rule. However, the most important of all three rules remains the first: even before odd. Why does CHP refuse to donate one electron to Mo^+ in $BzMoCHP^+$? Because the Bz has to accept it! In other words, ionization of CHP is linked to acceptance by BZ because the bond linking the two π ligands to the metal is a REL bond, not a covalent bond (see, by analogy the $BzFeCp^+$ complex in Figure 33.5). Bz is intrinsically a poor acceptor, and this is why the REL bond is directed from the even to the odd ligand rather than the other way around.

33.4 The Diborane Cleavage by Nucleophiles

Diborane is cleaved by weak nucleophiles to produce either symmetrical or unsymmetrical products.[14] Intuitively, one expects the first mode to dominate, yielding H_3BN (N = nucleophile). Yet the second mode, leading to

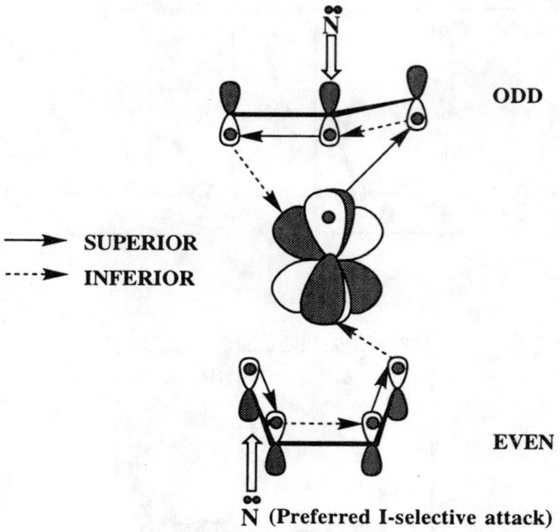

Figure 33.8 Rationalization of second Green rule.

thermodynamically inferior products, is frequently observed. Why does the $H_2BN—H—BH_3$ intermediate favor nucleophilic attack at the boron atom that already has a nucleophile attached? The answer is given in Figure 33.9, assuming a reactant-like transition state and I selectivity. The preferred nucleophilic attack is the one the conserves arrow codirectionality.

There have been many attempts to compute intermediates and transition states of ionic addition and substitution reactions occurring in solution. A commonplace error is one pointed out recently by Siegbahn: Neglect of the counterion of the ionic electrophile or nucleophile is catastrophic![15] Of course,

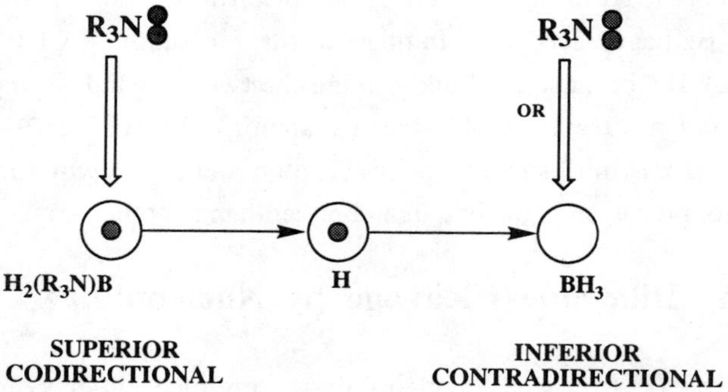

Figure 33.9 Orientational selectivity of the nucleophilic attack of $H_2(R_3N)B—H—BH_3$.

this is well-known to the thinking experimentalist. For example, the chemistry undergraduate learns that electrophilic addition of HCl to a double bond occurs in solution via the intermediacy of a high energy carbenium ion, which results from protonation of the double bond. On the other hand, an alkene has positive proton affinity in the gas phase. Proton plus olefin, that is, has higher energy than the resulting cationic complex (a carbenium ion or a nonclassical isomer) in the gas phase: The cationic complex is more stable than the reactants in the gas phase. This means that the correct model of the solution reaction is actually "H^+ coordinated to Cl^- plus olefin" (rather than "proton plus olefin") to yield a tight or solvent-separated ion pair. Siegbahn's computations are meaningful to the extent that they show that the counterion can influence the (metal vs. ligand) regioselectivity of nucleophilic attack on a transition metal complex.

References

1. (a) S.G. Davies, *Organotransition Metal Chemistry: Applications to Organic Synthesis*, Pergamon, Oxford, 1982, Chapter 4. (b) M. Rosenblum, *Acc. Chem. Res.* 7, 122 (1974).
2. (a) M.R. Baar and B.W. Roberts, *J. Chem. Soc., Chem. Commun.* 1129 (1979). (b) B.W. Roberts and J. Wong, *J. Chem. Soc., Chem. Commun.* 20 (1977).
3. O. Eisenstein and R. Hoffmann, *J. Am. Chem. Soc.* 103, 4308 (1981).
4. (a) A.J. Pearson, *Science* 223, 895 (1984). (b) R. Noyori, *Acc. Chem. Res.* 12, 61 (1979).
5. S.G. Davies, M.L.H. Green, and D.M.P. Mingos, *Tetrahedron* 34, 3047 (1976).
6. (a) M.F. Semmelhack and J.W. Herndon, *Organometallics* 2, 363 (1983). (b) M.F. Semmelhack and H.T.M. Le, *J. Am. Chem. Soc.* 106, 2715 (1984). (c) M.F. Semmelhack and T.M.L. Hahn, *J. Am. Chem. Soc.* 106, 2715 (1984); 107, 1455 (1985).
7. P. Sautet, O. Eisenstein, and K.M. Nicholas, *Organometallics* 6, 1845 (1987).
8. A.J. Pearson, *Metallo-organic Chemistry*, Wiley, New York, 1985.
9. J.H. Rigby, *Acc. Chem. Res.* 26, 579 (1993).
10. J.S. Ward and R. Pettit, *J. Am. Chem. Soc.* 93, 262 (1971).
11. P.A. Wender, N.C. Ihle, and C.R.D. Correia, *J. Am. Chem. Soc.* 110, 5904 (1988).
12. C.G. Kreiter, *Adv. Organomet. Chem.* 26, 297 (1986).

13. R. Boese, D.F. Harvey, M.J. Malaska, and K.P.C. Vollhardt, *J. Am. Chem. Soc.* 116, 11153 (1994).
14. K.F. Purcell and J.C. Kotz, *An Introduction to Inorganic Chemistry*, Saunders College Publishing, Philadelphia, 1980.
15. P.E.M. Siegbahn, *J. Am. Chem. Soc.* 117, 5409 (1995).

Chapter 34

Error-Free and Error-Full Organometallic Transition States

34.1 A Reexamination of Chemical Reactivity Across the Periodic Table

VB theory provides the basis for classifying reaction transition states:

Transition states are I selective, E selective, or frustrated.

The hallmark of a successful I-selective transition state is error-free arrow codirectionality.

One hallmark of a successful E-selective transition state is alternation of electropositive and electronegative atoms.

A frustrated transition state is capable neither of error-free arrow codirectionality nor of of hole–pair alternation. A lower energy path involving a T-bound diradical intermediate is now possible.

Our first goal is to search for error-free codirectional transition states. Ideally, a codirectional arrow train is initiated by one pair at one terminus and one pair at the other terminus of an extended linear array made up by the active AOs of the reacting molecules:

ACTIVATED ARROW TRAIN - HOLE-EXCESSIVE COMPLEX

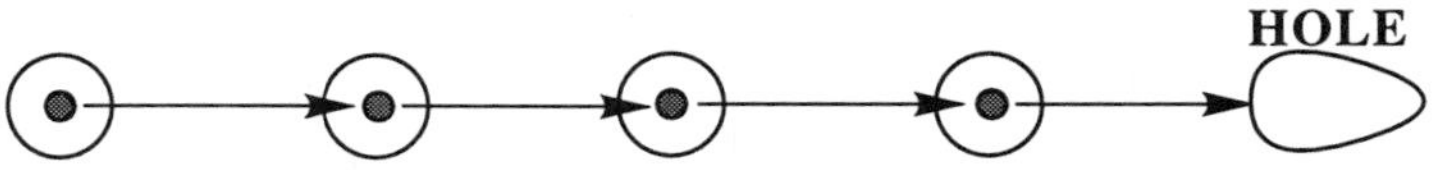

ACTIVATED ARROW TRAIN - PAIR-EXCESSIVE COMPLEX

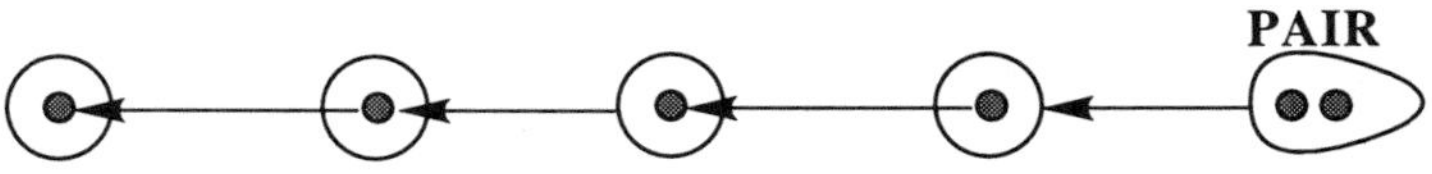

ACTIVATED ARROW TRAIN - HOLE/PAIR MATCHED COMPLEX

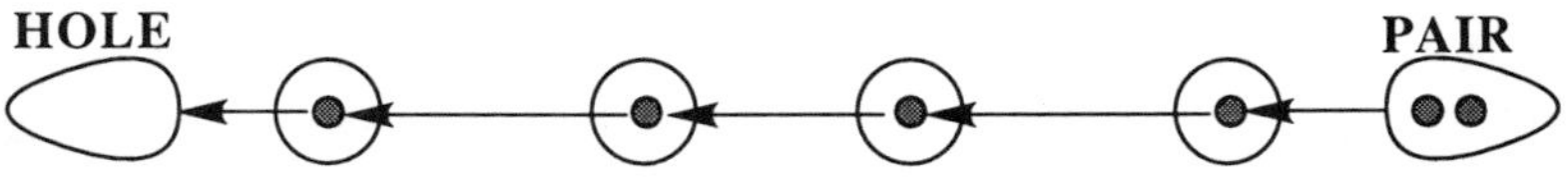

A transition state of this type is called *H/P matched.* Significant stability is also guaranteed if either a single hole or a pair initiates the codirectional arrow train. These transition states are called *H-excessive* or *P-excessive* states, respectively. To simplify things, we assume that orbital electronegativity does not restrict the degree of association at the transition state; that is, we assume that all transitions states are uniformly bound rather than bikratic. This is actually very likely whenever codirectional arrow trains are activated.

34.2 Error-Free, H/P-Matched Transition States

The names of Hofmann, Curtius, Schmidt, Lossen, and Wolff are associated with the early phase of the development of organic chemistry. Many a student had to memorize the corresponding "name reactions," which have one common denominator: They are rearrangements of an alkyl group attached to carbonyl to an adjacent true or incipient carbene or nitrene.[1]

A related problem is the 1,2-hydrogen shift in carbenes.[2] Rearrangements of the latter type occur via quantum mechanical tunneling[3] with a saddle point that can be formulated as an error-free, H/P matched complex of the following form.

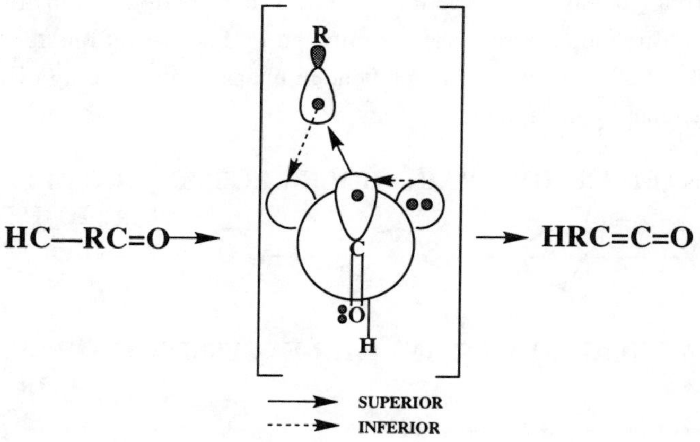

The cycloaddition of ketene plus alkene was once celebrated as a confirmation of the Woodward–Hoffmann rules. However, it has a nonaromatic transition state that can be formulated as an error-free, H/P-matched complex as shown (the O2p pair triggers the codirectional arrow train, which is propagated without errors because the oxygen atom is at the origin).

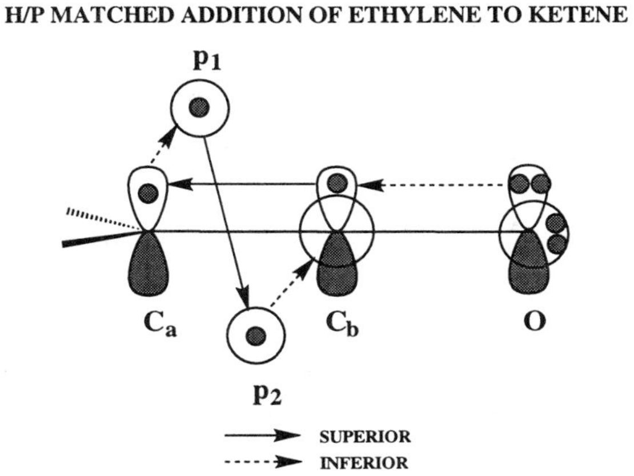

By following the arrows of the I formula, we conclude that the carbonyl carbon has carbenic character, as has been confirmed by ab initio computations.[4]

34.3 Error-Free, P-Excessive Transition States

The reductive fragmentation of bismuth polyphenyls places phenyls along a carbon chain on positions normally inaccessible by routine synthetic means. Barton[5] showed that under basic conditions, ketones, phenols, enols, and nitro compounds can be converted in high yield to very hindered arylated derivatives by $BiPh_3X_2$ reagents. The transition state can be formulated as a P-excessive complex as follows:

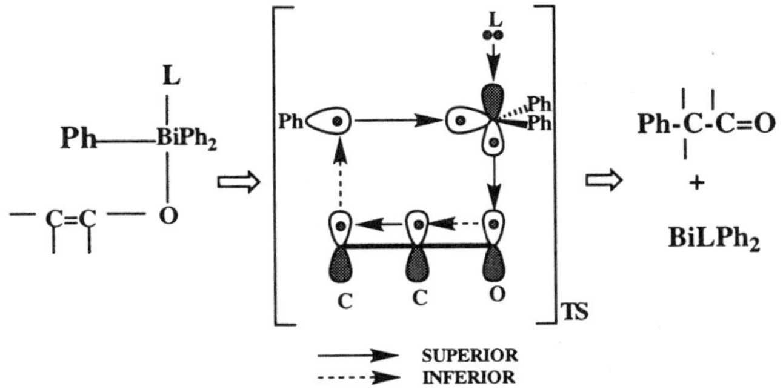

We expect that reductive fragmentation of U—M—R, where M is either a transition metal or a heavy p-block semimetal and U is a π-unsaturated fragment, will turn out to be a general, synthetically useful reaction. Already, in addition to bismuth, ligand coupling has been demonstrated for S(IV) and P(V) by Oae and co-workers,[6] for Te(IV) and Cu(III) by Barton and co-workers,[5] and for Pb(IV) by Pinhey and colleagues.[7]

34.4 Error-Free, H-Excessive Transition States

The I formula of the transition state of the addition of BH_3 to an olefin, an all-important reaction,[8] is as follows:

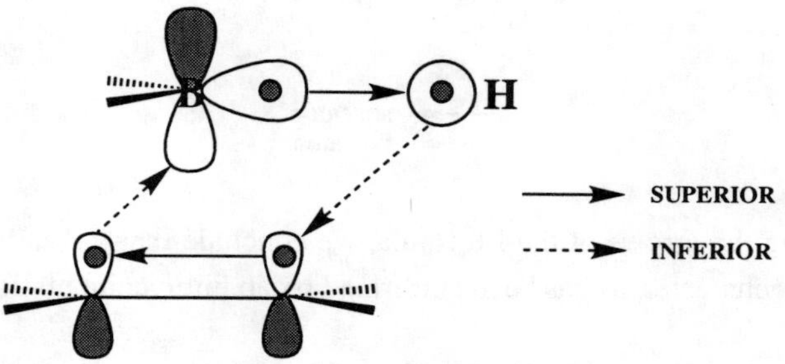

Since C and H have comparable electronegativities, no errors are incurred. Note the terminal B2p hole that triggers the codirectional arrow wave. An atomic hole can accept an arrow without penalty even if the arrow originates in a more electronegative atom. The same holds true for atom pairs.

The addition of BH_3 to ethylene is *not* a 2+2 cycloaddition! Rather, it is an H-excessive reaction of broad generality:

$$\text{B—Z} + \text{X—Y} \longrightarrow \text{B—X} + \text{Y—Z}$$

We expect that reactions of this type for different Z, X, and Y will be facile stereospecific reactions if no errors are involved. For instance, allylboranes undergo very fast [1,3] shifts in contrast to the [1,3] shifts of carbon centers. The very facility of these reactions limits their synthetic utility.[9,10]

$$R_2B\text{—}CHR'\text{—}CH{=}CH_2 \longrightarrow CHR'{=}CH\text{—}CH_2\text{—}BR_2$$

When R = allyl and R′ = H, the molecule is fluctional. Boranes may attack C—H bonds, liberating H_2 gas.[11] Alkyl-to-carbonyl migration on a transition metal is celebrated in monographs and journals.[12] By contrast, the corresponding boron reaction is practically unknown to most chemists. The presence of low-lying vacant orbitals thwarts isolation of the product acylborane. The mechanism is unknown.

34.5 Erroneous H-Excessive Transition States

The hallmark of an "erroneous reaction" is the presence of electronegative (electropositive) atoms flanked by electropositive (electronegative) atoms. *The presence of errors does not condemn a reaction complex.* Instead, it implies that this may have a very favorable E-, rather I-type, bonding. One example of a highly erroneous transition state is that of the Meerwein–Pondorff–Verley (MPV) reaction shown in Figure 34.1. We show the ideal I formula as well as the actual I formula, with the erroneous arrows labeled as such. The errors occur whenever arrows point from O to C or from O to Al. This implies that the complex is E- rather than I-bound. The hallmark of E-bonding is hole–pair alternation promoted by alternation of electronegative and electropositive (in a relative sense) atoms: —C—O—Al—O—C—H—. This is why Brown and co-workers have been successful in the synthesis of optically active alcohols by use of analogous boron reagents.[13] A related reaction is β-hydride transfer from R_2Mg to a ketone when normal Grignard action is blocked.[14] Other reactions (of yet undetermined mechanism) are known that effect cleavage of all three R—B bonds of trialkylboranes at 0°C![15]

MEERWEIN-PONDORFF-VERLEY REACTION

$$3\ R_2C{=}O + Al(i\text{-}PrO)_3 \longrightarrow (R_2CH\text{-}O)_3Al + 3\ CH_3COCH_3$$

IDEAL-CODIRECTIONAL

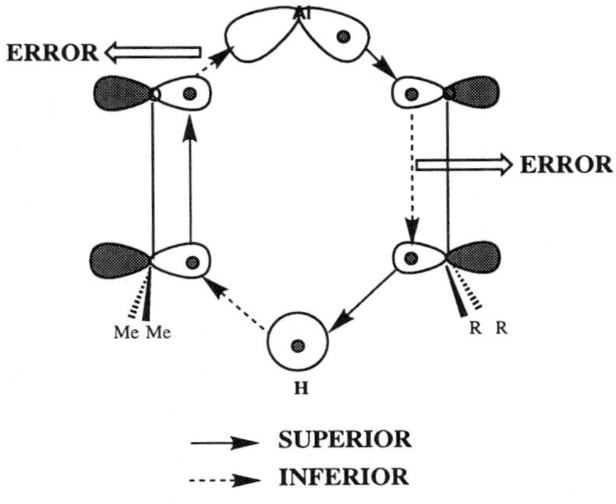

Figure 34.1 The two errors of the I-type transition state of the Meerwein–Pondorff–Verley reaction.

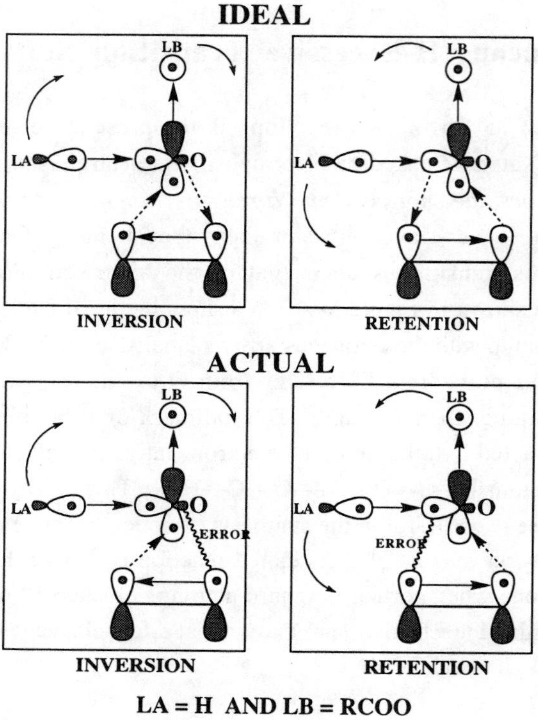

Figure 34.2 Oxygen atom transfer from RCOOOH to olefin.

34.6 Oxygen Insertion and the Heteroatom Exclusion Rule

The epoxidation of an olefin by peroxy acid is an example of oxygen insertion into a π bond. Oxo metals are also known to insert oxygen to π as well as σ bonds. What is the relationship of the two types of reaction? Assuming that the oxo metal has a formal M═O bond (but see Section 34.8), the I formulas speak for themselves.

In the case of epoxidation by peroxy acid, the inserting oxygen is flanked by carbon and by oxygen at the reactant stage. At the transition state, the transferring O is linked to one H, one O, and two C atoms. We represent this state of affairs by the notation (H,O)O(C,C), where the central atom is the transferring atom and the atoms in parentheses are those connected to it. Error avoidance requires that two flanking atoms act as donors and two as acceptors. It is readily seen that there is one error because the condition is not fulfilled: Three of the four atoms in parentheses, namely, C, C, and H, act as donors with respect to O. The I formulas for the two possible stereochemical modes of transfer confirm the presence of only one error (Figure 34.2). It involves donation from O to C. Since the electronegativity difference is not very large, the reaction is expected to be I selective but far from perfect. It is well known that the reaction is highly stereoretentive, and this implies a concerted process that can be viewed as nucleophilic attack by the olefin on the O—O bond coupled to "back donation" of one O2p lone pair to the olefin.[16]

The message is that molecules of the types X—O—O—Y, XYO—O, X—O, and even O═O (singlet oxygen) are capable of stereospecific oxygen transfer (epoxidation), provided X and Y are not much more electropositive than O. The transferring oxygen has oxenic character (singlet oxygen atom) and acts via an O2p pair and an O2p hole to make the new bonds, while the O2s lone pair is polarized away from the reaction centers by a hole borne by the atom connected to the oxenic oxygen. In other words, the transferring reagent must feature a doubly bonded oxygen with one active O2p lone pair. The existing data are consistent with this hypothesis.[17] In this regard, an interesting case is the behavior of N_2O, which is a resonance hybrid of the following structures:

$$\mathrm{N{=}N{=}O} \longleftrightarrow \mathrm{N{\equiv}N{-}O}$$

The "allenic" structure enables the molecule to transfer the oxygen atom to a saturated molecule.[18a,b] On the other hand, the "acetylenic" structure enables N_2O to transfer the terminal nitrogen atom to an unsaturated 3h/3e fragment acting with unstarred valence AOs, such as L_3Mo (L = NRAr),[18c] as follows:

$$\mathrm{L_3Mo + N{\equiv}N{-}O \longrightarrow L_3Mo{\equiv}N + NO}$$

As discussed in Chapter 27, the hallmark of L_3Mo is that it acts with three unstarred valence AOs of an sd^3-hybridized Mo to tie up the nitrogen atom by a ROX bond as illustrated:

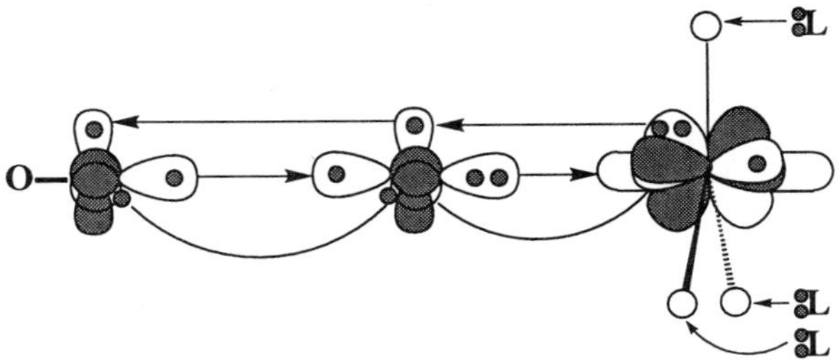

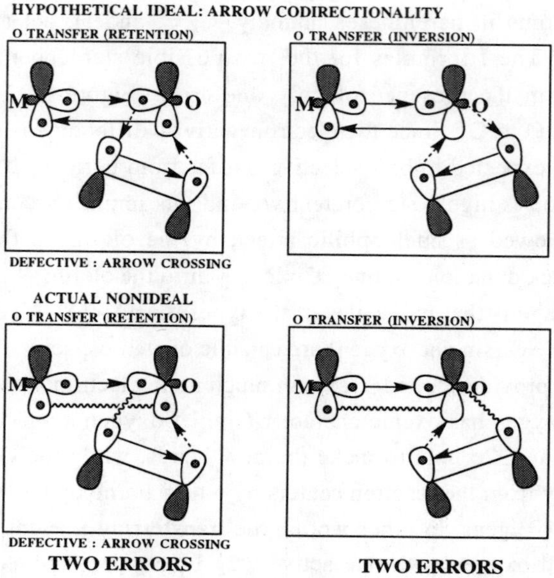

Figure 34.3 Oxygen atom transfer from $L_nM{=}O$ to olefin.

In the case of epoxidation by oxo metal, one can write analogous transition states as illustrated in Figure 34.3. Now, there is a major difference: We have an (M,M)O(C,C) array (M = metal), which means that there are two errors but we cannot tell whether these involve transfer from O to M or from O to C. To find the answer, we must construct the I formula and identify the errors. As shown in Figure 34.3, there is one major M/O and one minor C/O error. Hence, the oxygen transfer may well occur in a stepwise fashion via a diradical intermediate. The accumulated data suggest two points:

(a) A tremendous amount of effort continues to be expended in elucidating the mechanism of atom transfer from metal to olefin or alkane.[19,20] One striking aspect of the oxygen transfer reaction is the observation of stereospecificity with unactivated olefins and loss of stereochemistry in the case of olefins bearing radical-stabilizing substituents.

(b) Chromyl chloride reacts with alkanes in a stepwise fashion involving radical intermediates.[21]

The situation improves dramatically when oxygen is replaced by H_2C as the transferring group. Now, the M/O and C/O errors give way to a single M/C error. The reaction is now expected to be one-step. Furthermore, the I formulas reveal that the inversion path is superior for two reasons:

Crossing arrows indicate high contact repulsion and inferior I-bonding in the retention path.

Minimization of exchange repulsion requires that the donor metal must rotate away from the olefin in the process of bond making. Such rotation is possible only occur the inversion path.

The evidence points to one-step transfer of CX_2 from metal to olefin with inversion.[22] The thing to note here is that we now have a theory that differentiates CH_2 from NH and from O. In Chapter 24, we differentiated SiH_2 from PH and from S (see Section 24.4).

Two elements in the preceding discussion go contrary to conventional wisdom.

(a) A metal–oxygen "double bond" is, according to the popular oxidation state formalism, a doubly ionic bond ($M^{2+}O^{2-}$), and this implies that MO will add via ionic (rather than radical) intermediates. As a result, the observation of radical intermediates in reactions of metal oxides and organic molecules seems incongruous. The VB formulation of the MO bond as a REL bond removes the problem: Promotion to a diradical state from which further reaction can occur is feasible by a single (rather than double) electron transfer from oxygen to metal that effectively transforms a REL bond to an inferior T bond.

(b) What determines the reaction mechanism is not the strength of the reactant bonds that must be broken but, rather, whether the constituent atoms can support strong association at the transition state. The heteroatom exclusion rule says that an A—Z bond, where A is highly electropositive and Z highly electronegative, will fail to associate with K—K bonds, where K is an electroneutral atom (i.e., an atom with intermediate electronegativity). Two misdirected arrows must be eliminated, as indicated:

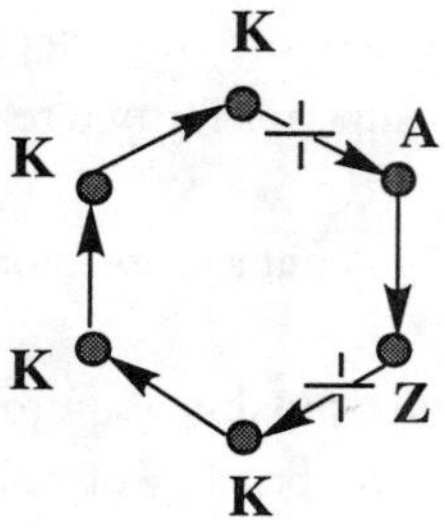

This means that M=O reagents are unlikely participants in pericyclic reactions. One implication is that cis-hydroxylation of olefins by permanganate may not proceed by initial 3+2 cycloaddition but, rather, by a 2+2 cycloaddition via a diradical intermediate that collapses to the 3+2 cycloadduct before bond rotation can occur.[23]

Bottom line: Combining an electropositive metal, an electronegative nonmetal (e.g., O), and electroneutral carbon and hydrogen is certain to produce errors in a pericyclic transition state no matter what its type (1+2, 2+2, 3+2, 4+2 cycloaddition, etc.). As a result, a pathway involving distinct T-bound diradical intermediates is always the most likely reaction mechanism.

34.7 Loop Reactions; The World's Best Transition State

The transition state of the addition of BH_3 to olefin, a prototypical *migratory insertion*, has low energy because of two physical conditions:

One hole activates a codirectional arrow train.

The string of AOs form a *loop* (fold of a thread) in which an acyclic AO array crosses over the same atom (*cross-atom*).

The two AOs of the cross-atom that partake in the loop are called *cross-AOs*, and they are subject to the condition of being degenerate or quasi-degenerate orthogonal AOs. In this way, the array is formally nonaromatic. The degeneracy condition tends to be met when the cross-atom is semimetallic (2s and 2p AOs of B and heavy p-block atoms) or metallic [(n – 1)d and ns AOs of transition metals]. At the same time, the atom is polarizable. As a result, overall I-bonding due to the activated codirectional arrow train becomes especially favorable.

In our example, the cross-atom is B and the two cross AOs are Bsp^x and B2p. CT delocalization involves, in part, one electron transfer from the former to the latter AO, and this can occur without charge separation (i.e., without the penalty of unfavorable Coulombic interaction). In other words, this electron hop is defined by two quasi-degenerate configurations, and we know that CT delocalization is facilitated by configurational degeneracy (Chapter 15).

Reactions of this type are called *hole-excessive loop reactions*. This is only one of three categories of loop reactions: H-excessive, P-excessive, and H/P-matched (H = hole, P = pair). The crucial point is that these formally nonpericyclic reactions have an attribute that the normal pericyclic reactions do not have: configurational degeneracy at the transition state because of the loop aspect. It follows that activated loop reactions must be competitive with conventional "allowed" pericyclic reactions. One look back in this chapter will indicate that all reactions discussed thus far have been loop reactions. For example, all group transfer reactions are loop reactions in which the transferring atom is the cross-atom. Indeed, the Woodward–Hoffmann rules have limited utility in organometallic chemistry precisely because the recurring reactions in this branch of chemistry are activated loop reactions!

What are the relative merits of the three categories of loop reactions? An E-selective and an I-selective transition state depend for stability on hole–pair alternation. This can be induced by means of an excess hole or an excess pair or a strategic placement of both. On this basis, an H/P-matched complex is superior to either an H- or a P-excessive complex. The situation changes once exchange repulsion of the pairs comes into play. Now, H-excessive is superior and P-excessive inferior among loop complexes. The H/P-matched species should be in between. This leads to a hypothesis: *The world's best transition state is an H-excessive, error-free loop transition state.* Support for this idea comes from awareness that organometallic chemistry is mostly the chemistry of β-elimination and migratory insertion.

A vast amount has been written about the mechanisms of olefin metathesis, alkyne metathesis, olefin oxidation, and olefin polymerization. All these can actually be formulated as loop reactions that have remained underappreciated because of the inability of MO theoretical models to expose their intrinsic superiority. Specifically, the frustration of trying to decipher a stack of one-electron MOs to find out what orbitals play a key role in a reaction involving the participation of two mutually orthogonal AOs at one reactive center is familiar to anyone who has tried to understand as simple a reaction as the butadiene-to-cyclobutene ring closure. In this last case, orthogonal AOs at one center play a key role, and all this is very hard to uncover from the pile of MOs in the computer printout.

The conceptual intractability of MO theory coupled with the wrong physics (i.e., the neglect of interelectronic repulsion at the EHMO level or the one-electron approximation of monodeterminantal SCFMO) has created the impression that there is nothing special about activated (H-excessive, P-excessive, or H/P-matched) loop reactions. VB theory paints a very different picture. The failure to go "orbital by orbital and electron by electron" in the analysis of reactions, and the failure to grasp the intrinsic superiority of an error-free, H-excessive loop reaction, have tangible consequences for the thinking of the organometallic chemist. For example, β-elimination of M—CH_2—CH_2—H is the reverse of migratory insertion of (H_2C=CH_2)M—H. This is an H-excessive loop reaction. But, what is α-elimination? Stimulated by some recent work by Schrock and co-workers,[24] let us consider the reaction paths open to X_3W—CH_2CH_3. Starting with the pseudo-octahedral metal, deleting three starred exo conical AOs, and assuming a deformation of the three equatorial ligands (indicated by arrows), we can compare the following transition states:

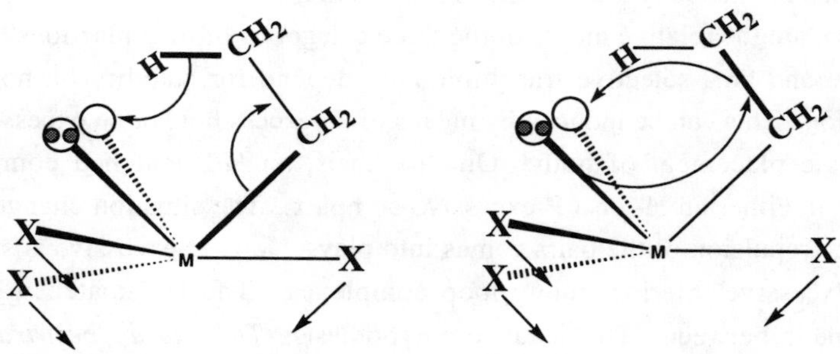

β-Elimination
HOLE-EXCESSIVE

α-Elimination
H/P-MATCHED

The β-elimination path yields $X_3WH(H_2C{=}CH_2)$ but the α-elimination path yields $X_3W{=}CH_2(CH_3)$. The active pairs involved in each process are shown in boldface. Actually, the intrinsic superiority of H over CH_3 as a migration group renders the alternative α-elimination to produce $X_3HW{=}CH(CH_3)$ superior. Now, the difference between the illustrative regioisomeric β- and α-elimination paths just shown is that the former is H-excessive but the latter H/P-matched. Hence, everything else being equal, the former must be preferred. The point now is that the meaning of the term "α-elimination" is obscured because the phenomenon itself can occur regardless of whether or not the axial metal pair is there. Deactivation of the axial metal pair can still lead to α-elimination producing $X_3HW[CH(CH_3)]$, where the entity in brackets is a singlet carbene.

The transition state is now different to the extent that it has two cross-atoms, rather than one, and it is H-excessive, rather than H/P-matched. Unless we can differentiate between reactions that are presently given the same label, no progress in the microelectronics of metal transition states can be made. Such distinctions can be achieved only by using specific VB formulas rather than delocalized MOs. By the way, in the case of Schrock's molecule, the three X atoms are effectively monovalent $R(SiR_3)N$ fragments of a triamidoamine ligand (a tetradentate nitrogen ligand). The fourth apical R_3N ligand has been omitted for simplicity. Because of the pair affinity of silicon (see Chapter 28), it is entirely possible that the tungsten pair is partly deactivated by the trimethylsilyl groups. This would explain why addition of H_2 to X_3WH is slow.

When loop reactions are not possible, acyclic as well as pericyclic reactions from the organic chemist's repertoire are the best alternatives. When either substituents or metal fragments act on the reactive centers, the dichotomy between "allowed" and "forbidden" pericyclic reactions disappears in a way that is easily explained by reference to the principle of arrow codirectionality. For example, the Hückel-antiaromatic transition state of the 2+2 all-carbon cycloaddition (and every homonuclear 2+2 addition) is defective in the sense that arrow codirectionality cannot be established (see Chapter 15). Strategic replacement of carbons by heteroatoms can (partly) restore arrow codirectionality and render the reaction favorable. As a result, the transition state of a nominally "forbidden" 2s+2s reaction is depressed relative to *all* T-bound species (reactants, products, and diradical intermediate), with the result that the reactions become effectively 2s+2s stereospecific even though an intermediate may be involved (ring closure of the intermediate fast relative to conformational equilibration). For example, codirectional arrow trains that associate, though imperfectly, all four reaction centers (Figure 34.4) rationalize four observations.

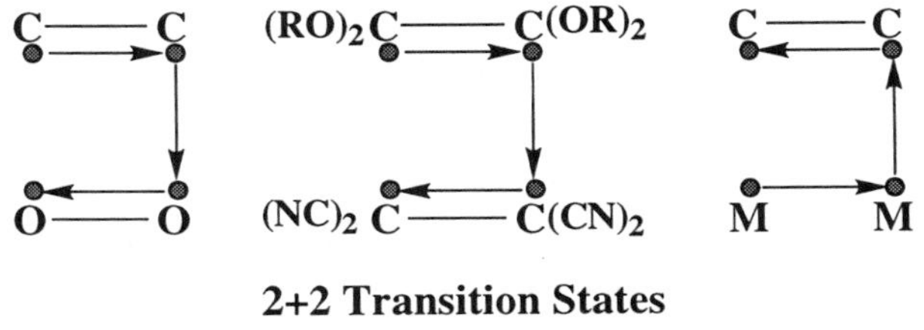

Figure 34.4 Codirectional arrow trains in "forbidden" 2+2 cycloadditions.

(a) Addition of donor olefins to singlet oxygen to produce a dioxetane is 2s+2s stereospecific.[17a] Isolation of products other than dioxetane (e.g., hydroperoxides) suggests the incursion of either a diradical or a perepoxide intermediate. Whatever the case, the system must ultimately go through a "2s+2s-like" transition state on the way to dioxetane product.

(b) One could conjecture that the highly 2s+2s stereoselective cycloaddition of 1,2-dicyclopropyl ethylene to TCNE involves a dipolar intermediate. If this were the case, one would expect to find cycloadducts other than the cyclobutane product, since the cyclopropylcarbinyl cationic end could rearrange to homoallyl and cyclobutyl cations, which could ultimately combine with the anionic end of the zwitterion to generate rearranged cycloadducts. It turns out that products derived from such ions are not found.[25a] Furthermore, tetramethoxyethylene reacts with 1,2-dicyanoethylene in a completely stereospecific manner.[25b]

(c) Cycloadditions and retrocycloadditions of metal dimers to alkenes and alkynes are known.[26]

(d) Ring closure of the $Fe(CO)_3$–cyclooctatetraene is disrotatory rather than conrotatory.[27] The I formula of the complex shows how the $Fe(CO)_3$ group acts as a donor–acceptor substituent to produce a codirectional arrow train within the reactive dienic fragment.

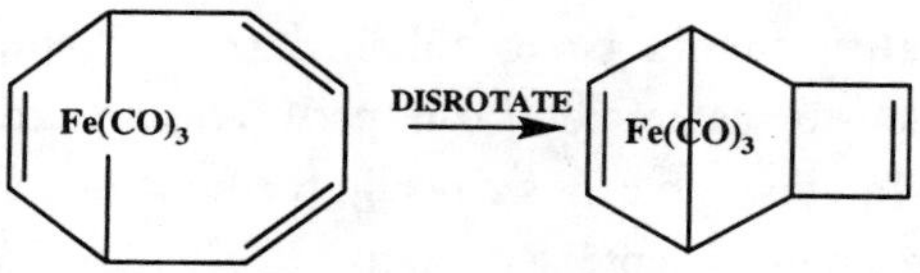

STRUCTURE OF METAL COMPLEX

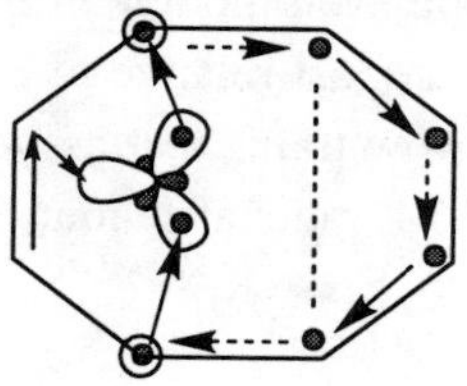

The synthetic chemist's paradise would be a world in which certain molecules would act as specific "transfer reagents" to donate some component to another target molecule. How do we design such molecules?

Figure 34.5a illustrates two types of arrow train. The characteristic feature of the two-tier arrow train is the generation of incipient carbenes or olefins as a consequence of CT delocalization over a cross-atom. In other words, a two-tier arrow train is a loop arrow train.

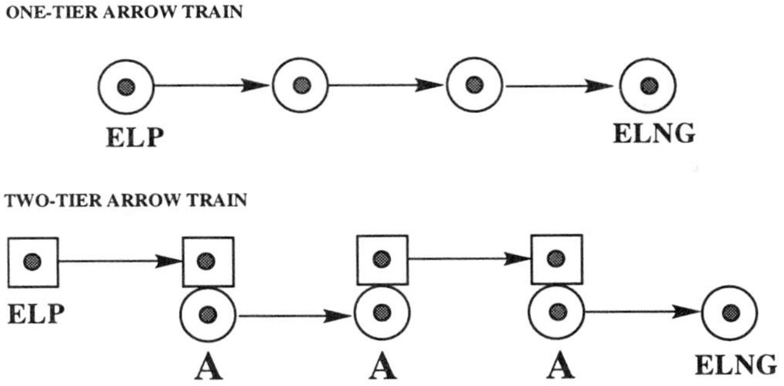

Figure 34.5 One- and two-tier arrow trains.

The simplest application of this concept is the cyclopropanation effected by the Simmons-Smith reagent (CH_2I_2 plus Zn/Hg amalgam) as illustrated in Figure 34.6a. This inspires the search for an analogous reaction that effects olefin, rather than carbene, addition to a second olefin as illustrated in Figure 34.6b. The way in which electron transfer effects bond metathesis will be revisited shortly.

Chemistry is the science of the transformation of matter. Hence, the main challenge for the theoretician is the design of new reactions that cannot be

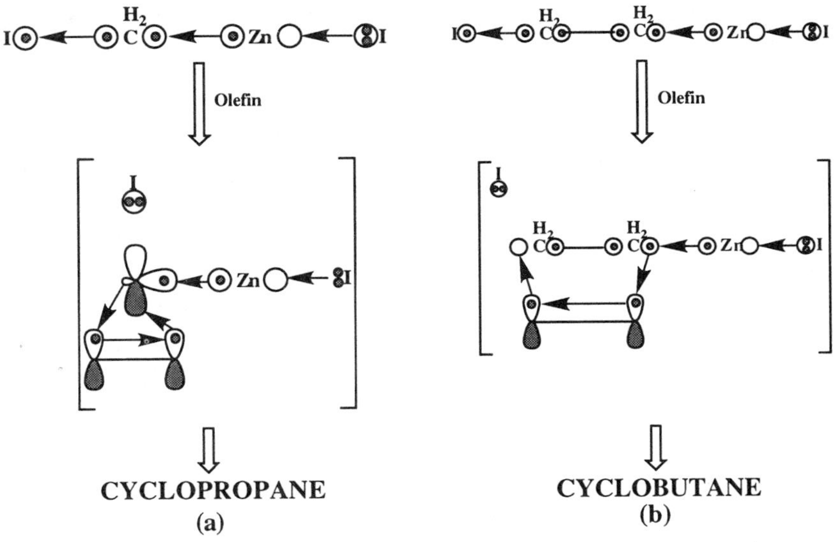

Figure 34.6 (a) Carbene and (b) olefin transfer induced by codirectional electron transfer.

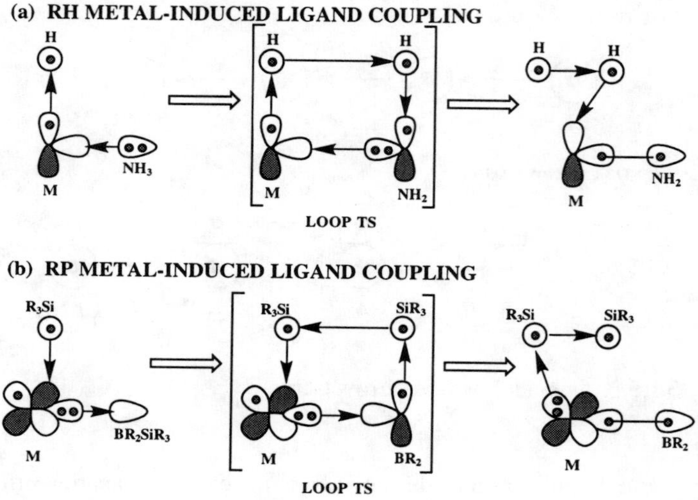

Figure 34.7 Two types of ligand coupling.

anticipated from mere extrapolation of known reactions. We now know how to do this, and Figure 34.7 shows two examples. The key features are:

(a) Each reaction has a favorable loop transition state.

(b) Each loop transition state has an error-free codirectional arrow train. The requirement of error-free arrow codirectionality defines the optimal reactants. When the metal acts as radical/hole (RH), the three additional reactive centers must be equally or more electronegative. When the metal acts as radical/pair (RP), the three additional reactive centers must be equally or more electropositive. Clearly, there are many other choices of atoms that fulfill these conditions. Which one does it best is the next question down the road.

(c) An RH metal must act with a "symmetric" hole to accommodate a donor product (Hückel array). By contrast, an RP metal must act with an "antisymmetric" hole to accommodate an acceptor product (Möbius array). The RH metal-induced ligand coupling of Figure 34.7a has actually been observed by Milstein and co-workers[28a] and we suspect that other analogues, unknown to us, have been reported or lie under cover in the literature.

Let us now combine the concepts of I bonding (Chapters 15–17) and the idea of the cross-cyclic reaction to shed light on a mystery: Why do reactants that are seemingly capable of entering pericyclic reactions bypass them in preference to paths leading to highly strained intermediates? For example, singlet O═O and diimide derivatives (XN═NX) can potentially act as ethylene analogues to partake in 2s+2a or 4s+2s cycloadditions. Nonetheless, the important work of Foote and Orphanopoulos[28b] suggests that they act by being promoted to a zwitterionic form, which adds to double bond by the cationic carbenic end to form epoxides or aziridines! Translated into our language, this implies that a (H/P-matched) loop reaction (e.g., carbene addition to a double bond) is intrinsically superior to a pericyclic reaction. Furthermore, recall the two ways of destabilizing an organic pericyclic transition state: Either replace all carbons by more electronegative atoms, or destroy the degeneracy of the association-promoting configurations by alternately replacing carbons by either more or less electronegative heteroatoms. This impairs CT, the basis of I bonding. The two heteroatomic analogues of ethylene mentioned above meet both criteria, as O and N are more electronegative than C. As a result, these reagents are not optimal participants in a pericyclic transition state.

We have compacted an endless subject, namely, loop transition states and loop molecules, in the space of one chapter section. Since loops can be combined to form knots, knot chemistry[29a] and knot theory[29b]become relevant to the orbital topology of molecules and transition states.[30]

34.8 The Mechanisms of Oxidation by Oxometals

VB theory provides a broad overview of oxidation because it allows us to see the reaction alternatives open to a given system. We illustrate our approach by focusing on $(YO)_3VO$ assuming d^3s-hybridization of V (Figure 25.6) and a trigonal pyramidal geometry, for simplicity. The three axial metal AOs make a formal triple bond with the oxygen atom and the three equatorial metal AOs make three formal covalent bonds with the three YOs. Each of the two π-type components of the axial VO triple bond can interact with π-type YO oxygen lone pairs [using the basal $(YO)_3V$ unit as our reference nodal plane]. When this complex or its isoelectronic analogues [e.g., $(O^-)_3Mn^{2+}O$] acts on molecule A—B, there are three possibilities:

(a) The VO triple bond acts as an "organic triple bond" to undergo a formally "forbidden" 2+2 addition either in a one-step fashion, as discussed later, or via the intermediacy of a diradical.

(b) One π component of the VO triple bond in conjunction with one π-type YO oxygen lone pair defines a VO(OY) system isoconjugate to an allyl anion that can add stereospecifically to A—B in a formally "allowed" 3+2 fashion. This explains cis-hydroxylation of olefins by permanganate ($M = Mn^{2+}$, Y =

electron) as well as the dependence of the catalytic activity of surface-dispersed vanadium oxide on the nature of the surface atoms (M = V, Y = Si or Ti of silica or titania).[31a,b] In general, the more Y is capable of tying up the lone pairs of the adjacent O in YO, the more unlikely the 3+2 reaction path or any other pathway in which the YO lone pairs play a crucial role.

(c) The third and most interesting possibility is that the initial addition of A—B to the metal is a hole-excessive reaction made possible by the localization of one of the three equatorial bond pairs to the more electronegative oxygen and subsequent migration of the YO^- as illustrated here. This, rather than possibility (a), may be actually the metalloxetane mechanism of cis-hydroxylation of olefins by permanganate and osmium tetroxide.[31c]

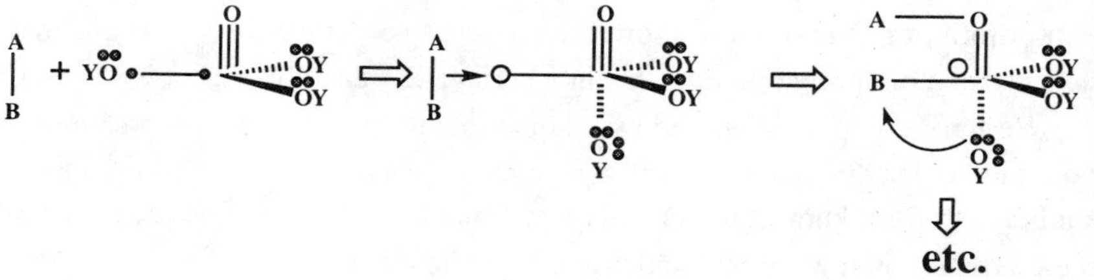

The Pt-catalyzed reduction of Re in PtRe/γ-Al_2O_3 reported by Augustine and Sachtlet[31d] may serve to illustrate how crucial for reactivity is the "untying" of metal holes. These authors proposed that the reduction studied involves migration of oxidic Re over the support surface to a zerovalent Pt center, where it encounters H_2. We can explain this result by assuming that pseudo-octahedral Pt acts as a 3h/0e piece, which acts to "unglue" a fraction of oxygen pairs coordinated to the equatorial holes of the Re center. In turn, the Re center is enabled to add H_2 by a hole-excessive loop reaction. Although we have used ReO_4^- as the substrate in the following VB scenario, an analogous mechanism can be written for ReO_2, and so on.

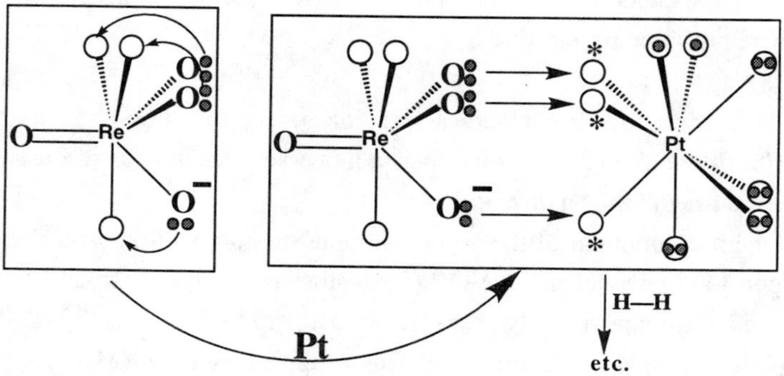

Let us now revisit the problem of *cis*-olefin hydroxylation by OsO_4 in a way that makes the disjunction between VB theory and conventional models transparent. Specifically, assuming that Os has six valence AOs (the five 5d orbitals and the 6s), the molecule may be described by any one of three different formulas depending on the bonding mechanism. As a result, Os can exist in any one of three different actual (rather than formal) oxidation states, namely, Os(II), Os(IV), and Os(VIII):

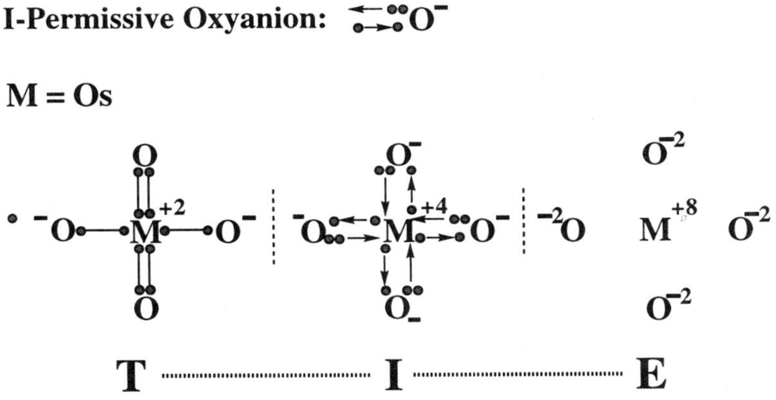

Let us focus on the I formula, recalling that an oxyanion is an I-permissive electrophilic radical; that is, it can use one dot (odd electron) and one pair to form a codirectional I bond. As a result, the formula representing optimal codirectional I-bonding has Os(IV) bonded to four O^- oxyanions. In this way, each one of the four Os(IV) dots associates with a pair and a dot of two different oxyanions. Each of the remaining two metal holes ties up two oxyanion lone pairs to form an inferior contradirectional I bond. As a result, all four oxygens are I-bound to Os(IV) primarily by utilization of only four singly occupied valence AOs. In other words, OsO_4 is effectively isoconjugate to four allyl anions! However, in a more detailed sense, all metal and oxygen electrons (except the exo oxygen lone pair) are involved in the bonding. Later on, we will add another piece of evidence in favor of our formulation of the bonding of OsO_4: Bridging implies I-bonding, and $(RO)_2OsO_2$ is known to be a dimer.

Recalling that the T, I, and E mechanisms of bonding fall on a continuum, the question can be asked: What is the actual formula of OsO_4? The answer is that it must approximate the preceding I formula. Sigma donor ligands (RED ligands), which form electrostatic bonds with the metals, enhance I-bonding by virtue of stabilizing the higher Os(IV) over the lower Os(II) oxidation state. For example, amines are expected to enhance the I over the T character of the Os—O bonds because the T formula features an Os(II) while the I formula features an Os(IV) metal. As a result, the effective electronic configuration of the metal changes upon amine coordination: Rather than six singly occupied AOs, there are four singly occupied and two unoccupied valence AOs in Os(IV). One of the two holes can easily become available to an incoming olefin in an H-excessive loop 2+2 reaction. As a result, amines are expected to enhance the rate of formal 2+2 cycloaddition yielding a metalloxetane, which ultimately leads to the osmate ester preceding the cis-hydroxylation product. This brings us inevitably to a key question: What is the mechanism of cis-hydroxylation by OsO_4?

Conventional wisdom says that a pericyclic reaction going through an aromatic transition state is the best possible mechanistic alternative. VB theory paints a different picture: An H- or P-excessive "nonaromatic" loop can be superior to a pericyclic "aromatic" transition state. Furthermore, if errors are present, a stepwise diradical path may be preferred. Two of the three possible

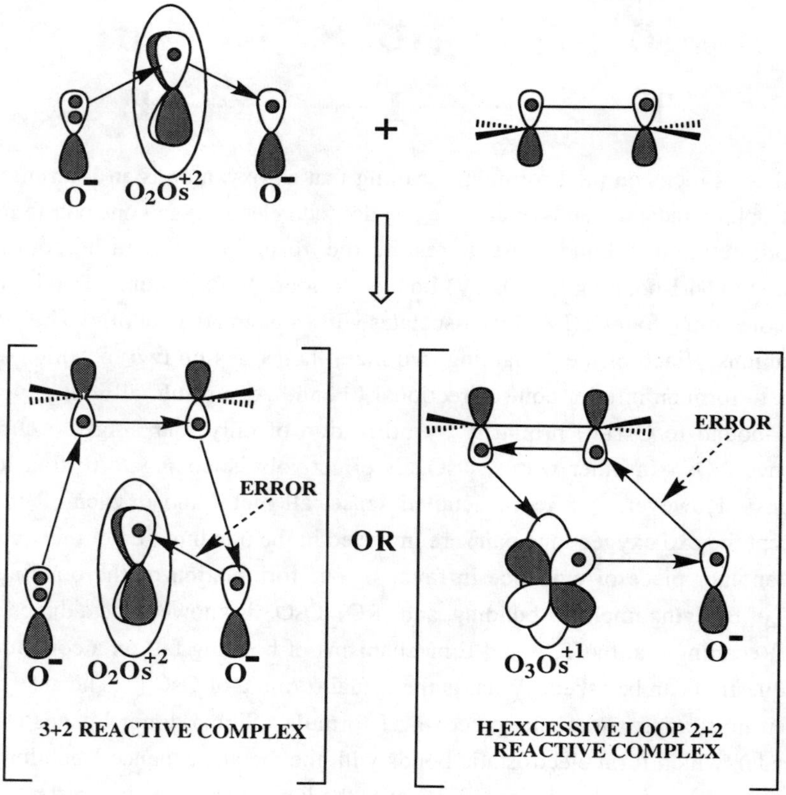

Figure 34.8 Two possibilities for the addition of OsO_4 to olefin.

scenarios of the addition of OsO_4 to olefin are shown in Figure 34.8. The 1,4-diradical mechanism is self-evident. The VB prediction is that a one-step "2+2" is superior to a one-step 3+2 addition, but a diradical mechanism may well turn out to be best because each of the one-step transition states has one error.

There are two pieces of data, reviewed in the monograph of Nugent and Mayer,[19] that point to a 2+2 mechanism, either one-step or stepwise:

Amines speed cis-hydroxylation, as noted earlier.

Replacing one oxygen by the isolobal cyclopentadienyl aborts cis-hydroxylation in the case of $CpReO_3$, because Cp is so large that the olefin cannot approach the metal. As a result, 3+2 addition is now observed.

Organometallic chemists avoid metal hybridization like the plague. They find comfort in the MO approach, where all that one sees are occupied and unoccupied symmetry MOs from H_2 to a metal solid. Some of the fear reflects worry about symmetry control. However, this is justified only in cyclic systems. In acyclic molecules, the atoms take care of themselves, so to speak. At the same time, this disaffection with metal hybridization points out the problem: Nonmetals hybridize to make strong T bonds, but metals hybridize to make strong I bonds! .No wonder that, lacking a distinction between these two mechanisms of chemical bonding, one feels uncomfortable with the notion of hybridization. Exactly how different VB thinking is from to the conventional models can be further illustrated by a relative of OsO_4, namely, the powerful oxidant $L_4OsO_2^{2+}$.[31e] Conventional wisdom says that this octahedral molecule involves an Os(VI) metal, which has one lone pair. This means that six two-electron ligands must be bound by only five valence holes. In turn, this means that the molecule is "hypervalent"! VB theory says both that the metal is not Os(VI) and that the molecule is not "hypervalent" (see Section 24.18). Rather, it is a normal I-bound complex in which a d^2s hybridized Os(IV) (see Figure 25.5) makes five I bonds in the following way:

Each one of the two equatorial holes ties up two trans-L ligands via contradirectional I bond.

Each of two axial d-π-type dots ties up an oxyanion dot and an oxyanion pair by a codirectional I bond.

Two oxyanionic σ-type lone pairs combine with the axial metal σ hole to define yet another contradirectional I bond.

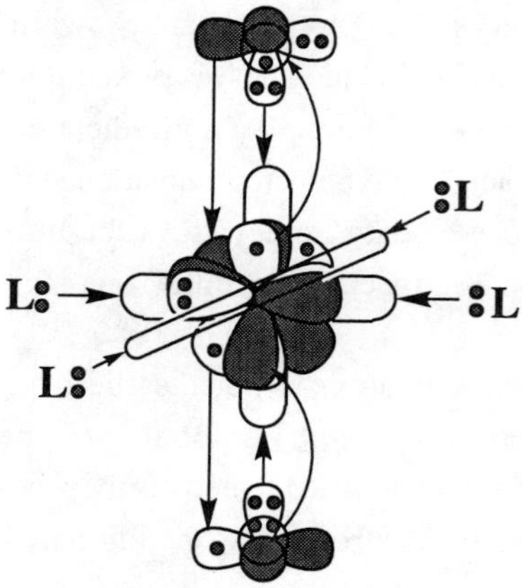

$L_4Os^{+4}(O^-)_2$

Bottom line: VB theory has drastically changed the way we view metal reactions. First, the concept of I-bonding reveals immediately the options of a molecule — whether, for example, it has holes that qualify it for a loop reaction. Then, instead of focusing on pericyclic, Woodward–Hoffmann-allowed reactions, we give first billing to loop reactions. Finally, diradical paths are indicators of errors in loop transition states. In this context, the demonstration of a diradical mechanism in the addition of $CrCl_2O_2$ to alkanes by Cook and Mayer[21] is important because it suggests that the error of the loop transition complex is serious enough to thwart the one-step "2+2" addition mechanism.

34.9 The Regioselectivity of Migratory Insertion of a π Bond into a Metal–Hydrogen Bond

A recent paper by Casey and Petrovich[31f] forced us to think through the issue of the regioselectivity of the reaction $L_nMH(H_2C{=}CH_2)$ to yield L_nMCH_2—CH_2—H by using a linear combination of four (!) new VB concepts: the hybridized metal, the starring procedure, the docking rule, and the loop reaction. The first point is that the H-excessive loop reaction shown in Figure 34.9 has a highly asymmetric transition state that involves AOs of five different types: two different metal AOs (one axial and one equatorial), one H1s AO, and two different C2p AOs, in the case of unsymmetrical olefins. This means that the four arrows representing CT delocalization carry different weights, depending on the details. More specifically, depending on the star character of the equatorial metal hole, there exist two limiting cases:

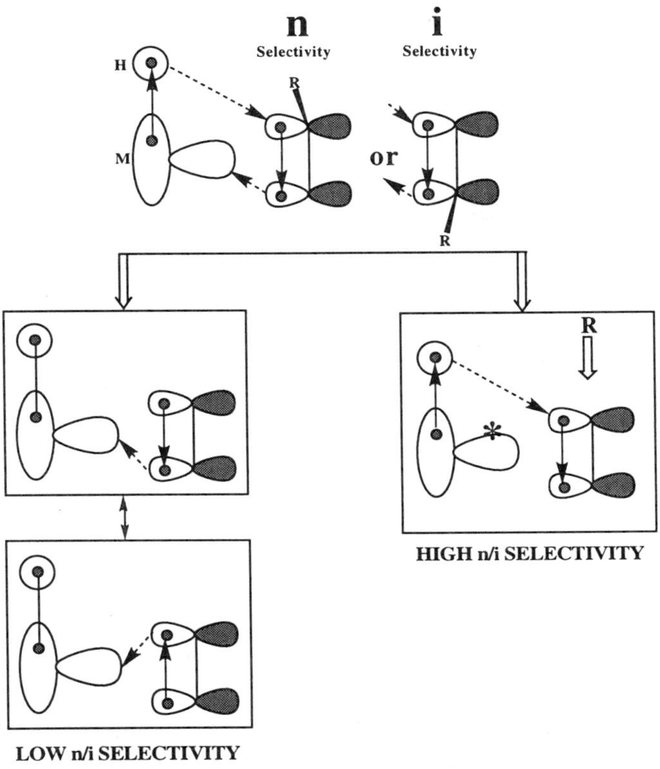

Figure 34.9 Selectivity of H-excessive loop reaction.

(a) When the metal hole is unstarred, the metal center acts as an effective electrophile. Therefore, the transition state resembles an aromatic two-electron complex isoconjugate to π-cyclopropenyl cation. As a result, the differentiation of the two ends of an unsymmetrical olefin is near zero.

(b) When the metal hole is starred, its action is effectively turned off and the metal center acts as an effective hydride donor. We now have a hydride–olefin complex that is isoconjugate to π-allyl anion. As a result, the differentiation of the two ends of an unsymmetrical olefin is high. The T formula of H_2C=)CH—CH_3, a model of an unsymmetrical olefin, reveals that the polarization of the π system is one in which the π dipole points toward the outer olefinic carbon. This predicts that the formal hydride will attack preferentially the interior olefinic carbon, giving rise to metal–(*n*-alkyl) product (*n*-selectivity).

Let us now apply these ideas to the generic rhodium complex JKLRhH(olefin), in which J, K, and L are three different ligands. No progress can be made unless we have a vivid description of the molecule's wavefunction.

This is easily attained by using the pentagonal d^3sp^3-hybridization model shown in Figure 25.9. By eliminating the action of the p_z metal AO, the scheme is transformed to d^3sp^2. There are three axial and five equatorial AOs. The crucial Rh AOs are the σ-type axial AO binding the hydrogen (plus one ligand) and one equatorial hole binding the olefin. The starring procedure reveals the following star-isomeric forms of the pentagonal base:

Star isomeric equatorial belts of $L_3HRh(Alkene)$

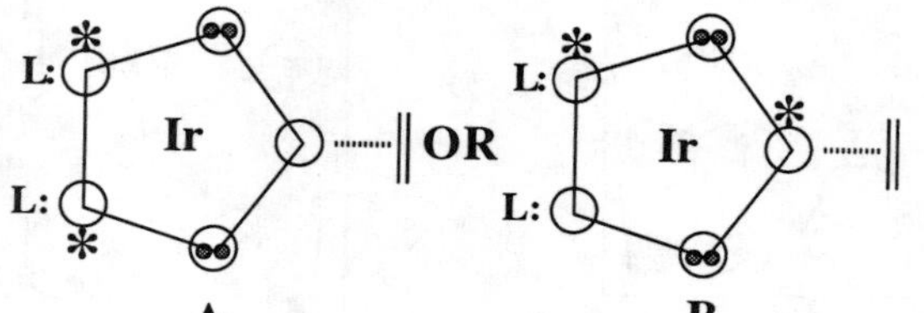

Which of the two forms is preferred depends on the identity of K and L. Form A is favored when L is a strong σ base (which docks on an unstarred hole). Form B is favored when L is a strong π acid (which docks on a starred hole). This predicts that constraining (by chelation) a phosphine into the equatorial belt and relegating a CO to an axial position should yield high n/i selectivity. Doing the opposite should tend to annihilate the selectivity (but never making it to the limit). We diagram the argument by focusing on the star character of the metal fragment coordinated to the olefin:

$L(R_3P)CORhH^-$

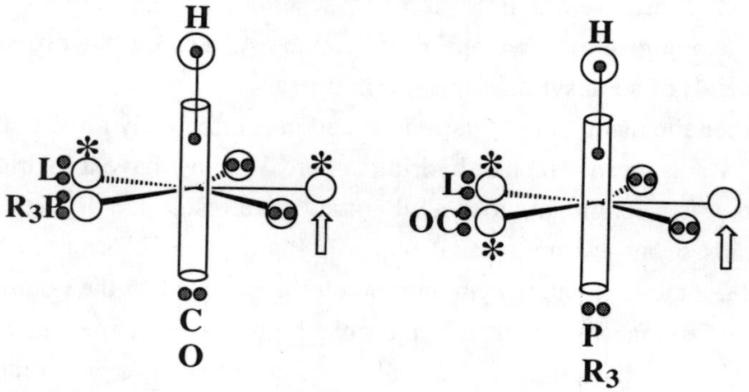

High n/i selectivity **Low n/i selectivity**

Bottom line: The observer ligands control the star character of the active metal AOs, and this determines the selectivity of metal reaction.

The paper of Casey and Petrovich[31f] is thorough both mechanistically and in an interpretational sense. Specifically, the authors used molecular mechanics calculations to investigate the hypothesis that the n/i selectivity is due to steric effects. They interpreted a quantitative disagreement between computation and experiment as evidence against a steric interpretation. Thus an electronic basis for n/i selectivity is suggested, and this is what our model defines. However, given the uncertainty of how to compute "steric effects," the issue remains open. What we have offered here is simply one more demonstration of the great difference between VB thinking and conventional models of chemical reactivity.

34.10 Fragmentation Induced by Electron Transfer

The classification of chemical reactions has been dictated by the evolution of experimental chemistry. However, as new reactions are invented, new common denominators become apparent. One theme that has been implicit until now is that a wide variety of chemical reactions are actually inner sphere ET reactions in which a codirectional arrow train (i.e., the most favorable form of CT delocalization) acts to transfer one electron from the origin to the terminus of the wave train. A beautiful example is the reaction of formaldehyde radical anion with methyl chloride studied computationally by Sastry and Shaik.[32] It was found that the oxygen terminus promotes direct substitution (SUB) while the carbon terminus of the radical anion promotes direct electron transfer (ET). From the I formulas of these transition states, it is clear that the two apparently unrelated reactions are really related: Each is an electron transfer reaction from O to Cl. The only difference is that the codirectional arrow train spans an even number of electrons in SUB and an odd number of electrons in ET.

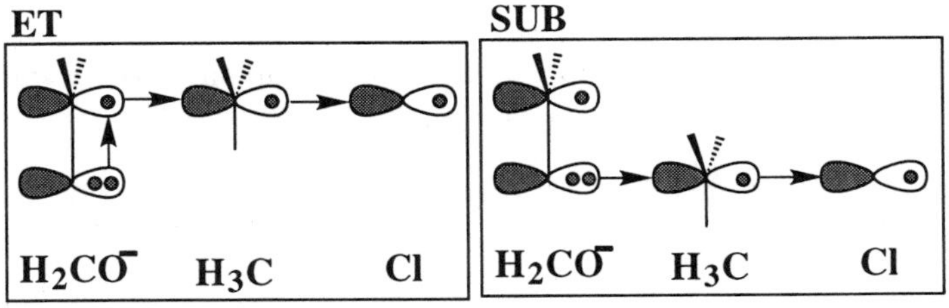

The next theme entails the recognition that the so-called ET reaction can be viewed as a bond switch, or fragmentation, reaction induced by electron transfer from one end to the other end of an AO open chain. We can have either even- or odd-electron ET fragmentations, as illustrated. The characteristic feature of each ET fragmentation transition state is a favorable (i.e., relatively error-free) arrow train.

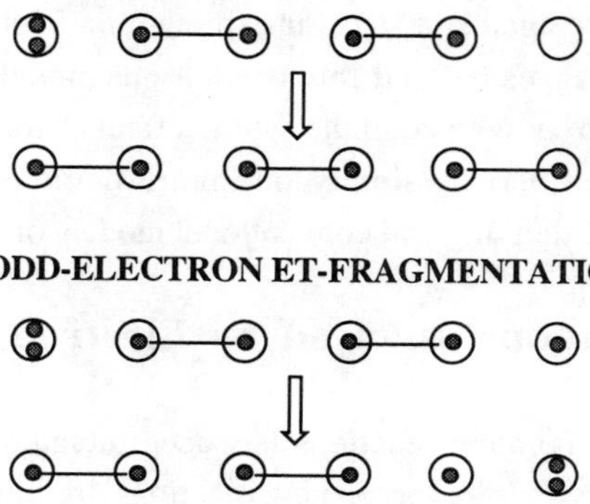

Even-electron ET-fragmentations are well-known in organic chemistry. Odd-electron ET fragmentations may occur in the mass spectrometer, one example being the McLafferty rearrangement.[33]

We can now see an alternative formulation of H-excessive loop reactions such as the addition of BH_3 to alkene. Specifically, these can be formulated as bond metatheses triggered by intra-atomic ET. For example, intra-atomic promotion of one electron from the singly occupied sp^2 to the vacant 2p of boron triggers transfer of H to one olefinic carbon and linkage of the second olefinic carbon to B. This series of events suggests that atom transfer reactions can be effected photochemically.

34.11 The Mechanism of Action of Cytochrome P 450

The following "oxygen-rebound" mechanism of alkane hydroxylation by cytochrome P 450 has been embraced by many.[34]

$$\mathrm{Fe{=}O + R{-}H \longrightarrow Fe{-}O{-}H + R \longrightarrow Fe + R{-}O{-}H}$$

On the other hand, the experimental data are permissive of other mechanistic schemes.[35] Armed with specific and detailed VB formulas, we can now suggest a plausible scenario, electron by electron and orbital by orbital. The reactive oxo complex of P 450 may exist as one of the two forms shown in Figure 34.10.

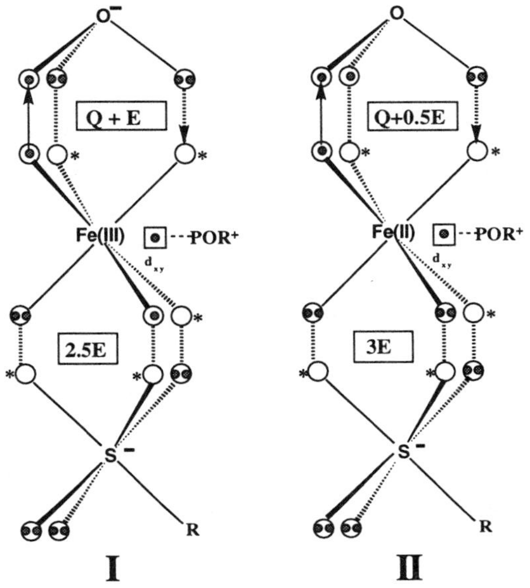

Figure 34.10 Two possible oxo iron derivatives of cytochrome P 450.

Note the hallmarks:

Because the proximal RS ligand enters as a 3h/2e piece (and, conceivably, because of the action of the protein environment), each species has three unpaired electrons.

Neither I nor II involves an Fe═O unit with a formal double bond.

Intermediate II can react via the oxygen-rebound mechanism, while intermediate I can react with an A—H bond by odd-electron ET fragmentation via a transition state of the following type.

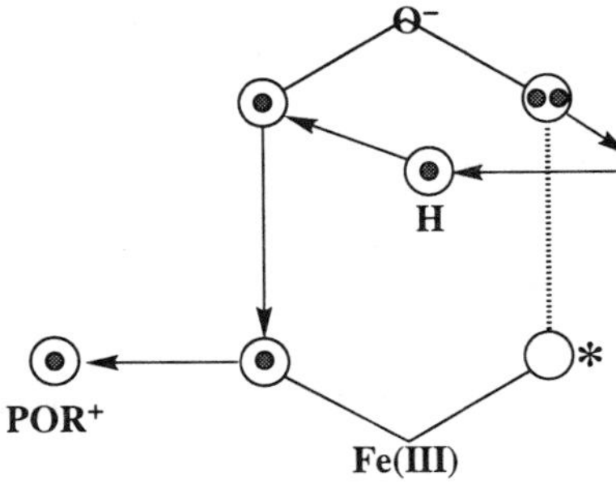

The driving force is now electron transfer from oxo anion to porphinato π-radical cation with concurrent bond metathesis. This mechanism parallels that deduced by Bruice and Ostovic for olefin epoxidation by oxo metalloporphyrins.[36] Furthermore, Figures 28.4 and 34.10 suggest an explanation of the difference between hemoglobin and cytochrome P 450 insofar as both dioxygen binding and the subsequent reaction are concerned. The difference lies in the nature of the proximal ligand (histidine vs. cysteinyl) that determines the allocation of the electrons in the upper and lower faces of the metalloporphyrin.

34.12 Theory and Synthesis

Let us reiterate three fundamental positions of this work:

(a) Molecular electronics is a "game of angstroms." In other words, we must be able to understand molecules "electron by electron, and orbital by orbital." This is evident to all familiar with the subtleties of chemical reactivity.

(b) In chemistry "everything goes." The real challenge is in deciphering the design, and this can be achieved only by falling back on a sound conceptual theory.

(c) The existing MO theoretical framework of chemistry is not adequate as a platform for exploiting the full slate of elements in synthesizing whatever target molecules one may wish. Most of the applications of VB theory presented in this work were deliberately selected to permit comparison of the MO interpretation (in the literature) with the VB argument. The process of understanding chemical bonding is just beginning.

The consequences of (c) have been implicitly realized by Trost,[37] who has clearly seen what many academics (who keep rehashing old prototypes) have failed to realize: Synthetic chemistry has not even begun to scratch the surface of what lies ahead, namely, synthesis via homogeneous metal catalysts involving reactions that cannot be conveniently realized in the absence of the metals. To put it bluntly, most academics who accept MO theory believe that it enables them to understand chemical bonding. Thus the creative search for new reactions is stifled simply because many imagine that the possibilities have been exhausted. This work argues the opposite: There are great opportunities for discovery, especially in "metal chemistry." Discoveries are now made by serendipity. Let us see whether VB theory can provide a rational plan!

Trost and co-workers found that two different metal systems, namely, CpCl(COD)Ru and $(AcO)_2Pd$, catalyze highly regioselective ene reactions.

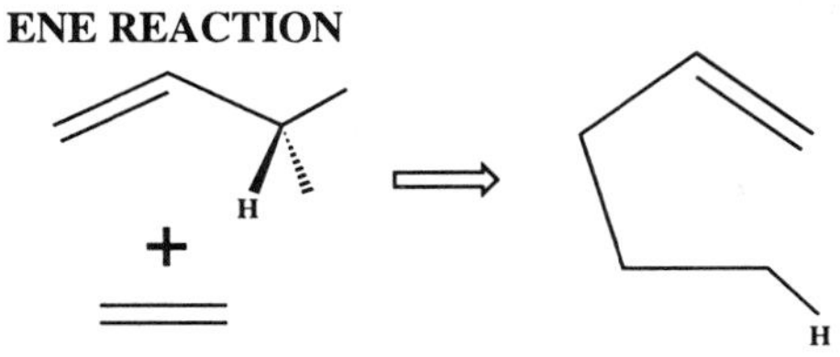

The first question is: What is the common denominator of these two apparently disparate complexes? Departure of the COD and Cl^- ligands from the ruthenium complex generate a 3h/0e $CpRu^+$ fragment with three equatorial lone pairs. The undissociated $(AcO)_2P$ has exactly the same valence configuration.

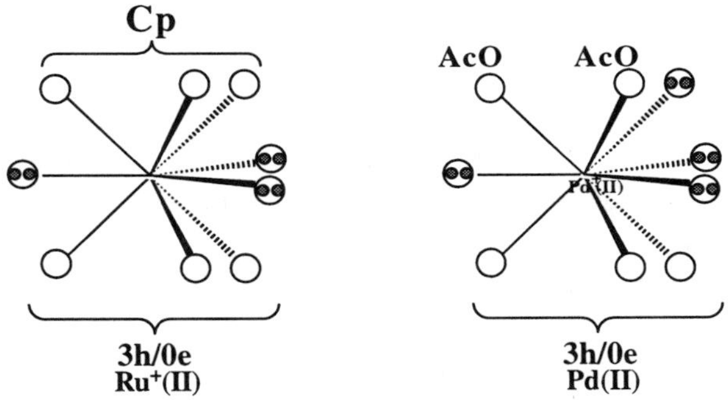

How do these fragments effect a regioselective ene reaction? The VB rules are straightforward:

The best transition state is an error-free loop transition state.
The best loop transition state is an H-excessive one.

These rules are obeyed as follows:

(a) One of the endo conical holes acts as a chelation hole to anchor one of the two olefinic reactants. If this is inaccessible, the metal can no longer effect intermolecular ene reactions. Each of the remaining two endo holes ties up one olefin by one codirectional I bond.

(b) Transfer of a pair from the equatorial to the endo set sets up a codirectional arrow train that leads to oxidative coupling that in turn creates a metallacyclopentane.

(c) The equatorial hole sets the stage for an H-excessive reaction: migratory insertion of the C—H bond into the M—C bond.

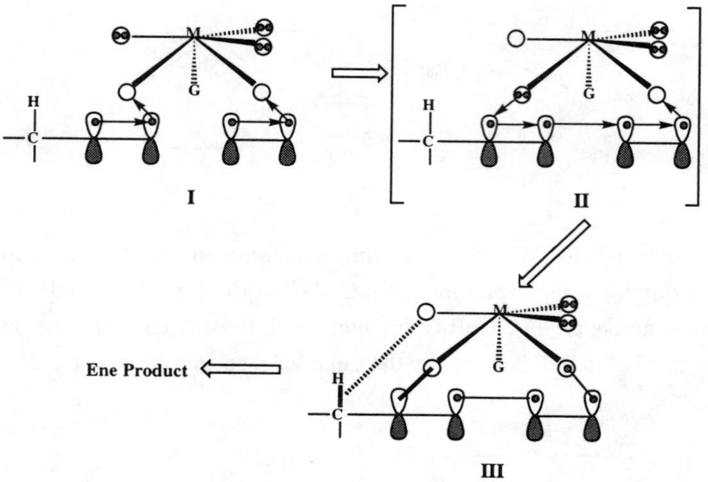

The key point is that all transition states are error-free loop transition states. The whole presentation is simple and transparent because VB theory gives us two invaluable weapons: specific formulas and clear concepts.

Since the ene reaction is a case of oxidative ligand coupling, let us look at the broader picture. Instead of using 3h/0e metal fragments (in which one hole is reserved for chelation and two are used to tie up two olefins) and requiring transfer of an equatorial pair to the conical set, we can replace the chelation hole by a pair in a 3h/2e metal fragment. A typical example is provided by the ability of CpCo, which is produced by dissociation of ligand L from CpCoL and is known to be an effective acetylene coupler, to produce aromatic molecules via a metallacyclopentadiene intermediate.[38] In summary, a metal fragment acts as a "push–pull" reagent to effect ligand coupling and, once the existence of the equatorial pairs has been recognized, either a 3h/0e or a 3h/2e metal can act as a coupler. Since acetylenes have the "extra" π bond that can indirectly stabilize the codirectional arrow train responsible for coupling, alkynes can be coupled with greater facility than alkenes.

What is the simplest example of ligand coupling? CpCo is isolobal to $(CO)_3Fe$. They both form stable diene complexes. This means that $(CO)_3Fe$ should also be an efficient coupler. Indeed, this is the case: The reaction of $(CO)_3$ Fe with 1,3-butadiene to produce the stable complex $(CO)_3$ Fe(1,3-butadiene) is the simplest illustrator of ligand coupling. To see this, consider the following case of 1,3-butadiene valence isomerization as an effective ligand coupling reaction in which two π bonds (two ligands) are coupled to a 1,4-π diradical or zwitterion.

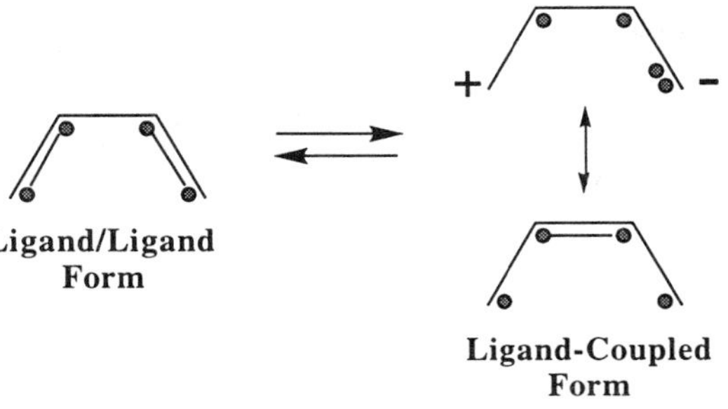

Since the $(CO)_3Fe$ acts as 3h/2e (two holes and one pair), it stabilizes preferentially the zwitterionic "ligand-coupled" form; that is, the zwitterion is tied up by one hole and one pair of Fe and the distal π bond is coordinated to the second hole of Fe. As a result, $(CO)_3Fe$(1,3-butadiene) is effectively a metallacyclopentene, and this is why nucleophilic attack occurs at C2 of the "diene fragment" (see Chapter 33, Figure 33.4). This is also why $(CO)_3Fe$(1,5-COD), which is incapable of ligand coupling because the two π bonds are constrained to be away from each other (two π bonds tied up by two Fe holes, two formal bonds), rearranges to $(CO)_3Fe$(1,3-COD), which is the superior "ligand-coupled" form (three formal bonds).[39] Finally, if the $(CO)_3Fe$(1,3-butadiene) contains a valence-isomerized 1,3-butadiene, collapse of the zwitterion to a covalent bond plus *disrotation* must lead, with only a small barrier, to the fully bonded valence isomer of 1,3-butadiene, namely, cyclobutene! Indeed, this is a textbook case: Coordination of $Fe(CO)_4$ to skeletally constrained cyclobutene leads to disrotatory ring opening after loss of one CO to produce the $(CO)_3Fe$(1,3-butadiene) complex. There is no violation of any rules here because both reactants and products have effectively the same spin coupling scheme. In other words, reactants and products are stereoisomeric forms of the same valence isomer. The crossing from one valence isomer to the other comes in the decomposition of $(CO)_3Fe$(1,3-butadiene) to $(CO)_3Fe$ plus 1,3-butadiene, and the metal took care of this.

If we can predict what metal *will* effect coupling, we should be able to predict what metal *will not* do so despite appearances. For example, in contrast to $(CO)_3Fe$, the 3h/0e fragment, $(CO)_3Cr$ does not show any coupling action. This is because $(CO)_3Cr$ requires the relocation of a pair from the equatorial belt, which is not possible, since the three equatorial lone pairs are tied up by the three carbonyls. This chemical stalemate highlights the condition for oxidative coupling by a 3h/0e metal: The equatorial pairs must be available. Also, we should be able to distinguish couplers from non couplers irrespective of complexity. For example, the "large" angular Cp_2Zr is an efficient coupler forming metallacyclopentane derivatives.[40] Starting with an all-eclipsed pseudo-octahedral Zr and recognizing that each Cp ties up three AOs leads to the realization that Zr will act as a 3h/2e fragment with three coplanar (rather than conical) AOs (see Figure 25.8b). This qualifies it as a ligand oxidative coupler.

The foregoing discussion is only a first-order treatment because we disregarded the effect on reactivity of the star character of the holes, assuming that the complex takes care of itself, so to speak. The next level involves explicit consideration of how the observer ligands (Cp or COs) determine the star character of the reactive holes and, as a result, the detailed features of this reaction. However, at this stage, such a treatment is superfluous simply because the right experiments involving incremental changes of metal, observer ligands, and reactive ligands have yet to be carried out.

Let us now use what transition metals have taught us to design heavy p-block ligand coupling reagents. What we need is initial attachment of two olefins to two holes followed by exchange of one hole by one pair, to spur the coupling. The strategy shown below is self-explanatory. It typifies what we hope to do full-time in the future, namely, use the experience of one region of the periodic table to develop analogous new reagents in another domain of the periodic table. With the current focus on transition metals (especially in the United States), a great deal of novel heavy p-block chemistry is there to be discovered by using the concepts of VB theory.

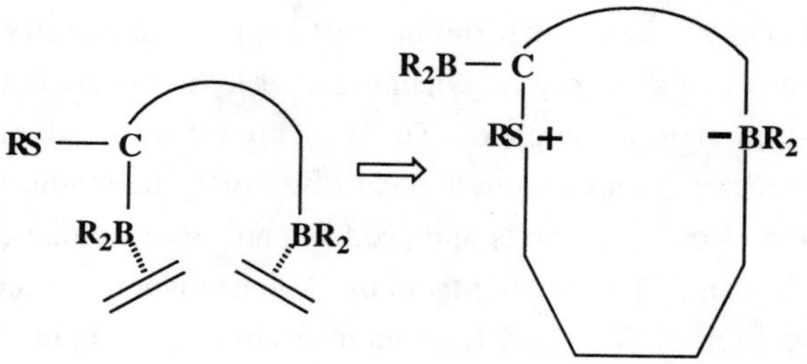

We now come to another key question: Is there any evidence in support of the proposition that the H-excessive loop transition state reigns supreme? Perthuisot and Jones discovered that biphenylene is catalytically hydrogenolyzed to biphenyl in the presence of a catalytic amount of $Cp^*(R_3P)RhH_2$ and H_2.[41] What is the active intermediate in this amazing transformation? Immediately, we look for a rearrangement of the initial complex capable of producing a vacant Rh AO to assist Rh—C bond making as well as Rh—C bond breaking. The first issue is the structure of the rhodium complex. Recently, Kelly et al.[42] reported the synthesis of a "molecular brake": A triptycene "wheel" is subjected to the braking action of a dimethoxybipyridyl group that upon coordination to Hg^{2+} acts as a "brake." The same situation on a more intimate scale can be identified in $Cp^*(R_3P)RhH_2$, as illustrated in Figure 34.11. A pseudo-octahedral Rh (see Chapter 25) can act with a lower conical three-tooth (3T) gear to bind the hydrogen and phosphine ligands under the influence of an equatorial 3T, gear which is either locked or free to rotate. It can be seen that a ROX ligand, like PR_3, can act as a lock of the two 3T gears, with the strength of the locking action being dependent on the star character (not shown explicitly) of the hole on which it is docked.

We can now envision a double-ligand rearrangement that will create a hole in the lower conical 3T gear: One hydrogen migrates to Me_5C_5 (designated Cp*), transforming it to Me_5C_5H (designated Cpe* in Figure 34.11). Since the former ties up three but the latter only two upper conical holes, there is now the

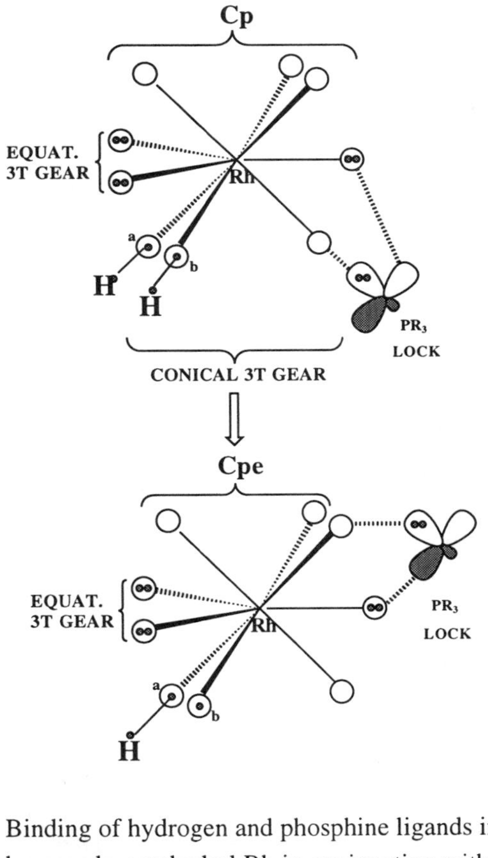

Figure 34.11 Binding of hydrogen and phosphine ligands in $Cp^*(R_3P)RhH_2$ by pseudo-octahedral Rh in conjunction with lower conical 3T gear.

opportunity for relocation of the phosphine ligand from the lower to the upper conical set in the way illustrated in the lower part of the figure.

The result is the generation of an intermediate, Cpe*(R_3P)*RhH, which is now capable of cleaving the σ C—C bond of biphenylene through an H-excessive loop reaction as follows (where the asterisk on the phosphine ligand indicates that it is tied by one upper conical AO, rather than one lower):

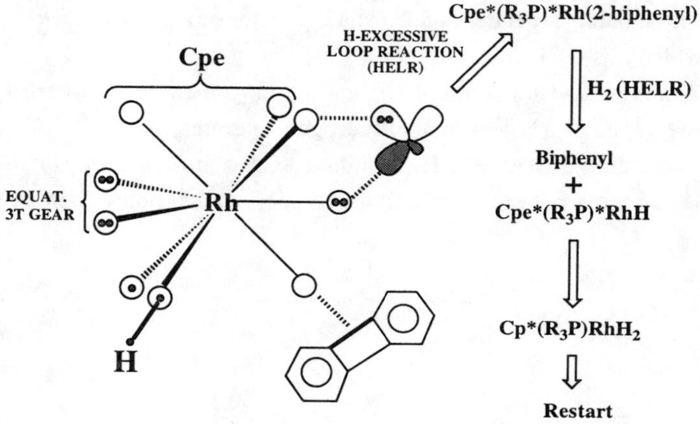

The product, Cpe*(R_3P)*Rh(2-biphenyl), is now all set up for cleavage by H_2 through another H-excessive loop reaction to generate biphenyl plus Cpe*(R_3P)*RhH, which rearranges to the original catalyst Cp*(R_3P)RhH_2, as illustrated above.

Here is now the key point: Perthuisot and Jones not only cited previous evidence[43] for the migration of hydrogen to Cp* in Cp*(R_3P)RhH_2 but also demonstrated that Cp*(R_3P)Rh(2,2′-biphenyl) cannot be the crucial intermediate because it is resistant to hydrogenolysis. Why? It cannot undergo hydrogenolysis by a loop reaction because a metal hole is no longer available (see the structure of the dihydride derivative in Figure 34.11), since the required carbon (C2 of biphenyl) migration to Cp is skeletally blocked. Hence, the entire catalytic cycle is a consequence of the ability of the initial metal complex to enter loop reactions by transmutation to an intermediate with one strategically placed metal hole.

Let us summarize the basic elements of our thinking. A ligand-coupling reaction is the "bond switch" valence isomerization that transforms two bonds to a 1,4-zwitterion, with the metal acting to stabilize both reactants and products. The bond-switched isomer (ligand-coupled isomer) is stabilized by matching the pair and the hole of the zwitterion with a metal hole and pair, respectively, so that each hole–pair couple is stabilized either by electrostatic or T-bonding. We can now see that there exist two limiting geometries of the reacting bonds compatible with the bonding capabilities of a transition metal:

The "linear" arrangement, typified by two olefins docking on two conical holes.

The "parallel" arrangement, typified by two carbonyls docking on two conical holes.

It is now evident that while ligand coupling of the linear type can be induced by the metal alone, bond switching of the parallel type requires the intervention of an external electrophile (nucleophile) and the flexible action of a metal pair (hole):

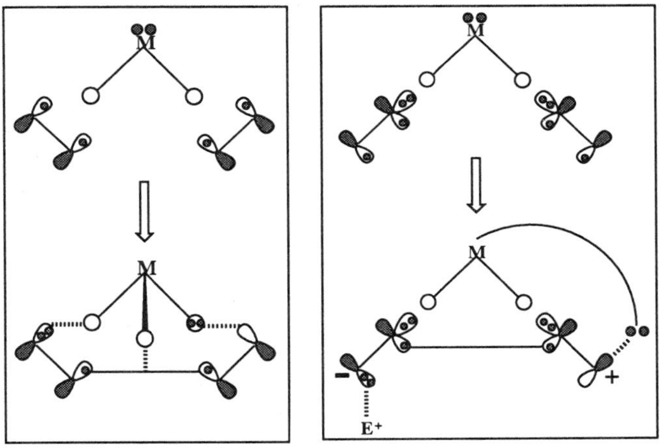

LINEAR COUPLING PARALLEL COUPLING

While many organometallic chemists are familiar with ligand couplings of the linear type (e.g., CpCo-catalyzed couplings), the corresponding processes of the parallel type have only recently surfaced in the work of Lippard and co-workers.[44]

The preparation of an oxo alkylidene complex $Cl_2L_2W(O)(CR_2)$ by Bryan and Mayer (see Section 31.4) suggests further interesting modalities of ligand coupling, such as the one indicated below. Here it is crucial for an initially saturated complex to make one hole available for stabilizing the "diradical" intermediate by one I bond. For this to happen, the complex must have two halides, which can fall back on electrostatic bonding by relinquishing a single metal hole that bound them by contradirectional I-bonding. The precursor of $Cl_2L_2W(O)(CR_2)$ seems to be $(R_2C\text{-}O)_2WL_2Cl_2$.

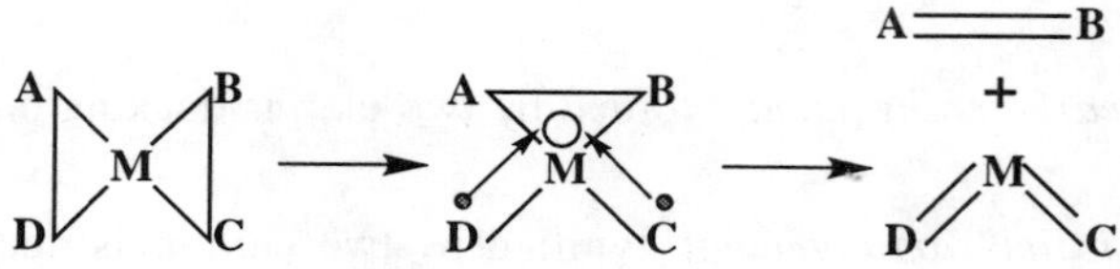

Assuming that the metal binds as d^4 and the chlorines as chlorides, the tungsten has two pairs and four holes for binding the ligands. Two pairs and two holes bind the two R_2C—O units, one hole binds the two ligands, and one hole binds the two chlorides. The latter can act as a stabilizer of the diradical intermediate.

Standing on different grounds are complexes of the type represented by nickelocene and bisallyl nickel. The latter is a key intermediate in the nickel-catalyzed diene polymerization. The hallmark is the presence of two three-electron bonds involving two Ni pairs and the terminal allyl carbons (Figure 34.12).

Two three-electron bonds can couple to produce one exchange bond plus two pairs. Thus, these complexes are ideally suited for effecting radical-like ligand coupling, which is distinct from ligand coupling effected by migratory insertion as in the case of BH_3 plus olefin addition. Not only do the formulas speak for

THE I-FORMULA OF BISALLYL NICKEL

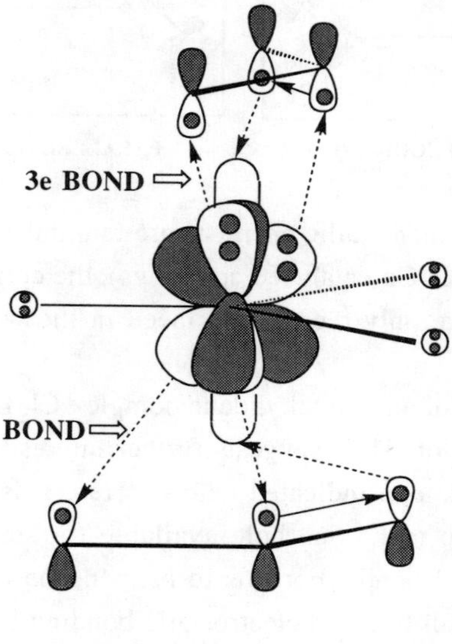

Figure 34.12 The electronic structure of bisallyl nickel.

themselves with respect to explaining data, they also suggest useful analogies. Thus, for example, we seek to make a p-block analogue of the transition metal allyl complexes in the way illustrated. The tuner substituent must be selected to effect the optimal three-electron bond.

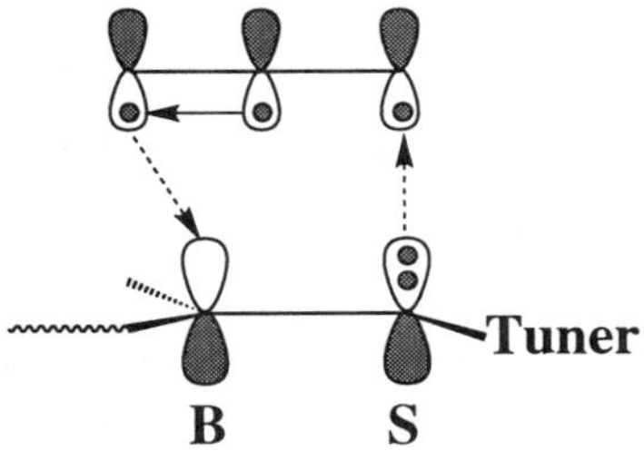

A clear discussion of the polymerization of 1,3-butadiene by nickel is given in the classic monograph by Collman and Hegedus.[45] The first step is the coupling of the two allyl fragments of bisallyl nickel to yield 1,5-hexadiene induced by two molecules of 1,3-butadiene (BU) that produce $NiBU_2$. The crucial step is the coupling of the two BU molecules into a bisallyl chain in the presence of one molecule of BU. What is the electronic mechanism of this coupling? To understand this, we must first understand how each BU is held by Ni in the parent $NiBU_2$ complex. The answer is simple: One BU is tied up to one π-type (d_{xz}) pair and the axial hole (q) of a d^3s Ni, and the other is tied up by the second π-type (d_{yz}) pair and the same axial hole as illustrated (for one BU) below:

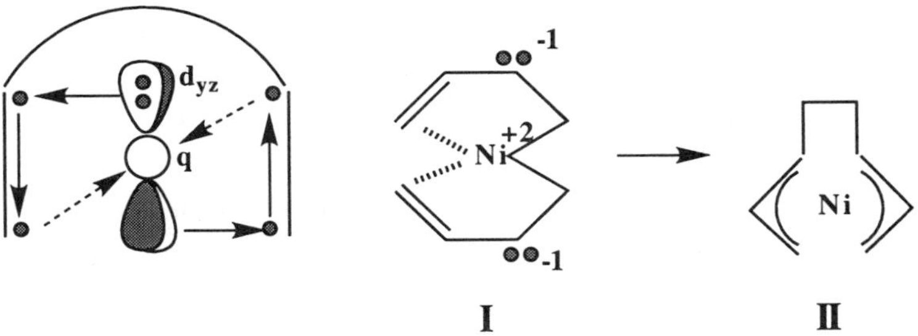

I II

Note how the association rule places a restriction on the number of arrows that can end up in the axial hole which, because it has 4s character, is higher lying (i.e., is daggered). The tetrahedral-like arrangement of the four olefinic fragments is a result of the orthogonality of the d_{xz} and d_{yz} AOs. One can easily produce the electronic structure of derivatives such as $PdBUCl_2$. We can obtain a conventional formula that is familiar to the chemist and comes close to the correct VB description by simply assuming that only one arrow can exit a pair. In this way, we obtain formula I which projects how reductive elimination yields II. Once again, formulas I and II speak for themselves. They reveal the regiochemistry of the coupling (assuming no steric effects), the effect of substituents, and so on.

34.13 Are Metallomimetic Complementary Catalytic Antibodies Viable?

Consider the following intriguing scenario. Upon coordination, $(CO)_3Fe$ effects ligand coupling (i.e., bond switching) in 1,3-butadiene.[46] Can we replace $(CO)_3Fe$ by a protein that does the same thing better? Since $(CO)_3Fe$ works as a 3h/2e fragment (two holes plus one pair), we want to build a protein with the same qualifications; that is, a protein that will have two acidic protons (simulating the two metal holes) plus one lone pair (simulating the one metal pair) in a tripodal geometry much like that of $(CO)_3Fe$. This means that we should elicit an antibody not with $(CO)_3$ Fe but with the complementary $(CO)_3$ Ni, which is 3h/4e and will direct complementary formation of a 3h/2e protein pocket. The whole issue is now the synthesis of a stable 3h/4e molecule that has C_{3v} geometry. One possibility is to chelate Cu^+ with a tribasic ligand and let electrostatic bonding stabilize what is formally an unsaturated 16-electron molecule. This can then be used to elicit a complementary 3h/2e antibody which can catalyze the bond switch of two π or σ bonds.

Is this strategy feasible? Theory takes us to the precipice and then we must rely on experiments. Given the ingenuity of synthetic organometallic chemists, one cannot rule out the creation of metallomimetic antibodies that may act as excellent catalysts of bond making and bond breaking. We can get a hint by trying the following investigation: If a catalytic antibody stabilizes a Diels–Alder transition state (three electron pairs), we have the implication that the same antibody may bind a 3h/6e metal fragment. While antibodies to stable metal complexes have been made,[47] studies of the type suggested here have not been carried out because of the obvious difficulty of using unstable metal fragments. Thus, we are left with a speculation that may ultimately lead to a pot of gold.

References

1. M.T.H. Liu, *Acc. Chem. Res.* 27, 287 (1994).
2. E.J. Dix, M.S. Herman, and J.L. Goodman, *J. Am. Chem. Soc.* 115, 10424 (1993).
3. S. Wierlacher, W. Sander, and M.T.H. Liu, *J. Am. Chem. Soc.* 115, 8943 (1993).
4. X. Wang and K.N. Houk, *J. Am. Chem. Soc.* 112, 1754 (1990).
5. (a) D.H.R. Barton, J.-P. Finet, C. Giannotti, and F. Halley, *J. Chem. Soc. Perkins Trans.* I, 241 (1987). (b) Review: J.-P. Finet, *Chem. Rev.* 89, 1487 (1989).
6. Y. Uchida, K. Onoue, N. Tada, F. Nagao, and S. Oae, *Tetrahedron Lett.* 30, 567 (1989).
7. M.G. Moloney, J.T. Pinhey, and E.G. Roche, *J. Chem. Soc. Perkins Trans.* 1333 (1989).
8. T. Onak, *Organoborane Chemistry*, Academic Press, New York, 1975.
9. A. Pelter, in *Rearrangements in Ground and Excited States*, Vol. 2, P. de Mayo, Ed., Academic Press, New York, 1980, Chapter 8.
10. B.M. Mikhailov and Y.N. Bubnov, *Organoboron Compounds in Organic Synthesis*, Harwood Academic Publishers, New York, 1984.
11. P. Koester, W. Larbig, and G.W. Rotermund, *J. Liebigs Ann. Chem.* 682, 21 (1965).
12. F. Calderazzo, *Angew. Chem. Int. Ed. Engl.* 16, 299 (1977).
13. H.C. Brown and P.V. Ramachandran, *Acc. Chem. Res.* 25, 16 (1992).
14. J.S. Rirtwistle, K. Lee, J.D. Morrison, W.A. Sanderson, and H.S. Mosher, *J. Org. Chem.* 29, 37 (1964).
15. L.H. Toporcer, R.E. Dessy, and S.I.E. Green, *J. Am. Chem. Soc.* 87, 1236 (1965).
16. (a) P.D. Bartlett, *Rec. Chem. Prog.* 18, 111 (1957). (b) Review: V.G. Dryuk, *Tetrahedron* 32, 2855 (1976).
17. (a) Singlet oxygen plus nucleophilic alkene: P.D. Bartlett and A.P. Schaap, *J. Am. Chem. Soc.* 92, 3223 (1970). *Active Oxygen: Reactive Oxygen Species in Chemistry*, C. Foote, J.S. Valentine, J.F. Liebman, and A. Greenberg, Eds., Chapman and Hall, New York, 1995. (b) Dioxiranes: R.W. Murray, *Chem. Rev.* 89, 1187 (1989). W. Adam, R. Curci, and J.O. Edwards, *Acc. Chem. Res.* 22, 205 (1989).
18. (a) P.T. Matsunaga, G.L. Hillhouse, and A.L. Rheingold, *J. Am. Chem. Soc.* 115, 2075 (1993). (b) W. Howard and G. Parkin, *J. Am. Chem. Soc.* 116, 606 (994). (c) C.E. Laplaza, A.L. Odom, W.M.

Davis, C.C. Cummins, and J.D. Protasiewicz, *J. Am. Chem. Soc.* 117, 4999 (1995).

19. W.A. Nugent and J.M. Mayer, *Metal–Ligand Multiple Bonds*, Wiley-Interscience, New York, 1988.
20. (a) Cyclopropanation: M.P. Doyle, *Chem. Rev.* 86, 919 (1986). (b) Aziridination: D.A. Evans, M.M. Faul, and M.T. Bilodeau, *J. Am. Chem. Soc.* 116, 2742 (1994). (c) Epoxidation: K.A. Jorgensen, *Chem. Rev.* 89, 431 (1989). (d) W. Zhang, N.H. Lee, and E.N. Jacobsen, *J. Am. Chem. Soc.* 116, 425 (1994), and references cited therein.
21. G.K. Cook and J.M. Mayer, *J. Am. Chem. Soc.* 116, 1855 (1994). See also: A.S. Goldstein, R.H. Beer, and R.S. Drago, *J. Am. Chem. Soc.* 116, 2424 (1994) and J.T. Groves, W.J. Kruper, and R.C. Haushalter, *J. Am. Chem. Soc.* 102, 6375 (1980).
22. M. Brookhart and Y. Liu, *J. Am. Chem. Soc.* 13, 939 (1991). Recent work suggests that inversion of an anionic center that is coordinated to a metal and leading to a cyclopropane can be favorable: P. Kocovsky, J. Srogl, M. Pour, and A. Gogoll, *J. Am. Chem. Soc.* 116, 186 (1994).
23. P.-O. Norrby, H.C. Kolb, and K.B. Sharpless, *J. Am. Chem. Soc.* 116, 8470 (1994).
24. R.R. Schrock, K.-Y. Shih, D.A. Dobbs, and W.M. Davis, *J. Am. Chem. Soc.* 117, 6609 (1995).
25. (a) S. Nishida, I. Moritani, and T. Teraji, *J. Org. Chem.* 38, 1878 (1973). (b) R.W. Hoffmann, U. Bressel, J. Gehlhaus, and H. Hauser, *Chem. Ber.* 104, 873 (1971).
26. (a) A.G. Brook, in *Heteroatom Chemistry*, E. Block, Ed., VCH, New York, 1990, p. 105. (b) A.G. Brook and A.R. Bassindale, in *Rearrangements in Ground and Excited States*, Vol. 2, P. de Mayo, Ed., Academic Press, New York, 1980, Chapter 9.
27. M. Cooke, J.A.K. Howard, C.R. Russ, F.G.A. Stone, and P. Woodward, *J. Organomet. Chem.* 78, C43 (1974); *J. Chem. Soc. Dalton Trans.* 70 (1976).
28. (a) R. Koelliker and D. Milstein, *J. Am. Chem. Soc.* 113, 8524 (1991). (b) I. Simonou, S. Khan, C.S. Foote, Y. Elemes, I.M. Mavridis, A. Pantidou, and M. Orfanopoulos, *J. Am. Chem. Soc.* 117, 7081 (1995), and references therein.
29. (a) J.-F. Nierengarten, C.O. Dietrich-Buchecker, and J.-P. Sauvage, *J. Am. Chem. Soc.* 116, 375 (1994). (b) C. Liang and K.M. Mislow, *J. Am. Chem. Soc.* 116, 11189 (1994).

30. At long last, it has been recognized that reactions other than pericyclic can have low energy transition states: D.M. Birney and P.E. Wagenseller, *J. Am. Chem. Soc.* 116, 6262 (1994).
31. (a) *Chem. Eng. News* 73(15), 37 (1995). (b) K. Tran, M.A. Hanning-Lee, A. Biswas, A.E. Stiegman, and G.W. Scott, *J. Am. Chem. Soc.* 117, 2618 (1995). (c) P.-O. Norrby, H.C. Kolb, and K.B. Sharpless, *J. Am. Chem. Soc.* 116, 8470 (1994). (d) S.M. Augustine and W.M.H. Sachtlet, *J. Catal.* 116, 184 (1989). (e) A. Dovletoglou and T.J. Meyer, *J. Am. Chem. Soc.* 116, 215 (1994). (f) C.P. Casey and L.M. Petrovich, *J. Am. Chem. Soc.* 117, 6007 (1995).
32. G.N. Sastry and S. Shaik, *J. Am. Chem. Soc.* 117, 3290 (1995).
33. F.W. McLafferty, *Anal. Chem.* 31, 82 (1959).
34. D.L. Harris and G.H. Lew, *J. Am. Chem. Soc.* 116, 11671 (1994).
35. M. Newcomb, M.-H. Le Tadic, D.A. Putt, and P.F. Hollenberg, *J. Am. Chem. Soc.* 117, 3312 (1995).
36. D. Ostovic and T.C. Bruice, *Acc. Chem. Res.* 25, 314 (1992). See also T.G. Traylor, C. Kim, J.L. Richards, F. Xu, and C.L. Perrin, *J. Am. Chem. Soc.* 117, 3468 (1995).
37. (a) *Chem. Eng. News* 73(25), 32 (1995). (b) B.M. Trost, *Science* 254, 1471 (1991). (c) B.M. Trost, *Angew. Chem. Int. Ed. Engl.* 34, 259 (1995). (d) B.M. Trost, A.F. Indolese, T.J.J. Mueller, and B. Treptow, *J. Am. Chem. Soc.* 117, 615 (1995). (e) B.M. Trost, G.J. Tanoury, M. Lautens, C. Chan, and D.T. MacPherson, *J. Am. Chem. Soc.* 116, 4255 (1994).
38. R. Boese, D.F. Harvey, M.J. Malaska, and K.P.C. Vollhardt, *J. Am. Chem. Soc.* 116, 11153 (1994), and references therein.
39. P. Powell, *Principles of Organometallic Chemistry*, 2nd ed., Chapman and Hall, New York, 1988.
40. H. Yasuda, K. Tatsumi, and A. Nakamura, *Acc. Chem. Res.* 18, 120 (1985).
41. C. Perthuisot and W.D. Jones, *J. Am. Chem. Soc.* 116, 3647 (1994).
42. T.R. Kelly, M.C. Bowyer, K.V. Bhaskar, D. Bebbington, A. Garcia, F. Lang, M.H. Kim, and M.P. Jette, *J. Am. Chem. Soc.* 116, 3657 (1994).
43. W.D. Jones, V.L. Kuykendall, and A.D. Selmeczy, *Organometallics* 10, 1577 (1991).
44. E. Carnahan, J.D. Protasiewicz, and S.J. Lippard, *Acc. Chem. Res.* 26, 90 (1993).

45. J.P. Collman, L.S. Hegedus, J.R. Norton, and R.G. Finke, *Principles and Applications of Organotransition Metal Chemistry*, University Science Books, Mill Valley, CA, 1987.
46. For pertinent structural data, see: A.D. Redhouse, "Structure of Organometallic Compounds" in *The Chemistry of the Metal–Carbon Bond*, F.R. Hartley, and S. Patai, Eds., Wiley, New York, 1986.
47. P.G. Schultz, R.A. Lerner, and S.J. Benkovic, *Chem. Eng. News* 68(22), 26 (1990).

Part IV

The Cluster I Bond

Chapter 35

The Many Important Lessons of the Li_4 Metal Cluster

35.1 Bare Metals: The End of Arrow Codirectionality

The combination of electropositive metals with electronegative nonmetals produces ionic solids that are bound principally by the E mechanism. The characteristic feature is the close-packing of anions and the deposition of the cations in the interstitial spaces. Pure metals are also solids and they have the same characteristic feature: close-packing of the metal atoms. We will now argue that the mechanism of bonding of homonuclear and heteronuclear metallic solids is essentially the same, namely, E-bonding, simply because arrow codirectionality has given way to arrow contradirectionality. In other words, the dominating configuration is an association-promoting alternant hole–pair configuration (ALT) which dictates principal E-bonding complemented by I-delocalization that is no longer of the REL type. We will argue that the simplest prototypes of metallic bonding, namely, the Li_4 rhombus and the Li_6 "raft," have crossed the line separating the organometals from the pure metals, as illustrated in Figure 35.1. Instead of having an activated codirectional arrow train, they have contradirectional I bonds.

35.2 Square Li_4 Foils One-Electron Concepts

Very few computational results are important in the development of new concepts simply because the experimentalist has been there first! One way for the computational chemist to make a worthwhile contribution is to calculate molecules that cannot be easily isolated because of the very nature of bonding. Indeed, ab initio calculations of this type have served to define the magnitude of the problem one faces when attempting to understand what is "covalent" and what is "metallic" bonding: According to ab initio calculations, rhombic H_4 lies about 150 kcal/mol above two isolated H_2 molecules[1] but rhombic Li_4 lies 13 kcal/mol below two isolated Li molecules.[2] Now, considering that many people interested in applications of quantum mechanics to molecular electronic structure typically deal with energy differences of the order of few kcal/mol (relative stability of conformational isomers, pericyclic vs. diradicaloid transition states, "classical" vs. "nonclassical" ions, etc.), the situation is clearly desperate: There

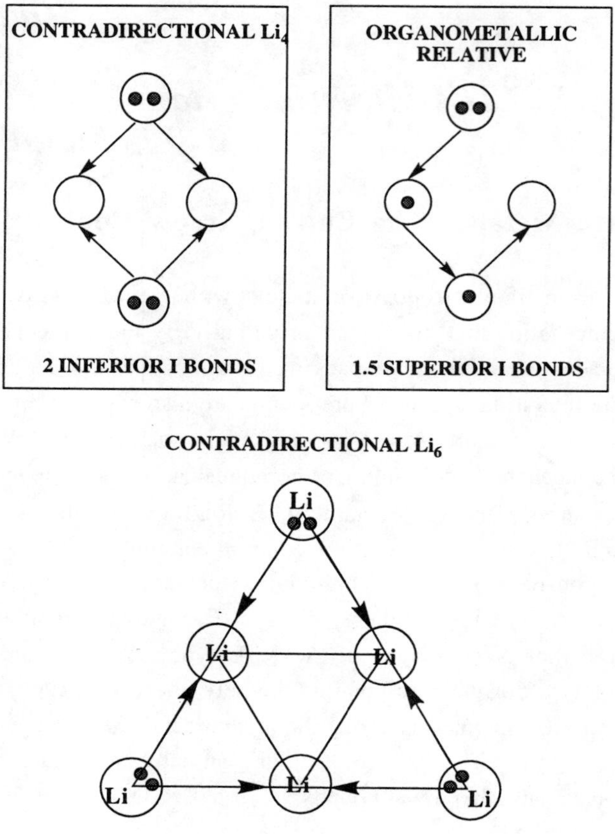

Figure 35.1 First-order electronic structure of Li_4 and Li_6.

is an estimated 163 kcal/mol swing in stability in going from H_4 to Li_4! This can only mean that these species are fundamentally different in their bonding. New formulas that provide a clear differentiation are needed. We have already done this with Li_2 and Li_2^+. However, it is Li_4 that deals the decisive blow to any illusion that metallic bonding resembles covalent bonding in the least.

There are two important papers regarding the structures of Li_4 and Li_6:

(a) McAdon and Goddard[3] predicted that Li_6 is not a regular hexagon, as one might have expected by analogy to π-benzene. Rather, it is an equilateral Li_3 triangle with overall D_{3h} symmetry, here each side is coordinated to a Li atom. This means that regular polygonal metal clusters distort, even when there is no apparent motivation for doing so, to be able to form a close-packed structure.

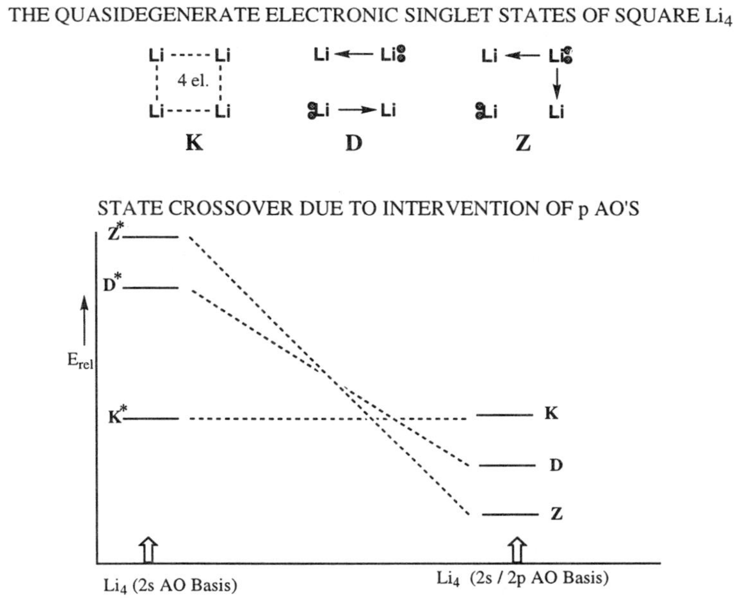

Figure 35.2 Li2p AOs induce crossing of the upper diradicaloid states of square Li_4.

(b) Maynau and Malrieu[4] showed that, while the lowest and highest energy singlet diradical states of square H_4 are B_{2g} and B_{1g}, respectively, the order switches around in square Li_4. What formerly was a highly unfavorable purely "ionic" state in H_4 (B_{1g}) becomes the ground state in Li_4. Equally important is the observation that the A_{1g} state also dips below the original B_{2g} ground state! This state of affairs is depicted in Figure 35.2, with $K^* = B_{2g}$, $D^* = A_{1g}$, and $Z^* = B_{1g}$.

The best way to understand what goes on with square Li_4 is to start with the simpler problem of the electronic structure of square Li, where each Li is assigned only one 2s AO. There are three singlet diradical states designated K* (B_{2g}), D*(A_{1g}), and Z*(B_{1g}). Examination of the corresponding VB wavefunctions (Table 12.1) reveals how symmetry controls the action of the critical association-promoting ALT and HRP configurations. Both are turned off in K while one (ALT) or both remain active in D and Z, respectively.

What are the I formulas of the D and Z states, and how many I bonds does each one have? With the aid of Table 12.1, we conclude that D and Z should be represented as follows:

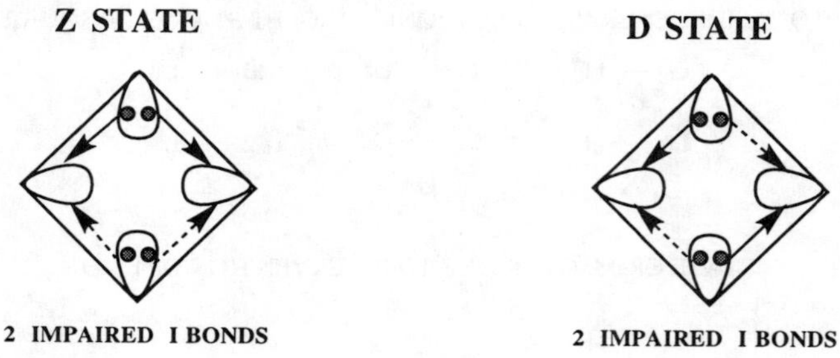

where a 2-e hop can occur by a combination of two arrows of the same type (solid or dashed), but a 2-e hop involving one solid and one dashed arrow is "symmetry forbidden":

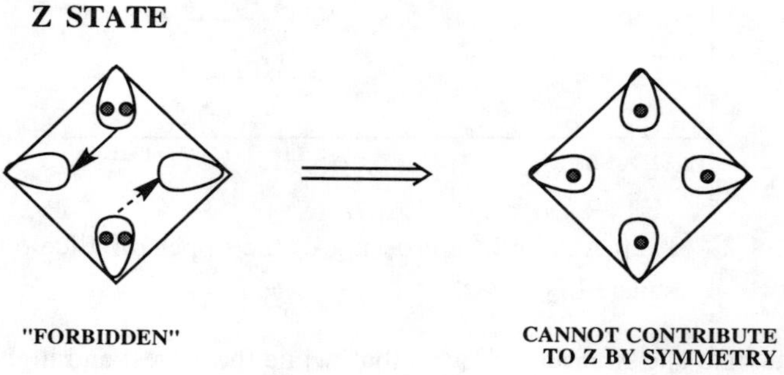

Finally, the "allowed" 2-e hop in D does not generate perfect pairing but, rather, diagonal covalent bonding.

Admitting now the tangential 2p AOs in the basis set amounts to a perturbation that has the following effect: It stabilizes the quadrupolar ALT configuration by allowing each pair of electrons to "dilate" and minimize intrapair Coulomb repulsion while still keeping away from the other electron pair. This is called *pair dilation.* In the absence of the ALT configuration, because of symmetry constraints, pair dilation is no longer possible because it has to originate in the nonalternant hole–pair configuration. Pair dilation would now cause one pair to run into another:

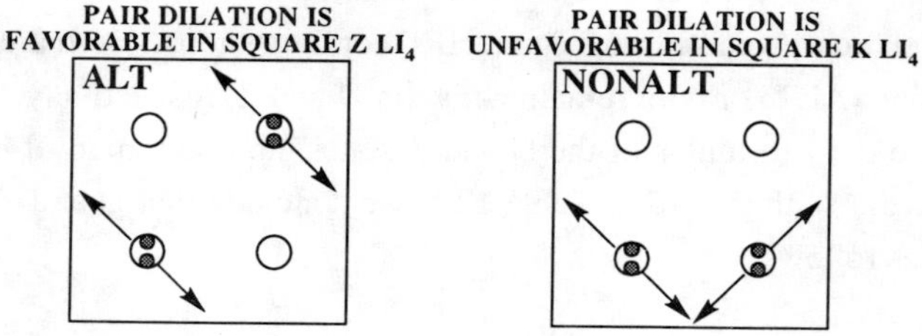

Because the ALT configuration can contribute only to D and Z but not to K, by symmetry, the effect of turning on the 2p (and higher) functions is to drive both D and Z below K. The tendency of Z to drop below K is greater than that of D because of the greater importance of the ALT configuration in the former, purely ionic, state.

In summary, the primary role of the tangential Li2p AOs is not that normally expected from a valence AO. Rather, it is to minimize interelectronic repulsion attending electron delocalization in the low energy radial Li2s AOs. This is produced by 2s/2p hybridization, which minimizes intrapair repulsion as well as interpair repulsion. The development of the wavefunction of Li_4 is shown in steps in Figure 35.3. The key point to note is that electron transfer to produce alternating ions would not have taken place in the absence of the Li2p AOs.

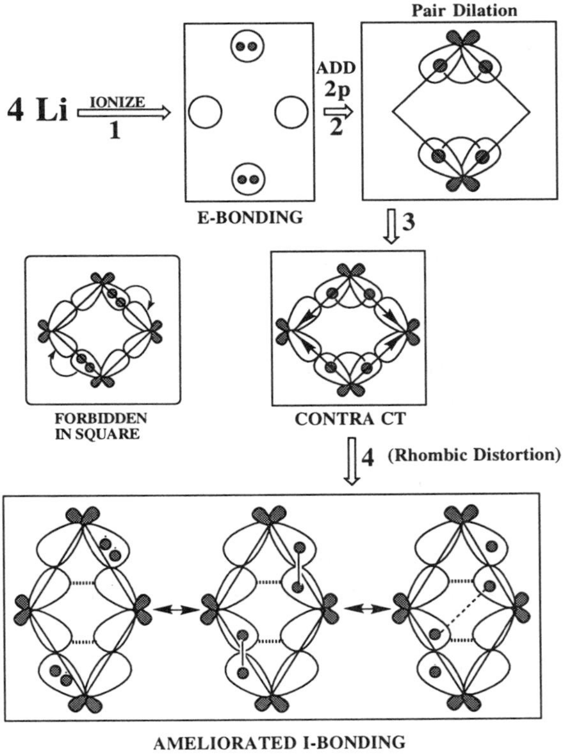

Figure 35.3 Step-by-step construction of the VB wavefunction of rhombic Li_4.

35.3 The Shell Model Interpretation of the Structure of Square Li_4; the Pivotal Role of Coulomb Exchange

Our goal is to replace the one-electron theory of chemistry by a model featuring many electrons. We hope to achieve this by providing explicit VB formulas while at the same time demonstrating what is wrong with manipulating orbitals EHMO style. We will now show how we arrive at the formulas of the K and Z states of square Li_4 in a formal sense by using the symmetry-adapted MOs of square Li_4 according to the shell model described in Chapter 4.

We begin by representing each Li by a single Li2s. The four MOs are designated by r_i. Allocating two electrons to r_1 and two electrons to the degenerate r_2 and r_3 creates three singlet states designated K, D, and Z having the following occupancies of the degenerate orbitals:

$$K = |r_2\bar{r}_3| + |r_3\bar{r}_2|$$

$$D = |r_2\bar{r}_2| - |r_3\bar{r}_3|$$

$$Z = |r_2\bar{r}_2| + |r_3\bar{r}_3|$$

In VB terms, the K state corresponds to the two interacting Kekulé structures. In MO terms, it has the lowest energy because the two odd electrons in r_2 and r_3 are localized in different regions of space: one at atoms a and c and the other at atoms b and d. On the other hand, in VB terms, the Z state is purely ionic, receiving a high contribution from the two (out-of-phase) quadrupolar ALT configurations. In MO terms, it has the highest energy largely because two electrons are forced to occupy the same MO, and this causes strong interelectronic repulsion. Borden and Davidson[5] have argued (analogously for π-cyclobutadiene) that K has lower energy than its triplet counterpart because there is CI involving promotion of one electron from r_1 to r_4. The resulting configuration compromises two triplet subsystems coupled into an overall singlet, with the result that electrons with the same spin occupy alternant sites: One triplet pair localizes at atoms a and c and the second triplet pair at atoms b and d. The resultant spin alternation is a characteristic feature of the in-phase combination of Kekulé structures. Hence, all that MO manipulations produce is a well-known result: Valence-space CI emphasizes the covalent Kekulé structures at the expense of the ionic structures, causing a reduction of interelectronic repulsion.

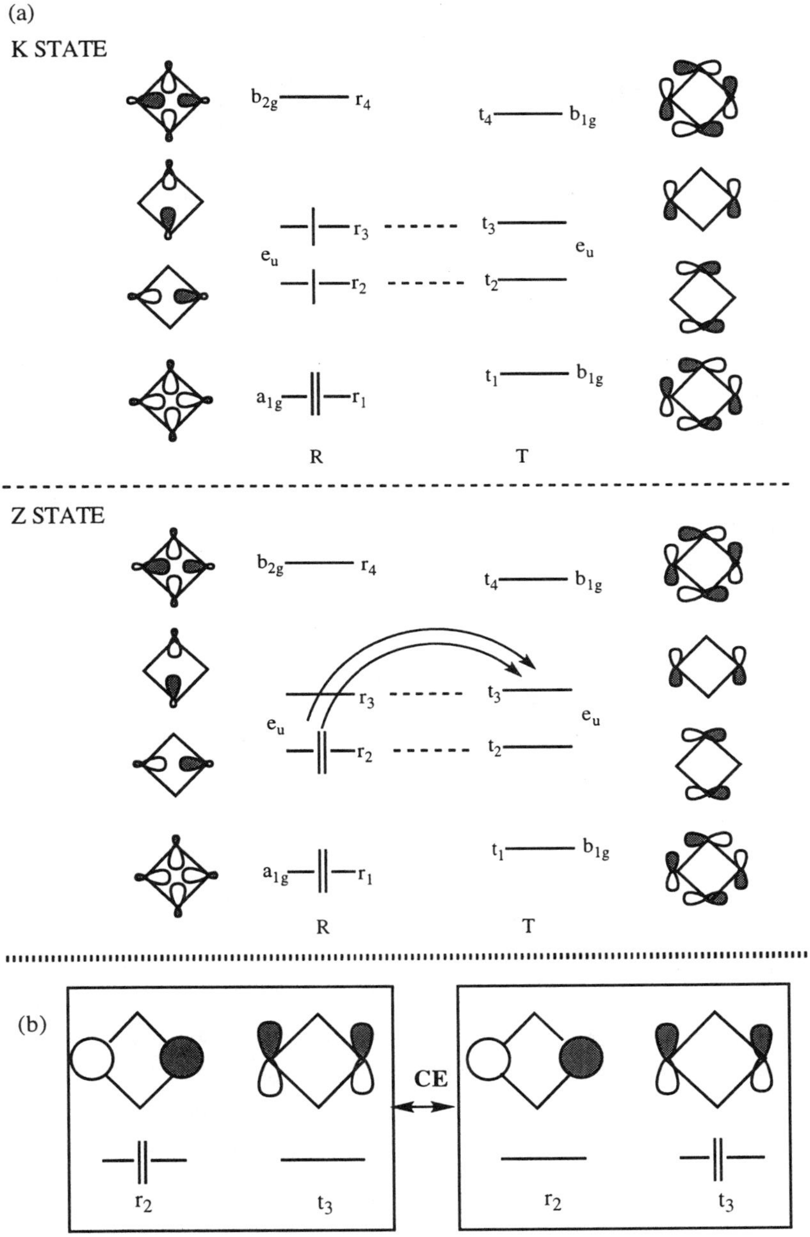

Figure 35.4 Structure of square Li$_4$: (a) Critical MOs and the electronic configurations of the spatially degenerate K and Z states. (b) Configuration interaction due to Coulomb exchange (CE).

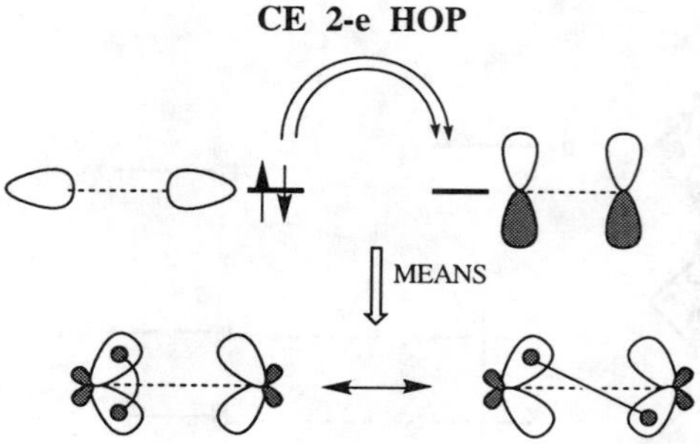

Figure 35.5 The physical significance of Coulomb exchange (CE).

When we augment the basis by admitting the tangential and π-type Li2p AOs, the situation changes. Proceeding as in Chapter 4, we build three MO manifolds, R, T, and P, starting from the radial n, tangential $2p_t$, and π-2p, AOs, respectively. Disregarding, for the moment, the P manifold, we construct the bond representations of the K and Z states as shown in Figure 35.4, where dashed lines indicate CT 2-e delocalization involving overlapping MOs and arrows indicate a Coulomb exchange (CE) 2-e hop involving orthogonal MOs. An electron pair is ideally CE-paired with a hole if both pair and hole MOs are diatom-localized because the corresponding CE hop effects intra-atomic pair correlation as well as interatomic dispersion as illustrated in Figure 35.5. Two fully delocalized MOs are not CE-paired.

We will now argue that the important orbital interactions that differentiate Li_4 from H_4 are not the ones indicated by the dashed lines, which are "contained" in EHMO or RHF theory, but, rather the CE hops present only in a CI treatment of the problem. To this end, consider the critical interactions in the K and Z states of Li_4 as depicted in Figure 35.4a. For brevity, we show only one of the two equivalent bond diagrams of Z. The second (not shown) has r_2 empty and r_3 doubly occupied. The key difference between K and Z lies in the way in which each of the e_u MOs are occupied: radical–radical (1,1) occupancy in K and pair–hole (2,0) occupancy in Z. As a result, a CE hop from r_2 to the nonoverlapping t_3, is possible only in Z and not in K. The crucial point is that this CE interaction, present in Z but absent in K, is strong because it meets the diatom localization condition spelled out in connection with Figure 35.4. The stabilization due to the CE hop is proportional to the bielectronic exchange integral $(r_2t_3|r_2t_3)$ and the CE hop itself is depicted in Figure 35.4b. The physical meaning of the bielectronic exchange integral has two aspects:

(a) An electron pair at one vertex of the square becomes correlated. This is because expansion of $(r_2t_3|r_2t_3)$ into bielectronic exchange integrals over AOs shows that one leading term is $(n_ap_{ta}|n_ap_{ta})$, where the subscript a is an index of the atomic Li center.

(b) Electrons along each diagonal become correlated by dispersion because of the action of a second leading term, $(n_ap_{ta}|n_cp_{tc})$.

The final conclusion is that since CE interaction occurs only in the "closed-shell" Z but not in the "open-shell" K state, Z will be stabilized relative to K as we go from H_4 to Li_4. In other words, intervention of the 2p AOs causes the state with the most unfavorable interelectronic repulsion in the absence of the 2p orbitals to become the state with the most favorable combination of kinetic energy and interelectronic repulsion in their presence.

35.4 Hollow, Boundary, and Vacuum Pairs and the Exceptional Stability of Odd Metal Polygons

According to the shell model, the valence electrons of a cluster can be classified as follows:

(a) Electrons occupying radial MOs that are not overlap-matched with tangential MOs are "inside electrons," and they are called *hollow electrons (H electrons).*

(b) Electrons occupying radial MOs that are overlap-matched with tangential MOs or vice versa are called *boundary electrons (B electrons).* If the radial partners have lower energy, the boundary electrons are displaced "in" (Bh electrons), while if the tangential partners have lower energy, they are displaced "out" (Bv electrons).

(c) Electrons occupying tangential MOs that are not overlap-matched with radial MOs are called *vacuum electrons (V electrons).*

The key points now are that Bh pairs can be simultaneously overlap-paired and CE-paired with tangential holes, and Bv pairs can be simultaneously overlap-paired and CE-paired with radial holes. This is not the case with either H or V electrons. To understand this important point, consider the reference configuration of the Z state of square Li_4 depicted in Figure 35.6. The two pairs are identified as follows:

(a) The r_1 pair is a *hollow pair*, which is only weakly CE-paired to the tangential holes.

(b) The r_2 pair is a *boundary pair*, because it is overlap-matched with t_2. However, at the same time, it is strongly CE-paired with t_3. Noting that r_2 and r_3 are components of an e_u-degenerate MO and so are t_2 and t_3, we see that the components of e-degenerate MOs are simultaneously overlap-paired and CE-paired as follows:

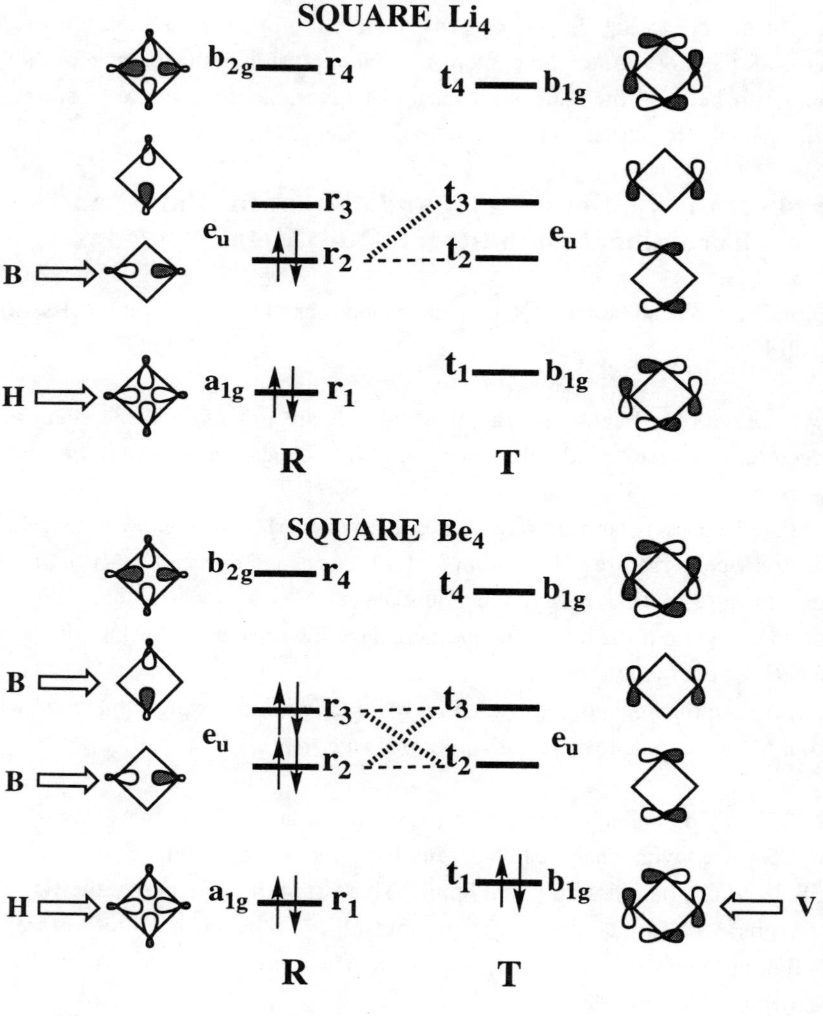

Figure 35.6 Hollow (H), boundary (B), and vacuum (V) electrons of square Li_4 and Be_4.

THE COUPLED CT AND CE INTERACTION OF e_u-DEGENERATE R AND T MOs

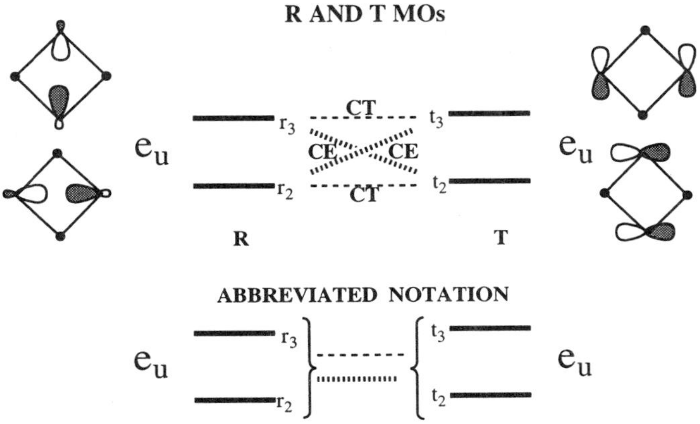

The electronic configuration of square Be_4 provides a further illustration of the classification of metal pairs into B, H, and V pairs.

We can use these concepts to understand how Be_n rings differ in stability. To do this, we construct the shell diagrams, identify the electron pair types, and determine the fraction of the pairs that are both overlap- and CE-stabilized. This fraction, the number of boundary pairs over the total number of pairs: B/(H + B + V), turns out to be equal to the number of e-degenerate radial MOs, N(e). For odd annulenes with n centers and n + 1 electrons, $N(e):N(t) = (n - 1)/n$. For even annulenes with n centers and n electrons, $N(e):N(t) = (n - 2)/n$. It follows that odd must be superior to even rings if indeed the coupled overlap/CE stabilization dominates the problem.

The energy change accompanying the transformation of linear triplet Be_n to regular polygonal singlet Be_n, has been calculated by ab initio techniques: It is positive for $n = 4$, zero when $n = 5$, and negative when $n = 3$.[6]

	ΔE (kcal/mol)
linear $Be_3 \longrightarrow Be_3\ (D_{3h})$	−6.3
linear $Be_4 \longrightarrow Be_4\ (D_{4h})$	+16.7
linear $Be_5 \longrightarrow Be_5\ (D_{5h})$	0

We predict that the reaction will be endothermic when $n = 6$.

The most stable form of Be_4 is not the square but the tetrahedral structure according to ab initio calculations.[6] The shell diagrams explain formally what one might have anticipated on the basis of knowing that Be is a close-packing-seeking metal: The fractions of B-pairs in the square and tetrahedral forms are 1/2 and 2/3, respectively.

Let us now put the results of the MOVB analysis in the context of the VB overview of rings. There are actually three "ring stories":

(a) The carbon rings are dominated by perfect pairing and additional contributions of characteristic VB structures which preserve electroneutrality but fall short of perfect pairing (Chapters 4 and 5).

(b) The inorganic rings made up of semimetals are actually bisaromatic R^xT^y species (where x and y are aromatic electron counts), and the odd members are superior because of activated CT delocalization (configurational degeneracy) (Chapter 24).

(c) The metal rings cannot be anything but relatives of the semimetal rings, with intra- becoming relatively more significant than interatomic CT delocalization because of the greater atom electropositivity. How can we express the difference between the semimetallic $(RP)_n$ and the metallic Be_n rings by simple VB formulas? The recipe turns out to be simple: Allocate the Be electron pairs to the R and T shells of each representative ring according to the shell model and seek to form the maximum number of codirectional I bonds linking one Be with its neighbor. An I bond of this type guarantees bonding by dispersion (a form of intra-atomic CT delocalization) at the limit of zero interatomic CT delocalization.

In discussing the results of this approach (Figure 35.7) we use the index b/n to specifies the number of I bonds by b and the number of metal centers by n. The larger the ratio, the lower the cohesive energy. It can be seen that, in each case, the b/n index is either $(n-1)/n$ in odd or $(n-2)/n$ in even rings, as predicted by the MOVB analysis. This makes sense: The number of I bonds (number of arrows divided by 2) must equal the number of B pairs because the latter support dispersion (by a 2-e CE hop) as well as interatomic CT delocalization (by an ordinary 2-e CT hop). With the exception of Be_6, these results have been obtained by using the simple recipe given above. The Be_6 case is interesting because it projects the importance of symmetry considerations, which are transparent in the MOVB but disguised in the VB method. Specifically, the simple recipe predicts that $i/n = 6/6$, while the MOVB shell diagram predicts a 4/6 fraction of B pairs. We interpret this discrepancy to imply that two of the six I bonds are effectively turned off by symmetry in the six-membered ring. The effect of symmetry can be incorporated in the following rule: Dispersive I bonds associate only an odd number of pairs and holes. In

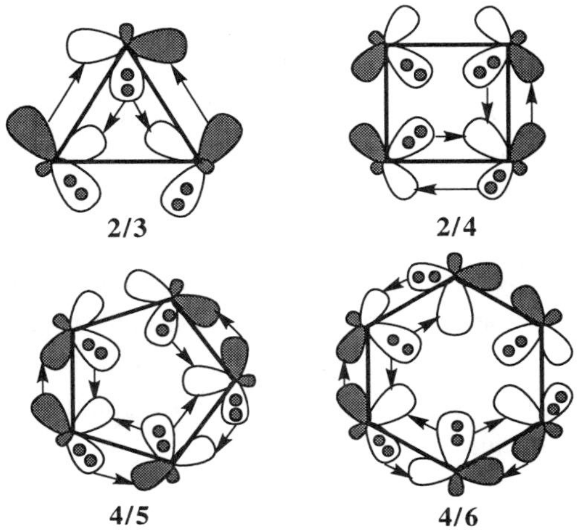

Figure 35.7 The ratio of dispersive I bonds over the number of ring vertices in the prototypical Be_n rings. Note the four missing arrows in the six-membered ring. In each case, only one of the *n* equivalent resonant I descriptors is shown.

this way, the b/n fraction becomes identical to the fraction of B pairs in all four prototypical rings.

We have arrived at an important result: Odd are superior to even rings on account of dispersion on account of the *b/n* ratios. Whether the three- or the five-membered ring is the best depends on the quality of the dispersion component of the I bonds. We can return to the MOVB analysis and ask the question: What ring has B pairs that enjoy maximal CE stabilization? The condition is that the boundary orbitals be diatom localized; furthermore, this diatom localization must involve neighboring atoms. This condition is met only by the three-membered ring, as illustrated. This means that, among odd rings, the three-membered ring is the best when dispersion prevails (electropositive metals), while the five-membered ring becomes superior when covalency is still somewhat important (semimetals).

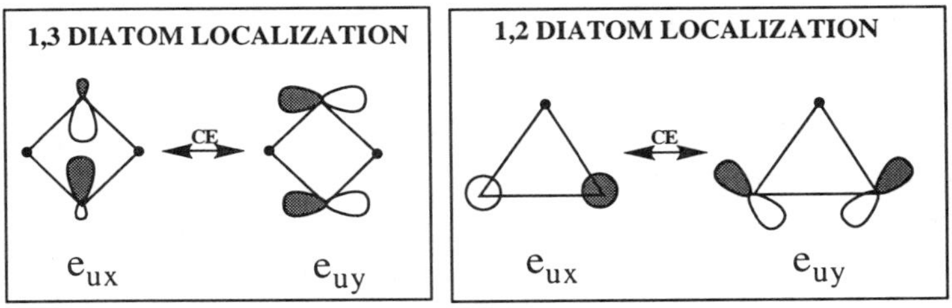

One of the most interesting trends of metal chemistry is the recurrent preference of diverse metal clusters for geometries having fivefold symmetry.[7] This seems to occur whether the metal involved is an "s^1" (e.g., Li, Cr) or an "s^2" (e.g., Be, Mn) metal. We have already exposed what is special about odd rings having three- or fivefold symmetry in the case of Be_n. The question remains: Why does the trend persist when a pair is replaced by a single valence electron? We suggest that Coulomb-correlated pairs and a corresponding number of exchange-correlated, high-spin-coupled odd electrons behave in an analogous manner. Instead of counting I bonds, we now count one-electron bonds, corresponding to individual arrows:

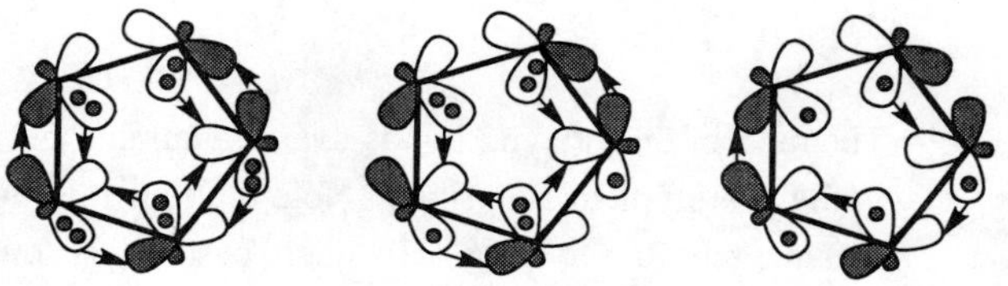

Be_5 $Be_5^{+2} = Li_5^{-3}$ TRIPLET Li_5 HEXAPLET

It can be seen that the b/n ratios of singlet Be_5 and hexaplet Li_5 are identical. Hence, what makes "s^2" metals prefer fivefold symmetry by staying low spin is also makes "s^1" metals prefer fivefold symmetry by going high spin. Of course, within an aperiodic solid, individual high-spin pentagonal bipyramids and icosahedra can couple either ferromagnetically or antiferromagnetically.

We end here by stressing the obvious: We are in the process of learning how to interpret orbital diagrams at the multideterminantal level. Testing the ensuing models requires information that is either unavailable or sporadic and accidental. The "pair–odd electron" analogy can be tested by a systematic computational investigation of the stereochemistry and bonding of low and high spin clusters. In practically every chapter (or even section) of this work, we must start anew.

35.5 Why Is Li_4 Rhombic?

Because rhombic distortion is effectively a *close-packing (CP) distortion*, our understanding of metal solids depends critically on our understanding of rhombic Li_4. The key issue is: Why do the two electron pairs of Li_4 stick together rather than flying apart like the two electron pairs in H_4? The VB answer is that rhombic distortion in Li_4 optimizes the two I bonds present in the square form. By contrast, H_4 undergoes rectangular distortion that causes segregation of the two T bonds present in the square form because their mutual exchange repulsion is annihilated by the distortion.

A contradirectional arrow train is less favorable than a codirectional arrow train, but it does have a saving grace: Individual codirectional 2-e CT hops can still occur, since we can always find two arrows that are codirectional. In other words, we can always transform a contra- to a codirectional arrow train by eliminating the "wrong" arrows. Of course, the penalty is a reduction of the number of I bonds. All this presupposes that symmetry does not "forbid" codirectional CT delocalization. It turns out that this is precisely what happens in the case of the Z state of *square* Li_4: Only contradirectional 2-e hops are "allowed" because of symmetry restrictions (see the VB four-electron wavefunction in Chapter 12). Hence, distortion of the square to a rhombus is the action necessary for ensuring the activation of both modes of 2-e CT delocalization that originate with the ALT configuration (which is selectively stabilized by the Li2p orbitals). This is equivalent to saying that rhombic distortion turns the dashed to solid arrows (see Section 35.2):

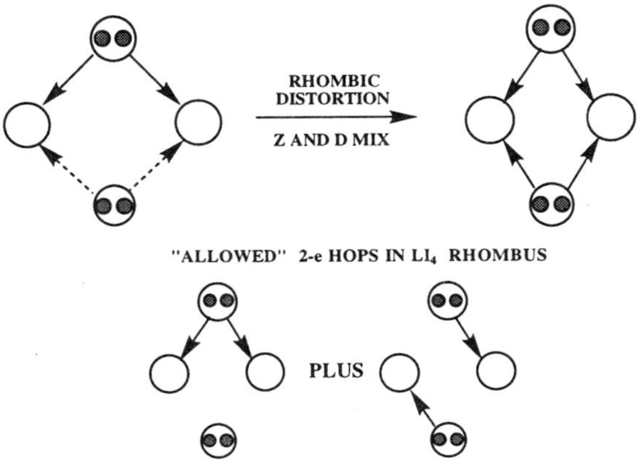

The physical meaning of this result is that two electrons can delocalize not only contradirectionally along two adjacent sides (as in the lowest Z state of the square) but also codirectionally along two opposite sides of the polygon. The final result is portrayed in Figure 35.3, which gives simple answers to key questions:

(a) What stabilizes the rhombus? In part, Coulomb attraction of the ions (already present in the square form) and, in part, ameliorated I-bonding. In other words, what is labeled I(w) in our bonding map of Chapter 3 has actually both an E and an I component.

(b) How can one visualize the CT delocalization in the rhombus? Rhombic distortion creates a four-electron conjugated system exactly analogous to the π-system of 1,3-butadiene, which is "trapped" by four Li cations. The physical meaning of "ameliorated I-bonding" is that interpair correlation is produced by two concomitant effects:

1. The two pairs are placed at the two ends of the *long* rhombus diagonal and their mutual repulsion is reduced.
2. Codirectional CT delocalization along opposite sides allows the two pairs to remain away from each other as they seek to minimize their kinetic energy. The dispersive interaction along the short diagonal of the rhombus (represented by the final resonance hybrid of Figure 35.3) transforms the square into a rhombus, and this is the signature of interpair correlation.

(c) Why does square cyclobutadiene distort to a rectangle but square Li_4 distort to a rhombus? Because the mechanism of bonding of the former is T while that of the latter is I.

The reader may ask: Why ionize the Li atoms in Li_4 and Li_6 instead of preserving covalent bonding in the reference configuration: To answer this question, we go back to Section 16.1, where we saw that Ω, the fraction of disengaged arrow pairs representing covalent bonds, is a function of atom electronegativity. In the case of electropositive Li, we can assume that $\Omega = 1$ (i.e., all three arrow pairs are disengaged). Now, this would produce codirectional CT delocalization (six codirectional arrows in descriptor B) at the price of net exchange repulsion (brought about by descriptor A). Assuming that the overlap of nonneighbor AOs is zero, we can annihilate exchange repulsion by replacing the Kekulé-type covalent VB configuration by the charge-alternant fully ionic configuration. Exchange repulsion is now zero, but (contradirectional) I-bonding is preserved. This is why the HP configuration becomes the reference configuration, which promotes further distortion to minimize Coulomb interpair repulsion.

There is an important lesson here: To achieve full association, the best option is to arrive at an ionic configuration in which pairs and holes alternate so that exchange repulsion goes to zero. At this limit, even contradirectional CT delocalization is sufficient to effect bonding. This is why metals and ionic compounds are solids. In other words, it is the exchange complement of I bonds that thwarts association.

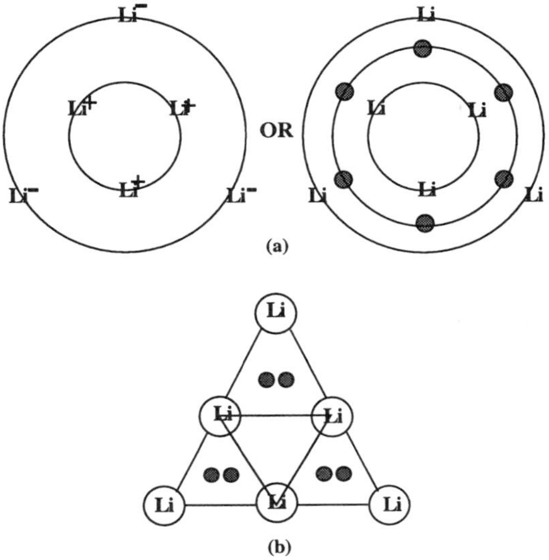

Figure 35.8 Two different ways of describing the electronic structure of Li_6.

We have analyzed the bonding of Li_4 and Li_6 because we seek to learn something about the electronic structure of solids. There are two equivalent interpretations of Figure 35.3 that lead to a physical model of close-packing in metal solids. For the first, we focus attention on the step responsible for ion formation, which produces the model shown in Figure 35.8a. A two-dimensional array of "s^1" metals is made up of alternating concentric circles of cations and anions or alternating concentric circles of cations and electrons. As we move away from the center circle in going from M_6 to M_n ($n > 6$), the cores making up outer circles are separated by longer distances, electron–cation coordination becomes inefficient, and the surface curls to form a spherical cluster. Assuming that the "limit" of a circle is a line (or, taking a cut through the surface comprising the concentric ion circles), the close-packing motif becomes obvious: Close-packed, alternating linear chains of anions and cations represent the best arrangement of the metal atoms.

By focusing attention on the rhombic distortion step, we obtain our second interpretation. In the model shown in Figure 35.8b, close-packing is seen as a consequence of interpair correlation. As we shall see, this model is best for visualizing the electronic structures of metal clusters.

35.6 Li_3 Ions and Radicals as Derivatives of the Li_4 Rhombus

Conceived for rationalizing the electronic structures of boron clusters and diborane itself,[8] the concept of the three-center/two-electron (3c-2e) bond has served as the "explanation" of the electronic structures of key prototypical systems from the simple H_3^+ to the "nonclassical" 2-norbornyl cation to the metallic Li_4. The idea seems eminently reasonable because it reasserts a conclusion of Hückel MO theory: The cyclopropenyl cation is a $4N + 2$ π-aromatic system. Extending the argument to Li_3^+ seems very natural. Indeed, most chemists accept the notion that H_3^+ and Li_3^+ are illustrators of the same bonding mechanism: covalent bonding. We will now argue that Li_3^+ as well as $Li_3\bullet$ and Li_3^- are derivatives of rhombic Li_4 and are bound by the same mechanism.

We begin by taking the normal route of considering Li_3^+ as an analogue of aromatic π-cyclopropenyl cation. We use the shell model, and we generate three MO manifolds, R, T, and P, subject to D_{3h} symmetry constraints (Figure 35.9a).

The electronic configuration is $R^2T^0P^0$. In a one-electron sense, this is a Hückel aromatic system. In a many-electron sense, the stability of this system is additionally due to Coulomb exchange corresponding to a 2-e hop that takes the two electrons occupying r_1 to q_1 and to $t_{1,2}$. The energy lowering associated with these 2-e hops depends on the magnitude of the corresponding repulsion exchange integral and the orbital energy gap. We interpret this interaction to mean that a resonating interstitial pair has formed and is aligned with the three Cartesian axes. Our formula is shown in Figure 35.9b. The derivation of Li_3^+ from rhombic Li_4 is shown in Figure 35.9c.

The Li_3 radical and Li_3^- anions must abandon the D_{3h} geometry because deformation is now required by the Jahn–Teller theorem. There are two possibilities:

A distortion producing an "acute" triangle. This geometrical form can be obtained by starting with a rhombus and removing one vertex located at the *long* diagonal.

A distortion producing an "obtuse" triangle. This geometrical form can be obtained by starting with a rhombus and removing one vertex located at the *short* diagonal.

Is there any way to predict the best choice?

Figure 35.10 shows the derivations of Li_3 and Li_3^- from Li_4. Focusing attention on the anion, the key difference between the "acute" and the "obtuse" form is that the former has one I and one T bond while the latter has two I bonds. As a result, the latter wins.

Gole et al.[9] have summarized computational and experimental data concerning the structures, electron affinities, and dissociation energies of neutral Li_3 and corresponding ion radicals. The trends can now be explained by the VB formulas of the Li trimers.

Experiment and computation indicate that the "acute" and "obtuse" forms of Li_3 are close in energy. By contrast, Li_3^- strongly prefers the "obtuse" form.

The electron affinity of Li_3 is greater than that of Li atom. Indeed, the best estimates are more than 1.0 eV for the former and only 0.6 eV for the latter.

Underlining the difference between nonmetals (T-bonding) and metals (I-bonding) are the following facts: H_3^+ is bound while H_3 and H_3^- are unbound relative to H_2 plus H and H_2 plus H^-, respectively. By contrast, Li_3^+, Li_3, and Li_3^- are all bound.

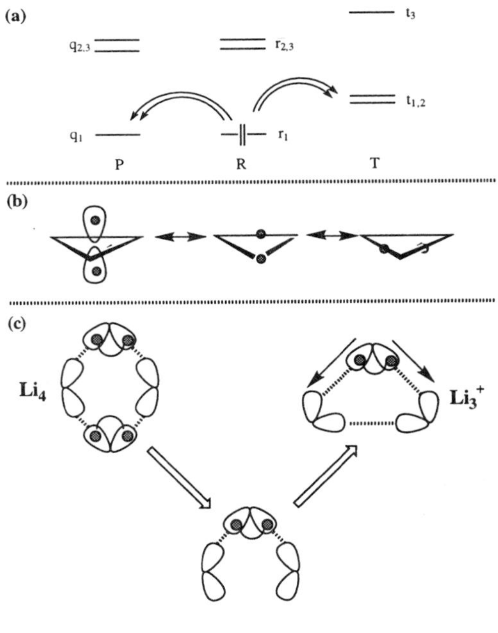

Figure 35.9 (a) Shell model of the bonding of Li_3^+; arrows indicate 2-e Coulomb exchange hop. (b) VB representation of the gas pair gluing three formal Li^+ cores. (c) Derivation of Li_3^+ from Li_4.

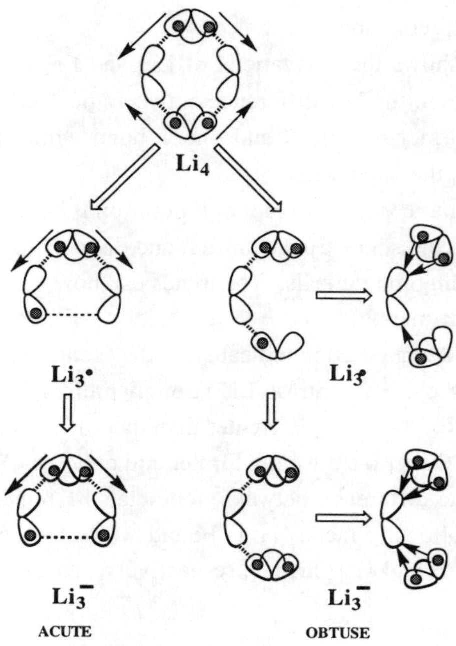

Figure 35.10 Derivation of Li_3^{-} and Li_3 from Li_4.

Li_3^{+} and Li_3^{-} ions have much larger dissociation energies (estimated 1.28 and 0.90 eV, respectively) than the Li_3 radical (0.40 eV).

35.7 Crossed-Antiaromaticity: The Quadrupolar Diradical and the Quadrupolar Zwitterion

Figure 35.11 identifies the difference between the homonuclear transition states of a (4*N* + 2)-electron "allowed" and a 4*N*-electron "forbidden" reaction, where "allowed" and "forbidden" refer to the thermal reaction. When a perturbation is turned on, the D or Z (or both) excited states can cross under the original ground K state in either the "allowed" or the "forbidden" mode. The "crossed" state represents an intermediate, which is viewed as a descendant of either the D or the Z state or both. "Allowed" and "forbidden" reactions are differentiated by the size of the energy gap separating K from D and Z. This creates the following conditions:

D(F) and Z(F) can easily cross below K(F). As a result, D(F) or Z(F) can appear as intermediates in "forbidden" reactions.

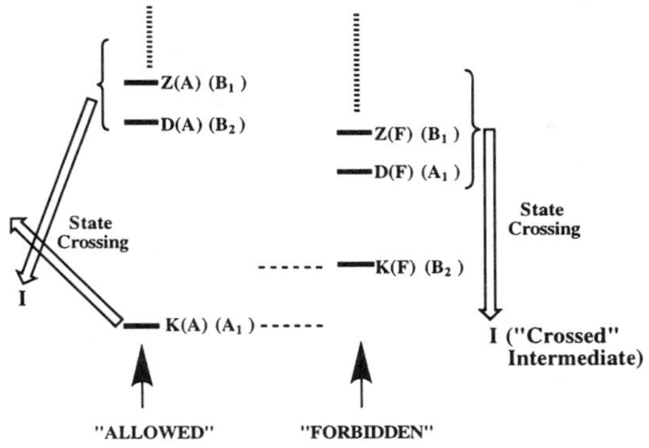

Figure 35.11 State manifolds of "allowed" (A) and "forbidden" (F) complexes. Symbols in parentheses are symmetry labels. Arrows indicate state crossing under the appropriate perturbation.

D(A) and Z(A) need a very large perturbation for crossing under K(A). As a result, they are observed as intermediates either in photochemical reactions or in high temperature pyrolysis.

The key message is that facile "forbidden" reactions can occur via the intermediacy of a species that corresponds to a "crossed" resonance hybrid of D(F) and Z(F). The condition for crossing is that D interact strongly with the close-lying Z. The distortion that produces such mixing (i.e., the distortion that turns the B_1 symmetry label of Z to A_1) can be easily found from group theoretical correlation tables. Once this has been done, the distortion mode itself identifies the substitution pattern, as shown below. However, the symmetry classification of the distortion (e.g., D_{2h}, C_{2v}) is by itself meaningless. For example, rectangular (symbolized by "re") D_{2h} fails but rhombic (symbolized by "rh") D_{2h} distortion succeeds in effecting the crossing of D below K by turning on its interaction with Z.

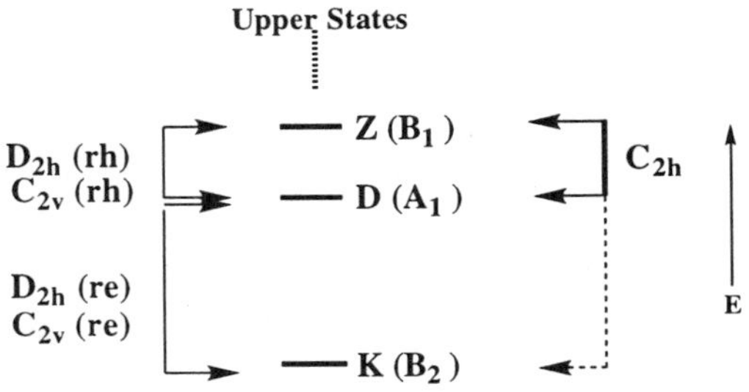

The stereochemical outcomes of the C_{2h} and D_{2h}(rh) distortion modes that couple D to Z are shown in Figure 35.12. Their effect is contrasted to the effect of D_{2h}(re) distortion . We conclude that D will cross *strongly* under K if the A atoms are replaced by substituents so that the complex attains either D_{2h}(rh) or C_{2h} symmetry. What is the best choice of substituents?

A VB state is made up of different subsets of equivalent VB structures, which are combined according to the symmetry constraints. For example, D has an in-phase combination of two equivalent alternant quadrupolar structures, ALT + ALT′, while Z has an out-of-phase combination of the same two structures, ALT – ALT′ (see Table 12.1). Two VB states interact strongly whenever the distortion causes a mixing of subsets of identical configurations that differ only in the nodal properties. The result of such a mixing is the projection of the VB structure that provides the best response to the applied perturbation. For example, placing alternating R and W groups will cause mixing of D and Z because this amounts to adding (ALT + ALT′) to (ALT – ALT′) so that ALT survives while ALT′ is annihilated. The result is that ALT, with the alternating

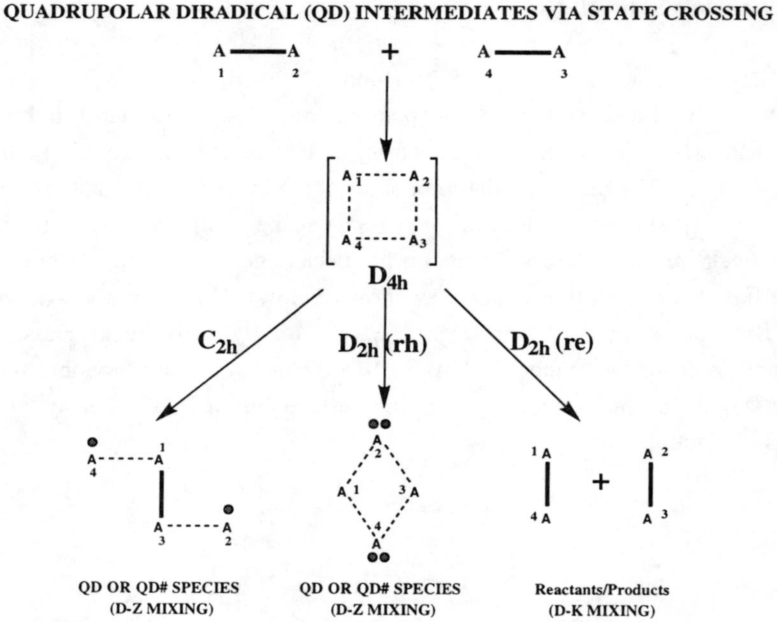

Figure 35.12 Formation of QD intermediates or QD# transition states by C_{2h} and D_{2h}(rh) distortion of a hypothetical square precursor (rh = rhombic, re = rectangular).

holes and pairs, is the optimal response to alternating R and W groups. In summary, the interaction of D and Z depends, in part, on the magnitude of the interaction of the symmetry-adapted quadrupolar structures which, in turn, depends on the electronegativity difference of the R and W atoms or groups that define the D_{2h} or C_{2h} complex.

The physical significance of the mixing of the in- (D) and out-of-phase (Z) quadrupolar structures via D_{2h} or C_{2h} distortion is schematically depicted in Figure 35.13. Assuming that the diagonal perfect pairing configuration promotes C_{2h} and the quadrupolar configuration promotes rhombic D_{2h}(rh) distortion, C_{2h} will give way to D_{2h}(rh) distortion x, as the electronegativity difference between R and W, increases. When x^- is modest, a 2+2 addition will proceed via the intermediacy of a quadrupolar diradical, while, when x^- is large, a

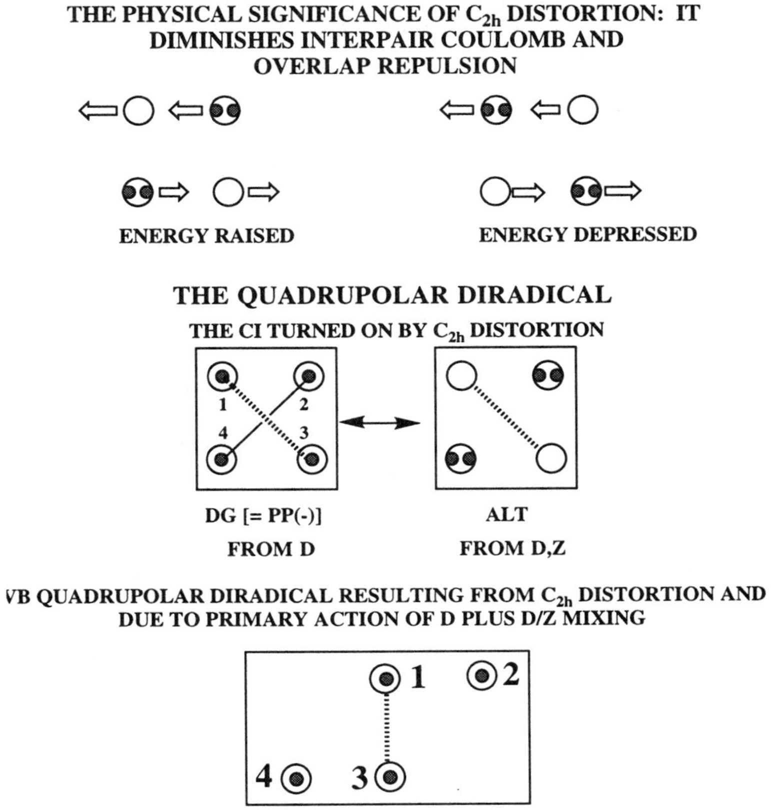

Figure 35.13 The electronic structure of a quadrupolar diradical (QD). Note how the 1—3 bond is formed where the quadrupolar structure has holes.

2+2 addition will occur via the intermediacy of a quadrupolar zwitterion .

The quadrupolar diradical transition state (QD#) leading to the formation of the quadrupolar diradical intermediate (QD) derives much of its stability from the CI depicted in Figure 35.13. The distortion of the diagonally bonded D establishes a single bond between the atoms that bear the holes in the stabilized quadrupolar VB structure because this produces minimization of interpair repulsion (of both the overlap and Coulomb types) . This leads to the prediction that the dimerization of two unsymmetrical olefins will occur in a preferential head-to-head fashion. The quadrupolar diradical represents a continuous evolution of the D state of the formal 2s+2s complex, which "parents" the transition state, QD#, as well as the final product, QD. The difference between QD# and QD lies in the relative weights of the DG and ALT configurations shown in Figure 35.13. Thus, QD# is expected to have tetraradical (because the difference in the two diagonal bonds is still small) and quadrupolar character, while QD is expected to have predominant diradical character (because one diagonal is now

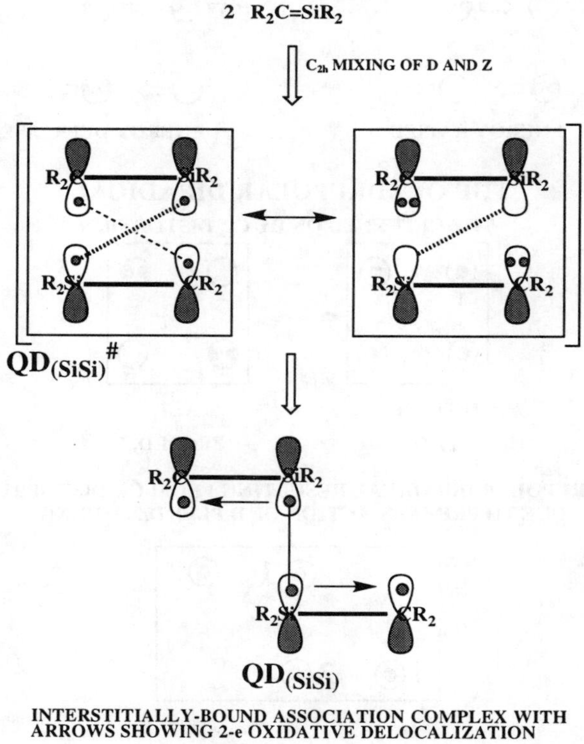

Figure 35.14 Counterintuitive formation of a quadrupolar diradical having an Si—Si bond in the cyclodimerization of silaethylene.

significantly shorter than the other and the DG configuration has dropped significantly below the ALT configuration).

A counterintuitive prediction follows: Head-to-head (HH) dimerization of R═W can occur via the intermediacy of two different QDs (depending on whether the single bond connects the Rs or the Ws), and the kinetically favored one has the two formal odd electrons localized on the two Ws; that is, the bond links the more electropositive Rs. The 2+2 cyclodimerization of silaethylene provide an excellent test of our theory. The reaction is "forbidden," but it may proceed via a QD intermediate because the Si3p AO has low electronegativity. The manner of formulating the cycloaddition is shown in Figure 35.14. We start with the C_{2h} "square" and let it evolve along the C_{2h} reaction coordinate, forming a bond (i.e., establishing the short diagonal) between the atoms that bear the holes in the quadrupolar structure. The preferred linkage of the atoms at the transition state is identified by the hatched lines.

Exactly how counterintuitive this scenario is can be appreciated by comparing the three QD diradicals shown in Figure 35.15. We expect the fastest formed diradical to be the one in which the bond connecting the reactants is weakest! Bernardi et al.[10] found that QD(SiSi)# is the lowest energy transition state. In a related work, Schaefer et al.[11] found that QD(SiSi) and QD(CC) have comparable stability even though the C—C bond is 14 kcal/mol stronger than

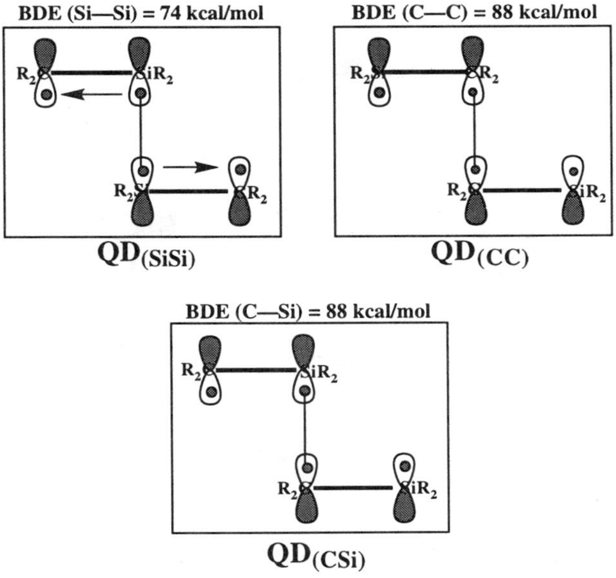

Figure 35.15 Three isomeric diradicals with varying I character.

the Si—Si bond. This makes sense: QD(SiSi) is entirely different from the organic analogues in which C replaces Si. Rather than being a T-bound organic molecule, it is an I-bound organosemimetallic association complex . Experimentally, it is found that 2+2 cyclodimerization of disilaethylene derivatives does not yield the more stable 1,3-disilacyclobutane as the major product. Rather, it yields the less stable head-to-head cycloadduct.[12] When either the electronegativity difference of R and W is very large or a homonuclear system is made up of highly electropositive metal atoms, the quadrupolar diradical is transformed to a quadrupolar zwitterion (QZ). Quadrupolar diradicals or zwitterions are implicated in diverse reactions of molecules containing semimetals and metals.[13–17]

We now return to the issue of the Cotton complexes featuring "multiple bonds." There are two alternatives: Either the metal–metal bonds are covalent and the complex is E/T-bound or the entire complex is bound by REL bonds

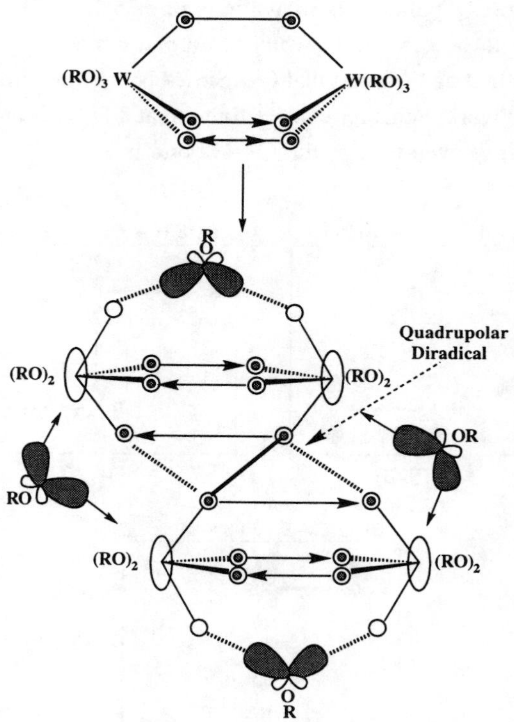

Figure 35.16 Quadrupolar diradical formation in the dimerization of $(RO)_6W_2$ suggests I-type metal–metal bonding in the reactants.

(i.e., codirectional CT delocalization). We can infer which of the two situations materializes in the following way. If a metal–metal "triple bond" is indeed covalent, the dimerization product must be an analogue of rectangular cyclobutadiene. However, if it is actually one I bond plus one T bond, dimerization could produce an I-bound C_{2h} homonuclear quadrupolar diradical in which two T bonds have been effectively turned into one I bond. Chisholm and co-workers[18] synthesized W_2L_6 (L = O—*i*-Pr), a molecule with a formal metal–metal triple bond. The complex dimerizes with a very low activation barrier to produce a C_{2h} metal species. The structures of the reactants and products are shown in Figure 35.16.

35.8 Organometallic Rhombuses Without Confusion

When bare metals are replaced by isolobal organometallic fragments, metal–metal bonding changes in the way indicated in Figure 35.1. Instead of contra-, we now have codirectional CT, with the same stereochemical consequences, because the ALT configuration remains an important, but not dominant, contributor.

The simplest example of a six-electron organometallic rhombus is $[Re(CO)_4]_4^{2-}$.[19] Assuming d^2s octahedral hybridization of Re, the $Re(CO)_4$ fragment is σ-isolobal to Li atom because it uses an ensemble of two valence AOs (one radial and one tangential), which contain one valence electron (Figure 35.17). Hence, the rhenium cluster is analogous to Li_4^{2-} and Be_2Li_2 with a codirectional replacing a contradirectional arrow train.

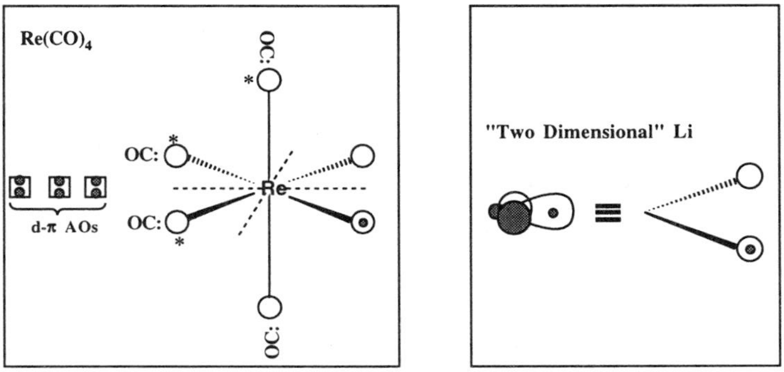

Figure 35.17 $Re(CO)_4$ is analogous to "Li" in which one tangential and one exo radial AO have been deleted.

Two 64-electron tetraruthenium carbonyl clusters with rhombic geometry have been described in the literature: $Ru_4(CO)_{10}(\mu\text{-}PPh_2)_4$ has s = 3.05 Å and d = 2.84 Å, where "s" symbolizes the side bond and "d" the shorter diagonal, and $Ru_4(CO)_{13}(\mu\text{-}PPh_2)_2$ with s = 2.91 and d(av) = 3.15 Å.[20] To understand how these rhombuses are put together, we begin with the individual fragments that make up the clusters.

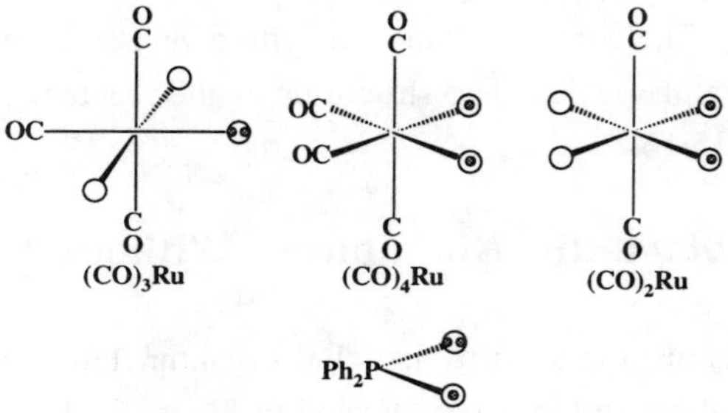

The assembly of the fragments to form the two clusters is shown in Figure 35.18. The VB formulas speak for themselves. With the concept of the I bond at hand, these representations become the clearest descriptors of the bonding, which is seen to be entirely different in the two complexes.

35.9 Why Are Physicists Indifferent to or Unaware of Hückel's Rule?

Ask any organic chemist about the utility of Hückel's rule and you will be assured that it is indispensable: The stability of cyclic π-conjugated systems and the stereochemistry of concerted reactions can be explained only by recourse to Hückel's rule. Next, pose the same question to inorganic or organometallic chemists and you will find that they have little use for this concept. In fact, many have observed with regret that the concept of aromaticity seems to have little relevance to organometallic chemistry. Finally, solid state or surface chemists are likely to be marginally, if at all, aware of Hückel's rule. Smalley puts it best: "I am not all that interested in the issue [i.e., the aromaticity of C_{60}] because I have yet to be in a situation where it mattered. Aromaticity is not a concept that comes readily to the lips of chemical physicists."[21] So, what lies behind the apparent failure of Hückel's rule to cut through interdisciplinary barriers?

A 6-ELECTRON METAL RHOMBUS (64-El Cluster)

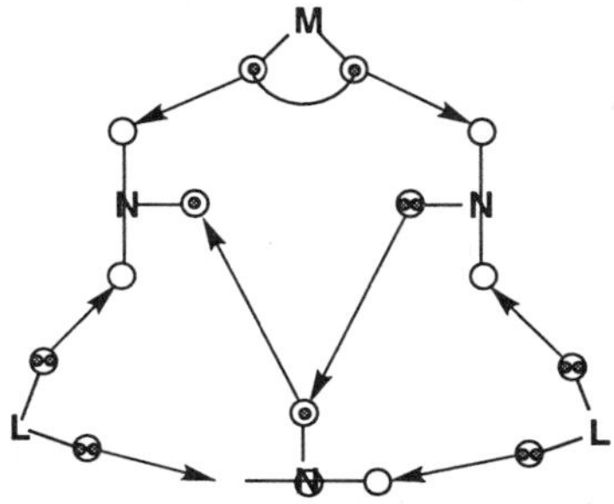

M = $Ru(CO)_4$, N = $Ru(CO)_3$, L = Ph_2P

A 4-ELECTRON METAL RHOMBUS (64-El Cluster)

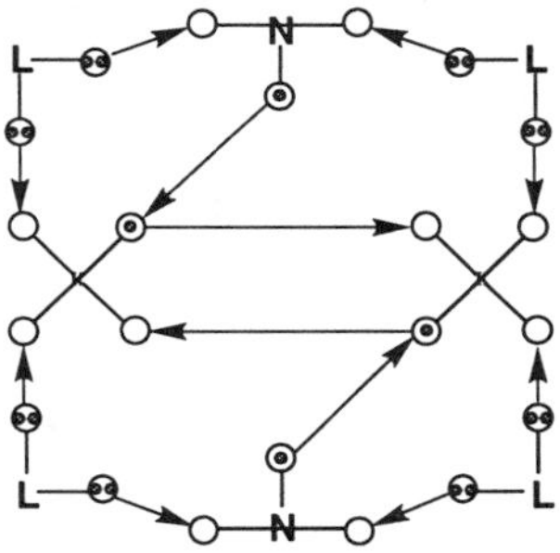

K = $Ru(CO)_2$, N = $Ru(CO)_3$, L = Ph_2P

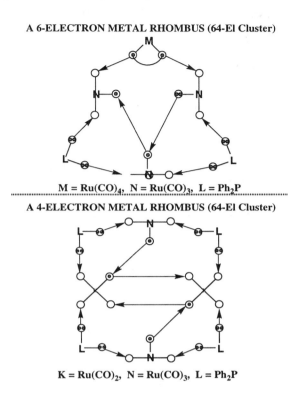

Figure 35.18 Construction of the clusters reported by Hogart et al.[20] Note that the fragments M and N differ in hapticity.

An aromatic complex is one in which all symmetry restrictions to CT delocalization have been lifted. If atom electronegativity is low, an aromatic complex affords the best possible I-bonding. This means that we expect aromaticity to reign supreme in molecules expected to be I-bound. These are metal-containing systems from organometallic complexes to metal solids. Why, then, do the very people who are supposed to be the beneficiaries of aromaticity find little utility in the concept or even ignore it? The answer is twofold:

(a) Aromaticity has been under the nose of organometallic chemists, who simply did not perceive it! The best example consists of the electronic structures of inorganic rings and Baudler's rule discussed in Chapter 24.

(b) For certain combinations of atoms, such as the combination of electronegative nonmetals with metals and the combination of pure metals, state crossing in a formal antiaromatic geometry can become responsible for the creation of a highly stabilized ground state, which can compete with the aromatic one. The bonding is now dictated by the ALT configuration, and aromatic systems become competitive with crossed antiaromatic ones. The stabilization of the ALT configuration by the low-lying unoccupied space represents atom ionization to form a highly stabilized multipolar array. As a result, static bonding annihilates the difference between "aromatic" and "antiaromatic." This is why the formally antiaromatic Li_4 is rhombic and the formally aromatic Li_6 is a "raft," with both being illustrators of "close packing" and bearing no apparent resemblance to what the organic chemist recognizes as aromatic (hexagonal Li_6) and antiaromatic (rectangular Li_4)!

35.10 Diborane Is the Codirectional CT Analogue of Rhombic Li_4

The famous diborane molecule is neither covalent (like ethane dication) nor I-bound by contradirectional CT delocalization and significant E-bonding character (like its aluminum derivative and Li_4). Rather, it is an illustrator of REL-bonding, as shown in Figure 35.19. We can easily distinguish these different cases:

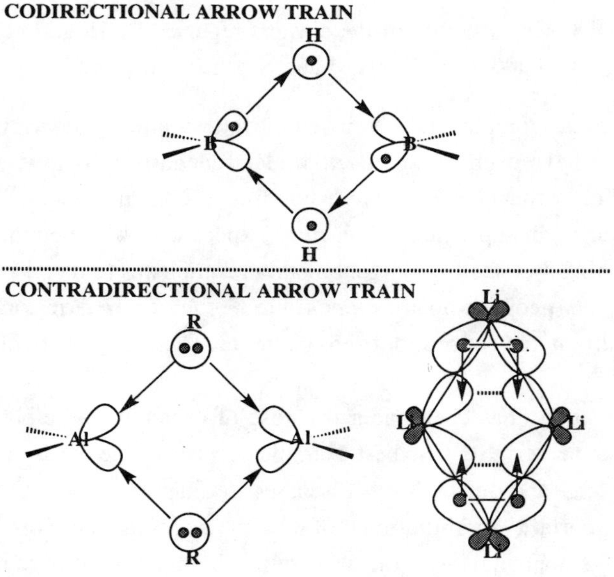

Figure 35.19 Isoelectronic molecules can differ in the gross and the detailed mechanism of chemical bonding.

(a) Everyone who has used $NaBH_4$ and $LiAlH_4$ in the laboratory knows that these are very different reagents. This is a consequence of REL-bonding of the former and E-bonding of the latter.

(b) Ethane dication and diborane are isoelectronic, and conventional theory expects them to be isostructural. By contrast, we predict that the replacement of two carbon cations by two borons will change the mechanism of bonding from T to I. Hence, we expect bona fide 3c-2e resonating T-bonding in ethane dication and I-bonding in diborane.

Schleyer et al.[22] determined the optimum structures of diborane and the isoelectronic ethane dication by ab initio calculations. They found that the two species have different structures (Figure 35.20). The ethane dication is T-delocalized and can be formulated as H_2 bound to a vacant C2p AO of ethylene dication in a true 3c-2e bonding sense. On the other hand, the bridged structure of diborane is a consequence of I-bonding, as discussed earlier.

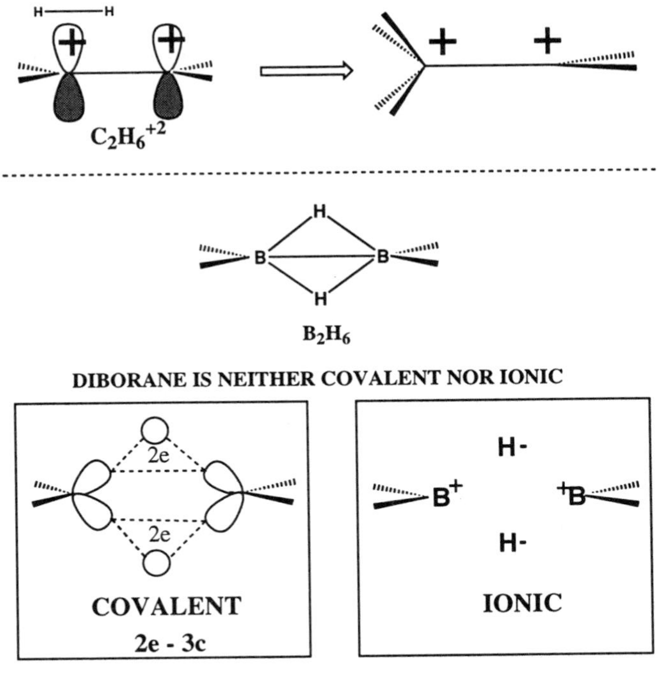

Figure 35.20 The structural difference between the isoelectronic $H_3CCH_3^{2+}$ and H_3BBH_3.

35.11 The Electronic Structures of Dications

We now make a small detour to revisit the world of organic molecules, where we shall expose the very different rules of bonding of carbocations. Recently, partly because of the inherent appeal of these fragments and partly because people have run out of monocations to compute, organic dications are attracting interest. In general, the best a hydrocarbon cation or polycation can do is form aromatic 3c-2e bonds (exemplified by H_3^+). These are resonating T bonds, which are differentiated from resonating I bonds in the sense that ionic structures are minor contributors to the total VB wavefunction. As a result, the rules for predicting the most stable form of a polycation are straightforward: Combine a stable polyene with one or more protons (because H is an excellent bridging atom), to form the maximum number of 3c-2e bonds. Each proton acts as a cap of a two-electron bond in the following order: π pairs are capped ahead of σ pairs and C—H are capped ahead of C—C σ pairs. Finally, the capping can occur either symmetrically or, if spatial overlap considerations so dictate, unsymmetrically. Here are some examples:

A simple carbenium ion $C_nH^+{}_{2n+1}$ is "C_nH_{2n} plus H^+," in which the proton caps the alkene π bond either symmetrically ("nonclassical" ion) or unsymmetrically.

An alkane dication $C_nH^{2+}_{2n+2}$ is "C_nH_{2n} plus H^+ plus H^+," in which one proton caps a C—C π bond and the other a C—H σ bond. This explains the computed structure of the ethane dication, as discussed later, as well as the computed structure of the propane dication.[23]

An alkene or cycloalkane dication $C_nH^{2+}_{2n}$ is "C_nH_{2n-2} plus H^+ plus H^+." An example is $^+H_2C—CH_2—CH_2—CH_2^+$ (unsymmetrical diprotonation of 1,3-butadiene).

35.12 What Li_4 and Be_4 Tell Us About the Structure of Solids

Despite the absence of close-packed layers in solid Li, rhombic Li_4 is a prototype of two-dimensional close-packed bonding. Solid Li forms non-close-packed layers, which are then stacked one on top of the other in a close-packing (CP) sense to produce a body centered cubic (bcc) architecture.[24] This means that something happens between the tetramer and the polymer that adversely affects CP bonding in two dimensions.

Recall the critical difference between the K and Z states of square Li_4: open-shell (1,1) arrangement of two electrons in the e_u MOs in the former and closed-shell (0,2) allocation in the latter. This configuration is responsible for superior 2-e delocalization in Z and its greater stability compared to K. In the rhombic Li tetramer, the local symmetry at every Li is C_{2v}. The construction of a close-packed layer of Li atoms can be formalized as a sequential distortion of an initial D_{nh} ring of atoms to a derivative with $D_{nh/2}$ symmetry until the process is exhausted. Some examples are as follows:

A $D_{4h} \longrightarrow D_{2h}$ is a CP transformation with n assuming sequentially the values $n = 1, 0$.

A $D_{6h} \longrightarrow D_{3h}$ is a CP transformation with n assuming sequentially the values $n = 1, 0$.

A $D_{8h} \longrightarrow D_{4h} \longrightarrow D_{2h}$ is a CP transformation with n assuming sequentially the values $n = 2, 1, 0$.

As a result of this transformation, local symmetry is reduced and states analogous to Z and K become actually resonance structures. In other words, Li_4 differs from Li solid to the extent that the former has zero while the latter has significant open-shell character. Since only the closed-shell configuration favors close packing, solid Li is much less predisposed than Li_4 to be close-packed. This is why Li and all alkalies adopt the bcc structure. By contrast, a closed-shell configuration is guaranteed when Li is replaced by Be, and this is why Be adopts the hexagonal close-packed (hcp) structure. In going from Be to Ba, the nd become quasi-degenerate with the np AOs (i.e., the unoccupied space is expanded). As a result, there is progressive enhancement of open-shell character, and this is why the alkaline earth structure changes from hcp (Be, Mg) to ccp (Ca, Sr) to bcc (Ba).

A simple explanation of why open-shell character grows in the close-packed alkalies (thus, favoring abandonment of close packing) as the number of Li atoms increases can be given by comparing the sequential CP distortion of a four- and an eight-atom D_{nh} ring and recalling that the driving force for close packing in Li_4 was the association-promoting, quadrupolar ALT configuration that dominates the Li_4 Z wavefunction. Figure 35.21a shows that in Li_4, sequential CP distortion retains the alternation of pairs and holes present in the square form. By contrast, in Li_8 (and larger rings), sequential CP distortion fails to preserve hole–pair alternation present in the octagonal D_{8h} form in the final D_{2h} rhomboidal cluster. As can be seen in Figure 35.21b, the electron pairs of the original ALT configuration end up at the periphery, leaving four holes to

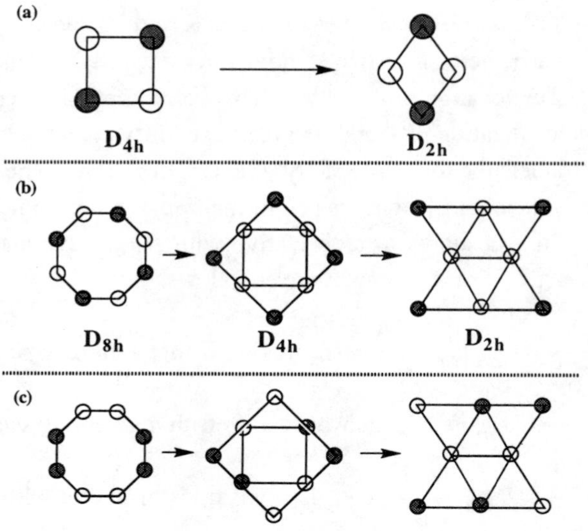

Figure 35.21 Sequential CP deformation in four- and eight-membered metal rings.

define the central rhombus. The physical meaning of this result is that electrons can minimize their mutual repulsion in a geometry incompatible with minimization of the total Coulombic energy (i.e., electron–electron plus electron–nucleus plus nucleus–nucleus interaction). The schematic representation of the D_{2h} Li_8 clearly shows that segregating the electrons "outside" causes severe diminution of nucleus–electron attraction. The same impasse is reached if we start with a nonalternant HP configuration in D_{8h} and end up with hole–pair alternation in the internal rhombus (Figure 35.21c). The transformation fails to abolish the pair adjacency, and severe interelectronic repulsion offsets the improvement in nucleus–electron attraction.

There is yet another way of looking at the problem. Close-packing of Li atoms in two dimensions produces an array of triangular voids, each of which is formed by the in-plane overlap of the two Lisp and one Li2p of each Li as depicted in Figure 35.22a. It should be noted that two triangular voids share one Li2p AO and, thus, interact. Furthermore, there exists the following relationship: Since the unit cell is a parallelogram that is subdivided into two triangles by the short diagonal, and since there is one Li atom per unit cell, each Li can deliver its electron to either one of two triangular voids. This is illustrated in Figure 35.22b. As a result, a close-packed Li layer has only half of the triangular hollows filled by one electron. The presence of unoccupied

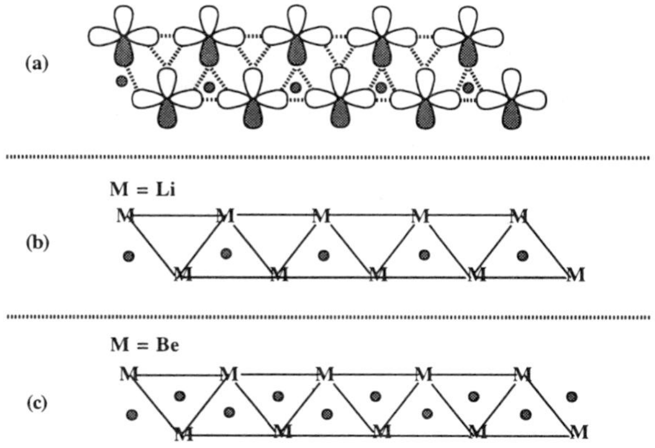

Figure 35.22 (a,b) The structure of metallic Li in two dimensions. Only half the triangular hollows are filled by a single electron. (c) The structure of metallic Be in two dimensions. Each triangular hollow is filled by a single electron.

triangular voids implies unfavorable nucleus–electron attraction and inferior net Coulomb interaction.

A drastic improvement can be made by arranging the Li atoms so that every void is filled by one electron. This condition is met by a square layer lattice, which is formed by the in-plane AO overlap of two Lisp and Li2p AOs as shown in Figure 35.23a. Now, the unit cell is a square with one Li per unit cell. This means that each square void is filled by one electron, as in Figure 35.23b. Therefore, Li_4 is close-packed rhombic while solid Li is neither hexagonal nor cubic close-packed but, rather, body-centered cubic.

The stereochemical disjunction between small cluster and solid persists in

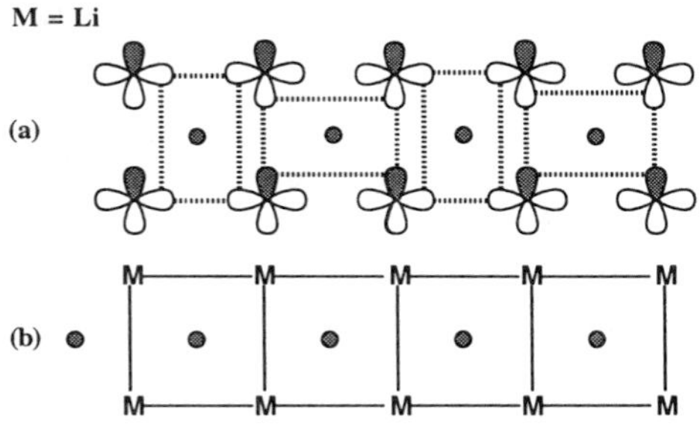

Figure 35.23 The structure of metallic Li in two dimensions. Each square hollow is filled by a single electron.

the case of Be. Square is superior to rhombic Be_4, but Be solid has an (hcp) structure.[24] Again, the most apt physical description would depict a close-packed layer of Be atoms having triangular voids, each of which was filled with one electron (Figure 35.22c). The difference between Li and Be is now crystal clear: The two metals adopt different layer-packing motifs because Li has one electron and Be has two, and the former can fill only half the triangular voids, but the latter can fill them all.

In this light, consider the relationship of K, an alkali metal, and Cu, a coinage metal, both of which are treated as univalent atoms. Copper differs from potassium in having a full 3d shell. However, these d electrons are seldom invoked in conventional bonding theories, and electron count schemes treat K and Cu on equal footing (i.e., each is seen as binding with one valence electron). On the other hand, K is bcc but Cu is ccp. We have already explained why K aquires open-shell character in the solid and thus avoids close-packing. How does a filled 3d shell modify the relative importance of open- and closed-shell spatially degenerate configurations? The Cu 3d pairs can be "dilated" only by Coulomb exchange with an upper 4s hole. This is possible only if the Cu4s electrons arrange themselves in a closed-shell (hole–pair rather than radical–radical) fashion. An equivalent physical explanation suggests that Cu forms a close-packed layer because this causes half the triangular voids to be filled by one valence electron each, while the remaining voids act as receptors of d electrons. In other words, Cu resembles Ca because of the action of the filled d shell.

35.13 The Size-Dependence of Metal–Ligand Bonding

What happens to the M—L bond in the cluster M_xL_n in which x varies from 1 to infinity? If we conceive of a small metal cluster as a "homonuclear ionic" species bound by the E mechanism and aided by contradirectional CT delocalization, we end up with the following simplified picture. The promotional energy necessary for converting atoms to ions becomes invested in E-bonding in an increasingly favorable way as the number of formal cations (anions) surrounding a central anion (cation) increases. This is why the cohesive energy of metal solids increases asymptotically with cluster size. This means that the relative importance of the static mechanism of bonding relative to contradirectional CT delocalization increases with size. How fast this shift in bonding mechanism occurs is revealed by ΔCE (in kcal/mol) which is defined as follows:

$$\Delta CE = \Delta H_f^\circ\,(\mathrm{M}) - BDE\,(\mathrm{M{-}M})$$

Table 35.1 ΔCE Values (kcal/mol) for the Categories of Transition Metals

High ($\Delta CE > 70$)	Borderline ($70 > \Delta CE > 60$)	Low ($60 > \Delta CE$)
Ti, Fe, Co	Sc, V	Cr, Mn, Ni, Cu, Zn
Zr, Ru	Y, Rh, Pd	Nb, Mo, Ag, Cd
Ta, W, Re, Os, Ir	Hf	Pt, Au, Hg

where ΔH_f° (M) is the heat of formation of the gaseous metal atom and *BDE* (M—M) is the bond dissociation energy of the metal dimer (taken from the Morse compilation[25]). The larger the ΔCE, the greater the difference in M_x—L bonding at the two extremes of x = small and $x = \infty$.

Transition metals can be classified in three categories, called "low," "borderline," and "high," depending on the ΔCE values shown in Table 35.1 for each transition series. According to our analysis, M_xL_n clusters in which M belongs to the "high" category should show a size dependence of M—L bonding. This would explain why dinitrogen chemisorbs on small iron clusters but only physisorbs on bulk iron.[26] If we assume that the borderline Pd leans toward the high domain, we also have explanations for three key observations:

(a) In $Pd_x(CO)_n$ clusters, the CO IR frequencies are different in the small and large species.[27]

(b) The CO frequency of supported Pd—CO clusters and of CO adsorbed on single-crystal faces *increases* as the n:x ratio (metal coverage) increases.[27] A metal cation causes an increase in the C2p character of the carbon σ lone pair and an increase in C2s character of the C—O bond (induction). As a result, the C—O bond becomes stronger and the frequency shifts to higher values as the number of carbonyls added to the metal cluster increases.

(c) Since only comparisons are physically meaningful, the most telling observation is that these trends fail to materialize in clusters that replace borderline Pd by low Cu.

35.14 Is the Map of Chemical Bonding "Cyclic"?

Consider what happens as atom electronegativity in the system A_6 decreases in going from A = H to A = Li. At the high limit, we have T-bonding and three H_2 molecules infinitely apart are the global minimum. As we go to the π system of benzene, orbital electronegativity is somewhat reduced (H1s to C2p), and we now stand between the T and I markers. The π system of benzene has impaired codirectional CT delocalization, and the T complement of the 1.5 I bonds is still sufficient to drive the three π bonds away from each other. Benzene is hexagonal because of the constraining action of the σ system. Finally, when A = Li, the system finds itself between the I and E markers. We can say that the continuum of chemical bonding is "linear," with T-bonding giving rise to codirectional I-bonding, which gives way to contradirectional I-bonding with a strong or dominant E-component:

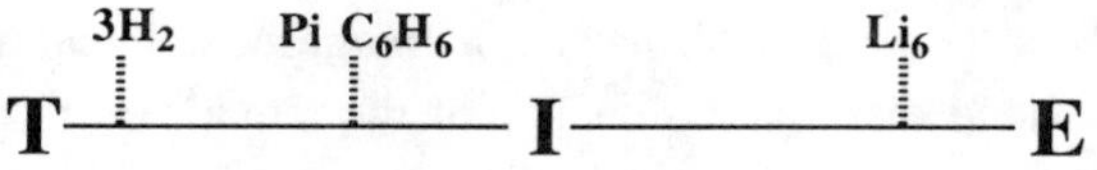

Consider now a different problem in which an atom A binds dihydrogen. At the high electronegativity limit (e.g., A = CH_2, NH, or O) we have T-bonding, and the molecule is the covalent "organic." In other words, the promotional energy of breaking the H—H bond is more than made up by forming two strong T bonds between A and H. When we shift to a neutral metal, the promotional energy expenditure is no longer justified. We now have I-bonding in which dihydrogen acts as OX ligand, and the crucial question becomes: What happens as we shift from neutral to cationic to biscationic metal? To find out, we note that the neutral (electropositive) metal is ideally suited for I-bonding, while the dicationic metal has two options:

By virtue of its charge, it binds dihydrogen by the E mechanism.

By virtue of its high electronegativity, it binds dihydrogen by the T mechanism; that is, a biscationic metal is analogous to a first-row nonmetal atom.

The implications of this scenario are clear-cut: When the metal becomes cationic, the dihydrogen becomes bound either by the I mechanism (as a ROX ligand) or by the E mechanism. In other words, the system falls between the I and E markers. However, when the metal is biscationic, the system can fall either toward the E marker or between the E and T markers. In other words, the continuum of chemical bonding is "cyclic":

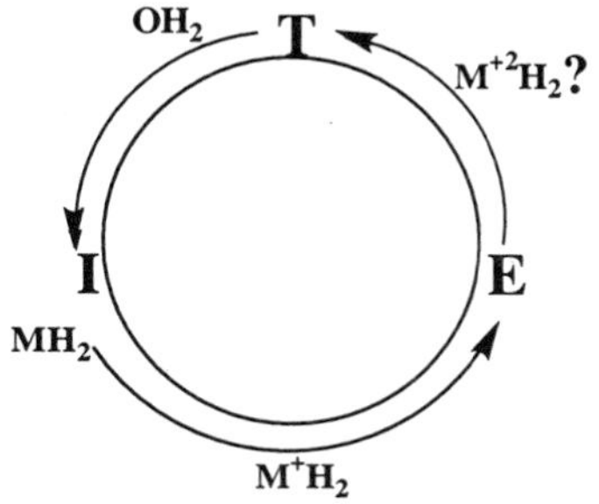

The experimental test is now straightforward: If the continuum is cyclic, we should be able to break the "E barrier" and observe a switch from "dihydride" to "dihydrogen" back to "dihydride" as the metal changes from neutral to monocationic and, finally, to biscationic. If the continuum is "linear," we should simply see a progressive decrease of the H—H distance in going from the neutral to the biscationic metal. The reason is that OX action of dihydrogen in combination with the neutral metal produces a formal H_2 radical anion, which has a three-electron bond that suffers from exchange repulsion. At the other extreme, RED action of dihydrogen in combination with the neutral metal produces a formal H_2 radical cation, which has a strong one-electron bond that is free of exchange repulsion. Alternatively, the dihydrogen binds by the E mechanism with essentially little geometry change. The "metal—H_2 puzzle" is defined as follows:

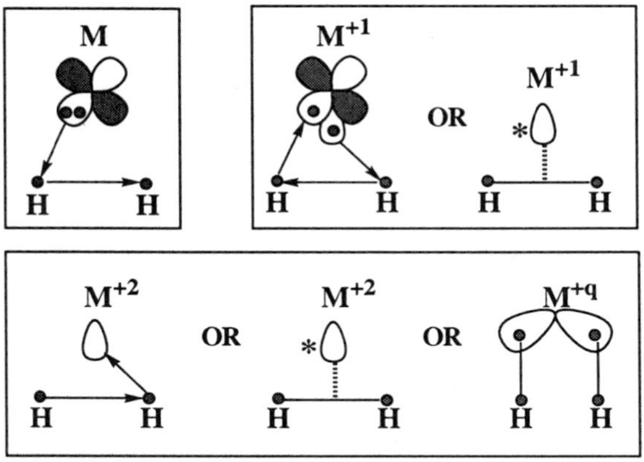

The work of Kubas, Hoff, and co-workers[28] leads to the final answer. These investigators have found that the $Re(R_3P)_2(CO)_3^{+}$ complex binds H_2O more strongly than N_2 or H_2, implying that we are in the realm of E-bonding, where a polar molecule is superior to a nonpolar one as a ligand. Will changing

Re^+ to Os^{2+} make T-bound dihydride competitive with E-bound water, or will it enhance the superiority of water over dihydrogen (both E-bound)? Work in progress by Heinekey, Radzewich, and co-workers may answer the question. It should be noted that the problem is tricky. In a metal complex of the type $L_n(CO)M^{2+}H_2$, there can be, at least, two T-bound isomers: the ketenic/dihydrogen form $L_nO{=}C{=}M^{2+}(H_2)$ and the carbonyl/dihydride form $L_n(CO)H{-}M^{2+}{-}H$. One or the other may be the global minimum, depending on M and its charge.

References

1. M. Rubinstein and I. Shavitt, *J. Chem. Phys.* 51, 2014 (1969).
2. H. Beckman, J. Koutecky, P. Botschwina, and W. Meyer, *Chem. Phys. Lett.* 67, 119 (1979).
3. M.H. McAdon and W.A. Goddard III, *J. Non-Cryst. Solids* 75, 149 (1985).
4. D. Maynau and J.P. Malrieu, *J. Chem. Phys.* 88, 3163 (1988).
5. W.T. Borden and E.R. Davidson, *Acc. Chem. Res.* 14, 69 (1981).
6. R.A. Whitesides, R. Krishnan, J.A. Pople, M.-B. Krogh-Jespersen, P.v.R. Schleyer, and G. Wenke, *J. Comput. Chem.* 1, 307 (1980).
7. (a) M.R. Hoare, *Adv. Chem. Phys.* 40, 49 (1979). (b) J.M. Basset and R. Ugo, in *Aspects in Homogeneous Catalysis*, R. Ugo, Ed., Reidel, Dordrecht, 1977. (c) J.G. Allpress and V. Saunders, *Surf. Sci.* 7, 1 (1967). (d) S. Mader, *J. Vac. Sci. Technol.* 8, 247 (1971). (e) E.B. Prestridge and D.J.C. Yates, *Nature* 234, 345 (1971). (f) I.B. Knight Jr., R.W. Woodward, R.J. van Zee, and W. Weltner Jr., *J. Chem. Phys.* 79, 5820 (1983).
8. (a) H.C. Longuet-Higgins, *Q. Rev.* 11, 121 (1957). (b) W.N. Lipscomb, *Boron Hydrides,* Benjamin, New York, 1963. (c) K. Wade, *Electron Deficient Compounds*, Nelson, London, 1971.
9. J.L. Gole, R.H. Childs, D.A. Dixon, and R.A. Eades, *J. Chem. Phys.* 72, 6368 (1980).
10. F. Bernardi, A. Bottoni, M. Olivucci, M.A. Robb, and A. Venturini, *J. Am. Chem. Soc.* 115, 3322 (1993).
11. E.T. Seidl, R.S. Grev, and H.F. Schaefer III, *J. Am. Chem. Soc.* 114, 3643 (1992).
12. A.G. Brook, in *Heteroatom Chemistry*, E.Block, Ed., VCH, New York, 1990, p. 105.

13. F.R. Jensen and B. Rickborn, *Electrophilic Substitution of Organomercurials*, McGraw-Hill, New York, 1968.
14. (a) D. Hanssgen and W. Roelle, *J. Organomet. Chem.* 63, 269 (1973). (b) H.W. Roesky, W. Schmieder, and K. Ambrosius, *Z. Naturforsch.* 34[B], 197 (1979)
15. (a) A.G. Brook, D.M. MacRae, and W.W. Limburg, *J. Am. Chem. Soc.* 89, 5493 (1967). (b) A.G. Brook and A.R. Bassindale, in *Rearrangements in Ground and Excited States,* Vol. 2, P. de Mayo, Ed., Academic Press, New York, 1980, Chapter 9.
16. J. Slutsky and H. Kwart, *J. Am. Chem. Soc.* 95, 8678 (1973).
17. M.T. Reetz, *Adv. Organomet. Chem.* 16, 33 (1977).
18. (a) M.H. Chisholm, D.L. Clark, K. Folting, J.C. Huffman, and M.J. Hampden-Smith, *J. Am. Chem. Soc.* 109, 7750 (1987). (b) M.H. Chisholm, D.L. Clark, and M.J. Hampden-Smith, *J. Am. Chem. Soc.* 111, 574 (1989).
19. M.R. Churchill and R. Bau, *Inorg. Chem.* 7, 2606 (1968).
20. G. Hogart, J.A. Phillips, F. VanGastel, N.J. Taylor, T.B. Marder, and A.J. Carty, *J. Chem. Soc. Chem. Commun.* 1570 (1988).
21. *Chem. Eng. News* 72(9), 40 (1994).
22 P.v.R. Schleyer, A.J. Kos, J.A. Pople, and A.T. Balaban, *J. Am. Chem. Soc.* 104, 3771 (1982).
23. G.A. Olah, N. Hartz, G. Rasul, G.K.S. Prakash, M. Burkhart, and K. Lammertsma, *J. Am. Chem. Soc.* 116, 3187 (1994).
24. J. Donohue, *The Structures of the Elements*, Wiley, New York, 1974.
25. M.D. Morse, *Chem. Rev.* 86, 1049 (1986).
26. R. Brill and J. Kurzidim, *Colloq. Int. CNRS* 187, 99 (1969).
27. M. Moskovits, *Acc. Chem. Res.* 12, 229 (1979). This is an important paper because it projects clearly the difficulties of applying conventional theory to the facts of surface science.
28. G.J. Kubas, C.J. Burns, G.R.K. Khalsa, L.S. Van der Sluys, G. Kiss, and C.D. Hoff, *Organometallics* 11, 3390 (1992), and previous papers.

Chapter 36

Rotated Molecules or Aromaticity Where Least Expected

Most chemists believe that π-benzene has an aromatic π system that would have retained a D_{6h} geometry if the σ bonds had been absent. H_6 dashes this popular hypothesis. VB theory provides the explanation: Aromaticity is a consequence of CT delocalization, and the H1s and C2p AOs are too electronegative to allow its full expression. On the other hand, no chemist thinks that multiple bonds have anything to do with aromaticity! Once again, VB theory paints a very different picture: Multiple bonds can form aromatic systems when the electron count is right and, most importantly, when the two atoms defining the multiple bond are metallic.

Two sp^2 hybrids constitute one angular bident and three sp^3 hybrids make up one conical trident. These two entities are thought of as two- and three-pronged *gears,* with a characteristic *bite angle*, which can interact in two different extreme modes as illustrated in Figures 36.1 and 36.2.

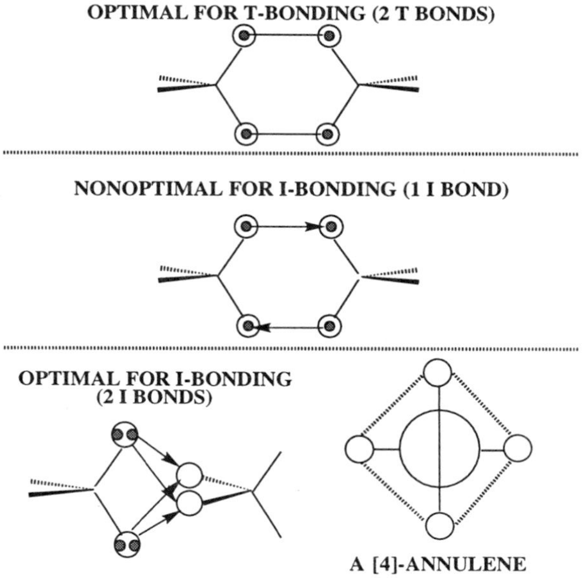

Figure 36.1 Different orbitals for different mechanisms of "double bonding."

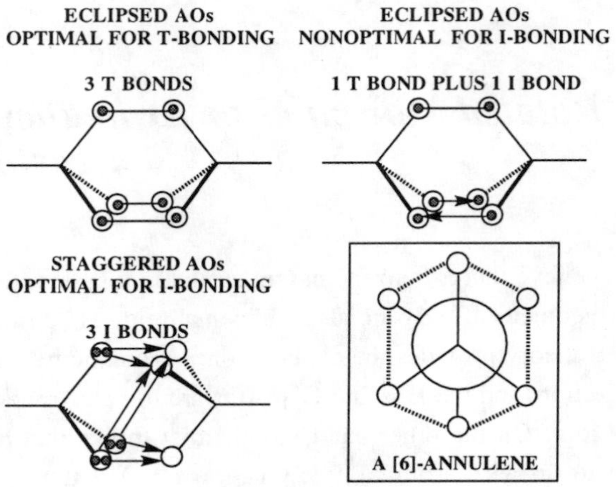

Figure 36.2 Different orbitals for different mechanisms of "triple bonding."

The *noninterlocking (NIN) mode*, which is suitable for T-bonding, defines pairs of interacting AOs. Two bidents produce a NIN2 and two tridents a NIN3 array. A four-electron NIN2 array, 4e-NIN2, is the conventional double bond and a six-electron NIN3 array, 6e-NIN3, is the conventional triple bond. Multiple bonds of these types connect first-row atoms H, C, N, O, and F and will not concern us further.

The *interlocking (IN) mode* is produced by rotation starting from the geometry characteristic of the NIN mode and is suitable for I-bonding. The meshing hybrids form cyclic arrays, which are isoconjugate to Hückel π-annulenes. Molecules having rotated multiple bonds are called *rotated molecules.* The characteristic feature is that the multiple bond connects either one metal (or semimetal) and one nonmetal or two metals (semimetals). Rotated molecules fall in the following categories:

Singlet aromatic species with $4N + 2$ electrons.

Triplet aromatic species with $4N$ electrons.

Singlet crossed-antiaromatic (see Chapter 35) species with $4N$ electrons, requiring that either the two metals of a homopolar multiple rotated bond be highly electropositive or that one be an electropositive metal and the other an electronegative nonmetal.

If aromatic metal-containing multiple bonds do exist, how is it that no one has seen them? The answer is that they are well disguised. Specifically, orbital rotation may convert an initial unrotated molecule to a rotated one belonging either to the same or a different point group: *invisible* and *visible* rotation, respectively. For example, planar D_{2h} disilaethylene can undergo visible rotation to adopt a D_{2d} geometry to produce an 4e-IN2 bond isoconjugate to an antiaromatic four-electron [4]-annulene. On the other hand, linear $D_{\infty h}$ disilaethyne can undergo invisible rotation that conserves the molecular shape to produce a 6e-IN3 bond isoconjugate to an aromatic six-electron [6]-annulene. It follows that while we can physically observe visible rotation, we can only compute invisible rotation. This limitation is a consequence of electron correlation.

One fundamental misconception is the notion that delocalization determines stability and bond strength. As we have discussed before, bond strength is determined by orbital electronegativity, which also determines the mechanism of delocalization, exchange versus CT. Bond strength and CT delocalization are antipodal consequences of orbital electronegativity: Strong bonds are consistent with exchange and weak bonds with CT delocalization. Since aromaticity has to do with CT delocalization, it should be sought in weakly bound systems such as metal-containing molecules and, especially, multiple bonds involving at least one metal atom. Thus, the notion of aromatic rotated multiple bonds is entirely consistent with the many-electron physics of chemical bonding and entirely inconsistent with the conventional one-electron concepts of chemistry. Finally, the concept of a "multiple bond" covers much more than, say, a double bond formed by the overlap of two sets of two sp^3 hybrids belonging to two carbon atoms. A set of two sp^3 hybrids can be replaced by two unhybridized p AOs of a dehybridized atom (e.g., a heavy p-block atom). Furthermore, one atom can be connected to two (three) atoms by a formal double (triple) bond. Thus, OH_2 can be regarded as $O{=}(H)_2$, and so on.

We now reopen the issue of multiple metal bonding by first listing some potential examples of invisible rotated molecules:

Sb_2 has a 6e-IN3 bond. It is a singlet-aromatic molecule.

Sn_2 has a 4e-IN3 bond and Te_2 has an 8e-IN3 bond. Each is a triplet-aromatic molecule.

Confinement of a metal (M) in the center of a porphyrin (P) ring and subsequent dimerization generates the metalloporphyrin dimer $[M(P)]_2$.

In the third case, the macrocycle hollow accommodates two δ-type d AOs, $d_{x^2-y^2}$ and d_{xy}, and an s-like toroidal sd_{z^2} hybrid. Two of the four nitrogen σ electron pairs combine with one δ-type hole, and the other two with the toroidal hole, to define two I bonds. Remember that one metal hole ties up two ligand pairs to form one I bond. When the metal is d^5 in its formal oxidation state, three electrons go to the three endo hybrids (in preparation for forming a formal six-electron triple bond upon dimerization) and two electrons go to the δ-type equatorial d_{xy} that points away from the nitrogen σ pairs. This is the case with $[Re(OEP)]_2$ (OEP = octaethylporphyrin).[1]

The situation changes when we go to the isoelectronic system [(OEP)MoRu(OEP)].[2] Here two nonbonding d_{xy} electrons of the more electropositive formal Mo^- are transferred to the endo space of the more electronegative formal Ru^+. This corresponds to internal reduction that converts a singlet six-electron to a triplet eight-electron triple bond. Since the multiple bonds involve metal atoms, we interpret these two results to imply the existence of a singlet aromatic 6e-IN3 bond in the homodimer and a triplet aromatic 8e-IN3 bond in the heterodimer. We suggest that [(OEP)MoZr(OEP)] will turn out to be a triplet aromatic 4e-IN3 derivative.

We now shift attention to visible rotation in which the geometry of the molecule itself reveals the aromaticity or crossed antiaromaticity of the multiple bond:

(a) The dication and the dianion of $R_2Sn{=}SnR_2$ should have D_{2d} geometry featuring singlet aromatic 2e-IN2 and 6e-IN2 bonds, respectively.

(b) The strategy for stabilizing planar methane discussed by Hoffmann[3] stimulated computational chemists to study organolithium molecules[4] and this, in turn, produced data in search of interpretation.[5,6] Planar dilithiomethane was computed to lie only about 10–20 kcal/mol above the "classical" tetrahedral structure. Our explanation is that *planar* H_2CLi_2 is a rotated molecule analogous to rhombic Li_4 and its R_2Li_2 derivative. It has a singlet crossed antiaromatic 4e-IN2 bond linking the carbon with the two lithiums, as shown in Figure 36.3.

If planar H_2CLi_2 is a rotated molecule, its stability must be sensitive to the carbon bite angle. This can be adjusted by replacing the H atoms by

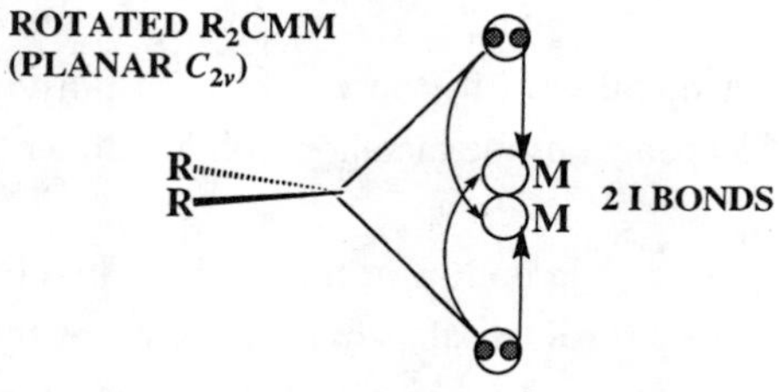

Figure 36.3 Planar C_{2v} dimetallamethane.

substituents. Ab initio computations indicate that the planar form is stabilized when H is replaced by F. This is expected to open up the bite angle of the carbon sp^3 AOs directed to the two lithium atoms. Since these two hybrids contain two electron pairs, this opening is exactly what we expect by analogy with Li_4 and R_2Li_2: The pairs go to the long diagonal of the rhombus.

(c) The final piece of the puzzle is the triplet form of planar H_2CLi_2, which is nearly equal in energy to the singlet form and can be formalized as a rotated molecule with a triplet aromatic 4N-IN2 bond.

The preceding discussion is yet another introduction to rethinking chemistry. We can pursue the implications by formulating additional problems. For example, the IN3 bond of Sb_2 can be isoconjugate to either a [6]-annulene with D_{6h} symmetry (π-benzene analogue) or to a [6]-annulene with D_{3h} symmetry (Li_6 analogue). For the latter to happen, the bite angle of one tripodal set of AOs at one atom must be greater than that at the second atom. This implies that a D_{3h} 6e-IN3 bond is best accommodated in a heterodimer. Does the [6]-annulene of the allegedly rotated Sb_2 have D_{6h} or D_{3h} geometry? The answer is given by the reaction of Bi_2 plus Sb_2 to produce two BiSb molecules. The high exothermicity of the metathesis suggests that Sb_2 and Bi_2 have each a rotated triple bond, which is analogous to D_{3h} Li_6, and that BiSb also has a rotated triple bond, which approaches analogy to $Li_3(NR_2)_3$.[7] The BiSb molecule can be represented approximately as shown in Figure 36.4.

While we have focused attention on metals, the same ideas can be extended to rotational isomerism of the nonmetallic alkenes. The first to recognize that the double bond of a 90°-rotated (perpendicular, P) ethylene is isoconjugate to the π system of cyclobutadiene was Mulder.[8] According to the analysis presented in Chapter 34, the P ethylene has three close-lying, K, D, and Z states. The D state corresponds to two singlet methylenes, or, equivalently, two methylenes

$$Sb_2 + Bi_2 \longrightarrow 2\ SbBi \quad \Delta E = -24\ \text{kcal/mol}$$

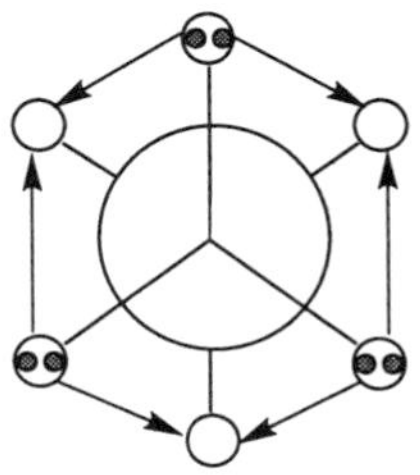

BiSb

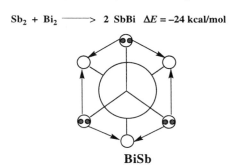

Figure 36.4 The BiSb molecule.

with a delocalized G bond. It follows that replacing the H_2C moieties of ethylene by Z_2A groups having a large singlet–triplet gap (e.g., F_2C or H_2Si) will produce a secondary minimum in the rotational hypersurface due to state crossing: The D state dips below K while interacting with both K and Z in a geometry that minimizes the exchange repulsion of the two carbenic electron pairs. This explains why, as is well known, it is much more favorable to break either the double or the single π bond of tetrafluoroethylene than of ethylene. We suggest that the best way of thinking about the rotational hypersurface of $R_2M{=}MR_2$, where M is Si or heavier congener, is by departing from the state manifold of the P form and considering how any applied perturbation affects the mixing of K, D, and Z. In this light, the twisted diradical-like secondary minimum and the trans-bent global minimum, which are frequently encountered in disilene chemistry, are manifestations of the key role of the low-lying D and Z states.

References

1. J.P. Collman, J.M. Garner, and L.K. Woo, *J. Am. Chem. Soc.* 111, 8141 (1989).
2. J.P. Collman, H.J. Arnold, K.J. Weissman, and J.M. Burton, *J. Am. Chem. Soc.* 116, 9761 (1994).
3. R. Hoffmann, R.W. Alder, and C.F. Wilcox Jr., *J. Am. Chem. Soc.* 92, 4992 (1970).
4. (a) J.B. Collins, J.D. Dill, E.D. Jemmis, Y. Apeloig, P.v.R. Schleyer, R. Seeger, and J.A. Pople, *J. Am. Chem. Soc.* 98, 5419 (1976). (b) W.D. Laidig and H.F. Schaefer, III *J. Am. Chem. Soc.* 100, 5972 (1978). (c) S.M. Bachrach and A. Streitwieser Jr., *J. Am. Chem. Soc.* 106, 5818 (1984).
5. P.v.R. Schleyer, *Pure Appl. Chem.* 56, 151(1984).
6. W.N. Setzer and P.v.R. Schleyer, *Adv. Organomet. Chem.* 24, 353 (1986).
7. R.E. Mulvey, *Chem. Soc. Rev.* 20, 167 (1991).
8. J.J.C. Mulder, *Nouv. J. Chim.* 4, 283 (1980).

Chapter 37

The Multicatenation Model of Polyhedral Metal Clusters

37.1 Matched Versus Unmatched Polyhedra

The story of borane clusters has been told over and again, and it is familiar to most inorganic chemists.[1] From our standpoint, what is remarkable about these species is not the aesthetic beauty of the closoboranes but rather the existence of nido and arachno derivatives that arise from a parent closoborane by removal of one or two vertices, respectively. In other words, the gross structure of the closoborane is retained even when vertices are lost but the number of skeletal electrons is conserved. Phosphorus and sulfur form rings, cages, and clusters that seem to be isoelectronic analogues of covalent organic molecules. For example, P_8 seems to be analogous to $(CH)_8$, cubane. However, in 1979, Gillespie[2] made an important discovery, the significance of which has yet to be fully appreciated. Specifically, he recognized that there exist heavy p-block clusters in which removal of one vertex produces a derivative with the same gross geometry. That is, the stereochemistry of the remnant cluster is retained, provided the number of skeletal electrons is conserved. One example presented by Gillespie is the shape relationship of Te_3S_3 with the parent P_4S_3 structure; discounting the inert ns pairs, these two molecules are valence-isoelectronic:

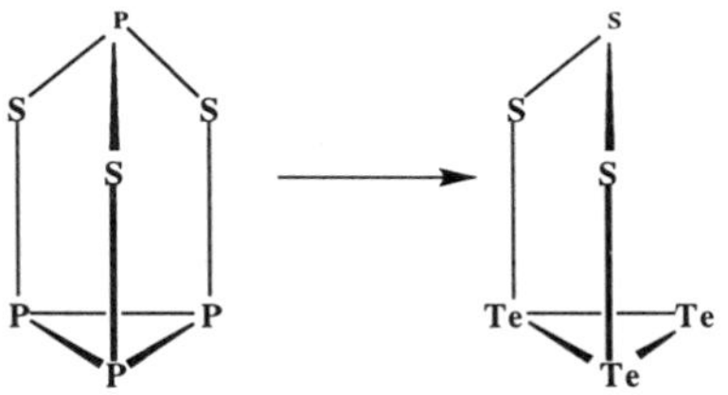

This suggests the following scenario: "Electron-deficient" as well as "electron-precise" clusters of metals or semimetals are actually polyhedra in which atom cores are Coulombically attached to a rigid three-dimensional network of interstitial electrons. The "rigidity" of the "electron polyhedron" is a direct consequence of minimization of interelectronic repulsion; that is, each electron is encompassed by a sphere defining the excluded volume which minimizes electron–electron repulsion. Metal clusters, thus, are miniature metal solids in which metal atom cores are attached on a correlated electron gas. In this chapter we develop a practical and predictive model of cluster stereochemistry while, in parallel, differentiating the VB ideas from the conventional notions of "electron deficiency" and "electron precision," which have dragged the chemist down the wrong path of "universal covalency." In other words, we want to teach how to make new clusters by falling back on the physics (rather than on the "orbital engineering") of the problem.

There are two key features of a homonuclear polyhedron M_n, where M is an atom or fragment:

(a) The fragment *synapticity*. This is the number of orbitals that can be used for polyhedral bonding, and it is symbolized by the label *j*h, where *j* is the number of available AOs. Examples of three- and four-hole fragments are shown in Figure 37.1.

(b) Each polyhedron is characterized by the total *number* of vertices as well as the *type* of vertices. We can identify the type of a vertex by the symbol *i*v, where *i* denotes the connectivity of the vertex and it is called the *degree* of the vertex (e.g., 3v means that the vertex is connected to three other vertices). From this definition it also follows that the number of edges that start at a given vertex is *i* and an *i*v vertex subtends *i* number of faces.

As an illustration, among the first dozen members, only three "pure" deltahedra have just one type of vertex: the tetrahedron, the octahedron, and the icosahedron. The rest have vertices of two different types, as shown in Table 37.1.

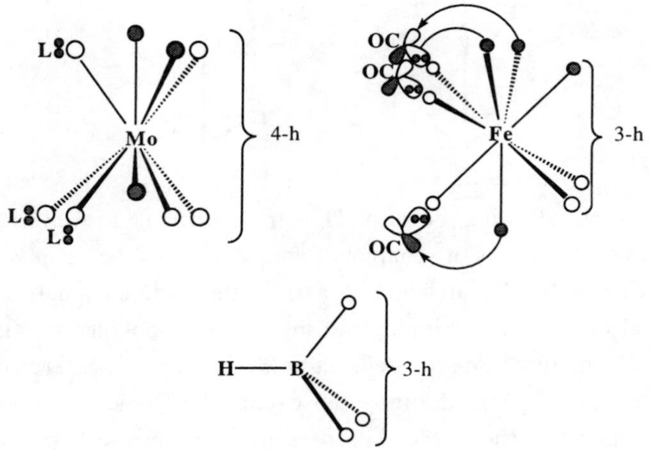

Figure 37.1 Organometallic fragments with different synapticities.

Table 37.1 Vertices and Faces of Representative Deltahedra

Deltahedron	Number of Vertices	Number of Faces	Vertex Types
Tetrahedron[a]	4	4	3v (4)
Trigonal bipyramid	5	6	3v(2), 4v(3)
Octahedron[a]	6	8	4v(6)
Pentagonal bipyramid	7	10	4v(5), 5v(2)
Triangular dodecahedron	8	12	4v(4), 5v(4)
Tricapped trigonal prism	9	14	4v(3), 5v(6)
Bi-end capped square antiprism	10	16	4v(2), 5v(8)
Icosahedron	12	20	5v(12)

[a]Pure deltahedron.

It is now apparent that depending on the nature of the fragment M and the nature of the polyhedron, we can have two situations:

(a) A j-h M fragment is matched with an iv vertex and $i = j$. If this condition holds for all pairings of a vertex and a fragment, we have a *matched polyhedron* ("orbital precise" polyhedron) to the extent that orbitals can be arranged so that they span exactly either all edges in pairs (two overlapping orbitals) or all faces in multiplets (three, four, five, etc. overlapping orbitals). For example, the tetrahedron is a *matched deltahedron* if we place four 3h fragments on the four 3v vertices, the trigonal bipyramid is a matched deltahedron if we place two 3h fragments on the two 3v vertices and three 4h fragments on the three 4v vertices, the octahedron is a matched deltahedron if we place six 4h fragments on the six 4v vertices, etc. Furthermore, two different orbital arrangements are possible for a matched tetrahedron: one involving "dual AO overlap" along each of the six edges and a second involving "triple AO overlap" on each of the four faces.

(b) When $i > j$ or $i < j$, for any one vertex-fragment pairing, we have an unmatched polyhedron. The case of greater interest is when $i > j$; here, there are not enough orbitals to span all edges in pairs or all faces in multiplets.

In a rule-dominated area such as cluster stereochemistry, it is prudent to make a sharp differentiation of our goals. Wade's rule systematizes a considerable body of cluster structural chemistry. On the other hand, it neither explains cluster bonding nor tells us how to make new systems, except by simple extrapolation from known structures. We will rederive Wade's rule as a subcase of a broader rule that has the correct physical underpinning. In doing so, we will always keep in mind that rules are terrible! Our aim is to represent every known molecule as an individual by a formula that all by itself defines the strategy for making a new molecule.

37.2 The Tire Chain Model and the VB Cluster Rule

We are now at an important crossroads. We know that borane clusters and organometallic analogues ("borane analogy") prefer deltahedral shapes. Our theory must take account of this fundamental observation on first principles. Indeed, the explanation is simple: *Cluster deltahedral preference is a consequence of I-bonding.* To see this important point, recall that a T bond can tie up two univalent atoms (two arrows in one interatomic overlap region) while an I bond (two arrows in two different interatomic overlap regions) can tie up either three or four univalent atoms:

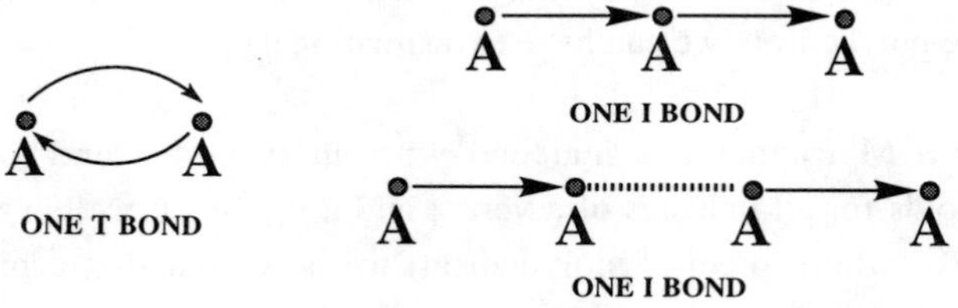

For four univalent atoms to become I-bonded, the two arrows must be codirectional, in which case two pairs of univalent atoms are held together by an overlap-independent quadrupolar Coulomb interaction. This is not the case for the three univalent atoms, which can all be held together by a single codirectional I bond in an overlap-dependent sense. It follows that the optimum I bond spans three AOs. In turn, this implies that in the absence of exchange repulsion, all metalloid and metal clusters will be bound by x-electron triangular AO arrays in which the value of x meets the condition of minimization of exchange repulsion. The optimum is $x = 2$. Now we have the implication that the optimal allocation of the electrons of electropositive elements is one pair per triangular AO array. Excess electrons will break up the cluster by exchange repulsion. Here is the surprising result: The rule just derived is based on the concept of the I bond, which has nothing to do with monodeterminantal MO theory and the concepts thereof! On the other hand, we can see how an empirical theory like EHMO theory contains, by cancellation, the right ingredients: EHMO theory is a "metallic" theory in which I-bonding is simulated by unrestrained CT delocalization (because of the neglect of interelectronic repulsion).

Recall two important results of Chapter 35:

(a) The preferred geometries of Li_4 and Li_6 are the consequence of I-bonding. Each can be visualized as an array of triangular hollows that accommodate two and three electron pairs, respectively. It follows that the best accommodation of T-bond pairs is along edges, but the best accommodation of I-bond pairs is within triangular hollows (catenae).

(b) The six-electron triple bond of trans-bent $H—B{\equiv}B—H^{2-}$ is an analogue of rhombic Be_2Li_2.

Since boron is an electropositive atom much like Li and Be, these findings can be extrapolated to the *tire chain model* of any cluster made up of metalloid or metal atoms. With respect to borane clusters, the model is as follows:

(a) A deltahedron approximates a sphere and the surface of a sphere approximates a two-dimensional surface. The approximation becomes increasingly satisfactory as the size of the deltahedron increases.

(b) Each HB unit of $(HB)_n$ makes use of three endo sp^3 boron hybrids and two electrons to effect cluster bonding. These orbitals form a *net*, which is made up of *n triangular catenae*. Each catena is made up of three AOs and contains two electrons.

(c) The net is fastened on the boron cores by one added electron pair, which plays the role of an intercatenal link (ICL) as in Be_2Li_2.

(d) The analogy of boranes and metal clusters is based on the observation that molecules of both types are I-bound, differing only in arrow directionality (e.g., see the comparison of diborane and Li_4 in Figure 35.19).

Although the author is averse to skiing, he finds the tire chain analogy very apt: The boron cores (i.e., the ensemble of HB^{2+} ions) define the surface of a sphere which, in turn, plays the role of a car tire. The net enshrouds the sphere and is fastened to it by one ICL. Thus, the net plays the role of a tire chain with one fastener. The latter is needed because the triangular faces of the deltahedron exceed the triangular catenae of the net. This is why we chose to name the VB model of borane clusters the tire chain model.

As an example, consider the classic case of the octahedral closo-hexaborane, $(BH)_6^{2-}$. According to Wade's rule, the magic number (MN) is:

$$MN = 2V + 2 = 14$$

where V is the number of polyhedral vertices. According to our approach, one can construct $3V/3 = V = 6$ triangular catenae (V = number of polyhedron vertices). Each BH fragment contributes a pair of electrons, and each pair is accommodated by one triangular catena. A net made up of six triangular catenae, each occupied by two electrons, may have to "stretch" to best fit a polyhedral surface having eight triangular faces. Thus, we conclude that the magic number of a closoborane is given by the following expression:

$$MN = 2\left(\frac{3V}{3}\right) + 2A = 2V + 2A$$

where A is the number of ICLs. For $A = 1$, we reproduce Wade's rule.

The tire chain model reveals that a stable deltahedral cluster may actually be formed when A is either zero or a positive or negative integer! This complies with one fundamental VB concept developed in Chapter 15: The very nature of the I bond guarantees multiple magic numbers. In this particular case, the ICLs act as "straps" to fasten the $2V$-electron triangular net on the atomic cores. We can expect to find deltahedral clusters that can fall in any one of three categories:

(a) Strap-negative cluster with $2V - 2A$ electrons. This is analogous to a chain that fits the tire so tightly that chain crosslinks must be removed for optimal fit.

(b) Strap-free cluster with $2V$ electrons. This is analogous to a chain that fits the tire perfectly and needs no straps (fasteners) for anchoring.

(c) Strapped cluster with $2V + 2A$ electrons. This is analogous to a chain that fits the tire so loosely that it must be held on with straps.

In summary, the physics of the situation suggests that the all-encompassing rule of cluster stability has the following form:

$$MN = 2V + 2A$$

where $A = \pm\, 0, 1, 2, 3$. This is referred to as the *VB cluster rule.*

Clusters of all three types have been known for a long time, and some pertinent data are collected in Table 37.2. Boron chlorides form strap-free $2V$-electron deltahedral clusters. Three examples are B_nCl_n with $n = 4$, 8, and 9.

Table 37.2 Strap-Negative ($V-A$), Strap-Free (V), and Strapped ($V+A$) Deltahedral Clusters

Molecule	Skeletal Pairs	Ref.
Trigonal bipyramid		
$Ni_5(CO)_{12}^{2-}$	$V+3$	(a)
$Rh_5(CO)_{14}^{2-}$	$V+3$	(b)
Compressed octahedron		
Tl_6^{6-}	V	(c)
Pentagonal bipyramid		
CpNiCpNiCp	$V+3$	(d)
Tricapped trigonal prism		
B_9Cl_9	V	(e)
Bi_9^{5+}	$V+2$	(f)
Dodecahedron		
B_8Cl_8	V	(g)
$(CpCO)_4B_4H_4$	V	(h)
$(CpNi)_4B_4H_4$	$V+2$	(i)
Elongated tricapped trigonal prism		
Tl_9^{9-}	V	(j)
Pentacapped trigonal prism		
Tl_{11}^{7-}	$V-2$	(k)

(a) G. Longoni, P. Chini, L.D. Lower, and L.F. Dahl, *J. Am. Chem. Soc.* 97, 5034 (1975).

(b) S. Martinengo, G. Ciani, and A. Sironi, *J. Chem. Soc. Chem. Commun.* 1059 (1979).

(c) Z. Dong and J.D. Corbett, *J. Am. Chem. Soc.* 115, 11299 (1993).

(d) A. Salzer and H. Werner, *Angew. Chem. Int. Ed. Engl.* 17, 869 (1978)

(e) M.B. Hursthouse, J. Kane, and A.G. Massey, *Nature (London)* 228, 659 (1970).

(f) R.M. Friedman and J.D. Corbett, *Inorg. Chem.* 12, 1134 (1973).

(g) R.A. Jacobson and W.N. Lipscomb, *J. Chem. Phys.* 31, 605 (1959).

(h) J.R. Pipal and R.N. Grimes, *Inorg. Chem.* 18, 257 (1979).

(i) J.R. Bowser, A. Bonny, J.R. Pipal, and R.N. Grimes, *J. Am. Chem. Soc.* 101, 6229 (1979).

(j) Z. Dong and J.D. Corbett, *J. Am. Chem. Soc.* 116, 3429 (1994).

(k) G. Cordier and V. Mueller, *Z. Kristallogr.* 198, 281 (1992).

The $n = 4$ species is a tetrahedron while the other two are isostructural with $B_8H_8^{2-}$ and $B_9H_9^{2-}$.[3,4] One additional example of strap-free deltahedral clusters comes from the category of alkyl lithiums: the octahedral 12-electron $(CyLi)_6(PhH)_2$.[5] Heavy p-block elements provide examples of strapped deltahedral clusters: Bi_9^{5+} (22 electrons), Sn_9^{3-}, and $B_9H_9^{2-}$ (20 electrons) are all tricapped trigonal prisms.[6a] A pentacapped trigonal prism, Tl_{11}^{7-}, is an example of a strap-negative deltahedral cluster.[6b] Examination of the literature of deltahedral clusters reveals that a single cage can be associated with diverse electron counts in which *A* ranges from positive to negative integral values. Time will tell what is the most frequent magic number (i.e., what is the most commonly occuring *A* value). For the time being, the two likely champions seem to be $A = 0$ and $A = 1$ (Wade's rule).

We will now indulge in some scientopsychology with, however, very specific chemical ramifications. A coordination chemist and an "EHMO theoretician" interested in developing chemical models view Wade's rule as applicable to "covalent" borane clusters and organometallic clusters. At the same time, they think that this rule has little or nothing to do with bare metal clusters in which the mysterious "correlation energy" is presumed to play a key role. The very same impression is shared by the "ab initio theoretician," who thinks that metals and semimetal clusters have little to do with the rules of organometallic chemistry. In contrast to all this, the VB cluster rule says that the preference for deltahedral cluster formation is a consequence of I-bonding spanning everything from boron to bare metal clusters, to a first approximation (see Section 24.12).

In this light, consider the case of $(Cu_2Se)_n$ clusters, the topic of a recent ab initio calculational study.[7a] According to our approach, the first-order guess of the cluster shape is made using the VB cluster rule with $A = 1$ (Wade's rule). When $n = 2$, there are 12 cluster electrons (Cu contributes one and Se four), and the predicted structure is a monocapped trigonal bipyramid corresponding to the structure found by the computations. When $n = 3$, the predicted structure is a monocapped dodecahedron rather than the calculated tricapped trigonal prism.

A popular topic among physical chemists and solid state physicists is the electronic structure of bare metal clusters. What is the shape? What is the spin? What are the reactivity properties? Many computational investigations have been reported in the literature, but all of them have been random. We now have the tools for at least suggesting how such investigations ought to be targeted. Specifically, coupling the concept of the sd^3 metal (Section 31.11) within the framework of the starring procedure (Chapter 31) to the VB cluster rule leads to automatic predictions:

Each metal uses the three axial AOs to effect cluster bonding.

The shape must be a deltahedron with a magic number consistent with the VB cluster rule. As an initial guess, we set $A = 0$ or 1. The $2V + 2A$ skeletal electrons are delocalized in the skeletal axial metal AOs.

Each metal has two unstarred equatorial AOs and one that is daggered.

The remaining electrons (i.e., those beyond the $2V + 2A$ count) are accommodated by the equatorial AOs with the provision that unstarred be occupied ahead of daggered AOs. These are the nonbonding electrons.

Odd electrons in the nonbonding AOs can be coupled in either a low or high spin sense.

We can now appreciate why the problem of the electronic structure of a bare metal cluster is a complex one. First, there is no singular deltahedral magic number. Second, occupancy of daggered nonbonding AOs can become justified if it relieves interelectronic repulsion. Third, there is no good way of estimating the relative advantage of low over high spin coupling because two nonbonding odd electrons on two different centers can be coupled antiferromagnetically by intervening nonbonding pairs, and the strength of this coupling depends on the (weak) overlap of the equatorial hybrids. However, the key point is that we now have a definite vantage point from which to view the complexities.

The analysis presented in Section 24.12 and the VB cluster rule tell a simple story. A deltahedron with $2V$ electrons may or may not be able to encapsulate an extra pair of electrons, depending on size. The small $(BH)_4$ tetrahedron fails. The large $(BH)_6$ octahedron succeeds. This means that either small or large clusters must be able to accommodate a single electron without significant distortion of the deltahedral shape. As a result, the VB cluster rule for odd-electron clusters is exactly analogous to that of even-electron clusters, with the most likely value of A being +1.

$$\mathrm{MN} = 2V + A$$

The $2V + 1$ rule of odd-electron metal (or semimetal) clusters rationalizes the results of van Zee et al., who found that the Ga_2As_3 is a trigonal bipyramid.[7b] Because of the inert pair effect, Ga acts as a 3h/1e and As as a 3h/3e atom. As a result, the cluster has $(2 \times 5) + 1$ electrons. It can be viewed as an As_3 ring with a three-electron π system. The odd π electron of As_3 couples to two apical Ga radicals to form an allyl-type array, which localizes the odd electron in the two Ga vertices in agreement with experiment. A related application can be made to ab initio calculations of the shape of Ga_3As_2.[7c] This can be viewed as an eight-electron Ga_2As_2 tetrahedron edge-capped by a ground Ga atom to produce the $^2A_1$ C_{2v} global minimum found by the computations. All this has direct implications for the abundance of Ga_xAs_y clusters determined by Smalley and co-workers.[7d] Finally, the odd-electron Tl_{13}^{10-} is also found to have the expected (slightly distorted) icosahedral structure.[7e]

37.3 The Difference Between the P_8 Cube and the $(CH)_8$ Cube

The traditional thinking is that P_8 is an analogue of the organic cubane because the 12 cube edges can accommodate the 24 valence electrons. VB theory paints a very different picture:

(a) Cubane has 12 edge pairs resulting from perfect pairing of 24 odd electrons contributed by the eight promoted CH units at the eight vertices. Cubane is T-bound.

(b) P_8 approaches the limit of I-bonding, and the difference between $(CH)_8$ and P_8 is equivalent to effecting orbital rotation at each cube vertex so that the 3 × 8 valence AOs originally directed along the edges (optimal arrangement for exchange delocalization because each AO overlaps significantly with only one neighbor AO) are now directed toward the centers of each face to form six square (SQ) catenae (optimal arrangement for 2-e CT delocalization because each AO overlaps with two neighbor AOs).

The "orbital rotation" of a 3h fragment on a vertex of a matched polyhedron that, in our example, converts 12 two-orbital edges to 6 four-orbital faces is shown in Figure 37.2. This means that the orbitals redirect themselves so that instead of optimal two-center T bonds, they can now form optimal I bonds in which each AO is connected to two neighbor AOs by two arrows directed in a way consistent with AO occupancy in the ground electronic configuration and the association rule. As a result, P_8 is a multicatenation cluster that can be represented by a resonance hybrid of the following form:

$$P_8\text{: } 3SQ^2 3SQ^6 \longleftrightarrow 6SQ^4 \longleftrightarrow 3SQ^6 3SQ^2$$

SQ identifies a square face spanned by a square catena, the number preceding it denotes the number of such faces, and the superscript denotes the number of electrons within the corresponding catena. The faces of every matched polyhedron are Hückel rings and each one can accommodate $4N + 2$, or $2Q$ (Q = odd) electrons. Hence, the magic number of a matched polyhedron depends on whether it has an even or odd number of faces. As a result, we obtain the following equation , where F symbolizes the number of polyhedral faces.

$$\mathrm{MN} = 2Q^{\mathrm{odd}} \times F$$

Stable matched polyhedra of metals or semimetals are multicatenated association complexes in which formally aromatic, nonaromatic or antiaromatic catenae span the polyhedron faces and comply with either a $4N + 2$ (F = odd) or $4N$ (F = even) total electron count. This is called the *multicatenation cluster model.*

The P_8 cube has an even number of faces; hence the magic number is $4N$ with $N = 6$. Because the faces interact as a result of the square catenae overlap, the formal electron count of each square face can be aromatic or antiaromatic (see hybrid above). Hence, each individual face retains memory of formal aromaticity

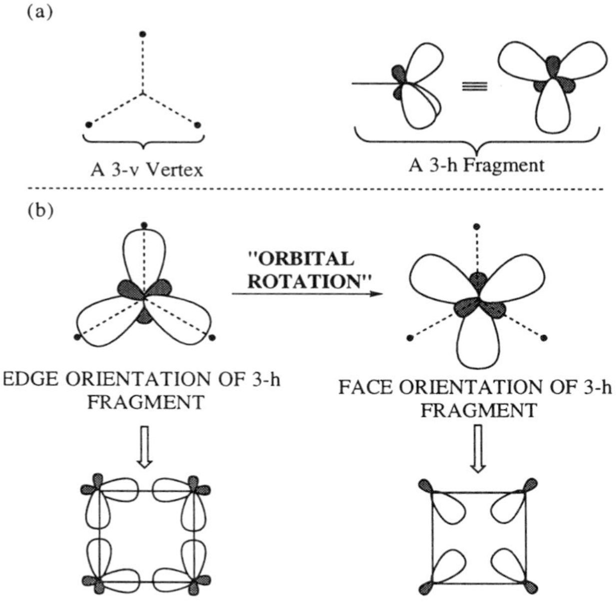

Figure 37.2 (a) Example of 3v vertex and 3h fragment. (b) Edge and face alignment of the hybrid AOs of the 3h fragment.

or antiaromaticity, and any perturbation predicted to stabilize a bona fide aromatic or antiaromatic complex is also expected to stabilize a P_8 face. Furthermore, the notion that a semimetallic cube can be formalized as six interacting facial antiaromatic [4]-annulenes leads to counterintuitive insights. Specifically, an antiaromatic [4]-annulene (e.g., square π-cyclobutadiene, square Li_4) has three low-lying energy states, namely, K, D, and Z (see Chapter 35). It follows that, according to the Multicatenation Cluster model, a cube can exist in three different forms:

A cube made up of six interacting K faces: the $(K)_6$ form.
A cube made up of six interacting D faces: the $(D)_6$ form.
A cube made up of six interacting Z faces: the $(Z)_6$ form.

This formulation assumes that the interaction of degenerate dominates the interaction of nondegenerate face states (i.e., K′–K″ is stronger than K′–D″ or K′–Z″ interaction, where the primes identify different cube faces). The lowest energy valence isomer is expected to be the $(K)_6$ form, but the other two may be thermally accessible if they do exist as distinct minima. The synthesis of a thermochromic $(RSi)_8$ by Matsumoto and co-workers (R = t-$BuMe_2Si$; see Chapter 4) suggests that this scenario is not unreasonable. The molecule is colorless at –196°C, yellow at 25°C, and orange at 280°C.

37.4 The VB View of the Platonic Solids

According to the multicatenation cluster model, a tetrahedron is viewed as a composite of four interacting faces, each of which is spanned by a triangular catena. This suggests two different magic numbers:

(a) Each catena accommodates two electrons and becomes formally aromatic. In actuality, we are approaching the situation represented by twice rhombic Li_4. The magic number is now 8. The tire chain becomes identical to the multicatenation cluster magic number for $A = 0$. In other words, the eight-electron tetrahedron needs no "fastening" by an ICL.

(b) Each catena accommodates three electrons and becomes formally nonaromatic. On the other hand, each pair of triangular catenae is linked by a covalent bond formed by the odd electrons. In actuality, we are approaching the situation represented by twice rhombic Be_2Li_2. The magic number is now 12. The tire chain becomes identical to the multicatenation cluster magic number for $A = 2$. In other words, the 12-electron tetrahedron has two ICLs. Both 8- and

12-electron stable tetrahedra are known. The simplest 8-electron clusters are Be_4, calculated to be tetrahedral,[8a] and the experimentally observed $(BCl)_4$.[3] The 12-electron magic number of the tetrahedron is exemplified by P_4.

A systematic and wide-ranging comparison of chemical predictions at the one-determinantal and many-determinantal level has never been attempted. Nonetheless, the writing on the wall is clear. Specifically, there exist some very useful, but isolated, computational studies, and these come mostly from Pople's extended group of students and co-workers. One example is a report of what happens when Be_4 is computed at different levels of sophistication.[8b] At the RHF level, the concept of I-bonding effectively vanishes; that is, Be_4 is effectively reproduced as if it were covalent. When the basis set is poor, the linear form turns out to be more stable than the tetrahedral. In other words, the calculation opts for perfect pairing. When the basis set is improved, *but always within the RHF framework*, the tetrahedral form becomes more stable. This means that the calculation now opts for "2e-3c bonding" of the type proposed by Longuet-Higgins: Each tetrahedral face is spanned by a triangular array of sp^3 hybrids containing two electrons (an analogue of cyclopropenyl cation). Finally, upon going to the multideterminantal level, the preference for the tetrahedral form is greatly enhanced. This means that the calculation has finally reproduced the right bonding, namely, I-bonding.

The four occupied MOs of tetrahedral Be_4 can be classified as "edge" or "face" or both. At the EHMO (or monodeterminantal SCFMO) level, it is immaterial whether we call the four pairs edge pairs or face pairs. Whether we talk about 2c-2e or 3c-2e bonding, we are still talking about covalent bonding! The distinction between edges and faces becomes meaningful only at the multideterminantal level where the physics of the problem is correctly reproduced. Now, as atom electronegativity decreases, the pairs move from the edges to the faces by trading T- for I-bonding. In other words, we go from bonding involving more interelectronic repulsion to bonding involving less.

It is instructive to compare the multicatenation cluster and the shell model of rings and clusters (see Chapter 4) and, to this extent, we show the shell electronic configurations consistent with the two different magic numbers of the tetrahedron indicating in parentheses whether the MO is edge (E)- or face (F)-localized.

$$Be_4 = a_1{}^2(E,F)t_2{}^6(E,F)$$

$$P_4 = a_1{}^2(E,F)t_2{}^6(E,F)e^4(E)$$

Because Be is a metal, face allocation wins over the edge variety when a choice exists, because faces and edges are optimal for I- and T-bonding, respectively. Hence, Be_4 has four face pairs, the stability of which is largely due to the action of higher lying unoccupied MOs (spanning 3d, 4s, and 4p nonvalence AOs), much as the stability of the two pairs of rhombic Li_4 was due to the action of the higher lying Li2p orbitals. P_4 can be formulated as having the Be_4 configuration plus two ICLs that are localized on opposite edges.

The validity of the VB cluster rule is further exposed by a systematic ab initio computation of the shape of Be_5.[8b] At the RHF level, a planar pentagon is found to be more stable than a trigonal bipyramid by 13 kcal/mol. Inclusion of correlation energy reverses the order: The deltahedral is now more stable than the planar structure by 38 kcal/mol! There are two messages here:

RHF and MO-CI calculations describe different types of bonding. The pentagon and the trigonal bipyramid is optimal for T- and I-bonding, respectively.

Stability of a deltahedral metal cluster is consistent with the electron count $2V + 2A$ with $A = 0$ (VB cluster rule).

This suggests the following scenario: The magic number of small deltahedral clusters in which one surface electron pair cannot delocalize in the sphere hollow because of the small size of the latter is $2V + 2A$ with $A = 0$. However, A becomes unity in the case of intermediate size deltahedra and may exceed unity in large deltahedra.

The multicatenation cluster model has various ramifications that are worth exploring. Thus, for example, let us reconsider the decision to regard the BH fragment as a trisynaptic unit in which B is sp^3-hybridized and binds by using three endo AOs and two valence electrons. The sp^3 hybridization is attained by paying the price of large 2s-to-2p excitation. If this amount is excessive, BH may opt for sp^2 hybridization, where the lower p character of each of the three sp^2 hybrids means lower promotional energy. Because of its high energy, the B2p can no longer be used for bonding. This leaves us with two endo sp^2 hybrids per vertex available for cluster bonding and an exo sp^2 hybrid that accommodates the hydrogen atom of the BH unit. By straddling each vertex with an sp^2 BH so that the two endo sp^2 hybrids orient themselves toward two nonadjacent triangular faces, we can form four triangular catenae spanning four alternant octahedral faces that are tetrahedrally disposed:

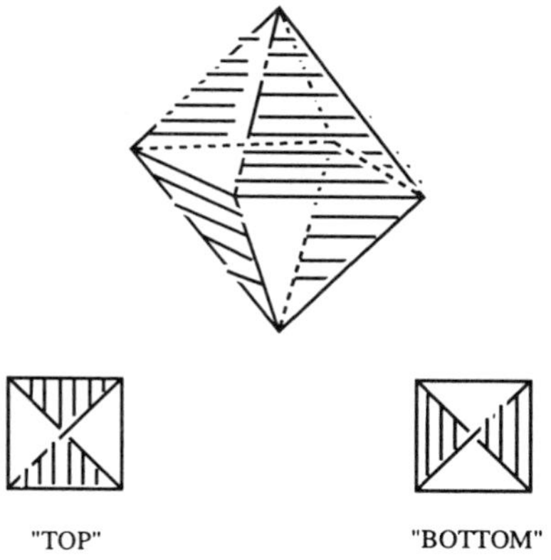

Thus, an octahedron with a magic number of eight electrons is expected whenever the promotional energy of the vertical fragment demands di- rather than trisynapticity. We can easily design eight-electron octahedra. One possible candidate is $SiLi_4Si$, assuming that Si acts in its divalent state with one lone pair placed in a 3s-like exo AO. Du Mont and Grenz found that $Sn_2(t\text{BuS})_4$ undergoes rapid intramolecular degenerate rearrangement via a transition state that presumably has an eight-electron octahedral $Sn(t\text{BuS})_4Sn$ structure.[9]

37.5 Uncapped Vertices and the Magic Numbers of Lithium Clusters

The magic number rules derived earlier were based on one key assumption: The exo AOs of the vertical fragments are either capped by ligands or accommodate lone pairs, hence are unavailable for accepting electrons. The situation changes drastically in the case of Li clusters. Each Li uses three sp^3 endo AOs to effect cluster bonding with one exo hybrid remaining vacant. When the Li atoms combine to form a polyhedron, application of the VB cluster rules is justified with one difference: The exo AOs combined in phase define a totally symmetric bonding exo MO which can serve as the receptacle of an additional pair. In such cases, a supplemental rule materializes in which the magic number of electrons increases by 2 relative to the one derived for the case of capped vertices:

$$MN\# = MN + 2$$

A simple example is CLi_6, which has a total of 10 electrons. As we have seen, the magic number of the octahedron when the promotional energy of the metal is disproportionately high (with respect to bond making) is 8. This means that CLi_6 is a cluster with two exo electrons, a phenomenon discovered by means ofab initio computations.[10a] In other words, MN# = 8 + 2. CLi_6 has been prepared and is unusually stable.[10b–d]

Obviously, there are many ways to test this proposal by carrying out simple ab initio computations. For example, CLi_4 is expected (and found by ab initio computations)[11] to be a global minimum. This can be viewed as an eight-electron Li tetrahedron with a guest carbon in compliance with the multicatenated cluster rule. Addition of two electrons is expected to generate yet another stable cluster with two exo electrons (MN# = 8 + 2). Thus, we expect that the CLi_4 dianion will be also a stable tetrahedron. Indeed, ab initio calculations of the isoelectronic Li_4O confirm these expectations.[11]

Schleyer and his group have made extensive calculational investigations of organolithium molecules guided by classical ideas of bonding. The results are summarized in a review entitled "Remarkable Structures of Lithium Compounds,"[12] which concludes that "lithium compounds are very different from their hydrogen counterparts and are characterized by polar character, electrostatic interactions, and multicenter covalent bonding." In other words, lithium compounds follow the ionic/covalent gospel but, somehow, turn out to be very different! This "understanding by computation" should now be contrasted to our conclusions: Lithium is different from hydrogen because the former is a metal and the latter a nonmetal. Metals support association bonding (I-bonding) and nonmetals covalent bonding (T-bonding), and the two mechanisms are distinct. This leads to a forecast: Since computational chemists think of organometals as metal derivatives of organic molecules, and since their choices of what to compute are guided by the covalent/ionic model, their results will be chemically meaningless when the global minimum does not belong to the set of structures they choose to investigate! We will now pursue this issue.

37.6 How to Predict Organolithium Global Minima

We saw that diborane and Li_4 can be regarded as I-bound rhombic structures that differ only in the quality of the I bonds: The former enjoys codirectional CT delocalization, but the latter is condemned to the contradirectional formwith the benefit, however, of E-bonding. It follows that to a first approximation the VB cluster rule should be valid across the board from borane to bare metal clusters simply because the CT delocalization mechanism of bonding persists throughout. In other words, we can assume that E-bonding does not produce sharp stereochemical distinctions and that, as a result, cluster stereoselectivity is due to CT delocalization irrespective of whether this is superior or inferior to E-bonding. As a specific example, consider the semimetallic $(ClB)_4$, the metallic Be_4, and the mixed $(CH_3Li)_4$ tetrahedral clusters. All are eight-electron clusters and all are expected to be tetrahedral even though the relative importance of E-bonding increases in the order given above.

In this light, let us now focus on organolithium molecules and extended structures. The central organizing principle is the VB cluster rule, with all the uncertainties imposed by the physics of the problem. Hence, the global minima of organolithium molecules must be *clusters* satisfying one of the magic electron counts prescribed by the VB cluster rule. In general, one can assume that the global minimum will be a $2V$-electron deltahedron, where V is the number of deltahedral vertices. The necessity of adding one electron pair as a "fastener" (arriving at a Wadian count) and a second pair to any available exo vacancies should be treated as likely but not obligatory.

"The seeming simplicity of 1,2-dilithioethene is deceptive": This is how Schaefer et al.[13] begin their recent description of ab initio computations of this molecule and cite no less than six previous studies. The 1995 paper reports a global minimum roughly 30 kcal/mol more stable than any one of the species studied earlier! This is a good illustration of our inability to guess the global minimum when it comes to organometals.

We now offer guidelines for predicting the structures of lithiated hydrocarbons. Specifically, the organolithium molecule ALi_n (A = nonmetal fragment, n = integer) is formalized as a composite of two types of radical taken in 1:1 ratio: Li atoms and hydrocarbon radicals. The latter are generated by splitting the A piece into R_1, R_2, etc., hydrocarbon radicals of maximum stability. As a result, the molecule is made up of n Li atoms and n hydrocarbon radicals. Since there are $2n$ monovalent units, the electron count is $2n$. This is the electron count of a deltahedron with n vertices that satisfies the VB cluster rule for $A = 0$. Hence, the n Li atoms and the n hydrocarbon radicals are assembled into a deltahedral cluster with n Li vertices and n R_i face or edge caps. When face-capping is stereochemically impossible, "internal capping" (i.e., placing one or more organic fragments inside the Li_n cage) becomes preferable. In short, an organolithium molecule cannot be anything but a deltahedral cluster obeying the VB cluster rule because this is the prescription for the best realization of I-bonding. When n Li atoms are matched by n hydrocarbon radicals, they produce the $2n$ magic number, which obeys the VB cluster rule with the assumption that no intercatenal links are necessary ($A = 0$).

We first apply our procedure to the case of "1,2-dilithioethene." The VB formula is $(C_2H_2)Li_2$. We decompose (C_2H_2) to HCC plus an H radical. The global minimum must be the four-electron cluster $(HCC, H)Li_2$. This cluster is the rhombus A, below. It is a bicapped [because one edge of a diatom (or any acyclic array) has two sides] Li_2 "collapsed deltahedron." The situation is entirely analogous to the "metallic" Li_4. To gain additional contact between one HCC π-bond pair and an Li2p hole, *A* distorts to produce the global minimum B found by the calculations. However, the computed energy difference between the two structures is practically zero.

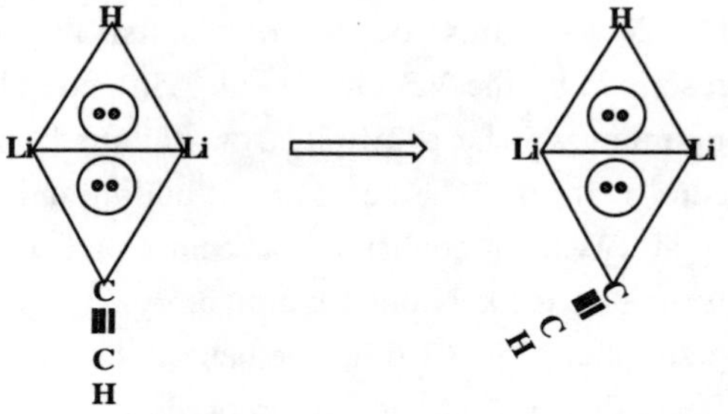

A B

Next, consider the case of "1,1,1-trilithioethane." The VB formula is $(C_2H_3)Li_3$. We decompose (C_2H_3) to HCC radical plus two H radicals. The global minimum must be the six-electron raft-type cluster (HCC, H, H)Li_3. This tricapped Li_3 triangle ("collapsed deltahedron"), representing a situation analogous to that of the metallic Li_6, is depicted as follows.

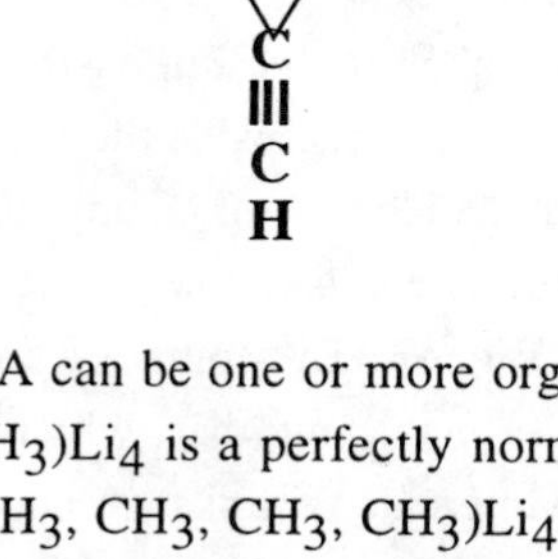

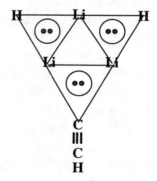

The organic fragment A can be one or more organic molecules. Thus, the species $(H_3CCH_3, H_3CCH_3)Li_4$ is a perfectly normal organolithium species. The global minimum is $(CH_3, CH_3, CH_3, CH_3)Li_4$, and the structure is an Li tetrahedron with each face capped by one CH_3. This is a well-known cluster.

What is the structure of "1,2-dilithioethane"? Since there are two Li atoms, this must be the rhombus $(H_3CCH_2, H)Li_2$. This is very different from the

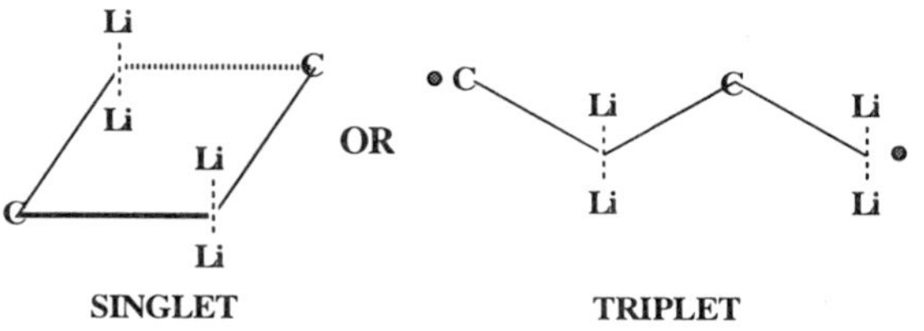

Figure 37.3 Predicted structures of C_2Li_4.

alleged C_{2h} "organic-like" global minimum computed by Schleyer and co-workers.

What is the structure of "$Li_2HCCH_2CH_2CH_2CH_2CHLi_2$"? The answer is: Combine 6 carbons and 10 hydrogens into the best 4 organic radicals and make a face-capped Li_4 tetrahedron! Now, there are many choices! Our "all-carbon" best scenario is ($H_2C{=}CH$, $HC{\equiv}C$, CH_3, CH_3)Li_4. In other words, we predict a distorted Li tetrahedron with the four faces capped by two methyl, one acetyleno, and one vinyl radical! Our "multihydrogen" best scenario is ($CH_3CH_2CH{=}$)$CHCH{=}CH$, H, H, H)Li_4. In other words, we predict a distorted Li tetrahedron with three faces capped by three hydrogens and the fourth face capped by a 1,3-hexadienyl radical.

Things become even more interesting when we come to permetallated carbon molecules. For example, consider the case of $Li_2C{=}CLi_2$, which has been computed many times as an ethylene derivative. One formulation is (C_2)Li_4, where C_2 is now a quintuplet tetraradical (i.e., a doubly excited carbon dimer). However, this C_2 tetraradical cannot cap each face of the Li_4 tetrahedron. Furthermore, it is too large to fit in the tetrahedral Li_4 hollow. The alternative formulation is (C,C)Li_4, where C is a triplet ground carbon. Again, the mere three-dimensional disposition of the odd electrons of each carbon atom prohibits effective capping of each Li_4 tetrahedral face by the two ground carbons. Thus, we have exhausted the utility of the guidelines given earlier. We must now think anew. One solution is to reduce the Li_n cluster size. Instead of one tetrahedral Li_4, we can use two Li_2 "clusters" and edge-bicap each one by a carbon atom. The predicted geometry is shown in Figure 37.3.

The difference between I-bonding and the ionic bonding model is exposed by a comparison of $(CH_3)_4Li_4$ and its derivatives. The cluster is an eight-electron tetrahedron. Replacing CH_3 by NH_2 changes the electron count from 8 to 16 because the amino group is a three-electron ligand. As a result, it is predicted that the tetrahedral structure will be destroyed, giving way to a planar rhombic or even a chain structure. Replacing methyl by hydroxy yields $(OH)_4Li_4$, in which Li is a 3h/1e and OH a 3h/5e ligand that may combine to form a 24-electron cube that can also be viewed as a tetracapped tetrahedron in which each of the four triangular faces is connected to a capping hydroxy by a formal triple bond. In short, I-bonding depends on electron count and predicts different structures for the tetramer $(LiZ)_4$ depending on the electron count of Z. By contrast, the ionic model predicts the same geometry irrespective of Z. These considerations emphasize:

The need to incorporate the lone pairs of heteroatomic fragments in the electron count.

The idea that loose bond pairs of a hydrocarbon fragment (e.g., the π system of the CCH fragment in "dilithioethene") are not incorporated in the electron count. This is an assumption that needs to be tested.

Bottom line: The organic formulas used to symbolize organolithium molecules are misleading because they are inspired by a bonding theory appropriate for nonmetals but not for organometals. They fail to project the global minimum and they should be abandoned. We have submitted a predictive recipe that is not intuitively obvious but can be easily tested by ab initio calculations.

37.7 The Electronic Structure of Ti_8C_{12}

The proposal that M_8C_{12} (M = Ti, Zr) has a *pentagonal dodecahedral* (PD) structure[14] leads to the question: Is there some remarkable feature about this cluster, or are there better alternatives? The multicatenation cluster model suggests the following strategy: Start with a Zr_8 polyhedron that is not necessarily a deltahedron, cap each face by carbon fragments to make formal multiple bonds, and examine the distortion potential for turning the polyhedron into a deltahedron, if it is not already one.

The most plausible scenario jumps on us: A Zr_8 cube has six faces and each can be capped by six (C_2^{2-}) fragments by six formal triple bonds, leaving eight nonbonding electrons, which are accommodated by the surface Zr AOs with the largest radial extension. Here are the details, with the aid of Figure 37.4.

(a) Each Zr adopts the d^3s^* hybridization mode (Figure 25.6) and acts as a trisynaptic piece. This hybridization mode can be derived from a pseudo-octahedral Zr by elimination of three starred exo conical hybrids. The three equatorial hybrids become AOs and resolve into one toroidal $s\text{-}d_{z^2}$ hybrid plus two δ-type d orbitals. The three endo conical AOs resolve into one $s\text{-}d_{z^2}$ hybrid

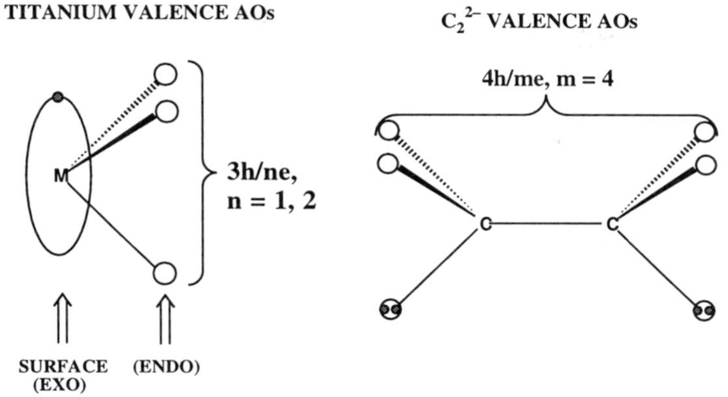

Figure 37.4 Constituent fragments of Ti_8C_{12}.

plus two π-type d AOs. Since the equatorial toroidal AO has the largest radial extension (because it has s character), this is the one that can best overlap with another toroidal AO of a neighboring Zr. Hence, this is the AO that accommodates "residual electrons" (i.e., electrons in excess of the cluster electrons), for the purpose of additional metal–metal bond making.

(b) Each carbon is reduced by one electron, and the 12 carbon monoanions are combined to form six C_2^{2-} fragments, each of which acts as a tetrasynaptic piece.

(c) The total number of electrons available for cluster bonding is now (4 × 8) – 12 zirconium electrons plus (4 × 6) dicarbon electrons (i.e., 44 cluster electrons).

(d) Each C_2^{2-} is tetrasynaptic and so is each Zr_8 cube face. Two tetrasynaptic pieces have the right ensemble of orbitals for forming a formal triple bond. That an acetylenic residue can act as a tetrasynaptic four-electron piece to form a triple bond with a cluster face will be further documented in the following chapter.

(e) The number of electrons required for making six triple bonds is 6 × 6 = 36. Each triple bond connects a C_2 cap to a Zr_8 cube face. This leaves 44 – 36 = 8 electrons, which go to the eight toroidal AOs of the 8 Zr atoms.

Will the C_2^{2-}-hexacapped Zr_8 cube distort? Since each cube vertex has one singly occupied toroidal AO, two types of distortion are predicted:

A D_{2d} distortion in which each of two trans tetragonal faces turns into a four-electron crossed-antiaromatic rhombus.

A T_d distortion that transforms the cube into a tetracapped eight-electron tetrahedron. In an all-too-rare instance of creative use of computations, Dance proposed such a T_d structure as the global minimum of Ti_8C_{12}.[15a]

Are there related clusters? Each of the six faces of a 24-electron cube can be capped by an organometallic fragment to generate a 36-electron cube with six radiating triple bonds. A case in point may very well be $[Ni(CO)]_8(PPh)_6$, which can be viewed in the following way: Six Ni(CO) fragments act like 3h/2e and two Ni(CO) fragments act like 3h/0e fragments to produce a cube with six two-electron faces, each of which is capped by a 3h/4e PhP fragment by a formal triple bond.[16]

The game is not over! The analysis of the bonding of inorganic rings in Chapter 24 suggests a second candidate for the global minimum of Ti_8C_{12}. Specifically, the average electronegativity of the 20 atoms of Ti_8C_{12} (2.0) is very close to the electronegativity of phosphorus (2.1). This implies that the shape of Ti_8C_{12} must be related to the shape of P_{20} (see Section 24.4). Any difference must be attributable to the exo AOs of the tetrahedral C and Ti, which are singly rather than doubly occupied. As a result, there are 10 weak exo bonds that may cause a deviation from the structure expected by analogy to the P_{20} structure. Recall that the predicted structure of P_{20} is *not* a dodecahedron!

The two proposed met-car structures are supposed to be the thermodynamically favored products of the carbon–titanium condensation and neither is a dodecahedron. However, the latter isomer has a characteristic feature: It is made up of five-membered rings that can act as entropy sinks as discussed in Chapter 5. Hence, a symmetrical dodecahedral Ti_8C_{12} may be the kinetically favored product at high temperatures. In other words, both icosahedral C_{60} and dodecahedral Ti_8C_{12} may be kinetic products, rather than thermodynamic, sharing a common feature: pentagonal rings that act as entropy sinks.

The treatment applied to Zr_8C_{12} rationalizes the distorted cubic structure of the Nb_4C_4 cluster.[17] This is formulated as $(Nb^+)_4(C^-)_4$ in which a d^3s^*-hybridized Nb^+ acts as 3e/3h with an additional equatorial electron (in the toroidal AO) and C^- as 3e/3h. The four equatorial electrons of $(Nb^+)_4$ define two formal bonds localized at opposite edges of the niobium tetrahedron. In an MO framework, these two electron pairs appear as an e^4 ensemble in T_d symmetry (much like the two electron pairs of the HOMO of the 12-electron tetrahedron). The electronic structure of the related Nb_4C_2 cluster, formulated as $(Nb)_2(Nb^+)_2(C^-)_2$, can be explained most conveniently by adopting the d^3s (instead of the d^3s^*) hybridization model.

37.8 The Counterintuitive VB View of the "Simple" Heavy p-Block Halides

SF_2, SF_4, and SF_6, are traditionally used as illustrators of apparently unrelated concepts: covalent bonding, "normal" versus "hypervalent" bonding, the theory of valence shell electron pair repulstion (VSEPR), and so on. By contrast, we have argued that heavy p-block and transition metal halides fall toward the I extremum of the T/I continuum. Treating a halogen atom as a one-electron ligand, as one semimetal, or as a metal pair can bind two electronegative halogen atoms by an I bond. Thus, what goes on in, say, SF_6, is qualitatively the same thing that goes on in deltahedral borane clusters, in metal clusters, and in metal solids! The only difference is that the directionality of the I bonds as well as the relative importance of E-bonding changes in going from homo- to heteronuclear systems. This leads to a formulation that is dictated by the physics of the bonding: *Mononuclear semimetal and metal halides are centered clusters.* If indeed the mononuclear heavy p-block and transition metal halides are bound by the same mechanism as the borane and organometallic clusters, two predictions follow:

They must be deltahedral; that is, the ligands must form a deltahedral cage, encapsulating the central atom.

They must have the same magic numbers as the boron and metal clusters discussed earlier. According to VB theory, these magic numbers are not rigid but flexible and they are given by the VB cluster expression (see Section 37.2).

In this light, we can now appreciate the following pattern: Trigonal AlF_3, tetrahedral SiF_4, trigonal bipyramidal PF_5, and octahedral SF_6 are all deltahedral clusters with a $2V$-electron count. This pattern repeats itself in the lower p-block rows. There is one deviant: XeF_6 is also a deltahedron, but its electron count is $2V + 2$. We will achieve a renewed appreciation of this "hypervalent" molecule shortly.

Let us now pursue the idea that metal fluorides and borane clusters are related. In the latter, the nido is derived from the closo structure by removing one polyhedral vertical core. Thus, the closo and nido forms are isoelectronic. By contrast, peeling off two fluorines from a closo-type metal fluoride generates a nido-type structure but reduces the electron count by two. This means that the metal fluorides define a closo/pseudonido/pseudoarachno (pseudo-CNA) concept that is entirely analogous to the familiar closo/nido/arachno concept of boron chemistry.

The implementation of the pseudo-CNA concept hinges on the definition of the closo metal fluoride. This is important because only this species qualifies as the reference with respect to which the geometries (pseudonido, pseudoarachno, etc.) are predicted. The closo metal fluoride is defined as one in which each metal pair is "saturated"; that is, each metal pair binds two fluorines by one I bond, and any remaining odd electron binds one fluorine by a T bond. Thus, the closo form of the metal fluoride MF_n is the one in which the metal is in its highest formal oxidation state. For example, the closo form of PF_n is PF_5. We express our conclusion as:

$$MF_{2n}\ (\text{CLOSO}) = M(n\ \text{pairs})F_{2n}$$

and

$$FMF_{2n}\ (\text{CLOSO}) = FM(n\ \text{pairs})F_{2n}$$

In this light, we can see several interrelations between three apparently unrelated molecules: SF_6, SF_4, and SF_2.

(a) SF_6 is an *associatively saturated* fluoride in which each sulfur pair binds two fluorines. Hence, it is a closo deltahedral (octahedral) cluster. The skeletal electron count is 12, and this is a $2V$ number because there are six octahedral vertices.

(b) Sequential removal of two fluorines at a time generates the experimental structures for SF_4 and SF_2. SF_4 is pseudonido and SF_2 is pseudoarachno SF_6.

(c) The geometry of SF_4 hinges on the decision of which I bond in SF_6 is the weakest. The first pair of fluorines comes off the sulfur pair of lowest energy, namely, the sp^2 pair.

(d) The geometry of SF_2 hinges on the decision of which I bond in SF_4 is the weakest. The second pair of fluorines (relative to SF_6) comes off as the higher-energy 3p pair. Remember: The higher the energy of the pair, the more favorable the REL bond formed by the interaction of one sulfur pair with two fluorine atoms.

The argument is illustrated pictorially in Figure 37.5.

Sulfur binds six fluorines by its three valence pairs to form octahedral SF_6, a closo cluster. In an exactly analogous fashion, tungsten binds six fluorines by its three valence pairs to form octahedral WF_6, also a closo cluster. This means that S and W have the same *oxidative valence* because they have the same

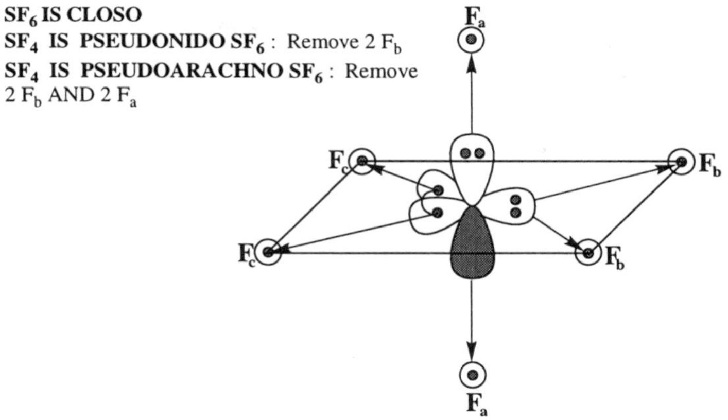

Figure 37.5 Electronic structure of SF_6.

number of valence pairs. In general, low spin transition metal fluorides are expected to be deltahedral clusters much like heavy p-block fluorides. The data given in Table 37.3 speak for themselves.

The famous ReH_9^{2-} is simply an analogue of $(BH)_9^{2-}$ minus two electrons: Both are tricapped trigonal prisms. By the way, 2*V*-electron octafluorometallate clusters are known, but the geometry deviates from that of an ideal deltahedron. Two examples are the square antiprismatic $[MoF_8]^{2-}$ and $[TaF_8]^{3-}$. The demonstration of quasi-dodecahedral symmetry in transition metal octafluorides has been the result of pioneering stereochemical studies by Hoard.[18]

Table 37.3 Electron Counts of Simple Deltahedral Transition Metal Fluorides and Derivatives

Molecule	2V Skeletal Electrons	Shape
ScF_3, AlF_3	6	Triangle
ZrF_4, SiF_4	8	Tetrahedron
VF_5, PF_5	10	Trigonal bipyramid
WF_6, TeF_6	12	Octahedron
ReF_7, IF_7	14	Pentagonal bipyramid
$Zr(O_4C_2)_4^{4-}$	16	Dodecahedron
ReH_9^{2-}	18	Tricapped trigonal prism
$Mo(CN)_8^{4-}$	18[a]	Dodecahedron

[a] $2V + 2$ skeletal electrons.

Table 37.4 Relation of Some p- and d-Block Fluorides

Fluorides		
p-Block	d-Block	Number of Bonds
AlF_3	LaF_3	I + T
SiF_4	ZrF_4	2I
PF_5	TaF_5	2I + T
SF_6	WF_6	3I
IF_7	ReF_7	3I + T

The pseudo-CNA principle can be further illustrated by reference to WF_6. Each of the three W pairs occupies a diagonal, σ-type sd^2 hybrid and associates with two trans fluorines. From this arrangement we can predict that sequential removal of pairs of fluorines will produce square WF_4 and linear WF_2. Recent ab initio calculations indicate that $W(OAr)_4$ and $Mo(OH)_4$ are square planar.[19] On the other hand, spectroscopic data have been interpreted to be consistent with tetrahedral MoF_4 and CrF_4.

In summary, the concept of the I bond and the closo deltahedron reveals common denominators where none could be seen. In every row of Table 37.4, a p- and a d-block fluoride are linked in three ways: They have the same *deltahedral* shape, the same number of bonds, and the same number of 2*V* valence electrons. All these molecules are elements of one and the same story: I-bonding. In terms of conventional theory, they represent three different types of species: "electron precise" (AlF_3, SiF_4), "hypervalent" (PF_5, SF_6, and IF_7) and "ionic" or "covalent/ionic" (d-block fluorides). Thus it is implied that isoelectronic p- and d-block deltahedra must have the same reactivity properties even though the valence orbitals of p-block atoms are ns and np while those of d-block transition metals are $(n-1)$d and ns!

37.9 Information Relay in Fe_4S_4 Cubes

There exist many 24-electron "push–pull" A_4W_4 cubes.[20–25] According to the multicatenation cluster model, a metal polyhedron is made up of AO catenae spanning the polyhedral faces. One strategy for stabilizing such a polyhedron is to cap each face with a fragment by a triple bond. For example, each face of the 24-electron cube can be formalized as an analogue of the K state of cyclobutadiene (Figure 37.6). To convert the K to Z faces and further stabilize

Figure 37.6 K and Z faces of a cube according to the multicatenation model.

the system, we must implement "push–pull" substitution (i.e., alternation of electronegative and electropositive atoms), to render each face crossed antiaromatic. These "push–pull" cubes can adopt two different geometries:

(a) Push-pull cubes with D_{2d} symmetry. We view this as a consequence of rhombic distortion of two of six K faces that causes crossing of Z below K. This distortion is preferred when the electronegativity difference between *A* and W is modest.

(b) Push-pull cubes with tetrahedral symmetry in which the four electropositive atoms form a tetrahedron that is face-capped by the four electronegative atoms. Recalling that a 12-electron tetrahedron has four three-electron faces, the "tetrahedral cube" is equivalent to a 12-electron tetrahedron in which one triple bond links each face with one capping atom or fragment. The total number of electrons is 24, the ordinary magic number of the cube. This is equivalent to converting each of the six K faces into Z faces. This distortion is preferred when the electronegativity difference between *A* and W is large.

In summary, D_{2d} distortion turns two of the K faces of a binary cube into crossed-antiaromatic Z faces and T_d distortion turns all six; the form of distortion that is preferred depends on the electronegativity difference. This is the VB rationalization of the D_{2d} shapes of the $[Fe_4S_4]^{-q}$ cubes.[26]

The next issue is: How do "innocent" ligands fine-tune the electronic properties of a metal cluster? We begin with the symmetrical model system $[(RSFe_4)S_4]^{4-}$, symbolized as $[Fe_4S_4]^{4-}$, in which each formal neutral Fe is coordinated to RS^-, which can act as 3h/6e or 3h/4e or 3h/2e. One of the possible valence states of the RS^-Fe fragment is shown in Figure 37.7a; note in particular that the 3h/6e action of the proximal RS^- ligand dictates 3h/2e action by the iron center of the RS^-Fe fragment. Since each RS^-Fe can act as 3h/2e and each sulfur atom can act as 3h/4e, we have a 24-electron cube.

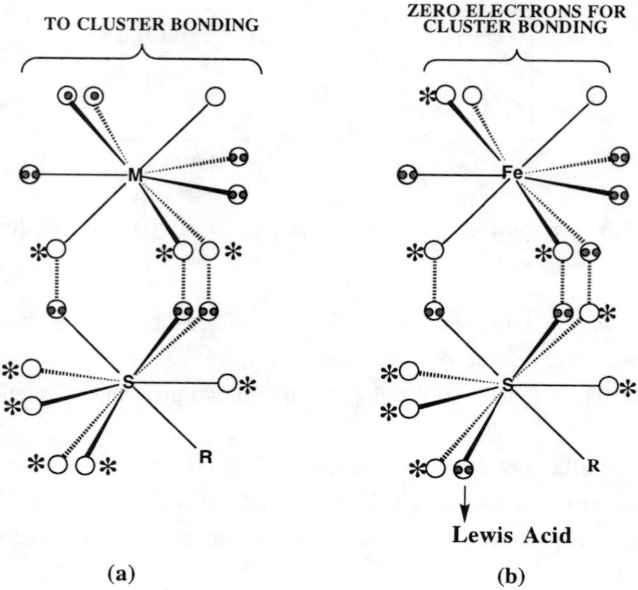

Figure 37.7 (a) The valence state of Fe in the Fe_4S_4 cube depends on the valence state of the RS^- proximal ligand. When the latter acts as 3h/6e, Fe binds as a 3h/2e fragment. (b) When the RS^- proximal ligand acts as 3h/4e, $(RS^-)Fe$ binds as a 3h/0e fragment.

How can we move two electrons from the cube to the outer envelope of the cluster? The answer is: Change the valence state of one RS^- from 3h/6e to 3h/4e. In turn, this changes the valence state of the iron center as depicted in Figure 37.7b. One motivation for the change is the presence of a weak Lewis acid that can tie up one exo RS^- pair. The number of skeletal electrons of the cube are now reduced from 24 to 22. This analysis raises the possibility of the following relay of information: The medium determines, via hydrogen bonding (or related coordination mechanisms), the valence state of RS^-, which, in turn, determines the number of skeletal cube electrons. This mechanism is not oxidation or reduction but rather a way of shifting electron density from the cube to the exo orbitals of the ligands which face the environment.

The biologically important $[(RSFe_4)S_4]^{-q}$ cubes have $q = 2$. The two electrons lost with respect to our model system come from equatorial Fe AOs. This creates two singly occupied equatorial Fe AOs, which can be antiferromagnetically coupled as illustrated in Figure 37.8. It is seen that, while

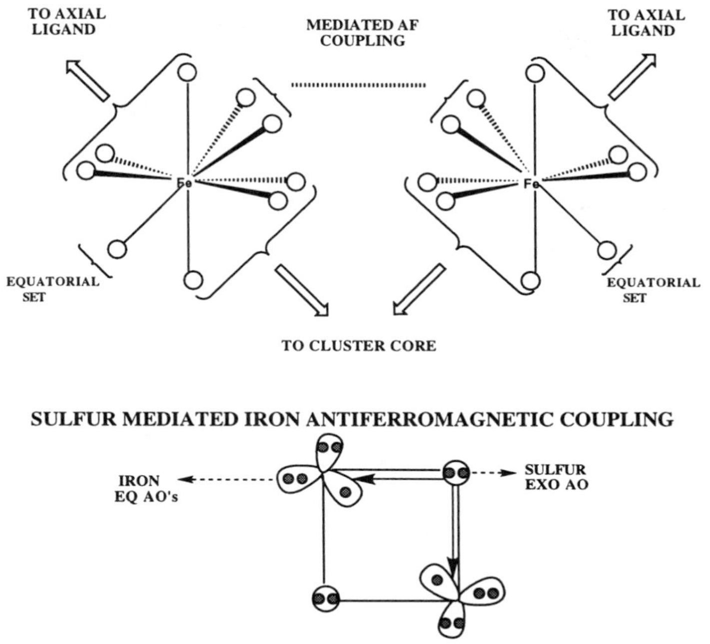

Figure 37.8 Bonding aspects of the Fe_4S_4 cube and illustration of antiferromagnetic coupling of odd equatorial electrons in oxidized species.

the endo radial electrons provide the glue for cube binding, the equatorial pairs interact via the exo radial sulfur pairs in such a way so that two equatorial odd electrons on two different irons are antiferromagnetically coupled by a single I bond. Thus, we have a highly intricate apparatus:

The endo AOs and electrons glue the cube together.

The equatorial iron electrons and the exo radial sulfur electrons lie on the surface of an idealized spherical cluster and are coupled to each other by I bonds.

The iron ligands respond to the environment by shifting electron density from the $[Fe_4S_4]^{-q}$ cube to the "outside world," or vice versa.

The way in which we do the electron bookkeeping and interpret the electronic structures of metal–cubanes can be illustrated by reference to two specific examples: The $(CpCrO)_4$ cube, prepared by Bottomley's group,[27] and the $[(PhSFe)Te]_4^{3-}$ cube, prepared by Bertini's group.[28]

The $(CpCrO)_4$ cube was found to involve antiferromagnetic coupling of the metal centers. As usual, each oxygen acts as a formal dianion and each Cr as a formal trication. This means that pseudo-octahedral Cr acts as 3h/0e and O as 3h/6e to generate a 24-electron cube. In addition, each Cr has one equatorial pair, one equatorial odd electron, and one (daggered) equatorial hole. Thus, we have four equatorial odd electrons that can couple in three different ways:

Low spin (antiferromagnetic coupling) to yield an $S = 0$ state.
Intermediate spin to yield an $S = 1$ state.
High spin to yield an $S = 2$ state.

The data suggest that the first possibility is the one that materializes.

The philosophy that first priority is given to abiding by the 24-electron count is justified by reports of low spin 24-electron metal–alkoxy cubes in which the metal has a full complement of six equatorial electrons. This configuration suggests an unwillingness on the part of the metal to transfer electrons from the equatorial to the endo conical set in order to diminish electron–electron repulsion by adopting an intermediate-spin configuration.

Our second example cube $[(PhSFe)Te]_4^{3-}$ cube was found to have three unpaired electrons. By following the procedure just used for the tetroxide, we expect a 24-electron cube with one odd equatorial electron. This is because each Fe acts as a formal 3h/0e and each Te as a formal 3h/6e piece. In addition, three of the four irons have six equatorial electrons but one has only five (i.e., one has one odd equatorial electron). To end up with three unpaired electrons, we need to transfer one electron from the equatorial to the exo conical set of Fe hybrids used for skeletal bonding. Since the exo electron must be accommodated in an exo unstarred Fe AO, an endo AO that was originally unstarred must now become starred. This results in the weakening of the cube bonding.

The central problem with "bioinorganic chemistry" is the same as that of "inorganic chemistry": a lack of molecular formulas that express the physics of bonding. As a result, extremely sophisticated electronic processes have been reduced to crude representations that miss completely the wide spectrum of possibilities open to a metal atom. Specifically, one and the same metal can bind from more than one valence state, and bioinorganic reactions are a challenge to understand how a metal shuttles between different valence states in serving a specific function. The VB model can be used to explain much of what is in the literature but, at the same time, it raises more questions than it answers. The hope is that experimentalists will rethink their data along these lines and devise new experiments to uncover the secrets of atom communication in clusters.

37.10 Iron–Sulfur Clusters, the Tire Chain Model, and the Triple-Bond Road

What is the global minimum of an M_xQ_y cluster where M is metal and Q a heavy p-block metalloid? A challenging subcase is that for which M = Fe and Q = S. As in the case of organolithium clusters, the answer is provided by the VB cluster rule, with the only difference being that the number of valence electrons far exceeds that needed for the preparation of a primitive deltahedron with $2V + 2A$ electrons. The resolution is straightforward: An Fe_xS_y cluster is an Fe_x deltahedron complying with the VB cluster rule having *y* S caps, each connected to a triangular catena by a formal triple bond.

$Fe_4S_3^-$ is the simplest application of the VB cluster model. Each Fe acts as a 3h/2e piece (with six equatorial nonbonding AOs) and each S acts as a 3h/4e piece. Fe_4 is an eight-electron tetrahedron with four triangular faces, each spanned by a two-electron triangular catena. In other words, we have a symmetrical array of four triangular catenae (forming the tire chain), each acting like a 3h/2e piece. Each of three triangular catenae is capped by one 3h/4e S via a formal triple bond. The fourth triangular catena (i.e., the fourth face of the tetrahedron) receives the extra odd electron to become a "nonaromatic" three-electron piece. This analysis predicts the structure forecast by Long and Holm, to be discussed shortly.

That was the easy part! Things get more challenging (and more revealing) in the case of $Fe_6S_y^-$. We now have an Fe_6 octahedron that has *eight* triangular faces but only *six* triangular catenae stretched over the polyhedron. This octahedron in the analogue of $(BH)_6^{2-}$ with $A = 0$ rather than $A = 2$. Unlike the case of the tetrahedron, in which the four triangular catenae are tetrahedrally disposed, matching the symmetry of the molecule, the six triangular catenae of the octahedron are asymmetrically disposed. That is, they are "stretched," to accommodate the eight-face polyhedral structure. The crucial point is that, in the case of $(BH)_6^{2-}$, octahedral (O_h) symmetry is imposed by the inter- and intra-atomic delocalization of 14 skeletal electrons among six asymmetrically disposed triangular catenae. (Remember: Atom-centered hybrids interact.) In VB theory, instead of using symmetry orbitals (or, more generally, symmetry functions) spanning eight octahedral faces) to build the total wavefunction of a symmetrical species, we use low symmetry local AOs or functions (e.g., six three-orbital arrays of six triangular catenae).

The problem is now apparent: We can use *y* of sulfurs to cap *y* of the six triangular catenae spanning the Fe_6 octahedron by *y* triple bonds; however, we do not know the geometry of these six asymmetrically disposed triangular catenae! In addition, we have no idea about the distortion of the metal deltahedron required for the best accommodation of the capping atoms as well as the best preservation of any uncapped triangular catenae. Despite all these shortcomings, the VB model places restrictions on the selection of the global minimum. Ideally, each Fe and each S must be tricoordinated. However, the requisite "stretching" of the triangular catenae in every deltahedron beyond the tetrahedron places a restriction: either Fe or S or, still better, both maintain tricoordination.

This section owes its genesis to a paper by Long and Holm[29] in which the authors suggest structures for laser-ablated iron–sulfur cluster anions of the type $Fe_xS_y^-$. Their proposal was based on a consideration of a cluster database with the assumption that terminal ligands can be removed and the hypothesis that a structure containing all three-coordinate irons is best. The result is that the predicted structures seem to have nothing to do with the undistorted deltahedra of boron chemistry. Our interpretation is that these very structures are actually sulfur-capped boron closo deltahedra in which BH has been replaced by Fe, each $(Fe_3)S$ unit involves a formal triple bond connecting three irons to one sulfur, and the number of excess electrons is one rather than two (i.e., $A = 0.5$ rather than $A = 1$). Moreover, the structure is unsymmetrical because the x sulfur-capped triangular catenae [defined by $3x$ iron valence AOs ($3x/3 = x$ triangular catenae)] are themselves unsymmetrically disposed on $x + b$, where b is an integer greater than zero, triangular faces of the metal $(2x + 1)$-electron deltahedron.

We can use the cluster $Fe_6S_6^-$ to illustrate these ideas. Since the Fe_6 octahedron has six triangular catenae, each having two electrons, the maximum number of capping sulfurs is six. Furthermore, the best structure must have tricoordinated Fe and S. This predicts an Fe_6S_6 adamantane-type structure. Adding the extra electron turns one formal two-electron bond to one formal three-electron bond. This is one of the possible $Fe_6S_6^-$ structures proposed by Long and Holm.[29] The key point is that a structure that seems to comply with the rules of organic chemistry (T-bonding: e.g., the adamantane-like $Fe_6S_6^-$), is actually a manifestation of entirely different metal-bonding principles (I-bonding). This is a recurring theme, which reaches its strongest form in Chapter 39: Nonmetal and metal atoms bind by different mechanisms. Sometimes this difference expresses itself in the differing molecular architectures of isoelectronic molecules. At other times, the difference hides behind apparent similarity.

37.11 Contact Repulsion as the Physical Basis of Bridging

The dimer of HC is the classical organic molecule acetylene, an illustrator of carbon–carbon triple bonding. On the basis of the isolobal analogy of EHMO theory (see Chapter 39), we expect to find $(CO)_3Co{\equiv}Co(CO)_3$ with a cobalt–cobalt triple bond. However, the dimer of $(CO)_3Co$ turns out to be bridged by two added carbonyls.[30a] We explain this by saying that metal bonding requires minimization of interelectronic repulsion and a "double-banana" or "triple-banana" bond suffers greatly from *contact repulsion* of closely spaced overlap densities, as illustrated:

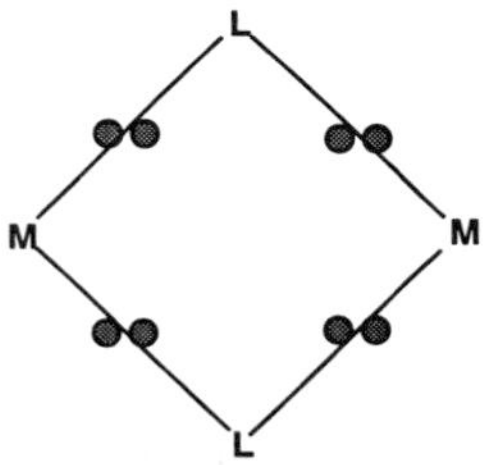

Bridging by two carbonyl ligands effectively transforms a two-membered ring into a four-membered ring. Coulomb exchange repulsion is now reduced as the overlap regions move away from each other.

To better appreciate this phenomenon, we consider two cases in which metal–metal bridging occurs despite the appearance of being unfavorable:

(a) In Figure 37.9, two carbonyls migrate from two vertices of a cluster to bridging positions to "support" what normally would be classified as a metal–metal double bond. This ligand rearrangement reduces the number of I bonds by one. If it were true that "all I bonds are equal," the process would be strongly disfavored. However, in our case, the three reactant I bonds are unfavorable because each is represented by two "parallel" arrows located in the same interatomic space, and this is the hallmark of a *single* I bond that involves large contact repulsion. This is not the case in the product I bonds. Thus, while the I-bond count favors the reactants, the quality of I bonds favors the products.

(b) In Figure 37.10a, a single ligand pair makes an I bond with a single metal hole. According to conventional theory, this is "allowed" because electron–electron repulsion is not explicitly considered. By contrast, such I-bond

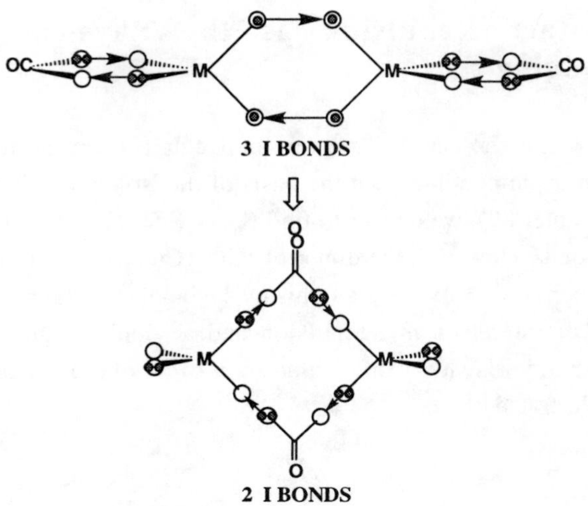

Figure 37.9 The change of the number of I bonds as ligands move from vertices to bridging sites.

is "forbidden" because it involves large contact repulsion. Bridging is the mechanism by which a "forbidden" I bond becomes "allowed." The analogous case of ligand hole interacting with metal pair is also shown in Figure 37.10b. Note the characteristic pattern: Bridging requires the extrusion of one ligand.

One thing that turned me off when I attempted a course in synthetic organometallic chemistry in the early 1970s was the "small" dinuclear cluster $(CO)_3Fe(CO)_3Fe(CO)_3$.[30b] A successful theory of metal bridging must explain

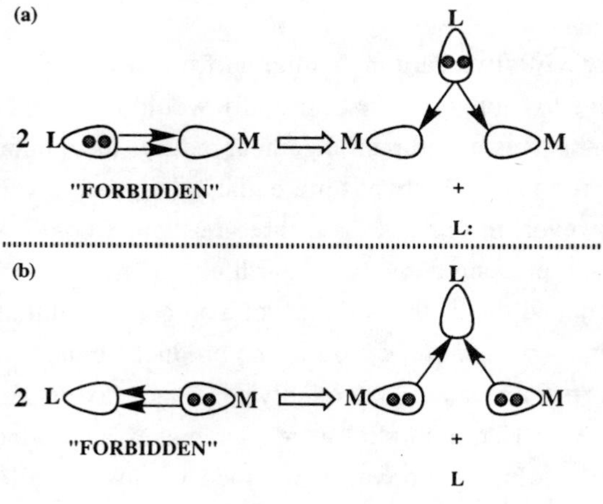

Figure 37.10 Bridging restores favorable I-bonding.

these key observations:

(a) Bridging carbonyls have lower CO stretching frequencies and this indicates that they act as better π acceptors than the nonbridging carbonyls.[31]

(b) Carbonyl migration on the cluster surface has a very low barrier.[32]

(c) Bridging ceases to exist as a more electropositive first transition series metal is replaced by a more electronegative third transition series metal. Among other examples of transition metal carbonyls, the reader is invited to compare the bridged $Co_4(CO)_{12}$ with the unbridged $Ir_4(CO)_{12}$.

The VB formula of diiron nonacarbonyl, formulated as $[(CO)_3Fe](CO)_3[Fe(CO)_3]$, is shown in Figure 37.11. The endo metal hybrids that accommodate the bridging CO fragments have greater star character (2/3)

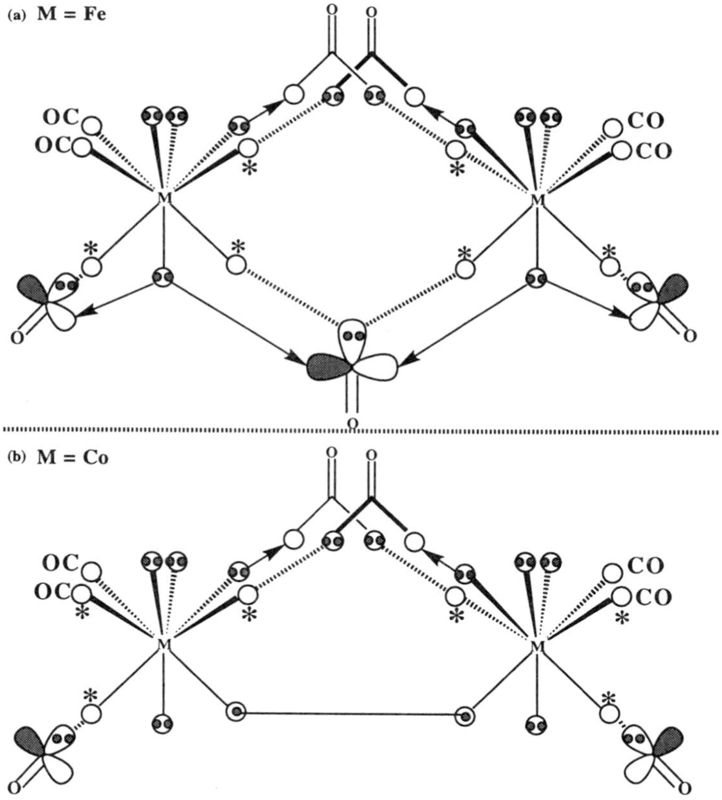

Figure 37.11 The VB formulas of the prototypical bridged dinuclear carbonyl clusters $Fe_2(CO)_9$ (a) and $Co_2(CO)_8$ (b).

than the exo hybrids. This means that the bridging carbonyls act predominantly as π-acceptor ligands, since σ donation to starred AOs is minimal. This simple model explains observations a to c.

Further support comes from the solid state structure of the binuclear cluster $[(CO)_3Co](CO)_2[Co(CO)_3]$,[33] which differentiates bridging from terminal carbonyls in the following ways:

Bridging carbonyls have longer Co—C and C—O bonds and lower CO stretching frequencies.

Terminal carbonyls have shorter Co—C and C—O bonds and higher CO stretching frequencies (ν).

In addition, terminal carbonyls make stronger bonds to the metal than the bridging form.[34] These trends can be explained by keeping in mind three points:

(a) Bonding effected by the radial σ pair causes an increase in ν_{CO} and a shortening of r_{CO} because the radial exo sp pair directed to the metal gains p character (to better overlap with the metal hole) and relinquishes s character to the endo sp AO dedicated to binding the oxygen. As a result, the σ component of the CO bond gets stronger.[35]

(b) Bonding effected by the tangential π hole causes a decrease in ν_{CO} and a lengthening of r_{CO} because acceptance of one electron by the π hole generates a three-electron π antibond (as a result of the presence of an "observer" π lone pair).

(c) A combination of σ-pair and π-hole action has an intermediate effect.

The structural trends can now be rationalized by reference to Figure 37.11 and by noting that the percentage of CO ligands docking on unstarred holes is greater for the terminal than for the bridging sites. In an overall sense, that is, terminal carbonyls have higher ROX character while bridging COs have higher OX character.

If bridging carbonyls act as OX ligands, their action must be turned off and bridging should be abolished as the metal electronegativity increases and OX gives way to ROX delocalization. Thus, for example, replacement of a transition metal by an isoelectronic species of higher electronegativity (e.g., replacing Co by Ir or Fe^- by Co) is expected to shift bridging ligands from edges or faces to the vertices of a cluster:

(a) $Fe_3(CO)_{12}$ has two carbonyls bridging two edges of the triangular metal framework.[36] The appropriate formulation is $Fe(CO)_4[Fe(CO)_3]_2(CO)_2$. By contrast, $Os_3(CO)_{12}$ is a triangular unbridged $Os(CO)_4$ trimer.[37]

(b) $Co_4(CO)_{12}$ has a tetrahedral metal framework with three edge-bridging carbonyls; that is, we have $Co(CO)_3[Co(CO)_2]_3(CO)_3$.[38,39] $Fe_4(CO)_{13}^{2-}$ has three edge- and one face-bridging carbonyl and is represented[40] by the formula $Fe(CO)_3[Fe(CO)_2](CO)_3(CO)^{2-}$. By contrast, $Ir_4(CO)_{12}$ is an unbridged tetrahedral tetramer of $Ir(CO)_3$.[41]

(c) Upon reaction with PPh_3, $Ir(CO)_4$ produces $Ir_4(CO)_{10}(PPh_3)_2$, which now has three edge-bridging carbonyls.[42,43] This is a consequence of the increase of electron density on two Ir centers produced by trading OX (OC—Ir—CO) for ROX (OC—M—PPh_3) delocalization on each of two centers.

(d) The following clusters have an octahedral metal skeleton and either edge- or face-bridging carbonyls:

1. $[Ni(CO)]_6(CO)(BH)_6^{2-}$. There are six edge-bridging carbonyls.[44]
2. $[Co(CO)]_6(CO)(BH)_8^{4-}$. There are eight face-bridging carbonyls.[45a]
3. $[Co(CO)]_3(CO)_3[Co(CO)_2]_3(CO)(BH)_3^{2-}$. There are three edge- and three face-bridging carbonyls.[45b]
4. $[Rh(CO)_2]_6(CO)_4$. There are four face-bridging carbonyls.[46]

In all cases the metal is either an electropositive first transition series metal or a second transition metal with an unusually low ionization potential (Co, Rh, Ir). There is increasing bridging with increasing anionic charge in the case of the cobalt octahedra. By contrast, $Os_6(CO)(BH)_{18}^{2-}$ is an unbridged hexamer of $Os(CO)_3$ plus two electrons.[47]

37.12 Contact Repulsion and the Electronic Structures of Inorganic Rings and Cages

Borazine, phosphazene, and tetrasulfur tetranitride are the main conceptual challenges of main group inorganic chemistry. Experimentalists working in this area have been suspicious that the many theoretical calculations and discussions of these molecules fall short of capturing the physics of the problem. Thus, many have recognized that borazine $(BHNH)_3$, widely advertised as the "inorganic benzene," is a very different beast. In discussing the electronic structure and reactivity of the related cycloboraphosphanes $(R_2PBH_2)_3$, Emsley and Hall point out that "there still remains the nagging doubt that we are missing something and that perhaps . . . the cycloboraphosphane system has

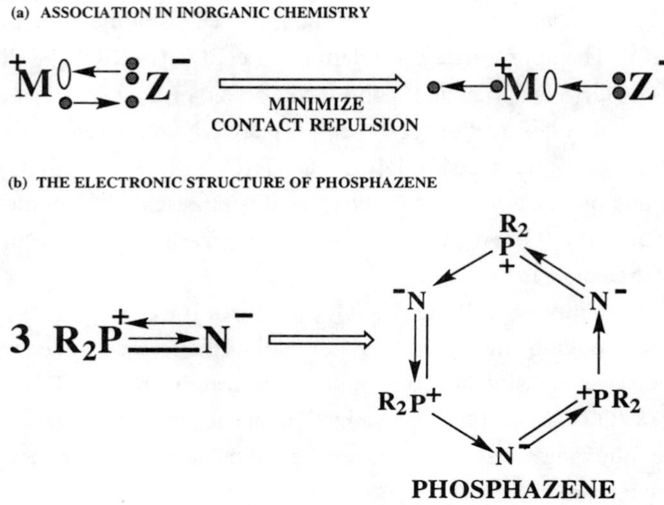

Figure 37.12 Ring formation represents association driven by minimization of contact repulsion.

some form of extrabonding"![48] The impossibility of understanding the bonding of S_4N_4 by conventional theory is projected by the bewildering reactivity of this compound. In a recent monograph, Woolins comments that "the diversity of reactions that S_4N_4 undergoes defies rationalization."[49]

We now know where the problem lies: These prototypical rings and cages are association complexes in which the components are connected by I bonds. In other words, the rings and cages are "hypervalent-like." Furthermore, the driving force for the association is that an inferior is replaced by a superior REL bond, which suffers from less contact repulsion. Thus, the reason for the polymerization of monomers to form rings and clusters is electron–electron repulsion.

We begin the story by displaying the formula of an MZ monomer in which one REL bond is engendered by an rr element of M and a divalent electronegative Z (Figure 37.12a). Two codirected arrows occupy the same interatomic region. Significant reduction of contact repulsion can occur if the arrows are assigned to different interatomic regions. Once again, contact repulsion dictates "bridging," which leads to formation of a ring as illustrated for the case of phosphazene in Figure 37.12b.

The trimerization of R_2PN to form phosphazene and the trimerization of RBNR to form borazine are motivated by the same factor (contact repulsion relief). The same story can be told for innumerable inorganic rings. For

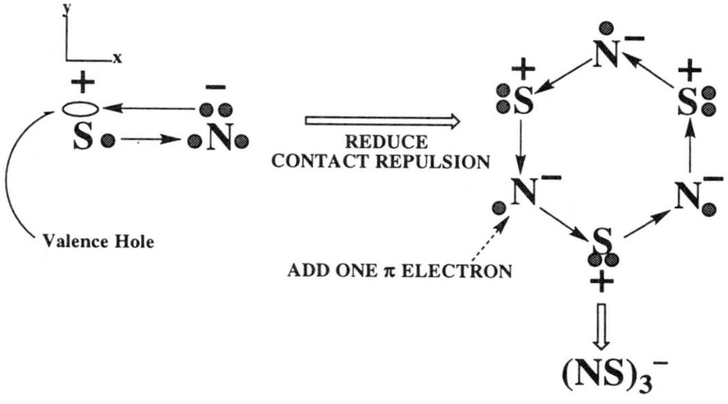

Figure 37.13 The σ electronic structure of $(SN)_3^-$.

example, the formula shown in Figure 37.13 makes clear that trimerization occurs in SN for the same reason as in borazine and phosphazene. Addition of an electron produces the well-known $(SN)(BH)_3^-$ ring, which has 10 π electrons.

The structure of S_4N_4 is derived from the organic cuneane, $(CH)_8$, by adding four valence (skeletal) electrons. It is rather amazing that no work in the literature poses the fundamental question: If the organic cage molecule cuneane is already highly unstable and if adding four electrons can only make the situation worse, how is it possible that S_4N_4 is such a stable molecule (which, however, does detonate to form S_2 plus N_2)?

Figure 37.14 argues that S_4N_4 is actually an association complex of two cis NSSN subunits each produced by dimerization of SN. This association is the result of turning intra-NSSN into inter-SNNS arrows. Again, this affords relaxation of contact repulsion because the arrows move further apart. The simple formalism presented in Figure 37.14 provides the theoretical rationalization of the observation that many phosphorus (or sulfur-isoelectronic) compounds can be derived from the P_4 tetrahedron by inserting fragments or atoms in the edges (e.g., P_4O_6). S_4N_4 can be regarded as $(S^+)_4$ tetrahedron into which one inserted four N^- anions into four edges.[50]

37.13 T Caps, I Caps, and the Stereochemistry of Clusters

The "organic" acetylene is a linear H—C≡C—H molecule that can be formulated as a composite of two carbon anions which combine to form C_2^{2-} and two protons: $C_2^{2-}(H^+)_2$. It is critical insofar as structure is concerned to identify the attachment mode of the two protons. The linear structure gives the answer:

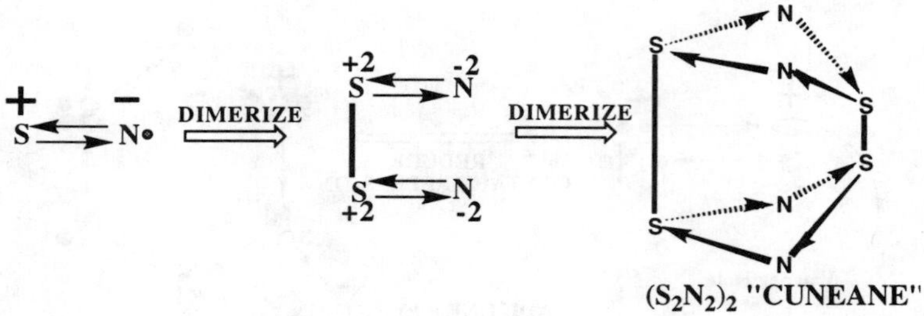

Figure 37.14 Electronic structure of the inorganic cuneane.

The two protons tie up the two "exo," sp-type lone pairs of C_2^{2-}. We say that each proton acts as a T cap to tie up a lone pair, to generate a T bond. This mode of capping is expected when the cap and the substrate atom are electronegative nonmetals, in which case the preferred bonding mechanism is T-bonding.

In one of these few ab initio calculations that "make a difference," Lischka and Koehler[51] computed the structure of the silicon analogue of acetylene and found it to be a "butterfly." This species can also be formulated as $Si_2^{2-}(H^+)_2$ with one important difference: Each proton acts as an I cap to tie up a skeletal cluster pair, to form an I bond. The fundamental difference in shape between C_2H_2 and Si_2H_2 is a consequence of a change in the bonding mechanism. In turn, this change is due to the replacement of an electronegative nonmetal by an electropositive semimetal (always in a relative sense).

We can now formulate a general procedure for predicting mixed cluster architecture:

(a) A molecule of the type M_xN_y, where M is metal or semimetal and N is a univalent nonmetal, is formulated as $(M_x)^{-y}(N^+)_y$.

(b) $(M_x)^{-y}$ is a cluster with magic numbers conforming to the VB cluster rule and made up of three types of "ingredients": face pairs, edge intercatenal links, and vertical exo lone pairs (or odd electrons).

(c) N^+ is a cap that can act either as a T cap (when the average electronegativity of M and N is high) to combine with exo lone pairs or as an I cap (when the average electronegativity of M and N is low) to combine with face or edge skeletal pairs.

Once again, the fundamental point is that *the structure of a mixed cluster is not unique; rather, it varies with atom electronegativity.* The more electronegative the atoms, the greater the cap preference for vertical attachment

(T caps), while the more electropositive the atoms, the greater the cap preference for face or edge attachment (I caps).

The organic molecule H_4C_4 can be formulated as $(C_4)^{4-}(H^+)_4$. The global minimum of the cluster $(C_4)^{4-}$ is the cumulene tetraanion $^{2-}C{=}C{=}C{=}C^{2-}$ because C is an electronegative atom and this means T bonding and electron pair segregation. Each H^+ will act as a T cap hence the global minimum is either $H_2C{=}C{=}C{=}CH_2$ or $H_2C{=}CH{-}C{\equiv}CH$.

The organometallic molecule H_4Sn_4 can also be formulated as $(Sn_4)^{4-}(H^+)_4$. The global minimum of the cluster $(Sn_4)^{4-}$ must be the same as that of the isoelectronic Sb_4 (i.e., a tetrahedron), having four exo lone pairs, four face pairs and two intercatenal links localized on two nonadjacent edges. The four protons will now act as I caps, and they will be distributed in a way consistent with the relative energies of the four face pairs and the two edge pairs. Since it is not always possible to predict whether face or edge pairs will be the preferred receptors of the I caps, it is safest to itemize the global minimum candidates:

(a) If the two edge pairs are destabilized relative to the face pairs, two protons will cap two opposite edges and two more will cap two faces.

(b) If the reverse condition is met, each face will be capped by a proton.

It turns out that the organometallic analogue of the face-capped tetrahedral H_4Sn_4 is known. This is the $H_4[Re(CO)_2]_4$ cluster, studied by Wilson and Bau.[52]

An interesting situation arises when we replace one Si by one C. Now, we still predict a tetrahedral $(CSn_3)^{4-}$ cluster, but the capping strategy changes: One proton functions as a T cap to tie up the exo C lone pair and three act as I caps to tie up either two edges and one face (geometry A) or three faces (geometry B). It turns out that $(CM_3)^{4-}(H^+)_4$, where M = $Os(CO)_3$, an organometallic fragment isolobal to Sn, is thought to be an intermediate in the conversion of an alkyne to an alkane; its likely geometry can be formulated as either A (in which one face cap has slipped to one edge to generate a higher symmetry C_{3v} structure) or B (with the face caps shifted toward the edges).[53]

37.14 The Stabilization of Radical Polymers

Experimentalists have been successful in stabilizing elementary reactive fragments like the diradicals R_2C, RN, and O by combining them with metals via an I bond due to the interaction of one metal pair (lp or rr) with two odd electrons of the oxygen atom. We now seek more far-fetched possibilities: If

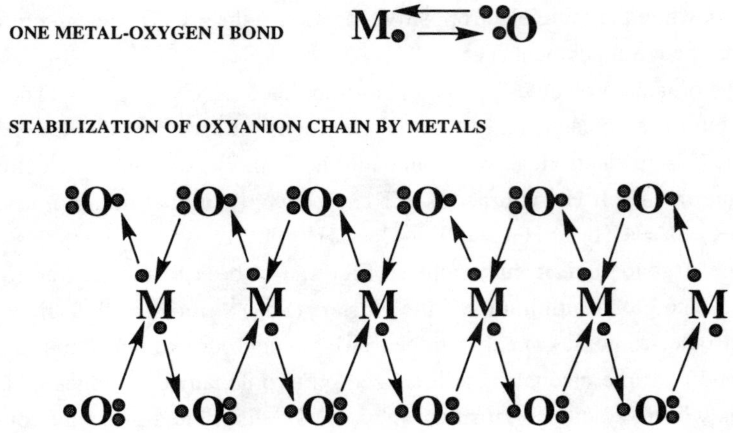

Figure 37.15 Polymeric metal–oxygen association by REL bonds.

one metal pair can bind one oxygen atom, many metal pairs can link many oxygen atoms. The situation is depicted schematically in Figure 37.15.

Are there any hints in the literature that this can indeed occur?

The cluster $[Zr_4(OH)_8(H_2O)_{16}]Cl_8$ studied by Vaughan and co-workers[54] can be formulated as $[Zr_4O_8(H^+)_8](H_2O)_{16}(Cl^-)_8$, and its three-dimensional structure is visualized as follows:

(a) Each Zr uses the square antiprismatic hybridization. There is one endo set of four hybrids directed to four oxygens, which occupy the vertices of one square face of an O_8 cube, and one exo set of four hybrids directed to four coordinated water molecules.

(b) At each O_8–cube vertex, three oxygen endo sp^3 hybrids are directed to the three faces spanned by the vertex. Two of the hybrids contain one electron and the third two electrons. The fourth hybrid is exo-directed and contains one pair.

(c) The oxygen endo hybrids are oriented to obtain two closed-shell trans faces denoted by CS and four open-shell cis faces denoted by OS (Figure 37.16a).

(d) Each OS face is capped by a Zr by two REL bonds, as illustrated in Figure 37.16b. Each CS face is coordinated to four protons by two REL-bonds as illustrated also in Figure 37.16b. The cluster structure (with the hydrogens being "invisible") is shown in Figure 37.16c).

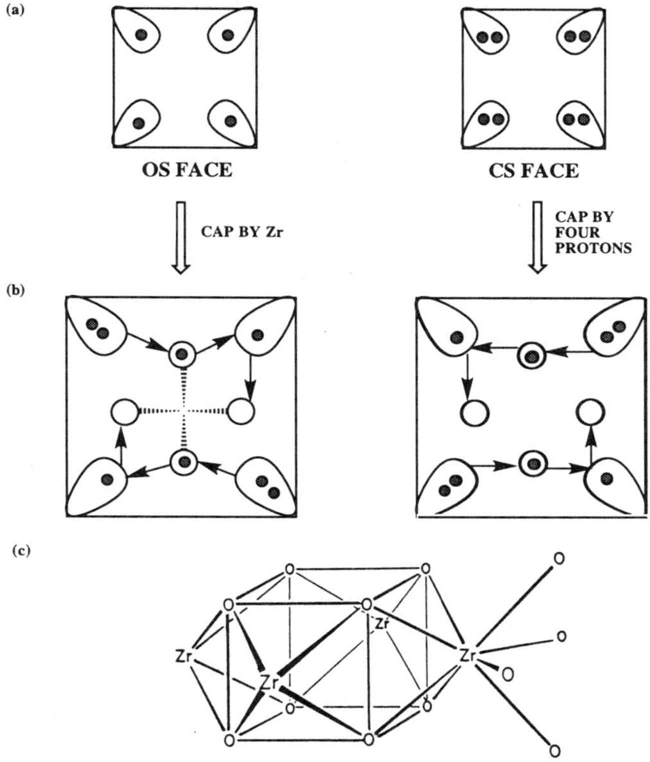

Figure 37.16 Assembly of Zr_4O_8 cube. Experimental structure adapted from ref. 54 with permission of the authors.

References

1. (a) K.Wade, *J. Chem. Soc. Chem. Commun.* 792 (1971); *Adv. Inorg. Chem. Radiochem.* 18, 1 (1976). (b) R.E. Williams, *Adv. Organomet. Chem.* 36, 1 (1994), and references cited therein. (c) R.W. Rudolph and W.R. Pretzer, *Inorg. Chem.* 11, 1974 (1972); R.W. Rudolph, *Acc. Chem. Res.* 9, 446 (1976). (d) D.M.P. Mingos and R.L. Johnston, *Struct. Bond.* 68, 29 (1987).
2. R.J. Gillespie, *Chem. Soc. Rev.* 8, 315 (1979).
3. J.A. Morrison, *Chem. Rev.* 91, 35 (1991).
4. P.R. LeBreton, S. Urano, M. Shahbaz, S.L. Emery, and J.A. Morrison, *J. Am. Chem. Soc.* 108, 3937 (1986).
5. R. Zerger, W. Rhine, and G. Stucky, *J. Am. Chem. Soc.* 96, 6048 (1974).

6. (a) J.D. Corbett, *Chem. Rev.* 85, 383 (1985). (b) G. Cordier and V. Mueller, *Z. Kristallogr.* 198, 281 (1992).

7. (a) A. Schaefer and R. Ahlrichs, *J. Am. Chem. Soc.* 116, 10686 (1994). (b) R.J. van Zee, S. Li, and W. Weltner Jr., *J. Chem. Phys.* 98, 4335 (1993). (c) M.Z. Liao, D. Dai, and K. Balasubramanian, *Chem. Phys. Lett.* 239, 124 (1995). (d) S.C. O'Brien, Y. Liu, Q.L. Zhang, F.K. Tittel, and R.E. Smalley, *J. Chem. Phys.* 84, 4074 (1986). (e) Z.-C. Dong and J.D. Corbett, *J. Am. Chem. Soc.* 117, 6447 (1995).

8. (a) R. Murphy and H.F. Schaefer III, in *Applied Quantum Chemistry*, V.H. Smith Jr., Ed., Reidel Publishing Company, New York, 1986, p. 431. (b) R.A. Whiteside, R. Krishnan, J.A. Pople, M.-B. Krogh-Jespersen, P.v.R. Schleyer, and G. Wenke, *J. Comput. Chem.* 1, 307 (1980).

9. W.-W. du Mont and M. Grenz, *Chem. Ber.* 118, 1045 (1985).

10. (a) P.v.R. Schleyer, E.-U. Würthwein, E. Kaufmann, and T. Clark, *J. Am. Chem. Soc.* 105, 5930 (1983). (b) H. Kudo, *Nature* 355, 432 (1992). (c) A. Märcker, *Angew. Chem. Int. Ed. Engl.* 31, 584 (1992). (d) Stable $PbNa_6$: C. Yeretzian, U. Röthlisberger, and E. Schumacher, *Chem. Phys. Lett.* 237, 334 (1995).

11. P.v.R. Schleyer, E.-U. Würthwein, and J.A. Pople, *J. Am. Chem. Soc.* 104, 5839 (1982).

12. P.v.R. Schleyer, *Pure Appl. Chem.* 56, 151 (1984). It should be noted that Schleyer's computations of lithium-containing molecules represent a welcome deviation from the usual route of using ab initio MO computations to reexamine the same old and tired hydrocarbons and their organic derivatives. Some of these molecules (e.g., $SiLi_4$) are marvelous examples of the VB concepts described in this work.

13. E.E. Bolton, W.D. Laidig, P.v.R. Schleyer, and H.F. Schaefer III, *J. Am. Chem. Soc.* 116, 9602 (1994). See also U. Röthlisberger and M.L. Klein, *J. Am. Chem. Soc.* 117, 42 (1995).

14. (a) B. Guo, K.P. Kerns, and A.W. Castleman Jr., *Science* 255, 1411 (1992) (b) C.H. Wei, B. Guo, H.T. Deng, K. Kerns, J. Purnell, S. Buzza, and A.W. Castleman Jr., *J. Am. Chem. Soc.* 116, 4475 (1994) and references therein.

15. (a) I. Dance, *J. Chem. Soc. Chem. Commun.* 1779 (1992). (b) M.-M. Rohmer, M. Benard, C. Bo, and J.-M. Poblet, *J. Am. Chem. Soc.* 117, 508 (1995). (c) For recent experimental developments, see C.S. Yeh, S. Afzaal, S.A. Lee, Y.G. Byun, and B.S. Freiser, *J. Am. Chem. Soc.* 116, 8806 (1994).

16. (a) L.D. Lower and L.F. Dahl, *J. Am. Chem. Soc.* 98, 5046 (1976). (b) D.F. Rieck, R.A. Montag, T.S. McKechnie, and L.F. Dahl, *J. Am. Chem. Soc.* 108, 1330 (1986), and references cited therein.
17. C.S. Yeh, Y.G. Byun, S. Afzaal, S.Z. Kan, S. Lee, B.S. Freiser, and P.J. Hay, *J. Am. Chem. Soc.* 117, 4042 (1995).
18. (a) J.L. Hoard and J.V. Silverton, *Inorg. Chem.* 2, 235 (1963). (b) G.L. Glenn, J.V. Silverton, and J.L. Hoard, *Inorg. Chem.* 2, 250 (1963). (c) Reviews: S.J. Lippard, *Prog. Inorg. Chem.* 8, 109 (1967); E.L. Muetterties and C.W. Wright, *Q. Rev.* 21, 109 (1967).
19. M.L. Listemann, J.C. Dewan, and R.R. Schrock, *J. Am. Chem. Soc.* 107, 7207 (1985).
20. $[Cr(CO)_5]_4(MeO)_4^{4-}$: T.J. McNeese, M.B. Cohen, and B.M. Foxman, *Organometallics* 3, 552 (1984).
21. $(R_3PCu)_4I_4$: G. Davies and M.A. El-Sayed, in *Copper Coordination Chemistry*, K.D. Karlin and J. Zubieta, Eds, Academic Press, New York, 1983. See also K.R. Kyle, C.K. Ryu, J.A. DiBenedetto, and P.C. Ford, *J. Am. Chem. Soc.* 113, 2954 (1991).
22. $(CpFe)_4(SR)_4$: C.H. Wei, G.R. Wilkes, P.M. Treichel, and L.F. Dahl, *Inorg. Chem.* 5, 900 (1966).
23. $Pb_4(OH)_4^{4+}$: F.C. Hentz and S.Y. Tyree Jr., *Inorg. Chem.* 3, 844 (1964).
24. Organophosphorus cubes: M. Regitz, in *Heteroatom Chemistry*, E. Block, Ed., VCH, New York, 1990, Chp. 17, p.295.
25. K. Isobe and A. Yagasaki, *Acc. Chem. Res.* 26, 524 (1993).
26. For a discussion of the structure and environment of the Fe_4S_4 cube of ferredoxin, see G. Backes, Y. Mino, T.M. Loehr, T.E. Meyer, M.A. Cusanovich, W.V. Sweeney, E.T. Adman, and J. Sanders-Loehr, *J. Am. Chem. Soc.* 113, 2055 (1991).
27. (a) F. Bottomley, D.E. Paez, and P.S. White, *J. Am. Chem. Soc.* 103, 5582 (1981). (b) F. Bottomley and L. Sutin, *Adv. Organomet. Chem.* 28, 339 (1988).
28. P. Barbaro, A. Bencini, I. Bertini, F. Briganti, and S. Midollini, *J. Am. Chem. Soc.* 112, 7238 (1990).
29. J.R. Long and R.H. Holm, *J. Am. Chem. Soc.* 116, 9987 (1994).
30. (a) G.G. Sumner, H.P. Klug, and L.E. Alexander, *Acta Crystallogr.* 17, 732 (1964). (b) F.A. Cotton and J.M. Troup, *J. Chem. Soc. Dalton Trans.* 800 (1974).

31. The characteristic IR frequencies of carbonyl, carbido, and hydrido clusters are tabulated in D.M.P. Mingos and D.J. Wales, *Introduction to Cluster Chemistry*, Prentice Hall, Englewood Cliffs, NJ, 1990, p. 205.
32. E. Band and E.L. Muetterties, *Chem. Rev.* 78, 639 (1978);
33. P.S. Braterman, *Metal Carbonyl Spectra*, Academic Press, New York, 1975, Chapter 7.
34. J.A. Connor, *Top. Curr. Chem.* 71, 71 (1977).
35. P.K. Hurlburt, J.J. Rack, J.S. Luck, S.F. Dec, J.D. Webb, O.P. Anderson, and S.H. Strauss, *J. Am. Chem. Soc.* 116, 10003 (1994).
36. C.H. Wei and L.F. Dahl, *J. Am. Chem. Soc.* 91, 1351 (1969).
37. M.R. Churchill and B.G. DeBoer, *Inorg. Chem.* 16, 878 (1977).
38. J.C. Calabrese, L.F. Dahl, P. Chini, G. Longoni, and S. Martinengo, *J. Am. Chem. Soc.* 96, 2614 (1974).
39. F.H. Carre, F.A. Cotton, and B.A. Frenz, *Inorg. Chem.* 15, 380 (1976).
40. R.J. Doedens and L.F. Dahl, *J. Am. Chem. Soc.* 88, 4847 (1966).
41. B.F.G. Johnson and R.E. Benfield, in *Transition Metal Clusters*, B.F.G. Johnson, Ed., Wiley, New York, 1980.
42. G.F. Stuntz and J.R. Shapley, *J. Am. Chem. Soc.* 99, 607 (1977).
43. V. Albano, P. Bellon, and V. Scatturin, *J. Chem. Soc. Chem. Commun.*, 730 (1967).
44. J.C. Calabrese, L.F. Dahl, A. Cavalieri, P. Chini, G. Longoni, and S. Martinengo, *J. Am. Chem. Soc.* 96, 2616 (1974).
45. (a) V.G. Albano, P.L. Bellon, P. Chini, and V. Scatturin, *J. Organomet. Chem.* 16, 461 (1969). (b) V. Albano, P. Chini, and V. Scatturin, *J. Organomet. Chem.* 15, 423 (1968).
46. E.R. Corey, L.F. Dahl, and W. Beck, *J. Am. Chem. Soc.* 85, 1202 (1963).
47. C.R. Eady, B.F.G. Johnson, J. Lewis, and M. McPartlin, *J. Chem. Soc. Chem. Commun.* 883 (1976).
48. J. Emsley and D. Hall, *The Chemistry of Phosphorus*, Halsted Press, Wiley, New York, 1976.
49. J.D. Woolins, *Nonmetal Rings, Cages and Clusters*, Wiley, New York, 1988.
50. For recent exposition of the apparently incomprehensible and bewildering architecture and reactivity of "heavy" main group molecules, see I. Haiduc and D.B. Sowerby, *The Chemistry of*

Inorganic Homo- and Heterocycles, Vols. I and II, Academic Press, London, 1987.

51. H. Lischka and H.-J. Koehler, *J. Am. Chem. Soc.* 105, 6646 (1983).
52. R.D. Wilson and R. Bau, *J. Am. Chem. Soc.* 98, 4687 (1976).
53. W. Bernhardt and H. Vahrenkamp, *Angew. Chem. Int. Ed. Engl.* 23, 141 (1984).
54. (a) A. Clearfield and P.A. Vaughan, *Acta Crystallogr.* 9, 555 (1956). (b) G.M. Muhta and P.A. Vaughan, *J. Chem. Phys.* 33, 194 (1960).
55. The analysis in section 37.4 predicts edge protonation of one intercatenal link (ICL) of P_4 as found by recent ab initio calculations (J.-L.M. Abboud, M. Herreros, R. Notario, M. Esseffar, O. Mo, and M. Yanez, *J. Am. Chem. Soc.* 118, 1126 (1996). Diprotonation should place two protons on opposite edges.

Chapter 38

The Toy Model

38.1 Rings as Triple-Deck Clusters

The aim of this chapter is to formulate a convenient procedure for the rationalization and prediction of medium and large clusters. We return to the comfort zone of the chemist: electron counts, formal bonds, and stability as opposed to the more refined differentiation among the T, I, and E forms of bonding. For example, what is the structure of P_6? Because heavy p-block atoms "hate" multiple bonding, one might conjecture that the most likely geometry of P_6 is a triangular prism. What about an octahedron or some lower symmetry derivative thereof? At first sight, this does not seem likely because the magic number of the octahedron according to Wade's rule is 14 and P_6 has four more valence electrons. We now outline a procedure for predicting the electron counts consistent with low energy pyramids and prisms using P_6 as the illustrator molecule.

Consider a P_6 square bipyramid made up of a square P_4 ring plus two apical P atoms. We begin with the P_4 ring in which four sp^2-hybridized P atoms combine to form four σ bonds and a four-electron π system. Next, we execute a 90° orbital rotation at each P atom as indicated in Figure 38.1a. The effect of this rotation is the creation of a P_4 *triple-deck cluster* in which the three decks are labeled L (lower), T (tangential), and U (upper) as in Figure 38.1b. Each of L and U is a *square face* formed by the overlap of sp^2 hybrids. T is a *square catena* formed by the overlap of the tangential 3p AOs. Each catena is tetrasynaptic (4h); that is, it has four available AOs with which to bind some other fragment, as illustrated in Figure 38.1c.

We can now allocate the 12 electrons of P_4 to U, T, and L to effect coordination of one apical P by the U face and a second apical P by the L face. Since the T catena is inactive, it must be itself stabilized; that is, it must have an aromatic electron count. To make the analogy to known chemistry as clear as possible, we transfer one electron from each apical P to the P_4 triple-deck cluster. We now have a total of 14 electrons in the dianionic P_4 cluster plus two cationic P caps with two electrons each. We distribute the 14 electrons to the P_4 cluster, to obtain the configuration $U^4T^6L^4$. As a result, each of U and L can make a *formal* triple bond with one P^+, which acts as a two-electron fragment. A formal triple bond has 4+2 electrons.

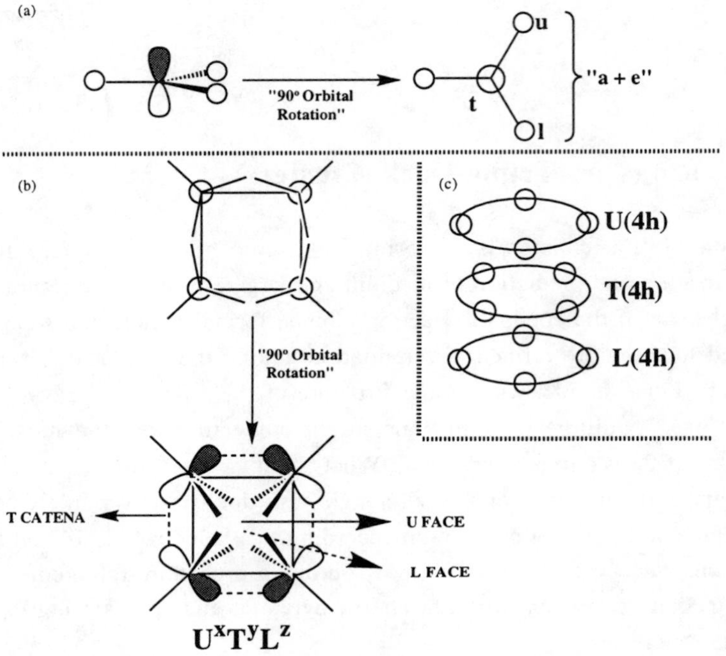

Figure 38.1 Generation of a triple-deck cluster from a planar ring constructed from sp^2 -hybridized atoms by rotation of the sp^2 ensemble.

The way in which a P^+ cap "fits" a P_4 face (U or L) by virtue of a perfect hole–pair match is shown in Figure 38.2. In this representation P^+ employs two of its three available P3p AOs. The third vacant P3p acts to relieve interelectronic repulsion engendered by CT delocalization in the active AOs. The important point is that even though we use only two of the three P3p AOs, we end up with a triple I bond (six arrows) that glues all four atoms of the face to the single apical P^+. The picture simply acquires more detail if the third P3p AO becomes active insofar as CT delocalization is concerned. The I-bond count remains unaltered, and the only difference is that the two pairs of the U face delocalize into two different vacant P3p AOs, instead of to a single one.

We formed a triple I bond by feeding six electrons into three P3p and four facial AOs. Having emphasized the I character of the bonding, we can now retrench in language familiar to the chemist by saying that the six-electron bond just described is formally equivalent to an "organic" triple bond, insofar as electron count is concerned. In MO terms, the orbitals of P^+ and one P_4 face allow the formation of a formal six-electron triple bond through the symmetry-

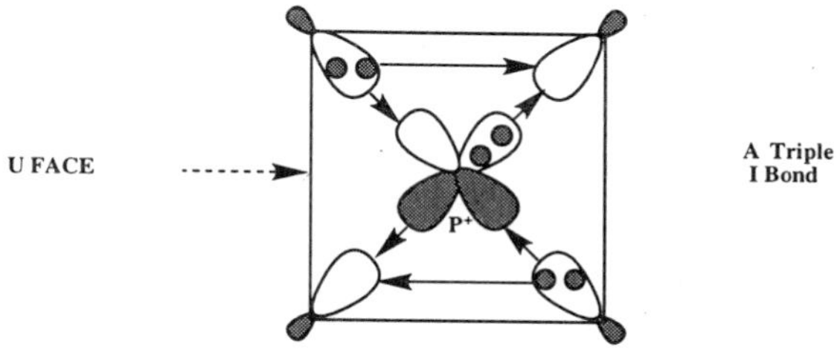

Figure 38.2 A triple I bond connecting an apical P to one face of a P_4 three-deck cluster. Two arrows count for one I bond.

allowed overlap of the three P3p and the three lowest energy facial P_4 MOs. We conclude that square bipyramidal P_6 is a viable candidate for the global minimum and that this scenario will be increasingly more plausible as P is replaced by a more electropositive congener and the system moves toward the I-bonding limit.

38.2 Pyramids and Prisms by Stacking Triple-Deck Clusters

Our analysis of square bipyramidal P_6 suggests that pyramids and prisms can be viewed as stacked triple-deck clusters and caps that are linked by formal multicenter multiple bonds. In our P_6 example, the P_4 cluster has a T catena with a Hückel aromatic electron count (six), and it is linked by two formal triple bonds to two capping P atoms. Thus, P_6 is a *cumulene* system that can be represented as follows:

$$\underbrace{(P) \equiv \overset{\vdots\vdots}{(P)_4} \equiv (P)}_{(P)_6}$$

The stability of a bipyramid depends on the number of electrons in the U and L faces plus the number of electrons contributed by the two apical atoms plus the number of the electrons that occupy the T catena, Z. In our case we have:

$$MN = 12 + Z$$

with $Z = 2$ or 6 (even T catena) or 4 (odd T catena).

When $Z = 2$, MN = 14, and this duplicates Wade's rule. More interesting is the $Z = 6$ family represented by the square bipyramidal (CpNi)[$(NiCp)_4$](CpNi) in which CpNi is analogous to P.[1] Hexagonal and pentagonal bipyramids have also been reported. Two examples are $CpMoAs_5MoCp$[2] and $Cp^*MoP_6MoCp^*$.[3] The first can be formulated as a 12+5 and the second as a 12+ 6 electron system. To better understand these formulations, we present Figure 38.3, which depicts the electronic structure of a cluster in the schematic fashion.

The synapticity of each U and L face, T catena, and cap is specified and next to it, in a separate square, we show the corresponding number of electrons. Thus, in the case of square bipyramidal P_6, one immediately sees that a trisynaptic cap and a tetrasynaptic face are connected by a six-electron formal triple bond and that the even T catena contains $4N + 2$ electrons. The representations shown in Figure 38.3, called *toy formulas*, answer in a very explicit fashion the following important questions:

(a) What is the difference between the 14-electron $(BH)_6^{2-}$ and the 18-electron P_6 square bipyramids? Four additional electrons in the T catena, because this preserves the aromaticity of T (since the electron count must be $4N + 2$).

(b) What is the difference between an 18-electron triangular prism and an 18-electron square bipyramid? The toy formulas speak for themselves.

(c) How can we prepare stable derivatives of these three different species? All we have to do is replace the trisynaptic two-electron P^+ cap by an isolobal fragment (i.e., a piece with the same synapticity and the same number of valence electrons).

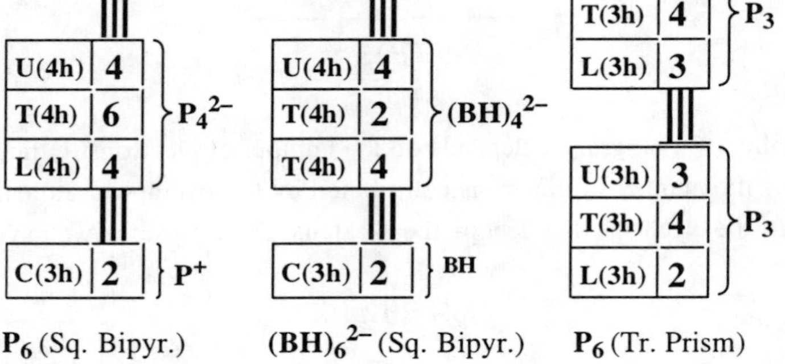

P_6 (Sq. Bipyr.) $(BH)_6^{2-}$ (Sq. Bipyr.) P_6 (Tr. Prism)

Figure 38.3 Toy diagrams for selected clusters.

In this light, we can now appreciate the electronic structures of $CpMoAs_5MoCp$ and $Cp^*MoP_6MoCp^*$ by noting that CpMo is a flexible "piece" that can act either as a 3h/1e or a 3h/0e fragment as illustrated in Figure 38.4. Much as ground divalent carbon responds to four hydrogen atoms by being promoted to the tetravalent state to make four bonds, CpMo attains the electronic configuration that leads to the generation of two interfacial triple bonds. As a result, the first is a 12+5 and the second a 12+6 bipyramid. Because the T catena of the $CpMoP_5MoCp$ cluster has an odd (rather than an even-aromatic) electron count, the As_5 undergoes Jahn–Teller distortion and ends up resembling the organic analogue "ethene plus allyl."

Recently, Scherer and co-workers synthesized an isoelectronic derivative of the P_5 sandwich in which the two capping organometallic fragments are different, one acting as 3h/0e $[(CO)_3Mo]$ and the other as 3h/1e (Cp^*Fe).[4] To establish two interfacial triple bonds, there is a transfer of one electron from the T catena to the face accommodating the $(CO)_3Mo$ fragment. As a result, we get not only two interfacial triple bonds but also a Möbius aromatic T catena. Therefore, the P_5 ring retains its integrity unlike the preceding case. Grimes has

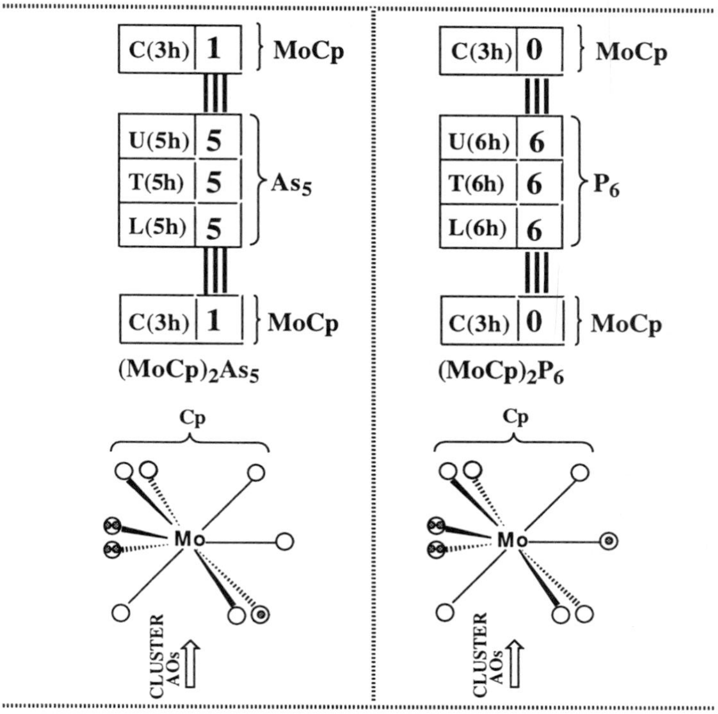

Figure 38.4 The two possible modes of action of the CpMo fragment.

also reported some interesting sandwiches to which one can apply the same techniques.[5]

38.3 Closoboranes Revisited

The tire chain model of borane clusters is revealing of the general physics of deltahedral clusters but is short on specifics, a deficiency now remedied by the toy model. The toy formulas of representative closoboranes are shown in Figure 38.5 and 38.6, and the corresponding compact formulas are shown below.

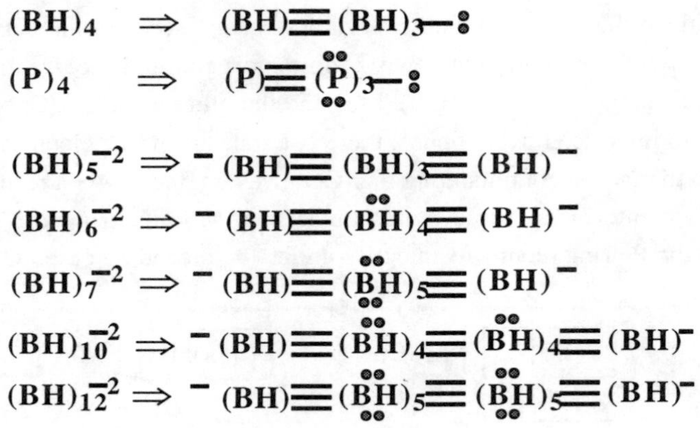

Since every deltahedron can be formulated as either a pyramid or a capped antiprism, the toy model is ideally suited for these systems. Note some key trends:

Each odd T catena has four (Möbius aromatic) electrons and each even T catena has two (Hückel aromatic) electrons.

A cap is connected to a face and a face is connected to a second face by a formal triple bond.

Electrons are allocated to U and L faces ahead of the T catena because the former two are made up of lower energy sp^2 hybrids while the latter is made up of higher energy 2p AOs.

We now have answers to two vexing questions:

(a) Why does the tetrahedron have two different magic numbers, neither of which obeys Wade's rule? Because the eight-electron species is a monocapped triangular ring in which there is one free face with an aromatic count, while the

T catena is unoccupied. Adding four electrons to the Möbius T catena generates the 12-electron tetrahedron.

(b) Why is it that closoboranes conforming to Wade's rule exist and yet are kinetically unstable? The $(BH)_5^{2-}$ closoborane is kinetically unstable because it has a vacant T catena. Delivery of four electrons is expected to generate a 16-electron, stable trigonal bipyramidal cluster.

A closoborane can be produced by starting with the neutral $(BH)_n$ molecule and replacing two BH by two CH fragments to form a carborane. Carborane derivatives, such as the dicarbollido group, $C_2B_9H_{11}^{2-}$ are often used as metal ligands. According to our approach, the electronic structure of this group is obtained by removing the upper (or lower) BH cap core of the icosahedral closododecaborane, $(BH)_{12}^{2-}$ and relinquishing the two cap electrons to the U (or L) face (Figure 38.6). This creates an uncapped face in which, additionally, two

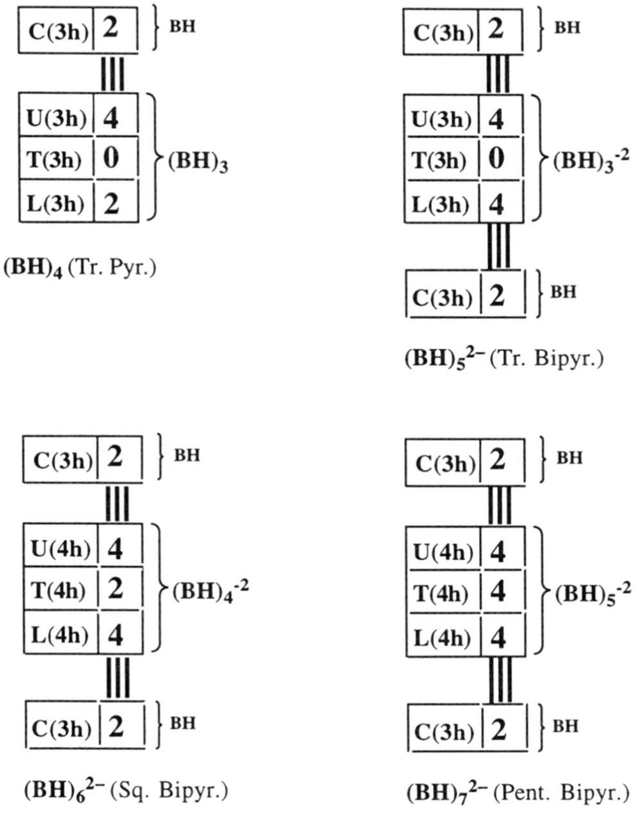

Figure 38.5 Toy diagrams for representative closoboranes.

BH groups have been replaced by two CH groups. This "open" face has an aromatic count of six electrons. Hence, the dicarbollido group is a 5h piece with six electrons analogous to the cyclopentadienyl anion. The binding mode of the dicarbollido group can be inferred from MO calculation of the frontier orbitals. With the toy model, not only can we identify the valence state of a group by falling back on first principles, but we can also tell what the rest of the electrons are doing, how they can become activated, and so on. The formula for the icosahedral closoborane in Figure 38.6, which takes up less than a page, says much more than volumes of MO outputs.

There is still a key question: According to the toy model of closoboranes, faces are connected by formal triple bonds. However, these are preserved in either prismatic or antiprismatic stacking of the rings that make up the cluster. Why is the antiprismatic arrangement better? The answer has been given before: Eclipsed interaction of two sp^3 sets is ideal for T-bonding while staggered interaction of two sp^3 sets is ideal for I-bonding. This is how the triple bond of acetylene differs from that of P_2. A multicenter antiprismatic is analogous to a staggered two-center triple bond, and a multicenter prismatic is analogous to a two-center eclipsed triple bond. In other words, the switch from T- to I-bonding is signaled by "orbital rotation," which can be either "invisible" (acetylene vs. P_2 case) or "visible" (prismatic vs. antiprismatic or deltahedral boron clusters).

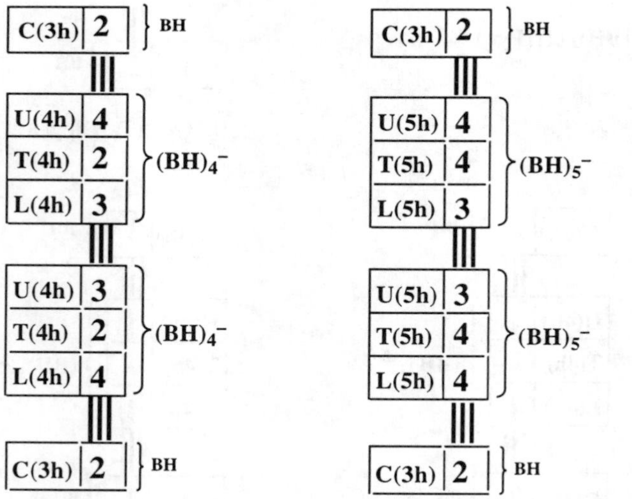

$(BH)_{10}^{2-}$ (Bicapped Sq. Antiprism) $(BH)_{12}^{2-}$ (Bicapped Pent. Antiprism)

Figure 38.6 Toy diagrams for representative closoboranes.

38.4 The Toy Game and the Magic Numbers of Metal Clusters

The characteristic feature of coordination chemistry is that individual "fragments," or "pieces," retain their gross identity within diverse species. Thus, for example, we can identify an $Fe(CO)_3$ unit in mononuclear organometallic complexes, in binuclear clusters, in high nuclearity clusters, and so on. Hence, every chemist is comfortable with the notion of transferability of pieces from one system to another. The key question is: What are the rules for deciding whether a given combination of pieces is successful? What we really want is not the post facto rationalization of the magic number of, say, ferrocene, but, rather, a new way of thinking about ferrocene that will permit us to design new ferrocene derivatives having different magic numbers.

The *toy game* leads to a rational design of new chemistry. Its conventions and rules are founded on MOVB theory in which MOs replace AOs but VB methodology is retained. The basic guidelines are as follows:

(a) Atoms, fragments as well as rings can be formalized as three-deck (TD), two-deck (BD), or one-deck (OD) "pieces."

(b) A TD piece has two faces, namely, the U and L faces, along with a T catena. The latter can be either a single catena or a multiplet of *interacting* catenae, the latter situation becoming commonplace in transition metal clusters. A BD piece has also two faces but no T catena. An OD piece has simply one face.

(c) A TD piece has an exo catena, denoted by R*. In most applications, this contains lone pairs or bond pairs and is full. Hence, it is not shown. When R* is "uncapped" (vacant), it can accommodate any number of electrons that U, T, and L cannot accept.

(d) Each face has a synapticity *jh* and an associated electron count. The symmetry-adapted MOs of a face have a key property: Each face with synapticity 3h or greater has one a-type plus one (or more) e-degenerate MOs available for cluster bonding. Typical examples are given below. Note that attention is focused on the a and e orbitals. For example, a tetrasynaptic face has four symmetry MOs but is regarded as "a + e" face by tentatively discounting the fourth b-type MO.

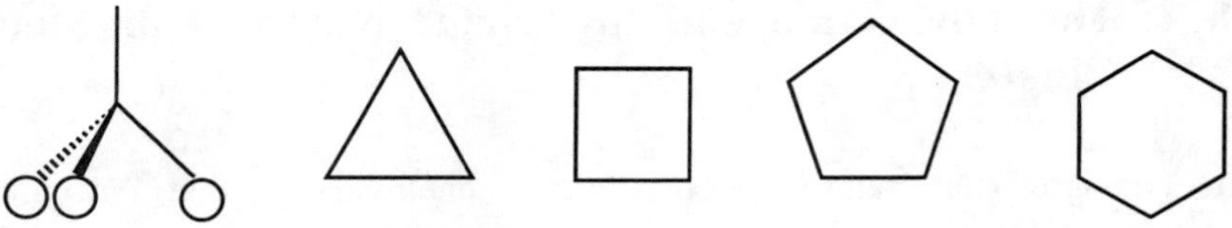

C_{3v}	C_{3v}	C_{4v}	C_{5v}	C_{6v}
"a + e"	"a + e"	"a + e"	"a + e" or "a + e + e"	"a + e" or "a + e + e"
3h	**3h**	**4h**	**5h**	**6h**

(e) The T catenae of TD pieces, which are formed by the overlap of tangential np AOs accommodate either 4N (odd catenae) or 4N + 2 (even catenae) electrons.

(f) Uncapped faces have a Hückel aromatic electron count. However, a Hückel antiaromatic count is also acceptable, provided a distortion exists that renders the face crossed antiaromatic.

(g) Pieces can be combined to form larger pieces by matching faces. This matching depends on synapticity and electron count, and the rules of combination are given in Table 38.1. In all case, two faces of diverse synapticities are connected by a formal triple (6 electrons) or quintuple (10 electrons) bond. Because each face is a Hückel aromatic system, a multiple multicenter bond connecting two faces (or one face and one cap) must have $4N + 2$ electrons, to ensure that each face can attain an aromatic electron count. Note that formal triple or quintuple bonding is guaranteed for faces of different synapticities that can successfully match in a one-to-one sense their "a" and "e" orbitals. For example, a 4h and a 3h face can combine to form a formal triple bond with six electrons.

Table 38.1 The Face-Matching Rules and the Triple-Bond Road

Matching Faces	Number of Electrons
F(3h)–F(3h)	6
F(4h)–F(3h)	6
F(4h)–F(4h)	6
F(5h)–F(3h)	6
F(5h)–F(4h)	6
F(5h)–F(5h)	6, 10

ANTIPRISMATIC OR PRISMATIC TUBE

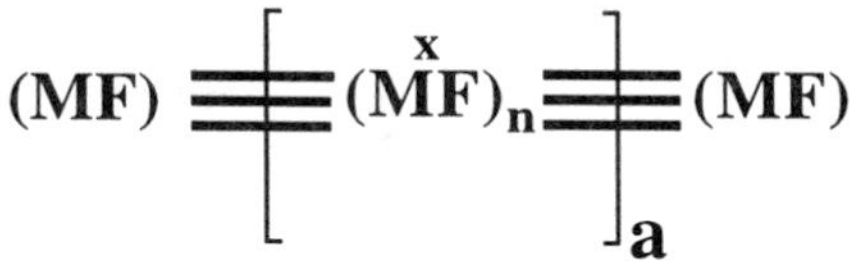

$$\mathbf{MN = 2n(a+1) + ax}$$

$$\mathbf{x = 4N+2 \ (n\ even)\ or\ 4N\ (n\ odd)}$$

Figure 38.7 The selection rules for stable tubular clusters: MF = metal fragment.

The compact formulas of the closoboranes and the toy game just defined lead directly to the formulation of the magic numbers of metal clusters spelled out in Figure 38.7. There is no longer a single magic number because there are three variables, a, n and x, with x, the number of electrons in the T catena of $(MF)_n$, being a function of n. As a simple example, consider how the Wade count of 14 skeletal electrons for $(BH)_6^{2-}$ comes about. The formula is:

$$MN = 6(a + 1) + ax$$

where $n = 4$, $a = 1$, and $x = 4N + 2$ with $N = 0$ because n is even. Indeed, this derivation has nothing to do with the conventional derivations of the magic numbers of borane clusters, and it forecasts the existence of multitudes of stable clusters, some of which have already been characterized.

38.5 The Rational Design of Metal Tubes; The Chevrel Phases

What is the strategy for making a triangular antiprismatic tube? We begin by considering the trigonal bipyramidal closopentaborane shown in Figure 38.8a as a model of a short "tube." The problem is immediately obvious: the tube will be kinetically unstable because the T catena is unfilled. The situation can be improved by replacing the three- by a four-membered boron ring to obtain the square bipyramidal "tube" shown in Figure 38.8b. Chevrel tubes are good examples of clusters that owe their stability to filled T catenae plus interfacial triple bonding, much like the closoboranes.

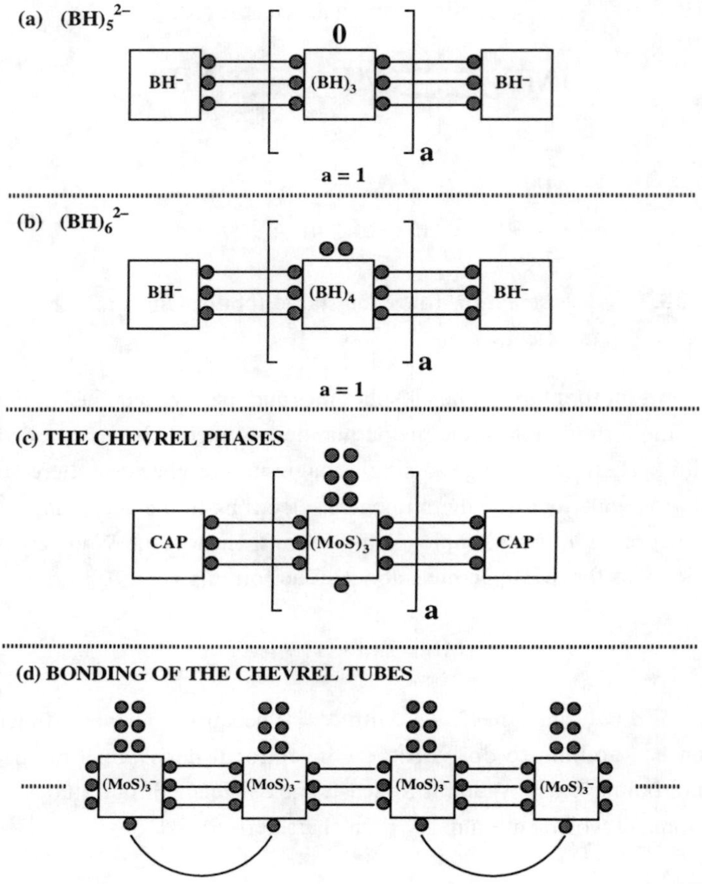

Figure 38.8 The analogy between closoboranes and Chevrel phases.

Mo_6X_8, Mo_9X_{11}, and $Mo_{12}X_{14}$ form a progression of structures in which (Mo_3X_3) rings, each being a triangular array of metal atoms edge-bridged by three X atoms (X = S, Se, Te), are stacked in an antiprismatic fashion with two X groups capping the terminal faces (Figure 38.9). These compounds are often referred to as the "Chevrel phases."[6–14]

The limiting case is represented by $M_2Mo_6X_6$, an infinite tube with $(Mo_3X_3^-)_{inf}$ stoichiometry (M = In, Tl, Na, K; X = S, Se, Te).[13,14] What is the electronic structure of the tube, and how can we modify it to obtain interesting electrical properties? This problem has been attacked by conventional MO band theoretical methods. The scenario always remains the same: Many orbitals, symmetry labels, and one-electron interactions, and few insights beyond what the experimentalist can obtain by extrapolation. By contrast, our interest lies in producing a formula that speaks for itself.

The toy formula for the infinite tube is given in Figure 38.8c. The analogy to closoborane clusters should be obvious. Nonetheless, since our aim is to teach a procedure that is applicable to every cluster, we will consider in detail the derivation of this formula. We begin by formulating each repeating unit as $(Mo^{2+})_3$ coordinated to $(S^{2-})_3$, and we seek to understand the structure of the metal triangle. We assume that each Mo acts with its five d and one s AO. The combination of the d_{z^2} with the s AO generates a radial orbital and a tangential totally symmetric orbital, r and t, respectively. The U and L faces are constructed by mixing the r with the π-type d_{yz}, and each face accommodates three electrons. This leaves out four tangential AOs per Mo: d_{xy}, d_{xz}, $d_{x^2-y^2}$ and t. As a result, we have four (rather than one) interacting tangential catenae:

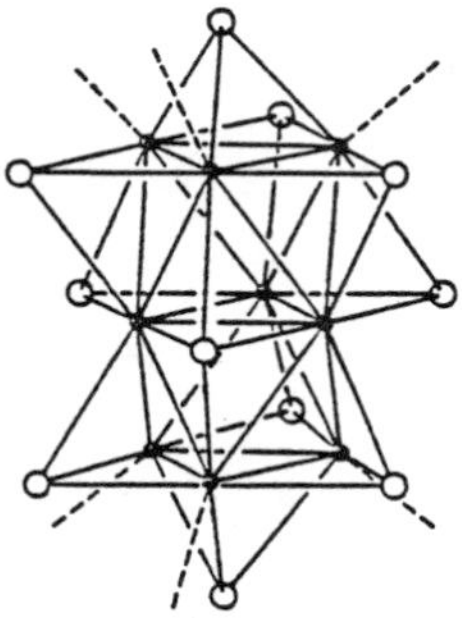

Mo_9Se_{11}

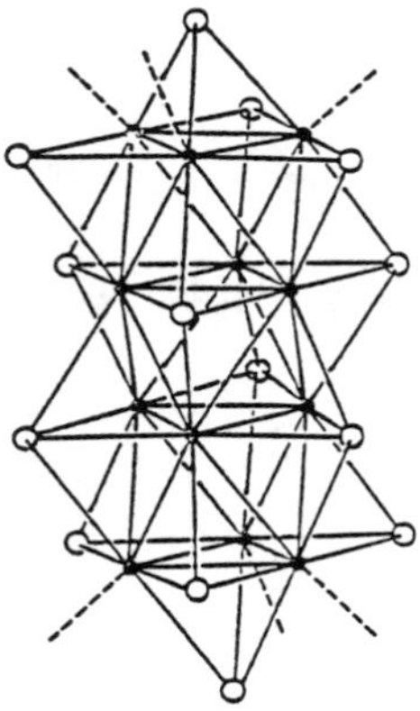

$Mo_{12}Se_{14}$

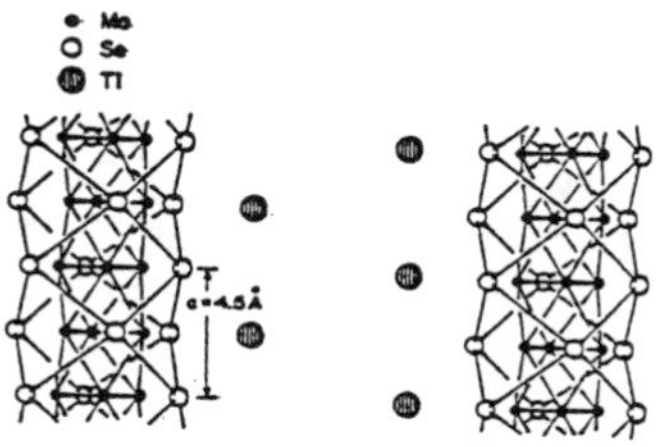

$Tl_2Mo_6Se_6$

Figure 38.9 Examples of Chevrel phases reproduced with permission from ref. 6. Copyright 1983 American Chemical Society.

T_M, $T_M^{\#}$, T_H, and T_H^* where the subscript indicates the type (i.e., Möbius or Hückel). An M-type catena accommodates 4*N* and an H-type catena 4*N* + 2 electrons. The four interacting catenae are said to define the T supercatena. The results are shown in Figure 38.10. Taking into consideration that the MOs of the T_H^* manifold are somewhat higher lying because of the predominant s character, we arrive at the following conclusions on the basis of first principles alone:

(a) The T supercatena of the Mo^{2+} ring can accommodate a *maximum* of 12 electrons because we have two Möbius (four electrons each) and two Hückel (two electrons each) catenae that make up the T supercatena. The "accommodating" low energy orbitals are enclosed in a square. In our case, only three of the six low energy T MOs are occupied. Hence, every T supercatena has six electrons. Thus, every Mo_3 subunit has 3+3 facial plus 6 T-catenal electrons, with one

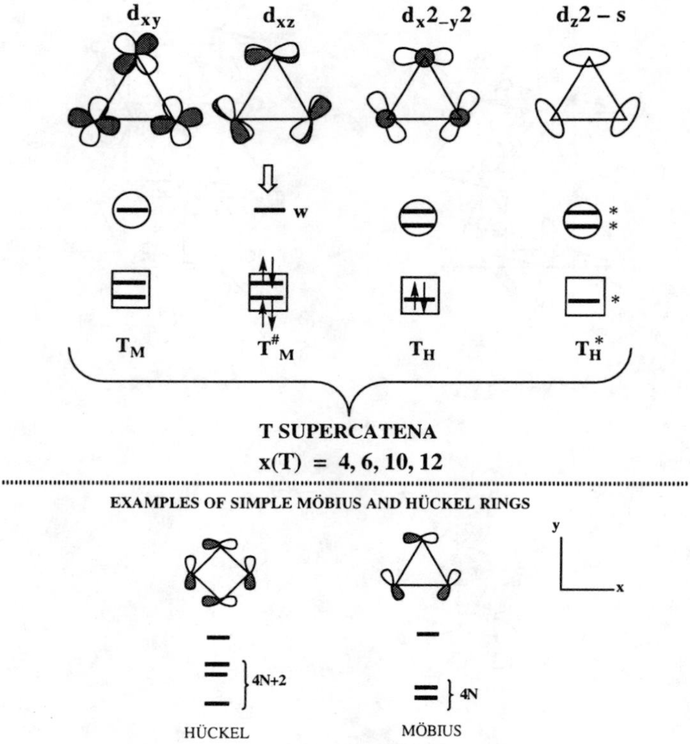

Figure 38.10 The T supercatena of the Mo_3 triangle is made up of four interacting tangential catenae: squares, bonding MOs; circles, MOs.

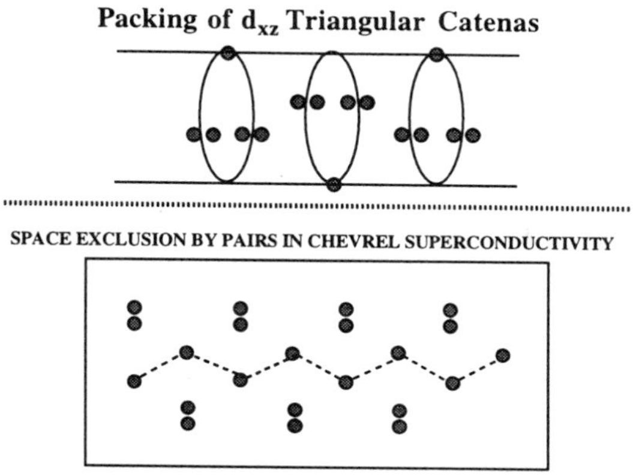

Figure 38.11 The critical tangential catena of the infinite chain $(Mo_3S_3)^{-1}$ and the distribution of the 5 electrons. Two pairs exclude space and induce superconductivity of the odd electrons.

additional odd electron awaiting assignment.

(b) The three sulfur dianions offer nine S3p electron pairs that, ideally, should be matched by nine holes of the T supercatena. This ideal is not met because the antibonding MO of $T_M^{\#}$ cannot match any one of them. This leaves out eight metal holes to accommodate eight sulfur pairs.

(c) The single unoccupied $T_M^{\#}$ MO can accept an additional odd electron because, although it is a high energy T-catenal MO, it is properly oriented for effecting *intercatenal* bonding between two adjacent metal rings. This point has been made in an EHMO paper on the subject.[6] As a result, the structure of the infinite tube can be represented as shown in Figure 38.8d.

(d) The electrical properties of the Chevrel tube can be understood by focusing on the $T_M^{\#}$ catenae, which contain the odd electron of each metal ring, and the way in which they define, in part, the surface of the tube. Our model is shown in Figure 38.11. The measured superconductivity of these phases is attributed to the exclusion by two neighboring bonding electron pairs of space into which the high energy, singlet-coupled odd electrons can be scattered. Thus, superconductivity involves the motion of odd electrons within "surface corridors" in which the $T_M^{\#}$ bonding pairs play the role of the "pillars."

In summary, the important thing here is that we now have a vivid picture of the infinite $(Mo_3S_{\overline{3}})$ chain. We have a structure–function theory, which dispels the illusion that there is one single magic number. Indeed, we expect to be able to prepare stable Chevrel tubes that differ only in the number of T-catenal electrons (see Figure 38.10). The infinite tube discussed earlier has $x(T) = 6$. A cursory examination of the literature reveals that species with $x(max) = 12$ is also known. Twelve is the maximum number of electrons the T supercatena can accommodate in the bonding MOs of its components (see Figure 38.10). Thus, $TlMo_3Se_3$ with 13 (= 3 + 3 + 6 + 1) is isostructural to $TlFe_3Te_3$ with 19 (= 3 + 3 + 12 + 1) metal electrons![15]

38.6 The Diamond–Square–Diamond Rearrangement and the Difference Between Si and BH

Wade's rules are empirical correlation and not theory. The recognition that known borane clusters have shapes analogous to those of metal carbonyl clusters under the assumption that HB is isolobal and isoelectronic to $(CO)_3Fe$ is a "connection" but not an "interpretation." A real understanding of how borane clusters are bound and, thus, why HB acts like $(CO)_3Fe$, should allow us to answer the following fundamental questions:

(a) Why are borane clusters fluctional? Why do they undergo with extreme facility rearrangements such as the diamond–square–diamond (DSD) interconversion when the latter is formally a "forbidden" reaction?

(b) If BH is isolobal with $Fe(CO)_3$ and Si, why are there no stable silicon deltahedra? Why are there no stable "alumane" clusters?

(c) Ab initio calculations by those who develop new computational methodology suggest that the models of bonding of mononuclear organometallic complexes are not supported by the calculational results.[16] Why should one expect the same models to be successful with polynuclear clusters other than by accident?

The trigonal bipyramidal (TBP) closopentaborane can be visualized either as a stack of "BH plus $(BH)_3^{2-}$ plus BH" or as a derivative of the square pyramid (SP) "BH plus $(BH)_4^{2-}$" as shown in Figure 38.12. The latter formulation exposes immediately why the SP mutates to the TBP form: The uncapped L-face is antiaromatic and undergoes a rhombic distortion much like square Li_4 does. This is the basis for the DSD rearrangement discovered by Lipscomb.[17] The process is facile simply because the uncapped square L face itself is a crossed-antiaromatic quadrupolar diradical (see Chapter 35), and the rearrangement is nothing other than a rhombic relaxation of four electropositive atoms connected by two I bonds.

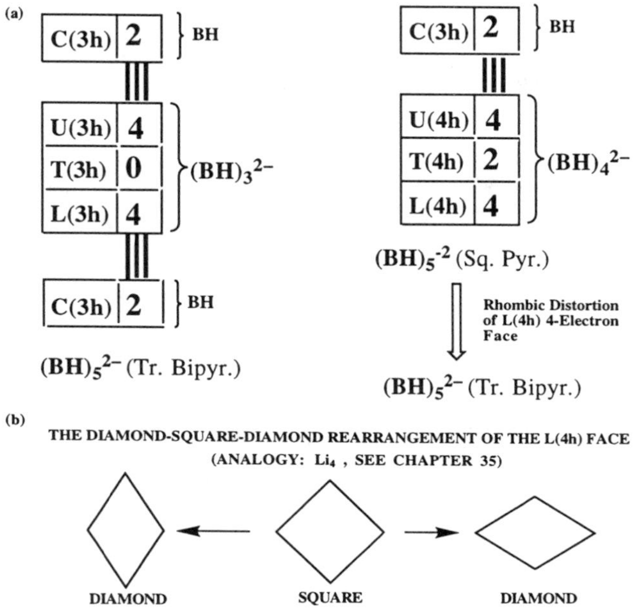

Figure 38.12 Distortion of a square bipyramid is driven by the rhombic distortion of the L(4h) exo antiaromatic face.

The story does not end here. To produce the TBP from the SP form, we need to follow up the in-plane rhombic distortion by a butterfly distortion, which amounts to lifting the two borons at the ends of the long rhombus diagonal above the plane. Since we are aiming at understanding the basic physics of this process, the simplest model is the transformation of planar square $(BH)_4$ to a tetrahedron.

Figure 38.13 shows the toy model of square planar $(BH)_4$, and it is immediately apparent that eight electrons can never be distributed in a way that will yield two aromatic faces and one aromatic T catena. One of these three elements must go antiaromatic. Once again, we see that one face (here, the L face) must accommodate four electrons and be antiaromatic, while the other face is aromatic. Rhombic distortion places the two electron pairs of the antiaromatic L face at the ends on the long diagonal. Here is the essential point: Because the sp^2 hybrids that make up the faces have strong spatial overlap, the butterfly distortion sets in, resulting in (a) the diminution of interpair exchange repulsion in the L face and (b) the enhancement of the aromaticity of the U face. This is why the preferred structure of $(BH)_4$ is a tetrahedron and not a rhombus.

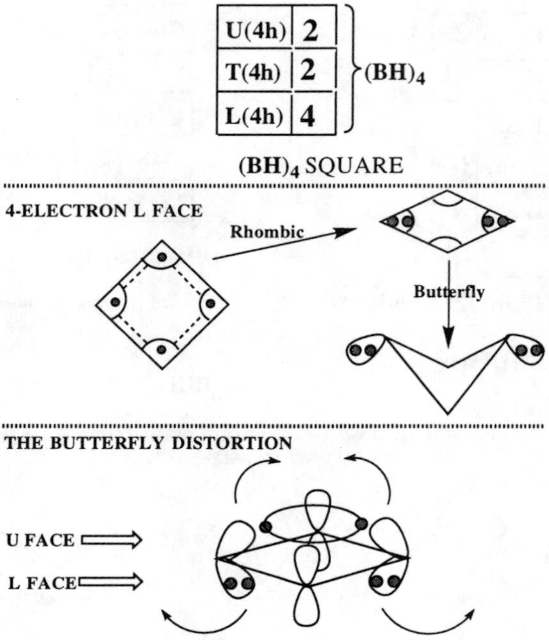

Figure 38.13 Rhombic distortion coupled to butterfly distortion relieves interpair repulsion and transforms square into tetrahedral $(BH)_4$. This is the simplest explanation of the physical basis of the Peierls distortion of metals.

How can we inactivate the butterfly distortion? If we replace the valence sp^2 AOs of HB by unhybridized B2p AOs, spatial overlap in the L and U faces will be diminished and there will no longer be a motivation for the butterfly distortion. Dehybridization (i.e., blockage of s-p mixing) occurs in neutral atoms that have inert ns AOs. These are all heavy congeners of B, C, N, O, and F. This is why, as discussed earlier, Si_4 prefers the rhombic geometry even though Si is isolobal and isoelectronic to BH,. And this is why there are no stable deltahedral silicon clusters. And this is why there are no "alumane" clusters.

38.7 Triple Bonds Are the Key to Metal Chemistry

The title of this section should be broadly descriptive: "The *n*-tuple bond road of metal chemistry, with *n* odd." However, we here restrict our attention to *formal* triple bonds. A vast number of observations can be systematized by assuming

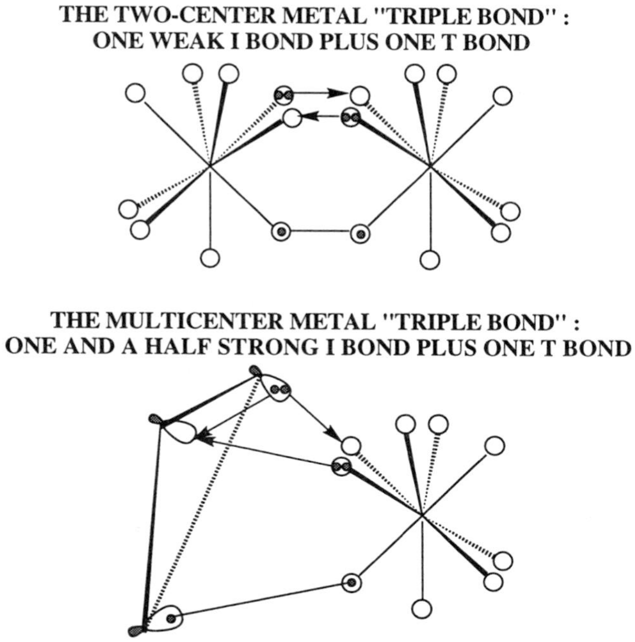

THE TWO-CENTER METAL "TRIPLE BOND" :
ONE WEAK I BOND PLUS ONE T BOND
THE MULTICENTER METAL "TRIPLE BOND" :
ONE AND A HALF STRONG I BOND PLUS ONE T BOND

Figure 38.14 The difference between a triple bond with two centers and one with multiple centers.

that the most effective way of assembling inorganic or organometallic pieces is to link them with formal multicenter triple bonds. The underpinning of this concept is presented schematically in Figure 38.14: A two-center "triple bond" formed by the overlap of singular metal AOs is horrendously unstable because it is effectively not a triple bond. Rather, it is actually one T bond plus one I bond. Perfect pairing is unfavorable in metallic systems (high electron–electron repulsion), and contact repulsion within the I bond is also high. By contrast, a *multicenter* "triple bond" formed by the overlap of fragment MOs is superior because it approaches the limit of one T bond plus two I bonds. In addition, the orientation of the overlap densities diminishes their mutual Coulomb repulsion (reduced contact repulsion).

Exactly where the "triple-bond road" is taking us can be best appreciated by a specific example. The octahedral molecule $(PR)_2[Fe(CO)_3]_2[Fe(CO)_2]_2(CO)$,[18] shown in Figure 38.15a in the way the chemist writes it, is "unsaturated" according to the rule for noble-gases. However, the localized unsaturation implied by the metal–metal double bond is not borne out by the cluster bond lengths. How is this cluster put together? The presence of the two 3h/4e PR caps tells us that the square cluster of four

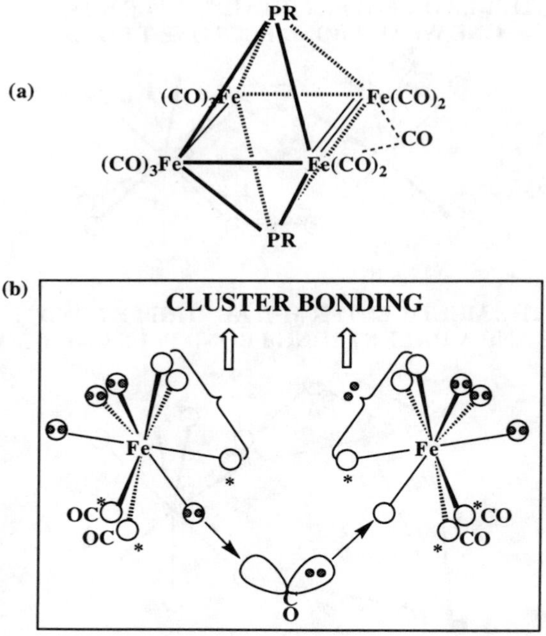

Figure 38.15 (a) The $(PR)_2[Fe(CO)_3]_2$ $[Fe(CO)_2]_2$ CO molecule. (b) Cluster bonding of two carbonyl-bridged $Fe(CO)_2$ fragments.

irons must provide two pairs, to be able to generate two triple bonds with the two PR caps, plus an aromatic count of electrons to fill the T catena of the Fe_4 ring. This count is either 2 or 6. Adopting the former configuration, we predict that the square base of the square bipyramid contributes a total of six cluster electrons. Added to the eight electrons contributed by the two PR caps, we have a total of 14 electrons, a number conforming to Wade's rule.

We now ask: Exactly how are the electrons allocated, and how does bridging occur? Each $Fe(CO)_3$ acts as 3h/2e and each $Fe(CO)_2$ acts as 3h/1e. Furthermore, the two $Fe(CO)_2$ units are connected by an exo Fe—Fe bond, which is "supported" by the bridging carbonyl in the fashion shown in Figure 38.15b. This leads to the toy representation in which two PR caps are connected to the square Fe_4 triple-deck toy cluster by two formal triple bonds, with one additional pair occupying the T catena of Fe_4. Each PR contributes four electrons and the $[Fe(CO)_3]_2[Fe(CO)_2]_2(CO)$ toy cluster contributes six [two electrons by each $Fe(CO)_3$ and one by each $Fe(CO)_2$]. The toy formula is as follows:

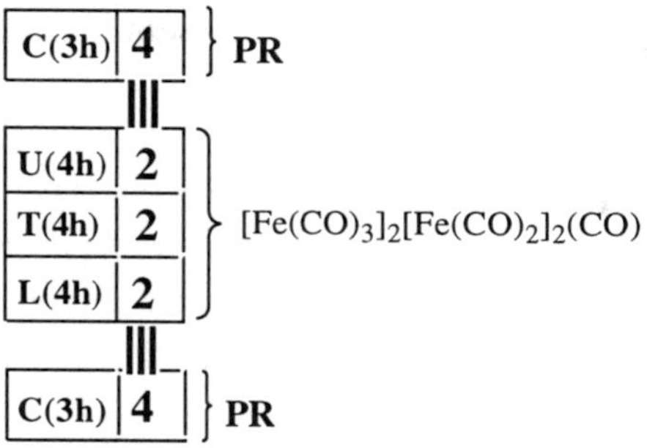

The octahedral cluster just described is known to undergo nucleophilic addition to yield $(PR)_2[Fe(CO)_3]_3[Fe(CO)_2L]$. This means that we must now accommodate an extra pair. There are two choices: either we disrupt one of the two triple bonds or we add the pair to the T catena, turning it to an antiaromatic system. The latter choice is more favorable because metals can cope with "antiaromaticity" through distortion. In summary, the "triple-bond road," in connection with the toy model, reveals where the electrons are in the cluster and where they go upon nucleophilic attack.

A second example is the 12-electron trigonal pyramidal ("tetrahedral") cluster $(PR)[Fe(CO)_3]_3[MnCp(CO)_2]$[19] in which $Fe(CO)_3$ acts as 3h/2e and $MnCp(CO)_2$ as 3h/4e (use the pseudo-octahedral hybridization model to confirm this). The toy formula of the 12-electron trigonal pyramid features a triple bond connecting the PR cap to the triangular $MnFe_2$ base. There are also a four-electron T catena and a two-electron uncapped L face of the triangular Fe_2Mn base (see the toy formula of a 12-electron trigonal pyramid in Figure 38.5). The six electrons in the T catena and the L face represent three formal bond pairs of the triangular $MnFe_2$ base (see the description of cyclopropane in Chapter 4). Experiment indicates that this cluster is in equilibrium with a valence isomeric species that has a disrupted formal Fe—Mn bond and adds a nucleophile L. Ring opening can produce three different localized zwitterions as follows:

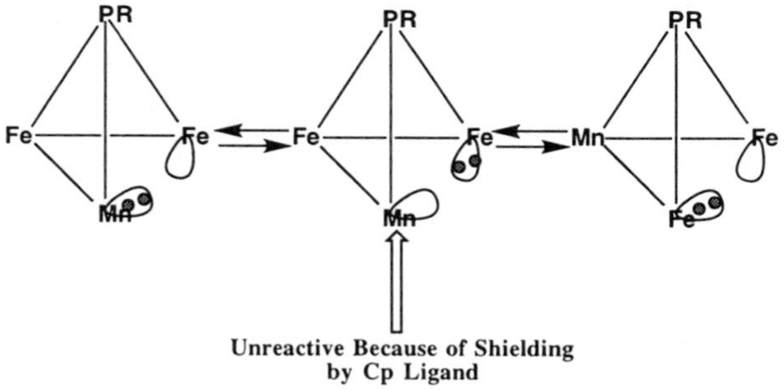

Nucleophilic attack occurs preferentially on the least sterically shielded hole (i.e., the iron site).

Casey and co-workers uncovered a paramagnetic $H(CoCp^*)_3H_3$ that is converted to a diamagnetic $CH_3C(CoCp^*)_3H_3$ upon reaction with acetylene.[20a] According to the toy model, the first has a pair of triplet coupled electrons in the Möbius T catena, a single bond linking one H with the lower face, and a triple bond linking the three H atoms with the upper face of the metal triangle. The second has one "lower" and one "upper" triple bond and a vacant T catena. The mechanism of the reaction can be rationalized and further refined by using the toy model and recognizing the pivotal effect of the T catena of the metal cluster. We also note that the diamagnetic cluster is a 12-electron octahedron, consistent with the VB cluster rule but not with Wade's rule. Finally, the cationic tetramer $(AuPR_3)_4^+$ can act as a three-electron piece to combine with a carbynic RC fragment, which also acts as a three-electron piece, by a formal multicenter triple bond.[20b]

38.8 Metal Cluster Stereoisomerism

Koutecky and his group have carried out extensive investigations of metal clusters.[21a] In one paper, they posed a key question: Is Li_{13} icosahedral (i.e., a deltahedral cluster) or is it a piece of a hexagonal close-packed (hcp) lattice? [21b] The toy model allows us to compare these two geometries on equal footing: Is a 1:5:5:1 (icosahedron) superior to a 3:6:3 (hcp) Li_{12} cluster with a guest Li at the center? The notation a:b:c:d specifies the number of vertices of the axially packed rings that make up the cluster. For example, the icosahedron is viewed as a composite of the following elements: 1 cap: one 5-ring: one 5-ring: 1 cap, with the two pentagonal Li_5 rings arranged in an antiprismatic fashion.

According to the toy model, the icosahedral $(BH)_{12}^{2-}$ closoborane is a 1:5:5:1 antiprismatic stack in which the four layers are connected by three triple bonds and each (Möbius aromatic) pentagonal T catena accommodates four electrons. Hence, the magic number is $(3 \times 6) + (2 \times 4) = 26$, which is identical to that predicted by Wade's rule. Li_{12} has 14 fewer electrons and cannot form a 12-electron trigonal bipyramid with each face capped by one Li core simply because it falls one atom short. This suggests a different view of clusters having electrons fewer than $2V + 2$. Specifically, the cluster is visualized as a stack of caps and rings connected by a combination of the following elements:

A six-electron *singlet* triple bond, symbolized by t(6).

A four-electron *triplet* triple bond, symbolized by t(4).

Aromatic uncapped faces, symbolized by f(4N + 2).

In this way, we can generate the toy diagrams of the various Li_{12} species shown in Figures 38.16 and 38.17. Bold lines indicate electron pair bonds, dashed lines one-electron bonds, and a single hatched line a σ hole. The thirteenth Li at the center of each polyhedron is not shown but is assumed to couple with the Li_{12} main frame in either a high or low spin mode. The latter is favored in a promoted icosahedral structure in which a σ hole can pefectly accommodate the odd electron of the thirteenth encapsulated Li atom. The four Li_{12} isomers can be represented as follows:

Bicapped pentagonal antiprism: t(4) + t(4) + t(4).
Promoted icosahedron: t(4) + t(4) + t(4).
Low spin hcp: t(6) + t(6).
High spin hcp: f(2) + t(4) + t(4) + f(2).

Here is now the physics of the problem. Each Li_{12} cluster can be thought of as a stack of slabs representing the atomic cores of the rings and sandwiching the valence electrons in slab gaps as illustrated in Figure 38.18. Only slabs that

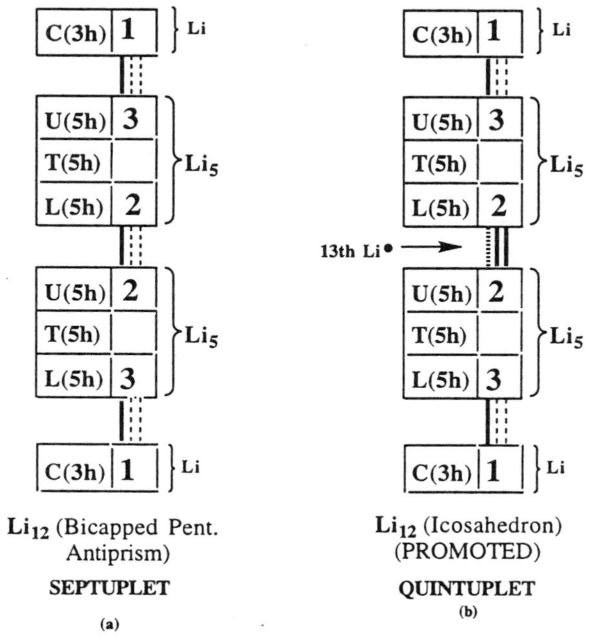

Figure 38.16 Toy formulas for two spin isomers of icosahedral Li_{12}.

correspond to three or more rings can be used. That is, a two-ring slab is "forbidden" for reasons discussed earlier (no e-degenerate orbitals). There exist many different isomeric stacking patterns, and we must ask: What is the optimum number of slabs, and what is the best type of slab to use to sandwich the valence electrons?

Here are some (but not all) possible isomers of Li_{12}:
Two-slab isomers: 6:6.
Three-slab isomers: 3:6:3, 4:5:3.
Four-slab isomers: 1:5:5:1, 2:4:4:2, 3:3:3:3.

The 2:4:4:2 isomer is rejected, but the rest remain viable candidates for the global minimum. The next thing is to identify the "optimum slab." This is an odd-membered ring because an n ring with n = odd has $(n - 1)/n$ e-degenerate character while an n ring with n = even has $(n - 2)/n$ e-degenerate character (see Section 35.4). We can now eliminate the 6:6, 3:6:3, 4:5:3, and 2:4:4:2 isomers from competition. This leaves the 1:5:5:1 and the 3:3:3:3 antiprismatic isomers as contenders. The toy formulas for these two species reveal that both have the same number of four-electron triple bonds. This identifies the problem: Two cluster stereoisomers with the same number of slabs (and with a total number of

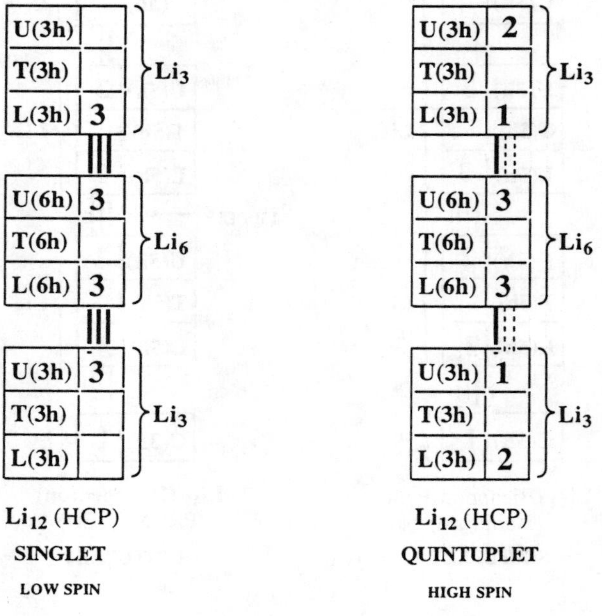

Figure 38.17 Toy formulas for low and high spin isomers of hcp Li_{12}.

electrons that falls short of saturating all interfacial bonding orbitals) cannot be differentiated by a mere electron counting because interslab bonding remains constant!

We are now stuck with a major problem, the dimensions of which can be appreciated by looking at a related but simpler case of cluster stereoisomerism. Specifically, M_7 pentagonal bipyramids (M = Li, Na, K) have been observed in rare-gas matrices. Furthermore, pentagonal symmetry is ubiquitous in clusters of alkalies and transition metals with a ground $ns^1(n-1)d^m$ configuration. For this reason, we expect the icosahedral 1:5:5:1 to be more stable than the 3:3:3:3 isomer. Thus, our problem is reasonably well defined: Why is the 1:5:1 form more stable than the 1:3:3 M_7 isomer? The answer lies in the organizing principle discussed in Chapter 35: Occupancy of e-degenerate MOs in rings is ideal for I-bonding.

The toy formulas of clusters do not project explicitly an important element, namely, the delocalization of the face electrons to the T catenae. This means that the constituent rings are, partly, preserved in the cluster. For example, the

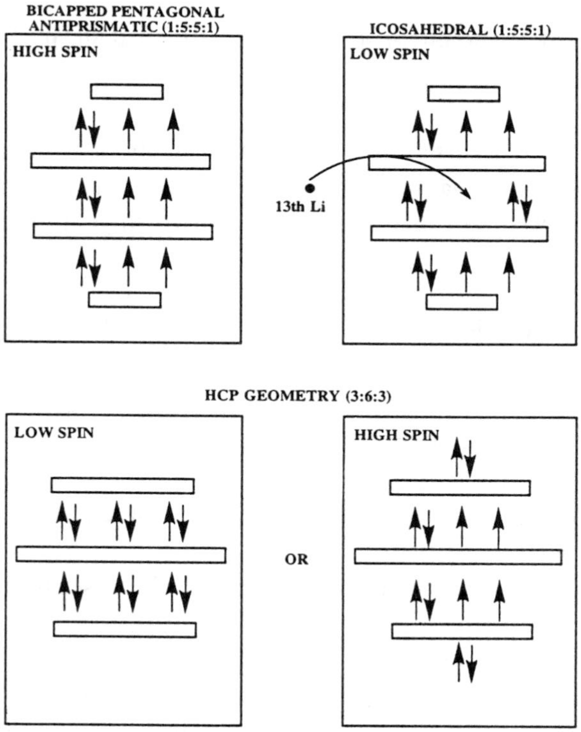

Figure 38.18 Physical interpretation of the structures of the Li_{12} clusters.

1:5:1 M_7 isomer has partial "M_5-ring character" and the 1:3:3 M_7 isomer has partial "double-M_3-ring character." What makes a metal ring stable? The presence of e pairs (i.e., pairs in e-degenerate MOs). How many e pairs are there potentially in the M_5 ring of the 1:5:1 isomer? Because of the resonance structure $M^+(M_5^-)M$, M_5 has M_5^- character and the latter has one totally symmetric pair and two e pairs. By contrast, the resonance structure $M^+(M_3^-)(M_3)$ assigns a total of three e-electrons to the two rings; that is, we have only one e pair. The reason is that the 1:3:3 partition requires that two pairs go to totally symmetric ring MOs, and this leaves out only three electrons for deposition into the e-type ring MOs. By contrast, the 1:5:1 partition requires that only one pair go to a totally symmetric ring MO, and this leaves out five electrons for deposition in the e-type MOs.

We have now obtained a selection rule for predicting the relative energies of stereoisomeric alkali clusters (and their transition metal analogues) within the context of our formulation of a cluster as stacked rings: The more stable stereoisomer is an a:b:c: composite, in which a, b, c, etc., are chosen as follows:

Odd are preferred over even values.

Unsymmetrical are preferred over symmetrical combinations of odd rings (e.g., 1:5 is superior to 3:3).

The important point is that this selection rule is entirely based on many-electron theory and the notion that interelectronic repulsion is the key determinant of metal cluster structures. This is why the e-degenerate MOs of rings have special significance.

38.9 The Icosahedral Motif in Coinage Metal Clusters

The toy formulation of an icosahedral M_{12} cluster reveals the existence of multiple magic numbers, two of which have been discussed:

The 26-electron magic number is exemplified by the singlet icosahedral $(BH)_{12}^{2-}$ closoborane.

The 12-electron magic number is exemplified by the septuplet icosahedral Li_{12}.

We now add an important third icosahedral member which has an exceedingly simple electronic structure: three formal triple bonds connect two BH caps and

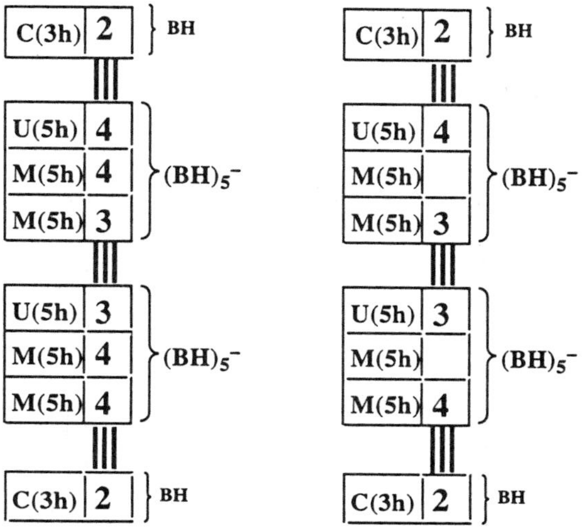

$(BH)_{12}^{2-}$ (Bicapped Pent. Antiprism) — **26 ELECTRONS**

$(BH)_{12}^{6+}$ (Bicapped Pent. Antiprism) — **18 ELECTRONS**

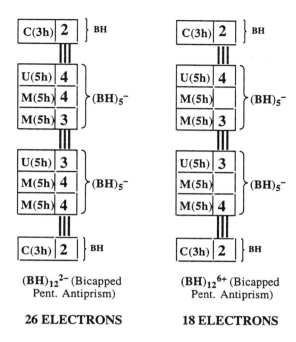

Figure 38.19 Two different magic numbers of the icosahedron. Note how the difference of eight electrons is due to differential T catenal occupation.

two $(BH)_5$ rings in the way indicated in Figure 38.19. The magic number is 18 electrons.

Stable 18- and 16-electron clusters, which can be thought of as derivatives of a gold icosahedron centered by a Pt atom, have been prepared by the groups of Steggerda and Pignolet.[22,23] $M(CO)(AuPPh_3)_8^{2+}$ is an example of an 18-electron and $M(AuPPh_3)_6(PPh_3)^{2+}$ and $M(AuPPh_3)_8^{2+}$ are examples of 16-electron clusters (M = Pt, Pd). That only the 16-electron clusters undergo facile oxidative addition is testimonial for the 18-electron magic number. The deviation of the 16-electron clusters from the magic number of 18 electrons is in the spirit of the VB cluster rule of Chapter 37.

38.10 The Reactions of Organometallic Clusters

The diversity of cluster reactivity is exhibited by a reaction studied by Huttner and co-workers[24] which we now treat using the toy formulas shown in Figure 38.20. The important point is that this photochemically induced reaction

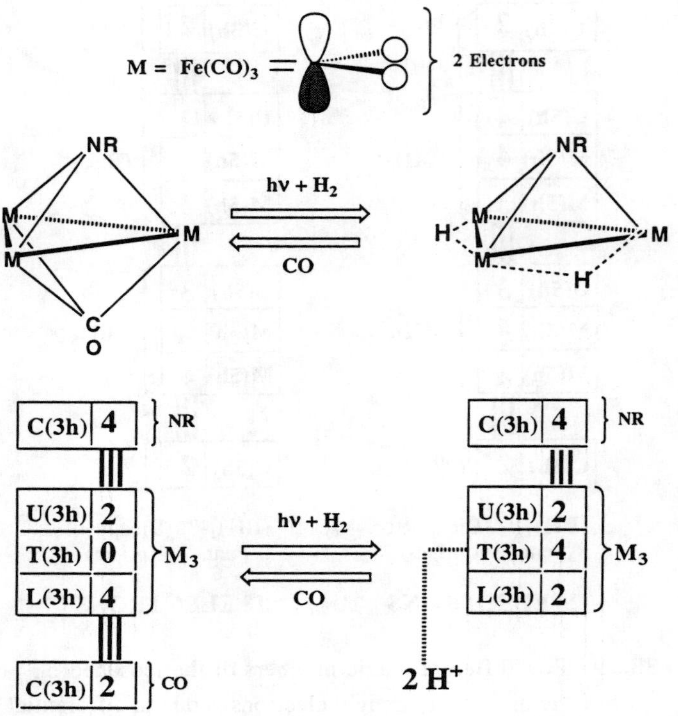

Figure 38.20 Photochemical addition of H_2 to an iron cluster formulated by the toy model.

involves a reshuffling of four electrons. Specifically, two electrons from the L face and two electrons from the H—H reactant must find their way to the originally vacant T catena. In principle, this rearrangement can come about by as many as four distinct steps, each involving a one-electron relocation. The mechanism is presently unknown.

38.11 Clusters as Catalysts

Adams and co-workers were able to prepare the complex $Pt_3Ru_6(CO)_{21}(\mu\text{-}H)_3(\mu_3\text{-}H)$ (I) shown here (structures reproduced with permission from ref. 25. Copyright 1992 American Chemical Society), along with a derivative $Pt_3Ru_6(CO)_{20}(\mu_3\text{-}PhCCPh)(\mu_3\text{-}H)(\mu\text{-}H)$ (II) which exhibits 100% selectivity in the hydrogenation of diphenylacetylene to Z-stilbene.[25]

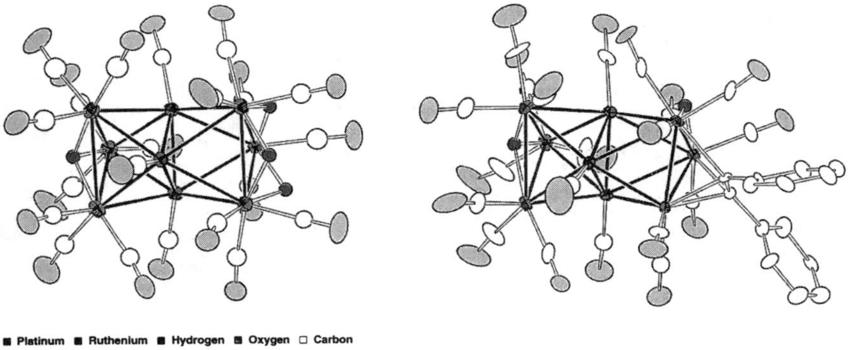

To understand how this happens, we must first understand where the electrons are in the two complexes. The toy model again comes to our rescue, and we proceed in three steps.

First, we dissect each cluster to its metal–carbonyl components, write the appropriate VB formula for each one (choosing a hybridization scheme that best accommodates the geometry of the fragment within the cluster), and identify the active orbitals and electrons. The results for molecule I are shown in Figure 38.21.

Next, we envision the complex as a stack of three triangular toy pieces as shown schematically in Figure 38.22.

The third step is to produce the toy formula of each complex by following the standard procedure. The final result is shown in Figure 38.23.

Because of space limitations, we simply invite the attention of the reader to what has been a recurring theme: When given the opportunity, metal atoms and π ligands make multicenter triple-I bonds. This is how the acetylene coordinates to the L face of the MM*K disk! The precise match of the orbitals can be represented as follows:

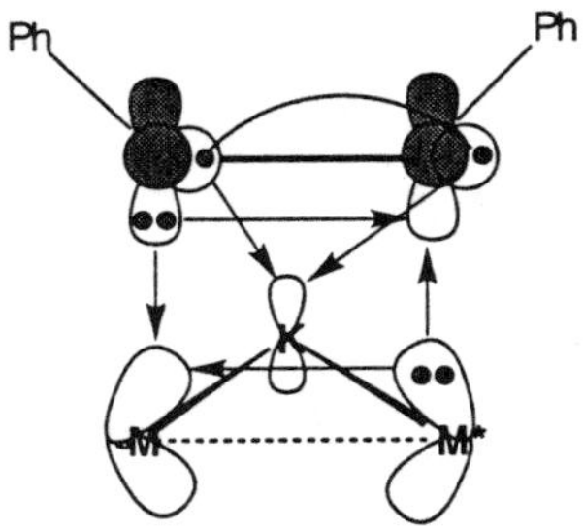

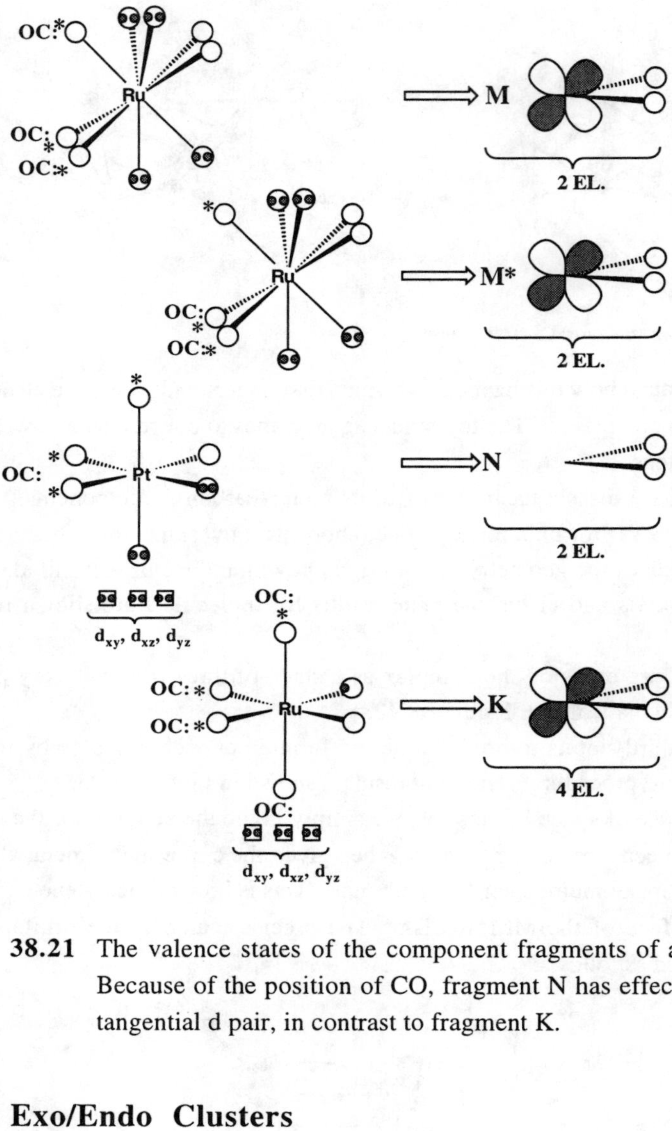

Figure 38.21 The valence states of the component fragments of a cluster. Because of the position of CO, fragment N has effectively no tangential d pair, in contrast to fragment K.

38.12 Exo/Endo Clusters

The toy game has revealed a secret: Stable clusters are formed by the union of toy pieces, as accomplished by formal triple bonds. We can exploit this finding to design clusters in which a pseudo-octahedrally hybridized metal utilizes the exo set of hybrids to make a triple bond with an organic π ligand and the endo set for making a metal ring or deltahedral cluster. In addition, the number of equatorial electrons is restricted to zero to accommodate σ pairs of the organic π

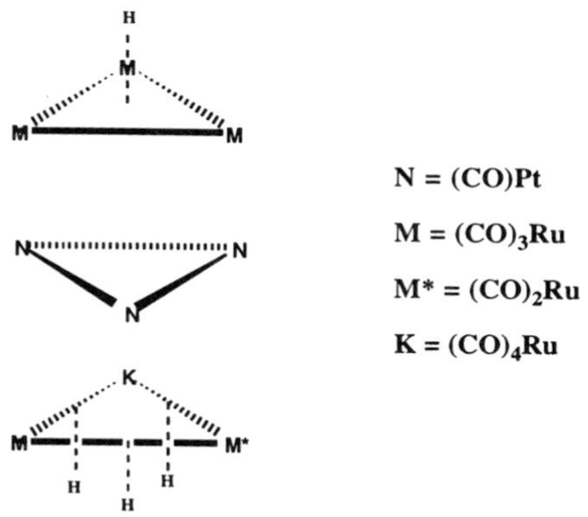

Figure 38.22 Cluster generated by the stacking of triple-deck clusters according to the toy model.

ligand. For example, the successful electron count of an even-electron metal is the one that assigns two endo electrons plus two, four, or six exo electrons that can complement a four, two-, or zero-electron π system of an organic fragment. The possible successful combinations are those generated by the following algorithm.

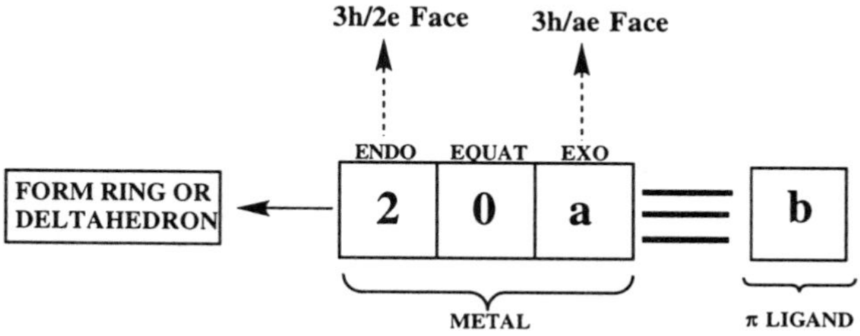

$$\mathbf{a + b = 6}$$

Recently, there has been an observation of stable (MoC_4)n clusters.[26] These meet the 2:0:4+2 condition if we assume that the C_4 fragment can be induced to act as a two-π-electron rhombus (analogous to Si_4).

38.13 Theory as a "Listening Device"

Organometallic chemistry, which has never shaken off the preoccupation with "electron counts," has yet to advance to the "formula stage." The literature is replete with quantum chemical calculations of all types which, however, fail to

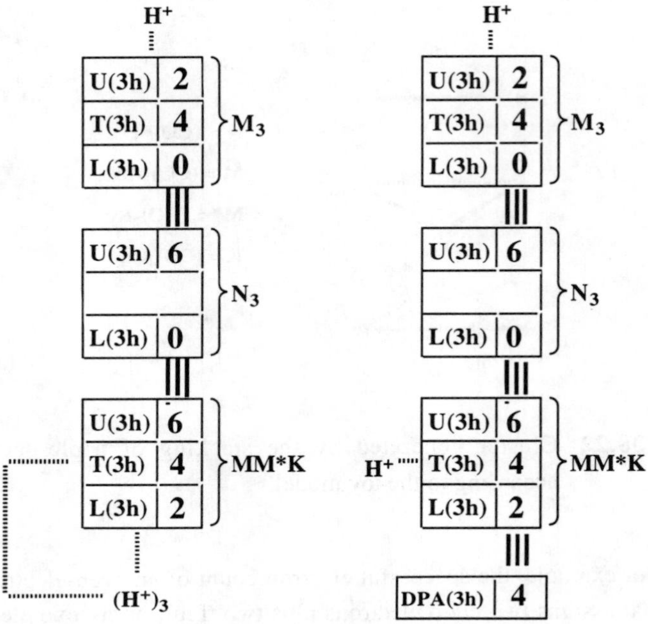

Figure 38.23 Toy formula of precursor I [$Pt_3Ru_6(CO)_{21}(\mu\text{-}H)_3(\mu_3\text{-}H)$] (left) and the derivative II [$Pt_3Ru_6(CO)_{20}(\mu_3\text{-}PhCCPh)(\mu_3\text{-}H)(\mu\text{-}H)$] (right).

answer the fundamental question of the experimentalist: What should I do next (beyond extrapolation from experience) to produce a new reagent, new catalysts, and so on? We now have the concepts, the electron-counting schemes, and, above all, the formulas to change the situation. The best approach is to use these tools as a *listening device*. Specifically, we examine the structure and reactivity of one complex, we write its VB formula and, then, we seek to identify and understand the unique features that differentiate this reference molecule from a second related system. Organometallic chemistry is a "game of angstroms," and our only hope of being creative lies in understanding how small differences in the molecular electronics translate to respectable structure and reactivity differences. Arriving at such an understanding is both entertaining and instructive because we let nature (i.e., the experimental results) speak for itself. Two specific examples will serve to illustrate this modus operandi.

Many organometallic chemists have sought to "activate" alkanes by homogeneous catalysts. One approach is to have the metal atom of a mononuclear complex add oxidatively to a C—C or a C—H bond in an

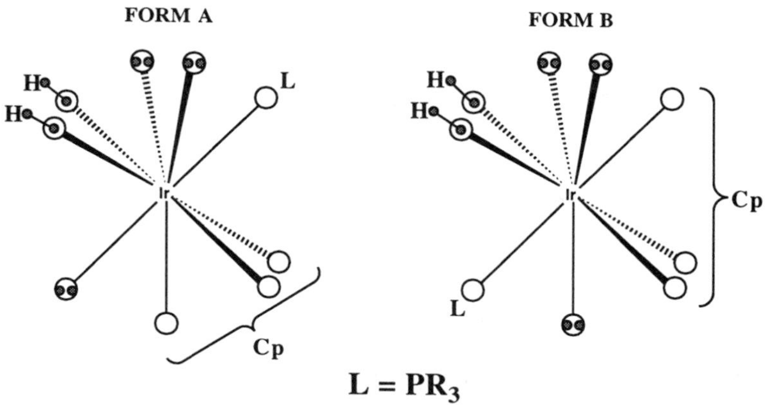

Figure 38.24 Two valence isomeric forms of $CpIrH_2L$.

intermolecular fashion. In other words, we want to have the metal attack a C—H bond of another molecule rather than a C—H bond belonging to one of its ligands. Bergman, Crabtree, Graham and their collaborators discovered an iridium complex of the type Cp^*LIrH_2 (L = phosphine) that acts in this way even when intramolecular oxidative addition can occur when L = triphenylphosphine.[27,28] We seek to understand why this is so by writing the pseudo-octahedral formula of the complex as shown in Figure 38.24.

There exist two different stereochemical arrangements that differ in the way the lone pairs are interlaced between the ligands although, in both, the L and the H atoms are "cis." In form A, the equatorial metal electrons can act as a shield to prevent the alkyl groups of the phosphine from coming close to the reactive conical hybrids that accommodate the H atoms. That this is no longer the case in B suggests that one key requirement of an active "intermolecular" organometallic reagent is the shielding of the reactive orbitals from the rest of the ligands by intervening metal lone pairs!

Having presented an example of using theory as a listening device, we proceed to the deliberate design of a preselected reagent. I have been intrigued by the possibility of mounting on the same metal or cluster a singlet and a triplet oxygen, which could selectively respond to different organic molecules in a way controlled by "observer" metal ligands. How might we do this? The toy model gives a straightforward answer: Take a triangular cluster with a 3h/4e U face and a 3h/2e L face, and make an eight-electron triplet triple bond using the U face

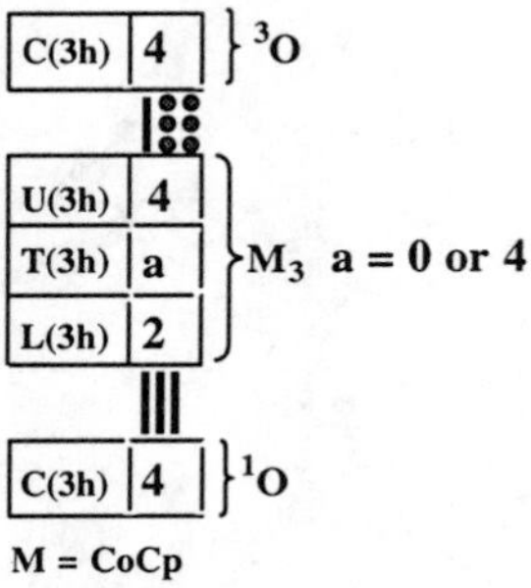

Figure 38.25 Toy formula of triplet $O(CpM)_3O$.

plus triplet O and a six-electron singlet triple bond using the L face and singlet O. The toy formula of the sought-after molecule is shown in Figure 38.25.

Does the literature cite any six-electron triangular clusters that are capped by nonmetals? Three different clusters[29] and their corresponding toy formulas are shown in Figure 38.26. First, note that one can be considered to be a 10-electron trigonal pyramid and two can be regarded as 12-electron trigonal bipyramids. *The important thing is that they all differ in the allocation of the skeletal electrons.* Furthermore, of the three, only the cobalt cluster has the property we desire, namely, four and two electrons in the U and L face, respectively. Indeed, this grouping succeeds where simple mononuclear complexes fail: It places CO trans to singlet O! Hence, the sought after complex is $(CoCp)_3(^1O)(^3O)$.

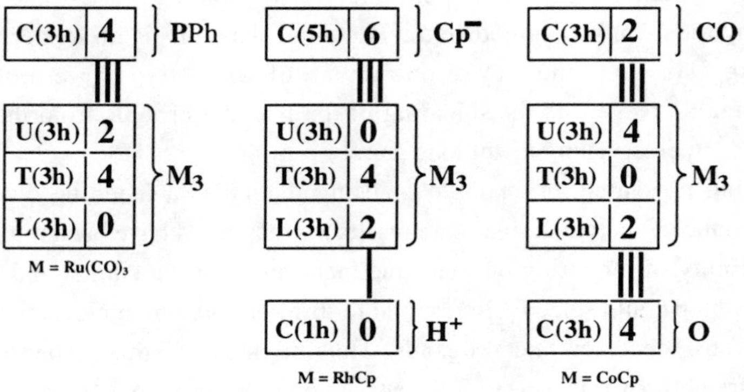

Figure 38.26 Triangular metal clusters capped by diverse ligands via triple bonds.

References

1. M.S. Paquette and L.F. Dahl, *J. Am. Chem. Soc.* 102, 6621 (1980).
2. A.L. Rheingold, M.J. Foley, and P.J. Sullivan, *J. Am. Chem. Soc.* 104, 4727 (1982).
3. O.J. Scherer, H. Sitzmann, and G. Wolmershäuser, *Angew. Chem. Int. Ed. Engl.* 24, 351 (1985).
4. O.J. Scherer, C. Blath, J. Braun, B. Hoebel, K. Pfeiffer, B. Rink, H. Slodzyk, P. Walther, B. Werner, and R. Winter, in *The Chemistry of Inorganic Ring Systems*, R. Steudel, Ed., Elsevier, Amsterdam, 1992, Chapter 11.
5. R.N. Grimes, in *Electron Deficient Boron and Carbon Clusters*, G.A. Olah, K. Wade, and R.E. Williams, Eds., Wiley, New York, 1991, Chapter 11.
6. T. Hughbanks and R. Hoffmann, *J. Am. Chem. Soc.* 105, 1150 (1983).
7. R. Chevrel, M. Sergent, and J. Prigent, *J. Solid State Chem.* 3, 515 (1971).
8. R. Chevrel, in *Superconductor Materials Science: Metallurgy, Fabrication and Applications*, S. Foner and B.B. Schwartz, Eds., Plenum, New York, 1981, Chapter 10.
9. Ø. Fischer, B. Seeber, M. Decroux, R. Chevrel, M. Potel, and M. Sergent, in *Superconductivity in d- and f- Band Metals*, H. Sohl and M.B. Maple, Eds., Academic Press, New York, 1980, p. 485.
10. A. Gruettner, K. Yvon, R. Chevrel, M. Potel, M. Sergent, and B. Seeker, *Acta Crystallogr.* B35, 285 (1979).
11. M. Potel, R. Chevrel,and M. Sergent, *Acta Crystallogr.* B36, 1319 (1980).
12. R. Chevrel, M. Potel, M. Sergent, M. Decroux, and Ø. Fischer, *J. Solid State Chem.* 35, 286 (1980).
13. M. Potel, R. Chevrel, and M. Sergent, *Acta Crystallogr.* B36, 1545 (1980).
14. M. Potel, R. Chevrel, and M. Sergent, *J. Solid State Chem.* 35, 286 (1980).
15. K. Klepp, and M. Boller, *Monatsh. Chem.* 110, 677 (1979).
16. E.R. Davidson, K.L. Kunze, F.B.C. Machado, and S.J. Chakravorty, *Acc. Chem. Res.* 26, 628 (1993).
17. W.N. Lipscomb, *Science* 153, 373 (1966).
18. H. Lang, L. Zsolnai, and G. Huttner, *J. Organomet. Chem.* 23, 282 (1985).

19. (a) G. Huttner, J. Schneider, H.-D. Mueller, G. Mohr, J.v. Seyerl, and L. Wohlfahrt, *Angew. Chem. Int. Ed. Engl.* 18, 76 (1979). (b) J. Schneider and G. Huttner, *Chem. Ber.* 116, 917 (1983).
20. (a) C.P. Casey, S.L. Hallenbeck, and R.A. Widenhoefer, *J. Am. Chem. Soc.* 117, 4607 (1995). (b) J. Vicente, M.T. Chicote, R. Guerrero, and P.G. Jones, *J. Am. Chem. Soc.* 118, 699 (1996).
21. (a) J. Koutecky and P. Fantucci, *Chem. Rev.* 86, 539 (1986). (b) G. Pacchioni and J. Koutecky, *J. Chem. Phys.* 81, 3588 (1989).
22. (a) R.P.F. Kanters, P.P.J. Schlebos, J.J. Bour, W.P. Bosman, H.J. Behm, and J.J. Steggerda, *Inorg. Chem.* 27, 4034 (1988). (b) L.N. Ito, J.D. Sweet, A.M. Mueting, L.H. Pignolet, M.F.J. Schoondergang, and J.J. Steggerda, *Inorg. Chem.* 28, 3696 (1989). (c) L.N. Ito, A.M.P. Felicissimo, and L.H. Pignolet, *Inorg. Chem.* 30, 387 (1991). (d) J.J. Steggerda, J.J. Bour, and J.W.A. van der Velden, *Recl. Trav. Chim. Pays-Bas* 101, 164 (1982).
23. Review: L.H. Pignolet, M.A. Aubart, K.L. Craighead, R.A.T. Gould, D.A. Krogstad, and J.S. Wiley, in press.
24. G. Huttner and K. Knoll, *Angew. Chem. Int. Ed. Engl.* 26, 743 (1987), and references cited therein.
25. R.D. Adams, Z. Li, P. Swepstone, W. Wu, and J. Yamamoto, *J. Am. Chem. Soc.* 114, 10657 (1992).
26. R.L. Hettich, C. Jin, R.E. Haufler, CM. Barshick, A.A. Tuinman, A.A. Puretzky, and A.V. Dem'yanenko, *Science* 263, 68 (1994).
27. (a) R.G. Bergman, *Science* 223, 902 (1984). (b) R.H. Schultz, A.A. Bengali, M.J. Tauber, B.H. Weiller, E.P. Wasserman, K.R. Kyle, C.B. Moore, and R.G. Bergman, *J. Am. Chem. Soc.* 116, 7369 (1994).
28. (a) R. Crabtree, *Chem. Revs.* 85, 235 (1985). (b) W.A.G. Graham, *J. Organomet. Chem.* 300, 81 (1986).
29. D.M.P. Mingos and D.J. Wales, *Introduction to Cluster Chemistry*, Prentice Hall, Englewood Cliffs, NJ (1990).

Chapter 39

The Collapse of the Isoelectronic (Isolobal) Analogy

39.1 The Borane Analogy and the Lessons of the Metalloboranes

During the years of the lean cows, there still have been two interrelated important ideas:

Wade's rule, which systematizes, in part, the structures of boron and organometallic clusters.[1]

The isolobal analogy developed by Hoffmann, according to which an RB fragment, a transition organometallic fragment, and an organic fragment have the same type of frontier symmetry orbitals.[2]

These two ideas define the so-called *borane analogy*. Since this concept has been repeatedly reviewed in the literature, we simply summarize the situation as seen from the vantage point of VB theory by saying that a bad formula is better than no formula at all. By "bad formula" we mean that CpCo is analogous neither to nonmetallic HC^+ nor to the semimetallic HB. The first aphorism should be, by now, self-evident, namely, that C^+ is an electronegative nonmetal while Fe is an electropositive metal, and the two fragments will bind by two different mechanisms. On the other hand, the second aphorism needs exposition.

In a perceptive review article, Grimes[3] summarized the contrasting behavior of borane and metalloborane clusters. According to Wade's rule, a CpCo is isolobal to a BH group and these fragments must be interchangeable without affecting geometry or stability. Actually, it is found that replacement of the latter by the former generates a metalloborane with no structural counterpart in borane or carborane clusters. For example, B_5H_9 is oxidized violently by air, but 2-$CpCoB_4H_8$ is an air-stable crystalline solid. The 1-$CpCoB_4H_8$ isomer is also stable.[4]

The VB story is entirely different. Because of our ability to write down specific formulas, we can see that the formally isolobal HB, $(CO)_3Fe$ and CpCo are chemically different in these respects.

HB is a two-electron trisynaptic fragment denoted by 3h/2e.

$(CO)_3Fe$ is also a 3h/2e fragment, but only because the three equatorial lone pairs are tied up by the vacant MOs of the three carbonyls.

CpCo has the flexibility of acting as either a 3h/2e piece with three equatorial lone pairs or as a 4h/4e piece with two equatorial pairs. The latter choice is particularly favorable when CpCo is called to match a square face with four electrons, in which case it can engender a formal quadruple bond.

We can now formulate a rule: An L_nM fragment in which M is formally d^8 and which is formally classified as a 3h/2e piece by adopting the pseudo-octahedral hybridization mode can upgrade a triple to a quadruple formal bond by mutating to a 4h/4e piece, provided L_n are donor ligands and the fragment matches an even metal-annulene with a total of four electrons. This rule explains the following trends.

(a) The relative instability of B_5H_9 compared to $CpCoB_4H_8$. The toy formulas are given in Figure 39.1.

(b) The statistical identity, despite the larger covalent radius of boron compared to carbon, of fact that the average Co—B distance in 1-$CpCoB_4H_8$[5] with the corresponding Co—C bond length in $CpCo(C_4H_4)$[6]. This reflects the difference between the carbon and boron square bases. Only the latter can promote formation of a quadruple multicenter bond with CpCo because the participation of all four MOs of the square face is required. When overlap is strong (nonmetal case), the MOs are strongly split and the single antibonding MO is highly destabilized. As a result, the π "face" of cyclobutadiene can make

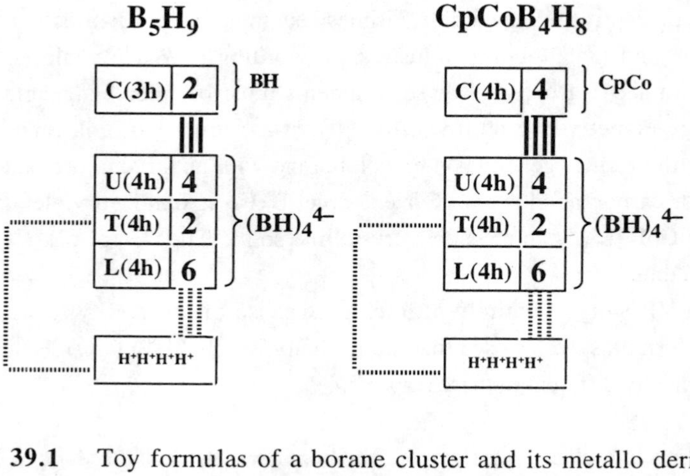

Figure 39.1 Toy formulas of a borane cluster and its metallo derivative. Note the difference in the synapticity of HB and CpCo.

only a formal triple bond with a 3h/2e `fragment. When overlap is weak (metal case), the four MOs tend to be quasi-degenerate. As a result, one face of the $(BH)_4$ square base can now make a formal quadruple bond (actually a bona fide double I bond) with a 4h/4e fragment.

(c) Although P and CpNi are isolobal, the P_6 square bipyramid (SBP) is unknown, whereas the $(CpNi)_6$ SBP has been prepared (see Chapter 38). The comparison of P and CpNi is entirely analogous to that of HB versus CpCo. P acts as 3h/3e but CpNi acts as 4h/5e.

The ability of BH and isolobal organometallic fragments to act as polyhedral caps is well recognized. However, the two types of cap are not interchangeable. Hence, the isolobal analogy fails. For example, in $(CpCo)_3B_4H_4$, one BH unit caps a Co_3 octahedral face.[7] By contrast, the corresponding borane, B_7H_7, is unknown in any geometry. More interestingly, boron-capped polyhedra are known only in the metal-containing systems. The VB interpretation is that the capping action of BH is, once again, due to the presence of equatorial pairs in the CpCo unit, which now acts as a 3h/2e piece with three equatorial lone pairs. The BH cap relinquishes its two electrons to the cluster cavity to produce the closo-octahedral $(CpCo)_3B_3H_3^{2-}$. The resulting HB^{2+} unit coordinates to three equatorial pairs which are contributed by each of the three Co atoms in the way illustrated in Figure 39.2.

Wade's rule assigns a magic count of 18 to the dodecahedral $(BH)_8^{2-}$ closoborane. Nonetheless, replacement of four BH by either four CpCo or four CpNi fragments in $(BH)_8$ generates stable dodecahedra with electron counts of 16 and 20 electrons, respectively.[8]

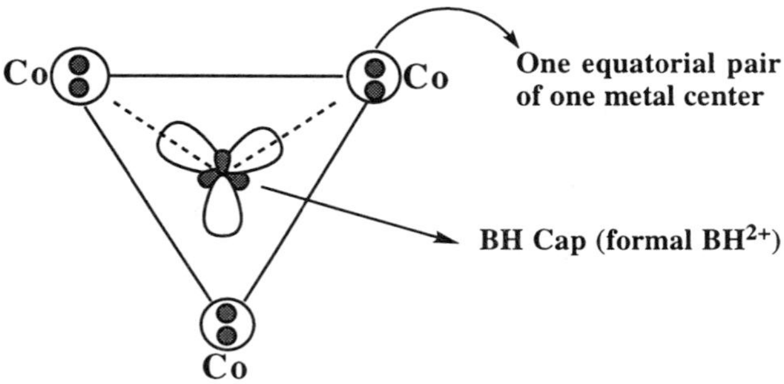

Figure 39.2 A BH^{2+} cap meets equatorial pairs upon coordination to a trigonal face of the octahedral Co_3B_3 cage of $(CpCo)_3B_4H_4$.

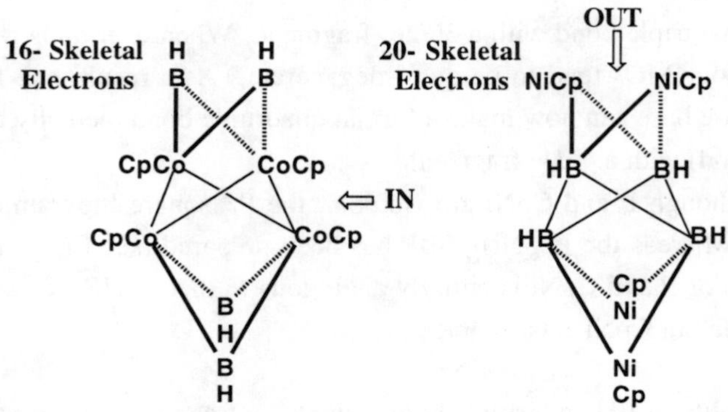

We have already seen that the dodecahedron is derived from the square prism by rhombic distortion of two opposite square faces. In the toy model formula, such a distortion turns two antiaromatic to two crossed-antiaromatic exo faces. The stability of the latter increases when there is alternation of electropositive and electronegative atoms, and this is exactly what is accomplished by replacing alternant BH by metal fragments. A key question is whether the metal fragments move "in" or "out" as the square prism undergoes the rhombic distortion. The experimental result is that CoCp fragments move "in" while NiCp groups move "out" as depicted schematically:

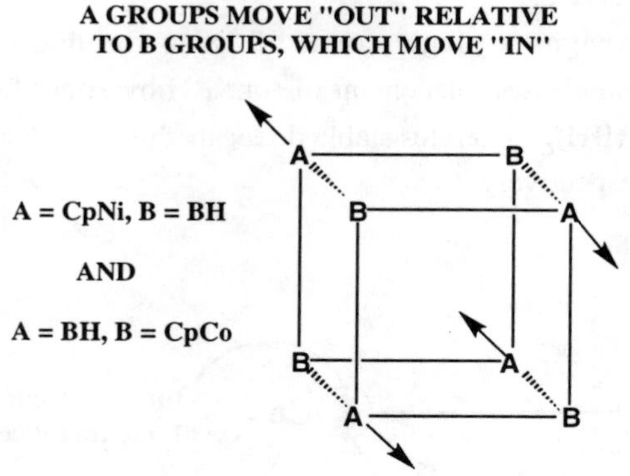

The toy formulas of the parent $(BH)_8^{2-}$ closoborane and the derivative $M_4(BH)_4$ (M = CpCo, CpNi) clusters are shown in Figure 39.3. The reason behind the stability of the 16- and 20-electron dodecahedra is the establishment of intercatenal linkages by either two-electron oxidation or two-electron reduction of the T catena of the 18-electron species.

Why do Co atoms move "in" but Ni atoms move "out" when a square prism is transformed to a deltahedron by rhombic distortion? The toy formulas of

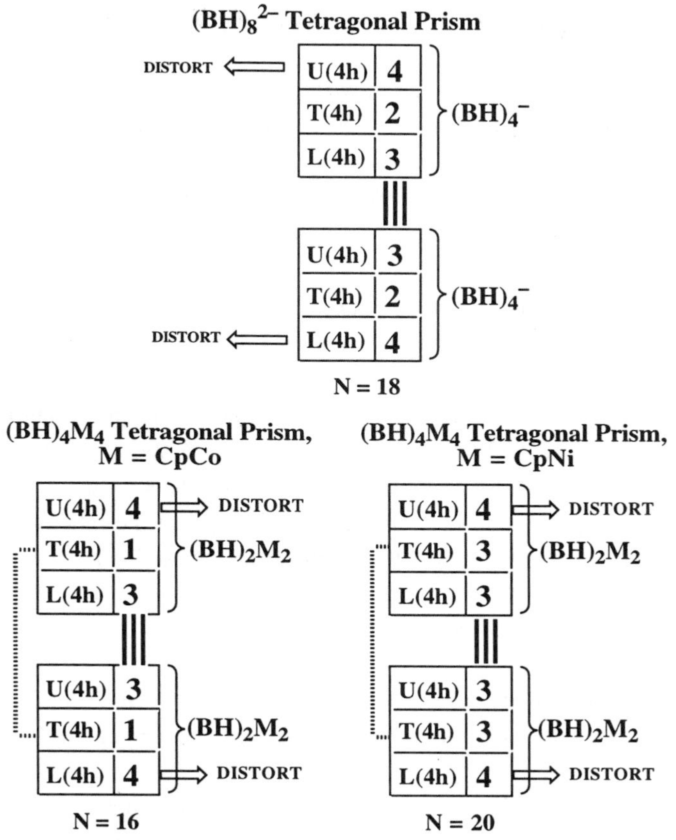

Figure 39.3 Toy diagrammatic derivation of the electronic structure of the $(BH)_8^{2-}$ dodecahedron and its derivatives, starting from the tetragonal prism.

Figure 39.3 suggest that the difference lies in the occupancy of the T catenae of the $(BH)_2M_2$ clusters: one per catena when M = CoCp and three per T catena when M = NiCp. The result is the following:

(a) When M = CoCp, the single odd electron of each catena localizes on the more electronegative Co atoms. As a result, intercatenal bonding involves the Co atoms which move "in."

(b) When M = NiCp, two of the three electrons of each catena localize on the more electronegative Ni atoms, leaving the single odd electron to localize on the less electronegative B atoms. As a result, intercatenal bonding involves the BH fragments, which move "in," while the NiCp fragments move "out."

39.2 The Problem with the Borane Analogy

The point is simple: Unless we can write explicit formulas showing the three-dimensional orientation of valence AOs and their occupancy, and unless we can interpret these formulas by falling back on a theory that captures the physics of the bonding, there can only be many happy accidents. Real progress in understanding organometallic chemistry will not take place. Here are some examples.

$(BH)_6^{2-}$ is an octahedral closoborane made up of BH units that act as 3h/2e fragments. The magic number is 14 electrons. $[Ni(CO)]_6(CO)_6^{2-}$ is also octahedral.[9] Hence, Ni(CO) is taken to be a 3h/0e fragment, for the sake of compliance with the 14-electron magic number. This interpretation of the bonding of $[Ni(CO)]_6(CO)_6^{2-}$ is contradicted by our analysis simply because, as shown earlier, OCNi is not 3h/0e but, rather, 4h/2e. This means that the borane analogy works by accident in rationalizing the shape of the $[Ni(CO)]_6(CO)_6^{2-}$, and this is evident by the failure of the same model to rationalize the following:

The planar dibridged structures of $LPdX_2PdL$ complexes that involve a metal–metal bond.[10]

The trigonal prismatic structure of the isoelectronic $[Pt(CO)]_6(CO)_6^{2-}$.[11]

The trigonal prismatic Pt tubes discovered by Chini and co-workers.[12]

A similar story can be told for the CpMo fragment, which has been found to form stable complexes with phosphorus rings. As illustrated below, this can act as a 3h/1e or 3h/0e fragment depending on the circumstances. When interfacial triple bonding (total of six electrons) calls for a contribution of one electron by CpMo, this adopts its 3h/1e configuration as in the $(CpMo)_2As_5$ complex. When it calls for a contribution of zero electrons, CpMo adopts a 3h/0e configuration as in the $(Cp^*Mo)_2As_6$ complex.

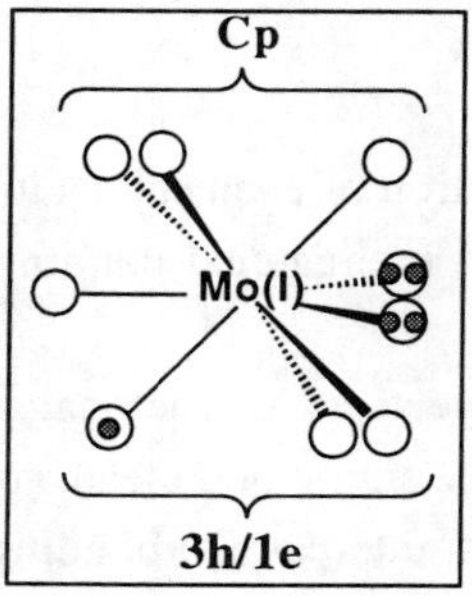

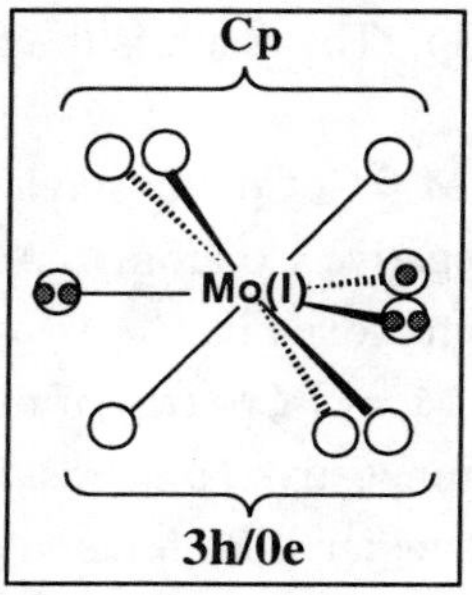

The observation to result from the presentation thus far in this section is that as we enlist an increasing number of bonding electrons by placing them in the endo conical AOs, we open up holes in the equatorial belt. For example, CpMn (and the isoelectronic $CpCr^-$) can act as 3h/0e with three equatorial pairs, as 3h/2e with two equatorial pairs and one equatorial hole, or as 3h/4e with one equatorial pair and two equatorial holes, and so on. Acting as 3h/2e, 3h/4e, or 3h/6e, this fragment can produce formally unsaturated clusters. However, this presumed unsaturation will not actually be present if the equatorial holes are tied up by pairs contributed by substituents of Cp. For example, one strategy for enforcing a 3h/2e valency is to hang a —(CH_2)—NH_2 group on the Cp, and so on.

Recently, Deck et al. prepared the formally unsaturated cluster shown below.[13] This a 14-electron octahedron in which a formal Cp^*Cr^- acts as a 3h/2e piece with one equatorial hole. The toy formula is shown along with the structure reproduced with permission from ref. 13 (copyright 1994 American Chemical Society). It may very well be that the single Cr^- equatorial hole is effectively tied up to two C—H bond pairs, one belonging to a methyl group borne by one Cp* and the other belonging to a methyl group borne by the second Cp*. The conformation of the Cp* units should provide a hint.

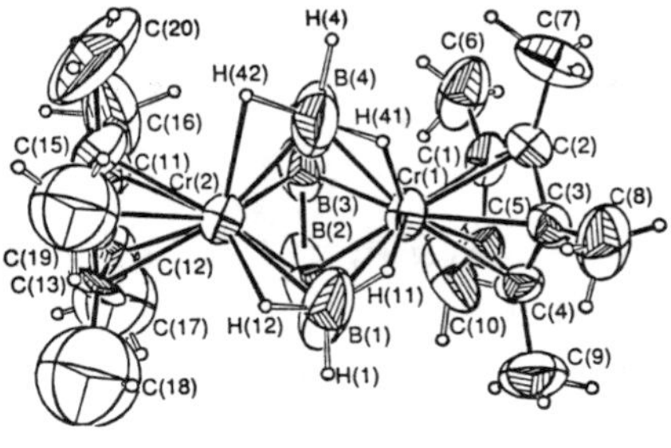

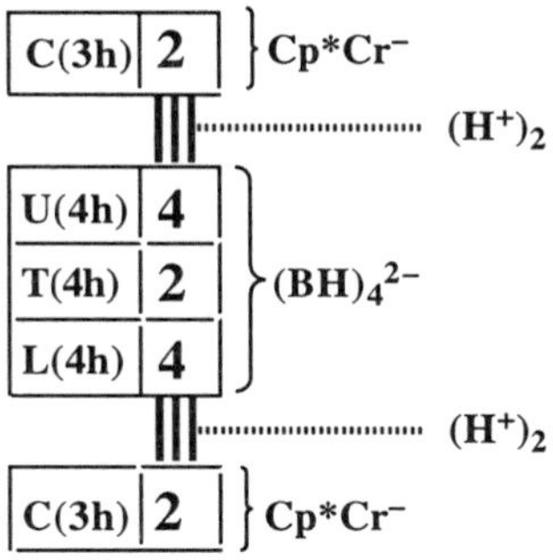

The electron-counting schemes devised by Wade and Hoffmann say that $(CO)_3Co$ is a three-electron fragment. VB theory shows an explicit formula that agrees but says more: $(CO)_3Co$ can act as 3h/3e with three equatorial pairs, as 3h/4e with two equatorial pairs and one equatorial odd electron, or as 3h/5e with one equatorial pair and two equatorial holes, and so on. How would $(RO^-)_3W^{3+}$ act? Making the usual assumption of treating alkoxide as a two-electron ligand, the conventional answer would be that this must be a minus-three-electron fragment within an organometallic cluster. By contrast, we can easily see that

$(RO^-)_3W^{3+}$ may very well act as a 3h/3e piece entirely analogous to $(CO)_3Co$, as found by Chisholm et al.[14]

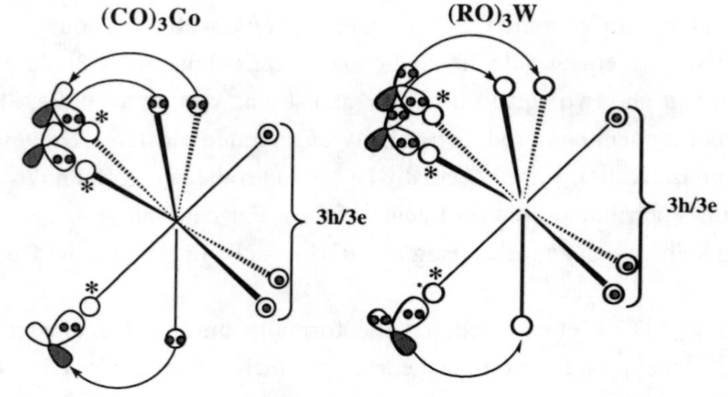

Organometallic chemists know that fragments "adjust" in order to bind. The problem is that this realization comes after the fact. MO computations also come after the fact. The VB formulas speak for themselves and suggest many strategies for realizing new structures and reactions. Even when we cannot make definite predictions, we can see the potentialities. If organic chemists had had to rely on MO computations to obtain the frontier orbitals of hydrocarbon fragments, organic chemistry would be in the dark ages. It is time that organometallic chemists are awakened to the inevitability of the chemical formula.

39.3 Prototypical Clusters of Inorganic Chemistry Do Not Conform to Wade's Rules

Many discussions of inorganic clusters[15,16] begin by showing the prototypical $Mo_6Cl_8^{4+}$ face-coordinated octahedron[17] and the $Nb_6Cl_{12}^{2+}$ edge-coordinated octahedron[18] or related structures. An octahedron has 8 faces and 12 edges, and a face or an edge can accommodate two electrons. Hence, the magic number of a polyhedron of this type is 40 electrons. This is exactly the electron count of each of the aforementioned clusters. The $Mo_6Cl_8^{4+}$ octahedron is formalized as $(Mo^{2+})_6(Cl^-)_8$. We start with a 4:2:3 hybridized Mo, eliminate the three exo-starred hybrids, and end up with a set of four endo conical and two equatorial AOs. The latter are so oriented that each spans two opposite triangular faces of the octahedron. Mo^{2+} acts as a 4h/4e piece (Figure 39.4). Six of these pieces assemble to form 12 formal two-electron bonds along the edges of an octahedron.

This leaves 12 interacting equatorial Mo holes with which to span 8 faces, each of which accommodates one chloride. The $Nb_6Cl_{12}^{2+}$ octahedron is formalized as $[(Nb^{2+})_6(Cl^-)_{12}]^{2+}$. Each Nb^{2+} now acts as a 4h/3e piece, and six of them assemble to form eight facial three-center/two-electron bonds. This leaves 12 interacting equatorial holes to accommodate 12 edge chlorides.[19]

Pt_6Cl_{12}, an edge-capped metal octahedron[20] that can be formulated as $(Pt^{2+})_6(Cl^-)_{12}$, is a more challenging problem. Again, the metal enters with 4:2:3 hybridization, the three exo starred hybrids are deleted, and each formal Pt^{2+} acts as a 4h/4e piece (Figure 39.4). The formal chlorides act as 2h/2e pieces in which one hybrid is starred. As a result, the 12 edges combine with the 12 chlorides to form 12 Q bonds. Thus, this cluster is an illustrator of the preferred 3h/2e action of chloride as well as a demonstrator of the Q bond.

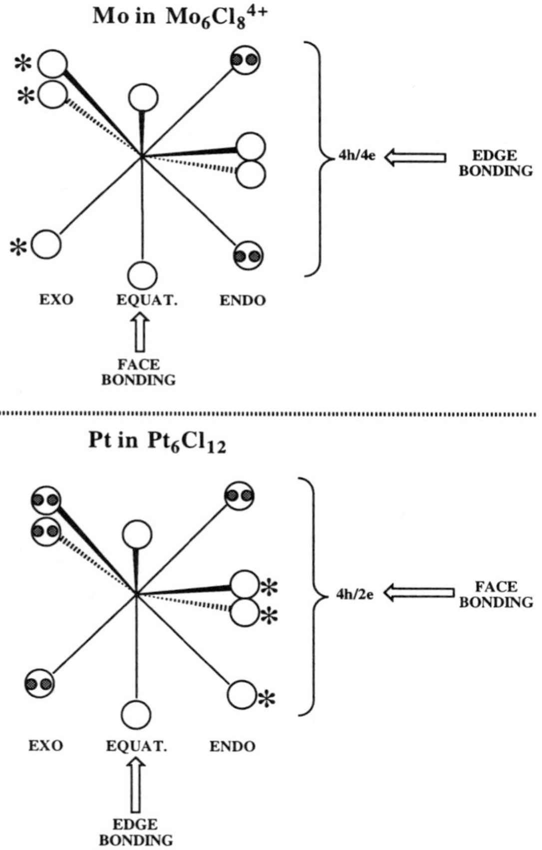

Figure 39.4 The electronic structures of Mo and Pt dications prepared for making an octahedral cluster.

Why do the three different clusters just discussed disobey Wade's rules? For the same reason that WF_6 does so: Because the rule covers only a subcase of I-bonding. Furthermore, since I-bonding implies bridging as well as multiplicity of magic numbers, there exist a variety of face-capped octahedra with differing numbers of skeletal electrons. Examples are $Nb_6I_8^{3+}$, $Re_6S_8^{3+}$, $Co_6S_8(PEt_3)_6$,[23] and $Fe_6S_8(PEt_3)_6^{2+}$, as reported, respectively, by Simon,[21] Spangerberg,[22] and Cecconi et al.[23,24]

Since we now understand that Wade's rule is a subcase of the VB cluster rule, which, in turn, is based on the concept of the I bond, we can deliberately design clusters that will disobey Wade's rules. For example, we have seen that the $(BH)_8^{2-}$ dodecahedron has two stable crossed-antiaromatic uncapped faces, and we have recognized that a subclass of closoboranes attain deltahedral structures as a result of "rhombic plus butterfly" distortion of regular polygonal faces. This is the characteristic distortion that transforms an antiaromatic to a crossed-antiaromatic species, which ends up having stability comparable to that of an aromatic species. We now put these ideas to work to design nido clusters that fall short of the Wade electron count by two electrons.

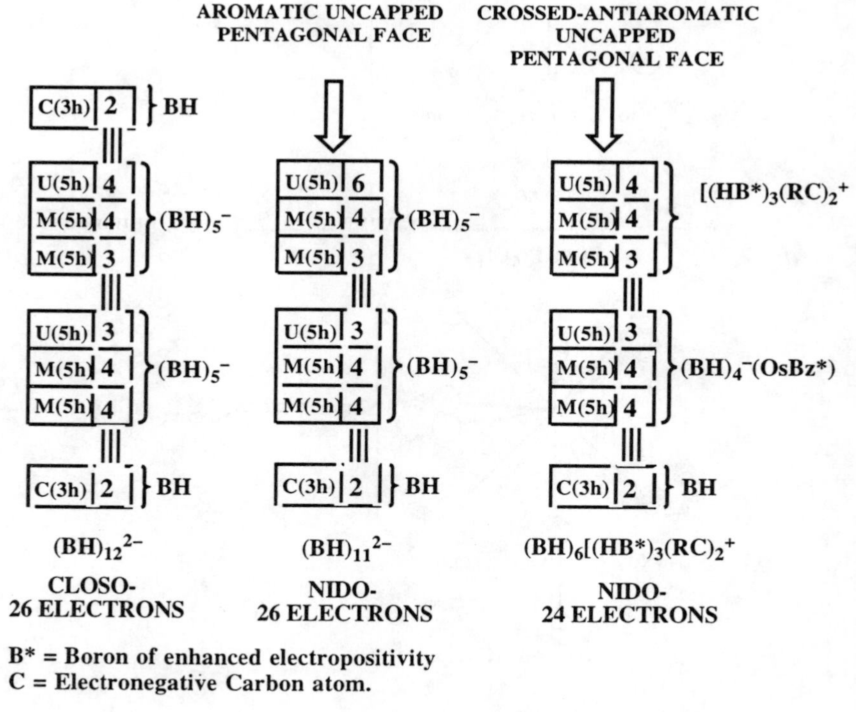

Figure 39.5 Crossed-antiaromatic nido-carboranes violate Wade's rule.

Consider the prototypical icosahedral closoborane $(BH)_{12}^{2-}$ with the magic number of 26 skeletal electrons. Removing one vertical BH core opens up a pentagonal face. We now have nido $(BH)_{11}^{4-}$, which still has MN = 26. Removing two electrons turns nido into closo $(BH)_{11}^{2-}$. Here is the problem: What is the perturbation that can cause $(BH)_{11}^{2-}$ to abandon the expected closo in favor of the nido structure? The answer is: Turn one aromatic pentagonal face of $(BH)_{11}^{4-}$ into a crossed-antiaromatic one by replacing the borons by alternating electronegative and electropositive atoms to the extent possible and removing two electrons. The strategy, always in the context of the toy game, is illustrated in Figure 39.5. The hallmark of the crossed-antiaromatic pentagonal face is familiar: hole–pair alternation (in part), which allows the formation of two I bonds.

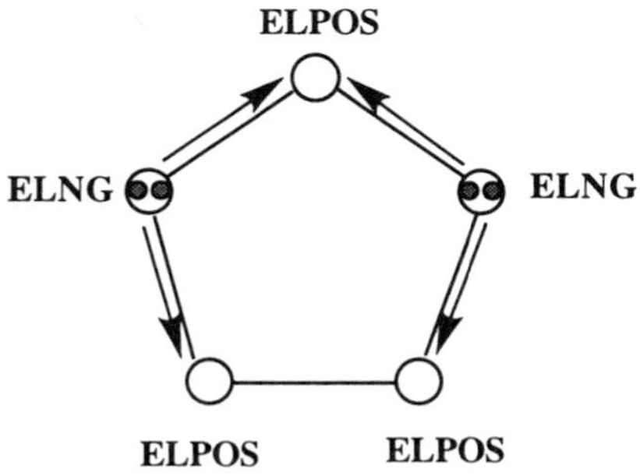

By turning aromatic into stabilized-antiaromatic faces with two less electrons via alternation of borons bearing π donors (to enhance their electropositivity) and electronegative atoms within a square or pentagonal polyhedral face, we can produce many exceptions to Wade's rules in the sense that closo will give way to nido preference. This trend was discovered (and its significance first fully appreciated) by Greenwood and co-workers in their studies of metalloboranes.[25] The carborane I shown in Scheme 39.1 adopts the nido rather than the closo structure. Not only does the open pentagonal face alternate boron and carbon, but at least one of the boron atoms has reduced electronegativity. The reason is that $Os(C_6Me_6)$, isolobal to BH, is present in the underlying bicapped pentagonal face, from which it acts to lower the electronegativity of the overlying boron in the open pentagonal face via its equatorial pairs.

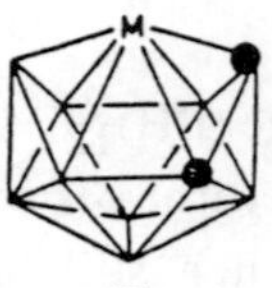

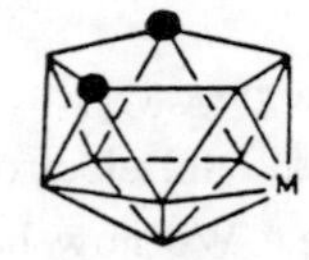

I **II**

1-C_6Me_6-2,4-Me_2-1,2,4-$OsC_2B_8H_8$ ⟹ 2-C_6Me_6-8,10-Me_2-1,2,4-$OsC_2B_8H_8$

CLOSO **NIDO**

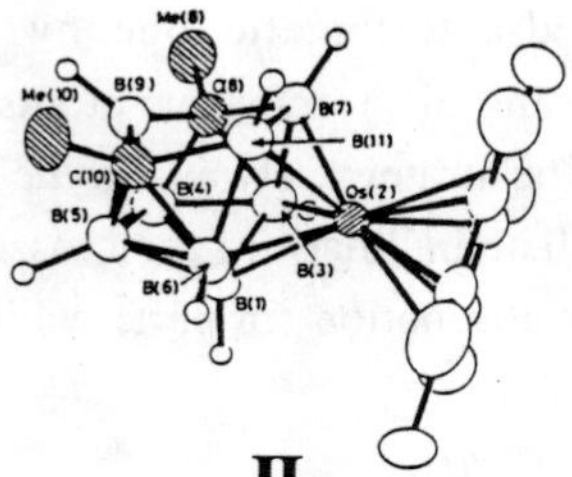

II

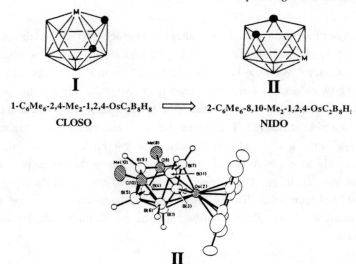

Scheme 39.1 Closo and nido $OsC_2B_8H_8$ clusters.

39.4 The Dependence of Cluster Structure on Electronegativity

Since there are two extreme mechanisms of bonding, T and I, it follows that two isoelectronic molecules can be fundamentally different if the constituent atoms have very different electronegativities. Electropositive (metallic) atoms favor polyhedra with a high ratio of triangular (TR) to square (SQ) faces. As atom electronegativity increases, the TR:SQ ratio of the optimum polyhedron decreases. Ultimately, any polyhedron containing triangular and/or square faces is destabilized relative to dimers infinitely apart from each other. We can find no better illustrator of the principles developed here than the contrasting structures of $[Ni(CO)]_6(CO)_6^{2-}$ (antiprismatic)[9] and $[Pt(CO)]_6(CO)_6^{2-}$ (prismatic).[11] According to the borane analogy, MCO (M = Ni, Pd, Pt) is a zero-electron fragment. Thus, one expects an octahedral structure with seven skeletal electron pairs irrespective of whether M is Ni, Pd, or Pt. The electronic structure of the MCO structure prepared for cluster bonding was shown in Figure 31.1b. This immediately explains the bonding of the triangular $(MCO)_3(CO)_3$ fragment (three bridging carbonyls) which, upon dimerization and addition of two electrons, yields the desired $[M(CO)]_6(CO)_6^{2-}$ complex. The assembly of two triangular $(MCO)_3(CO)_3$ fragments to the final complex is effected by axial hole–pair association as follows:

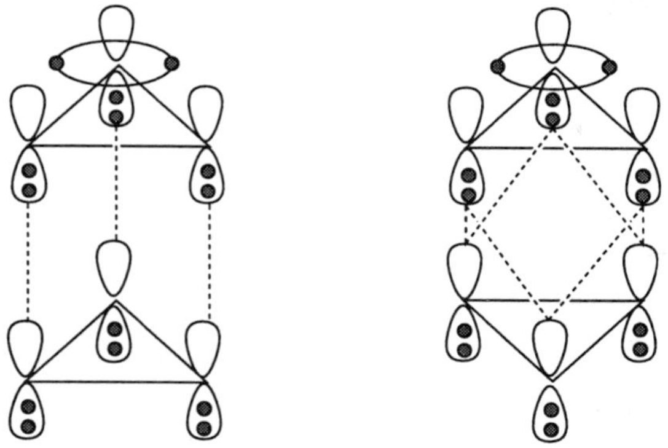

The staggered arrangement is intrinsically favorable for I-bonding because it permits the formation of three I bonds. One pair from the upper disk is connected by two arrows to two holes of the lower disk. On the other hand, the eclipsed arrangement is intrinsically favorable for T-bonding because it permits the formation of three T bonds by turning the eclipsed hole–pair couples into bond pairs. As atom electronegativity increases, I bonds are converted to T bonds and the prismatic is stabilized relative to the antiprismatic structure. This is why replacing Ni by the more electronegative Pt changes the geometry of the cluster from octahedral (antiprismatic) to prismatic. In other words, the Pt cluster is displaced toward the T end of the bonding continuum relative to the Ni cluster.

Calabrese et al.[26] synthesized a high nuclearity prismatic tube that can be viewed as the result of stacking $[Pt(CO)]_3(CO)_3$ disks so that axial pairs of one match axial holes of the other. There are two possible stacking patterns: prismatic and antiprismatic. The first mode is favored by Pt on account of its relatively high ionization potential. The process is completed by adding a pair to the open triangular face at one terminus.

By the way, the one-dimensional version of the prismatic tubes is the unbridged array of $L_2(CO)_3Os—Os(CO)_4—W(CO)_5$ prepared by Pomeroy and his co-workers.[27] A related theme is the construction of clusters by stacking alternately rings and metal atoms. Grimes and co-workers prepared a hexadecker sandwich in which cobalt atoms or cobalt hydrides are sandwiched by C_2B_3 carborane rings.[28] This species is a derivative of the diamagnetic cluster in which pseudo-octahedral metals with three equatorial pairs and three exo (upper) and three endo (lower) holes make delocalized triple bonds with six-π-electron rings. The result is a stack such as the one shown in Figure 39.6 in which each metal atom is connected to two rings (one above and one below) by three I bonds

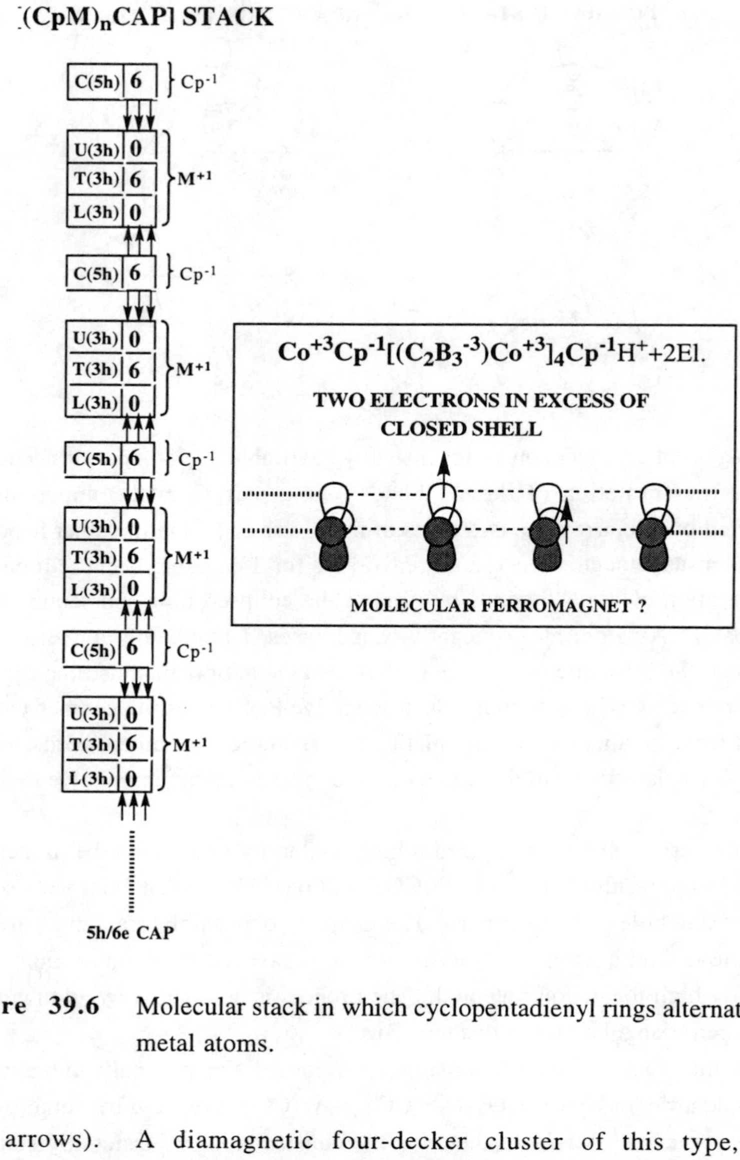

Figure 39.6 Molecular stack in which cyclopentadienyl rings alternate with metal atoms.

(six arrows). A diamagnetic four-decker cluster of this type, with carbon/boron/sulfur rings, was first prepared by Siebert and co-workers.[29] Actually, the Grimes cluster, $Cp^*Co(Et_2C_2B_3H_2Me)Co(Et_2C_2B_3H_3)$—$Co(Et_2C_2B_3H_3)CoH(Et_2C_2B_3H_2Me)CoCp^*$, is paramagnetic. It has one four-electron triplet–triple bond, which delocalizes into neighboring six-electron singlet–triple bonds.

39.5 The Isoelectronic and Isolobal Analogies Are Stoichiometric Guidelines Rather than Bonding Principles

Oxygen has one 2s and three 2p AOs and six valence electrons. Sulfur has one 3s and three 4p AOs and six valence electrons. Oxygen and sulfur are isoelectronic. This explains the similarity of the shapes of HOH and HSH (both molecules are bent). By contrast, we say that the very similarity hides a fundamental difference in bonding mechanism that is due to sulfur's superiority to oxygen with respect to electropositivity. As a result, HOH and HSH lie toward the T- and the I-delocalization limit, respectively. We expect that in two other isoelectronic systems, the difference in bonding mechanism will cause them to have radically different shapes.

Analogous to the isoelectronic principle is the isolobal principle.[30] For example, one may visualize BH as having one radial n and two tangential p AOs of a and e symmetry (in C_{3v}) and two valence electrons. The C_{3v} $Fe(CO)_3$ fragment also has three frontier MOs of a and e symmetry and a total of two electrons. Hence, BH and $Fe(CO)_3$ are said to be isolobal. The analogies can be extended. All the following fragments are either isolobal or isoelectronic, having one a and two e orbitals containing two valence electrons: C, Si, HC^+, HB, $Fe(CO)_3$, $Os(CO)_3$, $Co(CO)_3^+$, $Ir(CO)_3^+$. According to EHMO theory, replacing any one of these groups by another one should not result in a change in the architecture of the molecule unless there is some problem of orbital mismatching or steric repulsion. The notion of "isolobality" was developed by Hoffmann and his school,[30] and it motivated much fruitful research.[31–33] Nonetheless, our conclusions are at odds with the EHMO scenario: The shape of a molecule depends on the mechanism of bonding (i.e., whether T or I delocalization predominates). In turn, the type of delocalization depends on the electronegativity of the central atom. Thus, for example, we expect that despite their isolobality, HC^+ and HB will combine with a given fixed fragment in ways which are stereochemically different. The aspect of chemistry that makes the science itself interesting and challenging is the astounding diversity of molecular shapes and properties. Indeed, the central problem of conceptual chemistry is to make diversity understandable and predictable. It is in this sense that bonding theory starts after symmetry considerations have ended.

39.6 Isoelectronic and Isolobal Molecules of Different Colors Have Different Shapes and Properties

A whole book could be written just on the differences of isoelectronic or isolobal species. Therefore we restrict our attention to the fragment series HC, N, HSi, P, $(CO)_3Co$, and $(CO)_3Ir$, making it unambiguously clear that the "color" of atoms determines the bonding mechanism, hence, their shapes and properties. We begin with some facts.

(a) Nitrogen forms a strong triple bond to a second nitrogen, while phosphorus shows a preference to form rings, cages, and even deltahedral clusters (e.g., P_4).

(b) C_2H_2 is linear, with a carbon–carbon triple bond, while Si_2H_2 is computed to be "butterfly."[34]

(c) HC═CH is a stable organic molecule, but $(CO)_3Co{=}Co(CO)_3$ has not yet been made.

(d) Two acetylenes are more stable than tetrahedrane.[35] Similarly, tetrahedral N_4, irrespective of whether it is a secondary minimum, lies much higher than two N_2 fragments.[36] By contrast, tetrahedral P_4 is more stable than two P_2 fragments by 55 kcal/mol! Similarly, the tetramer $[Co(CO)_3]_4$ is stable and has a bridged tetrahedral structure.[37]

(e) $[Co(CO)_3]_4$ is tetrahedral, involving carbonyls as bridging groups. By contrast, $[Ir(CO)_3]_4$ is tetrahedral with no bridging carbonyls.[38]

39.7 The Isosynaptic Principle in Organometallic Chemistry

One of the major impulses for the development of the theory described in this work was the observation that isoelectronic or isolobal species are not necessarily isostructural. For example, C and Si are isoelectronic and both are isolobal to $Fe(CO)_3$. Yet, the structures of C-containing molecules are different from those of Si-containing isoelectronic molecules. By contrast, replacing Si by $Fe(CO)_3$ does not cause a change of gross molecular architecture. I concluded that the difference between C and Si and between C and $Fe(CO)_3$ must be due to a difference in the binding mechanism: C supports T-bonding, but the more electropositive Si and Fe support I-bonding. On the other hand, Si and $Fe(CO)_3$ must be essentially interchangeable within a molecule without causing a change of shape. We say that these two groups are *isosynaptic*. Two groups are isosynaptic if they are isolobal or isoelectronic and have the same color (see

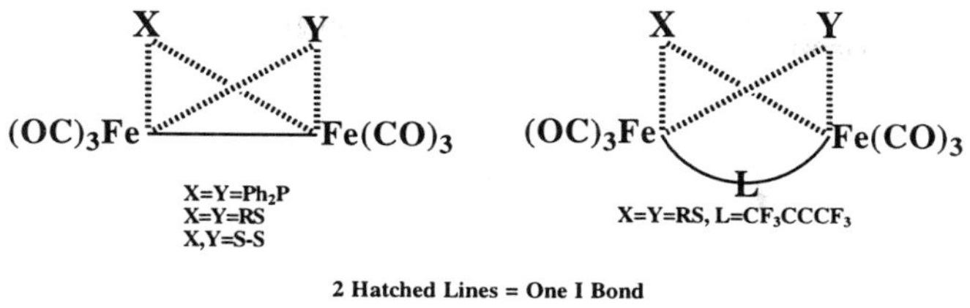

Figure 39.7 Organometallic analogues of the SiHHSi butterfly. Two hatched lines equal one I bond. One solid line equals one T bond.

Chapter 24). Green and red isolobal groups are regarded, to a first approximation, as isosynaptic. For example, red $Fe(CO)_3$ is isolobal to green Si. Hence, these groups should also be isosynaptic. Si_2H_2 was discovered computationally to have a butterfly structure.[34] The contrasting structure of the well-known acetylene molecule to that of disilaacetylene was one of the factors that led us to the hypothesis that stereochemical differences are consequences of different bonding mechanisms. A monograph by Marko and Marko-Monostory played a key role.[39a] If disilaacetylene is a butterfly, then $Fe(CO)_3$ organometallics with butterfly structures must be abundant. Mere perusal of the aforementioned monograph makes it evident that this is indeed the case. Typical structures are shown in Figure 39.7. The key point is that in these structures, replacement of $Fe(CO)_3$ by the isolobal but not isosynaptic C is predicted to produce a derivative having a completely different structure. The disjunction between carbon, on one hand, and silicon or $Fe(CO)_3$ structures, on the other, is projected in Figure 39.8. For example, starting with the SiHHSi butterfly and making the substitutions Si = $Fe(CO)_3$, Si = $Ru(CO)_3$, H = $(PPh_3)Au$, and H = $Co[(Co)_3]$═S, we obtain the butterfly structure reported by Fischer et al.[40c]

Going one step further, it is clear that Si, $Fe(CO)_3$, and $Re(CO)_3^-$ are isosynaptic only to a first approximation. Obviously, Re^- is much more electropositive than Si and, thus, we expect it to lie further along toward the I-delocalization end of the bonding pantheon (see Chapter 3). This is clearly illustrated by the comparison of Si_4 and the isolobal $[Re(CO)_3^-]_4(H^+)_4$ (formal structure). The former is computed to be a rhombus.[41a] By contrast, the rhenium species is an eight-electron, face-protonated tetrahedron.[41b]

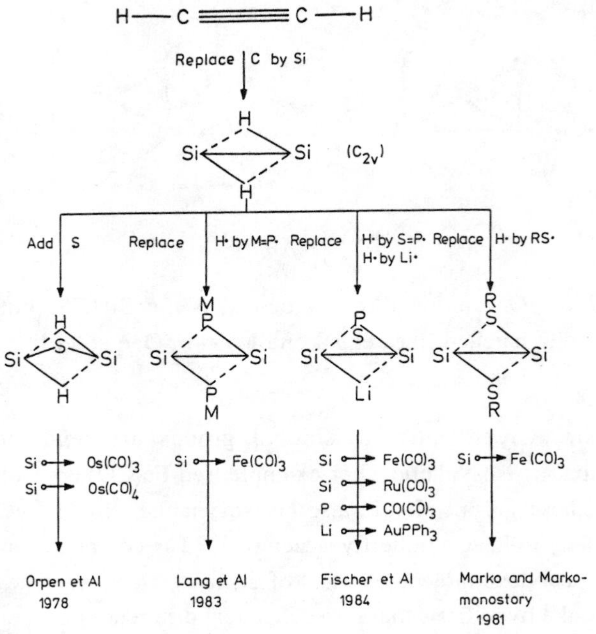

Figure 39.8 Stereochemistry depends on atom color [literature cited in ref. 40].

39.8 T Forms and I Forms of Molecules

We have built sufficient background to see the night-and-day difference of the thinking suggested by conventional models and by VB theory. As a specific case, we consider the VB approach to the elementary molecules CH_4 and SiH_4. In the case of methane, we accept maximal excitation of the central atom in exchange for maximum T-bond making. The predicted structure is the familiar carbon-centered tetrahedron. In the case of silane, however, the strategy changes completely:

(a) According to the colored periodic table (see Chapter 24), the central atom can opt for either T- or I-bonding because it is borderline.

(b) In the T form, Si enters in its ground configuration and makes two covalent bonds to form the conventional SiH_2 as illustrated in Figure 39.9a. Higher excitation to the tetravalent Si configuration produces a second T form, namely, the conventional tetrahedral SiH_4.

(c) In the I form, T bonds are replaced by REL bonds, or, more generally, codirectional arrow trains. The REL bonds must have configurational degeneracy, and they must be complementary so that one activates the other.

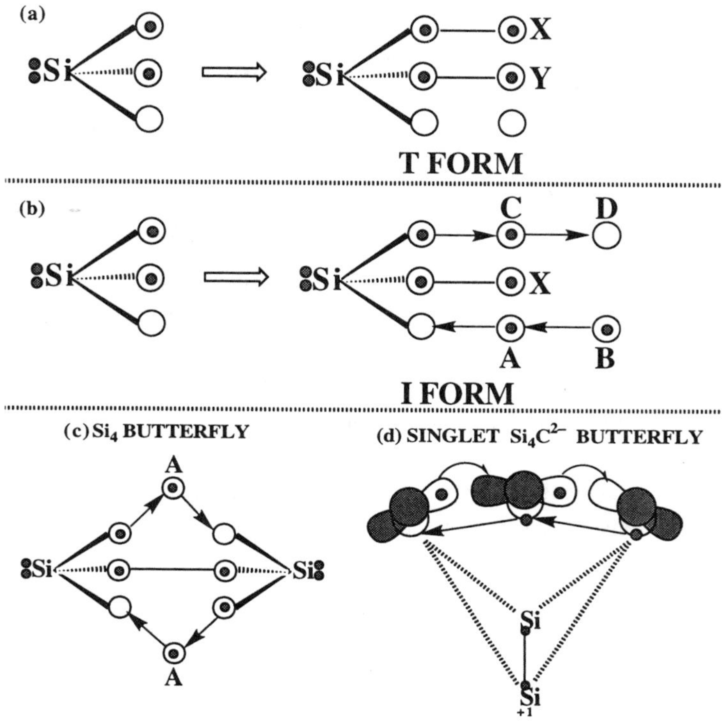

Figure 39.9 T- and I-binding modes initiated by Si atom. The I mode is realized in the silicon butterflies.

Figure 39.9b shows that starting with ground Si, one dot must become coordinated to a radical cation CD, the second dot must be linked covalently to X, and the remaining hole must become attached to a diatomic AB. Furthermore, to ensure configurational degeneracy of the REL bonds, C and A must be efficient bridging atoms (like hydrogen), and B and D must have the same electronegativity as Si. Indeed, most stable silicon compounds belong to the butterfly type shown in Figure 39.9c:

(a) When A = H, we have the Lischka–Koehler molecule. When A = Si, we get the triplet Si_4 butterfly.

(b) The triplet Si_4 butterfly can be coupled to ground triplet C^{2-} by two complementary REL bonds as shown in Figure 39.9d. The entire molecule (with the exception of a single Si—Si covalent bond) is now held together by REL bonds.

(c) Replacing Si by the isosynaptic $Fe(CO)_3$ generates the observed $[Fe(CO)_3]_4C^{2-}$ and the analogous $[Fe(CO)_3]_4COC$.[42] The preferred geometry of the isoelectronic C_5^{2-} is predicted to be the entirely different linear allenic structure *:C═C═C═C═C:*, where the stars are singlet- or triplet-coupled dots.

We can generate transition metal complexes by starting with heavy p-block analogues. Thus, for example, we can start with the Si_4C^{2-} analogue of $[Fe(CO)_3]_4C^{2-}$ and we can replace three Si's by three $Fe(CO)_3$, one wingtip by $Mn(CO)_3^-$, and the C^{2-} by O. This yields the known $[Mn(CO)_3]^-[Fe(CO)_3]_3O$ which, unlike the carbido complex, seems to be unwilling to be protonated.[43] Figure 39.10 shows an illustrative set of data highlighting the implications of "color" for molecular stereochemistry.[44]

The type of analysis just presented leads fast to the discovery of many stable species whose existence had not been suspected until now. For example, Figure

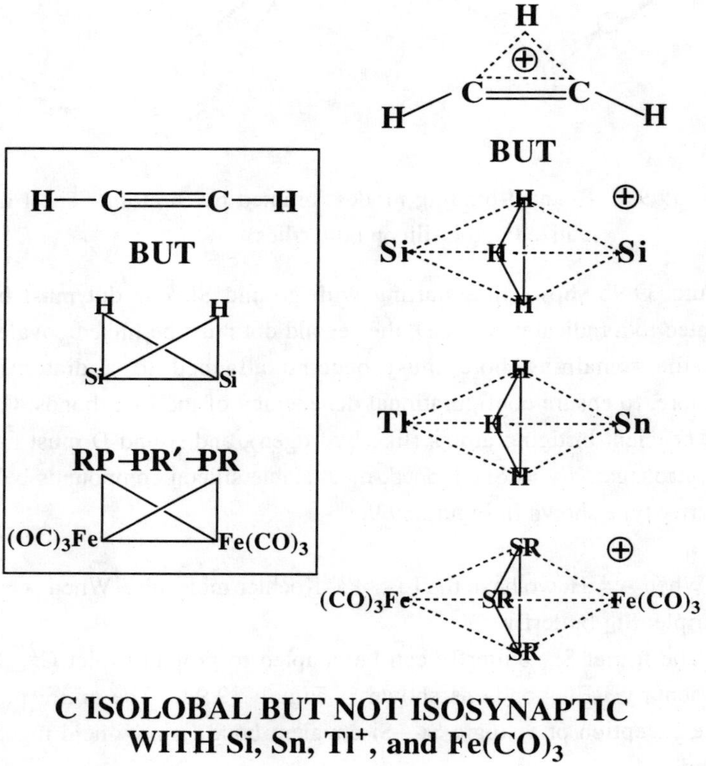

Figure 39.10 Examples of color-dependent stereochemistry.

39.11 shows the isoelectronic Si and BH fragments and formulates sulfur as a relative of silicon in the sense that a hole has been replaced by a pair. While the conventional $H_2B—BH_2$ is the T form of the BH dimer, the butterfly I form is yet another candidate for the global minimum. Things get much more interesting when we go to sulfur. We now see that the conventional HSSH isomer is the T form, while recognizing that the HSSH butterfly represents the I form. The latter differs from the silicon butterfly analogue only in the sense that two holes have been replaced by two pairs. Hence, we expect that the HSSH butterfly could be competitive with the HSSH gauche chain.

Far more important is the realization that replacing H by S produces four-membered rings of the types shown in Figure 39.12. One is the conventional cyclic form, but the second is a triplet butterfly. All we have to do now is make selective substitutions to stabilize the I-forms, as we saw in the case of the silicon molecules [e.g., replace S by $Ni(CO)_3$, etc.]. This opens up an unlimited territory of new molecules where new experiments and calculations

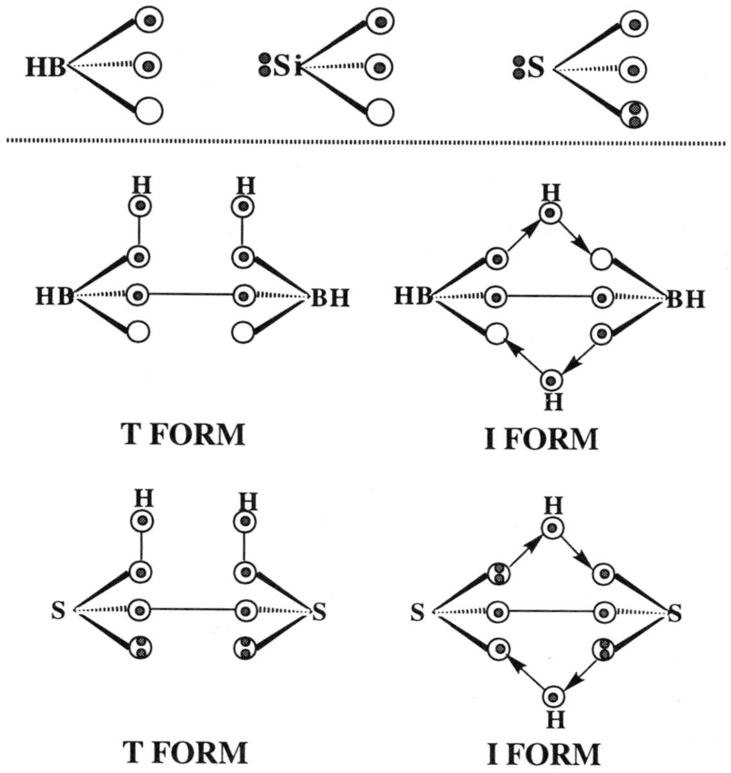

Figure 39.11 Examples of T and I forms of typical semimetals.

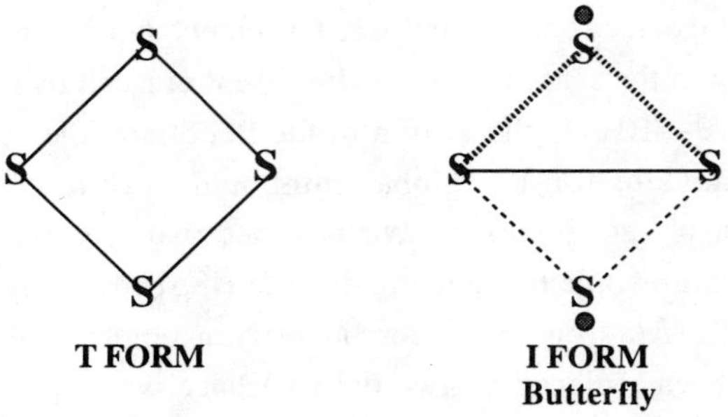

Figure 39.12 T and I forms of S_4.

will serve to forever banish notions of universal covalency to the dustbin of history.

An example of how a molecule comprising heavy main group atoms may inspire the discovery of an organometallic analogue is taken from the work of Einstein and Pomeroy and their associates, who prepared a planar C_{2v} $[Os(CO)_4]_3Os(CO)_3$ cluster.[45] The isosynaptic concept predicts the existence of a corresponding SiS_3 cluster in which divalent Si connected to a chain of three ground S atoms; such a cluster should have a gross geometry similar to that of the Pomeroy cluster. By contrast, the first-row analogue of SiS_3 should have a $O_2C{=}O$ global minimum featuring an O_2C three-membered ring. That is, the existence of the $[Os(CO)_4]_3Os(CO)_3$ organometallic cluster could have been predicted from existing data regarding heavy main group clusters and usage of the isosynaptic concept. Specifically, S_4^{2+} and related M_4^{2+} (M = S, Se, Te) species are known to have a planar square geometry.[46] Since Si and S are heavy main group atoms, the isoelectronic SiS_3 must have the same gross structure (distorted square) and this fact, plus the isosynaptic analogies, recommend the Pomeroy cluster as a feasible synthetic target.

Bottom line: The ways we think about methane and silane and their derivatives are entirely different. That polysilanes often resemble alkanes is an accident: Because Si is a borderline atom, the requisite high T complement of the REL bonds promotes structures that are analogous to those predicted by the construct of the covalent bond. That this is no more than coincidence can be fast appreciated: SiH_4 is tetrahedral but Si_2H_2 is butterfly. Figure 39.9 gives the algorithm for the prediction of countless stable silicon-containing molecules. The predicted global minima (especially as Si is replaced by heavier congeners) are very different from what one would predict by falling back on standard concepts. The knowledge that CH_4 and SiH_4 are isoelectronic and isostructural put chemists on the wrong track of seeking additional similarities between fundamentally unlike molecules: those that contain metal and those that do not.[47] If two classes of molecules have similarities as well as differences this does not mean they are not different, only that their similarities are accidental! On the other hand, the isoelectronic analogy remains valid when one is comparing molecules made up exclusively of nonmetal atoms. The challenge now is to develop new ideas for recognizing topological similarities in large "organic" molecules and macromolecules. Some novel mathematical concepts have been suggested by Mezey.[48]

References

1. (a) K. Wade, *J .Chem. Soc. Chem. Commun.* 792 (1971). (b) K. Wade, *Adv. Inorg. Chem. Radiochem.* 18, 1 (1976).
2. R. Hoffmann, *Angew. Chem. Int. Ed. Engl.* 21, 711 (1982).
3. R.N. Grimes, *Acc. Chem. Res.* 16, 22 (1983)
4. V.R. Miller, R. Weiss and R.N. Grimes *J. Am. Chem. Soc.* 99, 5646 (1977).
5. T.L. Venable, E. Sinn, and R.N. Grimes, cited in ref. 3.
6. P.E. Riley and R.E. Davis, *J. Organomet. Chem.* 113, 157 (1976).
7. J.R. Pipal and R.N. Grimes, *Inorg. Chem.* 16, 3255 (1977).
8. (a) J.R. Pipal and R.N. Grimes, *Inorg. Chem.* 18, 257 (1979). (b) J.R. Bowser, A. Bonny, J.R. Pipal, and R.N. Grimes, *J. Am. Chem. Soc.* 101, 6229 (1979).
9. J.C. Calabrese, L.F. Dahl, A. Cavalieri, P. Chini, G. Longoni, and S. Martinengo, *J. Am. Chem. Soc.* 96, 2616 (1974).
10. R.G. Holloway, B.R. Penfold, R. Colton, and M.J. McCormick, *J. Chem. Soc. Chem. Commun.* 485 (1976).
11. J.C. Calabrese, L.F. Dahl, P. Chini, G. Longoni, and S. Martinengo, *J. Am. Chem. Soc.* 96, 2614 (1974).
12. P. Chini, *J. Organomet. Chem.* 200, 37 (1981).
13. K.J. Deck, Y. Nishihara, M. Shang, and T.P. Fehlner, *J. Am. Chem. Soc.* 116, 8408 (1994).
14. M.C. Chisholm, D.L. Clark, M.J. Hampden-Smith, and D.H. Hoffman, *Angew. Chem. Int. Ed. Engl.* 28, 432 (1989).
15. E.L. Muetterties, *Chem. Eng. News* 28 (August 30, 1982).
16. D.M.P. Mingos and D.J. Wales, *Introduction to Cluster Chemistry*, Prentice-Hall, Englewood Cliffs, NJ, 1990.
17. P.A. Vaughan, *Proc. Natl. Acad. Sci. USA* 36, 461 (1950).

18. A. Simon, H.G. von Schnering, H. Wohrle, and H. Schaefer, *Z. Anorg. Allg. Chem.* 339, 155 (1965).
19. For early calculations, see F.A. Cotton and T.E. Haas, *Inorg. Chem.* 3, 10 (1964).
20. K. Brodersen, G. Thiele, and H.G. von Schnering, *Z. Anorg. Allg. Chem.* 337, 120 (1965).
21. A. Simon, H.-G. von Schnering, and H. Schaefer, *Z. Anorg. Allg. Chem.* 355, 295 (1967); H. Imoto and A. Simon, *Inorg. Chem.* 21, 308 (1982).
22. M. Spangerberg and W. Bronger, *Angew. Chem. Int. Ed. Engl.* 17, 368 (1978).
23. F. Cecconi, C.A. Ghilardi, and S. Middolini, *Inorg. Chim. Acta* 64, L-47 (1982).
24. F. Cecconi, C.A. Ghilardi, and S. Middolini, *J. Chem. Soc. Chem. Commun.* 640 (1981).
25. N.C. Greenwood, in *The Chemistry of Inorganic Ring Systems*, R. Steudel, Ed., Elsevier, Amsterdam, 1992, Chapter 1.
26. See ref. 11.
27. R.J. Batchelor, H.B. Davis, F.W.B. Einstein, and R.K. Pomeroy, *J. Am. Chem. Soc.* 112, 2036 (1990).
28. X. Wang, M. Sabat, and R.N. Grimes, *J. Am. Chem. Soc.* 116, 2687 (1994).
29. W. Siebert, C. Bohle, and C. Kruger, *Angew. Chem. Int. Ed. Engl.* 19, 746 (1980).
30. R. Hoffmann, *Angew. Chem. Int. Ed. Engl.* 21, 711 (1982).
31. F.G.A. Stone, *Angew. Chem. Int. Ed. Engl.* 23, 89 (1984).
32. W.A. Herrmann, *Angew. Chem. Int. Ed. Engl.* 25, 56 (1986).
33. G. Huttner, *Angew. Chem. Int. Ed. Engl.* 26, 743 (1987).
34. H. Lischka and H.-J. Koehler, *J. Am. Chem. Soc.* 105, 6646 (1983). The SiHHSi butterfly has been observed in the laboratory: M. Bogey, H. Bolvin, C. Demuynck, and J.-L. Destombes, *Phys. Rev. Lett.* 66, 413 (1991).
35. H. Kollmar, F. Carrion, M.J.S. Dewar, and R.C. Bingham, *J. Am. Chem. Soc.* 103, 5292 (1981).
36. P. Saxe and H.F. Schaefer III, *J. Am. Chem. Soc.* 105, 1760 (1983).
37. J. Lewis and B.F.G. Johnson, *Adv. Inorg. Chem. Radiochem.* 24, 225 (1981).
38. Transition metal clusters have been reviewed ad nauseum. A brief overview can be found in the first two chapters of ref. 16.

39. (a) L. Marko and B. Marko-Monostory, in *The Organic Chemistry of Iron*, Vol. 2, E.A.K. von Gustorf, F.-W. Grevels, and I. Fischler, Eds., Academic Press, New York, 1981. (b) P. Chini, in *The Organic Chemistry of Iron*, ibid., p. 89.

40. (a) A.G. Orpen, A.V. Rivera, E.G. Bryan, D. Pippard, G.M. Sheldrick, and K.D. Rouse, *J. Chem. Soc. Chem. Commun.* 723 (1978). (b) H. Lang, L. Zsolnai, and G. Huttner, *Angew. Chem. Int. Ed. Engl.* 22, 976 (1983). (c) K. Fischer, M. Mueller, and H. Vahrenkamp, *Angew. Chem. Int. Ed. Engl.* 23, 140 (1984). (d) Ref. 39a.

41. (a) G. Pacchioni and J. Koutecky, *Ber. Bunsenges. Phys. Chem.* 88, 242 (1984). (b) R.D. Wilson and R. Bau, *J. Am. Chem. Soc.* 98, 4687 (1976).

42. J.S. Bradley, *Adv. Organomet. Chem.* 1983, 22, 1.

43. M.A. Beno, J.M. Williams, M. Tachikawa, and E.L. Muetterties, *J. Am. Chem. Soc.* 103, 1485 (1981).

44. (a) C_2H_2 (exp.): W.J. Lafferty and R.J. Thibault, *J. Mol. Spectrosc.* 14, 79 (1964). (b) Si_2H_2 (calc.) H. Lischka and H.-J. Koehler, *J. Am. Chem. Soc.* 105, 6646 (1983). (c) $[Fe(CO)_3]_2(RPH)_2$ (exp.): J. Borm., L. Zsolnai, and G. Huttner, *Z. Naturforsch.* 41B, 532 (1986). (d) $C_2H_3^+$ (calc.): H. Lischka and H.-J. Koehler, *J. Am. Chem. Soc.* 100, 5297 (1978). (e) $Si_2H_3^+$ (calc.): H.-J. Koehler and H. Lischka, *Chem. Phys. Lett.* 112, 33 (1984). (f) $SnTl(RO)_3$ (exp.): M. Veith and R. Roesler, *Angew. Chem. Int. Ed. Engl.* 21, 858 (1982). (g) $[Fe(CO)_3]_2(RS)_3^+$ (exp.): A.J. Schultz and R. Eisenberg, *Inorg. Chem.* 12, 518 (1973).

45. V.J. Johnston, F.W.B. Einstein, and R.K. Pomeroy, *J. Am. Chem. Soc.* 109, 7220 (1987).

46. R.J. Gillespie, *Chem. Soc. Rev.* 8, 315 (1979).

47. A.A. Aradi and T.P. Fehlner, *Adv. Organomet. Chem.* 30, 189 (1990).

48. P.G. Mezey, *Shape in Chemistry. An Introduction to Molecular Shape and Topology*, VCH, New York, 1993.

Part V

Chemoelectricity, Chemomagnetism, and Beyond

Chapter 40

The Representation of the Electrical Properties of Solids by the I Formulas

40.1 Is Band Theory Chemically Useful?

One stringent test of a modern conceptual theory of chemical bonding is the interpretation of superconductivity at normal and high critical temperatures (T_c) on the microelectronic level. The discovery of high T_c superconductivity challenges us to develop an understanding of the structure–function relationship that exists in solid state physics. This type of problem is at the heart of chemistry itself. To be more specific, we wish to understand not only why metals can be superconductors but also why Sc and Cu are only conductors with widely differing resistivities while V is a superconductor, why V has a higher T_c than Ti, why $La_{1.85}Sr_{0.15}CuO_4$ has a higher T_c than any pure metal, why $YBa_2Cu_3O_7$ has a higher T_c than $La_{1.85}Sr_{0.15}CuO_4$, and so on. The ultimate question is: How do we go about raising the T_c of a superconductor?

It is impossible to understand superconductivity without first understanding conductivity at the microelectronic level. It is impossible to understand conductivity without first understanding the electronic structure of metals. It is impossible to understand the electronic structure of metals without prior understanding the electronic structures of metal dimers. It is impossible to appreciate the electronic structures of metals without comparing them to those of nonmetals. Because we have graduated from the chemistry of nonmetals *and* metals, we can now turn our attention to conductivity and superconductivity.

We have seen that the VB formulas of H_2 and Li_2 are entirely different. Since our formula of a molecule as simple as Li_2 is justified only at the level of "perfect theory" at which electron–electron repulsion plays a key role, since the formula is very different from that of H_2, and since the popular models of chemical bonding, such as EHMO theory, neglect electron–electron repulsion, the conclusion is inescapable: One-electron molecular theory stands as an obstacle to our understanding of metal-containing molecules. In an exactly analogous fashion, one-electron band theory stands as an obstacle to our understanding of metal-containing solids.

Because of the element's respectable electronegativity, atoms of hydrogen rely for bonding on exchange, rather than CT, delocalization. As a result, association is thwarted and even hexagonal H_6 (formally "aromatic") distorts to three H_2 infinitely apart from each other. The solid state analogue of this distortion is called *Peierls distortion*. Exactly the same feature of the H atom (i.e., modestly high electronegativity) renders the hypothetical H_{2n} polymer an insulator. Interatomic CT to produce H^+ $H:^-$ is energetically unfavorable because the electronegative, hence contracted, H1s becomes responsible for high on-site interelectronic Coulomb repulsion in $H:^-$. In solid state theory, this is referred to as *Mott insulation*.

Hydrogen is a gas because two H atoms form a covalent bond by the exchange mechanism, the covalent bonds of two H_2 molecules repel each other, a Peierls distortion transforms H_{2n} to nH_2, and we have an insulator. But the very presence of covalent bond in H_2 implies that the ionic excited state, which could promote both association (I-bonding) and conduction, is too high in energy. This means that Peierls distortion and Mott insulation are two faces of the same coin. So, hydrogen is an insulator (at room temperature) because H_2 is T-bound. That lithium behaves in exactly the opposite way is a fundamental consequence of the I-boundedness of Li_2. Two Li cations bridge a pair of electrons called the gas pair (or vice versa) with Coulomb forces holding the four pieces together. In solid Li, the atomic nuclei are embedded in an electron gas, and Peierls instability and Mott insulation disappear together. The Li atoms are indirectly glued together via attractive Coulomb interaction with interstitial electrons that keep away from each other to bring interelectronic repulsion to a minimum. In short, Li_2 units stick together while H_2 molecules repel each other because Li_2 and H_2 have little in common insofar as bonding is concerned. Since this difference shows up only at a high level of theory, trying to understand solid state metal bonding by using one-electron models is an exercise in futility.

Is there something useful in one-electron band theory? Could it not be that the type of band occupancy (quarter-filled, half-filled, three-quarters filled, etc.) is a useful index of solid properties? It is precisely this area of "minimum expectations" that exposes the illusory nature of one-electron band theory, for we have already seen examples of a magic electron count being a necessary but not sufficient condition for the manifestation of a phenomenon: D_{6h} H_6 has an aromatic count of $4N + 2$ electrons but is unstable relative to three H_2 molecules. By the same token, a band may be partly full and there may still be no conductivity. In an excellent monograph, Duffy points out that the electrical properties of solid (rock salt) transition metal oxides cannot be rationalized by band theory.[1] The octahedral metal atom has a set of three t_{2g} AOs that are used to construct the solid band. Transition metals with fewer than six electrons in the t_{2g} AOs (which fill before the higher e_g AOs) are all expected to be

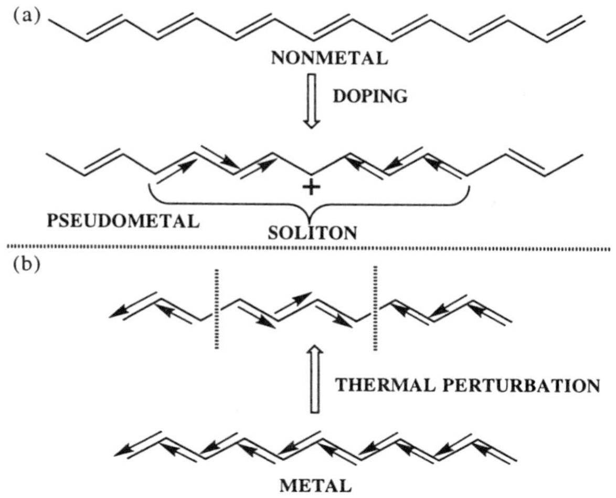

Figure 40.1 Arrow formulas of nonmetallic, pseudometallic, and metallic "polyenes."

conductors because they have partly filled t_{2g} bands. The prediction succeeds with TiO and VO but fails with CrO, MnO, FeO, and CoO. Once again, as in the case of aromaticity, the second condition is that the metal atoms be electropositive. In general, electropositivity decreases as we sweep from left to right along a d-block row.

40.2 I Formulas and Conductivity

The I formulas are based on the concept of the I bond and the association rule, and they are applicable to all metal-containing systems. Because they capture the physics of the problem, they reveal directly and in a pictorial and nonmathematical fashion the electronic structures of solid insulators and conductors, suggesting new designs inspired by the chemistry of molecules to which we have applied them. We illustrate our approach by reference to a one-dimensional π-conjugated system. In the case of carbon atomic centers, the appropriate representation is given by the T formula shown in Figure 40.1. The system is turned into a conductor by oxidation, which produces a soliton that is represented by an I/T formula. When the atomic centers are metal atoms, the formula changes to an I formula in which the arrows comply with the association rule. The hallmarks of the formula are two:

The presence of two dangling arrows implies instability. This is the Peierls instability explained physically in a fraction of a page!

Arrow codirectionality. This minimizes resistivity.

The I formula directly suggests two ways of raising the resistivity of the system:

By effecting a perturbation, such as thermal excitation, which destroys arrow codirectionality.

By applying a polarization perturbation that causes differentiation of the arrows to superior and inferior. The limit is a nonconductive ionic solid:

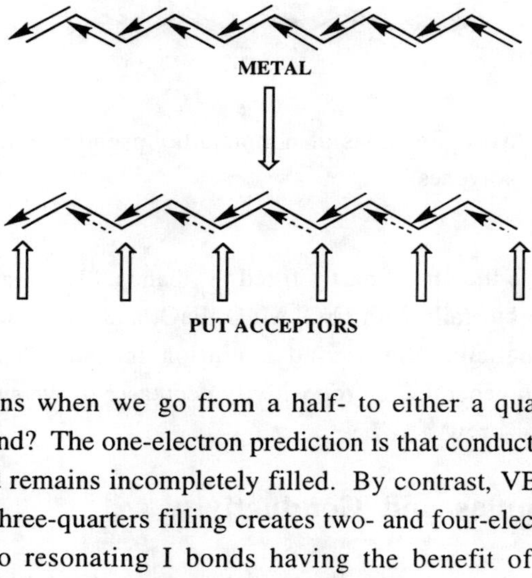

What happens when we go from a half- to either a quarter- or a three-quarters-filled band? The one-electron prediction is that conductivity is preserved because the band remains incompletely filled. By contrast, VB theory suggests that quarter and three-quarters filling creates two- and four-electron solitons that are equivalent to resonating I bonds having the benefit of configurational degeneracy. These considerations can be illustrated as follows:

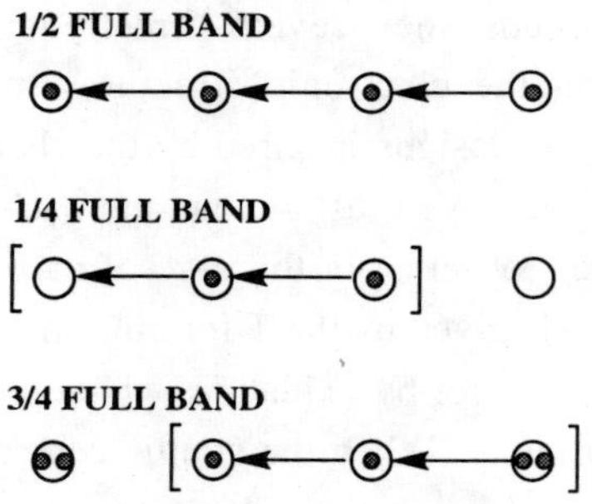

Peierls instability of metal solids, organometallic reactions, and electrical conductivity have one common denominator: the dangling arrows, which imply instability. To see this point, consider a ground organometallic complex as a "metastable" species carrying dangling arrows that can readily ascend to a transition state, where they coalesce to allow full CT delocalization (i.e., conduction) at the expense of atom promotion. The best example is the reaction of the type shown below, which appears variously in the organometallic literature under the headings "α- or β-elimination," "migratory insertion," and "ligand coupling." The fundamental nature of the process is a consequence of its very essence: the transformation of "nonconducting reactants" into "conducting transition state." The transition state is a mere "transition state" (and not a global minimum) only because of the promotional energy requirement for conduction.

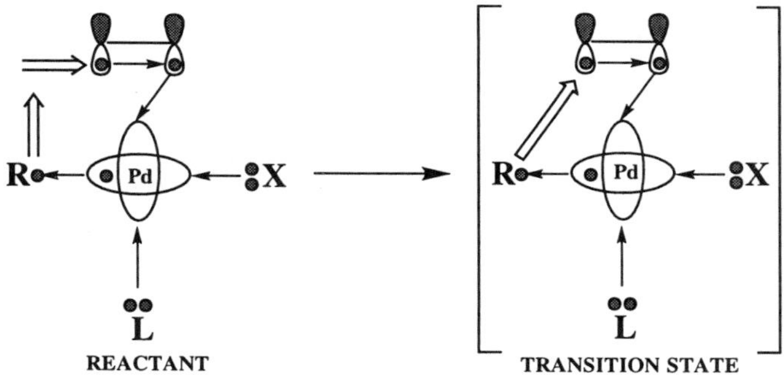

PRODUCT: R—CH_2—CH_2—PdLX

40.3 Anchored Prometals and Support Systems

We have argued that the π bonds of "aromatic" benzene repel each other. The problem is accentuated in the "nonaromatic" polyacetylene. In both systems, the σ frame constrains the π bonds to stay in mutual proximity. As a result, we can think of polyacetylene as made up of two components:

The constraining σ system.
The constrained π system.

From the work of Heeger, we know that polyacetylene, normally an insulator, can be doped to become a conductor.[2] Hence, we can envision the π system as a *prometal*, an insulator or semiconductor subsystem that can be turned into a conductor by some appropriate treatment. In the case of polyacetylene, the prometal π subsystem is anchored by a *supporting frame*, which is the σ subsystem. These definitions can be used to classify diverse materials that differ with respect to the nature of the prometal subsystem and the nature of the supporting frame. In this light, we can depict polyacetylene as follows:

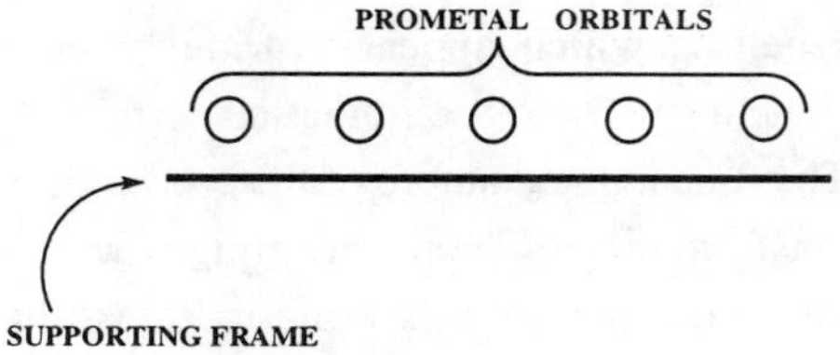

where the solid line denotes the supporting subsystem and the circles the orbitals of the linear prometal. As an additional example, consider the electronic structure of tetracyanoplatinate (TCP).[3]

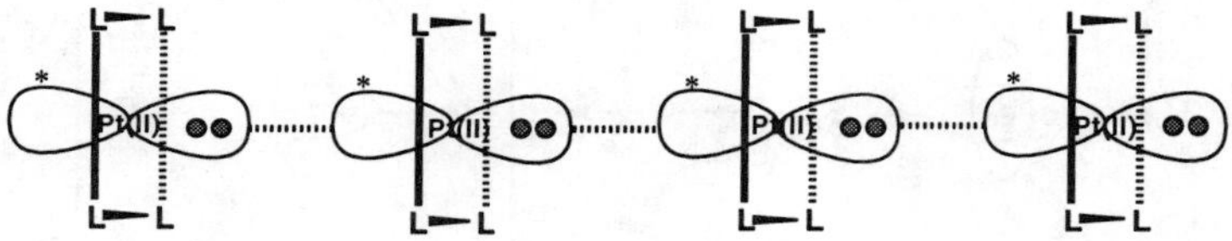

L = CN⁻

In recent years a great amount of activity and interest has centered on so-called one-dimensional "organic metals."[4–9] These are chains or stacks of organic fragments or molecules that show significant conductivity despite a temperature dependence different from that of a pure metal. We now take a look at the TCNQ (tetracyanoquinodimethane) and TTF (tetrathiafulvene) segregated stacks in the TTF-TCNQ solid.

What holds a TCNQ radical anion stack together in TTF-TCNQ where *complete* electron transfer from TTF to TCNQ has taken place? Here is an example of a problem that cannot be approached productively by the conventional MO models. By contrast, the formulas of VB theory speak for themselves: The supporting frame is defined by the subsystem of orbitals and electrons, which engender an attractive static interaction of the individual units as shown in Figure 40.2. The stack of TCNQ anion radicals is now visualized as a linear "pseudometallic polyene" which is held together by E bonds (like TCP) rather than by T bonds (like polyacetylene). Depending on the strength of the E bonds, we can have any one of three situations:

The unstable stack decomposes to individual TCNQ anion radicals or to corresponding dimers.

The stack dimerizes but retains its integrity; that is, we now have a stack with alternating short–long inter-TCNQ distances.

The stack remains uniform, with equidistant TCNQ units.

For many π donors, the ground state turns out to be the dimerized stack with a conducting uniform stack being a thermally accessible excited state.[10]

Incomplete electron transfer, say, half-electron transfer, creates a situation in which the static interaction now binds alternating ground TCNQ molecules and TCNQ anion radicals. Figure 40.3 shows how the binding of the supporting frame comes about and lists the electrons defining the prometal subsystem. Analogous considerations apply to the companion TTF stacks.

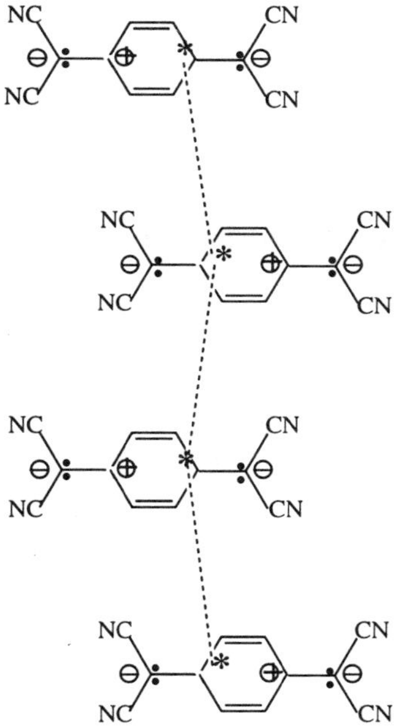

Figure 40.2 The bonding and the conduction channel (dashed line) of a stack of TCNQ anion radicals; the electrons contained in the AOs of the linear prometal subsystem are indicated by asterisks.

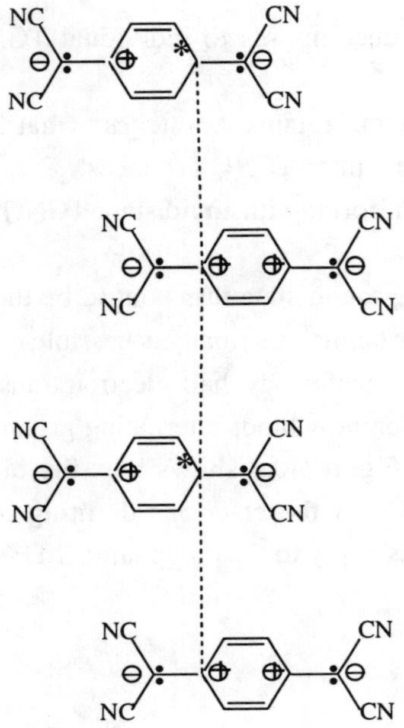

Figure 40.3 Bonding and conduction channel of a stack of TCNQs, half of which are reduced.

40.4 The Design of Organic Conductors

An organic conductor is produced by doping an insulator. How should we choose the substituents to bring about a reduction in the resistivity? Since electron–electron repulsion is steep, the substituents flanking the conducting strip must be π acceptors (V), as illustrated:

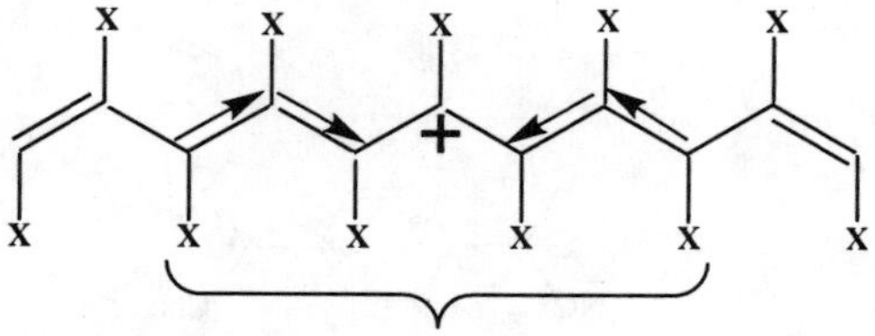

Some options are shown schematically in Figure 40.4. Interestingly, a recently reported "organic metal" champion of conductivity,[11] shown in Figure 40.5, is one of the designed systems.

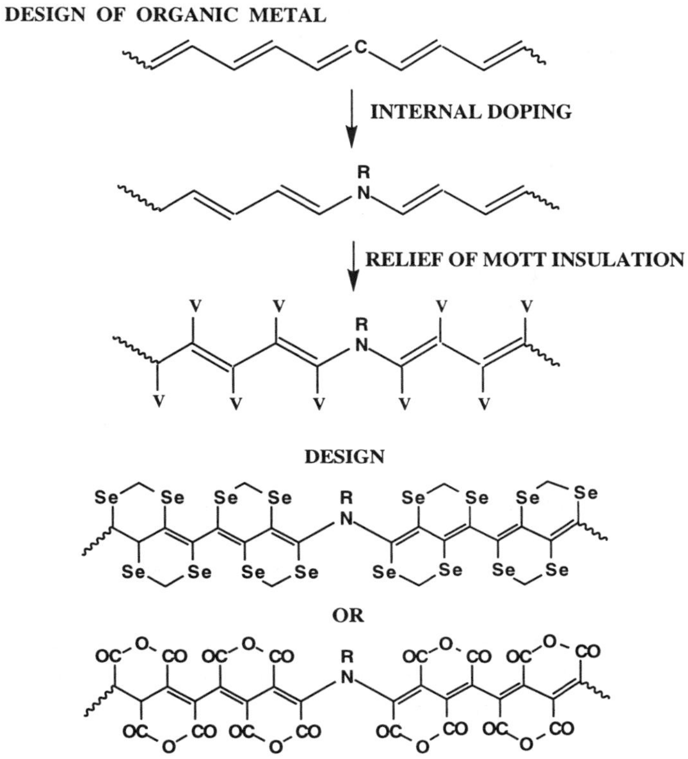

Figure 40.4 Design of an organic metal.

The analogy between our model organic conductor (Figure 40.4) and polycroconaine (Figure 40.5) is made clear by drawing the molecular skeleton of the latter in a way that exposes the repeating subunit as an "internally doped" derivative of polyacetylene-bearing V groups.

We can now return to the problem of conducting salts in which TTF acts as the donor. These compounds conduct electricity along the TTF stack because the sulfur or selenium atoms lend their nonvalence unoccupied space to the conducting π electrons. Thus, a TTF acceptor conducting system is, in a way, analogous to the croconaine system! Conductivity increases by replacing S by Se because the assisting vacuum space is improved as we go toward the metallic limit. The same thing is expected to happen when one replaces S by Se in the croconaine system.

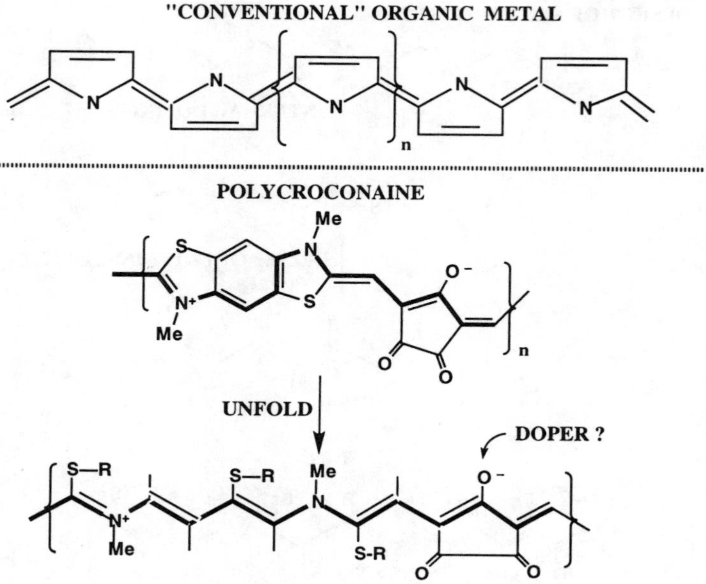

Figure 40.5 Reformulation of the organic conductor polycroconaine, exposing the substitution of the "conduction chain" by groups having low-lying unoccupied orbitals.

40.5 The Conductivity of Pure Metals

We have left the "conventional" for the last. What determines the conductivity of pure metals? Each metal atom within a solid binds primarily from its s^2d^{n-2} or s^1d^{n-1} configuration, whichever is lower. A one-dimensional metal can be visualized as a cylinder with a diameter equal to the radius of maximum electron density of the metal s AO (R_s) and with the surface of the cylinder occupied by s electrons. These are the metallic electrons. For constant R_s (and cylinder surface), and assuming that correlation effects are comparable (a point to which we will return shortly), the greater the number of metallic electrons, the lower their mobility (because of the enhanced probability of repulsive interaction of one with the remaining electrons), and the greater the tendency of the metal to be an insulator. According to this model, the resistivity of a solid should correlate with the promotional energy P required for taking the atom from the s^2d^{m-2} to the s^1d^{m-1} configuration.

$$P = ns^2\,(n-1)d^{m-2} \longrightarrow ns^1\,(n-1)d^{m-1}$$

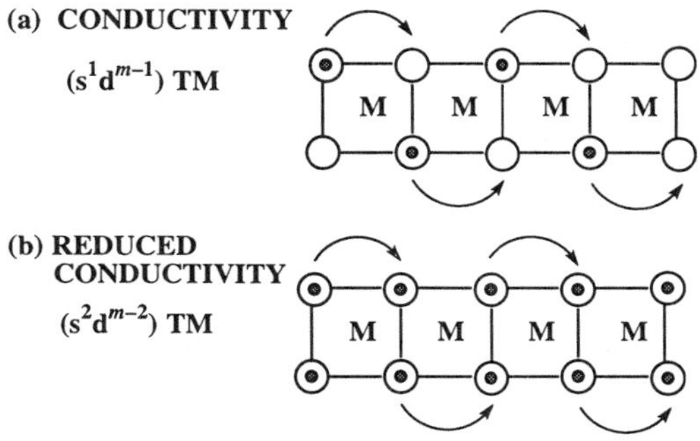

Figure 40.6 Transition metal conductivity depends on the relative stability of the $s^2d^{(m-2)}$ configuration.

When the metal is in the ns^1 $(n-1)d^{m-1}$ configuration, interstitial vacant AOs allow electrons to move without running into the repulsive field of another electron, as illustrated in Figure 40.6. By contrast, vacancies disappear when the metal is in the s^2 configuration, interelectronic repulsion thwarts mobility, and the solid tends to become an insulator. In summary, the higher the *P*, the greater the ns AO occupancy and the lower the conductivity.

It is too much to expect that a single parameter will correlate data across an entire row of metal elements. Yet, if one focuses on the trends of resistivity and *P* within each transition series, there are only two deviations (Ru/Rh and Ir/Pt). This is an amazing result, which is too good to be coincidental. Illustrative data for the first transition series are given in Table 40.1. It is seen that resistivity decreases from a high value in Sc all the way to Cr, and then peaks at Mn, which has the largest negative *P* of all first series transition elements. Then, it drops and levels off in sweeping from Fe to Zn.

Table 40.1 Promotional Energy *P* and Resistivity ρ for Metals in the First Transition Series

	Sc	Ti	V	Cr	Mn
P, eV	1.43	0.81	0.25	−1.00	2.14
ρ, $10^{-8}\Omega \cdot m$	61.0	42.0	25.0	12.9	185.0
	Fe	**Co**	**Ni**	**Cu**	**Zn**
P, eV	0.87	0.42	0.00	−1.49	
ρ, $10^{-8}\Omega \cdot m$	9.7	6.2	6.8	1.7	5.9

It should be pointed out that intrinsic resistivity is commonly written as a sum of one temperature-dependent and one temperature-independent term. The former is thought to reflect the action of the lattice and the latter the action of impurities. By contrast, according to our approach, the temperature-independent term is due to interference of conduction electrons by other conduction electrons, and this depends on *P*. Furthermore, an argument can be made that both the impurity effect and electron–electron repulsion have the same origin and a common index, namely, *P*. To see this point, consider that the difference in intrinsic resistivity between Cr and Mn is due primarily to the difference in the temperature-independent component of the same metals.[12] Thus impurities act much more deleteriously in Mn. In other words, Mn is such a bad conductor compared to Cr because of the impurity content and/or the impurity action. Impurities are much more likely to poison Mn than Cr because Mn has a much larger *P*, which, in itself, is one reason for the exceptionally low bond dissociation energy of the metal dimer and the cohesion of the solid. The more positive *P*, the less capable the metal to effect deshielding and engender strong bonding, as discussed earlier. The lower the cohesive energy of the metal, the more it will tend to be receptive to impurities.

We now sharpen the analysis and recognize the important effect of the vacuum space:

(a) The highly electropositive alkali and rare earth metals must have an expanded vacuum space. We indicate this symbolically by adding a second concentric cylinder on which electrons can be delocalized as illustrated in Figures 40.7a and 40.7b. This is equivalent to saying that the effective R_s of these metals is greater than that of transition metals because of their higher electropositivity.

(b) According to ab initio computations as well as inferences drawn from spectroscopic data, the ratio of the s and d radii of maximum electron density, called the $R_{ns}/R_{(n-1)d}$ ratio, increases along a row of the periodic table as one sweeps from the alkali metals to the inert gases. By contrast, the R_{np}/R_{ns} ratio decreases in the same direction. Finally, the $R_{ns}/R_{(n-1)d}$ ratio decreases down a column of the periodic table. These data help us to differentiate K from Cu and Ca from Zn in the following way: Because the 4p is much more contracted relative to 4s in Cu and Zn, these AOs can be mixed (at one and the same center) much more efficiently than the corresponding AOs of K and Ca. As a result, they can correlate the motion of electrons by the dispersion mechanism. The dispersion term involving correlated intra-atomic 1-e hops depends on a bielectronic exchange integral of the type (sp|sp). In turn, this depends on the

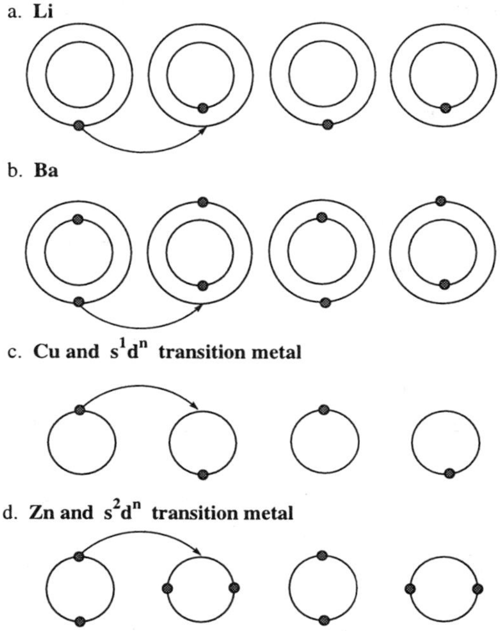

Figure 40.7 Electron transport in representative metals.

relative size of the s and p AOs: the more similar in size, the greater the dispersion stabilization.

Figure 40.7 shows the critical differences due to correlation among metal solids of different types. The alkalies and rare earths are placed on the uncorrelated limit and the late transition Cu and Zn families at the correlated limit.

40.6 Ferromagnets

There exist two conventional strategies for making a ferromagnetic system.

(a) The method of the "covalently anchored spins," according to which atoms with unpaired electrons are covalently bonded to a nonmetal chain that because of the mere skeletal constraints, prohibits the approach of the radical centers and the singlet-coupling of the electrons.

(b) The method of "caged spins," according to which odd electrons in contracted AOs are prevented from singlet coupling by the action of longer range valence AOs, which actually form the bonds:

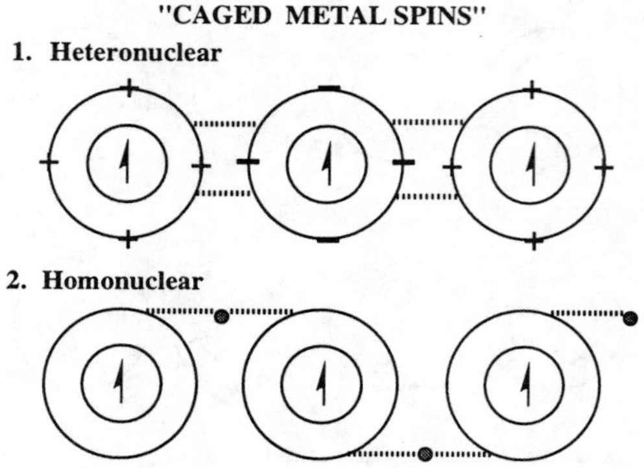

The simplest example of the homonuclear type is the ferromagnetic metal triad: Fe, Co, Ni.

Though interest in "new materials" has peaked in recent years, new ideas have not been forthcoming. Thus, we shall mention simple variations of the themes defined earlier. For example, experimental evidence has been obtained for ferromagnetic behavior in the charge transfer salt of decamethylferrocene (DMF) with tetracyanoethylene (TCNE).[13] This is an example of the "caged spins" theme: The odd electron on Fe cation radical is prevented from singlet-coupling with the odd electron of TCNE anion radical by the cyclopentadienyl rings. These rings are bound to the central metal by interstitialized electrons (as described earlier for neutral ferrocene), and this is partly why they are immune to attack by the TCNE radical anion.

VB theory says that the best strategy for making a ferromagnet rests on the concept of overlap dispersion, according to which two adjacent atoms bind by a combination of 2-e CT and dispersion if high-spin-coupled odd electrons are matched with either holes or pairs. Two different ferromagnetic chains bound by overlap dispersion are shown.

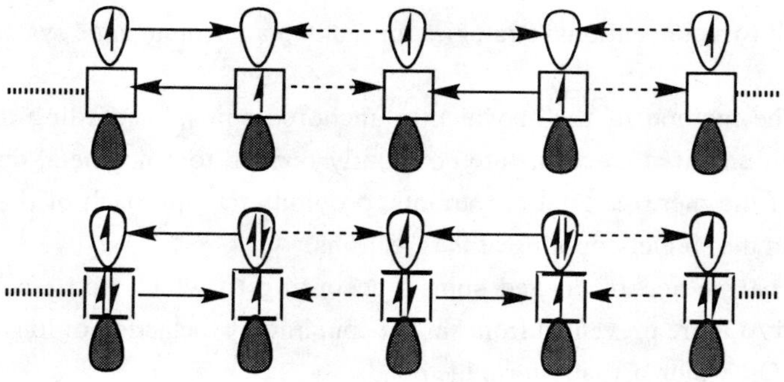

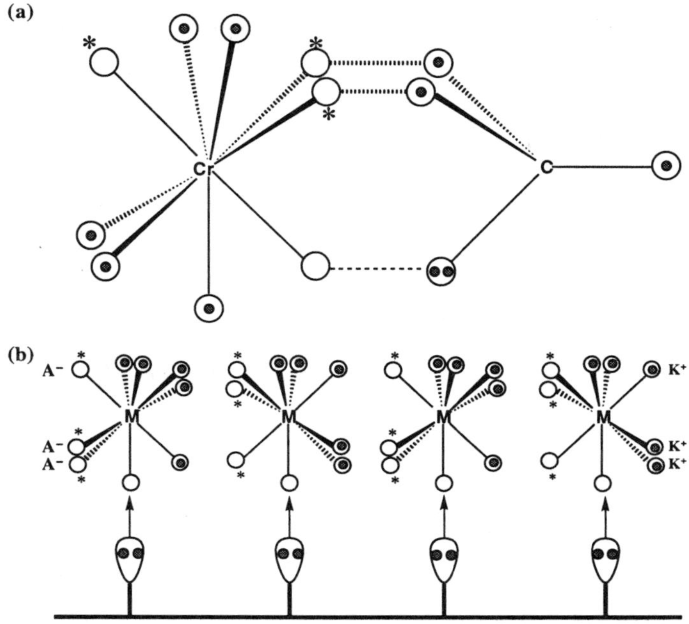

Figure 40.8 (a) CrC as a molecular ferromagnet. (b) Strategy for the design of a ferromagnet by coordinating odd electrons to deshielded atoms.

We can now appreciate why such materials are hard to make: They require combination with radicals of either Lewis acids or Lewis bases. They require, that is, combination of reactive chemical species that, unless constrained, are bound to react to produce singlet products.

There exists an alternative strategy that is related to the "caged spins" model and is inspired by the structure of CrC. By translating MO calculations[14] to VB language, we obtain the picture shown in Figure 40.8a. The physics is as follows. Ferromagnetic coupling of high spin atoms or fragments is possible if the odd electrons of one atom can coordinate with a deshielded core of its neighbor. The starring procedure reveals the operational principle, and the strategy is illustrated in Figure 40.8b: High spin atoms are ordered along a chain by coordination to "spacers," and coordination of the odd electrons of one atom with the deshielded core of the next is secured by the electric field of counterions, A^- and K^+.

Much as chemistry is dominated by the concept of the Lewis electron pair bond, the thinking of researchers interested in ferromagnetic materials is dominated by the McConnell model and the Heisenberg Hamiltonian.[15] When the ideas of McConnell were published, precious little was known about the magnetic properties of "small" molecules. Today, VB theory tells us that these properties can be explained by going all the way back to the Lewis concepts. For example, the union of π-methyl and π-allyl can lead to either an "antiferromagnetic" 1,3-butadiene-type chain or a "ferromagnetic" trimethylenemethane-type array because perfect pairing of all spins is possible only in the former. Today, we know that 1,3-butadiene has a singlet but trimethylenemethane has a triplet ground state. In the latter two spins are forced to occupy nonoverlapping AOs:

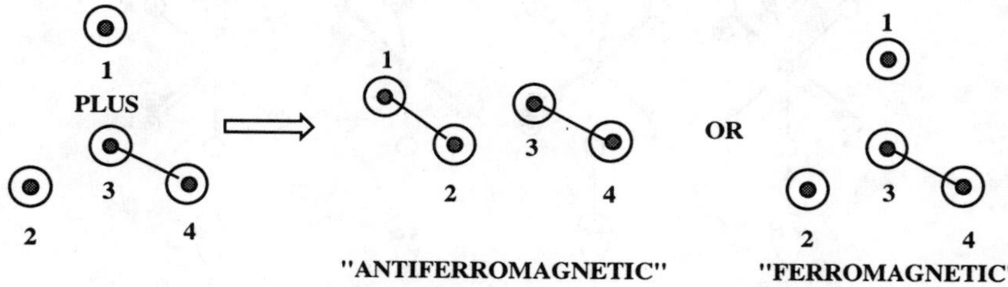

Now, reducing the symmetry of trimethylenemethane from D_{3h} to C_{2v} by bringing two nonbonded atoms together will convert it to singlet. Elongating the CC bonds in the D_{3h} geometry will ultimately convert it to a quintuplet. In this light, the McConnell model simply states that to make a ferromagnet, one must constrain spins to distances at which overlap is zero but Coulombic interaction is nonzero.

Thus, the problem of ferromagnetism is a problem in "constrained molecular architecture." The very nature of this work says that doing chemistry using the Heisenberg Hamiltonian is not justified when it comes to semimetals or metals: The very notion of the I bond, which is formed when a pair "stretches" into two adjacent radicals or holes, is incompatible with such a Hamiltonian because the bonding depends on CT. Furthermore, it means that when two spins (adjacent or nonadjacent) are exposed to the action of a proximal pair with which they can overlap, antiferromagnetism is almost always guaranteed. The rational design of ferromagnets is hard because the constraint is sharp: Two spins must not only be kept away from each other (but not so far that the Coulombic interaction vanishes) but they also must be protected from nearby electron pairs, which are prone to CT delocalization.[16]

Recent advances in the preparation and characterization of high nuclearity clusters sets up a different intriguing problem: Could we use the VB concepts to predict the spin state of a polynuclear cluster given its geometry. Lippard,

Sessoli, and co-workers synthesized recently decanuclear manganese clusters of the type $[Mn_{10}O_4(biphen)_4X_{12}]^{4-}$, where X is halogen.[17] The metal core can be formulated as an $[Mn(II)]_2[Mn(II)]_4$ octahedron with four triangular faces capped in a tetrahedral fashion by four Mn(II) atoms. Assuming pseudo-octahedral hybridization (see Chapter 25) of each Mn ion, we see that each metal enters with three odd equatorial electrons plus either one [Mn(II)] or two [Mn(II)] endo conical electrons for metal–metal bonding. The exo conical AOs are all starred, and they are used to accommodate the various ligands. This means that there are 16 $[(2 \times 6) + (1 \times 4)]$ skeletal electrons and 30 (3×10) odd, high-spin-coupled equatorial electrons. If we assume that metal orbitals interact either directly or indirectly (via the ligands) and that the magic number of a tetracapped octahedron corresponds to the Wadian count of 14 electrons (i.e., it has $2V + 2A$ electrons with $A = 1$), we must transfer two electrons from the endo conical to the equatorial AOs to get the right skeletal electron count. As a result, we now have 32 equatorial electrons in 30 equatorial hybrids. This predicts a total spin of $28 \times 0.5 = 14$. The experimental results indicate that the total spin lies somewhere between 12 and 14.

We end by reasserting that VB theory is ideally suited for the discussion of chemistry in general and spin problems in particular. Unfortunately, as with most everything else, the investigators use MO theory, and this turns simple problems to obscure riddles. For example, *m*-phenylene acts as a ferromagnetic coupler (i.e., triplet lies below singlet).[18] The explanation is straightforward: The molecule receives contributions from polyenyl-type (P) and trimethylenemethane-type (T) resonance structures:

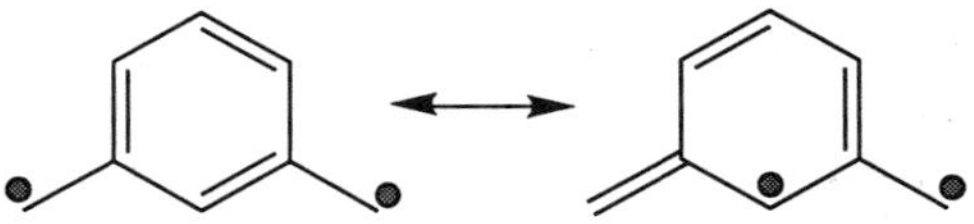

P-type **T-type**

The T-type structures are what makes the molecule triplet. What makes trimethylenemethane triplet has been explained before, using covalent-only VB theory.[19] As the percentage contribution of T-type structures is diminished, we expect that the singlet will dip below the triplet. Hence, all we have to do is increase the size of the annulene on which the two odd electrons are affixed in metal fashion. This can be done in two ways:

(a) Replace the benzene [6] by a larger [*n*]-annulene nucleus: for example, [10].

(b) Twist the two methylenes 90°, to now conjugate with the benzene σ frame. As we saw in Chapter 4, this is a "6+6" formally bisaromatic bisannulene. Rassat[20] and Iwamura[21] found that *m*-phenylene derivatives, where the skeletal constraints that promote such a deformation have singlet ground states.

Bottom line: To keep electrons triplet-coupled, one must impose a special type of delocalization (configuration aromaticity or overlap dispersion) that achieves the goal on the basis of the Pauli exchange principle (see the real VB theory of "aromaticity" in Chapters 12, 13). By now, we recognize from experience molecular fragments that do so (e.g., trimethylenemethane and ground O_2, respectively). If these effects are diluted or abolished, the two spins can almost always find a way to couple in an antiferromagnetic mode (e.g., by forming an I bond with an intervening pair or hole).[22]

References

1. J.A. Duffy, *Bonding, Energy Levels and Bands in Inorganic Solids*, Longman Scientific and Technical, Essex, England 1990.
2. C.K. Chiang, C.R. Fincher, Y.W. Park, A.J. Heeger, H. Shirakawa, E.J. Louis, S.C. Gau, and A.G. McDiarmid, *Phys. Rev. Lett.* 39, 1098 (1977).
3. K. Krogmann, *Angew. Chem. Int. Ed. Engl.* 8, 35 (1969).
4. V.V. Walatka, M.M. Labes, and J.H. Perlstein, *Phys. Rev. Lett.* 31, 1139 (1973); R.L. Greene, G.B. Street, and L.J. Suter, *Phys. Rev. Lett.* 34, 577 (1975).
5. *Chemistry of Oxide Superconductors*, C.N.R. Rao, Ed., Blackwell Scientific Publications, London, 1988.
6. For review, see J.B. Torrance, *Acc. Chem. Res.* 12, 79 (1979).
7. J. Ferraris, D.O. Cowan, V.V. Walatka Jr., and J.H. Perlstein, *J. Am. Chem. Soc.* 95, 948 (1973); L.B. Coleman, M.J. Cohen, D.J. Sandman, F.G. Yamagishi, A.F. Garito, and A.J. Heeger, *Solid State Commun.* 12, 1125 (1973).
8. D. Jerome, A. Mazaud, M. Ribault, and K. Bechgaard, *J. Phys. Lett.* 41, L195 (1980); K. Bechgaard, C.S. Jacobsen, K. Mortensen, H.J. Pedersen, and N. Thorup, *Solid State Commun.* 33, 1119 (1980).

9. For review of conductive polymers, see M.G. Kanatzidis, *Chem. Eng. News* 68(49), 36 (1990).

10. (a) M. Konno and Y. Saito, *Acta Crystallogr. B*, 30, 1294 (1974); 31, 2007 (1975). (b) M. Konno, T. Ishii, and Y. Saito, *Acta Crystallogr. B*, 33, 763 (1977).

11. *Chem. Eng. News* 70(35), 8 (1992). See also the intriguing work of Tour's group: T.W. Brockmann and J.M. Tour, *J. Am. Chem. Soc.* 117, 4437 (1995).

12. (a) G.T. Meaden, *Electrical Resistance of Metals*, Plenum Press, New York, 1965. (b) T.L. Martin and W.F. Leonard, *Electrons and Crystals*, Brooks/Cole, Belmont, CA, 1970, p. 485.

13. J.S. Miller, J.C. Calabrese, H. Rommelmann, S.R. Chittapeddi, R.W. Zhang, W.M. Reiff, and A.J. Epstein, *J. Am. Chem. Soc.* 109, 769 (1987) and references therein.

14. I. Shim and K.A. Gingerich, *Int. J. Quantum Chem.* 23, 409 (1989).

15. H. Kollmar and O. Kahn, *Acc. Chem. Res.* 26, 259 (1993).

16. High spin hydrocarbons are discussed by D.J. Klein, C.J. Nelin, S. Alexander, and F.A. Matsen, *J. Chem. Phys.* 77, 3101 (1982).

17. D.P. Goldberg, A. Caneschi, C.D. Delfs, R. Sessoli, and S.J. Lippard, *J. Am. Chem. Soc.* 117, 5789 (1995).

18. For data review, see S. Fang, M.S. Lee, D.A. Hrovat, and W.T. Borden, *J. Am. Chem. Soc.* 117, 6727 (1995).

19. N.D. Epiotis, "Unified Valence Bond Theory of Electronic Structure. Applications," in *Lecture Notes in Chemistry*, Vol. 34, Springer-Verlag, New York, 1983, Chapter 17.

20. M. Dvolaitzky, R. Chiarelli, and A. Rassat, *Angew. Chem. Int. Ed. Engl.* 31, 180 (1992).

21. F. Kanno, K. Inoue, N. Koga, and H. Iwamura, *J. Am. Chem. Soc.* 115, 847 (1993).

22. Nice presentations of the current thinking on molecular magnets can be found in O. Kahn, *Molecular Magnetism*, VCH, New York, 1993, and in J.S. Miller and A.J. Epstein, *Chem. Eng. News* 73 (40), 34 (1995).

Chapter 41

Normal and High T_c Superconductors

41.1 Connection of the BCS Superconductivity Model to the VB Theory of Chemical Bonding

Superconductivity is the pairing of conduction electrons below a critical temperature T_c that becomes responsible for zero resistivity.[1] Much has been written about the mechanism of "normal" and "high T_c" superconductivity.[2] All explanations, including the Bardeen–Cooper–Schrieffer (BCS) theory of "normal" superconductivity,[3] are founded on one-electron theory of electronic structure. In other words, all theories make one key assumption: Interelectronic repulsion does not upset pairing. By contrast, the main protagonist of the VB theory of chemical bonding is electron–electron interaction. This is what differentiates T- from I-bonding. We can now depict by specific formulas how individual atoms cope with interelectronic repulsion. This leads to a new overview of the subject.

Why is vanadium such a good superconductor whereas neither potassium nor copper exhibits superconductivity? This is the first question we must answer before we attempt to understand why doped La_3CuO_4 is a high T_c superconductor. Since the "gas electrons" of a metal are loosely bound, their mutual interaction is small, as is their interaction with the atomic cores. Hence, these electrons cannot sense the lattice vibrations and we have a good conductor. As the effective ionization potential (IP) is raised, we go toward the extreme of an insulator as the electrons are drawn closer to the nuclei. However, at the same time, the electrons approach nearer to the core and become capable of synchronizing their motion with the lattice vibrations. This is electron–phonon (e-ph) coupling. According to the BCS theory, e-ph coupling is the mechanism of pairing responsible for superconductivity in metal solids. Further increase of the IP leads to a nometallic state in which the now tightly bound electrons pair into covalent bonds and the system undergoes a Peierls distortion. Thus, the strategy is clear: To make a superconductor, start from a low resistivity conductor and increase slowly the effective IP of the metal atom. At some point, a superconductor is formed. Further on, the material is destroyed by Peierls distortion.

What is the physical underpinning of e-ph coupling? It is the favorable interaction of a gas pair with the lattice as portrayed schematically in Figure 41.1. The two electrons of the gas pair have opposite momenta. One can

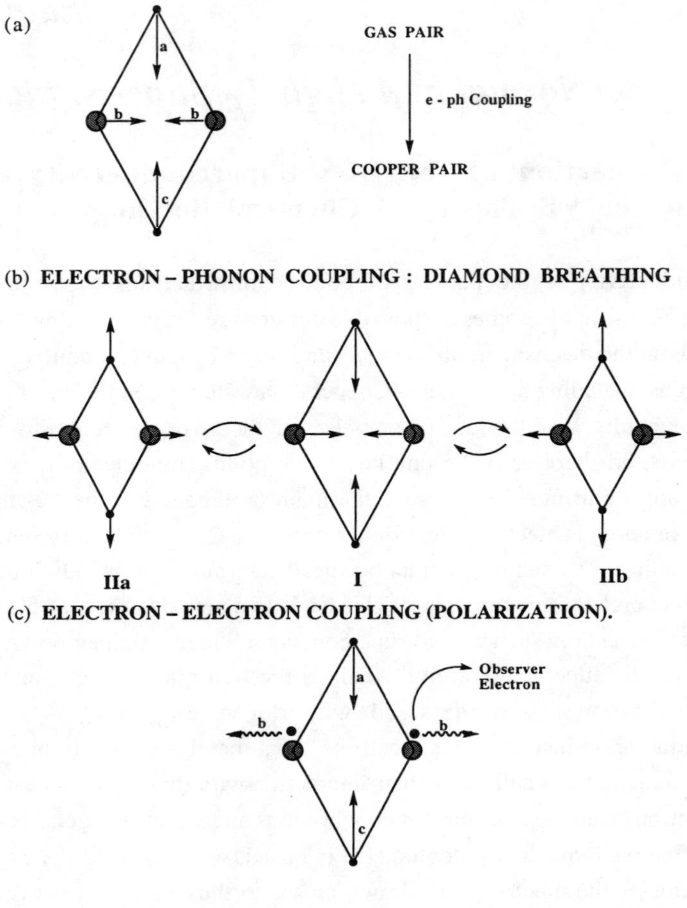

Figure 41.1 VB equivalent of the BCS model.

visualize the process in three steps (Figure 41.1a): One gas electron moves to the void between two atomic cores (stage a), the two cores move toward each other (stage b), and the second gas electron moves toward the original void (stage c). Thus, we have formation of a Cooper pair because of the action of the lattice. The coupled motion of the pair and the lattice is shown schematically in Figure 41.1b. Note the important feature of electrons moving with opposite momenta.

Once pairing has occurred, the mechanism of superconduction is as depicted in Figure 41.2a. In a linear metal chain segment, each pair glides on a cylinder surrounding the core while "vibrating" along the cylinder diameter. It is the "opposite momenta" condition of the BCS theory that defines the disposition of the gas pair with respect to the vibrating core. Note that the two electrons in an

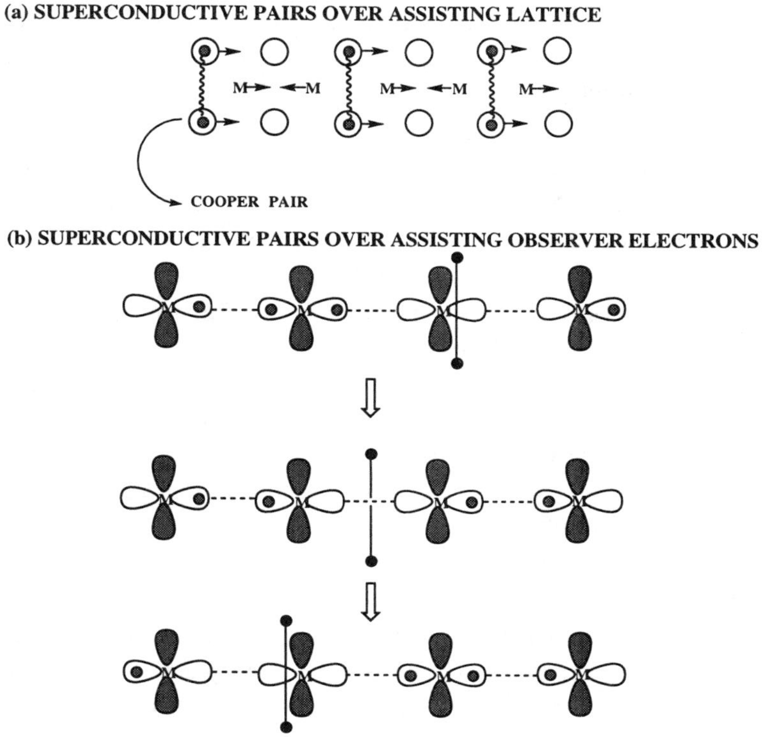

Figure 41.2 The two mechanisms of formation of Cooper pairs.

ordinary Lewis bond pair also have opposite momenta.[4] However, this does not mean that such a pair can be efficiently coupled to the lattice. This will depend on whether the pair sees an exposed core. The core exposure must be large enough for coupling to occur but not large enough to throw the system into a Peierls distortion.

One implicit assumption of BCS theory is that for strong superconduction to occur, the motion of Cooper pairs must be unimpeded by the "observer" electrons, which are tightly bound on the atomic cores. In other words, the metal s electrons must be able to avoid the metal d electrons in the case of transition metals. This brings us to the VB model of normal superconductivity, where the spotlight now falls on electron–electron repulsion:

(a) Each transition metal binds from the close-lying s^2d^{m-2}, s^1d^{m-1}, and s^0d^m configurations. *Bonding requires core deshielding.* This is brought about primarily by the mixing of these three configurations. In an orbital sense, this means that the ns mixes with a σ-type $(n-1)$d AO at every atom. The result is that one electron ends up in a conical and the other in an equatorial AO of a pseudo-octahedrally hybridized atom. One of these two electrons (the one in the conical set) becomes the *gas electron* and the other electron (the one in the equatorial set) becomes the *observer electron.* By extending this argument, we recognize that metal hybridization produces electrons of two types: gas electrons in the more diffuse "s-like" metal AOs and "observer" electrons in the tighter "d-like" AOs. When deshielding is effective, the gas electrons of one metal atom see an exposed core of the neighboring atom.

(b) Superconductivity is a two-phase problem: The gas electrons couple with the lattice to become Cooper pairs, provided the observer electrons are correlated with the gas electrons. Since this correlation is equivalent to deshielding and since deshielding determines the bond dissociation energy of a metal dimer and the heat of formation of a gaseous metal atom, we obtain the following connection: *Superconductivity depends on the BDE of the metal dimer.* The larger the BDE, the higher the T_c.

(c) The mechanism of superconduction falls between two extremes. The pairing mechanism is either the result of e-ph coupling with near-zero correlation of gas and observer electrons or the result of gas electron–observer electron correlation, called e-e coupling, with near-zero e-ph coupling. The maximum T_c is obtained when both mechanisms are activated. The argument is presented schematically in Figures 41.2a and 41.2b.

41.2 What Makes a Pure Metal a Good Superconductor?

Consider the first transition metal series. Which one is superconductor and which one has the lowest T_c? The critical index of core deshielding is the energy gap bracketing the three key metal atom configurations: s^2d^{m-2}, s^1d^{m-1}, and s^0d^m. This quantity, denoted by RAN, correlates with the dimer BDE's. We now expect it to correlate with the superconductor T_c, everything else being equal! So, we are proposing that dimer bonding (i.e., dimer BDE) and superconductivity are interrelated. The data in Table 41.1 hint that we are on the right track. In particular, two things stand out:

(a) Metals that have strong intermetallic bonds at the dimer level (because the gas pair "meshes" with the d electrons) are also the best superconductors: V and Nb have both the strongest metal–metal bonds and the highest T_c within the first and second transition series. Ta is also one of the best of the third transition series with regard to the same properties.

Table 41.1 Bond Dissociation Energies (kcal/mol)[a], RAN Values (ev)[b], and T_c (K)

	Sc	Ti	V	Cr	Mn
BDE	25.9	28.4	57.4	41.0	18.4
RAN	4.19	3.35	2.47	4.40	5.59
T_c		0.40	5.40		

	Fe	Co	Ni	Cu	Zn
BDE	20.8	21.9	47.7	46.3	1.3
RAN	4.07	3.36	1.71	—	—
T_c					0.85

	Y	Zr	Nb	Mo	Tc
BDE	37.3	73.8	115.2	101.2	73.1
RAN	3.63	2.66	1.32	3.18	—
T_c		0.61	9.25	0.92	7.80

	Ru	Rh	Pd	Ag	Cd
BDE	75.8	67.3	25.8	38.0	0.9
RAN	1.09	1.63	5.81	—	—
T_c	0.49				

	La	Hf	Ta	W	Re
BDE	57.6	78.4	92.2	115.2	92.2
P	0.36	1.69	1.04	–0.19	1.76
T_c	4.88	0.13	4.47	0.02	1.70
	6.00				

	Os	Ir	Pt	Au	Hg
BDE	99.1	85.3	85.5	52.8	1.6
P	0.75	0.40	–0.64	–1.74	
T_c	0.66	0.11			4.15
					3.95

[a] M.D. Morse, *Chem. Rev.* 86, 1049 (1986).

[b] *Handbook of Chemistry and Physics*, 46th Ed., The Chemical Rutter Co., Cleveland, OH, 1965.

(b) In moving down the column from V to Nb, two things happen: The lowering of the ns/(n – 1)d ratio of radii of maximum electron density decreases because the (n – 1)d values expand relative to the ns AOs and RAN decreases. Both these effects increase the effectiveness of deshielding and the strength of intermetallic bonding. As a result, the T_c is raised from 5.40 K to 9.25 K.

There is a second important trend: Gradually decreasing conductivity leads to superconductivity. Here are the facts.

(a) Alkali metals have only one valence electron. Because of the large (n – 1)d to ns promotional energy, Cu, Ag, and Au can also be regarded as one-electron atoms. As a result, these metals, being devoid of observer electrons, are not superconductors. That is, the single electron of Li sees a screened Li^+ core while the gas electron of a metal that has more than one electron sees an $M^{+(1+d)}$ core because of the gas electron–observer electron correlation. Electron–phonon coupling requires strong core deshielding.

(b) Superconductivity appears extremely weak at Ti (near zero T_c) and strong at V right between the two extremes of Sc, a poor conductor, and Cr, a good conductor. Conductivity depends on P (the s^2d^{m-2} to s^1d^{m-1} promotion) and superconductivity on RAN. It is the latter index that differentiates Tc from Mn and to a lesser extent from Re and what differentiates La from Sc and Y insofar as T_c is concerned (Table 41.1).

(c) A strong heavy isotope effect implies prime dependence of superconductivity on e-ph coupling. A weak heavy isotope effect implies prime dependence of superconductivity on e-e coupling. This satisfies the objection that the BCS mechanism cannot explain the absence of a heavy isotope effect in certain metals.

41.3 The Strategy for High T_c Superconductivity

High T_c superconductivity[5] poses two critical questions:

What is the mechanism of the pairing in high T_c superconductors?

Why did it take such a long time to discover the high T_c superconductors? (Normal superconductivity was discovered by Cammerlingh Onnes in 1890.[1])

The VB theory of chemical bonding makes straightforward suggestions. We begin by realizing that the early and still popular models of solid bonding and electrical properties are too simple and unphysical. For example, once we know that there exist three mechanisms of bonding, T, I, and E, the notion of one- and two-dimensional instability of solids according to the Peierls model vanishes. To see this point, consider first how the physics of the Peierls distortion is captured by the simple VB formula of a one-dimensional metal.

PEIERLS DISTORTION IN ONE DIMENSION

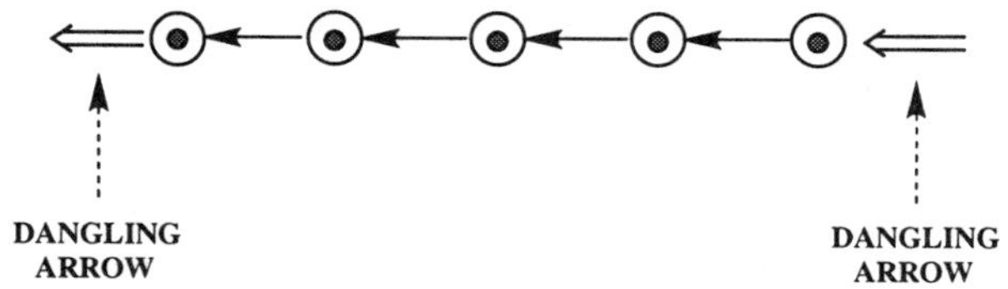

According to the Association rule, one arrow can enter and a second arrow can exit a radical site. This produces two dangling arrows at the two ends of the chain and predicts that the chain will be transformed to a cyclic structure. The concept of I-bond resonance predicts that there will be further distortion of the cyclic form to a three-dimensional form.

All this is too simple to be useful. We now know that labor can be subdivided among the valence electrons of molecules and solids. Some fraction effects E-bonding and another fraction effects additional I-bonding responsible for the electrical properties. As a result, there is nothing surprising about one-dimensional metals. A one-dimensional conductor can exist if E-bonding prevents the "coalescence" of the dangling arrows.

With these ideas, we now come to confront the superconductivity problem. We recognize that, in principle, a one-dimensional superconductor can be produced by the following strategy:

(a) Starting with an insulator chain, we activate it by doping. In this way, we initiate a codirectional arrow train and establish conduction.

(b) We flank the chain by systolic substituents (e.g., F, OR), to limit the space into which excitation of the conduction electrons can occur:

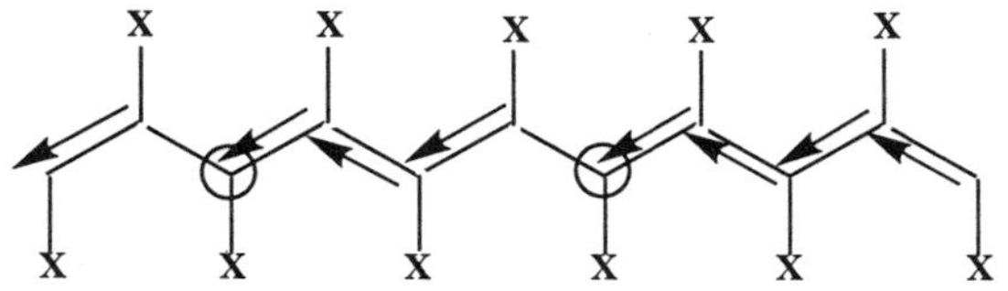

(c) The best realization of a codirectional arrow train occurs in aromatic systems. Hence, doping required for turning an original insulator into a superconductor must either preserve or impose aromaticity. This can be done only in two dimensions or three. Hence, the starting insulator chain must be replaced by an insulator plane.

We have reached an important conclusion: Strong superconductivity requires at least two dimensions because the strongest mechanism of electron coupling, namely, aromaticity does so! Does this pass the commonsense criterion? We have already seen that e-e coupling is equivalent to core deshielding and can assist e-ph coupling in normal metal superconductors. What other fundamental mechanism of electron coupling have we encountered in the course of this work? There is only one answer, and it was given in Chapter 13: Cooperative n-e delocalization, or, in more familiar terms, aromaticity.

What then has prevented speculation that that aromaticity is behind high T_c superconductivity? The answer is that there is no direct cyclic overlap in the repeating CuO_2 unit of the CuO_2 planes of cuprate superconductors (i.e., one copper and two adjacent oxygens do not form a cyclic array). However, cyclic overlap of the $Cu3d_{xy}$ and two adjacent $O2p_z$ valence AOs can be brought about indirectly by the vacuum space of the formal cations. Thus, what seems to be "acyclic" can become effectively "cyclic."

41.4 The Structure of Square Planar Complexes of Cu^{2+}

The critical structural features of a cuprate superconductor are summarized in Figure 41.3. One Cu^{2+} is coordinated in square planar fashion to four oxygen dianions, each of which is sp-hybridized, and directs one O2p doubly occupied AO toward the copper center. In addition, the pseudoatom M represents the action of the alkaline earth cations (typically, Ba^{2+}, Sr^{2+}, and Ca^{2+}) located above and below M.

The conventional one-electron approach focuses on the occupied AOs and neglects the action of the unoccupied Cu4s and Mns, Mnp, and M(n – 1)d AOs. Since we have already shown that the very facts of chemistry (e.g., the inorganic skeleton closet in Chapter 24) argue strongly against such an assumption, we must be prepared to assign important activity to these "forgotten" orbitals. Fortunately, there exist "small molecule" data that leave no doubt about the validity of this approach.

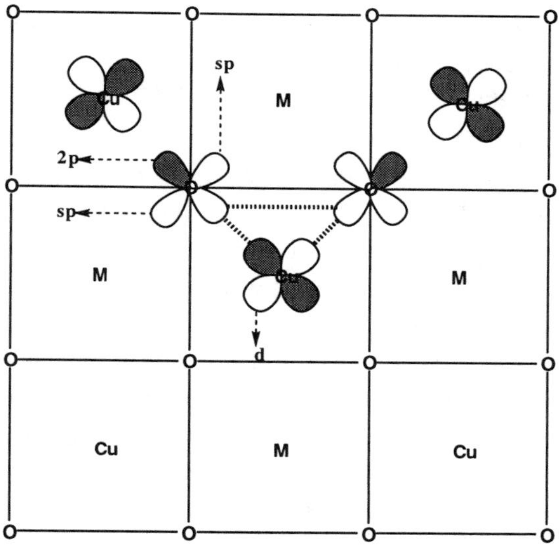

Figure 41.3 CuO_2 plane in typical high T_c superconductors. M simulates the action of alkaline earth cations.

We first focus on square planar L_4Cu^{2+}, where L = O^{2-}, in Figure 41.3. The key feature is that the odd electron goes to the Cu^{2+} d AO that has maximum overlap with the ligand MOs. This is done because the occupied O2p AOs destabilize the copper 3d AOs in an overlap sense. Hence, one should place two electrons in the least destabilized 3d AOs. This imposes single occupancy on the most destabilized (i.e., maximally overlapping) copper $3d_{xy}$ AO, which points straight toward the O2p AOs. There are two important points:

(a) The presence of an odd electron in the $3d_{xy}$ AO implies appreciable overlap of ligand and metal AOs.

(b) If the interaction of the (lower) ligand filled AOs with the (upper) copper d AOs is net destabilizing, what holds together planar CuO_4^{2-} is not overlap but static interaction. Hence, a polymer of the type $(CuO_2)_n$, where Cu has a square planar coordination, is a prometal in which an array of critical conduction AOs is held in place by a support system glued together by E-bonding.

We now come to the crucial argument, which is best presented in the form a hypothesis: *The vacuum space of the alkaline earth dications (and, in particular, the vacant ns AOs) is responsible for effective in-phase overlap of cis O2p AOs in the CuO_2 planes.* One should take these apparently nonoverlapping oxygen AOs as effectively overlapping.

41.5 Aromaticity as the Elusive Mechanism of Pairing in Cuprate Superconductors

The implications of the action of the vacuum space of the alkaline earths are now self-evident: *One singly oxidized CuO_2 unit in the CuO_2 plane of a cuprate superconductor is a four-electron Möbius aromatic system (4e-Möbius system).* The situation is illustrated in Figure 41.4. In VB terms, the aromaticity is the result of the cooperative action of 1-e (solid arrow) and 2-e (wavy double arrow) CT delocalization depicted in Figure 41.5. The net effect is that two odd electrons become singlet-coupled via the action of an observer pair. In the process, the identity of the electrons is lost. In the corresponding 2e-Hückel system, where the observer pair is replaced by an observer hole, the same thing happens.

Pearson[6] has provided a lucid discussion of how molecular vibrations cause the mixing of a ground and an excited state of a molecule and lead to geometrical distortion of the ground state. An exactly analogous situation obtains in the case of the mixing of lower and upper configurations (rather than states) to generate a VB aromatic system. The cooperative delocalization distributes the electron density symmetrically along the perimeter of the cyclic array of atomic centers. In the case of the 4e-Möbius system, the distribution has D_{3h} symmetry when all AOs are equivalent. A vibration that destroys D_{3h} symmetry (or a vibration that lowers the symmetry of an aromatic complex, in general) effectively destroys the aromatic coupling of the electrons. This means that aromatic delocalization is a temperature-dependent effect: Below a certain critical temperature, electron pairs with random spins couple via the action of neighboring holes or pairs, provided the orbitals have the right symmetry.

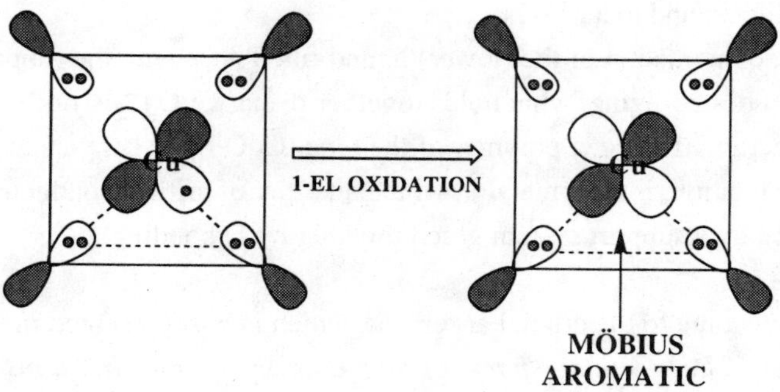

Figure 41.4 One-electron oxidation creates a Möbius aromatic system in the cuprate superconductors.

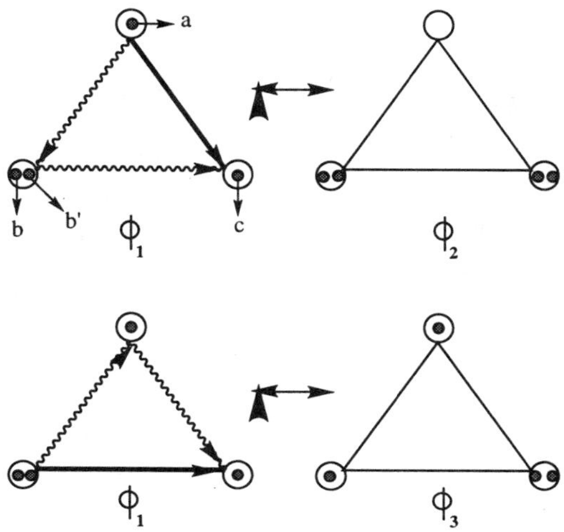

Figure 41.5 Two electrons become coupled by the action of an observer pair.

Recall now the central problem: Unless orbital electronegativity is low, a nominally "aromatic" system is unstable relative to the component bond pairs (e.g., hexagonal H_6 is unstable relative to three H_2 groups). Furthermore, even when orbital electronegativity is low, the competing influence of "observer pairs" may thwart an "aromatic" complex. In Figure 41.6, the realization of a 4e-Möbius system as a global minimum, or even as a transition state, fails because of lone pair exchange repulsion engendered by the surrounding pairs. A lower energy nonaromatic alternative is observed. The message is: *Realizing a stable Möbius aromatic system as a global minimum is as hard as making a high T_c superconductor!* The two phenomena are now linked by their very rarity.

The strategy is apparent: To generate a ground state 4e-Möbius system, use a supporting framework to anchor such a system. One way of implementing

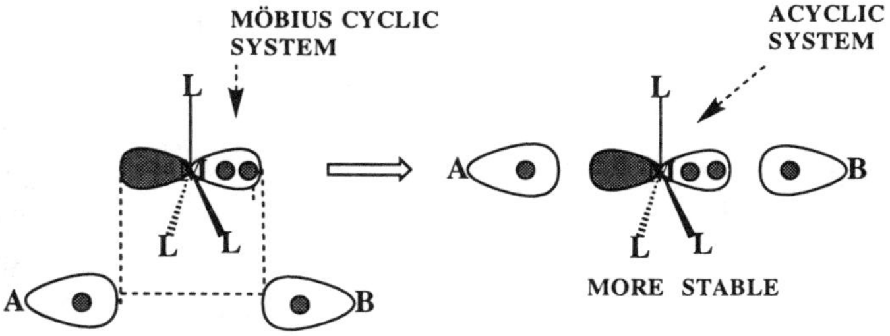

Figure 41.6 Externally destabilized Möbius aromatic system.

this strategy is to anchor a nonaromatic prometal on a supporting system and then proceed to turn it into an aromatic metal by means of doping. This aromatic system will now exhibit strong electron coupling that would manifest itself in superconductivity below a certain temperature at which the vibrations disturb the CT-induced electron pairing.

How does one choose the prometal and the supporting system? Note that the prometal is a subsystem of AOs and electrons and that the atoms defining the prometal have some orbitals and electrons belonging to the prometal and others belonging to the supporting framework. The most common mode of binding of the supporting system itself is either σ-bonding (T-bonding) or static interaction (E-bonding). The latter choice is best realized in "ionic solids." The stage is set: If aromaticity determines high T_c superconductivity, the CuO_2 layers must meet these conditions in a superb manner.

41.6 What Is Special About the High T_c Superconductors?

The unfolding discussion is based on experimental evidence as summarized in recent reviews and monographs.[7–10] Interpreted by the VB apparatus, the evidence taken in toto suggests the following scenario: The common denominator of all high T_c p-type superconductors is the presence of CuO_2 layers in which a fraction of the copper atoms are in the Cu^{3+} formal oxidation state, and their presence satisfies the condition for CuO_2 Möbius aromaticity.

We illustrate our approach by reference to the $La_{2-x}Ba_xCuO_4$ system, which has the structure shown in Figure 41.7a. If we look along one superconducting CuO_2 layer, replace La_2 by a dummy atom M [= $(La^{3+})_2$] located at the midpoint of the line connecting the two La atoms and omit, for the moment, the two oxygens that are coordinated to copper perpendicular to the plane, we obtain the structure shown in Figure 41.3. As discussed before, each formal O^{2-} is sp-hybridized, and each AO is doubly occupied. The sp hybrids point to the M hollows, and one 2p points toward the copper $3d_{xy}$. Because of the action of M, the most destabilized oxygen lone pair is the one that points toward the copper $3d_{xy}$ AO. Hence, oxidation is expected to remove one electron from this doubly occupied oxygen AO. However, whether this electron comes from O^{2-} or from Cu^{2+} is not of essence in what follows. For the sake of convenience in electron counting, it is preferable to regard the superconducting (doped) form as containing formal Cu^{3+} ions. Finally, the bridging of two copper ions by an O2p pair can lead to antiferromagnetic coupling of the Cu^{2+} $3d_{xy}$ odd electrons.

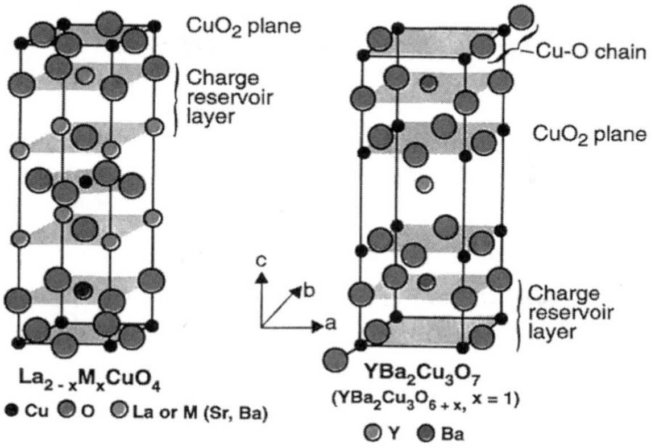

Figure 41.7 Representative high T_C superconductors reproduced with permission from ref. 26. Copyright 1992 American Chemical Society.

With this preamble, we now appreciate the unique qualifications of the cuprate superconductors. Specifically, removal of one electron from one Cu^{2+} generates the sought-after 4e-Möbius system. Note the critical feature: Copper furnishes a singly occupied antisymmetric d AO (Figure 41.4). The designation of the pdp array of orbitals as "cyclic" is based on the hypothesis that M acts to couple the nonbonded oxygen atoms. Furthermore, high T_C superconduction requires the absence of low energy unoccupied valence AOs, which can promote the generation of a nonaromatic system, or, in the physicist's language, to which the superconduction electrons can be scattered. This condition is admirably met by two features:

The formal Cu^{3+} has all valence 3d AOs completely full except the critical vacant $3d_{xy}$ AO.

The formal O^{2-} has a noble gas electronic configuration and no low-lying vacant nonvalence AOs.

If this scenario is correct, we can automatically predict how to destroy or impair the superconduction apparatus of $La_{2-x}Ba_xCuO_4$:

(a) By changing the symmetry of the critical d orbital. This can be effected by replacing Cu^{2+} by Ga^{2+}. In this case, the antisymmetric $3d_{xy}$ is replaced by the symmetric 4s AO, and this turns the GaO_2 repeating unit into an antiaromatic system upon removal of one electron.

(b) By introducing low-lying vacant AOs (e.g., by replacing O by S).

(c) By introducing metal d-holes (e.g. by replacing the Cu atom by a metal atom with 10 or fewer electrons).

41.7 Interpretation of High T_C Superconductivity

If the physical phenomenon is fundamental, it must be simple. High T_C superconductivity is fundamental. If the physical phenomenon is simple, its underpinning must be a fundamental law of nature. In chemistry, there is nothing more fundamental than VB aromaticity: Aromaticity is a manifestation of electron indistinguishability (Pauli principle). It represents the extreme at which the Pauli principle is consistent with strong (exchange and CT) delocalization. As a result, an aromatic system affords the best opportunity for the expression of I-bonding.

One extensively studied superconductor that provides a stringent test of our model is $YBa_2Cu_3O_{6+d}$. This metal oxide has the structure shown in Figure 41.7b, and it typifies most of the known p-type high T_C superconductors: It is made up of stacked layers that differ in atomic composition and execute different functions. The stacking pattern and the terminology we shall use are shown in Figure 41.8.

Note the three key components:

The superconductor layer is where the aromatic CuO_2 Möbius system responsible for superconductivity is formed.

The charge reservoir is the layer that contains the variable oxygens that control the formal oxidation states of the copper atoms.

The O-coupling layer contains the heavy ion that acts to couple the oxygen atoms of the CuO_2 layer, ensuring that the aromatic Möbius array does materialize.

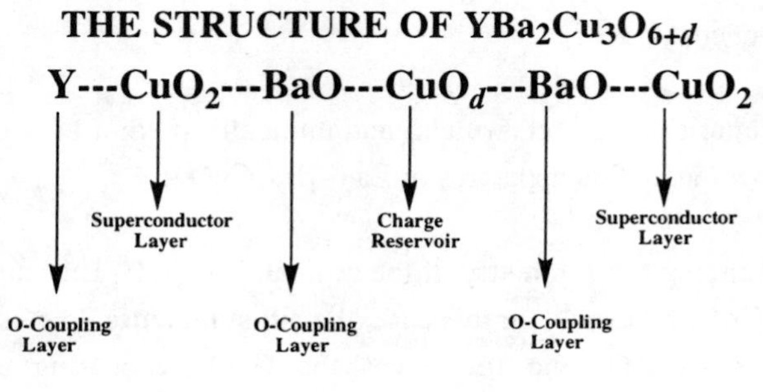

Figure 41.8 Stacking of layers in the $YBa_2Cu_3O_{6+d}$ superconductor.

The story begins by considering a derivative of the organic molecule cyclopentadienyl radical (C_5H_5) in which a first added electron goes to the π domain to transform it into a Hückel aromatic π system and a second added electron goes to the σ system causing a rupturing of a C—C σ bond and the indirect destruction of π aromaticity. This trivial example illustrates that reduction or oxidation can have two effects:

(a) It can generate a "modular subsystem" (i.e., a subsystem that has special chemical, electrical and magnetic properties). An aromatic system is an example of a modular system: It is strongly diamagnetic (aromatic ring current), it has strong predisposition toward substitution (as compared to addition) reactions, and so on.

(b) It can destroy a "structural subsystem" (i.e., a subsystem that provides structural support of the "modular subsystem").

Consider now what happens in the hypothetical case in which π and σ bonds converge to an intermediate state of bonding. Now, the consequences of the delivery of the first and the second electron to the simple cyclopentadienyl derivative would cease to be clear. However, the experimental manifestations would remain unaltered: Initial establishment of a special phenomenon will be followed by destruction of the phenomenon!

According to the amassed data, one can envision three different types of $YBa_2Cu_3O_{6+d}$ species, as shown in Figure 41.9. Each of the three forms can be written as a composite of an invariant part M and a variable part $(Cu^{+q})(O^{2-})_d$ in which the allowed q/d values are 1/0, 2/0.5, and 3/1. This leads to the following interpretation: $YBa_2Cu_3O_{6+d}$ becomes superconducting when the variable copper is either Cu^{3+} (d = 1) or Cu^{2+} (d = 0.5) because Cu^{2+} (by analogy to Ag^{2+}) can disproportionate to Cu^+ and Cu^{3+}. On the other hand, superconductivity is inconsistent with Cu^+. The key points are that superconductivity has to do with Cu^{3+}, and the existence of two different superconductor forms (the "high" T_c = 98 K and the "low" T_c= 58 K form) is a reflection of the mode of formation of Cu^{3+}: direct ("high" form) or indirect by disproportionation ("low" form).

This now begs the question: What do the *two* invariant Cu^{2+} ions [incorporated within M in the formula $M(Cu^{+q})(O^{2-})_d$] do? They form *one* symmetrical I bond by combining with *one* intervening O^{2-}, which is responsible for holding the structure together. This is the source of the antiferromagnetism of the insulator. In short, the idea is that $YBa_2Cu_3O_{6+d}$ has one modular and two structural copper ions, with the former determining

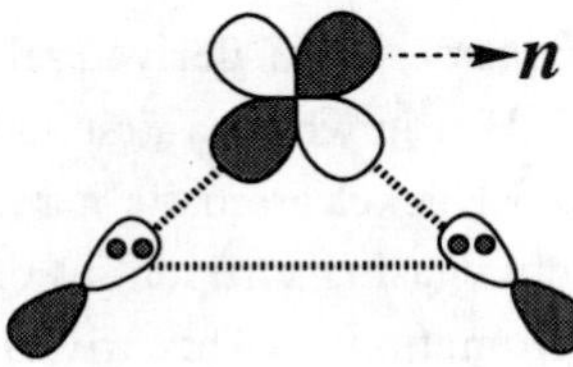

n VALUE

d=0	$Y^{3+}(Ba^{2+})_2(Cu^{2+})_2(Cu^{+})(O^{2-})_6$ **Antiferromagnetic Insulator**	**2**
d=0.5	$Y^{3+}(Ba^{2+})_2(Cu^{2+})_2(Cu^{2+})(O^{2-})_{6.5}$ T_c **= 58 K Superconductor**	**1**
d=1	$Y^{3+}(Ba^{2+})_2(Cu^{2+})_2(Cu^{3+})(O^{2-})_7$ = T_c **= 98 K Superconductor**	**0**

Each can be written as $M(Cu^{+q})(O^{2-})_d$ with

$M = Y^{3+}(Ba^{2+})_2(Cu^{2+})_2(O^{2-})_6$

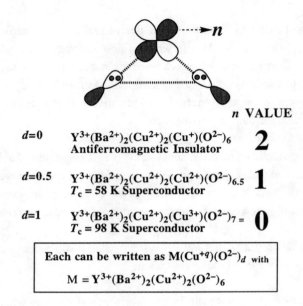

Figure 41.9 The VB formulation of the $YBa_2Cu_3O_{6+d}$ superconductor.

directly the electrical properties and the latter determining the structural integrity of the system. The modular copper shuttles among the Cu(III), Cu(II), and Cu(I) valence states, of which only the first (directly) and the second (indirectly) can fulfill the condition for CuO_2 Möbius aromaticity in the CuO_2 layers.

The preceding scenario explains why superconductivity is a sensitive function of oxygen stoichiometry and crystal disorder: Doping is effective up to the point that turns the modular copper from Cu^{2+} to Cu^{3+}. Beyond this point, it begins destroying the structural copper, namely, Cu^{2+} . This turns a symmetric into an asymmetric CuOCu I bond which, in turn, deforms the superconducting CuO_2 planes. Since aromaticity is maximized in high symmetry and since aromaticity is the basis of superconductivity, the dependence of T_c on doping should be parabolic, as illustrated by the phase diagram of the $La_{2-x}M_xCuO_4$ system shown in Figure 41.10. It is seen that the original insulator gives way to superconductor which upon further oxidation becomes a metal.

Let us now return to the "high" $YBa_2Cu_3O_{6+d}$ superconductor with $d = 1$. In this form, Y^{3+} and Ba^{2+} have interacted with geminal oxygen 2p electron pairs (geminal oxygens are those attached on the same Cu atom) in the process of coupling them to form an effectively cyclic Möbius aromatic $O^{2-}Cu^{3+}O^{2-}$ system. This means that the Y and BaO layers (i.e., the O-coupling layers) have electron density derived from the CuO_2 superconducting planes. Since aromaticity is a cooperative phenomenon in which CT delocalization in one promotes CT delocalization in a second interatomic space, the action of the Y and BaO layers will be turned off when aromaticity in the CuO_2 layers is no longer possible. This happens when the modular copper becomes Cu^+ (i.e., when $d = 0$). As d changes from 0.5 to 0, therefore, we must observe a shift of electron density from the Y and BaO layers to the CuO_2 layer accompanied by a sharp drop of T_c. It turns out that this is exactly what neutron diffraction experiments imply. The importance of the nonlinearity of the reduction of the superconducting CuO_2 planes has been stressed by Cava.[7]

Why are the oxygens lost from the CuO chains (CuO_d reservoir) and not from the CuO_2 planes? This is a question of obvious importance to which a straightforward answer can again be given: OCuO Möbius aromaticity can exist only in copper–oxygen layers, not in copper–oxygen chains. This is like saying that benzene is aromatic but 1,3,5-hexatriene is nonaromatic! This is why the first oxygens to go from $YBa_2Cu_3O_7$ are the less stabilized chain oxygens.

Bottom line: High T_c superconductivity has to do with "resonating aromaticity," which is embedded in a metallic system in which the rules of bond

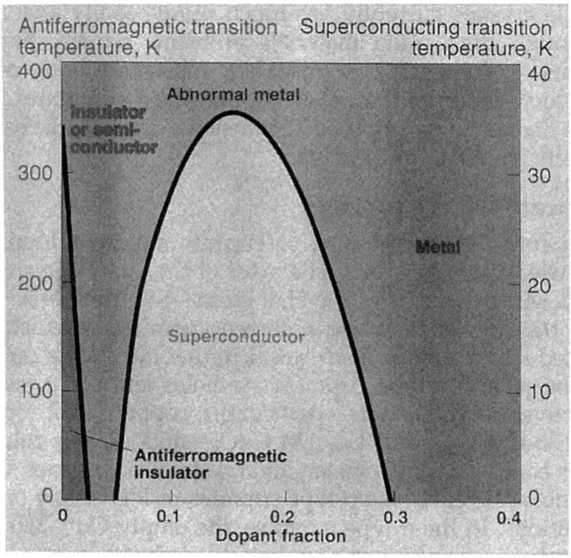

Figure 41.10 Phase diagram of $La_{2-x}M_xCuO_4$ reproduced with permission from ref. 26. Copyright 1992 American Chemical Society.

making (I-bonding) are not the rules of the covalent/ionic world of conventional theory. The interplay of modular and structural coppers is complex, and it is unrelated to the traditional indices used in correlating atomic composition with molecular structure and reactivity. Torrance and co-workers attempted to define similarities between high T_c superconductors that may shed light on the mechanism of the phenomenon.[8] By focusing on conventional chemical structural indices, one is befuddled by the observation that solids having structures "similar" to some high T_c superconductors are themselves either poor superconductors or no superconductors at all! Raveau et al.[9] conclude that the crystal chemistry of mixed-valence copper oxides is very complex and suggest that the presence of domains and defects may influence the superconducting properties of these oxides. Our model says that a transition metal other than d^9 is not "similar" to Cu(II), S is not "similar" to O, and so on.

41.8 Biology and High T_c Superconductivity

Our mechanism of high T_c superconductivity of cuprates rests on the concept of aromaticity. This may seem surprising and even unlikely to inorganic chemists and physicists who view metal–oxygen solids as "ionic" species having little to do with aromatic CT delocalization. On the other hand, we have seen in Chapter 24 that the place to look for aromaticity, as distinct from pseudoaromaticity, is in semimetal rings. In other words, we expect aromaticity in cyclic systems made up of atoms with low average electronegativity. This implies that transition metal–oxygen rings may also be "bisaromatic" species made up of aromatic R and T shells according to the shell model. This immediately suggests one test of our ideas: Four-membered rings of the type $(MO)_2$, where M is a transition metal and, in particular, copper, not only should exist but will bear the signature of bisaromaticity.

The eight-electron rhombic $(CuO)_2^{2+}$ can be formulated as a bisaromatic ring which, according to the shell model, has the electronic configuration R^2T^6 shown in Figure 41.11. Each formal Cu^{2+} contributes a 4s AO to complete the four-center R shell. As seen in Figure 41.11, the 2-e CT hop within the R shell (indicated by the two arrows) engenders a singlet oxygen atom, provided the Cu4s AO is active. Thus, if our superconductivity model is viable, rhombic $(CuO)_2^{2+}$ not only must be stable but also must behave either as a source of radical O^- or as a source of electrophilic oxenic oxygen. This scenario is supported by the experimental results of Karlin and co-workers[10] and Kitajima and Moro-oka.[11]

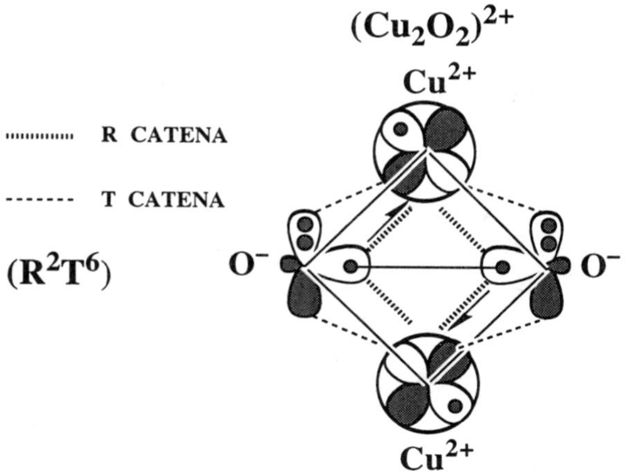

Figure 41.11 Shell model of $Cu_2O_2^{2+}$. Arrows indicate the dominant 2-e CT hop in the R catena.

The electronic structure of the rhombic $(CuO)_2^{2+}$ predicts the potential distortion modes such as those shown in Figure 41.12. We see that the transformation of the side-on to an end-on $(CuO)_2^{2+}$ eliminates the action of the Cu4s vacant AOs on the formal σ O—O bond. As a result, the dioxygen within the latter complex will be incapable of acting as an electrophile. Hence, the final conclusion is that either nucleophilic or radical oxygen reactivity is expected of the end-on complex and either electrophilic or radical oxygen reactivity is expected of the side-on complex. Clearly, the "innocent" ligands surrounding the copper atoms will play an important role in diverting a side-on complex toward either radical or electrophilic reactivity. Consistent with these expectations, the Karlin side-on $(CuO)_2^{2+}$ complex is not readily protonated. By contrast, the end-on complex (Figure 41.12) seems to readily give H_2O_2 upon protonation, and the peroxo fragment appears to be nucleophilic in character. Furthermore, the Karlin (butterfly) and Kitajima (planar) side-on $(CuO)_2^{2+}$ complexes show significantly different reactivities. The two structures differ with respect to the metal ligands, but each complex is diamagnetic. PPh_3 displaces O_2 from the butterfly but not from the planar complex.[10] More importantly, the Kitajima complex appears to react via radical intermediates resulting from homolytic cleavage of the side-on complex.[11,12]

The planar $CuBaO_2$ species shown in Figure 41.13 is a fair representation of the CuO_2 layer in La_2CuO_4. We now recall the distinguishing feature of heavy alkaline earths: The np and nd valence AOs tend to become degenerate,

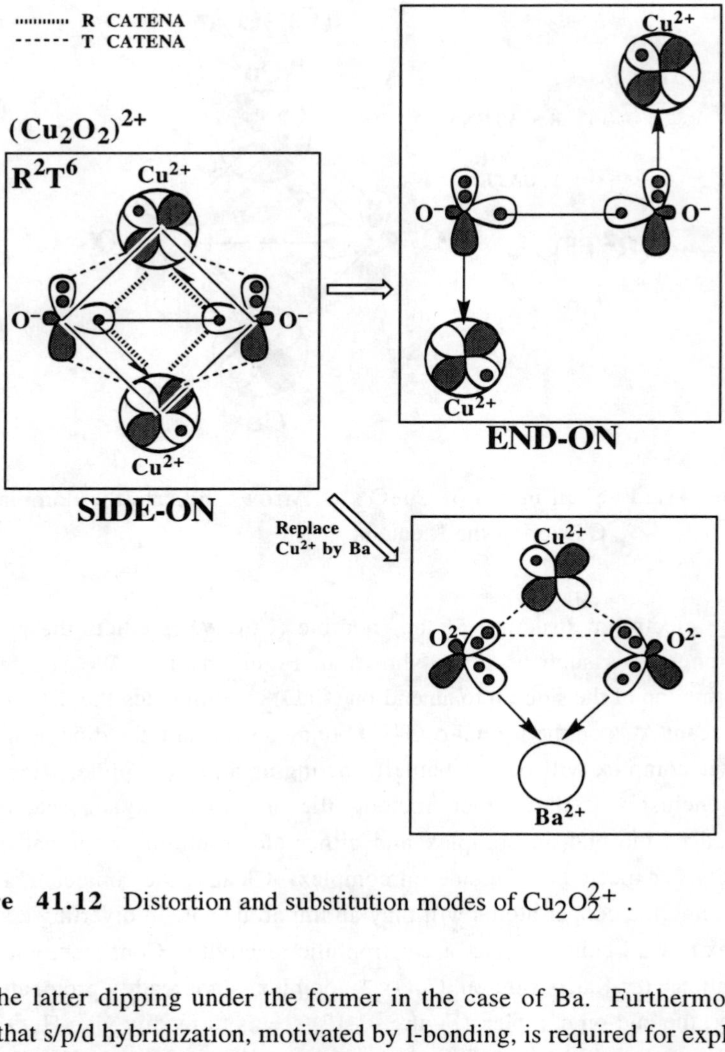

Figure 41.12 Distortion and substitution modes of $Cu_2O_2^{2+}$.

with the latter dipping under the former in the case of Ba. Furthermore, we recall that s/p/d hybridization, motivated by I-bonding, is required for explaining the angular shape of BaF_2. With this assumption, it is easily seen that Ba can act to couple nonbonded oxygens and generate, upon 1-e oxidation, a four-electron Möbius system.

In summary, hemocyanin, tyrosinase, and various oxidases containing one or more copper atoms can interact with dioxygen.[13] Awareness of this behavior has prompted intensive investigations of copper–oxygen complexes, and the results are consistent with our cuprate superconductor model: The difference between the side-on (rhombic) and end-on peroxo complexes as well as the difference between the Karlin and Kitajima complexes can be taken as indicators of the different requirements of I-bonding (side-on) and E-bonding (end-on),

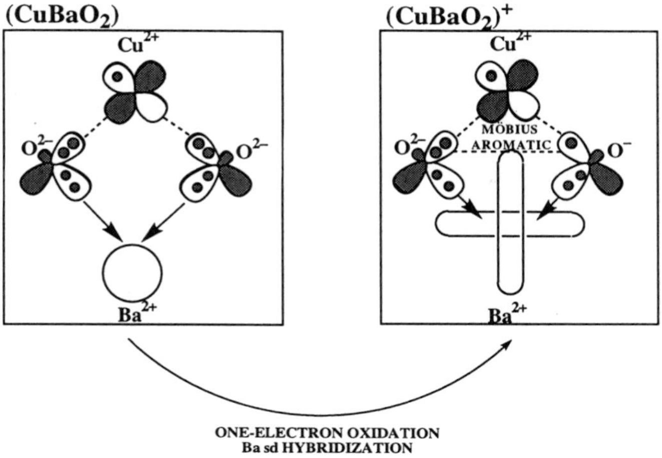

Figure 41.13 One-electron oxidation of $Cu_2O_2^{2+}$ establishes an aromatic array.

always in a relative sense. Furthermore, the preference for the side-on over the end-on complex can be taken as an indicator of the capability of the corresponding solid polymer to act as a high T_c superconductor.

41.9 Design of High T_c Superconductors

The first step is to visualize a solid or a polymer in a way consistent with the concepts developed thus far. The cuprate superconductors provide a good lesson. Specifically, we can envision the CuO_2 plane in its native state as a composite of a *prometal chain* involving alternating Cu and O coordinated to *effector* oxygens as illustrated in Figure 41.14. In this particular case, the prometal chain is made up by repeating prometal units (PU) and the effector unit (EU) is made up of a single atom, namely, oxygen atom. As a result, the formula of the polymer is $[(PU)(EU)]_n$. Since each PU has an odd and each EU an even number of electrons, the polymer is a polyradical with n radical sites. Removal of x electrons from the prometal chain constitutes "activation," which achieves two goals simultaneously:

It turns the insulator into a conductor.

It creates an aromatic subsystem.

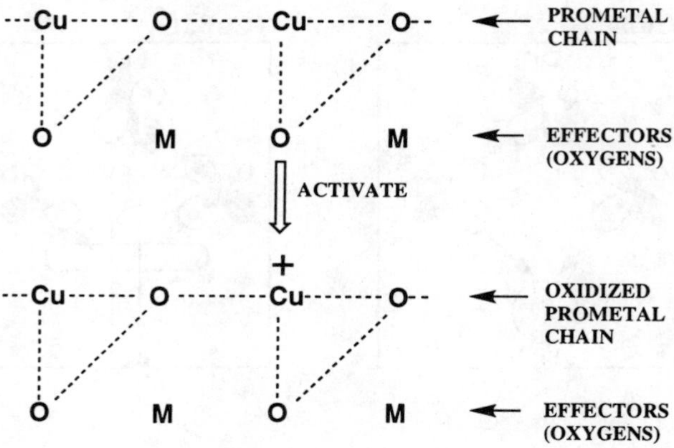

Figure 41.14 The assembly of prometal chain plus effectors.

The formula is now $[(PU)(EU)]_{n-x}$ $[(PU^+)(EU)]_x$. This means that x radical subsystems have been replaced by x aromatic closed-shell systems, as illustrated in Figure 41.15.

The strategy for promoting doping-induced superconductivity, as illustrated in Figure 41.16, can be summarized as follows: Couple the "doped orbital" to its nearest neighbor by an effector so that an aromatic system is produced. The overall plan is:

(a) A prometal chain is established.

(b) The effectors can be either single atoms or polyatomic fragments, and each effector may contribute an even or odd number of electrons toward the

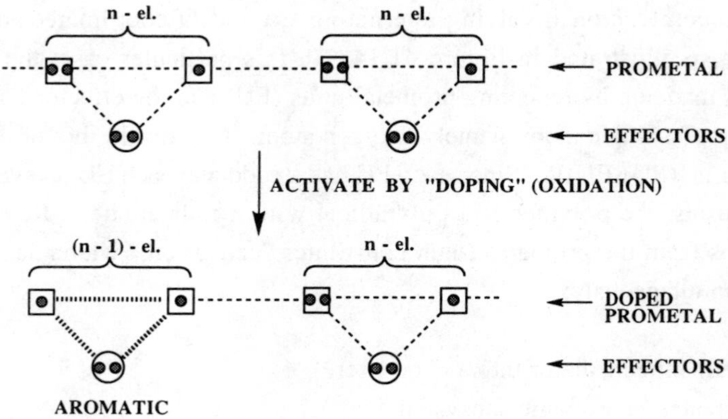

Figure 41.15 Doping establishes conductivity and creates islands of aromaticity.

realization of an aromatic subsystem. In each case, there exist two alternative effector electron counts that can generate either a Möbius or a Hückel aromatic subsystem.

(c) The atoms constituting the prometal chain as well as the effectors can be glued together either by static interaction (inorganic high T_C superconductors) or by σ bonds (organic high T_C superconductors).

(d) Activation is implemented either by direct oxidation or reduction by an extrinsic agent or by replacement of an atom by one with fewer or more electrons.

(e) Cyclic aromatic orbital interaction can be either direct (i.e., overlapping neighboring AOs forming a closed loop) or mediated indirectly by the vacuum space of one of the component atoms, as in the case of the cuprate superconductors.

To apply the approach, we start with an RR insulator like polyacetylene and activate it by direct or indirect (replacing C by N) doping that adds one electron to the prometal orbitals, followed by the addition of an effector with an odd number of electrons. When the insulator is polyacetylene and the effector is an allyl residue, we have a linear polymer of cyclopentadienyl (Cp) radicals Adding electrons converts a fraction of the rings to Hückel aromatic cyclopentadienide anions. Superconductivity arises from electron hopping, which constantly destroys and reconstitutes aromatic Cp rings. C_{60} fullerene is made up of 12 Cp

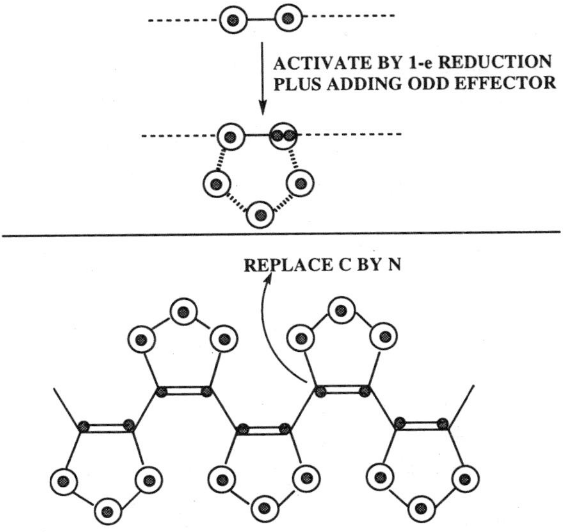

Figure 41.16 Design of an organic high T_C superconductor.

rings, and $C_{60}ALK_3$ (ALK = alkali metal) is a superconductor.[14] We suggest that this is a consequence of Cp^- aromaticity.

The last illustration immediately projects the difficulty of constructing high T_c superconductors. Specifically, if the number of dopant holes or electrons is x and the polymer is made up of n odd-electron prometal units and n even-electron effector units, there will be x singlet aromatic subsystems and $n - x$ radical nonaromatic subsystems. Note that an odd electron PU requires the presence of an even EU for aromaticity in the subsystem resulting from doping. An analogous situation occurs if the number of dopant holes or electrons is still x but now the polymer is made up of n even-electron prometal units and n odd-electron effector units. Again, there will still be x singlet aromatic subsystems and $n - x$ radical nonaromatic subsystems.

So, here is the key point: When starting with an insulator $(PU)_n$ chain, appendage of an effector partner (EU) that acts in conjunction with the dopant action to produce x aromatic subsystems requires that the polymer have $n - x$ radical sites. Hence, *the whole trick involved in making high T_c superconductors is to manufacture stabilized radical sites.* In the case of the cuprate superconductors, this condition is fulfilled because the formal Cu^{2+} has an odd electron that, being shielded by its surroundings, is relatively unreactive.

Bottom line: If high T_c superconduction involves an electronic mechanism of pairing, there can be no better candidate than aromaticity as formulated by VB theory. The many different theories of high T_c superconductivity are founded on the conventional covalent/ionic model of chemical bonding. By contrast, our analysis is founded on the concept of the I bond. *For optimal atom electronegativity*, the best I bond is realized in an aromatic system. The synthesis of high T_c superconductors is hindered by the instability of aromatic arrays, which persists unless the atom electronegativity is optimal (see Chapter 16). This is why success depends on our ability to constrain an aromatic array either by T-bonding (e.g., σ bonds constraining the quasi-aromatic "π system" of fullerenes) or by E-bonding (e.g., cuprate superconductors).

References

1. A.C. Rose-Innes and E.H. Rhoderick, *Introduction to Superconductivity*, Pergamon Press, New York, 1969.
2. (a) *Chemistry of High-Temperature Superconductors*, D.L. Nelson, M.S. Whittingham, and T.F. George, Eds., American Chemical

Society, Washington DC, 1987. (b) An excellent review: F.J. Adrian and D.O. Cowan, *Chem. Eng. News* 70 (51), 24 (1992).

3. J. Bardeen, L.N. Cooper, and J.R. Schrieffer, *Phys. Rev.* 108, 1175 (1957).
4. J.C. Slater, *Phys. Rev.* 35, 509 (1930).
5. (a) J.G. Bednorz and K.A.Z. Muller, *Phys. B: Condens. Matter* 84,189 (1986). (b) M.K. Wu, J.R. Ashburn, C.J. Torng, P.H. Hor, R.L. Meng, L. Gao, Z.J. Huang, Y.Q. Wang, and C.W. Chu, *Phys. Rev. Lett.* 58, 908 (1987). (c) P.H. Hor, L. Gao, R.L. Meng, Z.J. Huang, Y.Q. Wang, K. Forster, J. Vassiliou, and C.W. Chou, *Phys. Rev. Lett.* 58, 911 (1987).
6. R.G. Pearson, *Symmetry Rules for Chemical Reactions*, Wiley-Interscience, New York, 1976.
7. R.J. Cava, *Science* 247, 656 (1990).
8. J.B. Torrance, Y. Tokura, A. Nazzal, and S.S.P. Arkin, *Phys. Rev. Lett.* 60, 542 (1988).
9. B. Raveau, C. Michel, and M. Hervieu, in *Chemistry of High-Temperature Superconductors*, D.L. Nelson, M.S. Whittingham, and T.F. George, Eds., American Chemical Society, Washington DC, 1987, p. 122.
10. P.P. Paul, Z. Tyeklar, R.R. Jacobson, and K.D. Karlin, *J. Am. Chem. Soc.* 113, 5322 (1991).
11. N. Kitajima and Y. Moro-oka, *Chem. Rev.* 94, 737 (1994).
12. A similar story can be told for nitrogen (rather than oxygen) side-on coordination. For example, see M.D. Fryzuk, T.S. Haddad, and S.J. Rettig, *J. Am. Chem. Soc.*112, 8185 (1990), and W.J. Evans, T.A. Ulibarri, and J.W. Ziller, *J. Am. Chem. Soc.* 110, 6877 (1988).
13. (a) E.I. Solomon, F. Tuczek, D.E. Root, and C.A. Brown, *Chem. Rev.* 94, 827 (1994). (b) An insightful VB model of bridged copper dimers has been published: F. Tuczek and E.I. Solomon, *J. Am. Chem. Soc.* 116, 6916 (1994).
14. R.C. Haddon, *Acc. Chem. Res.* 25, 127 (1992).

Chapter 42

Is There Hyperbonding and Hyperchemistry?

We now come to consider the implications of VB theory. Specifically, we saw that there exist two mechanisms of chemical bonding, each having some advantages and some disadvantages. The T mechanism has the benefit of strong kinetic energy reduction, but only at the expense of strong interelectronic repulsion. At this limit, the covalent and association-promoting HRP and ALT configurations are separated by a large energy gap. Strong bonding is the result of exchange delocalization described by the covalent structures. On the other hand, the I mechanism effects smaller kinetic energy reduction, but with the benefit of minimization of interelectronic repulsion. At this limit, the covalent and association-promoting HRP and ALT configurations are separated by a small energy gap but the absolute magnitude of the matrix element is small and is appreciable only because electron–electron repulsion is minimized by CT delocalization. This is the "reality" as represented schematically in Figure 42.1.

Is there a way of combining the positive aspects of the two mechanisms in a single unique bonding mechanism? This corresponds to actualizing the "dream" also shown schematically in Figure 42.1. If so, what are the conditions for the observation of superbonding?

Closoboranes, $B_nH_n^{2-}$, are beautiful, but the real conceptual challenge comes in the form of the nido (B_nH_{n+4}), arachno (B_nH_{n+6}), and hypho (B_nH_{n+8}) boranes. The shapes of these molecules are those predicted by assuming that the much more electronegative hydrogens donate their electrons to the boron atoms to form closoborane cages with missing vertices. Thus, for example, the nidoborane B_5H_{5+4} has the shape expected from $(B_5H_5W^{2-})^{2-}(H^+)_4$, where W is a dummy HB^{2+} unit. In other words, it is a six-closoborane with one vertex missing and four formal protons coordinated to the cage electron pairs. Had we not known the answer, we would have guessed that the operationally significant formula ought to be the one derived by the standard formal oxidation state procedure, namely, $(B_5H_5)^{4+}(H^-)_4$. What is the physical basis for this remarkable reversal of intuition?

We can interpret the electronic structures of the nido, arachno, and hypho boranes as follows: The more electronegative hydrogens borrow the unoccupied space of the boron atoms in order to delocalize their electrons, which are then able to achieve two goals:

They are guaranteed strong kinetic energy reduction via delocalization as a result of exposure to the unscreened nuclei of the electronegative hydrogens.

At the same time, they avoid each other because they can use the boron vacant orbitals to correlate their motions.

Thus, we conclude that the nidoboranes may represent the first step toward attainment of the "dream" of *strong* binding via 2-e CT. This mechanism is inactivated when B is replaced by the more electropositive Al and the "vacuum borrowing" by the hydrogens becomes counterproductive (i.e., too much promotional energy to transfer density from H1s to Al3p).

In this light, we can envision the following scenario for *hyperbonding*, that is, bonding that effects strong kinetic energy reduction while, at the same time, minimizing interelectronic repulsion. This is achieved by embedding two (or more) electronegative nonmetal atoms (e.g., two deuterium atoms) in a metal solid. The two deuteriums relinquish their electrons to the metal solid and proceed to combine to form a "superatom," which is intermediate between two deuteriums and helium dication. The superatom $(D—D)^{2+}$ "borrows" the

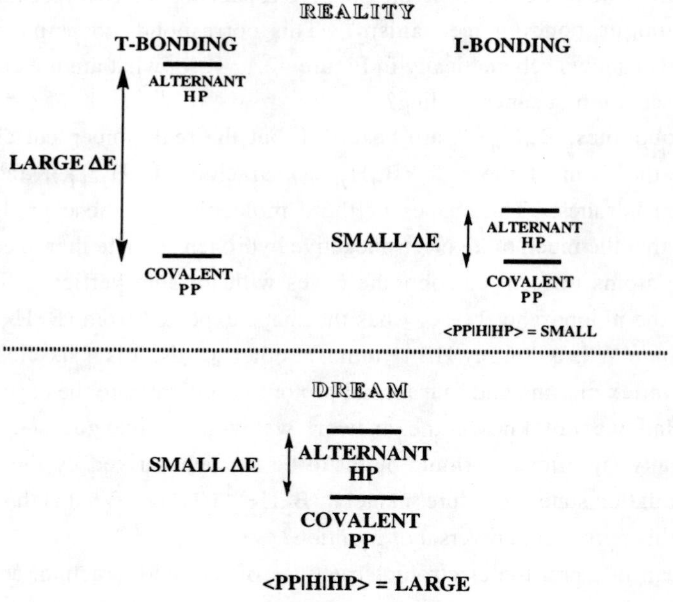

Figure 42.1 The interaction of the covalent (perfect pairing, PP) and association-promoting (hole–pair, HP) configurations as it occurs at the limits of T- and I-bonding and as it may occur under some ideal scenario.

vacuum space of the metal atoms to define an aromatic $4N + 2$ array of "plasma electrons" (i.e., electrons derived from the metal atoms and the original deuteriums). Aromatic delocalization now achieves two goals:

Kinetic energy reduction is sharp because of the extremely high electronegativity of the superatom.

Avoidance of interelectronic repulsion is made possible because the electrons can maximize their interparticle distances while staying within the "borrowed" metal AOs.

This scenario requires the presence of two partly fused deuterium atoms into a superatom, as illustrated in Figure 42.2, and it fails at the two extremes:

(a) At the "two deuteriums plus metal solid" extreme, we have organometallic bonding.

(b) At the "helium dication" extreme, the extraordinarily high electronegativity of He^{2+} causes the recapture of two "plasma electrons" to form He, which interacts with the metal solid in the ordinary chemical way (i.e., by dispersion).

We take a two-dimensional metal lattice where metal cores lie on the triangular vertices and interstitial (metallic) electrons lie in the triangular

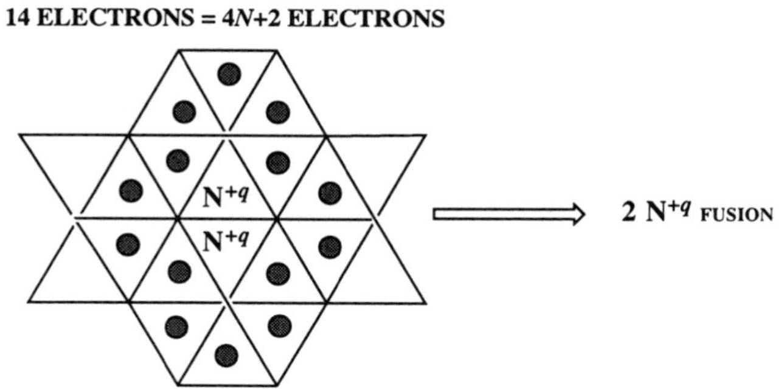

Figure 42.2 Aggregation of two metal cations creates a field that supports "aromatic" delocalization of surrounding gas pairs. The field exerted by the associated ions is such that kinetic energy reduction not only is strong, it is accompanied by minimization of interelectronic repulsion.

hollows (see Figure 42.2). Two nonmetal atoms, N, which shed their electrons to become N^{+q}, are placed in adjacent triangles sharing a side. This creates a field, which becomes responsible for strong kinetic reduction as a result of delocalization. Surrounding the two N^{+q} are now an odd number of pairs of (metallic) electrons, which are delocalized in a cyclic fashion around the N^{+q}—N^{+q} dimer within overlapping interstitial AOs. Thus, we have a microsystem effectively comprising atoms of very high electronegativity in which, nonetheless, $4N + 2$ electrons can delocalize via 2-e CT hops within a Hückel array of interstitial AOs. Thus, if there is hyperbonding, it must be occurring in the "blind spot" of chemical bonding, namely, at the transition state of a nuclear fusion reaction.

Thus, hyperbonding may be related to the recent controversy regarding "cold fusion." Because of the blending of the positive attributes of T and I delocalization, the logical extension of the VB concepts presented here suggests but does not require that the barrier to nuclear fusion of nonmetals (which have shed their electrons in the Fermi sea of metallic electrons and now appear as cations or polycations) be lowered in a "metal bath." In other words, nuclear reactions can be catalyzed by plasma electrons that are delocalized in metal AOs.

Chapter 43

Computational Chemistry: Curse or Panacea?

43.1 Can We Design Novel Chemistry by Carrying out Computations?

The biologist has the great stories of "Evolution" and "The Genetic Code" to tell the broad public. The physicist can spin the tale of "The Structure of Matter" or "The Structure of the Universe." In a world made up of molecules, the chemist, who has the best story of all — "The Society of Molecules" — has not chosen to recite it. Rather, the chemist steps forward with esoteric, boring, telephone-book itemizations of data: "The Chemistry of Alkenes," "Carbenium Ions," "Diradicals," . . . . No wonder the public finds little fascination in chemistry. Who can become excited about the physical and chemical properties of an apparently endless list of invisible entities symbolized by cryptic formulas? In what is the supreme irony, instead of being king (because "everything is molecules"), the chemist is the ugly duckling of the pond.[1]

Will computers help eliminate this unfair and absurd perception? The very essence of this work argues quite the contrary: The problem with chemistry is overspecialization, which up to a point is inevitable. It is bound to get worse, however, as long as computers are used mostly as number crunchers rather than as tools for the development of new ideas.

In the last 30 years, there has been fantastic progress in the accurate computation of small (especially organic) molecules. By comparison, virtually no progress has been made since Pauling's time in the conceptual domain of chemical theory. In the preface, I suggested that the accepted conceptual framework of chemistry is a fantasy with inevitable elements of reality. Figure 43.1 projects why: The chemical concepts of today cover indiscriminantly three different domains of chemical bonding.

For example, everything from hexagonal H_6 (T-bound) to cyclopropane (T-bound) to $(PH)_5$ (I-bound) to Li_6 (E/I-bound) is classified as "π aromatic" or "σ aromatic." The blame for this misconception must be laid at the feet of one-electron MO interaction diagrams. An MO diagram showing how symmetry orbitals interact can be constructed for *every* molecule. However, the nature and the stability of the molecule cannot be inferred from the MO diagram. H_6, π-benzene, and Li_6 all have three bonding MOs but as implied by their very different geometries, they are entirely different beasts. One must go beyond simple orbital overlap consideration to understand the bonding of these

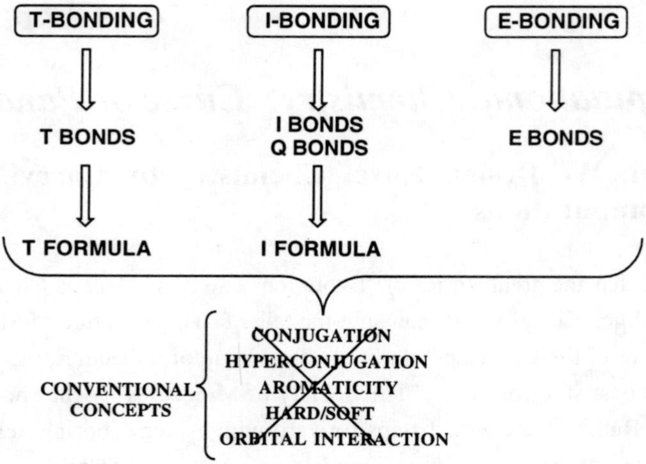

Figure 43.1 The three types of bond according to nonorthogonal VB theory.

molecules. Unless the grip of "orbital symmetry" is loosened, the chemist is condemned to the quagmire of "orbital engineering." For the time being, there seems to be no end to the monographs purporting to interpret chemistry through the concept of "orbital interaction." The idea is too simple to be resisted. But, it has no connection with reality!

The nonexistence of a true conceptual framework (i.e., the failure to differentiate between T-, I-, and E-bonding) is due to the seemingly reasonable assumption that electron–electron repulsion, at least in homonuclear systems, is qualitatively unimportant. Nothing could be further from the truth. Chemistry can be defined as the study of the *selective* transformations of molecules. Selectivity cannot be understood without explicit consideration of electron–electron repulsion. Without the latter, the distinction vanishes between T-, I-, and E-bonding (and all the selectivity principles developed in this work)! In this work, we discussed specific cases illustrating that even isoelectronic neighbors of one and the same row (e.g., Si, PH and S) have entirely different selectivities (Section 24.4).

There is one great irony: A Mulliken overlap population can be formulated as a useful computational tool only after one understands that chemical bonding is due to overlap. The point is simple: One cannot devise the indices for analyzing molecular wavefunctions without understanding chemical bonding. The problem is clearly illustrated by the apparent agreements and essential disagreements that lie hidden in the theoretical literature. For example, the

EHMO theorist and the ab initio theorist believe that their computations describe the same physical reality, in a qualitative sense, and that what separates them (aside from numerical accuracy) is mere style. Hoffmann says: "Why should I write my theory the way Bill Goddard, a theorist I admire, does, anymore than you expect Karl-Heinz Stockhausen and Pierre Boulez to write piano pieces that sound alike?"[2] Goddard agrees in the sense that he interprets his calculations to be a variation of the Pauling themes. For example, in discussing the bonding of $RuCH_2^+$, he says that "the result of a covalent double bond between a metal atom and CH_2 is in direct contradiction with the literal interpretation of the popular oxidation state formalism."[3] This statement bypasses serious discrepancies in computational results: $Ru{=}(CH_2)^+$ has a bond dissociation energy of –12 kcal/mol at the HF level, 27.6 kcal/mol at the perfect pairing GVB level and, finally, 68 kcal/mol at the GVB-CI level! All this apparent harmony (i.e., suggestions that the EHMO and the ab initio GVB stories are essentially the same) is contradicted by the applied mathematician who was not seduced by the apparent successes of the simple bonding models. Davidson notes: "Accurate ab initio calculations (on organometallics) remain an elusive goal. . . . The HF approximation is in fact hopelessly in error. . . . It is customary in this field to ignore these problems and carry out calculations which are marginally justified."[4] For the student of the history of science, we provide as a reference a listing of the seminal papers that detail the beginnings of the calculation of correlated wavefunctions.[5]

This work offers a resolution of the contradictory viewpoints:

(a) An EHMO calculation and a "perfect" MO-CI or VB-CI calculation capture two different physical realities. The difference is not a matter of style but a matter of substance. T-, I-, and E-bonding all differ in their physical and chemical consequences.

(b) The concept of the error-free codirectional arrow train renders polydeterminantal theory conceptually simple, reconfirming the motto of the mathematician: "To simplify, one must complexify." Thus we put in electron–electron repulsion (complexification) and we ended up differentiating between T and I bonds as well as between inferior and superior T and I bonds (simplification).

(c) A qualitative, leave alone a a quantitative, account of chemistry rests on electron–electron repulsion. Polydeterminantal calculations cannot be interpreted with tools tailor-made for monodeterminantal theory. This is why the mechanism of chemical bonding has remained elusive.

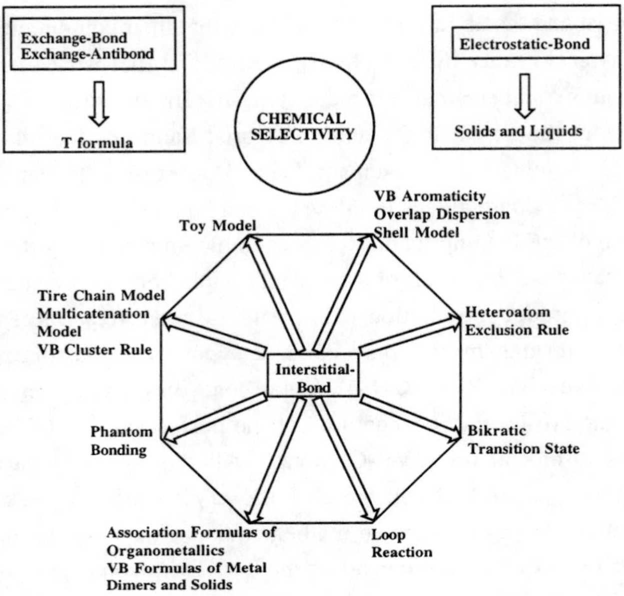

Figure 43.2 Some of the new VB concepts.

(d) Unlike the scenario often found in physics, one experiment does not yield a theory of chemical bonding. One computation of a molecule or a reaction path fails in the same way. What makes a "theory of chemical bonding" is an ocean of experimental data. Furthermore, not all drops of the ocean are equally significant, and this is why theory is not driven by those who "know the literature." To make the right choices, it is necessary to select the critical points that define and solve the problem; unfamiliarity with the entire ocean is not required. Physical organic chemistry came to being around the 1950s through the efforts of Ingold, Winstein, Bartlett, and Doering. Carbocations, carbanions, carbon-centered radicals, "no-mechanism" reactions came to center stage. Forty years later it is possible to compute with some reliability properties of reactants and intermediates, and to ascertain organic reaction paths. But, most of this work is done after the fact. In the 1990s, center stage is occupied by new materials, molecular recognition (directly relevant to the AIDS and cancer problems), high temperature superconductors, and so on. One may be able to "compute" a high T_C superconductor in the future but, by then, the experimentalist will have moved elsewhere.

Bottom line: The theory described in this work is not the result of computing. Quite the contrary. Ab initio computations have been interpreted by some practitioners in a way that is antithetical to our conclusions. An incomplete synopsis of the VB concepts that have no precedents in the literature is given in Figure 43.2. These are the product of the confluence of basic quantum mechanical notions and the phenomenology of chemistry. They mark the sharp contrast between the philosophy of this work and the philosophy of "canned computation," which now dominates the chemical literature.

43.2 A Prophecy of (Now Avoidable) Controversies

The periodic table has currently 109 elements, and all but C, N, O, F, and H obey the rules of I-bonding. T-bonding is equivalent to "covalent bonding" and then it is not! The concepts of Chapters 4 to 11 have little or nothing to do with what is in the literature. This defines the "spiraling confusion" scenario: Failure to extricate oneself from the labyrinth of conventional concepts and computations will condemn one to an ever-increasing level of confusion and frustration (unless, of course, all one wants is to get a paper published). Specifically, as isolation and characterization techniques constantly improve, new molecules, once thought to be inaccessible, will be made. Many will rush to "explain" their electronic structures by carrying out ab initio calculations, and soon an impasse will be reached: Stability will be found to be associated with either "localization" or "delocalization," and the argument will degenerate into discussions of "how much delocalized" a novel species is. This prophecy is already reality, judging from an article in *Chemical and Engineering News* entitled "Preparation of Stable Divalent Species Raises Issues of Electronic Structure."[6] In essence, three different groups prepared three different stable carbene derivatives, with the Arduengo-carbene researchers claiming "localization," the Denk-silylene researchers claiming "delocalization," and some warning against disagreements more semantic than real[7,8]:

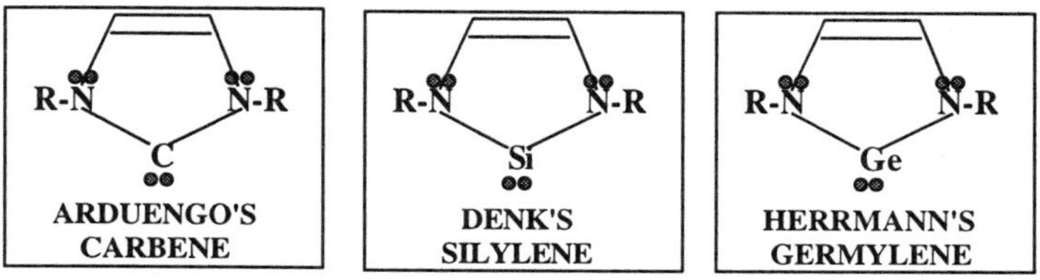

By now, the reader will anticipate our explanation: The Arduengo carbene is a T-bound molecule that conforms to the rules stated in Chapters 4 to 9, and it is representable by a *T formula.* On the other hand, the Denk silylene and Herrmann germylene are I-bound molecules conforming to a very different set of valence rules, and they are representable by *I formulas.* This is why ab initio computations predict that diaminosilylenes can actually dimerize to form the dibridged structures well known to p-block inorganic chemists.

Thus, both sides are right in the interpretation of their experiments (after all, they describe what they observe!), but neither can relate to the other because of the prevailing bonding models. Of course, in the context of VB theory, saying that silylene (and even more so germylene and stannylene) is "divalent" *or* "tetravalent" is wrong: Since a hole can bind either two radicals or two pairs, and since a pair can bind either two radicals or two holes, the maximal valency of silicon is eight! The compounds that chemists have produced represent different compromises between intrinsic valence and inevitable ligand repulsion in the coordination sphere.

Bottom line: There is nothing "semantic" in the looming controversy. One can either lump H and Li together as "univalent" or put them apart by saying that one is "nonmetallic" and the other "metallic." Arguing that such a categorization is arbitrary flies against any sense of chemical logic. Metals are fundamentally different from nonmetals. "The underlying physical laws for the mathematical theory of a large part of physics and the whole of chemistry are thus completely known and the difficulty is only that the exact applications of these laws leads to equations much too complicated to be soluble." We amend this well-known statement of Dirac to read that ". . . these laws lead to equations much too complicated to be chemically useful even if they were soluble." The experimental chemist asks: What should I do next that is new, important, and has practical applications? Only a global conceptual theory can yield an answer.

Beyond the "delocalization controversy," I can think of at least five other topics that may engender controversy.

The "p orbital controversy" in transition metals. The prediction is that p AOs of transition metals will turn out to act as polarization functions in ab initio computations, much like d AOs in heavy p-block elements. This will destroy the classical theoretical basis of the 18-electron rule and make many an organometallic chemist unhappy.

The "metal oxidation number controversy."[9a] The association formula of square planar $Cu(CF_3)_4^-$ is $(CF_3^-)_2Cu^+(CF_3)_2$ in which is CF_3^- is connected to the metal by one ROX bond and the two remaining CF_3 radicals are linked by one REL bond. The structure is entirely analogous to that of $(PR_3)_2$Pt(ethylene) discussed in Chapter 24. This predicts a nonformal Cu(I). Actually, the stability of anionic organometallics is a consequence of ensuring that a maximum number of ROX and REL bonds is compatible with a relatively low metal oxidation state. On the other hand, the classical oxidation formalism assumes ionic bonding and predicts a formal Cu(III). We expect future ab initio computations of metal charges to demonstrate clearly that the coordinate bond is neither covalent nor ionic but something else, namely, an I bond.

The "metallic bond controversy."[9b] Because chemists (either implicitly or explicitly) analyze all wavefunctions (from EHMO to ab initio MO-CI) by falling back on EHMO-type concepts, the temptation will be to postulate universal covalency. This runs head on against VB conclusions and the innumerable experimental and calculational facts that say quite the opposite: Things change spectacularly as we replace nonmetals by metals.

The "magic number controversy."[9c] The concept of aromaticity as presented by HMO theory fixated chemists on magic numbers. This preconception was further enhanced by EHMO-type models of cluster structure. The concept of the I bond says exactly the opposite: Metal complexes and clusters have multiple magic numbers. The issue of cluster electron count will become a source of controversy if chemists continue with the unphysical EHMO-type models. One indication of coming trends is the recent work of Puddephatt and co-workers,[9c] who prepared a octahedral Pt_6 cluster with fewer electrons than the Wadian magic number. This should be added to the growing list of clusters obeying the VB cluster rule rather than Wade's rule.

The "hypervalent controversy".[9d] Because one I bond connecting three atoms can have anywhere from two to four electrons, transition metal chemists will start discovering (much like their main group colleagues) the existence of what they will term "hypervalent" complexes (with respect to the 18-electron rule). The work of Geiger is an early indication that this is already happening.

The perception of "exceptional molecular stability" on the basis of conventional theory where, in fact, the stability is precisely what is expected by our VB theory of chemical bonding. For example, one typically expects isomers of the type $A{=}AZ_2$ and A=AHZ, where A is N or a heavier congener and Z an equally or more electronegative group, to be very unstable because of the required zwitterionic character of the A=A unit.[9e] On the other hand, we have seen in chapters 6–8 that this atomic arrangement is favorable because intrabond CT delocalization occurs with minimization of exchange repulsion; that is, $A{=}AZ_2$ has zero and A=AHZ has only one error in its T formula. As a result, when A becomes relatively electropositive (and zwitterionic character is no longer a severe problem), isomers of this type can easily become global minima.

43.3 Molecular Engineering

In recent years, there has been an explosion of physicochemical knowledge as a result of an ever-increasing ability to perform more discerning experiments and more accurate calculations. This has set up two unavoidable questions:

What do we choose to teach from the ocean of chemical facts, experimental and computational?

What is a legitimate and preeminent research goal?

VB theory answers these questions as follows.

In principle, everything ought to be taught because all chemistry is interrelated. For example, in a few lectures, one can cover organic rings (Chapter 4), inorganic rings (Chapter 24), metal solids (Chapter 35), and boron clusters (Chapters 37–39). Normally, an organic chemist hardly ever hears of boron deltahedra, an inorganic chemist hardly ever goes beyond the trivial aspects of carbocycles, and a transition metal chemist is acquainted hardly at all with the intricacies silicon and phosphorus rings.

Should we teach the facts or the theories of chemistry? This is an age-old debate, and this work gives a direct answer: Teach the facts of chemistry (rather than the facts of "organic," "inorganic," etc., chemistry) through the theory of chemistry (rather than the theory of "organic," "inorganic," etc., chemistry). This is possible because chemical formulas and reactions that were uninterpretable and, thus, had to be committed to memory, can now be explained in a self-consistent way across the periodic table.

In "anything-goes research" (euphemistically called "curiosity-driven research") all one has to do is dream of some combination of atoms and molecules and exploit modern experimental or computational techniques to collect data, publish papers, and get grants and awards. This process is ugly, unscientific and, from the economic standpoint, highly inefficient. Although we are only at the beginning, we could already produce specific and straightforward recipes for making new molecules or new materials. Some of these possibilities are listed (with the pertinent cross-references identified), to impress our motto: "No recipe, no theory."

Destrained rings: Sections 4.5–4.8
Stable organic molecules and polymers: Section 6.6; Chapters 7, 8.
Crowded rotational isomers: Section 11.6
Dearomatized molecules and transition states: Sections 16.3, 17.7

Successful and unsuccessful pericyclic transition states: Sections 15.7, 17.2–17.3
Stabilized π complexes: Section 28.9
Fast reactions: Section 34.7
Unusual diradicals: Section 35.7
Rotated molecules: Chapter 36
Bare metal clusters: Section 37.2
Organolithium clusters: Section 37.6
Metal tubes: Section 38.4–38.5
Isoelectronic but heterostructural molecules: Section 39.6
Organic conductors: Section 40.5
High T_c superconductors: Section 41.9
Metal-catalyzed nuclear reactions (?): Chapter 42

The most important aspect of this work is the delineation of the fundamentals of chemical selectivity. After all, chemistry is an exercise in chemical selectivity (i.e., everything is relative). We hope to further improve our understanding of chemical selectivity (the T/I/E disjunction) to be able to design selective molecules and materials. We call this enterprise *molecular engineering*, and we use the term to define a specific four-step process:

A chemical system (molecule, polymer, molecular assembly, etc.) is designed for executing a specific function. The design is based on conceptual theory.

Ab initio computations test and refine the design.

Synthetic chemistry and spectroscopy combine to make and characterize the system.

The successful system is passed on to the technologist for practical implementation.

We envision this to be the blueprint for a new Ph.D. curriculum, which promises the applicant a creative experience, a varied background, and employment flexibility. A specific example will serve to illustrate how "basic research" is differentiated from "applied research" in this model.

Many a computational chemist strives to calculate accurately the barrier of the prototypical H^- + H—H bond exchange, the simplest model of the crucial S_N2 reaction of organic chemistry. As computational capabilities improve, H^- is replaced by other nucleophiles, H—H is replaced by methane and its derivatives and, finally, solvent molecules are included in the reactive ensemble. While useful, these studies have a small impact on chemistry because the experimentalist has already been there. It is much more important to lead experimentalists to explorations they normally would be reluctant to undertake.

For instance, experimental organic chemistry exists as we know it today because molecules made up of H, C, O, N, and F are inert to solvolysis and, in particular, to hydrolysis and alcoholysis. As a result, organic molecules can be isolated and manipulated in ordinary solvents by exploiting two physical facts: Water is a poor nucleophile, and a first-row atom or fragment is a poor leaving group. Thus, for example, the S_N2 reaction of HOH plus H_3C—OR is effectively "forbidden." Now, the classical S_N2 reactions *in protic solvents* are reactions in which the optimal choice of nucleophile and leaving group is the one that produces an I-selective transition state. Recall that the order of nucleophilicity and leaving group ability is $I^- > Br^- > Cl^- > F^-$ because of the indirect effect of anion stabilization by the protic solvent. Thus, when thinking of S_N2 reactions, the organic chemist most frequently thinks about half of the story, namely, S_N2 reactions in protic solvents, which are facile because weak reactant and product solvation is compatible with a trigonal bipyramidal transition state that comes near the I-bonding ideal.

What about the other half of the story? This is an S_N2 reaction with a transition state that comes near the E-bonding ideal. Indeed, the very reversal of the order of nucleophilicity in aprotic dipolar solvents means that the intrinsic selectivity of the organic S_N2 transition state is E selectivity! The strategy is now apparent: Because of the small size of O (and first-row atoms, in general), HOH and OR can be turned into effective components of an E-selective S_N2 transition state by a perturbation that stabilizes a negatively charged water oxygen and an alkoxy anion. The obvious perturbation is coordination of metal cations devoid of valence electrons. A carbon substrate that will do the "impossible" (i.e., undergo facile hydrolysis in which the leaving group is a poor leaving group) can be designed as follows:

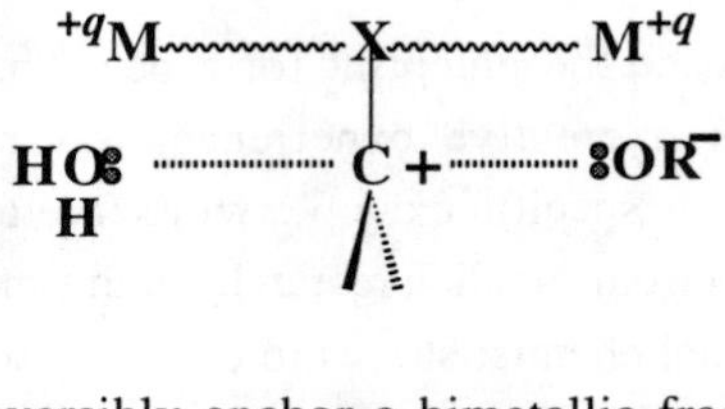

Our goal now is to reversibly anchor a bimetallic fragment, $M^{+q}XM^{+q}$, on carbon. The best strategy is to go to the literature to find whether the $M^{+q}XM^{+q}$ fragment has been synthesized, to get ideas for the modification of X to reversibly attach it to the carbon, and so on. Only after we have exhausted the goldmine of the literature do we turn to computations, as a "fast" substitute for experiment, to refine the design.

Every good chemist knows that there is nothing impossible when it comes to molecules. Quite the contrary: Everything is possible, provided we select the right pieces. Nature employs DNA as the genetic material partly because the phosphodiester linkage is resistant to hydrolysis. HOH is a poor nucleophile and OR is a poor leaving group. In a recent important paper, Tsubauchi and Bruice[10] report a remarkable 10^{13} rate enhancement in the hydrolysis of a phosphonate ester upon catalysis by two lanthanum cations, one of which is borne by the leaving group and one by the phosphorus center. This is a variation of the design just given with $M^{+q} = La^{3+}$, and it is is only a small indication of what is certain to come. While most of known chemistry is I-selective chemistry, E-selectivity holds the promise of enabling us to do what we once thought to be impossible.

An area of academic interest promising important practical applications is the utilization of metals to dismantle "resistant," stable molecules that have adverse environmental impact. We can use theory as a "listening device" to solve such problems. This approach can be illustrated by reference to the recent success of Groves and co-workers[11] in activating N_2O by a ruthenium porphyrin. These studies imply the formation of an intermediate PORRuN═N═ORuPOR that breaks up to PORRuN_2L plus PORRuOL, where L is THF. We want to find out why the dinuclear complex is formed, how it breaks up, and how the process can be tuned. The answer is suggested by one component of the π system of PORRuN═N═ORuPOR:

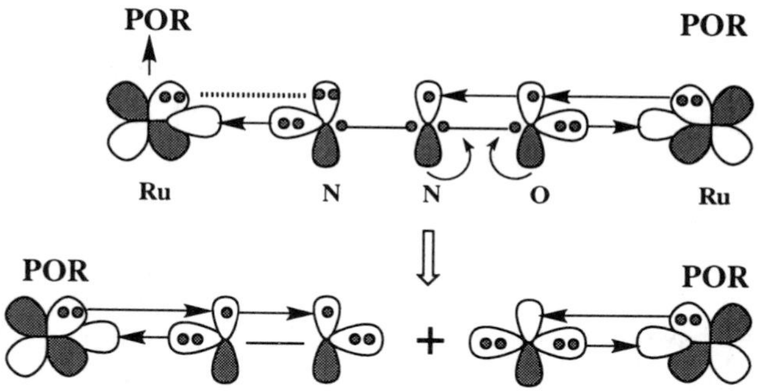

Here one nitrogen and one oxygen lone pair engender a four-electron antibond (hatched line) with one metal π-type d pair. Complex stability requires that the antibond be annihilated. This is effectively accomplished by delocalization of the Ru(II) d pair to the porphyrin. Turning off this delocalization promotes a two-electron reduction of the σ N—O bond (curved arrows) and product formation.

We now have a recipe: To stabilize the bisruthenium intermediate, put π-electron acceptors on two *trans*-pyrroles of the porphyrin. To turn on the σ-bond cleavage, turn off the action of the π acceptors by some effector base. In fact, we can install two molecular wheels on two *trans*-pyrroles and create a "molecular factory" dedicated to the processing of stable molecules. We end by noting a lesson of our "one orbital at a time" approach: A metal cannot tie up a ligand if ligand lone pairs run into metal pairs. The cure is to delocalize the metal pairs to an appropriate reservoir. The porphyrin macrocycle plays exactly this role.

What can be more interesting than the problem of life and death! We live by utilizing oxygen, but we get poisoned by inhaling cyanide. Problems of this nature are tackled at what is now the frontier of experimental chemical science: "bioinorganic chemistry." Cytochrome *c* oxidases catalyze the following exergonic process:

$$O_2 + 4H^+ + 4e^- \text{ (from cytochrome } c\text{)} \longrightarrow 2\,H_2O$$

The active site is believed to be an Fe(III)—Cu (II) heterodimer, which model studies by the groups of Karlin[12] and Holm[13] have shown to act (a) by trapping an O^{2-} dianion by *antiferromagetic* coupling of the two metal centers and (b) by trapping a CN^- anion by *ferromagnetic* coupling of the two metal centers.

The solution of the problem illustrates the "electron by electron, orbital by orbital" VB approach, and it hinges on our ability to represent the electronic structures of the participants in a compact and informative fashion by VB formulas. The L(POR)Fe(III) center can be represented by a ds-hybridized iron (see Figure 25.4) and the L_4Cu(II) center by a dsp^3 (trigonal bipyramidal hybridization) copper (Figure 25.10). L is a two-electron σ donor. By recognizing that cyanide is effectively analogous to a dipositive oxygen, we obtain the electronic formulas of the Fe(III)O(–II)Cu(II) and the model Fe(III)O(II)Cu(II) complexes shown in Figure 43.3. For maximal simplicity, π-type d orbitals have been replaced by the isolobal π-type p orbitals. The crucial point is now transparent: Cyanide poisons by its low-lying π LUMOs, which are responsible for a reorganization of the electrons in the Fe(III) orbitals that causes at least one π-type iron pair to match one cyanide π-type hole. This electronic rearrangement allows the formation of one strong ROX bond connecting Fe(III) and CN^-. As a result, CN^- ties up the metal (and especially the iron) centers and makes them unavailable for the processing of oxygen. In the conventional language of the chemist, cyanide poisons because it is a good nucleophile (i.e., a good ROX ligand).

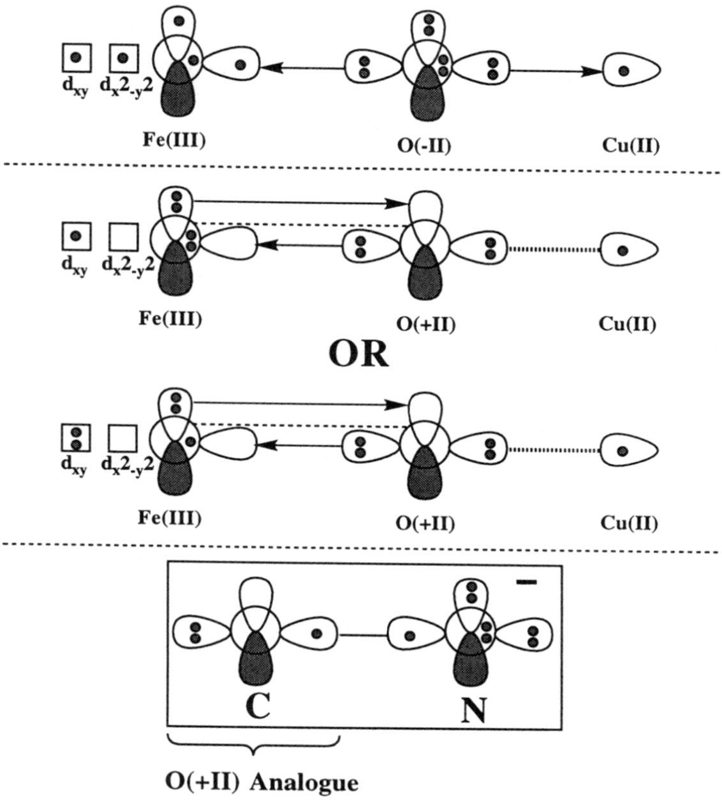

Figure 43.3 The difference between O (–2) and O (+2) and the analogy of the latter to CN (–1) when acting as metal-bridging ligands.

The O(–II) binding is due to a *contradirectional* OX bond, which is the cause of the antiferromagnetic coupling. This implies that the O^{2-} is essentially bound by the electrostatic mechanism and, as a result, can be readily protonated to form water. By contrast, the CN^- binding, simulated by the O(II) binding, is due principally to a superior codirectional iron-carbon ROX bond. The net result is that the two odd electrons end up in orthogonal AOs and can no longer be coupled by an intervening pair or hole into an OX or RED bond. Hence, we now have ferromagnetic coupling. The essence of this highly abridged presentation of a "life and death" problem is that with the right VB concepts at hand, we can use simple formulas to probe the mysteries of nature.

The point is simple: A computer cannot make a bald man grow hair. A computer cannot make the ugly handsome. A computer cannot stop aging. A computer cannot produce insights when the operator has none. To quote

Davidson, "Computers do not solve problems, people do."[14] The challenge for those aspiring to explain chemistry is simple: Can one go beyond the Pauling/Hückel world of ideas?[15] A recent review of pericyclic reactions by Houk[16] further defines the problem: What have we learned about the electronics of transition states after so much "computational blood" has been spilled? What physical principles of general applicability have ab initio calculations produced? A comparison of the Houk review with Chapter 17 of this work is left to the reader. Suffice to say that the Huisgen/Firestone controversy is highlighted, but the "physical meaning" of the affair is not even suspected.

This work breaks the conceptual impasse that has existed since the 1950s by offering concepts and accompanying formulas that are not only new but even antithetical to the ideas of Pauling and Hückel, which have been accorded the status of gospel. The T and I formulas sing a song entirely different from that heard in journals and conferences. This antithesis has nothing to do with formalistic details but, rather, with the very physics of the problem. As everyone knows, the Pauling/Hückel ideas (and subsequent applications by Lipscomb, Hoffmann, and Fukui) are based on EHMO or monodeterminantal SCFMO wavefunctions. By contrast, the ideas described in this work are the first to do justice to the correct (polydeterminantal MO or VB) wavefunction. At the same time, these ideas were not inspired by a computer printout. Quite the contrary. My experience has been that the more one tries to analyze MO wavefunctions, the more confusion is generated simply because new theory cannot be developed by calculating old indices. After a page touting "electronic wizardry," "awesome technology," "information superhighways," and so on, the editor of *Chemical and Engineering News* advises us that "it is okay to disconnect from the electronic world once in a while and take time to think."[17] By contrast, this work espouses that it is okay to disconnect from thinking once in a while to take time to see how ideas become quantitated by computation and how the resulting numbers stack up against experiment.

43.4 "New Theory" Means "New Gambles"

As technology improves, the information glut will keep on increasing. Soon, it will be clear that the problem is not *whether* to experiment or compute but, rather, *what* to investigate and *what* to calculate. This work provides a VB methodology of thinking about molecules in lieu of narrow rules and crude generalizations. This methodology is based on two elements. First, a reappreciation of the conceptual power of VB theory and, second, the device of new conceptual procedures. Two examples drawn from the very recent literature illustrate the point.

The big conceptual advantage of VB theory is that it makes the promotion/bond-making interplay transparent. In general, the higher the ionization potential (IP) of an atom, the more prone it is to be promoted if by doing so it can form either T or I bonds. Interpreted by the starring concept (Section 25.2), this means that unoccupied AOs have increasingly unstarred character as atom IP increases. For example, nonmetallic carbon is predominantly tetravalent binding from the $2s^1 2p^3$ electronic configuration while metallic tin (and lead) is frequently divalent binding from the $5s^2 5p^2$. This means that a C2p hole is an unstarred valence AO while a Sn5p hole tends to be a starred polarization hole. The general rule is that np AOs aquire more valence character (unstarred character) as we move up a column of the p-block, down a column in the d-block and from left to right in any p- or d-row.

In this light, consider now two possible mechanisms of oxidative addition of a transition metal, acting via a pair and a hole (of variable star character), to an A—B bond. The first is the one-step process expected when A and B have compable electronegativity and high IP (e.g., A—B is a C—H bond). This is expected when the metal hole is unstarred. The second is the stepwise process in which the metal acts as a nucleophile using its pair to attack A displacing B in an S_N2 fashion in the first step. The expelled B^- coordinates to the metal hole in the second step. This is expected when the metal hole is starred and when A and B have different electronegativities, with B being a good leaving group, e.g., A—B is a C—Cl bond. It is now clear that when a metal fragment is given a choice of oxidative addition to H_3C—H and H_3C—Cl, the preference will depend on the identity of the metal. Early (electropositive) transition metals will preferentially attack the latter while late (electronegative) transition metals will preferentially attack the former.[18]

As a second example of how the primordial concepts of VB theory combine with the new VB concepts developed here, consider the combination of the "simple" diatomic C_2 with SiH_2 and its relatives. Naively, one expects to form the silicon analogue of cyclopropyne, a covalent molecule. The VB story is entirely different: Since Si is an electropositive semimetal, it will combine with carbon by I bonds which are generated with minimum fragment excitation. Now, elementary physical chemistry texts often display the sequence of the valence MOs of a diatomic. One unappreciated but characteristic feature is the near-degeneracy of the $2\sigma_g$ (symmetric σ) and $1\pi_u$ (symmetric π) valence MOs. This means that ground C_2 can easily sustain a two electron promotion from its $2\sigma_g$ HOMO to its $1\pi_u$ LUMO to prodce a species which, in VB terms, is a π

bis-acid; that is, it has two 2p holes in addition to a double C═C bond and two exo carbon σ lone pairs. The cyclic molecule C_2SiH_2, characterized by Maier and coworkers,[19] can be simply described as promoted (bis-acid) C_2 linked to ground SiH_2 by one I bond, as illustrated below. The I bond is expected to be strong because CT delocalization occurs from the electropositive Si to the (more) electronegative C. This predicts a very low barrier to SiH_2 rotation, as found by ab initio calculations.[19]

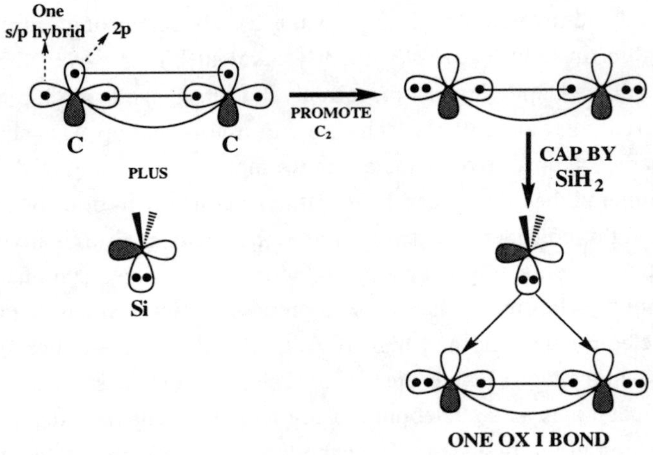

Since effectively only one of the two p holes of the promoted C_2 is tied up by one SiH_2 lone pair, it can be predicted that the cyclic molecule C_2SH_2 should also exist as a stable molecule. In this, two S lone pairs tie up the two C2p holes of promoted C_2. Bottom line: C_2SiH_2 is neither a covalent molecule nor a complex. It is a prototypical I-bound organometallic. It is easy to show that adding two hydrogens to the dicarbon fragment generates $H_2C_2SiH_2$ in which two I bonds (one s- and the other p-type) link the C_2H_2 and SiH_2 fragments in such a way so as to steeply enhance the barrier of rotation of SiH_2.

New theory shakes loose the albatross of incremental research by suggesting new and risky avenues of investigation. Can we convince nonmetals to behave like metalloids or metals? In particular, can we trick carbon to form deltahedral clusters much as boron does? We now have the theory that can guide us to this goal. How hard it is to get there is projected by the theory itself.

Closoboranes are deltahedral clusters made up of BH subunits with the addition of one extra pair of electrons. Each BH fragment acts as a 3h/2e piece. So, the first task is to define the strategy for producing a stabilized carbon 3h/2e fragment. One approach is to start with the acetylene derivative X—CC—SUB and derive the sought-after fragment by loss of X and appropriate choice of the substituent SUB. The latter should be a group that simultaneously turns one π component of the formal carbon–carbon triple bond into a diradical by acting as a radical stabilizer and turns the second π component into a zwitterion by acting as a dipolar, through-space stabilizer of the anionic part of the zwitterion. The strategy is as follows:

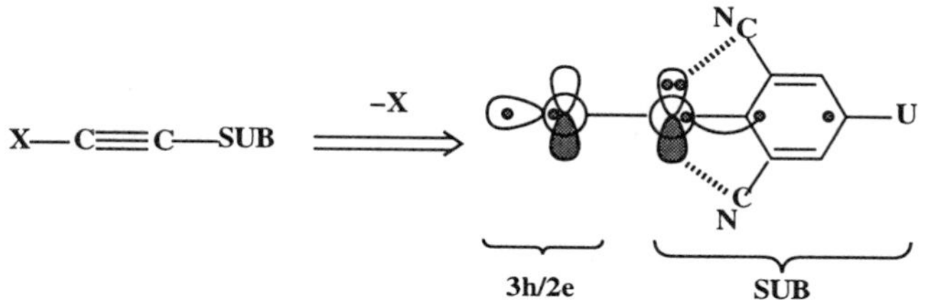

We now have the hope that *n* CC-SUB fragments can aggregate to a stable $(CC\text{-}SUB)_n$ deltahedron obeying the VB cluster rule of $2V + 2A$ electrons, where *A* is flexible and, in fact, unpredictable.

VB theory shows us what the risks are and tells us why we can still hope. For example, boron is fundamentally different from carbon. Because of their respective electronegativities, the former is predisposed toward I- and the latter toward T-bonding. T-bonding is incompatible with deltahedral cluster formation. Ab initio calculations reveal that H_2B binds differently from the isoelectronic H_2C^+: B_2H_6 and $C_2H_6^{2+}$ have different shapes. Countering these very real fears is the hope that C^+ will be convinced by perfect I-bonding (represented by the deltahedral cluster) to behave like B. In other words, carbon has not yet been given its best chance to act like its neighbor boron. We can now turn to ab initio calculations as a tool for a fast, preliminary test of our proposal. Thus our modus operandi is to use ab initio calculations to open up new fields rather than to eulogize what is already known.

Bottom line: So what if one has the best computational facilities? So what if someone else has the "perfect" computer programs? After all the computational and experimental data have been heaped on the desk, the inexorable task remains: Think and synthesize! Because life and resources are finite, one must answer the question: What do I select to do next from an ocean of possibilities? One solution is to throw darts at the map of conceivable choices. Another is to rely on basic chemical concepts. We advocate the latter approach. Ab initio computations can be used to disguise the lack of ideas. Ab initio computations can also be used to develop new ideas. The issue is not the tools but rather how one uses them.

The dedication of this work makes it evident that the heroes of the author are not found among scientists. However, a case for greatness can be made for those who demonstrated for the first time the doability of a problem formerly thought to be unassailable. Woodward has legendary status among organic chemists because he was the first to demonstrate that any organic molecule can be made regardless of complexity ("doability of complex synthesis"). Applied mathematicians deserve a similar status because they demonstrated the doability of computing molecules. This work argues that with the right concepts at hand, it is doable (and relatively easy) to express the physics of molecules by formulas based on polydeterminantal theory.

References

1. See the review of *The Third Culture: Beyond the Scientific Revolution,* J. Brockman, Ed., Simon & Schuster, New York, 1995, by J. Emsley in *Chem. Eng. News* 73(40), 42 (1995).
2. R. Hoffmann, *Angew. Chem. Int. Ed. Engl.* 27, 1593 (1988).
3. E.A. Carter and W.A. Goddard III, *J. Am. Chem. Soc.* 108, 2180 (1987).
4. (a) E.R. Davidson, in *The Challenge of the d and f Electrons*, American Chemical Society, Washington, DC, 1989, p. 153.
5. (a) E.A. Hylleraas, *Z. Phys.* 48, 469 (1928). (b) D.R. Hartree, W. Hartree, and B. Swirles, *Philos. Trans. R. Soc.* A238, 229 (1939). (c) P.O. Loewdin, *Adv. Chem. Phys.* 2, 207 (1959). (d) R.K. Nesbet, *J. Chem. Phys.* 43, 311 (1965). (e) C.F. Bender and E.R. Davidson, *J. Chem. Phys.* 70, 2675 (1966). (f) The systematic work of Pople and co-workers, starting with semiempirical SCFMO methods and culminating with ab initio perturbation MO-CI calculations of molecules of chemical interest, has been described in monographs on computational chemistry. See T. Clark, *Handbook of Computational Chemistry*, Wiley, New York, 1985.
6. *Chem. Eng. News* 72(18), 20 (1994).
7. (a) A. Arduengo III, R.L. Harlow, and M. Kline, *J. Am. Chem. Soc.* 113, 361 (1991). (b) A.J. Arduengo III, H.V. Rasika Dias, D.A. Dixon, R.L. Harlow, W.T. Klooster, and T.F. Koetzle, *J. Am. Chem. Soc.* 116, 6812 (1994).
8. (a) M. Denk, R. Lennon, R. Hayashi, R. West, A.V. Belyakov, H.P. Verne, A. Haaland, M. Wagner, and N. Metzler, *J. Am. Chem. Soc.*

116, 2691 (1994). (b) Y. Apeloig and T. Mueller, *J. Am. Chem. Soc.* 117, 5363 (1995). (c) W.A. Herrmann, M. Denk, J. Behm, W. Scherer, F.-R. Klingan, H. Bock, B. Solouki, and M. Wagner, *Angew. Chem. Int. Ed. Engl.* 31, 1485 (1992).

9. (a) M. Kaupp and H.G. von Schnering, *Angew. Chem. Int. Ed. Engl.* 34, 986 (1995); J.P. Snyder, *Angew. Chem. Int. Ed. Engl.* 34, 986 (1995). (b) J. Schoen, *Angew. Chem. Int. Ed. Engl.* (1995), in press. (c) L. Hao, G.J. Spivak, J. Xiao, J.J. Vittal, and R.J. Puddephatt, *J. Am. Chem. Soc.* 117, 7011 (1995). (d) W.E.Geiger, *Acc. Chem. Res.* 28, 351 (1995). (e) E. A. Salter , R. Z. Hinrichs, and C. Salter, *J. Am. Chem. Soc.* 118, 227 (1996).

10. A. Tsubouchi and T.C. Bruice, *J. Am. Chem. Soc.* 116, 11614 (1994).

11. J.T. Groves and J.S. Roman, *J. Am. Chem. Soc.* 117, 5594 (1995).

12. K.D. Karlin, A. Nanthakumar, S. Fox, N.N. Murthy, N. Ravi, B.H. Huynh, R.D. Orosz, and E.P. Day, *J. Am. Chem. Soc.* 116, 4753 (1994).

13. (a) S.C. Lee and R.H. Holm, *J. Am. Chem. Soc.* 115, 5833, 11789 (1993). (b) S.C. Lee, M.J. Scott, K. Kauffmann, E. Muenck, and R.H. Holm, *J. Am. Chem. Soc.* 116, 401 (1994).

14. E.R. Davidson, "Perpectives in Ab Initio Calculations," in *Reviews in Computational Chemistry*, Vol. 00, K.B. Lipkowitz and D.B. Boyd, Eds., VCH, New York, 1990, Chapter 11.

15. The Pauling philosophy, which resurfaces in this work, is nicely brought out in anecdotal form in *Chem. Eng. News* 73(19), 28 (1995).

16. K.N. Houk, J. Gonzalez, and Y. Li, *Acc. Chem. Res.* 28, 81 (1995). The authors close with an excerpt from the popular aria of the duke of Mantova from Verdi's *Rigoletto*. "La donna è mobile" expresses the reckless abandon and the hedonistic attitude that, in tandem with creative outbursts, are characteristic of the great Mediterranean cultures. The Houk article is the antithesis of this: a conservative restatement of old ideas dressed in ab initio computations.

17. *Chem. Eng. News* 73(13), 5 (1995).

18. This analysis seems to be supported by a comparison of recent results of Arndtsen and Bergman (ligated Ir cation attacks C—H in preference to C—Cl bonds) with unpublished work of Heinekey's group (ligated Re cation does the reverse). B.A. Arndtsen and R. G. Bergman, *Science* 270, 1970 (1995); D. M. Heinekey, private communication.

19. G. Maier, H. Pacl, H. P. Reisenauer, A. Meudt, and R. Janoschek, *J. Am. Chem. Soc.* 117, 12712 (1995).

Epilogue

Science is a constrained discipline in which nature is the arbiter of what is "right" and what is "wrong." Because of the very self-consistency of science, all who practice it are connected by the same matrix of ideas. In turn, this discourages adventurism and discovery, save by accident. The result is that there are many "useful scientists" but very few stars. Who remembers who received the Nobel Prize in chemistry the year —? The characteristic feature of a scientist is nonuniqueness. On the other hand, the unconstrained nature of art has a diametrically opposite consequence: Without any rigid matrix to keep them together, artists can express their individuality. Scientists cite one another but artists cite only themselves. Society celebrates the uniqueness of great artists (and even great athletes) but denies star status to (replaceable) scientists. The one or two exceptions (e.g., Einstein) confirm the rule, and no public awareness campaign can change the situation. The dedication of this work shows where my sympathies lie and explains the very content of it.

Chemical bonding is expressed by pairs of arrows. Two arrows can be oriented either in head-to-tail or head-to-head fashion. In each case, they can be either linearly or cyclically disposed. The *four* possibilities define the *three* types of chemical bonding. A tail-to-tail arrangement implies failed CT delocalization and, thus, E-bonding.

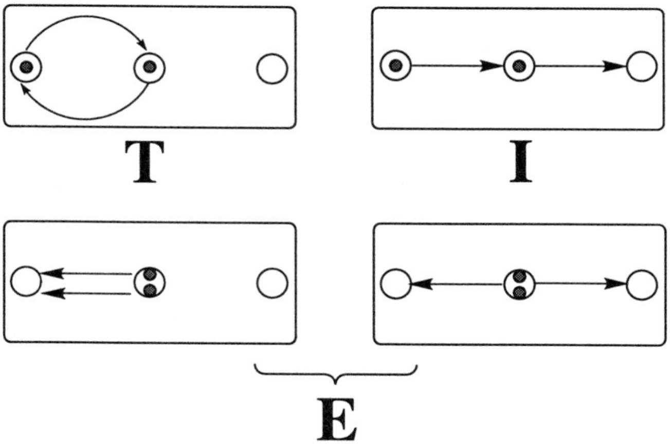

These ideas were conceived in the unique environment where Apollo, Neptune, and Aeolus reign supreme: the Greek islands of the Aegean sea, which lie near the birthplace of Demokritos, the first atomicist. Many formulas were scribbled at the Hotel Grotta, the Hotel Ikaros, and the patisserie "Rendezvous"

of Naxos. Going south, the island of Patmos and the archipelago of Lipsi, laboratories with blue skies and blue sea, offered antidotes to mind-numbing incremental science. By the time I arrived at the kingdom of Mrs. Irini at Megalo Horio of Tilos, the jewel of the Aegean, I had almost become convinced that I had made the wrong choice of profession. The Hotel "Annika," with Annika, Mitsos, Greek coffee and figs, and the endless beach of Kardamena drew the curtain down, and then it was back to the role of university professor.

I have left the most important for last. The development of the concepts described in this work has been akin to an oft-interrupted train journey that lasted many years and was not derailed because of Linda. Without her, there might still have been T and I formulas but no legend of love and devotion to go along.

Author Index

Subject Index

Note 4/14/97 MCL

Could we get objective pictures of T, E, & I bonding via electron density difference plots? perhaps look at ~~topology~~ of grad$(\rho - \rho_0)$

where ρ is the electronic charge density and ρ_0 is the sum of the (non-interacting) atomic charge densities? How do we deal with hybridization in this case. e.g. ρ_0 for C in CH_4? Is there a good way to use topological techniques (akin to Bader's) to elucidate the phenomena in this book? how does $\nabla^2 \rho$ relate to the ideas here?